Mathematical Excursions

INSTRUCTOR'S ANNOTATED EDITION

Richard N. Aufmann
Palomar College, California

Joanne S. Lockwood
Plymouth State College, New Hampshire

Richard D. Nation
Palomar College, California

Daniel K. Clegg
Palomar College, California

HOUGHTON MIFFLIN COMPANY
Boston   New York

Publisher: Jack Shira
Development Manager: Maureen Ross
Sponsoring Editor: Lauren Schultz
Assistant Editor: Lisa Pettinato
Senior Project Editor: Tamela Ambush
Senior Production/Design Coordinator: Carol Merrigan
Editorial Assistant: Lisa Sullivan
Manufacturing Manager: Florence Cadran
Senior Marketing Manager: Ben Rivera
Marketing Associate: Alexandra Shaw

Cover Photographer: © 2002 Yann Arthus-Bertrand/Altitude

Photo credits are found immediately after the
Answer section in the back of the book.

Printed in the U.S.A.

Library of Congress Catalog Card Number: 2002109357

ISBNs:
Student Text: 0-395-72779-0
Instructor's Annotated Edition: 0-618-30254-9

123456789-VHP-07 06 05 04 03

CONTENTS

CHAPTER

Logic   108

APPLICATIONS

Computer 199, 202, 212
Encryption 225
Mathematics 182, 190, 192, 201, 214, 223, 224, 234
Music 196
Postal Service 202

CHAPTER

Numeration Systems and Topics from Number Theory   *172*

CHAPTER

Applications of Equations *238*

CHAPTER

6

Applications of Functions *322*

CHAPTER

Mathematical Systems *404*

CHAPTER

Geometry *446*

CHAPTER

The Mathematics of Graphs *546*

CHAPTER

Statistics *766*

APPLICATIONS

Business 860, 889, 892, 900, 903, 904
Computer science 861
Construction 901
Criminology 896
Education 854, 856, 858, 859, 860, 861, 879,
    880, 884, 900, 901, 903, 904
Entertainment 878, 880, 881, 885
Essay contest 902
Film competition 883
Food preferences 864, 868, 869, 870, 872,
    877, 878, 885, 903, 904
Government 847, 849, 857, 858, 862, 867,
    877, 883, 890, 892, 893, 897, 898, 901
Health science 853, 859
Music 896
Populations 853, 863
Recreation 856, 878, 880, 881
Scholarships 884, 902
Security services 860
Social science 861
Sports 869, 871, 879, 882, 896, 897, 902
Wireless phones 879

CHAPTER

Mathematical Excursions is about mathematics as a system of knowing or understanding our surroundings. It is similar to an English literature textbook, an Introduction to Philosophy textbook, or perhaps an Introductory Psychology textbook. Each of those books provide glimpses into the thoughts and perceptions of some of the world's greatest writers, philosophers, and psychologists. Reading and studying their thoughts enables us to better understand the world we inhabit.

In a similar way, Mathematical Excursions provides glimpses into the nature of mathematics and how it is used to understand our world. This understanding, in conjunction with other disciplines, contributes to a more complete portrait of our world. Our contention is that ancient Greek architecture is quite dramatic but even more so when the "Golden Ratio" is considered. That I. M. Pei's work becomes even more interesting with a knowledge of elliptical shapes. That the challenges of sending information across the Internet is better understood by examining prime numbers. That the perils of radioactive waste take on new meaning with a knowledge of exponential functions. That generally, a knowledge of mathematics strengthens the way we know, perceive, and understand our surroundings.

One theme around which this book is written is, "What if you wanted to know how to . . . , what would you need to know?" Using this strategy, a contemporary problem is introduced and then the relevant mathematics needed to solve that problem is developed. With the mathematics in place, the solution to the problem is presented and additional applications of the mathematics are illustrated. A second theme is to have you explore a concept from different perspectives so that you can develop an appreciation for the diversity of problems that can be solved from a single concept.

Math Matters and Excursions are two features we have incorporated in the text. Math Matters are vignettes of interesting applications of the topic being discussed. Each section of the text ends with an Excursion, which is an extension of one of the topics of that section.

The exercise sets of Mathematical Excursions have been carefully selected to reinforce and extend the concepts developed in each section. The exercises range from drill and practice to interesting challenges. Some of the exercise sets include outlines for further explorations, suggestions for essays, critical thinking, and cooperative learning activities. In all cases, the exercises were chosen to illustrate the many facets of the topic under discussion.

The purpose of this book is to be a brief excursion into the castle of mathematics with all its myriad of rooms. Although we assume that the reader has a intermediate algebra background, each topic is carefully developed and appropriate material reviewed whenever necessary. When deciding on the depth of coverage, our singular criteria was to make mathematics accessible.

Chapter Opening Features

Chapter Opening Photos

Each chapter begins with photos and captions that are related to an Excursion in the chapter.

Web Icon

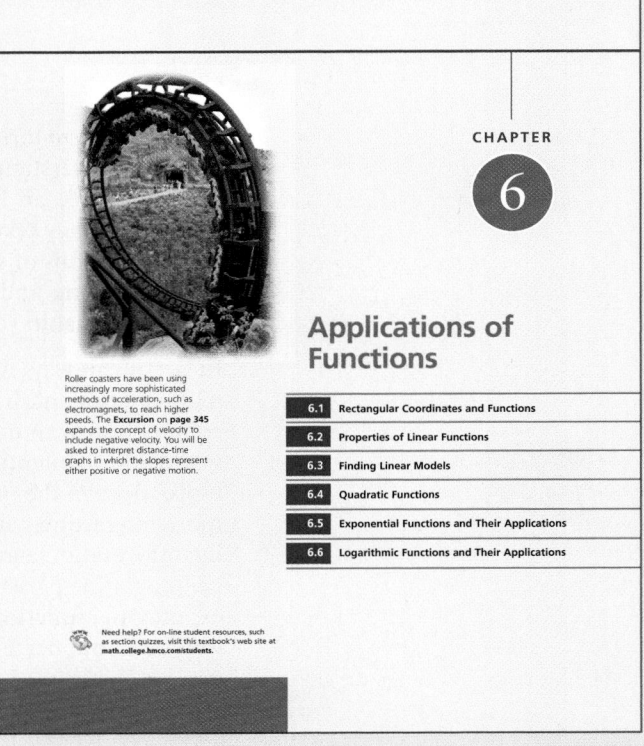

The web icon on this opening page lets students know of additional online resources at **math.college.hmco.com/students**.

Roller coasters have been using increasingly more sophisticated methods of acceleration, such as electromagnets, to reach higher speeds. The **Excursion** on **page 345** expands the concept of velocity to include negative velocity. You will be asked to interpret distance-time graphs in which the slopes represent either positive or negative motion.

6.1	Rectangular Coordinates and Functions
6.2	Properties of Linear Functions
6.3	Finding Linear Models
6.4	Quadratic Functions
6.5	Exponential Functions and Their Applications
6.6	Logarithmic Functions and Their Applications

Need help? For on-line student resources, such as section quizzes, visit this textbook's web site at **math.college.hmco.comstudents.**

page 322

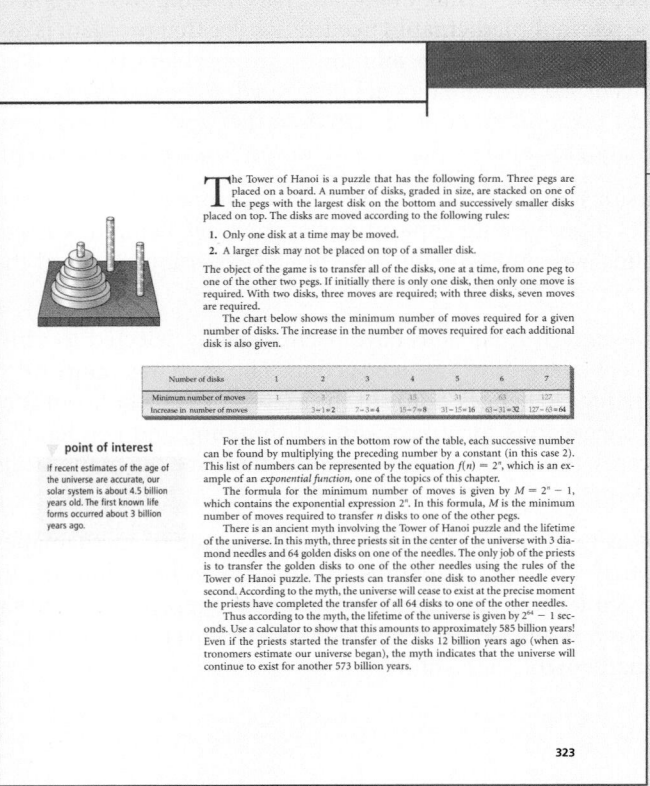

The Tower of Hanoi is a puzzle that has the following form. Three pegs are placed on a board. A number of disks, graded in size, are stacked on one of the pegs with the largest disk on the bottom and successively smaller disks placed on top. The disks are moved according to the following rules:

1. Only one disk at a time may be moved.
2. A larger disk may not be placed on top of a smaller disk.

The object of the game is to transfer all of the disks, one at a time, from one peg to one of the other two pegs. If initially there is only one disk, then only one move is required. With two disks, three moves are required; with three disks, seven moves are required.

The chart below shows the minimum number of moves required for a given number of disks. The increase in the number of moves required for each additional disk is also given.

Number of disks	1	2	3	4	5	6	7
Minimum number of moves	1	3	7	15	31	63	127
Increase in number of moves		$3-1=2$	$7-3=4$	$15-7=8$	$31-15=16$	$63-31=32$	$127-63=64$

▼ **point of interest**

If recent estimates of the age of the universe are accurate, our solar system is about 4.5 billion years old. The first known life forms occurred about 3 billion years ago.

For the list of numbers in the bottom row of the table, each successive number can be found by multiplying the preceding number by a constant (in this case 2). This list of numbers can be represented by the equation $f(n) = 2^n$, which is an example of an *exponential function*, one of the topics of this chapter.

The formula for the minimum number of moves is given by $M = 2^n - 1$, which contains the exponential expression 2^n. In this formula, M is the minimum number of moves required to transfer n disks to one of the other pegs.

There is an ancient myth involving the Tower of Hanoi puzzle and the lifetime of the universe. In this myth, three priests sit in the center of the universe with 3 diamond needles and 64 golden disks on one of the needles. The only job of the priests is to transfer the golden disks to one of the other needles using the rules of the Tower of Hanoi puzzle. The priests can transfer one disk to another needle every second. According to the myth, the universe will cease to exist at the precise moment the priests have completed the transfer of all 64 disks to one of the other needles.

Thus according to the myth, the lifetime of the universe is given by $2^{64} - 1$ seconds. Use a calculator to show that this amounts to approximately 585 billion years! Even if the priests started the transfer of the disks 12 billion years ago (when astronomers estimate our universe began), the myth indicates that the universe will continue to exist for another 573 billion years.

323

page 323

Chapter Opener Subject Matter

The second page of each chapter opener presents a motivational topic, an application from the chapter, or a new mathematical concept.

Interactive Method

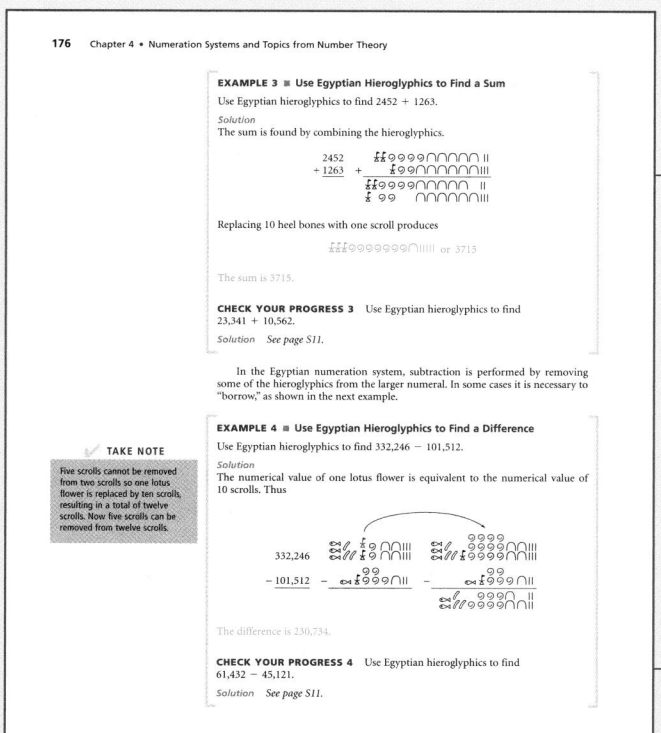

page 176

An Interactive Approach

Mathematical Excursions is written in a style that encourages the student to interact with the textbook. Each section contains a variety of worked examples. Each example is given a title so that the student can see at a glance the type of problem that is being solved. Most examples include annotations that assist the student in moving from step to step, and the final answer is in color in order to be readily identifiable.

Check Your Progress Exercises

Following each worked example is a Check Your Progress exercise for the student to work. By solving this exercise, the student actively practices concepts as they are presented in the text. For each Check Your Progress exercise, there is a detailed solution in the Solutions appendix.

Question/Answer Feature

At various places throughout the text, a Question is posed about the topic that is being developed. This question encourages students to pause, think about the current discussion, and answer the question. Students can immediately check their understanding by referring to the Answer to the question provided in a footnote on the same page. This feature creates another opportunity for the student to interact with the textbook.

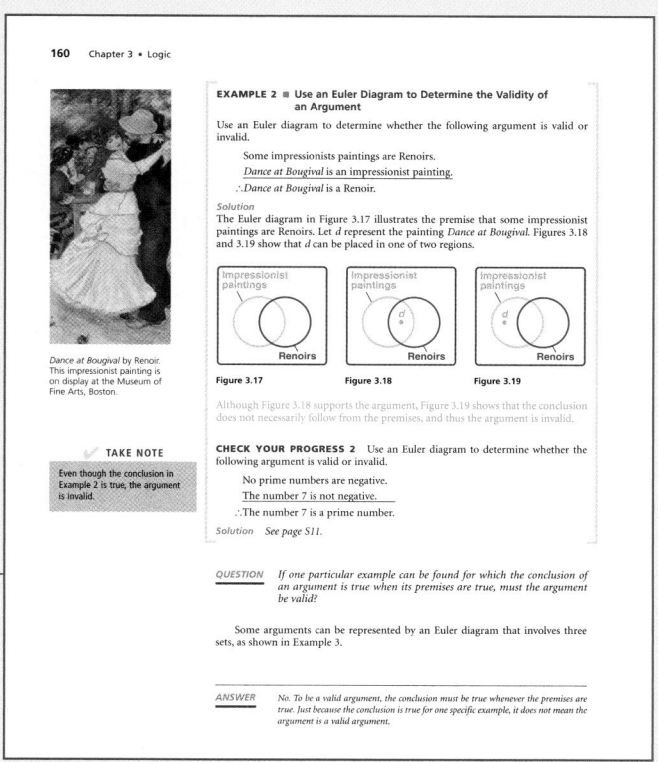

page 160

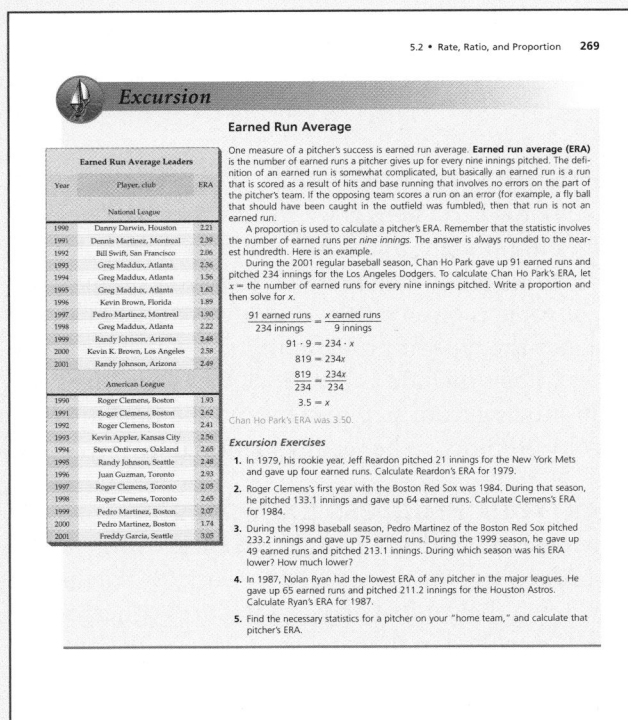

page 269

Interactive Method, *continued*

Excursions

Each section ends with an Excursion along with corresponding Excursion Exercises. These activities engage students in the mathematics of the section. Some Excursions are designed as in-class cooperative learning activities that lend themselves to a hands-on approach. They can also be assigned as projects or extra credit assignments. The Excursions are a unique and important feature of this text. They provide opportunities for students to take an active role in the learning process. The photos on the first page of a chapter opener relate to one of the Excursions in that chapter.

AIM for Success Student Preface

This 'how to use this text' preface explains what is required of a student to be successful and how this text has been designed to foster student success. AIM for Success can be used as a lesson on the first day of class or as a project for students to complete to strengthen their study skills.

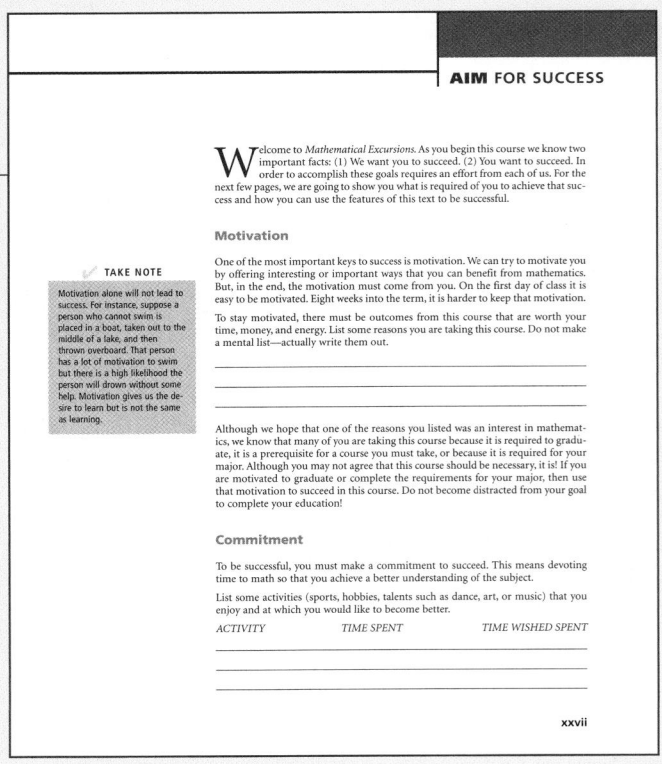

page xxvii

Math Matters and Margin Notes

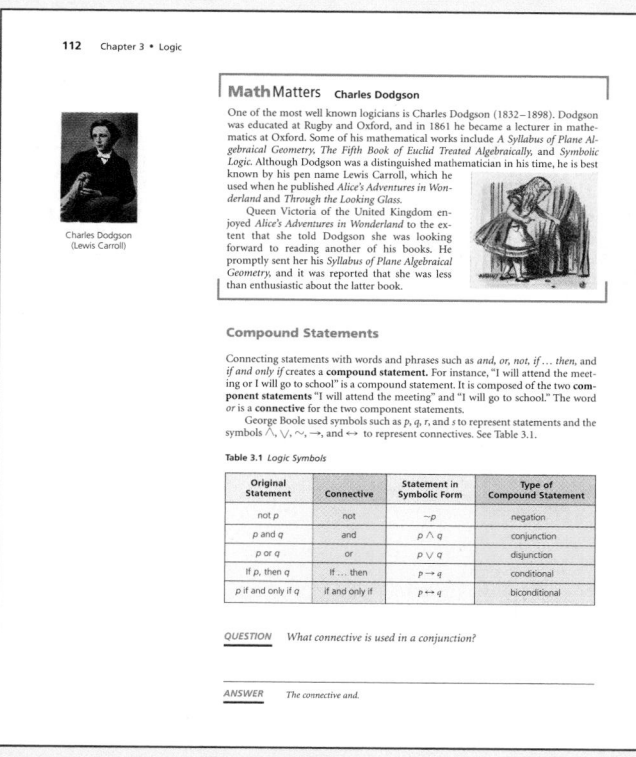

Math Matters

This feature of the text typically contains an interesting sidelight about mathematics, its history, or its applications.

Historical Note

These margin notes provide historical background information related to the concept under discussion or vignettes of individuals who were responsible for major advancements in their fields of expertise.

Calculator Note

These notes provide information about how to use the various features of a calculator.

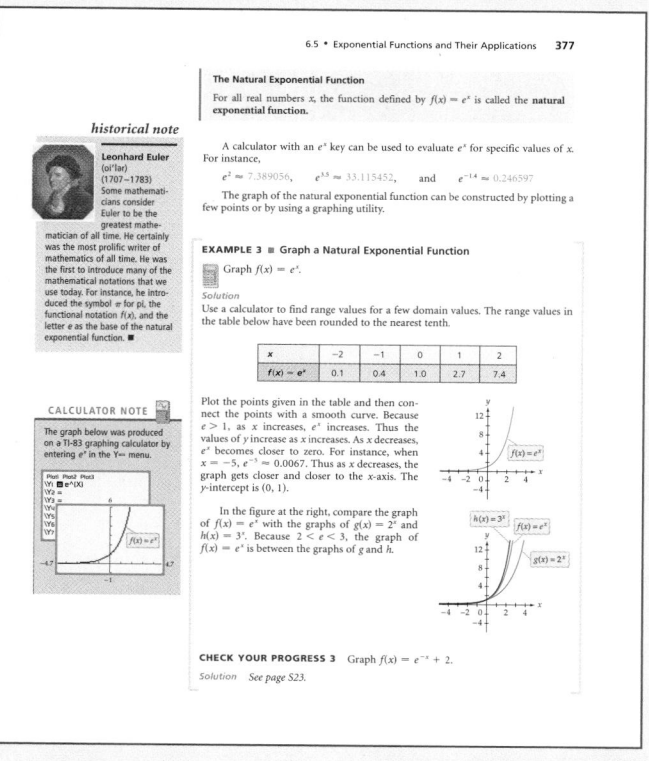

Margin Notes, *continued*

Point of Interest

These notes provide interesting information related to the topics under discussion. Many of these are of a contemporary nature and, as such, they provide students with the needed motivation for studying concepts that may at first seem abstract and obscure without this information.

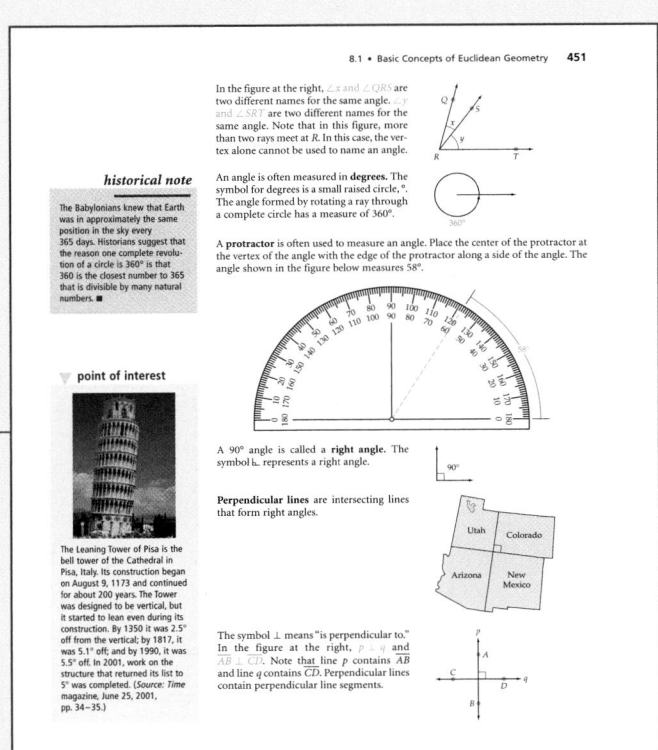

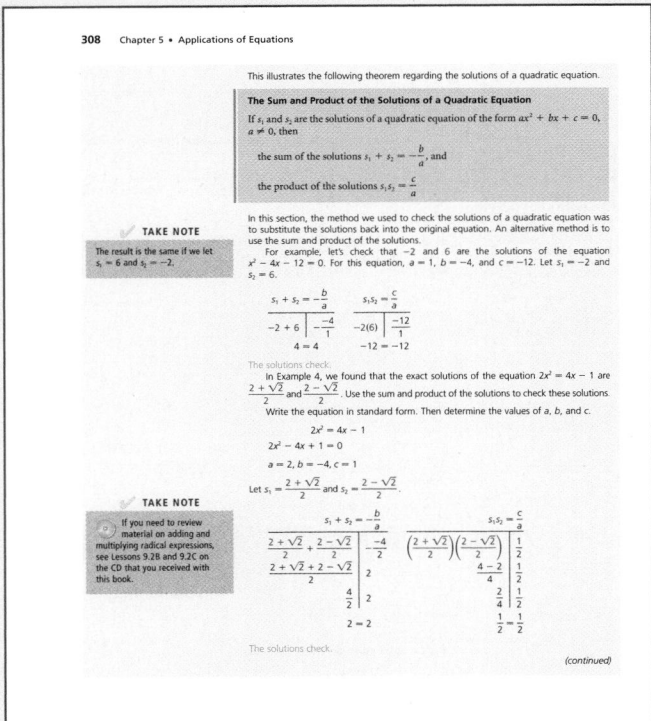

Take Note

These margin notes alert students to a point requiring special attention or are used to amplify the concepts that are currently being developed. Some Take Notes, identified by [CD icon], reference the student CD. A student who needs to review a prequisite skill or concept can find the needed material on this CD.

Exercises

Exercise Sets

The exercise sets of *Mathematical Excursions* were carefully written to provide a wide variety of exercises that range from drill and practice to interesting challenges. Exercise sets emphasize skill building, skill maintenance, concepts, and applications, when they are appropriate. Icons are used to identify various types of exercise.

Writing exercises

Data analysis exercises

Graphing calculator exercises

Internet exercises

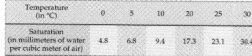

---

384 Chapter 6 • Applications of Functions

  b. Use the application to predict the number of ATMs in 2010.

37. The table below shows the saturation of water in air at various air temperatures.

Temperature (in °C)	0	5	10	20	25	30
Saturation (in milliliters of water per cubic meter of air)	4.8	6.8	9.4	17.3	23.1	30.4

  a. Find an exponential regression equation for these data. Round to the nearest thousandth.

  b. Use the equation to predict the number of milliliters of water per cubic meter of air at a temperature of 15°C. Round to the nearest tenth. *Hint:* The function is of the form $f(x) = k \cdot 2^{(cx)}$, where k and c are constants. Also $f(0) = 440$ and $f(12) = 880$.

38. Artificial snow is made at a ski resort by combining air and water in a ratio that depends on the outside air temperature. The table below shows the rate of air flow needed for various temperatures.

Temperature (in °F)	0	5	10	15	20
Air flow (in cubic feet per minute)	3.0	3.6	4.7	6.1	9.9

  a. Find an exponential regression equation for these data. Round to the nearest hundredth.

  b. Use the equation to predict the air flow needed when the temperature is 25°F. Round to the nearest tenth.

Extensions
CRITICAL THINKING

An exponential model for population growth or decay can be accurate over a short period of time. However, this model begins to fail because it does not account for the natural resources necessary to support growth, nor does it account for death within the population. Another model, called the *logistic model*, can account for some of these effects. The logistic model is given by $P(t) = \dfrac{mP_0}{P_0 + (m - P_0)e^{-kt}}$, where $P(t)$ is the population at time t, m is the maximum population that can be sup-

ported, P_0 is the population when $t = 0$, and k is a positive constant that is related to the growth of the population.

39. One model of Earth's population is given by $P(t) = \dfrac{280}{4 + 66e^{-0.021t}}$. In this equation, $P(t)$ is the population in billions and t is the number of years after 1980. Round answers to the nearest hundred million.

  a. According to this model, what was Earth's population in the year 2000?

  b. According to this model, what will be Earth's population in the year 2010?

  c. If t is very large, say greater than 500, then $e^{-0.021t} \approx 0$. What does this suggest about the maximum population that Earth can support?

40. Biologists have determined that the maximum wolf population in a certain preserve is 1000 wolves. Suppose the population of wolves in the preserve in the year 2000 was 500, and that k is estimated to be 0.025.

  a. Find a logistic function for the number of wolves in the preserve in year t, where t is the number of years after 2000.

  b. Find the estimated wolf population in 2015.

EXPLORATIONS

41. The formula used to calculate a monthly lease payment or a monthly car payment (for a purchase rather than a lease) is given by $P = \dfrac{Ar(1 + r)^n - Vr}{(1 + r)^n - 1}$, where P is the monthly payment, A is the amount of the loan, r is the *monthly* interest rate as a decimal, n is the number of months of the loan or lease, and V is the residual value of the car at the end of the lease. For a car purchase, $V = 0$.

  a. If the annual interest rate for a loan is 9%, what is the monthly interest rate as a decimal?

  b. Write the formula for a monthly car payment when the car is purchased rather than leased.

  c. Suppose you lease a car for 5 years. Find the monthly lease payment if the lease amount is $10,000, the residual value is $6000, and the annual interest rate is 6%.

  d. Suppose you purchase a car and secure a 5-year loan for $10,000 at an annual interest rate of 6%. Find the monthly payment.

  e. Why are the answers to parts c and d different?

page 384

Extensions

Extension exercises are placed at the end of each exercise set. As the name implies, these exercises are designed to extend concepts. In most cases these exercises are more challenging and require more time and effort than the preceding exercises. The Extension exercises always include at least two of the following types of exercises:

> *Critical Thinking*
> *Cooperative Learning*
> *Explorations*

Some Critical Thinking exercises require the application of two or more procedures or concepts.

The Cooperative Learning exercises are designed for small groups of 2 to 4 students.

Many of the Exploration exercises require students to search on the Internet or through reference materials in a library.

1.3 • Problem-Solving Strategies **41**

43. An airplane left Los Angeles at 8:20 A.M. and flew to Boston. The flying time was 6 hours 20 minutes. Boston is on Eastern Standard Time (EST) and Los Angeles is on Pacific Standard Time (PST), which is 3 hours behind EST. After the plane was on the ground for 1 hour it flew to Chicago, which is on Central Standard Time (CST). CST is 1 hour behind EST. The flying time from Boston to Chicago was 2 hours 20 minutes.

  a. What time, EST, did the plane arrive in Boston?

  b. What time, CST, did the plane arrive in Chicago?

44. a. List the four steps in Polya's problem-solving strategy.

  b. List eight problem-solving procedures that one might use in Polya's second step.

Extensions
CRITICAL THINKING

45. What is the 100th decimal digit in the decimal representation of $\frac{1}{7}$?

46. a. How many times larger is $3^{(3^3)}$ than $(3^3)^3$?

  b. How many times larger is $4^{(4^4)}$ than $(4^4)^4$? *Note:* Most calculators will not display the answer to this problem because it is too large. However, the answer can be determined in exponential form by applying the following properties of exponents.

$$(a^m)^n = a^{mn} \quad \text{and} \quad \frac{a^m}{a^n} = a^{m-n}$$

47. The mathematician Augustus De Morgan once wrote that he had the distinction of being x years old in the year x^2. He was 43 in the year 1849.

  a. Explain why people born in the year 1980 might share the distinction of being x years old in the year x^2. *Note:* Assume x is a natural number.

  b. What is the next year after 1980 for which people born in that year might be x years old in the year x^2?

48. Select a two-digit number between 50 and 100. Add 83 to your number. From this number form a new number by adding the digit in the hundreds place to the number formed by the other two digits (the digits in the tens place and the ones place). Now subtract this newly formed number from your original number. Your final result is 16. Use a deductive approach to show that the final result is always 16 regardless of which number you start with.

49. How many digits does it take in total to number a book from page 1 to page 240?

50. Consider a checkerboard with two red squares on opposite corners removed, as shown in the accompanying figure. Determine whether it is possible to completely cover the checkerboard with 31 dominoes if each domino is placed horizontally or vertically and each domino covers exactly two squares. If it is possible, show how to do it. If it is not possible, explain why it cannot be done.

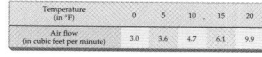

COOPERATIVE LEARNING

51. The object of this exercise is to create mathematical expressions that use exactly four 4's and that simplify to a counting number from 1 to 20, inclusive. You are allowed to use the following mathematical symbols: $+, -, \times, \div, \sqrt{\ }$, (, and). For example,

$$\frac{4}{4} + \frac{4}{4} = 2, \quad 4^{(4-4)} + 4 = 5, \quad \text{and}$$
$$4 - \sqrt{4} + 4 \times 4 = 18$$

52. The following puzzle is a famous *cryptarithm.*

```
  SEND
+ MORE
-------
 MONEY
```

Each letter in the cryptarithm represents one of the digits 0 through 9. The leading digits, represented by

page 41

End of Chapter

Chapter Summary

At the end of each chapter there is a Chapter Summary that includes *Key Terms* and *Essential Concepts* that were covered in the chapter. These chapter summaries provide a single point of reference as the student prepares for an examination. Each key word references the page number of the chapter where the word was first introduced.

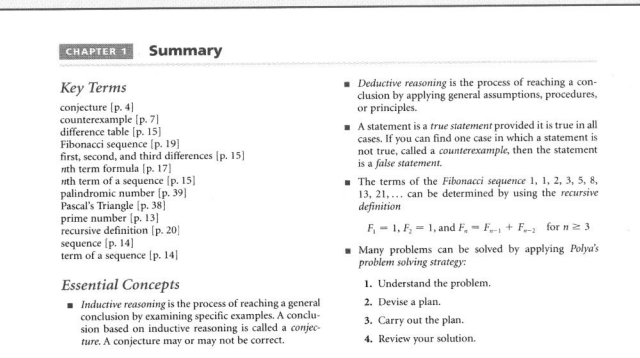

page 42

Chapter Review

Review exercises are found near the end of each chapter. These exercises were selected to help the student integrate the major topics presented in the chapter. The answers to all of the Chapter Review exercises appear in the answer section along with a section reference for each exercise. These section references indicate the section or sections where a student can locate the concepts needed to solve each exercise.

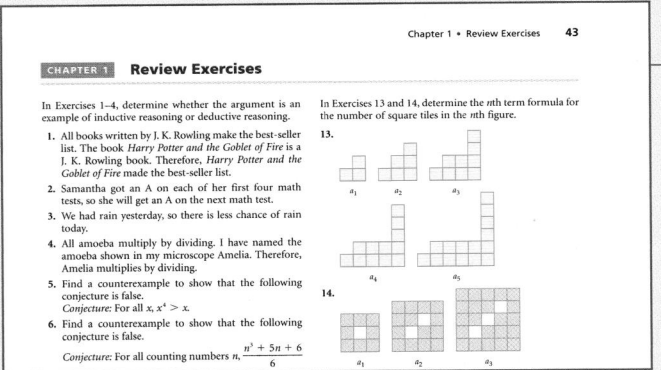

page 43

Chapter Test

The Chapter Test exercises are designed to simulate a possible test of the material in the chapter. The answers to all of the Chapter Test exercises appear in the answer section along with a section reference for each exercise. The section references indicate the section or sections where a student can locate the concepts needed to solve each exercise.

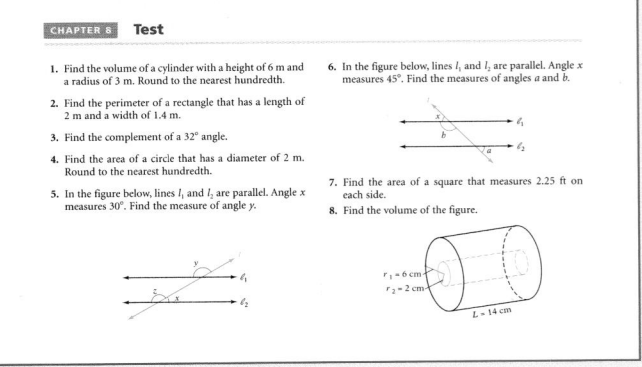

page 543

Supplements for the Instructor

Mathematical Excursions has an extensive support package for the instructor that includes:

Instructor's Annotated Edition (IAE): The *Instructor's Annotated Edition* is an exact replica of the student textbook with the following additional text-specific items for the instructor: answers to *all* of the end-of-section and end-of-chapter exercises, answers to *all* Excursion and Exploration exercises, Instructor Notes, Suggested Assignments, and P icons denoting tables and art that appear in Power-Point® slides. (The slides are available on the Instructor ClassPrep with HM Testing 6.0 CD-ROM and/or the files can be downloaded from our web site at math.college.hmco.com/instructors).

Instructor's Resource Manual: The *Instructor's Resource Manual* offers worked-out solutions to *all* of the exercises in each exercise set as well as answers to the Excursion and Exploration exercises. The manual contains a variety of ready-to-use printed Chapter Tests (two formats: free response and multiple choice). These tests can also be downloaded from our web site at math.college.hmco.com/instructors. In addition to the ready-to-use Chapter Tests, a *Printed Test Bank* is also available in the manual. The *Printed Test Bank* provides a printout of one example of each of the algorithmic items in the HM Testing 6.0 (See the description of HM Testing 6.0 below, under HM ClassPrep with HM Testing 6.0 CD-ROM). Also included in the manual are suggested syllabi that provide instructors with options for sequencing the course.

HM ClassPrep with HM Testing CD-ROM: This CD-ROM is a combination of two course management tools.

- HM Testing 6.0 computerized testing software provides instructors with an array of algorithmic test items, allowing for the creation of an unlimited number of tests for each chapter, including cumulative tests and final exams. HM Testing also offers online testing via a Local Area Network (LAN) or the Internet, as well as a grade book function.

- HM ClassPrep features supplements and text-specific resources, such as: suggested syllabi that provide instructors with options for sequencing the course, as well as an Index of Applications, Chapter Tests, PowerPoint® slides, Microsoft® Excel spreadsheets, Graphing Calculator Guide, and an Excel Guide.

Instructor Text-Specific Web Site: The companion web site provides additional teaching resources such as: suggested syllabi that provide instructors with options for sequencing the course as well as an Index of Applications, Chapter Tests, Excursions from the text, PowerPoint® slides, Excel spreadsheets, Graphing Calculator Guide, and an Excel Guide. Visit math.college.hmco.com/instructors and choose *Mathematical Excursions* from the list provided on the site. Appropriate items will be password-protected. Instructors have access to the student web site as well.

eduSpace®: eduSpace® is a text-specific online learning environment that combines algorithmic tutorials with homework capabilities and classroom management functions. Please contact your Houghton Mifflin sales representative for detailed information about the course content available for this text.

Two levels of service are provided for instructors.

- **Electronic grading** allows the instructor to complete their grades electronically and record students' results on the quizzes provided on eduSpace®.

■ **Course Management** allows the instructor to manage the course on a lecture-basis or manage a distance-learning course online.

Supplements for the Student

Mathematical Excursions has an extensive support package for the student that includes:

Student Solutions Manual: The *Student Solutions Manual* contains complete, worked-out solutions to *all* odd-numbered exercises and *all* of the solutions to the Chapter Reviews and Chapter Tests in the text.

CLAST Preparation Student Guide: The CLAST Preparation Student Guide is a competency-based study guide that reviews and offers preparatory material for the CLAST (College Level Academic Skills Test) objectives required by the State of Florida for mathematics. The guide includes a correlation of the CLAST objectives to the *Mathematical Excursions* text, worked-out examples, practice examples, cumulative reviews, and sample diagnostic tests with grading sheets.

HMmathSpace™ Tutorial CD ROM: . This new tutorial CD ROM allows students to practice skills and review concepts as many times as necessary by using algorithmically generated exercises and step-by-step solutions for practice. Among the many features of the CD-ROM, there is a Prerequisite Algebra Review, Graphing Calculator Guide, Excel Guide, and Excel spreadsheets that are referred to in the text.

SMARTTHINKING™ Live, On-line Tutoring: Houghton Mifflin has partnered with SMARTTHINKING™ to provide an easy-to-use, effective, online tutorial service. Through state-of-the-art tools and a two-way whiteboard, students communicate in real-time with qualified e-structors who can help the students understand difficult concepts and guide them through the problem-solving process while studying or completing homework.

Four levels of service are offered to the students.

■ **Live, online tutoring support** is available Sunday–Thursday 2pm–5pm and 9pm–1pm EST (hours are subject to change).

■ **Question submission** allows students to submit questions to the tutor outside the scheduled hours and receive a response within 24 hours.

■ **Pre-scheduled time** allows students to schedule tutoring with an e-structor in advance.

■ **Review past online sessions** allows students to access and review their progress from previous sessions on a personal academic home page.

Houghton Mifflin Instructional Videos/DVD's: These text-specific Videos and DVD's, professionally produced by Dana Mosely, provide explanations of key concepts, examples, and exercises in a lecture-based format. They offer students a valuable resource for further instruction and review.

Student Text-Specific Web Site: This textbook has a companion web site that provides additional learning resources. Visit math.college.hmco.com/students and choose *Mathematical Excursions* from the list provided on the site.

eduSpace®: eduSpace® is a text-specific online learning environment that combines algorithmic tutorials with homework capabilities. Text-specific content is available to help you understand the mathematics covered in this textbook.

Four levels of service are offered to the students.

- **Tutorials** help the student to review concepts that he or she may miss because of an absence from class. The students can also use the tutorials to review material for upcoming quizzes and tests.

- **Practice exercises** allow the student to reinforce skills and concepts, not yet mastered, by completing different types of exercises.

- **Homework assignments** can be accessed, completed, and submitted online if the instructor assigns these assignments.

- **Quizzes** can be used for practice or taken for a grade if your instructor assigns the quizzes.

Acknowledgments

The authors would like to thank the people who have reviewed this manuscript and provided many valuable suggestions.

Randall Allbritton
Daytona Beach Community College

Isali Alsina
Kean University

Bernadette Antkoviak
Harrisburg Area Community College

Charles N. Baker
West Liberty State College

Linda A. Bastian
Portland Community College

Carole A. Bauer
Triton College

Dr. Joan E. Bell
Northeastern State University

Brian Bradie
Christopher Newport University

Shelley Brooks
Baylor University

Jesse W. Bryne, Ph.D.
University of Central Oklahoma

Dr. J. Robert Buchanan
Millersville University

Thomas R. Caplinger
University of Memphis

Elizabeth Carrico
Illinois Central College

Penelope A. Coe
Central Connecticut State University

Dr. Donna Ericksen
Central Michigan University

Kenny Fister
Murray State University

Linda L. Galloway
Macon State College

Carolyn H. Goldberg
Niagara County Community College

Tracy Dawn Hamilton
Western Illinois University

Robert V. High
Hofstra University

Elaine Klett
Brookdale Community College

Denise LeGrand
University of Arkansas at Little Rock

Elaine M. Lytton
Sandhills Community College

Roger Marty
Cleveland State University

Dr. Pat Mower
Washburn University

Kathleen Offenholley
Brookdale Community College

Diana Pagel
The Victoria College

Dr. Anne Quinn
Edinboro University of Pennsylvania

Robert B. Sackett
Erie Community College

Mary Lee Seitz
Erie Community College—City Campus

Aaron Keith Trautwein
Carthage College

Susan Williford
Columbia State Community College

The authors would also like to give special thanks to Delaney Carrier, Tim Hempleman, Gina Sanders, and Lauri Semarne for their extra help with the preparation of this manuscript.

INSTRUCTOR NOTE
See the *Class Prep CD* for teaching tools and resources for this lesson.

Welcome to *Mathematical Excursions*. As you begin this course we know two important facts: (1) We want you to succeed. (2) You want to succeed. In order to accomplish these goals, an effort is required from each of us. For the next few pages, we are going to show you what is required of you to achieve that success and how you can use the features of this text to be successful.

Motivation

One of the most important keys to success is motivation. We can try to motivate you by offering interesting or important ways that you can benefit from mathematics. But, in the end, the motivation must come from you. On the first day of class it is easy to be motivated. Eight weeks into the term, it is harder to keep that motivation.

To stay motivated, there must be outcomes from this course that are worth your time, money, and energy. List some reasons you are taking this course. Do not make a mental list—actually write them out.

_____

_____

_____

✓ **TAKE NOTE**

Motivation alone will not lead to success. For instance, suppose a person who cannot swim is placed in a boat, taken out to the middle of a lake, and then thrown overboard. That person has a lot of motivation to swim but there is a high likelihood the person will drown without some help. Motivation gives us the desire to learn but is not the same as learning.

Although we hope that one of the reasons you listed was an interest in mathematics, we know that many of you are taking this course because it is required to graduate, it is a prerequisite for a course you must take, or because it is required for your major. Although you may not agree that this course should be necessary, it is! If you are motivated to graduate or complete the requirements for your major, then use that motivation to succeed in this course. Do not become distracted from your goal to complete your education!

Commitment

To be successful, you must make a commitment to succeed. This means devoting time to math so that you achieve a better understanding of the subject.

List some activities (sports, hobbies, talents such as dance, art, or music) that you enjoy and at which you would like to become better.

ACTIVITY	*TIME SPENT*	*TIME WISHED SPENT*

Thinking about these activities, put the number of hours that you spend each week practicing these activities next to the activity. Next to that number, indicate the number of hours a week you would like to spend on these activities.

Whether you listed surfing or sailing, aerobics or restoring cars, or any other activity you enjoy, note how many hours a week you spend on each activity. To succeed in math, you must be willing to commit the same amount of time. Success requires some sacrifice.

The "I Can't Do Math" Syndrome

There may be things you cannot do, for instance, lift a two-ton boulder. You can, however, do math. It is much easier than lifting the two-ton boulder. When you first learned the activities you listed above, you probably could not do them well. With practice, you got better. With practice, you will be better at math. Stay focused, motivated, and committed to success.

It is difficult for us to emphasize how important it is to overcome the "I Can't Do Math Syndrome." If you listen to interviews of very successful athletes after a particularly bad performance, you will note that they focus on the positive aspect of what they did, not the negative. Sports psychologists encourage athletes to always be positive—to have a "Can Do" attitude. You need to develop this attitude toward math.

Strategies for Success

Know the Course Requirements   To do your best in this course, you must know exactly what your instructor requires. Course requirements may be stated in a *syllabus,* which is a printed outline of the main topics of the course, or they may be presented orally. When they are listed in a syllabus or on other printed pages, keep them in a safe place. When they are presented orally, make sure to take complete notes. In either case, it is important that you understand them completely and follow them exactly. Be sure you know the answer to each of the following questions.

1. What is your instructor's name?

2. Where is your instructor's office?

3. At what times does your instructor hold office hours?

4. Besides the textbook, what other materials does your instructor require?

5. What is your instructor's attendance policy?

6. If you must be absent from a class meeting, what should you do before returning to class? What should you do when you return to class?

7. What is the instructor's policy regarding collection or grading of homework assignments?

8. What options are available if you are having difficulty with an assignment? Is there a math tutoring center?

9. If there is a math lab at your school, where is it located? What hours is it open?

10. What is the instructor's policy if you miss a quiz?

11. What is the instructor's policy if you miss an exam?

12. Where can you get help when studying for an exam?

Remember: Your instructor wants to see you succeed. If you need help, ask! Do not fall behind. If you were running a race and fell behind by 100 yards, you may be able to catch up but it will require more effort than had you not fallen behind.

Time Management   We know that there are demands on your time. Family, work, friends, and entertainment all compete for your time. We do not want to see you receive poor job evaluations because you are studying math. However, it is also true that we do not want to see you receive poor math test scores because you devoted too much time to work. When several competing and important tasks require your time and energy, the only way to manage the stress of being successful at both is to manage your time efficiently.

Instructors often advise students to spend twice the amount of time outside of class studying as they spend in the classroom. Time management is important if you are to accomplish this goal and succeed in school. The following activity is intended to help you structure your time more efficiently.

List the name of each course you are taking this term, the number of class hours each course meets, and the number of hours you should spend studying each subject outside of class. Then fill in a weekly schedule like the one printed below. Begin by writing in the hours spent in your classes, the hours spent at work (if you have a job), and any other commitments that are not flexible with respect to the time that you do them. Then begin to write down commitments that are more flexible, including hours spent studying. Remember to reserve time for activities such as meals and exercise. You should also schedule free time.

	Monday	Tuesday	Wednesday	Thursday	Friday	Saturday	Sunday
7–8 a.m.							
8–9 a.m.							
9–10 a.m.							
10–11 a.m.							
11–12 p.m.							
12–1 p.m.							
1–2 p.m.							
2–3 p.m.							
3–4 p.m.							
4–5 p.m.							
5–6 p.m.							
6–7 p.m.							
7–8 p.m.							
8–9 p.m.							
9–10 p.m.							
10–11 p.m.							
11–12 a.m.							

We know that many of you must work. If that is the case, realize that working 10 hours a week at a part-time job is equivalent to taking a three-unit class. If you must work, consider letting your education progress at a slower rate to allow you to be successful at both work and school. There is no rule that says you must finish school in a certain time frame.

Schedule Study Time   As we encouraged you to do by filling out the time management form above, schedule a certain time to study. You should think of this time like being at work or class. Reasons for "missing study time" should be as compelling as reasons for missing work or class. "I just didn't feel like it" is not a good reason to miss your scheduled study time. Although this may seem like an obvious exercise, list a few reasons you might want to study.

_____

_____

_____

Of course we have no way of knowing the reasons you listed, but from our experience one reason given quite frequently is "To pass the course." There is nothing wrong with that reason. If that is the most important reason for you to study, then use it to stay focused.

One method of keeping to a study schedule is to form a **study group**. Look for people who are committed to learning, who pay attention in class, and who are punctual. Ask them to join your group. Choose people with similar educational goals but different methods of learning. You can gain from seeing the material from a new perspective. Limit groups to four or five people; larger groups are unwieldy.

There are many ways to conduct a study group. Begin with the following suggestions and see what works best for your group.

1. Test each other by asking questions. Each group member might bring two or three sample test questions to each meeting.

2. Practice teaching each other. Many of us who are teachers learned a lot about our subject when we had to explain it to someone else.

3. Compare class notes. You might ask other students about material in your notes that is difficult for you to understand.

4. Brainstorm test questions.

5. Set an agenda for each meeting. Set approximate time limits for each agenda item and determine a quitting time.

And now, probably the most important aspect of studying is that it should be done in relatively small chunks. If you can only study three hours a week for this course (probably not enough for most people), do it in blocks of one hour on three separate days, preferably after class. Three hours of studying on a Sunday is not as productive as three hours of paced study.

Features of This Text That Promote Success

Preparing for Class   Before the class meeting in which your professor begins a new section, you should read the title of each section. Next, browse through the chapter material, being sure to note each word in bold type. These words indicate important concepts that you must know to learn the material. Do not worry about trying to understand all the material. Your professor is there to assist you with that endeavor. The purpose of browsing through the material is so that your brain will be prepared to accept and organize the new information when it is presented to you. Turn to page 768. Write down the title of Section 12.1.

Under the title of the section, write down the words in the section that are in bold print. It is not necessary for you to understand the meaning of these words. You are in this class to learn their meaning.

_____

_____

_____

Math is Not a Spectator Sport   To learn mathematics you must be an active participant. Listening and watching your professor do mathematics is not enough. Mathematics requires that you interact with the lesson you are studying. If you have been writing down the things we have asked you to do, you were being interactive. There are other ways this textbook has been designed so that you can be an active learner.

Check Your Progress   One of the key instructional features of this text is a completely worked-out example followed by a *Check Your Progress*.

Order is important.

$$P(20, 3) = \frac{20!}{(20 - 3)!}$$

$$= \frac{20!}{17!}$$

$$= \frac{20 \cdot 19 \cdot 18 \cdot 17!}{17!}$$

$$= 6840$$

There are 6840 different ways to award the three prizes.

11.2 • Permutations and Combinations   **707**

EXAMPLE 3 ▪ **Counting Permutations**

In 2001, the Kentucky Derby had 17 horses entered in the race. How many different finishes of first, second, third, and fourth place were possible?

Solution
Because the order in which the horses finish the race is important, the number of possible finishes of first, second, third, and fourth place is $P(17, 4)$.

$$P(17, 4) = \frac{17!}{(17 - 4)!} = \frac{17!}{13!} = \frac{17 \cdot 16 \cdot 15 \cdot 14 \cdot 13!}{13!}$$
$$= 17 \cdot 16 \cdot 15 \cdot 14 = 57{,}120$$

There were 57,120 possible finishes of first, second, third, and fourth place.

CHECK YOUR PROGRESS 3   A 10-K marathon has 20 people entered. In how many different ways can the first, second, and third place prizes be awarded?

Solution   See page S40.

---

page 707

Note that each Example is completely worked out and the *Check Your Progress* following the example is not. Study the worked-out example carefully by working through each step. Your should do this with paper and pencil.

Now work the *Check Your Progress*. If you get stuck, refer to the page number following the word *Solution* which directs you to the page on which the *Check Your*

Progress is solved—a complete worked-out solution is provided. Try to use the given solution to get a hint for the step you are stuck on. Then try to complete your solution.

When you have completed the solution, check your work against the solution we provide.

CHECK YOUR PROGRESS 3, *page 707*

The order in which the runners finish is important, so the number of ways to place first, second, and third is

$$P(20, 3) = \frac{20!}{(20-3)!} = \frac{20!}{17!} = \frac{20 \cdot 19 \cdot 18 \cdot 17!}{17!} = 6840$$

There are 6840 different ways to award the first, second, and third place prizes.

page S40

Be aware that frequently there is more than one way to solve a problem. Your answer, however, should be the same as the given answer. If you have any question as to whether your method will "always work," check with your instructor or with someone in the math center.

Browse through the textbook and write down the page numbers where two other paired example features occur.

Remember: Be an active participant in your learning process. When you are sitting in class watching and listening to an explanation, you may think that you understand. However, until you actually try to do it, you will have no confirmation of the new knowledge or skill. Most of us have had the experience of sitting in class thinking we knew how to do something only to get home and realize we didn't.

Rule Boxes Pay special attention to rules placed in boxes. These rules give you the reasons certain types of problems are solved the way they are. When you see a rule, try to rewrite the rule in your own words.

✓ **TAKE NOTE**

If a rule has more than one part, be sure to make a notation to that effect.

Arguments and Truth Tables

The following truth table procedure can be used to determine whether an argument is valid or invalid.

> **Truth Table Procedure to Determine the Validity of an Argument**
>
> 1. Write the argument in symbolic form.
> 2. Construct a truth table that shows the truth value of each premise and the truth value of the conclusion for all combinations of truth values of the component statements.
> 3. If the conclusion is true in every row of the truth table in which all the premises are true, the argument is valid. If the conclusion is false in any row in which all of the premises are true, the argument is invalid.

We will now use the above truth table procedure to determine the validity of the argument about Aristotle.

1. Once again we let h represent the statement "Aristotle was human" and m represent the statement "Aristotle was mortal." In symbolic form the argument is

$h \rightarrow m$	First premise
h	Second premise
$\therefore m$	Conclusion

page 147

Chapter Exercises   When you have completed studying a section, do the section exercises. Math is a subject that needs to be learned in small sections and practiced continually in order to be mastered. Doing the exercises in each exercise set will help you master the problem-solving techniques necessary for success. As you work through the exercises, check your answers to the odd-numbered exercises with those in the back of the book.

Preparing for a Test   There are important features of this text that can be used to prepare for a test.

- Chapter Summary

- Chapter Review Exercises

- Chapter Test

After completing a chapter, read the Chapter Summary. (See page 103 for the Chapter 2 Summary.) This summary highlights the important topics covered in the chapter. The page number following each topic refers you to the page in the text on which you can find more information about the concept.

Following the Chapter Summary are Chapter Review Exercises (see page 103). Doing the review exercises is an important way of testing your understanding of the chapter. The answer to each review exercise is given at the back of the book, along with, in brackets, the section reference from which the question was taken (see page A5). After checking your answers, restudy any section from which a question you missed was taken. It may be helpful to retry some of the exercises for that section to reinforce your problem-solving techniques.

Each chapter ends with a Chapter Test (see page 105). This test should be used to prepare for an exam. We suggest that you try the Chapter Test a few days before your actual exam. Take the test in a quiet place and try to complete the test in the same amount of time you will be allowed for your exam. When taking the Chapter Test, practice the strategies of successful test takers: 1) scan the entire test to get a feel for the questions; 2) read the directions carefully; 3) work the problems that are easiest for you first; and perhaps most importantly, 4) try to stay calm.

When you have completed the Chapter Test, check your answers (see page A6). Next to each answer is, in brackets, the reference to the section from which the question was taken. If you missed a question, review the material in that section and rework some of the exercises from that section. This will strengthen your ability to perform the skills in that section.

Your career goal goes here. →

Is it difficult to be successful? YES! Successful music groups, artists, professional athletes, teachers, sociologist, chefs, and _____ have to work very hard to achieve their goals. They focus on their goals and ignore distractions. The things we ask you to do to achieve success take time and commitment. We are confident that if you follow our suggestions, you will succeed.

Mathematical Excursions

INSTRUCTOR'S ANNOTATED EDITION

Problem Solving

Many games, such as chess, require problem-solving and reasoning skills. People use these skills to devise winning strategies. The **Excursion** on **page 10** illustrates how inductive and deductive reasoning is used to analyze and devise a winning strategy for the game of *Sprouts*.

Need help? For on-line student resources, such as section quizzes, visit this textbook's web site at **math.college.hmco.com/students.**

Two goals of this chapter are to help you become a better problem solver and to demonstrate that problem solving can be an enjoyable experience. One problem that many people enjoy is known as the Monte Hall (host of the game show *Let's Make a Deal*) problem. Here is a statement of this problem.

The grand prize in the game show *Let's Make a Deal* is behind one of three curtains. Behind each of the other two curtains is a less desirable prize. For instance, a goat may be behind one curtain and a box of candy behind the other. You select one of the curtains, say curtain A. To add drama, the host, Monte Hall, reveals one of the less desirable prizes behind one of the curtains. At this time you are given the opportunity to cancel your original choice and choose the remaining closed curtain, or to stay with your original choice.

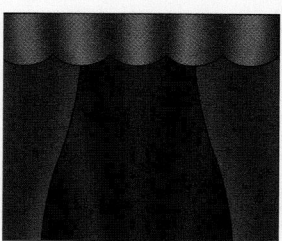 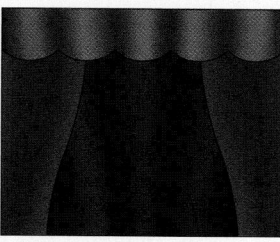 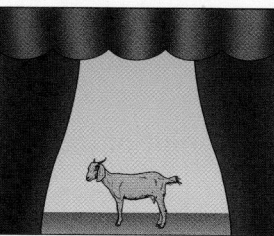

Curtain A   Curtain B   Curtain C

Marilyn vos Savant, author of the "Ask Marilyn" column featured in *Parade Magazine,* gave an analysis of this problem.[1] She claimed that you *double* your chances of winning the grand prize by switching to the other closed curtain. This analysis created a great deal of debate. In fact, many readers, including some mathematicians, responded with arguments that contradicted Marilyn's analysis.

What do you think? Do you have a better chance of winning the grand prize by

- switching to the other closed curtain or
- staying with your original choice?

Of course there is also the possibility that it does not matter, if the chances of winning are the same with either strategy.

Discuss the Monte Hall problem with some of your friends and classmates. Is everyone in agreement? Additional information on this problem is given in Exploration Exercise 50 on page 14.

Marilyn vos Savant

---

1. "Ask Marilyn," *Parade Magazine,* September 9, 1990, p. 15.

SECTION 1.1 | Inductive and Deductive Reasoning

Inductive Reasoning

The type of reasoning that forms a conclusion based on the examination of specific examples is called *inductive reasoning*. The conclusion formed by using inductive reasoning is often called a **conjecture,** since it may or may not be correct.

> **Inductive reasoning** is the process of reaching a general conclusion by examining specific examples.

When you examine a list of numbers and predict the next number in the list according to some pattern you have observed, you are using inductive reasoning.

EXAMPLE 1 ▪ Use Inductive Reasoning to Predict a Number

Use inductive reasoning to predict the most probable next number in each of the following lists.

a. 3, 6, 9, 12, 15, ?   **b.** 1, 3, 6, 10, 15, ?

Solution

a. Each successive number is 3 larger than the preceding number. Thus we predict that the most probable next number in the list is 3 larger than 15, which is 18.

b. The first two numbers differ by 2. The second and the third numbers differ by 3. It appears that the difference between any two numbers is always 1 more than the preceding difference. Since 10 and 15 differ by 5, we predict that the next number in the list will be 6 larger than 15, which is 21.

CHECK YOUR PROGRESS 1   Use inductive reasoning to predict the most probable next number in the following lists.

a. 5, 10, 15, 20, 25, ?   **b.** 2, 5, 10, 17, 26, ?

Solution   *See page S1.*

▼ **point of interest**

"Water boils down to nothing … snow boils down to nothing … ice boils down to nothing … everything boils down to nothing."

An example of inductive reasoning.

Inductive reasoning is not used just to predict the next number in a list. In Example 2 we use inductive reasoning to make a conjecture about an arithmetic procedure.

EXAMPLE 2 ▪ Use Inductive Reasoning to Make a Conjecture

Consider the following procedure: Pick a number. Multiply the number by 8, add 6 to the product, divide the sum by 2, and subtract 3.

Complete the above procedure for several different numbers. Use inductive reasoning to make a conjecture about the relationship between the size of the resulting number and the size of the original number.

✔ **TAKE NOTE**

In Example 5, we will use a deductive method to verify that the procedure in Example 2 always yields a result that is four times larger than the original number.

Solution

Suppose we pick 5 as our original number. Then the procedure would produce the following results:

Original number:	5
Multiply by 8:	$8 \times 5 = 40$
Add 6:	$40 + 6 = 46$
Divide by 2:	$46 \div 2 = 23$
Subtract 3:	$23 - 3 = 20$

We started with 5 and followed the procedure to produce 20. Starting with 6 as our original number produces a final result of 24. Starting with 10 produces a final result of 40. Starting with 100 produces a final result of 400. In each of these cases the resulting number is four times the original number. We *conjecture* that following the given procedure will produce a resulting number that is four times the original number.

CHECK YOUR PROGRESS 2   Consider the following procedure: Pick a number. Multiply the number by 9, add 15 to the product, divide the sum by 3, and subtract 5.

Complete the above procedure for several different numbers. Use inductive reasoning to make a conjecture about the relationship between the size of the resulting number and the size of the original number.

Solution   *See page S1.*

historical note

Galileo Galilei
(gǎl′-ə-lā′ē′) entered the University of Pisa to study medicine at the age of 17, but he soon realized that he was more interested in the study of astronomy and the physical sciences. Galileo's study of pendulums assisted in the development of pendulum clocks. ■

Scientists often use inductive reasoning. For instance, Galileo Galilei (1564–1642) used inductive reasoning to discover that the time required for a pendulum to complete one swing, called the *period* of the pendulum, depends on the length of the pendulum. Galileo did not have a clock, so he measured the periods of pendulums in "heartbeats." The following table shows some results obtained for pendulums of various lengths. For the sake of convenience, a length of 10 inches has been designated as 1 unit.

Length of pendulum, in units	Period of pendulum, in heartbeats
1	1
4	2
9	3
16	4

The period of a pendulum is the time it takes for the pendulum to swing from left to right and back to its original position.

EXAMPLE 3 ■ Use Inductive Reasoning to Solve an Application

Use the data in the table and inductive reasoning to answer each of the following.

a. If a pendulum has a length of 25 units, what is its period?

b. If the length of a pendulum is quadrupled, what happens to its period?

Solution

a. In the table on the previous page, each pendulum has a period that is the square root of its length. Thus we conjecture that a pendulum with a length of 25 units will have a period of 5 heartbeats.

b. In the table, a pendulum with a length of 4 units has a period that is twice that of a pendulum with a length of 1 unit. A pendulum with a length of 16 units has a period that is twice that of a pendulum with a length of 4 units. It appears that quadrupling the length of a pendulum doubles its period.

CHECK YOUR PROGRESS 3   A tsunami is a sea wave produced by an underwater earthquake. The velocity of a tsunami as it approaches land depends on the height of the tsunami. Use the table at the left and inductive reasoning to answer each of the following questions.

a. What happens to the height of a tsunami when its velocity is doubled?

b. What should be the height of a tsunami if its velocity is 30 feet per second?

Solution   See page S1.

Height of tsunami, in feet	Velocity of tsunami, in feet per second
4	6
9	9
16	12
25	15
36	18
49	21
64	24

Conclusions based on inductive reasoning may be incorrect. As an illustration, consider the circles shown below.

INSTRUCTOR NOTE
To produce the maximum number of regions, the points on the circle must be placed so that no three lines intersect at a single point.

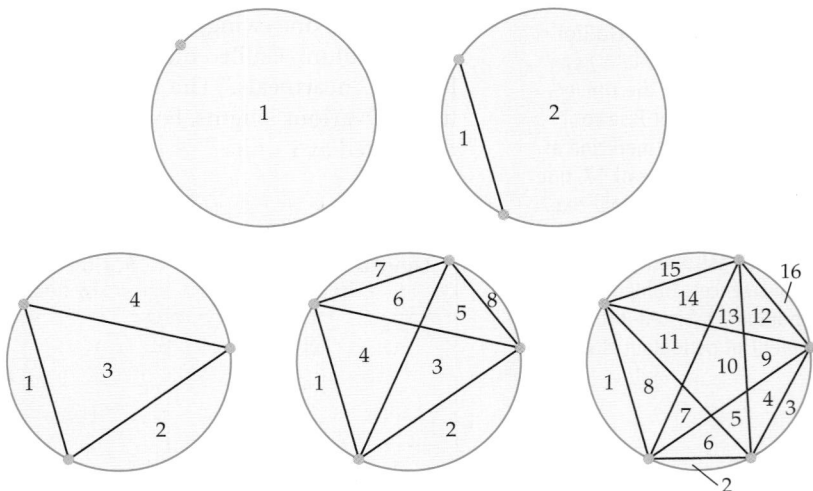

The maximum numbers of regions formed by connecting points on a circle.

For each circle, count the number of regions formed by the chords that connect the points on the circle. Your results should agree with the results in the following table.

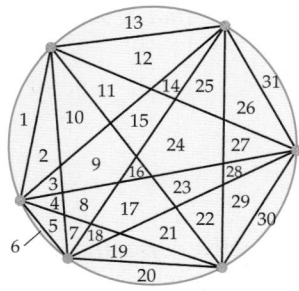

The line segments connecting six points on a circle yield a maximum of 31 regions.

Number of points	1	2	3	4	5	6
Maximum number of regions	1	2	4	8	16	?

There appears to be a pattern. Each additional point seems to double the number of regions. Guess the maximum number of regions you expect for a circle with six points. Check your guess by counting the maximum number of regions formed by the chords that connect six points on a *large* circle. Your drawing will show that for six points, the maximum number of regions is 31 (see the figure at the left), not 32 as you may have guessed. With seven points the maximum number of regions is 57. This is a good example to keep in mind. Just because a pattern holds true for a few cases, it does not mean the pattern will continue. When you use inductive reasoning you have no guarantee that your conclusion is correct.

Counterexamples

A statement is a true statement if and only if it is true in all cases. If you can find *one case* for which a statement is not true, called a **counterexample,** then the statement is a false statement. In Example 4 we verify that each statement is a false statement by finding a counterexample for each.

EXAMPLE 4 ■ Find a Counterexample

Verify that each of the following statements is a false statement by finding a counterexample.
For all x:

a. $|x| > 0$ **b.** $x^2 > x$ **c.** $\sqrt{x^2} = x$

Solution
A statement may have many counterexamples, but we need only find one counterexample to verify that the statement is false.

a. Let $x = 0$. Then $|0| = 0$. Because 0 is not greater than 0, we have found a counterexample. Thus for all x, $|x| > 0$ is a false statement.

b. For $x = 1$ we have $1^2 = 1$. Since 1 is not greater than 1, we have found a counterexample. Thus for all x, $x^2 > x$ is a false statement.

c. Consider $x = -3$. Then $\sqrt{(-3)^2} = \sqrt{9} = 3$. Since 3 is not equal to -3, we have found a counterexample. Thus for all x, $\sqrt{x^2} = x$ is a false statement.

CHECK YOUR PROGRESS 4   Verify that each of the following statements is a false statement by finding a counterexample for each.
For all x:

a. $\dfrac{x}{x} = 1$ **b.** $\dfrac{x+3}{3} = x + 1$ **c.** $\sqrt{x^2 + 16} = x + 4$

Solution   *See page S1.*

QUESTION *How many counterexamples are needed to prove that a statement is false?*

Deductive Reasoning

Another type of reasoning is called *deductive reasoning*. Deductive reasoning is distinguished from inductive reasoning in that it is the process of reaching a conclusion by applying general principles and procedures.

> **Deductive reasoning** is the process of reaching a conclusion by applying general assumptions, procedures, or principles.

EXAMPLE 5 ■ Use Deductive Reasoning to Establish a Conjecture

Use deductive reasoning to show that the following procedure produces a number that is four times the original number.

Procedure: Pick a number. Multiply the number by 8, add 6 to the product, divide the sum by 2, and subtract 3. *Note:* This is the same procedure as defined in Example 2.

Solution

Let n represent the original number.

Multiply the number by 8: $8n$

Add 6 to the product: $8n + 6$

Divide the sum by 2: $\dfrac{8n + 6}{2} = 4n + 3$

Subtract 3: $4n + 3 - 3 = 4n$

We started with n and ended with $4n$. The procedure given in this example produces a number that is four times the original number.

CHECK YOUR PROGRESS 5 Use deductive reasoning to show that the following procedure produces a number that is three times the original number.

Procedure: Pick a number. Multiply the number by 6, add 10 to the product, divide the sum by 2, and subtract 5. *Hint:* Let n represent the original number.

Solution See page S1.

ANSWER   One

Math Matters   The Game of MYST™ and Inductive Reasoning

Most games require that the players use a combination of deductive and inductive reasoning. However, the computer adventure game MYST™ has no specific rules and thus the player's only option is to explore and make use of inductive reasoning.

> Imagine your mind as a blank slate, like the pages of this journal. You must let Myst become your world. The land will offer up the answers you seek, if only you have eyes to see, ears to hear ... and wits to remember.[2]

Inductive Reasoning versus Deductive Reasoning

In Example 6 we analyze arguments to determine whether inductive or deductive reasoning is used.

EXAMPLE 6 ■ Determine Types of Reasoning

Determine whether each of the following arguments is an example of inductive reasoning or deductive reasoning.

a. During the past 10 years a tree has produced plums every other year. Last year the tree did not produce plums, so this year the tree will produce plums.

b. All home improvements cost more than the estimate. The contractor estimated my home improvement will cost $35,000. Thus my home improvement will cost more than $35,000.

Solution

a. This argument reaches a conclusion based on specific examples, so it is an example of inductive reasoning.

b. Because the conclusion is a specific case of a general assumption, this argument is an example of deductive reasoning.

CHECK YOUR PROGRESS 6 Determine whether each of the following arguments is an example of inductive reasoning or deductive reasoning.

a. All Danielle Steel novels are worth reading. The novel *A Perfect Stranger* is a Danielle Steel novel. Thus *A Perfect Stranger* is worth reading.

b. I know I will win a jackpot on this slot machine in the next 10 tries, because it has not paid out any money during the last 45 tries.

Solution   See page S1.

Excursion

A spot with 3 lives.

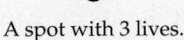

A spot with 2 lives.

A spot with 1 life.

A dead spot.

The status of a spot in the game of *Sprouts*.

The Game of Sprouts by John H. Conway

The mathematician John H. Conway has created several games that are easy to play but complex enough to provide plenty of mental stimulation. In 1967, Conway, along with Michael Paterson, created the two-person, paper-and-pencil game of *Sprouts*. After more than 30 years, the game of *Sprouts* has not been completely analyzed. Here are the rules for *Sprouts*.

Rules for Sprouts

1. A few spots (dots) are drawn on a piece of paper.

2. Players alternate turns. A turn consists of drawing a curve between two spots or drawing a curve that starts at a spot and ends at the same spot. The active player then places a new spot on the new curve. No curve may pass through a previously drawn spot. No curve may cross itself or a previously drawn curve.

3. A spot with no rays emanating from it has three lives. A spot with one ray emanating from it has two lives. A spot with two rays emanating from it has one life. A spot is dead and cannot be used when it has three rays emanating from it. See the figure at the left.

4. The winner is the last player who is able to draw a curve.

If *Sprouts* is played with just one spot (called *1-Spot Sprouts*), then the second player always wins. See the following analysis of *1-Spot Sprouts*.

(continued)

An Analysis of 1-Spot Sprouts

Curve drawn by the first player.   Curve drawn by the second player.

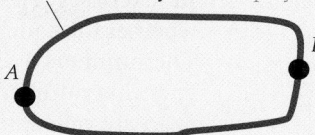   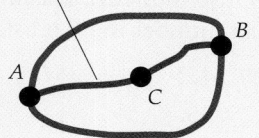

A game of *1-Spot Sprouts* starts with a single spot with three lives.

The first player has no options. The only possibility is to draw a curve from spot *A* back to spot *A*. After the curve is drawn, the first player completes the turn by placing another spot (spot *B*) on the curve. The game can continue because two lives remain.

The second player must now draw a curve between spot *A* and spot *B*. The curve can be drawn inside the previous curve as shown or outside the curve. In either case spot *C* is placed on the new curve. Only one life remains. The game is complete and the second player is the winner because the first player cannot draw another curve.

The game of *2-Spot Sprouts* (which starts with two spots) is more exciting than *1-Spot Sprouts* because there are several different games that can develop, depending on what move the first player makes and how the second player responds.

Excursion Exercises

1. With a partner, play a few games of *2-Spot Sprouts*. Alternate so that each player gets a chance to be the first player in several games.

   a. In a *2-Spot Sprouts* game, which player can guarantee a win?

   b. Did you use inductive or deductive reasoning to answer question 1a?

2. With a partner, play a few games of *3-Spot Sprouts*. Alternate so that each player gets a chance to be the first player in several games.

   a. In a *3-Spot Sprouts* game, which player can guarantee a win?

   b. Did you use inductive or deductive reasoning to answer question 2a?

3. Every *n-Spot Sprouts* game has a maximum of $3n - 1$ turns. To verify this, we use the following deductive argument.

   At the start of a game each spot has three lives. Thus an *n-Spot Sprouts* game starts with $3n$ lives. Each time a turn is completed, two lives are killed and one life is created. Thus each turn decreases the number of lives by one. Since each turn requires a curve to be drawn between two spots, the game cannot continue when only one life remains. Thus the maximum number of turns is $3n - 1$.

   It can also be shown that an *n-Spot Sprouts* game must have at least $2n$ turns. Play a few *4-Spot Sprouts* games. Did each *4-Spot Sprouts* game you played have at least $2(4) = 8$ turns and at most $3(4) - 1 = 11$ turns?

Exercise Set 1.1 (Suggested Assignment: 1–47 odds)

In Exercises 1–10, use inductive reasoning to predict the most probable next number in each list.

1. 4, 8, 12, 16, 20, 24, ?
2. 5, 11, 17, 23, 29, 35, ?
3. 3, 5, 9, 15, 23, 33, ?
4. 1, 8, 27, 64, 125, ?
5. 1, 4, 9, 16, 25, 36, 49, ?
6. 80, 70, 61, 53, 46, 40, ?
7. $\dfrac{3}{5}, \dfrac{5}{7}, \dfrac{7}{9}, \dfrac{9}{11}, \dfrac{11}{13}, \dfrac{13}{15}, ?$
8. $\dfrac{1}{2}, \dfrac{2}{3}, \dfrac{3}{4}, \dfrac{4}{5}, \dfrac{5}{6}, \dfrac{6}{7}, ?$
9. 2, 7, −3, 2, −8, −3, −13, −8, −18, ?
10. 1, 5, 12, 22, 35, ?

In Exercises 11–16, use inductive reasoning to decide whether the conclusion for each argument is correct. Note: The numbers 1, 2, 3, 4, ... are called natural numbers or counting numbers. The numbers ..., −3, −2, −1, 0, 1, 2, 3, ... are called integers.

11. The sum of any two even numbers is an even number.
12. If a number with three or more digits is divisible by 4, then the last two digits of the number are divisible by 4.
13. The product of an odd integer and an even integer is always an even number.
14. The cube of an odd integer is always an odd number.
15. Pick any counting number. Multiply the number by 6. Add 8 to the product. Divide the sum by 2. Subtract 4 from the quotient. The resulting number is twice the original number.
16. Pick any counting number. Multiply the number by 8. Subtract 4 from the product. Divide the difference by 4. Add 1 to the quotient. The resulting number is twice the original number.

Galileo used inclines similar to the one shown below to measure the distance balls of various weights would travel in equal time intervals. The conclusions that Galileo reached from these experiments were contrary to the prevailing Aristotelian theories on the subject, and he lost his post at the University of Pisa because of them.

An experiment with an incline and three balls produced the following results. The three balls are each the same size; however, ball A has a mass of 20 grams, ball B has a mass of 40 grams, and ball C has a mass of 80 grams.

Time, in seconds	Distance traveled, in inches		
	Ball A (20 grams)	Ball B (40 grams)	Ball C (80 grams)
1	6	6	6
2	24	24	24
3	54	54	54
4	96	96	96

In Exercises 17–24, use the above table and inductive reasoning to answer each question.

17. If the weight of a ball is doubled, what effect does this have on the distance it rolls in a given time interval?
18. If the weight of a ball is quadrupled, what effect does this have on the distance it rolls in a given time interval?
19. How far will ball A travel in 5 seconds?
20. How far will ball A travel in 6 seconds?
21. If a particular time is doubled, what effect does this have on the distance a ball travels?
22. If a particular time is tripled, what effect does this have on the distance a ball travels?

23. How much time is required for one of the balls to travel 1.5 inches?

24. How far will one of the balls travel in 1.5 seconds?

In Exercises 25–32, state whether the argument is an example of inductive reasoning or deductive reasoning.

25. I really liked Tom Clancy's novel *Executive Orders*, so I know I will like his next novel.

26. All pentagons have exactly five sides. Figure A is a pentagon. Therefore, Figure A has exactly five sides.

27. Every English setter likes to hunt. Duke is an English setter, so Duke likes to hunt.

28. Cats don't eat tomatoes. Scat is a cat. Therefore, Scat does not eat tomatoes.

29. A number is a "neat" number if the sum of the cubes of its digits equals the number. Therefore, 153 is a "neat" number.

30. The Atlanta Braves have won five games in a row. Therefore, the Atlanta Braves will win their next game.

31. Since

$$11 \times (1)(101) = 1111$$
$$11 \times (2)(101) = 2222$$
$$11 \times (3)(101) = 3333$$
$$11 \times (4)(101) = 4444$$
$$11 \times (5)(101) = 5555$$

we know that the product of 11 and a multiple of 101 is a number in which every digit is the same.

32. The following equations show that $n^2 - n + 11$ is a prime number for all counting numbers $n = 1, 2, 3, 4, \ldots$.

$$(1)^2 - 1 + 11 = 11 \qquad n = 1$$
$$(2)^2 - 2 + 11 = 13 \qquad n = 2$$
$$(3)^2 - 3 + 11 = 17 \qquad n = 3$$
$$(4)^2 - 4 + 11 = 23 \qquad n = 4$$

Note: A **prime number** is a counting number greater than 1 that has no counting number factors other than itself and 1. The first 10 prime numbers are 2, 3, 5, 7, 11, 13, 17, 19, 23, and 29.

In Exercises 33–42, find a counterexample to show that the statement is false.

33. $x > \dfrac{1}{x}$

34. $x + x > x$

35. $x^3 \geq x$

36. $|x + y| = |x| + |y|$

37. $-x < x$

38. $\dfrac{(x + 1)(x - 1)}{(x - 1)} = x + 1$ *Hint:* Division by zero is undefined.

39. If the sum of two natural numbers is even, then the product of the two natural numbers is even.

40. If the product of two natural numbers is even, then both of the numbers are even numbers.

41. Pick any three-digit counting number. Reverse the digits of the original number. The difference of these two numbers has a tens digit of 9.

42. If a counting number with two or more digits remains the same with its digits reversed, then the counting number is a multiple of 11.

43. Use deductive reasoning to show that the following procedure always produces a number that is equal to the original number.
Procedure: Pick a number. Multiply the number by 6 and add 8. Divide the sum by 2, subtract twice the original number, and subtract 4.

44. Use deductive reasoning to show that the following procedure always produces the number 5.
Procedure: Pick a number. Add 4 to the number and multiply the sum by 3. Subtract 7 and then decrease this difference by the triple of the original number.

Extensions
CRITICAL THINKING

45. Use inductive reasoning to predict the next letter in the following list.

O, T, T, F, F, S, S, E, …

Hint: Look for a pattern that involves letters from words used for counting.

46. Use inductive reasoning to predict the next symbol in the following list.

Ⴌ, ♄, 8, Ⴄ, ౙ, …

Hint: Look for a pattern that involves counting numbers and symmetry about a line.

47. For the World's Fair in 1850, the French physicist Jean Bernard Foucault (*foo-ko*) installed a pendulum in the Pantheon in Paris. Foucault's pendulum had a period of about 16.4 seconds. If a pendulum with a length of 0.25 meter has a period of 1 second, determine which of the following lengths best approximates the length of Foucault's pendulum. *Hint:* Use the results of Example 3b.

  a. 7 meters   **b.** 27 meters

  c. 47 meters   **d.** 67 meters

48. Find a counterexample to prove that the inductive argument in

  a. Exercise 31 is incorrect.

  b. Exercise 32 is incorrect.

EXPLORATIONS

49. When Galileo was teaching in Padua, he used inductive reasoning to discover some physical laws that govern the time of descent for a ball that rolls on a *cycloidial* path from point A to point B and for a ball that rolls on a straight incline from point A to point B, as shown in the photograph at the right. Do an Internet search for information about Galileo's experiment with a cycloid and write a short report on what Galileo discovered.

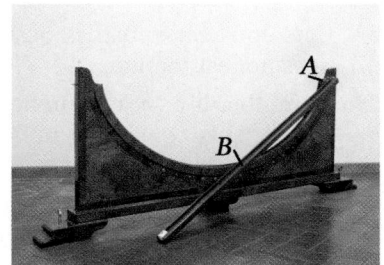

50. You can use the Internet to perform an experiment to determine the best strategy for playing the Monte Hall problem, which was stated in the Chapter 1 opener on page 3. Here is the procedure.

  a. Use a search engine to find a web site that provides a simulation of the Monte Hall problem. This problem is also known as the three-door problem, so search under both of these titles. Once you locate a site that provides a simulation, play the simulation 30 times using the strategy of not switching. Record the number of times you win the grand prize. Now play the simulation 30 times using the strategy of switching. How many times did you win the grand prize by not switching? How many times did you win the grand prize by switching?

  b. On the basis of this experiment, which strategy seems to be the best strategy for winning the grand prize? What type of reasoning have you used?

SECTION 1.2 | ## Problem Solving with Patterns

Terms of a Sequence

An ordered list of numbers such as

    5, 14, 27, 44, 65, . . .

is called a **sequence.** The numbers in a sequence that are separated by commas are the **terms** of the sequence. In the above sequence, 5 is the first term, 14 is the second term, 27 is the third term, 44 is the fourth term, and 65 is the fifth term. The three dots "…" indicate that the sequence continues beyond 65, which is the last written term. It is customary to use the subscript notation a_n to designate the **nth term of a sequence.** That is,

a_1 represents the first term of a sequence.

a_2 represents the second term of a sequence.

a_3 represents the third term of a sequence.

.
.
.

a_n represents the nth term of a sequence.

In the sequence 2, 6, 12, 20, 30, ..., $n^2 + n$, ...

$a_1 = 2$, $a_2 = 6$, $a_3 = 12$, $a_4 = 20$, $a_5 = 30$, and $a_n = n^2 + n$.

When we examine a sequence, it is natural to ask:

- What is the next term?
- What formula or rule can be used to generate the terms?

To answer these questions we often construct a **difference table,** which shows the differences between successive terms of the sequence. The following table is a difference table for the sequence 2, 5, 8, 11, 14, ...

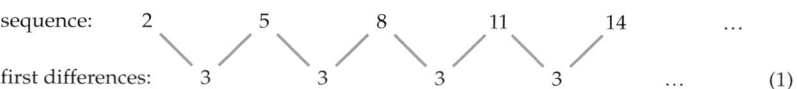

Each of the numbers in row (1) of the table is the difference between the two closest numbers just above it (upper right number minus upper left number). The differences in row (1) are called the **first differences** of the sequence. In this case the first differences are all the same. Thus, if we use the above difference table to predict the next number in the sequence, we predict that $14 + 3 = 17$ is the next term of the sequence. This prediction might be wrong; however, the pattern shown by the first differences seems to indicate that each successive term is 3 larger than the preceding term.

The following table is a difference table for the sequence 5, 14, 27, 44, 65, ...

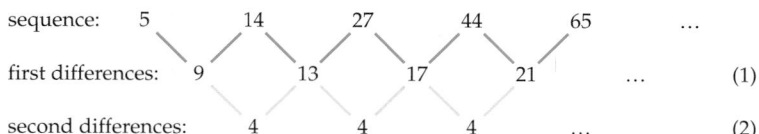

In this table the first differences are *not* all the same. In such a situation it is often helpful to compute the successive differences of the first differences. These are shown in row (2). These differences of the first differences are called the **second differences.** The differences of the second differences are called the **third differences.**

To predict the next term of a sequence, we often look for a pattern in a row of differences. For instance, in the following table, the second differences shown in blue are all the same constant, namely 4. If the pattern continues, then a 4 would also be the next second difference, and we can extend the table to the right as shown.

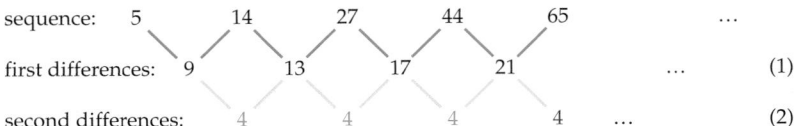

Now we work upward. That is, we add 4 to the first difference 21 to produce the next first difference, 25. We then add this difference to the fifth term, 65, to predict that 90 is the next term in the sequence. This process can be repeated to predict additional terms of the sequence.

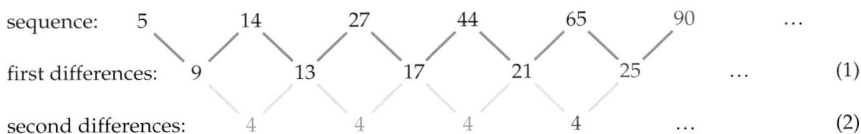

EXAMPLE 1 ■ Predict the Next Term of a Sequence

Use a difference table to predict the next term in the sequence.

2, 7, 24, 59, 118, 207, ...

Solution
Construct a difference table as shown below.

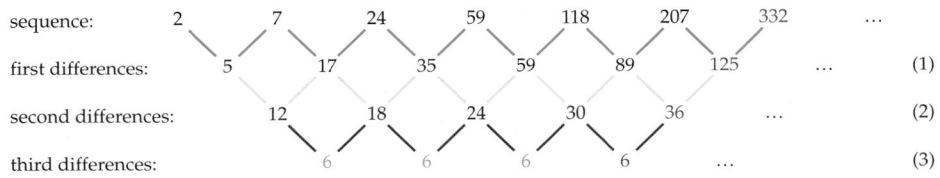

The third differences, shown in blue, are all the same constant, 6. Extending this row so that it includes an additional 6 enables us to predict that the next second difference will be 36. Adding 36 to the first difference 89 gives us the next first difference, 125. Adding 125 to the sixth term 207 yields 332. Using the method of extending the difference table, we predict that 332 is the next term in the sequence.

CHECK YOUR PROGRESS 1   Use a difference table to predict the next term in the sequence.

1, 14, 51, 124, 245, 426, ...

Solution   See page S1.

QUESTION   *Must the fifth term of the sequence 2, 4, 6, 8, ... be 10?*

ANSWER   *No. The fifth term could be any number. However, if you used the method shown in Example 1, then you would predict that the fifth term is 10.*

*n*th Term Formula for a Sequence

In Example 1 we used a difference table to predict the next term of a sequence. In some cases we can use patterns to predict a formula, called an ***n*th term formula,** that generates the terms of a sequence. As an example, consider the formula $a_n = 3n^2 + n$. This formula defines a sequence and provides a method for finding any term of the sequence. For instance, if we replace n with 1, 2, 3, 4, 5, and 6, then the formula $a_n = 3n^2 + n$ generates the sequence 4, 14, 30, 52, 80, 114. To find the 40th term, replace each n with 40.

$$a_{40} = 3(40)^2 + 40 = 4840$$

In Example 2 we make use of patterns to determine an *n*th term formula for a sequence given by geometric figures.

EXAMPLE 2 ■ Find an *n*th Term Formula

Assume the pattern shown by the square tiles in the following figures continues.

a. What is the *n*th term formula for the number of tiles in the *n*th figure of the sequence?

b. How many tiles are in the eighth figure of the sequence?

c. Which figure will consist of exactly 320 tiles?

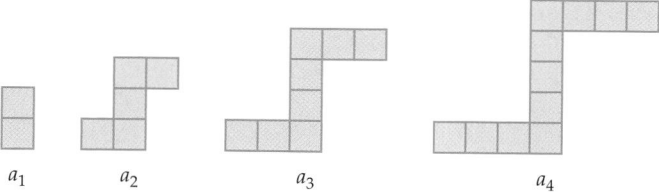

a_1 a_2 a_3 a_4

Solution

a. Examine the figures for patterns. Note that the second figure has two tiles on each of the horizontal sections and one tile between the horizontal sections. The third figure has three tiles on each horizontal section and two tiles between the horizontal sections. The fourth figure has four tiles on each horizontal section and three tiles between the horizontal sections.

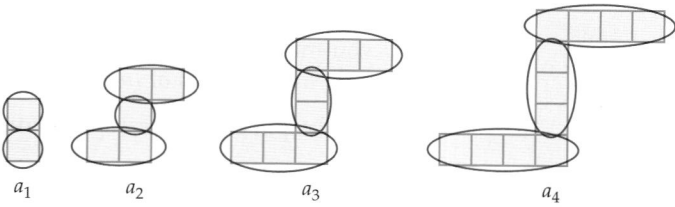

a_1 a_2 a_3 a_4

Thus the number of tiles in the *n*th figure is given by two groups of *n* plus a group of *n* less one. That is,

$$a_n = 2n + (n - 1)$$
$$a_n = 3n - 1$$

TAKE NOTE

Recall that the numbers 1, 2, 3, 4, ... are called *natural numbers* or *counting numbers.* We will often use the letter *n* to represent an arbitrary natural number.

TAKE NOTE

The method of grouping used in Example 2a is not unique. Do you see a different way to group the tiles? Does your method of grouping still produce the *n*th term formula $a_n = 3n - 1$?

b. The number of tiles in the eighth figure of the sequence is $3(8) - 1 = 23$.

c. To determine which figure in the sequence will have 320 tiles, we solve the equation $3n - 1 = 320$.

$$3n - 1 = 320$$
$$3n = 321$$
$$n = 107$$

The 107th figure is composed of 320 tiles.

CHECK YOUR PROGRESS 2   Assume the pattern shown by the square tiles in the following figure continues.

a. What is the nth term formula for the number of tiles in the nth figure of the sequence?

b. How many tiles are in the tenth figure of the sequence?

c. Which figure will consist of exactly 419 tiles?

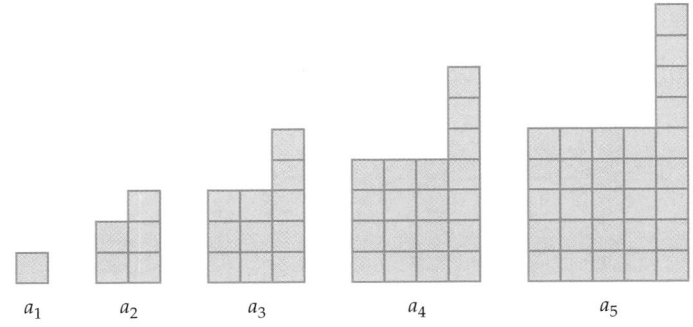

a_1 a_2 a_3 a_4 a_5

Solution   *See page S1.*

Math Matters   **Sequences on the Internet**

If you find it difficult to determine how the terms of a sequence are being generated, you might be able to find a solution on the Internet. One resource is *Sloane's On-Line Encyclopedia of Integer Sequences*© at:

http://www.research.att.com/~njas/sequences/

Here are two sequences from Sloane's web site.

- 1, 2, 3, 5, 9, 12, 21, 22, 23, 25, 29, 31, 32, 33, 35, 39, 41, 42, 43, 45, 49, 51, 52, ...
  nth term formula: *The natural numbers whose names, in English, end with vowels.*

- 5, 5, 5, 3, 4, 4, 4, 2, 5, 5, 5, 3, 6, 6, 6, 5, 10, 10, 10, 8, ...
  nth term formula: *Beethoven's Fifth Symphony;* 1 stands for the first note in the C minor scale, and so on.

The Fibonacci Sequence

Leonardo of Pisa, also known as Fibonacci (fē′bə-nä′chē) (c. 1170–1250), is one of the best known mathematicians of medieval Europe. In 1202, after a trip that took him to several Arab and Eastern countries, Fibonacci wrote the book *Liber Abaci*. In this book Fibonacci explained why the Hindu-Arabic numeration system that he had learned about during his travels was a more sophisticated and efficient system than the Roman numeration system. This book also contains a problem created by Fibonacci that concerns the birth rate of rabbits. Here is a statement of Fibonacci's rabbit problem.

Fibonacci

> At the beginning of a month, you are given a pair of newborn rabbits. After a month the rabbits have produced no offspring; however, every month thereafter, the pair of rabbits produces another pair of rabbits. The offspring reproduce in exactly the same manner. If none of the rabbits dies, how many pairs of rabbits will there be at the start of each succeeding month?

The solution of this problem is a sequence of numbers that we now call the **Fibonacci sequence.** The following figure shows the numbers of pairs of rabbits for the first 5 months. The larger rabbits represent mature rabbits that produce another pair of rabbits each month. The numbers in the blue region—1, 1, 2, 3, 5, 8—are the first six terms of the Fibonacci sequence.

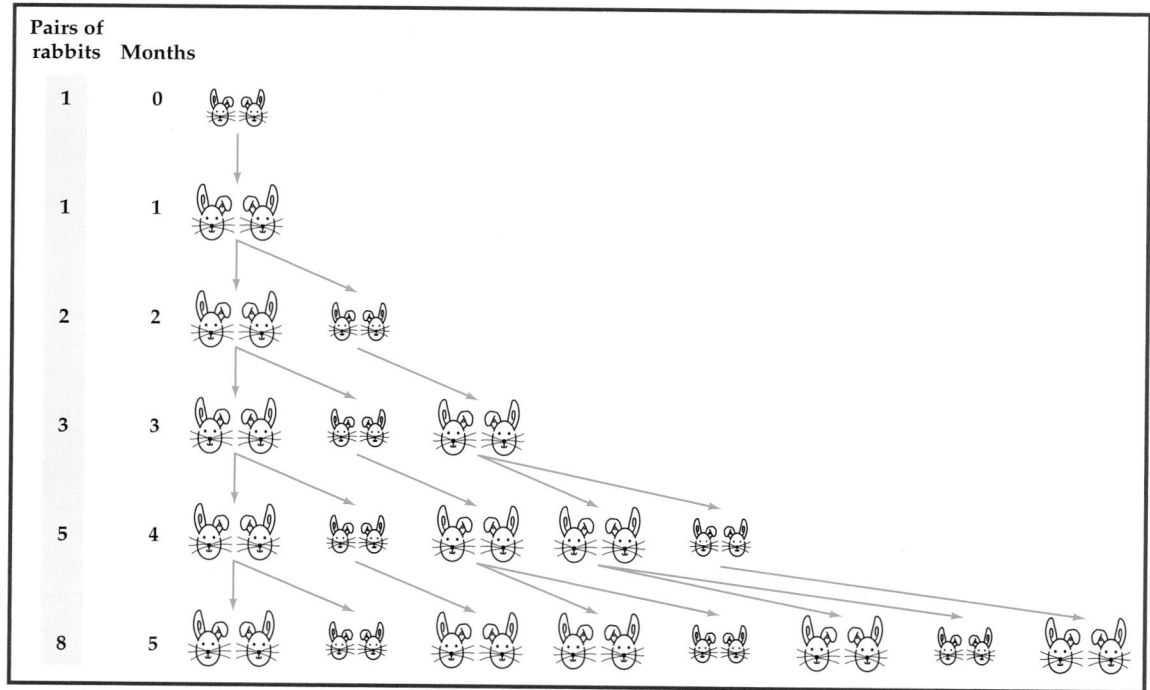

Fibonacci discovered that the number of pairs of rabbits for any month after the first two months can be determined by adding the numbers of pairs of rabbits in each of the *two previous* months. For instance, the number of pairs after 5 months is $3 + 5 = 8$.

INSTRUCTOR NOTE
To determine whether your
students understand the concept
of a recursive definition, ask them
to find a_3, a_4, and a_5 for the
sequence defined by $a_1 = 1$,
$a_2 = 2$, and

$$a_n = 3a_{n-1} + 2a_{n-2}$$

for $n \geq 3$.
They should find that $a_3 = 8$,
$a_4 = 28$, and $a_5 = 100$.

A **recursive definition** for a sequence is one in which each successive term of the sequence is defined by using some of the preceding terms. If we use the mathematical notation F_n to represent the nth Fibonacci number, then the numbers in the Fibonacci sequence are given by the following recursive definition.

The Fibonacci Numbers

$$F_1 = 1, \quad F_2 = 1, \quad \text{and} \quad F_n = F_{n-1} + F_{n-2} \quad \text{for } n \geq 3$$

EXAMPLE 3 ■ Find a Fibonacci Number

Use the definition of Fibonacci numbers to find the seventh and eighth Fibonacci numbers.

Solution
The first six Fibonacci numbers are 1, 1, 2, 3, 5, and 8. The seventh Fibonacci number is the sum of the two previous Fibonacci numbers. Thus,

$$F_7 = F_6 + F_5$$
$$= 8 + 5$$
$$= 13$$

The eighth Fibonacci number is

$$F_8 = F_7 + F_6$$
$$= 13 + 8$$
$$= 21$$

CHECK YOUR PROGRESS 3   Use the definition of Fibonacci numbers to find the ninth Fibonacci number.

Solution   See page S2.

▼ **point of interest**

A pineapple with eight spirals that rotate diagonally upward to the left and 13 spirals that rotate diagonally upward to the right. Many pine cones exhibit a pattern of five spirals in one direction and eight spirals in the other. The numbers 5 and 8 are also consecutive Fibonacci numbers.

The numbers of the Fibonacci sequence often occur in nature. For instance, pineapples have spirals formed by their hexagonal scales. The pineapple at the left has eight spirals that start at the bottom and rotate diagonally upward to the left, and 13 spirals that start at the bottom and rotate diagonally upward to the right. The numbers 8 and 13 are consecutive Fibonacci numbers.

We can find any term after the second term of the Fibonacci sequence by computing the sum of the previous two terms. However, this procedure of adding the previous two terms can be tedious. For instance, what is the 100th term or the 1000th term of the Fibonacci sequence? To find the 100th term we need to know the 98th and 99th terms. To find the 1000th term we need to know the 998th and 999th terms. Many mathematicians tried to find a nonrecursive nth term formula for the Fibonacci sequence without success, until a formula was discovered by Jacques Binet in 1843. Binet's formula is given in Exercise 23 of this section.

QUESTION   *What happens if you try to use a difference table to determine Fibonacci numbers?*

EXAMPLE 4 ■ Determine Properties of Fibonacci Numbers

Determine whether each of the following statements about Fibonacci numbers is true or false. *Note:* The first 10 terms of the Fibonacci sequence are 1, 1, 2, 3, 5, 8, 13, 21, 34, and 55.

a. F_{2n} is an odd number.

b. $2F_n - F_{n-2} = F_{n+1}$ for $n \geq 3$

Solution

a. The notation F_{2n} represents the second, fourth, sixth, eighth, ... Fibonacci numbers. An examination of Fibonacci numbers shows that the second Fibonacci number, 1, is odd and the fourth Fibonacci number, 3, is odd, but the sixth Fibonacci number, 8, is even. Thus the statement "F_{2n} is an odd number" is false.

b. Experiment to see whether $2F_n - F_{n-2} = F_{n+1}$ for several values of n. For instance, for $n = 7$, we get

$$2F_n - F_{n-2} = F_{n+1}$$
$$2F_7 - F_{7-2} = F_{7+1}$$
$$2F_7 - F_5 = F_8$$
$$2(13) - 5 = 21$$
$$26 - 5 = 21$$
$$21 = 21$$

which is true. Evaluating $2F_n - F_{n-2}$ for several additional values of n, $n \geq 3$, we find that in each case $2F_n - F_{n-2} = F_{n+1}$. Thus, by inductive reasoning, we conjecture $2F_n - F_{n-2} = F_{n+1}$ for $n \geq 3$ is a true statement. *Note:* This property of Fibonacci numbers can also be established using deductive reasoning. See Exercise 32 of this section.

> ✔ **TAKE NOTE**
>
> Pick any Fibonacci number larger than 1. The equation
>
> $$2F_n - F_{n-2} = F_{n+1}$$
>
> merely states that for numbers in the Fibonacci sequence
>
> 1, 1, 2, 3, 5, 8, 13, 21, ...
>
> the double of a Fibonacci number, F_n, less the Fibonacci number two to its left, is the Fibonacci number just to the right of F_n.

CHECK YOUR PROGRESS 4   Determine whether each of the following statements about Fibonacci numbers is true or false.

a. $2F_n > F_{n+1}$ for $n \geq 3$

b. $2F_n + 3 = F_{n+2}$

Solution   See page S2.

ANSWER   *The difference table for the numbers in the Fibonacci sequence does not contain a row of differences that are all the same constant.*

Excursion

Pythagoras
(c. 580 B.C. –520 B.C.)
The ancient Greek
philosopher and
mathematician
Pythagoras
(pĭ-thăg′ər-əs)
formed a secret brotherhood that
investigated topics in music, astronomy, philosophy, and mathematics. The Pythagoreans
believed that the nature of the
universe was directly related to
mathematics, and that whole
numbers and the ratios formed
by whole numbers could be used
to describe and represent all natural events.

The Pythagoreans were
particularly intrigued by the number 5 and the shape of a pentagon. They used the following
figure, which is a five-pointed
star inside of a regular pentagon,
as a secret symbol that could be
used to identify other members
of the brotherhood.

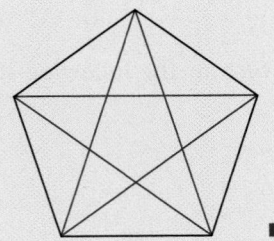

Polygonal Numbers

The ancient Greek mathematicians were interested in the geometric shapes associated
with numbers. For instance, they noticed that triangles can be constructed using 1, 3, 6,
10, or 15 dots, as shown in Figure 1.1. They called the numbers 1, 3, 6, 10, 15, . . . the
triangular numbers. The Greeks called the numbers 1, 4, 9, 16, 25, . . . the *square numbers* and the numbers 1, 5, 12, 22, 35, . . . the *pentagonal numbers*.

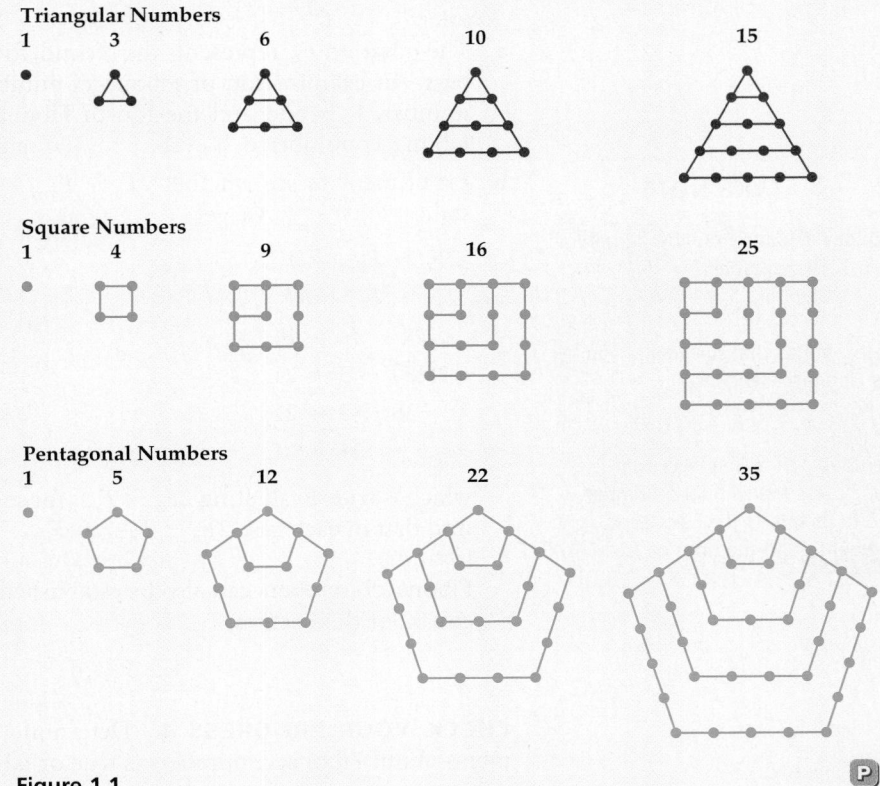

Figure 1.1

An *n*th term for the triangular numbers is:

$$Triangular_n = \frac{n(n+1)}{2}$$

The square numbers have an *n*th term formula of $Square_n = n^2$. The *n*th term formula for the pentagonal numbers is

$$Pentagonal_n = \frac{n(3n-1)}{2}$$

(continued)

Excursion Exercises

1. Extend Figure 1.1, page 22, by constructing drawings of the sixth triangular number, the sixth square number, and the sixth pentagonal number.

2. The figure below shows that the fourth triangular number, 10, added to the fifth triangular number, 15, produces the fifth square number, 25.

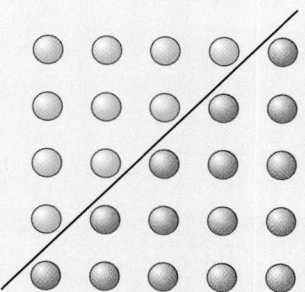

   a. Use a drawing to show that the fifth triangular number added to the sixth triangular number is the sixth square number.

   b. Verify that the 50th triangular number added to the 51st triangular number is the 51st square number. *Hint:* Use a numerical approach; don't use a drawing.

   c. Use nth term formulas to verify that the sum of the nth triangular number and the $(n + 1)$st triangular number is always the square number $(n + 1)^2$.

3. Construct a drawing of the fourth hexagonal number.

Exercise Set 1.2   (Suggested Assignment: 1–29 odds)

In Exercises 1–6, construct a difference table to predict the next term of each sequence.

1. $1, 7, 17, 31, 49, 71, \ldots$

2. $10, 10, 12, 16, 22, 30, \ldots$

3. $-1, 4, 21, 56, 115, 204, \ldots$

4. $0, 10, 24, 56, 112, 190, \ldots$

5. $9, 4, 3, 12, 37, 84, \ldots$

6. $17, 15, 25, 53, 105, 187, \ldots$

In Exercises 7–10, use the given nth term formula to compute the first five terms of the sequence.

7. $a_n = \dfrac{n(2n + 1)}{2}$

8. $a_n = \dfrac{n}{n + 1}$

9. $a_n = 5n^2 - 3n$

10. $a_n = 2n^3 - n^2$

In Exercises 11–14, determine the nth term formula for the number of square tiles in the nth figure.

11.

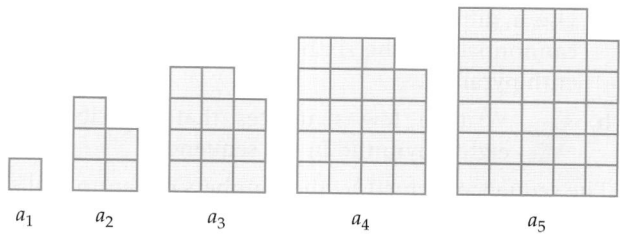

12.

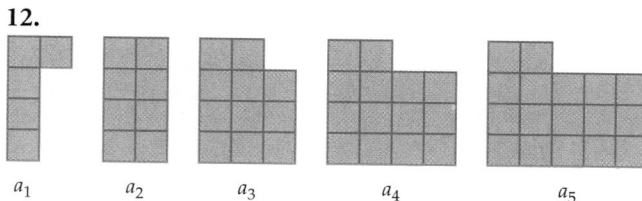

13.

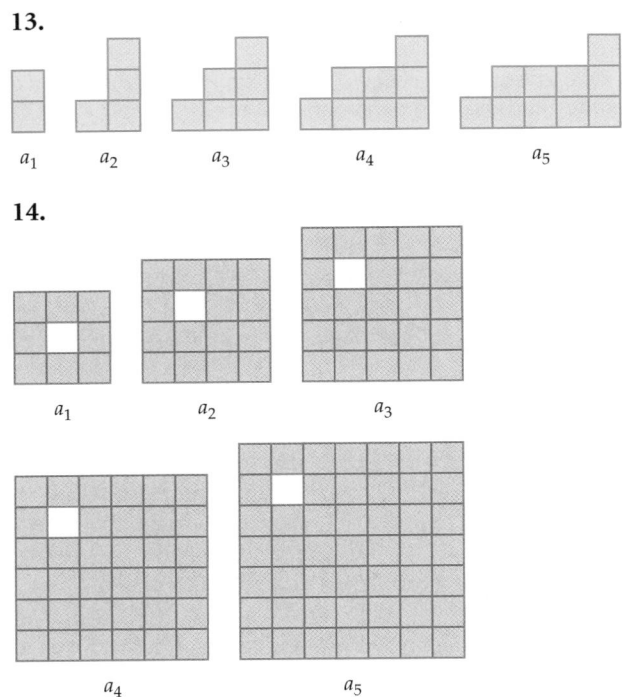

a_1 a_2 a_3 a_4 a_5

14.

a_1 a_2 a_3

a_4 a_5

Cannonballs can be stacked to form a pyramid with a triangular base. Five of these pyramids are shown below. Use these figures in Exercises 15 and 16.

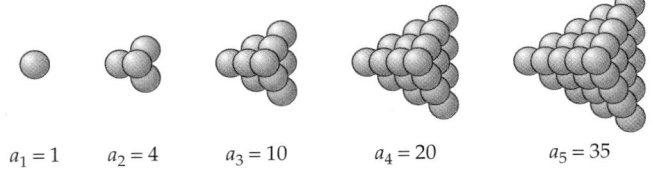

$a_1 = 1$ $a_2 = 4$ $a_3 = 10$ $a_4 = 20$ $a_5 = 35$

15. a. Use a difference table to predict the number of cannonballs in the sixth pyramid and in the seventh pyramid.

  b. Write a few sentences that describe the eighth pyramid in the sequence.

16. The sequence formed by the numbers of cannonballs in the above pyramids is called the *tetrahedral sequence.* The nth term formula for the tetrahedral sequence is

$$tet_n = \frac{1}{6}n(n + 1)(n + 2)$$

Find tet_{10}.

17. One cut of a stick of licorice produces two pieces. Two cuts produce three pieces. Three cuts produce four pieces.

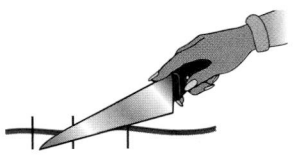

  a. How many pieces are produced by five cuts and by six cuts? Assume the cuts are made in the manner shown above.

  b. Predict the nth term formula for the number of pieces of licorice that are produced by n cuts, made in the manner shown above, of a stick of licorice.

18. One straight cut across a pizza produces two pieces. Two cuts can produce a maximum of four pieces. Three cuts can produce a maximum of seven pieces. Four cuts can produce a maximum of 11 pieces.

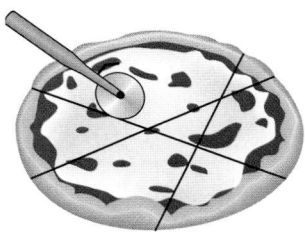

  a. Use a difference table to predict the maximum number of pieces that can be produced with seven cuts.

  b. How are the pizza-slicing numbers related to the triangular numbers, which are defined by

$$Triangular_n = \frac{n(n + 1)}{2}?$$

19. One straight cut through a thick piece of cheese produces two pieces. Two straight cuts can produce a maximum of four pieces. Three straight cuts can produce a maximum of eight pieces. You might be inclined to think that every additional cut doubles the previous number of pieces. However, for four straight cuts, you will find that you get a maximum of 15 pieces. An nth term formula for the maximum number of pieces, P, that can be produced by n straight cuts is

$$P(n) = \frac{n^3 + 5n + 6}{6}$$

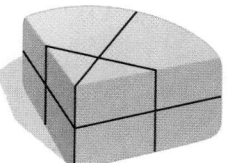

a. Use the formula at the bottom right of the previous page to determine the maximum number of pieces that can be produced by five straight cuts.

b. What is the fewest number of straight cuts that you can use if you wish to produce at least 60 pieces? *Hint:* Use the formula and experiment with larger and larger values of n.

20. The Fibonacci sequence has many unusual properties. Experiment to decide which of the following properties are valid. *Note:* F_n represents the nth Fibonacci number.

a. $3F_n - F_{n-2} = F_{n+2}$ for $n \geq 3$

b. $F_n F_{n+3} = F_{n+1} F_{n+2}$

c. F_{3n} is an even number.

d. $5F_n - 2F_{n-2} = F_{n+3}$ for $n \geq 3$

21. Find the third, fourth, and fifth terms of the sequence defined by $a_1 = 3$, $a_2 = 5$, and $a_n = 2a_{n-1} - a_{n-2}$ for $n \geq 3$.

22. Find the third, fourth, and fifth terms of the sequence defined by $a_1 = 2$, $a_2 = 3$, and $a_n = (-1)^n a_{n-1} + a_{n-2}$ for $n \geq 3$.

23. The following formula is known as *Binet's Formula* for the nth Fibonacci number.

$$F_n = \frac{1}{\sqrt{5}}\left[\left(\frac{1 + \sqrt{5}}{2}\right)^n - \left(\frac{1 - \sqrt{5}}{2}\right)^n\right]$$

The advantage of this formula over the recursive formula $F_n = F_{n-1} + F_{n-2}$ is that you can determine the nth Fibonacci number without finding the two preceding Fibonacci numbers.

Use Binet's Formula and a calculator to find the 20th, 30th, and 40th Fibonacci numbers.

24. Binet's Formula (see Exercise 23) can be simplified if you round your calculator results to the nearest integer. In the following formula, *nint* is an abbreviation for "the nearest integer of."

$$f_n = nint\left\{\frac{1}{\sqrt{5}}\left(\frac{1 + \sqrt{5}}{2}\right)^n\right\}$$

If you use $n = 8$ in the above formula, a calculator will show 21.0095194943 for the value inside the braces. Rounding this number to the nearest integer produces 21 as the eighth Fibonacci number.

Use the above form of Binet's Formula and a calculator to find the 16th, 21st, and 32nd Fibonacci numbers.

Extensions

CRITICAL THINKING

25. The ancient Greeks often discovered mathematical relationships by using geometric drawings. Study the accompanying drawing to determine what needs to be put in place of the question mark to make the equation a true statement.

$$1 + 3 + 5 + 7 + \cdots + (2n - 1) = ?$$

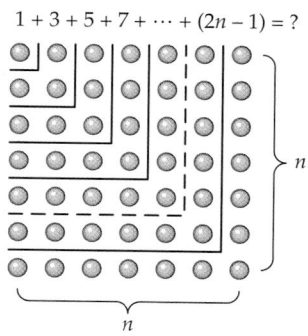

26. The nth term formula

$$a_n = \frac{n(n - 1)(n - 2)(n - 3)(n - 4)}{4 \cdot 3 \cdot 2 \cdot 1} + 2n$$

generates 2, 4, 6, 8, 15 for $n = 1, 2, 3, 4, 5$. Make minor changes to the above formula to produce an nth term formula that will generate the following finite sequences.

a. 2, 4, 6, 8, 20

b. 2, 4, 6, 8, 30

COOPERATIVE LEARNING

Near the end of the eighteenth century, the German astronomers J. Daniel Titus and John Ehlert Bode discovered a pattern concerning the distances of the planets from the sun. This pattern is now known as *Bode's rule*. To understand Bode's rule, examine the following table, which shows the *average* distances from the sun of planets Mercury, Venus, Earth, Mars, Jupiter, and Saturn. In the 1770s these were the only known planets. The distances are measured in astronomical units (AU), where 1 AU is the average distance of Earth from the sun. Bode noticed that he could approximate these distances by the following rule.

The sun and nine known planets of our solar system. The distances from the sun are not drawn to scale.

   Start with the numbers 0 and 0.3, and then continue to double 0.3 to produce the sequence 0, 0.3, 0.6, 1.2, 2.4, 4.8, 9.6, 19.2. To each of these numbers add 0.4. This produces Bode's numbers: 0.4, 0.7, 1.0, 1.6, 2.8, 5.2, 10.0, 19.6. Most astronomers did not pay much attention to Bode's numbers until Uranus was discovered in 1781. The actual distance from Uranus to the sun is 19.18 AU. The fact that this distance is close to Bode's eighth number inspired astronomers to search for a planet with an orbit between the orbits of Mars and Jupiter. No such planet has ever been discovered; however, in 1801, the asteroid Ceres, with a diameter of 480 miles, was discovered in this vicinity.

Planets	Mercury	Venus	Earth	Mars	?	Jupiter	Saturn	?	?	?
Actual distance from the sun (AU)	0.38	0.72	1.00	1.52	?	5.20	9.54	?	?	?
Bode's number	0.4	0.7	1.0	1.6	2.8	5.2	10.0	19.6	?	?

27. The next planet after Uranus is Neptune.

    a. Use Bode's ninth number to predict the average distance from the sun to Neptune.

    b. How does your answer in part a compare to Neptune's actual average distance from the sun, which is 30.60 AU? *Note:* Some people think that Bode's ninth number does not produce accurate results for Neptune because Pluto's orbit is such that Pluto is sometimes closer to the sun than Neptune.

28. The next planet after Neptune is Pluto. How does Bode's 10th number compare to Pluto's actual average distance from the sun of 39.52 AU?

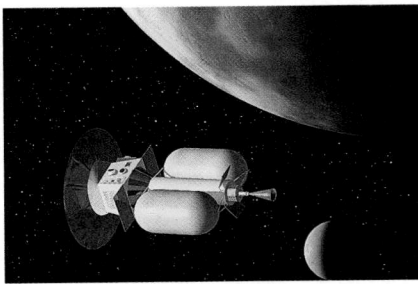

A drawing of Pluto, its satellite Charon, and the Pluto-Kuiper Express.

NASA has scheduled the Pluto-Kuiper Express for a robotic mission to Pluto in the year 2010.

29. As of 2002, no planet in our solar system has been discovered beyond Pluto. However, in 1992, astronomers discovered a vast population of small bodies that orbit the sun in a path that varies from as little as 30 AU up to 50 AU from the sun. These bodies are now known as the Kuiper Belt.

   a. According to Bode's 11th number, the next planet after Pluto would have what average distance from the sun?

   b. Is it likely that the bodies in the Kuiper Belt are the remnants of a planet beyond Pluto predicted by Bode's 11th number? Explain.

30. In 1995, three planets were discovered that orbit the pulsar known as PSR B1257+12. The planets have orbits that measure 0.19 AU, 0.36 AU, and 0.47 AU.

   a. If the largest orbit of 0.47 AU is thought of as 1 unit, then determine the sizes in units of the smaller orbits. (Round to the nearest 0.01.)

   b. How do the results in part a compare with Bode's first two numbers?

   c. On the basis of the results in part b, can we conclude that there is a universal mechanism for the placement of planets around stars, and that Bode's rule is an accurate model of this mechanism?

EXPLORATIONS

31. The *Tower of Hanoi* is a puzzle invented by Edouard Lucas in 1883. The puzzle consists of three pegs and a number of disks of distinct diameters stacked on one of the pegs such that the largest disk is on the bottom, the next largest is placed on the largest disk, and so on as shown by the figure in the next column.

   The object of the puzzle is to transfer the tower to one of the other pegs. The rules require that *only one disk be moved at a time* and that *a larger disk may not be placed on a smaller disk*. All pegs may be used.

Determine the *minimum* number of moves required to transfer all of the disks to another peg for each of the following situations.

   a. You start with only one disk.

   b. You start with two disks.

   c. You start with three disks. (*Note:* You can use a stack of various size coins to simulate the puzzle, or you can use one of the many web sites that provide a JavaScript simulation of the puzzle.)

   d. You start with four disks.

   e. You start with five disks.

   f. You start with n disks.

   g. Lucas included with the Tower puzzle a legend about a tower that had 64 gold disks on one of three diamond needles. A group of priests had the task of transferring the 64 disks to one of the other needles using the same rules as the Tower of Hanoi puzzle. When they had completed the transfer, the tower would crumble and the universe would cease to exist. Assuming that the priests could transfer one disk to another needle every second, how many years would it take them to transfer all of the 64 disks to one of the other needles?

32. Use the recursive definition for Fibonacci numbers and *deductive* reasoning to verify that, for Fibonacci numbers, $2F_n - F_{n-2} = F_{n+1}$ for $n \geq 3$. *Hint:* By definition $F_{n+1} = F_n + F_{n-1}$ and $F_n = F_{n-1} + F_{n-2}$.

Polya's Problem-Solving Strategy

historical note

George Polya
After a brief stay at Brown University, George Polya (pōl'yə) moved to Stanford University in 1942 and taught there until his retirement. While at Stanford, he published 10 books and a number of articles for mathematics journals. Of the books Polya published, *How to Solve it* (1945) is one of his best known. In this book, Polya outlines a strategy for solving problems from virtually any discipline.

"A great discovery solves a great problem but there is a grain of discovery in the solution of any problem. Your problem may be modest; but if it challenges your curiosity and brings into play your inventive faculties, and if you solve it by your own means, you may experience the tension and enjoy the triumph of discovery." ∎

Ancient mathematicians such as Euclid and Pappus were interested in solving mathematical problems, but they were also interested in *heuristics*, the study of the methods and rules of discovery and invention. In the seventeenth century, the mathematician and philosopher René Descartes (1596–1650) contributed to the field of heuristics. He tried to develop a universal problem-solving method. Although he did not achieve this goal, he did publish some of his ideas in *Rules for the Direction of the Mind* and his better known work *Discourse de la Methode*.

Another mathematician and philosopher, Gottfried Wilhelm Leibnitz (1646–1716), planned to write a book on heuristics titled *Art of Invention*. Of the problem-solving process, Leibnitz wrote: "Nothing is more important than to see the sources of invention which are, in my opinion, more interesting than the inventions themselves."

One of the foremost recent mathematicians to make a study of problem solving was George Polya (1887–1985). He was born in Hungary and moved to the United States in 1940. The basic problem-solving strategy that Polya advocated consisted of the following four steps.

Polya's Four-Step Problem-Solving Strategy

1. Understand the problem.
2. Devise a plan.
3. Carry out the plan.
4. Review the solution.

Polya's four steps are deceptively simple. To become a good problem solver, it helps to examine each of these steps and determine what is involved.

Understand the Problem   This part of Polya's four-step strategy is often overlooked. You must have a clear understanding of the problem. To help you focus on understanding the problem, consider the following questions.

- Can you restate the problem in your own words?
- Can you determine what is known about these types of problems?
- Is there missing information that, if known, would allow you to solve the problem?
- Is there extraneous information that is not needed to solve the problem?
- What is the goal?

Devise a Plan   Successful problem solvers use a variety of techniques when they attempt to solve a problem. Here are some frequently-used procedures.

- Make a list of the known information.
- Make a list of information that is needed.

- Draw a diagram.
- Make an organized list that shows all the possibilities.
- Make a table.
- Work backwards.
- Try to solve a similar but simpler problem.
- Look for a pattern.
- Write an equation. If necessary, define what each variable represents.
- Perform an experiment.
- Guess at a solution and then check your result.
- Use indirect reasoning.

Carry Out the Plan   Once you have devised a plan, you must carry it out.

- Work carefully.
- Keep an accurate and neat record of all your attempts.
- Realize that some of your initial plans will not work and that you may have to devise another plan or modify your existing plan.

Review the Solution   Once you have found a solution, check the solution.

- Ensure that the solution is consistent with the facts of the problem.
- Interpret the solution in the context of the problem.
- Ask yourself whether there are generalizations of the solution that could apply to other problems.

In Example 1 we apply Polya's four-step problem-solving strategy to solve a problem involving the number of routes between two points.

EXAMPLE 1 ■ Apply Polya's Strategy

Consider the map shown in Figure 1.2. Allison wishes to walk along the streets from point A to point B. How many direct routes can Allison take?

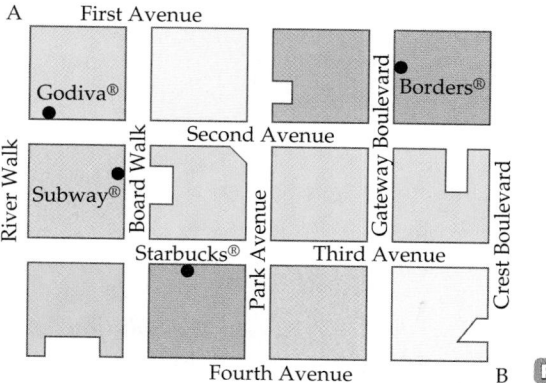

Figure 1.2 *City Map*

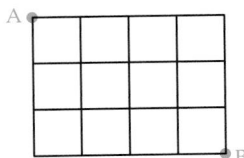

A simple diagram of the street map in Figure 1.2.

Solution

Understand the Problem   We would not be able to answer the question if Allison retraced her path or traveled away from point B. Thus we assume that on a direct route, she always travels along a street in a direction that gets her closer to point B.

Devise a Plan   The map in Figure 1.2 has many extraneous details. Thus we make a diagram that allows us to concentrate on the essential information. See the figure at the left.

   Because there are many routes, we consider the similar but simpler diagrams shown below. The number at each street intersection represents the number of routes from point A to that particular intersection.

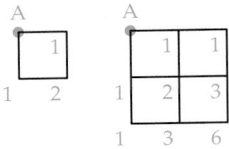

Simple Street Diagrams

Look for patterns. It appears that the number of routes to an intersection is the *sum* of the number of routes to the adjacent intersection to its left and the number of routes to the intersection directly above. For instance, the number of routes to the intersection labeled 6 is the sum of the number of routes to the intersection to its left, which is three, and the number of routes to the intersection directly above, which is also three.

Carry Out the Plan   Using the pattern discovered above, we see from the following figure that the number of routes from point A to point B is $20 + 15 = 35$.

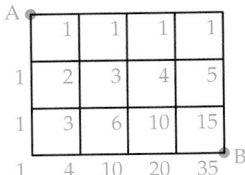

A street diagram with the number of routes to each intersection labeled.

Review the Solution   Ask yourself if a result of 35 seems reasonable. If you were required to draw each route, could you devise a scheme that would enable you to draw each route without missing a route or duplicating a route?

CHECK YOUR PROGRESS 1   Consider the street map in Figure 1.2. Allison wishes to walk directly from point A to point B. How many different routes can she take if she wants to go past Starbucks on Third Avenue?

Solution   See page S2.

Example 2 illustrates the technique of using an organized list.

EXAMPLE 2 ■ Apply Polya's Strategy

A baseball team won two out of their last four games. In how many different orders could they have two wins and two losses in four games?

Solution

Understand the Problem   There are many different orders. The team may have won two straight games and lost the last two (WWLL). Or maybe they lost the first two games and won the last two (LLWW). Of course there are other possibilities, such as (WLWL).

Devise a Plan   We will make an *organized list* of all the possible orders. An organized list is a list that is produced using a system that ensures that each of the different orders will be listed once and only once.

Carry Out the Plan   Each entry in our list must contain two W's and two L's. We will use a strategy that makes sure each order is considered, with no duplications. One such strategy is to always write a W unless doing so will produce too many W's or a duplicate of one of the previous orders. If it is not possible to write a W, then and only then do we write an L. This strategy produces the six different orders shown below.

1. WWLL (Start with two wins)
2. WLWL (Start with one win)
3. WLLW
4. LWWL (Start with one loss)
5. LWLW
6. LLWW (Start with two losses)

Review the Solution   We have made an organized list. The list has no duplicates and the list considers all possibilities, so we are confident that there are six different orders in which a baseball team can win two out of four games.

CHECK YOUR PROGRESS 2   A true-false quiz contains five questions. In how many ways can a student answer the questions if the student answers two of the questions with "false" and the other three with "true"?

Solution   See page S2.

INSTRUCTOR NOTE
Inform your students that the type of problem illustrated in Example 2 is a counting problem. Counting problems are covered in more detail in Chapter 11.

In Example 3 we make use of several problem solving strategies to solve a problem involving the total number of games to be played.

EXAMPLE 3 ■ Apply Polya's Strategy

In a basketball league consisting of 10 teams, each team plays each of the other teams exactly three times. How many league games will be played?

Solution

Understand the Problem   There are 10 teams in the league and each team plays exactly three games against each of the other teams. The problem is to determine the total number of league games that will be played.

Devise a Plan   Try the strategy of working a similar but simpler problem. Consider a league with only four teams (denoted by A, B, C, and D) in which each team plays each of the other teams only once. The diagram at the left illustrates that the games can be represented by line segments that connect the points A, B, C, and D.

    Since each of the four teams will play a game against each of the other three, we might conclude that this would result in $4 \cdot 3 = 12$ games. However, the diagram shows only six line segments. It appears that our procedure has counted each game twice. For instance, when team A plays team B, team B also plays team A. To produce the correct result, we must divide our previous result, 12, by 2. Hence, four teams can play each other once in $\frac{4 \cdot 3}{2} = 6$ games.

Carry Out the Plan   Using the process developed above, we see that 10 teams can play each other *once* in a total of $\frac{10 \cdot 9}{2} = 45$ games. Since each team plays each opponent exactly three times, the total number of games is $45 \cdot 3 = 135$.

Review the Solution   We could check our work by making a diagram that includes all 10 teams represented by dots labeled A, B, C, D, E, F, G, H, I, and J. Because this diagram would be somewhat complicated, let's try the method of making an organized list. The figure at the left shows an organized list in which the notation BC represents a game between team B and team C. The notation CB is not shown because it also represents a game between team B and team C. This list shows that 45 games are required for each team to play each of the other teams once. Also notice that the first row has nine items, the second row has eight items, the third row has seven items, and so on. Thus 10 teams require

$$9 + 8 + 7 + 6 + 5 + 4 + 3 + 2 + 1 = 45$$

games if each team plays every other team once and $45 \cdot 3 = 135$ games if each team plays exactly three games against each opponent.

CHECK YOUR PROGRESS 3   If six people greet each other at a meeting by shaking hands with one another, how many handshakes will take place?

Solution   See page S2.

In Example 4 we make use of a table to solve a problem.

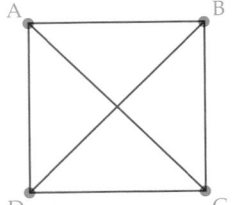

The possible pairings of a league with only four teams.

AB AC AD AE AF AG AH AI AJ
BC BD BE BF BG BH BI BJ
CD CE CF CG CH CI CJ
DE DF DG DH DI DJ
EF EG EH EI EJ
FG FH FI FJ
GH GI GJ
HI HJ
IJ

An organized list of all possible games.

EXAMPLE 4 ■ Apply Polya's Strategy

Determine the digit 100 places to the right of the decimal point in the decimal representation $\frac{7}{27}$.

Solution

Understand the Problem   Express the fraction $\frac{7}{27}$ as a decimal and look for a pattern that will enable us to determine the digit 100 places to the right of the decimal point.

Devise a Plan   Dividing 27 into 7 by long division or by using a calculator produces the decimal 0.259259259.... Since the decimal representation repeats the digits 259 over and over forever, we know that the digit located 100 places to the right of the decimal point is either a 2, a 5, or a 9. A table may help us to see a pattern and enable us to determine which one of these digits is in the 100th place. Since the decimal digits repeat every three digits, we use a table with three columns.

The first 15 decimal digits of $\dfrac{7}{27}$

Column 1 Location	Digit	Column 2 Location	Digit	Column 3 Location	Digit
1st	2	2nd	5	3rd	9
4th	2	5th	5	6th	9
7th	2	8th	5	9th	9
10th	2	11th	5	12th	9
13th	2	14th	5	15th	9
⋮		⋮		⋮	

Carry Out the Plan   Only in column 3 is each of the decimal digit *locations* evenly divisible by 3. From this pattern we can tell that the 99th decimal digit (because 99 is evenly divisible by 3) must be a 9. Since a 2 always follows a 9 in the pattern, the 100th decimal digit must be a 2.

Review the Solution   The above table illustrates additional patterns. For instance, if each of the location numbers in column 1 is divided by 3, a remainder of 1 is produced. If each of the location numbers in column 2 is divided by 3, a remainder of 2 is produced. Thus we can find the decimal digit in any location by dividing the location number by 3 and examining the remainder. For instance, to find the digit in the 3200th decimal place of $\frac{7}{27}$, merely divide 3200 by 3 and examine the remainder, which is 2. Thus, the digit 3200 places to the right of the decimal point is a 5.

CHECK YOUR PROGRESS 4   Determine the ones digit of 4^{200}.

Solution   See page S3.

Example 5 illustrates the method of working backwards. In problems in which you know a final result, this method may require the least effort.

✔ TAKE NOTE

Example 5 can also be worked by using algebra. Let A be the amount of money Stacy had just before she purchased the hotel. Then

$$\frac{1}{2} \cdot \left[\frac{1}{2} \cdot (A - 800) + 200 \right] = 2500$$

$$\frac{1}{2} \cdot (A - 800) + 200 = 5000$$

$$A - 800 + 400 = 10{,}000$$

$$A - 400 = 10{,}000$$

$$A = 10{,}400$$

Which do you prefer, the algebraic method or the method of working backwards?

EXAMPLE 5 ■ Apply Polya's Strategy

In consecutive turns of a Monopoly game, Stacy first paid $800 for a hotel. She then lost half her money when she landed on Boardwalk. Next, she collected $200 for passing GO. She then lost half her remaining money when she landed on Illinois Avenue. Stacy now has $2500. How much did she have just before she purchased the hotel?

Solution

Understand the Problem   We need to determine the number of dollars that Stacy had just prior to her $800 hotel purchase.

Devise a Plan   We could guess and check, but we might need to make several guesses before we found the correct solution. An algebraic method might work, but setting up the necessary equation could be a challenge. Since we know the end result, let's try the method of working backwards.

Carry Out the Plan   Stacy must have had $5000 just before she landed on Illinois Avenue; $4800 just before she passed GO; and $9600 prior to landing on Boardwalk. This means she had $10,400 just before she purchased the hotel.

Review the Solution   To check our solution we start with $10,400 and proceed through each of the transactions. $10,400 less $800 is $9600. Half of $9600 is $4800. $4800 increased by $200 is $5000. Half of $5000 is $2500.

CHECK YOUR PROGRESS 5   Melody picks a number. She doubles the number, squares the result, divides the square by 3, subtracts 30 from the quotient, and gets 18. What are the possible numbers that Melody could have picked? What operation does Melody perform that prevents us from knowing with 100% certainty which number she picked?

Solution   *See page S3.*

Some problems are best solved by making guesses and checking.

EXAMPLE 6 ■ Apply Polya's Strategy

The cube of a natural number is 648 more than its square. What is the natural number?

Solution

Understand the Problem   The natural numbers are 1, 2, 3, 4, ... If we call the number we are seeking n, then $n^3 - n^2$ must equal 648.

Devise a Plan   Solving the equation $n^3 - n^2 = 648$ could prove to be a difficult task, so we consider making a guess and checking to see if our guess is a solution.

Carry Out the Plan   If we guess that the number is 5, then the cube of 5 is 125 and the square of 5 is 25. The difference between 125 and 25 is only 100. Thus the number we seek must be larger than 5.

Try a larger number, say 10. The cube of 10 is 1000 and the square of 10 is 100. The difference between 1000 and 100 is 900. Thus 10 is larger than the number we are looking for.

Next we try a natural number between 5 and 10, say 9. The cube of 9 is 729 and the square of 9 is 81. The difference between 729 and 81 is 648. Thus 9 is a solution.

Review the Solution   We know that 9 is a solution, but is it the only solution? Increasing the size of n increases the difference between n^3 and n^2. Decreasing the size of n decreases the difference between n^3 and n^2. Thus 9 is the only natural number solution.

CHECK YOUR PROGRESS 6   Nothing is known about the personal life of the ancient Greek mathematician Diophantus except for the information in the following epigram. "Diophantus passed $\frac{1}{6}$ of his life in childhood, $\frac{1}{12}$ in youth, and $\frac{1}{7}$ more as a bachelor. Five years after his marriage was born a son who died four years before his father, at $\frac{1}{2}$ his father's (final) age."

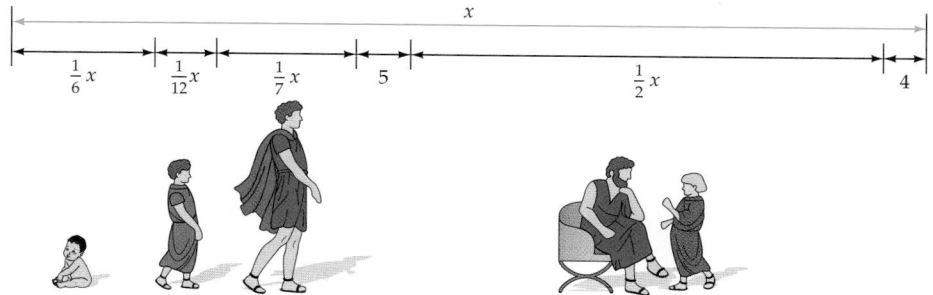

A diagram of the data, where *x* represents the age of Diophantus when he died.

How old was Diophantus when he died? (*Hint:* Although an equation can be used to solve this problem, the method of guessing and checking will probably require less effort. Also assume that his age, when he died, is a natural number.)

Solution   *See page S3.*

QUESTION   *Is the process of guessing at a solution and checking your result one of Polya's problem-solving strategies?*

ANSWER   *Yes*

Karl Friedrich Gauss
(1777–1855)

Math Matters   A Mathematical Prodigy

Karl Friedrich Gauss (gous) was a German scientist and mathematician. His work encompassed several disciplines, including number theory, differential geometry, analysis, astronomy, geodesy, and optics. He is often called the "Prince of Mathematicians" and is known for having shown remarkable mathematical prowess as early as age three. It is reported that soon after Gauss entered elementary school, his teacher assigned an arithmetic problem designed to keep the students occupied for a lengthy period of time. The problem consisted of finding the sum of the first 100 natural numbers. Gauss was able to determine the correct sum in a matter of a few seconds. The following solution shows the thought process that Gauss applied as he solved the problem.

Understand the Problem   The sum of the first 100 natural numbers is represented by

$$1 + 2 + 3 + \cdots + 98 + 99 + 100$$

Devise a Plan   Adding the first 100 natural numbers from left to right would produce the desired sum, but would be time consuming and laborious. Gauss considered another method. He added 1 and 100 to produce 101. He noticed that 2 and 99 have a sum of 101, and that 3 and 98 have a sum of 101. Thus the 100 numbers could be thought of as 50 pairs, each with a sum of 101.

$$1 + 2 + 3 + \cdots + 98 + 99 + 100$$

```
            ⌊── 101 ──⌋
      ⌊──────── 101 ────────⌋
⌊──────────── 101 ────────────⌋
```

Carry Out the Plan   To find the sum of the 50 pairs, each with a sum of 101, Gauss computed $50 \cdot 101$ and arrived at 5050 as the solution.

Review the Solution   Because the addends in an addition problem can be placed in any order without changing the sum, Gauss was confident that he had the correct solution.

An Extension   The solution of one problem often leads to solutions of additional problems. For instance, the sum $1 + 2 + 3 + \cdots + (n - 2) + (n - 1) + n$ can be found by using the following formula.

A summation formula for the first *n* natural numbers:

$$1 + 2 + 3 + \cdots + (n - 2) + (n - 1) + n = \frac{n(n + 1)}{2}$$

Trick Problems

Some problems can be classified as "trick" problems. They may involve misleading wording, as shown in Example 7.

EXAMPLE 7 ■ Solve a Trick Problem

Rearrange the letters of "new door" to make one word.

Solution

Understand the Problem   It appears that we need to rearrange the letters to make a single word.

Devise a Plan   After many unsuccessful attempts to find a single word, we consider that the wording of the problem may be misleading. We decide to look for a trick.

Carry Out the Plan   The letters can be arranged to form the words "one word."

Review the Solution   We could continue to search for a single word formed by rearranging the letters of "new door"; however, the directions said to form one word. We conclude that the problem was a trick problem and that "one word" is the solution.

CHECK YOUR PROGRESS 7   Two U.S. coins have a total value of 35¢. One of the coins is not a quarter. What are the two coins?

Solution   See page S3.

Excursion

Routes on a Probability Demonstrator

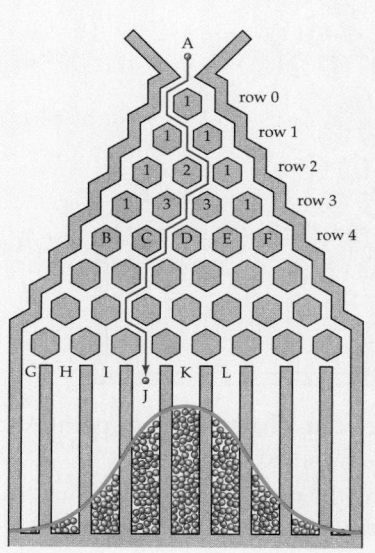

The object shown at the left is called a *Galton board*. It was invented by the English statistician Francis Galton (1822–1911). This particular board has 256 small red balls that are released so that they fall through an array of hexagons. The board is designed such that when a ball falls on a vertex of one of the hexagons, it is equally likely to fall to the left or to the right. As the ball continues its downward path it strikes a vertex of a hexagon in the next row, where the process of falling to the left or to the right is repeated. After the ball passes through all the rows of hexagons it falls into one of the bins at the bottom. In most cases the balls will form a *bell shape*, as shown by the green curve. Examine the numbers displayed in the hexagons in rows 0 through 3. Each number indicates the number of different routes a ball can take from point A to the top of that particular hexagon.

Excursion Exercises

1. How many routes can a ball take as it travels from point A to point B, from A to C, from A to D, from A to E, and from A to F? Hint: This problem is similar to Example 1 on page 29.

2. How many routes can a ball take as it travels from point A to points G, H, I, J, and K?

3. Explain how you know that the number of routes from point A to point J is the same as the number of routes from point A to point L.

4. Explain why the greatest number of balls tend to fall into the center bin.

Exercise Set 1.3   (Suggested Assignment: 1–47 odds)

Use Polya's four-step problem-solving strategy and the problem-solving procedures presented in this lesson to solve each of the following exercises.

1. There are 364 first-grade students in Park Elementary School. If there are 26 more girls than boys, how many girls are there?

2. If two ladders are placed end to end, their combined height is 31.5 feet. One ladder is 6.5 feet shorter than the other ladder. What are the heights of the two ladders?

3. How many squares are in the following figure?

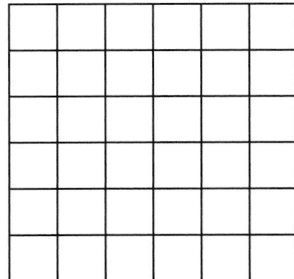

4. What is the 44th decimal digit in the decimal representation of $\frac{1}{11}$?

$$\frac{1}{11} = 0.09090909\ldots$$

5. A shirt and a tie together cost $30. The shirt costs $20 more than the tie. What is the cost of the shirt?

6. In a basketball league consisting of 12 teams, each team plays each of the other teams exactly twice. How many league games will be played?

7. Consider the following map. Tyler wishes to walk along the streets from point A to point B. How many direct routes (no backtracking) can Tyler take?

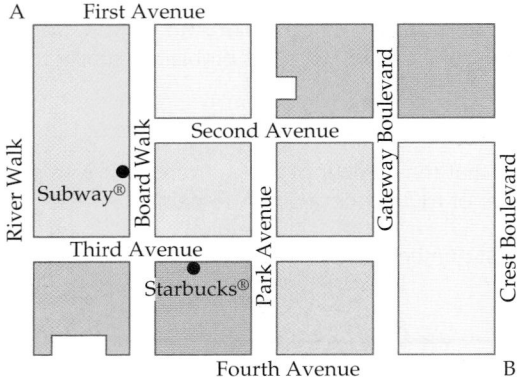

8. Use the map in Exercise 7 to answer each of the following.

   a. How many direct routes are there from A to B if you want to pass by Starbucks?

   b. How many direct routes are there from A to B if you want to stop at Subway for a sandwich?

   c. How many direct routes are there from A to B if you want to stop at Starbucks and at Subway?

9. In how many ways can you answer a 12-question true-false test if you answer each question with either a "true" or a "false"?

10. A frog is at the bottom of a 17-foot well. Each time the frog leaps it moves up 3 feet. If the frog has not reached the top of the well, then the frog slides back 1 foot before it is ready to make another leap. How many leaps will the frog need to escape the well?

11. Consider the following probability demonstrator. Note that this probability demonstrator has one additional row of hexagons compared with the probability demonstrator on page 37.

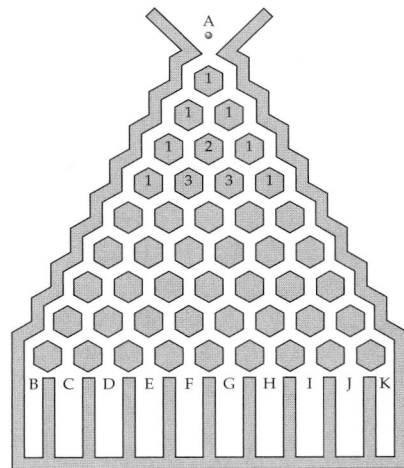

   a. How many routes can a ball take from point A to points B, C, D, E, F, G, and H?

   b. ✎ Explain why the number of routes from A to F is the same as the number from A to G.

12. The triangular pattern shown on the following page is known as **Pascal's triangle.** Pascal's triangle has intrigued mathematicians for hundreds of years. Although it is named after the mathematician Blaise Pascal (1623–1662), there is evidence that it was first developed in China in the 1300s. The numbers in Pascal's

triangle are created in the following manner. Each row begins and ends with the number 1. Any other number in a row is the sum of the two closest numbers above it. For instance, the first 10 in row 5 is the sum of the first 4 and the 6 above it in row 4.

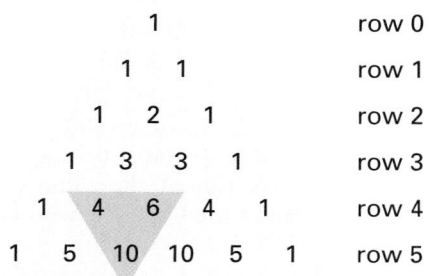

row 0
row 1
row 2
row 3
row 4
row 5

Pascal's Triangle

There are many patterns that can be discovered in Pascal's triangle.

a. Find the sum of the numbers in each row of the portion of Pascal's triangle shown above. What pattern do you observe concerning these sums? Predict the sum of the numbers in row 9 of Pascal's triangle.

b. How are the numbers in Pascal's triangle related to the number of routes found for the probability demonstrator in Exercise 11?

c. The numbers $1, 3, 6, 10, 15, \ldots, \dfrac{n(n+1)}{2}, \ldots$ are called *triangular numbers*. Where do the triangular numbers appear in Pascal's triangle?

13. If eight people greet each other at a meeting by shaking hands with one another, how many handshakes take place?

14. Twenty-four points are placed around a circle. A line segment is drawn between each pair of points. How many line segments are drawn?

15. The number of ducks and pigs in a field totals 35. The total number of legs among them is 98. Assuming each duck has exactly two legs and each pig has exactly four legs, determine how many ducks and how many pigs are in the field.

16. A runner runs around a circular track at a constant rate. There are 10 markers evenly spaced around the track. If the runner takes 12 seconds to run from the first marker to the fourth marker, determine the time it takes the runner to make one complete trip around the track.

17. How many ways can you make change for 25¢ using dimes, nickels, and/or pennies?

18. A room measures 12 feet by 15 feet. How many 3-foot by 3-foot squares of carpet are needed to cover the floor of this room?

19. Determine the units digit of 4^{7022}.

20. Determine the units digit of 2^{6543}.

21. Determine the units digit of $3^{11,707}$.

22. Determine the units digit of 8^{8985}.

23. Find the following sums without using a calculator or a formula. *Hint:* Apply the procedure used by Gauss. (See the *Math Matters* on page 36.)

a. $1 + 2 + 3 + 4 + \cdots + 397 + 398 + 399 + 400$

b. $1 + 2 + 3 + 4 + \cdots + 547 + 548 + 549 + 550$

c. $2 + 4 + 6 + 8 + \cdots + 80 + 82 + 84 + 86$

24. Explain how you could modify the procedure used by Gauss (see the *Math Matters* on page 36) to find the following sum.

$$1 + 2 + 3 + 4 + \cdots + 62 + 63 + 64 + 65$$

25. A **palindromic number** is a natural number that reads the same from left to right as it reads from right to left. For instance, 23,732 is a palindromic number. Find the smallest palindromic number greater than 1000 whose digits have a sum that is larger than 5.

26. The following powers of 11 are all palindromic numbers (see Exercise 25).

$$11^1 = 11$$
$$11^2 = 121$$
$$11^3 = 1331$$

Is 11^n a palindromic number for all natural numbers n? Explain your answer.

27. Find all three-digit palindromic numbers that are also the square of a natural number.

28. Find all four-digit palindromic numbers that are also the cube of a natural number.

29. Three volumes of the series *Mathematics: Its Content, Methods, and Meaning* are on a shelf with no space between the volumes. Each volume is 1 inch thick without its covers. Each cover is $\frac{1}{8}$ inch thick. See the figure at the top of the next page. A bookworm bores horizontally from the first page of Volume 1 to the last page of Volume III. How far does the bookworm travel?

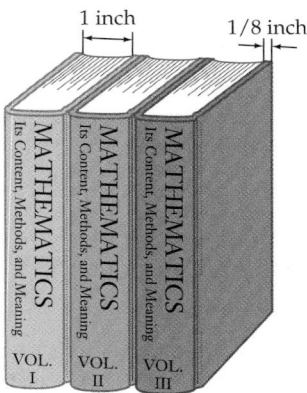

1 inch   1/8 inch

30. Nine dots are arranged as shown. Is it possible to connect the nine dots with exactly four lines if you are not allowed to retrace any part of a line and you are not allowed to remove your pencil from the paper? If it can be done, demonstrate with a drawing.

31. Oak Hill Cemetery in San Marcos, California will not bury anyone living outside of California? Why?

32. If you take 22 pennies from a pile of 57 pennies, how many pennies do you have?

33. The bacteria in a petri dish grow so that each day the number of bacteria doubles. On what day will the number of bacteria be half of the number present on the 12th day?

34. How long will it take to cut an 8-foot board into 1 foot lengths if each cut takes exactly 15 seconds? Assume no time is required between each cut.

35. Four people on one side of a river need to cross the river in a boat that can carry a maximum load of 180 pounds. The weights of the people are 80, 100, 150, and 170 pounds.

    a. Explain how the people can use the boat to get every one to the opposite side of the river.

    b. What is the minimum number of crossings that must be made by the boat?

36. On three examinations Dana received scores of 82, 91, and 76. What score does Dana need on the fourth examination to raise his average to 85?

37. Is it legal for a man to marry his widow's sister? Explain.

38. What is the largest number that can be formed using only two digits?

39. Two athletes race each other in a 100-yard dash. The faster runner finishes the race 10 yards ahead of the slower runner. Assume that each of the runners runs at a constant rate. In a second 100-yard dash, the faster runner starts 10 yards behind the starting line. If both runners run at the same constant rate at which they ran in the first race, who will win the second race?

40. One of the problems in George Polya's book *How to Solve It* reads as follows. "Bob wants a piece of land, exactly level, which has four boundary lines. Two boundary lines run exactly north-south, the others exactly east-west, and each boundary line measures exactly 100 feet. Can Bob buy such a piece of land in the U.S.?"[3]

41. In the movie *Die Hard: With A Vengeance* (1995) Bruce Willis and Samuel L. Jackson are given a 5-gallon jug and a 3-gallon jug and they must put *exactly* 4 gallons of water on a scale to keep a bomb from exploding. Explain how they could accomplish this feat. Take your time, you have 2 minutes.

42. You have eight coins. They all look identical, but one is a fake and is slightly lighter than the others. Explain how you can use a balance scale to determine which coin is the fake in exactly

    a. three weighings.

    b. two weighings.

---

3. From George Polya, *How to Solve It, 2e.* Copyright © 1957 by Princeton University Press. Reprinted with permission of Princeton University Press.

43. An airplane left Los Angeles at 8:20 A.M. and flew to Boston. The flying time was 6 hours 20 minutes. Boston is on Eastern Standard Time (EST) and Los Angeles is on Pacific Standard Time (PST), which is 3 hours behind EST. After the plane was on the ground for 1 hour it flew to Chicago, which is on Central Standard Time (CST). CST is 1 hour behind EST. The flying time from Boston to Chicago was 2 hours 20 minutes.

   a. What time, EST, did the plane arrive in Boston?

   b. What time, CST, did the plane arrive in Chicago?

44. a. List the four steps in Polya's problem-solving strategy.

   b. List eight problem-solving procedures that one might use in Polya's second step.

Extensions

CRITICAL THINKING

45. What is the 100th decimal digit in the decimal representation of $\frac{1}{7}$?

46. a. [calc] How many times larger is $3^{(3^3)}$ than $(3^3)^3$?

   b. How many times larger is $4^{(4^4)}$ than $(4^4)^4$? *Note:* Most calculators will not display the answer to this problem because it is too large. However, the answer can be determined in exponential form by applying the following properties of exponents.

$$(a^m)^n = a^{mn} \quad \text{and} \quad \frac{a^m}{a^n} = a^{m-n}$$

47. The mathematician Augustus De Morgan once wrote that he had the distinction of being x years old in the year x^2. He was 43 in the year 1849.

   a. Explain why people born in the year 1980 might share the distinction of being x years old in the year x^2. *Note:* Assume x is a natural number.

   b. What is the next year after 1980 for which people born in that year might be x years old in the year x^2?

48. Select a two-digit number between 50 and 100. Add 83 to your number. From this number form a new number by adding the digit in the hundreds place to the number formed by the other two digits (the digits in the tens place and the ones place). Now subtract this newly formed number from your original number. Your final result is 16. Use a deductive approach to show that the final result is always 16 regardless of which number you start with.

49. How many digits does it take in total to number a book from page 1 to page 240?

50. Consider a checkerboard with two red squares on opposite corners removed, as shown in the accompanying figure. Determine whether it is possible to completely cover the checkerboard with 31 dominoes if each domino is placed horizontally or vertically and each domino covers exactly two squares. If it is possible, show how to do it. If it is not possible, explain why it cannot be done.

COOPERATIVE LEARNING

51. The object of this exercise is to create mathematical expressions that use exactly four 4's and that simplify to a counting number from 1 to 20, inclusive. You are allowed to use the following mathematical symbols: $+, -, \times, \div, \sqrt{}, (, \text{ and })$. For example,

$$\frac{4}{4} + \frac{4}{4} = 2, \quad 4^{(4-4)} + 4 = 5, \quad \text{and}$$
$$4 - \sqrt{4} + 4 \times 4 = 18$$

52. The following puzzle is a famous *cryptarithm*.

$$\begin{array}{r} \text{SEND} \\ + \text{MORE} \\ \hline \text{MONEY} \end{array}$$

Each letter in the cryptarithm represents one of the digits 0 through 9. The leading digits, represented by

S and M, are not zero. Determine which digit is represented by each of the letters so that the addition is correct. *Note:* A letter that is used more than once, such as M, represents the same digit in each position in which it appears.

EXPLORATIONS

53. There are many unsolved problems in mathematics. One famous unsolved problem is known as the *Collatz problem,* or the $3x + 1$ problem. This problem was created by L. Collatz in 1937. Although the procedures in the Collatz problem are easy to understand, the problem remains unsolved. Search the Internet or a library to find information on the Collatz problem.

   a. Write a short report that explains the Collatz problem. In your report explain the meaning of a "hailstone" sequence.

   b. Show that for each of the natural numbers 2, 3, 4, ..., 10, the Collatz procedure does generate a sequence that "returns" to 1.

54. Paul Erdos (1913–1996) was a mathematician known for his elegant solutions of problems in number theory, combinatorics, discrete mathematics, and graph theory. He loved to solve mathematical problems and for those problems he could not solve, he offered financial rewards, up to $10,000, to the person who could provide a solution.

   Write a report on the life of Paul Erdos. In your report include information about the type of problem that Erdos considered the most interesting. What did Erdos have to say about the Collatz problem mentioned in Exploration Exercise 53?

CHAPTER 1   ## Summary

Key Terms

conjecture [p. 4]
counterexample [p. 7]
difference table [p. 15]
Fibonacci sequence [p. 19]
first, second, and third differences [p. 15]
nth term formula [p. 17]
nth term of a sequence [p. 15]
palindromic number [p. 39]
Pascal's Triangle [p. 38]
prime number [p. 13]
recursive definition [p. 20]
sequence [p. 14]
term of a sequence [p. 14]

Essential Concepts

- *Inductive reasoning* is the process of reaching a general conclusion by examining specific examples. A conclusion based on inductive reasoning is called a *conjecture*. A conjecture may or may not be correct.

- *Deductive reasoning* is the process of reaching a conclusion by applying general assumptions, procedures, or principles.

- A statement is a *true statement* provided it is true in all cases. If you can find one case in which a statement is not true, called a *counterexample*, then the statement is a *false statement*.

- The terms of the *Fibonacci sequence* 1, 1, 2, 3, 5, 8, 13, 21, ... can be determined by using the *recursive definition*

$$F_1 = 1, F_2 = 1, \text{ and } F_n = F_{n-1} + F_{n-2} \quad \text{for } n \geq 3$$

- Many problems can be solved by applying *Polya's problem solving strategy:*

   1. Understand the problem.

   2. Devise a plan.

   3. Carry out the plan.

   4. Review your solution.

Review Exercises

In Exercises 1–4, determine whether the argument is an example of inductive reasoning or deductive reasoning.

1. All books written by J. K. Rowling make the best-seller list. The book *Harry Potter and the Goblet of Fire* is a J. K. Rowling book. Therefore, *Harry Potter and the Goblet of Fire* made the best-seller list.

2. Samantha got an A on each of her first four math tests, so she will get an A on the next math test.

3. We had rain yesterday, so there is less chance of rain today.

4. All amoeba multiply by dividing. I have named the amoeba shown in my microscope Amelia. Therefore, Amelia multiplies by dividing.

5. Find a counterexample to show that the following conjecture is false.
   Conjecture: For all x, $x^4 > x$.

6. Find a counterexample to show that the following conjecture is false.
   Conjecture: For all counting numbers n, $\dfrac{n^3 + 5n + 6}{6}$ is an even number.

7. Find a counterexample to show that the following conjecture is false.
   Conjecture: For all x, $(x + 4)^2 = x^2 + 16$.

8. Find a counterexample to show that the following conjecture is false.
   Conjecture: For numbers a and b, $(a + b)^3 = a^3 + b^3$.

9. Use a difference table to predict the next term of each sequence.
   a. $-2, 2, 12, 28, 50, 78, ?$
   b. $-4, -1, 14, 47, 104, 191, 314, ?$

10. Use a difference table to predict the next term of each sequence.
    a. $5, 6, 3, -4, -15, -30, -49, ?$
    b. $2, 0, -18, -64, -150, -288, -490, ?$

11. A sequence has an nth term formula of
    $$a_n = 4n^2 - n - 2$$
    Use the nth term formula to determine the first five terms of the sequence and the 20th term of the sequence.

12. A sequence has an nth term formula of
    $$a_n = -2n^3 + 5n$$
    Use the nth term formula to determine the first five terms of the sequence and the 25th term of the sequence.

In Exercises 13 and 14, determine the nth term formula for the number of square tiles in the nth figure.

13.

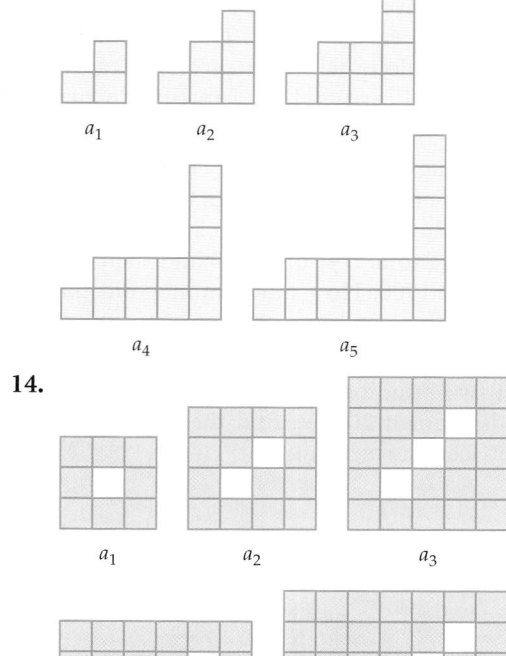

14.

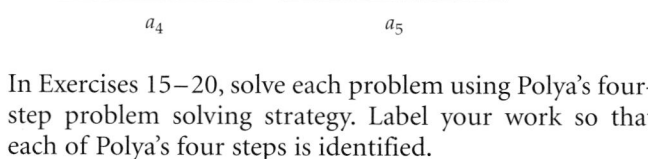

In Exercises 15–20, solve each problem using Polya's four-step problem solving strategy. Label your work so that each of Polya's four steps is identified.

15. A rancher decides to enclose a rectangular region by using an existing fence along one side of the region and 2240 feet of new fence on the other three sides. The rancher wants the length of the rectangular region to be five times longer than its width. What will be the dimensions of the rectangular region?

16. In how many ways can you answer a 15-question test if you answer each question with either a "true," a "false," or an "always false"?

17. You have purchased five new tires for your car. Each tire is designed to provide exactly 40,000 miles of use.
    a. What is the maximum mileage you can get from this set of five tires?

b. If you had purchased six of the tires, what would be the maximum mileage you could get?

18. A rancher needs to get a dog, a rabbit, and a basket of carrots across a river. The rancher has a small boat that will only stay afloat carrying the rancher and one of the critters or the rancher and the carrots. The rancher cannot leave the dog alone with the rabbit because the dog will eat the rabbit. The rancher cannot leave the rabbit alone with the carrots because the rabbit will eat the carrots. How can the rancher get across the river with the critters and the carrots?

19. An investor bought 20 shares of stock for a total cost of $1200 and then sold all the shares for $1400. A few months later the investor bought 25 shares of the same stock for a total cost of $1800 and then sold all the shares for $1900. How much money did the investor earn on these investments?

20. The ends of a 60-foot rope are attached to the upper corners of two buildings that are the same height. If the rope dips down 30 feet, find the horizontal distance x between the buildings.

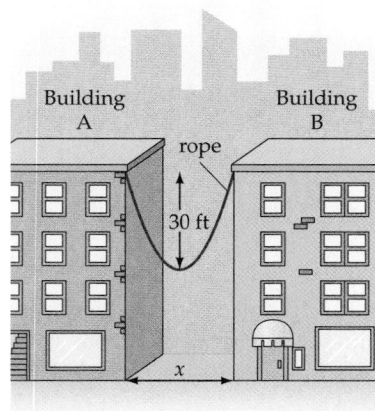

21. The fourth power of a natural number is 1372 more than three times its cube. What is the natural number?

22. If 15 people greet each other at a meeting by shaking hands with one another, how many handshakes will take place?

23. List five strategies that are included in Polya's second step (devise a plan).

24. List three strategies that are included in Polya's fourth step (review the solution).

25. Consider the following figures.

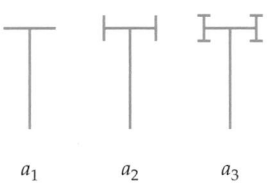

Figure a_1 consists of two line segments and figure a_2 consists of four line segments. If the pattern of adding a smaller line segment to each end of the shortest line segments continues, how many line segments will be in

a. figure a_{10}? **b.** figure a_{30}?

26. In the following addition problem each letter represents one of the digits 0, 1, 2, 3, 4, 5, 6, 7, 8, or 9. The leading digits represented by A and B are nonzero digits. What digit is represented by each letter?

27. In how many different ways can change be made for a dollar using only quarters and/or nickels?

28. In how many different orders can a basketball team win three out of their last five games?

29. What is the units digit of 7^{56}?

30. What is the units digit of 23^{85}?

31. Use deductive reasoning to show that the following procedure always produces a number that is twice the original number.

Procedure: Pick a number. Multiply the number by 4, add 12 to the product, divide the sum by 2, and subtract 6.

32. How many rectangles are in the following figure?

33. Explain why 2004 nickels are worth more than 100 dollars.

34. Bob goes to a home improvement store. Bob asks a clerk the cost of 1. The clerk answers fifty cents. Bob asks the cost of 2, and the clerk answers fifty cents. Bob asks the cost of 6 and the clerk answers fifty cents. If Bob pays a total of $1.50 for 126, what items did Bob purchase?

35. A *perfect number* is any natural number that is equal to the sum of all its natural number divisors except itself. For instance, 6 is a perfect number because its divisors other than itself are 1, 2, and 3, and the sum of these divisors is 6. Find a perfect number that is greater than 20 but less than 30.

36. Jennifer's parents have three children. They named their first child Ellen and their second child Elle. What did they name their third child?

37. Recall that palindromic numbers read the same from left to right as they read from right to left. For instance, 37,573 is a palindromic number. Find the smallest palindromic number larger than 1000 that is a multiple of 5.

38. A **narcissistic number** is a two-digit natural number that is equal to the sum of the squares of its digits. Find all narcissistic numbers.

39. Two different lines can intersect in at most one point. Three different lines can intersect in at most three points and four different lines can intersect in at most six points.

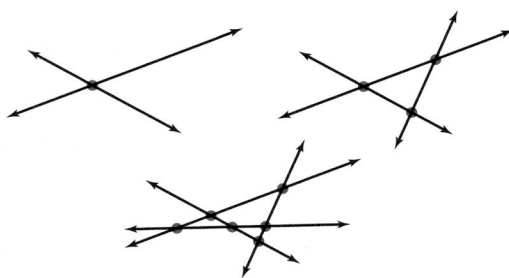

a. Determine the maximum number of intersections for five different lines.

b. Does it appear, by inductive reasoning, that the maximum number of intersection points I_n for n different lines is given by $I_n = \dfrac{n(n-1)}{2}$?

40. A student has noticed the following pattern.

$$9^1 = 9 \text{ has one digit.}$$
$$9^2 = 81 \text{ has two digits.}$$
$$9^3 = 729 \text{ has three digits.}$$
$$\vdots$$
$$9^{10} = 3{,}486{,}784{,}401 \text{ has ten digits.}$$

a. Find a natural number n such that the number of digits in the decimal expansion of 9^n is *not* equal to n.

b. A professor indicates that you can receive five extra-credit points if you write all of the digits in the decimal expansion of $9^{(9^9)}$. Is this a worthwhile project? Explain.

CHAPTER 1   **Test**

In Exercises 1–4, determine whether the problem is an example of inductive reasoning or deductive reasoning.

1. All novels by Sidney Sheldon are gruesome. The novel *The Sky is Falling* was written by Sidney Sheldon. Therefore, *The Sky is Falling* is a gruesome novel.

2. Judith Michael's last novel made the top-ten list, so her next novel will also make the top-ten list.

3. Two computer programs, a *bubble sort* and a *shell sort*, are used to sort data. In each of 50 experiments the shell sort program took less time to sort the data then did the bubble sort program. Thus the shell sort program is the faster of the two sorting programs.

4. If a figure is a rectangle, then it is a parallelogram. Figure A is a rectangle. Therefore, Figure A is a parallelogram.

5. Use a difference table to predict the next term in the sequence 1, 0, 9, 32, 75, 144, 245, ...

6. List the first 10 terms of the Fibonacci sequence.

In Exercises 7 and 8, determine the *n*th term formula for the number of square tiles in the *n*th figure.

7.

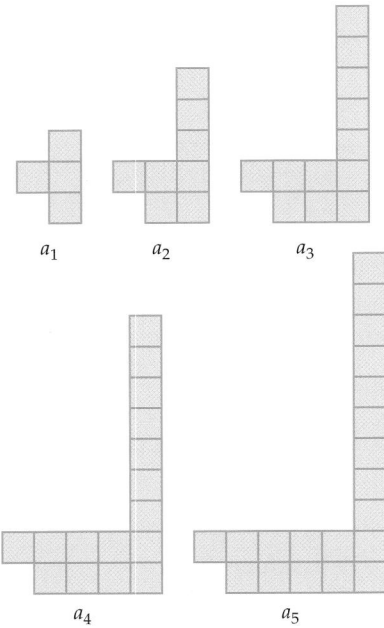

a_1 a_2 a_3

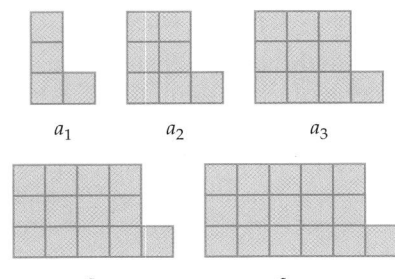

a_4 a_5

8.

a_1 a_2 a_3

a_4 a_5

9. A sequence has an *n*th term formula of

$$a_n = (-1)^n \left(\frac{n(n-1)}{2} \right)$$

Use the *n*th term formula to determine the first five terms and the 105th term in the sequence.

10. In the following sequence, each term after the second term is the sum of the two preceding terms.

$$2, 5, 7, 12, 19, 31, 50, 81, \ldots$$

Find the next three terms of the sequence.

11. State the four steps of Polya's four-step problem-solving strategy.

12. How many different ways can change be made for a dollar using only half-dollars, quarters, and/or dimes?

13. In how many different orders can a basketball team win four out of their last six games?

14. What is the units digit of 3^{4513}?

15. Shelly has saved some money for a vacation. Shelly spends half of her vacation money on an airline ticket; she then spends $50 for sunglasses, $22 for a taxi, and one-third of her remaining money for her a room with a view. After her sister repays her a loan of $150, Shelly finds that she has $326. How much vacation money did Shelly have at the start of her vacation?

16. How many different direct routes are there from point A to point B in the following figure?

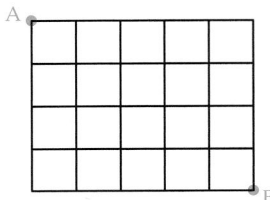

17. In a league of nine football teams, each team plays every other team in the league exactly once. How many league games will take place?

18. The cube of a natural number is 1089 more than twice its square. What is the natural number?

19. Palindromic numbers read the same from left to right as they read from right to left. For instance, 286,682 is a palindromic number. Find the smallest palindromic number larger than 500 that is a multiple of 3.

20. Find a counterexample to show that the following conjecture is false.

Conjecture: For all *x*, $\dfrac{(x-4)(x+3)}{x-4} = x + 3$.

Sets

The Hoover Dam is the largest dam in the United States. The dam uses the walls of the Black Canyon and its own mass to hold back the water from the Colorado River. The main purpose of the dam is to create hydroelectric power for use in Nevada, Arizona, and California. The dam generates more than four billion kilowatt-hours of electric power a year. Concepts involving fuzzy sets are often used in the control of valves and dam gates at power plants. In the **Exercursion exercises** on **page 55,** you will use fuzzy sets to define vague terms such as cold, hot, warm, old, and young.

Need help? For on-line student resources, such as section quizzes, visit this textbook's web site at **math.college.hmco.com/students.**

In mathematics any group or collection of objects is called a *set*. A simple application of sets occurs when you use a search engine (such as Yahoo, AltaVista, Google, or Lycos) to find a topic on the Internet. You merely enter a few words describing what you are searching for and click the "Search" button. The search engine then creates a list (set) of web sites that contain a match for the words you submitted.

For instance, suppose you wish to make a dessert. You decide to search the Internet for a chocolate cake recipe. You do a search for the words "chocolate cake" and you obtain a set containing 300,108 matches. This is a very large number, so you decide to narrow your search. One method of narrowing your search is to use the AND option found in the Advanced Search link of some search engines. An AND search is an all-words search. That is, an AND search finds only those sites that contain all of the words submitted. An AND search for "chocolate cake" produces a set containing 74,400 matches. This is a more reasonable number, but it is still quite large.

Search for:

chocolate cake| Search

Advanced Search

You attempt to narrow the search even further by using an AND search for the words "chocolate cake recipe." This search returns 17,945 matches. An AND search for "flourless chocolate cake recipe" returns only 913 matches. The second of these sites provides you with a recipe and states that it is fabulous and foolproof.

Sometimes it is helpful to perform a search using the OR option. An OR search is an any-words search. That is, an OR search finds all those sites that contain any of the words you submitted.

Many additional applications of sets are given in this chapter.

| # Basic Properties of Sets

Sets

The constellation Scorpius is a set of stars.

Human beings share the desire to organize and classify objects and concepts. For instance, ancient astronomers classified certain groups of stars as constellations. Modern day chemists classify solutions as acidic, neutral, or basic. Geologists organize the history of the Earth into *eras,* such as the Cenozoic, Mesozoic, Paleozoic, and Archeozoic, to name a few.

In mathematics, any group or collection of objects is called a **set.** The objects that belong to a set are the **elements** or **members** of the set. Sets are often denoted by placing braces around the elements of the set. The following sets of numbers are used extensively in many areas of mathematics.

> **Basic Number Sets**
>
> **Natural Numbers or Counting Numbers** $N = \{1, 2, 3, 4, 5, \ldots\}$
>
> **Whole Numbers** $W = \{0, 1, 2, 3, 4, 5, \ldots\}$
>
> **Integers** $I = \{\ldots, -4, -3, -2, -1, 0, 1, 2, 3, 4, \ldots\}$
>
> **Rational Numbers** $Q = \{$all terminating or repeating decimals$\}$
>
> **Irrational Numbers** $\mathcal{I} = \{$all nonterminating, nonrepeating decimals$\}$
>
> **Real Numbers** $R = \{$all rational or irrational numbers$\}$ **P**

✔ **TAKE NOTE**

In this chapter the letters *N, W, I, Q, $\mathcal{I}$,* and *R* will often be used to represent the basic number sets defined at the right.

The set of natural numbers is also called the set of counting numbers. The three dots ... are called an **ellipsis** and indicate that the elements of the set continue in a manner suggested by the elements that are listed. Commas are used to separate the elements of a set.

The integers ..., $-4, -3, -2, -1$ are **negative integers.** The integers 1, 2, 3, 4, ... are **positive integers.** Note that the natural numbers and the positive integers are the same set of numbers. The integer zero is neither a positive nor a negative integer.

If a number in decimal form terminates or repeats a block of digits, then the number is a rational number. Rational numbers can also be written as a fraction p/q, where p and q are integers and $q \neq 0$. For example,

$$\frac{1}{4} = 0.25 \quad \text{and} \quad \frac{3}{11} = 0.\overline{27}$$

are rational numbers. The bar over the 27 means that the block of digits 27 repeats without end; that is, $0.\overline{27} = 0.27272727\ldots$.

A decimal that neither terminates nor repeats is an irrational number. For instance, $0.35335333\ldots$ is a nonterminating, nonrepeating decimal and thus is an irrational number. Irrational numbers cannot be written as a fraction p/q, where p and q are integers.

Every real number is either a rational number or an irrational number. If a real number is written as a decimal, it is a terminating decimal, a repeating decimal, or a nonterminating, nonrepeating decimal.

The **roster method** of writing a set consists of listing the set's elements inside braces.

TAKE NOTE

When listing the elements of a set, the order is not important. Thus {1, 2, 3, 4} and {2, 4, 1, 3} both represent the same set.

INSTRUCTOR NOTE

Students should be reminded that a listing such as 1, 2, 3 is not a set. With the roster method the elements need to be placed inside of braces to form a set.

EXAMPLE 1 ■ The Roster Method

Use the roster method to write each of the given sets.

a. The set of natural numbers less than 5

b. The solution set of $x + 5 = -1$

c. The set of negative integers greater than -4

Solution

a. The set of *natural numbers* or *counting numbers* is given by {1, 2, 3, 4, 5, 6, 7, ...}. The natural numbers less than 5 are 1, 2, 3, and 4. Using the roster method we write this set as {1, 2, 3, 4}.

b. Adding -5 to each side of the equation produces $x = -6$. The solution set of $x + 5 = -1$ is {−6}.

c. The set of negative integers greater than -4 is {−3, −2, −1}.

CHECK YOUR PROGRESS 1 Use the roster method to write each of the given sets.

a. The set of whole numbers less than 4

b. The set of counting numbers larger than 11 and less than or equal to 19

c. The set of negative integers between -5 and 7

Solution   *See page S3.*

Definitions Regarding Sets

A set is **well defined** if it is possible to determine whether any given item is an element of the set. For instance, the set of letters of the English alphabet is well defined. The set of great songs is not a well-defined set. It is not possible to determine whether any given song is an element of the set of great songs, because "great" is a subjective term.

The statement "4 is an element of the set of natural numbers" can be written using mathematical notation as $4 \in N$. The symbol $\in$ is read "is an element of." To state that "-3 is not an element of the set of natural numbers," we use the "is not an element of" symbol $\notin$, and write $-3 \notin N$.

TAKE NOTE

Recall that *N* denotes the set of natural numbers, *I* denotes the set of integers, and *W* denotes the set of whole numbers.

EXAMPLE 2 ■ True or False

Determine whether each statement is true or false.

a. $4 \in \{2, 3, 4, 7\}$ **b.** $-5 \in N$ **c.** $\frac{1}{2} \notin I$

d. The set of nice cars is a well-defined set.

Solution

a. Because 4 is an element of the given set, the statement is true.

b. There are no negative natural numbers, so the statement is false.

c. Because $\frac{1}{2}$ is not an integer, the statement is true.

d. The word *nice* is not precise, so the statement is false.

CHECK YOUR PROGRESS 2   Determine whether each statement is true or false.

a. $5.2 \in \{1, 2, 3, 4, 5, 6\}$

b. $-101 \in I$

c. $2.5 \notin W$

d. The set of all integers larger than π is a well-defined set.

Solution   See page S3.

The **empty set,** or **null set,** is the set that contains no elements. The symbol $\varnothing$ or $\{\ \}$ is used to represent the empty set. As an example of the empty set, consider the set of natural numbers that are negative integers.

Another method of representing a set is **set-builder notation.** Set-builder notation is especially useful when describing infinite sets. For instance, in set-builder notation, the set of natural numbers greater than 7 is written as follows.

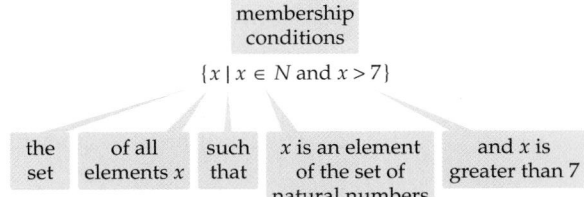

The above set-builder notation is read as "the set of all x such that x is an element of the set of natural numbers and x is greater than 7." It is impossible to list all the elements of the set, but set-builder notation defines the set by describing its elements.

EXAMPLE 3 ■ **Set-Builder Notation**

Use set-builder notation to write the following sets.

a. The set of integers greater than -3

b. The set of whole numbers less than 1000

Solution
a. $\{x \,|\, x \in I \text{ and } x > -3\}$ **b.** $\{x \,|\, x \in W \text{ and } x < 1000\}$

CHECK YOUR PROGRESS 3   Use set-builder notation to write the following sets.

a. The set of integers less than 9

b. The set of natural numbers greater than 4

Solution   See page S4.

A set is **finite** if the number of elements in the set is a whole number. The **cardinal number** of a finite set is the number of elements in the set. The cardinal number of a finite set A is denoted by the notation $n(A)$. For instance, if $A = \{1, 4, 6, 9\}$, then $n(A) = 4$. In this case, A has a cardinal number of 4, which is sometimes stated as "A has a *cardinality* of 4."

EXAMPLE 4 ■ **The Cardinality of a Set**

Find the cardinality of each of the following sets.

a. $J = \{2, 5\}$ **b.** $S = \{3, 4, 5, 6, 7, \ldots, 31\}$ **c.** $T = \{3, 3, 7, 51\}$

Solution

a. Set J contains exactly two elements, so J has a cardinality of 2. Using mathematical notation we state this as $n(J) = 2$.

b. Only a few elements are actually listed. The number of natural numbers from 1 to 31 is 31. If we omit the numbers 1 and 2, then the number of natural numbers from 3 to 31 must be $31 - 2 = 29$. Thus $n(S) = 29$.

c. Elements that are listed more than once are counted only once. Thus $n(T) = 3$.

CHECK YOUR PROGRESS 4   Find the cardinality of the following sets.

a. $C = \{-1, 5, 4, 11, 13\}$ **b.** $D = \{0\}$ **c.** $E = \varnothing$

Solution   *See page S4.*

The following definitions play an important role in our work with sets.

Definition of Equal Sets

Set A is **equal** to set B, denoted by $A = B$, if and only if A and B have exactly the same elements.

For instance $\{d, e, f\} = \{e, f, d\}$.

Definition of Equivalent Sets

Set A is **equivalent** to set B, denoted by $A \sim B$, if and only if A and B have the same number of elements.

QUESTION   *If two sets are equal, must they also be equivalent?*

EXAMPLE 5 ■ **Equal Sets and Equivalent Sets**

State whether each of the following pairs of sets are equal, equivalent, both, or neither.

a. $\{a, e, i, o, u\}, \{3, 7, 11, 15, 19\}$ **b.** $\{4, -2, 7\}, \{3, 4, 7, 9\}$

Solution

a. The sets are not equal. However, each set has exactly five elements, so the sets are equivalent.

b. The first set has three elements and the second set has four elements, so the sets are not equal and are not equivalent.

ANSWER   *Yes. If the sets are equal, then they have exactly the same elements; therefore, they also have the same number of elements.*

CHECK YOUR PROGRESS 5 State whether each of the following pairs of sets are equal, equivalent, both, or neither.

a. $\{x\,|\,x \in W \text{ and } x \leq 5\}$, $\{\alpha, \beta, \Gamma, \Delta, \delta, \varepsilon\}$

b. $\{5, 10, 15, 20, 25, 30, \ldots, 80\}$, $\{x\,|\,x \in N \text{ and } x < 17\}$

Solution   *See page S4.*

Math Matters   Georg Cantor

Georg Cantor

Georg Cantor (kǎn′tər) (1845–1918) was a German mathematician who developed many new concepts regarding the theory of sets. Cantor studied under the famous mathematicians Karl Weirstrass and Leopold Kronecker at the University of Berlin. Although Cantor demonstrated a talent for mathematics, his professors were unaware that Cantor would produce extraordinary results that would cause a major stir in the mathematical community.

Cantor never achieved his lifelong goal of a professorship at the University of Berlin. Instead he spent his active career at the undistinguished University of Halle. It was during this period, when Cantor was between the ages of 29 and 39, that he produced his best work. Much of this work was of a controversial nature. One of the simplest of the controversial concepts concerned

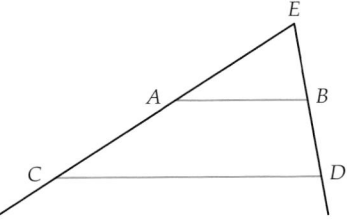

points on a line segment. For instance, consider the line segment $\overline{AB}$ and the line segment $\overline{CD}$ in the figure above. Which of these two line segments do you think contains the most points? Cantor was able to prove that they both contain the same number of points. In fact, he was able to prove that any line segment, no matter how short, contains the same number of points as a line, or a plane, or all of three-dimensional space. We will take a closer look at some of the mathematics developed by Cantor in the last section of this chapter.

Excursion

Fuzzy Sets

In traditional set theory, an element either belongs to a set or does not belong to the set. For instance, let $A = \{x\,|\,x \text{ is an even integer}\}$. Given $x = 8$, we have $x \in A$. However, if $x = 11$, then $x \notin A$. For any given integer, we can decide whether x belongs to A.

Now consider the set $B = \{x\,|\,x \text{ is a number close to 10}\}$. Does 8 belong to this set? Does 9.9 belong to the set? Does 10.001 belong to the set? Does 10 belong to the set? Does −50 belong to the set? Given the imprecision of the words "close to," it is impossible to know which numbers belong to set B.

(continued)

historical note

Lofti Zadeh

Lotfi Zadeh's work in the area of fuzzy sets has led to a new area of mathematics called *soft computing*. On the topic of soft computing, Zadeh stated, "The essence of soft computing is that unlike the traditional, hard computing, soft computing is aimed at an accommodation with the pervasive imprecision of the real world. Thus the guiding principle of soft computing is: Exploit the tolerance for imprecision, uncertainty and partial truth to achieve tractability, robustness, low solution cost and better rapport with reality. In the final analysis, the role model for soft computing is the human mind." ■

In 1965, Lotfi A. Zadeh of the University of California, Berkeley, published a paper titled *Fuzzy Sets* in which he described the mathematics of fuzzy set theory. This theory proposed that "to some degree," many of the numbers 8, 9.9, 10.001, 10 and −50 belong to set B defined in the previous paragraph. Zadeh proposed giving each element of a set a *membership grade* or *membership value*. This value is a number from 0 to 1. The closer the membership value is to 1, the more the certainty that an element belongs to the set. The closer the membership value is to 0, the less the certainty that an element belongs to the set. Elements of fuzzy sets are written in the form (element, membership value). Here is an example of a fuzzy set.

$$C = \{(8, 0.4), (9.9, 0.9), (10.001, 0.999), (10, 1), (-50, 0)\}$$

An examination of the membership values suggests that we are certain that 10 belongs to C (membership value is 1) and we are certain that −50 does not belong to C (membership value is 0). Every other element belongs to the set "to some degree."

The concept of a fuzzy set has been used in many real-world applications. Here are a few examples.

- Control of heating and air conditioning systems
- Compensation against vibrations in camcorders
- Voice recognition by computers
- Control of valves and dam gates at power plants
- Control of robots
- Control of subway trains
- Automatic camera focusing

A Fuzzy Heating System

Typical heating systems are controlled by a thermostat that turns a furnace on when the room temperature drops below a set point and turns the furnace off when the room temperature exceeds the set point. The furnace either runs at full force or it shuts down completely. This type of heating system is inefficient, and the frequent off and on changes can be annoying. A fuzzy heating system makes use of "fuzzy" definitions such as cold, warm, and hot to direct the furnace to run at low, medium, or full force. This results in a more efficient heating system and fewer temperature fluctuations.

Excursion Exercises

1. Mark, Erica, Larry, and Jennifer have each defined a fuzzy set to describe what they feel is a "good" grade. Each person paired the letter grades A, B, C, D, and F with a membership value. The results are as follows.

   Mark: $M = \{(A, 1), (B, 0.75), (C, 0.5), (D, 0.5), (F, 0)\}$

   Erica: $E = \{(A, 1), (B, 0), (C, 0), (D, 0), (F, 0)\}$

   Larry: $L = \{(A, 1), (B, 1), (C, 1), (D, 1), (F, 0)\}$

   Jennifer: $J = \{(A, 1), (B, 0.8), (C, 0.6), (D, 0.1), (F, 0)\}$

   a. Which of the four people considers an A grade to be the only good grade?
   b. Which of the four people is most likely to be satisfied with a grade of D or better?

(continued)

   c. Write a fuzzy set that you would use to describe the set of good grades. Consider only the letter grades A, B, C, D, and F.

2. In some fuzzy sets, membership values are given by a *membership graph* or by a *formula*. For instance, the following figure is a graph of the membership values of the fuzzy set *OLD*.

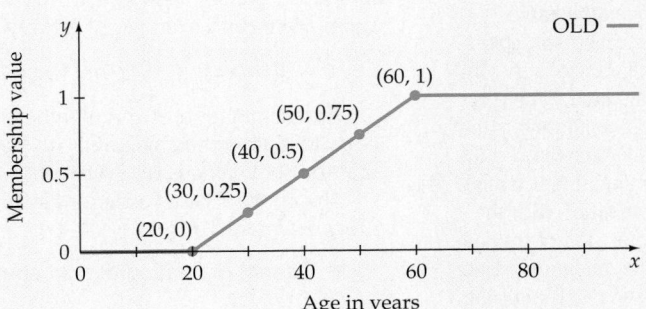

Use the membership graph of *OLD* to determine the membership value of each of the following.
a. $x = 15$ **b.** $x = 50$ **c.** $x = 65$
d. Use the graph of *OLD* to determine the age x with a membership value of 0.25.

An ordered pair (x, y) of a fuzzy set is a **crossover point** if its membership value is 0.5.
e. Find the crossover point for *OLD*.

3. The following membership graph provides a definition of real numbers x that are "about" 4.

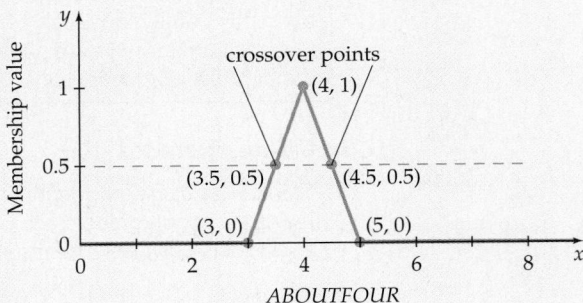

Use the graph of *ABOUTFOUR* to determine the membership value of:
a. $x = 2$ **b.** $x = 3.5$ **c.** $x = 7$
d. Use the graph of *ABOUTFOUR* to determine its crossover points.

4. The membership graphs in the following figure provide definitions of the fuzzy sets *COLD* and *WARM*.

(continued)

INSTRUCTOR NOTE
The process of creating a fuzzy definition is called **fuzzification**. The graphs in Excursion Exercise 4 are fuzzifications of the terms *cold* and *warm*. Fuzzification is not a unique process. For instance, "cold" can be defined in many different ways. Have your students create membership graphs that provide fuzzifications of terms such as *middle aged, short, tall, rich, poor, good test scores,* and *poor test scores.* It can be fun to compare their fuzzifications.

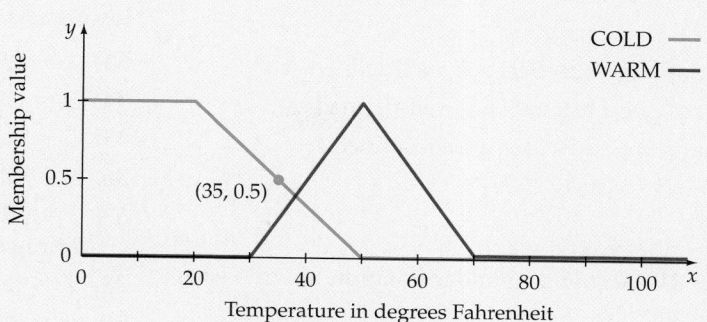

The point (35, 0.5) on the membership graph of *COLD* indicates that the membership value for $x = 35$ is 0.5. Thus, by this definition, 35°F is 50% cold. Use the above graphs to estimate

a. the *WARM* membership value for $x = 40$.
b. the *WARM* membership value for $x = 50$.
c. the crossover points of *WARM*.

5. The membership graph in Excursion Exercise 2 shows one person's idea of what ages are "old." Use a grid similar to the following to draw a membership graph that you feel defines the concept of being "young" in terms of a person's age in years.

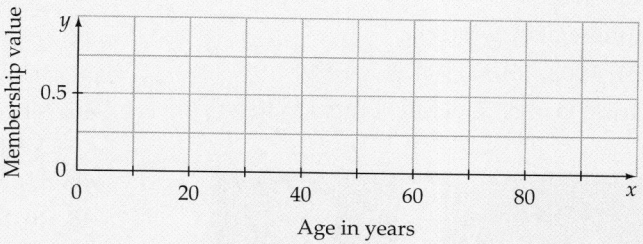

Show your membership graph to a few of your friends. Do they concur with your definition of "young?"

Exercise Set 2.1   (Suggested Assignment: 1–65 every other odd)

In Exercises 1–10, use the roster method to write each of the given sets. For some exercises you may need to consult a reference, such as the Internet or an encyclopedia.

1. The set of negative integers greater than −6
2. The set of whole numbers less than 12
3. The set of integers x that satisfy $x - 4 = 3$
4. The set of integers x that satisfy $2x - 1 = -11$
5. The set of counting numbers that satisfy $x + 4 = 1$
6. The set of whole numbers that satisfy $x - 1 < 4$
7. The set of the seven dwarfs

8. The set of oceans on Earth
9. The set of U.S. presidents that have served after Richard M. Nixon
10. The set of months with exactly 30 days

In Exercises 11–20, determine whether each statement is true or false. If the statement is false, give the reason.

11. $b \in \{a, b, c\}$
12. $0 \notin N$
13. $\{b\} \in \{a, b, c\}$

14. $\{1, 5, 9\} \sim \{\Psi, \Pi, \Sigma\}$

15. $\{0\} \sim \varnothing$

16. The set of large numbers is a well-defined set.

17. The set of good teachers is a well-defined set.

18. The set $\{x \mid 2 \leq x \leq 3\}$ is a well-defined set.

19. $\{x^2 \mid x \in I\} = \{x^2 \mid x \in N\}$

20. $0 \in \varnothing$

In Exercises 21–32, use set builder notation to write each of the following sets.

21. $\{1, 2, 3, 4, 5, 6, 7, 8, 9, 10, 11, 12\}$

22. $\{45, 55, 65, 75\}$

23. $\{5, 10, 15\}$

24. $\{1, 4, 9, 16, 25, 36, 49, 64, 81\}$

25. {January, March, May, July, August, October, December}

26. {Iowa, Ohio, Utah}

27. {Arizona, Alabama, Arkansas, Alaska}

28. {Mexico, Canada}

29. {spring, summer, fall, winter}

30. $\{1900, 1901, 1902, 1903, 1904, \ldots, 1999\}$

31. {Rupert Grint, Daniel Radcliffe, Emma Watson}

32. {George Walker Bush, Dick Cheney}

In Exercises 33–42, find the cardinality of each of the following sets.

33. $A = \{2, 4, 6, 8, 10, 12, 14, 16, 18, 20, 22\}$

34. $B = \{7, 14, 21, 28, 35, 42, 49, 56\}$

35. $D = \{$all dogs that can spell "elephant"$\}$

36. $S = \{$all states in the United States$\}$

37. $J = \{$all states in the United States that border Minnesota$\}$

38. $T = \{$all stripes on the U.S. flag$\}$

39. $N = \{$all baseball teams in the National League$\}$

40. $C = \{$all chess pieces on a chess board at the start of a chess game$\}$

41. $\{3, 6, 9, 12, 15, \ldots, 363\}$

42. $\{7, 11, 15, 19, 23, 27, \ldots, 407\}$

In Exercises 43–52, state whether each of the given pairs of sets are equal, equivalent, both, or neither.

43. The set of U.S. senators; the set of U.S. representatives

44. The set of single-digit counting numbers; the set of pins used in a regulation bowling game

45. The set of negative whole numbers; the set of integers between 4 and 5

46. The set of counting numbers; the set of two-digit integers

47. $\{1, 2, 3\}$; $\{$I, II, III$\}$

48. $\{6, 8, 10, 12\}$; $\{1, 2, 3, 4\}$

49. $\{a, b, c\}$; $\{a, b, b, c\}$

50. $\{2, 5\}$; $\{0, 1\}$

51. $\{t, e, n\}$; $\{ten\}$

52. $\{\ \}$; $\{0\}$

In Exercises 53–64, determine whether each of the sets is a well-defined set.

53. The set of good foods

54. The set of the six most populated cities in the United States

55. The set of tall buildings in the city of Chicago

56. The set of states that border Colorado

57. The set of even integers

58. The set of rational numbers of the form $\frac{1}{p}$, where p is a counting number

59. The set of former presidents of the United States that are alive at the present time

60. The set of real numbers larger than 89,000

61. The set of small countries

62. The set of great cities in which to live

63. The set consisting of the best soda drink

64. The set of fine wines

Extensions

CRITICAL THINKING

In Exercises 65–68, determine whether the given sets are equal. Recall that W represents the set of whole numbers and N represents the set of natural numbers.

65. $A = \{2n + 1 \mid n \in W\}$
$\quad B = \{2n - 1 \mid n \in N\}$

66. $A = \left\{16\left(\frac{1}{2}\right)^{n-1} \,\middle|\, n \in N\right\}$
$\quad B = \left\{16\left(\frac{1}{2}\right)^{n} \,\middle|\, n \in W\right\}$

67. $A = \{2n - 1 \mid n \in N\}$
$\quad B = \left\{\dfrac{n(n + 1)}{2} \,\middle|\, n \in N\right\}$

68. $A = \{3n + 1 \mid n \in W\}$
$\quad B = \{3n - 2 \mid n \in N\}$

69. Give an example of a set that cannot be written using the roster method.

EXPLORATIONS

70. In this section we have introduced the concept of *cardinal numbers*. Use the Internet or a mathematical textbook to find information about *ordinal numbers* and *nominal numbers*. Write a few sentences that explain the differences between these three types of numbers.

SECTION 2.2   **Subsets**

The Universal Set and the Complement of a Set

In complex problem solving situations and even in routine daily activities, we need to understand the set of all elements that are under consideration. For instance, when an instructor assigns letter grades, the possible choices may include A, B, C, D, F, and I. In this case the letter H is not a consideration. When you place a telephone call, you know that the area code is given by a natural number with three digits. In this instance a rational number such as $\frac{2}{3}$ is not a consideration. The set of all elements that are being considered is called the **universal set.** We will use the letter U to denote the universal set.

> **The Complement of a Set**
>
> The **complement** of a set A, denoted by A', is the set of all elements of the universal set U that are not elements of A.

EXAMPLE 1 ■ **Find the Complement of a Set**

Let $U = \{1, 2, 3, 4, 5, 6, 7, 8, 9, 10\}$, $S = \{2, 4, 6, 7\}$, and $T = \{x \mid x < 10 \text{ and } x \in \text{the odd counting numbers}\}$. Find

a. S' **b.** T'

Solution

a. The elements of the universal set are 1, 2, 3, 4, 5, 6, 7, 8, 9, and 10. From these elements we wish to exclude the elements of S, which are 2, 4, 6, and 7. Therefore $S' = \{1, 3, 5, 8, 9, 10\}$.

b. $T = \{1, 3, 5, 7, 9\}$. Excluding the elements of T from U gives us $T' = \{2, 4, 6, 8, 10\}$.

CHECK YOUR PROGRESS 1   Let $U = \{0, 2, 3, 4, 6, 7, 17\}$, $M = \{0, 4, 6, 17\}$, and $P = \{x \mid x < 7 \text{ and } x \in \text{the even natural numbers}\}$. Find

a. M' **b.** P'

Solution   *See page S4.*

There are two fundamental results concerning the universal set and the empty set. Because the universal set contains all elements under consideration, the complement of the universal set is the empty set. Conversely, the complement of the empty set is the universal set, because the empty set has no elements and the universal set contains all the elements under consideration. Using mathematical notation, we state these fundamental results as follows.

The Complement of the Universal Set and the Complement of the Empty Set

$U' = \varnothing$ and $\varnothing' = U$

Subsets

Consider the set of letters in the alphabet and the set of vowels $\{a, e, i, o, u\}$. Every element of the set of vowels is an element of the set of letters in the alphabet. The set of vowels is said to be a *subset* of the set of letters in the alphabet. We will often find it useful to examine subsets of a given set.

A Subset of a Set

Set A is a **subset** of set B, denoted by $A \subseteq B$, if and only if every element of A is also an element of B.

Here are two fundamental subset relationships.

Subset Relationships

$A \subseteq A$, for any set A
$\varnothing \subseteq A$, for any set A

To convince yourself that the empty set is a subset of any set, consider the following. We know that a set is a subset of a second set provided every element of the first set is an element of the second set. Pick an arbitrary set A. Because every element of the empty set (*there are none*) is an element of A, we know that $\varnothing \subseteq A$.

The notation $A \not\subseteq B$ is used to denote that A is *not* a subset of B. To show that A is not a subset of B, it is necessary to find at least one element of A that is not an element of B.

TAKE NOTE

Recall that W represents the set of whole numbers and N represents the set of natural numbers.

EXAMPLE 2 ■ True or False

Determine whether each statement is true or false.

a. $\{5, 10, 15, 20\} \subseteq \{0, 5, 10, 15, 20, 25, 30\}$

b. $W \subseteq N$

c. $\{2, 4, 6\} \subseteq \{2, 4, 6\}$

d. $\varnothing \subseteq \{1, 2, 3\}$

Solution

a. True; every element of the first set is an element of the second set.

b. False; 0 is a whole number, but 0 is not a natural number.

c. True; every set is a subset of itself.

d. True; the empty set is a subset of every set.

CHECK YOUR PROGRESS 2   Determine whether each statement is true or false.

a. $\{1, 3, 5\} \subseteq \{1, 5, 9\}$

b. The set of counting numbers is a subset of the set of natural numbers.

c. $\varnothing \subseteq U$

d. $\{-6, 0, 11\} \subseteq I$

Solution   See page S4.

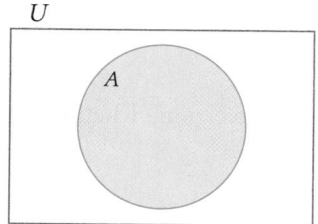

A Venn diagram.

The English logician John Venn (1834–1923) developed diagrams, which we now refer to as *Venn diagrams,* that can be used to illustrate sets and relationships between sets. In a **Venn diagram,** the universal set is represented by a rectangular region and subsets of the universal set are generally represented by oval or circular regions drawn inside the rectangle. The Venn diagram at the left shows a universal set and one of its subsets, labeled as set A. The size of the circle is not a concern. The region outside of the circle, but inside of the rectangle, represents the set A'.

QUESTION   *What set is represented by $(A')'$?*

ANSWER   *The set A' contains the elements of U that are not in A. By definition the set $(A')'$ contains only the elements of U that are elements of A. Thus $(A')' = A$.*

Proper Subsets of a Set

> **Proper Subset**
>
> Set A is a **proper subset** of set B, denoted by $A \subset B$, if every element of A is an element of B, and $A \neq B$.

To illustrate the difference between subsets and proper subsets, consider the following two examples.

1. Let $R = \{$Mars, Venus$\}$ and $S = \{$Mars, Venus, Mercury$\}$. The first set R is a subset of the second set S, because every element of R is an element of S. In addition, R is also a proper subset of S, because $R \neq S$.

2. Let $T = \{$Europe, Africa$\}$ and $V = \{$Africa, Europe$\}$. The first set T is a subset of the second set V; however, T is *not* a proper subset of V because $T = V$.

Venn diagrams can be used to represent proper subset relationships. For instance, if a set B is a proper subset of a set A, then we illustrate this relationship in a Venn diagram by drawing a circle labeled B inside of a circle labeled A. See the Venn diagram at the left.

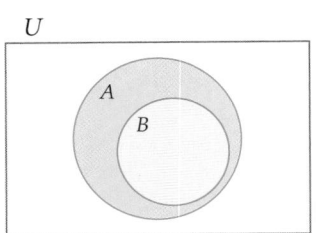

B is a proper subset of A.

EXAMPLE 3 ■ Proper Subsets

For each of the following, determine whether the first set is a proper subset of the second set.

a. $\{a, e, i, o, u\}, \{e, i, o, u, a\}$ **b.** N, I

Solution

a. Because the sets are equal, the first set is not a proper subset of the second set.

b. Every natural number is an integer, so the set of natural numbers is a subset of the set of integers. The set of integers contains elements that are not natural numbers, such as -3. Thus the set of natural numbers is a proper subset of the set of integers.

CHECK YOUR PROGRESS 3   For each of the following, determine whether the first set is a proper subset of the second set.

a. N, W **b.** $\{1, 4, 5\}, \{5, 1, 4\}$

Solution   See page S4.

Some counting problems in the study of probability require that we find all of the subsets of a given set. One way to find all the subsets of a given set is to use the method of making an organized list. First list the empty set, which has no elements. Next list all the sets that have exactly one element, followed by all the sets that contain exactly two elements, followed by all the sets that contain exactly three elements, and so on. This process is illustrated in the following example.

EXAMPLE 4 ■ **List all the Subsets of a Set**

List all the subsets of {1, 2, 3, 4}.

Solution

An organized list produces the following subsets.

{ }	• Subsets with 0 elements
{1}, {2}, {3}, {4}	• Subsets with 1 element
{1, 2}, {1, 3}, {1, 4}, {2, 3}, {2, 4}, {3, 4}	• Subsets with 2 elements
{1, 2, 3}, {1, 2, 4}, {1, 3, 4}, {2, 3, 4}	• Subsets with 3 elements
{1, 2, 3, 4}	• Subsets with 4 elements

CHECK YOUR PROGRESS 4   List all of the subsets of {a, b, c, d, e}.

Solution   *See page S4.*

Number of Subsets of a Set

INSTRUCTOR NOTE
Ask your students the following question:

If a set has n elements, then how many proper subsets does it have?

Answer: $2^n - 1$

The counting techniques developed in Section 1.3 can be used to produce the following result.

> **The Number of Subsets of a Set**
>
> A set with n elements has 2^n subsets.

EXAMPLE 5 ■ **The Number of Subsets of a Set**

Find the number of subsets of each set.

a. {1, 2, 3, 4, 5, 6}
b. {4, 5, 6, 7, 8, ..., 15}
c. ∅

Solution

a. {1, 2, 3, 4, 5, 6} has six elements. It has $2^6 = 64$ subsets.
b. {4, 5, 6, 7, 8, ..., 15} has 12 elements. It has $2^{12} = 4096$ subsets.
c. The empty set has zero elements. It has only one subset ($2^0 = 1$).

CHECK YOUR PROGRESS 5   Find the number of subsets of each set.

a. {Mars, Jupiter, Pluto}
b. $\{x \mid x \leq 15, x \in N\}$
c. {2, 3, 4, 5, ..., 11}

Solution   *See page S4.*

CALCULATOR NOTE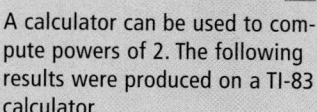

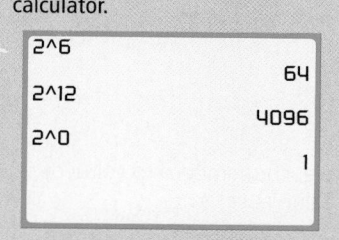

A calculator can be used to compute powers of 2. The following results were produced on a TI-83 calculator.

2^6	
	64
2^12	
	4096
2^0	
	1

Math Matters   The Barber's Paradox

Some problems that concern sets have led to paradoxes. For instance, in 1902, the mathematician Bertrand Russell developed the following paradox. "Is the set *A* of all sets that are not elements of themselves an element of itself?" Both the assumption that *A* is an element of *A* and the assumption that *A* is not an element of *A* lead to a contradiction. Russell's paradox has been popularized as follows.

> The town barber shaves all males who do not shave themselves, and he shaves only those males. The town barber is a male who shaves. Who shaves the barber?

The assumption that the barber shaves himself leads to a contradiction, and the assumption that the barber does not shave himself also leads to a contradiction.

Excursion

Subsets and Complements of Fuzzy Sets

✓ **TAKE NOTE**

A set such as {3, 5, 9} is called a *crisp set,* to distinguish it from a fuzzy set.

This excursion extends the concept of fuzzy sets that was developed in the Excursion in Section 2.1. Recall that the elements of a fuzzy set are ordered pairs. For any ordered pair (x, y) of a fuzzy set, the membership value y is a real number such that $0 \leq y \leq 1$.

The set of all x-values that are being considered is called the **universal set for the fuzzy set** and it is denoted by X.

Definition of a Fuzzy Subset

If the fuzzy sets $A = \{(x_1, a_1), (x_2, a_2), (x_3, a_3), \ldots\}$ and $B = \{(x_1, b_1), (x_2, b_2), (x_3, b_3), \ldots\}$ are both defined on the universal set $X = \{x_1, x_2, x_3, \ldots\}$, then $A \subseteq B$ if and only if $a_i \leq b_i$ for all i.

A fuzzy set A is a **subset** of a fuzzy set B if and only if the membership value of each element of A *is less than or equal to* its corresponding membership value in set B. For instance, in Excursion Exercise 1 in Section 2.1, Mark and Erica used fuzzy sets to describe the set of good grades as follows:

Mark: $M = \{(A, 1), (B, 0.75), (C, 0.5), (D, 0.5), (F, 0)\}$

Erica: $E = \{(A, 1), (B, 0), (C, 0), (D, 0), (F, 0)\}$

In this case fuzzy set E is a subset of fuzzy set M because each membership value of set E *is less than or equal to* its corresponding membership value in set M.

Definition of the Complement of a Fuzzy Set

Let A be the fuzzy set $\{(x_1, a_1), (x_2, a_2), (x_3, a_3), \ldots\}$ defined on the universal set $X = \{x_1, x_2, x_3, \ldots\}$. **Then the complement** of A is the fuzzy set $A' = \{(x_1, b_1), (x_2, b_2), (x_3, b_3), \ldots\}$, where each $b_i = 1 - a_i$.

(continued)

Each element of the fuzzy set A' has a membership value that is 1 minus its membership value in the fuzzy set A. For example, the complement of

$$S = \{(math, 0.8), (history, 0.4), (biology, 0.3), (art, 0.1), (music, 0.7)\}$$

is the fuzzy set

$$S' = \{(math, 0.2), (history, 0.6), (biology, 0.7), (art, 0.9), (music, 0.3)\}.$$

The membership values in S' were calculated by subtracting the corresponding membership values in S from 1. For instance, the membership value of math in set S is 0.8. Thus the membership value of math in set S' is $1 - 0.8 = 0.2$.

Excursion Exercises

1. Let $K = \{(1, 0.4), (2, 0.6), (3, 0.8), (4, 1)\}$ and
$J = \{(1, 0.3), (2, 0.6), (3, 0.5), (4, 0.1)\}$ be fuzzy sets defined on $X = \{1, 2, 3, 4\}$. Is $J \subseteq K$? Explain.

2. Consider the following membership graphs of *YOUNG* and *ADOLESCENT* defined on $X = \{x \mid 0 \leq x \leq 50\}$, where x is age in years.

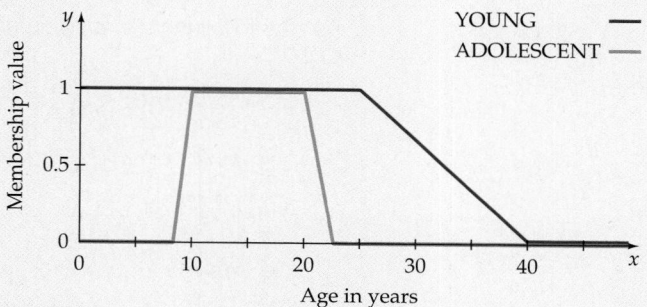

Is the fuzzy set *ADOLESCENT* a subset of the fuzzy set *YOUNG*? Explain.

3. Let the universal set be $\{A, B, C, D, F\}$ and let

$$G = \{(A, 1), (B, 0.7), (C, 0.4), (D, 0.1), (F, 0)\}$$

be a fuzzy set defined by Greg to describe what he feels is a good grade. Determine G'.

4. Let $C = \{(Ferrari, 0.9), (Ford Mustang, 0.6), (Dodge Neon, 0.5), (Hummer, 0.7)\}$ be a fuzzy set defined on the universal set $\{Ferrari, Ford Mustang, Dodge Neon, Hummer\}$. Determine C'.

Consider the following membership graph.

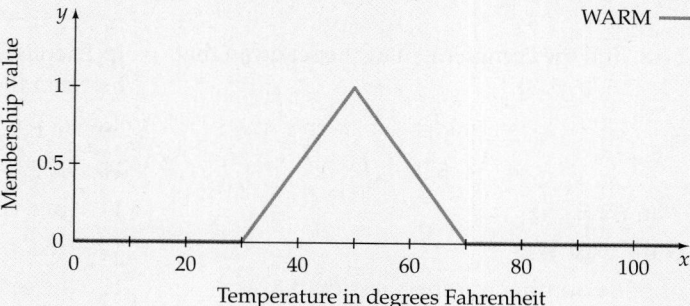

(continued)

The membership graph of *WARM'* can be drawn by *reflecting* the graph of *WARM* about the graph of the line $y = 0.5$, as shown in the following figure.

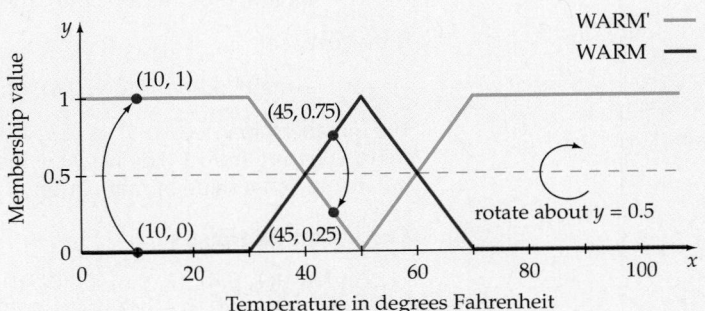

Note that when the membership graph of *WARM* is at a height of 0, the membership graph of *WARM'* is at a height of 1, and vice versa. In general, for any point (x, a) on the graph of *WARM*, there is a corresponding point $(x, 1 - a)$ on the graph of *WARM'*.

5. Use the following membership graph of *COLD* to draw the membership graph of *COLD'*.

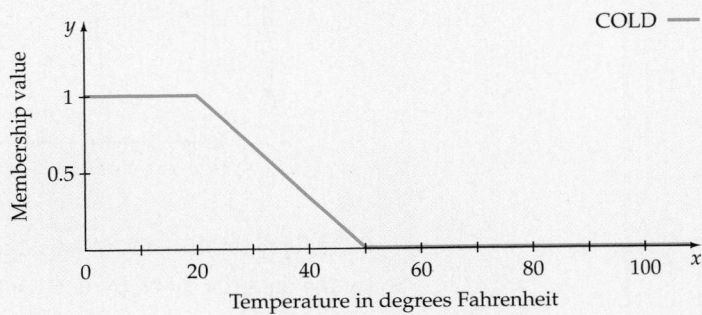

Exercise Set 2.2   (Suggested Assignment: 1–57 odds)

In Exercises 1–8, find the complement of the set given that $U = \{0, 1, 2, 3, 4, 5, 6, 7, 8\}$.

1. $\{2, 4, 6, 7\}$ **2.** $\{3, 6\}$

3. $\varnothing$ **4.** $\{4, 5, 6, 7, 8\}$

5. $\{x \mid x < 7 \text{ and } x \in N\}$

6. $\{x \mid x < 6 \text{ and } x \in W\}$

7. The set of odd counting numbers less than 8

8. The set of even counting numbers less than 10

In Exercises 9–18, insert either $\subseteq$ or $\nsubseteq$ in the blank space between the sets to make a true statement.

9. $\{a, b, c, d\}$ $\{a, b, c, d, e, f, g\}$

10. $\{3, 5, 7\}$ $\{3, 4, 5, 6\}$

11. $\{big, small, little\}$ $\{large, petite, short\}$

12. $\{red, white, blue\}$ $\{$the colors in the American flag$\}$

13. the set of integers   the set of rational numbers

14. the set of real numbers   the set of integers

15. ∅   {a, e, i, o, u}
16. {all sandwiches}   {all hamburgers}
17. {2, 4, 6, ..., 5000}   the set of even whole numbers
18. $\{x \mid x < 10 \text{ and } x \in Q\}$ the set of integers

In Exercises 19–36, let $U = \{p, q, r, s, t\}$, $D = \{p, r, s, t\}$, $E = \{q, s\}$, $F = \{p, t\}$, and $G = \{s\}$. Determine whether each statement is true or false.

19. $F \subseteq D$ 20. $D \subseteq F$
21. $F \subset D$ 22. $E \subset F$
23. $G \subset E$ 24. $E \subset D$
25. $G' \subset D$ 26. $E = F'$
27. $\varnothing \subset D$ 28. $\varnothing \subset \varnothing$
29. $D' \subseteq E$ 30. $G \in E$
31. $F \in D$ 32. $G \not\subseteq F$

33. D has exactly eight subsets and seven proper subsets.
34. U has exactly 32 subsets.
35. F and F' each have exactly four subsets.
36. $\{0\} = \varnothing$

37. A class of 16 students has 2^{16} subsets. Use a calculator to determine how long (to the nearest hour) it would take you to write all the subsets, assuming you can write each subset in 1 second.

38. A class of 32 students has 2^{32} subsets. Use a calculator to determine how long (to the nearest year) it would take you to write all the subsets, assuming you can write each subset in 1 second.

In Exercises 39–42, list all subsets of the given set.

39. $\{\alpha, \beta\}$ 40. $\{\alpha, \beta, \Gamma, \Delta\}$
41. $\{I, II, III\}$ 42. $\varnothing$

In Exercises 43–50, find the number of subsets of the given set.

43. $\{2, 5\}$ 44. $\{1, 7, 11\}$
45. $\{x \mid x \text{ is an even counting number between 7 and 21}\}$
46. $\{x \mid x \text{ is an odd integer between } -4 \text{ and } 8\}$
47. The set of eleven players on a football team
48. The set of all letters of our alphabet
49. The set of all negative whole numbers
50. The set of all single-digit natural numbers

51. Suppose you have a nickel, two dimes, and a quarter. One of the dimes was minted in 1976, and the other one was minted in 1992.

a. Assuming you choose at least one coin, how many different sets of coins can you form?

b. Assuming you choose at least one coin, how many different sums of money can you produce?

c. Explain why the answers in part a and part b are not the same.

52. The number of subsets of a set with n elements is 2^n.

a. Use a calculator to find the exact value of 2^{18}, 2^{19}, and 2^{20}.

b. What is the largest integer power of 2 for which your calculator will display the exact value?

53. Elementary school teachers use plastic pieces called *attribute pieces* to illustrate subset concepts and to determine whether a student has learned to distinguish between different shapes, sizes, and colors. The following figure shows 12 attribute pieces.

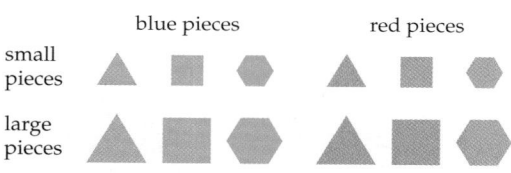

A set of 12 attribute pieces

A student has been asked to form the following sets. Determine the number of elements the student should have in each set.

a. The set of red attribute pieces
b. The set of red squares
c. The set of hexagons
d. The set of large blue triangles

54. A delicatessen makes a roast-beef-on-sour-dough sandwich for which you can choose from eight condiments.

a. How many different types of roast-beef-on-sourdough sandwiches can the delicatessen prepare?

b. What is the minimum number of condiments the delicatessen must have available if it wished to offer at least 2000 different types of roast-beef-on-sour-dough sandwiches?

55. A restaurant provides a brunch where the omelets are individually prepared. Each guest is allowed to choose from 10 different ingredients.

a. How many different types of omelets can the restaurant prepare?

b. What is the minimum number of ingredients that must be available if the restaurant wants to advertise that it offers over 4000 different omelets?

56. A truck company makes a pickup truck with 12 upgrade options. Some of the options are air conditioning, chrome wheels, and a CD player.

a. How many different versions of this truck can the company produce?

b. What is the minimum number of upgrade options the company must be able to provide if it wishes to offer at least 14,000 different versions of this truck?

Extensions

CRITICAL THINKING

57. **a.** Explain why $\{2\} \notin \{1, 2, 3\}$.

   b. Explain why $1 \not\subseteq \{1, 2, 3\}$.

   c. Consider the set $\{1, \{1\}\}$. Does this set have one or two elements? Explain.

58. **a.** A set has 1024 subsets. How many elements are in the set?

   b. A set has 255 proper subsets. How many elements are in the set?

   c. Is it possible for a set to have an odd number of subsets? Explain.

59. Five people, designated A, B, C, D, and E, serve on a committee. To pass a motion, at least three of the committee members must vote for the motion. In such a situation any set of three or more voters is called a **winning coalition** because if this set of people votes for a motion, the motion will pass. Any non-empty set of two or fewer voters is called a **losing coalition**.

   a. List all the winning coalitions.

   b. List all the losing coalitions.

EXPLORATIONS

60. Following is a list of all the subsets of $\{a, b, c, d\}$. Subsets with

   0 elements: $\{\ \}$
   1 element: $\{a\}, \{b\}, \{c\}, \{d\}$
   2 elements: $\{a, b\}, \{a, c\}, \{a, d\}, \{b, c\}, \{b, d\}, \{c, d\}$
   3 elements: $\{a, b, c\}, \{a, b, d\}, \{a, c, d\}, \{b, c, d\}$
   4 elements: $\{a, b, c, d\}$

There is 1 subset with zero elements and there are 4 subsets with exactly one element, 6 subsets with exactly two elements, 4 subsets with exactly three elements, and 1 subset with exactly four elements. Note that the numbers 1, 4, 6, 4, 1 are the numbers in row 4 of Pascal's triangle, which is shown below. Recall that the numbers in Pascal's triangle are created in the following manner. Each row begins and ends with the number 1. Any other number in a row is the sum of the two closest numbers above it. For instance, the first 10 in row 5 is the sum of the first 4 and the 6 in row 4.

					1						row 0
				1		1					row 1
			1		2		1				row 2
		1		3		3		1			row 3
	1		4		6		4		1		row 4
1		5		10		10		5		1	row 5

Pascal's Triangle

a. Use Pascal's triangle to make a conjecture about the numbers of subsets of {a, b, c, d, e} that have: zero elements, exactly one element, exactly two elements, exactly three elements, exactly four elements, and exactly five elements. Use your work from Check Your Progress 4 on page 63 to verify that your conjecture is correct.

b. Extend Pascal's triangle to show row 6. Use row 6 of Pascal's triangle to make a conjecture about the number of subsets of {a, b, c, d, e, f} that have exactly three elements. Make a list of all the subsets of {a, b, c, d, e, f} that have exactly three elements to verify that your conjecture is correct.

SECTION 2.3 | **Set Operations**

Intersection and Union of Sets

In Section 2.2 we defined the operation of finding the complement of a set. In this section we define the set operations *intersection* and *union*. In everyday usage, the word "intersection" refers to the *common region* where two streets cross. See the figure at the left. The intersection of two sets is defined in a similar manner.

> **Intersection of Sets**
>
> The **intersection** of sets A and B, denoted by $A \cap B$, is the set of elements common to both A and B.
>
> $$A \cap B = \{x \mid x \in A \quad \text{and} \quad x \in B\}$$

U

$A \cap B$

In the figure at the left, the region shown in blue represents the intersection of sets A and B.

TAKE NOTE

It is a mistake to write

$\{1, 5, 9\} \cap \{3, 5, 9\} = 5, 9$

The intersection of two sets is a set. Thus

$\{1, 5, 9\} \cap \{3, 5, 9\} = \{5, 9\}$

EXAMPLE 1 ■ Find Intersections

Let $A = \{1, 4, 5, 7\}$, $B = \{2, 3, 4, 5, 6\}$, and $C = \{3, 6, 9\}$. Find

a. $A \cap B$ **b.** $A \cap C$

Solution

a. The elements common to both sets are 4 and 5.

$$A \cap B = \{1, 4, 5, 7\} \cap \{2, 3, 4, 5, 6\}$$
$$= \{4, 5\}$$

b. Sets A and C have no common elements. Thus $A \cap C = \varnothing$.

CHECK YOUR PROGRESS 1   Let $D = \{0, 3, 8, 9\}$, $E = \{3, 4, 8, 9, 11\}$, and $F = \{0, 2, 6, 8\}$. Find

a. $D \cap E$ **b.** $D \cap F$

Solution   See page S4.

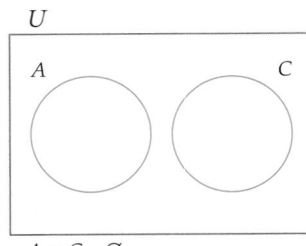

U

$A \cap C = \varnothing$

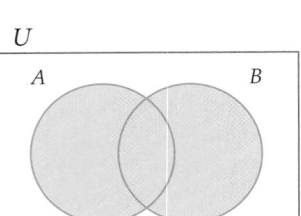

U

$A \cup B$

Two sets are **disjoint** if their intersection is the empty set. The sets A and C in Example 1b are disjoint. The Venn diagram at the left illustrates two disjoint sets.

In everyday usage, the word "union" refers to the act of uniting or joining together. The union of two sets has a similar meaning.

> ### Union of Sets
>
> The **union** of sets A and B, denoted by $A \cup B$, is the set that contains all the elements that belong to A or to B or to both.
>
> $$A \cup B = \{x \mid x \in A \quad \text{or} \quad x \in B\}$$

In the figure at the left, the region shown in blue represents the union of sets A and B.

EXAMPLE 2 ■ Find Unions

Let $A = \{1, 4, 5, 7\}$, $B = \{2, 3, 4, 5, 6\}$, and $C = \{3, 6, 9\}$. Find

a. $A \cup B$ **b.** $A \cup C$

Solution

a. List all the elements of set A, which are 1, 4, 5, and 7. Then add to your list the elements of set B that have not already been listed—in this case 2, 3, and 6. Enclose all elements with a pair of braces. Thus

$$A \cup B = \{1, 4, 5, 7\} \cup \{2, 3, 4, 5, 6\}$$
$$= \{1, 2, 3, 4, 5, 6, 7\}$$

b. $A \cup C = \{1, 4, 5, 7\} \cup \{3, 6, 9\}$
$$= \{1, 3, 4, 5, 6, 7, 9\}$$

CHECK YOUR PROGRESS 2   Let $D = \{0, 4, 8, 9\}$, $E = \{1, 5, 4, 7\}$, and $F = \{2, 6, 8\}$. Find

a. $D \cup E$ **b.** $D \cup F$

Solution   See page S4.

In mathematical problems that involve sets, the word "and" is interpreted to mean *intersection*. For instance, the phrase "the elements of A and B" means the elements of $A \cap B$. Similarly, the word "or" is interpreted to mean *union*. The phrase "the elements of A or B" means the elements of $A \cup B$.

EXAMPLE 3 ■ Describe Sets

Write a sentence that describes the set.

a. $A \cup (B \cap C)$ **b.** $J \cap K'$

Solution

a. The set $A \cup (B \cap C)$ can be described as "the set of all elements that are in A, or are in B and C."

b. The set $J \cap K'$ can be described as "the set of all elements that are in J and are not in K."

✔ **TAKE NOTE**

Would you like soup or salad?

In a sentence, the word "or" can mean one or the other, but not both. For instance, if a menu states that you can have soup or salad with your meal, this generally means that you can have either soup or salad for the price of the meal, but not both. In this case the word "or" is said to be an *exclusive or*. In the mathematical statement "*A* or *B*" the "or" is an *inclusive or*. It means *A* or *B*, or both.

Write a sentence that describes the set.

a. $D \cap (E' \cup F)$ **b.** $L' \cup M$

Solution   See page S4.

Venn Diagrams and Equality of Sets

The equality $A \cup B = A \cap B$ is true for some sets A and B, but not for all sets A and B. For instance, if $A = \{1, 2\}$ and $B = \{1, 2\}$, then $A \cup B = A \cap B$. However, we can prove that, in general, $A \cup B \neq A \cap B$ by finding an example for which the expressions are not equal. One such example is $A = \{1, 2, 3\}$ and $B = \{2, 3\}$. In this case $A \cup B = \{1, 2, 3\}$, whereas $A \cap B = \{2, 3\}$. This example is called a *counter-example*. The point to remember is that if you wish to show that two set expressions are *not equal*, then you need to find just one counterexample.

In the next example, we present a technique that uses Venn diagrams to determine whether two set expressions are equal.

EXAMPLE 4 ■ Equality of Sets

Use Venn diagrams to determine whether $(A \cup B)' = A' \cap B'$ for all sets A and B.

Solution

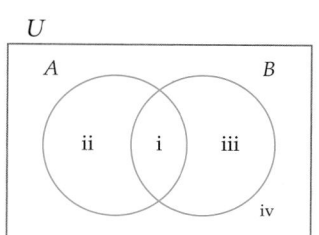

Draw a Venn diagram that shows the two sets A and B, as in the figure at the left. Label the four regions as shown. To determine what region(s) represents $(A \cup B)'$, first note that $A \cup B$ consists of regions i, ii, and iii. Thus $(A \cup B)'$ is represented by region iv. See Figure 2.1.

Draw a second Venn diagram. To determine what region(s) represents $A' \cap B'$, we shade A' (regions iii and iv) with a diagonal up pattern and we shade B' (regions ii and iv) with a diagonal down pattern. The intersection of these shaded regions, which is region iv, represents $A' \cap B'$. See Figure 2.2.

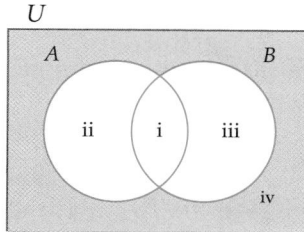

Figure 2.1

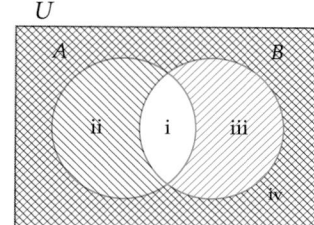

Figure 2.2

Because both $(A \cup B)'$ and $A' \cap B'$ are represented by the same region, we know that $(A \cup B)' = A' \cap B'$ for all sets A and B.

Use Venn diagrams to determine whether $(A \cap B)' = A' \cup B'$ for all sets A and B.

Solution   See page S4.

The properties that were verified in Example 4 and Check Your Progress 4 are known as **De Morgan's laws.**

De Morgan's Laws

For all sets A and B,

$(A \cup B)' = A' \cap B'$ and $(A \cap B)' = A' \cup B'$

De Morgan's law $(A \cup B)' = A' \cap B'$ can be stated as "the complement of the union of two sets is the intersection of the complements of the sets. De Morgan's law $(A \cap B)' = A' \cup B'$ can be stated as "the complement of the intersection of two sets is the union of the complements of the sets."

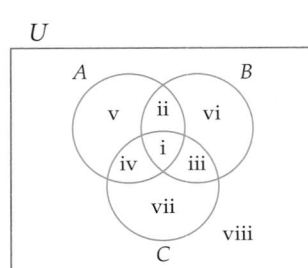

Augustus De Morgan

Math Matters   **Augustus De Morgan (1806–1871)**

Augustus De Morgan (dĭ môr′gən) was the first professor of mathematics at University College London and a cofounder of the London Mathematical Society. De Morgan formulated his laws during his study of symbolic logic. De Morgan's laws have applications in the areas of set theory, mathematical logic, and the design of electrical circuits.

Venn Diagrams Involving Three Sets

In the next example, we extend the Venn diagram procedure illustrated in the previous examples to expressions that involve three sets.

EXAMPLE 5 ■ **Equality of Set Expressions**

Use Venn diagrams to determine whether $A \cup (B \cap C) = (A \cup B) \cap C$ for all sets A, B, and C.

Solution

Draw a Venn diagram that shows the sets A, B, and C and the eight regions they form. See the figure at the left. To determine what region(s) represents $A \cup (B \cap C)$, we first consider $(B \cap C)$, represented by regions i and iii, because it is in parentheses. Set A is represented by the regions i, ii, iv, and v. Thus $A \cup (B \cap C)$ is represented by all of the listed regions (namely i, ii, iii, iv, and v), as shown in Figure 2.3 on the following page.

Draw a second Venn diagram showing the three sets A, B, and C. To determine what region(s) represents $(A \cup B) \cap C$, we first consider $(A \cup B)$ (regions i, ii, iii, iv, v, and vi) because it is in parentheses. Set C is represented by regions i, iii, iv, and vii. Therefore, the intersection of $(A \cup B)$ and C is represented by the overlap, or regions i, iii, and iv. See Figure 2.4 on the following page.

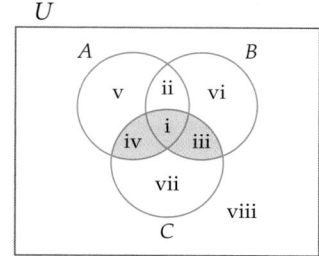

Figure 2.3   **Figure 2.4**

Because the sets $A \cup (B \cap C)$ and $(A \cup B) \cap C$ are represented by different regions, we conclude that $A \cup (B \cap C) \neq (A \cup B) \cap C$.

CHECK YOUR PROGRESS 5   Use Venn diagrams to determine whether $A \cup (B \cap C) = (A \cup B) \cap (A \cup C)$ for all sets A, B, and C.

Solution   *See page S5.*

Venn diagrams can be used to verify each of the following properties.

Properties of Sets

For all sets A and B,
   Commutative Properties
      $A \cap B = B \cap A$ Commutative property of intersection
      $A \cup B = B \cup A$ Commutative property of union

For all sets A, B, and C,
   Associative Properties
      $(A \cap B) \cap C = A \cap (B \cap C)$ Associative property of intersection
      $(A \cup B) \cup C = A \cup (B \cup C)$ Associative property of union
   Distributive Properties
      $A \cap (B \cup C) = (A \cap B) \cup (A \cap C)$ Distributive property of intersection over union
      $A \cup (B \cap C) = (A \cup B) \cap (A \cup C)$ Distributive property of union over intersection

P

QUESTION   *Does $(B \cup C) \cap A = (A \cap B) \cup (A \cap C)$?*

ANSWER   *Yes. The commutative property of intersection allows us to write $(B \cup C) \cap A$ as $A \cap (B \cup C)$, and $A \cap (B \cup C) = (A \cap B) \cup (A \cap C)$ by the distributive property of intersection over union.*

Application: Blood Groups and Blood Types

historical note

The Nobel Prize is an award granted to people who have made significant contributions to society. Nobel Prizes are awarded annually for achievements in physics, chemistry, physiology or medicine, literature, peace, and economics. The prizes were first established in 1901 by the Swedish industrialist Alfred Nobel, who invented dynamite. ■

Karl Landsteiner won a Nobel prize in 1930 for his discovery of the four different human blood groups. He discovered that the blood of each individual contains exactly one of the following combinations of antigens.

- Only A antigens (blood group A)
- Only B antigens (blood group B)
- Both A and B antigens (blood group AB)
- No A antigens and no B antigens (blood group O)

These four blood groups are represented by the Venn diagram at the left below.

In 1941, Landsteiner and Alexander Wiener discovered that human blood may or may not contain an Rh, or rhesus, factor. Blood with this factor is called Rh-positive and denoted by Rh+. Blood without this factor is called Rh-negative and is denoted by Rh−.

The Venn diagram in Figure 2.5 illustrates the eight blood types (A+, B+, AB+, O+, A−, B−, AB−, O−) that are possible if we consider antigens and the Rh factor.

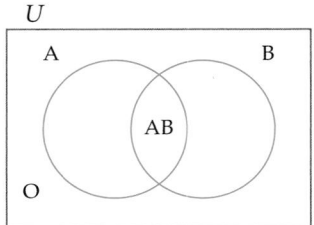

The four blood groups

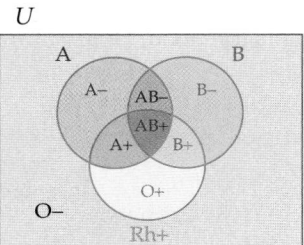

Figure 2.5 The eight blood types

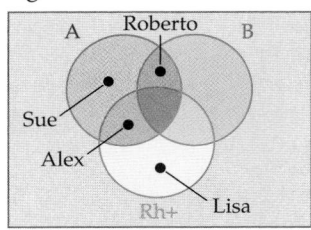

Figure 2.6

EXAMPLE 6 ■ Venn Diagrams and Blood Type

Use the Venn diagrams in Figures 2.5 and 2.6 to determine the blood type of each of the following people.

a. Sue   **b.** Lisa

Solution

a. Because Sue is in blood group A, not in blood group B, and not Rh+, her blood type is A−.

b. Lisa is in blood group O and she is Rh+, so her blood type is O+.

CHECK YOUR PROGRESS 6   Use the Venn diagrams in Figures 2.5 and 2.6 to determine the blood type of each of the following people.

a. Alex   **b.** Roberto

Solution   See page S5.

The following table shows the blood types that can safely be given during a blood transfusion to persons of each of the eight blood types.

Blood Transfusion Table

Recipient Blood Type	Donor Blood Type
A+	A+, A−, O+, O−
B+	B+, B−, O+, O−
AB+	A+, A−, B+, B−, AB+, AB−, O+, O−
O+	O+, O−
A−	A−, O−
B−	B−, O−
AB−	A−, B−, AB−, O−
O−	O−

Source: American Red Cross

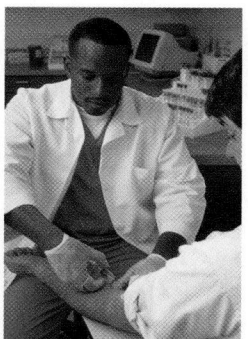

EXAMPLE 7 ▪ Applications of the Blood Transfusion Table

Use the blood transfusion table and Figures 2.5 and 2.6 to answer the following questions.

a. Can Sue safely be given a type O+ blood transfusion?

b. Why is a person with type O− blood called a *universal donor?*

Solution

a. Sue's blood type is A−. The blood transfusion table shows that she can safely receive blood only if it is type A− or type O−. Thus it is not safe for Sue to receive type O+ blood in a blood transfusion.

b. The blood transfusion table shows that all eight blood types can safely receive type O− blood. Thus a person with type O− blood is said to be a universal donor.

CHECK YOUR PROGRESS 7   Use the blood transfusion table and Figures 2.5 and 2.6 to answer the following questions.

a. Is it safe for Alex to receive type A− blood in a blood transfusion?

b. What blood type do you have if you are classified as a *universal recipient?*

Solution   See page S7.

Excursion

Union and Intersection of Fuzzy Sets

This Excursion extends the concepts of fuzzy sets that were developed in Sections 2.1 and 2.2.

There are a number of ways in which the *union of two fuzzy sets* and the *intersection of two fuzzy sets* can be defined. The definitions we will use are called the **standard union operator** and the **standard intersection operator.** These standard operators preserve many of the set relations that exist in standard set theory.

> ### Union and Intersection of Two Fuzzy Sets
>
> Let $A = \{(x_1, a_1), (x_2, a_2), (x_3, a_3), \ldots\}$ and $B = \{(x_1, b_1), (x_2, b_2), (x_3, b_3), \ldots\}$ Then
>
> $$A \cup B = \{(x_1, c_1), (x_2, c_2), (x_3, c_3), \ldots\}$$
>
> where c_i is the *maximum* of the two numbers a_i and b_i and
>
> $$A \cap B = \{(x_1, c_1), (x_2, c_2), (x_3, c_3), \ldots\}$$
>
> where c_i is the *minimum* of the two numbers a_i and b_i.

Each element of the fuzzy set $A \cup B$ has a membership value that is the *maximum* of its membership value in the fuzzy set A and its membership value in the fuzzy set B. Each element of the fuzzy set $A \cap B$ has a membership value that is the *minimum* of its membership value in fuzzy set A and its membership value in the fuzzy set B. In the following example we form the union and intersection of two fuzzy sets. Let P and S be defined as follows.

Paul: $P = \{(\text{math}, 0.2), (\text{history}, 0.5), (\text{biology}, 0.7), (\text{art}, 0.8), (\text{music}, 0.9)\}$

Sally: $S = \{(\text{math}, 0.8), (\text{history}, 0.4), (\text{biology}, 0.3), (\text{art}, 0.1), (\text{music}, 0.7)\}$

Then

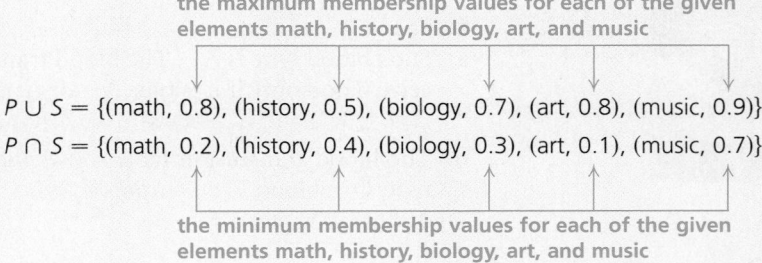

the maximum membership values for each of the given elements math, history, biology, art, and music

$P \cup S = \{(\text{math}, 0.8), (\text{history}, 0.5), (\text{biology}, 0.7), (\text{art}, 0.8), (\text{music}, 0.9)\}$

$P \cap S = \{(\text{math}, 0.2), (\text{history}, 0.4), (\text{biology}, 0.3), (\text{art}, 0.1), (\text{music}, 0.7)\}$

the minimum membership values for each of the given elements math, history, biology, art, and music

Excursion Exercises

In Excursion Exercise 1 of Section 2.1, we defined the following fuzzy sets.

Mark: $M = \{(A, 1), (B, 0.75), (C, 0.5), (D, 0.5), (F, 0)\}$

Erica: $E = \{(A, 1), (B, 0), (C, 0), (D, 0), (F, 0)\}$

Larry: $L = \{(A, 1), (B, 1), (C, 1), (D, 1), (F, 0)\}$

Jennifer: $J = \{(A, 1), (B, 0.8), (C, 0.6), (D, 0.1), (F, 0)\}$

(continued)

Use these fuzzy sets to find each of the following.

1. $M \cup J$　　**2.** $M \cap J$　　**3.** $E \cup J'$　　**4.** $J \cap L'$　　**5.** $J \cap (M' \cup L')$

Consider the following membership graphs.

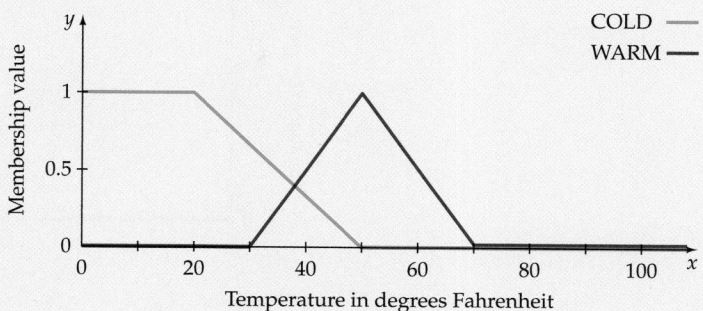

> ✔ **TAKE NOTE**
>
> The following lyrics from the old Scottish song *Loch Lomond* provide an easy way to remember how to draw the graph of the union or intersection of two membership graphs.
>
> > Oh! ye'll take the high road and I'll take the low road, and I'll be in Scotland afore ye.
>
> The graph of the union of two membership graphs takes the "high road" provided by the graphs and the graph of the intersection of two membership graphs takes the "low road" provided by the graphs.

The membership graph of *COLD* ∪ *WARM* is shown in purple in the following figure. The membership graph of *COLD* ∪ *WARM* lies on either the membership graph of *COLD* or

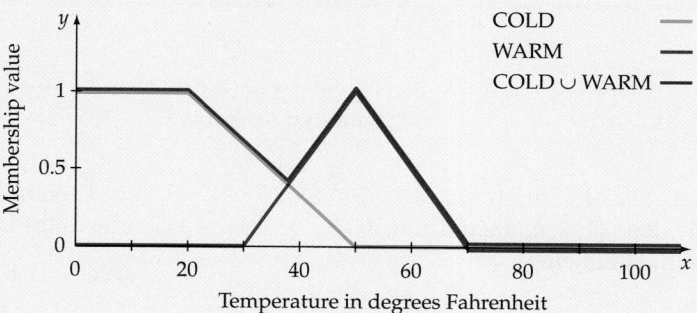

the membership graph of *WARM*, depending on which of these graphs is *highest* at any given temperature *x*.

The membership graph of *COLD* ∩ *WARM* is shown in green in the following figure. The membership graph of *COLD* ∩ *WARM* lies on either the membership graph of *COLD*

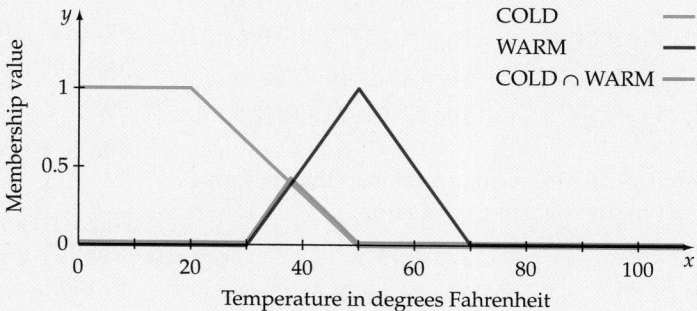

(continued)

or the membership graph of *WARM*, depending on which of these graphs is *lowest* at any given temperature *x*.

6. Use the following graphs to draw the membership graph of *WARM* ∪ *HOT*.

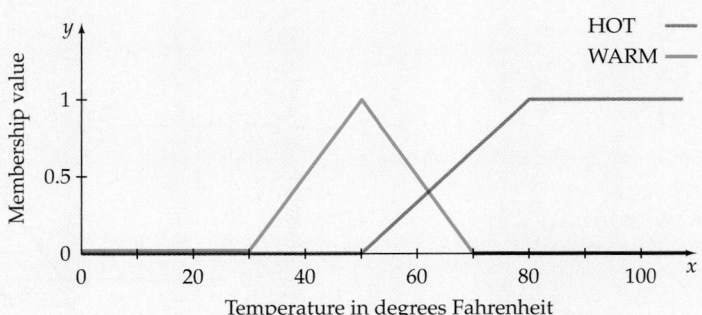

7. Let $X = \{a, b, c, d, e\}$ be the universal set. Determine whether De Morgan's Law $(A \cap B)' = A' \cup B'$ holds true for the fuzzy sets
$A = \{(a, 0.3), (b, 0.8), (c, 1), (d, 0.2), (e, 0.75)\}$ and
$B = \{(a, 0.5), (b, 0.4), (c, 0.9), (d, 0.7), (e, 0.45)\}.$

Exercise Set 2.3   (Suggested Assignment: 1–77 every other odd)

In Exercises 1–20, let $U = \{1, 2, 3, 4, 5, 6, 7, 8\}$, $A = \{2, 4, 6\}$, $B = \{1, 2, 5, 8\}$, and $C = \{1, 3, 7\}$. Find each of the following.

1. $A \cup B$ **2.** $A \cap B$

3. $A \cap B'$ **4.** $B \cap C'$

5. $(A \cup B)'$ **6.** $(A' \cap B)'$

7. $A \cup (B \cup C)$ **8.** $A \cap (B \cup C)$

9. $A \cap (B \cap C)$ **10.** $A' \cup (B \cap C)$

11. $B \cap (B \cup C)$ **12.** $A \cap A'$

13. $B \cup B'$ **14.** $(A \cap (B \cup C))'$

15. $(A \cup C') \cap (B \cup A')$ **16.** $(A \cup C') \cup (B \cup A')$

17. $(C \cup B') \cup \varnothing$ **18.** $(A' \cup B) \cap \varnothing$

19. $(A \cup B) \cap (B \cap C')$ **20.** $(B \cap A') \cup (B' \cup C)$

 In Exercises 21–28, write a sentence that describes the given mathematical expression.

21. $L' \cup T$ **22.** $J' \cap K$

23. $A \cup (B' \cap C)$ **24.** $(A \cup B) \cap C'$

25. $T \cap (J \cup K')$ **26.** $(A \cap B) \cup C$

27. $(W \cap V) \cup (W \cap Z)$ **28.** $D \cap (E \cup F)'$

In Exercises 29–36, draw a Venn diagram to show each of the following sets.

29. $A \cap B'$ **30.** $(A \cap B)'$

31. $(A \cup B)'$ **32.** $(A' \cap B) \cup B'$

33. $A \cap (B \cup C')$ **34.** $A \cap (B' \cap C)$

35. $(A \cup C) \cap (B \cup C')$ **36.** $(A' \cap B) \cup (A \cap C')$

In Exercises 37–40, draw two Venn diagrams to determine whether the following expressions are equal for all sets *A* and *B*.

37. $A \cap B'$; $A' \cup B$

38. $A' \cap B$; $A \cup B'$

39. $A \cup (A' \cap B)$; $A \cup B$

40. $A' \cap (B \cup B')$; $A' \cup (B \cap B')$

In Exercises 41–46, draw two Venn diagrams to determine whether the following expressions are equal for all sets *A*, *B*, and *C*.

41. $(A \cup C) \cap B'$; $A' \cup (B \cup C)$

42. $A' \cap (B \cap C)$, $(A \cup B') \cap C$

43. $(A' \cap B) \cup C$, $(A' \cap C) \cap (A' \cap B)$

44. $A' \cup (B' \cap C)$, $(A' \cup B') \cap (A' \cup C)$

45. $((A \cup B) \cap C)'$, $(A' \cap B') \cup C'$

46. $(A \cap B) \cap C$, $(A \cup B \cup C)'$

Computers and televisions make use of *additive color mixing*. The following figure shows that when the *primary colors* red R, green G, and blue B are mixed together using additive color mixing, they produce white, W. Using set notation, we state this as $R \cap B \cap G = W$. The colors yellow Y, cyan C, and magenta M are called *secondary colors*. A secondary color is produced by adding exactly two of the primary colors.

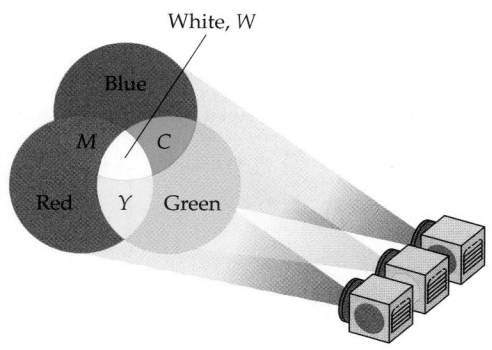

The ExploreScience web site (**www.explore-science.com**) has an interactive experiment called *Additive Colors* that enables you to mix red, blue, and green to produce the results shown above.

In Exercises 47–49, determine which color is represented by each of the following. Assume the colors are being mixed using additive color mixing. (Use R for red, G for green, and B for blue.)

47. $R \cap G \cap B'$

48. $R \cap G' \cap B$

49. $R' \cap G \cap B$

Artists that paint with pigments use *subtractive color mixing* to produce different colors. In a subtractive color mixing system, the primary colors are cyan C, magenta M, and yellow Y. The following figure shows that when the three primary colors are mixed in equal amounts, using subtractive color mixing, they form black, K. Using set nota-

tion, we state this as $C \cap M \cap Y = K$. In subtractive color mixing the colors red R, blue B, and green G are the secondary colors. As mentioned previously, a secondary color is produced by mixing equal amounts of exactly two of the primary colors.

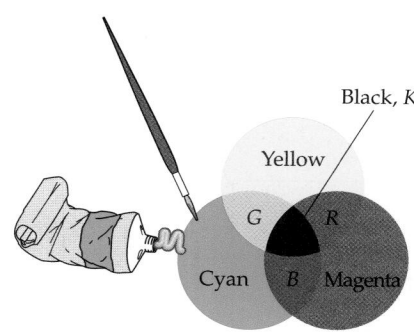

The ExploreScience web site (**www.explore-science.com**) has an interactive experiment called *Subtractive Colors* that enables you to mix cyan, yellow, and magenta to produce the results shown above.

In Exercises 50–52, determine which color is represented by each of the following. Assume the colors are being mixed using subtractive color mixing. (Use C for cyan, M for magenta, and Y for yellow.)

50. $C \cap M \cap Y'$

51. $C' \cap M \cap Y$

52. $C \cap M' \cap Y$

In Exercises 53–62, use set notation to describe the shaded region. You may use any of the following symbols: A, B, C, $\cap$, $\cup$, and $'$. Keep in mind that each shaded region has more than one set description.

53.

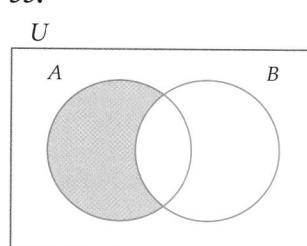

54.

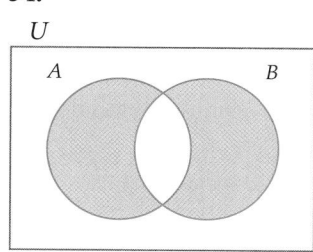

55.

56.

57.

58.

59.

60.

61.

62.

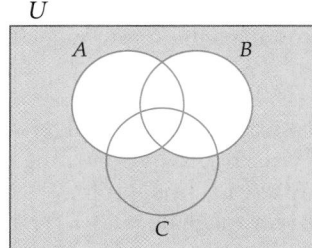

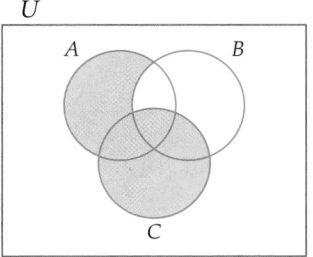

In Exercises 63–66, draw a Venn diagram with each of the given elements placed in the correct region.

63. $U = \{-1, 0, 1, 2, 3, 4, 5, 6, 7, 8, 9\}$
$A = \{1, 3, 5\}$
$B = \{3, 5, 7, 8\}$
$C = \{-1, 8, 9\}$

64. $U = \{2, 4, 6, 8, 10, 12, 14\}$
$A = \{2, 10, 12\}$
$B = \{4, 8\}$
$C = \{6, 8, 10\}$

65. $U = \{$Sue, Bob, Al, Jo, Ann, Herb, Eric, Mike, Sal$\}$
$A = \{$Sue, Herb$\}$
$B = \{$Sue, Eric, Jo, Ann$\}$
$Rh+ = \{$Eric, Sal, Al, Herb$\}$

66. $U = \{$Hal, Marie, Rob, Armando, Joel, Juan, Melody$\}$
$A = \{$Marie, Armando, Melody$\}$
$B = \{$Rob, Juan, Hal$\}$
$Rh+ = \{$Hal, Marie, Rob, Joel, Jaun, Melody$\}$

In Exercises 67 and 68, use two Venn diagrams to verify the following properties for all sets A, B, and C.

67. The associative property of intersection

$$(A \cap B) \cap C = A \cap (B \cap C)$$

68. The distributive property of intersection over union

$$A \cap (B \cup C) = (A \cap B) \cup (A \cap C)$$

Extensions

CRITICAL THINKING

Another operation that can be defined on sets A and B is the **difference of the sets,** denoted by $A - B$. Here is a formal definition of the difference of sets A and B.

$$A - B = \{x \,|\, x \in A \quad \text{and} \quad x \notin B\}$$

Thus $A - B$ is the set of elements that belong to A but not to B. For instance, let $A = \{1, 2, 3, 7, 8\}$ and $B = \{2, 7, 11\}$. Then $A - B = \{1, 3, 8\}$.

In Exercises 69–74, determine each difference given that $U = \{1, 2, 3, 4, 5, 6, 7, 8, 9\}$, $A = \{2, 4, 6, 8\}$, and $B = \{2, 3, 8, 9\}$.

69. $B - A$

70. $A - B$

71. $A - B'$

72. $B' - A$

73. $A' - B'$

74. $A' - B$

75. Write a few paragraphs about the life of John Venn and his work in the area of mathematics.

The Venn diagram on the following page illustrates that four sets can partition the universal set into 16 different regions.

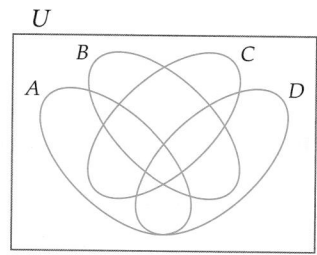

In Exercises 76 and 77, use a Venn diagram similar to the one above to shade in the region represented by the given expression.

76. $(A \cap B) \cup (C' \cap D)$

77. $(A \cup B)' \cap (C \cap D)$

SECTION 2.4   ## Applications of Sets

Surveys: An Application of Sets

Counting problems occur in many areas of applied mathematics. To solve these counting problems we often make use of a Venn diagram and the inclusion-exclusion principle, which will be presented in this section.

EXAMPLE 1 ■ A Survey of Preferences

A movie company is making plans for future movies it wishes to produce. The company has done a random survey of 1000 people. The results of the survey are shown below.

> 695 people like action adventures.
>
> 340 people like comedies.
>
> 180 people like both action adventures and comedies.

Of the people surveyed, how many people:

a. like action adventures but not comedies?

b. like comedies but not action adventures?

c. do not like either of these types of movies?

---

1. Anthony W. F. Edwards, "Venn diagrams for many sets," *New Scientist*, 7 January 1989, pp. 51–56.

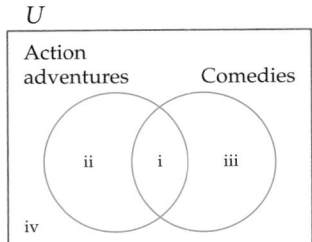

Figure 2.7

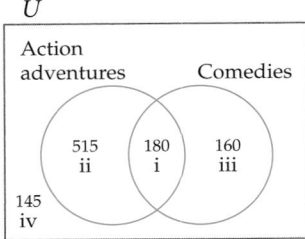

Figure 2.8

Solution

A Venn diagram can be used to illustrate the results of the survey. We use two overlapping circles (see Figure 2.7). One circle represents the set of people who like action adventures and the other represents the set of people who like comedies. The region i where the circles intersect represents the set of people who like both types of movies.

We start with the information that 180 people like both types of movies and write 180 in region i. See Figure 2.8.

a. Regions i and ii have a total of 695 people. So far we have accounted for 180 of these people in region i. Thus the number of people in region ii, which is the set of people who like action adventures but do not like comedies, is $695 - 180 = 515$.

b. Regions i and iii have a total of 340 people. Thus the number of people in region iii, which is the set of people who like comedies but do not like action adventures, is $340 - 180 = 160$.

c. The people that do not like action adventure movies or comedies is represented by region iv. The number of people in region iv must be the total number of people, which is 1000, less the number of people accounted for in regions i, ii, and iii, which is 855. Thus the number of people who do not like either type of movie is $1000 - 855 = 145$.

CHECK YOUR PROGRESS 1   The athletic director of a school has surveyed 200 students. The survey results are shown below.

　　140 students like volleyball.

　　120 students like basketball.

　　85 students like both volleyball and basketball.

Of the students surveyed, how many students:

a. like volleyball but not basketball?

b. like basketball but not volleyball?

c. do not like either of these sports?

Solution   See page S5.

In the next example we consider a more complicated survey that involves three types of music. It is helpful to use the following techniques to solve the counting problems in this section.

Counting Techniques with Venn Diagrams

　1. Draw a Venn diagram that shows all of the subsets involved in the problem.

　2. Use the information in the problem to label each subset with its cardinal number. If possible, start with the innermost regions.

　3. Use the cardinal numbers from Step 2 to answer the questions posed in the problem.

EXAMPLE 2 ■ A Music Survey

A music teacher has surveyed 495 students. The results of the survey are listed below.

> 320 students like rap music.
>
> 395 students like rock music.
>
> 295 students like heavy metal music.
>
> 280 students like both rap music and rock music.
>
> 190 students like both rap music and heavy metal music.
>
> 245 students like both rock music and heavy metal music.
>
> 160 students like all three.

How many students:

a. like exactly two of the three types of music?

b. like only rock music?

c. like only one of the three types of music?

Solution

The Venn diagram at the left shows three overlapping circles. Region i represents the set of students who like all three types of music. Each of the regions v, vi, and vii represent the students who like only one type of music.

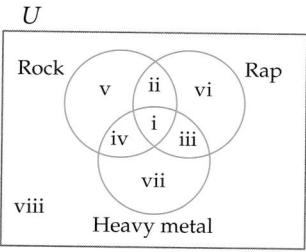

a. The survey shows that 245 students like rock and heavy metal music, so the numbers we place in regions i and iv must have a sum of 245. Since region i has 160 students, we see that region iv must have 245 − 160 = 85 students. In a similar manner we can determine that region ii has 120 students and region iii has 30 students. Thus 85 + 120 + 30 = 235 students like exactly two of the three types of music.

b. The sum of the students represented by regions i, ii, iv, and v must be 395. The number of students in region v must be the difference between this total and the sum of the numbers of students in region i, ii, and iv. Thus the number of students who like only rock music is 395 − (160 + 120 + 85) = 30. See the Venn diagram at the left.

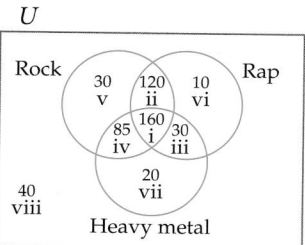

c. Using the same reasoning as in part b, we find that region vi has 10 students and region vii has 20 students. To find the number of students who like only one type of music, find the sum of the numbers of students in regions v, vi, and vii, which is 30 + 10 + 20 = 60. See the Venn diagram at the left.

CHECK YOUR PROGRESS 2

An activities director for a cruise ship has surveyed 240 passengers. Of the 240 passengers,

> 135 like swimming.   80 like swimming and dancing.
>
> 150 like dancing.   40 like swimming and games.
>
> 65 like games.   25 like dancing and games.
>
>   15 like all three activities.

How many passengers:

a. like exactly two of the three types of activities?

b. like only swimming?

c. like none of these activities?

Solution   See page S5.

Grace Chisholm Young

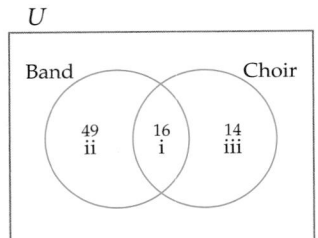

MathMatters   **Grace Chisholm Young (1868–1944)**

Grace Chisholm Young studied mathematics at Girton College, which is part of Cambridge University. In England at that time, women were not allowed to earn a university degree, so she decided to continue her mathematical studies at the University of Göttingen in Germany, where her advisor was the renowned mathematician Felix Klein. She excelled while at Göttingen and at the age of 27 earned her doctorate in mathematics, magna cum laude. She was the first woman to officially earn a doctorate degree from a German university. Shortly after her graduation she married the mathematician William Young. Together they published several mathematical papers and books, one of which was the first textbook on set theory.

The Inclusion-Exclusion Principle

A music director wishes to take the band and the choir on a field trip. There are 65 students in the band and 30 students in the choir. The number of students in both the band and the choir is 16. How many students should the music director plan on taking on the field trip?

Using the process developed in the previous examples, we find that the number of students that are in only the band is $65 - 16 = 49$. The number of students that are in only the choir is $30 - 16 = 14$. See the Venn Diagram at the left. Adding the numbers of students in regions i, ii, and iii gives us a total of $49 + 16 + 14 = 79$ students that might go on the field trip.

Although we can use Venn diagrams to solve counting problems, it is more convenient to make use of the following technique. First add the number of students in the band to the number of students in the choir. Then subtract the number of students that are in both the band and the choir. This technique gives us a total of $(65 + 30) - 16 = 79$ students, the same result as above. The reason we subtract the 16 students is we have counted each of them twice. Note that first we include the students that are in both the band and the choir twice, and then we exclude them once. This procedure leads us to the following result.

The Inclusion-Exclusion Principle

For all finite sets A and B.

$$n(A \cup B) = n(A) + n(B) - n(A \cap B)$$

QUESTION   *What must be true of the finite sets A and B if*
$n(A \cup B) = n(A) + n(B)$?

ANSWER   *A and B must be disjoint sets.*

EXAMPLE 3 ■ An Application of the Inclusion-Exclusion Principle

A school finds that 430 of its students are registered in chemistry, 560 are registered in mathematics, and 225 are registered in both chemistry and mathematics. How many students are registered in chemistry or mathematics.

Solution

Let C = {students registered in chemistry} and let M = {students registered in mathematics}.

$$n(C \cup M) = n(C) + n(M) - n(C \cap M)$$
$$= 430 + 560 - 225$$
$$= 765$$

CHECK YOUR PROGRESS 3   A high school has 80 athletes who play basketball, 60 athletes who play soccer, and 24 athletes who play both basketball and soccer. How many athletes play either basketball or soccer?

Solution   See page S5.

The inclusion-exclusion principle can be used provided we know the number of elements in any three of the four sets in the formula.

EXAMPLE 4 ■ An Application of the Inclusion-Exclusion Principle

Given $n(A) = 15$, $n(B) = 32$, and $n(A \cup B) = 41$, find $n(A \cap B)$.

Solution

Substitute the given information in the inclusion-exclusion formula and solve for the unknown.

$$n(A \cup B) = n(A) + n(B) - n(A \cap B)$$
$$41 = 15 + 32 - n(A \cap B)$$
$$41 = 47 - n(A \cap B)$$

Thus

$$n(A \cap B) = 47 - 41$$
$$n(A \cap B) = 6$$

CHECK YOUR PROGRESS 4   Given $n(A) = 785$, $n(B) = 162$, and $n(A \cup B) = 852$, find $n(A \cap B)$.

Solution   See page S6.

The inclusion-exclusion formula can be adjusted and applied to problems that involve percents. In the following formula we denote "the percent in set A" by the notation $p(A)$.

The Percent Inclusion-Exclusion Formula

For all finite sets A and B, $p(A \cup B) = p(A) + p(B) - p(A \cap B)$.

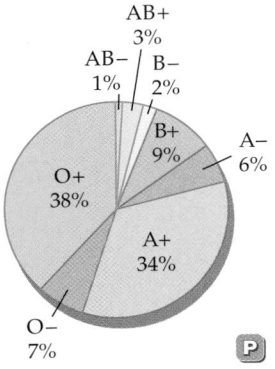

Percentage of U.S. Population with Each Blood Type

Source: American Association of Blood Banks, **http://www.aabb.org/ All_About_Blood/FAQs/ aabb_faqs.htm**

EXAMPLE 5 ■ An Application of the Percent Inclusion-Exclusion Formula

 The American Association of Blood Banks reports that about

44% of the U.S. population has the A antigen.

15% of the U.S. population has the B antigen.

4% of the U.S. population has both the A and the B antigen.

(*Source:* **http://www.aabb.org/All_About_Blood/FAQs/aabb_faqs.htm**)

Use the percent inclusion-exclusion formula to estimate the percent of the U.S. population that has the A antigen or the B antigen.

Solution

We are given $p(A) = 44\%$, $p(B) = 15\%$, and $p(A \cap B) = 4\%$. Substituting in the percent inclusion-exclusion formula gives

$$p(A \cup B) = p(A) + p(B) - p(A \cap B)$$
$$= 44\% + 15\% - 4\%$$
$$= 55\%$$

Thus about 55% of the U.S. population has the A antigen or the B antigen. Notice that this result checks with the data given in the pie chart at the left (1% + 3% + 2% + 9% + 6% + 34% = 55%).

CHECK YOUR PROGRESS 5   The American Association of Blood Banks reports that about

44% of the U.S. population has the A antigen.

84% of the U.S. population is Rh+.

91% of the U.S. population either has the A antigen or is Rh+.

(*Source:* **http://www.aabb.org/All_About_Blood/FAQs/aabb_faws.html**)

Use the percent inclusion-exclusion formula to estimate the percent of the U.S. population that has the A antigen *and* is Rh+.

Solution   *See page S6.*

In the next example the data is provided in a table. The number in column *A* and row *M* represents the number of elements in $A \cap M$. The sum of all the numbers in column *A* and column *L* represents the number of elements in $A \cup L$.

EXAMPLE 6 ■ A Survey Presented in Tabular Form

A survey of men *M*, women *W*, and children *C* concerning the use of the Internet search engines Alta Vista *A*, Yahoo *Y*, and Lycos *L* shows the following results.

	Alta Vista (*A*)	Yahoo (*Y*)	Lycos (*L*)
Men (*M*)	440	310	275
Women (*W*)	390	280	325
Children (*C*)	140	410	40

Use the data in the table to find each of the following.

a. $n(W \cap Y)$ **b.** $n(A \cap C')$ **c.** $n(M \cap (A \cup L))$

Solution

a. The table shows that 280 of the women surveyed use Yahoo as a search engine. Thus, $n(W \cap Y) = 280$.

b. The set $A \cap C'$ is the set of surveyed Alta Vista users who are men or women. The number in this set is $440 + 390 = 830$.

c. The number of men in the survey that use either Alta Vista or Lycos is $440 + 275 = 715$.

CHECK YOUR PROGRESS 6 Use the table in Example 6 to find each of the following.

a. $n(Y \cap C)$ **b.** $n(L \cap M')$ **c.** $n((A \cap M) \cup (A \cap W))$

Solution *See page S6.*

Excursion

Voting Systems

There are many types of voting systems. When voting for or against a resolution, a one-person, one-vote *majority system* is often used to decide the outcome. In this type of voting, each voter receives one vote and the resolution passes only if it receives *most* of the votes.

In any voting system, the number of votes that is required to pass a resolution is called the **quota**. A **coalition** is a set of voters each of whom votes the same way, either for or against a resolution. A **winning coalition** is a set of voters the sum of whose votes is greater than or equal to the quota. A **losing coalition** is a set of voters the sum of whose votes is less than the quota.

You can find all the winning coalitions in a voting process by making an organized list. For instance, consider the committee consisting of Alice, Barry, Cheryl, and Dylan. To decide on any issues, they use a one-person, one-vote majority voting system. Since each of the four voters has a single vote, the quota for this majority voting system is 3. The winning coalitions consist of all subsets of the voters that have three or more people. We list these winning coalitions in the table below, where A represents Alice, B represents Barry, C represents Cheryl, and D represents Dylan.

Winning Coalition	Sum of the Votes
{A, B, C}	3
{A, B, D}	3
{A, C, D}	3
{B, C, D}	3
{A, B, C, D}	4

(continued)

historical note

An ostrakon

In ancient Greece, the citizens of Athens adopted a procedure that allowed them to vote for the expulsion of any prominent person. The purpose of this procedure, known as an *ostracism,* was to limit the political power that any one person could attain.

In an ostracism, each voter turned in a *potsherd,* a piece of pottery fragment, on which was inscribed the name of the person the voter wished to ostracize. The pottery fragments used in the voting process became known as *ostrakon.*

The person who received the majority of the votes, above some set minimum, was exiled from Athens for a period of 10 years. ■

A **weighted voting system** is one in which some voters' votes carry more weight regarding the outcome of an election. As an example, consider a selection committee that consists of four people designated by A, B, C, and D. Voter A's vote has a weight of 2 and the vote of each other member of the committee has a weight of 1. The quota for this weighted voting system is 3. A winning coalition must have a weighted voting sum of at least 3. The winning coalitions are listed in the table below.

Winning Coalition	Sum of the Weighted Votes
{A, B}	3
{A, C}	3
{A, D}	3
{B, C, D}	3
{A, B, C}	4
{A, B, D}	4
{A, C, D}	4
{A, B, C, D}	5

A **minimal winning coalition** is a winning coalition that has no proper subset that is a winning coalition. In a minimal winning coalition each voter is said to be a **critical voter,** because if any of the voters leaves the coalition, the coalition will then become a losing coalition. In the above table, the minimal winning coalitions are {A, B}, {A, C}, {A, D}, and {B, C, D}. If any single voter leaves one of these coalitions, then the coalition will become a losing coalition. The coalition {A, B, C, D} is not a minimal winning coalition, because it contains at least one proper subset, for instance {A, B, C}, that is a winning coalition.

Excursion Exercises

1. A selection committee consists of Ryan, Susan, and Trevor. To decide on issues, they use a one-person, one-vote majority voting system.
   a. Find all winning coalitions.
   b. Find all losing coalitions.

2. A selection committee consists of three people designated by M, N, and P. M's vote has a weight of 3, N's vote has a weight of 2, and P's vote has a weight of 1. The quota for this weighted voting system is 4. Find all winning coalitions.

3. Determine the minimal winning coalitions for the voting system in Excursion Exercise 2.

Additional information on the applications of mathematics to voting systems is given in Chapter 13.

Exercise Set 2.4

(Suggested Assignment: 1–25 odds)

In Exercises 1–8, let $U = \{$English, French, History, Math, Physics, Chemistry, Psychology, Drama$\}$,
$A = \{$English, History, Psychology, Drama$\}$,
$B = \{$Math, Physics, Chemistry, Psychology, Drama$\}$, and
$C = \{$French, History, Chemistry$\}$.
Find each of the following.

1. $n(B \cup C)$
2. $n(A \cup B)$
3. $n(B) + n(C)$
4. $n(A) + n(B)$
5. $n(A \cup B \cup C)$
6. $n(A \cap B)$
7. $n(A) + n(B) + n(C)$
8. $n(A \cap B \cap C)$

9. Verify that for A and B as defined in Exercises 1–8, $n(A \cup B) = n(A) + n(B) - n(A \cap B)$.

10. Verify that for A and C as defined in Exercises 1–8, $n(A \cup C) = n(A) + n(C) - n(A \cap C)$.

11. Given $n(J) = 245$, $n(K) = 178$, and $n(J \cup K) = 310$, find $n(J \cap K)$.

12. Given $n(L) = 780$, $n(M) = 240$, and $n(L \cap M) = 50$, find $n(L \cup M)$.

13. Given $n(A) = 1500$, $n(A \cup B) = 2250$, and $n(A \cap B) = 310$, find $n(B)$.

14. Given $n(A) = 640$, $n(B) = 280$, and $n(A \cup B) = 765$, find $n(A \cap B)$.

In Exercises 15 and 16, use the given information to find the number of elements in each of the regions labeled with a question mark.

15. $n(A) = 28$, $n(B) = 31$, $n(C) = 40$, $n(A \cap B) = 15$, $n(U) = 75$

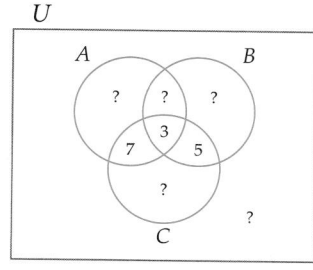

16. $n(A) = 610$, $n(B) = 440$, $n(C) = 1000$, $n(U) = 2900$

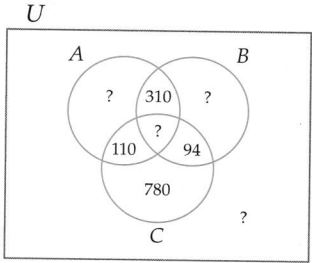

17. In a survey of 600 investors, it was reported that 380 had invested in stocks, 325 had invested in bonds, and 75 had not invested in either stocks or bonds.

   a. How many investors had invested in both stocks and bonds?

   b. How many investors had invested only in stocks?

18. A survey of 1500 commuters in New York City showed that 1140 take the subway, 680 take the bus, and 120 do not take either the bus or the subway.

   a. How many commuters take both the bus and the subway?

   b. How many commuters take only the subway?

19. A team physician has determined that of all the athletes who were treated for minor back pain, 72% responded to an analgesic, 59% responded to a muscle relaxant, and 44% responded to both forms of treatment.

   a. What percent of the athletes who were treated responded to the muscle relaxant but not the analgesic?

   b. What percent of the athletes who were treated did not respond to either form of treatment?

20. The management of a hotel conducted a survey. It found that of the 2560 guests who were surveyed,

1785 tip the wait staff.
1219 tip the luggage handlers.
831 tip the maids.
275 tip only the maids and the luggage handlers.
700 tip only the wait staff and the maids.
755 tip only the wait staff and the luggage handlers.
245 tip all three services.
210 do not tip these services.

How many of the surveyed guests tip:

a. exactly two of the three services?

b. only the wait staff?

c. only one of the three services?

21. A computer company advertises its computers in *BYTE* magazine, in *PC Magazine,* and on television. A survey of 770 customers finds that the numbers of customers who are familiar with the company's computers because of the different forms of advertising are as follows.

305, *BYTE* magazine
290, *PC Magazine*
390, television
110, *BYTE* magazine and *PC Magazine*
135, *PC Magazine* and television
150, *BYTE* magazine and television
85, all three sources

How many of the surveyed customers know about the computers because of:

a. exactly one of these forms of advertising?

b. exactly two of these forms of advertising?

c. *BYTE* magazine and neither of the other two forms of advertising?

22. A report from the American Association of Blood Banks shows that in the United States:

44% of the population has the A antigen.
15% of the population has the B antigen.
84% of the population has the Rh+ factor.
34% of the population is blood type A+.
9% of the population is blood type B+.
4% of the population has the A antigen and the B antigen.
3% of the population is blood type AB+.

(*Source:* **http://www.aabb.org/All_About_Blood/ FAQs/aabb_faqs.htm.**) Find the percent of the U.S. population that is:

a. A−   **b.** O+   **c.** O−

Extensions

CRITICAL THINKING

23. Given that set A has 47 elements and set B has 25 elements, determine each of the following.

a. The maximum possible number of elements in $A \cup B$.

b. The minimum possible number of elements in $A \cup B$.

c. The maximum possible number of elements in $A \cap B$.

d. The minimum possible number of elements in $A \cap B$.

24. Given that set A has 16 elements, set B has 12 elements, and set C has 7 elements, determine each of the following.

a. The maximum possible number of elements in $A \cup B \cup C$.

b. The minimum possible number of elements in $A \cup B \cup C$.

c. The maximum possible number of elements in $A \cap (B \cup C)$.

d. The minimum possible number of elements in $A \cap (B \cup C)$.

The following Venn diagram displays U parceled into 16 distinct regions by four sets.

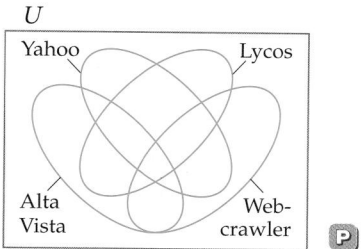

Use the Venn diagram to answer Exercise 25.

25. A survey of 1250 Internet users shows the following results concerning the use of the search engines Alta Vista, Yahoo, Lycos, and Webcrawler. Of the 1250 people surveyed,

> 585 use Alta Vista.
> 620 use Yahoo.
> 560 use Lycos.
> 450 use Webcrawler.
> 100 use only Alta Vista, Yahoo, and Lycos.
> 41 use only Alta Vista, Yahoo, and Webcrawler.
> 50 use only Alta Vista, Lycos, and Webcrawler.
> 80 use only Yahoo, Lycos, and Webcrawler.
> 55 use only Alta Vista and Yahoo.
> 34 use only Alta Vista and Lycos.
> 45 use only Alta Vista and Webcrawler.
> 50 use only Yahoo and Lycos.
> 30 use only Yahoo and Webcrawler.
> 45 use only Lycos and Webcrawler.
> 60 use all four.

How many of the Internet users:
a. use only Alta Vista?
b. use exactly three of the four search engines?
c. do not use any of the four search engines?

26. Exactly one of the following equations is a valid inclusion-exclusion formula for the union of three finite sets. Which equation do you think is the valid formula? *Hint:* Use the data in Example 2 to check your choice.

a. $n(A \cup B \cup C) = n(A) + n(B) + n(C)$

b. $n(A \cup B \cup C) = n(A) + n(B) + n(C) - n(A \cap B \cap C)$

c. $n(A \cup B \cup C) = n(A) + n(B) + n(C) - n(A \cap B) - n(A \cap C) - n(B \cap C)$

d. $n(A \cup B \cup C) = n(A) + n(B) + n(C) - n(A \cap B) - n(A \cap C) - n(B \cap C) + n(A \cap B \cap C)$

SECTION 2.5 | Infinite Sets

One-To-One Correspondences

Much of Georg Cantor's work with sets concerned infinite sets. Some of Cantor's work with infinite sets was so revolutionary that it was not readily accepted by his contemporaries. Today, however, his work is generally accepted and it provides unifying ideas in several diverse areas of mathematics.

Much of Cantor's set theory is based on the simple concept of a *one-to-one correspondence*.

One-to-One Correspondence

A **one-to-one correspondence** (or 1−1 correspondence) between two sets A and B is a rule or procedure that pairs each element of A with exactly one element of B and each element of B with exactly one element of A.

Many practical problems can be solved by applying the concept of a one-to-one correspondence. For instance, consider a concert hall that has 890 seats. During a performance the manager of the concert hall observes that every person occupies exactly one seat and that every seat is occupied. Thus, without doing any counting, the manager knows that there are 890 people in attendance. During a different performance the manager notes that all but six seats are filled, and thus there are $890 - 6 = 884$ people in attendance.

Recall that two sets are equivalent if and only if they have the same number of elements. One method of showing that two sets are equivalent is to establish a one-to-one correspondence between the elements of the sets.

One-to-One Correspondence and Equivalent Sets

Two sets A and B are equivalent, denoted by $A \sim B$, if and only if A and B can be placed in a one-to-one correspondence.

Set {a, b, c, d, e} is equivalent to set {1, 2, 3, 4, 5} because we can show that the elements of each set can be placed in a one-to-one correspondence. One method of establishing this one-to-one correspondence is shown in the following figure.

$$
\begin{array}{ccccc}
\{a, & b, & c, & d, & e\} \\
\updownarrow & \updownarrow & \updownarrow & \updownarrow & \updownarrow \\
\{1, & 2, & 3, & 4, & 5\}
\end{array}
$$

Each element of {a, b, c, d, e} has been paired with exactly one element of {1, 2, 3, 4, 5}, and each element of {1, 2, 3, 4, 5} has been paired with exactly one element of {a, b, c, d, e}. This is not the only one-to-one correspondence that we can establish. The figure at the left shows another one-to-one correspondence between the sets. In any case, we know that both sets have the same number of elements because we have established a one-to-one correspondence between the sets.

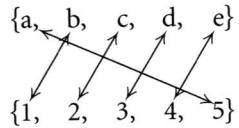

Sometimes a set is defined by including a general element. For instance, in the set {3, 6, 9, 12, 15, ..., 3n, ...}, the 3n (where n is a natural number) indicates that all the elements of the set are multiples of 3.

Some sets can be placed in a one-to-one correspondence with a proper subset of themselves. Example 1 illustrates this concept for the set of natural numbers.

EXAMPLE 1 ■ Establish a One-to-One Correspondence

Establish a one-to-one correspondence between the set of natural numbers $N = \{1, 2, 3, 4, 5, \ldots, n, \ldots\}$ and the set of even natural numbers $E = \{2, 4, 6, 8, 10, \ldots, 2n, \ldots\}$.

Solution

Write the sets so that one is aligned below the other. Draw arrows to show how you wish to pair the elements of each set. One possible method is shown in the following figure.

$$N = \{1, 2, 3, 4, \ldots, n, \ldots\}$$

$$\uparrow \uparrow \uparrow \uparrow \qquad \uparrow$$
$$\downarrow \downarrow \downarrow \downarrow \qquad \downarrow$$

$$E = \{2, 4, 6, 8, \ldots, 2n, \ldots\}$$

In the above correspondence, each natural number $n \in N$ is paired with the even number $(2n) \in E$. The *general correspondence* $n \leftrightarrow (2n)$ enables us to determine exactly which element of E will be paired with any given element of N, and vice versa. For instance, under this correspondence, $19 \in N$ is paired with the even number $2 \cdot 19 = 38 \in E$, and $100 \in E$ is paired with the natural number $\frac{1}{2} \cdot 100 = 50 \in N$. The general correspondence $n \leftrightarrow (2n)$ establishes a one-to-one correspondence between the sets.

CHECK YOUR PROGRESS 1 Establish a one-to-one correspondence between the set of natural numbers $N = \{1, 2, 3, 4, 5, \ldots, n, \ldots\}$ and the set of odd natural numbers $D = \{1, 3, 5, 7, 9, \ldots, 2n - 1, \ldots\}$.

Solution See page S6.

Infinite Sets

Definition of an Infinite Set

A set is an **infinite set** if it can be placed in a one-to-one correspondence with a proper subset of itself.

We know that the set of natural numbers N is an infinite set because in Example 1 we were able to establish a one-to-one correspondence between the elements of N and the elements of one of its proper subsets, E.

QUESTION   *Can the set {1, 2, 3} be placed in a one-to-one correspondence with one of its proper subsets?*

ANSWER   *No. The set {1, 2, 3} is a finite set with three elements. Every proper subset of {1, 2, 3} has two or fewer elements.*

✔ **TAKE NOTE**

The solution shown in Example 2 is not the only way to establish that S is an infinite set. For instance, $R = \{10, 15, 20, \ldots 5n + 5, \ldots\}$ is also a proper set of S, and the sets S and R can be placed in a one-to-one correspondence as follows.

$S = \{\, 5,\ \ 10,\ \ 15,\ \ 20,\ \ \ldots,\ \ \ \ 5n,\ \ \ \ldots\}$

$R = \{10,\ 15,\ 20,\ 25,\ \ldots,\ 5n + 5, \ldots\}$

This one-to-one correspondence between S and one of its proper subsets R also establishes that S is an infinite set.

EXAMPLE 2 ■ Verify that a Set is an Infinite Set

Verify that $S = \{5, 10, 15, 20, \ldots, 5n, \ldots\}$ is an infinite set.

Solution

One proper subset of S is $T = \{10, 20, 30, 40, \ldots, 10n, \ldots\}$, which was produced by deleting the odd numbers in S. To establish a one-to-one correspondence between set S and set T, consider the following diagram.

$S = \{\, 5,\ \ 10,\ 15,\ 20,\ \ldots,\ \ 5n,\ \ldots\}$

$T = \{10,\ 20,\ 30,\ 40,\ \ldots,\ 10n,\ \ldots\}$

In the above correspondence, each $(5n) \in S$ is paired with $(10n) \in T$. The *general correspondence* $(5n) \leftrightarrow (10n)$ establishes a one-to-one correspondence between S and one of its proper subsets, namely T. Thus S is an infinite set.

CHECK YOUR PROGRESS 2   Verify that $V = \{40, 41, 42, 43, \ldots, 40 + n, \ldots\}$ is an infinite set.

Solution   *See page S6.*

The Cardinality of Infinite Sets

The symbol $\aleph_0$ is used to represent the cardinal number for the set N of natural numbers. ($\aleph$ is the first letter of the Hebrew alphabet and is pronounced *aleph*. $\aleph_0$ is read as "aleph-null.") Using mathematical notation, we write this concept as $n(N) = \aleph_0$. Since $\aleph_0$ represents a cardinality larger than any finite number, it is called a **transfinite number**. Many infinite sets have a cardinality of $\aleph_0$. In Example 3, for instance, we show that the cardinality of the set of integers is $\aleph_0$ by establishing a one-to-one correspondence between the elements of the set of integers and the elements of the set of natural numbers.

EXAMPLE 3 ■ Establish the Cardinality of the Set of Integers

Show that the set of integers $I = \{\ldots, -4, -3, -2, -1, 0, 1, 2, 3, 4, \ldots\}$ has a cardinality of $\aleph_0$.

Solution

First we try to establish a one-to-one correspondence between I and N, with the elements in each set arranged as shown below. No general method of pairing the elements of N with the elements of I seems to emerge from this figure.

$N = \{1, 2, 3, 4, 5, 6, 7, 8, 9, 10, 11, \ldots\}$

?

$I = \{\ldots, -5, -4, -3, -2, -1, 0, 1, 2, 3, 4, 5, \ldots\}$

If we arrange the elements of I as shown in the figure below, then *two* general correspondences, shown by the blue arrows and the red arrows, can be identified.

$$N = \{1, 2, 3, 4, 5, 6, 7, 8, 9, 10, 11, \ldots, 2n-1, 2n, \ldots\}$$

$$I = \{0, 1, -1, 2, -2, 3, -3, 4, -4, 5, -5 \ldots, -n+1, n, \ldots\}$$

- Each even natural number $2n$ of N is paired with the integer n of I. This correspondence is shown by the blue arrows.
- Each odd natural number $2n-1$ of N is paired with the integer $-n+1$ of I. This correspondence is shown by the red arrows.

Together the two general correspondences $(2n) \leftrightarrow n$ and $(2n-1) \leftrightarrow (-n+1)$ establish a one-to-one correspondence between the elements of I and the elements of N. Thus the cardinality of the set of integers must be the same as the cardinality of the set of natural numbers, which is $\aleph_0$.

CHECK YOUR PROGRESS 3   Show that $M = \left\{\frac{1}{2}, \frac{1}{3}, \frac{1}{4}, \frac{1}{5}, \ldots, \frac{1}{n+1}, \ldots\right\}$ has a cardinality of $\aleph_0$.

Solution   See page S6.

Cantor was also able to show that the set of positive rational numbers is equivalent to the set of natural numbers. Recall that a rational number is a number that can be written as a fraction p/q where p and q are integers and $q \neq 0$. Cantor's proof used an array of rational numbers similar to the array shown below.

Theorem   The set $Q+$ of positive rational numbers is equivalent to the set N of natural numbers.

Proof   Consider the following array of positive rational numbers.

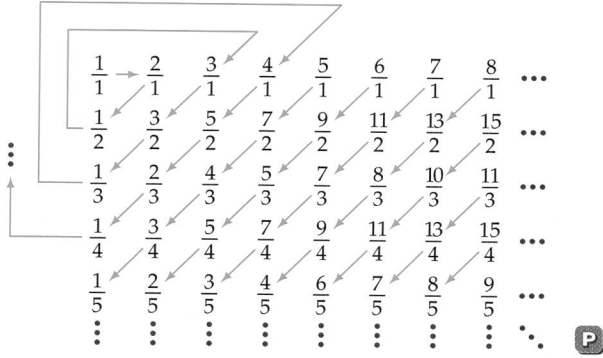

An array of all the positive rational numbers

The first row of the above array contains, in order from smallest to largest, all the positive rational numbers *which when expressed in lowest terms* have a denominator of 1. The second row contains the positive rational numbers *which when*

> **TAKE NOTE**
>
> The rational number $\frac{2}{2}$ is not listed in the second row because $\frac{2}{2} = 1 = \frac{1}{1}$, which is already listed in the first row.

expressed in lowest terms have a denominator of 2. The third row contains the positive rational numbers *which when expressed in lowest terms* have a denominator of 3. This process continues indefinitely.

Cantor reasoned that every positive rational number appears once and only once in this array. Note that $\frac{3}{5}$ appears in the fifth row. In general, if $\frac{p}{q}$ is in lowest terms, then it appears in row q.

At this point Cantor used a numbering procedure that establishes a one-to-one correspondence between the natural numbers and the positive rational numbers in the array. The numbering procedure starts in the upper left corner with $\frac{1}{1}$. Cantor considered this to be the first number in the array, so he assigned the natural number 1 to this rational number. He then moved to the right and assigned the natural number 2 to the rational number $\frac{2}{1}$. From this point on, he followed the diagonal paths shown by the arrows and assigned each number he encountered to the next consecutive natural number. When he reached the bottom of a diagonal, he moved up to the top of the array and continued to number the rational numbers in the next diagonal. The following table shows the first 10 rational numbers Cantor numbered using this scheme.

Rational number in the array	$\frac{1}{1}$	$\frac{2}{1}$	$\frac{1}{2}$	$\frac{3}{1}$	$\frac{3}{2}$	$\frac{1}{3}$	$\frac{4}{1}$	$\frac{5}{2}$	$\frac{2}{3}$	$\frac{1}{4}$
Corresponding natural number	1	2	3	4	5	6	7	8	9	10

This numbering procedure shows that each element of $Q+$ can be paired with exactly one element of N, and each element of N can be paired with exactly one element of $Q+$. Thus $Q+$ and N are equivalent sets. ∎

The negative rational numbers $Q-$ can also be placed in a one-to-one correspondence with the set of natural numbers in a similar manner.

QUESTION   *Using Cantor's numbering scheme, which rational numbers in the array shown on page 95 would be assigned the natural numbers 11, 12, 13, 14, and 15?*

Definition of a Countable Set

A set is a **countable set** if and only if it is a finite set or an infinite set that is equivalent to the set of natural numbers.

Every infinite set that is countable has a cardinality of $\aleph_0$. Every infinite set that we have considered up to this point is countable. You might think that all infinite sets are countable; however, Cantor was able to show that this is not the case. Con-

ANSWER   *The rational numbers in the next diagonal, namely $\frac{5}{1}, \frac{7}{2}, \frac{4}{3}, \frac{3}{4},$ and $\frac{1}{5},$ would be assigned to the natural numbers 11, 12, 13, 14, and 15, respectively.*

sider for example $A = \{x \mid x \in R \text{ and } 0 < x < 1\}$. To show that A is *not* a countable set, we use a *proof by contradiction,* where we assume that A is countable and then proceed until we arrive at a contradiction.

To better understand the concept of a proof by contradiction, consider the situation in which you are at a point where a road splits into two roads. See the figure at the left. Assume you know that only one of the two roads leads to your desired destination. If you can show that one of the roads cannot get you to your destination, then you know, without ever traveling down the other road, that it is the road that leads to your destination. In the following proof we know that set A is either a countable set or set A is not a countable set. To establish that A is *not* countable, we show that the assumption that A is countable leads to a contradiction. In other words, our assumption that A is countable must be incorrect and we are forced to conclude that A is not countable.

A is countable   *A is not countable*

Theorem   The set $A = \{x \mid x \in R \text{ and } 0 < x < 1\}$ is not a countable set.

Proof by contradiction   Either A is countable or A is not countable. Assume A is countable. Then we can place the elements of A, which we will represent by a_1, a_2, a_3, $a_4, \ldots$, in a one-to-one correspondence with the elements of the natural numbers as shown below.

$$N = \{1, \quad 2, \quad 3, \quad 4, \quad \ldots, \quad n, \quad \ldots\}$$

$$A = \{a_1, \ a_2, \ a_3, \ a_4, \ \ldots, \ a_n, \ \ldots\}$$

For example, the numbers a_1, a_2, a_3, $a_4, \ldots, a_n, \ldots$ could be as shown below.

$$1 \leftrightarrow a_1 = 0\,.\,\boxed{3}\,5\ 7\ 3\ 4\ 8\ 5\ldots$$
$$2 \leftrightarrow a_2 = 0\,.\,0\,\boxed{6}\,5\ 2\ 8\ 9\ 1\ldots$$
$$3 \leftrightarrow a_3 = 0\,.\,6\ 8\,\boxed{2}\,3\ 5\ 1\ 4\ldots$$
$$4 \leftrightarrow a_4 = 0\,.\,0\ 5\ 0\,\boxed{0}\,3\ 1\ 0\ldots$$
$$\vdots$$
$$n \leftrightarrow a_n = 0\,.\,3\ 1\ 5\ 5\ 7\ 2\ 8 \ldots\boxed{5}\ldots$$
$$\vdots$$

— *n*th decimal digit of a_n **P**

At this point we use a "diagonal technique" to construct a real number d that is greater than 0 and less than 1, and is not in the above list. We construct d by writing a decimal that *differs* from a_1 in the first decimal place, differs from a_2 in the second decimal place, differs from a_3 in the third decimal place, and, in general, differs from a_n in the nth decimal place. For instance, in the above list, a_1 has 3 as its first decimal digit. The first decimal digit of d can be any digit other than 3, say 4. The real number a_2 has 6 as its second decimal digit. The second decimal digit of d can be any digit other than 6, say 7. The real number a_3 has 2 as its third decimal digit. The third decimal digit of d can be any digit other than 2, say 3. Continue in this manner to determine the decimal digits of d. Now $d = 0.473\ldots$ must be in A because $0 < d < 1$. However, d is not in A, because d differs from each of the numbers in A in at least one decimal place.

INSTRUCTOR NOTE
You may wish to explain that the procedure used in the proof at the right cannot be used on the set of rational numbers between 0 and 1 because it is not possible to ensure that the number d created by the procedure is a rational number.

We have reached a contradiction. Our assumption that the elements of A could be placed in a one-to-one correspondence with the elements of the natural numbers must be false. Thus A is not a countable set.   ■

An infinite set that is not countable is said to be **uncountable.** Since A is uncountable, the cardinality of A is not $\aleph_0$. Cantor used the letter c, which is the first letter of the word *continuum,* to represent the cardinality of A. Cantor was also able to show that set A is equivalent to the set of all real numbers R. Thus the cardinality of R is also c. Cantor was able to prove that $c > \aleph_0$.

A Comparison of Transfinite Cardinal Numbers

$c > \aleph_0$

Up to this point, all of the infinite sets we have considered have a cardinality of either $\aleph_0$ or c. The following table lists several infinite sets and the transfinite cardinal number that is associated with each set.

The Cardinality of Some Infinite Sets	
Set	**Cardinal Number**
Natural numbers, N	$\aleph_0$
Integers, I	$\aleph_0$
Rational numbers, Q	$\aleph_0$
Irrational Numbers, $\mathscr{I}$	c
Any set of the form $\{x \mid a \leq x \leq b\}$, where a and b are real numbers and $a \neq b$.	c
Real numbers, R	c

Your intuition may suggest that $\aleph_0$ and c are the only two cardinal numbers associated with infinite sets; however, this is not the case. In fact, Cantor was able to show that no matter how large the cardinal number of a set, we can find a set that has a larger cardinal number. Thus there are infinitely many transfinite numbers. Cantor's proof of this concept is now known as *Cantor's theorem.*

Cantor's Theorem

Let S be any set. The set of all subsets of S has a cardinal number that is larger than the cardinal number of S.

The set of all subsets of S is called the **power set** of S, and is denoted by $P(S)$. We can see that Cantor's theorem is true for the finite set $S = \{a, b, c\}$ because the cardinality of S is 3 and S has $2^3 = 8$ subsets. The interesting part of Cantor's theorem is that it also applies to infinite sets.

Some of the following theorems can be established by using the techniques illustrated in the Excursion that follows.

Transfinite Arithmetic Theorems

- For any whole number a, $\aleph_0 + a = \aleph_0$ and $\aleph_0 - a = \aleph_0$
- $\aleph_0 + \aleph_0 = \aleph_0$ and, in general, $\underbrace{\aleph_0 + \aleph_0 + \aleph_0 + \cdots + \aleph_0}_{\text{a finite number of aleph nulls}} = \aleph_0$
- $c + c = c$ and, in general, $\underbrace{c + c + c + \cdots + c}_{\text{a finite number of } c\text{'s}} = c$
- $\aleph_0 + c = c$
- $\aleph_0 c = c$

Math Matters   Criticism and Praise

Georg Cantor's work in the area of infinite sets was not well received by some of his colleagues. For instance, the mathematician Leopold Kronecker tried to stop the publication of some of Cantor's work. He felt many of Cantor's theorems were ridiculous and asked, "How can one infinity be greater than another?" The following quote illustrates that Cantor was aware that his work would attract harsh criticism.

> ...I realize that in this undertaking I place myself in a certain opposition to views widely held concerning the mathematical infinite and to opinions frequently defended on the nature of numbers.[2]

A few mathematicians were willing to show support for Cantor's work. For instance, the famous mathematician David Hilbert stated that Cantor's work was

> ...the finest product of mathematical genius and one of the supreme achievements of purely intellectual human activity.[3]

 *Excursion*

Transfinite Arithmetic

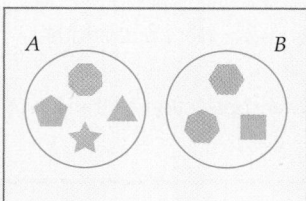

Disjoint sets are often used to explain addition. The sum $4 + 3$, for example, can be deduced by selecting two disjoint sets, one with exactly four elements and one with exactly three elements. See the Venn diagram at the left. Now form the union of the two sets. The union of the two sets has exactly seven elements; thus, $4 + 3 = 7$. In mathematical notation, we write

$$n(A) + n(B) = n(A \cup B)$$
$$4 + 3 = 7$$

(continued)

---

2. *Source:* **http://www-groups.dcs.st-and.ac.uk/%7Ehistory/Mathematicians/Cantor.html**
3. See note 2 above.

Cantor extended this idea to infinite sets. He reasoned that the sum $\aleph_0 + 1$ could be determined by selecting two disjoint sets, one with cardinality of $\aleph_0$ and one with cardinality of 1. In this case the set N of natural numbers and the set $Z = \{0\}$ are appropriate choices. Thus

$$n(N) + n(Z) = n(N \cup Z)$$
$$= n(W)$$
$$\aleph_0 + 1 = \aleph_0$$

and, in general, for any whole number a, $\aleph_0 + a = \aleph_0$.

To find the sum $\aleph_0 + \aleph_0$, use two disjoint sets, each with cardinality of $\aleph_0$. The set E of even natural numbers and the set D of odd natural numbers satisfy the necessary conditions. Since E and D are disjoint sets, we know

$$n(E) + n(D) = n(E \cup D)$$
$$= n(W)$$
$$\aleph_0 + \aleph_0 = \aleph_0$$

Thus $\aleph_0 + \aleph_0 = \aleph_0$ and, in general,

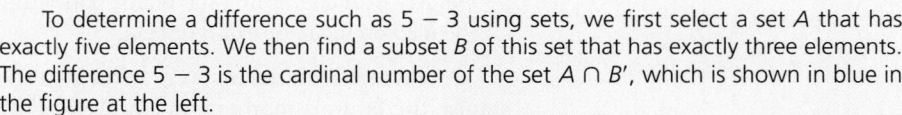

$$\underbrace{\aleph_0 + \aleph_0 + \aleph_0 + \cdots + \aleph_0}_{\text{a finite number of aleph-nulls}} = \aleph_0$$

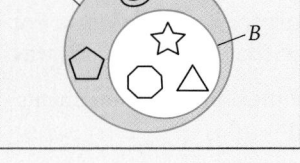

To determine a difference such as $5 - 3$ using sets, we first select a set A that has exactly five elements. We then find a subset B of this set that has exactly three elements. The difference $5 - 3$ is the cardinal number of the set $A \cap B'$, which is shown in blue in the figure at the left.

To determine $\aleph_0 - 3$, select a set with $\aleph_0$ elements, such as N, and then select a subset of this set that has exactly three elements. One such subset is $C = \{1, 2, 3\}$. The difference $\aleph_0 - 3$ is the cardinal number of the set $N \cap C' = \{4, 5, 6, 7, 8, \ldots\}$. Since $N \cap C'$ is a countable set, we can conclude that $\aleph_0 - 3 = \aleph_0$. This procedure can be generalized to show that for any whole number a, $\aleph_0 - a = \aleph_0$.

Excursion Exercises

1. Use two disjoint sets to show that $\aleph_0 + 1 = \aleph_0$.

2. Use two disjoint sets other than the set of even natural numbers and the set of odd natural numbers to show that $\aleph_0 + \aleph_0 = \aleph_0$.

3. Use sets to show that $\aleph_0 - 6 = \aleph_0$.

4. **a.** Find two sets that can be used to show that $\aleph_0 - \aleph_0 = \aleph_0$. Now find another two sets that can be used to show that $\aleph_0 - \aleph_0 = 1$.

   b. Use the results of Excursion Exercise 4a to explain why subtraction of transfinite numbers is an undefined operation.

Exercise Set 2.5   (Suggested Assignment: 1–29 odds)

1. a. Use arrows to establish a one-to-one correspondence between $V = \{a, e, i\}$ and $M = \{3, 6, 9\}$.

   b. How many different one-to-one correspondences between V and M can be established?

2. Establish a one-to-one correspondence between the set of natural numbers $N = \{1, 2, 3, 4, 5, \ldots, n, \ldots\}$ and $F = \{5, 10, 15, 20, \ldots, 5n, \ldots\}$ by stating a general rule that can be used to pair the elements of the sets.

3. Establish a one-to-one correspondence between $D = \{1, 3, 5, \ldots, 2n - 1, \ldots\}$ and $M = \{3, 6, 9, \ldots, 3n, \ldots\}$ by stating a general rule that can be used to pair the elements of the sets.

In Exercises 4–10, state the cardinality of each set.

4. $\{2, 11, 19, 31\}$

5. $\{2, 9, 16, \ldots, 7n - 5, \ldots\}$, where n is a natural number

6. The set Q of rational numbers

7. The set R of real numbers

8. The set $\mathscr{I}$ of irrational numbers

9. $\{x \mid 5 \leq x \leq 9\}$

10. The set of subsets of $\{1, 5, 9, 11\}$

In Exercises 11–14, determine whether the given sets are equivalent.

11. The set of natural numbers and the set of integers

12. The set of whole numbers and the set of real numbers

13. The set of rational numbers and the set of integers

14. The set of rational numbers and the set of real numbers

In Exercises 15–18, show that the given set is an infinite set by placing it in a one-to-one correspondence with a proper subset of itself.

15. $A = \{5, 10, 15, 20, 25, 30, \ldots, 5n, \ldots\}$

16. $B = \{11, 15, 19, 23, 27, 31, \ldots, 4n + 7, \ldots\}$

17. $C = \left\{\dfrac{1}{2}, \dfrac{3}{4}, \dfrac{5}{6}, \dfrac{7}{8}, \dfrac{9}{10}, \ldots, \dfrac{2n - 1}{2n}, \ldots\right\}$

18. $D = \left\{\dfrac{1}{2}, \dfrac{1}{3}, \dfrac{1}{4}, \dfrac{1}{5}, \dfrac{1}{6}, \ldots, \dfrac{1}{n + 1}, \ldots\right\}$

In Exercises 19–26, show that the given set has a cardinality of $\aleph_0$ by establishing a one-to-one correspondence between the elements of the given set and the elements of N.

19. $\{50, 51, 52, 53, \ldots, n + 49, \ldots\}$

20. $\{10, 5, 0, -5, -10, -15, \ldots, -5n + 15, \ldots\}$

21. $\left\{1, \dfrac{1}{3}, \dfrac{1}{9}, \dfrac{1}{27}, \ldots, \dfrac{1}{3^{n-1}}, \ldots\right\}$

22. $\{-12, -18, -24, -30, \ldots, -6n - 6, \ldots\}$

23. $\{10, 100, 1000, \ldots, 10^n, \ldots\}$

24. $\left\{1, \dfrac{1}{2}, \dfrac{1}{4}, \dfrac{1}{8}, \ldots, \dfrac{1}{2^{n-1}}, \ldots\right\}$

25. $\{1, 8, 27, 64, \ldots, n^3, \ldots\}$

26. $\{0.1, 0.01, 0.001, 0.0001, \ldots, 10^{-n}, \ldots\}$

Extensions
CRITICAL THINKING

27. a. Place the set $M = \{3, 6, 9, 12, 15, \ldots\}$ of positive multiples of 3 in a one-to-one correspondence with the set K of all natural numbers that are not multiples of 3. Write a sentence or two that explains the general rule you used to establish the one-to-one correspondence.

   b. Use your rule to determine what number from K is paired with the number 606 from M.

   c. Use your rule to determine what number from M is paired with the number 899 from K.

In the figure below every point on line segment AB corresponds to a real number from 0 to 1 and every real number from 0 to 1 corresponds to a point on line segment AB.

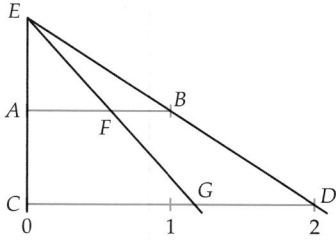

The line segment CD represents the real numbers from 0 to 2. Note that any point F on line segment AB can be paired with a unique point G on line segment CD by drawing a line from E through F. Also, any arbitrary point G on line segment CD can be paired with a unique point F on line segment AB by drawing the line EG. This geometric procedure establishes a one-to-one correspondence between the set $\{x \mid 0 \le x \le 1\}$ and the set $\{x \mid 0 \le x \le 2\}$. Thus $\{x \mid 0 \le x \le 1\} \sim \{x \mid 0 \le x \le 2\}$.

28. Draw a figure that can be used to verify each of the following.

   a. $\{x \mid 0 \le x \le 1\} \sim \{x \mid 0 \le x \le 5\}$
   b. $\{x \mid 2 \le x \le 5\} \sim \{x \mid 1 \le x \le 8\}$

29. Consider the semicircle with arc length 1 and center C and the line L_1 in the following figure. Each point on the semicircle, *other than the endpoints,* represents a unique real number between 0 and 1. Each point on line L_1 represents a unique real number.

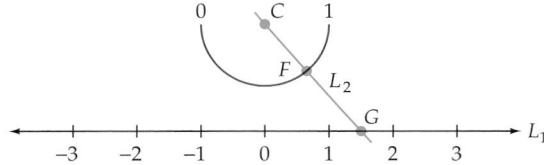

Any line through C that intersects the semicircle at a point other than one of its endpoints will intersect line L_1 at a unique point. Also, any line through C that intersects line L_1 will intersect the semicircle at a unique point that is not an endpoint of the semicircle. What can we conclude from this correspondence?

30. Explain how to use the following figure to verify that the set of all points on the circle is equivalent to the set of all points on the square.

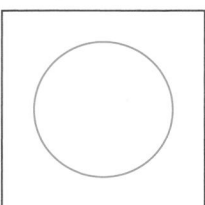

31. The Hilbert Hotel is an imaginary hotel created by the mathematician David Hilbert (1862–1943). The hotel has an infinite number of rooms. Each room is numbered with a natural number; room 1, room 2, room 3, and so on. Search the Internet for information on Hilbert's Hotel. Write a few paragraphs that explain some of the interesting questions that arise when guests arrive to stay at the hotel.

Mary Pat Campbell has written a song about a hotel with an infinite number of rooms. Her song is titled *Hotel Aleph Null—yeah.* Here are the lyrics for the chorus of her song, which is to be sung to the tune of *Hotel California* by the Eagles. (*Source:* **http://www.marypat.org/mathcamp/doc2001/ hellrelays.html#hotel**)[4]

> **Hotel Aleph Null—yeah**
> Welcome to the Hotel Aleph Null—yeah
> What a lovely place (what a lovely place)
> Got a lot of space
> Packin' em in at the Hotel Aleph Null—yeah
> Any time of year
> You can find space here

32. Cantor conjectured that no set can have a cardinality larger than $\aleph_0$ but smaller than c. This conjecture has become known as the *Continuum Hypothesis.* Search the Internet for information on the Continuum Hypothesis and write a short report that explains how the Continuum Hypothesis was resolved.

---

4. Reprinted by permission of Mary Pat Campbell.

Summary

Key Terms

aleph-null [p. 94]
cardinal number [p. 52]
complement of a set [p. 59]
countable set [p. 96]
counting number [p. 50]
disjoint sets [p. 70]
element (member) of a set [p. 50]
empty set or null set [p. 52]
equal sets [p. 53]
equivalent sets [p. 53]
finite set [p. 52]
infinite set [p. 93]
integer [p. 50]
intersection [p. 69]
irrational number [p. 50]
natural number [p. 50]
one-to-one correspondence [p. 91]
power set [p. 98]
proper subset [p. 62]
rational number [p. 50]
real number [p. 50]
roster method [p. 50]
set [p. 50]
set-builder notation [p. 52]
subset [p. 60]
transfinite number [p. 94]
uncountable set [p. 98]
union [p. 70]
universal set [p. 59]
Venn diagram [p. 61]
well-defined set [p. 51]
whole number [p. 50]

Essential Concepts

- A set with n elements has 2^n subsets.
- The *intersection* of sets A and B, denoted by $A \cap B$, is the set of elements common to both A and B.
$$A \cap B = \{x \mid x \in A \quad \text{and} \quad x \in B\}$$
- The *union* of sets A and B, denoted by $A \cup B$, is the set that contains all the elements that belong to A or to B or to both.
$$A \cup B = \{x \mid x \in A \quad \text{or} \quad x \in B\}$$
- **De Morgan's Laws**
$$(A \cap B)' = A' \cup B' \qquad (A \cup B)' = A' \cap B'$$
- **Commutative Properties of Sets**
$$A \cap B = B \cap A \qquad A \cup B = B \cup A$$
- **Associative Properties of Sets**
$$(A \cap B) \cap C = A \cap (B \cap C)$$
$$(A \cup B) \cup C = A \cup (B \cup C)$$
- **Distributive Properties of Sets**
$$A \cap (B \cup C) = (A \cap B) \cup (A \cap C)$$
$$A \cup (B \cap C) = (A \cup B) \cap (A \cup C)$$
- **The Inclusion-Exclusion Principle**
For any finite sets A and B,
$$n(A \cup B) = n(A) + n(B) - n(A \cap B)$$
- **Cantor's Theorem**
Let S be any set. The set of all subsets of S has a larger cardinal number than the cardinal number of S.
- **Transfinite Arithmetic Theorems**
For any whole number a,
$$\aleph_0 + a = \aleph_0, \quad \aleph_0 - a = \aleph_0, \quad \aleph_0 + \aleph_0 = \aleph_0,$$
$$\aleph_0 + c = c, \quad \aleph_0 c = c, \quad c + c = c$$

Review Exercises

In Exercises 1–4, use the roster method to write each set.

1. The set of whole numbers less than 8
2. The set of integers that satisfy $x^2 = 64$
3. The set of natural numbers that satisfy $x + 3 \leq 7$
4. The set of counting numbers larger than -3 and less than or equal to 6

In Exercises 5–8, use set-builder notation to write each set.

5. The set of integers greater than −6
6. {April, June, September, November}
7. {Kansas, Kentucky}
8. {1, 8, 27, 64, 125}

In Exercises 9–12, determine whether the statement is true or false.

9. {3} ∈ {1, 2, 3, 4}
10. −11 ∈ I
11. {a, b, c} ∼ {1, 5, 9}
12. The set of small numbers is a well-defined set.

In Exercises 13–20, let $U = \{2, 6, 8, 10, 12, 14, 16, 18\}$, $A = \{2, 6, 10\}$, $B = \{6, 10, 16, 18\}$, and $C = \{14, 16\}$. Find each of the following.

13. $A \cap B$
14. $A \cup B$
15. $A' \cap C$
16. $B \cup C'$
17. $A \cup (B \cap C)$
18. $(A \cup C)' \cap B'$
19. $(A \cap B')'$
20. $(A \cup B \cup C)'$

In Exercises 21–24, determine whether the first set is a proper subset of the second set.

21. The set of natural numbers; the set of whole numbers
22. The set of integers; the set of real numbers
23. The set of counting numbers; the set of natural numbers
24. The set of real numbers; the set of rational numbers

In Exercises 25–28, list all the subsets of the given set.

25. {I, II}
26. {s, u, n}
27. {penny, nickel, dime, quarter}
28. {A, B, C, D, E}

In Exercises 29–32, find the number of subsets of the given set.

29. The set of the four musketeers
30. The set of the letters of the English alphabet
31. The set of the letters of "uncopyrightable," which is the longest English word with no repeated letters
32. The set of the seven dwarfs

In Exercises 33–36, draw a Venn diagram to represent the given set.

33. $A \cap B'$
34. $A' \cup B'$
35. $(A \cup B) \cup C'$
36. $A \cap (B' \cup C)$

In Exercises 37–40, draw Venn diagrams to determine whether the expressions are equal for all sets A, B, and C.

37. $A' \cup (B \cup C)$; $(A' \cup B) \cup (A' \cup C)$
38. $(A \cap B) \cap C'$; $(A' \cup B') \cup C$
39. $A \cap (B' \cap C)$; $(A \cup B') \cap (A \cup C)$
40. $A \cap (B \cup C)$; $A' \cap (B \cup C)$

In Exercises 41 and 42, use set notation to describe the shaded region.

41.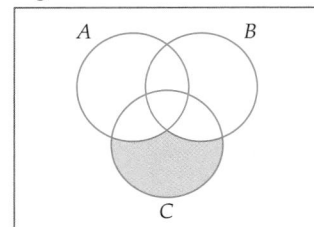

42.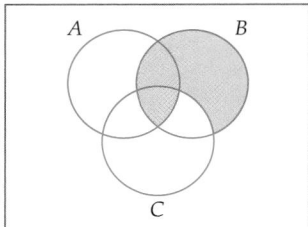

In Exercise 43 and 44, draw a Venn diagram with each of the given elements placed in the correct region.

43. $U = \{e, h, r, d, w, s, t\}$
    $A = \{t, r, e\}$
    $B = \{w, s, r, e\}$
    $C' = \{s, r, d, h\}$
44. $U = \{\alpha, \beta, \Gamma, \gamma, \Delta, \delta, \varepsilon, \theta\}$
    $A' = \{\beta, \Delta, \theta, \gamma\}$
    $B = \{\delta, \varepsilon\}$
    $C = \{\beta, \varepsilon, \Gamma\}$

45. In a survey at a health club, 208 members indicated that they enjoy aerobic exercises, 145 indicated that they enjoy weight training, 97 indicated that they enjoy both aerobics and weight training, and 135 indicated that they do not enjoy either of these types of exercise. How many members were surveyed?

46. A gourmet coffee bar conducted a survey to determine the preferences of its customers. Of the customers surveyed,

    221 like espresso.
    127 like cappuccino and chocolate-flavored coffee.
    182 like cappuccino.
    136 like espresso and chocolate-flavored coffee.
    209 like chocolate-flavored coffee.
    96 like all three types of coffee.
    116 like espresso and cappuccino.
    82 like none of these types of coffee.

How many of the customers in the survey:

a. like only chocolate-flavored coffee?

b. like cappuccino and chocolate-flavored coffee but not espresso?

c. like espresso and cappuccino but not chocolate-flavored coffee?

d. like exactly one of the three types of coffee?

In Exercises 47–50, establish a one-to-one correspondence between the sets.

47. $\{1, 3, 6, 10\}$; $\{1, 2, 3, 4\}$

48. $\{x \mid x > 10 \text{ and } x \in N\}$; $\{2, 4, 6, 8, \ldots, 2n, \ldots\}$

49. $\{3, 6, 9, 12, \ldots, 3n, \ldots\}$; $\{10, 100, 1000, \ldots, 10^n, \ldots\}$

50. $\{x \mid 0 \leq x \leq 1\}$; $\{x \mid 0 \leq x \leq 4\}$ (*Hint:* Use a drawing.)

In Exercises 51 and 52, show that the given set is an infinite set.

51. $A = \{6, 10, 14, 18, \ldots, 4n + 2, \ldots\}$

52. $B = \left\{1, \dfrac{1}{2}, \dfrac{1}{4}, \dfrac{1}{8}, \ldots, \dfrac{1}{2^{n-1}}, \ldots\right\}$

In Exercises 53–60, state the cardinality of each set.

53. $\{5, 6, 7, 8, 6\}$

54. $\{4, 6, 8, 10, 12, \ldots, 22\}$

55. $\{0, \varnothing\}$

56. The set of all states in the U.S. that border the Gulf of Mexico

57. The set of integers less than 1,000,000

58. The set of rational numbers between 0 and 1

59. The set of irrational numbers between 0 and 1

60. The set of real numbers between 0 and 1

In Exercises 61–68, find each of the following, where $\aleph_0$ and c are transfinite cardinal numbers.

61. $\aleph_0 - 700$ **62.** $\aleph_0 + 4100$

63. $\aleph_0 + (\aleph_0 + \aleph_0)$ **64.** $\aleph_0 + c$

65. $c - 7$ **66.** $c + (c + c)$

67. $5\aleph_0$ **68.** $15c$

CHAPTER 2	**Test**

In Exercises 1–6, let $U = \{1, 2, 3, 4, 5, 6, 7, 8, 9, 10\}$ $A = \{3, 5, 7, 8\}$, $B = \{2, 3, 8, 9, 10\}$, and $C = \{1, 4, 7, 8\}$. Use the roster method to write each of the following sets.

1. $A \cup B$ **2.** $A' \cap B$

3. $(A \cap B)'$ **4.** $(A \cup B')'$

5. $A' \cup (B \cap C')$ **6.** $A \cap (B' \cup C)$

In Exercises 7 and 8, use set-builder notation to write each of the given sets.

7. $\{0, 1, 2, 3, 4, 5, 6\}$ **8.** $\{-3, -2, -1, 0, 1, 2\}$

In Exercises 9 and 10, state the cardinality of the given set.

9. a. The set of whole numbers less than 4

    b. The set of rational numbers between 7 and 8

10. a. The set of natural numbers

    b. The set of real numbers

In Exercises 11 and 12, state whether the given sets are equal, equivalent, both, or neither.

11. a. the set of natural numbers; the set of integers

    b. the set of whole numbers; the set of positive integers

12. a. the set of rational numbers; the set of irrational numbers

    b. the set of real numbers; the set of irrational numbers

13. List all of the subsets of {a, b, c, d}.

14. Determine the number of subsets of a set with 21 elements.

15. State whether each statement is true or false.

    a. $\{4\} \in \{1, 2, 3, 4, 5, 6, 7\}$

    b. The set of rational numbers is a well-defined set.

    c. $A \subset A$

    d. The set of positive even whole numbers is equivalent to the set of natural numbers.

16. Draw a Venn diagram to represent the given set.

    a. $(A \cup B') \cap C$

    b. $(A' \cap B) \cup (A \cap C')$

17. Use set notation to describe the shaded region.

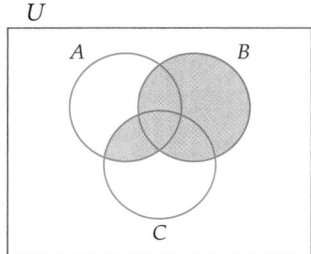

18. In the town of LeMars, 385 families have a CD player, 142 families have a DVD player, 41 families have both a CD player and a DVD player, and 55 families do not have a CD player or a DVD player. How many families live in LeMars?

19. Show a method that can be used to establish a one-to-one correspondence between the elements of the following sets.

$$\{5, 10, 15, 20, 25, \ldots, 5n, \ldots\}; \quad W$$

20. Prove that the following set is an infinite set by illustrating a one-to-one correspondence between the elements of the set and the elements of one of the set's proper subsets.

$$\{3, 6, 9, 12, \ldots, 3n \ldots\}$$

CHAPTER

3

Logic

Logic gates are an important component of modern digital computers. The **Excursion** on **page 136** illustrates how logic gates are used to implement some of the mathematical procedures used by modern computers.

Need help? For on-line student resources, such as section quizzes, visit this textbook's web site at **math.college.hmco.com/students.**

In the study of logic, English sentences are often represented by a letter. For instance, in this chapter opener,

e represents "All men are created equal."

t represents "I am trading places."

a represents "I get Abe's place."

g represents "I get George's place."

The following table shows the connective symbols that are used in this chapter.

Original Statement	Connective	Statement in Symbolic Form
not *p*	not	$\sim p$
p and *q*	and	$p \wedge q$
p or *q*	or	$p \vee q$
If *p*, then *q*	If … then	$p \rightarrow q$
p if and only if *q*	if and only if	$p \leftrightarrow q$

Note that $\sim t$ is read, "I am not trading places."

Use the above information to translate the dialogue in the following speech bubbles.

Logic Statements and Quantifiers

Logic Statements

One of the first mathematicians to make a serious study of symbolic logic was Gottfried Wilhelm Leibniz (1646–1716). Leibniz tried to advance the study of logic from a merely philosophical subject to a formal mathematical subject. Leibniz never completely achieved this goal; however, several mathematicians, such as Augustus De Morgan (1806–1871) and George Boole (1815–1864), contributed to the advancement of symbolic logic as a mathematical discipline.

Boole published *The Mathematical Analysis of Logic* in 1848. In 1854 he published the more extensive work *An Investigation of the Laws of Thought.* Concerning this document, the mathematician Bertrand Russell stated, "Pure mathematics was discovered by Boole in a work which is called *The Laws of Thought.*" Although some mathematicians feel this is an exaggeration, the following paragraph extracted from *An Investigation of the Laws of Thought* gives some insight into the nature and scope of this document.

The design of the following treatise is to investigate the fundamental laws of those operations of the mind by which reasoning is performed; to give expression to them in the language of a Calculus, and upon this foundation to establish the science of Logic and construct its method; to make the method itself the basis of a general method for the application of the mathematical doctrine of probabilities; and finally, to collect from the various elements of truth brought to view in the course of these inquiries some probable intimations concerning the nature and constitution of the human mind.[1]

Every language contains different types of sentences, such as statements, questions, and commands. For instance,

"Is the test today?" is a question.

"Go get the newspaper" is a command.

"This is a nice car" is an opinion.

"Denver is the capital of Colorado" is a statement of fact.

The symbolic logic that Boole was instrumental in creating applies only to sentences that are *statements* as defined below.

> **Definition of a Statement**
>
> A **statement** is a declarative sentence that is either true or false, but not both true and false.

---

1. Bell, E. T. *Men of Mathematics.* New York: Simon and Schuster, Inc., Touchstone Books, Reissue edition, 1986.

It may not be necessary to determine whether a sentence is true or false to determine whether it is a statement. For instance, the following sentence is either true or false:

Every even number greater than 2 can be written as the sum of two prime numbers.

At this time mathematicians have not determined whether the sentence is true or false, but they do know that it is either true or false and that it is not both true and false. Thus the sentence is a statement.

EXAMPLE 1 ■ Identify Statements

Determine whether each sentence is a statement.

a. Florida is a state in the United States.

b. The word *dog* has four letters.

c. How are you?

d. $9^{(9^9)} + 2$ is a prime number.

e. $x + 1 = 5$

Solution

a. Florida is one of the 50 states in the United States, so this sentence is true and it is a statement.

b. The word *dog* consists of exactly three letters, so this sentence is false and it is a statement.

c. The sentence "How are you?" is a question; it is not a declarative sentence. Thus it is not a statement.

d. You may not know if $9^{(9^9)} + 2$ is a prime number or not; however, you do know that it is a whole number larger than 1, so it is either a prime number or it is not a prime number. The sentence is either true or it is false, and it is not both true and false, so it is a statement.

e. $x + 1 = 5$ is a statement. It is known as an *open statement*. It is true for $x = 4$, and it is false for any other values of x. For any given value of x, it is true or false but not both.

CHECK YOUR PROGRESS 1   Determine whether each sentence is a statement.

a. Open the door.

b. 7055 is a large number.

c. $4 + 5 = 8$

d. In the year 2009, the president of the United States will be a woman.

e. $x > 3$

Solution   *See page S6.*

Charles Dodgson
(Lewis Carroll)

Math Matters   Charles Dodgson

One of the most well known logicians is Charles Dodgson (1832–1898). Dodgson was educated at Rugby and Oxford, and in 1861 he became a lecturer in mathematics at Oxford. Some of his mathematical works include *A Syllabus of Plane Algebraical Geometry, The Fifth Book of Euclid Treated Algebraically,* and *Symbolic Logic.* Although Dodgson was a distinguished mathematician in his time, he is best

known by his pen name Lewis Carroll, which he used when he published *Alice's Adventures in Wonderland* and *Through the Looking Glass.*

Queen Victoria of the United Kingdom enjoyed *Alice's Adventures in Wonderland* to the extent that she told Dodgson she was looking forward to reading another of his books. He promptly sent her his *Syllabus of Plane Algebraical Geometry,* and it was reported that she was less than enthusiastic about the latter book.

Compound Statements

Connecting statements with words and phrases such as *and, or, not, if … then,* and *if and only if* creates a **compound statement.** For instance, "I will attend the meeting or I will go to school" is a compound statement. It is composed of the two **component statements** "I will attend the meeting" and "I will go to school." The word *or* is a **connective** for the two component statements.

George Boole used symbols such as *p, q, r,* and *s* to represent statements and the symbols $\wedge$, $\vee$, $\sim$, $\rightarrow$, and $\leftrightarrow$ to represent connectives. See Table 3.1.

Table 3.1 *Logic Symbols*

Original Statement	Connective	Statement in Symbolic Form	Type of Compound Statement
not *p*	not	$\sim p$	negation
p and *q*	and	$p \wedge q$	conjunction
p or *q*	or	$p \vee q$	disjunction
If *p*, then *q*	If … then	$p \rightarrow q$	conditional
p if and only if *q*	if and only if	$p \leftrightarrow q$	biconditional

QUESTION   *What connective is used in a conjunction?*

ANSWER   *The connective and.*

*The Truth Table
for ~p*

p	*~p*
T	F
F	T

> **Truth Value and Truth Tables**
>
> The **truth value** of a statement is true (T) if the statement is true and false (F) if the statement is false. A **truth table** is a table that shows the truth value of a statement for all possible truth values of its components.

The *negation* of the statement "Today is Friday" is the statement "Today is not Friday." In symbolic logic, the tilde symbol ~ is used to denote the negation of a statement. If a statement p is true, its negation $~p$ is false, and if a statement p is false, its negation $~p$ is true. See the table at the left. The negation of the negation of a statement is the original statement. Thus, $~(~p)$ can be replaced by p in any statement.

EXAMPLE 2 ■ Write the Negation of a Statement

Write the negation of each statement.

a. Bill Gates has a yacht.

b. The number 10 is a prime number.

c. The Dolphins lost the game

Solution

a. Bill Gates does not have a yacht.

b. The number 10 is not a prime number.

c. The Dolphins did not lose the game.

CHECK YOUR PROGRESS 2   Write the negation of each statement.

a. 1001 is divisible by 7.

b. 5 is an even number.

c. The fire engine is not red.

Solution   *See page S6.*

We will often find it useful to write compound statements in symbolic form.

EXAMPLE 3 ■ Write Compound Statements in Symbolic Form

Consider the following statements.

    p: Today is Friday.

    q: It is raining.

    r: I am going to a movie.

    s: I am not going to the basketball game.

Write the following compound statements in symbolic form.

a. Today is Friday and it is raining.

b. It is not raining and I am going to a movie.

c. I am going to the basketball game or I am going to a movie.

d. If it is raining, then I am not going to the basektball game.

Solution

a. $p \wedge q$ **b.** $~q \wedge r$ **c.** $~s \vee r$ **d.** $q \rightarrow s$

CHECK YOUR PROGRESS 3   Use p, q, r, and s as defined in Example 3 to write the following compound statements in symbolic form.

a. Today is not Friday and I am going to a movie.

b. I am going to the basketball game and I am not going to a movie.

c. I am going to the movie if and only if it is raining.

d. If today is Friday, then I am not going to a movie.

Solution   *See page S7.*

In the next example, we translate symbolic logic statements into English sentences.

EXAMPLE 4 ■ Translate Symbolic Statements

Consider the following statements.

p: The game will be played in Atlanta.

q: The game will be shown on CBS.

r: The game will not be shown on ESPN.

s: The Dodgers are favored to win.

Write each of the following symbolic statements in words.

a. $q \wedge p$ **b.** $\sim r \wedge s$ **c.** $s \leftrightarrow \sim p$

Solution

a. The game will be shown on CBS and the game will be played in Atlanta.

b. The game will be shown on ESPN and the Dodgers are favored to win.

c. The Dodgers are favored to win if and only if the game will not be played in Atlanta.

CHECK YOUR PROGRESS 4   Consider the statements given in Example 4. Write each of the following symbolic statements in words.

a. $q \vee \sim r$ **b.** $r \rightarrow q$ **c.** $q \leftrightarrow r$

Solution   *See page S7.*

The Truth Table for $p \wedge q$

p	q	$p \wedge q$
T	T	T
T	F	F
F	T	F
F	F	F

If you order cake *and* ice cream in a restaurant, the waiter will bring *both* cake and ice cream. In general, the **conjunction** $p \wedge q$ is true if both p and q are true and the conjunction is false if either p or q is false. The truth table at the left shows the four possible cases that arise when we form a conjunction of two statements.

Truth Value of a Conjunction

The conjunction $p \wedge q$ is true if and only if both p and q are true.

Sometimes the word *but* is used in place of the connective *and* to form a conjunction. For instance, "My local phone company is Pacific Bell, but my long distance carrier is Sprint" is equivalent to the conjunction "My local phone company is Pacific Bell and my long distance carrier is Sprint."

The Truth Table for p ∨ q

p	*q*	*p ∨ q*
T	T	T
T	F	T
F	T	T
F	F	F

Any **disjunction** $p \lor q$ is true if p is true or q is true or both p and q are true. The truth table at the left shows that the disjunction p or q is false if both p and q are false; however, it is true in all other cases.

> **Truth Value of a Disjunction**
>
> The disjunction $p \lor q$ is true if p is true, if q is true, or if both p and q are true.

EXAMPLE 5 ■ Determine the Truth Value of a Statement

Determine whether each statement is true or false.

a. $7 \geq 5$

b. 5 is a whole number and 5 is an even number.

c. 2 is a prime number and 2 is an even number.

Solution

a. $7 \geq 5$ means $7 > 5$ or $7 = 5$. Because $7 > 5$ is true, the statement $7 \geq 5$ is a true statement.

b. This is a false statement because 5 is not an even number.

c. This is a true statement because each component statement is true.

CHECK YOUR PROGRESS 5   Determine whether each statement is true or false.

a. 21 is a rational number and 21 is a natural number.

b. $4 \leq 9$

c. $-7 \geq -3$

Solution   See page S7.

Truth tables for the conditional and biconditional are given in Section 3.3.

Quantifiers and Negation

In a statement, the word *some* and the phrases *there exists* and *at least one* are called **existential quantifiers.** Existential quantifiers are used as prefixes to assert the existence of something.

In a statement, the words *none, no, all,* and *every* are called **universal quantifiers.** The universal quantifiers *none* and *no* deny the existence of something, whereas the universal quantifiers *all* and *every* are used to assert that every element of a given set satisfies some condition.

Recall that the negation of a false statement is a true statement and the negation of a true statement is a false statement. It is important to remember this fact when forming the negation of a quantified statement. For instance, what is the negation of the false statement, "All dogs are mean"? You may think that the negation is "No dogs are mean," but this is also a false statement. Thus, the statement "No dogs are mean" is not the negation of "All dogs are mean." The negation of "All dogs are mean," which is a false statement, is in fact "Some dogs are not mean," which is a true statement. The statement "Some dogs are not mean" can also be stated as "At least one dog is not mean" or "There exists a dog that is not mean."

What is the negation of the false statement "No doctors write in a legible manner"? Whatever the negation is, we know it must be a true statement. The negation cannot be "All doctors write in a legible manner," because this is also a false statement. The negation is "Some doctors write in a legible manner." This can also be stated as "There exists at least one doctor who writes in a legible manner."

Table 3.2 summarizes the concepts needed to write the negations of statements that contain one of the quantifiers *all*, *none*, or *some*.

Table 3.2 *The Negation of a Statement that Contains a Quantifier*

Original Statement	Negation
All _____ are _____ .	Some _____ are not _____ .
No(ne) _____ .	Some _____ .
Some _____ are not _____ .	All _____ are _____ .
Some _____ .	No(ne) _____ .

EXAMPLE 6 ■ Write the Negation of a Quantified Statement

Write the negation of each of the following statements.

a. Some baseball players are worth a million dollars.

b. All movies are worth the price of admission.

c. No odd numbers are divisible by 2.

Solution

a. No baseball player is worth a million dollars.

b. Some movies are not worth the price of admission.

c. Some odd numbers are divisible by 2.

CHECK YOUR PROGRESS 6   Write the negation of the following statements.

a. All bears are brown.

b. No math class is fun.

c. Some vegetables are not green.

Solution   See page S7.

Excursion

Switching Networks

Claude E. Shannon

In 1939, Claude E. Shannon (1916–2001) wrote a thesis on an application of symbolic logic to *switching networks*. A switching network consists of wires and switches that can open or close. Switching networks are used in many electrical appliances, telephone equipment, and computers. Figure 3.1 shows a switching network that consists of a single switch P that connects two terminals. An electric current can flow from one terminal to the other terminal provided the switch P is in the closed position. If P is in the open position, then the current cannot flow from one terminal to the other. If a current can flow between the terminals we say that a network is closed, and if a current cannot flow between the terminals we say that the network is open. We designate this network by the letter P. There exists an analogy between a network P and a statement p in that a network is either open or it is closed, and a statement is either true or it is false.

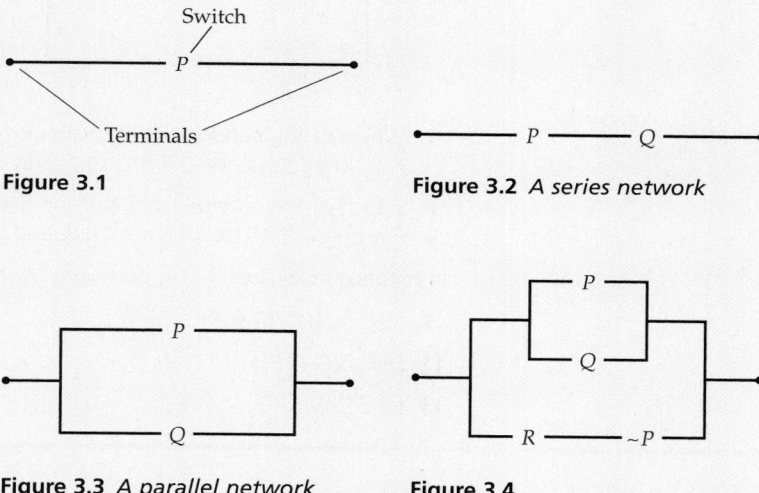

Figure 3.1

Figure 3.2 *A series network*

Figure 3.3 *A parallel network*

Figure 3.4

Figure 3.2 shows two switches P and Q connected in **series.** This series network is closed if and only if both switches are closed. We will use $P \wedge Q$ to denote this series network because it is analogous to the logic statement $p \wedge q$, which is true if and only if both p and q are true.

Figure 3.3 shows two switches P and Q connected in **parallel.** This parallel network is closed if either P or Q is closed. We will designate this parallel network by $P \vee Q$ because it is analogous to the logic statement $p \vee q$, which is true if p is true or if q is true.

Series and parallel networks can be combined to produce more complicated networks, as shown in Figure 3.4.

The network shown in Figure 3.4 is closed provided P or Q is closed or provided both R and $\sim P$ are closed. Note that the switch $\sim P$ is closed if P is open, and $\sim P$ is open if P is closed. We use the symbolic statement $(P \vee Q) \vee (R \wedge \sim P)$ to represent this network.

If two switches are always open at the same time and always closed at the same time, then we will use the same letter to designate both switches.

(continued)

Excursion Exercises

Write a symbolic statement to represent each of the networks in Excursion Exercises 1–6.

1.

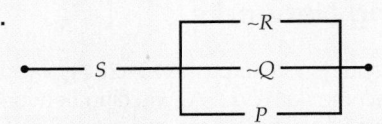

2.

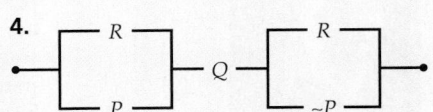

3.

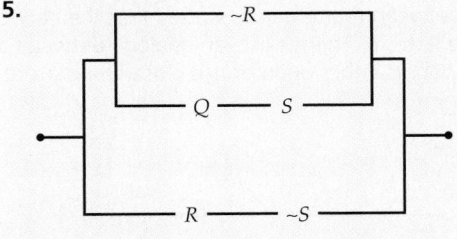

4.

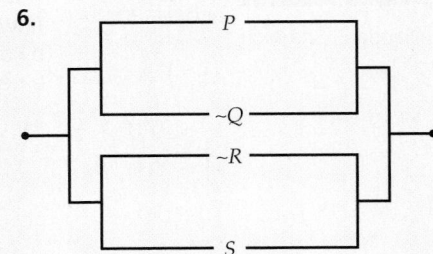

5.

6.

7. Which of the networks in Excursion Exercises 1–6 are closed networks, given that *P* is closed, *Q* is open, *R* is closed, and *S* is open?

8. Which of the networks in Excursion Exercises 1–6 are closed networks, given that *P* is open, *Q* is closed, *R* is closed, and *S* is closed?

In Excursion Exercises 9–14, draw a network to represent each statement.

9. $(\sim P \vee Q) \wedge (R \wedge P)$

10. $P \wedge [(Q \wedge \sim R) \vee R]$

11. $[\sim P \wedge Q \wedge R] \vee (P \wedge R)$

12. $(Q \vee R) \vee (S \vee \sim P)$

13. $[(\sim P \wedge R) \vee Q] \vee (\sim R)$

14. $(P \vee Q \vee R) \wedge S \wedge (\sim Q \vee R)$

Exercise Set 3.1   (Suggested Assignment: 1–73 every other odd)

In Exercises 1–10, determine whether each sentence is a statement.

1. West Virginia is west of the Mississippi river.

2. 1031 is a prime number.

3. The area code for Storm Lake, Iowa is 512.

4. Some negative numbers are rational numbers.

5. Have a fun trip.

6. Do you like to read?

7. All hexagons have exactly five sides.

8. If x is a negative number, then x^2 is a positive number.

9. Mathematics courses are better than history courses.

10. Every real number is a rational number.

In Exercises 11–18, determine the components of each compound statement.

11. The principal will attend the class on Tuesday or Wednesday.

12. 5 is an odd number and 6 is an even number.

13. A triangle is an acute triangle if and only if it has three acute angles.

14. Some birds can swim and some fish can fly.

15. I ordered a salad and a cola.

16. If this is Saturday, then tomorrow is Sunday.

17. $5 + 2 \geq 6$

18. $9 - 1 \leq 8$

In Exercises 19–22, write the negation of each statement.

19. The Giants lost the game.

20. The lunch was served at noon.

21. The game did not go into overtime.

22. The game was not shown on ABC.

In Exercises 23–32, write each sentence in symbolic form. Represent each component of the sentence with the letter indicated in parentheses. Also state whether the sentence is a conjunction, a disjunction, a negation, a conditional, or a biconditional.

23. If today is Wednesday (w), then tomorrow is Thursday (t).

24. It is not true that Sue took the tickets (t).

25. All squares (s) are rectangles (r).

26. I went to the post office (p) and the bookstore (s).

27. A triangle is an equilateral triangle (l) if and only if it is an equiangular triangle (a).

28. A number is an even number (e) if and only if it has a factor of 2 (t).

29. If it is a dog (d), it has fleas (f).

30. Polynomials that have exactly three terms (p) are called trinomials (t).

31. I will major in mathematics (m) or computer science (c).

32. All pentagons (p) have exactly five sides (s).

In Exercises 33–38, write each symbolic statement in words. Use p, q, r, s, t, and u as defined below.

> p: The tour goes to Italy.
> q: The tour goes to Spain.
> r: We go to Venice.
> s: We go to Florence.
> t: The hotel fees are included.
> u: The meals are not included.

33. $p \land \sim q$

34. $r \lor s$

35. $r \to \sim s$

36. $p \to r$

37. $s \leftrightarrow \sim r$

38. $\sim t \land u$

In Exercises 39–50, use the definitions presented in Table 3.2, page 116, to write the negation of each quantified statement.

39. Some cats do not have claws.

40. Some dogs are not friendly.

41. All classic movies were first produced in black and white.

42. Everybody enjoyed the dinner.

43. None of the numbers were even numbers.

44. At least one student received an A.

45. No irrational number can be written as a terminating decimal.

46. All cameras use film.

47. All cars run on gasoline.

48. None of the students took my advice.

49. Every item is on sale.

50. All of the telephone lines are not busy.

In Exercises 51–64, determine whether each statement is true or false.

51. $7 < 5$ or $3 > 1$.

52. $3 \leq 9$

53. $(-1)^{50} = 1$ and $(-1)^{99} = -1$.

54. $7 \neq 3$ or 9 is a prime number.

55. $-5 \geq -11$

56. $4.5 \leq 5.4$

57. 2 is an odd number or 2 is an even number.

58. 5 is a natural number and 5 is a rational number.

59. There exists an even prime number.

60. The square of any real number is a positive number.

61. Some real numbers are irrational.

62. All irrational numbers are real numbers.

63. Every integer is a rational number.

64. Every rational number is an integer.

Extensions
CRITICAL THINKING

In Exercises 65–70, translate each quote into symbolic form. For each component, indicate what letter you used to represent the component.

65. If music must be the food of love, play on.   *William Shakespeare*

66. If you aren't fired with enthusiasm, then you will be fired with enthusiasm. *Vince Lombardi*

67. Those who do not learn from history are condemned to repeat it. *George Santayana*

68. Until he extends the circle of his compassion to all living things, man will not himself find peace. *Albert Schweitzer*

69. If people concentrated on the really important things in life, there'd be a shortage of fishing poles. *Doug Larson*

70. If you're killed, you've lost a very important part of your life. *Brooke Shields*

In Exercises 71–76, translate each mathematical statement into symbolic form. For each component, indicate what letter you used to represent the component.

71. An angle is a right angle if and only if its measure is 90°.

72. Any angle inscribed in a semicircle is a right angle.

73. If two sides of a triangle are equal in length, the angles opposite those sides are congruent.

74. The sum of the measures of the three angles of any triangle is 180°.

75. All squares are rectangles.

76. If the corresponding sides of two triangles are proportional, then the triangles are similar.

EXPLORATIONS

77. Raymond Smullyan is a logician, a philospher, a professor, and an author of many books on logic and puzzles. Some of his fans rate his puzzle books as the best ever written. Search the Internet to find information on the life of Smullyan and his work in the area of logic. Write a few paragraphs that summarize your findings.

SECTION 3.2 | **Truth Tables and Applications**

Truth Tables

In Section 3.1, we defined truth tables for the negation of a statement, the conjunction of two statements, and the disjunction of two statements. Each of these truth tables is shown below for review purposes.

Negation

p	$\sim p$
T	F
F	T

Conjunction

p	q	$p \wedge q$
T	T	T
T	F	F
F	T	F
F	F	F

Disjunction

p	q	$p \vee q$
T	T	T
T	F	T
F	T	T
F	F	F

p	q	Given Statement
T	T	
T	F	
F	T	
F	F	

Standard truth table form for a given statement that involves only the two simple statements p and q.

In this section, we consider methods of constructing truth tables for a statement that involves a combination of conjunctions, disjunctions, and/or negations. If the given statement involves only the two simple statements p and q, then start with a table with four rows (see the table at the left), called the **standard truth table form**, and proceed as shown in Example 1.

EXAMPLE 1 ■ Truth Tables

a. Construct a table for $\sim(\sim p \vee q) \vee q$.

b. Use the truth table from part a to determine the truth value of $\sim(\sim p \vee q) \vee q$, given that p is true and q is false.

Solution

a. Start with the standard truth table form and then include a $\sim p$ column.

p	q	~p
T	T	F
T	F	F
F	T	T
F	F	T

Now use the truth values from the $\sim p$ and q columns to produce the truth values for $\sim p \vee q$, as shown in the following table.

p	q	~p	~p ∨ q
T	T	F	T
T	F	F	F
F	T	T	T
F	F	T	T

Negate the truth values in the $\sim p \vee q$ column to produce the following.

p	q	~p	~p ∨ q	~(~p ∨ q)
T	T	F	T	F
T	F	F	F	T
F	T	T	T	F
F	F	T	T	F

As our last step we form the disjunction of $\sim(\sim p \vee q)$ and q and place the results in the rightmost column of the table. The shaded column is the truth table for $\sim(\sim p \vee q) \vee q$.

p	q	~p	~p $\vee$ q	~(~p $\vee$ q)	~(~p $\vee$ q) $\vee$ q	
T	T	F	T	F	T	Row 1
T	F	F	F	T	T	Row 2
F	T	T	T	F	T	Row 3
F	F	T	T	F	F	Row 4

b. In row 2 of the above truth table, we see that p is true, q is false, and the statement ~($\sim p \vee q$) $\vee q$ in the rightmost column is true.

CHECK YOUR PROGRESS 1

a. Construct a truth table for $(p \wedge \sim q) \vee (\sim p \vee q)$.

b. Use the truth table from part a to determine the truth value of $(p \wedge \sim q) \vee (\sim p \vee q)$, given that p is true and q is false.

Solution   See page S7.

Compound statements that involve exactly three simple statements require a standard truth table form with $2^3 = 8$ rows, as shown at the left.

p	q	r	Given Statement
T	T	T	
T	T	F	
T	F	T	
T	F	F	
F	T	T	
F	T	F	
F	F	T	
F	F	F	

Standard truth table form for a statement that involves the three simple statements p, q, and r.

EXAMPLE 2 ■ Truth Tables

a. Construct a truth table for $(p \wedge q) \wedge (\sim r \vee q)$.

b. Use the truth table from part a to determine the truth value of $(p \wedge q) \wedge (\sim r \vee q)$, given that p is true, q is true, and r is false.

Solution

a. Using the procedures developed in Example 1 we can produce the following table. The shaded column is the truth table for $(p \wedge q) \wedge (\sim r \vee q)$. The numbers in the squares below the columns denote the order in which the columns were constructed.

p	q	r	p $\wedge$ q	~r	~r $\vee$ q	(p $\wedge$ q) $\wedge$ (~r $\vee$ q)	
T	T	T	T	F	T	T	Row 1
T	T	F	T	T	T	T	Row 2
T	F	T	F	F	F	F	Row 3
T	F	F	F	T	T	F	Row 4
F	T	T	F	F	T	F	Row 5
F	T	F	F	T	T	F	Row 6
F	F	T	F	F	F	F	Row 7
F	F	F	F	T	T	F	Row 8
			⬛ 1	⬛ 2	⬛ 3	⬛ 4	

b. In row 2 of the above truth table we see that $(p \wedge q) \wedge (\sim r \vee q)$ is true when p is true, q is true, and r is false.

Plugger caller I.D.

Plugger Logic

CHECK YOUR PROGRESS 2

a. Construct a truth table for $(\sim p \wedge r) \vee (q \wedge \sim r)$.

b. Use the truth table from part a to determine the truth value of $(\sim p \wedge r) \vee (q \wedge \sim r)$, given that p is false, q is true, and r is false.

Solution   *See page S7.*

Example 3 shows a *generalized procedure* that can be used to construct a truth table. This generalized procedure requires less writing than the procedure used in Examples 1 and 2.

A Generalized Procedure for the Construction of a Truth Table

If the given statement has n simple statements, then start with a standard form that has 2^n rows.

1. Enter the truth value for each simple statement p, q, r, and their negations, as needed.

2. Use the truth values from Step 1 to enter the truth value under each connective within parentheses. If some parentheses are nested inside other parentheses, work from the inside out.

3. Use the truth values from Step 2 to enter the truth values under any remaining connectives, working from left to right.

EXAMPLE 3 ■ **Truth Tables**

Use the generalized procedure given above to construct a truth table for $p \vee [\sim(p \wedge \sim q)]$.

Solution

The given statement $p \vee [\sim(p \wedge \sim q)]$ has the two simple statements p and q. Thus we start with a standard form that has $2^2 = 4$ rows.

1. First enter the truth values for the statements p and $\sim q$, as shown in the columns numbered 1, 2, and 3 of the following table.

p	q	p	$\vee$	$[\sim$	$(p$	$\wedge$	$\sim q)]$
T	T	T	T	T	T	F	F
T	F	T	T	F	T	T	T
F	T	F	T	T	F	F	F
F	F	F	T	T	F	F	T
		1	6	5	2	4	3

2. Use the truth values in the columns numbered 1, 2, and 3 to determine which truth values to enter under the connectives. See the columns numbered 4, 5, and 6. The shaded column is the truth table for $p \vee [\sim(p \wedge \sim q)]$.

CHECK YOUR PROGRESS 3 Construct a truth table for $\sim p \vee (p \wedge q)$.

Solution *See page S8.*

Math Matters A Three-Valued Logic

historical note

Jan Lukasiewicz (lōō-kä-shā-vēch) (1878–1956) was the Polish Minister of Education in 1919 and served as a professor of mathematics at Warsaw University from 1920–1939. Most of Lukasiewicz's work was in the area of logic. He is well known for developing *polish notation,* which was first used in logic to eliminate the need for parentheses in symbolic statements. Today *reverse polish notation* is used by many computers and calculators to perform computations without the need to enter parentheses. ■

In traditional logic a statement is either true or it is false. Many mathematicians have tried to extend traditional logic so that sentences that are *partially* true are assigned a truth value other than T or F. Jan Lukasiewicz was one of the first mathematicians to consider a three-valued logic in which a statement is either true, false, or "somewhere between true and false." In his three-valued logic, Lukasiewicz classified the truth value of a statement as either true (T), false (F), or maybe (M). The following table shows truth values for negation, conjunction, and disjunction in this three-valued logic.

p	q	Negation $\sim p$	Conjunction $p \wedge q$	Disjunction $p \vee q$
T	T	F	T	T
T	M	F	M	T
T	F	F	F	T
M	T	M	M	T
M	M	M	M	M
M	F	M	F	M
F	T	T	F	T
F	M	T	F	M
F	F	T	F	F

Equivalent Statements

✔ TAKE NOTE

In the remaining sections of this chapter, the ≡ symbol will often be used to denote that two statements are equivalent.

Two statements are **equivalent** if they both have the same truth value for all possible truth values of their component statements. Equivalent statements have identical truth values in the final columns of their truth tables. The notation $p \equiv q$ is used to indicate that the statements p and q are equivalent.

EXAMPLE 4 ■ Verify that Two Statements Are Equivalent

Show that $\sim(p \vee \sim q)$ and $\sim p \wedge q$ are equivalent statements.

Solution
Construct two truth tables and compare the results. The truth tables on the following page show that $\sim(p \vee \sim q)$ and $\sim p \wedge q$ have the same truth values for all possible truth values of their component statements. Thus the statements are equivalent.

p	q	~(p ∨ ~q)
T	T	F
T	F	F
F	T	T
F	F	F

p	q	~p ∧ q
T	T	F
T	F	F
F	T	T
F	F	F

└──────identical truth values──────┘
Thus ~(p ∨ ~q) ≡ ~p ∧ q

CHECK YOUR PROGRESS 4   Show that $p \vee (p \wedge \sim q)$ and p are equivalent.

Solution   *See page S8.*

The truth tables in Table 3.3 show that $\sim(p \vee q)$ and $\sim p \wedge \sim q$ are equivalent statements. The truth tables in Table 3.4 show that $\sim(p \wedge q)$ and $\sim p \vee \sim q$ are equivalent statements.

Table 3.3

p	q	~(p ∨ q)	~p ∧ ~q
T	T	F	F
T	F	F	F
F	T	F	F
F	F	T	T

Table 3.4

p	q	~(p ∧ q)	~p ∨ ~q
T	T	F	F
T	F	T	T
F	T	T	T
F	F	T	T

These equivalences are known as **De Morgan's laws for statements.**

De Morgan's Laws for Statements

For any statements p and q,

$$\sim(p \vee q) \equiv \sim p \wedge \sim q$$
$$\sim(p \wedge q) \equiv \sim p \vee \sim q$$

De Morgan's laws can be used to restate certain English sentences in an equivalent form.

EXAMPLE 5 ■ State an Equivalent Form

Use one of De Morgan's laws to restate the following sentence in an equivalent form.

It is not the case that I graduated or I got a job.

Solution

Let p represent the statement "I graduated." Let q represent the statement "I got a job." In symbolic form, the original sentence is $\sim(p \vee q)$. One of De Morgan's laws states that this is equivalent to $\sim p \wedge \sim q$. Thus a sentence that is equivalent to the original sentence is "I did not graduate and I did not get a job."

CHECK YOUR PROGRESS 5   Use one of De Morgan's laws to restate the following sentence in an equivalent form.

It is not true that I am going to the dance and I am going to the game.

Solution   *See page S8.*

Tautologies and Self-Contradictions

A **tautology** is a statement that is always true. A **self-contradiction** is a statement that is always false.

EXAMPLE 6 ▨ Verify Tautologies and Self-Contradictions

Show that $p \vee (\sim p \vee q)$ is a tautology.

Solution
Construct a truth table as shown below.

p	q	p	$\vee$	$(\sim p$	$\vee$	$q)$
T	T	T	T	F	T	T
T	F	T	T	F	F	F
F	T	F	T	T	T	T
F	F	F	T	T	T	F
		1	5	2	4	3

The table shows that $p \vee (\sim p \vee q)$ is always true. Thus $p \vee (\sim p \vee q)$ is a tautology.

CHECK YOUR PROGRESS 6   Show that $p \wedge (\sim p \wedge q)$ is a self-contradiction.

Solution   *See page S8.*

QUESTION   *Is the statement $x + 2 = 5$ a tautology or a self-contradiction?*

ANSWER   *Neither. The statement is not true for all values of x, and it is not false for all values of x.*

Excursion

Switching Networks—Part II

The Excursion in Section 3.1 introduced the application of symbolic logic to switching networks. This Excursion makes use of *closure tables* to determine under what conditions a switching network is open or closed. In a closure table, we use a 1 to designate that a switch or switching network is closed and a 0 to indicate that it is open.

Figure 3.5 shows a switching network that consists of the single switch *P* and a second network that consists of the single switch ~*P*. The table below shows that the switching network ~*P* is open when *P* is closed and closed when *P* is open.

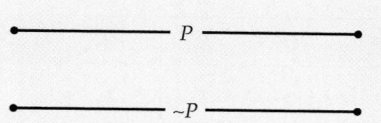

Negation Closure Table

P	~P
1	0
0	1

Figure 3.5

Figure 3.6 shows switches *P* and *Q* connected to form a series network. The table below shows that this series network is closed if and only if both *P* and *Q* are closed.

Series Network Closure Table

P	Q	P ∧ Q
1	1	1
1	0	0
0	1	0
0	0	0

Figure 3.6 *A series network*

Figure 3.7 shows switches *P* and *Q* connected to form a parallel network. The table below shows that this parallel network is closed if *P* is closed or if *Q* is closed.

Parallel Network Closure Table

P	Q	P ∨ Q
1	1	1
1	0	1
0	1	1
0	0	0

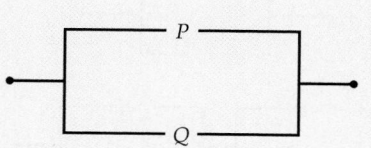

Figure 3.7 *A parallel network*

(continued)

Now consider the network shown in Figure 3.8. To determine the required conditions under which the network is closed, we first write a symbolic statement that represents the network and we then construct a closure table.

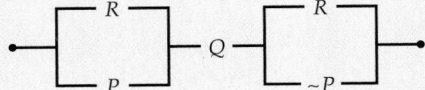

Figure 3.8

A symbolic statement that represents the network in Figure 3.8 is

$$((R \vee P) \wedge Q) \wedge (R \vee {\sim}P)$$

The closure table for this network is shown below.

P	Q	R	[(**R**	∨	**P**)	∧	**Q**)]	∧	(**R**	∨	**~P**)	
1	1	1	1	1	1	1	1	1	1	1	0	Row 1
1	1	0	0	1	1	1	1	0	0	0	0	Row 2
1	0	1	1	1	1	0	0	0	1	1	0	Row 3
1	0	0	0	1	1	0	0	0	0	0	0	Row 4
0	1	1	1	1	0	1	1	1	1	1	1	Row 5
0	1	0	0	0	0	0	1	0	0	1	1	Row 6
0	0	1	1	1	0	0	0	0	1	1	1	Row 7
0	0	0	0	0	0	0	0	0	0	1	1	Row 8

$$\boxed{1} \quad \boxed{6} \quad \boxed{2} \quad \boxed{7} \quad \boxed{3} \quad \boxed{9} \quad \boxed{4} \quad \boxed{8} \quad \boxed{5}$$

Rows 1 and 5 of the above table show that the network is closed whenever

- *P* is closed, *Q* is closed, and *R* is closed, or
- *P* is open, *Q* is closed, and *R* is closed.

Thus the switching network in Figure 3.8 is closed provided *Q* is closed and *R* is closed. The switching network is open under all other conditions.

Excursion Exercises

Construct a closure table for each of the following switching networks. Use the closure table to determine under what conditions the network will be closed.

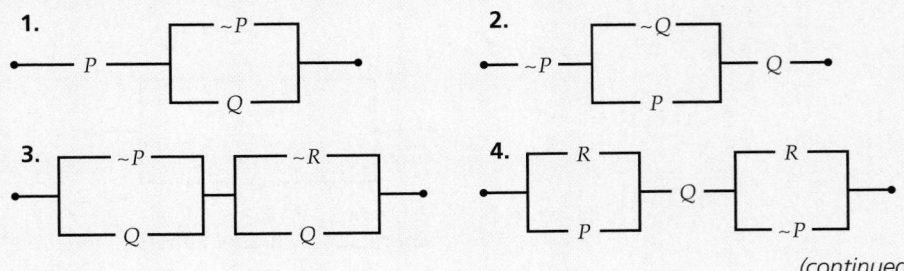

(continued)

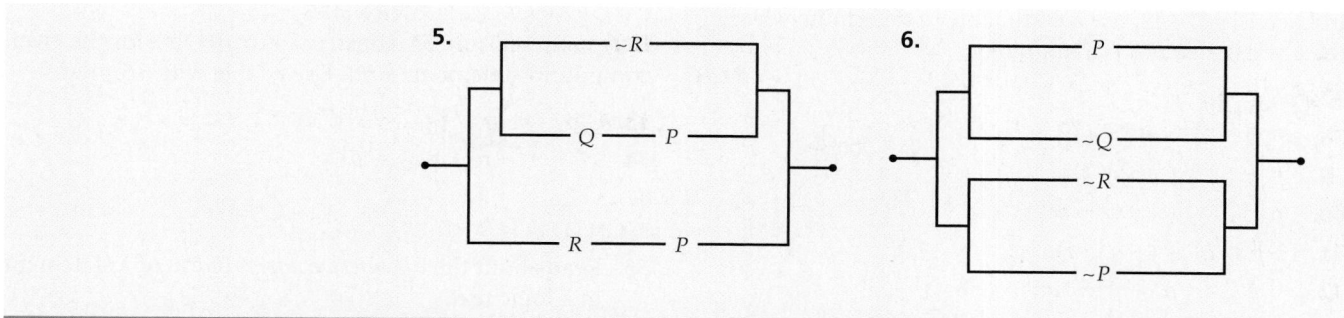

Exercise Set 3.2   (Suggested Assignment: 1–53 odds)

In Exercises 1–8, determine the truth value of the compound statement given that p is a false statement, q is a true statement, and r is a true statement.

1. $p \vee (\sim q \vee r)$

2. $r \wedge \sim (p \vee r)$

3. $(p \wedge q) \vee (\sim p \wedge \sim q)$

4. $(p \wedge q) \vee [(\sim p \wedge \sim q) \vee q]$

5. $[\sim (p \wedge \sim q) \vee r] \wedge (p \wedge \sim r)$

6. $(p \wedge \sim q) \vee [(p \wedge \sim q) \vee r]$

7. $[(p \wedge \sim q) \vee \sim r] \wedge (q \wedge r)$

8. $(\sim p \wedge q) \wedge [(p \wedge \sim q) \vee r]$

9. a. Given that p is a false statement, what can be said about $p \wedge (q \vee r)$?

  b. Explain why it is not necessary to know the truth values of q and r to determine the truth value of $p \wedge (q \vee r)$ in part a above.

10. a. Given that q is a true statement, what can be said about $q \vee \sim r$?

  b. Explain why it is not necessary to know the truth value of r to determine the truth value of $q \vee \sim r$ in part a above.

In Exercises 11–24, construct a truth table for each compound statement.

11. $\sim p \vee q$

12. $(q \wedge \sim p) \vee \sim q$

13. $p \wedge \sim q$

14. $p \vee [\sim (p \wedge \sim q)]$

15. $(p \wedge \sim q) \vee [\sim (p \wedge q)]$

16. $(p \vee q) \wedge [\sim (p \vee \sim q)]$

17. $\sim (p \vee q) \wedge (\sim r \vee q)$

18. $[\sim (r \wedge \sim q)] \vee (\sim p \vee q)$

19. $(p \wedge \sim r) \vee [\sim q \vee (p \wedge r)]$

20. $[r \wedge (\sim p \vee q)] \wedge (r \vee \sim q)$

21. $[(p \wedge q) \vee (r \wedge \sim p)] \wedge (r \vee \sim q)$

22. $(p \wedge q) \wedge \{[\sim (\sim p \vee r)] \wedge q\}$

23. $q \vee [\sim r \vee (p \wedge r)]$

24. $\{[\sim (p \vee \sim r)] \wedge \sim q\} \vee r$

In Exercises 25–30, use a truth table to show that each pair of compound statements is equivalent.

25. $p \vee (p \wedge r); p$

26. $q \wedge (q \vee r); q$

27. $p \wedge (q \vee r); (p \wedge q) \vee (p \wedge r)$

28. $p \vee (q \wedge r); (p \vee q) \wedge (p \vee r)$

29. $p \vee (q \wedge \sim p); p \vee q$

30. $\sim [p \vee (q \wedge r)]; \sim p \wedge (\sim q \vee \sim r)$

In Exercises 31–36, make use of one of De Morgan's laws to write the given statement in an equivalent form.

31. It is not the case that it rained or it snowed.

32. I did not pass the test and I did not complete the course.

33. She did not visit France and she did not visit Italy.

34. It is not true that I bought a new car and I moved to Florida.

35. It is not true that she received a promotion or that she received a raise.

36. It is not the case that the students cut classes or took part in the demonstration.

In Exercises 37–42, use a truth table to determine whether the given statement is a tautology.

37. $p \vee \sim p$
38. $q \vee [\sim(q \wedge r) \wedge \sim q]$
39. $(p \vee q) \vee (\sim p \vee q)$
40. $(p \wedge q) \vee (\sim p \vee \sim q)$
41. $(\sim p \vee q) \vee (\sim q \vee r)$
42. $\sim[p \wedge (\sim p \vee q)] \vee q$

In Exercises 43 to 48, use a truth table to determine whether the given statement is a self-contradiction.

43. $\sim r \wedge r$
44. $\sim(p \vee \sim p)$
45. $p \wedge (\sim p \wedge q)$
46. $\sim[(p \vee q) \vee (\sim p \vee q)]$
47. $[p \wedge (\sim p \vee q)] \vee q$
48. $\sim[p \vee (\sim p \vee q)]$
49. Explain why the statement $7 \le 8$ is a disjunction.
50. a. Why is the statement $5 \le 7$ true?
    b. Why is the statement $7 \le 7$ true?

Extensions

CRITICAL THINKING

51. How many rows are needed to construct a truth table for the statement $[p \wedge (q \vee \sim r)] \vee (s \wedge \sim t)$?
52. Explain why no truth table can have exactly 100 rows.

COOPERATIVE LEARNING

In Exercises 53 and 54, construct a truth table for the given compound statement. *Hint:* Use a table with 16 rows.

53. $[(p \wedge \sim q) \vee (q \wedge \sim r)] \wedge (r \vee \sim s)$
54. $s \wedge [\sim(\sim r \vee q) \vee \sim p]$

EXPLORATIONS

55. Read about the *disjunctive normal form* of a statement in a logic text.

    a. What is the disjunctive normal form of a statement that has the following truth table?

p	q	r	given statement
T	T	T	T
T	T	F	F
T	F	T	T
T	F	F	F
F	T	T	F
F	T	F	F
F	F	T	T
F	F	F	F

    b. Explain why the disjunctive normal form is a valuable concept.

56. Read about the *conjunctive normal form* of a statement in a logic text. What is the conjunctive normal form of the statement defined by the truth table in Exercise 55a?

SECTION 3.3   **The Conditional and the Biconditional**

Conditional Statements

If you don't get in that plane, you'll regret it. Maybe not today, maybe not tomorrow, but soon, and for the rest of your life.

The above quote is from the movie *Casablanca*. Rick, played by Humphrey Bogart, is trying to convince Ilsa, played by Ingrid Bergman, to get on the plane with

Humphrey Bogart and
Ingrid Bergman star in
Casablanca (1942).

Laszlo. The sentence "If you don't get in that plane, you'll regret it" is a *conditional statement*. **Conditional statements** can be written in *if p, then q* form or in *if p, q* form. For instance, all of the following are conditional statements.

If we order pizza, then we can have it delivered.

If you go to the movie, you will not be able to meet us for dinner.

If *n* is a prime number greater than 2, then *n* is an odd number.

In any conditional statement represented by "If *p*, then *q*" or by "If *p*, *q*", the *p* statement is called the **antecedent** and the *q* statement is called the **consequent**.

EXAMPLE 1 ▪ Identify the Antecedent and Consequent of a Conditional

Identify the antecedent and consequent in the following statements.

a. If our school was this nice, I would go there more than once a week.
— *The Basketball Diaries*

b. If you don't stop and look around once in a while, you could miss it.
—Ferris in *Ferris Bueller's Day Off*

c. If you strike me down, I shall become more powerful than you can possibly imagine.—Obi-Wan Kenobi, Star Wars, Episode IV, *A New Hope*

Solution

a. *Antecedent:* our school was this nice
Consequent: I would go there more than once a week

b. *Antecedent:* you don't stop and look around once in a while
Consequent: you could miss it

c. *Antecedent:* you strike me down
Consequent: I shall become more powerful than you can possibly imagine

---

CHECK YOUR PROGRESS 1   Identify the antecedent and consequent in each of the following conditional statements.

a. If I study for at least 6 hours, then I will get an A on the test.

b. If I get the job, I will buy a new car.

c. If you can dream it, you can do it.

Solution   See page S8.

Arrow Notation

The conditional statement "If *p*, then *q*" can be written using the **arrow notation** $p \rightarrow q$. The arrow notation $p \rightarrow q$ is read as "if *p*, then *q*" or as "*p* implies *q*."

The Truth Table for the Conditional $p \rightarrow q$

To determine the truth table for $p \rightarrow q$, consider the advertising slogan for a web authoring software product that states "If you can use a word processor, you can create a web page." This slogan is a conditional statement. The antecedent is *p*, "you can use a word processor," and the consequent is q, "you can create a web page." Now consider the truth value of $p \rightarrow q$ for each of the following four possibilities.

Table 3.5

p: you can use a word processor	q: you can create a web page	$p \rightarrow q$	
T	T	?	Row 1
T	F	?	Row 2
F	T	?	Row 3
F	F	?	Row 4

Row 1: Antecedent T, consequent T You can use a word processor, and you can create a web page. In this case the truth value of the advertisement is true. To complete Table 3.5, we place a T in place of the question mark in row 1.

Row 2: Antecedent T, consequent F You can use a word processor, but you cannot create a web page. In this case the advertisement is false. We put an F in place of the question mark in row 2 of Table 3.5.

Row 3: Antecedent F, consequent T You cannot use a word processor, but you can create a web page. Because the advertisement does not make any statement about what you might or might not be able to do if you cannot use a word processor, we cannot state that the advertisement is false, and we are compelled to place a T in place of the question mark in row 3 of Table 3.5.

Row 4: Antecedent F, consequent F You cannot use a word processor, and you cannot create a web page. Once again we must consider the truth value in this case to be true because the advertisement does not make any statement about what you might or might not be able to do if you cannot use a word processor. We place a T in place of the question mark in row 4 of Table 3.5.

The truth table for the conditional $p \rightarrow q$ is given in Table 3.6.

Table 3.6 *The Truth Table for* $p \rightarrow q$

p	q	$p \rightarrow q$
T	T	T
T	F	F
F	T	T
F	F	T

Truth Value of the Conditional $p \rightarrow q$

The conditional $p \rightarrow q$ is false if p is true and q is false. It is true in all other cases.

EXAMPLE 2 ■ Find the Truth Value of a Conditional

Determine the truth value of each of the following.

a. If 2 is an integer, then 2 is a rational number.

b. If 3 is a negative number, then $5 > 7$.

c. If $5 > 3$, then $2 + 7 = 4$.

Solution

a. Because the consequent is true, this is a true statement.

b. Because the antecedent is false, this is a true statement.

c. Because the antecedent is true and the consequent is false, this is a false statement.

CHECK YOUR PROGRESS 2   Determine the truth value of each of the following.

a. If $4 \geq 3$, then $2 + 5 = 6$.

b. If $5 > 9$, then $4 > 9$.

c. If Tuesday follows Monday, then April follows March.

Solution   *See page S8.*

CALCULATOR NOTE

TI-83

Program FACTOR

```
0→dim L₁
Prompt N
1→S: 2→F:0→E
√N→M
While F≤M
While fPart (N/F)=0
E+1→E:N/F→N
End
If E>0
Then
F→L₁(S)
E→L₁(S+1)
S+2→S:0→E
√N→M
End
If F=2
Then
3→F
Else
F+2→F
End: End
If N≠1
Then
N→L₁(S)
1→L₁(S+1)
End
L₁
```

EXAMPLE 3 ■ Construct a Truth Table for a Statement Involving a Conditional

Construct a truth table for $[p \wedge (q \vee {\sim}p)] \to {\sim}p$.

Solution

Using the generalized procedure for truth table construction, we produce the following table.

p	*q*	[*p*	∧	(*q*	∨	∼*p*)]	→	∼*p*
T	T	T	T	T	T	F	F	F
T	F	T	F	F	F	F	T	F
F	T	F	F	T	T	T	T	T
F	F	F	F	F	T	T	T	T
		☐1	☐6	☐2	☐5	☐3	☐7	☐4

CHECK YOUR PROGRESS 3   Construct a truth table for $[p \wedge (p \to q)] \to q$.

Solution   *See page S8.*

Math Matters   Use Conditional Statements to Control a Calculator Program

Computer and calculator programs use conditional statements to control the flow of a program. For instance, the "If…Then" instruction in a TI-83 calculator program directs the calculator to execute a group of commands if a condition is true and to skip to the End statement if the condition is false. See the program steps below.

:If *condition*

:Then (skip to End if *condition* is false)

:*command* if *condition* is true

:*command* if *condition* is true

:End

:*command*

The TI-82/83 program FACTOR shown at the left factors a number N into its prime factors. Note the use of the "If…Then" instructions highlighted in red.

An Equivalent Form of the Conditional

Table 3.7 *The Truth Table for* $\sim p \vee q$

p	*q*	$\sim p \vee q$
T	T	T
T	F	F
F	T	T
F	F	T

The truth table for $\sim p \vee q$ is shown in Table 3.7. The truth values in this table are identical to the truth values in Table 3.6. Hence, the conditional $p \rightarrow q$ is equivalent to the disjunction $\sim p \vee q$.

An Equivalent Form of the Conditional $p \rightarrow q$

$$p \rightarrow q \equiv \sim p \vee q$$

EXAMPLE 4 ■ Write a Conditional in Its Equivalent Disjunctive Form

Write each of the following in its equivalent disjunctive form.

a. If I could play the guitar, I would join the band.

b. If Arnold cannot play, then the Dodgers will lose.

Solution
In each case we write the disjunction of the negation of the antecedent and the consequent.

a. I cannot play the guitar or I would join the band.

b. Arnold can play or the Dodgers will lose.

CHECK YOUR PROGRESS 4   Write each of the following in its equivalent disjunctive form.

a. If I don't move to Georgia, I will live in Houston.

b. If the number is divisible by 2, then the number is even.

Solution   See page S8.

The Negation of the Conditional

Because $p \rightarrow q \equiv \sim p \vee q$, an equivalent form of $\sim(p \rightarrow q)$ is given by $\sim(\sim p \vee q)$, which, by one of De Morgan's laws, can be expressed as the conjunction $p \wedge \sim q$.

The Negation of $p \rightarrow q$

$$\sim(p \rightarrow q) \equiv p \wedge \sim q$$

EXAMPLE 5 ■ Write the Negation of a Conditional Statement

Write the negation of each conditional statement.

a. If they pay me the money, I will sign the contract.

b. If the lines are parallel, then they do not intersect.

Solution
In each case, we write the conjunction of the antecedent and the negation of the consequent.

a. They paid me the money and I did not sign the contract.
b. The lines are parallel and they intersect.

CHECK YOUR PROGRESS 5   Write the negation of each conditional statement.

a. If I finish the report, I will go to the concert.
b. If the square of n is 25, then n is 5 or -5.

Solution   *See page S8.*

The Biconditional

The statement $(p \rightarrow q) \wedge (q \rightarrow p)$ is called a **biconditional** and is denoted by $p \leftrightarrow q$, which is read as "p if and only if q."

> **Definition of the Biconditional $p \leftrightarrow q$**
> $$p \leftrightarrow q \equiv [(p \rightarrow q) \wedge (q \rightarrow p)]$$

Table 3.8 *The Truth Table for $p \leftrightarrow q$*

p	q	$p \leftrightarrow q$
T	T	T
T	F	F
F	T	F
F	F	T

Truth table 3.8 shows that $p \leftrightarrow q$ is true only when the components p and q have the same truth value.

EXAMPLE 6 ■ Determine the Truth Value of a Biconditional

State whether each biconditional is true or false.

a. $x + 4 = 7$ if and only if $x = 3$.
b. $x^2 = 36$ if and only if $x = 6$.

Solution
a. Both components are true when $x = 3$ and both are false when $x \neq 3$. Both components have the same truth value for any value of x, so this is a true statement.
b. If $x = -6$, the first component is true and the second component is false. Thus this is a false statement.

CHECK YOUR PROGRESS 6   State whether each biconditional is true or false.

a. $x > 7$ if and only if $x > 6$.
b. $x + 5 > 7$ if and only if $x > 2$.

Solution   *See page S8.*

Excursion

Logic Gates

Modern digital computers use *gates* to process information. These gates are designed to receive two types of electronic impulses, which are generally represented as a 1 or a 0. Figure 3.9 shows a *NOT gate*. It is constructed so that a stream of impulses that enter the gate will exit the gate as a stream of impulses in which each 1 is converted to a 0 and each 0 is converted to a 1.

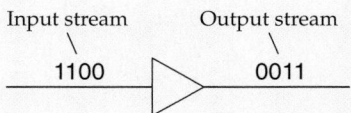

Figure 3.9 *NOT gate*

Note the similarity between the logical connective *not* and the logic gate NOT. The *not* connective converts the sequence of truth values T F to F T. The NOT gate converts the input stream 1 0 to 0 1. If the 1's are replaced with T's and the 0's with F's, then the NOT logic gate yields the same results as the *not* connective.

Many gates are designed so that two input streams are converted to one output stream. For instance, Figure 3.10 shows an *AND gate*. The AND gate is constructed so that a 1 is the output if and only if both input streams have a 1. In any other situation a 0 is produced as the output.

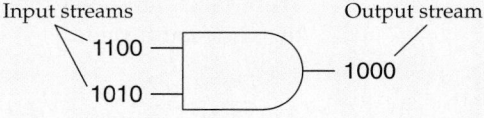

Figure 3.10 *AND gate*

Note the similarity between the logical connective *and* and the logic gate AND. The *and* connective combines the sequence of truth values T T F F with the truth values T F T F to produce T F F F. The AND gate combines the input stream 1 1 0 0 with the input stream 1 0 1 0 to produce 1 0 0 0. If the 1's are replaced with T's and the 0's with F's, then the AND logic gate yields the same result as the *and* connective.

The *OR gate* is constructed so that its output is a 0 if and only if both input streams have a 0. All other situations yield a 1 as the output. See Figure 3.11.

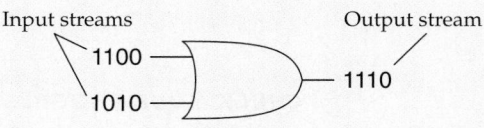

Figure 3.11 *OR gate*

(continued)

Figure 3.12 shows an network that consists of a NOT gate and an AND gate.

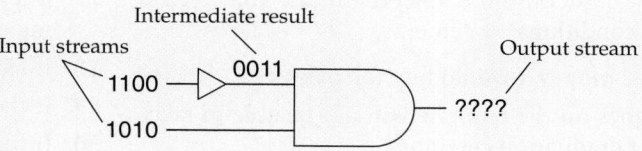

Figure 3.12

QUESTION   *What is the output stream for the network in Figure 3.12?*

Excursion Exercises

1. For each of the following, determine the output stream for the given input streams.

   a.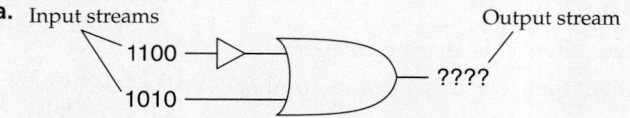

   b.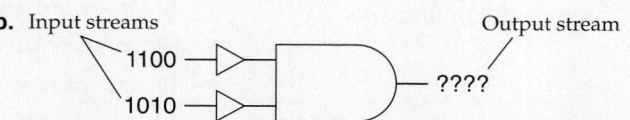

   c.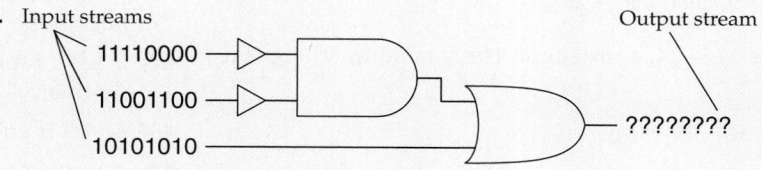

2. Construct a network using NOT, AND, and OR gates as needed that accepts the two input streams 1 1 0 0 and 1 0 1 0 and produces the output stream 0 1 1 1.

ANSWER   *0 0 1 0*

Exercise Set 3.3   (Suggested Assignment: 1–61 odds)

In Exercises 1–6, identify the antecedent and the consequent of each conditional statement.

1. If I had the money, I would buy the painting.

2. If Shelly goes on the trip, she will not be able to take part in the graduation ceremony.

3. If they had a guard dog, then no one would trespass on their property.

4. If I don't get to school before 7:30, I won't be able to find a parking place.

5. If I change my major, I must reapply for admission.

6. If your blood type is type O−, then you are classified as a universal blood donor.

In Exercises 7–14, determine the truth value of the given statement.

7. If x is an even integer, then x^2 is an even integer.

8. If x is a prime number, then $x + 2$ is a prime number.

9. If all frogs can dance, then today is Monday.

10. If all cats are black, then I am a millionaire.

11. If $4 < 3$, then $7 = 8$.

12. If $x < 2$, then $x + 5 < 7$.

13. If $|x| = 6$, then $x = 6$.

14. If $\pi = 3$, then $2\pi = 6$.

In Exercises 15–24, construct a truth table for the given statement.

15. $(p \wedge \sim q) \rightarrow [\sim(p \wedge q)]$

16. $[(p \rightarrow q) \wedge p] \rightarrow p$

17. $[(p \rightarrow q) \wedge p] \rightarrow q$

18. $(\sim p \vee \sim q) \rightarrow \sim(p \wedge q)$

19. $[r \wedge (\sim p \vee q)] \rightarrow (r \vee \sim q)$

20. $[(p \rightarrow \sim r) \wedge q] \rightarrow \sim r$

21. $[(p \rightarrow q) \vee (r \wedge \sim p)] \rightarrow (r \vee \sim q)$

22. $\{p \wedge [(p \rightarrow q) \wedge (q \rightarrow r)]\} \rightarrow r$

23. $[\sim(p \rightarrow \sim r) \wedge \sim q] \rightarrow r$

24. $[p \wedge (r \rightarrow \sim q)] \rightarrow (r \vee q)$

In Exercises 25–30, write each conditional statement in its equivalent disjunctive form.

25. If she could sing, she would be perfect for the part.

26. If he does not get frustrated, he will be able to complete the job.

27. If x is an irrational number, then x is not a terminating decimal.

28. If Mr. Hyde had a brain, he would be dangerous.

29. If the fog does not lift, our flight will be cancelled.

30. If the Yankees win the pennant, Carol will be happy.

In Exercises 31–36, write the negation of each conditional statement in its equivalent conjunctive form.

31. If they offer me the contract, I will accept.

32. If I paint the house, I will get the money.

33. If pigs had wings, pigs could fly.

34. If we had a telescope, we could see that comet.

35. If she travels to Italy, she will visit her relatives.

36. If Paul could play better defense, he could be a professional basketball player.

In Exercises 37–46, state whether the given biconditional is true or false. Assume that x and y are real numbers.

37. $x^2 = 9$ if and only if $x = 3$.

38. x is a positive number if and only if $x > 0$.

39. $|x|$ is a positive number if and only if $x \neq 0$.

40. $|x + y| = x + y$ if and only if $x + y > 0$.

41. A number is a rational number if and only if the number can be written as a terminating decimal.

42. $0.\overline{3}$ is a rational number if and only if $\frac{1}{3}$ is a rational number.

43. $4 = 7$ if and only if $2 = 3$.

44. x is an even number if and only if x is not an odd number.

45. Triangle ABC is an equilateral triangle if and only if triangle ABC is an equiangular triangle.

46. Today is March 1 if and only if yesterday was February 28.

In Exercises 47–52, let v represent "I will take a vacation," let p represent "I get the promotion," and let t represent "I am transferred." Write each of the following statements in symbolic form.

47. If I get the promotion, I will take a vacation.

48. If I am not transferred, I will take a vacation.

49. If I am transferred, then I will not take a vacation.

50. If I will not take a vacation, then I will not be transferred and I get the promotion.

51. If I am not transferred and I get the promotion, then I will take a vacation.

52. If I get the promotion, then I am transferred and I will take a vacation.

In Exercises 53–58, construct a truth table for each statement to determine if the statements are equivalent.

53. $p \rightarrow \sim r$; $r \vee \sim p$

54. $p \rightarrow q$; $q \rightarrow p$

55. $\sim p \rightarrow (p \vee r)$; r

56. $p \rightarrow q$; $\sim q \rightarrow \sim p$

57. $p \rightarrow (q \vee r)$; $(p \rightarrow q) \vee (p \rightarrow r)$

58. $\sim q \rightarrow p$; $p \vee q$

Extensions

CRITICAL THINKING

The statement "All squares are rectangles" can be written as "If a figure is a square, then it is a rectangle." In Exercises 59–64, write each statement given in "All p are q" form in the form "If it is a p, then it is a q."

59. All rational numbers are real numbers.

60. All whole numbers are integers.

61. All repeating decimals are rational numbers.

62. All multiples of 5 end with a 0 or with a 5.

63. All Sauropods are herbivorous.

64. All paintings by Vincent van Gogh are valuable.

EXPLORATIONS

65. If you have access to a TI-82 or a TI-83 calculator, enter the program FACTOR on page 133 into the calculator and demonstrate the program to a classmate.

66. Many calculator programs are available on the Internet. One source for Texas Instruments calculator programs is **ticalc.org.** Search the Internet and write a few paragraphs about the programs you found to be the most interesting.

SECTION 3.4   The Conditional and Related Statements

Equivalent Forms of the Conditional

Every conditional statement can be stated in many equivalent forms. It is not even necessary to state the antecedent before the consequent. For instance, the conditional "If I live in Boston, then I must live in Massachusetts" can also be stated as

I must live in Massachusetts, if I live in Boston.

Table 3.9 lists some of the various forms that may be used to write a conditional statement.

Table 3.9 *Common Forms of* $p \rightarrow q$

Every conditional statement $p \rightarrow q$ can be written in the following equivalent forms	
If p, then q.	Every p is a q.
If p, q.	q, if p.
p only if q.	q provided p.
p implies q.	q is a necessary condition for p.
Not p or q.	p is a sufficient condition for q.

EXAMPLE 1 ■ Write a statement in an Equivalent Form

Write each of the following in "If p, then q" form.

a. The number is an even number provided it is divisible by 2.

b. Today is Friday, only if yesterday was Thursday.

Solution

a. The statement "The number is an even number provided it is divisible by 2" is in "q provided p" form. The antecedent is "it is divisible by 2," and the consequent is "the number is an even number." Thus its "If p, then q" form is

   If it is divisible by 2, then the number is an even number.

b. The statement "Today is Friday, only if yesterday was Thursday" is in "p only if q" form. The antecedent is "today is Friday." The consequent is "yesterday was Thursday." Its "If p, then q" form is

   If today is Friday, then yesterday was Thursday.

CHECK YOUR PROGRESS 1   Write each of the following in "If p, then q" form.

a. Every square is a rectangle.

b. Being older than 30 is sufficient to show I am at least 21.

Solution   *See page S8.*

The Converse, the Inverse, and the Contrapositive

Every conditional statement has three related statements. They are called the *converse*, the *inverse*, and the *contrapositive*.

Statements Related to the Conditional Statement

The **converse** of $p \rightarrow q$ is $q \rightarrow p$.

The **inverse** of $p \rightarrow q$ is $\sim p \rightarrow \sim q$.

The **contrapositive** of $p \rightarrow q$ is $\sim q \rightarrow \sim p$.

The above definitions show the following:

- The converse of $p \rightarrow q$ is formed by interchanging the antecedent p with the consequent q.

- The inverse of $p \rightarrow q$ is formed by negating the antecedent p and negating the consequent q.

- The contrapositive of $p \rightarrow q$ is formed by negating both the antecedent p and the consequent q and interchanging these negated statements.

EXAMPLE 2 ■ Write the Converse, Inverse, and Contrapositive of a Conditional

Write the converse, inverse, and contrapositive of

If I get the job, then I will rent the apartment.

Solution
Converse: If I rent the apartment, then I get the job.
Inverse: If I do not get the job, then I will not rent the apartment.
Contrapositive: If I do not rent the apartment, then I did not get the job.

CHECK YOUR PROGRESS 2   Write the converse, inverse, and contrapositive of

If we have a quiz today, then we will not have a quiz tomorrow.

Solution   *See page S8.*

Table 3.10 shows that any conditional statement is equivalent to its contrapositive, and that the converse of a conditional is equivalent to the inverse of the conditional.

Table 3.10 *Truth Tables for the Conditional and Related Statements*

p	q	Conditional $p \rightarrow q$	Converse $q \rightarrow p$	Inverse $\sim p \rightarrow \sim q$	Contrapositive $\sim q \rightarrow \sim p$
T	T	T	T	T	T
T	F	F	T	T	F
F	T	T	F	F	T
F	F	T	T	T	T

$$q \rightarrow p \equiv \sim p \rightarrow \sim q$$
$$p \rightarrow q \equiv \sim q \rightarrow \sim p$$

EXAMPLE 3 ■ Determine Whether Related Statements are Equivalent

Determine whether the given statements are equivalent.

a. If a number ends with a 5, then the number is divisible by 5.
If a number is divisible by 5, then the number ends with a 5.
b. If two lines in a plane do not intersect, then the lines are parallel.
If two lines in a plane are not parallel, then the lines intersect.

Solution
a. The second statement is the converse of the first. The statements are not equivalent.
b. The second statement is the contrapositive of the first. The statements are equivalent.

CHECK YOUR PROGRESS 3   Determine whether the given statements are equivalent.

a. If $a = b$, then $a \cdot c = b \cdot c$.
  If $a \neq b$, then $a \cdot c \neq b \cdot c$.
b. If I live in Nashville, then I live in Tennessee.
  If I do not live in Tennessee, then I do not live in Nashville.

Solution   *See page S9.*

In mathematics, it is often necessary to prove statements that are in "If p, then q" form. If a proof cannot be readily produced, mathematicians often try to prove the contrapositive "If $\sim q$, then $\sim p$." Because a conditional and its contrapositive are equivalent statements, a proof of either statement also establishes the proof of the other statement.

QUESTION   *A mathematician wishes to prove the following statement about the integer x.*

If x^2 is an odd integer, then x is an odd integer.   (I)

If the mathematician is able to prove the statement, "If x is an even integer, then x^2 is an even integer," does this also prove statement (I)?

EXAMPLE 4 ■ Use the Contrapositive to Determine a Truth Value

Write the contrapositive of each statement and use the contrapositive to determine whether the original statement is true or false. In part b, assume x is an integer.

a. If $a + b$ is not divisible by 5, then a and b are not both divisible by 5.
b. If x^3 is an odd integer, then x is an odd integer.
c. If a geometric figure is not a rectangle, then it is not a square.

Solution
a. If a and b are both divisible by 5, then $a + b$ is divisible by 5. This is a true statement, so the original statement is also true.
b. If x is an even integer, then x^3 is an even integer. This is a true statement, so the original statement is also true.
c. If a geometric figure is a square, then it is a rectangle. This is a true statement, so the original statement is also true.

CHECK YOUR PROGRESS 4   Write the contrapositive of each statement and use the contrapositive to determine whether the original statement is true or false.

a. If $3 + x$ is an odd integer, then x is an even integer. (Assume x is an integer.)
b. If two triangles are not similar triangles, then they are not congruent triangles. *Note:* Similar triangles have the same shape. Congruent triangles have the same size and shape.
c. If today is not Wednesday, then tomorrow is not Thursday.

Solution   *See page S9.*

ANSWER   *Yes, because the second statement is the contrapositive of (I).*

Math Matters   Grace Hopper

Rear Admiral Grace Hopper

Grace Hopper (1906–1992) was a visionary in the field of computer programming. She was a mathematics professor at Vassar from 1931 to 1943, but retired from teaching to start a career in the U.S. Navy at the age of 37.

The Navy assigned Hopper to the Bureau of Ordnance Computation at Harvard University. It was here that she was given the opportunity to program computers. It has often been reported that she was the third person to program the world's first large-scale digital computer. Grace Hopper had a passion for computers and computer programming. She wanted to develop a computer language that would be user-friendly and enable people to use computers in a more productive manner.

Grace Hopper had a long list of accomplishments. She designed some of the first computer compilers, she was one of the first to introduce English commands into computer languages, and she wrote the precursor to the computer language COBOL.

Grace Hopper retired from the Navy (for the first time) in 1966. In 1967 she was recalled to active duty and continued to serve in the Navy until 1986, at which time she was the nation's oldest active duty officer.

In 1951, the UNIVAC I computer that Grace Hopper was programming started to malfunction. The malfunction was caused by a moth that had become lodged in one of the computer's relays. Grace Hopper pasted the moth into the UNIVAC I logbook with a label that read, "computer bug." Since then computer programmers have used the word *bug* to indicate any problem associated with a computer program. Modern computers use logic gates instead of relays to process information, so actual bugs are not a problem; however, bugs such as the "Year 2000 bug" can cause serious problems.

 *Excursion*

Sheffer's Stroke and the NAND Gate

Table 3.11 *Sheffer's stroke*

p	q	$p\,|\,q$
T	T	F
T	F	T
F	T	T
F	F	T

In 1913, the logician Henry M. Sheffer created a connective that we now refer to as *Sheffer's stroke* (or *NAND*). This connective is often denoted by the symbol $|$. Table 3.11 shows that $p\,|\,q$ is equivalent to $\sim(p \wedge q)$. Sheffer's stroke $p\,|\,q$ is false when both p and q are true and it is true in all other cases.

Any logic statement can be written using only Sheffer's stroke connectives. For instance, Table 3.12 shows that $p\,|\,p \equiv \sim p$ and $(p\,|\,p)\,|\,(q\,|\,q) \equiv p \vee q$.

Figure 3.13 shows a logic gate called a NAND gate. This gate models the Sheffer's stroke connective in that its output is 0 when both input streams are 1 and its output is 1 in all other cases.

(continued)

Table 3.12

p	q	p|p	(p|p)|(q|q)
T	T	F	T
T	F	F	T
F	T	T	T
F	F	T	F

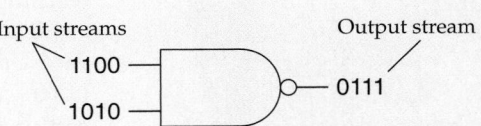

Figure 3.13 *NAND gate*

Excursion Exercises

1. a. Complete a truth table for $p|(q|q)$.

   b. Use the results of Excursion Exercise 1a to determine an equivalent statement for $p|(q|q)$.

2. a. Complete a truth table for $(p|q)|(p|q)$.

   b. Use the results of Excursion Exercise 2a to determine an equivalent statement for $(p|q)|(p|q)$.

3. a. Determine the output stream for the following network of NAND gates. Note: In a network of logic gates, a solid circle • is used to indicate a connection. A symbol such as ⟜ is used to indicate "no connection."

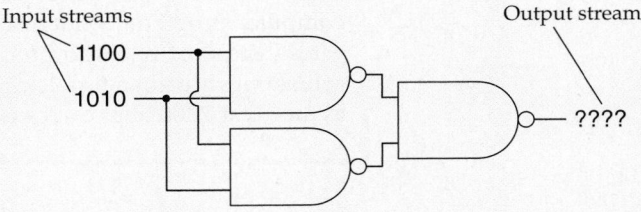

Figure 3.14

   b. What logic gate is modeled by the network in Figure 3.14?

4. NAND gates are functionally complete in that any logic gate can be constructed using only NAND gates. Construct a network of NAND gates that would produce the same output stream as an OR gate.

Exercise Set 3.4   (Suggested Assignment: 1–47 odds)

In Exercises 1–10, write each statement in "If p, then q" form.

1. We will be in good shape for the ski trip provided we take the aerobics class.

2. We can get a dog only if we install a fence around the back yard.

3. Every odd prime number is greater than 2.

4. The triangle is a 30-60-90 triangle, if the length of the hypotenuse is twice the length of the shorter leg.

5. He can join the band, if he has the talent to play a keyboard.

6. Every theropod is carnivorous.

7. I will be able to prepare for the test only if I have the textbook.

8. I will be able to receive my credential provided Education 147 is offered in the spring semester.

9. Being in excellent shape is a necessary condition for running the Boston marathon.

10. If it is an ankylosaur, it is quadrupedal.

In Exercises 11–24, write the **a.** converse, **b.** inverse, and **c.** contrapositive of the given statement.

11. If I were rich, I would quit this job.

12. If we had a car, then we would be able to take the class.

13. If she does not return soon, we will not be able to attend the party.

14. I will be in the talent show only if I can do the same comedy routine I did for the banquet.

15. Every parallelogram is a quadrilateral.

16. If you get the promotion, you will need to move to Denver.

17. I would be able to get current information about astronomy provided I had access to the Internet.

18. You need four-wheel drive to make the trip to Death Valley.

19. We will not have enough money for dinner, if we take a taxi.

20. If you are the president of the United States, then your age is at least 35.

21. She will visit Kauai only if she can extend her vacation for at least two days.

22. In a right triangle, the acute angles are complementary.

23. Two lines perpendicular to a given line are parallel.

24. If $x + 5 = 12$, then $x = 7$.

In Exercises 25–30, determine whether the given statements are equivalent.

25. If Kevin wins, we will celebrate.
    If we celebrate, then Kevin will win.

26. If I save \$1000, I will go on the field trip.
    If I go on the field trip, then I saved \$1000.

27. If she attends the meeting, she will make the sale.
    If she does not make the sale, then she did not attend the meeting.

28. If you understand algebra, you can remember algebra.
    If you do not understand algebra, you cannot remember algebra.

29. If $a > b$, then $ac > bc$.
    If $a \leq b$, then $ac \leq bc$.

30. If $a < b$, then $\dfrac{1}{a} > \dfrac{1}{b}$.
    If $\dfrac{1}{a} \leq \dfrac{1}{b}$, then $a \geq b$.
    (Assume $a \neq 0$ and $b \neq 0$.)

In Exercises 31–36, write the contrapositive of the statement and use the contrapositive to determine whether the given statement is true or false.

31. If $3x - 7 = 11$, then $x \neq 7$.

32. If $x \neq 3$, then $5x + 7 \neq 22$.

33. If $a \neq 3$, then $|a| \neq 3$.

34. If $a + b$ is divisible by 3, then a is divisible by 3 and b is divisible by 3.

35. If $\sqrt{a + b} \neq 5$, then $a + b \neq 25$.

36. Assume x is an integer. If x^2 is an even integer, then x is an even integer.

37. What is the converse of the inverse of the contrapositive of $p \rightarrow q$?

38. What is the inverse of the converse of the contrapositive of $p \rightarrow q$?

Extensions

CRITICAL THINKING

39. Give an example of a true conditional statement whose

    a. converse is true.   **b.** converse is false.

40. Give an example of a true conditional statement whose

    a. inverse is true.   **b.** inverse is false.

In Exercises 41–44, determine the original statement if the given statement is related to the original in the manner indicated.

41. *Converse:* If you can do it, you can dream it.

42. *Inverse:* If I did not have a dime, I would not spend it.

43. *Contrapositive:* If I were a singer, I would not be a dancer.

44. *Negation:* Pigs have wings and pigs cannot fly.

45. Explain why it is not possible to find an example of a true conditional statement whose contrapositive is false.

46. If a conditional statement is false, must its converse be true? Explain.

47. Lewis Carroll (Charles Dodgson) wrote many puzzles, many of which he recorded in his diaries. Solve the following puzzle, which appears in one of his diaries.

The Dodo says that the Hatter tells lies.
The Hatter says that the March Hare tells lies.
The March Hare says that both the Dodo and the Hatter tell lies.
Who is telling the truth?[2]

Hint: Consider the three different cases in which only one of the characters is telling the truth. In only one of these cases can all three of the statements be true.

EXPLORATIONS

48. Use a library or the Internet to find puzzles created by Lewis Carroll. For the puzzle that you think is most interesting, write an explanation of the puzzle and give its solution.

SECTION 3.5 Arguments

historical note

Aristotle
(ăr'ĭ-stŏt'l)
(384–322 B.C.)
was an ancient
Greek philosopher
who studied
under Plato. He
wrote about many subjects, including logic, biology, politics, astronomy, metaphysics, and ethics.
His ideas about logic and the
reasoning process have had a
major impact on mathematics
and philosophy. ■

Arguments

In this section we consider methods of analyzing arguments to determine whether they are *valid* or *invalid*. For instance, consider the following argument.

If Aristotle was human, then Aristotle was mortal. Aristotle was human. Therefore, Aristotle was mortal.

To determine whether the above argument is a valid argument, we must first define the terms *argument* and *valid argument*.

> **Definition of an Argument and a Valid Argument**
>
> An **argument** consists of a set of statements called **premises** and another statement called the **conclusion**. An argument is **valid** if the conclusion is true whenever all the premises are assumed to be true. An argument is **invalid** if it is not a valid argument.

In the argument about Aristotle, the two premises and the conclusion are shown below. It is customary to place a horizontal line between the premises and the conclusion.

First Premise:	If Aristotle was human, then Aristotle was mortal.
Second Premise:	Aristotle was human.
Conclusion:	Therefore, Aristotle was mortal.

---

2. The above puzzle is from *Lewis Carroll's Games and Puzzles,* newly compiled and edited by Edward Wakeling. New York: Dover Publications, Inc., copyright 1992, p. 11, puzzle 9, "Who's Telling the Truth?"

Arguments can be written in **symbolic form.** For instance, if we let h represent the statement "Aristotle was human" and m represent the statement "Aristotle was mortal," then the argument can be expressed as

$$h \rightarrow m$$
$$\underline{h \qquad\quad}$$
$$\therefore m$$

The three dots $\therefore$ are a symbol for "therefore."

EXAMPLE 1 ■ **Write an Argument in Symbolic Form**

Write the following argument in symbolic form.

The fish is fresh or I will not order it. The fish is fresh. Therefore I will order it.

Solution
Let f represent the statement "The fish is fresh." Let o represent the statement "I will order it." The symbolic form of the argument is

$$f \vee \sim o$$
$$\underline{f \qquad\quad}$$
$$\therefore o$$

CHECK YOUR PROGRESS 1   Write the following argument in symbolic form.

If she doesn't get on the plane, she will regret it. She does not regret it. Therefore, she got on the plane.

Solution   *See page S9.*

Arguments and Truth Tables

The following truth table procedure can be used to determine whether an argument is valid or invalid.

> **Truth Table Procedure to Determine the Validity of an Argument**
>
> **1.** Write the argument in symbolic form.
>
> **2.** Construct a truth table that shows the truth value of each premise and the truth value of the conclusion for all combinations of truth values of the component statements.
>
> **3.** If the conclusion is true in every row of the truth table in which all the premises are true, the argument is valid. If the conclusion is false in any row in which all of the premises are true, the argument is invalid.

We will now use the above truth table procedure to determine the validity of the argument about Aristotle.

1. Once again we let h represent the statement "Aristotle was human" and m represent the statement "Aristotle was mortal." In symbolic form the argument is

$$h \rightarrow m \qquad \text{First premise}$$
$$\underline{h \qquad\quad} \qquad \text{Second premise}$$
$$\therefore m \qquad\qquad \text{Conclusion}$$

2. Construct a truth table as shown below.

h	m	First premise $h \rightarrow m$	Second premise h	Conclusion m	
T	T	T	T	T	Row 1
T	F	F	T	F	Row 2
F	T	T	F	T	Row 3
F	F	T	F	F	Row 4

3. Row 1 is the only row in which all the premises are true, so it is the only row that we examine. Because the conclusion is true in row 1, the argument is valid.

In Example 2, we use the truth table method to determine the validity of a more complicated argument.

EXAMPLE 2 ■ Determine the Validity of an Argument

Determine whether the following argument is valid or invalid.

> If it rains, then the game will not be played. It is not raining. Therefore, the game will be played.

Solution

If we let r represent "it rains" and g represent "the game will be played," then the symbolic form is

$$r \rightarrow \sim g$$
$$\underline{\sim r}$$
$$\therefore g$$

The truth table for this argument is as follows.

r	g	First premise $r \rightarrow \sim g$	Second premise $\sim r$	Conclusion g	
T	T	F	F	T	Row 1
T	F	T	F	F	Row 2
F	T	T	T	T	Row 3
F	F	T	T	F	Row 4

QUESTION *Why do we need to examine only rows 3 and 4?*

Because the conclusion in row 4 is false and the premises are both true, we know the argument is invalid.

ANSWER *Rows 3 and 4 are the only rows in which all of the premises are true.*

CHECK YOUR PROGRESS 2   Determine the validity of the following argument.

If the stock market rises, then the bond market will fall.

The bond market did not fall.

∴The stock market did not rise.

Solution   *See page S9.*

The argument in Example 3 involves three statements. Thus we use a truth table with $2^3 = 8$ rows to determine the validity of the argument.

EXAMPLE 3 ■ Determine the Validity of an Argument

Determine whether the following argument is valid or invalid.

If I am going to run the marathon, then I will buy new shoes.

If I buy new shoes, then I will not buy a television.

∴If I buy a television, I will not run the marathon.

Solution
Label the statements

m: I am going to run the marathon.

s: I will buy new shoes.

t: I will buy a television.

The symbolic form of the argument is

$m \rightarrow s$

$s \rightarrow \sim t$

$\therefore t \rightarrow \sim m$

The truth table for this argument is as follows.

m	*s*	*t*	First premise $m \rightarrow s$	Second premise $s \rightarrow \sim t$	Conclusion $t \rightarrow \sim m$	
T	T	T	T	F	F	Row 1
T	T	F	T	T	T	Row 2
T	F	T	F	T	F	Row 3
T	F	F	F	T	T	Row 4
F	T	T	T	F	T	Row 5
F	T	F	T	T	T	Row 6
F	F	T	T	T	T	Row 7
F	F	F	T	T	T	Row 8

The only rows in which both premises are true are rows 2, 6, 7, and 8. Because the conclusion is true in each of these rows, the argument is valid.

CHECK YOUR PROGRESS 3   Determine whether the following argument is valid or invalid.

> If I arrive before 8 A.M., then I will make the flight.
>
> If I make the flight, then I will give the presentation.
>
> ∴If I arrive before 8 A.M., then I will give the presentation.

Solution   *See page S9.*

Standard Forms

Some arguments can be shown to be valid if they have the same symbolic form as an argument that is known to be valid. For instance, we have shown that the argument

$$h \rightarrow m$$
$$h$$
$$\therefore m$$

is valid. This symbolic form is known as **modus ponens** or the **law of detachment.** All arguments that have this symbolic form are valid. Table 3.13 shows four symbolic forms and the name used to identify each form. Any argument that has a symbolic form identical to one of these symbolic forms is a valid argument.

Table 3.13 *Standard Forms of Four Valid Arguments*

Modus ponens	Modus tollens	Law of syllogism	Disjunctive syllogism
$p \rightarrow q$	$p \rightarrow q$	$p \rightarrow q$	$p \vee q$
p	$\sim q$	$q \rightarrow r$	$\sim p$
$\therefore q$	$\therefore \sim p$	$\therefore p \rightarrow r$	$\therefore q$

The law of syllogism can be extended to include more than two conditional premises. For example, if the premises of an argument are $a \rightarrow b$, $b \rightarrow c$, $c \rightarrow d$, ..., $y \rightarrow z$, then a valid conclusion for the argument is $a \rightarrow z$. We will refer to any argument of this form with more than two conditional premises as the **extended law of syllogism.**

Table 3.14 shows two symbolic forms associated with invalid arguments. Any argument that has one of these symbolic forms is invalid.

Table 3.14 *Standard Forms of Two Invalid Arguments*

Fallacy of the converse	Fallacy of the inverse
$p \rightarrow q$	$p \rightarrow q$
q	$\sim p$
$\therefore p$	$\therefore \sim q$

EXAMPLE 4 ■ Use Standard Forms to Determine the Validity of an Argument

Use standard forms to determine whether the following argument is valid or invalid.

> The program is interesting or I will watch the basketball game.
> The program is not interesting.
> ∴I will watch the basketball game.

Solution

Label the statements

> *i:* The program is interesting.
> *w:* I will watch the basketball game.

In symbolic form the argument is:

$$i \vee w$$
$$\sim i$$
$$\therefore w$$

This symbolic form matches the standard form known as disjunctive syllogism. Thus the argument is valid.

CHECK YOUR PROGRESS 4   Use standard forms to determine whether the following argument is valid or invalid.

> If I go to Florida for spring break, then I will not study.
> I did not go to Florida for spring break.
> ∴I studied.

Solution   *See page S10.*

Consider an argument with the following symbolic form.

$q \rightarrow r$	Premise 1
$r \rightarrow s$	Premise 2
$\sim t \rightarrow \sim s$	Premise 3
q	Premise 4
$\therefore t$	

To determine whether the argument is valid or invalid using a truth table, we would require a table with $2^4 = 16$ rows. It would be time consuming to construct such a table and, with the large number of truth values to be determined, we might make an error. Thus we consider a different approach that makes use of a sequence of valid arguments to arrive at a conclusion.

$q \rightarrow r$	Premise 1
$r \rightarrow s$	Premise 2
$\therefore q \rightarrow s$	Law of syllogism

Waterfall by M. C. Escher

M. C. Escher (1898–1972) created many works of art that defy logic. In this lithograph, the water completes a full cycle even though the water is always traveling downward.

$$q \rightarrow s \qquad \text{The previous conclusion}$$

$$\underline{s \rightarrow t} \qquad \text{Premise 3 expressed in an equivalent form}$$

$$\therefore q \rightarrow t \qquad \text{Law of syllogism}$$

$$q \rightarrow t \qquad \text{The previous conclusion}$$

$$\underline{q \qquad\qquad} \qquad \text{Premise 4}$$

$$\therefore t \qquad \text{Modus ponens}$$

This sequence of valid arguments shows that *t* is a valid conclusion for the original argument.

EXAMPLE 5 ■ Determine the Validity of an Argument

Determine whether the following argument is valid.

> If the movie was directed by Steven Spielberg (*s*), then I want to see it (*w*). The movie's production costs must exceed 50 million dollars (*c*) or I do not want to see it. The movie's production costs were less than 50 million dollars. Therefore, the movie was not directed by Steven Spielberg.

Solution

In symbolic form the argument is

$$s \rightarrow w \qquad \text{Premise 1}$$

$$c \lor \sim w \qquad \text{Premise 2}$$

$$\underline{\sim c \qquad\qquad} \qquad \text{Premise 3}$$

$$\therefore \sim s \qquad \text{Conclusion}$$

Premise 2 can be written as $\sim w \lor c$, which is equivalent to $w \rightarrow c$. Applying the law of syllogism to Premise 1 and this equivalent form of Premise 2 produces

$$s \rightarrow w \qquad \text{Premise 1}$$

$$\underline{w \rightarrow c} \qquad \text{Equivalent form of Premise 2}$$

$$\therefore s \rightarrow c \qquad \text{Law of syllogism}$$

Combining the above conclusion $s \rightarrow c$ with Premise 3 gives us

$$s \rightarrow c \qquad \text{Conclusion from above}$$

$$\underline{\sim c \qquad\qquad} \qquad \text{Premise 3}$$

$$\therefore \sim s \qquad \text{Modus tollens}$$

This sequence of valid arguments has produced the desired conclusion, $\sim s$. Thus the original argument is valid.

CHECK YOUR PROGRESS 5   Determine whether the following argument is valid.

> I start to fall asleep if I read a math book. I drink soda whenever I start to fall asleep. If I drink a soda, then I must eat a candy bar. Therefore, I eat a candy bar whenever I read a math book.

Hint: p whenever *q* is equivalent to $q \rightarrow p$.

Solution   See page S10.

In the next example, we use standard forms to determine a valid conclusion for an argument.

EXAMPLE 6 ■ Determine a Valid Conclusion for an Argument

Use all of the premises to determine a valid conclusion for the following argument.

We will not go to Japan ($\sim j$) or we will go to Hong Kong (h). If we visit my uncle (u), then we will go to Singapore (s). If we go to Hong Kong, then we will not go to Singapore.

Solution

In symbolic form the argument is

$$\sim j \vee h \qquad \text{Premise 1}$$
$$u \rightarrow s \qquad \text{Premise 2}$$
$$\underline{h \rightarrow \sim s} \qquad \text{Premise 3}$$
$$\therefore ?$$

The first premise can be written as $j \rightarrow h$. The second premise can be written as $\sim s \rightarrow \sim u$. Therefore the argument can be written as

$$j \rightarrow h$$
$$\sim s \rightarrow \sim u$$
$$\underline{h \rightarrow \sim s}$$
$$\therefore ?$$

Interchanging the second and third premises yields

$$j \rightarrow h$$
$$h \rightarrow \sim s$$
$$\underline{\sim s \rightarrow \sim u}$$
$$\therefore ?$$

An application of the extended law of syllogism produces

$$j \rightarrow h$$
$$h \rightarrow \sim s$$
$$\underline{\sim s \rightarrow \sim u}$$
$$\therefore j \rightarrow \sim u$$

Thus a valid conclusion for the original argument is "If we go to Japan (j), then we will not visit my uncle ($\sim u$)."

CHECK YOUR PROGRESS 6 Use all of the premises to determine a valid conclusion for the following argument.

$$\sim m \vee t$$
$$t \rightarrow \sim d$$
$$e \vee g$$
$$\underline{e \rightarrow d}$$
$$\therefore ?$$

Solution *See page S10.*

> ✔ **TAKE NOTE**
>
> In Example 6 we are rewriting and reordering the statements so that the extended law of syllogism can be applied.

Math Matters   The Paradox of the Unexpected Hanging

The following paradox, known as "the paradox of the unexpected hanging," has proved difficult to analyze.

The man was sentenced on Saturday. "The hanging will take place at noon," said the judge to the prisoner, "on one of the seven days of next week. But you will not know which day it is until you are so informed on the morning of the day of the hanging."

The judge was known to be a man who always kept his word. The prisoner, accompanied by his lawyer, went back to his cell. As soon as the two men were alone the lawyer broke into a grin. "Don't you see?" he exclaimed. "The judge's sentence cannot possibly be carried out."

"I don't see," said the prisoner.

"Let me explain. They obviously can't hang you next Saturday. Saturday is the last day of the week. On Friday afternoon you would still be alive and you would know with absolute certainty that the hanging would be on Saturday. You would know this *before* you were told so on Saturday morning. That would violate the judge's decree."

"True," said the prisoner.

"Saturday, then, is positively ruled out," continued the lawyer. "This leaves Friday as the last day they can hang you. But they can't hang you on Friday because by Thursday afternoon only two days would remain: Friday and Saturday. Since Saturday is not a possible day, the hanging would have to be on Friday. Your knowledge of that fact would violate the judge's decree again. So Friday is out. This leaves Thursday as the last possible day. But Thursday is out because if you're alive Wednesday afternoon, you'll know that Thursday is to be the day."

"I get it," said the prisoner, who was beginning to feel much better. "In exactly the same way I can rule out Wednesday, Tuesday, and Monday. That leaves only tomorrow. But they can't hang me tomorrow because I know it today!"

In brief, the judge's decree seems to be self-refuting. There is nothing logically contradictory in the two statements that make up his decree; nevertheless, it cannot be carried out in practice.

He [the prisoner] is convinced, by what appears to be unimpeachable logic, that he cannot be hanged without contradicting the conditions specified in his sentence. Then, on Thursday morning, to his great surprise, the hangman arrives. Clearly he did not expect him. What is more surprising, the judge's decree is now seen to be perfectly correct. The sentence can be carried out exactly as stated.[3]

3. Reprinted with the permission of Simon & Schuster from *The Unexpected Hanging and Other Mathematical Diversions* by Martin Gardner. Copyright © 1969 by Martin Gardner.

Excursion

Fallacies

Any argument that is not valid is called a **fallacy.** Ancient logicians enjoyed the study of fallacies and took pride in their ability to analyze and categorize different types of fallacies. In this Excursion we consider the four fallacies known as *circulus in probando,* the fallacy of experts, the fallacy of equivocation, and the fallacy of accident.

Circulus in Probando

A fallacy of *circulus in probando* is an argument that uses a premise as the conclusion. For instance, consider the following argument.

> The Chicago Bulls are the best basketball team because there is no basketball team that is better than the Chicago Bulls.

The fallacy of *circulus in probando* is also know as *circular reasoning* or *begging the question.*

Fallacy of Experts

A fallacy of experts is an argument that uses an expert (or a celebrity) to lend support to a product or an idea. Often the product or idea is outside the expert's area of expertise. The following endorsements may qualify as fallacy of experts arguments.

> Tiger Woods for Rolex watches
>
> Lindsey Wagner for Ford Motor Company

Fallacy of Equivocation

A fallacy of equivocation is an argument that uses a word with two interpretations in two different ways. The following argument is an example of a fallacy of equivocation.

> The highway sign read $268 fine for littering,
>
> so I decided fine, for $268, I will litter.

Fallacy of Accident

The following argument is an example of a fallacy of accident.

> Everyone should visit Europe.
>
> Therefore, prisoners on death row should be allowed to visit Europe.

Using more formal language, we can state the argument as follows.

> If you are a prisoner on death row (d), then you are a person (p).
>
> If you are a person (p), then you should be allowed to visit Europe (e).
>
> ∴If you are a prisoner on death row, then you should be allowed to visit Europe.

The symbolic form of the argument is

$$d \rightarrow p$$
$$p \rightarrow e$$
$$\therefore d \rightarrow e$$

(continued)

This argument appears to be a valid argument because it has the standard form of the law of syllogism. Common sense tells us the argument is not valid, so where have we gone wrong in our analysis of the argument?

The problem occurs with the interpretation of the word "everyone." Often, when we say "everyone," we really mean "most everyone." A fallacy of accident may occur whenever we use a statement that is often true in place of a statement that is always true.

Excursion Exercises

1. Write an argument that is an example of *circulus in probando*.

2. Give an example of an argument that is a fallacy of experts.

3. Write an argument that is an example of a fallacy of accident.

4. Write an argument that is an example of a fallacy of equivocation.

5. Algebraic arguments often consist of a list of statements. In a valid algebraic argument, each statement (after the premises) can be deduced from the previous statements. The following argument that $1 = 2$ contains exactly one step that is not valid. Identify the step and explain why it is not valid.

Let		
	$a = b$	• Premise
	$a^2 = ab$	• Multiply each side by a.
	$a^2 - b^2 = ab - b^2$	• Subtract b^2 from each side.
	$(a + b)(a - b) = b(a - b)$	• Factor each side.
	$a + b = b$	• Divide each side by $(a - b)$.
	$b + b = b$	• Substitute b for a.
	$2b = b$	• Collect like terms.
	$2 = 1$	• Divide each side by b.

Exercise Set 3.5

(Suggested Assignment: 1–45 odds)

In Exercises 1–6, use the indicated letters to write each argument in symbolic form.

1. If you can read this bumper sticker (r), you're too close (c). You can read the bumper sticker. Therefore, you're too close.

2. If Lois Lane marries Clark Kent (m), then superman will get a new uniform (u). Superman does not get a new uniform. Therefore, Lois Lane did not marry Clark Kent.

3. If the price of gold rises (g), the stock market will fall (s). The price of gold did not rise. Therefore, the stock market did not fall.

4. I am going shopping (s) or I am going to the museum (m). I went to the museum. Therefore, I did not go shopping.

5. If we search the Internet (s), we will find information on logic (i). We searched the Internet. Therefore, we found information on logic.

6. If we check the sports results on the Excite channel (c), we will know who won the match (w). We know who won the match. Therefore, we checked the sports results on the Excite channel.

In Exercises 7–18, use a truth table to determine whether the argument is valid or invalid.

7. $p \lor \sim q$
$\underline{\sim q}$
$\therefore p$

8. $\sim p \land q$
$\underline{\sim p}$
$\therefore q$

9. $p \rightarrow \sim q$
$\underline{\sim q}$
$\therefore p$

10. $p \rightarrow \sim q$
$\underline{p}$
$\therefore \sim q$

11. $\sim p \to \sim q$
$\underline{\sim p}$
$\therefore \sim q$

12. $\sim p \to q$
$\underline{p \qquad}$
$\therefore \sim q$

13. $(p \to q) \wedge (\sim p \to q)$
$\underline{q \qquad\qquad\qquad}$
$\therefore p$

14. $(p \vee q) \wedge (p \wedge q)$
$\underline{p \qquad\qquad\qquad}$
$\therefore q$

15. $(p \wedge \sim q) \vee (p \to q)$
$\underline{q \vee p \qquad\qquad}$
$\therefore \sim p \wedge q$

16. $(p \wedge \sim q) \to (p \vee q)$
$\underline{q \to \sim p \qquad\qquad}$
$\therefore p \to q$

17. $(p \wedge \sim q) \vee (p \vee r)$
$\underline{r \qquad\qquad\qquad}$
$\therefore p \vee q$

18. $(p \to q) \to (r \to \sim q)$
$\underline{p \qquad\qquad\qquad}$
$\therefore \sim r$

In Exercises 19–24, use the indicated letters to write the argument in symbolic form. Then use a truth table to determine whether the argument is valid or invalid.

19. If you finish your homework (h), you may attend the reception (r). You did not finish your homework. Therefore, you cannot go to the reception.

20. The X Games will be held in Oceanside (o) if and only if the city of Oceanside agrees to pay $100,000 in prize money ($a$). If San Diego agrees to pay $200,000 in prize money ($s$), then the city of Oceanside will not agree to pay $100,000 in prize money. Therefore, if the X Games were held in Oceanside, then San Diego did not agree to pay $200,000 in prize money.

21. If I can't buy the house ($\sim b$), then at least I can dream about it (d). I can buy the house or at least I can dream about it. Therefore, I can buy the house.

22. If the winds are from the east (e), then we will not have a big surf ($\sim s$). We do not have a big surf. Therefore, the winds are from the east.

23. If I master college algebra (c), then I will be prepared for trigonometry (t). I am prepared for trigonometry. Therefore, I mastered college algebra.

24. If it is a blot (b), then it is not a clot ($\sim c$). If it is a zlot (z), then it is a clot. It is a blot. Therefore, it is not a zlot.

In Exercises 25–34, determine whether the argument is valid or invalid by comparing its symbolic form with the standard symbolic forms given in Tables 3.13 and 3.14. For each valid argument, state the name of its standard form.

25. If you take Art 151 in the fall, you will be eligible to take Art 152 in the spring. You were not able to take Art 152 in the spring. Therefore, you did not take Art 151 in the fall.

26. He will attend Stanford or Yale. He did not attend Yale. Therefore, he attended Stanford.

27. If I had a nickel for every logic problem I have solved, then I would be rich. I have not received a nickel for every logic problem I have solved. Therefore, I am not rich.

28. If it is a dog, then it has fleas. It has fleas. Therefore, it is a dog.

29. If we serve salmon, then Vicky will join us for lunch. If Vicky joins us for lunch, then Marilyn will not join us for lunch. Therefore, if we serve salmon, Marilyn will not join us for lunch.

30. If I go to college, then I will not be able to work for my Dad. I did not go to college. Therefore, I went to work for my Dad.

31. If my cat is left alone in the apartment, then she claws the sofa. Yesterday I left my cat alone in the apartment. Therefore, my cat clawed the sofa.

32. If I wish to use the new software, then I cannot continue to use this computer. I don't wish to use the new software. Therefore, I can continue to use this computer.

33. If Rita buys a new car, then she will not go on the cruise. Rita went on the cruise. Therefore, Rita did not buy a new car.

34. If Hideo Nomo pitches, then I will go to the game. I did not go to the game. Therefore, Hideo Nomo did not pitch.

In Exercises 35–40, use a sequence of valid arguments to show that each argument is valid.

35. $\sim p \to r$
$r \to t$
$\underline{\sim t \qquad}$
$\therefore p$

36. $r \to \sim s$
$s \vee \sim t$
$\underline{r \qquad}$
$\therefore \sim t$

37. If we sell the boat (s), then we will not go to the river ($\sim r$). If we don't go to the river, then we will go camping (c). If we do not buy a tent ($\sim t$), then we will not go camping. Therefore, if we sell the boat, then we will buy a tent.

38. If it is an ammonite (a), then it is from the Cretaceous period (c). If it is not from the Mesozoic era ($\sim m$), then it is not from the Cretaceous period. If it is from the Mesozoic era, then it is at least 65 million years old (s). Therefore, if it is an ammonite, then it is at least 65 million years old.

39. If the computer is not operating ($\sim o$), then I will not be able to finish my report ($\sim f$). If the office is closed (c), then the computer is not operating. Therefore, if I am able to finish my report, then the office is open.

40. If he reads the manuscript (r), he will like it (l). If he likes it, he will publish it (p). If he publishes it, then you will get royalties (m). You did not get royalties. Therefore, he did not read the manuscript.

Extensions
CRITICAL THINKING

In Exercises 41–44, use all of the premises to determine a valid conclusion for the given argument.

41.
$$\sim(p \land \sim q)$$
$$\underline{\quad p \quad}$$
$$\therefore ?$$

42.
$$\sim s \to q$$
$$\sim t \to \sim q$$
$$\underline{\quad \sim t \quad}$$
$$\therefore ?$$

43. If it is a theropod, then it is not herbivorous. If it is not herbivorous, then it is not a sauropod. It is a sauropod. Therefore, _____.

44. If you buy the car, you will need a loan. You do not need a loan or you will make monthly payments. You buy the car. Therefore, _____.

COOPERATIVE LEARNING

45. The following argument is from *Symbolic Logic* by Lewis Carroll, written in 1896. Determine whether the argument is valid or invalid.

> Babies are illogical.
> Nobody is despised who can manage a crocodile.
> Illogical persons are despised.
> Hence, babies cannot manage crocodiles.

EXPLORATIONS

46. Consult a logic text or search the Internet for information on fallacies. Write a report that includes examples of at least three of the following fallacies.

Ad hominem
Ad populum
Ad baculum
Ad vercundiam
Non sequitur
Fallacy of false cause
Pluriam interrogationem

SECTION 3.6 **Euler Diagrams**

Euler Diagrams

Many arguments involve sets whose elements are described using the quantifiers *all, some,* and *none.* The mathematician Leonhard Euler (laônhärt oi′lər) used diagrams to determine whether arguments that involved quantifiers were valid or invalid. The following figures show Euler diagrams that illustrate the four possible relationships that can exist between two sets.

All P s are Q s.   No P s are Q s.   Some P s are Q s.   Some P s are not Q s.

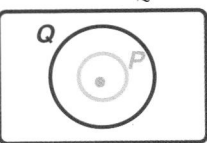 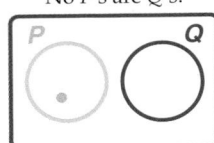 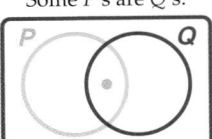 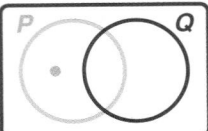

Euler diagrams

Euler used diagrams to illustrate logic concepts. Some 100 years later, John Venn extended the use of Euler's diagrams to illustrate many types of mathematics. In this section, we will construct diagrams to determine the validity of arguments. We will refer to these diagrams as Euler diagrams.

EXAMPLE 1 ■ Use an Euler Diagram to Determine the Validity of an Argument

Use an Euler diagram to determine whether the following argument is valid or invalid.

> All college courses are fun.
> This course is a college course.
> ∴This course is fun.

Solution

The first premise indicates that the set of college courses is a subset of the set of fun courses. We illustrate this subset relationship with an Euler diagram, as shown in Figure 3.15. The second premise tells us that "this course" is an element of the set of college courses. If we use c to represent "this course," then c must be placed inside the set of college courses, as shown in Figure 3.16.

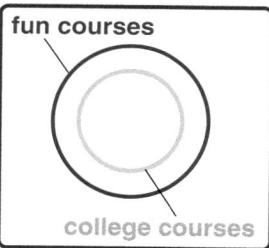

Figure 3.15

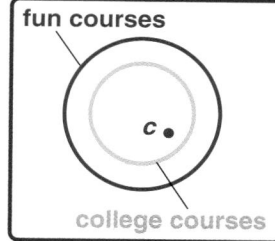

Figure 3.16

Figure 3.16 illustrates that c must also be an element of the set of fun courses. Thus the argument is valid.

CHECK YOUR PROGRESS 1 Use an Euler diagram to determine whether the following argument is valid or invalid.

> All lawyers drive BMW's.
> Susan is a lawyer.
> ∴Susan drives a BMW.

Solution See page S10.

If an Euler diagram can be drawn so that the conclusion does not necessarily follow from the premises, then the argument is invalid. This concept is illustrated in the next example.

Dance at Bougival by Renoir. This impressionist painting is on display at the Museum of Fine Arts, Boston.

EXAMPLE 2 ■ Use an Euler Diagram to Determine the Validity of an Argument

Use an Euler diagram to determine whether the following argument is valid or invalid.

> Some impressionists paintings are Renoirs.
> *Dance at Bougival* is an impressionist painting.
> ∴*Dance at Bougival* is a Renoir.

Solution

The Euler diagram in Figure 3.17 illustrates the premise that some impressionist paintings are Renoirs. Let *d* represent the painting *Dance at Bougival*. Figures 3.18 and 3.19 show that *d* can be placed in one of two regions.

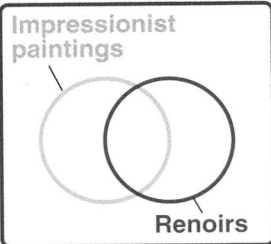

Figure 3.17

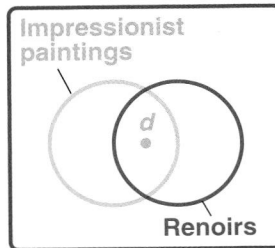

Figure 3.18

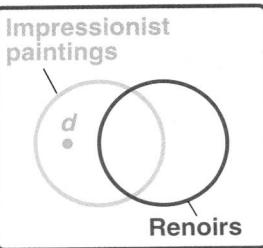

Figure 3.19

Although Figure 3.18 supports the argument, Figure 3.19 shows that the conclusion does not necessarily follow from the premises, and thus the argument is invalid.

CHECK YOUR PROGRESS 2   Use an Euler diagram to determine whether the following argument is valid or invalid.

> No prime numbers are negative.
> The number 7 is not negative.
> ∴The number 7 is a prime number.

Solution   *See page S11.*

✔ **TAKE NOTE**

Even though the conclusion in Example 2 is true, the argument is invalid.

QUESTION   *If one particular example can be found for which the conclusion of an argument is true when its premises are true, must the argument be valid?*

Some arguments can be represented by an Euler diagram that involves three sets, as shown in Example 3.

ANSWER   *No. To be a valid argument, the conclusion must be true whenever the premises are true. Just because the conclusion is true for one specific example, it does not mean the argument is a valid argument.*

EXAMPLE 3 ■ Use an Euler Diagram to Determine the Validity of an Argument

Use an Euler diagram to determine whether the following argument is valid or invalid.

> No psychologist can juggle.
> All clowns can juggle.
> ∴No psychologist is a clown.

Solution

The Euler diagram in Figure 3.20 shows that the set of psychologists and the set of jugglers are disjoint sets. Figure 3.21 shows that because the set of clowns is a subset of the set of jugglers, no psychologists *p* are elements of the set of clowns. Thus the argument is valid.

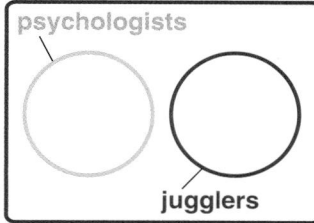

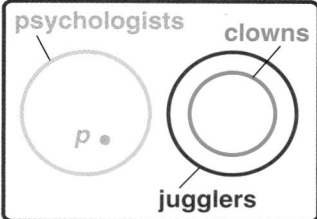

Figure 3.20   **Figure 3.21**

CHECK YOUR PROGRESS 3   Use an Euler diagram to determine whether the following argument is valid or invalid.

> No mathematics professors are good-looking.
> All good-looking people are models.
> ∴No mathematics professor is a model.

Solution   *See page S11.*

Math Matters   A Famous Puzzle

Three men decide to rent a room for one night. The regular room rate is $25; however, the desk clerk charges the men $30 because it will be easier for each man to pay one-third of $30 then it would be for each man to pay one-third of $25. Each man pays $10 and the porter shows them to their room.

After a short period of time the desk clerk starts to feel guilty and gives the porter $5, along with instructions to return the $5 to the three men.

On the way to the room the porter decides to give each man $1 and pocket $2. After all, the men would find it difficult to split $5 evenly.

Thus each man has paid $10 and received a refund of $1. After the refund, the men have paid a total of $27. The porter has $2. The $27 added to the $2 equals $29.

QUESTION   *Where is the missing dollar? (See Answer on the following page.)*

Euler Diagrams and the Extended Law of Syllogism

Example 4 uses Euler diagrams to visually illustrate the extended law of syllogism from Section 3.5.

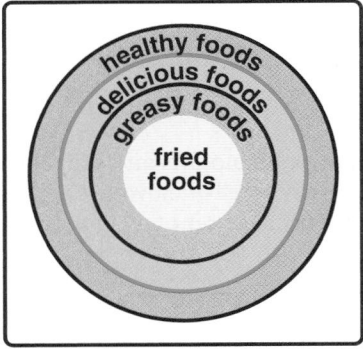

✔ **TAKE NOTE**

Although the conclusion in Example 4 is false, the argument in Example 4 is valid.

EXAMPLE 4 ■ **Use an Euler Diagram to Determine the Validity of an Argument**

Use an Euler diagram to determine whether the following argument is valid or invalid.

> All fried foods are greasy.
> All greasy foods are delicious.
> All delicious foods are healthy.
> ∴ All fried foods are healthy.

Solution

The figure at the left illustrates that every fried food is an element of the set of healthy foods, so the argument is valid.

CHECK YOUR PROGRESS 4   Use an Euler diagram to determine whether the following argument is valid or invalid.

> All squares are rhombi.
> All rhombi are parallelograms.
> All parallelograms are quadrilaterals.
> ∴ All squares are quadrilaterals.

Solution   See page S11.

Using Euler Diagrams to Form Conclusions

In Example 5, we make use of an Euler diagram to determine a valid conclusion for an argument.

ANSWER   *The $2 the porter kept was added to the $27 the men spent to produce a total of $29. The fact that this amount just happens to be close to $30 is a coincidence. All the money can be located if we total the $2 the porter has, the $3 that was returned to the men, and the $25 the desk clerk has, to produce $30.*

EXAMPLE 5 ▧ **Use an Euler Diagram to Determine the Conclusion for an Argument**

Use an Euler diagram and all of the premises in the following argument to determine a valid conclusion for the argument.

> All Ms are Ns.
> No Ns are Ps.
> ∴?

Solution

The first premise indicates that the set of Ms is a subset of the set of Ns. The second premise indicates that the set of Ns and the set of Ps are disjoint sets. The following Euler diagram illustrates these set relationships. An examination of the Euler diagram allows us to conclude that no Ms are Ps.

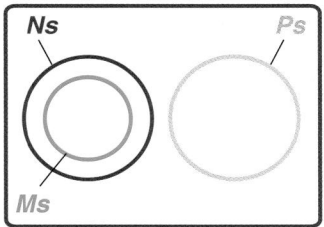

CHECK YOUR PROGRESS 5   Use an Euler diagram and all of the premises in the following argument to determine a valid conclusion for the argument.

> Some rabbits are white.
> All white animals like tomatoes.
> ∴?

Solution   *See page S11.*

Excursion

Using Logic to Solve Puzzles

Many puzzles can be solved by making an assumption and then checking to see if the assumption is supported by the conditions (premises) associated with the puzzle. For instance, consider the following addition problem in which each letter represents a digit from 0 through 9, and different letters represent different digits.

✔ **TAKE NOTE**

When working with cryptarithms, we assume that the leading digit of each number is a nonzero digit.

```
  T A
+ B T
-----
T E E
```

(continued)

Note that the T in T E E is a carry from the middle column. Because the sum of any two single digits plus a previous carry of at most 1 is 19 or less, the T in T E E must be a 1. Replacing all the T's with 1's produces:

$$
\begin{array}{r}
1\,A \\
+\,B\,1 \\
\hline
1\,E\,E
\end{array}
$$

Now B must be an 8 or a 9, because these are the only digits that would produce a carry into the leftmost column.

Case 1: Assume B is a 9. Then A must be an 8 or smaller, and A + 1 does not produce a carry into the middle column. The sum of the digits in the middle column is 10 thus E is a 0. This presents a dilemma because the units digit of A + 1 must also be a 0, which requires A to be a 9. The assumption that B is a 9 is not supported by the conditions of the problem; thus we reject the assumption that B is a 9.

Case 2: Assume B is an 8. To produce the required carry into the leftmost column there must be a carry from the column on the right. Thus A must be a 9 and we have the result shown below.

$$
\begin{array}{r}
1\,9 \\
+\,8\,1 \\
\hline
1\,0\,0
\end{array}
$$

A check shows that this solution satisfies all the conditions of the problem.

Excursion Exercises

Solve the following cryptarithms. Assume that no leading digit is a 0.
(*Source:* **http://www.geocities.com/Athens/Agora/2160/puzzles.html**)[4]

1.
$$
\begin{array}{r}
S\,O \\
+\,S\,O \\
\hline
T\,O\,O
\end{array}
$$

2.
$$
\begin{array}{r}
U\,S \\
+\,A\,S \\
\hline
A\,L\,L
\end{array}
$$

3.
$$
\begin{array}{r}
C\,O\,C\,A \\
+\,C\,O\,L\,A \\
\hline
O\,A\,S\,I\,S
\end{array}
$$

4.
$$
\begin{array}{r}
A\,T \\
E\,A\,S\,T \\
+\,W\,E\,S\,T \\
\hline
S\,O\,U\,T\,H
\end{array}
$$

Exercise Set 3.6   (Suggested Assignment: 1–27 odds)

In Exercises 1–20, use an Euler diagram to determine whether the argument is valid or invalid.

1. All frogs are poetical.
   Kermit is a frog.

   ∴Kermit is poetical.

2. All Oreo cookies have a filling.
   All Fig Newtons have a filling.

   ∴All Fig Newtons are Oreo cookies.

3. Some plants have flowers.
   All things that have flowers are beautiful.

   ∴Some plants are beautiful.

4. No squares are triangles.
   Some triangles are equilateral.

   ∴No squares are equilateral.

5. No rocker would do the Mariachi.
   All baseball fans do the Mariachi.

   ∴No rocker is a baseball fan.

6. Nuclear energy is not safe.
   Some electric energy is safe.

   ∴No electric energy is nuclear energy.

7. Some birds bite.
   All things that bite are dangerous.

   ∴Some birds are dangerous.

8. All fish can swim.
   That barracuda can swim.

   ∴That barracuda is a fish.

9. All men behave badly.
   Some hockey players behave badly.

   ∴Some hockey players are men.

10. All grass is green.
    That ground cover is not green.

    ∴That ground cover is not grass.

11. Most teenagers drink soda.
    No CEOs drink soda.

    ∴No CEO is a teenager.

12. Some students like history.
    Vern is a student.

    ∴Vern likes history.

13. No mathematics test is fun.
    All fun things are worth your time.

    ∴No mathematics test is worth your time.

14. All prudent people shun sharks.
    No accountant is imprudent.

    ∴No accountant fails to shun sharks.

15. All candidates without a master's degree will not be considered for the position of director.
    All candidates who are not considered for the position of director should apply for the position of assistant.

    ∴All candidates without a master's degree should apply for the position of assistant.

16. Some whales make good pets.
    Some good pets are cute.
    Some cute pets bite.

    ∴Some whales bite.

17. All prime numbers are odd.
    2 is a prime number.

    ∴2 is an odd number.

18. All Lewis Carroll arguments are valid.
    Some valid arguments are syllogisms.

    ∴Some Lewis Carroll arguments are syllogisms.

19. All aerobics classes are fun.
    Jan's class is fun.

    ∴Jan's class is an aerobics class.

20. No sane person takes a math class.
    Some students that take a math class can juggle.

    ∴No sane person can juggle.

In Exercises 21–26, use all of the premises in each argument to determine a valid conclusion for the argument.

21. All Reuben sandwiches are good.
    All good sandwiches have pastrami.
    All sandwiches with pastrami need mustard.

    ∴?

22. All cats are strange.
    Boomer is not strange.

    ∴?

23. All multiples of 11 end with a 5.
    1001 is a multiple of 11.

    ∴?

24. If it isn't broken, then I do not fix it.
    If I do not fix it, then I do not get paid.

    ∴?

25.   Some horses are frisky.
     All frisky horses are grey.
     ∴?

26.   If we like to ski, then we will move to Vail.
     If we move to Vail, then we will not buy a house.
     If we do not buy a condo, then we will buy a house.
     ∴?

Extensions
CRITICAL THINKING

In the following *crossnumber puzzle* each square holds a single digit from 0 through 9. Use the clues under the *Across* and *Down* headings to solve the puzzle.

27. *Across*

  1. One-fourth of 3 across

  3. Two more than 1 down with its digits reversed

  Down

  1. Larger than 20 and less than 30

  2. Half of 1 down

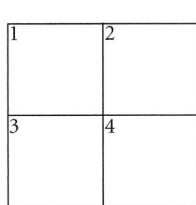

EXPLORATIONS

28. Lewis Carroll (Charles Dodgson) devised a *bilateral diagram* (two-part board) to analyze syllogisms. His method has some advantages over Euler diagrams and Venn diagrams. Use a library or the Internet to find information on Carroll's method of analyzing syllogisms. Write a few paragraphs that explain his method and its advantages.

CHAPTER 3   **Summary**

Key Terms

antecedent [p. 131]
argument [p. 146]
arrow notation [p. 131]
biconditional [p. 135]
component statement [p. 112]
compound statement [p. 112]
conclusion [p. 146]
conditional [p. 131]
conjunction [p. 114]
connective [p. 112]
consequent [p. 131]
contrapositive [p. 140]
converse [p. 140]
disjunction [p. 114]
disjunctive form of the conditional [p. 134]
disjunctive syllogism [p. 150]
equivalent statements [p. 124]
Euler diagram [p. 158]
existential quantifier [p. 115]

fallacy of the converse [p. 150]
fallacy of the inverse [p. 150]
invalid argument [p. 146]
inverse [p. 140]
law of syllogism [p. 150]
modus ponens [p. 150]
modus tollens [p. 150]
negation [p. 112]
premise [p. 146]
quantifier [p. 115]
self-contradiction [p. 126]
standard forms of arguments [p. 150]
standard truth table form [p. 122]
statement [p. 110]
symbolic form [p. 112]
tautology [p. 126]
truth table [p. 113]
truth value [p. 113]
universal quantifier [p. 115]
valid argument [p. 146]

Essential Concepts

- **Truth Values**

  $\sim p$ is true if and only if p is false.

  $p \wedge q$ is true if and only if both p and q are true.

  $p \vee q$ is true if and only if p is true, q is true, or both p and q are true.

- **De Morgan's Laws for Statements**

  $\sim(p \wedge q) \equiv \sim p \vee \sim q$ and $\sim(p \vee q) \equiv \sim p \wedge \sim q$

- **Equivalent Forms**

  $p \to q \equiv \sim p \vee q$

  $\sim(p \to q) \equiv p \wedge \sim q$

  $p \leftrightarrow q \equiv [(p \to q) \wedge (q \to p)]$

- **Statements Related to the Conditional Statement**

  The *converse* of $p \to q$ is $q \to p$.

  The *inverse* of $p \to q$ is $\sim p \to \sim q$.

  The *contrapositive* of $p \to q$ is $\sim q \to \sim p$.

- **Valid Arguments**

Modus ponens	Modus tollens	Law of syllogism	Disjunctive syllogism
$p \to q$	$p \to q$	$p \to q$	$p \vee q$
p	$\sim q$	$q \to r$	$\sim p$
$\therefore q$	$\therefore \sim p$	$\therefore p \to r$	$\therefore q$

- **Invalid Arguments**

Fallacy of the converse	Fallacy of the inverse
$p \to q$	$p \to q$
q	$\sim p$
$\therefore p$	$\therefore \sim q$

CHAPTER 3	**Review Exercises**

In Exercises 1–6, determine whether each sentence is a statement. Assume that *a* and *b* are real numbers.

1. How much is a ticket to London?
2. 91 is a prime number.
3. $a > b$
4. $a^2 \geq 0$
5. Lock the car.
6. Clark Kent is Superman.

In Exercises 7–10, write each sentence in symbolic form. Represent each component of the sentence with the letter indicated in parentheses. Also state whether the sentence is a conjunction, a disjunction, a negation, a conditional, or a biconditional.

7. Today is Monday (*m*) and it is my birthday (*b*).
8. If *x* is divisible by 2 (*d*), then *x* is an even number (*e*).
9. I am going to the dance (*g*) if and only if I have a date (*d*).
10. All triangles (*t*) have exactly three sides (*s*).

In Exercises 11–16, write the negation of each quantified statement.

11. Some dogs bite.
12. Every dessert at the Cove restaurant is good.
13. All winners receive a prize.
14. Some cameras do not use film.
15. None of the students received an A.
16. At least one person enjoyed the story.

In Exercises 17–22, determine whether each statement is true or false.

17. $5 > 2$ or $5 = 2$.
18. $3 \neq 5$ and 7 is a prime number.
19. $4 \leq 7$
20. $-3 < -1$
21. Every repeating decimal is a rational number.
22. There exists a real number that is not positive and not negative.

In Exercises 23–28, determine the truth value of the statement given that p is true, q is false, and r is false.

23. $(p \wedge q) \vee (\sim p \vee q)$

24. $(p \rightarrow \sim q) \leftrightarrow \sim(p \vee q)$

25. $(p \wedge \sim q) \wedge (\sim r \vee q)$

26. $(r \wedge \sim p) \vee [(p \vee \sim q) \leftrightarrow (q \rightarrow r)]$

27. $[p \wedge (r \rightarrow q)] \rightarrow (q \vee \sim r)$

28. $(\sim q \vee \sim r) \rightarrow [(p \leftrightarrow \sim r) \wedge q]$

In Exercises 29–36, construct a truth table for the given statement.

29. $(\sim p \rightarrow q) \vee (\sim q \wedge p)$

30. $\sim p \leftrightarrow (q \vee p)$

31. $\sim(p \vee \sim q) \wedge (q \rightarrow p)$

32. $(p \leftrightarrow q) \vee (\sim q \wedge p)$

33. $(r \leftrightarrow \sim q) \vee (p \rightarrow q)$

34. $(\sim r \vee \sim q) \wedge (q \rightarrow p)$

35. $[p \leftrightarrow (q \rightarrow \sim r)] \wedge \sim q$

36. $\sim(p \wedge q) \rightarrow (\sim q \vee \sim r)$

In Exercises 37–40, make use of De Morgan's laws to write the given statement in an equivalent form.

37. It is not true that Bob failed the English proficiency test and he registered for a speech course.

38. Ellen did not go to work this morning and she did not take her medication.

39. Wendy will go to the store this afternoon or she will not be able to fix her fettuccine al pesto recipe.

40. Gina enjoyed the movie, but she did not enjoy the party.

In Exercises 41–44, use a truth table to show that the given pairs of statements are equivalent.

41. $\sim p \rightarrow \sim q; p \vee \sim q$

42. $\sim p \vee q; \sim(p \wedge \sim q)$

43. $p \vee (q \wedge \sim p); p \vee q$

44. $p \leftrightarrow q; (p \wedge q) \vee (\sim p \wedge \sim q)$

In Exercises 45–48, use a truth table to determine whether the given statement is a tautology or a self-contradiction.

45. $p \wedge (q \wedge \sim p)$

46. $(p \wedge q) \vee (p \rightarrow \sim q)$

47. $[\sim(p \rightarrow q)] \leftrightarrow (p \wedge \sim q)$

48. $p \vee (p \rightarrow q)$

In Exercises 49–52, identify the antecedent and the consequent of each conditional statement.

49. If he has talent, he will succeed.

50. If I had a credential, I could get the job.

51. I will follow the exercise program provided I join the fitness club.

52. I will attend only if it is free.

In Exercises 53–56, write each conditional statement in its equivalent disjunctive form.

53. If she were tall, she would be on the volleyball team.

54. If he can stay awake, he can finish the report.

55. Rob will start provided he is not ill.

56. Sharon will be promoted only if she closes the deal.

In Exercises 57–60, write the negation of each conditional statement in its equivalent conjunctive form.

57. If I get my paycheck, I will purchase a ticket.

58. The tomatoes will get big only if you provide plenty of water.

59. If you entered Cleggmore University, then you had a high score on the SAT exam.

60. If Ryan enrolls at a university, then he will enroll at Yale.

In Exercises 61–66, determine whether the given statement is true or false. Assume that x and y are real numbers.

61. $x = y$ if an only if $|x| = |y|$.

62. $x > y$ if an only if $x - y > 0$.

63. If $x + y = 2x$, then $y = x$.

64. If $x > y$, then $\frac{1}{x} > \frac{1}{y}$.

65. If $x^2 > 0$, then $x > 0$.

66. If $x^2 = y^2$, then $x = y$.

In Exercises 67–70, write each statement in "If p, then q" form.

67. Every nonrepeating, nonterminating decimal is an irrational number.

68. Being well known is a necessary condition for a politician.

69. I could buy the house provided I could sell my condominium.

70. Being divisible by 9 is a sufficient condition for being divisible by 3.

In Exercises 71–76, write the **a.** converse, **b.** inverse, and **c.** contrapositive of the given statement.

71. If $x + 4 > 7$, then $x > 3$.

72. All recipes in this book can be prepared in less than 20 minutes.

73. If a and b are both divisible by 3, then $(a + b)$ is divisible by 3.

74. If you build it, they will come.

75. Every trapezoid has exactly two parallel sides.

76. If they like it, they will return.

77. What is the inverse of the contrapositive of $p \rightarrow q$?

78. What is the converse of the contrapositive of the inverse of $p \rightarrow q$?

In Exercises 79–82, determine the original statement if the given statement is related to the original statement in the manner indicated.

79. *Converse:* If $x > 2$, then x is an odd prime number.

80. *Negation:* The senator will attend the meeting and she will not vote on the motion.

81. *Inverse:* If their manager will not contact me, then I will not purchase any of their products.

82. *Contrapositive:* If Ginny can't rollerblade, then I can't rollerblade.

In Exercises 83–86, use a truth table to determine whether the argument is valid or invalid.

83. $(p \wedge \sim q) \wedge (\sim p \rightarrow q)$
$$\frac{p}{\therefore \sim q}$$

84. $p \rightarrow \sim q$
$$\frac{q}{\therefore \sim p}$$

85. r
$p \rightarrow \sim r$
$$\frac{\sim p \rightarrow q}{\therefore p \wedge q}$$

86. $(p \vee \sim r) \rightarrow (q \wedge r)$
$$\frac{r \wedge p}{\therefore p \vee q}$$

In Exercises 87–92, determine whether the argument is valid or invalid by comparing its symbolic form with the symbolic forms in Tables 3.13 and 3.14.

87. We will serve either fish or chicken for lunch. We did not serve fish for lunch. Therefore, we served chicken for lunch.

88. If Mike is a CEO, then he will be able to afford to make a donation. If Mike can afford to make a donation, then he loves to ski. Therefore, if Mike does not love to ski, he is not a CEO.

89. If we wish to win the lottery, we must buy a lottery ticket. We did not win the lottery. Therefore, we did not buy a lottery ticket.

90. Robert can charge it on his MasterCard or his Visa. Robert does not use his MasterCard. Therefore, Robert charged it to his Visa.

91. If we are going to have a caesar salad, then we need to buy some eggs. We did not buy eggs. Therefore, we are not going to have a caesar salad.

92. If we serve lasagna, then Eva will not come to our dinner party. We did not serve lasagna. Therefore, Eva came to our dinner party.

In Exercises 93–96, use an Euler diagram to determine whether the argument is valid or invalid.

93. No wizard can yodel.
All lizards can yodel.
∴No wizard is a lizard.

94. Some dogs have tails.
Some dogs are big.
∴Some big dogs have tails.

95. All Italian villas are wonderful. It is not wise to invest in expensive villas. Some wonderful villas are expensive. Therefore, it is not wise to invest in Italian villas.

96. All logicians like to sing "It's a small world after all." Some logicians have been presidential candidates. Therefore, some presidential candidates like to sing "It's a small world after all."

CHAPTER 3 **Test**

1. Determine whether each sentence is a statement.
   a. Look for the cat.
   b. Clark Kent is afraid of the dark.

2. Write the negation of each statement.
   a. Some trees are not green.
   b. None of the kids had seen the movie.

3. Determine whether each statement is true or false.
   a. $5 \leq 4$
   b. $-2 \geq -2$

4. Determine the truth value of each statement given that p is true, q is false, and r is true.
   a. $(p \lor \sim q) \land (\sim r \land q)$
   b. $(r \lor \sim p) \lor [(p \lor \sim q) \leftrightarrow (q \to r)]$

In Exercises 5 and 6, construct a truth table for the given statement.

5. $\sim(p \land \sim q) \lor (q \to p)$
6. $(r \leftrightarrow \sim q) \land (p \to q)$

7. Use one of De Morgan's laws to write the following in an equivalent form.

   Elle did not eat breakfast and she did not take a lunch break.

8. ✎ What is a tautology?

9. Write $p \to q$ in its equivalent disjunctive form.

10. Determine whether the given statement is true or false. Assume that x, y, and z are real numbers.
    a. $x = y$ if $|x| = |y|$.
    b. If $x > y$, then $xz > yz$.

11. Write the **a.** converse, **b.** inverse, and **c.** contrapositive of the following statement.

    If $x + 7 > 11$, then $x > 4$.

12. Write the standard form known as modus ponens.

13. Write the standard form known as the law of syllogism.

In Exercises 14 and 15, use a truth table to determine whether the argument is valid or invalid.

14. $(p \land \sim q) \land (\sim p \to q)$
    $\underline{p \hspace{4cm}}$
    $\therefore \sim q$

15. r
    $p \to \sim r$
    $\underline{\sim p \to q}$
    $\therefore p \land q$

In Exercises 16 to 20, determine whether the argument is valid or invalid. Explain how you made your decision.

16. If we wish to win the talent contest, we must practice. We did not win the contest. Therefore, we did not practice.

17. Gina will take a job in Atlanta or she will take a job in Kansas City. Gina did not take a job in Atlanta. Therefore, Gina took a job in Kansas City.

18. No wizard can glow in the dark.
    Some lizards can glow in the dark.
    $\therefore$ No wizard is a lizard.

19. Some novels are worth reading.
    War and Peace is a novel.
    $\therefore$ *War and Peace* is worth reading.

20. If I cut my night class, then I will go to the party. I went to the party. Therefore, I cut my night class.

Numeration Systems and Topics from Number Theory

Archeological exploration can provide today's world with valuable information about past civilizations such as their written languages and numeration systems. In the **Excursion exercises** on **page 180,** you will use your knowledge of the Roman numeration system in order to discover information about the traditional Chinese numeration system.

Need help? For on-line student resources, such as section quizzes, visit this textbook's web site at **math.college.hmco.com/students.**

In this chapter we will examine several numeration systems. As each numeration system is presented, try to answer the following questions.

- How many symbols must be memorized to use this system?
- How do you perform arithmetic computations in this system?
- What are the advantages of this particular system compared to the other numeration systems that have been presented?
- What are the disadvantages of this system?

The numeration system that most of us use today is called the Hindu-Arabic numeration system. In Europe, the adoption of the Hindu-Arabic system took place over a period of several hundred years. Most people were reluctant to change from their familiar Roman numeration system to the Hindu-Arabic system. The illustration below is from the 1504 text *Margarita Philosophica* by Gregor Reisch. It shows a mathematical contest between Boethius (a philosopher and author of a math text that was widely used in his time) performing a computation with Hindu-Arabic numerals and Pythagoras computing with the aid of a counting board. The figure in the background is the Spirit of Arithmeticae.

SECTION 4.1 Early Numeration Systems

The Egyptian Numeration System

In mathematics, symbols that are used to represent numbers are called **numerals.** A number can be represented by many different numerals. For instance, the concept of "eightness" is represented by each of the following.

Hindu-Arabic: 8   Tally: 𝗟𝗛𝗧 ||| Roman: VIII

Chinese: ﾉＬ   Egyptian: |||||||| Babylonian: 𝖸𝖸𝖸𝖸𝖸𝖸𝖸𝖸 Ⓟ

A **numeration system** consists of a set of numerals and a method of arranging the numerals to represent numbers. The numeration system that most people use today is known as the *Hindu-Arabic numeration system.* It makes use of the 10 numerals 0, 1, 2, 3, 4, 5, 6, 7, 8, and 9. Before we examine the Hindu-Arabic numeration system in detail, it will be helpful to study some of the earliest numeration systems that were developed by the Egyptians, the Romans, and the Chinese.

The Egyptian numeration system uses pictorial symbols called **hieroglyphics** as numerals. The Egyptian hieroglyphic system is an **additive system** because any given number is written by using numerals whose sum equals the number. Table 4.1 gives the Egyptian hieroglyphics for powers of 10 from one to one million.

Table 4.1 *Egyptian Hieroglyphics for Powers of 10*

Hindu-Arabic Numeral	Egyptian Hieroglyphic	Description of Hieroglyphic
1			stroke
10	∩	heel bone
100	୭	scroll
1000	⨍	lotus flower
10,000	╱	pointing finger
100,000	⋈	fish
1,000,000	ⵖ	astonished person

To write the number 300, the Egyptians wrote the scroll hieroglyphic three times: ୭୭୭. In the Egyptian hieroglyphic system, the order of the hieroglyphics is of no importance. Each of the following Egyptian numerals represents 321.

୭୭୭∩∩|, ∩∩୭|୭୭, ୭|∩୭∩୭, ∩୭୭∩

EXAMPLE 1 ■ Write a Numeral Using Egyptian Hieroglyphics

Write 3452 using Egyptian hieroglyphics.

Solution

3452 = 3000 + 400 + 50 + 2. Thus the Egyptian numeral for 3452 is
𓆼𓆼𓆼𓍢𓍢𓍢𓍢𓎆𓎆𓎆𓎆𓎆𓏺𓏺

CHECK YOUR PROGRESS 1 Write 201,473 using Egyptian hieroglyphics.

Solution *See page S11.*

QUESTION *Do the Egyptian hieroglyphics* 𓍢𓍢𓎆𓏺 *and* 𓎆𓏺𓍢𓍢 *represent the same number?*

EXAMPLE 2 ■ Evaluate a Numeral Written Using Egyptian Hieroglyphics

Write 𓆐𓆐𓂭𓂭𓂭𓆼𓆼𓍢𓍢𓍢𓍢𓎆𓏺𓏺𓏺 as a Hindu-Arabic numeral.

Solution

(2 × 100,000) + (3 × 10,000) + (2 × 1000) + (4 × 100) + (1 × 10) + (3 × 1) = 232,413

CHECK YOUR PROGRESS 2 Write 𓁨𓆐𓆐𓆐𓂭𓆼𓆼𓆼𓆼𓍢𓍢𓍢𓎆𓎆𓏺𓏺𓏺𓏺𓏺𓏺 as a Hindu-Arabic numeral.

Solution *See page S11.*

historical note

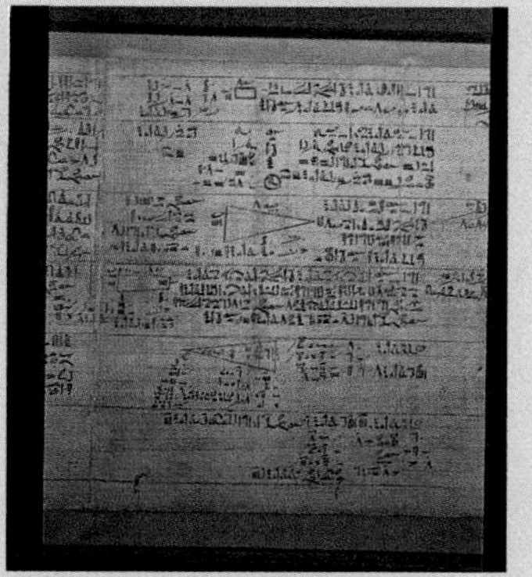

A portion of the Rhind papyrus.

The Rhind papyrus is named after Alexander Henry Rhind, who purchased the papyrus in Egypt in 1858 A.D. Today the Rhind papyrus is preserved in the British Museum in London. ■

One of the earliest written documents of mathematics is the Rhind papyrus (see the figure at the left). This tablet was found in Egypt in 1858 A.D., but it is estimated that the writings date back to 1650 B.C. The Rhind papyrus contains 85 mathematical problems. Studying these problems has enabled mathematicians and historians to understand some of the mathematical procedures used in the early Egyptian numeration system.

The operation of addition with Egyptian hieroglyphics is a simple grouping process. In some cases the final sum can be simplified by replacing a group of hieroglyphics by a single hieroglyphic with a larger numeric value. This technique is illustrated in Example 3.

ANSWER *Yes, they both represent the number 211.*

EXAMPLE 3 ■ Use Egyptian Hieroglyphics to Find a Sum

Use Egyptian hieroglyphics to find 2452 + 1263.

Solution

The sum is found by combining the hieroglyphics.

$$
\begin{array}{r}
2452 \\
+\ 1263 \\
\end{array}
$$

Replacing 10 heel bones with one scroll produces

꤫꤫꤫999999∩||||| or 3715

The sum is 3715.

CHECK YOUR PROGRESS 3   Use Egyptian hieroglyphics to find 23,341 + 10,562.

Solution   See page S11.

In the Egyptian numeration system, subtraction is performed by removing some of the hieroglyphics from the larger numeral. In some cases it is necessary to "borrow," as shown in the next example.

✔ **TAKE NOTE**

Five scrolls cannot be removed from two scrolls so one lotus flower is replaced by ten scrolls, resulting in a total of twelve scrolls. Now five scrolls can be removed from twelve scrolls.

EXAMPLE 4 ■ Use Egyptian Hieroglyphics to Find a Difference

Use Egyptian hieroglyphics to find 332,246 − 101,512.

Solution

The numerical value of one lotus flower is equivalent to the numerical value of 10 scrolls. Thus

$$
\begin{array}{r}
332,246 \\
-\ 101,512 \\
\end{array}
$$

The difference is 230,734.

CHECK YOUR PROGRESS 4   Use Egyptian hieroglyphics to find 61,432 − 45,121.

Solution   See page S11.

Math Matters   Early Egyptian Fractions

Evidence gained from the Rhind papyrus shows that the Egyptian method of calculating with fractions was much different from the methods we use today. All Egyptian fractions $\left(\text{except for } \frac{2}{3}\right)$ were represented in terms of unit fractions, which are fractions of the form $\frac{1}{n}$, for some natural number $n > 1$. The Egyptians wrote these unit fractions by placing an oval over the numeral that represented the denominator. For example,

$$\underset{\text{|||}}{\bigcirc} = \frac{1}{3} \qquad \underset{\cap\text{|||||}}{\bigcirc} = \frac{1}{15}$$

If a fraction was not a unit fraction, then the Egyptians wrote the fraction as the sum of *distinct* unit fractions. For instance,

$$\frac{2}{5} \text{ was written as the sum of } \frac{1}{3} \text{ and } \frac{1}{15}.$$

Of course, $\frac{2}{5} = \frac{1}{5} + \frac{1}{5}$, but (for some mysterious reason) the early Egyptian numeration system didn't allow repetitions. The Rhind papyrus includes a table that shows how to write fractions of the form $\frac{2}{k}$, where k is an odd number from 5 to 101, in terms of unit fractions. Some of these are listed below.

$$\frac{2}{7} = \frac{1}{4} + \frac{1}{28} \qquad \frac{2}{11} = \frac{1}{6} + \frac{1}{66} \qquad \frac{2}{19} = \frac{1}{12} + \frac{1}{76} + \frac{1}{114}$$

Table 4.2 *Roman Numerals*

Hindu-Arabic Numeral	Roman Numeral
1	I
5	V
10	X
50	L
100	C
500	D
1000	M

P

INSTRUCTOR NOTE
Inform your students that the C used in Roman numerals is from the Latin word for hundred, which is *centum*. The word *century,* which means a period of 100 years, is a derivation of *centum.* The M is from the Latin word for thousand, which is *mille.* We use the word *millennium* to designate 1000 years.

The Roman Numeration System

The Roman numeration system was used in Europe during the reign of the Roman Empire. Today we still make limited use of Roman numerals on clock faces, on the cornerstones of buildings, and in numbering the volumes of periodicals and books. Table 4.2 shows the numerals used in the Roman numeration system. If the Roman numerals are listed so that each numeral has a larger value than the numeral to its right, then the value of the Roman numeral is found by adding the values of each numeral. For example,

$$\text{CLX} = 100 + 50 + 10 = 160$$

If a Roman numeral is repeated two or three times in succession, we add to determine its numerical value. For instance, $\text{XX} = 10 + 10 = 20$ and $\text{CCC} = 100 + 100 + 100 = 300$. Each of the numerals I, X, C, and M may be repeated up to three times. The numerals V, L, and D are not repeated.

Although the Roman numeral system is an additive system, it also incorporates a subtraction property. In the Roman numeral system, the value of a numeral is determined by adding the values of the numerals from left to right. However, if the value of a numeral is less than the value of the numeral to its right, the smaller value is subtracted from the next larger value. For instance, $\text{VI} = 5 + 1 = 6$; however, $\text{IV} = 5 - 1 = 4$. In the Roman system the only numerals whose values can be subtracted from the value of the numeral to the right are I, X, and C. Also, the

The Roman numeral system evolved over a period of several years, and thus some Roman numerals displayed on ancient structures do not adhere to the basic rules given at the right. For instance, in the Colosseum in Rome (c. 80 A.D.), the numeral XXVIIII appears above archway 29 instead of the numeral XXIX.

subtraction of these values is allowed only if the value of the numeral to the right is within two rows as shown in Table 4.2. That is, the value of the numeral to be subtracted must be *no less than* one-tenth of the value of the numeral it is to be subtracted from. For instance, XL = 40 and XC = 90, but XD does not represent 490 because the value of X is less than one-tenth the value of D. To write 490 using Roman numerals, we write CDXC.

A Summary of the Basic Rules Employed in the Roman Numeral System

$$I = 1, \quad V = 5, \quad X = 10, \quad L = 50, \quad C = 100, \quad D = 500, \quad M = 1000$$

1. If the numerals are listed so that each numeral has a larger value than the numeral to the right, then the value of the Roman numeral is found by adding the values of the numerals.

2. Each of the numerals, I, X, C, and M may be repeated up to three times. The numerals V, L, and D are not repeated. If a numeral is repeated two or three times in succession, we add to determine its numerical value.

3. The only numerals whose values can be subtracted from the value of the numeral to the right are I, X, and C. The value of the numeral to be subtracted must be no less than one-tenth of the value of the numeral to its right.

EXAMPLE 5 ■ Evaluate a Roman Numeral

Write DCIV as a Hindu-Arabic numeral.

Solution
Because the value of D is larger than the value of C, we add their numerical values. The value of I is less than the value of V, so we subtract the smaller value from the larger value. Thus

$$DCIV = (DC) + (IV) = (500 + 100) + (5 - 1) = 600 + 4 = 604$$

CHECK YOUR PROGRESS 5   Write MCDXLV as a Hindu-Arabic numeral.

Solution   *See page S11.*

The spreadsheet program Excel has a function that converts Hindu-Arabic numerals to Roman numerals. In the Edit Formula dialogue box, type = ROMAN(n), where n is the number you wish to convert to a Roman numeral.

EXAMPLE 6 ■ Write a Hindu-Arabic Numeral as a Roman Numeral

Write 579 as a Roman numeral.

Solution
$$579 = 500 + 50 + 10 + 10 + 9$$

In Roman numerals 9 is written as IX. Thus 579 = DLXXIX.

CHECK YOUR PROGRESS 6   Write 473 as a Roman numeral.

Solution   *See page S11.*

In the Roman numeral system a bar over a numeral is used to denote a value 1000 times the value of the numeral. For instance,

$$\overline{V} = 5 \times 1000 = 5000 \qquad \overline{IV}LXX = (4 \times 1000) + 70 = 4070$$

TAKE NOTE

The method of writing a bar over a numeral should be used only to write Roman numerals that cannot be written using the basic rules. For instance, the Roman numeral for 2003 is MMIII, not $\overline{\text{IIIII}}$.

EXAMPLE 7 ■ **Convert Between Roman Numerals and Hindu-Arabic Numerals**

a. Write $\overline{\text{IV}}$DLXXII as a Hindu-Arabic numeral.

b. Write 6125 as a Roman numeral.

Solution

a. $\overline{\text{IV}}\text{DLXXII} = (\overline{\text{IV}}) + (\text{DLXXII})$
$$= (4 \times 1000) + (572)$$
$$= 4572$$

b. The Roman numeral 6 is written VI and 125 is written as CXXV. Thus in Roman numerals 6125 is $\overline{\text{VI}}$CXXV.

CHECK YOUR PROGRESS 7

a. Write $\overline{\text{VII}}$CCLIV as a Hindu-Arabic numeral.

b. Write 8070 as a Roman numeral.

Solution   *See page S11.*

Excursion

A Rosetta Tablet for the Traditional Chinese Numeration System

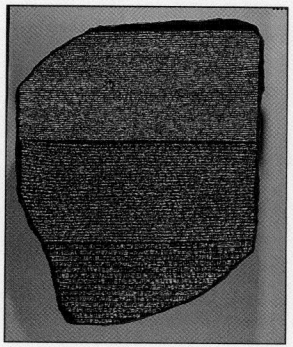

The Rosetta Stone

Most of the knowledge we have gained about early numeration systems has been obtained from inscriptions found on ancient tablets or stones. The information provided by these inscriptions has often been difficult to interpret. For several centuries archeologists had little success in interpreting the Egyptian hieroglyphics they had discovered. Then, in 1799, a group of French military engineers discovered a basalt stone near Rosetta in the Nile delta. This stone, which we now call the Rosetta Stone, has an inscription in three scripts: Greek, Egyptian Demotic, and Egyptian hieroglyphic. It was soon discovered that all three scripts contained the same message. The Greek script was easy to translate and, from its translation, clues were uncovered that enabled scholars to translate many of the documents that up to that time had been unreadable.

Pretend that you are an archeologist. Your team has just discovered an old tablet from 1998. This tablet displays Roman numerals and traditional Chinese numerals. It also provides hints in the form of a crossword puzzle about the traditional Chinese numeration system. Study the inscriptions on the following tablet and then complete the Excursion Exercises that follow.

(continued)

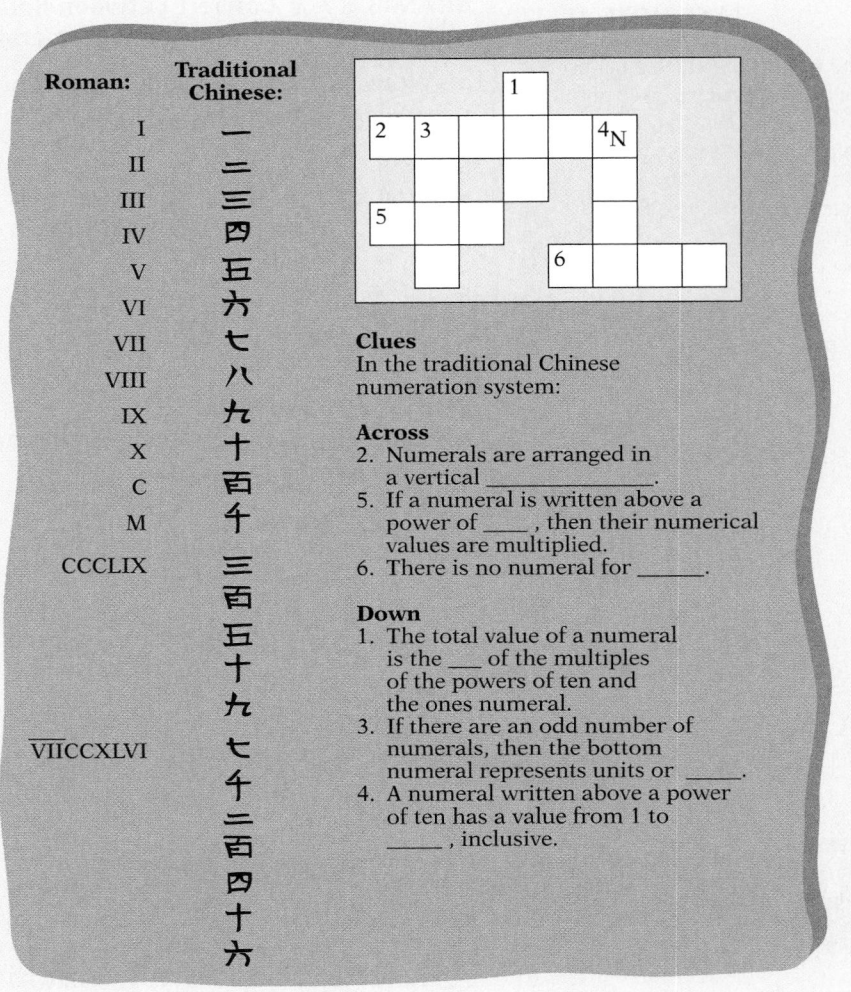

Roman:	Traditional Chinese:
I	一
II	二
III	三
IV	四
V	五
VI	六
VII	七
VIII	八
IX	九
X	十
C	百
M	千
CCCLIX	三百五十九
$\overline{VII}$CCXLVI	七千二百四十六

Clues

In the traditional Chinese numeration system:

Across
2. Numerals are arranged in a vertical _____.
5. If a numeral is written above a power of ____ , then their numerical values are multiplied.
6. There is no numeral for _____.

Down
1. The total value of a numeral is the ___ of the multiples of the powers of ten and the ones numeral.
3. If there are an odd number of numerals, then the bottom numeral represents units or _____.
4. A numeral written above a power of ten has a value from 1 to _____ , inclusive.

Excursion Exercises

1. Complete the crossword puzzle shown on the above tablet.

2. Write 26 as a traditional Chinese numeral.

3. Write 357 as a traditional Chinese numeral.

4. Write the Hindu-Arabic numeral given by each of the following traditional Chinese numerals.

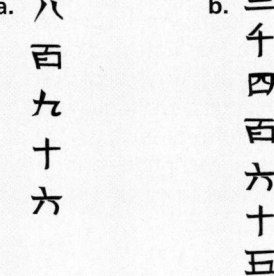

a.   八
     百
     九
     十
     六

b.   二
     千
     四
     百
     六
     十
     五

(continued)

5. a. How many Hindu-Arabic numerals are required to write four thousand five hundred twenty-eight?

  b. How many traditional Chinese numerals are required to write four thousand five hundred twenty-eight?

6. The traditional Chinese numeration system is no longer in use. Give a reason that may have contributed to its demise.

Exercise Set 4.1   (Suggested Assignment: 1–49 odds)

In Exercises 1–8, write each Hindu-Arabic numeral using Egyptian hieroglyphics.

1. 46

2. 82

3. 103

4. 157

5. 2568

6. 3152

7. 1,405,203

8. 653,271

In Exercises 9–16, write each Egyptian numeral as a Hindu-Arabic numeral.

9. 9999∩∩III
9999∩∩II

10. 9∩∩∩∩IIII
9∩∩∩∩III

11. ⚡99∩∩II

12. ⫻⫻ 99∩I
⊠⫻⚡⚡9∩∩II

13. ⫻⫻⫻⚡⚡⚡999∩∩∩IIIII
⫻⫻⫻⚡⚡9999∩∩∩IIII

14. ⌇⊠⫻⫻⚡⚡⚡9999∩
⌇⊠⫻ ⚡⚡⚡999∩III

15. ⌇⌇⌇⋈⚡99III
⌇⌇⫻⫻ ⚡99III

16. ⌇⋈⋈⚡99
⌇⋈⋈99

In Exercises 17–24, use Egyptian hieroglyphics to find each sum or difference.

17. 51 + 43

18. 67 + 58

19. 231 + 435

20. 623 + 124

21. 83 − 51

22. 94 − 23

23. 254 − 198

24. 640 − 278

In Exercises 25–32, write each Roman numeral as a Hindu-Arabic numeral.

25. DCL

26. MCX

27. MCDIX

28. MDCCII

29. $\overline{\text{IX}}$XLIV

30. $\overline{\text{VII}}$DXVII

31. $\overline{\text{XI}}$CDLXI

32. $\overline{\text{IV}}$CCXXI

In Exercises 33–40, write each Hindu-Arabic numeral as a Roman numeral.

33. 157

34. 231

35. 787

36. 1343

37. 683

38. 959

39. 6898

40. 4357

Egyptian Multiplication The Rhind papyrus contains problems that show a *doubling procedure* used by the Egyptians to find the product of two whole numbers. The following examples illustrate this doubling procedure. In the examples we have used Hindu-Arabic numerals so that you can concentrate on the doubling procedure and not be distracted by the Egyptian hieroglyphics. The first example determines the product 5 × 27 by computing two successive doublings of 27 and then forming the sum of the blue numbers in the rows marked with a check. Note that the rows marked with a check show that one 27 is 27 and four 27's is 108. Thus five 27's is the sum of 27 and 108, or 135.

$$
\begin{array}{ll}
\checkmark\ 1 & 27 \quad \text{double} \\
2 & 54 \quad \text{double} \\
\checkmark\ 4 & 108 \\
\overline{5} & \overline{135} \quad \longleftarrow \text{This sum is the} \\
& \qquad\qquad \text{product of 5 and 27.}
\end{array}
$$

In the next example, we use the Egyptian doubling procedure to find the product of 35 and 94. Because the sum of 1, 2, and 32 is 35, we add only the blue numbers in the rows marked with a check to find that $35 \times 94 = 94 + 188 + 3008 = 3290$.

$$
\begin{array}{ll}
\checkmark\ 1 & 94 \quad \text{double} \\
\checkmark\ 2 & 188 \quad \text{double} \\
4 & 376 \quad \text{double} \\
8 & 752 \quad \text{double} \\
16 & 1504 \quad \text{double} \\
\checkmark\ 32 & 3008 \\
\overline{35} & \overline{3290} \quad \longleftarrow \text{This sum is the} \\
& \qquad\qquad \text{product of 35 and 94.}
\end{array}
$$

In Exercises 41–48, use the Egyptian doubling procedure to find each product.

41. 8×63

42. 4×57

43. 7×29

44. 9×33

45. 17×35

46. 26×43

47. 23×108

48. 72×215

Extensions

CRITICAL THINKING

49. a. State a reason why you might prefer to use the Egyptian hieroglyphic numeration system rather than the Roman numeration system.

b. State a reason why you might prefer to use the Roman numeration system rather than the Egyptian hieroglyphic numeration system.

50. What is the largest number that can be written using Roman numerals without using the bar over a numeral or the subtraction property?

EXPLORATIONS

51. The Ionic Greek numeration system assigned numerical values to the letters of the Greek alphabet. Research the Ionic Greek numeration system and write a report that explains this numeration system. Include information about some of the advantages and disadvantages of this system compared with our present Hindu-Arabic numeration system.

52. Some clock faces display the Roman numeral IV as IIII. Research this topic and write a few paragraphs that explain at least three possible reasons for this variation.

53. The Rhind papyrus (see page 175) contained solutions to several mathematical problems. Some of these solutions made use of a procedure called the *method of false position*. Research the method of false position and write a report that explains this method. In your report, include a specific mathematical problem and its solution by the method of false position.

Place-Value Systems

Expanded Form

historical note

Abu Ja'far Muhammad ibn Musa al'Khwarizmi (ca. 790–850 A.D.) Al'Khwarizmi (ăl′khwăhr′ĭz mee) produced two important texts. One of these texts advocated the use of the Hindu-Arabic numeration system. The twelfth century Latin translation of this book is called *Liber Algoritmi de Numero Indorum,* or *Al-Khwarizmi on the Hindu Art of Reckoning.* In Europe, the people who favored the adoption of the Hindu-Arabic numeration system became known as *algorists* because of Al'Khwarizmi's *Liber Algoritmi de Numero Indorum* text. The Europeans who opposed the Hindu-Arabic system were called *abacists.* They advocated the use of Roman numerals and often performed computations with the aid of an abacus. ■

The most common numeration system used by people today is the Hindu-Arabic numeration system. It is called the Hindu-Arabic system because it was first developed in India (around 800 A.D.) and then refined by the Arabs. It makes use of the 10 symbols 0, 1, 2, 3, 4, 5, 6, 7, 8, and 9. The reason for the 10 symbols, called *digits,* is related to the fact that we have 10 fingers. The Hindu-Arabic numeration system is also called the *decimal system,* where the word *decimal* is a derivation of the Latin word *decem,* which means "ten."

One important feature of the Hindu-Arabic numeration system is that it is a *place-value* or *positional-value system.* This means that the numerical value of each digit in a Hindu-Arabic numeral depends on its *place* or *position* in the numeral. For instance, the 3 in 31 represents 3 tens, whereas the 3 in 53 represents 3 ones. The Hindu-Arabic numeration system is a **base ten numeration system** because the place values are the powers of 10:

$$\ldots, 10^5, 10^4, 10^3, 10^2, 10^1, 10^0$$

The place value associated with the nth digit of a numeral (counting from right to left) is 10^{n-1}. For instance, in the numeral 7532, the 7 is the fourth digit from the right and is in the $10^{4-1} = 10^3$, or thousands', place. The numeral 2 is the first digit from the right and is in the $10^{1-1} = 10^0$, or ones', place. The *indicated sum* of each digit of a numeral multiplied by its respective place value is called the **expanded form** of the numeral.

EXAMPLE 1 ■ Write a Numeral in its Expanded Form

Write 4672 in expanded form.

Solution

$$4672 = 4000 + 600 + 70 + 2$$
$$= (4 \times 1000) + (6 \times 100) + (7 \times 10) + (2 \times 1)$$

The above expanded form can also be written as $(4 \times 10^3) + (6 \times 10^2) + (7 \times 10^1) + (2 \times 10^0)$.

CHECK YOUR PROGRESS 1   Write 17,325 in expanded form.

Solution   *See page S12.*

If a number is written in expanded form, it can be simplified to its ordinary decimal form by performing the indicated operations. The *Order of Operations Agreement* states that we should first perform the exponentiations, then perform the multiplications, and finish by performing the additions.

EXAMPLE 2 ■ **Simplify a Number Written in Expanded Form**

Simplify: $(2 \times 10^3) + (7 \times 10^2) + (6 \times 10^1) + (3 \times 10^0)$

Solution

$$(2 \times 10^3) + (7 \times 10^2) + (6 \times 10^1) + (3 \times 10^0)$$
$$= (2 \times 1000) + (7 \times 100) + (6 \times 10) + (3 \times 1)$$
$$= 2000 + 700 + 60 + 3$$
$$= 2763$$

CHECK YOUR PROGRESS 2 Simplify:

$$(5 \times 10^4) + (9 \times 10^3) + (2 \times 10^2) + (7 \times 10^1) + (4 \times 10^0)$$

Solution *See page S12.*

In the next few examples, we make use of the expanded form of a numeral to compute sums and differences. An examination of these examples will help you better understand the computational algorithms used in the Hindu-Arabic numeration system.

EXAMPLE 3 ■ **Use Expanded Form to Find a Sum**

Use expanded forms of 26 and 31 to find their sum.

Solution

$$26 = (2 \times 10) + 6$$
$$+\ 31 = (3 \times 10) + 1$$
$$\overline{(5 \times 10) + 7 = 50 + 7 = 57}$$

CHECK YOUR PROGRESS 3 Use expanded forms to find the sum of 152 and 234.

Solution *See page S12.*

If the expanded form of a sum contains one or more powers of 10 that have multipliers larger than 9, then we simplify by rewriting the sum with multipliers that are less than or equal to 9. This process is known as *carrying*.

EXAMPLE 4 ■ **Use Expanded Forms to Find a Sum**

Use expanded forms of 85 and 57 to find their sum.

Solution

$$85 = (8 \times 10) + 5$$
$$+\ 57 = (5 \times 10) + 7$$
$$\overline{(13 \times 10) + 12}$$
$$(10 + 3) \times 10 + 10 + 2$$
$$100 + 30 + 10 + 2 = 100 + 40 + 2 = 142$$

TAKE NOTE

From the expanded forms in Example 4, note that 12 is 1 ten and 2 ones. The 1 ten is added to the 13 tens, resulting in a total of 14 tens. When we add columns of numbers, this is shown as "carry a 1." Because the 1 is placed in the tens column, we are actually adding 10.

$$\begin{array}{r} 1 \\ 85 \\ +\ 57 \\ \hline 142 \end{array}$$

CHECK YOUR PROGRESS 4 Use expanded forms to find the sum of 147 and 329.

Solution *See page S12.*

In the next example, we use the expanded forms of numerals to analyze the concept of "borrowing" in a subtraction problem.

EXAMPLE 5 ■ Use Expanded Forms to Find a Difference

Use the expanded forms of 457 and 283 to find 457 − 283.

Solution

$$457 = (4 \times 100) + (5 \times 10) + 7$$
$$- 283 = (2 \times 100) + (8 \times 10) + 3$$

At this point, this example is similar to Example 4 in Section 4.1. We cannot remove 8 tens from 5 tens so 1 hundred is replaced by 10 tens.

$$457 = (4 \times 100) + (5 \times 10) + 7 \qquad \bullet\, 4 \times 100 = 3 \times 100 + 100$$
$$= (3 \times 100) + (10 \times 10) + (5 \times 10) + 7 \qquad = 3 \times 100 + 10 \times 10$$
$$= (3 \times 100) + (15 \times 10) + 7$$

We can now remove 8 tens from 15 tens.

$$457 = (3 \times 100) + (15 \times 10) + 7$$
$$- 283 = (2 \times 100) + (8 \times 10) + 3$$
$$= (1 \times 100) + (7 \times 10) + 4 = 100 + 70 + 4 = 174$$

CHECK YOUR PROGRESS 5 Use expanded forms to find the difference 382 − 157.

Solution *See page S12.*

✔ **TAKE NOTE**

From the expanded forms in Example 5, note that we "borrowed" 1 hundred as 10 tens. This explains how we show borrowing when numbers are subtracted using place value form.

$$\begin{array}{r} 3 \\ \cancel{4}{}^{1}57 \\ -\ 2\ 83 \\ \hline 1\ 74 \end{array}$$

The Babylonian Numeration System

The Babylonian numeration system uses a base of 60. The place values in the Babylonian system are given in the following table.

Table 4.3 *Place Values in the Babylonian Numeration System*

...	60^3 = 216,000	60^2 = 3600	60^1 = 60	60^0 = 1

The Babylonians recorded their numerals on damp clay using a wedge-shaped stylus. A vertical wedge shape represented one unit and a sideways "vee" shape represented 10 units.

Ⅰ 1

⟨ 10

To represent a number smaller than 60, the Babylonians used an *additive* feature similar to that used by the Egyptians. For example, the Babylonian numeral for 32 is

$$\text{《《《YY}$$

For the number 60 and larger numbers, the Babylonians left a small space between groups of symbols to indicate a different place value. This procedure is illustrated in the following example.

EXAMPLE 6 ■ **Write a Babylonian Numeral as a Hindu-Arabic Numeral**

Write ❮ ❮❮❮❮ ❮❮❮❮❮❮❮❮ as a Hindu-Arabic numeral.

Solution

❮	《《《Y	《《YYYYY
1 group of 60^2	31 groups of 60	25 ones

$$= (1 \times 60^2) + (31 \times 60) + (25 \times 1)$$
$$= 3600 + 1860 + 25$$
$$= 5485$$

CHECK YOUR PROGRESS 6 Write 《《Y YYYYY 《《《YYYY as a Hindu-Arabic numeral.

Solution See page S12.

QUESTION In the Babylonian numeration system, does $YY = Y\ Y$?

In the next example we illustrate a division process that can be used to convert Hindu-Arabic numerals to Babylonian numerals.

EXAMPLE 7 ■ **Write a Hindu-Arabic Numeral as a Babylonian Numeral**

Write 8503 as a Babylonian numeral.

Solution

The Babylonian numeration system uses place values of

$$60^0, 60^1, 60^2, 60^3, \ldots.$$

Evaluating the powers produces

$$1, 60, 3600, 216{,}000, \ldots$$

The largest of these powers that is contained in 8503 is 3600. One method of finding how many groups of 3600 are in 8503 is to divide 3600 into 8503. Refer to the

ANSWER *No.* $YY = 2$, *whereas* $Y\ Y = (1 \times 60) + (1 \times 1) = 61$.

first division shown below. Now divide to determine how many groups of 60 are contained in the remainder 1303.

$$
\begin{array}{r}
2 \\
3600{\overline{\smash{\big)}\,8503}} \\
\underline{7200} \\
1303
\end{array}
\qquad
\begin{array}{r}
21 \\
60{\overline{\smash{\big)}\,1303}} \\
\underline{120} \\
103 \\
\underline{60} \\
43
\end{array}
$$

The above computations show that 8503 consists of 2 groups of 3600 and 21 groups of 60, with 43 left over. Thus

$$8503 = (2 \times 60^2) + (21 \times 60^1) + (43 \times 60^0)$$

As a Babylonian numeral, 8503 is written

$$\text{YY} \quad \text{≪Y} \quad \text{≪≪≪YYY}$$

CHECK YOUR PROGRESS 7   Write 12,578 as a Babylonian numeral.

Solution   *See page S12.*

Math Matters   Zero as a Placeholder and as a Number

When the Babylonian numeration system first began to develop around 1700 B.C., it did not make use of a symbol for zero. The Babylonians merely used an empty space to indicate that a place value was missing. This procedure of "leaving a space" can be confusing. How big is an empty space? Is that one empty space or two empty spaces? Around 300 B.C., the Babylonians started to use the symbol ▲ to indicate that a particular place value was missing. For instance, YY ▲ ◁Y represented $(2 \times 60^2) + (11 \times 1) = 7211$. In this case the zero placeholder indicates that there are no 60's. There is evidence that although the Babylonians used the zero placeholder, they did not use the number zero.

The Mayan Numeration System

The Mayan civilization existed in the Yucatan area of southern Mexico and in Guatemala, Belize, and parts of El Salvador and Honduras. It started as far back as 9000 B.C. and reached its zenith during the period from 200 A.D. to 900 A.D. Among their many accomplishments, the Maya are best known for their complex hieroglyphic writing system, their sophisticated calendars, and their remarkable numeration system.

The Maya used three calendars—the solar calendar, the ceremonial calendar, and the Venus calendar. The solar calendar consisted of about 365.24 days. Of these, 360 days were divided into 18 months, each with 20 days. The Mayan numeration system was strongly influenced by this solar calendar, as evidenced by the use of the numbers 18 and 20 in determining place values. See Table 4.4

Table 4.4 *Place Values in the Mayan Numeration system*

	18×20^3	18×20^2	18×20^1	20^1	20^0
⋯	$= 144{,}000$	$= 7200$	$= 360$	$= 20$	$= 1$

The Mayan numeration system was one of the first systems to use a symbol for zero as a placeholder. The Mayan numeration system used only three symbols. A dot was used to represent 1, a horizontal bar represented 5, and a conch shell represented 0. The following table shows how the Maya used a combination of these three symbols to write the whole numbers from 0 to 19. Note that each numeral contains at most four dots and at most three horizontal bars.

Table 4.5 *Mayan Numerals*

To write numbers larger than 19, the Maya used a vertical arrangement with the largest place value at the top. The following example illustrates the process of converting a Mayan numeral to a Hindu-Arabic numeral.

EXAMPLE 8 ■ Write a Mayan Numeral as a Hindu-Arabic Numeral

Write each of the following as a Hindu-Arabic numeral.

a.   b.

Solution

a.
$$10 \times 360 = 3600$$
$$8 \times 20 = 160$$
$$11 \times 1 = +11$$
$$\overline{3771}$$

b.
$$5 \times 7200 = 36{,}000$$
$$0 \times 360 = 0$$
$$12 \times 20 = 240$$
$$3 \times 1 = + \ 3$$
$$\overline{36{,}243}$$

CHECK YOUR PROGRESS 8   Write each of the following as a Hindu-Arabic numeral.

a.
```
  ·
  ══
  ⟨●⟩
  ·
```

b.
```
····
  ·
══
····
```

Solution   *See page S12.*

In the next example, we illustrate how the concept of place value is used to convert Hindu-Arabic numerals to Mayan numerals.

EXAMPLE 9 ■ Write a Hindu-Arabic Numeral as a Mayan Numeral

Write 7495 as a Mayan numeral.

Solution

The place values used in the Mayan numeration system are

$$20^0, 20^1, 18 \times 20^1, 18 \times 20^2, 18 \times 20^3, \ldots$$

or

$$1, 20, 360, 7200, 144{,}000, \ldots$$

Removing 1 group of 7200 from 7495 leaves 295. No groups of 360 can be obtained from 295, so we divide 295 by the next smaller place value of 20 to find that 295 equals 14 groups of 20 with 15 left over.

$$
\begin{array}{r}
1 \\
7200\overline{)7495} \\
7200 \\
\hline
295
\end{array}
\qquad
\begin{array}{r}
14 \\
20\overline{)295} \\
20 \\
\hline
95 \\
80 \\
\hline
15
\end{array}
$$

Thus

$$7495 = (1 \times 7200) + (0 \times 360) + (14 \times 20) + (15 \times 1)$$

In Mayan numbers, 7495 is written as

```
      ·

    ⟨●⟩

    ····

    ══
```

CHECK YOUR PROGRESS 9   Write 11,480 as a Mayan numeral.

Solution   *See page S12.*

Excursion

Subtraction via the Nines Complement and the End-Around Carry

INSTRUCTOR NOTE
The concepts in this Excursion are extended in the Excursion in Section 4.4, page 212.

In the subtraction $5627 - 2564 = 3063$, the number 5627 is called the *minuend*, 2564 is called the *subtrahend*, and 3063 is called the *difference.* In the Hindu-Arabic base ten system, subtraction can be performed by a process that involves addition and the *nines complement* of the subtrahend. The **nines complement** of a single digit n is the number $9 - n$. For instance, the nines complement of 3 is 6, the nines complement of 1 is 8, and the nines complement of 0 is 9. The nines complement of a number with more than one digit is the number that is formed by taking the nines complement of each digit. The nines complement of 25 is 74 and the nines complement of 867 is 132.

> **Subtraction by Using the Nines Complement and the End-Around Carry**
>
> To subtract by using the nines complement:
>
> **1.** Add the nines complement of the subtrahend to the minuend.
>
> **2.** Take away 1 from the leftmost digit and add 1 to the units digit. This is referred to as the end-around carry procedure.

The following example illustrates the process of subtracting 2564 from 5627 by using the nines complement.

$$
\begin{array}{r}
5627 \\
- \ 2564 \\
\end{array}
\quad
\begin{array}{l}
\text{Minuend} \\
\text{Subtrahend}
\end{array}
$$

$$
\begin{array}{r}
5627 \\
+ \ 7435 \\
\hline
13062 \\
\end{array}
\quad
\begin{array}{l}
\text{Minuend} \\
\text{Replace the subtrahend with the nines complement} \\
\text{of the subtrahend and add.}
\end{array}
$$

$$
\begin{array}{r}
13062 \\
+ \quad\ \ 1 \\
\hline
3063 \\
\end{array}
\quad
\begin{array}{l}
\text{Take away 1 from the leftmost digit and add 1 to} \\
\text{the units digit. This is the end-around carry procedure.}
\end{array}
$$

Thus

$$
\begin{array}{r}
5627 \\
- \ 2564 \\
\hline
3063 \\
\end{array}
$$

If the subtrahend has fewer digits than the minuend, leading zeros should be inserted in the subtrahend so that it has the same number of digits as the minuend. This process is illustrated below for $2547 - 358$.

$$
\begin{array}{r}
2547 \\
- \ \ 358 \\
\end{array}
\quad
\begin{array}{l}
\text{Minuend} \\
\text{Subtrahend}
\end{array}
$$

$$
\begin{array}{r}
2547 \\
- \ 0358 \\
\end{array}
\quad
\text{Insert a leading zero.}
$$

(continued)

$$\begin{array}{r} 2547 \\ + \ \underline{9641} \\ 12188 \end{array}$$
Minuend
Nines complement of subtrahend

$$\begin{array}{r} 12188 \\ + \ \underline{\quad 1} \\ 2189 \end{array}$$
Take away 1 from the leftmost digit and add 1 to the units digit.

Verify that 2189 is the correct difference.

Excursion Exercises

For Exercises 1−6, use the nines complement of the subtrahend to find the indicated difference.

1. 724 − 351

2. 2405 − 1608

3. 91,572 − 7824

4. 214,577 − 48,231

5. 3,156,782 − 875,236

6. 54,327,105 − 7,678,235

7. Explain why the nines complement and the end-around carry procedure produce the correct answer to a subtraction problem.

Exercise Set 4.2   (Suggested Assignment: 1−55 odds)

In Exercises 1−8, write each numeral in its expanded form.

1. 48

2. 93

3. 420

4. 501

5. 6803

6. 9045

7. 10,208

8. 67,482

In Exercises 9−16, simplify each expansion.

9. $(4 \times 10^2) + (5 \times 10^1) + (6 \times 10^0)$

10. $(7 \times 10^2) + (6 \times 10^1) + (3 \times 10^0)$

11. $(5 \times 10^3) + (7 \times 10^1) + (6 \times 10^0)$

12. $(3 \times 10^3) + (1 \times 10^2) + (2 \times 10^1) + (8 \times 10^0)$

13. $(3 \times 10^4) + (5 \times 10^3) + (4 \times 10^2) + (7 \times 10^0)$

14. $(2 \times 10^5) + (3 \times 10^4) + (6 \times 10^2) + (7 \times 10^1) + (5 \times 10^0)$

15. $(6 \times 10^5) + (8 \times 10^4) + (3 \times 10^3) + (0 \times 10^2) + (4 \times 10^1)$

16. $(5 \times 10^7) + (3 \times 10^6) + (7 \times 10^3) + (9 \times 10^2) + (2 \times 10^0)$

In Exercises 17−20, use expanded forms to find each sum.

17. 35 + 41

18. 42 + 56

19. 257 + 138

20. 352 + 461

In Exercises 21−24, use expanded forms to find each difference.

21. 62 − 35

22. 193 − 157

23. 4725 − 1362

24. 85,381 − 64,156

In Exercises 25−32, write each Babylonian numeral as a Hindu-Arabic numeral.

25. ⟨⟨𝌆𝌆𝌆

26. ⟨⟨⟨⟨𝌆𝌆𝌆𝌆𝌆

27. 𝌆 ⟨⟨⟨𝌆𝌆𝌆𝌆𝌆𝌆𝌆

28. ⟨𝌆𝌆 ⟨⟨𝌆𝌆𝌆𝌆𝌆𝌆

29. ⟨⟨ 𝌆𝌆 ⟨𝌆𝌆𝌆

30. ⟪𒁹 ⟨𒁹 ⟨𒁹𒁹

31. ⟨ 𒁹𒁹𒁹 ⟨𒁹 𒁹𒁹𒁹𒁹𒁹𒁹

32. ⟪𒁹 ⟨𒁹 𒁹 ⟪⟨𒁹𒁹𒁹𒁹

In Exercises 33–40, write each Hindu-Arabic numeral as a Babylonian numeral.

33. 42   **34.** 57

35. 128   **36.** 540

37. 5678   **38.** 7821

39. 10,584   **40.** 12,687

In Exercises 41–48, write each Mayan numeral as a Hindu-Arabic numeral.

41.   **42.**

43.   **44.**

45.   **46.**

47.   **48.**

In Exercises 49–56, write each Hindu-Arabic numeral as a Mayan numeral.

49. 137   **50.** 253

51. 948   **52.** 1265

53. 1693   **54.** 2728

55. 7432   **56.** 8654

Extensions

CRITICAL THINKING

57. a. State a reason why you might prefer to use the Babylonian numeration system instead of the Mayan numeration system.

    b. State a reason why you might prefer to use the Mayan numeration system instead of the Babylonian numeration system.

58. Explain why it might be easy to mistake the number 122 for the number 4 when 122 is written as a Babylonian numeral.

EXPLORATIONS

59. A student has created a *base three* numeration system. The student has named this numeration system ZUT because Z, U, and T are the symbols used in this system: Z represents 0, U represents 1, and T represents 2. The place values in this system are: $\ldots, 3^3 = 27, 3^2 = 9, 3^1 = 3, 3^0 = 1$.

Write each ZUT numeral as a Hindu-Arabic numeral.

    a. TU   **b.** TZT   **c.** UZTT

Write each Hindu-Arabic numeral as a ZUT numeral.

    d. 37   **e.** 87   **f.** 144

SECTION 4.3 | ## Different Base Systems

Converting Non-Base-Ten Numerals to Base Ten

✔ **TAKE NOTE**

Recall that in the expression

$$b^n$$

b is the *base*, *n* is the *exponent*, and b^n is the *power*.

Recall that the Hindu-Arabic numeration system is a base ten system because its place values

$$\ldots, 10^5, 10^4, 10^3, 10^2, 10^1, 10^0$$

all have 10 as their base. The Babylonian numeration system is a base sixty system because its place values

$$\ldots, 60^5, 60^4, 60^3, 60^2, 60^1, 60^0$$

all have 60 as their base. In general, a base b (where b is a natural number greater than 1) numeration system has place values of

$$\ldots, b^5, b^4, b^3, b^2, b^1, b^0$$

Many people think that our base ten numeration system was chosen because it is the easiest to use, but this is not the case. In reality most people find it easier to use our base ten system only because they have had a great deal of experience with the base ten system and have not had much experience with non-base-ten systems. In this section, we examine some non-base-ten numeration systems. To reduce the amount of memorization that would be required to learn new symbols for each of these new systems, we will (as far as possible) make use of our familiar Hindu-Arabic symbols. For instance, if we discuss a base four numeration system that requires four basic symbols, then we will use the four Hindu-Arabic symbols 0, 1, 2, and 3 and the place values

$$\ldots, 4^5, 4^4, 4^3, 4^2, 4^1, 4^0$$

The base eight, or **octal,** numeration system uses the Hindu-Arabic symbols 0, 1, 2, 3, 4, 5, 6, and 7 and the place values

$$\ldots, 8^5, 8^4, 8^3, 8^2, 8^1, 8^0$$

To differentiate between bases, we will label each non-base-ten numeral with a subscript that indicates the base. For instance, 23_{four} represents a base four numeral. If a numeral is written without a subscript, then it is understood that the base is ten. Thus 23 written without a subscript is understood to be the base ten numeral 23.

To convert a non-base-ten numeral to base ten, we write the numeral in its expanded form, as shown in the following example.

EXAMPLE 1 ■ Convert to Base Ten

Convert 2314_{five} to base ten.

Solution

In the base five numeration system, the place values are

$$\ldots, 5^4, 5^3, 5^2, 5^1, 5^0$$

The expanded form of 2314_{five} is

$$
\begin{aligned}
2314_{five} &= (2 \times 5^3) + (3 \times 5^2) + (1 \times 5^1) + (4 \times 5^0) \\
&= (2 \times 125) + (3 \times 25) + (1 \times 5) + (4 \times 1) \\
&= 250 + 75 + 5 + 4 \\
&= 334
\end{aligned}
$$

Thus $2314_{five} = 334$.

CHECK YOUR PROGRESS 1   Convert 3156_{seven} to base ten.

Solution   See page S12.

QUESTION   *Does the notation 26_{five} make sense?*

ANSWER   *No. The expression 26_{five} is a meaningless expression because there is no 6 in base five.*

In base two, which is called the **binary numeration system,** the place values are the powers of two.

$$\dots, 2^7, 2^6, 2^5, 2^4, 2^3, 2^2, 2^1, 2^0$$

The binary numeration system uses only the two digits 0 and 1. These *bi*nary dig*its* are often called **bits.** To convert a base two numeral to base ten, write the numeral in its expanded form and then evaluate the expanded form.

EXAMPLE 2 ■ Convert to Base Ten

Convert 10110111_{two} to base ten.

Solution
$$\begin{aligned}
10110111_{two} &= (1 \times 2^7) + (0 \times 2^6) + (1 \times 2^5) + (1 \times 2^4) + (0 \times 2^3) \\
&\quad + (1 \times 2^2) + (1 \times 2^1) + (1 \times 2^0) \\
&= (1 \times 128) + (0 \times 64) + (1 \times 32) + (1 \times 16) + (0 \times 8) \\
&\quad + (1 \times 4) + (1 \times 2) + (1 \times 1) \\
&= 128 + 0 + 32 + 16 + 0 + 4 + 2 + 1 \\
&= 183
\end{aligned}$$

CHECK YOUR PROGRESS 2   Convert 111000101_{two} to base ten.

Solution   *See page S12.*

> ✔ **TAKE NOTE**
>
> The base twelve numeration system is called the **duodecimal system.** A group called the Dozenal Society of America advocates the replacement of our base ten decimal system with the duodecimal system. If you wish to find out more about this organization, you can contact them at: The Dozenal Society of America, Nassau Community College, Garden City, New York. Not surprisingly, the dues are $12 per year and $144 ($12^2 = 144$) for a lifetime membership.

The base twelve numeration system requires 12 distinct symbols. We will use the symbols 0, 1, 2, 3, 4, 5, 6, 7, 8, 9, A, and B as our base twelve numeration system symbols. The symbols 0 through 9 have their usual meaning; however, A is used to represent 10 and B to represent 11.

EXAMPLE 3 ■ Convert to Base Ten

Convert $B37_{twelve}$ to base ten.

Solution
In the base twelve numeration system the place values are

$$\dots, 12^4, 12^3, 12^2, 12^1, 12^0$$

Thus
$$\begin{aligned}
B37_{twelve} &= (11 \times 12^2) + (3 \times 12^1) + (7 \times 12^0) \\
&= 1584 + 36 + 7 \\
&= 1627
\end{aligned}$$

CHECK YOUR PROGRESS 3   Convert $A5B_{twelve}$ to base ten.

Solution   *See page S12.*

Computer programmers often write programs that use the base sixteen numeration system, which is also called the **hexadecimal system.** This system uses the symbols 0, 1, 2, 3, 4, 5, 6, 7, 8, 9, A, B, C, D, E, and F. Table 4.6 shows that A represents 10, B represents 11, C represents 12, D represents 13, E represents 14, and F represents 15.

Table 4.6 *Decimal and Hexadecimal Equivalents*

Base Ten Decimal	Base Sixteen Hexadecimal
0	0
1	1
2	2
3	3
4	4
5	5
6	6
7	7
8	8
9	9
10	A
11	B
12	C
13	D
14	E
15	F

EXAMPLE 4 ■ **Convert to Base Ten**

Convert $3E8_{sixteen}$ to base ten.

Solution

In the base sixteen numeration system the place values are

$$\ldots, 16^4, 16^3, 16^2, 16^1, 16^0$$

Thus

$$3E8_{sixteen} = (3 \times 16^2) + (14 \times 16^1) + (8 \times 16^0)$$
$$= 768 + 224 + 8$$
$$= 1000$$

CHECK YOUR PROGRESS 4   Convert $C24F_{sixteen}$ to base ten.

Solution   *See page S12.*

Converting from Base Ten to Another Base

The most efficient method of converting a number written in base ten to another base makes use of a *successive division process*. For example, to convert 219 to base four, divide 219 by 4 and write the quotient 54 and the remainder 3, as shown below. Now divide the quotient 54 by the base to get a new quotient of 13 and a new remainder of 2. Continuing the process, divide the quotient 13 by 4 to get a new quotient of 3 and a remainder of 1. Because our last quotient, 3, is less than the base, 4, we stop the division process. The answer is given by the last quotient, 3, and the remainders, shown in red in the following diagram. That is, $219 = 3123_{four}$.

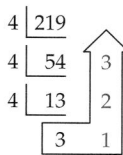

You can understand how the successive division process converts a base ten numeral to another base by analyzing the process. The first division shows there are 54 fours in 219, with **3 ones** left over. The second division shows that there are 13 sixteens (two successive divisions by 4 is the same as dividing by 16) in 219, and the remainder 2 indicates that there are **2 fours** left over. The last division shows that there are 3 sixty-fours (three successive divisions by 4 is the same as dividing by 64) in 219, and the remainder 1 indicates that there is **1 sixteen** left over. In mathematical notation these results are written as follows.

$$219 = (3 \times 64) + (1 \times 16) + (2 \times 4) + (3 \times 1)$$
$$= (3 \times 4^3) + (1 \times 4^2) + (2 \times 4^1) + (3 \times 4^0)$$
$$= 3123_{four}$$

EXAMPLE 5 ■ **Convert a Base Ten Numeral to Another Base**

Convert 5821 to **a.** base three and **b.** base sixteen.

Solution

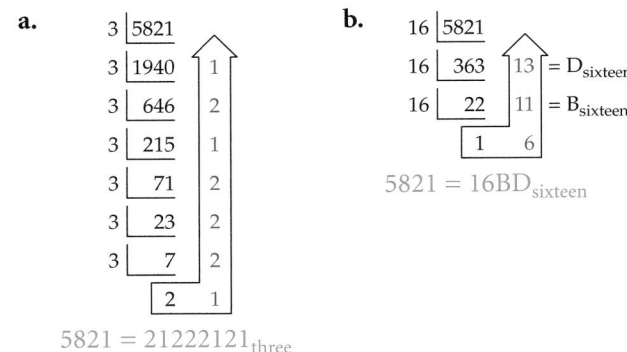

a.

$$
\begin{array}{r|r}
3 & 5821 \\
\hline
3 & 1940 \quad 1 \\
\hline
3 & 646 \quad 2 \\
\hline
3 & 215 \quad 1 \\
\hline
3 & 71 \quad 2 \\
\hline
3 & 23 \quad 2 \\
\hline
3 & 7 \quad 2 \\
\hline
& 2 \quad 1 \\
\end{array}
$$

$$5821 = 21222121_{\text{three}}$$

b.

$$
\begin{array}{r|r}
16 & 5821 \\
\hline
16 & 363 \quad 13 = D_{\text{sixteen}} \\
\hline
16 & 22 \quad 11 = B_{\text{sixteen}} \\
\hline
& 1 \quad 6 \\
\end{array}
$$

$$5821 = 16BD_{\text{sixteen}}$$

CHECK YOUR PROGRESS 5   Convert 1952 to **a.** base five and **b.** base twelve.

Solution   *See page S13.*

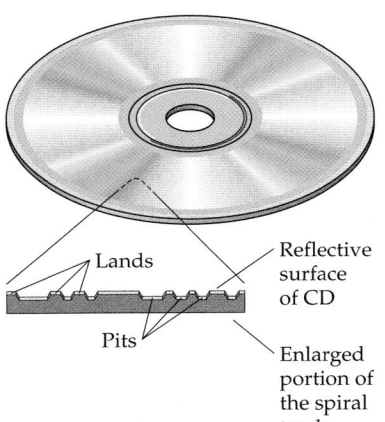

Lands
Pits
Reflective surface of CD
Enlarged portion of the spiral track

Table 4.7 *Octal and Binary Equivalents*

Octal	Binary
0	000
1	001
2	010
3	011
4	100
5	101
6	110
7	111

Math Matters   Music by the Numbers

The binary numeration system is used to encode music on a CD (compact disc). The figure at the left shows the surface of a CD, which consists of flat regions called *lands* and small indentations called *pits*. As a laser beam tracks along a spiral path, the beam is reflected to a sensor when it shines on a land, but is not reflected to the sensor when it shines on a pit. The sensor interprets a reflection as a 1 and no reflection as a 0. As the CD is playing, the sensor receives a series of 1's and 0's, which the CD player converts to music. On a typical CD, the spiral path that the laser follows loops around the disc over 20,000 times and contains about 650 megabytes of data. A **byte** is eight bits, so this amounts to 5,200,000,000 bits, each of which is represented by a pit or a land.

Converting Directly Between Computer Bases

Although computers compute internally by using base two (binary system), humans generally find it easier to compute with a larger base. Fortunately, there are easy conversion techniques that can be used to convert a base two numeral directly to a base eight (octal) numeral or a base sixteen (hexadecimal) numeral. Before we explain the techniques, it will help to become familiar with the information in Table 4.7, which shows the eight octal symbols and their binary equivalents.

To convert from octal to binary, just replace each octal symbol with its three-bit binary equivalent.

EXAMPLE 6 ■ **Convert Directly from Base Eight to Base Two**

Convert 5724_{eight} directly to binary form.

Solution

$$
\begin{array}{cccc}
5 & 7 & 2 & 4_{eight} \\
\| & \| & \| & \| \\
101 & 111 & 010 & 100_{two}
\end{array}
$$

$5724_{eight} = 101111010100_{two}$

CHECK YOUR PROGRESS 6   Convert 63210_{eight} directly to binary form.

Solution   *See page S13.*

Because every group of three binary bits is equivalent to an octal symbol, we can convert from binary directly to octal by breaking a binary numeral into groups of three (from right to left) and replacing each group with its octal equivalent.

EXAMPLE 7 ■ **Convert Directly from Base Two to Base Eight**

Convert 11100101_{two} directly to octal form.

Solution

Starting from the right, break the binary numeral into groups of three. Then replace each group with its octal equivalent.

┌─── This zero was inserted to
│   make a group of three.
↓

$$
\begin{array}{ccc}
011 & 100 & 101_{two} \\
\| & \| & \| \\
3 & 4 & 5_{eight}
\end{array}
$$

$11100101_{two} = 345_{eight}$

CHECK YOUR PROGRESS 7   Convert 111010011100_{two} directly to octal form.

Solution   *See page S13.*

Table 4.8 shows the hexadecimal symbols and their binary equivalents. To convert from hexadecimal to binary, replace each hexadecimal symbol with its four-bit binary equivalent.

EXAMPLE 8 ■ **Convert Directly from Base Sixteen to Base Two**

Convert $BAD_{sixteen}$ directly to binary form.

Solution

$$
\begin{array}{ccc}
B & A & D_{sixteen} \\
\| & \| & \| \\
1011 & 1010 & 1101_{two}
\end{array}
$$

$BAD_{sixteen} = 101110101101_{two}$

Table 4.8 *Hexadecimal and Binary Equivalents*

Hexadecimal	Binary
0	0000
1	0001
2	0010
3	0011
4	0100
5	0101
6	0110
7	0111
8	1000
9	1001
A	1010
B	1011
C	1100
D	1101
E	1110
F	1111

CHECK YOUR PROGRESS 8   Convert C5A$_{sixteen}$ directly to binary form.

Solution   *See page S13.*

Because every group of four binary bits is equivalent to a hexadecimal symbol, we can convert from binary to hexadecimal by breaking the binary numeral into groups of four (from right to left) and replacing each group with its hexadecimal equivalent.

EXAMPLE 9 ■ Convert Directly from Base Two to Base Sixteen

Convert 10110010100011$_{two}$ directly to hexadecimal form.

Solution
Starting from the right, break the binary numeral into groups of four. Replace each group with its hexadecimal equivalent.

Insert two zeros to make a group of four.

$$0010 \quad 1100 \quad 1010 \quad 0011_{two}$$
$$\| \qquad \| \qquad \| \qquad \|$$
$$2 \qquad C \qquad A \qquad 3$$

10110010100011$_{two}$ = 2CA3$_{sixteen}$

CHECK YOUR PROGRESS 9   Convert 101000111010010$_{two}$ directly to hexadecimal form.

Solution   *See page S13.*

The Double-Dabble Method

There is a short cut that can be used to convert a base two numeral to base ten. The advantage of this short cut, called the *double-dabble method,* is that you can start at the left of the numeral and work your way to the right without first determining the place value of each bit in the base two numeral.

EXAMPLE 10 ■ Apply the Double-Dabble Method

Use the double-dabble method to convert 1011001$_{two}$ to base ten.

Solution
Start at the left with the first 1 and move to the right. Every time you pass by a 0, double your current number. Every time you pass by a 1, dabble. Dabbling is accomplished by doubling your current number and adding 1.

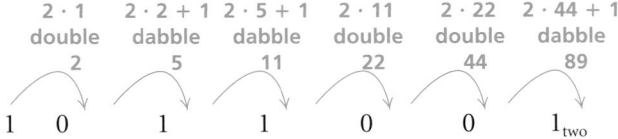

$2 \cdot 1$	$2 \cdot 2 + 1$	$2 \cdot 5 + 1$	$2 \cdot 11$	$2 \cdot 22$	$2 \cdot 44 + 1$
double	dabble	dabble	double	double	dabble
2	5	11	22	44	89

1   0   1   1   0   0   1$_{two}$

As we pass by the final 1 in the units place, we dabble 44 to get 89. Thus 1011001$_{two}$ = 89.

CHECK YOUR PROGRESS 10 Use the double-dabble method to convert 1110010_{two} to base ten.

Solution *See page S13.*

Excursion

Information Retrieval via a Binary Search

To complete this Excursion, you must first construct a set of 31 cards that we refer to as a deck of *binary cards*. Templates for constructing the cards are available at our web site, **www.hmco.com,** under the file name binarycards. Use a computer to print the templates onto a medium-weight card stock similar to that used for playing cards. Specific directions are provided with the templates.

We are living in the information age, but information is not useful if it cannot be retrieved when you need it. The binary numeration system is vital to the retrieval of information. To illustrate the connection between retrieval of information and the binary system, examine the card in the following figure. The card is labeled with the base ten numeral 20, and the holes and notches at the top of the card represent 20 in binary notation. A hole is used to indicate a 1 and a notch is used to indicate a 0. In the figure, the card has holes in the third and fifth binary-place-value positions (counting from right to left) and notches cut out of the first, second, and fourth positions.

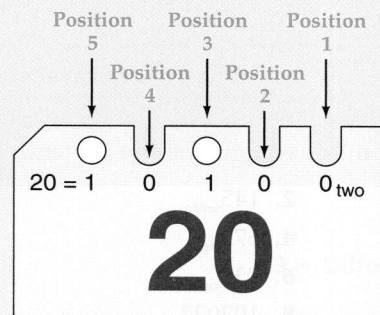

After you have constructed your deck of binary cards, take a few seconds to shuffle the deck. To find the card labeled with the numeral 20, complete the following process.

1. Use a thin dowel (or the tip of a sharp pencil) to lift out the cards that have a hole in the fifth position. *Keep* these cards and set the other cards off to the side.

2. From the cards that are *kept,* use the dowel to lift out the cards with a hole in the fourth position. Set these cards off to the side.

3. From the cards that are *kept,* use the dowel to lift out the cards that have a hole in the third position. *Keep* these cards and place the others off to the side.

(continued)

(without first converting to base ten) to the indicated base.

41. 352_{eight} to base two

42. $A4_{sixteen}$ to base two

43. 11001010_{two} to base eight

44. 111011100101_{two} to base sixteen

a. Numbers written in base two are divisible by 2 if and only if the number ends with a 0.

b. In base six, the next counting number after 55_{six} is 100_{six}.

c. In base sixteen, the next counting number after $3BF_{sixteen}$ is $3C0_{sixteen}$.

Extensions

CRITICAL THINKING

In the computer game *Riven*, a D'ni numeration system is used. Although the D'ni numeration system is a base twenty-five numeration system with 25 distinct numerals, you really need to memorize only the first five numerals, which are shown below.

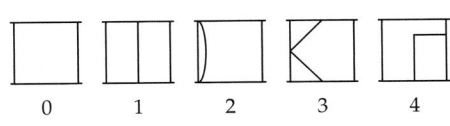

The basic D'ni numerals

If two D'ni numerals are placed side-by-side, then the numeral on the left is in the twenty-fives' place and the numeral on the right is in the ones' place. Thus

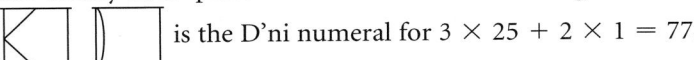 is the D'ni numeral for $3 \times 25 + 2 \times 1 = 77$.

51. Convert the following D'ni numeral to base ten.

52. Convert the following D'ni numeral to base ten.

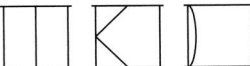

4. From the cards that are *kept,* use the dowel to lift out the cards with a hole in the second position. Set these cards off to the side.

5. From the cards that are *kept,* use the dowel to lift out the card that has a hole in the first position. Set this card off to the side.

The card that remains is the card labeled with the numeral 20. You have just completed a binary search.

Binary Sort

The binary numeration system can also be used to implement a *sorting* procedure. To illustrate, shuffle your deck of cards. Use the dowel to lift out the cards that have a hole in the rightmost position. Place these cards, *face up,* in the back of the other cards. Now use the dowel to lift out the cards that have a hole in the next position to the left. Place these cards, face up, in back of the other cards. Continue this process of lifting out the cards in the next position to the left and placing them in back of the other cards until you have completed the process for all five positions.

Excursion Exercises

1. Examine the numerals on the cards. What do you notice about the order of the numerals? Explain why they are in this order.

2. If you wanted to sort 1000 cards from smallest to largest value by using the binary sort procedure, how many positions (where each position is either a hole or a notch) would be required at the top of each card? How many positions are needed to sort 10,000 cards?

3. Explain why the above sorting procedure cannot be implemented with base three cards.

Exercise Set 4.3   (Suggested Assignment: 1–59 odds)

In Exercises 1–10, convert the given numeral to base ten.   In Exercises 21–24, use expanded forms to convert the given base two numeral to base ten.

Rotating any of the D'ni numerals for 1, 2, 3, and 4 by a 90° counterclockwise rotation produces a numeral with a value five times its original value. For instance, rotating the numeral for 1 produces ⬚, which is the D'ni numeral for 5, and

rotating the numeral for 2 produces ⬚, which is the D'ni numeral for 10.

53. Write the D'ni numeral for 15.   **54.** Write the D'ni numeral for 20.

In the D'ni numeration system, explained above, many numerals are obtained by rotating a basic numeral and then overlaying it on one of the basic numerals. For instance, if you rotate the D'ni numeral for 1, you get the numeral for 5. If you then overlay the numeral for 5 on the numeral for 1, you get the numeral for 5 + 1 = 6.

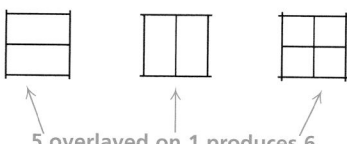

5 overlayed on 1 produces 6.

55. Write the D'ni numeral for 8.   **56.** Write the D'ni numeral for 22.
57. Convert the following D'ni numeral to base ten.   **58.** Convert the following D'ni numeral to base ten.

59. a. State one advantage of the hexadecimal numeration system over the decimal numeration system.

 b. State one advantage of the decimal numeration system over the hexadecimal numeration system.

60. a. State one advantage of the D'ni numeration system over the decimal numeration system.

 b. State one advantage of the decimal numeration system over the D'ni numeration system.

EXPLORATIONS

61. ASCII, pronounced *ask-key,* is an acronym for the American Standard Code for Information Interchange. In this code, each of the characters that can be typed on a computer keyboard is represented by a number. For instance, the letter A is assigned the number 65, which when written as an 8-bit binary numeral is 01000001. Research the topic of ASCII. Write a report about ASCII and its applications.

62. The U.S. Postal Service uses a *Postnet code* to write zip codes + 4 on envelopes. The Postnet code is a bar code that is based on the binary numeration system. Postnet code is very useful because it can be read by a machine. Write a few paragraphs that explain how to convert a zip code + 4 to its Postnet code. What is the Postnet code for your zip code + 4?

Erin Q. Smith
1836 First Avenue
Escondido, CA
92027-4405

SECTION 4.4 | ## Arithmetic in Different Bases

Addition in Different Bases

Most computers and calculators make use of the base two (binary) numeration system to perform arithmetic computations. For instance, if you use a calculator to find the sum of 9 and 5, the calculator first converts the 9 to 1001_{two} and the 5 to 101_{two}. The calculator uses electronic circuitry called *binary adders* to find the sum of 1001_{two} and 101_{two} as 1110_{two}. The calculator then converts 1110_{two} to base ten and displays the sum 14. All of the conversions and the base two addition are done internally in a fraction of a second, which gives the user the impression that the calculator performed the addition in base ten.

The following examples illustrate how to perform arithmetic in different bases. We first consider the operation of addition in the binary numeration system. Table 4.9 is an addition table for base two. It is similar to the base ten addition table that you memorized in elementary school, except that it is much smaller because base two involves only the bits 0 and 1. The numbers shown in red in Table 4.9 illustrate that $1_{two} + 1_{two} = 10_{two}$.

Table 4.9 *A Binary Addition Table*

Second addend

+	0	1
0	0	1
1	1	10

First addend

Sums

EXAMPLE 1 ■ Add Base Two Numerals

Find the sum of 11110_{two} and 1011_{two}.

Solution

Arrange the numbers vertically, keeping the bits of the same place value in the same column.

$$
\begin{array}{r}
\,1\;\;1\;\;1\;\;1\;\;0_{two} \\
+\;\;\;\;\,1\;\;0\;\;1\;\;1_{two} \\
\hline
1_{two}
\end{array}
$$

Start by adding the bits in the ones' column: $0_{two} + 1_{two} = 1_{two}$. Then move left and add the bits in the twos' column. When the sum of the bits in a column exceeds 1, the addition will involve carrying, as shown below.

$$
\begin{array}{r}
{}^{1} \\
\,1\;\;1\;\;1\;\;1\;\;0_{two} \\
+\;\;\;\;\,1\;\;0\;\;1\;\;1_{two} \\
\hline
0\;\;1_{two}
\end{array}
$$

Add the bits in the twos' column.
$1_{two} + 1_{two} = 10_{two}$
Write the 0 in the twos' column and carry the 1 to the fours' column.

$$
\begin{array}{r}
{\scriptstyle 1 \quad 1}\\
1\ 1\ 1\ 1\ 0_{two}\\
+\qquad 1\ 0\ 1\ 1_{two}\\
\hline
0\ 0\ 1_{two}
\end{array}
$$

Add the bits in the fours' column.
$(1_{two} + 1_{two}) + 0_{two} = 10_{two} + 0_{two} = 10_{two}$
Write the 0 in the fours' column and carry the 1 to the eights' column.

$$
\begin{array}{r}
{\scriptstyle 1 \quad 1 \quad 1 \quad 1}\\
1\ 1\ 1\ 1\ 0_{two}\\
+\qquad 1\ 0\ 1\ 1_{two}\\
\hline
1\ 0\ 1\ 0\ 0\ 1_{two}
\end{array}
$$

Add the bits in the eights' column.
$(1_{two} + 1_{two}) + 1_{two} = 10_{two} + 1_{two} = 11_{two}$
Write a 1 in the eights' column and carry a 1 to the sixteens' column. Continue to add the bits in each column to the left of the eights' column.

The sum of 11110_{two} and 1011_{two} is 101001_{two}.

CHECK YOUR PROGRESS 1 Find the sum of 11001_{two} and 1101_{two}.

Solution *See page S13.*

Table 4.10 *A Base Four Addition Table*

+	0	1	2	3
0	0	1	2	3
1	1	2	3	10
2	2	3	10	11
3	3	10	11	12

There are four symbols in base four, namely 0, 1, 2, and 3. Table 4.10 shows a base four addition table that lists all the sums that can be produced by adding two base four digits. The numbers shown in red in Table 4.10 illustrate that $2_{four} + 3_{four} = 11_{four}$.

In the next example we compute the sum of two numbers written in base four.

EXAMPLE 2 ■ Add Base Four Numerals

Find the sum of 23_{four} and 13_{four}.

Solution
Arrange the numbers vertically, keeping the digits of the same place value in the same column.

$$
\begin{array}{r}
\text{\tiny SIXTEENS}\ \ \text{\tiny FOURS}\ \ \text{\tiny ONES}\\
{\scriptstyle 1}\\
2\ \ 3_{four}\\
+\quad 1\ \ 3_{four}\\
\hline
2_{four}
\end{array}
$$

Add the digits in the ones' column.
Table 4.10 shows that $3_{four} + 3_{four} = 12_{four}$.
Write the 2 in the ones' column and carry the 1 to the fours' column.

$$
\begin{array}{r}
{\scriptstyle 1 \quad 1}\\
2\ \ 3_{four}\\
+\quad 1\ \ 3_{four}\\
\hline
1\ 0\ 2_{four}
\end{array}
$$

Add the digits in the fours' column:
$(1_{four} + 2_{four}) + 1_{four} = 3_{four} + 1_{four} = 10_{four}$.
Write the 0 in the fours' column and carry the 1 to the sixteens' column. Bring down the 1 that was carried to the sixteens' column to form the sum 102_{four}.

The sum of 23_{four} and 13_{four} is 102_{four}.

CHECK YOUR PROGRESS 2   Find $42_{\text{five}} + 23_{\text{five}}$.

Solution   *See page S13.*

In the previous examples we used a table to determine the necessary sums. However, it is generally quicker to find a sum by computing the base ten sum of the digits in each column and then converting each base ten sum back to its equivalent in the given base. The next two examples illustrate this summation technique.

EXAMPLE 3 ■ **Add Base Six Numerals**

Find $25_{\text{six}} + 32_{\text{six}} + 42_{\text{six}}$.

Solution
Arrange the numbers vertically, keeping the digits of the same place value in the same column.

Add the digits in the ones' column:
$5_{\text{six}} + 2_{\text{six}} + 2_{\text{six}} = 5 + 2 + 2 = 9$.
Convert 9 to base six. ($9 = 13_{\text{six}}$)
Write the 3 in the ones' column and carry the 1 to the sixes' column.

Add the digits in the sixes' column and convert the sum to base six.
$1_{\text{six}} + 2_{\text{six}} + 3_{\text{six}} + 4_{\text{six}} = 1 + 2 + 3 + 4 = 10 = 14_{\text{six}}$
Write the 4 in the sixes' column and carry the 1 to the thirty-sixes' column. Bring down the 1 that was carried to the thirty-sixes' column to form the sum 143_{six}.

$25_{\text{six}} + 32_{\text{six}} + 42_{\text{six}} = 143_{\text{six}}$

CHECK YOUR PROGRESS 3   Find $35_{\text{seven}} + 46_{\text{seven}} + 24_{\text{seven}}$.

Solution   *See page S13.*

In the next example, we solve an addition problem that involves a base greater than ten.

EXAMPLE 4 ■ Add Base Twelve Numerals

Find $A97_{twelve} + 8BA_{twelve}$.

Solution

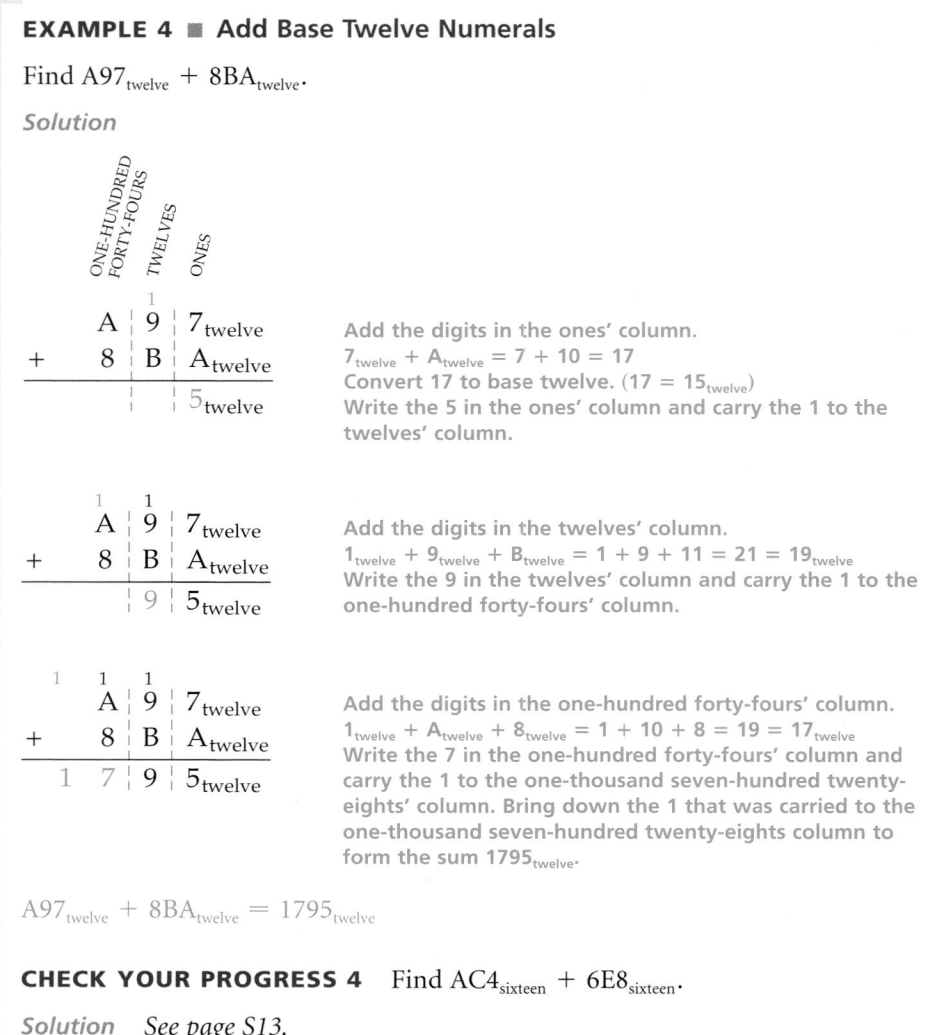

Add the digits in the ones' column.
$7_{twelve} + A_{twelve} = 7 + 10 = 17$
Convert 17 to base twelve. $(17 = 15_{twelve})$
Write the 5 in the ones' column and carry the 1 to the twelves' column.

Add the digits in the twelves' column.
$1_{twelve} + 9_{twelve} + B_{twelve} = 1 + 9 + 11 = 21 = 19_{twelve}$
Write the 9 in the twelves' column and carry the 1 to the one-hundred forty-fours' column.

Add the digits in the one-hundred forty-fours' column.
$1_{twelve} + A_{twelve} + 8_{twelve} = 1 + 10 + 8 = 19 = 17_{twelve}$
Write the 7 in the one-hundred forty-fours' column and carry the 1 to the one-thousand seven-hundred twenty-eights' column. Bring down the 1 that was carried to the one-thousand seven-hundred twenty-eights column to form the sum 1795_{twelve}.

$A97_{twelve} + 8BA_{twelve} = 1795_{twelve}$

CHECK YOUR PROGRESS 4   Find $AC4_{sixteen} + 6E8_{sixteen}$.

Solution   *See page S13.*

Subtraction in Different Bases

To subtract two numbers written in the same base, begin by arranging the numbers vertically, keeping digits that have the same place value in the same column. It will be necessary to borrow whenever a digit in the subtrahend is greater than its corresponding digit in the minuend. Every number that is borrowed will be a power of the base.

EXAMPLE 5 ■ Subtract Base Seven Numerals

Find $463_{seven} - 124_{seven}$.

Solution
Arrange the numbers vertically, keeping the digits of the same place value in the same column.

TAKE NOTE

In this section assume that the small numerals, used to illustrate the borrowing process, are written in the same base as the numerals in the given subtraction problem.

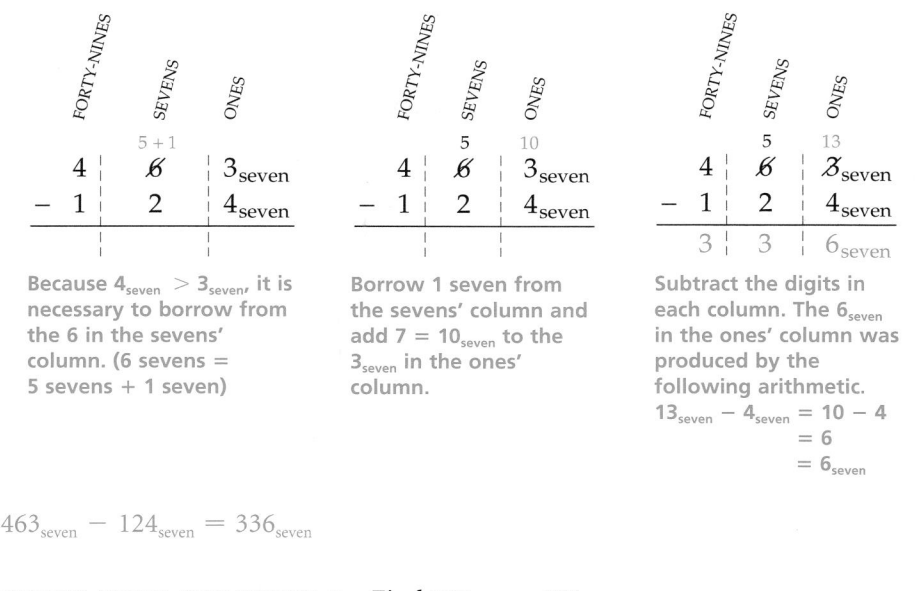

Because $4_{seven} > 3_{seven}$, it is necessary to borrow from the 6 in the sevens' column. (6 sevens = 5 sevens + 1 seven)

Borrow 1 seven from the sevens' column and add $7 = 10_{seven}$ to the 3_{seven} in the ones' column.

Subtract the digits in each column. The 6_{seven} in the ones' column was produced by the following arithmetic.
$$13_{seven} - 4_{seven} = 10 - 4$$
$$= 6$$
$$= 6_{seven}$$

$463_{seven} - 124_{seven} = 336_{seven}$

CHECK YOUR PROGRESS 5   Find $365_{nine} - 183_{nine}$.

Solution   *See page S13.*

Table 4.11 *Decimal and Hexadecimal Equivalents*

Base Ten Decimal	Base Sixteen Hexadecimal
0	0
1	1
2	2
3	3
4	4
5	5
6	6
7	7
8	8
9	9
10	A
11	B
12	C
13	D
14	E
15	F

EXAMPLE 6 ■ Subtract Base Sixteen Numerals

Find $7AB_{sixteen} - 3E4_{sixteen}$.

Solution

Table 4.11 shows the hexadecimal digits and their decimal equivalents. Because $B_{sixteen}$ is greater than $4_{sixteen}$, there is no need to borrow to find the difference in the ones' column. However, $A_{sixteen}$ is less than $E_{sixteen}$, so it is necessary to borrow to find the difference in the sixteens' column.

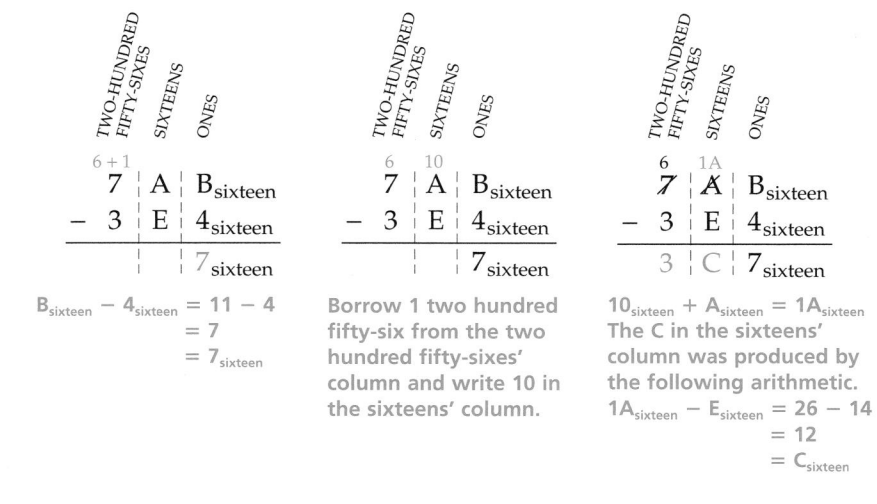

$B_{sixteen} - 4_{sixteen} = 11 - 4$
$= 7$
$= 7_{sixteen}$

Borrow 1 two hundred fifty-six from the two hundred fifty-sixes' column and write 10 in the sixteens' column.

$10_{sixteen} + A_{sixteen} = 1A_{sixteen}$
The C in the sixteens' column was produced by the following arithmetic.
$1A_{sixteen} - E_{sixteen} = 26 - 14$
$= 12$
$= C_{sixteen}$

$7AB_{sixteen} - 3E4_{sixteen} = 3C7_{sixteen}$

CHECK YOUR PROGRESS 6 Find $83A_{\text{twelve}} - 467_{\text{twelve}}$.

Solution See page S13.

Multiplication in Different Bases

Table 4.12 *A Base Four Multiplication Table*

×	0	1	2	3
0	0	0	0	0
1	0	1	2	3
2	0	2	10	12
3	0	3	12	21

To perform multiplication in bases other than base ten, it is often helpful to first write a multiplication table for the given base. Table 4.12 shows a multiplication table for base four. The numbers shown in red in the table illustrate that $2_{\text{four}} \times 3_{\text{four}} = 12_{\text{four}}$. You can verify this result by converting the numbers to base ten, multiplying in base ten, and then converting back to base four. Here is the actual arithmetic.

$$2_{\text{four}} \times 3_{\text{four}} = 2 \times 3 = 6 = 12_{\text{four}}$$

QUESTION What is $5_{\text{six}} \times 4_{\text{six}}$?

EXAMPLE 7 ▪ Multiply Base Four Numerals

Use the base four multiplication table to find $3_{\text{four}} \times 123_{\text{four}}$.

Solution
Arrange the numbers vertically, keeping the digits of the same place value in the same column. Use Table 4.12 to multiply 3_{four} times each digit in 123_{four}. If any of these multiplications produces a two-digit product, then write down the digit on the right and carry the digit on the left.

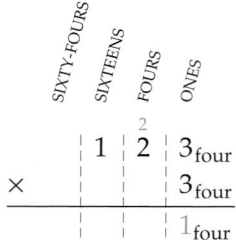

$3_{\text{four}} \times 3_{\text{four}} = 21_{\text{four}}$
Write the 1 in the ones' column and carry the 2.

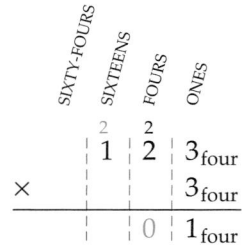

$3_{\text{four}} \times 2_{\text{four}} = 12_{\text{four}}$
$12_{\text{four}} + 2_{\text{four}}$ (the carry) $= 20_{\text{four}}$
Write the 0 in the fours' column and carry the 2.

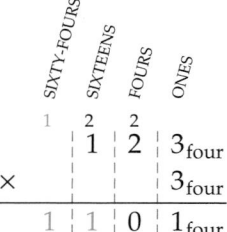

$3_{\text{four}} \times 1_{\text{four}} = 3_{\text{four}}$
$3_{\text{four}} + 2_{\text{four}}$ (the carry) $= 11_{\text{four}}$
Write a 1 in the sixteens' column and carry a 1 to the sixty-fours' column. Bring down the 1 that was carried to the sixty-fours' column to form the product 1101_{four}.

$123_{\text{four}} \times 3_{\text{four}} = 1101_{\text{four}}$

CHECK YOUR PROGRESS 7 Find $213_{\text{four}} \times 2_{\text{four}}$.

Solution See page S13.

ANSWER $5_{\text{six}} \times 4_{\text{six}} = 20 = 32_{\text{six}}$

Writing all of the entries in a multiplication table for a large base such as base twelve can be time-consuming. In such cases you may prefer to multiply in base ten and then convert each product back to the given base. The next example illustrates this multiplication method.

EXAMPLE 8 ■ Multiply Base Twelve Numerals

Find $53_{twelve} \times 27_{twelve}$.

Solution
Arrange the numbers vertically, keeping the digits of the same place value in the same column. Start by multiplying each digit of the multiplicand by the ones digit of the multiplier.

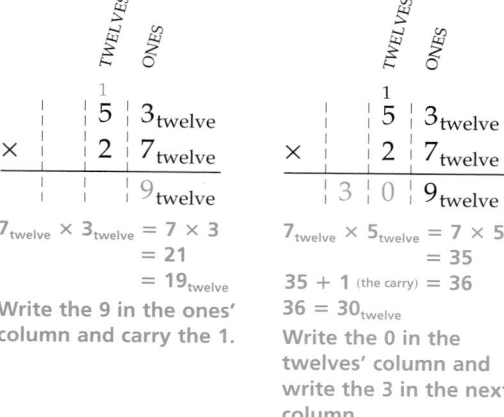

$7_{twelve} \times 3_{twelve} = 7 \times 3$
$\qquad = 21$
$\qquad = 19_{twelve}$
Write the 9 in the ones'
column and carry the 1.

$7_{twelve} \times 5_{twelve} = 7 \times 5$
$\qquad = 35$
$35 + 1 \text{ (the carry)} = 36$
$36 = 30_{twelve}$
Write the 0 in the twelves' column and write the 3 in the next column.

Now multiply each digit of the multiplicand by the twelves digit of the multiplier.

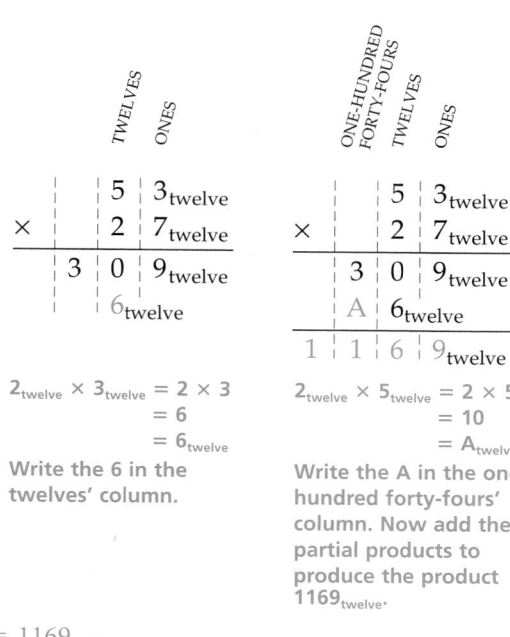

$2_{twelve} \times 3_{twelve} = 2 \times 3$
$\qquad = 6$
$\qquad = 6_{twelve}$
Write the 6 in the twelves' column.

$2_{twelve} \times 5_{twelve} = 2 \times 5$
$\qquad = 10$
$\qquad = A_{twelve}$
Write the A in the one hundred forty-fours' column. Now add the partial products to produce the product 1169_{twelve}.

$53_{twelve} \times 27_{twelve} = 1169_{twelve}$

CHECK YOUR PROGRESS 8 Find $34_{eight} \times 25_{eight}$.

Solution *See page S14.*

Math Matters   The Fields Medal

The Fields Medal

A Nobel Prize is awarded each year in the categories of chemistry, physics, physiology, medicine, literature, and peace. However, no award is given in mathematics. Why Alfred Nobel chose not to provide an award in the category of mathematics is unclear. There has been some speculation that Nobel had a personal conflict with the mathematician Gosta Mittag-Leffler. Other historians have suggested that Nobel didn't think of mathematics as a practical discipline that provided great benefits to society.

The Canadian mathematician John Charles Fields (1863–1932) felt that a prestigious award should be given in the area of mathematics. Fields helped establish the Fields Medal, which was first given to Lars Valerian Ahlfors and Jesse Douglas in 1936. The International Congress of Mathematicians had planned to give two Fields Medals every four years after 1936, but because of World War II, the next Fields Medals were not given until 1950.

It was Fields's wish that the Fields Medal recognize both existing work and the promise of future achievement. Because of this concern for future achievement, the International Congress of Mathematicians decided to restrict those eligible for the Fields Medal to mathematicians under the age of 40.

Division in Different Bases

To perform a division in a base other than base ten, it is helpful to first make a list of a few multiples of the divisor. This procedure is illustrated in the following example.

EXAMPLE 9 ■ Divide Base Seven Numerals

Find $253_{seven} \div 3_{seven}$.

Solution
First list a few multiples of 3_{seven}.

$$3_{seven} \times 0_{seven} = 3 \times 0 = 0 = 0_{seven} \qquad 3_{seven} \times 4_{seven} = 3 \times 4 = 12 = 15_{seven}$$
$$3_{seven} \times 1_{seven} = 3 \times 1 = 3 = 3_{seven} \qquad 3_{seven} \times 5_{seven} = 3 \times 5 = 15 = 21_{seven}$$
$$3_{seven} \times 2_{seven} = 3 \times 2 = 6 = 6_{seven} \qquad 3_{seven} \times 6_{seven} = 3 \times 6 = 18 = 24_{seven}$$
$$3_{seven} \times 3_{seven} = 3 \times 3 = 9 = 12_{seven}$$

Because $3_{seven} \times 6_{seven} = 24_{seven}$ is slightly less than 25_{seven}, we pick 6 as our first numeral in the quotient when dividing 25_{seven} by 3_{seven}.

$$
\begin{array}{r}
6 \\
3_{seven}\overline{)2\,5\,3}_{seven} \\
2\,4 \\
\hline
1
\end{array}
$$

$3_{seven} \times 6_{seven} = 24_{seven}$
Subtract 24_{seven} from 25_{seven}.

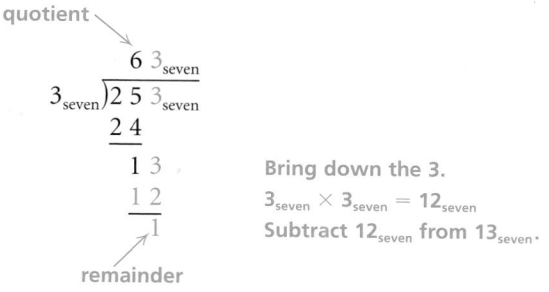

quotient

$$3_{seven}\overline{)2\,5\,3_{seven}}$$ quotient $6\,3_{seven}$

$$\begin{array}{r} 6\ 3_{seven} \\ 3_{seven}\overline{)2\ 5\ 3_{seven}} \\ \underline{2\ 4} \\ 1\ 3 \\ \underline{1\ 2} \\ 1 \end{array}$$

Bring down the 3.

$3_{seven} \times 3_{seven} = 12_{seven}$

Subtract 12_{seven} from 13_{seven}.

remainder

Thus $253_{seven} \div 3_{seven} = 63_{seven}$ with a remainder of 1_{seven}.

CHECK YOUR PROGRESS 9   Find $324_{five} \div 3_{five}$.

Solution   *See page S14.*

In a base two division problem, the only multiples of the divisor that are used are zero times the divisor and one times the divisor.

EXAMPLE 10 ■ Divide Base Two Numerals

Find $101011_{two} \div 11_{two}$.

Solution

The divisor is 11_{two}. The multiples of the divisor that may be needed are $11_{two} \times 0_{two} = 0_{two}$ and $11_{two} \times 1_{two} = 11_{two}$. Also note that because $10_{two} - 1_{two} = 2 - 1 = 1 = 1_{two}$, we know that

$$\begin{array}{r} 10_{two} \\ -\ 1_{two} \\ \hline 1_{two} \end{array}$$

$$\begin{array}{r} 1\ 1\ 1\ 0_{two} \\ 11_{two}\overline{)1\ 0\ 1\ 0\ 1\ 1_{two}} \\ \underline{1\ 1} \\ 1\ 0\ 0 \\ \underline{1\ 1} \\ 1\ 1 \\ \underline{1\ 1} \\ 0\ 1 \\ \underline{0} \\ 1 \end{array}$$

Therefore $101011_{two} \div 11_{two} = 1110_{two}$ with a remainder of 1_{two}.

CHECK YOUR PROGRESS 10   Find $1110011_{two} \div 10_{two}$.

Solution   *See page S14.*

Excursion

Subtraction in Base Two via the Ones Complement and the End-Around Carry

Computers and calculators are often designed so that the number of required circuits is minimized. Instead of using separate circuits to perform addition and subtraction, engineers make use of an *end-around carry procedure* that uses addition to perform subtraction. The end-around carry procedure also makes use of the ones complement of a number. In base two, the ones complement of 0 is 1 and the ones complement of 1 is 0. Thus the ones complement of any base two number can be found by changing each 1 to a 0 and each 0 to a 1.

Subtraction Using the Ones Complement and the End-Around Carry

To subtract a base two number from a larger base two number:

1. Add the ones complement of the subtrahend to the minuend.

2. Take away 1 from the leftmost bit and add 1 to the units bit.

The following example illustrates the process of subtracting 1001_{two} from 1101_{two} using the ones complement and the end-around carry procedure.

$$\begin{array}{rl} 1101_{two} & \text{Minuend} \\ -\ 1001_{two} & \text{Subtrahend} \end{array}$$

$$\begin{array}{rl} 1101_{two} & \\ +\ 0110_{two} & \text{Ones complement of subtrahend} \\ \hline 10011_{two} & \end{array}$$

$$\begin{array}{rl} 10011_{two} & \text{Take away 1 from the leftmost bit} \\ +\qquad 1_{two} & \text{and add 1 to the ones bit.} \\ \hline 100_{two} & \end{array}$$

$$1101_{two} - 1001_{two} = 100_{two}$$

If the subtrahend has fewer bits than the minuend, leading zeros should be inserted in the subtrahend so that it has the same number of bits as the minuend. This process is illustrated below for the subtraction $1010110_{two} - 11001_{two}$.

$$\begin{array}{rl} 1010110_{two} & \text{Minuend} \\ -\quad 11001_{two} & \text{Subtrahend} \end{array}$$

$$\begin{array}{rl} 1010110_{two} & \\ -\ 0011001_{two} & \text{Insert two leading zeros.} \end{array}$$

$$\begin{array}{rl} 1010110_{two} & \\ +\ 1100110_{two} & \text{Ones complement of subtrahend} \\ \hline 10111100 & \end{array}$$

(continued)

$$10111100_{two}$$
$$+\qquad 1_{two}$$
$$\overline{111101_{two}}$$

Take away 1 from the leftmost bit
and add 1 to the ones bit.

$$1010110_{two} - 11001_{two} = 111101_{two}$$

Excursion Exercises

Use the ones complement of the subtrahend and the end-around carry method to find each difference.

1. $1110_{two} - 1001_{two}$

2. $101011_{two} - 100010_{two}$

3. $101001010_{two} - 1011101_{two}$

4. $111011100110_{two} - 101010100_{two}$

5. $1111101011_{two} - 1001111_{two}$

6. $1110010101100_{two} - 100011110_{two}$

Exercise Set 4.4

(Suggested Assignment: 1–49, every other odd, and 51)

In Exercises 1–9, find each sum in the same base as the addends.

1. $204_{five} + 123_{five}$

2. $323_{four} + 212_{four}$

3. $5625_{seven} + 634_{seven}$

4. $1011_{two} + 101_{two}$

5. $110101_{two} + 10011_{two}$

6. $11001010_{two} + 1100111_{two}$

7. $8B5_{twelve} + 578_{twelve}$

8. $379_{sixteen} + 856_{sixteen}$

9. $C489_{sixteen} + BAD_{sixteen}$

In Exercises 10–18, find each difference.

10. $534_{six} - 241_{six}$

11. $7325_{eight} - 563_{eight}$

12. $6148_{nine} - 782_{nine}$

13. $11010_{two} - 1011_{two}$

14. $111001_{two} - 10101_{two}$

15. $11010100_{two} - 1011011_{two}$

16. $9C5_{sixteen} - 687_{sixteen}$

17. $43A7_{twelve} - 289_{twelve}$

18. $BAB2_{twelve} - 475_{twelve}$

In Exercises 19–30, find each product.

19. $212_{three} \times 2_{three}$

20. $4132_{five} \times 4_{five}$

21. $7354_{eight} \times 5_{eight}$

22. $11011_{two} \times 11_{two}$

23. $101010_{two} \times 10_{two}$

24. $110100_{two} \times 101_{two}$

25. $453_{eight} \times 25_{eight}$

26. $1254_{six} \times 43_{six}$

27. $1323_{four} \times 132_{four}$

28. $895_{twelve} \times 43_{twelve}$

29. $BAD_{sixteen} \times 5_{sixteen}$

30. $798_{sixteen} \times 23_{sixteen}$

In Exercises 31–39, find each quotient and remainder.

31. $231_{four} \div 3_{four}$

32. $672_{eight} \div 5_{eight}$

33. $5341_{six} \div 4_{six}$

34. $11011_{two} \div 10_{two}$

35. $101010_{two} \div 11_{two}$

36. $1011011_{two} \div 100_{two}$

37. $457_{twelve} \div 5_{twelve}$

38. $832_{\text{sixteen}} \div 7_{\text{sixteen}}$

39. $234_{\text{five}} \div 12_{\text{five}}$

40. If $232_{\text{base } x} = 92$, find base x.

41. If $143_{\text{base } x} = 10200_{\text{three}}$, find base x.

42. If $46_{\text{base } x} = 101010_{\text{two}}$, find base x.

43. Consider the addition $384 + 245$.

   a. Use base ten addition to find the sum.

   b. Convert 384 and 245 to base two.

   c. Find the base two sum of the base two numbers you found in part b.

   d. Convert the base two sum from part c to base ten.

   e. How does the answer to part a compare with the answer to part d?

44. Consider the subtraction $457 - 318$.

   a. Use base ten subtraction to find the difference.

   b. Convert 457 and 318 to base two.

   c. Find the base two difference of the base two numbers you found in part b.

   d. Convert the base two difference from part c to base ten.

   e. How does the answer to part a compare with the answer to part d?

45. Consider the multiplication 247×26.

   a. Use base ten multiplication to find the product.

   b. Convert 247 and 26 to base two.

   c. Find the base two product of the base two numbers you found in part b.

   d. Convert the base two product from part c to base ten.

   e. How does the answer to part a compare with the answer to part d?

Extensions

CRITICAL THINKING

46. Explain the error in the following base eight subtraction.

$$\begin{array}{r} 751_{\text{eight}} \\ -\ 126_{\text{eight}} \\ \hline 625_{\text{eight}} \end{array}$$

47. Determine the base used in the following multiplication.

$$314_{\text{base } x} \times 24_{\text{base } x} = 11202_{\text{base } x}$$

48. The base ten number 12 is an even number. In base seven, 12 is written as 15_{seven}. Is 12 an odd number in base seven?

49. Explain why there is no numeration system with a base of 1.

50. In the following base four addition problem, each letter represents one of the numerals 0, 1, 2, or 3. No two different letters represent the same numeral. Determine which digit is represented by each letter.

$$\begin{array}{r} \textbf{N O}_{\text{ four}} \\ +\ \textbf{A T}_{\text{ four}} \\ \hline \textbf{N O T}_{\text{ four}} \end{array}$$

51. In the following base six addition problem, each letter represents one of the numerals 0, 1, 2, 3, 4, or 5. No two different letters represent the same numeral. Determine which digit is represented by each letter.

$$\begin{array}{r} \textbf{M A}_{\text{ six}} \\ +\ \textbf{A S}_{\text{ six}} \\ \hline \textbf{M O M}_{\text{ six}} \end{array}$$

EXPLORATIONS

It is possible to use a negative number as the base of a numeration system. For instance, the negative base four numeral $32_{\text{negative four}}$ represents the number $3 \times (-4)^1 + 2 \times (-4)^0 = -12 + 2 = -10$.

52. a. Convert each of the following negative base numerals to base 10:

    $143_{\text{negative five}}$

    $74_{\text{negative nine}}$

    $10110_{\text{negative two}}$

   b. Write -27 as a negative base five numeral.

   c. Write 64 as a negative base three numeral.

   d. Write 112 as a negative base ten numeral.

Prime Numbers and Selected Topics from Number Theory

Prime Numbers

Number theory is a mathematical discipline that is primarily concerned with the properties that are exhibited by the natural numbers. The mathematician Carl Friedrich Gauss established many theorems in number theory. Concerning this branch of mathematics, Gauss once stated, "Mathematics is the Queen of the Sciences and number theory is the Queen of Mathematics." Many topics in number theory involve the concept of a *divisor* or *factor*.

> **Definition of Divisor**
>
> The natural number a is a **divisor** or **factor** of the natural number b provided there exists a natural number j such that $aj = b$.

In less formal terms, a natural number a is a divisor of the natural number b provided $b \div a$ has a remainder of 0. For instance, 10 has divisors of 1, 2, 5, and 10 because each of these numbers divides into 10 with a remainder of 0.

EXAMPLE 1 ■ Find Divisors

Determine all of the natural number divisors of each number.

a. 6   **b.** 42   **c.** 17

Solution

a. Divide 6 by 1, 2, 3, 4, 5, and 6. The division of 6 by 1, 2, 3, and 6 each produces a natural number quotient and a remainder of 0. Thus 1, 2, 3, and 6 are divisors of 6. Dividing 6 by 4 and 6 by 5 does not produce a remainder of 0. Therefore 4 and 5 are not divisors of 6.

b. The only natural numbers from 1 to 42 that divide into 42 with a remainder of 0 are 1, 2, 3, 6, 7, 14, 21, and 42. Thus the divisors of 42 are 1, 2, 3, 6, 7, 14, 21, and 42.

c. The only divisors of 17 are 1 and 17.

CHECK YOUR PROGRESS 1   Determine all of the natural number divisors of each number.

a. 9   **b.** 11   **c.** 24

Solution   *See page S14.*

It is worth noting that every natural number greater than 1 has itself as a factor and 1 as a factor. If a natural number greater than 1 has only 1 and itself as factors, then it is a very special number known as a *prime number*.

Definition of a Prime Number and a Composite Number

A **prime number** is a natural number greater than 1 that has exactly two factors (divisors): itself and 1.

A **composite number** is a natural number greater than 1 that is not a prime number.

The ten smallest prime numbers are 2, 3, 5, 7, 11, 13, 17, 19, 23, and 29. Each of these numbers has only itself and 1 as factors. The ten smallest composite numbers are 4, 6, 8, 9, 10, 12, 14, 15, 16, and 18.

EXAMPLE 2 ■ Classify a Number as a Prime Number or a Composite Number

Determine whether each number is a prime number or a composite number.

a. 41   **b.** 51

Solution

a. The only divisors of 41 and 1 are 41. Thus 41 is a prime number.

b. The divisors of 51 are 1, 3, 17, and 51. Thus 51 is a composite number.

CHECK YOUR PROGRESS 2   Determine whether each number is a prime number or a composite number.

a. 47   **b.** 171

Solution   See page S14.

QUESTION   *Are all prime numbers odd numbers?*

Divisibility Tests

To determine whether one number is divisible by a smaller number, we often apply a **divisibility test,** which is a procedure that enables one to determine whether the smaller number is a divisor of the larger number without actually dividing the smaller number into the larger number. Table 4.13 provides divisibility tests for the numbers 2, 3, 4, 5, 6, 8, 9, 10, and 11.

ANSWER   *No. The even number 2 is a prime number.*

Table 4.13 *Base Ten Divisibility Tests*

A number is divisible by the following divisor if:	Divisibility Test	Example
2	The number is an even number.	846 is divisible by 2 because 846 is an even number.
3	The sum of the digits of the number is divisible by 3.	531 is divisible by 3 because $5 + 3 + 1 = 9$ is divisible by 3.
4	The last two digits of the number form a number that is divisible by 4.	1924 is divisible by 4 because the last two digits form the number 24, which is divisible by 4.
5	The number ends with a 0 or a 5.	8785 is divisible by 5 because it ends with 5.
6	The number is divisible by 2 and by 3.	972 is divisible by 6 because it is divisible by 2 and also by 3.
8	The last three digits of the number form a number that is divisible by 8.	19,168 is divisible by 8 because the last three digits form the number 168, which is divisible by 8.
9	The sum of the digits of the number is divisible by 9.	621,513 is divisible by 9 because the sum of the digits is 18, which is divisible by 9.
10	The last digit is 0.	970 is divisible by 10 because it ends with 0.
11	Start at one end of the number and compute the sum of every other digit. Next compute the sum of the remaining digits. If the difference of these sums is divisible by 11, then the original number is divisible by 11.	4807 is divisible by 11 because the difference of the sum of the digits shown in blue $(8 + 7 = 15)$ and the sum of the remaining digits shown in red $(4 + 0 = 4)$ is $15 - 4 = 11$, which is divisible by 11.

EXAMPLE 3 ■ Apply Divisibility Tests

Use divisibility tests to determine whether 16,278 is divisible by the following numbers.

a. 2   **b.** 3   **c.** 5   **d.** 8   **e.** 11

Solution

a. Because 16,278 is an even number, it is divisible by 2.

b. The sum of the digits of 16,278 is 24, which is divisible by 3. Therefore, 16,278 is divisible by 3.

c. The number 16,278 does not end with a 0 or a 5. Therefore, 16,278 is not divisible by 5.

d. The last three digits of 16,278 form the number 278, which is not divisible by 8. Thus 16,278 is not divisible by 8.

e. The sum of the digits with even place-value powers is $1 + 2 + 8 = 11$. The sum of the digits with odd place-value powers is $6 + 7 = 13$. The difference of these sums is $13 - 11 = 2$. This difference is not divisible by 11, so 16,278 is not divisible by 11.

CHECK YOUR PROGRESS 3 Use divisibility tests to determine whether 341,565 is divisible by each of the following numbers.

a. 3   **b.** 4   **c.** 10   **d.** 11

Solution   *See page S14.*

Prime Factorization

The **prime factorization** of a composite number is a factorization that contains only prime numbers. Many proofs in number theory make use of the following important theorem.

> **The Fundamental Theorem of Arithmetic**
>
> Every composite number can be written as a unique product of prime numbers (disregarding the order of the factors).

To find the prime factorization of a composite number, rewrite the number as a product of two smaller natural numbers. If these smaller numbers are both prime numbers, then you are finished. If either of the smaller numbers is not a prime number, then rewrite it as a product of smaller natural numbers. Continue this procedure until all factors are primes. In Example 4 we make use of a *tree diagram* to organize the factorization process.

EXAMPLE 4 ■ **Find the Prime Factorization of a Number**
Determine the prime factorization of the following numbers.

a. 84   **b.** 495

Solution

a. The following tree diagrams show two different ways of finding the prime factorization of 84, which is $2 \cdot 2 \cdot 3 \cdot 7 = 2^2 \cdot 3 \cdot 7$. Each number in the tree is equal to the product of the two smaller numbers below it. The numbers (in red) at the extreme ends of the branches are the prime factors.

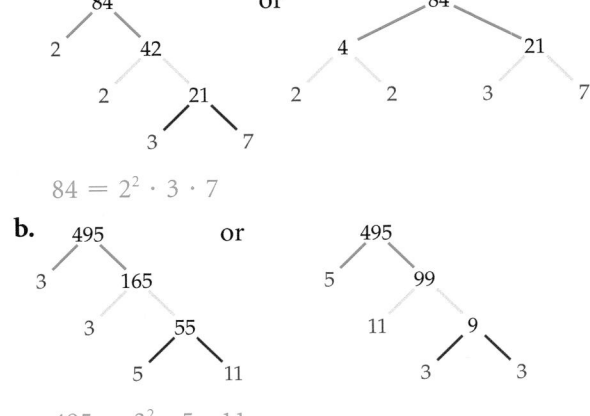

$$84 = 2^2 \cdot 3 \cdot 7$$

$$495 = 3^2 \cdot 5 \cdot 11$$

▼ point of interest

Frank Nelson Cole (1861–1926) concentrated his mathematical work in the areas of number theory and group theory. He is well known for his 1903 presentation to the American Mathematical Society. Without speaking a word, he went to a chalkboard and wrote

$$2^{67} - 1 =$$
$$147573952589676412927$$

Many mathematicians considered this large number to be a prime number. Then Cole moved to a second chalkboard and computed the product

$$761838257287 \cdot 193707721$$

When the audience saw that this product was identical to the result on the first chalkboard, they gave Cole the only standing ovation ever given during an American Mathematical Society presentation. They realized that Cole had factored $2^{67} - 1$, which was a most remarkable feat, considering that no computers existed at that time.

CHECK YOUR PROGRESS 4   Determine the prime factorization of the following.

a. 315   **b.** 273

Solution   *See page S14.*

Srinivasa Ramanujan
(1887–1920)

INSTRUCTOR NOTE
The number 1729 is also a Carmichael number, a number that is not prime but satisfied Fermat's Little Theorem (see Exercise 31, Section 4.6). Thus, 1729 provides a counterexample to the converse of Fermat's Little Theorem.

Math Matters   Srinivasa Ramanujan

On January 16, 1913, the young 26-year-old Srinivasa Ramanujan (Rä-mä′noo-jŭn) sent a letter from Madras, India, to the illustrious English mathematician G. H. Hardy. The letter requested that Hardy give his opinion about several mathematical ideas that Ramanujan had developed. In the letter Ramanujan explained, "I have not trodden through the conventional regular course which is followed in a University course, but I am striking out a new path for myself." Much of the mathematics was written using unconventional terms and notation; however, Hardy recognized (after many detailed readings and with the help of other mathematicians at Cambridge University) that Ramanujan was "a mathematician of the highest quality, a man of altogether exceptional originality and power."

On March 17, 1914, Ramanujan set sail for England, where he joined Hardy in a most unusual collaboration that lasted until Ramanujan returned to India in 1919. The following famous story is often told to illustrate the remarkable mathematical genius of Ramanujan.

After Hardy had taken a taxicab to visit Ramanujan, he made the remark that the license plate number for the taxi was "1729, a rather dull number." Ramanujan immediately responded by saying that 1729 was a most interesting number, because it is the smallest natural number that can be expressed in two different ways as the sum of two cubes.

$$1^3 + 12^3 = 1729 \quad \text{and} \quad 9^3 + 10^3 = 1729$$

An interesting biography of the life of Srinivasa Ramanujan is given in *The Man Who Knew Infinity: A Life of the Genius Ramanujan* by Robert Kanigel.[1]

TAKE NOTE

To determine whether a natural number is a prime number, it is necessary to consider only divisors from 2 up to the square root of the number, because every composite number n has at least one divisor less than or equal to $\sqrt{n}$. The proof of this statement is outlined in Exercise 67 of this section.

It is possible to determine whether a natural number n is a prime number by checking each natural number from 2 up to the largest integer not greater than $\sqrt{n}$ to see whether each is a divisor of n. If none of these numbers is a divisor of n, then n is a prime number. For large values of n, this division method is generally time-consuming and tedious. The Greek astronomer and mathematician Eratosthenes (about 276–192 B.C.) recognized that multiplication is generally easier than division, and he devised a method that makes use of multiples to determine every

1. Kanigel, Robert. *The Man Who Knew Infinity: A Life of the Genius Ramanujan.* New York: Simon & Schuster, 1991.

prime number in a list of natural numbers. Today we call this method the *Sieve of Eratosthenes.*

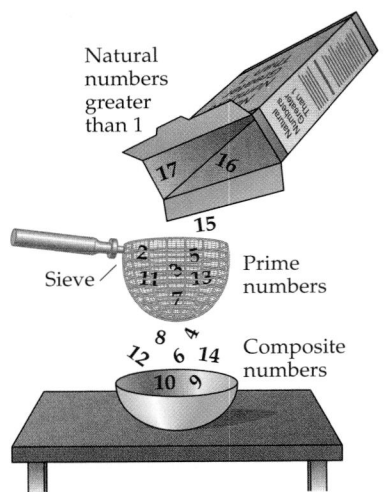

▼ **point of interest**

In article 329 of *Disquisitiones Arithmeticae,* Gauss wrote:

"The problem of distinguishing prime numbers from composite numbers and of resolving the latter into their prime factors is known to be one of the most important and useful in arithmetic... The dignity of the science itself seems to require that every possible means be explored for the solution of a problem so elegant and so celebrated." (*Source: The Little Book of Big Primes* by Paulo Ribenboim. New York: Springer-Verlag, 1991.)

To sift prime numbers, first make a list of consecutive natural numbers. In Table 4.14 we have listed the consecutive counting numbers from 2 to 100.

- Cross out every multiple of 2 larger than 2. The next smallest remaining number in the list is 3. Cross out every multiple of 3 larger than 3.
- Call the next smallest remaining number in the list k. Cross out every multiple of k larger than k. Repeat this step for all $k < \sqrt{100}$.

Table 4.14 *The Sieve Method of Finding Primes*

	2	3	4	5	6	7	8	9	10
11	12	13	14	15	16	17	18	19	20
21	22	23	24	25	26	27	28	29	30
31	32	33	34	35	36	37	38	39	40
41	42	43	44	45	46	47	48	49	50
51	52	53	54	55	56	57	58	59	60
61	62	63	64	65	66	67	68	69	70
71	72	73	74	75	76	77	78	79	80
81	82	83	84	85	86	87	88	89	90
91	92	93	94	95	96	97	98	99	100

The numbers in blue that are not crossed out are prime numbers. Table 4.14 shows that there are 25 prime numbers less than 100.

Over 2000 years ago, Euclid proved that the set of prime numbers is an infinite set. Euclid's proof is an *indirect proof* or a *proof by contradiction.* Essentially his proof shows that for any finite list of prime numbers, we can create a number T, as described at the top of the following page, such that any prime factor of T can be shown to be a prime number that is not in the list. Thus there must be an infinite number of primes because it is not possible for all of the primes to be in any finite list.

INSTRUCTOR NOTE
It may be helpful to give a specific example to illustrate that

$$T = (p_1 \cdot p_2 \cdot p_3 \cdots p_r) + 1$$

is not divisible by any of the primes $p_1, p_2, p_3, \ldots, p_r$. For instance, if $T = (2 \cdot 3 \cdot 5 \cdot 7) + 1$, then T is not divisible by 2, 3, 5, or 7. In fact, T divided by any of the primes 2, 3, 5, or 7 yields a remainder of 1.

Euclid's Proof Assume *all* of the prime numbers are contained in the list $p_1, p_2, p_3, \ldots, p_r$. Let $T = (p_1 \cdot p_2 \cdot p_3 \cdots p_r) + 1$. Either T is a prime number or T has a prime divisor. If T is a prime then it is a prime that is not in our list and we have reached a contradiction. If T is not prime, then one of the primes $p_1, p_2, p_3, \ldots, p_r$ must be a divisor of T. However, the number T is not divisible by any of the primes $p_1, p_2, p_3, \ldots, p_r$ because each p_i divides $p_1 \cdot p_2 \cdot p_3 \cdots p_r$ but does not divide 1. Hence any prime divisor of T, say p, is a prime number that is not in the list $p_1, p_2, p_3, \ldots, p_r$. So p is yet another prime number, and $p_1, p_2, p_3, \ldots, p_r$ is not a complete list of all the prime numbers. ■

Excursion

The Distribution of the Primes

Many mathematicians have searched without success for a mathematical formula that can be used to generate the sequence of prime numbers. We know that the prime numbers form an infinite sequence. However, the distribution of the prime numbers within the sequence of natural numbers is very complicated. The ratio of prime numbers to composite numbers appears to become smaller and smaller as larger and larger numbers are considered. In general, the number of consecutive composite numbers that come between two prime numbers tends to increase as the size of the numbers becomes larger; however, this increase is erratic and appears to be unpredictable.

In this Excursion we refer to a list of two or more consecutive composite numbers as a **prime desert**. For instance, 8, 9, 10 is a prime desert because it consists of three consecutive composite numbers. The longest prime desert shown in Table 4.14 on the previous page is the seven consecutive composite numbers 90, 91, 92, 93, 94, 95, and 96. A formula that involves *factorials* can be used to form prime deserts of any finite length.

Definition of *n* factorial

If n is a natural number, then $n!$, which is read "n factorial," is defined as

$$n! = n \cdot (n - 1) \cdot \cdots \cdot 3 \cdot 2 \cdot 1$$

As an example of a factorial, consider $4! = 4 \cdot 3 \cdot 2 \cdot 1 = 24$.

The sequence

$$4! + 2, \, 4! + 3, \, 4! + 4$$

is a prime desert of the three composite numbers 26, 27, and 28. Figure 4.1 on the following page shows a prime desert of 10 consecutive composite numbers. Figure 4.2

(continued)

shows a procedure that can be used to produce a prime desert of n composite numbers, where n is any natural number greater than 2.

$$
\left.\begin{aligned}
&11! + 2 \\
&11! + 3 \\
&11! + 4 \\
&11! + 5 \\
&11! + 6 \\
&\quad\vdots \\
&11! + 10 \\
&11! + 11
\end{aligned}\right\} \quad
\begin{aligned}
&\text{A prime desert of} \\
&\text{10 consecutive} \\
&\text{composite} \\
&\text{numbers}
\end{aligned}
\qquad
\left.\begin{aligned}
&(n + 1)! + 2 \\
&(n + 1)! + 3 \\
&(n + 1)! + 4 \\
&(n + 1)! + 5 \\
&(n + 1)! + 6 \\
&\quad\vdots \\
&(n + 1)! + n \\
&(n + 1)! + (n + 1)
\end{aligned}\right\} \quad
\begin{aligned}
&\text{A prime desert of} \\
&n \text{ consecutive} \\
&\text{composite} \\
&\text{numbers}
\end{aligned}
$$

Figure 4.1   **Figure 4.2**

A procedure for generating a prime desert

A prime desert of length one million is shown by the sequence

$$1{,}000{,}001! + 2; \; 1{,}000{,}001! + 3; \; 1{,}000{,}001! + 4; \ldots; 1{,}000{,}001! + 1{,}000{,}001$$

It appears that the distribution of prime numbers is similar to the situation wherein a mathematical gardener plants an infinite number of grass seeds on a windy day. Many of the grass seeds fall close to the gardener, but some are blown down the street and into the next neighborhood. There are gaps where no grass seeds are within 1 mile of each other. Farther down the road there are gaps where no grass seeds are within 10 miles of each other. No matter how far the gardener travels and how long it has been since the last grass seed was spotted, the gardener knows that more grass seeds will appear.

Excursion Exercises

1. Explain how you know that each of the numbers

$$1{,}000{,}001! + 2; \; 1{,}000{,}001! + 3; \; 1{,}000{,}001! + 4; \ldots; 1{,}000{,}001! + 1{,}000{,}001$$

is a composite number.

2. Use factorials to generate the numbers in a prime desert of 12 consecutive composite numbers. Now use a calculator to evaluate each number in this prime desert.

3. Use factorials and "..." notation to represent a prime desert of

   a. 20 consecutive composite numbers.

   b. 500,000 consecutive composite numbers.

   c. 7 billion consecutive composite numbers.

Exercise Set 4.5   (Suggested Assignment: 1–37 odds; 38; 39–63 odds)

In Exercises 1–8, determine all natural number divisors of the given number.

1. 20

2. 32

3. 65

4. 75

5. 41

6. 79

7. 110

8. 150

In Exercises 9–16, determine whether each number is a prime number or a composite number.

9. 21

10. 31

11. 37

12. 39

13. 101

14. 81

15. 79

16. 161

In Exercises 17–24, use the divisibility tests in Table 4.13 to determine whether the given number is divisible by each of the following: 2, 3, 4, 5, 6, 8, 9, and 10.

17. 210

18. 314

19. 51

20. 168

21. 2568

22. 3525

23. 4190

24. 6123

In Exercises 25–36, write the prime factorization of the number.

25. 18

26. 48

27. 120

28. 380

29. 425

30. 625

31. 1024

32. 1410

33. 6312

34. 3155

35. 18,234

36. 19,345

37. Use the Sieve of Eratosthenes procedure to find all prime numbers from 2 to 200. *Hint:* Because $\sqrt{200} \approx 14.1$, you need to continue the sieve procedure up to $k = 13$. Note: You do not need to consider $k = 14$ because 14 is not a prime number.

38. Use your list of prime numbers from Exercise 37 to find the number of prime numbers from:

  a. 2 to 50

  b. 51 to 100

  c. 101 to 150

  d. 151 to 200

39. If the natural numbers n and $n + 2$ are both prime numbers, then they are said to be **twin primes.** For example, 11 and 13 are twin primes. It is not known whether the set of twin primes is an infinite set or a finite set. Use the list of primes from Exercise 37 to write all twin primes less than 200.

40. If the natural numbers n, $n + 2$, and $n + 4$ are all prime numbers, then they are said to be **prime triplets.** Write a few sentences that explain why the prime triplets 3, 5, and 7 are the only prime triplets.

41. In 1742, Christian Goldbach conjectured that every even number greater than 2 can be written as the sum of two prime numbers. Many mathematicians have tried to prove or disprove this conjecture without succeeding. Show that *Goldbach's conjecture* is true for each of the following even numbers.

  a. 24

  b. 50

  c. 144

  d. 210

42. The square of a natural number is called a **perfect square.** Pick six perfect squares. For each perfect square determine the number of distinct natural-number factors of the perfect square. Make a conjecture about the number of distinct natural-number factors of any perfect square.

Every prime number has a divisibility test. Many of these divisibility tests are slight variations of the following divisibility test for 7.

A divisibility test for 7   To determine whether a given base ten number is divisible by 7, double the ones digit of the given number. Find the difference between this number and the number formed by omitting the ones digit from the given number. If necessary, repeat this procedure until you obtain a small final difference.

If the final difference is divisible by 7, then the given number is also divisible by 7. If the final difference is not divisible by 7, then the given number is not divisible by 7.

Example   Use the above divisibility test to determine whether 301 is divisible by 7.

Solution   The double of the ones digit is 2. Subtracting 2 from 30, which is the number formed by omitting the ones digit from the original number, yields 28. Because 28 is divisible by 7, the original number 301 is divisible by 7.

In Exercises 43–50, use the above divisibility test for 7 to determine whether each number is divisible by 7.

43. 182

44. 203

45. 1001

46. 2403

47. 11,561

48. 13,842

49. 204,316

50. 789,327

A divisibility test for 13 To determine whether a given base ten number is divisible by 13, multiply the ones digit of the given number by 4. Find the sum of this multiple of 4 and the number formed by omitting the ones digit from the given number. If necessary, repeat this procedure until you obtain a small final sum.

If the final sum is divisible by 13, then the given number is divisible by 13.

If the final sum is not divisible by 13, then the given number is not divisible by 13.

Example Use the above divisibility test to determine whether 1079 is divisible by 13.

Solution Four times the ones digit is 36. The number formed by omitting the ones digit is 107. The sum of 36 and 107 is 143. Now repeat the procedure on 143. Four times the ones digit is 12. The sum of 12 and 14, which is the number formed by omitting the ones digit, is 26. Because 26 is divisible by 13, the original number 1079 is divisible by 13.

In Exercises 51–58, use the above divisibility test for 13 to determine whether each number is divisible by 13.

51. 91   **52.** 273   **53.** 1885   **54.** 8931

55. 14,507   **56.** 22,184   **57.** 13,351   **58.** 85,657

Extensions

CRITICAL THINKING

59. Determine a divisibility test for 17. Hint: One divisibility test for 17 is similar to the divisibilty test on the previous page for 7 in that it involves the last digit of the given number and the operation of subtraction.

60. Determine a divisibility test for 19. Hint: One divisibility test for 19 is similar to the above divisibility test for 13 in that it involves the last digit of the given number and the operation of addition.

Number of Divisors of a Composite Number The following method can be used to determine the number of divisors of a composite number. First find the prime factorization (in exponential form) of the composite number. Add 1 to each exponent in the prime factorization and then compute the product of these exponents. This product is equal to the number of divisors of the composite number. To illustrate that this procedure yields the correct result, consider the composite number 12, which has the six divisors 1, 2, 3, 4, 6, and 12. The prime factorization of 12 is $12 = 2^2 \cdot 3^1$. Adding 1 to each of the exponents produces the numbers 3 and 2. The product of 3 and 2 is 6, which agrees with the result obtained by listing all of the divisors.

In Exercises 61–64, determine the number of divisors of each composite number.

61. 60   **62.** 84

63. 288   **64.** 360

EXPLORATIONS

65. In the 1870s, the mathematician Eduard Kummer used a proof similar to the following to show that there exist an infinite number of prime numbers. Supply the missing reasons in parts a and b.

a. *Proof* Assume there exist only a finite number of prime numbers, say $p_1,\ p_2,\ p_3, \ldots,\ p_r$. Let $N = p_1 p_2 p_3 \cdots p_r > 2$. The natural number $N - 1$ has at least one common prime factor with N. Why?

b. Call the common prime factor from part a p_i. Now p_i divides N and p_i divides $N - 1$. Thus p_i divides their difference: $N - (N - 1) = 1$. Why?

This leads to a contradiction, because no prime number is a divisor of 1. Hence Kummer concluded that the original assumption was incorrect, and there must exist an infinite number of prime numbers.

66. ***Theorem*** If a number of the form 111...1 is a prime number, then the number of 1's in the number is a prime number. For instance,

$$11 \text{ and } 1{,}111{,}111{,}111{,}111{,}111{,}111$$

are prime numbers, and the number of 1's in each number (two in the first number and 19 in the second number) is a prime number.

a. What is the converse of the above theorem? Is the converse of a theorem always true?

b. The number $111 = 3 \cdot 37$, so 111 is not a prime number. Explain why this does not contradict the above theorem.

67. State the missing reasons in parts a, b, and c of the following proof.

a. ***Theorem*** Every composite number n has at least one divisor less than or equal to $\sqrt{n}$.

Proof Assume a is a divisor of n. Then there exists a natural number j such that $aj = n$. Why?

b. Now a and j cannot both be greater than $\sqrt{n}$, because this would imply that $aj > \sqrt{n}\sqrt{n}$. However, $\sqrt{n}\sqrt{n}$ simplifies to n, which equals aj. What contradiction does this lead to?

c. Thus either a or j must be less than or equal to $\sqrt{n}$. Because j is also a divisor of n, the proof is complete. How do we know that j is a divisor of n?

68. In 1977, Ron Rivest, Adi Shamir, and Leonard Adleman invented a method for encrypting information. Their method is known as the RSA algorithm. Today the RSA algorithm is used by both the Microsoft Internet Explorer and Netscape Navigator web browsers, as well as by VISA and MasterCard to ensure secure electronic credit card transactions. The RSA algorithm involves large prime numbers. Research the RSA algorithm and write a report about some of the reasons why this algorithm has become one of the most popular of all the encryption algorithms.

SECTION 4.6 | ## Additional Topics from Number Theory

Perfect, Deficient, and Abundant Numbers

The ancient Greek mathematicians personified the natural numbers. For instance, they considered the odd natural numbers as male and the even natural numbers as female. They also used the concept of a *proper factor* to classify a natural number as *perfect, deficient,* or *abundant.* The **proper factors** of a natural number n include all natural number factors of n except for the number n itself. For instance, the proper factors of 10 are 1, 2, and 5. The proper factors of 16 are 1, 2, 4, and 8.

> **Perfect, Deficient, and Abundant Numbers**
>
> A natural number is
>
> - **perfect** if it is equal to the sum of its proper factors.
> - **deficient** if it is greater than the sum of its proper factors.
> - **abundant** if it is less than the sum of its proper factors.

EXAMPLE 1 ■ **Classify a Number as Perfect, Deficient, or Abundant**

Determine whether the following numbers are perfect, deficient, or abundant.

a. 6   **b.** 20   **c.** 25

Solution

a. The proper factors of 6 are 1, 2, and 3. The sum of these proper factors is $1 + 2 + 3 = 6$. Because 6 is equal to the sum of its proper divisors, 6 is a perfect number.

b. The proper factors of 20 are 1, 2, 4, 5, and 10. The sum of these proper factors is $1 + 2 + 4 + 5 + 10 = 22$. Because 20 is less than the sum of its proper factors, 20 is an abundant number.

c. The proper factors of 25 are 1 and 5. The sum of these proper factors is $1 + 5 = 6$. Because 25 is greater than the sum of its proper factors, 25 is a deficient number.

CHECK YOUR PROGRESS 1 Determine whether the following numbers are perfect, deficient, or abundant.

a. 24   **b.** 28   **c.** 35

Solution *See page S14.*

Mersenne Numbers and Perfect Numbers

As a French monk in the religious order known as the Minims, Marin Mersenne (mər-sĕn′) devoted himself to prayer and his studies, which included topics from number theory. Mersenne took it upon himself to collect and disseminate mathematical information to scientists and mathematicians through Europe. Mersenne was particularly interested in prime numbers of the form $2^n - 1$, where n is a prime number. Today numbers of the form $2^n - 1$ are known as **Mersenne numbers.**

Some Mersenne numbers are prime and some are composite. For instance, the Mersenne numbers $2^2 - 1$ and $2^3 - 1$ are prime numbers, but the Mersenne number $2^{11} - 1 = 2047$ is not prime because $2047 = 23 \cdot 89$.

Marin Mersenne
(1588–1648)

EXAMPLE 2 ■ **Determine Whether a Mersenne Number is a Prime Number**

Determine whether the Mersenne number $2^5 - 1$ is a prime number.

Solution
$2^5 - 1 = 31$ and 31 is a prime number. Thus $2^5 - 1$ is a Mersenne prime.

CHECK YOUR PROGRESS 2 Determine whether the Mersenne number $2^7 - 1$ is a prime number.

Solution *See page S15.*

The ancient Greeks knew that the first four perfect numbers were 6, 28, 496, and 8128. In fact, proposition 36 from Volume IX of Euclid's *Elements* states a procedure that uses Mersenne primes to produce a perfect number.

> **Euclid's Procedure for Generating a Perfect Number**
>
> If n and $2^n - 1$ are both prime numbers, then
>
> $$2^{n-1}(2^n - 1)$$
>
> is a perfect number.

Euclid's procedure shows how every Mersenne prime can be used to produce a perfect number. For instance,

If $n = 2$, then $2^{2-1}(2^2 - 1) = 2(3) = 6$.

If $n = 3$, then $2^{3-1}(2^3 - 1) = 4(7) = 28$.

If $n = 5$, then $2^{5-1}(2^5 - 1) = 16(31) = 496$.

If $n = 7$, then $2^{7-1}(2^7 - 1) = 64(127) = 8128$.

The fifth perfect number was not discovered until the year 1461. It is $2^{12}(2^{13} - 1) = 33,550,336$. The sixth and seventh perfect numbers were discovered in 1588 by P. A. Cataldi. In exponential form, they are $2^{16}(2^{17} - 1)$ and $2^{18}(2^{19} - 1)$. It is interesting to observe that the ones digits of the first five perfect numbers alternate: 6, 8, 6, 8, 6. Evaluate $2^{16}(2^{17} - 1)$ and $2^{18}(2^{19} - 1)$ to determine if this alternating pattern continues for the first seven perfect numbers.

EXAMPLE 3 ■ Use a Given Mersenne Prime Number to Write a Perfect Number

In 1750, Leonhard Euler proved that $2^{31} - 1$ is a Mersenne prime. Use Euclid's theorem to write the perfect number associated with this prime.

Solution
Euler's theorem states that if n and $2^n - 1$ are both prime numbers, then $2^{n-1}(2^n - 1)$ is a perfect number. In this example $n = 31$, which is a prime number. We are given that $2^{31} - 1$ is a prime number, so the perfect number we seek is $2^{30}(2^{31} - 1)$.

CHECK YOUR PROGRESS 3   In 1883, I. M. Pervushin proved that $2^{61} - 1$ is a Mersenne prime. Use Euclid's theorem to write the perfect number associated with this prime.

Solution   See page S15.

QUESTION   *Must a perfect number produced by Euclid's perfect-number-generating procedure be an even number?*

The search for Mersenne primes and their associated perfect numbers still continues. As of March 2002, only 39 Mersenne primes (and 39 perfect numbers) had

---

ANSWER   *Yes. The 2^{n-1} factor of $2^{n-1}(2^n - 1)$ ensures that this product will be an even number.*

been discovered. The largest known Mersenne prime is $(2^{13466917} - 1)$. See Table 4.15. This gigantic prime number, which has 4,053,946 digits, was discovered on November 14, 2001, by Michael Cameron, a 20-year-old student from Owen Sound, Ontario, Canada. Cameron used his personal computer and a program developed by George Woltman to discover this prime. More information about this program is given in the *Math Matters* on page 229.

Table 4.15 *Some Mersenne Primes and Their Associated Perfect Numbers*

	Mersenne Prime	Perfect Number	Date	Discoverer
1	$2^2 - 1$	$2^1(2^2 - 1) = 6$	B.C.	
2	$2^3 - 1$	$2^2(2^3 - 1) = 28$	B.C.	
3	$2^5 - 1$	$2^4(2^5 - 1) = 496$	B.C.	
4	$2^7 - 1$	$2^6(2^7 - 1) = 8128$	B.C.	
5	$2^{13} - 1$	$2^{12}(2^{13} - 1) = 33{,}550{,}336$	1461	Unknown
6	$2^{17} - 1$	$2^{16}(2^{17} - 1)$	1588	Cataldi
7	$2^{19} - 1$	$2^{18}(2^{19} - 1)$	1588	Cataldi
$\vdots$	$\vdots$	$\vdots$	$\vdots$	$\vdots$
30	$2^{132049} - 1$	$2^{132048}(2^{132049} - 1)$	1983	Slowinski
31	$2^{216091} - 1$	$2^{216090}(2^{216091} - 1)$	1985	Slowinski
32	$2^{756839} - 1$	$2^{756838}(2^{756839} - 1)$	1992	Slowinski and Gage
33	$2^{859433} - 1$	$2^{859432}(2^{859433} - 1)$	1994	Slowinski and Gage
34	$2^{1257787} - 1$	$2^{1257786}(2^{1257787} - 1)$	1996	Slowinski and Gage
35	$2^{1398269} - 1$	$2^{1398268}(2^{1398269} - 1)$	1996	Armengaud and Woltman, et al. (GIMPS)
?	$2^{2976221} - 1$	$2^{2976220}(2^{2976221} - 1)$	1997	Spence and Woltman, et al. (GIMPS)
?	$2^{3021377} - 1$	$2^{3021376}(2^{3021377} - 1)$	1998	Clarkson, Woltman, and Kurowski, et al. (GIMPS, PrimeNet)
?	$2^{6972593} - 1$	$2^{6972592}(2^{6972593} - 1)$	1999	Hajratwala, Woltman, and Kurowski, et al. (GIMPS, PrimeNet)
?	$2^{13466917} - 1$	$2^{13466916}(2^{13466917} - 1)$	2001	Cameron, Woltman, and Kurowski, et al. (GIMPS, PrimeNet)

Source: *The Little Book of Big Primes* by Paulo Ribenboim and the GIMPS homepage (**http://www.mersenne.org**)

The numbers in the bottom four rows of Table 4.15 are the 36th, 37th, 38th, and 39th Mersenne primes to have been discovered. However, they may not be the 36th, 37th, 38th, and 39th Mersenne primes, since there are many smaller numbers that have yet to be tested.

Math Matters   The Great Internet Mersenne Prime Search

From 1952 to 1996, large-scale computers were used to find Mersenne primes. However, during the past few years, small personal computers working in parallel have joined in the search. This search using personal computers has been organized by a fast-growing Internet organization known as the Great Internet Mersenne Prime Search (GIMPS). Members of this group use the Internet to download a Mersenne prime program that runs on their personal computers. Each member is assigned a range of numbers to check for Mersenne primes. A search over a specified range of numbers can take several weeks, but because the program is designed to run in the background, you can still use your computer to perform its regular duties. As of August 2002, five Mersenne primes had been discovered by the members of GIMPS (see the bottom five rows of Table 4.15). One of the current goals of GIMPS is to find a prime number that has more than 10 million digits. In fact, the Electronic Frontier Foundation is offering a $100,000 reward to the person or group to discover a 10-million-digit prime number. If you are using your computer just to run a screen saver, why not join in the search? You can get the needed program and additional information at

> **http://www.mersenne.org**

Who knows, maybe you will discover a new Mersenne prime and share in the reward money. Of course, we know that you really just want to have your name added to Table 4.15.

The Number of Digits in b^x

To determine the number of digits in a Mersenne number, mathematicians make use of the following formula, which involves finding the greatest integer of a number. **The greatest integer** of a number k is the greatest integer less than or equal to k. For instance, the greatest integer of 5 is 5 and the greatest integer of 7.8 is 7.

The Number of Digits in b^x

The number of digits in the number b^x, where b is a natural number less than 10 and x is a natural number, is the greatest integer of $(x \log b) + 1$.

INSTRUCTOR NOTE
A complete coverage of logarithms is given in Chapter 6.

✔ **TAKE NOTE**

If 2^{19} were a power of 10, such as 100,000, then $2^{19} - 1$ would equal 99,999 and it would have one less digit than 2^{19}. However, we know 2^{19} is not a power of 10 and thus $2^{19} - 1$ has the same number of digits as 2^{19}.

EXAMPLE 4 ■ **Determine the Number of Digits in a Mersenne Number**

Find the number of digits in the Mersenne prime number $2^{19} - 1$.

Solution

First consider just 2^{19}. The base b is 2. The exponent x is 19.

$$(x \log b) + 1 = (19 \log 2) + 1$$
$$\approx 5.72 + 1$$
$$= 6.72$$

The greatest integer of 6.72 is 6. Thus 2^{19} has six digits. The Mersenne number $2^{19} - 1$ has six digits.

CALCULATOR NOTE

To evaluate 19 log 2 on a TI-83 calculator press

19 | LOG | 2 |) | | ENTER |

CHECK YOUR PROGRESS 4   Find the number of digits in the Mersenne prime number $2^{2976221} - 1$.

Solution   *See page S15.*

Euclid's perfect-number-generating formula produces only *even* perfect numbers. Do odd perfect numbers exist? As of August 2002, no odd perfect number had been discovered and no one had been able to prove that odd perfect numbers exist or do not exist. The question of the existence of an odd perfect number is one of the major unanswered questions of number theory.

Fermat's Last Theorem

historical note

Pierre de Fermat (1601–1665) In the mathematical community, Pierre de Fermat (fĕhr′məh) is known as the Prince of Amateurs because although he spent his professional life as a councilor and judge, he spent his leisure time working on mathematics. As an amateur mathematician, Fermat did important work in analytic geometry and calculus, but he is remembered today for his work in the area of number theory. Fermat stated theorems he had developed, but seldom did he provide the actual proofs. He did not want to waste his time showing the details required of a mathematical proof, and then spend additional time defending his work once it was scrutinized by other mathematicians.

By the twentieth century all but one of Fermat's proposed theories had been proved by other mathematicians. The remaining unproved theorem became known as Fermat's Last Theorem. ■

In 1637, the French mathematician Pierre de Fermat wrote in the margin of a book:

> It is impossible to divide a cube into two cubes, or a fourth power into two fourth powers, or in general any power greater than the second into two like powers, and I have a truly marvelous demonstration of it. But this margin will not contain it.

This problem, which became known as **Fermat's Last Theorem,** can also be stated in the following manner.

Fermat's Last Theorem

There are no natural numbers x, y, z, and n that satisfy $x^n + y^n = z^n$, where n is greater than 2.

Fermat's Last Theorem has attracted a great deal of attention over the last three centuries. The theorem has become so well known that it is simply called "FLT." Some of the reasons for its popularity include the following.

- Very little mathematical knowledge is required to understand the statement of FLT.
- FLT is an extension of the well-known Pythagorean theorem $x^2 + y^2 = z^2$, which has several natural number solutions. Two such solutions are $3^2 + 4^2 = 5^2$ and $5^2 + 12^2 = 13^2$.
- It seems so simple. After all, while reading the text *Arithmetica* by Diophantus, Fermat wrote that he had discovered a *truly marvelous proof* of FLT. The only reason that Fermat gave for not providing his proof was that the margin of *Arithmetica* was too narrow to contain it.

Many famous mathematicians have worked on FLT. Some of these mathematicians tried to disprove FLT by searching for natural numbers x, y, z, and n that satisfied $x^n + y^n = z^n$, $n > 2$.

EXAMPLE 5 ■ **Check a Possible Solution to Fermat's Last Theorem**

Determine whether $x = 6$, $y = 8$, and $n = 3$ satisfies the equation $x^n + y^n = z^n$ where z is a natural number.

Solution

Substituting 6 for x, 8 for y, and 3 for n in $x^n + y^n = z^n$ yields

$$6^3 + 8^3 = z^3$$
$$216 + 512 = z^3$$
$$728 = z^3$$

The real solution of $z^3 = 728$ is $\sqrt[3]{728} \approx 8.99588289$, which is not a natural number. Thus $x = 6$, $y = 8$, and $n = 3$ does not satisfy the equation $x^n + y^n = z^n$ where z is a natural number.

CHECK YOUR PROGRESS 5   Determine whether $x = 9$, $y = 11$, and $n = 4$ satisfies the equation $x^n + y^n = z^n$ where z is a natural number.

Solution   *See page S15.*

▼ **point of interest**

Andrew Wiles

Fermat's Last Theorem had been labeled by some mathematicians as the world's hardest mathematical problem. After solving Fermat's Last Theorem, Andrew Wiles made the following remarks: "Having solved this problem there's certainly a sense of loss, but at the same time there is this tremendous sense of freedom. I was so obsessed by this problem that for eight years I was thinking about it all the time—when I woke up in the morning to when I went to sleep at night. That's a long time to think about one thing. That particular odyssey is now over. My mind is at rest."

In the eighteenth century, the great mathematician Leonhard Euler was able to make some progress on a proof of FLT. He adapted a technique that he found in Fermat's notes about another problem. In this problem, Fermat gave an outline of how to prove that the special case $x^4 + y^4 = z^4$ has no natural number solutions. Using a similar procedure, Euler was able to show that $x^3 + y^3 = z^3$ also has no natural number solutions. Thus all that was left was to show that $x^n + y^n = z^n$ has no solutions with n greater than 4.

In the nineteenth century, additional work on FLT was done by Sophie Germain, Augustin Louis Cauchy, and Gabriel Lame. Each of these mathematicians produced some interesting results, but FLT still remained unsolved.

In 1983, the German mathematician Gerd Faltings used concepts from differential geometry to prove that the number of solutions to FLT must be finite. Then, in 1988, the Japanese mathematician Yoichi Miyaoka claimed he could show that the number of solutions to FLT was not only finite, but that the number of solutions was zero, and thus he had proved FLT. At first Miyaoka's work appeared to be a valid proof, but after a few weeks of examination a flaw was discovered. Several mathematicians looked for a way to repair the flaw, but eventually they came to the conclusion that although Miyaoka had developed some interesting mathematics, he had failed to establish the validity of FLT.

In 1993, Andrew Wiles of Princeton University made a major advance toward a proof of FLT. Wiles first became familiar with FLT when he was only 10 years old. At his local public library, Wiles first learned about FLT in the book *The Last Problem* by Eric Temple Bell. In reflecting on his first thoughts about FLT, Wiles recalled

> It looked so simple, and yet all the great mathematicians in history couldn't solve it. Here was a problem that I, a ten-year-old, could understand and I knew from that moment that I would never let it go. I had to solve it.[2]

---

2. Singh, Simon. *Fermat's Enigma: The Quest to Solve the World's Greatest Mathematical Problem.* New York: Walker Publishing Company, Inc., 1997, p. 6.

Wiles took a most unusual approach to solving FLT. Whereas most contemporary mathematicians share their ideas and coordinate their efforts, Wiles decided to work alone. After 7 years of working in the attic of his home, Wiles was ready to present his work. In June of 1993 Wiles gave a series of three lectures at the Isaac Newton Institute in Cambridge, England. After showing that FLT was a corollary of his major theorem, Wiles's concluding remark was "I think I'll stop here." Many mathematicians in the audience felt that Wiles had produced a valid proof of FLT, but a formal verification by several mathematical referees was required before Wiles's work could be classified as an official proof. The verification process was lengthy and complex. Wiles's written work was about 200 pages in length, it covered several different areas of mathematics, and it used hundreds of sophisticated logical arguments and a great many mathematical calculations. Any mistake could result in an invalid proof. Thus it was not too surprising when a flaw was discovered in late 1993. The flaw did not necessarily imply that Wiles's proof could not be repaired, but it did indicate that it was not a valid proof in its present form.

It appeared that once again the proof of FLT had eluded a great effort by a well-known mathematician. Several months passed and Wiles was still unable to fix the flaw. It was a most depressing period for Wiles. He felt that he was close to solving one of the world's hardest mathematical problems, yet all of his creative efforts failed to turn the flawed proof into a valid proof. Several mathematicians felt that it was not possible to repair the flaw, but Wiles did not give up. Finally, in late 1994, Wiles had an insight that eventually led to a valid proof. The insight required Wiles to seek additional help from Richard Taylor, who had been a student of Wiles. On October 15, 1994, Wiles and Taylor presented to the world a proof that has now been judged to be a valid proof of FLT. Their proof is certainly not the "truly marvelous proof" that Fermat said he had discovered. But it has been deemed a wonderful proof that makes use of several new mathematical procedures and concepts.

INSTRUCTOR NOTE
NOVA has produced a wonderful video called *The Proof* that chronicles Wiles's efforts to prove Fermat's Last Theorem. This video can be ordered at **http://www.pbs.org/wgbh/nova/teachers/programs/**

Excursion

A Sum of the Divisors Formula

Consider the numbers 10, 12, and 28 and the sums of the proper factors of these numbers.

$$10: 1 + 2 + 5 = 8 \qquad 12: 1 + 2 + 3 + 4 + 6 = 16$$
$$28: 1 + 2 + 4 + 7 + 14 = 28$$

From the above sums we see that 10 is a deficient number, 12 is an abundant number, and 28 is a perfect number. The goal of this Excursion is to find a method that will enable us to determine whether a number is deficient, abundant, or perfect without having to first find all of its proper factors and then compute their sum.

(continued)

In the following example, we use the number 108 and its prime factorization $2^2 \cdot 3^3$ to illustrate that every factor of 108 can be written as a product of powers of its prime factors. Table 4.16 includes all the proper factors of 108 (the numbers in blue) plus the factor $2^2 \cdot 3^3$, which is 108 itself. The sum of each column is shown at the bottom (the numbers in red).

Table 4.16 *Every Factor of 108 Expressed as a Product of Powers of its Prime Factors*

	1	$1 \cdot 3$	$1 \cdot 3^2$	$1 \cdot 3^3$
	2	$2 \cdot 3$	$2 \cdot 3^2$	$2 \cdot 3^3$
	2^2	$2^2 \cdot 3$	$2^2 \cdot 3^2$	$2^2 \cdot 3^3$
Sum	7	$7 \cdot 3$	$7 \cdot 3^2$	$7 \cdot 3^3$

The sum of *all* the factors of 108 is the sum of the numbers in the bottom row.

$$\text{Sum of all factors of } 108 = 7 + 7 \cdot 3 + 7 \cdot 3^2 + 7 \cdot 3^3$$
$$= 7(1 + 3 + 3^2 + 3^3) = 7(40) = 280$$

To find the sum of just the *proper* factors, we must subtract 108 from 280, which gives us 172. Thus 108 is abundant.

We now look for a pattern for the sum of all the factors. Note that

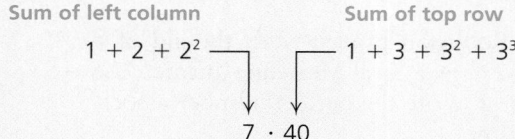

Sum of left column   Sum of top row
$1 + 2 + 2^2$ $1 + 3 + 3^2 + 3^3$

$7 \cdot 40$

This result suggests that the sum of the factors of a number can be found by finding the sum of all the prime power factors of each prime factor and then computing the product of those sums. Because we are interested only in the sum of the proper factors, we subtract the original number. Although we have not proved this result, it is a true statement and can save much time and effort. For instance, the sum of the proper factors of 3240 can be found as follows.

$$3240 = 2^3 \cdot 3^4 \cdot 5$$

Compute the sum of *all* the prime power factors of each prime factor.

$$1 + 2 + 2^2 + 2^3 = 15 \qquad 1 + 3 + 3^2 + 3^3 + 3^4 = 121 \qquad 1 + 5 = 6$$

The sum of the proper factors of 3240 is $(15)(121)(6) - 3240 = 7650$.

Excursion Exercises

Use the above technique to find the sum of the proper factors of each number and then state whether the number is deficient, abundant, or perfect.

1. 200   **2.** 262   **3.** 325   **4.** 496

5. Use deductive reasoning to prove that every prime number is deficient.

6. Use inductive reasoning to decide whether every multiple of 6 greater than 6 is abundant.

Exercise Set 4.6   (Suggested Assignment: 1–31 odds)

In Exercises 1–12, determine whether each number is perfect, deficient, or abundant.

1. 18   **2.** 32

3. 91   **4.** 51

5. 19   **6.** 144

7. 204   **8.** 128

9. 610   **10.** 508

11. 291   **12.** 1001

In Exercises 13–16, determine whether each Mersenne number is a prime number.

13. $2^3 - 1$ **14.** $2^5 - 1$

15. $2^7 - 1$ **16.** $2^{13} - 1$

17. In 1876, E. Lucas proved, without the aid of a computer, that $2^{127} - 1$ is a Mersenne prime. Use Euclid's theorem to write the perfect number associated with this prime.

18. In 1952, R. M. Robinson proved, with the aid of a computer, that $2^{521} - 1$ is a Mersenne prime. Use Euclid's theorem to write the perfect number associated with this prime.

In Exercises 19–22, determine the number of digits in the given Mersenne prime.

19. $2^{17} - 1$ **20.** $2^{132049} - 1$

21. $2^{1398269} - 1$ **22.** $2^{3021377} - 1$

In Exercises 23–26, determine the number of digits in the given perfect number. *Hint:* Use the theorem $b^p \cdot b^q = b^{p+q}$ and the formula for the number of digits in b^x from the section.

23. $2^{18}(2^{19} - 1)$ **24.** $2^{216090}(2^{216091} - 1)$

25. $2^{1257786}(2^{1257787} - 1)$ **26.** $2^{2976220}(2^{2976221} - 1)$

27. Verify that $x = 9$, $y = 15$, and $n = 5$ do not yield a solution to the equation $x^n + y^n = z^n$ where z is a natural number.

28. Verify that $x = 7$, $y = 19$, and $n = 6$ do not yield a solution to the equation $x^n + y^n = z^n$ where z is a natural number.

Extensions
CRITICAL THINKING

29. Prove that $231^n + 455^n = 1347^n$ cannot be a solution to the equation $x^n + y^n = z^n$ where z is a natural number. *Hint:* Examine the ones digits of the powers.

30. Prove that $4078^n + 3433^n = 12{,}046^n$ cannot be a solution to the equation $x^n + y^n = z^n$ where z is a natural number. *Hint:* Examine the ones digits of the powers.

31. A theorem known as *Fermat's Little Theorem* states, "If n is a prime number and a is any natural number, then $a^n - a$ is divisible by n." Verify Fermat's Little Theorem for

a. $n = 7$ and $a = 12$.

b. $n = 11$ and $a = 8$.

32. The Greeks considered the pair of numbers 220 and 284 to be *amicable* or *friendly* numbers because the sum of the proper divisors of one of the numbers is the other number.

The sum of the proper factors of 220 is

$$1 + 2 + 4 + 5 + 10 + 11 + 20 + 22 +$$
$$44 + 55 + 110 = 284$$

The sum of the proper factors of 284 is

$$1 + 2 + 4 + 71 + 142 = 220$$

Determine whether

a. 60 and 84 are amicable numbers.

b. 1184 and 1210 are amicable numbers.

EXPLORATIONS

33. Determine the smallest odd abundant number. *Hint:* It is greater than 900 but less than 1000.

34. Numbers of the form $2^{2^m} + 1$, where m is a whole number, are called *Fermat numbers*. Fermat believed that all Fermat numbers were prime. Prove that Fermat was wrong.

CHAPTER 4 **Summary**

Key Terms

additive system [p. 174]
amicable numbers [p. 234]
Babylonian numeration system [p. 185]
base ten [p. 183]
binary adders [p. 203]
binary numeration system [p. 194]
bit [p. 194]
decimal system [p. 183]
digits [p. 183]
divisibility test [p. 216]
double-dabble method [p. 198]
duodecimal system [p. 194]
Egyptian multiplication procedure [p. 181]
expanded form [p. 183]
Fermat numbers [p. 234]
Fermat's Last Theorem [p. 230]
Fermat's Little Theorem [p. 219]
Goldbach's conjecture [p. 223]
hexadecimal system [p. 194]
hieroglyphics [p. 174]
Hindu-Arabic numeration system [p. 183]
indirect proof [p. 220]
Mayan numeration system [p. 187]
Mersenne number [p. 226]
Mersenne prime [p. 226]
number theory [p. 215]
numeral [p. 174]
numeration system [p. 174]
octal numeration system [p. 193]
perfect square [p. 223]
place-value or positional-value system [p. 183]
prime factorization [p. 218]

prime triplets [p. 223]
proof by contradiction [p. 220]
proper factor [p. 225]
Roman numeration system [p. 177]
Sieve of Eratosthenes [p. 220]
successive division process [p. 195]
twin primes [p. 223]
zero as a place holder [p. 187]

Essential Concepts

- A *base b numeration system* (where b is a natural number greater than 1) has place values of $\ldots, b^5, b^4, b^3, b^2, b^1, b^0$.

- The natural number a is a *divisor,* or *factor,* of the natural number b provided there exists a natural number j such that $aj = b$.

- A *prime number* is a natural number greater than 1 that has exactly two factors (divisors), itself and 1. A *composite number* is a natural number greater than 1 that is not a prime number.

- **The Fundamental Theorem of Arithmetic**
  Every composite number can be written as a unique product of prime numbers (disregarding the order of the factors).

- A natural number is *perfect* if it is equal to the sum of its proper factors. It is *deficient* if it is greater than the sum of its proper factors. It is *abundant* if it is less than the sum of its proper factors.

- **Euclid's Perfect-Number-Generating Procedure**
  If n and $2^n - 1$ are both prime numbers, then $2^{n-1}(2^n - 1)$ is a perfect number.

CHAPTER 4   **Review Exercises**

1. Write 4,506,325 using Egyptian hieroglyphics.
2. Write 3,124,043 using Egyptian hieroglyphics.
3. Write the Egyptian hieroglyphic

   as a Hindu-Arabic numeral.

4. Write the Egyptian hieroglyphic

   as a Hindu-Arabic numeral.

In Exercises 5–8, write each Roman numeral as a Hindu-Arabic numeral.

5. CCCXLIX   **6.** DCCLXXIV

7. $\overline{\text{IX}}$DCXL   **8.** $\overline{\text{XCII}}$CDXLIV

In Exercises 9–12, write each Hindu-Arabic numeral as a Roman numeral.

9. 567   **10.** 823

11. 2489   **12.** 1335

In Exercises 13 and 14, write each Hindu-Arabic numeral in expanded form.

13. 432   **14.** 456,327

In Exercises 15 and 16, simplify each expanded form.

15. $(5 \times 10^6) + (3 \times 10^4) + (8 \times 10^3) + (2 \times 10^2) + (4 \times 10^0)$

16. $(3 \times 10^5) + (8 \times 10^4) + (7 \times 10^3) + (9 \times 10^2) + (6 \times 10^1)$

In Exercises 17–20, write each Babylonian numeral as a Hindu-Arabic numeral.

17. ⟨𝍠𝍠𝍠 ⟨⟨𝍠

18. ⟨⟨𝍠𝍠𝍠𝍠𝍠𝍠 ⟨⟨⟨⟨𝍠𝍠𝍠

19. ⟨⟨𝍠 ⟨𝍠𝍠𝍠𝍠 𝍠

20. ⟨⟨𝍠𝍠𝍠𝍠 ⟨𝍠𝍠𝍠𝍠𝍠𝍠 ⟨⟨⟨𝍠𝍠𝍠

In Exercises 21–24, write each Hindu-Arabic numeral as a Babylonian numeral.

21. 721   **22.** 1080

23. 12,543   **24.** 19,281

In Exercises 25–28, write each Mayan numeral as a Hindu-Arabic numeral.

25.

26.

27.

28.

In Exercises 29–32, write each Hindu-Arabic numeral as a Mayan numeral.

29. 522   **30.** 346

31. 1862   **32.** 1987

In Exercises 33–36, convert each numeral to base ten.

33. 45_{six} **34.** 172_{nine}

35. $\text{E3}_{\text{sixteen}}$ **36.** $\text{1BA}_{\text{twelve}}$

In Exercises 37–40, convert each numeral to the indicated base.

37. 346_{nine} to base six   **38.** 1532_{six} to base eight

39. 275_{twelve} to base nine   **40.** $\text{67A}_{\text{sixteen}}$ to base twelve

In Exercises 41–48, convert each numeral directly (without first converting to base ten) to the indicated base.

41. 11100_{two} to base eight

42. 1010100_{two} to base eight

43. 1110001101_{two} to base sixteen

44. 11101010100_{two} to base sixteen

45. 25_{eight} to base two

46. 1472_{eight} to base two

47. $\text{4A}_{\text{sixteen}}$ to base two

48. $\text{C72}_{\text{sixteen}}$ to base two

In Exercises 49–56, perform the indicated operation.

49. $235_{\text{six}} + 144_{\text{six}}$ **50.** $673_{\text{eight}} + 345_{\text{eight}}$

51. $672_{\text{nine}} - 135_{\text{nine}}$ **52.** $1332_{\text{four}} - 213_{\text{four}}$

53. $542_{\text{eight}} \times 25_{\text{eight}}$ **54.** $3421_{\text{five}} \times 43_{\text{five}}$

55. $1010101_{\text{two}} \div 11_{\text{two}}$ **56.** $321_{\text{four}} \div 12_{\text{four}}$

In Exercises 57–60, determine the prime factorization of the given number.

57. 45   **58.** 54

59. 153   **60.** 285

In Exercises 61–64, determine whether the given number is a prime number or a composite number.

61. 501   **62.** 781

63. 689   **64.** 1003

In Exercises 65–68, determine whether the given number is perfect, deficient, or abundant.

65. 28   **66.** 81

67. 144   **68.** 200

In Exercises 69 and 70, use Euclid's perfect-number-generating procedure to write the perfect number associated with the given Mersenne prime.

69. $2^{61} - 1$ **70.** $2^{1279} - 1$

In Exercises 71–74, use the Egyptian doubling procedure to find the given product.

71. 8×46 **72.** 9×57

73. 14×83 **74.** 21×143

In Exercises 75–78, use the double-dabble method to convert each base two numeral to base ten.

75. 110011010_{two} **76.** 100010101_{two}

77. 10000010001_{two} **78.** 11001010000_{two}

79. State the (base ten) divisibility test for 3.

80. State the (base ten) divisibility test for 6.

81. State the Fundamental Theorem of Arithmetic.

82. How many odd perfect numbers had been discovered as of March 2002?

83. Find the number of digits in the Mersenne number $2^{132049} - 1$.

84. Find the number of digits in the Mersenne number $2^{2976221} - 1$.

CHAPTER 4 **Test**

---

1. Write 3124 using Egyptian hieroglyphics.

2. Write the Egyptian hieroglyphic

as a Hindu-Arabic numeral.

3. Write the Roman numeral MCDXLVII as a Hindu-Arabic numeral.

4. Write 2609 as a Roman numeral.

5. Write 67,485 in expanded form.

6. Simplify:

$$(5 \times 10^5) + (3 \times 10^4) + (2 \times 10^2) + (8 \times 10^1) + (4 \times 10^0)$$

7. Write the Babylonian numeral

as a Hindu-Arabic numeral.

8. Write 9675 as a Babylonian numeral.

9. Write the Mayan numeral

as a Hindu-Arabic numeral.

10. Write 502 as a Mayan numeral.

11. Convert 3542_{six} to base ten.

12. Convert 2148 to **a.** base eight and **b.** base twelve.

13. Convert 4567_{eight} to binary form.

14. Convert 101010110111_{two} to hexadecimal form.

In Exercises 15–18, perform the indicated operation.

15. $34_{five} + 23_{five}$

16. $462_{eight} - 147_{eight}$

17. $101110_{two} \times 101_{two}$

18. $431_{seven} \div 5_{seven}$

19. Determine the prime factorization of 230.

20. Determine whether 1001 is a prime number or a composite number.

21. Use divisibility tests to determine whether 1,737,285,147 is divisible by **a.** 2, **b.** 3, or **c.** 5.

22. Use divisibility tests to determine whether 19,531,333,276 is divisible by **a.** 4, **b.** 6, or **c.** 11.

23. Determine whether 96 is perfect, deficient, or abundant.

24. Use Euclid's perfect-number-generating procedure to write the perfect number associated with the Mersenne prime $2^{17} - 1$.

The Earned Run Average (ERA) is an important statistical measure of a baseball pitcher's performance. Mariano Rivera, a relief pitcher for the New York Yankees, is considered by many baseball fans to be the top relief pitcher in baseball. From 1995 to 2001, Mariano's career ERA was 2.61. In 1999 he won his third World Series with the Yankees and was named the Most Valuable Player of the Series after pitching 12 scoreless innings and posting a 0.38 ERA (2 runs in $47\frac{1}{3}$ innings), which is ranked as the lowest post-season ERA in baseball history. The **Excursion exercises** on **page 269** demonstrate how to calculate a pitcher's ERA.

CHAPTER

5

Applications of Equations

5.1	**First-Degree Equations**
5.2	**Rate, Ratio, and Proportion**
5.3	**Percent**
5.4	**Second-Degree Equations**

Need help? For on-line student resources, such as section quizzes, visit this textbook's web site at **math.college.hmco.com/students.**

A s you progress in your study of mathematics, you have probably noticed that the problems became less concrete and more abstract. Problems that are concrete provide information pertaining to a specific instance. Abstract problems are theoretical; they are stated without reference to a specific instance. Here is an example of a concrete problem.

How many cents are in 5 dollars?

You know that there are 100 cents in 1 dollar. To find the number of cents in 5 dollars, multiply 5 by 100.

$$100 \cdot 5 = 500$$

There are 500 cents in 5 dollars.

Here is a related abstract problem.

How many cents are in d dollars?

Use the same procedure to find the number of cents in d dollars: Multiply d by 100.

$$100 \cdot d = 100d$$

There are $100d$ cents in d dollars.

This problem can be taken a step further.

If one pen costs c cents, how many pens can be purchased with d dollars?

Consider the same problem using numbers in place of the variables.

If one pen costs 25 cents, how many pens can be purchased with 2 dollars?

To solve this problem, you need to calculate the number of cents in 2 dollars (multiply 2 by 100), and divide the result by the cost per pen (25 cents).

$$\frac{100 \cdot 2}{25} = \frac{200}{25} = 8$$

If one pen costs 25 cents, 8 pens can be purchased with 2 dollars.

Use the same procedure to solve the related abstract problem. Calculate the number of cents in d dollars (multiply d by 100), and divide the result by the cost per pen (c cents).

$$\frac{100 \cdot d}{c} = \frac{100d}{c}$$

If one pen costs c cents, $\dfrac{100d}{c}$ pens can be purchased with d dollars.

At the heart of the study of algebra is the use of variables. It is the variables in the above problems that make the problems abstract. Variables enable us to generalize situations and state rules about mathematics. These rules are often stated in the form of equations and formulas. In this chapter, we will be using equations and formulas to solve problems.

TAKE NOTE

Recall that the Order of Operations Agreement states that when simplifying a numerical expression you should perform the operations in the following order:

1. Perform operations inside parentheses.

2. Simplify exponential expressions.

3. Do multiplication and division from left to right.

4. Do addition and subtraction from left to right.

Solving First-Degree Equations

The fuel economy, in miles per gallon, of a particular car traveling at a speed of v miles per hour can be calculated using the variable expression $-0.02v^2 + 1.6v + 3$. For example, suppose the speed of a car is 30 miles per hour. We can calculate the fuel economy by substituting 30 for v in the variable expression and then using the Order of Operations Agreement to evaluate the resulting numerical expression.

$$-0.02v^2 + 1.6v + 3$$
$$-0.02(30)^2 + 1.6(30) + 3 = -0.02(900) + 1.6(30) + 3$$
$$= -18 + 48 + 3$$
$$= 33$$

The fuel economy is 33 miles per gallon.

The **terms** of a variable expression are the addends of the expression. The expression $-0.02v^2 + 1.6v + 3$ has three terms. The terms $-0.02v^2$ and $1.6v$ are **variable terms** because each contains a variable. The term 3 is a **constant term;** it does not contain a variable.

Each variable term is composed of a **numerical coefficient** and a **variable part** (the variable or variables and their exponents). For the variable term $-0.02v^2$, -0.02 is the coefficient and v^2 is the variable part.

Like terms of a variable expression are terms with the same variable part. Constant terms are also like terms. Examples of like terms are:

$4x$ and $7x$
$9y$ and y
$5x^2y$ and $6x^2y$
8 and -3

An **equation** expresses the equality of two mathematical expressions. Each of the following is an equation.

$8 + 5 = 13$
$4y - 6 = 10$
$x^2 - 2x + 1 = 0$
$b = 7$

Each of the equations below is a **first-degree equation in one variable.** *First degree* means that the variable has an exponent of 1.

$x + 11 = 14$
$3z + 5 = 8z$
$2(6y - 1) = 34$

▼ **point of interest**

One of the most famous equations is $E = mc^2$. This equation, stated by Albert Einstein (īn'stīn), shows that there is a relationship between mass m and energy E. In this equation, c is the speed of light.

TAKE NOTE

It is important to note the difference between an expression and an equation. An equation contains an equal sign; an expression doesn't.

historical note

Finding solutions of equations has been a principal aim of mathematics for thousands of years. However, the equal sign did not occur in any text until 1557, when Robert Recorde (c. 1510–1558) used the symbol in *The Whetstone of Witte*. ∎

INSTRUCTOR NOTE
Students have a tendency to think of equations and expressions as the same thing. Emphasize that equations always have equal signs. Also, we *solve* equations and *simplify* expressions.

✔ **TAKE NOTE**

In the Multiplication Property, it is necessary to state $c \neq 0$ so that the solutions of the equation are not changed. For example, if $\frac{1}{2}x = 4$, then $x = 8$. But if we multiply each side of the equation by 0, we have

$$0 \cdot \frac{1}{2}x = 0 \cdot 4$$

$$0 = 0$$

The solution $x = 8$ is lost.

QUESTION *Which of the following are first-degree equations in one variable?*

   a. $5y + 4 = 9 - 3(2y + 1)$ *b.* $\sqrt{x} + 9 = 16$

   c. $p = -14$ *d.* $2x - 5 = x^2 - 9$

   e. $3y + 7 = 4z - 10$

A **solution** of an equation is a number that, when substituted for the variable, results in a true equation.

   3 is a solution of the equation $x + 4 = 7$ because $3 + 4 = 7$.

   9 is not a solution of the equation $x + 4 = 7$ because $9 + 4 \neq 7$.

To **solve an equation** means to find all solutions of the equation. The following properties of equations are often used to solve equations.

Properties of Equations

Addition Property
The same number can be added to each side of an equation without changing the solution of the equation.

   If $a = b$, then $a + c = b + c$.

Subtraction Property
The same number can be subtracted from each side of an equation without changing the solution of the equation.

   If $a = b$, then $a - c = b - c$.

Multiplication Property
Each side of an equation can be multiplied by the same nonzero number without changing the solution of the equation.

   If $a = b$ and $c \neq 0$, then $ac = bc$.

Division Property
Each side of an equation can be divided by the same nonzero number without changing the solution of the equation.

   If $a = b$ and $c \neq 0$, then $\dfrac{a}{c} = \dfrac{b}{c}$.

In solving a first-degree equation in one variable, the goal is to rewrite the equation with the variable alone on one side of the equation and a constant term on the other side of the equation.

ANSWER *The equations in **a** and **c** are first-degree equations in one variable. The equation in **b** is not a first-degree equation in one variable because it contains the square root of a variable. The equation in **d** contains a variable with an exponent other than 1. The equation in **e** contains two variables.*

✔ **TAKE NOTE**

You should always check the solution of an equation. The check for the example at the right is shown below.

$$t + 9 = -4$$
$$\overline{-13 + 9 \ \big| \ -4}$$
$$-4 = -4$$

This is a true equation. The solution checks.

For example, to solve the equation $t + 9 = -4$, use the Subtraction Property to subtract the constant term (9) from each side of the equation.

$$t + 9 = -4$$
$$t + 9 - 9 = -4 - 9$$
$$t = -13$$

Now the variable (t) is alone on one side of the equation and a constant term (-13) is on the other side. The solution is -13.

To solve the equation $-5q = 120$, use the Division Property. Divide each side of the equation by the coefficient -5.

$$-5q = 120$$
$$\frac{-5q}{-5} = \frac{120}{-5}$$
$$q = -24$$

Now the variable (q) is alone on one side of the equation and a constant (-24) is on the other side. The solution is -24.

✔ **TAKE NOTE**

An equation has some properties that are similar to those of a balance scale. For instance, if a balance scale is in balance and equal weights are added to each side of the scale, then the balance scale remains in balance. If an equation is true, then adding the same number to each side of the equation produces another true equation.

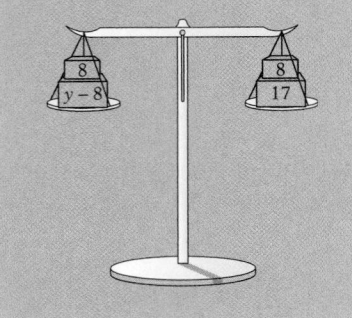

EXAMPLE 1 ■ **Solve a First-Degree Equation Using One of the Properties of Equations**

Solve.

a. $y - 8 = 17$ **b.** $4x = -2$ **c.** $-5 = 9 + b$ **d.** $-a = -36$

Solution

a. Because 8 is subtracted from y, use the Addition Property to add 8 to each side of the equation.

$$y - 8 = 17$$
$$y - 8 + 8 = 17 + 8$$
$$y = 25 \qquad \text{• A check will show that 25 is a solution.}$$

The solution is 25.

b. Because x is multiplied by 4, use the Division Property to divide each side of the equation by 4.

$$4x = -2$$
$$\frac{4x}{4} = \frac{-2}{4}$$
$$x = -\frac{1}{2} \qquad \text{• A check will show that } -\tfrac{1}{2} \text{ is a solution.}$$

The solution is $-\dfrac{1}{2}$.

c. Because 9 is added to b, use the Subtraction Property to subtract 9 from each side of the equation.

$$-5 = 9 + b$$
$$-5 - 9 = 9 - 9 + b$$
$$-14 = b$$

The solution is -14.

d. The coefficient of the variable is -1. Use the Multiplication Property to multiply each side of the equation by -1.

$$-a = -36$$
$$-1(-1a) = -1(-36)$$
$$a = 36$$

The solution is 36.

CHECK YOUR PROGRESS 1   Solve.

a. $c - 6 = -13$ **b.** $4 = -8z$ **c.** $22 + m = -9$ **d.** $5x = 0$

Solution   See page S15.

When solving more complicated first-degree equations in one variable, use the following sequence of steps.

Steps for Solving a First-Degree Equation in One Variable

1. If the equation contains fractions, multiply each side of the equation by the least common multiple (LCM) of the denominators to clear the equation of fractions.

2. Use the Distributive Property to remove parentheses.

3. Combine any like terms on the right side of the equation and any like terms on the left side of the equation.

4. Use the Addition or Subtraction Property to rewrite the equation with only one variable term and only one constant term.

5. Use the Multiplication or Division Property to rewrite the equation with the variable alone on one side of the equation and a constant term on the other side of the equation.

If one of the above steps is not needed to solve a given equation, proceed to the next step. Remember that the goal is to rewrite the equation with the variable alone on one side of the equation and a constant term on the other side of the equation.

EXAMPLE 2 ■ **Solve a First-Degree Equation Using the Properties of Equations**

Solve.

a. $5x + 9 = 23 - 2x$ **b.** $8x - 3(4x - 5) = -2x + 6$

c. $\dfrac{3x}{4} - 6 = \dfrac{x}{3} - 1$

Solution

a. There are no fractions (Step 1) or parentheses (Step 2). There are no like terms on either side of the equation (Step 3). Use the Addition Property to rewrite the equation with only one variable term (Step 4). Add $2x$ to each side of the equation.

$$5x + 9 = 23 - 2x$$
$$5x + 2x + 9 = 23 - 2x + 2x$$
$$7x + 9 = 23$$

Use the Subtraction Property to rewrite the equation with only one constant term (Step 4). Subtract 9 from each side of the equation.

$$7x + 9 - 9 = 23 - 9$$
$$7x = 14$$

Use the Division Property to rewrite the equation with the x alone on one side of the equation (Step 5). Divide each side of the equation by 7.

$$\frac{7x}{7} = \frac{14}{7}$$
$$x = 2$$

The solution is 2.

b. There are no fractions (Step 1). Use the Distributive Property to remove parentheses (Step 2).

$$8x - 3(4x - 5) = -2x + 6$$
$$8x - 12x + 15 = -2x + 6$$

Combine like terms on the left side of the equation (Step 3). Then rewrite the equation with the variable alone on one side and a constant on the other.

$$-4x + 15 = -2x + 6 \qquad \text{• Combine like terms.}$$
$$-4x + 2x + 15 = -2x + 2x + 6 \qquad \text{• The Addition Property}$$
$$-2x + 15 = 6$$
$$-2x + 15 - 15 = 6 - 15 \qquad \text{• The Subtraction Property}$$
$$-2x = -9$$
$$\frac{-2x}{-2} = \frac{-9}{-2} \qquad \text{• The Division Property}$$
$$x = \frac{9}{2}$$

The solution is $\dfrac{9}{2}$.

historical note

The letter x is used universally as the standard letter for a single unknown, which is why x-rays were so named. The scientists who discovered them did not know what they were, and so labeled them the "unknown rays," or x-rays. ■

✔ **TAKE NOTE**

Recall that the least common multiple (LCM) of two numbers is the smallest number that both numbers divide into evenly. For the example at the right, the LCM of the denominators 4 and 3 is 12.

c. The equation contains fractions (Step 1); multiply each side of the equation by the LCM of the denominators. Then rewrite the equation with the variable alone on one side and a constant on the other.

$$\frac{3x}{4} - 6 = \frac{x}{3} - 1$$

$$12\left(\frac{3x}{4} - 6\right) = 12\left(\frac{x}{3} - 1\right)$$ • The Multiplication Property

$$12 \cdot \frac{3x}{4} - 12 \cdot 6 = 12 \cdot \frac{x}{3} - 12 \cdot 1$$ • The Distributive Property

$$9x - 72 = 4x - 12$$

$$9x - 4x - 72 = 4x - 4x - 12$$ • The Subtraction Property

$$5x - 72 = -12$$

$$5x - 72 + 72 = -12 + 72$$ • The Addition Property

$$5x = 60$$

$$\frac{5x}{5} = \frac{60}{5}$$ • The Division Property

$$x = 12$$

The solution is 12.

CHECK YOUR PROGRESS 2 Solve.

a. $4x + 3 = 7x + 9$ **b.** $7 - (5x - 8) = 4x + 3$ **c.** $\frac{3x - 1}{4} + \frac{1}{3} = \frac{7}{3}$

Solution   *See page S15.*

Math Matters   **The Hubble Space Telescope**

The Hubble Space Telescope missed the stars it was targeted to photograph during the second week of May, 1990, because it was pointing in the wrong direction. The telescope was off by about one-half of one degree as a result of an arithmetic error—an addition instead of a subtraction.

Applications

In some applications of equations, we are given a formula that can be used to solve the problem. This is illustrated in Example 3.

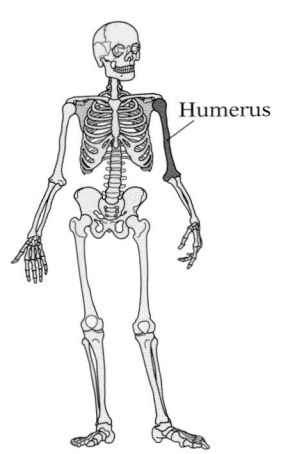

Humerus

EXAMPLE 3 ■ **Solve an Application**

Forensic scientists have determined that the equation $H = 2.9L + 78.1$ can be used to approximate the height H, in centimeters, of an adult based on the length L, in centimeters, of the adult's humerus (the bone extending from the shoulder to the elbow).

a. Use this formula to approximate the height of an adult whose humerus measures 36 centimeters.

b. According to this formula, what is the length of the humerus of an adult whose height is 168 centimeters?

Solution

a. Substitute 36 for L in the given formula. Solve the resulting equation for H.

$$H = 2.9L + 78.1$$
$$H = 2.9(36) + 78.1$$
$$H = 104.4 + 78.1$$
$$H = 182.5$$

The adult's height is approximately 182.5 centimeters.

b. Substitute 168 for H in the given formula. Solve the resulting equation for L.

$$H = 2.9L + 78.1$$
$$168 = 2.9L + 78.1$$
$$168 - 78.1 = 29L + 78.1 - 78.1$$
$$89.9 = 2.9L$$
$$\frac{89.9}{2.9} = \frac{2.9L}{2.9}$$
$$31 = L$$

The length of the adult's humerus is approximately 31 centimeters.

CHECK YOUR PROGRESS 3   The amount of garbage generated by each person living in the United States has been increasing and is approximated by the equation $P = 0.05Y - 95$, where P is the number of pounds of garbage generated per person per day and Y is the year.

a. Find the amount of garbage generated per person per day in 1990.

b. According to the equation, in what year will 5.25 pounds of garbage be generated per person per day?

Solution   *See page S15.*

In some applications of equations, we are not given a formula to use to solve the problem. Instead, we must use the given information to write an equation whose solution answers the question stated in the problem. This is illustrated in Examples 4 and 5.

EXAMPLE 4 ■ Solve an Application of First-Degree Equations

The cost of electricity in a certain city is $.08 for each of the first 300 kWh (kilowatt-hours) and $.13 for each kilowatt-hour over 300 kWh. Find the number of kilowatt-hours used by a family that receives a $51.95 electric bill.

Solution
Let k = the number of kilowatt-hours used by the family. Write an equation and then solve the equation for k.

$.08 for each of the first 300 kWh + $.13 for each kilowatt-hour over 300	=	$51.95

$$0.08(300) + 0.13(k - 300) = 51.95$$
$$24 + 0.13k - 39 = 51.95$$
$$0.13k - 15 = 51.95$$
$$0.13k - 15 + 15 = 51.95 + 15$$
$$0.13k = 66.95$$
$$\frac{0.13k}{0.13} = \frac{66.95}{0.13}$$
$$k = 515$$

The family used 515 kWh of electricity.

> ✔ **TAKE NOTE**
>
> If the family uses 500 kilowatt-hours of electricity, they are billed $.13 per kilowatt-hour for 200 kilowatt-hours (500 − 300). If they use 650 kilowatt-hours, they are billed $.13 per kilowatt-hour for 350 kilowatt-hours (650 − 300). If they use k kilowatt-hours, $k > 300$, they are billed $.13 per kilowatt-hour for (k − 300) kilowatt-hours.

CHECK YOUR PROGRESS 4

For a classified ad, a newspaper charges by the word: $7.50 for the first 25 words and 25¢ for each additional word after the first 25. You don't want to spend more than $14 on the ad you want to place in the newspaper. Determine the number of words you can place in the ad for $14.

Solution   *See page S16.*

EXAMPLE 5 ■ Solve an Application of First-Degree Equations

In 1990, the population of New Mexico was 1,515,100 and the population of West Virginia was 1,793,500. During the 1990s, New Mexico's population increased at an average rate of 21,235 people per year while West Virginia's population decreased at an average rate of 15,600 people per year. If these rate changes remained stable, in what year would the populations of New Mexico and West Virginia have been the same? Round to the nearest year.

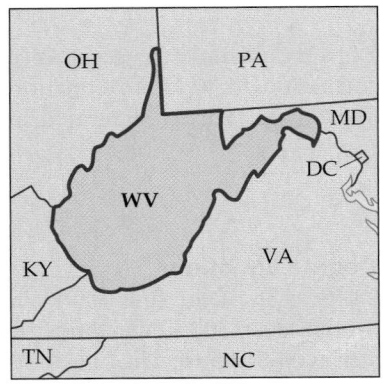

▼ **point of interest**

Is the population of your state increasing or decreasing? You can find out by checking a reference such as the *Information Please Almanac*, which was the source for the data in Example 5 and Check Your Progress 5.

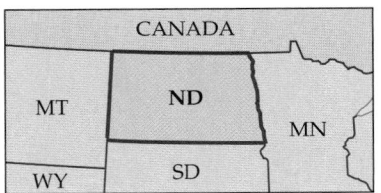

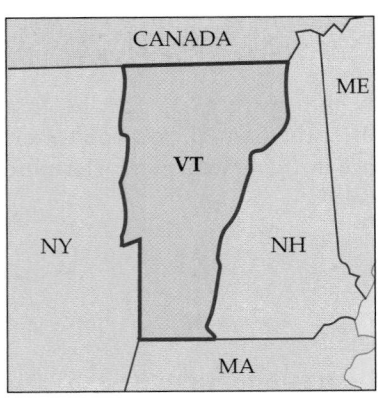

Solution

Let $n =$ the number of years until the populations are the same. Write an equation and then solve the equation for n.

The 1990 population of New Mexico plus the annual increase times n	$=$	the 1990 population of West Virginia minus the annual decrease times n

$$1{,}515{,}100 + 21{,}235n = 1{,}793{,}500 - 15{,}600n$$
$$1{,}515{,}100 + 21{,}235n + 15{,}600n = 1{,}793{,}500 - 15{,}600n + 15{,}600n$$
$$1{,}515{,}100 + 36{,}835n = 1{,}793{,}500$$
$$1{,}515{,}100 - 1{,}515{,}100 + 36{,}835n = 1{,}793{,}500 - 1{,}515{,}100$$
$$36{,}835n = 278{,}400$$
$$\frac{36{,}835n}{36{,}835} = \frac{278{,}400}{36{,}835}$$
$$n \approx 8$$

The variable n represents the number of years after 1990. Add 8 to the year 1990.

$$1990 + 8 = 1998$$

The populations would have been the same in 1998.

CHECK YOUR PROGRESS 5   In 1990, the population of North Dakota was 638,800 and the population of Vermont was 562,576. During the 1990s, North Dakota's population decreased at an average rate of 1370 people per year while Vermont's population increased at an average rate of 5116 people per year. If these rate changes remained stable, in what year would the populations of North Dakota and Vermont be the same? Round to the nearest year.

Solution   *See page S16.*

Literal Equations

A **literal equation** is an equation that contains more than one variable. Examples of literal equations are:

$$2x + 3y = 6$$
$$4a - 2b + c = 0$$

A **formula** is a literal equation that states a rule about measurement. Examples of formulas are shown below. These formulas are taken from physics, mathematics, and business.

$$\frac{1}{R_1} + \frac{1}{R_2} = \frac{1}{R}$$
$$s = a + (n - 1)d$$
$$A = P + Prt$$

QUESTION   *Which of the following are literal equations?*

a. $5a - 3b = 7$

b. $a^2 + b^2 = c^2$

c. $a_1 + (n - 1)d$

d. $3x - 7 = 5 + 4x$

The addition, subtraction, multiplication, and division properties of equations are used to solve a literal equation for one of the variables. In solving a literal equation for one of the variables, the goal is to rewrite the equation so that the letter being solved for is alone on one side of the equation and all numbers and other variables are on the other side. This is illustrated in Example 6.

EXAMPLE 6 ■ Solve a Literal Equation

a. Solve $A = P(1 + i)$ for i.

b. Solve $I = \dfrac{E}{R + r}$ for R.

Solution

a. The goal is to rewrite the equation so that i is alone on one side of the equation and all other numbers and letters are on the other side. We will begin by using the Distributive Property on the right side of the equation.

$$A = P(1 + i)$$
$$A = P + Pi$$

Subtract P from each side of the equation.

$$A - P = P - P + Pi$$
$$A - P = Pi$$

Divide each side of the equation by P.

$$\frac{A - P}{P} = \frac{Pi}{P}$$
$$\frac{A - P}{P} = i$$

b. The goal is to rewrite the equation so that R is alone on one side of the equation and all other variables are on the other side of the equation. Because the equation contains a fraction, we will first multiply both sides of the equation by the denominator $R + r$ to clear the equation of fractions.

$$I = \frac{E}{R + r}$$
$$(R + r)I = (R + r)\frac{E}{R + r}$$
$$RI + rI = E$$

INSTRUCTOR NOTE

Some students will want to simplify the expression $\dfrac{A - P}{P}$.
Remind them that we can divide by common *factors* only.

Many students will find it helpful if you provide additional examples of solving literal equations for an indicated variable. Here are a few possibilities.

1. Solve $PV = nRT$ for T.

$$T = \frac{PV}{nR}$$

2. Solve $V = \frac{1}{3}s^2h$ for h.

$$h = \frac{3V}{s^2}$$

3. Solve $S = C + rC$ for r.

$$r = \frac{S - C}{C}$$

ANSWER   *a and b are literal equations. c is not an equation. d does not have more than one variable.*

Subtract from the left side of the equation the term that does not contain a capital R.

$$RI + rI - rI = E - rI$$
$$RI = E - rI$$

Divide each side of the equation by I.

$$\frac{RI}{I} = \frac{E - rI}{I}$$

$$R = \frac{E - rI}{I}$$

CHECK YOUR PROGRESS 6

a. Solve $s = \dfrac{A + L}{2}$ for L.

b. Solve $L = a(1 + ct)$ for c.

Solution   See page S16.

Excursion

Body Mass Index

Body mass index, or **BMI,** expresses the relationship between a person's height and weight. It is a measurement for gauging a person's weight-related level of risk for high blood pressure, heart disease, and diabetes. A BMI value of 25 or less indicates a very low to low risk; a BMI value of 25 to 30 indicates a low to moderate risk; a BMI of 30 or more indicates a moderate to very high risk.

The formula for body mass index is

$$B = \frac{705W}{H^2}$$

where B is the BMI, W is weight in pounds, and H is height in inches.

(continued)

To determine how much a woman who is 5′4″ should weigh in order to have a BMI of 24, first convert 5′4″ to inches.

$$5′4″ = 5(12)″ + 4″ = 60″ + 4″ = 64″$$

Substitute 24 for B and 64 for H in the body mass index formula. Then solve the resulting equation for W.

$$B = \frac{705W}{H^2}$$

$$24 = \frac{705W}{64^2}$$ • B = 24, H = 64

$$24 = \frac{705W}{4096}$$

$$4096(24) = 4096\left(\frac{705W}{4096}\right)$$ • Multiply each side of the equation by 4096.

$$98{,}304 = 705W$$

$$\frac{98{,}304}{705} = \frac{705W}{705}$$ • Divide each side of the equation by 705.

$$139 \approx W$$

A woman who is 5′4″ should weigh about 139 pounds in order to have a BMI of 24.

Excursion Exercises

1. Amy is 140 pounds and 5′8″ tall. Calculate Amy's BMI. Round to the nearest tenth. Would you rank Amy as a low, moderate, or high risk for weight-related disease?

2. Carlos is 6′1″ and weighs 225 pounds. Calculate Carlos's BMI. Round to the nearest tenth. Would you rank Carlos as a low, moderate, or high risk for weight-related disease?

3. Roger is 5′11″. How much should he weigh in order to have a BMI of 25? Round to the nearest whole number.

4. Brenda is 5′3″. How much should she weigh in order to have a BMI of 24? Round to the nearest whole number.

5. Bohdan weighs 185 pounds and is 5′9″. How many pounds must Bohdan lose in order to reach a BMI of 23? Round to the nearest whole number.

6. Pat is 6′3″ and weighs 245 pounds. Calculate the number of pounds Pat must lose in order to reach a BMI of 22. Round to the nearest whole number.

7. Zack weighs 205 pounds and is 6′0″. He would like to lower his BMI to 20.

   a. By how many points must Zack lower his BMI? Round to the nearest tenth.

   b. How many pounds must Zack lose in order to reach a BMI of 20? Round to the nearest whole number.

8. Felicia weighs 160 pounds and is 5′7″. She would like to lower her BMI to 20.

   a. By how many points must Felicia lower her BMI? Round to the nearest tenth.

   b. How many pounds must Felicia lose in order to reach a BMI of 20? Round to the nearest whole number.

Exercise Set 5.1   (Suggested Assignment: 5–41, every other odd; 43–91, odds)

1. What is the difference between an expression and an equation? Provide an example of each.

2. What is the solution of the equation $x = 8$? Use your answer to explain why the goal in solving

an equation is to get the variable alone on one side of the equation.

3. Explain how to check the solution of an equation.

In Exercises 4–41, solve the equation.

4. $x + 7 = -5$

5. $9 + b = 21$

6. $-9 = z - 8$

7. $b - 11 = 11$

8. $-3x = 150$

9. $-48 = 6z$

10. $-9a = -108$

11. $-\dfrac{3}{4}x = 15$

12. $\dfrac{5}{2}x = -10$

13. $-\dfrac{x}{4} = -2$

14. $\dfrac{2x}{5} = -8$

15. $4 - 2b = 2 - 4b$

16. $4y - 10 = 6 + 2y$

17. $5x - 3 = 9x - 7$

18. $10z + 6 = 4 + 5z$

19. $3m + 5 = 2 - 6m$

20. $6a - 1 = 2 + 2a$

21. $5x + 7 = 8x + 5$

22. $2 - 6y = 5 - 7y$

23. $4b + 15 = 3 - 2b$

24. $2(x + 1) + 5x = 23$

25. $9n - 15 = 3(2n - 1)$

26. $7a - (3a - 4) = 12$

27. $5(3 - 2y) = 3 - 4y$

28. $9 - 7x = 4(1 - 3x)$

29. $2(3b + 5) - 1 = 10b + 1$

30. $2z - 2 = 5 - (9 - 6z)$

31. $4a + 3 = 7 - (5 - 8a)$

32. $5(6 - 2x) = 2(5 - 3x)$

33. $4(3y + 1) = 2(y - 8)$

34. $2(3b - 5) = 4(6b - 2)$

35. $3(x - 4) = 1 - (2x - 7)$

36. $\dfrac{2y}{3} - 4 = \dfrac{y}{6} - 1$

37. $\dfrac{x}{8} + 2 = \dfrac{3x}{4} - 3$

38. $\dfrac{2x - 3}{3} + \dfrac{1}{2} = \dfrac{5}{6}$

39. $\dfrac{2}{3} + \dfrac{3x + 1}{4} = \dfrac{5}{3}$

40. $\dfrac{1}{2}(x + 4) = \dfrac{1}{3}(3x - 6)$

41. $\dfrac{3}{4}(x - 8) = \dfrac{1}{2}(2x + 4)$

The monthly car payment on a 60-month car loan at a 9 percent rate is calculated by using the formula $P = 0.02076L$, where P is the monthly car payment and L is the loan amount. Use this formula for Exercises 42 and 43.

42. If you can afford a maximum monthly car payment of $300, what is the maximum loan amount you can afford? Round to the nearest cent.

43. If the maximum monthly car payment you can afford is $350, what is the maximum loan amount you can afford? Round to the nearest cent.

The music you hear when listening to a magnetic cassette tape is the result of the magnetic tape passing over magnetic heads, which read the magnetic information on the tape. The length of time a tape will play depends on the length of the tape and the operating speed of the tape player. The formula is $T = \frac{L}{S}$, where T is the time in seconds, L is the length of the tape in inches, and S is the operating speed in inches per second. Use this formula for Exercises 44–46.

44. How long a tape does a marine biologist need to record 16 seconds of whale songs at an operating speed of $1\frac{7}{8}$ inches per second?

45. How long a tape does a police officer need to record a 3-minute confession at an operating speed of $7\frac{1}{2}$ inches per second?

46. A U2 cassette tape takes 50 minutes to play on a system with an operating speed of $3\frac{3}{4}$ inches per second. Find the length of the tape.

The pressure on a diver can be calculated using the formula $P = 15 + \frac{1}{2}D$, where P is the pressure in pounds per square inch and D is the depth in feet. Use this formula for Exercises 47 and 48.

47. Find the depth of a diver when the pressure on the diver is 45 pounds per square inch.

48. Find the depth of a diver when the pressure on the diver is 55 pounds per square inch.

The world-record time for a 1-mile race can be approximated by $t = 17.08 - 0.0067y$, where y is the year of the race and t is the time, in minutes, of the race. Use this formula for Exercises 49 and 50.

49. Approximate the year in which the first "4-minute mile" was run. The actual year was 1954.

50. In 1985, the world-record time for a 1-mile race was 3.77 minutes. For what year does the equation predict this record time?

Black ice is an ice covering on roads that is especially difficult to see and therefore extremely dangerous for motorists. The distance a car traveling at 30 miles per hour will slide after its brakes are applied is related to the outside temperature by the formula $C = \frac{1}{4}D - 45$, where C is the Celsius temperature and D is the distance, in feet, the car will slide. Use this formula for Exercises 51 and 52.

51. Determine the distance a car will slide on black ice when the outside air temperature is $-3°C$.

52. How far will a car slide on black ice when the outside air temperature is $-11°C$?

The formula $N = 7C - 30$ approximates N, the number of times per minute a cricket chirps when the air temperature is C degrees Celsius. Use this formula for Exercises 53 and 54.

53. What is the approximate air temperature when a cricket chirps 100 times per minute? Round to the nearest tenth.

54. Determine the approximate air temperature when a cricket chirps 140 times per minute. Round to the nearest tenth.

In order to equalize the chances of winning, some players in a bowling league are given a handicap, or a bonus of extra points. Some leagues use the formula $H = 0.8(200 - A)$, where H is the handicap and A is the bowler's average score in past games. Use this formula for Exercises 55 and 56.

55. A bowler has a handicap of 20. What is the bowler's average score?

56. Find the average score of a bowler who has a handicap of 25.

In Exercises 57–69, write an equation as part of solving the problem.

57. As a result of depreciation, the value of a car is now $13,200. This is three-fifths of its original value. Find the original value of the car.

58. The operating speed of a personal computer is 100 megahertz. This is one-third the speed of a newer model. Find the speed of the newer personal computer.

59. The chart below shows some average starting salaries for college graduates in a recent year. (*Source:* Michigan State University)

a. The average starting salary for a college graduate with a chemical engineering major is $431 more than twice the average starting salary for a telecommunications major. Find the average starting salary for a telecommunications major.

b. The average starting salary for a college graduate with an electrical engineering major is $7989 less than twice the average starting salary for a liberal arts major. Find the average starting salary for a liberal arts major.

College major	1998 Salary
Chemical Engineering	$44,557
Electrical Engineering	$41,167
Computer Science	$38,741
Physics	$36,692
Chemistry	$35,227
Mathematics	$33,180
Nursing	$31,802
Journalism	$24,588

60. The table on the following page shows 1998's top earners, based on income from entertainment work alone. (*Source: Forbes* magazine)

a. Steven Spielberg's 1998 income was $15 million more than four times Stephen King's 1998 income. Find Stephen King's 1998 income from entertainment.

b. Tim Allen's 1998 income was $18 million less than twice Eddie Murphy's 1998 income. Find Eddie Murphy's 1998 income from entertainment.

1998 Ranking	Star	1998 Income (in millions)
1	Jerry Seinfeld	225
2	Larry David	200
3	Stephen Spielberg	175
4	Oprah Winfrey	125
5	James Cameron	115
6	Tim Allen	77
7	Michael Crichton	65
8	Harrison Ford	58
9	Rolling Stones	57
10	Master P	56.5
11	Robin Williams	56
12	Celine Dion	55

61. The purchase price of a big-screen TV, including finance charges, was $3276. A down payment of $450 was made, and the remainder was paid in 24 equal monthly installments. Find the monthly payment.

62. The cost to replace a water pump in a sports car was $600. This included $375 for the water pump and $45 per hour for labor. How many hours of labor were required to replace the water pump?

63. The table below provides statistics on the space shuttle Discovery, on which John Glenn was a crew member in 1998. His previous space flight was on the Friendship 7 in 1962. (*Source: Time* magazine, August 17, 1998)

a. The lift-off thrust of the Discovery was 85,000 kilograms less than 20 times the lift-off thrust of the Friendship 7. Find the lift-off thrust of the Friendship 7.

b. The weight of the Discovery was 290 kilograms greater than 36 times the weight of the Friendship 7. Find the weight of the Friendship 7.

The Space Shuttle *Discovery*	
Crew size	7
Crew work area	66 cubic meters
Windows	5
Computers	5
Toggle switches	856
Lift-off thrust	3,175,000 kilograms
Weight	69,770 kilograms

64. A university employs a total of 600 teaching assistants and research assistants. There are three times as many teaching assistants as research assistants. Find the number of research assistants employed by the university.

65. A service station attendant is paid time-and-a-half for working over 40 hours per week. Last week the attendant worked 47 hours and earned $631.25. Find the attendant's regular hourly wage.

66. An investor deposited $5000 in two accounts. Two times the smaller deposit is $1000 more than the larger deposit. Find the amount deposited in each account.

67. A computer screen consists of tiny dots of light called *pixels*. In a certain graphics mode, there are 640 horizontal pixels. This is 40 more than three times the number of vertical pixels. Find the number of vertical pixels.

68. An overnight mail service charges $5.60 for the first 6 ounces and $.85 for each additional ounce or fraction of an ounce. Find the weight, in ounces, of a package that cost $10.70 to deliver.

69. The charges for a long-distance telephone call are $1.42 for the first 3 minutes and $.65 for each additional minute or fraction of a minute. If charges for a call were $10.52, for how many minutes did the phone call last?

In Exercises 70–87, solve the formula for the indicated variable.

70. $A = \dfrac{1}{2}bh$; h (Geometry)

71. $P = a + b + c$; b (Geometry)

72. $d = rt$; t (Physics)

73. $E = IR$; R (Physics)

74. $PV = nRT$; R (Chemistry)

75. $I = Prt$; r (Business)

76. $P = 2L + 2W$; W (Geometry)

77. $F = \dfrac{9}{5}C + 32$; C (Temperature conversion)

78. $P = R - C$; C (Business)

79. $A = P + Prt$; t (Business)

80. $S = V_0t - 16t^2$; V_0 (Physics)

81. $T = fm - gm$; f (Engineering)

82. $P = \dfrac{R - C}{n}$; R (Business)

83. $R = \dfrac{C - S}{t}$; S (Business)

84. $V = \dfrac{1}{3}\pi r^2 h$; h (Geometry)

85. $A = \dfrac{1}{2}h(b_1 + b_2)$; b_2 (Geometry)

86. $a_n = a_1 + (n - 1)d$; d (Mathematics)

87. $S = 2\pi r^2 + 2\pi rh$; h (Geometry)

In Exercises 88 and 89, solve the equation for y.

88. $2x - y = 4$ **89.** $4x + 3y = 6$

In Exercises 90 and 91, solve the equation for x.

90. $ax + by + c = 0$ **91.** $y - y_1 = m(x - x_1)$

Extensions

CRITICAL THINKING

In Exercises 92–95, solve the equation.

92. $3(4x + 2) = 7 - 4(1 - 3x)$

93. $9 - (6 - 4x) = 3(1 - 2x)$

94. $4(x + 5) = 30 - (10 - 4x)$

95. $7(x - 20) - 5(x - 22) = 2x - 30$

96. Use the numbers 5, 10, and 15 to make equations by filling in the boxes: $x + \square = \square - \square$. Each equation must use all three numbers.

    a. What is the largest possible solution of these equations?

    b. What is the smallest possible solution of these equations?

97. Solve the equation $ax + b = cx + d$ for x. Is your solution valid for all numbers a, b, c, and d? Explain.

COOPERATIVE LEARNING

98. Arrange the numbers

1, 1, 1, 1, 2, 2, 2, 2, 3, 3, 3, 3, 4, 4, 4, 4

in the grid so that the sum of each row, each column, each of the two diagonals, the four corner boxes, and the set of four boxes at the corners is the same.

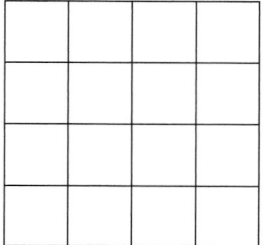

EXPLORATIONS

99. Recall that formulas state rules about relationships. When we know there is an explicit relationship between two variables, we can write a formula to express the relationship.

For example, the Hooksett tollbooth on Interstate 93 in New Hampshire collects a toll of $.75 from each car that passes through the toll. We can write a formula that describes the amount of money collected from passenger cars on any given day.

Let A be the total amount of money collected, and let c be the number of cars that pass through the tollbooth on a given day. Then

$$A = \$.75c$$

is a formula that expresses the total amount of money collected from passenger cars on any given day.

a. How much money is collected from passenger cars on a day on which 5500 passenger cars pass through the toll?

b. How many passenger cars passed through the toll on a day on which $3243.75 was collected in tolls from passenger cars?

Write formulas for each of the following situations. Include as part of your answer a list of variables that were used, and state what each variable represents.

c. Write a formula to represent the total cost to rent a copier from a company that charges $225 per month plus $.03 per copy made.

d. Suppose you buy a used car with 30,000 miles on it. You expect to drive the car about 750 miles per month. Write a formula to represent the total number of miles the car has been driven after you have owned it for m months.

e. A parking garage charges $2.50 for the first hour and $1.75 for each additional hour. Write a formula to represent the parking charge for parking in this garage for h hours.

f. Write a formula to represent the total cost to rent a car from a company that rents cars for $19.95 per day plus 25¢ for every mile driven over 100 miles.

g. Think of a mathematical relationship that can be modeled by a formula. Identify the variables in the relationship and write a formula that models the relationship.

SECTION 5.2   **Rate, Ratio, and Proportion**

Rates

The word *rate* is used frequently in our everyday lives. It is used in such contexts as unemployment rate, tax rate, interest rate, hourly rate, infant mortality rate, school dropout rate, inflation rate, and postage rate.

A **rate** is a comparison of two quantities. A rate can be written as a fraction.

A car travels 135 miles on 6 gallons of gas. The miles-to-gallons rate is written

$$\frac{135 \text{ miles}}{6 \text{ gallons}}$$

Note that the units (miles and gallons) are written as part of the rate.

▼ **point of interest**

Unit rates are used in a wide variety of situations. One unit rate you may not be familiar with is used in the airline industry: cubic feet of air per minute per person. Typical rates are: economy class, 7 cubic feet/minute/person; first class, 50 cubic feet/minute/person; cockpit, 150 cubic feet/minute/person.

A **unit rate** is a rate in which the number in the denominator is 1. To find a unit rate, divide the number in the numerator of the rate by the number in the denominator of the rate. For the example above:

$$135 \div 6 = 22.5$$

The unit rate is $\dfrac{22.5 \text{ miles}}{1 \text{ gallon}}$.

This rate can be written 22.5 miles/gallon or 22.5 miles per gallon, where the word *per* has the meaning "for every."

Unit rates make comparisons easier. For example, if you travel 37 miles per hour and I travel 43 miles per hour, we know that I am traveling faster than you are. It is more difficult to compare speeds if we are told that you are traveling $\dfrac{111 \text{ miles}}{3 \text{ hours}}$ and I am traveling $\dfrac{172 \text{ miles}}{4 \text{ hours}}$.

EXAMPLE 1 ■ Calculate a Unit Rate

A dental hygienist earns $780 for working a 40-hour week. What is the hygienist's hourly rate of pay?

Solution
The hygienist's rate of pay is $\dfrac{\$780}{40 \text{ hours}}$.
To find the hourly rate of pay, divide 780 by 40.

$$780 \div 40 = 19.5$$

$$\frac{\$780}{40 \text{ hours}} = \frac{\$19.50}{1 \text{ hour}} = \$19.50/\text{hour}$$

The hygienist's hourly rate of pay is $19.50 per hour.

CHECK YOUR PROGRESS 1 You pay $4.92 for 1.5 pounds of hamburger. What is the cost per pound?

Solution See page S16.

EXAMPLE 2 ■ Solve an Application of Unit Rates

A teacher earns a salary of $34,200 per year. Currently the school year is 180 days. If the school year were extended to 220 days, as is proposed in some states, what annual salary should the teacher be paid if the salary is based on the number of days worked per year?

Solution
Find the current salary per day.

$$\frac{\$34,200}{180 \text{ days}} = \frac{\$190}{1 \text{ day}} = \$190/\text{day}$$

Multiply the salary per day by the number of days in the proposed school year.

$$\frac{\$190}{1 \text{ day}} \cdot 220 \text{ days} = \$190(220) = \$41,800$$

The teacher's annual salary should be $41,800.

CHECK YOUR PROGRESS 2 In September 1997, the federal minimum wage rose from $4.75 per hour to $5.15 per hour. Find the increase in pay per week for an employee working 35 hours per week and earning the federal minimum wage.

Solution   *See page S16.*

Grocery stores are required to provide customers with unit price information. The **unit price** of a product is its cost per unit of measure. Unit pricing is an application of unit rate.

The price of a 2-pound box of spaghetti is $1.89. The unit price of the spaghetti is the cost per pound. To find the unit price, write the rate as a unit rate.

The numerator is the price and the denominator is the quantity. Divide the number in the numerator by the number in the denominator.

$$\frac{\$1.89}{2 \text{ pounds}} = \frac{\$.945}{1 \text{ pound}}$$

The unit price of the spaghetti is $.945 per pound.

Unit pricing is used by consumers to answer the question, "Which is the better buy?" The answer is that the product with the lower unit price is the more economical purchase.

EXAMPLE 3 ■ Determine the More Economical Purchase

Which is the more economical purchase, an 18-ounce jar of peanut butter priced at $3.49 or a 12-ounce jar of peanut butter priced at $2.59?

Solution
Find the unit price for each item.

$$\frac{\$3.49}{18 \text{ ounces}} \approx \frac{\$.194}{1 \text{ ounce}} \qquad \frac{\$2.59}{12 \text{ ounces}} \approx \frac{\$.216}{1 \text{ ounce}}$$

Compare the two prices per ounce.

$$\$.194 < \$.216$$

The item with the lower unit price is the more economical purchase.
The more economical purchase is the 18-ounce jar priced at $3.49.

CHECK YOUR PROGRESS 3 Which is the more economical purchase, 32 ounces of detergent for $2.99 or 48 ounces of detergent for $3.99?

Solution   *See page S16.*

Rates such as crime statistics or data on fatalities are often written as rates per hundred, per thousand, per hundred thousand, or per million. For example, the table below shows bicycle deaths per million people in the states with the highest and the lowest rates. (*Source:* Environmental Working Group)

Rates of Bicycle Fatalities (Deaths per Million People)			
Highest		Lowest	
Florida	8.8	North Dakota	1.7
Arizona	7.0	Oklahoma	1.6
Louisiana	5.9	New Hampshire	1.4
South Carolina	5.4	West Virginia	1.2
North Carolina	4.5	Rhode Island	1.1

▼ **point of interest**

In 2001, the Centers for Disease Control reported that the teen pregnancy rate in the United States had fallen to 94.3 pregnancies per 1000 girls aged 15 to 19.

The rates in this table are easier to read than they would be if they were unit rates. As a comparison, consider that the bicycle fatalities in North Carolina would be written as 0.0000045 as a unit rate. Also, it is easier to understand that 7 out of every million people living in Arizona die in bicycle accidents, rather than to consider that 0.000007 out of every person in Arizona dies in a bicycle accident.

QUESTION   *What does the rate given for Oklahoma mean?*

Another application of rates is in the area of international trade. Suppose a company in France purchases a shipment of sneakers from an American company. The French company must exchange euros, which is France's currency, for U.S. dollars in order to pay for the order. The number of euros that are equivalent to one U.S. dollar is called the *exchange rate.*

The table below shows the exchange rates per U.S. dollar for four foreign countries on January 24, 2002. Use this table for Example 4 and Check Your Progress 4.

Exchange Rate per U.S. Dollar	
British Pound	0.7026
Canadian Dollar	1.602
Japanese Yen	134.5
Swiss Franc	1.676

EXAMPLE 4 ■ Solve an Application Using Exchange Rates

a. How many Japanese yen are needed to pay for an order costing $10,000?

b. Find the number of British pounds that would be exchanged for $5000.

Solution

a. Multiply the number of yen per $1 by 10,000.

10,000(134.5) = 1,345,000

1,345,000 Japanese yen are needed to pay for an order costing $10,000.

ANSWER   *Oklahoma's rate of 1.6 means that 1.6 out of every million people living in Oklahoma die in bicycle accidents.*

b. Multiply the number of pounds per $1 by 5000.

5000(0.7026) = 3513

3513 British pounds would be exchanged for $5000.

CHECK YOUR PROGRESS 4
a. How many Canadian dollars would be needed to pay for an order costing $20,000?
b. Find the number of Swiss francs that would be exchanged for $25,000.

Solution   See page S17.

Solution   See page S17.

Ratios

A **ratio** is the comparison of two quantities that have the same units. A ratio can be written in three different ways:

1. As a fraction $\dfrac{2}{3}$

2. As two numbers separated by a colon (:)   2 : 3

3. As two numbers separated by the word *to*   2 to 3

Although units, such as hours, miles, or dollars, are written as part of a rate, units are not written as part of a ratio.

According to the most recent census, there are 50 million married women in the United States, and 30 million of these women work in the labor force. The ratio of the number of married women who are employed in the labor force to the total number of married women in the country is calculated below. Note that the ratio is written in simplest form.

$$\frac{30,000,000}{50,000,000} = \frac{3}{5} \quad \text{or} \quad 3:5 \quad \text{or} \quad 3 \text{ to } 5$$

The ratio 3 to 5 tells us that 3 out of every 5 married women in the United States are part of the labor force.

Given that 30 million of the 50 million married women in the country work in the labor force, we can calculate the number of married women who do not work in the labor force.

50 million − 30 million = 20 million

The ratio of the number of married women who are not in the labor force to the number of married women who are is:

$$\frac{20,000,000}{30,000,000} = \frac{2}{3} \quad \text{or} \quad 2:3 \quad \text{or} \quad 2 \text{ to } 3$$

The ratio 2 to 3 tells us that for every 2 married women who are not in the labor force, there are 3 married women who are in the labor force.

EXAMPLE 5 ■ Determine a Ratio in Simplest Form

A survey revealed that, on average, eighth-graders watch approximately 21 hours of television each week. Find the ratio, as a fraction in simplest form, of the number of hours spent watching television to the total number of hours in a week.

▼ point of interest

It is believed that billiards was invented in France during the reign of Louis XI (1423–1483). In the United States, the standard billiard table is 4 feet 6 inches by 9 feet. This is a ratio of 1 : 2. The same ratio holds for carom and snooker tables, which are 5 feet by 10 feet.

INSTRUCTOR NOTE
Ratios have applications to many disciplines. Investors talk of price-earnings ratios. Accountants use the current ratio, which is the ratio of current assets to current liabilities. Metallurgists use ratios to make various grades of steel.

Solution

A ratio is the comparison of two quantities with the same units. In this problem we are given both hours and weeks. We must first convert one week to hours.

$$\frac{24 \text{ hours}}{1 \text{ day}} \cdot 7 \text{ days} = (24 \text{ hours})(7) = 168 \text{ hours}$$

Write in simplest form the ratio of the number of hours spent watching television to the number of hours in one week.

$$\frac{21 \text{ hours}}{1 \text{ week}} = \frac{21 \text{ hours}}{168 \text{ hours}} = \frac{21}{168} = \frac{1}{8}$$

The ratio is $\frac{1}{8}$.

CHECK YOUR PROGRESS 5

a. According to the National Low Income Housing Coalition, a minimum-wage worker ($5.15 per hour) living in New Jersey would have to work 120 hours per week to afford the rent on an average two-bedroom apartment and be within the federal standard of 30% of income for housing. Find the ratio, as a fraction in simplest form, of the number of hours a minimum-wage worker would spend working per week to the total number of hours in a week.

b. Although a minimum-wage worker in New Jersey would have to work 120 hours per week to afford a two-bedroom rental, the national average is 60 hours of work per week. For this "average" worker, find the ratio, written using the word *to,* of the number of hours per week spent working to the number of hours not spent working.

Solution See page S17.

A **unit ratio** is a ratio in which the number in the denominator is 1. One situation in which a unit ratio is used is student-faculty ratios.

The table below shows the number of full-time men and women undergraduates, as well as the number of full-time faculty, at two universities in the Pacific 10. Use this table for Example 6 and Check Your Progress 6. (*Source:* Barron's Profile of American Colleges, 24th edition, c. 2001)

University	Men	Women	Faculty
Oregon State University	6444	5386	1081
University of Oregon	5507	6221	777

EXAMPLE 6 ■ Determine a Unit Ratio

 Calculate the student-faculty ratio at Oregon State University. Round to the nearest whole number. Write the ratio using the word *to.*

Solution

Add the number of male undergraduates and the number of female undergraduates to determine the total number of students.

$$6444 + 5386 = 11{,}830$$

Write the ratio of the total number of students to the number of faculty. Divide the numerator and denominator by the denominator. Then round the numerator to the nearest whole number.

$$\frac{11{,}830}{1081} \approx \frac{10.9436}{1} \approx \frac{11}{1}$$

The ratio is approximately 11 to 1.

CHECK YOUR PROGRESS 6

 Calculate the student-faculty ratio at the University of Oregon. Round to the nearest whole number. Write the ratio using the word *to*.

Solution *See page S17.*

Math Matters   Scale Model Buildings

George E. Slye of Tuftonboro, New Hampshire has built scale models of more than 100 of the best-known buildings in America and Canada. From photographs, floor plans, roof plans, and architectural drawings, Slye has constructed a replica of each building using a scale of 1 inch per 200 feet.

The buildings and landmarks are grouped together on an 8-foot-by-8-foot base as if they were all erected in a single city. Slye's city includes such well-known landmarks as New York's Empire State Building, Chicago's Sears Tower, Boston's John Hancock Tower, Seattle's Space Needle, and San Francisco's Golden Gate Bridge.

Proportions

Thus far in this section we have discussed rates and ratios. Now that you have an understanding of rates and ratios, you are ready to work with proportions. A **proportion** is an equation that states the equality of two rates or ratios. The following are examples of proportions.

$$\frac{250 \text{ miles}}{5 \text{ hours}} = \frac{50 \text{ miles}}{1 \text{ hour}} \qquad \frac{3}{6} = \frac{1}{2}$$

The first example above is the equality of two rates. Note that the units in the numerators (miles) are the same and the units in the denominators (hours) are the same. The second example is the equality of two ratios. Remember that units are not written as part of a ratio.

The definition of a proportion can be stated as follows: If $\frac{a}{b}$ and $\frac{c}{d}$ are equal ratios or rates, then $\frac{a}{b} = \frac{c}{d}$ is a proportion.

Each of the four members in a proportion is called a **term.** Each term is numbered as shown below.

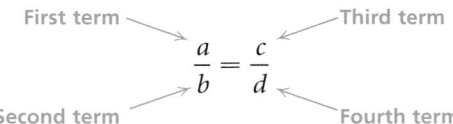

First term ⟍   ⟋ Third term
$$\frac{a}{b} = \frac{c}{d}$$
Second term ⟋   ⟍ Fourth term

The second and third terms of the proportion are called the **means** and the first and fourth terms are called the **extremes.**

If we multiply both sides of the proportion by the product of the denominators, we obtain the following result.

$$\frac{a}{b} = \frac{c}{d}$$
$$bd\left(\frac{a}{b}\right) = bd\left(\frac{c}{d}\right)$$
$$ad = bc$$

Note that ad is the product of the extremes and bc is the product of the means. In any proportion, the product of the means equals the product of the extremes. This is sometimes phrased, "the cross products are equal."

In the proportion $\frac{3}{4} = \frac{9}{12}$, the cross products are equal.

$$\frac{3}{4} \quad \frac{9}{12} \qquad \begin{array}{l} 4 \cdot 9 = 36 \longleftarrow \text{Product of the means} \\ 3 \cdot 12 = 36 \longleftarrow \text{Product of the extremes} \end{array}$$

QUESTION   *For the proportion $\frac{5}{8} = \frac{10}{16}$, **a.** name the first and third terms, **b.** write the product of the means, and **c.** write the product of the extremes.*

ANSWER   *a. The first term is 5. The third term is 10. **b.** The product of the means is $8(10) = 80$. **c.** The product of the extremes is $5(16) = 80$.*

Sometimes one of the terms in a proportion is unknown. In this case, it is necessary to solve the proportion for the unknown number. The **cross-products method,** which is based on the fact that the product of the means equals the product of the extremes, can be used to solve the proportion. Remember that the cross-products method is just a short cut for multiplying each side of the equation by the least common multiple of the denominators.

Cross-Products Method of Solving a Proportion

If $\frac{a}{b} = \frac{c}{d}$, then $ad = bc$.

EXAMPLE 7 ■ Solve a Proportion

Solve: $\dfrac{8}{5} = \dfrac{n}{6}$

Solution

Use the cross-products method of solving a proportion: the product of the means equals the product of the extremes. Then solve the resulting equation for n.

$$\frac{8}{5} = \frac{n}{6}$$

$$8 \cdot 6 = 5 \cdot n$$

$$48 = 5n$$

$$\frac{48}{5} = \frac{5n}{5}$$

$$9.6 = n$$

The solution is 9.6.

> ✔ **TAKE NOTE**
>
> Be sure to check the solution.
>
> $$\frac{8}{5} = \frac{9.6}{6}$$
>
> $$8 \cdot 6 = 5 \cdot 9.6$$
>
> $$48 = 48$$
>
> The solution checks.

CHECK YOUR PROGRESS 7 Solve: $\dfrac{42}{x} = \dfrac{5}{8}$

Solution *See page S17.*

Proportions are useful for solving a wide variety of application problems. Remember that when using the given information to write a proportion involving two rates, the units in the numerators of the rates need to be the same and the units in the denominators of the rates need to be the same. It is helpful to keep in mind that when we write a proportion, we are stating that two rates or ratios are equal.

EXAMPLE 8 ■ **Solve an Application Using a Proportion**

If you travel 290 miles in your car on 15 gallons of gasoline, how far can you travel in your car on 12 gallons of gasoline?

Solution
Let x = the unknown number of miles.
Write a proportion and then solve the proportion for x.

$$\frac{290 \text{ miles}}{15 \text{ gallons}} = \frac{x \text{ miles}}{12 \text{ gallons}}$$

$$\frac{290}{15} = \frac{x}{12}$$

$$290 \cdot 12 = 15 \cdot x$$

$$3480 = 15x$$

$$232 = x$$

You can travel 232 miles on 12 gallons of gasoline.

CHECK YOUR PROGRESS 8 On a map, a distance of 2 centimeters represents 15 kilometers. What is the distance between two cities that are 7 centimeters apart on the map?

Solution *See page S17.*

EXAMPLE 9 ■ **Solve an Application Using a Proportion**

The table below shows three of the universities in the Big Ten Conference and their student-faculty ratios. (*Source:* Barron's Profile of American Colleges, 24th edition, c. 2001) There are approximately 29,450 full-time undergraduate students at Michigan State University. Approximate the number of faculty at Michigan State.

University	Student-Faculty Ratio
Michigan State University	12 to 1
University of Illinios	16 to 1
University of Iowa	10 to 1

Solution

Let $F =$ the number of faculty members.

Write a proportion and then solve the proportion for F.

$$\frac{12 \text{ students}}{1 \text{ faculty}} = \frac{29{,}450 \text{ students}}{F \text{ faculty}}$$

$$12 \cdot F = 1(29{,}450)$$

$$12F = 29{,}450$$

$$\frac{12F}{12} = \frac{29{,}450}{12}$$

$$F \approx 2454$$

There are approximately 2454 faculty members at Michigan State University.

CHECK YOUR PROGRESS 9 The profits of a firm are shared by its two partners in the ratio 7:5. If the partner receiving the larger amount of this year's profits receives $28,000, what amount does the other partner receive?

Solution *See page S17.*

EXAMPLE 10 ■ Solve an Application Using a Proportion

In the United States, the average annual number of deaths per million people aged 5 to 34 from asthma is 3.5. Approximately how many people aged 5 to 34 die from asthma each year in this country? Use a figure of 150,000,000 for the number of U.S. residents who are 5 to 34 years old. (*Source:* National Center for Health Statistics)

Solution

Let $d =$ the number of people aged 5 to 34 who die each year from asthma in the United States. Write and solve a proportion. One rate is 3.5 deaths per million people.

$$\frac{3.5 \text{ deaths}}{1{,}000{,}000 \text{ people}} = \frac{d \text{ deaths}}{150{,}000{,}000 \text{ people}}$$

$$3.5(150{,}000{,}000) = 1{,}000{,}000 \cdot d$$

$$525{,}000{,}000 = 1{,}000{,}000d$$

$$\frac{525{,}000{,}000}{1{,}000{,}000} = \frac{1{,}000{,}000d}{1{,}000{,}000}$$

$$525 = d$$

In the United States, approximately 525 people aged 5 to 34 die each year from asthma.

CHECK YOUR PROGRESS 10

New York City has the highest death rate from asthma in the United States. The average annual number of deaths per million people aged 5 to 34 from asthma in New York City is 10.1. Approximately how many people aged 5 to 34 die from asthma each year in New York City? Use a figure of 4,000,000 for the number of residents 5 to 34 years old in New York City. Round to the nearest whole number. (*Source:* National Center for Health Statistics)

Solution *See page S17.*

MathMatters   **Scale Models for Special Effects**

Special-effects artists use scale models to create dinosaurs, exploding spaceships, and aliens. A scale model is produced by using ratios and proportions to determine the size of the scale model.

Excursion

Earned Run Average

Earned Run Average Leaders		
Year	Player, club	ERA
	National League	
1990	Danny Darwin, Houston	2.21
1991	Dennis Martinez, Montreal	2.39
1992	Bill Swift, San Francisco	2.06
1993	Greg Maddux, Atlanta	2.36
1994	Greg Maddux, Atlanta	1.56
1995	Greg Maddux, Atlanta	1.63
1996	Kevin Brown, Florida	1.89
1997	Pedro Martinez, Montreal	1.90
1998	Greg Maddux, Atlanta	2.22
1999	Randy Johnson, Arizona	2.48
2000	Kevin K. Brown, Los Angeles	2.58
2001	Randy Johnson, Arizona	2.49
	American League	
1990	Roger Clemens, Boston	1.93
1991	Roger Clemens, Boston	2.62
1992	Roger Clemens, Boston	2.41
1993	Kevin Appler, Kansas City	2.56
1994	Steve Ontiveros, Oakland	2.65
1995	Randy Johnson, Seattle	2.48
1996	Juan Guzman, Toronto	2.93
1997	Roger Clemens, Toronto	2.05
1998	Roger Clemens, Toronto	2.65
1999	Pedro Martinez, Boston	2.07
2000	Pedro Martinez, Boston	1.74
2001	Freddy Garcia, Seattle	3.05

One measure of a pitcher's success is earned run average. **Earned run average (ERA)** is the number of earned runs a pitcher gives up for every nine innings pitched. The definition of an earned run is somewhat complicated, but basically an earned run is a run that is scored as a result of hits and base running that involves no errors on the part of the pitcher's team. If the opposing team scores a run on an error (for example, a fly ball that should have been caught in the outfield was fumbled), then that run is not an earned run.

A proportion is used to calculate a pitcher's ERA. Remember that the statistic involves the number of earned runs per *nine innings.* The answer is always rounded to the nearest hundredth. Here is an example.

During the 2001 regular baseball season, Chan Ho Park gave up 91 earned runs and pitched 234 innings for the Los Angeles Dodgers. To calculate Chan Ho Park's ERA, let x = the number of earned runs for every nine innings pitched. Write a proportion and then solve for x.

$$\frac{91 \text{ earned runs}}{234 \text{ innings}} = \frac{x \text{ earned runs}}{9 \text{ innings}}$$

$$91 \cdot 9 = 234 \cdot x$$

$$819 = 234x$$

$$\frac{819}{234} = \frac{234x}{234}$$

$$3.5 = x$$

Chan Ho Park's ERA was 3.50.

Excursion Exercises

1. In 1979, his rookie year, Jeff Reardon pitched 21 innings for the New York Mets and gave up four earned runs. Calculate Reardon's ERA for 1979.

2. Roger Clemens's first year with the Boston Red Sox was 1984. During that season, he pitched 133.1 innings and gave up 64 earned runs. Calculate Clemens's ERA for 1984.

3. During the 1998 baseball season, Pedro Martinez of the Boston Red Sox pitched 233.2 innings and gave up 75 earned runs. During the 1999 season, he gave up 49 earned runs and pitched 213.1 innings. During which season was his ERA lower? How much lower?

4. In 1987, Nolan Ryan had the lowest ERA of any pitcher in the major leagues. He gave up 65 earned runs and pitched 211.2 innings for the Houston Astros. Calculate Ryan's ERA for 1987.

5. Find the necessary statistics for a pitcher on your "home team," and calculate that pitcher's ERA.

Exercise Set 5.2

(Suggested Assignment: 9–69, odds)

1. Provide two examples of situations in which unit rates are used.

2. Explain why unit rates are used to describe situations involving units such as miles per gallon.

3. What is the purpose of exchange rates in international trade?

4. Provide two examples of situations in which ratios are used.

5. Explain why ratios are used to describe situations involving information such as student-teacher ratios.

6. What does the phrase "the cross products are equal" mean?

7. Explain why the product of the means in a proportion equals the product of the extremes.

In Exercises 8–13, write the expression as a unit rate.

8. 582 miles in 12 hours

9. 138 miles on 6 gallons of gasoline

10. 544 words typed in 8 minutes

11. 100 meters in 8 seconds

12. $9100 for 350 shares of stock

13. 1000 square feet of wall covered with 2.5 gallons of paint

Solve Exercises 14–19.

14. A machinist earns $490 for working a 35-hour week. What is the machinist's hourly rate of pay?

15. The Saturn-5 rocket uses 534,000 gallons of fuel in 2.5 minutes. How much fuel does the rocket use in 1 minute?

16. During filming, an IMAX camera uses 65-mm film at a rate of 5.6 feet per second.
    a. At what rate per minute does the camera go through film?
    b. How quickly does the camera use a 500-foot roll of 65-mm film? Round to the nearest second.

17. Which is the more economical purchase, a 32-ounce jar of mayonnaise for $2.79 or a 24-ounce jar of mayonnaise for $2.09?

18. Which is the more economical purchase, an 18-ounce box of corn flakes for $2.89 or a 24-ounce box of corn flakes for $3.89?

19. You have a choice of receiving a wage of $34,000 per year, $2840 per month, $650 per week, or $16.50 per hour. Which pay choice would you take? Assume a 40-hour work week and 52 weeks of work per year.

20. Baseball statisticians calculate a hitter's at-bats per home run by dividing the number of times the player has been at bat by the number of home runs the player has hit.
    a. Calculate the at-bats per home run for each player in the table on the following page. Round to the nearest tenth.
    b. Which player has the lowest rate of at-bats per home run? Which player has the second lowest rate?
    c. Why is this rate used for comparison rather than the number of home runs a player has hit?

Year	Baseball Player	Number of Times at Bat	Number of Home Runs Hit	Number of At-Bats per Home Run
1921	Babe Ruth	540	59	_____
1927	Babe Ruth	540	60	_____
1930	Hack Wilson	585	56	_____
1932	Jimmie Foxx	585	58	_____
1938	Hank Greenberg	556	58	_____
1961	Roger Maris	590	61	_____
1961	Mickey Mantle	514	54	_____
1964	Willie Mays	558	52	_____
1977	George Foster	615	52	_____
1998	Mark McGwire	509	70	_____
1998	Sammy Sosa	643	66	_____
2001	Barry Bonds	476	73	_____

21. The table below shows the population and area of three countries. The population density of a country is the number of people per square mile.

a. Which country has the lowest population density?

b. How many more people per square mile are there in India than in the United States? Round to the nearest whole number.

Country	Population	Area (in square miles)
Australia	19,358,000	2,938,000
India	1,029,991,000	1,146,000
United States	281,422,000	3,535,000

22. Forrester Research, Inc., compiled the following estimates on consumer use of email in the United States.

a. Complete the last column of the table on the following page by calculating the estimated number of messages per day that each user receives. Round to the nearest tenth.

b. How many times greater is the predicted number of messages per person per day in 2005 than the estimated number in 1995?

Year	Number of Users (in millions)	Messages per Day (in millions)	Messages per Person per Day
1992	5	10	_____
1993	8	17	_____
1994	13	28	_____
1995	22	53	_____
1996	40	100	_____
1997	55	150	_____
1998	75	225	_____
1999	95	313	_____
2000	115	402	_____
2001	135	500	_____
2005	170	5000	_____

The table below shows the exchange rates per U.S. dollar for eight foreign countries on November 18, 2002. Use this table for Exercises 23–26.

Exchange Rate per U.S. Dollar	
British Pound	0.6325
Danish Krone	7.3576
Indian Rupee	48.1300
Japanese Yen	121

23. How many Danish krona are equivalent to $10,000?

24. Find the number of Indian rupees that would be exchanged for $45,000.

25. Find the cost, in British pounds, of an order of American computer hardware costing $38,000.

26. Calculate the cost, in Japanese yen, of an American car costing $24,000.

Researchers at the Centers for Disease Control and Prevention estimate that 5 million young people living today will die of tobacco-related diseases. Almost one-third of children who become regular smokers will die of a smoking-related illness such as heart disease or lung cancer. The table below shows, for eight states in our nation, the number of children under 18 who are expected to become smokers and the number who are projected to die of smoking-related illness. Use the table for Exercises 27 and 28.

State	Projected Number of Smokers	Projected Number of Deaths
Alabama	260,639	83,404
Alaska	56,246	17,999
Arizona	307,864	98,516
Arkansas	155,690	49,821
California	1,446,550	462,896
Colorado	271,694	86,942
Connecticut	175,501	56,160
Delaware	51,806	16,578

27. Find the ratio of the projected number of smokers in each state listed to the projected number of deaths. Round to the nearest thousandth. Write the ratio using the word *to*.

28. **a.** Did the researchers calculate different probabilities of death from smoking-related illnesses for each state?

   b. If the projected number of smokers in Florida is 928,464, what would you expect the researchers to project as the number of deaths from smoking-related illnesses in Florida?

The table below shows the number of full-time men and women undergraduates and the number of full-time faculty at universities in the Big East. Use this table for Exercises 29–32. Round ratios to the nearest whole number. (*Source: Barron's Profile of American Colleges, 24th edition, c. 2001*)

University	Men	Women	Faculty
Boston College	4339	4851	631
Georgetown University	2797	3288	647
Syracuse University	4828	5654	778
University of Connecticut	5348	6063	1043
University of Miami	3625	4201	578
Villanova University	3162	3212	499

29. Calculate the student-faculty ratio at Syracuse University. Write the ratio using a colon and using the word *to*. What does this ratio mean?

30. Which school listed has the lowest student-faculty ratio?

31. Which school listed has the highest student-faculty ratio?

32. Which two schools listed have the same student-faculty ratio?

33. A bank uses the ratio of a borrower's total monthly debt to total monthly income to determine eligibility for a loan. This ratio is called the debt-equity ratio. First National Bank requires that a borrower have a debt-equity ratio that is less than $\frac{2}{5}$. Would the homeowner whose monthly income and debt are given below qualify for a loan using these standards?

Monthly Income (in dollars)		Monthly Debt (in dollars)	
Salary	3400	Mortgage	1800
Interest	83	Property tax	104
Rent	640	Insurance	27
Dividends	34	Credit cards	354
		Car loan	199

Solve Exercises 34–49. Round to the nearest hundredth.

34. $\dfrac{3}{8} = \dfrac{x}{12}$

35. $\dfrac{3}{y} = \dfrac{7}{40}$

36. $\dfrac{7}{12} = \dfrac{25}{d}$

37. $\dfrac{16}{d} = \dfrac{25}{40}$

38. $\dfrac{15}{45} = \dfrac{72}{c}$

39. $\dfrac{120}{c} = \dfrac{144}{25}$

40. $\dfrac{65}{20} = \dfrac{14}{a}$

41. $\dfrac{4}{a} = \dfrac{9}{5}$

42. $\dfrac{0.5}{2.3} = \dfrac{b}{20}$

43. $\dfrac{1.2}{2.8} = \dfrac{b}{32}$

44. $\dfrac{0.7}{1.2} = \dfrac{6.4}{x}$

45. $\dfrac{2.5}{0.6} = \dfrac{165}{x}$

46. $\dfrac{y}{6.25} = \dfrac{16}{87}$

47. $\dfrac{y}{2.54} = \dfrac{132}{640}$

48. $\dfrac{1.2}{0.44} = \dfrac{m}{14.2}$

49. $\dfrac{12.5}{m} = \dfrac{102}{55}$

Solve Exercises 50–61.

50. The ratio of weight on the moon to weight on Earth is 1 : 6. How much would a 174-pound person weigh on the moon?

51. A management consulting firm recommends that the ratio of middle-management salaries to management trainee salaries be 5 : 4. Using this recommendation, what is the annual middle-management salary if the annual management trainee salary is $36,000?

52. A gardening crew uses 2 pounds of fertilizer for every 100 square feet of lawn. At this rate, how many pounds of fertilizer does the crew use on a lawn that measures 2500 square feet?

53. The dosage of a cold medication is 2 milligrams for every 80 pounds of body weight. How many milligrams of this medication are required for a person who weighs 220 pounds?

54. If your car can travel 70.5 miles on 3 gallons of gasoline, how far can the car travel on 14 gallons of gasoline?

55. The scale on the architectural plans for a new house is 1 inch equals 3 feet. Find the length and width of a room that measures 5 inches by 8 inches on the drawing.

56. The scale on a map is 1.25 inches equals 10 miles. Find the distance between Carlsbad and Del Mar, which are 2 inches apart on the map.

57. A pre-election survey showed that two out of every three eligible voters would cast ballots in the county election. There are 240,000 eligible voters in the county. How many people are expected to vote in the election?

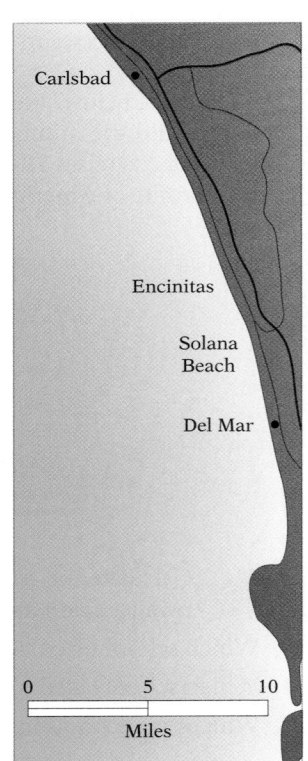

58. A paint manufacturer suggests using 1 gallon of paint for every 400 square feet of wall. At this rate, how many gallons of paint would be required to paint a room that has 1400 square feet of wall?

59. Amanda Chicopee bought a new car and drove 22,000 miles in the first four months. At the same rate, how many miles will Amanda drive in 3 years?

60. Three people put their money together to buy lottery tickets. The first person put in $25, the second person put in $30, and the third person put in $35. One of their tickets was a winning ticket. If they won $4.5 million, what was the first person's share of the winnings?

61. A pancake 4 inches in diameter contains 5 grams of fat. How many grams of fat are in a pancake 6 inches in diameter? Explain how you arrived at your answer.

Extensions

CRITICAL THINKING

62. For U.S. cities with populations over 100,000, those with the highest homicide rates in a recent year are listed below, along with the number of murders in each city and the homicide rate per 100,000 people. (*Source:* FBI Uniform Crime Reports)

a. Explain what the statistics for Atlanta mean.

b. What information not provided here was needed to calculate the homicide rates given for each city?

City	Total Number of Murders	Homicide Rate per 100,000 People
Washington	397	73
New Orleans	351	72
Richmond, VA	112	55
Atlanta	196	47
Baltimore	328	46

In Exercises 63 and 64, assume each denominator is a non-zero real number.

63. Determine whether the statement is true or false.

a. A quotient $a \div b$ is a ratio.

b. If $\dfrac{a}{b} = \dfrac{c}{d}$, then $\dfrac{b}{a} = \dfrac{d}{c}$.

c. If $\dfrac{a}{b} = \dfrac{c}{d}$, then $\dfrac{a}{c} = \dfrac{b}{d}$.

d. If $\dfrac{a}{b} = \dfrac{c}{d}$, then $\dfrac{a}{d} = \dfrac{c}{b}$.

64. If $\dfrac{a}{b} = \dfrac{c}{d}$, show that $\dfrac{a+b}{b} = \dfrac{c+d}{d}$.

COOPERATIVE LEARNING

65. Advertising rates during daytime television programs are lower than during prime time. In a recent year, 30 seconds of ad time during a soap opera cost about $20,000, whereas the same ad during prime time cost from $50,000 to $500,000. Ad rates for daytime shows are based on a cost per thousand viewers. Network daytime ad rates were around $10 per thousand women aged 18 to 49. By way of comparison, ad rates for the Oprah Winfrey Show and the Rosie O'Donnell Show were about $12 per thousand, and for the less popular talk shows ad rates were about $6.50 per thousand. (*Source: Working Woman* magazine)

a. Explain why different time slots and different shows demand different advertising rates.

b. Why are advertisers concerned about the gender and age of the audience watching a particular program?

c. How do Nielsen ratings help determine what advertising rates can be charged for an ad airing during a particular program?

66. According to a recent study, the crime rate against travelers in the United States is lower than that against the general population. The article reporting this study included a bar graph similar to the one at the right.

a. Why are the figures reported based on crime victims per 1000 adults per year?

b. Use the given figures for personal crimes and property crimes to set up a proportion. Is the proportion true?

c. Why might the crime rate against travelers be lower than that against the general population?

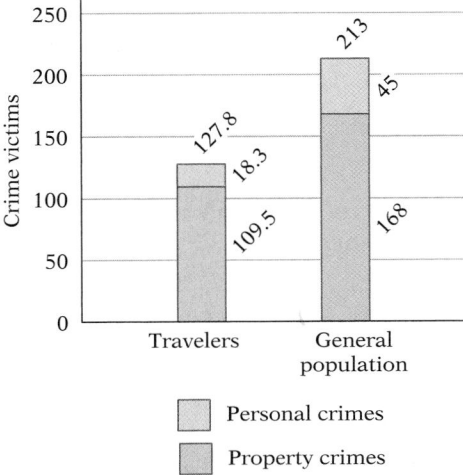

Crime Victims Per 1000 Adults Per Year
Source: Travel Industry Association of America

EXPLORATIONS

67. The 50 states in the United States are listed below by their USPS two-letter abbreviations. The number following each state is the figure provided by the U.S. Bureau of the Census for the population of that state in 2000. Figures are in millions and are rounded to the nearest hundred thousand.

Populations of the 50 States in the United States (in millions)				
AL: 4.4	HI: 1.2	MA: 6.3	NM: 1.8	SD: 0.8
AK: 0.6	ID: 1.3	MI: 9.9	NY: 19.0	TN: 5.7
AZ: 5.1	IL: 12.4	MN: 4.9	NC: 8.0	TX: 20.9
AR: 2.7	IN: 6.1	MS: 2.8	ND: 0.6	UT: 2.2
CA: 33.9	IA: 2.9	MO: 5.6	OH: 11.4	VT: 0.6
CO: 4.3	KS: 2.7	MT: 0.9	OK: 3.5	VA: 7.1
CT: 3.4	KY: 4.0	NE: 1.7	OR: 3.4	WA: 5.9
DE: 0.8	LA: 4.5	NV: 2.0	PA: 12.3	WV: 1.8
FL: 16.0	ME: 1.3	NH: 1.2	RI: 1.0	WI: 5.4
GA: 8.2	MD: 5.3	NJ: 8.4	SC: 4.0	WY: 0.5

The House of Representatives has a total of 435 members. These members represent the 50 states in proportion to each state's population. As stated in Article XIV, Section 2, of the Constitution of the United States, "Representatives shall be apportioned among the several states according to their respective numbers, counting the whole number of persons in each state." Based on the populations for the states given above, determine how many representatives each state should elect to Congress.

Compare your list against the actual number of representatives each state has. These numbers can be found in any almanac.

| ## Percent

Percents

An understanding of percent is vital to comprehending the events that take place in our world today. We are constantly confronted with phrases such as "unemployment of 5%," "annual inflation of 7%," "6% increase in fuel prices," "25% of the daily minimum requirement," and "increase in tuition and fees of 10%."

Percent means "for every 100." Therefore, unemployment of 5% means that 5 out of every 100 people are unemployed. An increase in tuition of 10% means that tuition has gone up $10 for every $100 it cost previously.

▼ **point of interest**

Of all the errors made on federal income tax returns, the four most common errors account for 76% of the mistakes. These errors include an omitted entry (30.7%), an incorrect entry (19.1%), an error in mathematics (17.4%), and an entry on the wrong line (8.8%).

QUESTION   *When adults were asked to name their favorite cookie, 52% said chocolate chip. What does this statistic mean? (Source: WEAREVER)*

A percent is a ratio of a number to 100. Thus $\frac{1}{100} = 1\%$, $\frac{50}{100} = 50\%$, and $\frac{99}{100} = 99\%$. Because $1\% = \frac{1}{100}$ and $\frac{1}{100} = 0.01$, we can also write 1% as 0.01.

$$1\% = \frac{1}{100} = \mathbf{0.01}$$

The equivalence $1\% = 0.01$ is used to write a percent as a decimal or to write a decimal as a percent.

To write 17% as a decimal:

$$17\% = 17(1\%) = 17(0.01) = 0.17$$

Note that this is the same as removing the percent sign and moving the decimal point two places to the left.

To write 0.17 as a percent:

$$0.17 = 17(0.01) = 17(1\%) = 17\%$$

Note that this is the same as moving the decimal point two places to the right and writing a percent sign at the right of the number.

ANSWER   *52 out of every 100 people surveyed responded that their favorite cookie was chocolate chip. (In the same survey, the following responses were also given: oatmeal raisin, 10%; peanut butter, 9%; oatmeal, 7%; sugar, 4%; molasses, 4%; chocolate chip oatmeal, 3%.)*

EXAMPLE 1 ■ Write a Percent as a Decimal

Write the percent as a decimal.

a. 24%   **b.** 183%   **c.** 6.5%   **d.** 0.9%

Solution

To write a percent as a decimal, remove the percent sign and move the decimal point two places to the left.

a. 24% = 0.24

b. 183% = 1.83

c. 6.5% = 0.065

d. 0.9% = 0.009

CHECK YOUR PROGRESS 1   Write the percent as a decimal.

a. 74%   **b.** 152%   **c.** 8.3%   **d.** 0.6%

Solution   *See page S17.*

▼ **point of interest**

According to a Pathfinder Research Group survey, more than 94% of adults have heard of the Three Stooges. The choices of a favorite among those who have one are:

Curly:	52%
Moe:	31%
Larry:	12%
Curly Joe:	3%
Shemp:	2%

EXAMPLE 2 ■ Write a Decimal as a Percent

Write the decimal as a percent.

a. 0.62   **b.** 1.5   **c.** 0.059   **d.** 0.008

Solution

To write a decimal as a percent, move the decimal point two places to the right and write a percent sign.

a. 0.62 = 62%

b. 1.5 = 150%

c. 0.059 = 5.9%

d. 0.008 = 0.8%

CHECK YOUR PROGRESS 2   Write the decimal as a percent.

a. 0.3   **b.** 1.65   **c.** 0.072   **d.** 0.004

Solution   *See page S17.*

The equivalence $1\% = \frac{1}{100}$ is used to write a percent as a fraction.
To write 16% as a fraction:

$$16\% = 16(1\%) = 16\left(\frac{1}{100}\right) = \frac{16}{100} = \frac{4}{25}$$

Note that this is the same as removing the percent sign and multiplying by $\frac{1}{100}$. The fraction is written in simplest form.

INSTRUCTOR NOTE
Example 3d and Check Your Progress 3d are difficult for students. Here is another in-class example to use.
Write $12\frac{1}{2}\%$ as a fraction.

Solution:

$$12\frac{1}{2}\% = \frac{25}{2}\%$$
$$= \left(\frac{25}{2}\right)\left(\frac{1}{100}\right)$$
$$= \frac{1}{8}$$

EXAMPLE 3 ■ Write a Percent as a Fraction

Write the percent as a fraction.

a. 25%   **b.** 120%   **c.** 7.5%   **d.** $33\frac{1}{3}\%$

Solution

To write a percent as a fraction, remove the percent sign and multiply by $\frac{1}{100}$. Then write the fraction in simplest form.

a. $25\% = 25\left(\frac{1}{100}\right) = \frac{25}{100} = \frac{1}{4}$

b. $120\% = 120\left(\frac{1}{100}\right) = \frac{120}{100} = 1\frac{20}{100} = 1\frac{1}{5}$

c. $7.5\% = 7.5\left(\frac{1}{100}\right) = \frac{7.5}{100} = \frac{75}{1000} = \frac{3}{40}$

d. $33\frac{1}{3}\% = \frac{100}{3}\% = \frac{100}{3}\left(\frac{1}{100}\right) = \frac{1}{3}$

CHECK YOUR PROGRESS 3   Write the percent as a fraction.

a. 8%   **b.** 180%   **c.** 2.5%   **d.** $66\frac{2}{3}\%$

Solution   *See page S17.*

To write a fraction as a percent, first write the fraction as a decimal. Then write the decimal as a percent.

✔ **TAKE NOTE**

To write a fraction as a decimal, divide the number in the numerator by the number in the denominator. For example,

$$\frac{4}{5} = 4 \div 5 = 0.8.$$

EXAMPLE 4 ■ Write a Fraction as a Percent

Write the fraction as a percent.

a. $\frac{3}{4}$ **b.** $\frac{5}{8}$ **c.** $\frac{1}{6}$ **d.** $1\frac{1}{2}$

Solution

To write a fraction as a percent, write the fraction as a decimal. Then write the decimal as a percent.

a. $\frac{3}{4} = 0.75 = 75\%$

b. $\frac{5}{8} = 0.625 = 62.5\%$

c. $\frac{1}{6} = 0.16\overline{6} = 16.\overline{6}\%$

d. $1\frac{1}{2} = 1.5 = 150\%$

CHECK YOUR PROGRESS 4 Write the fraction as a percent.

a. $\dfrac{1}{4}$ **b.** $\dfrac{3}{8}$ **c.** $\dfrac{5}{6}$ **d.** $1\dfrac{2}{3}$

Solution *See page S18.*

Math Matters   College Graduates' Job Expectations

The table below shows the expectations of a recent class of college graduates. (*Source:* Yankelovich Partners for Phoenix Home Life Mutual Insurance)

	Men	Women
Expect a job by graduation	31%	23%
Expect a job within 6 months	32%	23%
Expect a starting pay of $30,000 or more	55%	43%
Expect to be richer than their parents	64%	59%

Percent Problems: The Proportion Method

INSTRUCTOR NOTE
This objective explains the proportion method of solving the basic percent equation. Coverage of this material is optional.

Finding the solution of an application problem involving percent generally requires writing and solving an equation. Two methods of writing the equation will be developed in this section—the *proportion method* and the *basic percent equation*. We will present the proportion method first.

The proportion method of solving a percent problem is based on writing two ratios. One ratio is the percent ratio, written $\frac{\text{percent}}{100}$. The second ratio is the amount-to-base ratio, written $\frac{\text{amount}}{\text{base}}$. These two ratios form the proportion used to solve percent problems.

The Proportion Used to Solve Percent Problems

$$\frac{\text{Percent}}{100} = \frac{\text{amount}}{\text{base}}$$

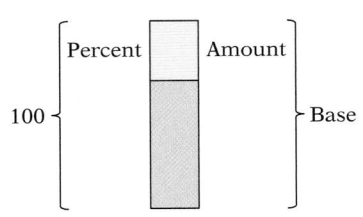

Diagram of the Proportion Method of Solving Percent Problems

The proportion method can be illustrated by a diagram. The rectangle at the left is divided into two parts. On the left, the whole rectangle is represented by 100 and the part by percent. On the right, the whole rectangle is represented by the *base* and the part by the *amount*. The ratio of percent to 100 is equal to the ratio of the amount to the base.

When solving a percent problem, first identify the percent, the base, and the amount. It is helpful to know that the base usually follows the phrase "percent of."

QUESTION   *In the statement "15% of 40 is 6," which number is the percent? Which number is the base? Which number is the amount?*

INSTRUCTOR NOTE
Some students find it easier to remember the proportion method by using the equation

$$\frac{is}{of} = \frac{n}{100}$$

EXAMPLE 5 ■ Solve a Percent Problem for the Base Using the Proportion Method

The average size of a house in 1997 was 2100 square feet. This is approximately 125% of the average size of a house in 1977. What was the average size of a house in 1977?

Solution
We want to answer the question, "125% of what number is 2100?" Write and solve a proportion. The percent is 125%. The amount is 2100. The base is the average size of a house in 1977.

$$\frac{\text{Percent}}{100} = \frac{\text{amount}}{\text{base}}$$

$$\frac{125}{100} = \frac{2100}{B}$$

$$125 \cdot B = 100(2100)$$

$$125B = 210,000$$

$$\frac{125B}{125} = \frac{210,000}{125}$$

$$B = 1680$$

The average size of a house in 1977 was 1680 square feet.

CHECK YOUR PROGRESS 5   A used Chevrolet Blazer was purchased for $22,400. This was 70% of the cost of the Blazer when new. What was the cost of the Blazer when it was new?

Solution   See page S18.

▼ **point of interest**

According to the U.S. Department of Agriculture, of the 356 billion pounds of food produced annually in the United States, about 96 billion pounds are wasted. This is approximately 27% of all the food produced in the United States.

EXAMPLE 6 ■ Solve a Percent Problem for the Percent Using the Proportion Method

During 1996, Texas suffered through one of its longest droughts in history. Of the $5 billion in losses caused by the drought, $1.1 billion was direct losses to ranchers. What percent of the total losses was direct losses to ranchers?

ANSWER   *The percent is 15. The base is 40. (It follows the phrase "percent of.") The amount is 6.*

Solution

We want to answer the question, "What percent of $5 billion is $1.1 billion?" Write and solve a proportion. The base is $5 billion. The amount is $1.1 billion. The percent is unknown.

$$\frac{\text{Percent}}{100} = \frac{\text{amount}}{\text{base}}$$

$$\frac{p}{100} = \frac{1.1}{5}$$

$$p \cdot 5 = 100(1.1)$$

$$5p = 110$$

$$\frac{5p}{5} = \frac{110}{5}$$

$$p = 22$$

Direct losses to ranchers represent 22% of the total losses.

CHECK YOUR PROGRESS 6

Of the approximately 1,300,000 enlisted women and men in the U.S. military, 416,000 are over the age of 30. What percent of the enlisted people are over the age of 30?

Solution   *See page S18.*

EXAMPLE 7 ■ Solve a Percent Problem for the Amount Using the Proportion Method

In a certain year, Blockbuster Video customers rented 24% of the approximately 3.7 billion videos rented that year. How many million videos did Blockbuster Video rent that year?

Solution

We want to answer the question, "24% of 3.7 billion is what number?" Write and solve a proportion. The percent is 24%. The base is 3.7 billion. The amount is the number of videos Blockbuster Video rented during the year.

$$\frac{\text{Percent}}{100} = \frac{\text{amount}}{\text{base}}$$

$$\frac{24}{100} = \frac{A}{3.7}$$

$$100(A) = 24(3.7)$$

$$100A = 88.8$$

$$\frac{100A}{100} = \frac{88.8}{100}$$

$$A = 0.888$$

The number 0.888 is in billions. We need to convert it to millions.

0.888 billion = 888 million

Blockbuster Video rented approximately 888 million videos that year.

CHECK YOUR PROGRESS 7   A General Motors buyers' incentive program offered a 3.5% rebate on the selling price of a new car. What rebate would a customer receive who purchased a $32,500 car under this program?

Solution   *See page S18.*

Percent Problems: The Basic Percent Equation

A second method of solving a percent problem is to use the basic percent equation.

> **The Basic Percent Equation**
>
> $PB = A$, where P is the percent, B is the base, and A is the amount.

When solving a percent problem using the proportion method, we have to first identify the percent, the base, and the amount. The same is true when solving percent problems using the basic percent equation. Remember that the base usually follows the phrase "percent of."

When using the basic percent equation, the percent must be written as a decimal or a fraction. This is illustrated in Example 8.

EXAMPLE 8 ■ Solve a Percent Problem for the Amount Using the Basic Percent Equation

A real estate broker receives a commission of 6% of the selling price of a house. Find the amount the broker receives on the sale of a $175,000 home.

Solution
We want to answer the question, "6% of $175,000 is what number?" Use the basic percent equation. The percent is 6% = 0.06. The base is 175,000. The amount is the amount the broker receives on the sale of the home.

$$PB = A$$
$$0.06(175,000) = A$$
$$10,500 = A$$

The real estate broker received a commission of $10,500 on the sale.

CHECK YOUR PROGRESS 8   New Hampshire public school teachers contribute 5% of their wages to the New Hampshire Retirement System. What amount is contributed during one year by a teacher whose annual salary is $32,685?

Solution   *See page S18.*

EXAMPLE 9 ■ **Solve a Percent Problem for the Base Using the Basic Percent Equation**

An investor received a payment of $480, which was 12% of the value of the investment. Find the value of the investment.

Solution
We want to answer the question, "12% of what number is 480?" Use the basic percent equation. The percent is 12% = 0.12. The amount is 480. The base is the value of the investment.

$$PB = A$$
$$0.12B = 480$$
$$\frac{0.12B}{0.12} = \frac{480}{0.12}$$
$$B = 4000$$

The value of the investment is $4000.

CHECK YOUR PROGRESS 9   A real estate broker receives a commission of 6% of the selling price of a house. Find the selling price of a home on whose sale the broker received a commission of $14,370.

Solution   *See page S18.*

EXAMPLE 10 ■ **Solve a Percent Problem for the Percent Using the Basic Percent Equation**

If you answer 96 questions correctly on a 120-question exam, what percent of the questions did you answer correctly?

Solution
We want to answer the question, "What percent of 120 questions is 96 questions?" Use the basic percent equation. The base is 120. The amount is 96. The percent is unknown.

$$PB = A$$
$$P \cdot 120 = 96$$
$$\frac{P \cdot 120}{120} = \frac{96}{120}$$
$$P = 0.8$$
$$P = 80\%$$

You answered 80% of the questions correctly.

CHECK YOUR PROGRESS 10   If you answer 63 questions correctly on a 90-question exam, what percent of the questions did you answer correctly?

Solution   *See page S18.*

The table below shows the average cost in the United States for five of the most popular home remodeling projects and the average percent of that cost recouped when the home is sold. Use this table for Example 11 and Check Your Progress 11. (*Source:* National Association of Home Builders)

Home Remodeling Project	Average Cost	Percent Recouped
Addition to the master suite	$36,472	84%
Attic bedroom	$22,840	84%
Major kitchen remodeling	$21,262	90%
Bathroom addition	$11,645	91%
Minor kitchen remodeling	$8,507	94%

EXAMPLE 11 ■ Solve an Application Using the Basic Percent Equation

 Find the difference between the cost of adding an attic bedroom to your home and the amount by which the addition increases the sale price of your home.

Solution

The cost of building the attic bedroom is $22,840, and the sale price increases by 84% of that amount. We need to find the difference between $22,840 and 84% of $22,840.

Use the basic percent equation to find 84% of $22,840. The percent is 84% = 0.84. The base is 22,840. The amount is unknown.

$$PB = A$$
$$0.84(22{,}840) = A$$
$$19{,}185.60 = A$$

Subtract 19,185.60 (the amount of the cost that is recouped when the home is sold) from 22,840 (the cost of building the attic bedroom).

$$22{,}840 - 19{,}185.60 = 3654.40$$

The difference between the cost of the addition and the increase in value of your home is $3654.40.

CHECK YOUR PROGRESS 11

Find the difference between the cost of a major kitchen remodeling in your home and the amount by which the remodeling increases the sale price of your home.

Solution   See page S18.

Percent Increase

When a family moves from one part of the country to another, they are concerned about the difference in the cost of living. Will food, housing, and gasoline cost more in that part of the country? Will they need a larger salary in order to make ends meet?

We can use one number to represent the increased cost of living from one city to another so that no matter what salary you make, you can determine how much you will need to earn in order to maintain the same standard of living. That one number is a percent.

For example, look at the information in the table below. (*Source:* **www.home-fair.com/homefair/cmr/salcalc.html**)

If you live in	and are moving to	you will need to make this percent of your current salary
Cincinnati, Ohio	San Francisco, California	236%
St. Louis, Missouri	Boston, Massachusetts	213%
Denver, Colorado	New York, New York	239%

A family in Cincinnati living on $30,000 per year would need 236% of their current income to maintain the same standard of living in San Francisco. Likewise, a family living on $75,000 per year would need 236% of their current income.

$$30,000(2.36) = 70,800 \qquad 75,000(2.36) = 177,000$$

The family from Cincinnati living on $30,000 would need an annual income of $70,800 in San Francisco to maintain their standard of living. The family living on $75,000 would need an annual income of $177,000 in San Francisco to maintain their standard of living.

We have used one number, 236%, to represent the increase in the cost of living from Cincinnati to San Francisco. No matter what a family's present income, they can use 236% to determine their necessary comparable income.

QUESTION *How much would a family in Denver, Colorado living on $55,000 per year need in New York City to maintain a comparable lifestyle? Use the table above.*

The percent used to determine the increase in the cost of living is a *percent increase*. **Percent increase** is used to show how much a quantity has increased over its original value. Statements that illustrate the use of percent increase include "sales volume increased by 11% over last year's sales volume" and "employees received an 8% pay increase."

The **federal debt** is the amount the government owes after borrowing the money it needs to pay for its expenses. It is considered a good measure of how much of the government's spending is financed by debt as opposed to taxation. The graph below shows the federal debt, according to the U.S. Department of the Treasury, at the end of the fiscal years 1980, 1985, 1990, 1995, and 2000. A fiscal year is the 12-month period that the annual budget spans, from October 1 to September 30. Use the graph for Example 12 and Check Your Progress 12.

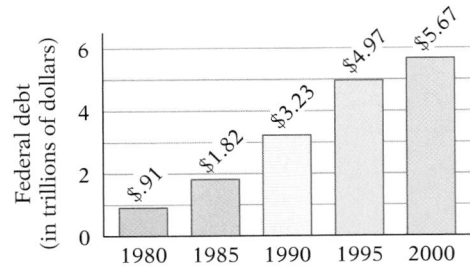

ANSWER *In New York City, the family would need $55,000(2.39) = $131,450 per year to maintain a comparable lifestyle.*

EXAMPLE 12 ∎ Solve an Application Involving Percent Increase

 Find the percent increase in the federal debt from 1980 to 1995. Round to the nearest tenth of a percent.

Solution

Calculate the amount of increase in the federal debt from 1980 to 1995.

$$4.97 - 0.91 = 4.06$$

We will use the basic percent equation. (The proportion method could also be used.) The base is the debt in 1980. The amount is the amount of increase in the debt. The percent is unknown.

$$PB = A$$
$$P \cdot 0.91 = 4.06$$
$$\frac{P \cdot 0.91}{0.91} = \frac{4.06}{0.91}$$
$$P \approx 4.462$$

The percent increase in the federal debt from 1980 to 1995 was 446.2%.

CHECK YOUR PROGRESS 12

 Find the percent increase in the federal debt from 1985 to 2000. Round to the nearest tenth of a percent.

Solution   *See page S18.*

Notice in Example 12 that the percent increase is a measure of the *amount of increase* over an *original value.* Therefore, in the basic percent equation, the amount A is the *amount of increase* and the base B is the *original value,* in this case the debt in 1980.

Percent Decrease

The federal debt is not the same as the federal deficit. The **federal deficit** is the amount by which government spending exceeds the federal budget. The table below shows that the federal deficit in May of 1997 was less than the federal deficit in May 1996. (*Source:* U.S. Treasury Department)

First 8 months of Fiscal Year	Federal Deficit
1996 (October '95 – May '96)	$108.4 billion
1997 (October '96 – May '97)	$65.4 billion

The decrease in the federal deficit can be expressed as a percent. First find the amount of decrease in the deficit.

$$108.4 - 65.4 = 43.0$$

We will use the basic percent equation to find the percent. The base is the deficit in May 1996. The amount is the amount of decrease.

$$PB = A$$
$$P \cdot 108.4 = 43.0$$
$$\frac{P \cdot 108.4}{108.4} = \frac{43.0}{108.4}$$
$$P \approx 0.397$$

The federal deficit decreased by 39.7% from May 1996 to May 1997.

The percent used to measure the decrease in the federal deficit is a *percent decrease*. **Percent decrease** is used to show how much a quantity has decreased from its original value. Statements that illustrate the use of percent decrease include "the president's approval rating has decreased 9% over last month" and "there has been a 15% decrease in the number of industrial accidents."

Note in the deficit example above that the percent decrease is a measure of the *amount of decrease* over an *original value*. Therefore, in the basic percent equation, the amount A is the *amount of decrease* and the base B is the *original value*, in this case the deficit in May 1996.

The table below shows the approximate values of three different cars in 1997. Use this table for Example 13 and Check Your Progress 13.

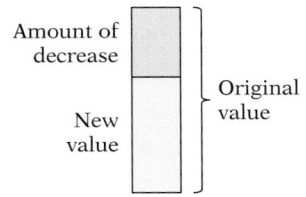

Amount of decrease

New value

Original value

Model	1997 Retail Value
Pontiac Firebird	$22,884
Chevrolet Camaro	$23,170
Ford Mustang	$23,985

EXAMPLE 13 ■ Solve an Application Involving Percent Decrease

In a recent year, *Money* magazine reported that from 1997 to 2002, the value of a Pontiac Firebird would decrease 45% from its 1997 value. Using this estimate, find the resale value of a Firebird in 2002.

Solution

We will write and solve a proportion. (The basic percent equation could also be used.) The percent is 45%. The base is $22,884. The amount is unknown.

$$\frac{\text{Percent}}{100} = \frac{\text{amount}}{\text{base}}$$
$$\frac{45}{100} = \frac{A}{22,884}$$
$$45(22,884) = 100(A)$$
$$1,029,780 = 100A$$
$$\frac{1,029,780}{100} = \frac{100A}{100}$$
$$10,297.80 = A$$

Subtract the decrease in value from the 1997 retail value.

$$22,884 - 10,297.80 = 12,586.20$$

The resale value of a Pontiac Firebird in 2002 was expected to be $12,586.20.

CHECK YOUR PROGRESS 13

 In a recent year, *Money* magazine reported that from 1997 to 2002, the value of a Ford Mustang would decrease 40% from its 1997 value. Using this estimate, find the resale value of a Mustang in 2002.

Solution   *See page S19.*

Excursion

Federal Income Tax

Income taxes are the chief source of revenue for the federal government. If you are employed, your employer probably withholds some money from each of your paychecks for federal income tax. At the end of each year, your employer sends you a **Wage and Tax Statement Form** (**W-2 form**), which states the amount of money you earned that year and how much was withheld for taxes.

Every employee is required by law to prepare an income tax return by April 15 of each year and send it to the Internal Revenue Service (IRS). On the income tax return, you must report your total income, or **gross income.** Then you subtract from the gross income any adjustments (such as deductions for charitable contributions or exemptions for people who are dependent on your income) to determine your **adjusted gross income.** You use your adjusted gross income and either a tax table or a tax rate schedule to determine your **tax liability,** or the amount of income you owe to the federal government.

After calculating your tax liability, compare it with the amount withheld for federal income tax, as shown on your W-2 form. If the tax liability is less than the amount withheld, you are entitled to a tax refund. If the tax liability is greater than the amount withheld, you owe the IRS money; you have a **balance due.**

The 2001 Tax Rate Schedules are shown on page 290. To use these schedules for the exercises that follow, first classify the taxpayer as single, married filing jointly, or married filing separately. Then determine into which range the adjusted gross income falls. Perform the calculations shown to the right of that range to determine the tax liability.

For example, consider a taxpayer who is single and has an adjusted gross income of $48,720. To find the taxpayer's tax liability, use Schedule X, the schedule for taxpayers whose filing status is single.

An income of $48,720 falls in the range $27,050 to $65,550. The tax is $4057.50 + 27.5% of the amount over $27,050. Find the amount over $27.050.

$$\$48,720 - \$27,050 = \$21,670$$

(continued)

Calculate the tax liability.

$$\$4057.50 + 27.5\%(21{,}670) = \$4057.50 + 0.275(\$21{,}670)$$
$$= \$4057.50 + \$5959.25$$
$$= \$10{,}016.75$$

The taxpayer's liability is $10,016.75.

2001
Tax Rate
Schedules

Schedule X—Use if your filing status is Single

If the amount on Form 1040, line 39, is: Over—	But not over—	Enter on Form 1040, line 40		of the amount over—
$0	$27,050	...	15%	$0
27,050	65,550	$4,057.50 +	27.5%	27,050
65,550	136,750	14,645.00 +	30.5%	65,550
136,750	297,350	36,361.00 +	35.5%	136,750
297,350	...	93,374.00 +	39.1%	297,350

Schedule Y-1—Use if your filing status is Married filing jointly or Qualifying widow(er)

If the amount on Form 1040, line 39, is: Over—	But not over—	Enter on Form 1040, line 40		of the amount over—
$0	$45,200	...	15%	$0
45,200	109,250	$6,780.00 +	27.5%	45,200
109,250	166,500	24,393.75 +	30.5%	109,250
166,500	297,350	41,855.00 +	35.5%	166,500
297,300	...	88,306.75 +	39.1%	297,350

Schedule Y-2—Use if your filing status is Married filing separately

If the amount on Form 1040, line 39, is: Over—	But not over—	Enter on Form 1040, line 40		of the amount over—
$0	$22,600	...	15%	$0
22,600	54,625	$3,390.00 +	27.5%	22,600
54,625	83,250	12,196.88 +	30.5%	54,625
83,250	148,675	20,927.50 +	35.5%	83,250
148,675	...	44,153.38 +	39.1%	148,675

(continued)

Excursion Exercises

Use the 2001 Tax Rate Schedules to solve Exercises 1–8.

1. Joseph Abruzzio is married and filing separately. He has an adjusted gross income of $63,850. Find Joseph's tax liability.

2. Angela Lopez is single and has an adjusted gross income of $31,680. Find Angela's tax liability.

3. Dee Pinckney is married and filing jointly. She has an adjusted gross income of $58,120. The W-2 form shows the amount withheld as $7124. Find Dee's tax liability and determine her tax refund or balance due.

4. Jeremy Littlefield is single and has an adjusted gross income of $72,800. His W-2 form lists the amount withheld as $18,420. Find Jeremy's tax liability and determine his tax refund or balance due.

5. Does a taxpayer in the 30.5% tax bracket pay 30.5% of his or her earnings in income tax? Explain your answer.

6. A single taxpayer has an adjusted gross income of $154,000. On what amount of the $154,000 does the taxpayer pay 35.5% to the Internal Revenue Service?

7. In the table for single taxpayers, how were the figures $4057.50 and $14,645 arrived at?

8. In the table for married persons filing jointly, how were the figures $6780 and $24,393.75 arrived at?

Exercise Set 5.3

(Suggested Assignment: 3–39, odds)

1. Name three situations in which percent is used.

2. Explain why multiplying a number by 100% does not change the value of the number.

3. Multiplying a number by 300% is the same as multiplying it by what whole number?

4. Describe each ratio in the proportion used to solve a percent problem.

5. Employee A had an annual salary of $32,000, Employee B had an annual salary of $38,000, and Employee C had an annual salary of $36,000 before each employee was given a 5% raise. Which of the three employees' annual salaries is now the highest? Explain how you arrived at your answer.

6. Each of three employees earned an annual salary of $35,000 before Employee A was given a 3% raise, Employee B was given a 6% raise, and Employee C was given a 4.5% raise. Which of the three employees' annual salaries is now the highest? Explain how you arrived at your answer.

Complete the table of equivalent fractions, decimals, and percents.

	Fraction	Decimal	Percent
7.	$\dfrac{1}{2}$		
8.		0.75	
9.			40%
10.	$\dfrac{3}{8}$		
11.		0.7	
12.			56.25%
13.	$\dfrac{11}{20}$		
14.		0.52	
15.			15.625%
16.	$\dfrac{9}{50}$		

17. In 1997, for the first time in major league baseball's history, interleague baseball games were played during the regular season. According to a Harris Poll, 73% of fans approved and 20% of fans disapproved of the change.

   a. How many fans, out of every 100 surveyed, approved of interleague games?

   b. Did more fans approve or disapprove of the change?

   c. Fans surveyed gave one of three responses: approve, disapprove, or don't know. What percent of the fans surveyed responded that they didn't know? Explain how you calculated the percent.

Solve Exercises 18–34.

18. In 1987, the average number of labor hours required to build a car was 36 hours. In 1997, the average number of labor hours required to build a car was approximately 70% of the average number of labor hours in 1987. What was the average number of labor hours required to build a car in 1997?

19. During a recent year, charitable contributions in the United States totaled $190 billion. The table at the right shows to whom this money was donated. Determine how much money was donated to educational organizations. (*Source:* Giving USA 2000/AAFRC Trust for Philanthropy)

Recipients	Percent of Total Charitable Contributions
Religion	43%
Education	14%
Health	9%
Other	34%

20. Of the 44 million people in the United States who do not have health insurance, 13.2 million are between the ages of 18 and 24. What percent of the people in the United States who do not have health insurance are between the ages of 18 and 24? (*Source:* U.S. Census Bureau)

21. A survey of 1236 adults nationwide asked, "What irks you most about the actions of other motorists?" The response "tailgaters" was given by 293 people. What percent of those surveyed were most irked by tailgaters? Round to the nearest tenth of a percent. (*Source:* Reuters/Zogby)

22. A survey by the *Boston Globe* questioned elementary and middle-school students about television. Sixty-eight students, or 42.5% of those surveyed, said that they had a television in their bedrooms at home. How many students were included in the survey?

23. a. Women who are planning a 7-day vacation budget an average of $290 for meals. This is 21.0% of the total amount budgeted for the trip. What is the total amount women budget for a week's vacation? Round to the nearest dollar.

b. Men who are planning a 7-day vacation budget an average of $438 for lodging. This is 26.9% of the total amount budgeted for the trip. What is the total amount men budget for a week's vacation? Round to the nearest dollar. (*Source:* American Express Travel Index)

24. The average costs associated with owning a dog over its average 11-year life span are shown in the graph at the right. These costs do not include the price of the puppy when purchased. The category labeled "Other" includes such expenses as fencing and repairing furniture damaged by the pet.

a. Calculate the total of all the expenses.

b. What percent of the total is each category? Round to the nearest tenth of a percent.

c. If the price of the puppy were included in these data, how would it affect the percents you calculated in part b?

d. What does it mean to say that these are average costs?

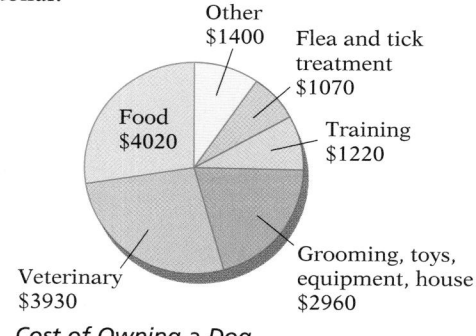

Cost of Owning a Dog

Source: American Kennel Club, *USA Today* research

25. The two circle graphs show how surveyed employees actually spend their time and how they would prefer to spend their time. Assume that employees have 112 hours a week that are not spent sleeping. Round answers to the nearest tenth of an hour. (*Source: Wall Street Journal* Supplement, Work & Family 3/31/97; from *Families and Work Institute*)

a. What is the actual number of hours per week that employees spend with family and friends?

b. What is the number of hours that employees would prefer to spend on their jobs or careers?

c. What is the difference between the number of hours an employee would prefer to spend on him- or herself and the actual amount of time the employee spends on him- or herself?

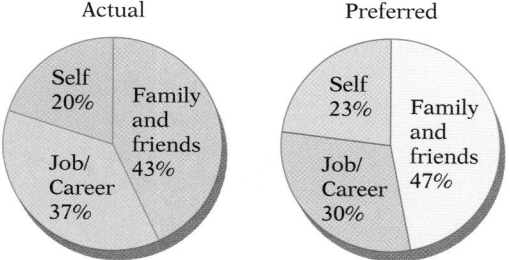

26. The graph below shows the number of prison inmates in the United States for the years 1990, 1992, 1994, and 1996.

   a. What percent of the number of state inmates was the number of federal inmates in 1990? In 1996? Round to the nearest tenth of a percent.

   b. If the ratio of federal inmates to state inmates in 1990 had remained constant, how many state inmates would there have been to the 105,544 federal inmates in 1996? Is this more or less than the actual number of state inmates in 1996? Does this mean that the number of federal inmates is growing at a more rapid rate or that the number of state inmates is growing at a more rapid rate?

   c. Explain how parts a and b are two methods of measuring the same change.

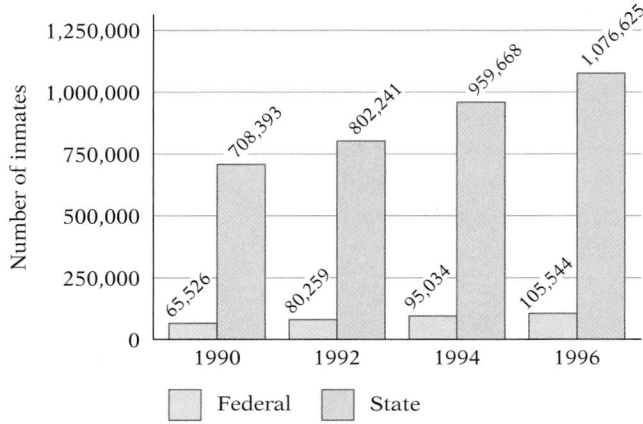

Number of Inmates in the United States

Source: U.S. Bureau of Justice Statistics

27. The graph below shows the projected growth in the number of telecommuters.

   a. During which 2-year period is the percent increase in the number of telecommuters the greatest?

   b. During which 2-year period is the percent increase in the number of telecommuters the lowest?

   c. Does the growth in telecommuting increase more slowly or more rapidly as we move from 1998 to 2006?

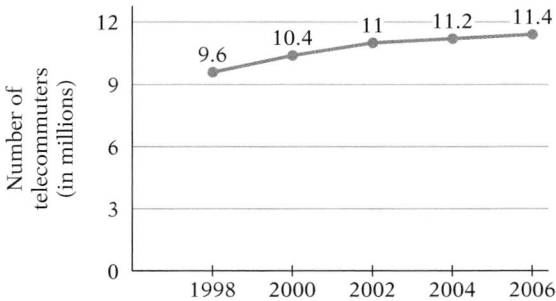

Projected Growth in Telecommuting

28. In a recent year, the states listed below had the highest rates of truck-crash fatalities. (*Source:* Citizens for Reliable and Safe Highways)

State	Number of Truck-Crash Fatalities	Number of Truck-Crash Fatalities per 100,000 People
Alabama	160	3.76
Arkansas	102	4.11
Idaho	38	3.27
Iowa	86	3.10
Mississippi	123	4.56
Montana	30	3.45
West Virginia	53	2.90
Wyoming	17	3.54

a. What state has the highest rate of truck-crash deaths?

For each of the following states, find the population to the nearest thousand people. Calculate the percent of the population of each state that died in truck accidents. Round to the nearest hundred thousandth of a percent.

b. Mississippi

c. West Virginia

d. Montana

e. Compare the percents in parts b, c, and d to the numbers of truck-crash fatalities per 100,000 people listed in the table. Based on your observations, what percent of the population of Idaho do you think was killed in truck accidents?

f. Explain the relationship between the rates in the table and the percents you calculated.

29. The table below lists the states with the highest rates of high-tech employees per 1000 workers. High-tech employees are those employed in the computer and electronics industries. Also provided is the civilian labor force for each of these states. (*Source:* American Electronics Association; U.S. Bureau of Labor Statistics)

State	High-Tech Employees per 1000 Workers	Labor Force
Arizona	53	2,229,300
California	62	10,081,300
Colorado	75	2,200,500
Maryland	51	2,798,400
Massachusetts	75	3,291,100
Minnesota	55	4,173,500
New Hampshire	78	663,000
New Jersey	55	4,173,500
Vermont	52	332,200
Virginia	52	3,575,000

a. Find the number of high-tech employees in each state listed. Round to the nearest whole number.

b. Which state has the highest rate of high-tech employment? Which state has the largest number of high-tech employees?

c. The computer and electronics industries employ approximately 4 million workers. Do more or less than half of the high-tech employees work in the 10 states listed in the table?

d. What percent of the state's labor force is employed in the computer and electronics industries in Massachusetts? in New Jersey? in Virginia? Round to the nearest tenth of a percent.

e. Compare your answers to part d with the numbers of high-tech employees per 100 workers listed in the table. Based on your observations, what percent of the labor force in Arizona is employed in the high-tech industries?

f. Explain the relationship between the rates in the table and the percents you calculated.

30. During the last 40 years, the consumption of eggs in the United States has decreased by 35%. Forty years ago, the average consumption was 400 eggs per person per year. What is the average consumption of eggs today?

31. The table at the right shows the estimated number of millionaire households in the United States for selected years.

a. What is the percent increase in the estimated number of millionaire households from 1975 to 1997?

b. Find the percent increase in the estimated number of millionaire households from 1997 to 2005.

c. Find the percent increase in the estimated number of millionaire households from 1975 to 2005.

d. Provide an explanation for the dramatic increase in the estimated number of millionaire households from 1975 to 2005.

Year	Number of Households Containing Millionaires
1975	350,000
1997	3,500,000
2005	5,600,000

32. The graph at the right shows the projected growth of the number of Americans aged 85 and older.

a. What is the percent increase in the population of this age group from 1995 to 2030?

b. What is the percent increase in the population of this age group from 2030 to 2050?

c. What is the percent increase in the population of this age group from 1995 to 2050?

d. How many times larger is the population in 2050 than in 1995? How could you determine this number from the answer to part c?

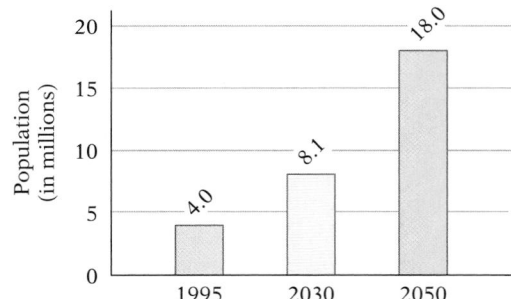

Projected Growth (in millions) of the Population of Americans Aged 85 and Older

Source: U.S. Census Bureau

33. The Bureau of Labor Statistics provides information on the fastest-growing occupations in this country. These are listed in the table on the following page, along with the predicted percent increase in employment from 1994 to 2005.

Occupation	Employment 1994	Percent Increase 1994 – 2005
Personal and home care aides	179,000	119%
Home health aides	420,000	102%
Systems analysts	483,000	92%
Computer engineers	195,000	90%
Physical and corrective therapy assistants and aides	78,000	83%
Electronic pagination systems workers	18,000	83%
Occupational therapy assistants and aides	16,000	82%
Physical therapists	102,000	80%
Residential counselors	165,000	76%
Human services workers	168,000	75%
Occupational therapists	54,000	72%
Manicurists	38,000	69%
Medical assistants	206,000	59%
Paralegals	110,000	58%
Medical record technicians	81,000	56%
Special education teachers	388,000	53%
Amusement and recreation attendants	267,000	52%
Correctional officers	310,000	51%
Operations research analysts	44,000	50%
Guards	867,000	48%
Speech-language pathologists and audiologists	85,000	46%
Private detective and investigators	55,000	44%
Surgical technologists	46,000	43%
Dental hygienists	127,000	42%
Dental assistants	190,000	42%
Adjustment clerks	373,000	40%
Teacher aides and educational assistants	932,000	39%
Data processing equipment repairers	75,000	38%
Nursery and greenhouse managers	19,000	37%
Securities and financial services sales representatives	246,000	37%

a. Which occupation employed the largest number of people in 1994?

b. Which occupation is expected to have the largest percent increase in employment?

c. What is the expected increase in the number of people employed as correctional officers from 1994 to 2005?

d. How many people are expected to be employed as physical therapists in 2005?

e. Is it expected that more people will be employed as home health aides or as guards in 2005?

f. Why can't the answer to part e be determined by simply comparing the percent increases from the two occupations?

g. From the list, choose an occupation that interests you. Calculate the projected number of people who will be employed in that job in 2005.

34. The voter turnout in the 1996 presidential election was lower in every state than it was in the 1992 election. The three states with the largest percent of voter turnout are shown below, along with the voter turnout in 1992 and the percent decrease in voter turnout from 1992 to 1996.

a. How many more people voted in the 1992 election than in the 1996 election in the state of Montana?

b. How many Maine voters turned out to vote in the 1996 presidential election?

c. How many people in Minnesota voted in the 1996 election?

State	1992 Voter Turnout	Percent Decrease in Voter Turnout in 1996 Election
Maine	650,000	7.9%
Minnesota	2,300,000	8.8%
Montana	430,000	7.4%

Extensions
CRITICAL THINKING

35. Your employer agrees to give you a 5% raise after one year on the job, a 6% raise the next year, and a 7% raise the following year. Is your salary after the third year greater than, less than, or the same as it would be if you had received a 6% raise each year?

36. Approximately 73% of Americans who work in large offices work on weekends, either at home or in the office. The table below shows the average number of hours these workers report they work on a weekend. Approximately what percent of Americans who work in large offices work 11 or more hours on weekends?

Numbers of Hours Worked on Weekends	Percent
0 – 1	3%
2 – 5	32%
6 – 10	42%
11 or more	23%

37. In a survey of consumers, approximately 43% said they would be willing to pay $1500 more for a new car if the car had an EPA rating of 80 miles per gallon. If your car currently gets 28 miles per gallon and you drive approximately 10,000 miles per year, in how many months would your savings on gasoline pay for the increased cost of such a car? Assume that the average cost of gasoline is $1.50 per gallon.

COOPERATIVE LEARNING

38. **Demography** is the statistical study of human populations. Many groups are interested in the sizes of certain segments of the population and projections of population growth. For example, public school administrators want estimates of the number of school-age children who will be living in their districts 10 years from now.

The U.S. government provides estimates of the future U.S. population. You can find these projections at the Census Bureau website at **www. census.gov.** Three different projections are provided: a lowest series, a middle series, and a highest series. These series reflect different theories on how fast the population of this country will grow.

The table below contains data from the Census Bureau website. The figures are from the middle series of projections.

	Age	Under 5	5 – 17	18 – 24	25 – 34	35 – 44	45 – 54	55 – 64	65 – 74	75 & older
2010	Male	9,712	26,544	13,338	18,535	22,181	18,092	11,433	8,180	6,165
	Female	9,274	25,251	12,920	18,699	22,478	18,938	12,529	9,956	10,408
2050	Male	13,877	35,381	18,462	24,533	23,352	21,150	20,403	16,699	19,378
	Female	13,299	33,630	17,871	24,832	24,041	22,344	21,965	18,032	24,751

For the following exercises, round all percents to the nearest tenth of a percent.

a. Which of the age groups listed are of interest to public school officials?

b. Which age groups are of interest to nursing home administrators?

c. Which age groups are of concern to accountants determining future benefits to be paid out by the Social Security Administration?

d. Which age group is of interest to manufacturers of disposable diapers?

e. Which age group is of primary concern to college and university admissions officers?

f. In which age groups do the males outnumber the females? In which do the females outnumber the males?

g. What percent of the projected population aged 75 and older in the year 2010 is female? Does this percent decrease in 2050? If so, by how much?

h. Find the difference between the percent of the population that will be 65 or over in 2010 and the percent that will be 65 or older in 2050.

i. Assume that the work force consists of people aged 25 to 64. What percent increase is expected in this population from 2010 to 2050?

j. What percent of the population is the work force expected to be in 2010? What percent of the population is the work force expected to be in 2050?

k. Why are the answers to parts h, i, and j of concern to the Social Security Administration?

l. Describe any patterns you see in the table.

m. Calculate a statistic based on the data in the table and explain why it would be of interest to an institution (such as a school system) or a manufacturer of consumer goods (such as disposable diapers).

39. Nielsen Media Research surveys television viewers to determine the numbers of people watching particular shows. There are an estimated 102.2 million U.S. households with televisions. Each **rating point** represents 1% of that number, or 1,022,000, households. Therefore, if *60 Minutes* received a rating of 9.6, then 9.6%, or $(0.096)(102,200,000) = 9,811,200$, households watched that program.

A rating point does not mean that 1,022,000 people are watching a program. A rating point refers to the number of households with TV sets tuned to that program; there may be more than one person watching a television set in the household.

Nielsen Media Research also describes a program's share of the market. **Share** is the percent of households with television sets in use that are tuned to a program. Suppose the same week that *60 Minutes* received 9.6 rating points, the show received a share of 18%. This would mean that 18% of all households with a television *turned on* were tuned to *60 Minutes,* whereas 9.6% of all households with a television were tuned to the program.

a. If *Law and Order* received a Nielsen rating of 10.8 and a share of 19, how many TV households watched the program that week? How many TV households were watching television during that hour? Round to the nearest hundred thousand.

b. Suppose *Primetime Thursday* received a rating of 16.2 and a share of 28. How many TV households watched the program that week? How many TV households were watching television during that hour? Round to the nearest hundred thousand.

c. Suppose *CSI: Crime Scene Investigation* received a rating of 13.6 during a week in which 34,750,000 people were watching the show. Find the average number of people per TV household who watched the program. Round to the nearest tenth.

Nielsen Media Research has a web site on the Internet. You can locate the site by using a search engine. The site does not list rating points or market share, but these statistics can be found on other web sites by using a search engine.

d. Find the top two prime-time television shows for last week. Calculate the number of TV households that watched the programs. Compare the figures with the top two sports programs for last week.

SECTION 5.4 | **Second-Degree Equations**

Second-Degree Equations in Standard Form

▼ **point of interest**

The word *quadratic* comes from the Latin word *quadratus,* which means "to make square." Note that the highest term in a quadratic equation contains the variable squared.

In Section 5.1, we introduced first-degree equations in one variable. The topic of this section is second-degree equations in one variable. A **second-degree equation in one variable** is an equation of the form $ax^2 + bx + c = 0$, where a and b are coefficients and c is a constant, and $a \neq 0$. An equation of this form is also called a **quadratic equation.** Here are three examples of second-degree equations in one variable.

$$4x^2 - 7x + 1 = 0 \qquad a = 4, b = -7, c = 1$$
$$3z^2 - 6 = 0 \qquad a = 3, b = 0, c = -6$$
$$t^2 + 10t = 0 \qquad a = 1, b = 10, c = 0$$

Note that although the value of a cannot be 0, the value of b or c can be 0.

A second-degree equation is in **standard form** when the expression $ax^2 + bx + c$ is in **descending order** (the exponents on the variables decrease from left to right) and set equal to zero. For instance, $2x^2 + 8x - 3 = 0$ is written in standard form; $x^2 = 4x - 8$ is not in standard form.

QUESTION   *Which of the following are second-degree equations written in standard form?*

a. $3y^2 + 5y - 2 = 0$ *b.* $8p - 4p^2 + 7 = 0$

c. $z^3 - 6z + 9 = 0$ *d.* $4r^2 + r - 1 = 6$

e. $v^2 - 16 = 0$

EXAMPLE 1 ■ Write a Quadratic Equation in Standard Form

Write the equation $x^2 = 3x - 8$ in standard form.

Solution

Subtract $3x$ from each side of the equation.

$$x^2 = 3x - 8$$
$$x^2 - 3x = 3x - 3x - 8$$
$$x^2 - 3x = -8$$

Then add 8 to each side of the equation.

$$x^2 - 3x + 8 = -8 + 8$$
$$x^2 - 3x + 8 = 0$$

CHECK YOUR PROGRESS 1   Write $2s^2 = 6 - 4s$ in standard form.

Solution   *See page S19.*

Solving Second-Degree Equations by Factoring

Recall that the Multiplication Property of Zero states that the product of a number and zero is zero.

If a is a real number, then $a \cdot 0 = 0$.

Consider the equation $a \cdot b = 0$. If this is a true equation, then either $a = 0$ or $b = 0$. This is summarized in the Principle of Zero Products.

ANSWER   *The equations in **a** and **e** are second-degree equations in standard form. The equation in **b** is not in standard form because $8p - 4p^2 + 7$ is not written in descending order. The equation in **c** is not a second-degree equation because there is an exponent of 3 on the variable. The equation in **d** is not in standard form because it is not equal to 0.*

> **Principle of Zero Products**
>
> If the product of two factors is zero, then at least one of the factors must be zero.
>
> If $ab = 0$, then $a = 0$ or $b = 0$.

The Principle of Zero Products is often used to solve equations. This is illustrated in Example 2.

EXAMPLE 2 ■ Solve an Equation Using the Principle of Zero Products

Solve: $(x - 4)(x + 6) = 0$

Solution

In the expression $(x - 4)(x + 6)$, we are multiplying two numbers. Because their product is 0, one of the numbers must be equal to zero. The number $x - 4 = 0$ or the number $x + 6 = 0$. Solve each of these equations for x.

$$(x - 4)(x + 6) = 0$$

$$x - 4 = 0 \qquad x + 6 = 0$$
$$x = 4 \qquad x = -6$$

Check:
$$\frac{(x - 4)(x + 6) = 0}{\begin{array}{c|c} (4 - 4)(4 + 6) & 0 \\ (0)(10) & 0 \\ 0 = 0 \end{array}} \qquad \frac{(x - 4)(x + 6) = 0}{\begin{array}{c|c} (-6 - 4)(-6 + 6) & 0 \\ (-10)(0) & 0 \\ 0 = 0 \end{array}}$$

The solutions are 4 and -6.

CHECK YOUR PROGRESS 2 Solve: $(n + 5)(2n - 3) = 0$

Solution See page S19.

✔ **TAKE NOTE**

Note that both 4 and -6 check as solutions. The equation $(x - 4)(x + 6) = 0$ has two solutions.

historical note

Chu Shih-chieh, a Chinese mathematician who lived around 1300, wrote the book *Ssu-yuan yu-chien*, which dealt with equations of degree as high as 14. ■

A second-degree equation can be solved by using the Principle of Zero Products when the expression $ax^2 + bx + c$ is factorable. This is illustrated in Example 3.

EXAMPLE 3 ■ Solve a Quadratic Equation by Factoring

Solve: $2x^2 + x = 6$

Solution

In order to use the Principle of Zero Products to solve a second-degree equation, the equation must be in standard form. Subtract 6 from each side of the given equation.

$$2x^2 + x = 6$$
$$2x^2 + x - 6 = 6 - 6$$
$$2x^2 + x - 6 = 0$$

Factor $2x^2 + x - 6$.

$$(2x - 3)(x + 2) = 0$$

CALCULATOR NOTE

These solutions can be checked on a scientific calculator. Many calculators require the following keystrokes.

2 [X] [(] 1.5 [INV] [X²] [)]
[+] 1.5 [=]

and

2 [X] [(] 2 [+/−] [INV] [X²] [)]
[+] 2 [+/−] [=]

On a graphing calculator, enter

2 [X] 1.5 [X²] [+] 1.5
[ENTER]

and

2 [X] [(] [(−)] 2 [)] [X²] [+]
[(−)] 2 [ENTER]

In each case, the display is 6.

Use the Principle of Zero Products. Set each factor equal to zero. Then solve each equation for x.

$$2x - 3 = 0 \qquad\qquad x + 2 = 0$$
$$2x = 3 \qquad\qquad\quad x = -2$$
$$x = \frac{3}{2}$$

Check:

$$2x^2 + x = 6 \qquad\qquad 2x^2 + x = 6$$

$$
\begin{array}{c|c}
2\left(\dfrac{3}{2}\right)^2 + \dfrac{3}{2} & 6 \\[2ex]
2\left(\dfrac{9}{4}\right) + \dfrac{3}{2} & 6 \\[2ex]
\dfrac{9}{2} + \dfrac{3}{2} & 6 \\[2ex]
6 & = 6
\end{array}
\qquad\qquad
\begin{array}{c|c}
2(-2)^2 + (-2) & 6 \\[2ex]
2(4) + (-2) & 6 \\[2ex]
8 + (-2) & 6 \\[2ex]
6 & = 6
\end{array}
$$

The solutions are $\dfrac{3}{2}$ and -2.

CHECK YOUR PROGRESS 3   Solve: $2x^2 = x + 1$

Solution   *See page S19.*

Note from Example 3 the steps involved in solving a second-degree equation by factoring. These are outlined below.

Solving a Second-Degree Equation by Factoring

1. Write the equation in standard form.

2. Factor the expression $ax^2 + bx + c$.

3. Use the Principle of Zero Products to set each factor equal to zero.

4. Solve each of the resulting equations for the variable.

Solving Second-Degree Equations by Using the Quadratic Formula

When using only integers, not all second-degree equations can be solved by factoring. Any equation that cannot be solved easily by factoring can be solved by using the *quadratic formula*, which is given on the following page.

historical note

The quadratic formula is one of the oldest formulas in mathematics. It is not known where it was first derived or by whom. However, there is evidence of its use as early as 2000 B.C. in ancient Babylonia (the area of the Middle East now known as Iraq). Hindu mathematicians were using it over a thousand years ago. ■

The Quadratic Formula

The solutions of the equation $ax^2 + bx + c = 0$, $a \neq 0$, are

$$x = \frac{-b + \sqrt{b^2 - 4ac}}{2a} \quad \text{and} \quad x = \frac{-b - \sqrt{b^2 - 4ac}}{2a}$$

The quadratic formula is frequently written in the form

$$x = \frac{-b \pm \sqrt{b^2 - 4ac}}{2a}$$

To use the quadratic formula, first write the second-degree equation in standard form. Determine the values of a, b, and c. Substitute the values of a, b, and c into the quadratic formula. Then evaluate the resulting expression.

✔ **TAKE NOTE**

If you need to review material on simplifying radical expressions, see Lesson 9.2A on the CD that you received with this book.

CALCULATOR NOTE

To find the decimal approximation of $\dfrac{2 + \sqrt{2}}{2}$ on a scientific calculator, use the following keystrokes.

(2 + 2 √) ÷ 2 =

Note that parentheses are used to ensure that the entire numerator is divided by the denominator.

On a graphing calculator, enter

(2 + 2nd √ 2))
÷ 2 ENTER

EXAMPLE 4 ■ **Solve a Quadratic Equation by Using the Quadratic Formula**

Solve the equation $2x^2 = 4x - 1$ by using the quadratic formula. Give exact solutions and approximate solutions to the nearest thousandth.

Solution

Write the equation in standard form by subtracting $4x$ from each side of the equation and adding 1 to each side of the equation. Then determine the values of a, b, and c.

$$2x^2 = 4x - 1$$
$$2x^2 - 4x + 1 = 0$$
$$a = 2, \; b = -4, \; c = 1$$

Substitute the values of a, b, and c into the quadratic formula. Then evaluate the resulting expression.

$$x = \frac{-b \pm \sqrt{b^2 - 4ac}}{2a}$$

$$x = \frac{-(-4) \pm \sqrt{(-4)^2 - 4(2)(1)}}{2(2)} = \frac{4 \pm \sqrt{16 - 8}}{4}$$

$$= \frac{4 \pm \sqrt{8}}{4} = \frac{4 \pm 2\sqrt{2}}{4} = \frac{2(2 \pm \sqrt{2})}{2(2)} = \frac{2 \pm \sqrt{2}}{2}$$

The exact solutions are $\dfrac{2 + \sqrt{2}}{2}$ and $\dfrac{2 - \sqrt{2}}{2}$.

$$\frac{2 + \sqrt{2}}{2} \approx 1.707 \qquad \frac{2 - \sqrt{2}}{2} \approx 0.293$$

To the nearest thousandth, the solutions are 1.707 and 0.293.

CHECK YOUR PROGRESS 4 Solve the equation $2x^2 = 8x - 5$ by using the quadratic formula. Give exact solutions and approximate solutions to the nearest thousandth.

Solution See page S19.

INSTRUCTOR NOTE
It may help to give examples of just determining a, b, and c for different equations. Here are some suggestions.

$$2x = x^2 - 3$$
$$4x - 3x^2 + 3 = 0$$
$$2x^2 - x = 0$$

Students will ask whether it matters to which side of the equation a term is moved. To convince these students that it does not matter, have them solve $3x = x^2 - 4$ by rewriting it as

$$x^2 - 3x - 4 = 0$$

and as

$$-x^2 + 3x + 4 = 0$$

One of the difficulties students have with the quadratic formula is correct substitution. Have them first make sure that the equation is in standard form. Then circling a, b, and c may help. Remind them that the sign that precedes a number is the sign of the number. For Example 4, we have

$$\overset{a}{②}x^2 \overset{b}{\ominus} 4x \overset{c}{\oplus} 1 = 0$$

✔ **TAKE NOTE**

The square root of a negative number is not a real number because there is no real number that, when squared, equals a negative number. Therefore $\sqrt{-19}$ is not a real number.

The exact solutions to Example 4 are irrational numbers. It is also possible for a quadratic equation to have no real number solution. This is illustrated in Example 5.

EXAMPLE 5 ■ **Solve a Quadratic Equation by Using the Quadratic Formula**

Solve by using the quadratic formula: $t^2 + 7 = 3t$

Solution
Write the equation in standard form by subtracting $3t$ from each side of the equation. Then determine the values of a, b, and c.

$$t^2 + 7 = 3t$$
$$t^2 - 3t + 7 = 0$$
$$a = 1, b = -3, c = 7$$

Substitute the values of a, b, and c into the quadratic formula. Then evaluate the resulting expression.

$$t = \frac{-b \pm \sqrt{b^2 - 4ac}}{2a}$$
$$t = \frac{-(-3) \pm \sqrt{(-3)^2 - 4(1)(7)}}{2(1)}$$
$$= \frac{3 \pm \sqrt{9 - 28}}{2} = \frac{3 \pm \sqrt{-19}}{2}$$

$\sqrt{-19}$ is not a real number.

The equation has no real number solutions.

CHECK YOUR PROGRESS 5   Solve by using the quadratic formula: $z^2 = -6 - 2z$

Solution   See page S19.

Math Matters   The Discriminant

In Example 5, the second-degree equation has no real number solutions. In the quadratic formula, the quantity $b^2 - 4ac$ under the radical sign is called the **discriminant**. When a, b, and c are real numbers, the discriminant determines whether or not a quadratic equation has real number solutions.

The Effect of the Discriminant on the Solutions of a Second-Degree Equation

1. If $b^2 - 4ac \geq 0$, the equation has real number solutions.

2. If $b^2 - 4ac < 0$, the equation has no real number solutions.

For example, for the equation $x^2 - 4x - 5 = 0$, $a = 1$, $b = -4$, and $c = -5$.

$$b^2 - 4ac = (-4)^2 - 4(1)(-5) = 16 + 20 = 36$$
$$36 > 0$$

The discriminant is greater than 0. The equation has real number solutions.

Applications of Second-Degree Equations

Second-degree equations have many applications to the real world. Examples 6 and 7 illustrate two such applications.

EXAMPLE 6 ▪ Solve an Application of Quadratic Equations by Factoring

An arrow is projected straight up into the air with an initial velocity of 48 feet per second. At what times will the arrow be 32 feet above the ground? Use the equation $h = 48t - 16t^2$, where h is the height, in feet, above the ground after t seconds.

Solution
We are asked to find the times when the arrow will be 32 feet above the ground, so we are given a value for h. Substitute 32 for h in the given equation and solve for t.

$$h = 48t - 16t^2$$
$$32 = 48t - 16t^2$$

This is a second-degree equation. Write the equation in standard form by adding $16t^2$ to each side of the equation and subtracting $48t$ from each side of the equation.

$$16t^2 - 48t + 32 = 0$$
$$16(t^2 - 3t + 2) = 0$$
$$t^2 - 3t + 2 = 0 \qquad \bullet \text{ Divide each side of the equation by 16.}$$
$$(t - 1)(t - 2) = 0$$
$$t - 1 = 0 \qquad t - 2 = 0$$
$$t = 1 \qquad\quad t = 2$$

The arrow will be 32 feet above the ground 1 second after its release and 2 seconds after its release.

CHECK YOUR PROGRESS 6   An object is projected straight up into the air with an initial velocity of 64 feet per second. At what times will the object be on the ground? Use the equation $h = 64t - 16t^2$, where h is the height, in feet, above the ground after t seconds.

Solution   *See page S20.*

> ✔ **TAKE NOTE**
>
> It would also be correct to subtract 32 from each side of the equation. However, many people prefer to have the coefficient of the squared term positive.

EXAMPLE 7 ▪ Solve an Application of Quadratic Equations by Using the Quadratic Formula

A baseball player hits a ball. The height of the ball above the ground after t seconds can be approximated by the equation $h = -16t^2 + 75t + 5$. When will the ball hit the ground? Round to the nearest hundredth.

Solution

We are asked to determine the number of seconds from the time the ball is hit until it is on the ground. When the ball is on the ground, its height above the ground is 0 feet. Substitute 0 for h and solve for t.

$$h = -16t^2 + 75t + 5$$
$$0 = -16t^2 + 75t + 5$$
$$16t^2 - 75t - 5 = 0$$

This is a second-degree equation. It is not easily factored. Use the quadratic formula to solve for t.

$$a = 16, b = -75, c = -5$$
$$t = \frac{-b \pm \sqrt{b^2 - 4ac}}{2a}$$
$$t = \frac{-(-75) \pm \sqrt{(-75)^2 - 4(16)(-5)}}{2(16)} = \frac{75 \pm \sqrt{5945}}{32}$$
$$t = \frac{75 + \sqrt{5945}}{32} \approx 4.75 \qquad t = \frac{75 - \sqrt{5945}}{32} \approx -0.07$$

The ball strikes the ground 4.75 seconds after the baseball player hits it.

CHECK YOUR PROGRESS 7 A basketball player shoots at a basket that is 25 feet away. The height h, in feet, of the ball above the ground after t seconds is given by $h = -16t^2 + 32t + 6.5$. How many seconds after the ball is released does it hit the basket? Note: The basket is 10 feet off the ground. Round to the nearest hundredth.

Solution See page S20.

> ✔ **TAKE NOTE**
>
> The time until the ball hits the ground cannot be a negative number. Therefore, -0.07 is not a solution of this application.

Excursion

The Sum and Product of the Solutions of a Quadratic Equation

The solutions of the equation $x^2 + 3x - 10 = 0$ are -5 and 2.

$$x^2 + 3x - 10 = 0$$
$$(x + 5)(x - 2) = 0$$
$$x + 5 = 0 \qquad x - 2 = 0$$
$$x = -5 \qquad\quad x = 2$$

Note that the sum of the solutions is equal to $-b$, the opposite of the coefficient of x.

$$-5 + 2 = -3$$

The product of the solutions is equal to c, the constant term.

$$-5(2) = -10$$

> ✔ **TAKE NOTE**
>
> Look closely at the example at the right, in which -5 and 2 are solutions of the quadratic equation $(x + 5)(x - 2) = 0$. Using variables, we can state that if s_1 and s_2 are solutions of a quadratic equation, then the quadratic equation can be written in the form $(x - s_1)(x - s_2) = 0$.

(continued)

This illustrates the following theorem regarding the solutions of a quadratic equation.

> **The Sum and Product of the Solutions of a Quadratic Equation**
>
> If s_1 and s_2 are the solutions of a quadratic equation of the form $ax^2 + bx + c = 0$, $a \neq 0$, then
>
> the sum of the solutions $s_1 + s_2 = -\dfrac{b}{a}$, and
>
> the product of the solutions $s_1 s_2 = \dfrac{c}{a}$

✔ **TAKE NOTE**

The result is the same if we let $s_1 = 6$ and $s_2 = -2$.

In this section, the method we used to check the solutions of a quadratic equation was to substitute the solutions back into the original equation. An alternative method is to use the sum and product of the solutions.

For example, let's check that -2 and 6 are the solutions of the equation $x^2 - 4x - 12 = 0$. For this equation, $a = 1$, $b = -4$, and $c = -12$. Let $s_1 = -2$ and $s_2 = 6$.

$$
\begin{array}{c|c}
s_1 + s_2 = -\dfrac{b}{a} & s_1 s_2 = \dfrac{c}{a} \\
\hline
-2 + 6 \;\Big|\; -\dfrac{-4}{1} & -2(6) \;\Big|\; \dfrac{-12}{1} \\
4 = 4 & -12 = -12
\end{array}
$$

The solutions check.

In Example 4, we found that the exact solutions of the equation $2x^2 = 4x - 1$ are $\dfrac{2 + \sqrt{2}}{2}$ and $\dfrac{2 - \sqrt{2}}{2}$. Use the sum and product of the solutions to check these solutions.

Write the equation in standard form. Then determine the values of a, b, and c.

$$2x^2 = 4x - 1$$
$$2x^2 - 4x + 1 = 0$$
$$a = 2, \; b = -4, \; c = 1$$

Let $s_1 = \dfrac{2 + \sqrt{2}}{2}$ and $s_2 = \dfrac{2 - \sqrt{2}}{2}$.

✔ **TAKE NOTE**

If you need to review material on adding and multiplying radical expressions, see Lessons 9.2B and 9.2C on the CD that you received with this book.

$$
\begin{array}{c|c}
s_1 + s_2 = -\dfrac{b}{a} & s_1 s_2 = \dfrac{c}{a} \\
\hline
\dfrac{2 + \sqrt{2}}{2} + \dfrac{2 - \sqrt{2}}{2} \;\Big|\; -\dfrac{-4}{2} & \left(\dfrac{2 + \sqrt{2}}{2}\right)\left(\dfrac{2 - \sqrt{2}}{2}\right) \;\Big|\; \dfrac{1}{2} \\
\dfrac{2 + \sqrt{2} + 2 - \sqrt{2}}{2} \;\Big|\; 2 & \dfrac{4 - 2}{4} \;\Big|\; \dfrac{1}{2} \\
\dfrac{4}{2} \;\Big|\; 2 & \dfrac{2}{4} \;\Big|\; \dfrac{1}{2} \\
2 = 2 & \dfrac{1}{2} = \dfrac{1}{2}
\end{array}
$$

The solutions check.

(continued)

If we divide both sides of the equation $ax^2 + bx + c = 0$, $a \neq 0$, by a, the result is the equation

$$x^2 + \frac{b}{a}x + \frac{c}{a} = 0$$

Using this model and the sum and products of the solutions of a quadratic equation, we can find a quadratic equation given its solutions. The method is given below.

A Quadratic Equation with Solutions s_1 and s_2

If the solutions of a quadratic equation are s_1 and s_2, then the quadratic equation is

$$x^2 - (s_1 + s_2)x + s_1 s_2 = 0$$

To write a quadratic equation that has solutions $\frac{2}{3}$ and 1, let $s_1 = \frac{2}{3}$ and $s_2 = 1$. Substitute these values into the equation $x^2 - (s_1 + s_2)x + s_1 s_2 = 0$ and simplify.

$$x^2 - (s_1 + s_2)x + s_1 s_2 = 0$$
$$x^2 - \left(\frac{2}{3} + 1\right)x + \left(\frac{2}{3} \cdot 1\right) = 0$$
$$x^2 - \frac{5}{3}x + \frac{2}{3} = 0$$

We want a, b, and c to be integers. Multiply each side of the equation by 3.

$$3\left(x^2 - \frac{5}{3}x + \frac{2}{3}\right) = 3(0)$$
$$3x^2 - 5x + 2 = 0$$

A quadratic equation with solutions $\frac{2}{3}$ and 1 is $3x^2 - 5x + 2 = 0$.

Excursion Exercises

In Exercises 1–8, solve the equation and then check the solutions using the sum and product of the solutions.

1. $x^2 - 10 = 3x$ **2.** $x^2 + 16 = 8x$

3. $3x^2 + 5x = 12$ **4.** $3x^2 + 8x = 3$

5. $x^2 = 6x + 3$ **6.** $x^2 = 2x + 5$

7. $4x + 1 = 4x^2$ **8.** $x + 1 = x^2$

In Exercises 9–16, write a quadratic equation that has integer coefficients and has the given pair of solutions.

9. -1 and 6   **10.** -5 and -4

11. 3 and $\frac{1}{2}$ **12.** $-\frac{3}{4}$ and 2

13. $\frac{1}{4}$ and $-\frac{3}{2}$ **14.** $\frac{2}{3}$ and $-\frac{2}{3}$

15. $2 + \sqrt{2}$ and $2 - \sqrt{2}$ **16.** $1 + \sqrt{3}$ and $1 - \sqrt{3}$

Exercise Set 5.4   (Suggested Assignment: 5–25, every other odd; 29–43, odds)

1. Explain the importance of writing a second-degree equation in standard form as the first step in solving the equation. Include in your explanation how the Principle of Zero Products is used to solve the equation.

2. Explain why the restriction $a \neq 0$ is given in the definition of a quadratic equation.

3. Write a second-degree equation that you can solve by factoring.

4. Write a second-degree equation that you can solve by using the quadratic formula but not by factoring.

Solve Exercises 5–28. First try to solve the equation by factoring. If you are unable to solve the equation by factoring, solve the equation by using the quadratic formula. For equations with solutions that are irrational numbers, give exact solutions and approximate solutions to the nearest thousandth.

5. $r^2 - 3r = 10$
6. $p^2 + 5p = 6$
7. $t^2 = t + 1$

8. $u^2 = u + 3$
9. $y^2 - 6y = 4$
10. $w^2 + 4w = 2$

11. $9z^2 - 18z = 0$
12. $4y^2 + 20y = 0$
13. $z^2 = z + 4$

14. $r^2 = r - 1$
15. $2s^2 = 4s + 5$
16. $3u^2 = 6u + 1$

17. $r^2 = 4r + 7$
18. $s^2 + 6s = 1$
19. $2x^2 = 9x + 18$

20. $3y^2 = 4y + 4$
21. $6x - 11 = x^2$
22. $-8y - 17 = y^2$

23. $4 - 15u = 4u^2$
24. $3 - 2y = 8y^2$
25. $6y^2 - 4 = 5y$

26. $6v^2 - 3 = 7v$
27. $y - 2 = y^2 - y - 6$
28. $8s - 11 = s^2 - 6s + 8$

Solve Exercises 29–43. Round answers to nearest hundredth.

29. The height h, in feet, of a golf ball t seconds after it has been hit is given by the equation $h = -16t^2 + 60t$. How many seconds after the ball is hit will the height of the ball be 36 feet?

30. The area A, in square meters, of a rectangle with a perimeter of 100 meters is given by the equation $A = 50w - w^2$, where w is the width of the rectangle in meters. What is the width of a rectangle if its area is 400 square meters?

31. In the diagram below, the total number of circles T when there are n rows is given by $T = 0.5n^2 + 0.5n$. Verify the formula for the four figures shown. Determine the number of rows when the total number of circles is 55.

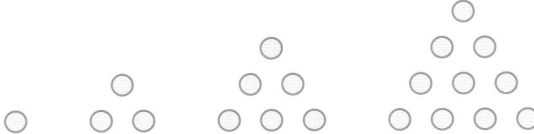

32. If an astronaut on the moon throws a ball upward with an initial velocity of 6 meters per second, its approximate height h, in meters, after t seconds is given by the equation $h = -0.8t^2 + 6t$. How long after it is released will the ball be 8 meters above the surface of the moon?

33. When a driver decides to stop a car, it takes time first for the driver to react and put a foot on the brake and then it takes additional time for the car to slow down. The total distance traveled during this period of time is called the *stopping distance* of the car. For some cars, the stopping distance d, in feet, is given by the equation $d = 0.05r^2 + r$, where r is the speed of the car in miles per hour.

    a. Find the distance needed to stop a car traveling at 60 miles per hour.

    b. If skid marks at an accident site are 75 feet long, how fast was the car traveling?

34. At La Quebrada in Acapulco, Mexico, cliff divers dive from a rock cliff that is 27 meters above the water. The equation $h = -x^2 + 2x + 27$ describes the height h, in meters, of the diver above the water when the diver is x feet away from the cliff.

    a. When the diver enters the water, how far is the diver from the cliff?

    b. Does the diver ever reach a height of 28 meters above the water? If so, how far is the diver from the cliff at that time?

    c. Does the diver ever reach a height of 30 meters above the water? If so, how far is the diver from the cliff at that time?

35. The hang time of a football that is kicked on the opening kickoff is given by $s = -16t^2 + 88t + 1$, where s is the height in feet of the football t seconds after leaving the kicker's foot. What is the hang time of a kickoff that hits the ground without being caught?

36. In a slow pitch softball game, the height of a ball thrown by a pitcher can be approximated by the equation $h = -16t^2 + 24t + 4$, where h is the height, in feet, of the ball and t is the time, in seconds, since it was released by the pitcher. If a batter hits the ball when it is 2 feet off the ground, for how many seconds has the ball been in the air?

37. The path of water from a hose on a fire tugboat can be approximated by the equation $y = -0.005x^2 + 1.2x + 10$, where y is the height, in feet, of the water above the ocean when the water is x feet from the tugboat. When the water from the hose is 5 feet above the ocean, at what distance from the tugboat is it?

38. An event in the Summer Olympics is 10-meter springboard diving. In this event, the height h, in meters, of a diver above the water t seconds after jumping is given by the equation $h = -4.9t^2 + 7.8t + 10$. What is the height above the water of a diver after 2 seconds?

39. A penalty kick in soccer is made from a penalty mark that is 36 feet from a goal that is 8 feet high. A possible equation for the flight of a penalty kick is $h = -0.002x^2 + 0.36x$, where h is the height, in feet, of the ball x feet from the goal. Assuming that the flight of the kick is toward the goal and that it is not touched by the goalie, will the ball land in the net?

40. In Germany there are no speed limits on some portions of the autobahn (the highway). Other portions have a speed limit of 180 kilometers per hour (approximately 112 miles per hour). The distance d, in meters, required to stop a car traveling v kilometers per hour is $d = 0.0056v^2 + 0.14v$. Approximate the maximum speed a driver can be going and still be able to stop within 150 meters.

41. A model rocket is launched with an initial velocity of 200 feet per second. The height h, in feet, of the rocket t seconds after the launch is given by $h = -16t^2 + 200t$. How many seconds after the launch will the rocket be 300 feet above the ground? Round to the nearest hundredth of a second.

42. The Water Arc is a fountain that shoots water across the Chicago River from a water cannon. The path of the water can be approximated by the equation $h = -0.006x^2 + 1.2x + 10$, where x is the horizontal distance, in feet, from the cannon and h is the height, in feet, of the water above the river. On one particular day, some people were walking along the opposite side of the river from the Water Arc when a pulse of water was shot in their direction. If the distance from the Water Arc to the people was 220 feet, did they get wet from the cannon's water?

German Autobahn System

43. The graph below shows the cost of a first-class postage stamp from the 1950s to 2002. A second-degree equation that approximately models these data is $y = 0.0087x^2 - 0.6125x + 9.9096$, where $x \geq 50$ and $x = 50$ for the year 1950, and y is the cost, in cents, of a first-class stamp. Using the model equation, predict the cost of a first-class stamp in the year 2020. Round to the nearest cent.

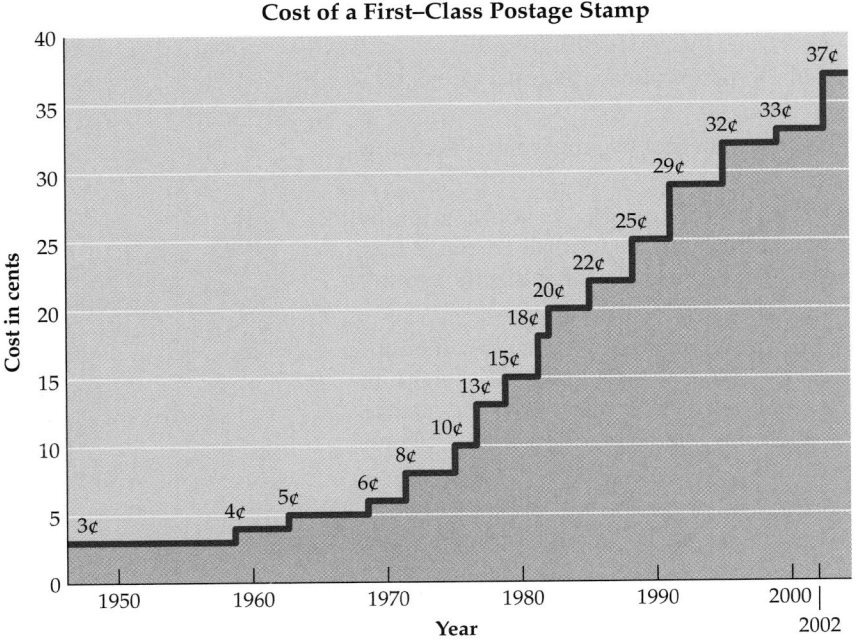

Cost of a First–Class Postage Stamp

Extensions

CRITICAL THINKING

44. Show that the solutions of the equation $ax^2 + bx = 0$, $a \neq 0$, are 0 and $-\dfrac{b}{a}$.

45. In a second-degree equation in standard form, why is the expression $ax^2 + bx + c$ factorable over the integers only when the discriminant is a perfect square? (See Math Matters, page 305.)

46. A wire 8 feet long is cut into two pieces. A circle is formed from one piece and a square is formed from the other. The total area of both figures is given by $A = \dfrac{1}{16}(8 - x)^2 + \dfrac{x^2}{4\pi}$. What is the length of each piece of wire if the total area is 4.5 square feet? Round to the nearest thousandth.

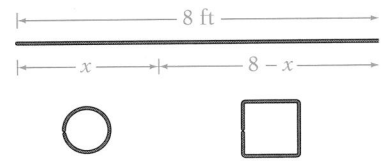

In Exercises 47–52, solve the equation for x.

47. $x^2 + 16ax + 48a^2 = 0$

48. $x^2 - 8bx + 15b^2 = 0$

49. $2x^2 + 3bx + b^2 = 0$

50. $3x^2 - 4cx + c^2 = 0$

51. $2x^2 - xy - 3y^2 = 0$

52. $x^2 - 4xy + 4y^2 = 0$

COOPERATIVE LEARNING

53. Show that the equation $x^2 + bx - 1 = 0$ always has real number solutions, regardless of the value of b.

54. Show that the equation $2x^2 + bx - 2 = 0$ always has real number solutions, regardless of the value of b.

CHAPTER 5 # Summary

Key Terms

constant term [p. 240]
descending order [p. 301]
equation [p. 240]
extremes of a proportion [p. 264]
first-degree equation in one variable [p. 240]
formula [p. 248]
like terms [p. 240]
literal equation [p. 248]
means of a proportion [p. 264]
numerical coefficient [p. 240]
percent [p. 277]
percent decrease [p. 288]
percent increase [p. 286]
proportion [p. 264]
quadratic equation [p. 300]
rate [p. 257]
ratio [p. 261]
second-degree equation in one variable [p. 300]
solution of an equation [p. 241]
solve an equation [p. 241]
standard form [p. 301]
term of a proportion [p. 264]
terms of a variable expression [p. 240]
unit price [p. 259]
unit rate [p. 258]
unit ratio [p. 262]
variable part [p. 240]
variable term [p. 240]

Essential Concepts

■ **Properties of Equations**

Addition Property
The same number can be added to each side of an equation without changing the solution of the equation.
If $a = b$, then $a + c = b + c$.

Subtraction Property
The same number can be subtracted from each side of an equation without changing the solution of the equation.
If $a = b$, then $a - c = b - c$.

Multiplication Property
Each side of an equation can be multiplied by the same nonzero number without changing the solution of the equation.

If $a = b$ and $c \neq 0$, then $ac = bc$.

Division Property
Each side of an equation can be divided by the same nonzero number without changing the solution of the equation.
If $a = b$ and $c \neq 0$, then $\frac{a}{c} = \frac{b}{c}$.

■ **Steps for Solving a First-Degree Equation in One Variable**

1. If the equation contains fractions, multiply each side of the equation by the least common multiple (LCM) of the denominators.

2. Use the Distributive Property to remove parentheses.

3. Combine any like terms on the right side of the equation and any like terms on the left side of the equation.

4. Use the Addition or Subtraction Property to rewrite the equation with only one variable term and only one constant term.

5. Use the Multiplication or Division Property to rewrite the equation with the variable alone on one side of the equation and a constant on the other side of the equation.

■ **Solve a Literal Equation for One of the Variables**
The goal is to rewrite the equation so that the letter being solved for is alone on one side of the equation and all numbers and other variables are on the other side.

■ **Calculate a Unit Rate**
Divide the number in the numerator of the rate by the number in the denominator of the rate.

■ **Write a Ratio**
A ratio can be written in three different ways: as a fraction, as two numbers separated by a colon (:), or as two numbers separated by the word *to*. Although units, such as hours, miles, or dollars, are written as part of a rate, units are not written as part of a ratio.

■ **Cross-Products Method of Solving a Proportion**
If $\frac{a}{b} = \frac{c}{d}$, then $bc = ad$.

■ **Write a Percent as a Decimal**
Remove the percent sign and move the decimal point two places to the left.

■ **Write a Decimal as a Percent**
Move the decimal point two places to the right and write a percent sign.

■ **Write a Percent as a Fraction**
Remove the percent sign and multiply by $\frac{1}{100}$.

■ **Write a Fraction as a Percent**
First write the fraction as a decimal. Then write the decimal as a percent.

■ **Proportion Used to Solve Percent Problems**
$\frac{\text{Percent}}{100} = \frac{\text{amount}}{\text{base}}$

■ **Basic Percent Equation**
$PB = A$, where P is the percent, B is the base, and A is the amount.

■ **Principle of Zero Products**
If the product of two factors is zero, then at least one of the factors must be zero.
If $ab = 0$, then $a = 0$ or $b = 0$.

■ **Solving a Second-Degree Equation by Factoring**

1. Write the equation in standard form.
2. Factor the polynomial $ax^2 + bx + c$.
3. Use the Principle of Zero Products to set each factor of the polynomial equal to zero.
4. Solve each equation for the variable.

■ **The Quadratic Formula**
The solutions of the equation $ax^2 + bx + c = 0$, $a \neq 0$, are $x = \dfrac{-b \pm \sqrt{b^2 - 4ac}}{2a}$.

CHAPTER 5   **Review Exercises**

In Exercises 1–8, solve the equation.

1. $5x + 3 = 10x - 17$

2. $3x + \dfrac{1}{8} = \dfrac{1}{2}$

3. $6x + 3(2x - 1) = -27$

4. $\dfrac{5}{12} = \dfrac{n}{8}$

5. $4y^2 + 9 = 0$

6. $x^2 - x = 30$

7. $x^2 = 4x - 1$

8. $x + 3 = x^2$

In Exercises 9 and 10, solve the formula for the given variable.

9. $4x + 3y = 12$; y

10. $f = v + at$; t

11. In June, the temperature at various elevations of the Grand Canyon can be approximated by the equation $T = -0.005x + 113.25$, where T is the temperature in degrees Fahrenheit and x is the elevation (distance above sea level) in feet. Use this equation to find the elevation at Inner Gorge, the bottom of the canyon, where the temperature is 101°F.

12. Find the time it takes for the velocity of a falling object to increase from 4 feet per second to 100 feet per second. Use the equation $v = v_0 + 32t$, where v is the final velocity of the falling object, v_0 is the initial velocity, and t is the time it takes for the object to fall.

13. A chemist mixes 100 grams of water at 80°C with 50 grams of water at 20°C. Use the formula $m_1(T_1 - T) = m_2(T - T_2)$ to find the final temperature of the water after mixing. In this equation, m_1 is the quantity of water at the hotter temperature, T_1 is the temperature of the hotter water, m_2 is the quantity of water at the cooler temperature, T_2 is the temperature of the cooler water, and T is the final temperature of water after mixing.

14. A computer bulletin board service charges $4.25 per month plus $.08 for each minute over 30 minutes that the service is used. For how many minutes did a subscriber use this service during a month in which the monthly charge was $4.97?

15. An automobile was driven 326.6 miles on 11.5 gallons of gasoline. Find the number of miles driven per gallon of gas.

16. A house with an original value of $120,000 increased in value to $160,000 in 5 years. Write, as a fraction in simplest form, the ratio of the increase in value to the original value of the house.

17. The table below shows the number of sexual harassment complaints from 1993 through 1997. (*Source:* Equal Employment Opportunity Commission) The number of sexual harassment complaints in 1997 was 2123 more than twice the number in 1991. Find the number of sexual harassment complaints in 1991.

Year	Number of Sexual Harassment Complaints
1993	11,908
1994	14,420
1995	15,549
1996	15,342
1997	15,889

18. The table below shows the population and area of the five most populous cities in the United States.

a. The cities are listed in the table according to population, from largest to smallest. Rank the cities according to population density, from largest to smallest.

b. How many more people per square mile are there in New York than in Houston? Round to the nearest whole number.

City	Population	Area (in square miles)
New York	8,008,000	321.8
Los Angeles	3,695,000	467.4
Chicago	2,896,000	228.469
Houston	1,954,000	594.03
Philadelphia	1,519,000	136

19. The table below shows the number of full-time men and women undergraduates and the number of full-time faculty at the seven colleges in Arizona. In parts **a.**, **b.**, and **c.**, round ratios to the nearest whole number. (*Source:* Barron's Profile of American Colleges, 24th edition, c. 2001)

University	Men	Women	Faculty
Arizona State University	12,451	13,952	1,248
DeVry Institute of Technology/Phoenix	2,192	710	81
Embry-Riddle Aeronautical University	1,210	238	61
Grand Canyon University	419	824	96
Northern Arizona University	4,990	6,800	690
Prescott College	309	425	53
University of Arizona	10,229	11,375	1,495

a.   Calculate the student-faculty ratio at Northern Arizona University. Write the ratio using a colon and using the word *to*. What does this ratio mean?

b. Which school listed has the lowest student-faculty ratio? the highest?

c. Which two schools listed have the same student-faculty ratio?

20. The Randolph Company spent $35,000 for advertising last year. Department A and Department B share the cost of advertising in the ratio 3:7. Find the amount allocated to each department.

21. Three tablespoons of a liquid plant fertilizer are to be added to every 4 gallons of water. How many tablespoons of fertilizer are required for 10 gallons of water?

22.   The table below shows how each dollar of projected spending by the federal government for a recent year was distributed. Social Security, interest payments, Medicare, Medicaid, and other entitlements are fixed expenditures. Nondefense discretionary and defense are the only discretionary spending by the federal government. The projected budget for the year was $1687.5 billion.

a. Is at least one-half of federal spending discretionary spending?

b. Find the ratio of fixed expenditures to discretionary spending.

c. Find the amount of the budget to be spent on fixed expenditures.

d. Find the amount of the budget to be spent on Social Security.

How Your Federal Tax Dollar is Spent	
Social Security	23 cents
Interest payments	15 cents
Medicare	12 cents
Medicaid	6 cents
Other entitlements	12 cents
Nondefense discretionary	17 cents
Defense	15 cents

23.   According to the U.S. Bureau of the Census, the projected population of males and females in the United States in 2025 and 2050 is as shown in the table below.

Year	Males	Females
2025	164,119,000	170,931,000
2050	193,234,000	200,696,000

a. What percent of the projected population in 2025 is female? Round to the nearest tenth of a percent.

b. Does the percent of the projected population that is female in 2050 differ by more or less than one percent from the percent that is female in 2025?

24.   According to the Scarborough Report, San Francisco is the city that has the highest percentage of people with a current U.S. passport: 38.6% of the population, or approximately 283,700 people, have a U.S. passport. Estimate the population of San Francisco. Round to the nearest hundred.

25. The graph at the right shows the number of people, in thousands, that are listed on the national patient waiting list for organ transplants. What percent of those listed are waiting for a kidney transplant? Round to the nearest tenth of a percent.

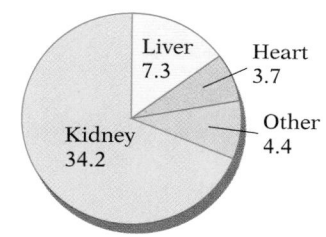

26. The table at the right shows the fat, saturated fat, cholesterol, and calorie content of a 90-gram ground-beef burger and a 90-gram soy burger.
    a. Compared to the beef burger, by what percent is the fat content decreased in the soy burger?
    b. What is the percent decrease in cholesterol in the soy burger compared to the beef burger?
    c. Calculate the percent decrease in calories in the soy burger compared to the beef burger.

Number of People on the National Waiting List for Organ Transplants (in thousands)

Source: United Network for Organ Sharing

	Beef Burger	Soy Burger
Fat	24 g	4 g
Saturated fat	10 g	1.5 g
Cholesterol	75 mg	0 mg
Calories	280	140

27. In January 2001, the Environmental Protection Agency (EPA) required all 54,000 community water systems in the United States to reduce the amount of arsenic in drinking water by 40 parts per billion from 50 parts per billion. What percent decrease does this represent?

28. The chart below shows, by age group, the number of girls playing youth soccer. Also shown is the percent of total youth soccer players who are girls. (*Source:* American Youth Soccer Organization)

Age group	5 – 6	7 – 8	9 – 10	11 – 12	13 – 14	15 – 16	17 – 18
Number of girls playing	23,805	45,181	46,758	39,939	26,157	11,518	4,430
Percent of all players	33%	33%	36%	40%	41%	42%	38%

   a. Which age group has the greatest number of girls playing youth soccer?
   b. Which age group has the largest percent of girls participating in youth soccer?
   c. What percent of the girls playing youth soccer are ages 7 to 10? Round to the nearest tenth of a percent. Is this more or less than half of all the girls playing?
   d. How many boys ages 17 to 18 play youth soccer? Round to the nearest ten.
   e. The girls playing youth soccer represent 36% of all youth soccer players. How many young people play youth soccer? Round to the nearest hundred.

29. A small rocket is shot from the edge of a cliff. The height h, in meters, of the rocket above the cliff is given by $h = 30t - 5t^2$, where t is the time in seconds after the rocket is shot. Find the times at which the rocket is 25 meters above the cliff.

30. The height h, in feet, of a ball t seconds after being thrown from a height of 6 feet is given by the equation $h = -16t^2 + 32t + 6$. After how many seconds is the ball 18 feet above the ground? Round to the nearest tenth.

CHAPTER 5   **Test**

In Exercises 1–5, solve the equation.

1. $\dfrac{x}{4} - 3 = \dfrac{1}{2}$

2. $x + 5(3x - 20) = 10(x - 4)$

3. $\dfrac{7}{16} = \dfrac{x}{12}$

4. $x^2 = 12x - 27$

5. $3x^2 - 4x = 1$

In Exercises 6 and 7, solve the formula for the given variable.

6. $x - 2y = 15;\ y$

7. $C = \dfrac{5}{9}(F - 32);\ F$

8. Old Faithful is a geyser in Yellowstone National Park. It is so named because of its regular eruptions for the past 100 years. An equation that can predict the approximate time until the next eruption is $T = 12.4L + 32$, where T is the time, in minutes, until the next eruption and L is the duration, in minutes, of the last eruption. Use this equation to determine the duration of the last eruption when the time between two eruptions is 63 minutes.

9. A library charges a fine for each overdue book. The fine is 15¢ for the first day plus 7¢ a day for each additional day the book is overdue. If the fine for a book is 78¢, how many days overdue is the book?

10. You drive 246.6 miles in 4.5 hours. Find your average rate in miles per hour.

11. The table below lists the largest city parks in the United States. The land acreage of Griffith Park in Los Angeles is three acres more than five times the acreage of New York's Central Park. What is the acreage of Central Park?

City Park	Land Acreage
Cullen Park (Houston)	10,534
Fairmont Park (Philadelphia)	8,700
Griffith Park (Los Angeles)	4,218
Eagle Creek Park (Indianapolis)	3,800
Pelham Bay Park (Bronx, NY)	2,764
Mission Bay Park (San Diego)	2,300

12. The table on the following page shows six Major League lifetime record holders for batting. (*Source: Information Please Almanac*)

a. Calculate the number of at-bats per home run for each player in the table. Round to the nearest thousandth.

b. The players are listed in the table alphabetically. Rank the players according to the number of at-bats per home run, starting with the best rate.

Baseball Player	Number of Times at Bat	Number of Home Runs Hit	Number of At-Bats per Home Run
Ty Cobb	11,429	4,191	_____
Billy Hamilton	6,284	2,163	_____
Rogers Hornsby	8,137	2,930	_____
Joe Jackson	4,981	1,774	_____
Tris Speaker	10,195	3,514	_____
Ted Williams	7,706	2,654	_____

13. In a recent year, the gross revenue from television rights, merchandise, corporate hospitality and tickets for the four major men's golf tournaments was as shown in the table below. What is the ratio, as a fraction in simplest form, of the gross revenue from the British Open to the gross revenue from the U.S. Open?

Golf Championship	Gross Revenue (in millions)
U. S. Open	$35
PGA Championship	$30.5
The Masters	$22
British Open	$20

14. The two partners in a partnership share the profits of their business in the ratio 5:3. Last year the profits were $180,000. Find the amount received by each partner.

15. The directions on a bag of plant food recommend one-half pound for every 50 square feet of lawn. How many pounds of plant food should be used on a lawn that measures 275 square feet?

16. The table below lists the U.S. cities with populations over 100,000 that had the highest number of violent crimes per 1000 residents per year. The violent crimes include murder, rape, aggravated assault, and robbery. (*Source:* FBI Uniform Crime Reports)

City	Violent Crimes per 1000 People
Baltimore	13.4
Baton Rouge	13.9
Gainesville, Florida	14.2
Lawton, Oklahoma	13.3
Los Angeles - Long Beach	14.2
Miami - Dade	18.9
New Orleans	13.3
New York	13.9

a. Which city has the highest rate of violent crimes?

b. The population of Baltimore is approximately 703,000. Estimate the number of violent crimes committed in that city. Round to the nearest whole number.

17. During a recent year, nearly 1.2 million dogs or litters were registered with the American Kennel Club. The most popular breed was the Labrador retriever, with 172,841 registered. What percent of the registrations were Labrador retrievers? Round to the nearest tenth of a percent. (*Source:* American Kennel Club)

18. The graph at the right shows the value of the personal computers (PCs) exported from Silicon Valley from 1993 through 1996.

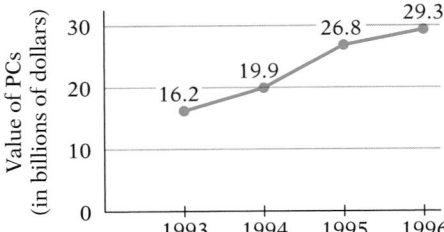

Value of PCs Exported from Silicon Valley (in billions of dollars)

Source: *Time* magazine, October 13, 1997

a. Which percent of the total value of the PCs exported from Silicon Valley during the 4 years is the value of the PCs exported in 1996? Round to the nearest tenth of a percent.

b. Between which two consecutive years shown on the graph is the percent increase the greatest?

c. What is the percent increase in the value of the exports from 1993 to 1996? Round to the nearest tenth of a percent.

19. The number of working farms in the United States in 1977 was 2.5 million. In 1987, there were 2.2 million working farms, and in 1997, there were 2.0 million working farms. (*Source:* CNN)

a. Find the percent decrease in the number of working farms from 1977 to 1997.

b. If the percent decrease in the number of working farms from 1997 to 2017 is the same as it was from 1977 to 1997, how many working farms will there be in the United States in 2017?

c. Provide an explanation for the decrease in the number of working farms in the United States.

20. The equation $h = -16t^2 + 28t + 6$ can be used to find the height h, in feet, of a shot t seconds after a shot putter has released it. After how many seconds is the shot put 10 feet above the ground? Round to the nearest tenth.

CHAPTER

6

Applications of Functions

Roller coasters have been using increasingly more sophisticated methods of acceleration, such as electromagnets, to reach higher speeds. The **Excursion** on **page 345** expands the concept of velocity to include negative velocity. You will be asked to interpret distance-time graphs in which the slopes represent either positive or negative motion.

Need help? For on-line student resources, such as section quizzes, visit this textbook's web site at **math.college.hmco.com/students.**

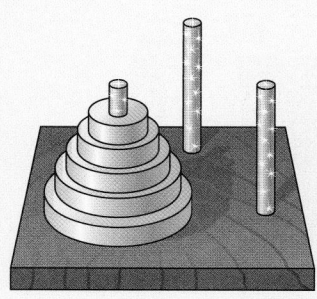

The Tower of Hanoi is a puzzle that has the following form. Three pegs are placed on a board. A number of disks, graded in size, are stacked on one of the pegs with the largest disk on the bottom and successively smaller disks placed on top. The disks are moved according to the following rules:

1. Only one disk at a time may be moved.

2. A larger disk may not be placed on top of a smaller disk.

The object of the game is to transfer all of the disks, one at a time, from one peg to one of the other two pegs. If initially there is only one disk, then only one move is required. With two disks, three moves are required; with three disks, seven moves are required.

The chart below shows the minimum number of moves required for a given number of disks. The increase in the number of moves required for each additional disk is also given.

Number of disks	1	2	3	4	5	6	7
Minimum number of moves	1	3	7	15	31	63	127
Increase in number of moves		$3-1=2$	$7-3=4$	$15-7=8$	$31-15=16$	$63-31=32$	$127-63=64$

For the list of numbers in the bottom row of the table, each successive number can be found by multiplying the preceding number by a constant (in this case 2). This list of numbers can be represented by the equation $f(n) = 2^n$, which is an example of an *exponential function*, one of the topics of this chapter.

The formula for the minimum number of moves is given by $M = 2^n - 1$, which contains the exponential expression 2^n. In this formula, M is the minimum number of moves required to transfer n disks to one of the other pegs.

There is an ancient myth involving the Tower of Hanoi puzzle and the lifetime of the universe. In this myth, three priests sit in the center of the universe with 3 diamond needles and 64 golden disks on one of the needles. The only job of the priests is to transfer the golden disks to one of the other needles using the rules of the Tower of Hanoi puzzle. The priests can transfer one disk to another needle every second. According to the myth, the universe will cease to exist at the precise moment the priests have completed the transfer of all 64 disks to one of the other needles.

Thus according to the myth, the lifetime of the universe is given by $2^{64} - 1$ seconds. Use a calculator to show that this amounts to approximately 585 billion years! Even if the priests started the transfer of the disks 12 billion years ago (when astronomers estimate our universe began), the myth indicates that the universe will continue to exist for another 573 billion years.

| # Rectangular Coordinates and Functions

Introduction to Rectangular Coordinate Systems

When archaeologists excavate a site, a *coordinate grid* is laid over the site so that records can be kept not only of what was found, but also of *where* it was found. The grid below is from an archaeological dig at Poggio Colla, a site in the Mugello about 20 miles northeast of Florence, Italy.

historical note

The concept of a coordinate system developed over time, culminating in 1637 with the publication of *Discourse on the Method for Rightly Directing One's Reason and Searching for Truth in the Sciences* by René Descartes (1596–1650) and *Introduction to Plane and Solid Loci* by Pierre de Fermat (1601–1665). Of the two mathematicians, Descartes is usually given more credit for developing the concept of a coordinate system. In fact, he became so famous in Le Haye, the town in which he was born, that the town was renamed Le Haye–Descartes. ∎

In mathematics we encounter a similar problem, that of locating a point in a plane. One way to solve the problem is to use a *rectangular coordinate system*.

A **rectangular coordinate system** is formed by two number lines, one horizontal and one vertical, that intersect at the zero point of each line. The point of intersection is called the **origin.** The two number lines are called the **coordinate axes,** or simply the **axes.** Frequently, the horizontal axis is labeled the *x*-axis and the vertical axis is labeled the *y*-axis. In this case, the axes form what is called the **xy-plane.**

The two axes divide the plane into four regions called **quadrants,** which are numbered counterclockwise, using Roman numerals, from I to IV, starting at the upper right.

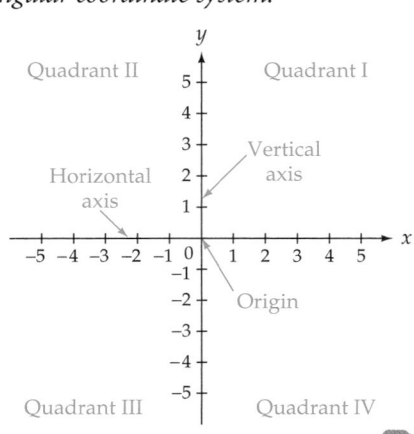

▼ **point of interest**

The word *abscissa* has the same root as the word *scissors*. When open, a pair of scissors looks like an *x*.

Each point in the plane can be identified by a pair of numbers called an **ordered pair.** The first number of the ordered pair measures a horizontal change from the *y*-axis and is called the **abscissa,** or ***x*-coordinate.** The second number of the ordered pair measures a vertical change from the *x*-axis and is called the **ordinate,** or ***y*-coordinate.** The ordered pair (x, y) associated with a point is also called the **coordinates** of the point.

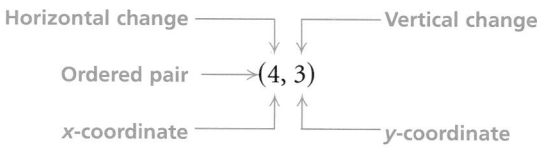

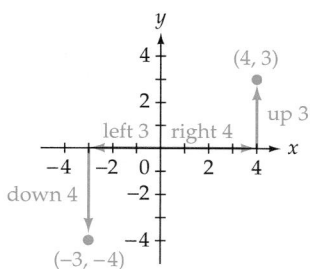

To **graph,** or **plot,** a point means to place a dot at the coordinates of the point. For example, to graph the ordered pair $(4, 3)$, start at the origin. Move 4 units to the right and then 3 units up. Draw a dot. To graph $(-3, -4)$, start at the origin. Move 3 units left and then 4 units down. Draw a dot.

The **graph of an ordered pair** is the dot drawn at the coordinates of the point in the plane. The graphs of the ordered pairs $(4, 3)$ and $(-3, -4)$ are shown above.

The graphs of the points whose coordinates are $(2, 3)$ and $(3, 2)$ are shown at the right. Note that they are different points. The order in which the numbers in an ordered pair are listed is important.

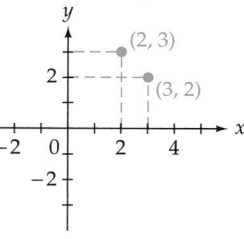

✔ **TAKE NOTE**

This is *very* important. An *ordered pair* is a pair of coordinates, and the *order* in which the coordinates are listed matters.

If the axes are labeled with letters other than *x* or *y*, then we refer to the ordered pair using the given labels. For instance, if the horizontal axis is labeled *t* and the vertical axis is labeled *d*, then the ordered pairs are written as (t, d). We sometimes refer to the first number in an ordered pair as the **first coordinate** of the ordered pair and to the second number as the **second coordinate** of the ordered pair.

One purpose of a coordinate system is to draw a picture of the solutions of an **equation in two variables.** Examples of equations in two variables are shown at the right.

$$y = 3x - 2$$
$$x^2 + y^2 = 25$$
$$s = t^2 - 4t + 1$$

A **solution of an equation in two variables** is an ordered pair that makes the equation a true statement. For instance, as shown below, $(2, 4)$ is a solution of $y = 3x - 2$ but $(3, -1)$ is not a solution of the equation.

$y = 3x - 2$		
4	$3(2) - 2$	• $x = 2, y = 4$
4	$6 - 2$	
$4 = 4$		• Checks.

$y = 3x - 2$		
-1	$3(3) - 2$	• $x = 3, y = -1$
-1	$9 - 2$	
$-1 \neq 7$		• Does not check.

QUESTION   Is $(-2, 1)$ a solution of $y = 3x + 7$?

ANSWER   Yes, because $3(-2) + 7 = 1$.

The **graph of an equation in two variables** is a drawing of all the ordered pair solutions of the equation. To create a graph of an equation, find some ordered-pair solutions of the equation, plot the corresponding points, and then connect the points with a smooth curve.

EXAMPLE 1 ■ Graph an Equation in Two Variables

Graph $y = 3x - 2$.

Solution

To find ordered-pair solutions, select various values of x and calculate the corresponding values of y. Plot the ordered pairs. After the ordered pairs have been graphed, draw a smooth curve through the points. It is convenient to keep track of the solutions in a table.

When choosing values of x, we often choose integer values because the resulting ordered pairs are easier to graph.

x	$3x - 2 = y$	(x, y)
-2	$3(-2) - 2 = -8$	$(-2, -8)$
-1	$3(-1) - 2 = -5$	$(-1, -5)$
0	$3(0) - 2 = -2$	$(0, -2)$
1	$3(1) - 2 = 1$	$(1, 1)$
2	$3(2) - 2 = 4$	$(2, 4)$
3	$3(3) - 2 = 7$	$(3, 7)$

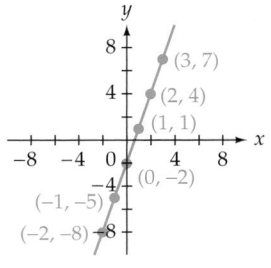

CHECK YOUR PROGRESS 1   Graph $y = -2x + 3$.

Solution   *See page S20.*

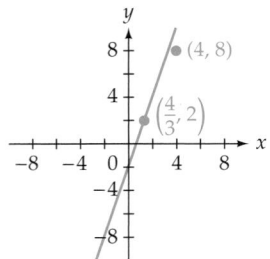

The graph of $y = 3x - 2$ is shown again at the left. Note that the ordered pair $\left(\frac{4}{3}, 2\right)$ is a solution of the equation and is a point on the graph. The ordered pair $(4, 8)$ is *not* a solution of the equation and is *not* a point on the graph. Every ordered-pair solution of the equation is a point on the graph and every point on the graph is an ordered-pair solution of the equation.

EXAMPLE 2 ■ Graph an Equation in Two Variables

Graph $y = x^2 + 4x$.

Solution

Select various values of x and calculate the corresponding values of y. Plot the ordered pairs. After the ordered pairs have been graphed, draw a smooth curve through the points. Here is a table showing some possible ordered pairs.

x	$x^2 + 4x = y$	(x, y)
-5	$(-5)^2 + 4(-5) = 5$	$(-5, 5)$
-4	$(-4)^2 + 4(-4) = 0$	$(-4, 0)$
-3	$(-3)^2 + 4(-3) = -3$	$(-3, -3)$
-2	$(-2)^2 + 4(-2) = -4$	$(-2, -4)$
-1	$(-1)^2 + 4(-1) = -3$	$(-1, -3)$
0	$(0)^2 + 4(0) = 0$	$(0, 0)$
1	$(1)^2 + 4(1) = 5$	$(1, 5)$

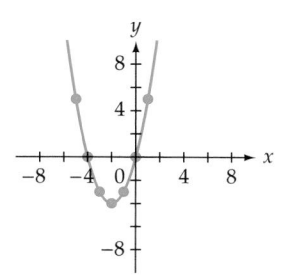

CHECK YOUR PROGRESS 2   Graph $y = -x^2 + 1$.

Solution   See page S20.

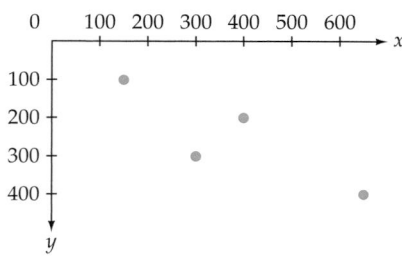

Math Matters   Computer Software Program Coordinate Systems

Some computer software programs use a coordinate system that is different from the *xy*-coordinate system we have discussed. For instance, in one particular software program, the origin $(0, 0)$ represents the top left point of a computer screen, as shown at the left. The points $(150, 100)$, $(300, 300)$, $(400, 200)$, and $(650, 400)$ are shown on the graph.

Introduction to Functions

An important part of mathematics is the study of the relationship between known quantities. Exploring relationships between known quantities frequently results in equations in two variables. For instance, as a car is driven, the fuel in the gas tank is burned. There is a correspondence between the number of gallons of fuel used and the number of miles traveled. If a car gets 25 miles per gallon, then the car consumes 0.04 gallon of fuel for each mile driven. For the sake of simplicity, we will assume that the car always consumes 0.04 gallon of gasoline for each mile driven. The equation $g = 0.04d$ defines how the number of gallons used, g, depends on the number of miles driven, d.

Distance traveled (in miles), d	25	50	100	250	300
Fuel used (in gallons), g	1	2	4	10	12

The ordered pairs in the table above are only some of the possible ordered pairs. Other possibilities are $(90, 3.6)$, $(125, 5)$, and $(235, 9.4)$. If all of the ordered pairs of the equation were drawn, the graph would appear as a line. The graph of the equation and the ordered pairs we have calculated are shown on the following page. Note that the graphs of all the ordered pairs are on the same line.

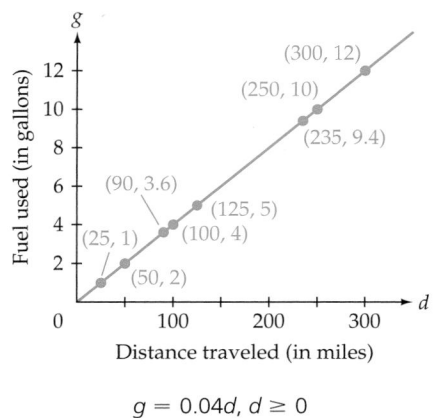

$$g = 0.04d, \ d \geq 0$$

QUESTION   *What is the meaning of the ordered pair (125, 5)?*

The ordered pairs, the graph, and the equation are all different ways of expressing the correspondence, or relationship, between the two variables. This correspondence, which pairs the number of miles driven with the number of gallons of fuel used, is called a *function*.

Here are some additional examples of functions, along with a specific example of each correspondence.

To each real number 5	there corresponds ⟶	its square 25
To each score on an exam 87	there corresponds ⟶	a grade B
To each student Alexander Sterling	there corresponds ⟶	a student identification number S18723519

An important fact about each of these correspondences is that each result is *unique*. For instance, for the real number 5, there is *exactly one* square, 25. With this in mind, we now state the definition of a function.

> **Definition of a Function**
>
> A **function** is a correspondence, or relationship, between two sets called the **domain** and **range** such that for each element of the domain there corresponds *exactly one* element of the range.

As an example of domain and range, consider the function that pairs a test score with a letter grade. The domain is the real numbers from 0 to 100. The range is the letters A, B, C, D, and F.

✓ **TAKE NOTE**

Because the square of any real number is a positive number or zero ($0^2 = 0$), the range of the function that pairs a number with its square contains the positive numbers and zero. Therefore, the range is the nonnegative real numbers. The difference between *nonnegative* numbers and *positive* numbers is that nonnegative numbers include zero; positive numbers do not include zero.

Test score	Grade
90–100	A
80–89	B
70–79	C
60–69	D
0–59	F

ANSWER   *A car that gets 25 miles per gallon can travel 125 miles on 5 gallons of fuel.*

Although a function can be described in terms of ordered pairs, by a graph, or by an equation, this text will focus on functions defined by equations in two variables. For instance, when gravity is the only force acting on a falling body, a function that describes the distance s, in feet, an object will fall in t seconds is given by $s = 16t^2$.

Given a value of t (time), the value of s (the distance the object falls) can be found. For instance, given $t = 3$,

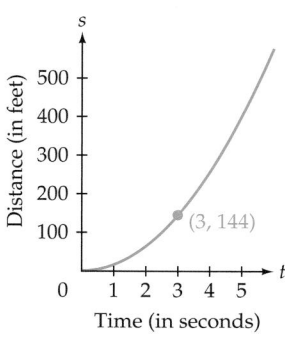

Distance (in feet) / **Time (in seconds)**

$$s = 16t^2$$
$$s = 16(3)^2 \quad \text{• Replace } t \text{ by } \mathbf{3}.$$
$$s = 16(9) \quad \text{• Simplify.}$$
$$s = 144$$

The object falls 144 feet in 3 seconds.

Because the distance the object falls *depends on* how long it has been falling, s is called the **dependent variable** and t is called the **independent variable**. Some of the ordered pairs of this function are (3, 144), (1, 16), (0, 0), and $\left(\frac{1}{4}, 1\right)$. The ordered pairs can be written as (t, s), where $s = 16t^2$. By substituting $16t^2$ for s, we can also write the ordered pairs as $(t, 16t^2)$. For the equation $s = 16t^2$, we say that "distance is a function of time." A graph of the function is shown at the left.

Not all equations in two variables define a function. For instance,

$$y^2 = x^2 + 9$$

is not an equation that defines a function because

$$5^2 = 4^2 + 9 \quad \text{and} \quad (-5)^2 = 4^2 + 9$$

The ordered pairs (4, 5) and (4, −5) both belong to the equation. Consequently, there are two ordered pairs with the same first coordinate, 4, but *different* second coordinates, 5 and −5. By definition, the equation does not define a function. The phrase "y is a function of x," or a similar phrase with different variables, is used to describe those equations in two variables that define functions.

Functional notation is frequently used for equations that define functions. Just as the letter x is commonly used as a variable, the letter f is commonly used to name a function.

To describe the relationship between a number and its square using functional notation, we can write $f(x) = x^2$. The symbol $f(x)$ is read "the value of f at x" or "f of x." The symbol $f(x)$ is the **value of the function** and represents the value of the dependent variable for a given value of the independent variable. We will often write $y = f(x)$ to emphasize the relationship between the independent variable, x, and the dependent variable, y. Remember: y and $f(x)$ are different symbols for the same number. Also, the *name* of the function is f, the *value* of the function at x is $f(x)$.

The letters used to represent a function are somewhat arbitrary. All of the following equations represent the same function.

✔ **TAKE NOTE**

The notation $f(x)$ does *not* mean "f times x." The letter f stands for the name of the function, and $f(x)$ is the value of the function at x.

$$\left.\begin{array}{l} f(x) = x^2 \\ g(t) = t^2 \\ P(v) = v^2 \end{array}\right\} \quad \text{Each of these equations represents the square function.}$$

The process of finding $f(x)$ for a given value of x is called **evaluating the function.** For instance, to evaluate $f(x) = x^2$ when $x = 4$, replace x by 4 and simplify.

$$f(x) = x^2$$
$$f(4) = 4^2 = 16 \quad \text{• Replace } x \text{ by } \mathbf{4}. \text{ Then simplify.}$$

The *value* of the function is 16 when $x = 4$. This means that an ordered pair of the function is $(4, 16)$.

EXAMPLE 3 ■ Evaluate a Function

Evaluate $s(t) = 2t^2 - 3t + 1$ when $t = -2$.

Solution

$$s(t) = 2t^2 - 3t + 1$$
$$s(-2) = 2(-2)^2 - 3(-2) + 1 \quad \text{• Replace } t \text{ by } -2. \text{ Then simplify.}$$
$$= 15$$

The value of the function is 15 when $t = -2$.

CHECK YOUR PROGRESS 3 Evaluate $f(z) = z^2 - z$ when $z = -3$.

Solution See page S21.

Any letter or combination of letters can be used to name a function. In the next example, the letters *SA* are used to name a *Surface Area* function.

EXAMPLE 4 ■ Application of Evaluating a Function

The surface area of a cube (the sum of the areas of each of the six faces) is given by $SA(s) = 6s^2$, where $SA(s)$ is the surface area of the cube and s is the length of one side of the cube. Find the surface area of a cube that has a side of length 10 centimeters.

Solution

$$SA(s) = 6s^2$$
$$SA(10) = 6(10)^2 \quad \text{• Replace } s \text{ by } 10.$$
$$= 6(100) \quad \text{• Simplify.}$$
$$= 600$$

The surface area of the cube is 600 square centimeters.

Diagonal

CHECK YOUR PROGRESS 4 A **diagonal** of a polygon is a line segment from one vertex to a nonadjacent vertex, as shown at the left. The total number of diagonals of a polygon is given by $N(s) = \dfrac{s^2 - 3s}{2}$, where $N(s)$ is the total number of diagonals and s is the number of sides of the polygon. Find the total number of diagonals of a polygon with 12 sides.

Solution See page S21.

Math Matters   The Special Theory of Relativity

In 1905, Albert Einstein published a paper that set the framework for relativity theory. This theory, now called the Special Theory of Relativity, explains, among other things, how mass changes for a body in motion. Basically the theory states that the mass of a body is a function of its velocity. That is, the mass of a body changes as its speed increases.

The function can be given by $M(v) = \dfrac{m_0}{\sqrt{1 - \dfrac{v^2}{c^2}}}$, where m_0 is the mass of the

body at rest (its mass when its speed is zero) and c is the velocity of light, which Einstein showed was the same for all observers. The table below shows how a 5-kilogram mass increases as its speed becomes closer and closer to the speed of light.

Speed	Mass (kilograms)
30 meters/second—speed of a car on an expressway	5
240 meters/second—speed of a commercial jet	5
3.0×10^7 meters/second—10% of the speed of light	5.025
1.5×10^8 meters/second—50% of the speed of light	5.774
2.7×10^8 meters/second—90% of the speed of light	11.471

Note that for speeds of everyday objects, such as a car or plane, the increase in mass is negligible. Physicists have verified these increases using particle accelerators that can accelerate a particle such as an electron to more than 99.9% of the speed of light.

historical note

Albert Einstein
(īn′stīn)
(1879–1955) was honored by *Time* magazine as Person of the Century. According to the magazine, "He was the pre-eminent scientist in a century dominated by science. The touchstones of the era—the Bomb, the Big Bang, quantum physics and electronics—all bear his imprint." ∎

Graphs of Functions

The graph of a function can be drawn by finding ordered pairs of the function, plotting the points corresponding to the ordered pairs, and then connecting the points with a curve.

For example, to graph $f(x) = x^3 + 1$, select several values of x and evaluate the function at each value. Recall that $f(x)$ and y are different symbols for the same quantity.

TAKE NOTE

We are basically creating the graph of an equation in two variables as we did earlier. The only difference is the use of functional notation.

x	$f(x) = x^3 + 1$	(x, y)
−2	$f(-2) = (-2)^3 + 1 = -7$	(−2, −7)
−1	$f(-1) = (-1)^3 + 1 = 0$	(−1, 0)
0	$f(0) = (0)^3 + 1 = 1$	(0, 1)
1	$f(1) = (1)^3 + 1 = 2$	(1, 2)
2	$f(2) = (2)^3 + 1 = 9$	(2, 9)

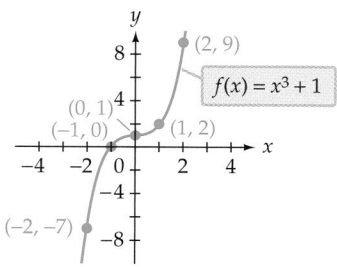

Plot the ordered pairs and draw a smooth curve through the points.

INSTRUCTOR NOTE
For the graph of $h(x) = x^2 - 3$ at the right, ask students the following questions.

1. Are there other values of y for which there are two different x-coordinates? If so, name one.

2. Are there values of y for which there is one unique x-coordinate? If so, name it.

3. Are there values of y for which there is no x-coordinate? If so, name one.

EXAMPLE 5 ■ **Graph a Function**

Graph $h(x) = x^2 - 3$.

Solution

x	$h(x) = x^2 - 3$	(x, y)
-3	$h(-3) = (-3)^2 - 3 = 6$	$(-3, 6)$
-2	$h(-2) = (-2)^2 - 3 = 1$	$(-2, 1)$
-1	$h(-1) = (-1)^2 - 3 = -2$	$(-1, -2)$
0	$h(0) = (0)^2 - 3 = -3$	$(0, -3)$
1	$h(1) = (1)^2 - 3 = -2$	$(1, -2)$
2	$h(2) = (2)^2 - 3 = 1$	$(2, 1)$
3	$h(3) = (3)^2 - 3 = 6$	$(3, 6)$

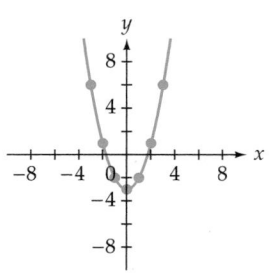

Plot the ordered pairs and draw a smooth curve through the points.

CHECK YOUR PROGRESS 5   Graph $f(x) = 2 - \dfrac{3}{4}x$.

Solution   *See page S21.*

Excursion

Dilations of a Geometric Figure

A **dilation** of a geometric figure changes the size of the figure by either enlarging it or reducing it. This is accomplished by multiplying the coordinates of the figure by a positive number called the **constant of dilation.** Examples of enlarging (multiplying the coordinates by a number greater than 1) and reducing (multiplying the coordinates by a number between 0 and 1) a geometric figure are shown at the top of the following page.

(continued)

▼ **point of interest**

Photocopy machines have reduction and enlargement features that function essentially as constants of dilation. The numbers are usually expressed as a percent. A copier selection of 50% reduces the size of the object being copied by 50%. A copier selection of 125% increases the size of the object being copied by 125%.

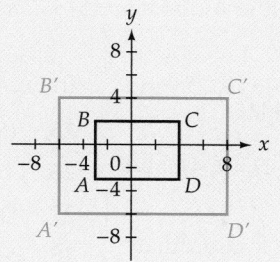

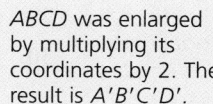

ABCD was enlarged by multiplying its coordinates by 2. The result is *A'B'C'D'*.

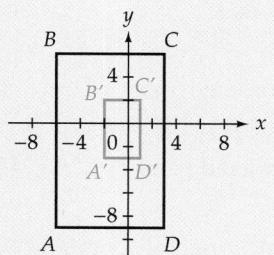

ABCD was reduced by multiplying its coordinates by $\frac{1}{3}$. The result is *A'B'C'D'*.

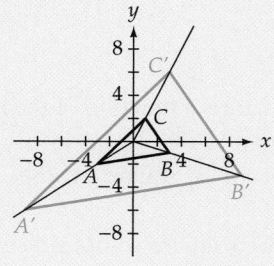

When each of the coordinates of a figure is multiplied by the same number in order to produce a dilatation, the **center of dilation** will be the origin of the coordinate system. For triangle *ABC* at the left, a constant of dilation of 3 was used to produce triangle *A'B'C'*. Note that lines through the vertices of the two triangles intersect at the origin, the center of dilation. The center of dilation, however, can be any point in the plane.

Excursion Exercises

1. A dilation is performed on the figure with vertices $A(-2, 0)$, $B(2, 0)$, $C(4, -2)$, $D(2, -4)$, and $E(-2, -4)$.

   a. Draw the original figure and a new figure using a constant of dilation of 2.

   b. Draw the original figure and a new figure using a constant of dilation of $\frac{1}{2}$.

2. Because each of the coordinates of a geometric figure is multiplied by a number, the lengths of the sides of the figure will change. It is possible to show that the lengths change by a factor equal to the constant of dilation. In this exercise, you will examine the effect of a dilation on the angles of a geometric figure. Draw some figures and then draw a dilation of each figure using the origin as the center of dilation. Using a protractor, determine whether the measures of the angles of the dilated figure are different from the measures of the corresponding angles of the original figure.

3. Graphic artists use centers of dilation to create three-dimensional effects. Consider the block letter A shown at the left. Draw another block letter A by changing the center of dilation to see how it affects the 3-D look of the letter. Programs such as PowerPoint use these methods to create various shading options for design elements in a presentation.

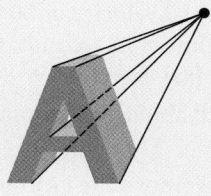

4. Draw an enlargement and a reduction of the figure at the left for the given center of dilation *P*.

5. On a blank piece of paper, draw a rectangle 4 inches by 6 inches with the center of the rectangle in the center of the paper. Make various photocopies of the rectangle using the reduction and enlargement settings on a copy machine. Where is the center of dilation for the copy machine?

Exercise Set 6.1   (Suggested Assignment: 1–55, odds, 57–62)

1. Graph the ordered pairs $(0, -1)$, $(2, 0)$, $(3, 2)$, and $(-1, 4)$.

2. Graph the ordered pairs $(-1, -3)$, $(0, -4)$, $(0, 4)$, and $(3, -2)$.

3. Draw a line through all points with an x-coordinate of 2.

4. Draw a line through all points with an x-coordinate of -3.

5. Draw a line through all points with a y-coordinate of -3.

6. Draw a line through all points with a y-coordinate of 4.

7. Graph the ordered-pair solutions of $y = x^2$ when $x = -2, -1, 0, 1,$ and 2.

8. Graph the ordered-pair solutions of $y = -x^2 + 1$ when $x = -2, -1, 0, 1,$ and 2.

9. Graph the ordered-pair solutions of $y = |x + 1|$ when $x = -5, -3, 0, 3,$ and 5.

10. Graph the ordered-pair solutions of $y = -2|x|$ when $x = -3, -1, 0, 1,$ and 3.

11. Graph the ordered-pair solutions of $y = -x^2 + 2$ when $x = -2, -1, 0, 1,$ and 2.

12. Graph the ordered-pair solutions of $y = -x^2 + 4$ when $x = -3, -1, 0, 1,$ and 3.

13. Graph the ordered-pair solutions of $y = x^3 - 2$ when $x = -1, 0, 1,$ and 2.

14. Graph the ordered-pair solutions of $y = -x^3 + 1$ when $x = -1, 0, 1,$ and $\frac{3}{2}$.

In Exercises 15–24, graph each equation.

15. $y = 2x - 1$

16. $y = -3x + 2$

17. $y = \frac{2}{3}x + 1$

18. $y = -\frac{x}{2} - 3$

19. $y = \frac{1}{2}x^2$

20. $y = \frac{1}{3}x^2$

21. $y = 2x^2 - 1$

22. $y = -3x^2 + 2$

23. $y = |x - 1|$

24. $y = |x - 3|$

In Exercises 25–32, evaluate the function for the given value.

25. $f(x) = 2x + 7$; $x = -2$

26. $y(x) = 1 - 3x$; $x = -4$

27. $f(t) = t^2 - t - 3$; $t = 3$

28. $P(n) = n^2 - 4n - 7$; $n = -3$

29. $v(s) = s^3 + 3s^2 - 4s - 2$; $s = -2$

30. $f(x) = 3x^3 - 4x^2 + 7$; $x = 2$

31. $T(p) = \dfrac{p^2}{p - 2}$; $p = 0$

32. $s(t) = \dfrac{4t}{t^2 + 2}$; $t = 2$

33. The perimeter P of a square is a function of the length s of one of its sides and is given by $P(s) = 4s$.

    a. Find the perimeter of a square whose side is 4 meters.

    b. Find the perimeter of a square whose side is 5 feet.

34. The area of a circle is a function of its radius and is given by $A(r) = \pi r^2$.

    a. Find the area of a circle whose radius is 3 inches. Round to the nearest tenth.

    b. Find the area of a circle whose radius is 12 centimeters. Round to the nearest tenth.

35. The height h, in feet, of a ball that is released 4 feet above the ground with an initial velocity of 80 feet per second is a function of the time t, in seconds, the ball is in the air and is given by $h(t) = -16t^2 + 80t + 4$.

    a. Find the height of the ball above the ground 2 seconds after it is released.

    b. Find the height of the ball above the ground 4 seconds after it is released.

36. The distance d, in miles, a forest fire ranger can see from an observation tower is a function of the height h, in feet, of the tower above level ground and is given by $d(h) = 1.5\sqrt{h}$.

    a. Find the distance a ranger can see whose eye level is 20 feet above level ground. Round to the nearest tenth.

    b. Find the distance a ranger can see whose eye level is 35 feet above level ground. Round to the nearest tenth.

37. The speed s, in feet per second, of sound in air depends on the temperature t of the air in degrees Celsius and is given by $s(t) = \dfrac{1087\sqrt{t + 273}}{16.52}$.

  a. What is the speed of sound in air when the temperature is 0°C (the temperature at which water freezes)? Round to the nearest foot per second.

  b. What is the speed of sound in air when the temperature is 25°C? Round to the nearest foot per second.

38. In a softball league in which each team plays every other team three times, the number of games N that must be scheduled depends on the number of teams n in the league and is given by $N(n) = \frac{3}{2}n^2 - \frac{3}{2}n$.

  a. How many games must be scheduled for a league that has five teams?

  b. How many games must be scheduled for a league that has six teams?

39. The percent concentration P of salt in a particular salt water solution depends on the number of grams x of salt that are added to the solution and is given by $P(x) = \dfrac{100x + 100}{x + 10}$.

  a. What is the original percent concentration of salt?

  b. What is the percent concentration of salt after 5 more grams of salt are added?

40. The time T, in seconds, it takes a pendulum to make one swing depends on the length of the pendulum and is given by $T(L) = 2\pi\sqrt{\frac{L}{32}}$, where L is the length of the pendulum in feet.

  a. Find the time it takes the pendulum to make one swing if the length of the pendulum is 3 feet. Round to the nearest hundredth.

  b. Find the time it takes the pendulum to make one swing if the length of the pendulum is 9 inches. Round to the nearest tenth.

Graph each of the following.

41. $f(x) = 2x - 5$ **42.** $f(x) = -2x + 4$

43. $f(x) = -x + 4$ **44.** $f(x) = 3x - 1$

45. $g(x) = \dfrac{2}{3}x + 2$ **46.** $h(x) = \dfrac{5}{2}x - 1$

47. $F(x) = -\dfrac{1}{2}x + 3$ **48.** $F(x) = -\dfrac{3}{4}x - 1$

49. $f(x) = x^2 - 1$ **50.** $f(x) = x^2 + 2$

51. $f(x) = -x^2 + 4$ **52.** $f(x) = -2x^2 + 5$

53. $g(x) = x^2 - 4x$ **54.** $h(x) = x^2 + 4x$

55. $P(x) = x^2 - x - 6$ **56.** $P(x) = x^2 - 2x - 3$

Extensions

CRITICAL THINKING

57. Find the area of the rectangle.

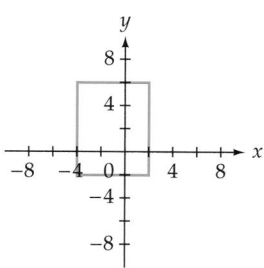

58. Find the area of the triangle.

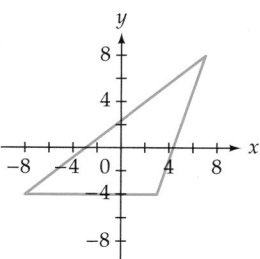

59. Suppose f is a function. Is it possible to have $f(2) = 4$ and $f(2) = 7$? Explain your answer.

60. Suppose f is a function and $f(a) = 4$ and $f(b) = 4$. Does this mean that $a = b$?

61. If $f(x) = 2x + 5$ and $f(a) = 9$, find a.

62. If $f(x) = x^2$ and $f(a) = 9$, find a.

63. Let $f(a, b) =$ the sum of a and b.
Let $g(a, b) =$ the product of a and b.
Find $f(2, 5) + g(2, 5)$.

64. Let $f(a, b) =$ the greatest common factor of a and b and let $g(a, b) =$ the least common multiple of a and b. Find $f(14, 35) + g(14, 35)$.

65. Given $f(x) = x^2 + 3$, for what value of x is $f(x)$ smallest?

66. Given $f(x) = -x^2 + 4x$, for what value of x is $f(x)$ greatest?

EXPLORATIONS

67. Consider the function given by

$$M(x, y) = \frac{x + y}{2} + \frac{|x - y|}{2}.$$

a. Complete the following table.

x	y	$M(x, y) = \dfrac{x + y}{2} + \dfrac{\lvert x - y \rvert}{2}$
−5	11	$M(-5, 11) = \dfrac{-5 + 11}{2} + \dfrac{\lvert -5 - 11 \rvert}{2} = 11$
10	8	
−3	−1	
12	−13	
−11	15	

b. Extend the table by choosing some additional values of x and y.

c. How is the value of the function related to the values of x and y? *Hint:* For x = −5 and y = 11, the value of the function was 11, the value of y.

d. The function $M(x, y)$ is sometimes referred to as the *maximum function*. Why is this a good name for this function?

e. Try to create a *minimum function*—that is, a function that yields the minimum of two numbers x and y.

SECTION 6.2 # Properties of Linear Functions

Intercepts

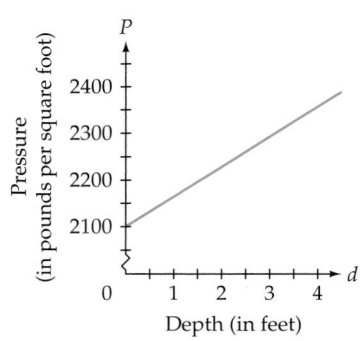

Pressure (in pounds per square foot)

P

2400
2300
2200
2100

0   1   2   3   4   d

Depth (in feet)

The graph at the left shows the pressure on a diver as the diver descends in the ocean. The equation of this graph can be represented by $P(d) = 64d + 2100$, where $P(d)$ is the pressure, in pounds per square foot, on a diver d feet below the surface of the ocean. By evaluating the function for various values of d, we can determine the pressure on the diver at those depths. For instance, when d = 2, we have

$$P(d) = 64d + 2100$$
$$P(2) = 64(2) + 2100$$
$$= 128 + 2100$$
$$= 2228$$

The pressure on a diver 2 feet below the ocean's surface is 2228 pounds per square foot.

The function $P(d) = 64d + 2100$ is an example of a *linear function*.

> **Linear Function**
>
> A **linear function** is one that can be written in the form $f(x) = mx + b$, where m is the coefficient of x and b is a constant.

For the linear function $P(d) = 64d + 2100$, m = 64 and b = 2100.

Here are some other examples of linear functions.

$f(x) = 2x + 5$ • $m = 2, b = 5$

$g(t) = \dfrac{2}{3}t - 1$ • $m = \dfrac{2}{3}, b = -1$

$v(s) = -2s$ • $m = -2, b = 0$

$h(x) = 3$ • $m = 0, b = 3$

$f(x) = 2 - 4x$ • $m = -4, b = 2$

Note that different variables can be used to designate a linear function.

QUESTION *Which of the following are linear functions?*

    a. $f(x) = 2x^2 + 5$ ***b.*** $g(x) = 1 - 3x$

Consider the linear function $f(x) = 2x + 4$. The graph of the function is shown below, along with a table listing some of its ordered pairs.

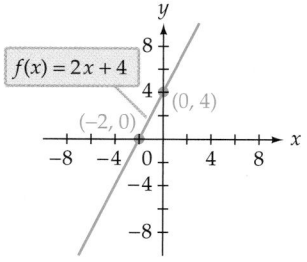

x	**f(x) = 2x + 4**	**(x, y)**
−3	$f(-3) = 2(-3) + 4 = -2$	(−3, −2)
−2	$f(-2) = 2(-2) + 4 = 0$	(−2, 0)
−1	$f(-1) = 2(-1) + 4 = 2$	(−1, 2)
0	$f(0) = 2(0) + 4 = 4$	(0, 4)
1	$f(1) = 2(1) + 4 = 6$	(1, 6)

✔ **TAKE NOTE**

Note that the graph of a *linear* function is a straight *line*. Observe that when the graph crosses the x-axis, the y-coordinate is 0. When the graph crosses the y-axis, the x-coordinate is 0. The table confirms these observations.

From the table and the graph we can see that when $x = -2$, $y = 0$, and the graph crosses the x-axis at (−2, 0). The point $(-2, 0)$ is called the **x-intercept** of the graph. When $x = 0$, $y = 4$, and the graph crosses the y-axis at (0, 4). The point $(0, 4)$ is called the **y-intercept** of the graph.

ANSWER   *a. Because $f(x) = 2x^2 + 5$ has an x^2 term, f is not a linear function. **b.** Because $g(x) = 1 - 3x$ can be written in the form $f(x) = mx + b$ as $g(x) = -3x + 1$ ($m = -3$ and $b = 1$), g is a linear function.*

EXAMPLE 1 ■ **Find the *x*- and *y*-intercepts of a Graph**

Find the *x*- and *y*-intercepts of the graph of $g(x) = -3x + 2$.

Solution

When a graph crosses the *x*-axis, the *y*-coordinate of the point is 0. Therefore, to find the *x*-intercept, replace $g(x)$ by 0 and solve the equation for *x*. [Recall that $g(x)$ is another name for *y*.]

$$g(x) = -3x + 2$$
$$0 = -3x + 2 \qquad \text{• Replace } g(x) \text{ by } \mathbf{0}.$$
$$-2 = -3x$$
$$\frac{2}{3} = x$$

The *x*-intercept is $\left(\frac{2}{3}, 0\right)$.

When a graph crosses the *y*-axis, the *x*-coordinate of the point is 0. Therefore, to find the *y*-intercept, evaluate the function when *x* is 0.

$$g(x) = -3x + 2$$
$$g(0) = -3(0) + 2 \qquad \text{• Evaluate } g(x) \text{ when } x = \mathbf{0}. \text{ Then simplify.}$$
$$= 2$$

The *y*-intercept is $(0, 2)$.

CHECK YOUR PROGRESS 1 Find the *x*- and *y*-intercepts of the graph of $f(x) = \frac{1}{2}x + 3$.

Solution *See page S21.*

TAKE NOTE

To find the *y*-intercept of $y = mx + b$ [we have replaced $f(x)$ by *y*], let $x = 0$. Then

$$y = mx + b$$
$$y = m(0) + b$$
$$= b$$

The *y*-intercept is $(0, b)$. This result is shown at the right.

In Example 1, note that the *y*-coordinate of the *y*-intercept of $g(x) = -3x + 2$ has the same value as *b* in the equation $f(x) = mx + b$. This is always true.

> **y-intercept**
>
> The *y*-intercept of the graph of $f(x) = mx + b$ is $(0, b)$.

If we evaluate the linear function that models pressure on a diver, $P(d) = 64d + 2100$, at 0, we have

$$P(d) = 64d + 2100$$
$$P(0) = 64(0) + 2100 = 2100$$

TAKE NOTE

We are working with the function $P(d) = 64d + 2100$. Therefore, the intercepts on the horizontal axis of a graph of the function are *d*-intercepts rather than *x*-intercepts, and the intercept on the vertical axis is a *P*-intercept rather than a *y*-intercept.

In this case, the *P*-intercept (the intercept on the vertical axis) is $(0, 2100)$. In the context of the application, this means that the pressure on a diver 0 feet below the ocean's surface is 2100 pounds per square foot. Another way of saying "zero feet below the ocean's surface" is "at sea level." Thus the pressure on the diver, or anyone else for that matter, at sea level is 2100 pounds per square foot.

Both the *x*- and *y*-intercept can have meaning in an application problem. This is demonstrated in the next example.

EXAMPLE 2 ■ Application of the Intercepts of a Linear Function

After a parachute is deployed, a function that models the height of the parachutist above the ground is $f(t) = -10t + 2800$, where $f(t)$ is the height, in feet, of the parachutist t seconds after the parachute is deployed. Find the intercepts on the vertical and horizontal axes and explain what they mean in the context of the problem.

Solution

To find the intercept on the vertical axis, evaluate the function when t is 0.

$$f(t) = -10t + 2800$$
$$f(0) = -10(0) + 2800 = 2800$$

The intercept on the vertical axis is $(0, 2800)$. This means that the parachutist is 2800 feet above the ground when the parachute is deployed.

To find the intercept on the horizontal axis, set $f(t) = 0$ and solve for t.

$$f(t) = -10t + 2800$$
$$0 = -10t + 2800$$
$$-2800 = -10t$$
$$280 = t$$

The intercept on the horizontal axis is $(280, 0)$. This means that the parachutist reaches the ground 280 seconds after the parachute is deployed. Note that the parachutist reaches the ground when $f(t) = 0$.

CHECK YOUR PROGRESS 2   A function that models the descent of a certain small airplane is given by $g(t) = -20t + 8000$, where $g(t)$ is the height, in feet, of the airplane t seconds after it begins its descent. Find the intercepts on the vertical and horizontal axes and explain what they mean in the context of the problem.

Solution   See page S21.

Slope of a Line

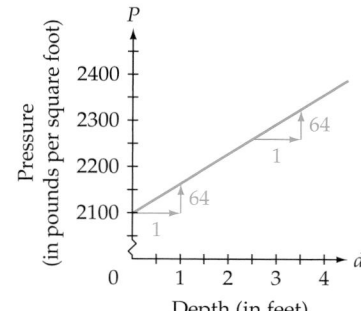

Consider again the linear function $P(d) = 64d + 2100$, which models the pressure on a diver as the diver descends below the ocean's surface. From the graph at the left, you can see that when the depth of the diver increases by 1 foot, the pressure on the diver increases by 64 pounds per square foot. This can be verified algebraically.

$P(0) = 64(0) + 2100 = 2100$ • Pressure at sea level
$P(1) = 64(1) + 2100 = 2164$ • Pressure after descending **1** foot
$2164 - 2100 = 64$ • Change in pressure

If we choose two other depths that differ by 1 foot, such as 2.5 feet and 3.5 feet (see the graph at the left), the change in pressure is the same.

$P(2.5) = 64(2.5) + 2100 = 2260$ • Pressure at **2.5** feet below the surface
$P(3.5) = 64(3.5) + 2100 = 2324$ • Pressure at **3.5** feet below the surface
$2324 - 2260 = 64$ • Change in pressure

The **slope** of a line is the change in the vertical direction caused by one unit of change in the horizontal direction. For $P(d) = 64d + 2100$, the slope is 64. In the context of the problem, the slope means that the pressure on a diver increases by 64 pounds per square foot for each additional foot the diver descends. Note that the slope (64) has the same value as the coefficient of d in $P(d) = 64d + 2100$. This connection between the slope and the coefficient of the variable in a linear function always holds.

> ### Slope
>
> For a linear function given by $f(x) = mx + b$, the slope of the graph of the function is m, the coefficient of the variable.

QUESTION   *What is the slope of each of the following?*

a. $y = -2x + 3$ **b.** $f(x) = x + 4$ **c.** $g(x) = 3 - 4x$

d. $y = \dfrac{1}{2}x - 5$

The slope of a line can be calculated by using the coordinates of any two distinct points on the line and the following formula.

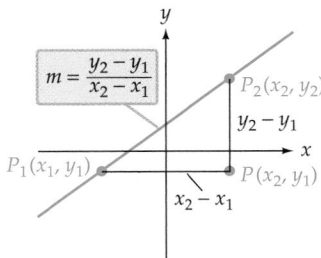

> ### Slope of a Line
>
> Let (x_1, y_1) and (x_2, y_2) be two points on a line. Then the **slope** of the line through the two points is the ratio of the change in the y-coordinates to the change in the x-coordinates.
>
> $$m = \frac{\text{change in } y}{\text{change in } x} = \frac{y_2 - y_1}{x_2 - x_1}, \quad x_1 \neq x_2$$

QUESTION   *Why is the restriction $x_1 \neq x_2$ required in the definition of slope?*

ANSWER   **a.** -2 **b.** 1 **c.** -4 **d.** $\dfrac{1}{2}$

ANSWER   *If $x_1 = x_2$, then the difference $x_2 - x_1 = 0$. This would make the denominator 0, and division by 0 is not defined.*

EXAMPLE 3 ■ Find the Slope of a Line Between Two Points

Find the slope of the line between the two points.

a. $(-4, -3)$ and $(-1, 1)$ **b.** $(-2, 3)$ and $(1, -3)$

c. $(-1, -3)$ and $(4, -3)$ **d.** $(4, 3)$ and $(4, -1)$

Solution

a. $(x_1, y_1) = (-4, -3), (x_2, y_2) = (-1, 1)$

$$m = \frac{y_2 - y_1}{x_2 - x_1} = \frac{1 - (-3)}{-1 - (-4)} = \frac{4}{3}$$

The slope is $\frac{4}{3}$. A *positive* slope indicates that the line slopes *upward* to the right. For this particular line, the value of *y increases* by $\frac{4}{3}$ when *x* increases by 1.

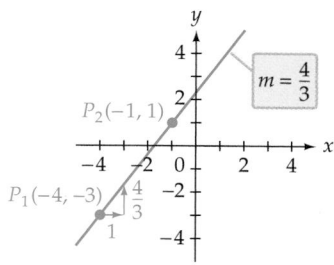

> **✔ TAKE NOTE**
>
> When we talk about *y* values increasing (as in part a) or decreasing (as in part b), we always mean as we move from left to right.

b. $(x_1, y_1) = (-2, 3), (x_2, y_2) = (1, -3)$

$$m = \frac{y_2 - y_1}{x_2 - x_1} = \frac{-3 - 3}{1 - (-2)} = \frac{-6}{3} = -2$$

The slope is -2. A *negative* slope indicates that the line slopes *downward* to the right. For this particular line, the value of *y decreases* by 2 when *x* increases by 1.

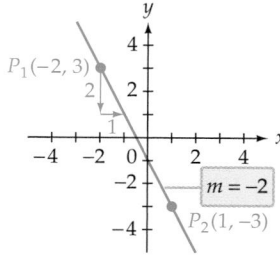

---

c. $(x_1, y_1) = (-1, -3), (x_2, y_2) = (4, -3)$

$$m = \frac{y_2 - y_1}{x_2 - x_1} = \frac{-3 - (-3)}{4 - (-1)} = \frac{0}{5} = 0$$

The slope is 0. A *zero* slope indicates that the line is *horizontal*. For this particular line, the value of *y stays the same* when *x* increases by any amount.

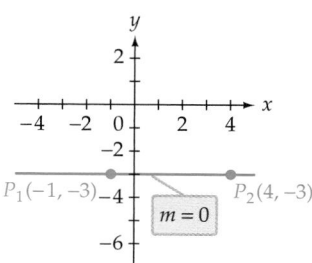

> **✔ TAKE NOTE**
>
> A horizontal line has zero slope. A line that has no slope, or whose slope is undefined, is a vertical line.

d. $(x_1, y_1) = (4, 3), (x_2, y_2) = (4, -1)$

$$m = \frac{y_2 - y_1}{x_2 - x_1} = \frac{-1 - 3}{4 - 4} = \frac{-4}{0} \quad \text{Division by 0 is undefined.}$$

If the denominator of the slope formula is zero, the line has *no slope*. Sometimes we say that the slope of the line is *undefined*.

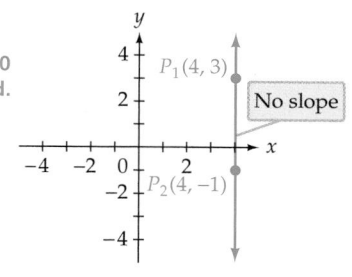

CHECK YOUR PROGRESS 3   Find the slope of the line between the two points.

a. $(-6, 5)$ and $(4, -5)$ **b.** $(-5, 0)$ and $(-5, 7)$

c. $(-7, -2)$ and $(8, 8)$ **d.** $(-6, 7)$ and $(1, 7)$

Solution   *See page S21.*

Suppose a jogger is running at a constant velocity of 6 miles per hour. Then the linear function $d = 6t$ relates the time t, in hours, spent running to the distance traveled d, in miles. A table of values is shown below.

Time, t, in hours	0	0.5	1	1.5	2	2.5
Distance, d, in miles	0	3	6	9	12	15

Because the equation $d = 6t$ represents a linear function, the slope of the graph of the equation is 6. This can be confirmed by choosing any two points on the graph shown below and finding the slope of the line between the two points. The points $(0.5, 3)$ and $(2, 12)$ are used here.

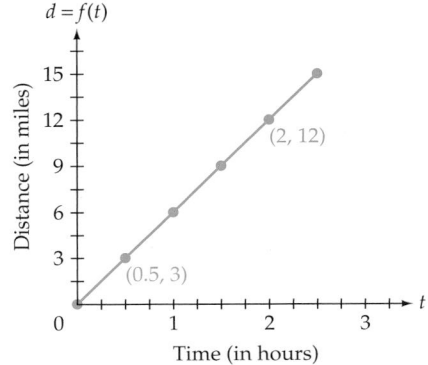

$$m = \frac{\text{change in } d}{\text{change in } t} = \frac{12 \text{ miles} - 3 \text{ miles}}{2 \text{ hours} - 0.5 \text{ hours}} = \frac{9 \text{ miles}}{1.5 \text{ hours}} = 6 \text{ miles per hour}$$

This example demonstrates that the slope of the graph of an object in uniform motion is the same as the velocity of the object. In a more general way, any time we discuss the velocity of an object, we are discussing the slope of the graph that describes the relationship between the distance the object travels and the time it travels.

EXAMPLE 4 ■ **Application of the Slope of a Linear Function**

The equation $T(x) = -6.5x + 20$ approximates the temperature $T(x)$, in degrees Celsius, at x kilometers above sea level. What is the slope of this function? Write a sentence that explains the meaning of the slope in the context of the problem.

Solution
For the linear function $T(x) = -6.5x + 20$, the slope is the coefficient of x. Therefore, the slope is -6.5. The slope means that the temperature is decreasing (because the slope is negative) 6.5°C for each 1 kilometer increase in height above sea level.

CHECK YOUR PROGRESS 4   The distance that a homing pigeon can fly can be approximated by $d(t) = 50t$, where $d(t)$ is the distance, in miles, flown by the pigeon in t hours. Find the slope of this function. What is the meaning of the slope in the context of the problem?

Solution   *See page S22.*

Math Matters

Galileo Galilei (găl-ĭ-lā-ē) (1564–1642) was one of the most influential scientists of his time. In addition to inventing the telescope, with which he discovered the moons of Jupiter, Galileo successfully argued that Aristotle's assertion that heavy objects drop at a greater velocity than lighter ones was incorrect. According to legend, Galileo went to the top of the Leaning Tower of Pisa and dropped two balls at the same time, one weighing twice the other. His assistant, standing on the ground, observed that both balls reached the ground at the same time.

There is no historical evidence that Galileo actually performed this experiment, but he did do something similar. Galileo correctly reasoned that if Aristotle's assertion were true, then balls of different weights should roll down a ramp at different speeds. Galileo did carry out this experiment and was able to show that, in fact, balls of different weights reached the end of the ramp at the same time. Galileo was not able to determine why this happened, and it took Issac Newton, born the same year that Galileo died, to formulate the first theory of gravity.

Slope-Intercept Form of a Straight Line

The value of the slope of a line gives the change in y for a *1 unit* change in x. For instance, a slope of -3 means that y changes by -3 as x changes by 1; a slope of $\frac{4}{3}$ means that y changes by $\frac{4}{3}$ as x changes by 1. Because it is difficult to graph a change of $\frac{4}{3}$, it is easier to think of a fractional slope in terms of integer changes in x and y. As shown at the right, for a slope of $\frac{4}{3}$ we have

$$m = \frac{\text{change in } y}{\text{change in } x} = \frac{4}{3}$$

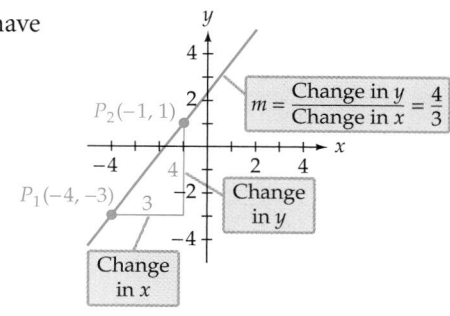

That is, for a slope of $\frac{4}{3}$, y changes by 4 as x changes by 3.

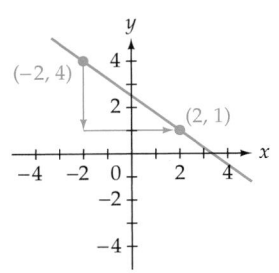

EXAMPLE 5 ■ Graph a Line Given a Point on the Line and the Slope

Draw the line that passes through $(-2, 4)$ and has slope $-\frac{3}{4}$.

Solution
Place a dot at $(-2, 4)$ and then rewrite $-\frac{3}{4}$ as $\frac{-3}{4}$. Starting from $(-2, 4)$, move 3 units down (the change in y) and then 4 units to the right (the change in x). Place a dot at that location and then draw a line through the two points.

CHECK YOUR PROGRESS 5 Draw the line that passes through $(2, 4)$ and has slope -1.

Solution *See page S22.*

INSTRUCTOR NOTE
It is important for students to associate the graph of a straight line with the equation $y = mx + b$. You might ask them to state whether the graph of each of the following equations is a straight line. (The graphs of equations b and d are straight lines.)
a. $y = x^2 + 1$
b. $y = -x$
c. $y = \dfrac{1}{x}$
d. $y = 2 - \dfrac{1}{2}x$
e. $y = \sqrt{x} - 1$

Because the slope and y-intercept can be determined directly from the equation $f(x) = mx + b$, this equation is called the *slope-intercept form* of a straight line.

> **Slope-Intercept Form of the Equation of a Line**
>
> The graph of $f(x) = mx + b$ is a straight line with slope m and y-intercept $(0, b)$.

When a function is written in this form, it is possible to create a quick graph of the function.

EXAMPLE 6 ■ Graph a Linear Function Using the Slope and y-intercept

Graph $f(x) = -\frac{2}{3}x + 4$ by using the slope and y-intercept.

Solution
From the equation, the slope is $-\frac{2}{3}$ and the y-intercept is $(0, 4)$. Place a dot at the y-intercept. We can write the slope as $m = -\frac{2}{3} = \frac{-2}{3}$. Starting from the y-intercept, move 2 units down and 3 units to the right and place another dot. Now draw a line through the two points.

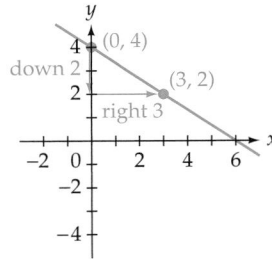

CHECK YOUR PROGRESS 6 Graph $y = \frac{3}{4}x - 5$ by using the slope and y-intercept.

Solution *See page S22.*

Excursion

Negative Velocity

We can expand the concept of velocity to include negative velocity. Suppose a car travels in a straight line starting at a given point. If the car is moving to the right, then we say that its velocity is positive. If the car is moving to the left, then we say that its velocity is negative. For instance, a velocity of −45 miles per hour means the car is moving to the left at 45 miles per hour.

If we were to graph the motion of an object on a distance-time graph, a positive velocity would be indicated by a positive slope; a negative velocity would be indicated by a negative slope.

The graph at the left represents a car traveling on a straight road. Answer the following questions on the basis of this graph.

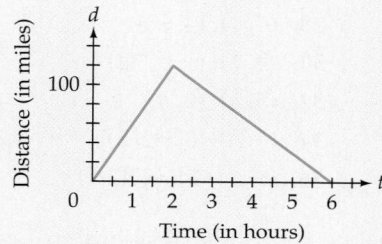

1. Between what two times is the car moving to the right?

2. Between what two times does the car have a positive velocity?

3. Between what two times is the car moving to the left?

4. Between what two times does the car have a negative velocity?

5. After 2 hours, how far is the car from its starting position?

6. How long after the car leaves its starting position does it return to its starting position?

7. What is the velocity of the car during its first 2 hours of travel?

8. What is the velocity of the car during its last 4 hours of travel?

The graph below represents another car traveling on a straight road, but this car's motion is a little more complicated. Use this graph for the questions below.

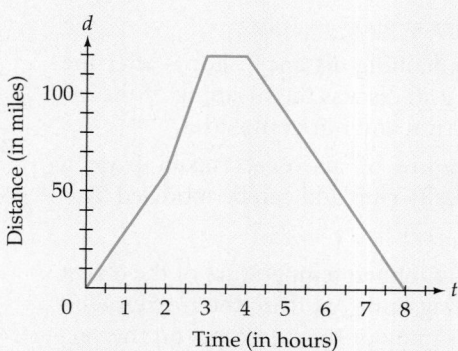

9. What is the slope of the line between hours 3 and 4?

10. What is the velocity of the car between hours 3 and 4? Is the car moving?

11. During which of the following intervals of time is the absolute value of the velocity greatest: 0 to 2 hours, 2 to 3 hours, 3 to 4 hours, or 4 to 8 hours?

Exercise Set 6.2   (Suggested Assignment: 1–51, odds; 53–55)

In Exercises 1–14, find the *x*- and *y*-intercepts of the graph of the equation.

1. $f(x) = 3x - 6$

2. $f(x) = 2x + 8$

3. $y = \dfrac{2}{3}x - 4$

4. $y = -\dfrac{3}{4}x + 6$

5. $y = -x - 4$

6. $y = -\dfrac{x}{2} + 1$

7. $3x + 4y = 12$

8. $5x - 2y = 10$

9. $2x - 3y = 9$

10. $4x + 3y = 8$

11. $\dfrac{x}{2} + \dfrac{y}{3} = 1$

12. $\dfrac{x}{3} - \dfrac{y}{2} = 1$

13. $x - \dfrac{y}{2} = 1$

14. $-\dfrac{x}{4} + \dfrac{y}{3} = 1$

15. ✎ There is a relationship between the number of times a cricket chirps per minute and the air temperature. A linear model of this relationship is given by
$$f(x) = 7x - 30$$
where *x* is the temperature in degrees Celsius and $f(x)$ is the number of chirps per minute. Find and discuss the meaning of the *x*-intercept.

16. ✎ An approximate linear model that gives the remaining distance, in miles, a plane must travel from Los Angeles to Paris is given by
$$s(t) = 6000 - 500t$$
where $s(t)$ is the remaining distance *t* hours after the flight begins. Find and discuss the meaning of the intercepts on the vertical and horizontal axes.

17. ✎ The temperature of an object taken from a freezer gradually rises and can be modeled by
$$T(x) = 3x - 15$$
where $T(x)$ is the Fahrenheit temperature of the object *x* minutes after being removed from the freezer. Find and discuss the meaning of the intercepts on the vertical and horizontal axes.

18. ✎ A retired biologist begins withdrawing money from a retirement account according to the linear model
$$A(t) = 100,000 - 2500t$$
where $A(t)$ is the amount, in dollars, remaining in the account *t* months after withdrawals begin. Find and

discuss the meaning of the intercepts on the vertical and horizontal axes.

In Exercises 19–36, find the slope of the line containing the two points.

19. $(1, 3), (3, 1)$

20. $(2, 3), (5, 1)$

21. $(-1, 4), (2, 5)$

22. $(3, -2), (1, 4)$

23. $(-1, 3), (-4, 5)$

24. $(-1, -2), (-3, 2)$

25. $(0, 3), (4, 0)$

26. $(-2, 0), (0, 3)$

27. $(2, 4), (2, -2)$

28. $(4, 1), (4, -3)$

29. $(2, 5), (-3, -2)$

30. $(4, 1), (-1, -2)$

31. $(2, 3), (-1, 3)$

32. $(3, 4), (0, 4)$

33. $(0, 4), (-2, 5)$

34. $(-2, 3), (-2, 5)$

35. $(-3, -1), (-3, 4)$

36. $(-2, -5), (-4, -1)$

37. ✎ The graph below shows the relationship between the distance traveled by a motorist and the time of travel. Find the slope of the line between the two points shown on the graph. Write a sentence that states the meaning of the slope.

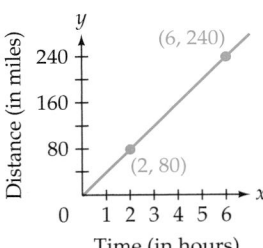

38. ✎ The graph below shows the relationship between the value of a building and the depreciation allowed for income tax purposes. Find the slope of the line between the two points shown on the graph. Write a sentence that states the meaning of the slope.

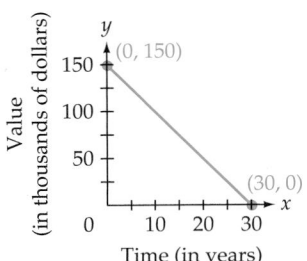

39. The graph below shows the relationship between the amount of tax and the amount of taxable income between $22,101 and $54,500. Find the slope of the line between the two points shown on the graph. Write a sentence that states the meaning of the slope.

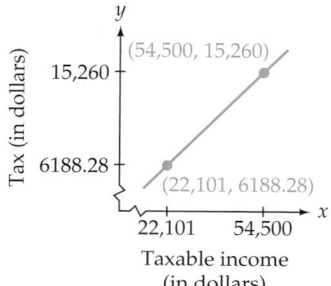

40. The graph below shows the relationship between the payment on a mortgage and the amount of the mortgage. Find the slope of the line between the two points shown on the graph. Write a sentence that states the meaning of the slope.

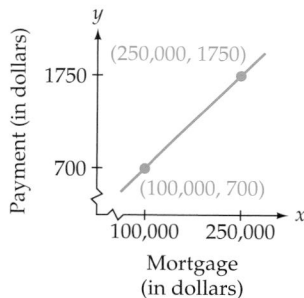

41. The graph below shows the relationship between distance and time for the world record for the 5000-meter run set by Said Aouita in 1987. (Assume Aouita ran the race at a constant rate.) Find the slope of the line between the two points shown on the graph. Write a sentence that states the meaning of the slope.

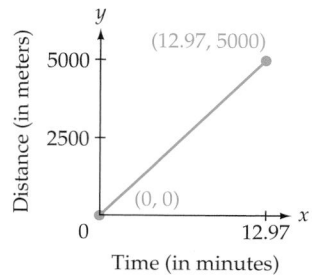

42. The graph below shows the relationship between distance and time for the world record for the 10,000-meter run set by Arturo Barrios in 1989. (Assume Barrios ran the race at a constant rate.) Find the slope of the line between the two points shown on the graph. Write a sentence that states the meaning of the slope.

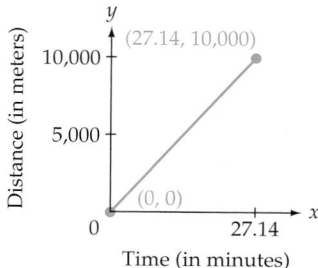

43. Graph the line that passes through the point $(-1, -3)$ and has slope $\frac{4}{3}$.

44. Graph the line that passes through the point $(-2, -3)$ and has slope $\frac{5}{4}$.

45. Graph the line that passes through the point $(-3, 0)$ and has slope -3.

46. Graph the line that passes through the point $(2, 0)$ and has slope -1.

In Exercises 47–52, graph using the slope and y-intercept.

47. $f(x) = \frac{1}{2}x + 2$

48. $f(x) = \frac{2}{3}x - 3$

49. $f(x) = -\frac{3}{2}x$

50. $f(x) = \frac{3}{4}x$

51. $f(x) = \frac{1}{3}x - 1$

52. $f(x) = -\frac{3}{2}x + 6$

Extensions

CRITICAL THINKING

53. Lois and Tanya start from the same place on a straight jogging course, at the same time, and jog in the same direction. Lois is jogging at 9 kilometers per hour, and Tanya is jogging at 6 kilometers per hour. The graphs on the following page show the distance each jogger has traveled in x hours and the distance between the joggers in x hours. Which graph represents the distance Lois has traveled in x hours? Which graph represents the distance Tanya has traveled in x hours? Which graph represents the distance between Lois and Tanya in x hours?

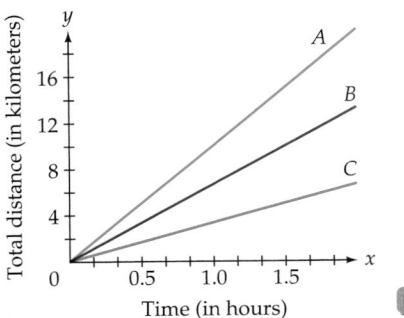

54. A chemist is filling two cans from a faucet that releases water at a constant rate. Can 1 has a diameter of 20 millimeters and can 2 has a diameter of 30 millimeters.

a. In the following graph, which line represents the depth of the water in can 1 after x seconds?

b. Use the graph to estimate the difference in the depths of the water in the two cans after 15 seconds.

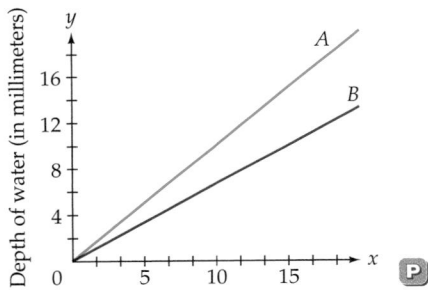

55. The American National Standards Institute (ANSI) states that the slope of a wheelchair ramp must not exceed $\frac{1}{12}$.

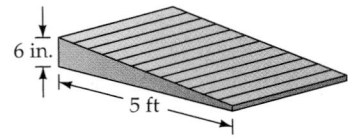

Does the ramp pictured above meet the requirements of ANSI?

56. A ramp for a wheelchair must be 14 inches high. What is the minimum length of this ramp so that it meets the ANSI requirements stated in Exercise 55?

57. If $(2, 3)$ are the coordinates of a point on a line that has slope 2, what is the y-coordinate of the point on the line at which $x = 4$?

58. If $(-1, 2)$ are the coordinates of a point on a line that has slope -3, what is the y-coordinate of the point on the line at which $x = 1$?

59. If $(1, 4)$ are the coordinates of a point on a line that has slope $\frac{2}{3}$, what is the y-coordinate of the point on the line at which $x = -2$?

60. If $(-2, -1)$ are the coordinates of a point on a line that has slope $\frac{3}{2}$, what is the y-coordinate of the point on the line at which $x = -6$?

61. What effect does increasing the coefficient of x have on the graph of $y = mx + b$?

62. What effect does decreasing the coefficient of x have on the graph of $y = mx + b$?

63. What effect does increasing the constant term have on the graph of $y = mx + b$?

64. What effect does decreasing the constant term have on the graph of $y = mx + b$?

65. Do the graphs of all straight lines have a y-intercept? If not, give an example of one that does not.

66. If two lines have the same slope and the same y-intercept, must the graphs of the lines be the same? If not, give an example.

EXPLORATIONS

67. When you climb a staircase, the flat part of a stair that you step on is called the *tread* of the stair. The *riser* is the vertical part of the stair. The slope of a staircase is the ratio of the length of the riser to the length of the tread. Because the design of a staircase may affect safety, most cities have building codes that give rules for the design of a staircase.

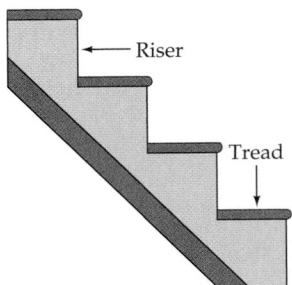

a. The traditional design of a staircase calls for a 9-inch tread and an 8.25-inch riser. What is the slope of this staircase?

b. A newer design for a staircase uses an 11-inch tread and a 7-inch riser. What is the slope of this staircase?

c. An architect is designing a house with a staircase that is 8 feet high and 12 feet long. Is the architect using the traditional design in part a or the newer design in part b? Explain your answer.

d. Staircases that have a slope between 0.5 and 0.7 are usually considered safer than those with a slope greater than 0.7. Design a safe staircase that goes from the first floor of a house to the basement, which is 9 feet below the first floor.

e. Measure the tread and riser for three staircases you encounter. Do these staircases match the traditional design in part a or the newer design in part b?

68. In the diagram at the right, lines l_1 and l_2 are perpendicular with slopes m_1 and m_2, respectively, and $m_1 > 0$ and $m_2 < 0$. Line segment $\overline{AC}$ has length 1.

a. Show that the length of $\overline{BC}$ is m_1.

b. Show that the length of $\overline{CD}$ is $-m_2$. Note that because m_2 is a negative number, $-m_2$ is a positive number.

c. Show that triangles ACB and ACD are similar right triangles. **Similar triangles** have the same shape; the corresponding angles are equal, and corresponding sides are in proportion. (*Suggestion:* Show that the measure of angle ADC equals the measure of angle BAC.)

d. Show that $\dfrac{m_1}{1} = \dfrac{1}{-m_2}$. Use the fact that the ratios of corresponding sides of similar triangles are equal.

e. Use the equation in part d to show that $m_1 m_2 = -1$.

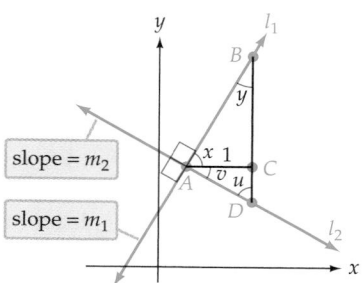

SECTION 6.3 **Finding Linear Models**

Finding Linear Models

Suppose that a car uses 0.04 gallon of gas per mile driven and that the fuel tank, which holds 18 gallons of gas, is full. Using this information, we can determine a linear model for the amount of fuel remaining in the gas tank after driving x miles.

Recall that a linear function is one that can be written in the form $f(x) = mx + b$, where m is the slope of the line and b is the y-intercept. The slope is the rate at which the car is using fuel, 0.04 gallon per mile. Because the car is consuming the fuel, the amount of fuel in the tank is decreasing. Therefore, the slope is negative and we have $m = -0.04$.

The amount of fuel in the tank depends on the number of miles, x, the car has been driven. Before the car starts (that is, when $x = 0$), there are 18 gallons of gas in the tank. The y-intercept is $(0, 18)$.

Using this information, we can create the linear function.

$$f(x) = mx + b$$
$$f(x) = -0.04x + 18 \qquad \bullet \text{ Replace } m \text{ by } -0.04 \text{ and } b \text{ by } 18.$$

> **✓ TAKE NOTE**
>
> When creating a linear model, the slope will be the quantity that is expressed using the word *per*. The car discussed at the right uses 0.04 gallon *per* mile. The slope is negative because the amount of fuel in the tank is decreasing.

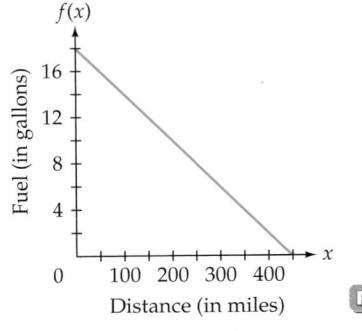

Fuel (in gallons)

$f(x)$

16

12

8

4

0   100 200 300 400

Distance (in miles)

The linear function that models the amount of fuel remaining in the tank is given by $f(x) = -0.04x + 18$, where $f(x)$ is the amount of fuel, in gallons, remaining after driving x miles. The graph of the function is shown at the left.

The x-intercept of a graph is the point at which $f(x) = 0$. For this application, $f(x) = 0$ when there are 0 gallons of fuel remaining in the tank. Thus, replacing $f(x)$ by 0 in $f(x) = -0.04x + 18$ and solving for x will give the number of miles the car can be driven before running out of gas.

$$f(x) = -0.04x + 18$$
$$0 = -0.04x + 18 \qquad \bullet \text{ Replace } f(x) \text{ by } \mathbf{0}.$$
$$-18 = -0.04x$$
$$450 = x$$

The car can travel 450 miles before running out of gas.

Recall that the domain of a function is all possible values of x, and the range of a function is all possible values of $f(x)$. For the function $f(x) = -0.04x + 18$, which was used above to model the fuel remaining in the gas tank of the car, the domain is $\{x \mid 0 \le x \le 450\}$ because the fuel tank is empty when the car has traveled 450 miles. The range is $\{y \mid 0 \le y \le 18\}$ because the tank can hold up to 18 gallons of fuel. Sometimes it is convenient to write the domain $\{x \mid 0 \le x \le 450\}$ as $[0, 450]$ and the range $\{y \mid 0 \le y \le 18\}$ as $[0, 18]$. The notation $[0, 450]$ and $[0, 18]$ is called **interval notation**. Using interval notation, a domain of $\{x \mid a \le x \le b\}$ is written $[a, b]$ and a range of $\{y \mid c \le y \le d\}$ is written $[c, d]$.

QUESTION   *Why does it not make sense for the domain of $f(x) = -0.04x + 18$ to exceed 450?*

EXAMPLE 1 ■ **Application of Finding a Linear Model Given the Slope and y-intercept**

Suppose a 20-gallon gas tank contains 2 gallons when a motorist decides to fill up the tank. If the gas pump fills the tank at a rate of 0.1 gallon per second, find a linear function that models the amount of fuel in the tank t seconds after fueling begins.

Solution
When fueling begins, at $t = 0$, there are 2 gallons of gas in the tank. Therefore, the y-intercept is $(0, 2)$. The slope is the rate at which fuel is being added to the tank. Because the amount of fuel in the tank is increasing, the slope is positive and we have $m = 0.1$. To find the linear function, replace m and b by their respective values.

$$f(t) = mt + b$$
$$f(t) = 0.1t + 2 \qquad \bullet \text{ Replace } m \text{ by } \mathbf{0.1} \text{ and } b \text{ by } \mathbf{2}.$$

The linear function is $f(t) = 0.1t + 2$, where $f(t)$ is the number of gallons of fuel in the tank t seconds after fueling begins.

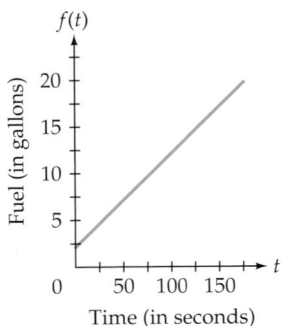

Fuel (in gallons)

$f(t)$

20

15

10

5

0   50   100 150

Time (in seconds)

ANSWER   *If $x > 450$, then $f(x) < 0$. This would mean that the tank has negative gallons of gas. For instance, $f(500) = -2$.*

CHECK YOUR PROGRESS 1   The boiling point of water at sea level is 100°C. The boiling point decreases 3.5°C per 1 kilometer increase in altitude. Find a linear function that gives the boiling point of water as a function of altitude.

Solution   *See page S22.*

For each of the previous examples, the known point on the graph of the linear function was the *y*-intercept. This information enabled us to determine *b* for the linear function $f(x) = mx + b$. In some cases, a point other than the *y*-intercept is given. In such a case, the *point–slope formula* is used to find the equation of the line.

✔ **TAKE NOTE**

Using parentheses may help when substituting into the point–slope formula.

$$y - y_1 = m(x - x_1)$$

$$y - (\ ) = (\ )[x - (\ )]$$

Point–Slope Formula of a Straight Line

Let (x_1, y_1) be a point on a line and let *m* be the slope of the line. Then the equation of the line can be found using the point–slope formula

$$y - y_1 = m(x - x_1)$$

EXAMPLE 2 ■ Find the Equation of a Line Given the Slope and a Point on the Line

Find the equation of the line that passes through $(1, -3)$ and has slope -2.

Solution

$$y - y_1 = m(x - x_1) \qquad \bullet \text{ Use the point–slope formula.}$$
$$y - (-3) = -2(x - 1) \qquad \bullet \ m = -2, (x_1, y_1) = (1, -3)$$
$$y + 3 = -2x + 2$$
$$y = -2x - 1$$

✔ **TAKE NOTE**

Recall that $f(x)$ and *y* are different symbols for the same quantity, the value of the function at *x*.

Note that we wrote the equation of the line as $y = -2x - 1$. We could also write the equation in functional notation as $f(x) = -2x - 1$.

CHECK YOUR PROGRESS 2   Find the equation of the line that passes through $(-2, 2)$ and has slope $-\frac{1}{2}$.

Solution   *See page S22.*

EXAMPLE 3 ■ Application of Finding a Linear Model Given a Point and the Slope

Based on data from the *Kelley Blue Book,* the value of a certain car decreases approximately $250 per month. If the value of the car 2 years after it was purchased was $14,000, find a linear function that models the value of the car after *x* months of ownership. Use this function to find the value of the car after 3 years of ownership.

Solution
Let V represent the value of the car after x months. Then $V = 14{,}000$ when $x = 24$ (2 years is 24 months). A solution of the equation is $(24, 14{,}000)$. The car is decreasing in value at a rate of \$250 per month. Therefore, the slope is -250. Now use the point–slope formula to find the linear equation that models the function.

$$V - V_1 = m(x - x_1)$$
$$V - 14{,}000 = -250(x - 24) \qquad \bullet \ x_1 = \mathbf{24}, \ V_1 = \mathbf{14{,}000}, \ m = \mathbf{-250}$$
$$V - 14{,}000 = -250x + 6000$$
$$V = -250x + 20{,}000$$

A linear function that models the value of the car after x months of ownership is $V(x) = -250x + 20{,}000$.

To find the value of the car after 3 years (36 months), evaluate the function when $x = 36$.

$$V(x) = -250x + 20{,}000$$
$$V(36) = -250(36) + 20{,}000 = 11{,}000$$

The value of the car is \$11,000 after 3 years of ownership.

CHECK YOUR PROGRESS 3 During a brisk walk, a person burns about 3.8 calories per minute. If a person has burned 191 calories in 50 minutes, determine a linear function that models the number of calories burned after t minutes.

Solution *See page S22.*

The next example shows how to find the equation of a line given two points on the line.

EXAMPLE 4 ■ **Find the Equation of a Line Given Two Points on the Line**

Find the equation of the line that passes through $P_1(6, -4)$ and $P_2(3, 2)$.

Solution
Find the slope of the line between the two points.

$$m = \frac{y_2 - y_1}{x_2 - x_1} = \frac{2 - (-4)}{3 - 6} = \frac{6}{-3} = -2$$

Use the point–slope formula to find the equation of the line.

$$y - y_1 = m(x - x_1)$$
$$y - (-4) = -2(x - 6) \qquad \bullet \ m = \mathbf{-2}, \ x_1 = \mathbf{6}, \ y_1 = \mathbf{-4}$$
$$y + 4 = -2x + 12$$
$$y = -2x + 8$$

CHECK YOUR PROGRESS 4 Find the equation of the line that passes through $P_1(-2, 3)$ and $P_2(4, 1)$.

Solution *See page S22.*

✔ **TAKE NOTE**

There are many ways to find the equation of a line. However, in every case, there must be enough information to determine a point on the line and to find the slope of the line. When you are doing problems of this type, look for different ways that information may be presented. For instance, in Example 4, even though the slope of the line is not given, knowing two points enables us to find the slope.

INSTRUCTOR NOTE
You might have students use $P_2(3, 2)$ in the point-slope formula at the right and verify that the same equation results.

Math Matters Perspective: Using Straight Lines in Art

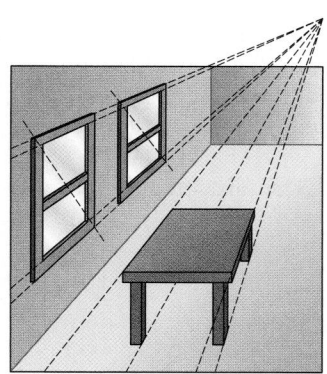

Many paintings we see today have a three-dimensional quality to them, even though they are painted on a flat surface. This was not always the case. It wasn't until the Renaissance that artists started to paint "in perspective." Using lines is one way to create this perspective. Here is a simple example.

Draw a dot, called the *vanishing point*, and a rectangle on a piece of paper. Draw windows as shown. To keep the perspective accurate, the lines through opposite corners of the windows should be parallel. A table in proper perspective is created in the same way.

This method of creating perspective was employed by Leonardo Da Vinci. Find a copy of his painting *The Last Supper* and, using a ruler, see whether you can find the vanishing point by drawing lines along the edges of the windows on the sides of the painting.

Regression Lines

There are many situations in which a linear function can be used to approximate collected data. For instance, the table below shows the maximum exercise heart rates for specific individuals of various ages who exercise regularly.

Age, x, in years	20	25	30	32	43	55	28	42	50	55	62
Heart rate, y, in beats per minute	160	150	148	145	140	130	155	140	132	125	125

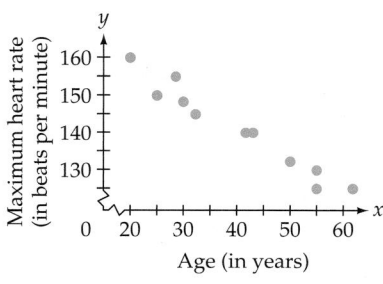

The graph at the left, called a **scatter diagram**, is a graph of the ordered pairs of the table. These ordered pairs suggest that the maximum exercise heart rate for an individual decreases as the person's age increases.

Although these points do not lie on one line, it is possible to find a line that *approximately fits* the data. One way to do this is to select two data points and then find the equation of the line that passes through the two points. To do this, we first find the slope of the line between the two points and then use the point–slope formula to find the equation of the line. Suppose we choose $(20, 160)$ as P_1 and $(62, 125)$ as P_2. Then the slope of the line between P_1 and P_2 is

$$m = \frac{y_2 - y_1}{x_2 - x_1} = \frac{125 - 160}{62 - 20} = -\frac{35}{42} = -\frac{5}{6}$$

Now use the point–slope formula.

$$y - y_1 = m(x - x_1)$$

$$y - 160 = -\frac{5}{6}(x - 20) \quad \bullet \ m = -\frac{5}{6}, x_1 = 20, y_1 = 160$$

$$y - 160 = -\frac{5}{6}x + \frac{50}{3} \quad \bullet \text{ Multiply by } -\frac{5}{6}.$$

$$y = -\frac{5}{6}x + \frac{530}{3} \quad \bullet \text{ Add 160 to each side of the equation.}$$

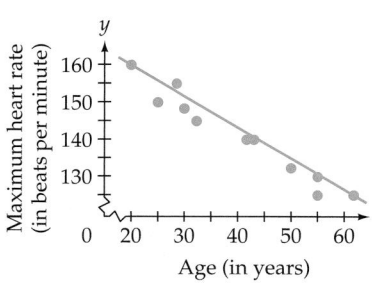

The graph of $y = -\frac{5}{6}x + \frac{530}{3}$

The graph of $y = -\frac{5}{6}x + \frac{530}{3}$ is shown at the left. This line *approximates* the data.

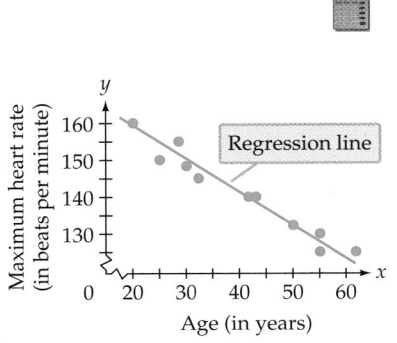

Maximum heart rate (in beats per minute) vs. Age (in years), with regression line.

The equation of the line we found by choosing two data points gives an approximate linear model for the data. If we had chosen different points, the result would have been a different equation. Among all the equations that might have been chosen, statisticians have determined the line of *best fit* is called the **regression line.** A graphing calculator can be used to find the equation of a regression line. For instance, the equation of the regression line for the maximum heart rate data is given by $y = -0.827x + 174$. The graph of this line is shown at the left. Using this model, an exercise physiologist can determine the recommended maximum exercise heart rate for an individual of any particular age.

For example, suppose an individual is 28 years old. The physiologist would replace x by 28 and determine the value of y.

$$y = -0.827x + 174$$
$$y = -0.827(28) + 174 \quad \bullet \text{ Replace } x \text{ by } \mathbf{28.}$$
$$= 150.844$$

The maximum exercise heart rate recommended for a 28-year-old person is approximately 151 beats per minute.

The calculation of a regression equation can be accomplished with a graphing calculator by using the STAT key. The table below shows the data collected by a chemistry student who is trying to determine a relationship between the temperature, in degrees Celsius, and volume, in liters, of 1 gram of oxygen at a constant pressure. Chemists refer to this relationship as Charles's Law.

Temperature, T, in degrees Celsius	−100	−75	−50	−25	0	25	50
Volume, V, in liters	0.43	0.5	0.57	0.62	0.7	0.75	0.81

To find the regression equation for these data, press the STAT key and then select EDIT from the menu. This will bring up a table into which you can enter data. Let L1 be the independent variable (temperature) and L2 be the dependent variable (volume).

Once the data have been entered, select the $\boxed{\text{STAT}}$ key again, highlight CALC, and arrow down to LinReg(ax+b) to see the linear regression equation. Selecting this option will paste LinReg(ax+b) onto the home screen. Now you can just press $\boxed{\text{ENTER}}$ and the values for a and b will appear on the screen. Entering LinReg(ax+b)L1,L2,Y1 not only shows the results on the home screen, but pastes the regression equation into Y1 in the Y= editor. This will enable you to easily graph or evaluate the regression equation.

For this set of data, the regression equation is $V = 0.0025286T + 0.68892857$. To determine the volume of 1 gram of oxygen when the temperature is $-30°C$, replace T by -30 and evaluate the expression. This can be done using your calculator. Use the Y-VARS menu to place Y1 on the screen ($\boxed{\text{VARS}} \triangleright \boxed{\text{ENTER}} \boxed{\text{ENTER}}$). Then enter the value -30, within parentheses, as shown at the left. After hitting $\boxed{\text{ENTER}}$, the volume will be displayed as approximately 0.61 liter.

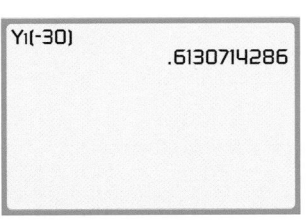

CALCULATOR NOTE

If r and r^2 do not appear on the screen of your TI-83, press [2nd] CATALOG D (above the [x⁻¹] key) and then scroll to DiagnosticsOn. Now press [ENTER] twice.

You may have noticed some additional results on the screen when the regression equation was calculated. The variable r is called the **correlation coefficient,** and r^2 is called the **coefficient of determination.** Statisticians use these numbers to determine how well the regression equation approximates the data. If $r = 1$, the data exactly fit a line of positive slope. If $r = -1$, the data exactly fit a line of negative slope. In general, the closer r^2 is to 1, the closer the data fit a linear model. For our purposes, we will assume that the given data are approximated by a linear function.

EXAMPLE 5 ■ Find a Linear Regression Equation

Sodium thiosulfate is used by photographers to develop some types of film. The amount of this chemical that will dissolve in water depends on the temperature of the water. The table below gives the number of grams of sodium thiosulfate that will dissolve in 100 milliliters of water at various temperatures.

Temperature, x, in degrees Celsius	20	35	50	60	75	90	100
Sodium thiosulfate dissolved, y, in grams	50	80	120	145	175	205	230

a. Find the linear regression equation for these data.

b. How many grams of sodium thiosulfate does the model predict will dissolve in 100 milliliters of water when the temperature is 70°C? Round to the nearest tenth.

Solution

a. Using a calculator, the regression equation is $y = 2.2517731x + 5.2482270$.

b. Evaluate the regression equation when $x = 70$.

$$y = 2.2517731x + 5.2482270$$
$$= 2.2517731(70) + 5.2482270 \qquad \text{• Replace } x \text{ by 70.}$$
$$= 162.872344$$

Approximately 162.9 grams of sodium thiosulfate will dissolve when the temperature is 70°C.

CHECK YOUR PROGRESS 5   The heights and weights of women swimmers on a college swim team are given in the table below.

Height, x, in inches	68	64	65	67	62	67	65
Weight, y, in pounds	132	108	108	125	102	130	105

a. Find the linear regression equation for these data.

b. Use your regression equation to estimate the weight of a woman swimmer who is 63 inches tall. Round to the nearest whole number.

Solution   *See page S22.*

Excursion

A Linear Business Model

Two people decide to open a business reconditioning toner cartridges for copy machines. They rent a building for $7000 per year and estimate that building maintenance, taxes, and insurance will cost $6500 per year. Each person wants to make $12 per hour in the first year and will work 10 hours per day for 260 days of the year. Assume that it costs $28 to restore a cartridge and that the restored cartridge can be sold for $45.

1. Write a linear function for the total cost C to operate the business and restore n cartridges during the first year, not including the hourly wage the owners wish to earn.

2. Write a linear function for the total revenue R the business will earn during the first year by selling n cartridges.

3. How many cartridges must the business restore and sell annually to break even, not including the hourly wage the owners wish to earn?

4. How many cartridges must the business restore and sell annually for the owners to pay all expenses and earn the hourly wage they desire?

5. Suppose the entrepreneurs are successful in their business and are restoring and selling 25 cartridges each day of the 260 days they are open. What would be their hourly wage for the year if all the profit is shared equally?

6. As the company becomes successful and is selling and restoring 25 cartridges each day of the 260 days it is open, the entrepreneurs decide to hire a part-time employee 4 hours per day and pay the employee $8 per hour. How many additional cartridges must be restored and sold each year just to cover the cost of the new employee? You can neglect employee costs such as social security, worker's compensation, and other benefits.

7. Suppose the company decides that it could increase its business by advertising. Answer Exercises 1, 2, 3, and 5 if the owners decide to spend $400 per month on advertising.

Exercise Set 6.3   (Suggested Assignment: 1–29, odds)

In Exercises 1 to 8, find the equation of the line that passes through the given point and has the given slope.

1. $(0, 5)$, $m = 2$

2. $(2, 3)$, $m = \dfrac{1}{2}$

3. $(-1, 7)$, $m = -3$

4. $(0, 0)$, $m = \dfrac{1}{2}$

5. $(3, 5)$, $m = -\dfrac{2}{3}$

6. $(0, -3)$, $m = -1$

7. $(-2, -3)$, $m = 0$

8. $(4, -5)$, $m = -2$

In Exercises 9–16, find the equation of the line that passes through the given points.

9. $(0, 2)$, $(3, 5)$

10. $(0, -3)$, $(-4, 5)$

11. $(0, 3)$, $(2, 0)$

12. $(-2, -3)$, $(-1, -2)$

13. $(2, 0)$, $(0, -1)$

14. $(3, -4)$, $(-2, -4)$

15. $(-2, 5)$, $(2, -5)$

16. $(2, 1)$, $(-2, -3)$

17. The operator of a hotel estimates that 500 rooms per night will be rented if the room rate per night is $75. For each $10 increase in the price of a room, six fewer rooms will be rented. Determine a linear function that will predict the number of rooms that will be rented per night for a given price per room. Use this model to predict the number of rooms that will be rented if the room rate is $100 per night.

18. A general building contractor estimates that the cost to build a new home is $30,000 plus $85 for each square foot of floor space in the house. Determine a linear function that will give the cost of building a house that contains a given number of square feet of floor space. Use this model to determine the cost to build a house that contains 1800 square feet of floor space.

19. A plane travels 830 miles in 2 hours. Determine a linear model that will predict the number of miles the plane can travel in a given interval of time. Use this model to predict the distance the plane will travel in $4\frac{1}{2}$ hours.

20. An account executive receives a base salary plus a commission. On $20,000 in monthly sales, an account executive would receive compensation of $1800. On $50,000 in monthly sales, an account executive would receive compensation of $3000. Determine a linear function that will yield the compensation of a sales executive for a given amount of monthly sales. Use this model to determine the compensation of an account executive who has $85,000 in monthly sales.

21. A manufacturer of pickup trucks has determined that 50,000 trucks per month can be sold at a price of $12,000 per truck. At a price of $11,000, the number of trucks sold per month would increase to 55,000. Determine a linear function that will predict the number of trucks that would be sold per month at a given price. Use this model to predict the number of trucks that would be sold at a price of $12,500.

22. A manufacturer of graphing calculators has determined that 10,000 calculators per week will be sold at a price of $95 per calculator. At a price of $90, it is estimated that 12,000 calculators would be sold. Determine a linear function that will predict the number of calculators that would be sold per week at a given price. Use this model to predict the number of calculators that would be sold at a price of $75.

23. A research hospital did a study on the relationship between stress and diastolic blood pressure. The results from eight patients in the study are given in the table below.

Stress test score, x	55	62	58	78	92	88	75	80
Blood pressure, y	70	85	72	85	96	90	82	85

   a. Find the linear regression equation for these data. Round to the nearest hundredth.

   b. Use the regression equation to estimate the diastolic blood pressure of a person whose stress test score was 85. Round to the nearest whole number.

24. The average hourly earnings, in dollars, of non-farm workers in the United States for the years 1995 to 2001 are given in the table below. (*Source:* U.S. Bureau of Labor Statistics)

Year, x	1995	1996	1997	1998	1999	2000	2001
Hourly wage, y	7.38	7.41	7.46	7.65	7.83	7.88	7.90

a. Using $x = 0$ to correspond to 1995, find the linear regression equation for these data. Round to the nearest hundredth.

b. Use the regression equation to estimate the expected average hourly wage, to the nearest cent, in 2005.

25. The table below shows the numbers of students, in thousands, who have graduated from or are projected to graduate from high school for the years 1996 to 2005. (*Source:* U.S. Bureau of Labor Statistics)

Year, x	1996	1997	1998	1999	2000	2001	2002	2003	2004	2005
Number of students (in thousands), y	2540	2633	2740	2786	2820	2837	2886	2929	2935	2944

a. Using $x = 0$ to correspond to 1996, find the linear regression equation for these data. Round to the nearest hundredth.

b. Use the regression equation to estimate the expected number of high school graduates in 2010. Round to the nearest thousand.

26. An automotive engineer studied the relationship between the speed of a car and the number of miles traveled per gallon of fuel consumed at that speed. The results of the study are shown in the table below.

Speed (in miles per hour), x	40	25	30	50	60	80	55	35	45
Consumption (in miles per gallon), y	26	27	28	24	22	21	23	27	25

a. Find the linear regression equation for these data. Round to the nearest hundredth.

b. Use the regression equation to estimate the expected number of miles traveled per gallon of fuel consumed for a car traveling at 65 miles per hour. Round to the nearest whole number.

27. A meteorologist studied the maximum temperatures at various latitudes for January of a certain year. The results of the study are shown in the table below.

Latitude (in °N), x	22	30	36	42	56	51	48
Maximum temperature (in °F), y	80	65	47	54	21	44	52

a. Find the linear regression equation for these data. Round to the nearest hundredth.

b. Use the regression equation to estimate the expected maximum temperature in January at a latitude of 45°N. Round to the nearest whole number.

28. A zoologist studied the running speeds of animals in terms of the animals' body lengths. The results of the study are shown in the table below.

Body length (in centimeters), x	1	9	15	16	24	25	60
Running speed (in meters per second), y	1	2.5	7.5	5	7.4	7.6	20

a. Find the linear regression equation for these data. Round to the nearest hundredth.

b. Use the regression equation to estimate the expected running speed of a deer mouse, whose body length is 10 centimeters. Round to the nearest tenth.

Extensions

CRITICAL THINKING

29. A line contains the points $(4, -1)$ and $(2, 1)$. Find the coordinates of three other points on the line.

30. If f is a linear function for which $f(1) = 3$ and $f(-1) = 5$, find $f(4)$.

31. The ordered pairs $(0, 1)$, $(4, 9)$ and $(3, n)$ are solutions of the same linear equation. Find n.

32. The ordered pairs $(2, 2)$, $(-1, 5)$ and $(3, n)$ are solutions of the same linear equation. Find n.

33. Is there a linear function that contains the ordered pairs $(2, 4)$, $(-1, -5)$, and $(0, 2)$? If so, find the function and explain why there is such a function. If not, explain why there is no such function.

34. Is there a linear function that contains the ordered pairs $(5, 1)$, $(4, 2)$, and $(0, 6)$? If so, find the function and explain why there is such a function. If not, explain why there is no such function.

35. Assume that the maximum speed your car will travel varies linearly with the steepness of the hill it is climbing or descending. If the hill is 5° up, your car can travel 77 kilometers per hour. If the hill is 2° down $(-2°)$, your car can travel 154 kilometers per hour. When your car's top speed is 99 kilometers per hour, how steep is the hill? State your answer in degrees and note whether the car is climbing or descending.

EXPLORATIONS

36. A person who can row at a rate of 3 miles per hour in calm water is trying to cross a river in which a current of 4 miles per hour runs perpendicular to the direction of rowing. See the figure at the right.

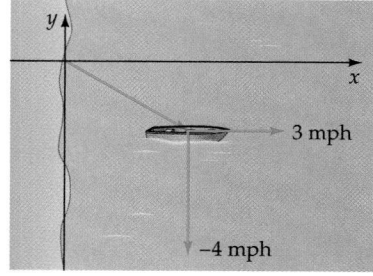

Because of the current, the boat is being pushed downstream at the same time that it is moving across the river. Because the boat is traveling at 3 miles per hour in the x direction, its horizontal position after t hours is given by $x = 3t$. The current is pushing the boat in the negative y direction at 4 miles per hour. Therefore, the boat's vertical position after t hours is given by $y = -4t$, where -4 indicates that the boat is moving downstream. The set of equations $x = 3t$ and $y = -4t$ are called **parametric equations,** and t is called the **parameter.**

a. What is the location of the boat after 15 minutes (0.25 hour)?

b. If the river is 1 mile wide, how far down the river will the boat be when it reaches the other shore? *Hint:* Find the time it takes the boat to cross the river by solving $x = 3t$ for t when $x = 1$. Then replace t by this value in $y = -4t$ and simplify.

c. For the parametric equations $x = 3t$ and $y = -4t$, write y in terms of x by solving $x = 3t$ for t and then substituting this expression into $y = -4t$.

37. In the diagram below, a plane flying at 5000 feet above sea level begins a gradual ascent.

a. Determine parametric equations for the path of the plane.

b. What is the altitude of the plane 5 minutes after it begins its ascent?

c. What is the altitude of the plane after it has traveled 12,000 feet in the positive x direction?

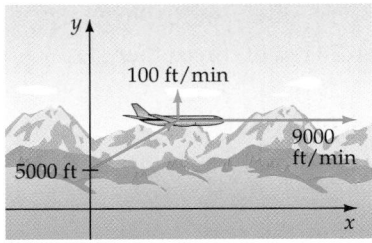

SECTION 6.4 | ## Quadratic Functions

Properties of Quadratic Functions

Recall that a linear function is a function of the form $f(x) = mx + b$. The graph of a linear function has certain characteristics. It is a straight line with slope m and y-intercept $(0, b)$.

A **quadratic function** in a single variable x is a function of the form $f(x) = ax^2 + bx + c$, $a \neq 0$. Examples of quadratic functions are given below.

$$f(x) = x^2 - 3x + 1 \qquad \bullet \; a = 1, b = -3, c = 1$$
$$g(t) = -2t^2 - 4 \qquad \bullet \; a = -2, b = 0, c = -4$$
$$h(p) = 4 - 2p - p^2 \qquad \bullet \; a = -1, b = -2, c = 4$$
$$f(x) = 2x^2 + 6x \qquad \bullet \; a = 2, b = 6, c = 0$$

The graph of a quadratic function in a single variable x is a **parabola.** The graphs of two of these quadratic functions are shown below.

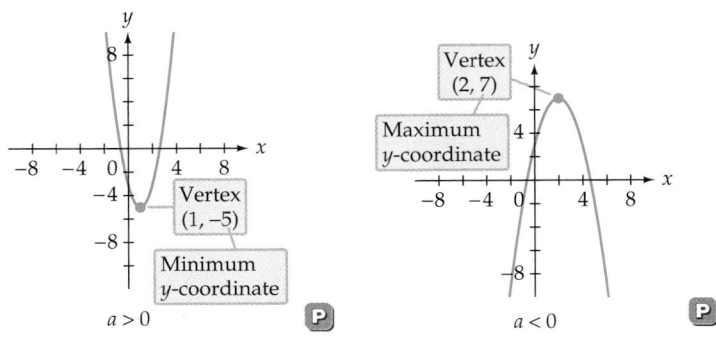

For the figure on the left above, $f(x) = 2x^2 - 4x - 3$. The value of a is *positive* $(a = 2)$ and the graph opens up. For the figure on the right above, $f(x) = -x^2 + 4x + 3$. The value of a is *negative* $(a = -1)$ and the graph opens down. The point at which the graph of a parabola has a minimum or a maximum is called the *vertex* of the parabola. The **vertex** of a parabola is the point with the smallest y-coordinate when $a > 0$ and the point with largest y-coordinate when $a < 0$.

The **axis of symmetry** of the graph of a quadratic function is a vertical line that passes through the vertex of the parabola. To understand the concept of the axis of symmetry of a graph, think of folding the graph along that line. The two portions of the graph will match up.

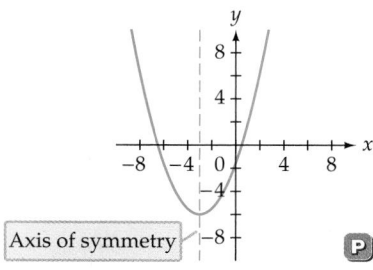

Axis of symmetry

The following formula enables us to determine the vertex of a parabola without having to graph the function.

Vertex of a Parabola

Let $f(x) = ax^2 + bx + c$ be the equation of a parabola. The coordinates of the vertex are $\left(-\dfrac{b}{2a}, f\left(-\dfrac{b}{2a} \right) \right)$.

EXAMPLE 1 ■ Find the Vertex of a Parabola

Find the vertex of the parabola whose equation is $y = -3x^2 + 6x + 1$.

Solution

$$x = -\frac{b}{2a} = -\frac{6}{2(-3)} = 1$$
 • Find the x-coordinate of the vertex. $a = -3$, $b = 6$

$$y = -3x^2 + 6x + 1$$
 • Find the y-coordinate of the vertex by replacing x by **1** and solving for y.

$$y = -3(1)^2 + 6(1) + 1$$
$$y = 4$$

The vertex is $(1, 4)$.

CHECK YOUR PROGRESS 1   Find the vertex of the parabola whose equation is $y = x^2 - 2$.

Solution   *See page S22.*

Math Matters   Paraboloids

The movie *Contact* was based on a novel by astronomer Carl Sagan. In the movie, Jodie Foster plays an astronomer who is searching for extraterrestrial intelligence. One scene from the movie takes place at the Very Large Array (VLA) in New Mexico. The VLA consists of 27 large radio telescopes whose dishes are paraboloids, the three-dimensional version of a parabola. A parabolic shape is used because of its unique reflective property: parallel rays of light or radio waves that strike the surface of a parabola are reflected to the same point. This point is called the *focus* of the parabola.

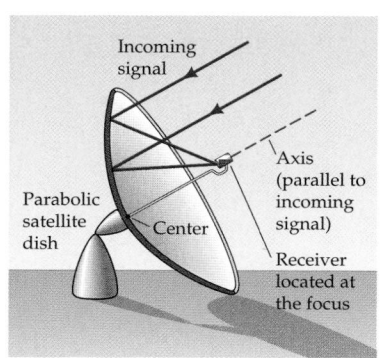

The photos above show the layout of the radio telescopes of the VLA and a more detailed picture of one of the telescopes. The figure at the far right shows the reflective property of a parabola. Note that all the incoming rays are reflected to the focus.

The reflective property of a parabola is also used in optical telescopes and headlights on a car. In the case of headlights, the bulb is placed at the focus and the light is reflected along parallel rays from the reflective surface of the headlight, thereby making a more concentrated beam of light.

x-Intercepts of Parabolas

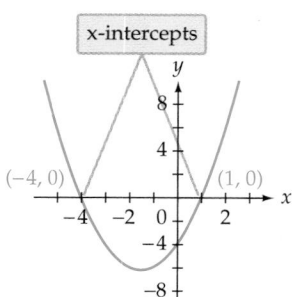

Recall that a point at which a graph crosses the *x*- or *y*-axis is called an *intercept* of the graph. The *x*-intercepts of the graph of an equation can be found by setting $y = 0$.

The graph of $y = x^2 + 3x - 4$ is shown at the left. The points whose coordinates are $(-4, 0)$ and $(1, 0)$ are *x*-intercepts of the graph. We can algebraically determine the *x*-intercepts by solving an equation.

EXAMPLE 2 ■ **Find the *x*-intercepts of a Parabola**

Find the *x*-intercepts of the graph of the parabola given by the equation.

a. $y = 4x^2 + 4x + 1$ **b.** $y = x^2 + 2x - 2$

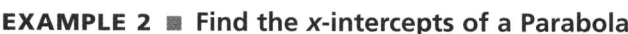

Solution

a. $y = 4x^2 + 4x + 1$

$0 = 4x^2 + 4x + 1$ • Let $y = \mathbf{0}$.

$0 = (2x + 1)(2x + 1)$ • Solve for x by factoring.

$2x + 1 = 0 \qquad\qquad 2x + 1 = 0$

$x = -\dfrac{1}{2} \qquad\qquad x = -\dfrac{1}{2}$

The x-intercept is $\left(-\dfrac{1}{2}, 0\right)$.

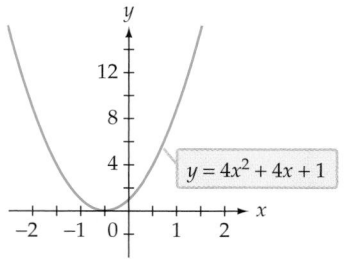

b. $y = x^2 + 2x - 2$

$0 = x^2 + 2x - 2$ • Let $y = \mathbf{0}$. The trinomial $x^2 + 2x - 2$ is nonfactorable over the integers.

$x = \dfrac{-b \pm \sqrt{b^2 - 4ac}}{2a}$ • Use the quadratic formula to solve for x. $a = \mathbf{1}$, $b = \mathbf{2}$, $c = \mathbf{-2}$

$= \dfrac{-(2) \pm \sqrt{(2)^2 - 4(1)(-2)}}{2(1)} = \dfrac{-2 \pm \sqrt{4 + 8}}{2}$

$= \dfrac{-2 \pm \sqrt{12}}{2} = \dfrac{-2 \pm 2\sqrt{3}}{2} = -1 \pm \sqrt{3}$

The x-intercepts are $\left(-1 + \sqrt{3}, 0\right)$ and $\left(-1 - \sqrt{3}, 0\right)$.

CHECK YOUR PROGRESS 2   Find the x-intercepts of the graph of the parabola given by the equation.

a. $y = 2x^2 - 5x + 2$ **b.** $y = x^2 + 4x + 4$

Solution   *See page S22.*

Minimum and Maximum of a Quadratic Function

Note that for the graphs below, when $a > 0$, the vertex is the point with the minimum y-coordinate. When $a < 0$, the vertex is the point with the maximum y-coordinate.

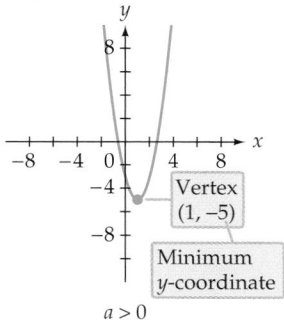

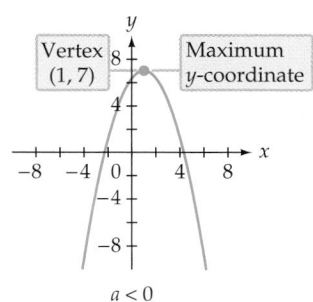

Finding the minimum or maximum value of a quadratic function is a matter of finding the vertex of the graph of the function.

EXAMPLE 3 ■ **Find the Minimum or Maximum Value of a Quadratic Function**

Find the maximum value of $f(x) = -2x^2 + 4x + 3$.

Solution

$$x = -\frac{b}{2a} = -\frac{4}{2(-2)} = 1$$

- Find the x-coordinate of the vertex.
  $a = -2, b = 4$

$$f(x) = -2x^2 + 4x + 3$$

- Find the y-coordinate of the vertex by replacing x by **1** and solving for y.

$$f(1) = -2(1)^2 + 4(1) + 3$$
$$f(1) = 5$$

The vertex is $(1, 5)$.

The maximum value of the function is 5, the y-coordinate of the vertex.

CHECK YOUR PROGRESS 3 Find the minimum value of $f(x) = 2x^2 - 3x + 1$.

Solution See page S23.

QUESTION *The vertex of a parabola that opens up is $(-4, 7)$. What is the minimum value of the function?*

Applications of Quadratic Functions

EXAMPLE 4 ■ **Application of Finding the Minimum of a Quadratic Function**

A mining company has determined that the cost c, in dollars per ton, of mining a mineral is given by $c(x) = 0.2x^2 - 2x + 12$, where x is the number of tons of the mineral that are mined. Find the number of tons of the mineral that should be mined to minimize the cost. What is the minimum cost?

Solution
To find the number of tons of the mineral that should be mined to minimize the cost and to find the minimum cost, find the x- and y-coordinates of the vertex of the graph of $c(x) = 0.2x^2 - 2x + 12$.

$$x = -\frac{b}{2a} = -\frac{-2}{2(0.2)} = 5$$

- Find the x-coordinate of the vertex.
  $a = 0.2, b = -2$

ANSWER *The minimum value of the function is 7, the y-coordinate of the vertex.*

To minimize the cost, 5 tons of the mineral should be mined.

$$c(x) = 0.2x^2 - 2x + 12$$ • Find the *y*-coordinate of the vertex by replacing *x* by **5** and solving for *y*.

$$c(5) = 0.2(5)^2 - 2(5) + 12$$
$$c(5) = 7$$

The minimum cost per ton is $7.

CHECK YOUR PROGRESS 4   The height *s*, in feet, of a ball thrown straight up is given by $s(t) = -16t^2 + 64t + 4$, where *t* is the time in seconds. Find the time it takes the ball to reach its maximum height. What is the maximum height?

Solution   See page S23.

EXAMPLE 5 ■ **Application of Finding the Maximum of a Quadratic Function**

A lifeguard has 600 feet of rope with buoys attached to lay out a rectangular swimming area on a lake. If the beach forms one side of the rectangle, find the dimensions of the rectangle that will enclose the greatest swimming area.

Solution

Let *l* represent the length of the rectangle, let *w* represent the width of the rectangle, and let *A* (which is unknown) represent the area of the rectangle. See the figure at the left. Use these variables to write expressions for the perimeter and area of the rectangle.

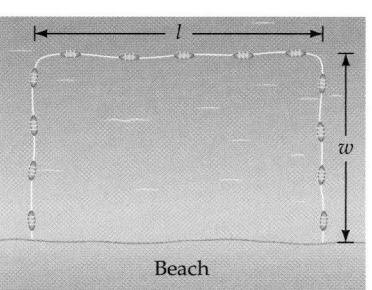

Perimeter: $w + l + w = 600$ • There are **600** feet of rope.
$$2w + l = 600$$

Area: $A = lw$

The goal is to maximize *A*. To do this, first write *A* in terms of a single variable. This can be accomplished by solving $2w + l = 600$ for *l* and then substituting into $A = lw$.

$$2w + l = 600$$ • Solve for *l*.
$$l = -2w + 600$$

$$A = lw$$
$$= (-2w + 600)w$$ • Substitute $-2w + 600$ for *l*.
$$A = -2w^2 + 600w$$ • Multiply. This is now a quadratic equation.
    $a = -2, b = 600$

Find the *w*-coordinate of the vertex.

$$w = -\frac{b}{2a} = -\frac{600}{2(-2)} = 150$$ • Find the *w*-coordinate of the vertex.
    $a = -2, b = 600$

The width is 150 feet. To find *l*, replace *w* by 150 in $l = -2w + 600$ and solve for *l*.

$$l = -2w + 600$$
$$l = -2(150) + 600 = -300 + 600 = 300$$

The length is 300 feet. The dimensions of the rectangle with maximum area are 150 feet by 300 feet.

CHECK YOUR PROGRESS 5   A mason is forming a rectangular floor for a storage shed. The perimeter of the rectangle is 44 feet. What dimensions will give the floor a maximum area?

Solution   *See page S23.*

Excursion

Reflective Properties of a Parabola

The fact that the graph of $y = ax^2 + bx + c$ is a parabola is based on the following geometric definition of a parabola.

> **Definition of a Parabola**
>
> A parabola is the set of points in the plane that are equidistant from a fixed line (the **directrix**) and a fixed point (the **focus**) not on the line.

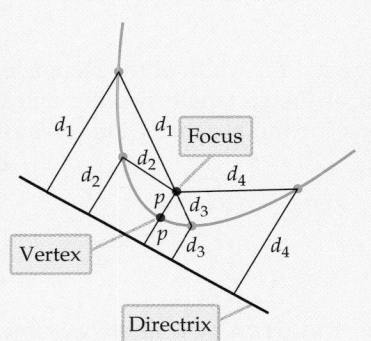

This geometric definition of a parabola is illustrated in the figure at the left. Basically, for a point to be on a parabola, the distance from the point to the focus must equal the distance from the point to the directrix. Note also that the vertex is halfway between the focus and the directrix. This distance is traditionally labeled p.

The equation of a parabola that opens up with vertex at the origin can be written in terms of the distance p between the vertex and focus as $y = \frac{1}{4p}x^2$. For this equation, the coordinates of the focus are $(0, p)$. For instance, to find the coordinates of the focus for $y = \frac{1}{4}x^2$, let $\frac{1}{4p} = \frac{1}{4}$ and solve for p.

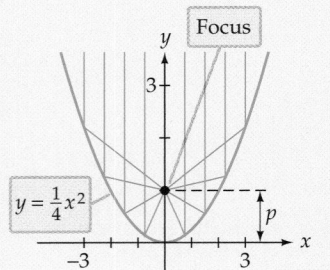

$$\frac{1}{4p} = \frac{1}{4}$$

$$1 = \frac{4p}{4} \qquad \text{• Multiply each side of the equation by } 4p.$$

$$1 = p \qquad \text{• Simplify the right side of the equation.}$$

The coordinates of the focus are $(0, 1)$.

1. Find the coordinates of the focus for the parabola whose equation is $y = 0.4x^2$.

Optical telescopes work on the same principle as radio telescopes (see Math Matters, p. 362) except that light hits a mirror that has been shaped into a paraboloid. The light is reflected to the focus, where another mirror reflects it through a lens to the observer. See the diagram at the top of the following page.

(continued)

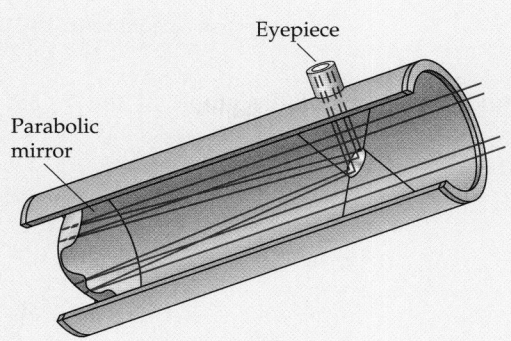

Eyepiece

Parabolic mirror

Palomar Observatory with the shutters open.

2. The telescope at the Palomar Observatory in California has a parabolic mirror. The diameter of the circle at the top of the parabolic mirror has a 200-inch diameter. An equation that approximates the parabolic surface of the mirror is $y = \frac{1}{2639}x^2$, where x and y are measured in inches. How far is the focus from the vertex of the mirror?

If a point on a parabola whose vertex is at the origin is known, then the equation of the parabola can be found. For instance, if $(4, 1)$ is a point on a parabola with vertex at the origin, then we can find the equation as follows.

$y = \dfrac{1}{4p}x^2$ • **Begin with the equation of the parabola.**

$1 = \dfrac{1}{4p}(4)^2$ • **The known point is (4, 1). Replace x by 4 and y by 1.**

$1 = \dfrac{4}{p}$ • **Solve for p.**

$p = 4$ • **$p = 4$ in the equation $y = \dfrac{1}{4p}x^2$.**

The equation of the parabola is $y = \frac{1}{16}x^2$.

Find a flashlight and measure the diameter of its lens cover and the depth of the reflecting parabolic surface. See the diagram below.

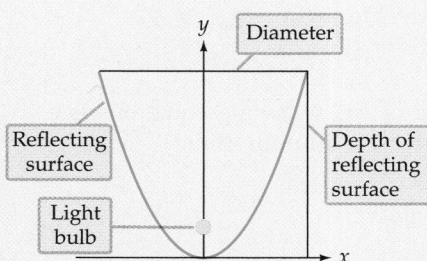

3. If coordinate axes are set up as shown, find the equation of the parabola.

4. Find the location of the focus. Explain why the light bulb should be placed at this point.

Exercise Set 6.4   (Suggested Assignment: 1–21, every other odd; 29–45, odds)

In Exercises 1–12, find the vertex of the graph of the equation.

1. $y = x^2 - 2$

2. $y = x^2 + 2$

3. $y = -x^2 - 1$

4. $y = -x^2 + 3$

5. $y = -\dfrac{1}{2}x^2 + 2$

6. $y = \dfrac{1}{2}x^2$

7. $y = 2x^2 - 1$

8. $y = x^2 - 2x$

9. $y = x^2 - x - 2$

10. $y = x^2 - 3x + 2$

11. $y = 2x^2 - x - 5$

12. $y = 2x^2 - x - 3$

In Exercises 13–24, find the x-intercepts of the parabola given by the equation.

13. $y = 2x^2 - 4x$

14. $y = 3x^2 + 6x$

15. $y = 4x^2 + 11x + 6$

16. $y = x^2 - 9$

17. $y = x^2 + 2x - 1$

18. $y = x^2 + 4x - 3$

19. $y = -x^2 - 4x - 5$

20. $y = -x^2 - 2x + 1$

21. $y = -x^2 + 4x + 1$

22. $y = x^2 + 6x + 10$

23. $y = 2x^2 - 5x - 3$

24. $y = x^2 - 2$

In Exercises 25–32, find the minimum or maximum value of the quadratic function. State whether the value is a minimum or a maximum.

25. $f(x) = x^2 - 2x + 3$

26. $f(x) = 2x^2 + 4x$

27. $f(x) = -2x^2 + 4x - 5$

28. $f(x) = 3x^2 + 3x - 2$

29. $f(x) = x^2 - 5x + 3$

30. $f(x) = 2x^2 - 3x$

31. $f(x) = -x^2 - x + 2$

32. $f(x) = -3x^2 + 4x - 2$

33. The graph of which of the following equations is a parabola with the largest minimum value?

    a. $y = x^2 - 2x - 3$

    b. $y = x^2 - 10x + 20$

    c. $y = 3x^2 - 1$

34. The height s, in feet, of a rock thrown upward at an initial speed of 64 feet per second from a cliff 50 feet above an ocean beach is given by the function

$$s(t) = -16t^2 + 64t + 50$$

where t is the time in seconds. Find the maximum height above the beach that the rock will attain.

35. The height s, in feet, of a ball thrown upward at an initial speed of 80 feet per second from a platform 50 feet high is given by

$$s(t) = -16t^2 + 80t + 50$$

where t is the time in seconds. Find the maximum height above the ground that the ball will attain.

36. A manufacturer of microwave ovens estimates that the revenue R, in dollars, the company receives is related to the price p, in dollars, of an oven by the function

$$R(p) = 125p - 0.25p^2$$

What price per oven will give the maximum revenue?

37. A manufacturer of camera lenses estimates that the average monthly cost C of a lens is given by the function

$$C(x) = 0.1x^2 - 20x + 2000$$

where x is the number of lenses produced each month. Find the number of lenses the company should produce to minimize the average cost.

38. A pool is treated with a chemical to reduce the number of algae. The number of algae in the pool t days after the treatment can be approximated by the function

$$A(t) = 40t^2 - 400t + 500$$

How many days after treatment will the pool have the least number of algae?

39. The suspension cable that supports a very small footbridge hangs in the shape of a parabola. The height h, in feet, of the cable above the bridge is given by the function

$$h(x) = 0.25x^2 - 0.8x + 25, 0 < x < 3.2$$

where x is the distance in feet from one end of the bridge. What is the minimum height of the cable above the bridge?

40. The net annual income I, in dollars, of a family physician can be modeled by the equation

$$I(x) = -290(x - 48)^2 + 148{,}000$$

where x is the age of the physician and $27 \le x \le 70$. Find (a) the age at which the physician's income will be a maximum and (b) the maximum income.

41. Karen is throwing an orange to her brother Saul, who is standing on the balcony of their home. The height h, in feet, of the orange above the ground t seconds after it is thrown is given by

$$h(t) = -16t^2 + 32t + 4, 0 \le t \le 2.118$$

If Saul's outstretched arms are 18 feet above the ground, will the orange ever be high enough so that he can catch it?

42. Some football fields are built in a parabolic-mound shape so that water will drain off the field. A model for the shape of such a field is given by

$$h(x) = -0.00023475x^2 + 0.0375x$$

where h is the height of the field in feet at a distance of x feet from the sideline and $0 \leq x \leq 159.744$. What is the maximum height of the field? Round to the nearest tenth.

43. The Clarence Buckingham Memorial Fountain in Chicago shoots water from a nozzle at the base of the fountain. The height h, in feet, of the water above the ground t seconds after it leaves the nozzle is given by

$$h(t) = -16t^2 + 90t + 15$$

What is the maximum height of the water spout? Round to the nearest tenth.

44. On wet concrete, the stopping distance s, in feet, of a car traveling v miles per hour is given by

$$s(v) = 0.055v^2 + 1.1v$$

At what maximum speed could a car be traveling on wet concrete and still stop at a stop sign 44 feet away?

45. The fuel efficiency of an average car is given by the equation

$$E(v) = -0.018v^2 + 1.476v + 3.4$$

where E is the fuel efficiency in miles per gallon and v is the speed of the car in miles per hour.

a. What speed will yield the maximum fuel efficiency?

b. What is the maximum fuel efficiency?

46. A rancher has 200 feet of fencing to build a rectangular corral alongside an existing fence. Determine the dimensions of the corral that will maximize the enclosed area.

Extensions

CRITICAL THINKING

In Exercises 47 and 48, find the value of k such that the graph of the equation contains the given point.

47. $f(x) = x^2 - 3x + k$; $(2, 5)$

48. $f(x) = 2x^2 + kx - 3$; $(4, -3)$

49. For $f(x) = 2x^2 - 5x + k$, we have $f\left(-\frac{3}{2}\right) = 0$. Find another value of x for which $f(x) = 0$.

50. For what values of k does the graph of $f(x) = 2x^2 - kx + 8$ just touch the x-axis without crossing it?

EXPLORATIONS

A real number x is called a **zero of a function** if the function evaluated at x is 0. That is, if $f(x) = 0$, then x is called a zero of the function. For instance, evaluating $f(x) = x^2 + x - 6$ when $x = -3$, we have

$$f(x) = x^2 + x - 6$$
$$f(-3) = (-3)^2 + (-3) - 6 \qquad \text{• Replace } x \text{ by } -3.$$
$$f(-3) = 9 - 3 - 6 = 0$$

For this function, $f(-3) = 0$, so -3 is a zero of the function.

51. Verify that 2 is a zero of $f(x) = x^2 + x - 6$ by showing that $f(2) = 0$.

The graph of $f(x) = x^2 + x - 6$ is shown below. Note that the graph crosses the x-axis at -3 and 2, the two zeros of the function. The points $(-3, 0)$ and $(2, 0)$ are x-intercepts of the graph.

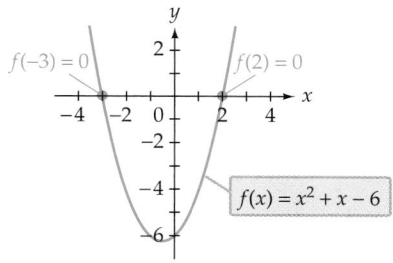

Consider the equation $0 = x^2 + x - 6$, which is $f(x) = x^2 + x - 6$ with $f(x)$ replaced by 0. By solving $0 = x^2 + x - 6$, we get

$$0 = x^2 + x - 6$$
$$0 = (x + 3)(x - 2) \qquad \text{• Solve by factoring.}$$

$$x + 3 = 0 \qquad x - 2 = 0$$
$$x = -3 \qquad x = 2$$

Observe that the solutions of the equation are the zeros of the function. This important connection between the real zeros of a function, the x-intercepts of the graph, and the real solutions of an equation is the basis of using a graphing calculator to solve an equation. The method discussed below of solving a quadratic equation using a graphing calculator is based on a TI-83/TI-83 Plus calculator. Other calculators will necessitate a slightly different approach.

Approximate the solutions of $x^2 + 4x = 6$ by using a graphing calculator.

i. Write the equation in standard form.

$$x^2 + 4x = 6$$
$$x^2 + 4x - 6 = 0$$

Press $\boxed{\text{Y=}}$ and enter $x^2 + 4x - 6$ for Y1.

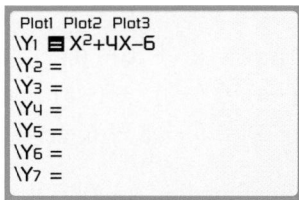

ii. Press $\boxed{\text{GRAPH}}$. If the graph does not appear on the screen, press $\boxed{\text{ZOOM}}$ 6.

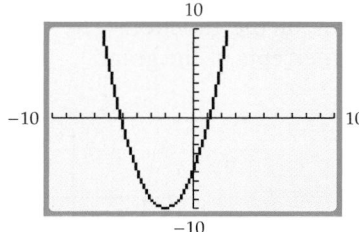

iii. Press $\boxed{\text{2nd}}$ CALC 2. Note that the second menu item is zero. This will begin the calculation of the zeros of the function, which are the solutions of the equation.

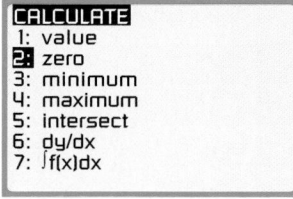

iv. At the bottom of the screen you will see LeftBound? This is asking you to move the blinking cursor so that it is to the *left* of the first x-intercept. Use the left arrow key to move the cursor to the left of the first x-intercept. The values of x and y that appear on your calculator may be different from the ones shown here. Just be sure you are to the left of the x-intercept. When you are done, press $\boxed{\text{ENTER}}$.

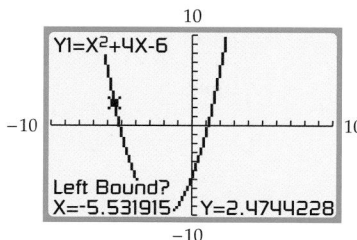

v. At the bottom of the screen you will see `RightBound?` This is asking you to move the blinking cursor so that it is to the *right* of the x-intercept. Use the right arrow key to move the cursor to the right of the x-intercept. The values of x and y that appear on your calculator may be different from the ones shown here. Just be sure you are to the right of the x-intercept. When you are done, press ENTER.

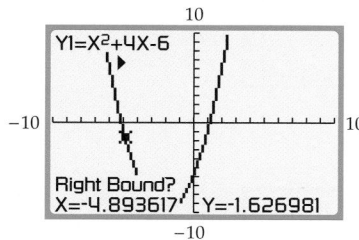

vi. At the bottom of the screen you will see `Guess?` This is asking you to move the blinking cursor so that it is close to the x-intercept. Use the arrow keys to move the cursor to the approximate x-intercept. The values of x and y that appear on your calculator may be different from the ones shown here. When you are done, press ENTER.

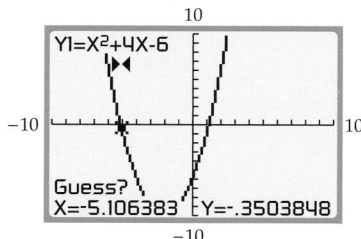

vii. The zero of the function is approximately -5.162278. Thus one approximate solution of $x^2 + 4x = 6$ is -5.162278. Note that the value of y is given as $Y1 = {}^-1E^-12$. This is how the calculator writes a number in scientific notation. We would normally write $Y1 = -1.0 \times 10^{-12}$. This number is very close to zero.

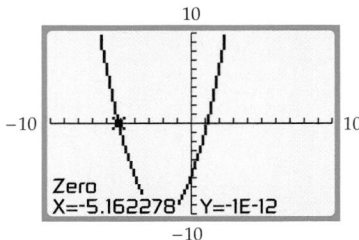

To find the other solution, repeat steps **iii** through **vi**. The screens are shown below.

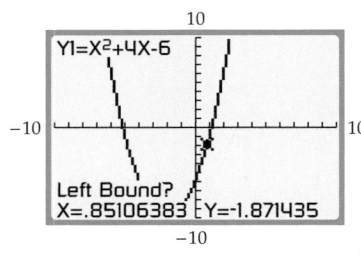

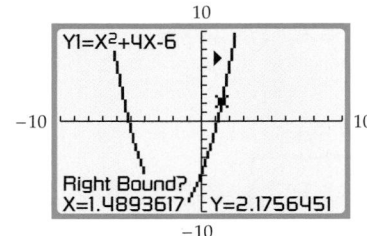

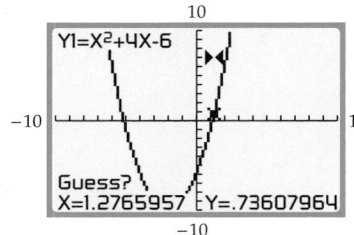

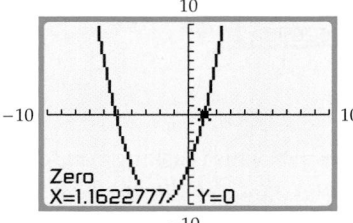

A second zero of the function is approximately 1.1622777. Thus the two solutions of $x^2 + 4x = 6$ are approximately -5.162278 and 1.1622777.

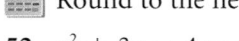 In Exercises 52–57, find the solutions of the equation. Round to the nearest hundredth.

52. $x^2 + 3x - 4 = 0$ **53.** $x^2 - 4x - 5 = 0$

54. $x^2 + 3.4x = 4.15$ **55.** $2x^2 - \dfrac{5}{9}x = \dfrac{3}{8}$

56. $\pi x^2 - \sqrt{17}x - 2 = 0$ **57.** $\sqrt{2}x^2 + x - \sqrt{7} = 0$

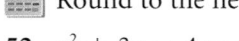 In Exercises 58–60, find the real solutions of the equation.

58. $x^3 - 2x^2 - 5x + 6 = 0$

59. $x^3 + 2x^2 - 11x - 12 = 0$

60. $x^4 + 2x^3 - 13x^2 - 14x + 24 = 0$

This method of finding solutions can be extended to find the real zeros, and therefore the real solutions, of any equation that can be graphed.

Exponential Functions and Their Applications

Introduction to Exponential Functions

In 1965, Gordon Moore, one of the cofounders of Intel Corporation, observed that the maximum number of transistors that could be placed on a microprocessor seemed to be doubling every 18 to 24 months. The table below shows how the maximum number of transistors on various Intel processors has changed over time. (*Source:* **www.intel.com/research/silicon/mooreslaw.htm**)

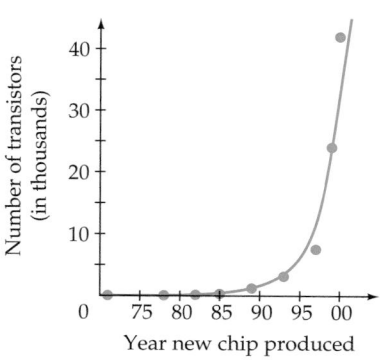

Figure 6.1

Year	1971	1978	1982	1985	1989	1993	1997	1999	2000
Number of transistors per microprocessor (in thousands)	2.3	29	120	275	1180	3100	7500	24,000	42,000

The curve in Figure 6.1 that approximately passes through the points is the graph of a mathematical model of the data. The model is an *exponential function*. Because y is increasing (growing), it is an *exponential growth function*.

When light enters water, the intensity of the light decreases with the depth of the water. The graph in Figure 6.2 shows a model, for Lake Michigan, of the decrease in the percent of light available as the depth of the water increases. This model is also an exponential function. In this case y is decreasing (decaying), and the model is an *exponential decay function*.

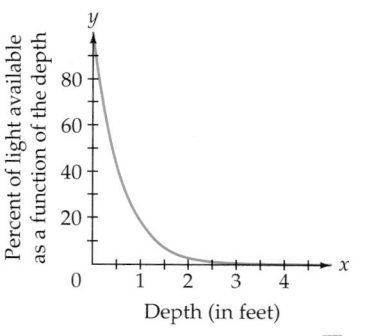

Figure 6.2

> The **exponential function** is defined by $f(x) = b^x$, where b is called the base, $b > 0$, $b \neq 1$, and x is any real number.

The base b of $f(x) = b^x$ must be positive. If the base were a negative number, the value of the function would not be a real number for some values of x. For instance, if $b = -4$ and $x = \frac{1}{2}$, then $f\left(\frac{1}{2}\right) = (-4)^{1/2} = \sqrt{-4}$, which is not a real number. Also, $b \neq 1$ because when $b = 1$, $b^x = 1^x = 1$, a constant function.

At the right we evaluate $f(x) = 2^x$ for $x = 3$ and $x = -2$.

$$f(x) = 2^x$$
$$f(3) = 2^3 = 8$$
$$f(-2) = 2^{-2} = \frac{1}{2^2} = \frac{1}{4}$$
$$f\left(\sqrt{2}\right) = 2^{\sqrt{2}} \approx 2^{1.4142} \approx 2.6651$$

✔ **TAKE NOTE**

If you need to review material on fractional exponents or the square root of a negative number, see Section 9.1 on the CD that you received with this book.

To evaluate the exponential function $f(x) = 2^x$ for an irrational number such as $x = \sqrt{2}$, we use a rational approximation of $\sqrt{2}$ (for instance, 1.4142) and a calculator to obtain an approximation of the function.

CALCULATOR NOTE

To evaluate $f(\pi)$ we used a calculator. For a scientific calculator, enter

3 $\boxed{y^x}$ π $\boxed{=}$

For the TI-83 graphing calculator, enter

3 $\boxed{\wedge}$ $\boxed{2nd}$ π $\boxed{ENTER}$

EXAMPLE 1 ■ Evaluate an Exponential Function

Evaluate $f(x) = 3^x$ at $x = 2$, $x = -4$, and $x = \pi$. If necessary, round to the nearest hundred thousandth.

Solution

$$f(2) = 3^2 = 9$$

$$f(-4) = 3^{-4} = \frac{1}{3^4} = \frac{1}{81}$$

$$f(\pi) = 3^\pi \approx 3^{3.1415927} \approx 31.54428$$

CHECK YOUR PROGRESS 1   Evaluate $g(x) = \left(\frac{1}{2}\right)^x$ when $x = 3$, $x = -1$, and $x = \sqrt{3}$. If necessary, round to the nearest thousandth.

Solution   See page S23.

Graphs of Exponential Functions

INSTRUCTOR NOTE
It is important for students to distinguish between $p(x) = x^2$ and $f(x) = 2^x$. The first is a polynomial function; the second is an exponential function. Polynomial functions have a variable base and a constant exponent, while exponential functions are characterized by a constant base and a variable exponent. You might graph $g(x) = x^2$ so that students can see the difference between the graph of $g(x) = x^2$ and the graph of $f(x) = 2^x$.

The graph of $f(x) = 2^x$ is shown below. The coordinates of some of the points on the graph are given in the table.

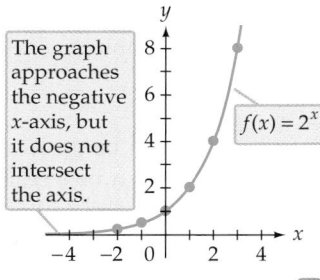

The graph approaches the negative x-axis, but it does not intersect the axis.

Figure 6.3

x	$f(x) = 2^x$	(x, y)
-2	$f(-2) = 2^{-2} = \dfrac{1}{4}$	$\left(-2, \dfrac{1}{4}\right)$
-1	$f(-1) = 2^{-1} = \dfrac{1}{2}$	$\left(-1, \dfrac{1}{2}\right)$
0	$f(0) = 2^0 = 1$	$(0, 1)$
1	$f(1) = 2^1 = 2$	$(1, 2)$
2	$f(2) = 2^2 = 4$	$(2, 4)$
3	$f(3) = 2^3 = 8$	$(3, 8)$

Observe that the values of *y increase* as *x* increases. This is an exponential growth function. This is typical of the graph of all exponential functions for which the base is *greater than* 1. For the function $f(x) = 2^x$, $b = 2$, which is greater than 1.

Now consider the graph of an exponential function for which the base is between 0 and 1. The graph of $f(x) = \left(\frac{1}{2}\right)^x$ is shown on the following page. The coordinates of some of the points on the graph are given in the table.

x	$f(x) = \left(\dfrac{1}{2}\right)^x$	(x, y)
−3	$f(-3) = \left(\dfrac{1}{2}\right)^{-3} = 8$	(−3, 8)
−2	$f(-2) = \left(\dfrac{1}{2}\right)^{-2} = 4$	(−2, 4)
−1	$f(-1) = \left(\dfrac{1}{2}\right)^{-1} = 2$	(−1, 2)
0	$f(0) = \left(\dfrac{1}{2}\right)^{0} = 1$	(0, 1)
1	$f(1) = \left(\dfrac{1}{2}\right)^{1} = \dfrac{1}{2}$	$\left(1, \dfrac{1}{2}\right)$
2	$f(2) = \left(\dfrac{1}{2}\right)^{2} = \dfrac{1}{4}$	$\left(2, \dfrac{1}{4}\right)$

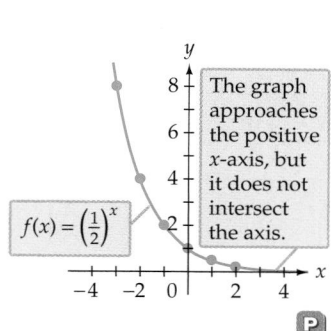

The graph approaches the positive x-axis, but it does not intersect the axis.

$f(x) = \left(\dfrac{1}{2}\right)^x$

Figure 6.4

Observe that the values of *y decrease* as *x* increases. This is an exponential decay function. This is typical of the graph of all exponential functions for which the positive base is *less than* 1. For the function $f(x) = \left(\dfrac{1}{2}\right)^x$, $b = \dfrac{1}{2}$, which is less than 1.

QUESTION   *Is $f(x) = 0.25^x$ an exponential growth or exponential decay function?*

EXAMPLE 2 ■ Graph an Exponential Function

State whether $g(x) = \left(\dfrac{3}{4}\right)^x$ is an exponential growth function or an exponential decay function. Then graph the function.

Solution
Because the base $\dfrac{3}{4}$ is less than 1, *g* is an exponential decay function. Because it is an exponential decay function, the *y*-values will decrease as *x* increases. The *y*-intercept of the graph is the point (0, 1), and the graph also passes through $\left(1, \dfrac{3}{4}\right)$. Plot a few additional points. (See the table on the following page.) Then draw a smooth curve through the points, as shown below.

ANSWER   *The base is 0.25, which is less than 1. It is an exponential decay function.*

x	$g(x) = \left(\dfrac{3}{4}\right)^x$	(x, y)
-3	$\left(\dfrac{3}{4}\right)^{-3} = \dfrac{64}{27}$	$\left(-3, \dfrac{64}{27}\right)$
-2	$\left(\dfrac{3}{4}\right)^{-2} = \dfrac{16}{9}$	$\left(-2, \dfrac{16}{9}\right)$
-1	$\left(\dfrac{3}{4}\right)^{-1} = \dfrac{4}{3}$	$\left(-1, \dfrac{4}{3}\right)$
0	$\left(\dfrac{3}{4}\right)^{0} = 1$	$(0, 1)$
1	$\left(\dfrac{3}{4}\right)^{1} = \dfrac{3}{4}$	$\left(1, \dfrac{3}{4}\right)$
2	$\left(\dfrac{3}{4}\right)^{2} = \dfrac{9}{16}$	$\left(2, \dfrac{9}{16}\right)$
3	$\left(\dfrac{3}{4}\right)^{3} = \dfrac{27}{64}$	$\left(3, \dfrac{27}{64}\right)$

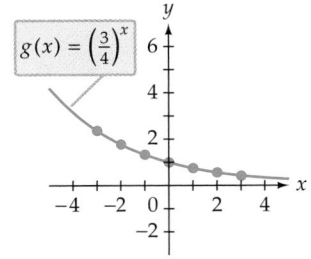

$g(x) = \left(\dfrac{3}{4}\right)^x$

CHECK YOUR PROGRESS 2   State whether $f(x) = \left(\dfrac{3}{2}\right)^x$ is an exponential growth function or an exponential decay function. Then graph the function.

Solution   See page S23.

The Natural Exponential Function

The irrational number π is often used in applications that involve circles. Another irrational number, denoted by the letter e, is useful in applications that involve growth or decay.

n	$\left(1 + \dfrac{1}{n}\right)^n$
10	2.59374246
100	2.70481383
1000	2.71692393
10,000	2.71814593
100,000	2.71826824
1,000,000	2.71828047

Definition of e

The number e is defined as the number that

$$\left(1 + \frac{1}{n}\right)^n$$

approaches as n increases without bound.

The letter e was chosen in honor of the Swiss mathematician Leonhard Euler. He was able to compute the value of e to several decimal places by evaluating $\left(1 + \frac{1}{n}\right)^n$ for large values of n, as shown at the left. The value of e accurate to eight decimal places is 2.71828183.

> **The Natural Exponential Function**
>
> For all real numbers x, the function defined by $f(x) = e^x$ is called the **natural exponential function**.

A calculator with an e^x key can be used to evaluate e^x for specific values of x. For instance,

$$e^2 \approx 7.389056, \qquad e^{3.5} \approx 33.115452, \qquad \text{and} \qquad e^{-1.4} \approx 0.246597$$

The graph of the natural exponential function can be constructed by plotting a few points or by using a graphing utility.

EXAMPLE 3 ■ Graph a Natural Exponential Function

 Graph $f(x) = e^x$.

Solution
Use a calculator to find range values for a few domain values. The range values in the table below have been rounded to the nearest tenth.

x	-2	-1	0	1	2
$f(x) = e^x$	0.1	0.4	1.0	2.7	7.4

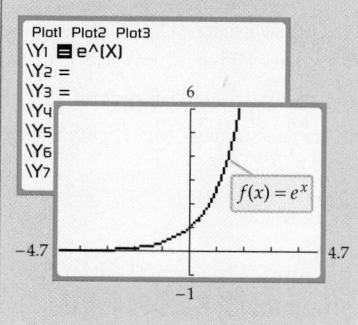
Plot the points given in the table and then connect the points with a smooth curve. Because $e > 1$, as x increases, e^x increases. Thus the values of y increase as x increases. As x decreases, e^x becomes closer to zero. For instance, when $x = -5$, $e^{-5} \approx 0.0067$. Thus as x decreases, the graph gets closer and closer to the x-axis. The y-intercept is $(0, 1)$.

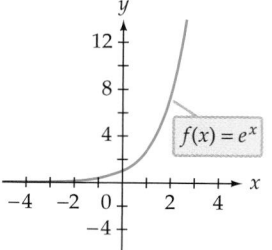

In the figure at the right, compare the graph of $f(x) = e^x$ with the graphs of $g(x) = 2^x$ and $h(x) = 3^x$. Because $2 < e < 3$, the graph of $f(x) = e^x$ is between the graphs of g and h.

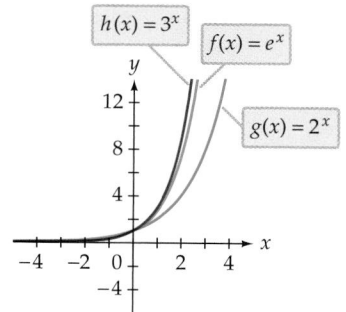

CHECK YOUR PROGRESS 3 Graph $f(x) = e^{-x} + 2$.

Solution See page S23.

Applications of Exponential Functions

Many applications can be effectively modeled by an exponential function. For instance, when money is deposited into a compound interest account, the value of the money can be represented by an exponential growth function. When physicians use a test that involves a radioactive element, the amount of radioactivity remaining in the patient's body can be modeled by an exponential decay function.

EXAMPLE 4 ■ Application of an Exponential Function

When an amount of money P is placed in an account that earns compound interest, the value A of the money after t years is given by the compound interest formula

$$A = P\left(1 + \frac{r}{n}\right)^{nt}$$

where r is the annual interest rate as a decimal and n is the number of compounding periods per year. Suppose $500 is placed in an account that earns 8% interest compounded daily. Find the value of the investment after 5 years.

Solution
Use the compound interest formula. Because interest is compounded daily, $n = 365$.

$$A = P\left(1 + \frac{r}{n}\right)^{nt}$$

$$= 500\left(1 + \frac{0.08}{365}\right)^{365(5)} \qquad \bullet\ P = 500,\ r = 0.08,\ n = 365,\ t = 5$$

$$\approx 500(1.491759) \approx 745.88$$

After 5 years, there is $745.88 in the account.

CHECK YOUR PROGRESS 4 The radioactive isotope iodine-131 is used to monitor thyroid activity. The number of grams N of iodine-131 in the body t hours after an injection is given by $N(t) = 1.5\left(\frac{1}{2}\right)^{t/193.7}$. Find the number of grams of the isotope in the body 24 hours after an injection. Round to the nearest ten-thousandth.

Solution See page S24.

The next example is based on Newton's Law of Cooling. This exponential function can be used to model the temperature of something that is being cooled.

EXAMPLE 5 ■ Application of an Exponential Function

A cup of coffee is heated to 160°F and placed in a room that maintains a temperature of 70°F. The temperature of the coffee after t minutes is given by $T = 70 + 90e^{-0.0485t}$. Find the temperature of the coffee 20 minutes after it is placed in the room. Round to the nearest degree.

Solution

Evaluate the function $T = 70 + 90e^{-0.0485t}$ for $t = 20$.

$$T = 70 + 90e^{-0.0485t}$$
$$= 70 + 90e^{-0.0485(20)} \quad \bullet \text{ Substitute } \mathbf{20} \text{ for } t.$$
$$\approx 70 + 34.1$$
$$\approx 104.1$$

After 20 minutes the temperature of the coffee is about 104°F.

CHECK YOUR PROGRESS 5 The function $A(t) = 200e^{-0.014t}$ gives the amount of aspirin, in milligrams, in a patient's bloodstream t minutes after the aspirin has been administered. Find the amount of aspirin in the patient's bloodstream after 45 minutes. Round to the nearest milligram.

Solution *See page S24.*

Math Matters   Marie Curie

In 1903, Marie Sklodowska-Curie (1867–1934) became the first woman to receive a Nobel prize in physics. She was awarded the prize along with her husband, Pierre Curie, and Henri Becquerel for their discovery of radioactivity. In fact, Marie Curie coined the word *radioactivity.*

In 1911, Marie Curie became the first person to win a second Nobel prize, this time alone and in chemistry, for the isolation of radium. She also discovered the element polonium, which she named after Poland, her birthplace. The radioactive phenomena that she studied are modeled by exponential functions.

The stamp at the left was printed in 1938 to commemorate Pierre and Marie Curie. When the stamp was issued, there was a surcharge added to the regular price of the stamp. The additional revenue was donated to cancer research. Marie Curie died in 1934 from leukemia, which was caused by her constant exposure to radiation from the radioactive elements she studied.

In 1935, one of Marie Curie's daughters, Irene Joloit-Curie, won a Nobel prize in chemistry.

Exponential Regression

Earlier in this chapter we examined linear regression models. In some cases, an exponential function may more closely model a set of data.

▼ **point of interest**

The Hope Diamond, shown below, is the world's largest deep-blue diamond. It weighs 45.52 carats. We should not expect the function $y \approx 4067.6(1.3816)^x$ in Example 6 to yield an accurate value of the Hope Diamond because the Hope Diamond is not the same type of diamond as the diamonds in Example 6, and its weight is much larger.

The Hope Diamond is on display at the Smithsonian Museum of Natural History in Washington, D.C.

EXAMPLE 6 ■ Find an Exponential Regression Equation

A diamond merchant has determined the values of several white diamonds that have different weights, measured in carats, but are similar in quality. See the table below.

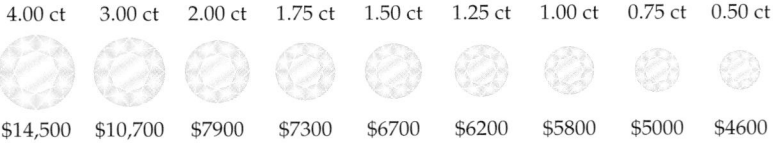

4.00 ct	3.00 ct	2.00 ct	1.75 ct	1.50 ct	1.25 ct	1.00 ct	0.75 ct	0.50 ct
$14,500	$10,700	$7900	$7300	$6700	$6200	$5800	$5000	$4600

Find an exponential function that models the values of the diamonds as a function of their weights and use the model to predict the value of a 3.5-carat diamond of similar quality.

Solution

Use a graphing calculator to find the regression equation. The calculator display below shows that the exponential regression equation is $y \approx 4067.6(1.3816)^x$, where x is the carat weight of the diamond and y is the value of the diamond.

```
ExpReg
 y = a*b^x
 a = 4067.641145
 b = 1.381644186
 r2 = .994881215
 r = .9974373238
```

To use the regression equation to predict the value of a 3.5-carat diamond of similar quality, substitute 3.5 for x and evaluate.

$$y \approx 4067.6(1.3816)^x$$
$$y \approx 4067.6(1.3816)^{3.5}$$
$$y \approx 12{,}609$$

According to the modeling function, the value of a 3.5-carat diamond of similar quality is $12,609.

Notice on the calculator screen above that the correlation coefficient is about 0.9974, which is very close to 1. This indicates that the equation $y \approx 4067.6(1.3816)^x$ provides a good fit for the data.

CHECK YOUR PROGRESS 6 The table below shows Earth's atmospheric pressure P at an altitude of a kilometers. Find an exponential function that models the atmospheric pressure as a function of altitude. Use the function to estimate, to the nearest tenth, the atmospheric pressure at an altitude of 24 kilometers.

Altitude, a, above sea level, in kilometers	Atmospheric pressure, P, in newtons per square centimeter
0	10.3
2	8.0
4	6.4
6	5.1
8	4.0
10	3.2
12	2.5
14	2.0
16	1.6
18	1.3

Solution   *See page S24.*

 **Excursion**

Chess and Exponential Functions

According to legend, when Sissa Ben Dahir of India invented the game of chess, King Shirham was so impressed with the game that he summoned the game's inventor and offered him the reward of his choosing. The inventor pointed to the chessboard and requested, for his reward, one grain of wheat on the first square, two grains of wheat on the second square, four grains on the third square, eight grains on the fourth square, and so on for all 64 squares on the chessboard. The king considered this a very modest reward and said he would grant the inventor's wish.

1. This portion of this Excursion will enable you to find a formula for the amount of wheat on the nth square of the chessboard. You may want to use the chart below as you answer the questions. It may help you to see a pattern.

Square number, n	1	2	3	4	5	6	7
Number of grains of wheat on square n	1	2	4				
Total number of grains of wheat on squares 1 through n	1	1 + 2 =	3 + 4 =				

a. How many grains of wheat are on each of the first seven squares?

(continued)

What is the total number (the sum) of grains of wheat on the first

   b. two squares?

   c. three squares?

   d. four squares?

   e. five squares?

   f. six squares?

   g. seven squares?

2. Use inductive reasoning to find a function that gives the total number (the sum) of grains of wheat on the first n squares of the chessboard. Test your function to ensure that it works for parts b through g.

3. If all 64 squares of the chessboard are piled with wheat as requested by Sissa Ben Dahir, how many grains of wheat are on the board?

4. A grain of wheat weighs approximately 0.000008 kilogram. Find the total weight of the wheat requested by Sissa Ben Dahir.

5. In a recent year, a total of 6.5×10^8 metric tons of wheat were produced in the world. At this level, how many years of wheat production would be required to fill the request of Sissa Ben Dahir? One metric ton equals 1000 kilograms.

Exercise Set 6.5   (Suggested Assignment: 1–39, odds)

1. Given $f(x) = 3^x$, evaluate:

   a. $f(2)$ **b.** $f(0)$ **c.** $f(-2)$

2. Given $H(x) = 2^x$, evaluate:

   a. $H(-3)$ **b.** $H(0)$ **c.** $H(2)$

3. Given $g(x) = 2^{x+1}$, evaluate:

   a. $g(3)$ **b.** $g(1)$ **c.** $g(-3)$

4. Given $F(x) = 3^{x-2}$, evaluate:

   a. $F(-4)$ **b.** $F(-1)$ **c.** $F(0)$

5. Given $G(r) = \left(\frac{1}{2}\right)^{2r}$, evaluate:

   a. $G(0)$ **b.** $G\left(\frac{3}{2}\right)$ **c.** $G(-2)$

6. Given $R(t) = \left(\frac{1}{3}\right)^{3t}$, evaluate:

   a. $R\left(-\frac{1}{3}\right)$ **b.** $R(1)$ **c.** $R(-2)$

7. Given $h(x) = e^{x/2}$, evaluate the following. Round to the nearest ten-thousandth.

   a. $h(4)$ **b.** $h(-2)$ **c.** $h\left(\frac{1}{2}\right)$

8. Given $f(x) = e^{2x}$, evaluate the following. Round to the nearest ten-thousandth.

   a. $f(-2)$ **b.** $f\left(-\frac{2}{3}\right)$ **c.** $f(2)$

9. Given $H(x) = e^{-x+3}$, evaluate the following. Round to the nearest ten-thousandth.

   a. $H(-1)$ **b.** $H(3)$ **c.** $H(5)$

10. Given $g(x) = e^{-x/2}$, evaluate the following. Round to the nearest ten-thousandth.

   a. $g(-3)$ **b.** $g(4)$ **c.** $g\left(\frac{1}{2}\right)$

11. Given $F(x) = 2^{x^2}$, evaluate the following. Round to the nearest ten-thousandth.

   a. $F(2)$ **b.** $F(-2)$ **c.** $F\left(\frac{3}{4}\right)$

12. Given $Q(x) = 2^{-x^2}$, evaluate:

   a. $Q(3)$ **b.** $Q(-1)$ **c.** $Q(-2)$

13. Given $f(x) = e^{-x^2/2}$, evaluate the following. Round to the nearest ten-thousandth.

   a. $f(-2)$ **b.** $f(2)$ **c.** $f(-3)$

14. Given $h(x) = e^{-2x} + 1$, evaluate the following. Round to the nearest ten-thousandth.

   a. $h(-1)$ **b.** $h(3)$ **c.** $h(-2)$

In Exercises 15–24, graph the function.

15. $f(x) = 2^x + 1$

16. $f(x) = 3^x - 2$

17. $g(x) = 3^{x/2}$

18. $h(x) = 2^{-x/2}$

19. $f(x) = 2^{x+3}$

20. $g(x) = 4^{-x} + 1$

21. $H(x) = 2^{2x}$

22. $F(x) = 2^{-x}$

23. $f(x) = e^{-x}$

24. $y(x) = e^{2x}$

In Exercises 25 and 26, use the compound interest formula $A = P\left(1 + \dfrac{r}{n}\right)^{nt}$, where P is the amount deposited, A is the value of the money after t years, r is the annual interest rate as a decimal, and n is the number of compounding periods per year.

25. A computer network specialist deposits $2500 into a retirement account that earns 7.5% annual interest, compounded daily. What is the value of the investment after 20 years?

26. A $10,000 certificate of deposit (CD) earns 5% annual interest, compounded daily. What is the value of the investment after 20 years?

27. Some banks now use continuous compounding of an amount invested. In this case, the equation that models the value of an initial investment of A dollars in t years at an annual interest rate of r is given by $P = Ae^{rt}$. Using this equation, find the value in 5 years of an investment of $2500 that earns 5% annual interest.

28. An isotope of technetium is used to prepare images of internal body organs. This isotope has a half-life of approximately 6 hours. If a patient is injected with 30 milligrams of this isotope, what will be the technetium level in the patient after 3 hours? Use the formula $A = 30\left(\frac{1}{2}\right)^{t/6}$, where A is the technetium level, in milligrams, in the patient after t hours. Round to the nearest tenth.

29. Iodine-131 is an isotope that is used to study the functioning of the thyroid gland. This isotope has a half-life of approximately 8 days. If a patient is given an injection that contains 8 micrograms of iodine-131, what will be the iodine level in the patient after 5 days? Use the formula $A = 8\left(\frac{1}{2}\right)^{t/8}$, where A is the amount of the isotope, in micrograms, in the patient after t days. Round to the nearest tenth.

30. The percent of correct welds that a student can make will increase with practice and can be approximated by the equation $P = 100[1 - (0.75)^t]$, where P is the

percent of correct welds and t is the number of weeks of practice. Find the percent of correct welds that a student will make after 4 weeks of practice. Round to the nearest percent.

31. The "concert A" note on a piano is the first A below middle C. When that key is struck, the string associated with the key vibrates 440 times per second. The next A above concert A vibrates twice as fast. An exponential function with a base of two is used to determine the frequency of the 11 notes between the two A's. Find this function. *Hint:* The function is of the form $f(x) = k \cdot 2^{(cx)}$, where k and c are constants. Also $f(0) = 440$ and $f(12) = 880$.

32. Atmospheric pressure changes as you rise above Earth's surface. At an altitude of h kilometers, where $0 < h < 80$, the pressure P in newtons per square centimeter is approximately modeled by the equation $P(h) = 10.13e^{-0.116h}$.

a. What is the approximate pressure at 40 kilometers above Earth?

b. What is the approximate pressure on Earth's surface?

c. Does atmospheric pressure increase or decrease as you rise above Earth's surface?

33. The number of automobiles in the United States in 1900 was around 8000. In the year 2000, the number of automobiles in the United States reached 200 million. Find an exponential model for the data and use the model to predict the number of automobiles in the United Sates in 2010. Use $t = 0$ to represent the year 1900. Round to the nearest hundred thousand.

34. One estimate gives the world panda population as 3200 in 1980 and 590 in 2000. Find an exponential model for the data and use the model to predict the panda population in 2040. Use $t = 0$ to represent the year 1980. Round to the nearest whole number.

35. An initial amount of 100 micrograms of polonium decays to 75 micrograms in approximately 34.5 days. Find an exponential model for the amount of polonium in the sample after t days. Round to the nearest hundredth.

36.   The table below shows the number of ATMs in the United States for the years 1995 through 2000.

Year	1995	1996	1997	1998	1999	2000
ATMs	122,700	139,100	165,000	187,000	227,000	280,000

a. Find an exponential regression equation for these data, using $t = 0$ to represent 1995. Round to the nearest hundredth.

b. Use the equation to predict the number of ATMs in 2010.

37. The table below shows the saturation of water in air at various air temperatures.

Temperature (in °C)	0	5	10	20	25	30
Saturation (in millimeters of water per cubic meter of air)	4.8	6.8	9.4	17.3	23.1	30.4

a. Find an exponential regression equation for these data. Round to the nearest thousandth.

b. Use the equation to predict the number of milliliters of water per cubic meter of air at a temperature of 15°C. Round to the nearest tenth.

38. Artificial snow is made at a ski resort by combining air and water in a ratio that depends on the outside air temperature. The table below shows the rate of air flow needed for various temperatures.

Temperature (in °F)	0	5	10	15	20
Air flow (in cubic feet per minute)	3.0	3.6	4.7	6.1	9.9

a. Find an exponential regression equation for these data. Round to the nearest hundredth.

b. Use the equation to predict the air flow needed when the temperature is 25°F. Round to the nearest tenth.

Extensions

CRITICAL THINKING

An exponential model for population growth or decay can be accurate over a short period of time. However, this model begins to fail because it does not account for the natural resources necessary to support growth, nor does it account for death within the population. Another model, called the *logistic model,* can account for some of these effects. The logistic model is given by

$$P(t) = \frac{mP_0}{P_0 + (m - P_0)e^{-kt}},$$ where $P(t)$ is the population at time t, m is the maximum population that can be sup-

ported, P_0 is the population when $t = 0$, and k is a positive constant that is related to the growth of the population.

39. One model of Earth's population is given by

$$P(t) = \frac{280}{4 + 66e^{-0.021t}}.$$ In this equation, $P(t)$ is the population in billions and t is the number of years after 1980. Round answers to the nearest hundred million.

a. According to this model, what was Earth's population in the year 2000?

b. According to this model, what will be Earth's population in the year 2010?

c. If t is very large, say greater than 500, then $e^{-0.021t} \approx 0$. What does this suggest about the maximum population that Earth can support?

40. Biologists have determined that the maximum wolf population in a certain preserve is 1000 wolves. Suppose the population of wolves in the preserve in the year 2000 was 500, and that k is estimated to be 0.025.

a. Find a logistic function for the number of wolves in the preserve in year t, where t is the number of years after 2000.

b. Find the estimated wolf population in 2015.

EXPLORATIONS

41. The formula used to calculate a monthly lease payment or a monthly car payment (for a purchase rather than a lease) is given by $P = \dfrac{Ar(1 + r)^n - Vr}{(1 + r)^n - 1}$, where P is the monthly payment, A is the amount of the loan, r is the *monthly* interest rate as a decimal, n is the number of months of the loan or lease, and V is the residual value of the car at the end of the lease. For a car purchase, $V = 0$.

a. If the annual interest rate for a loan is 9%, what is the monthly interest rate as a decimal?

b. Write the formula for a monthly car payment when the car is purchased rather than leased.

c. Suppose you lease a car for 5 years. Find the monthly lease payment if the lease amount is $10,000, the residual value is $6000, and the annual interest rate is 6%.

d. Suppose you purchase a car and secure a 5-year loan for $10,000 at an annual interest rate of 6%. Find the monthly payment.

e. Why are the answers to parts c and d different?

The total amount C that has been repaid on a loan or lease is given by $C = \dfrac{(P - Ar)[(1 + r)^n - 1]}{r}$.

f. Using the lease payment in part c, find the total amount that will be repaid in 5 years. How much remains to be paid?

g. Using the monthly payment in part d, find the total amount that will be repaid in 5 years. How much remains to be paid?

h.  Explain why the answers to parts f and g make sense.

SECTION 6.6 | **Logarithmic Functions and Their Applications**

Introduction to Logarithms

Consider the equation $2^x = 32$. By trial and error, we find that when $x = 5$, $2^5 = 32$. Therefore, the solution of the equation is 5. Now consider the equation $2^x = 13$. Using trial and error again, we find that

$$2^3 = 8 \qquad \text{and} \qquad 2^4 = 16$$

Because 13 is between 8 and 16, we can conclude that x is between 3 and 4. To find its exact value, it would help to have a function that is the opposite of raising a number to a power. This function would find the power of 2 that produces 13.

Around the mid-sixteenth century, mathematicians created such a function, which we now call a *logarithmic function*. We write the solution of $2^x = 13$ as $x = \log_2 13$. This is read "x equals the logarithm base 2 of 13" and it means "x equals the power of 2 that produces 13." When logarithms were first introduced, tables were used to find a numerical value of x. Today, a calculator is used. Using a calculator, as we will show later, we can approximate the value of x as 3.7. This means that $2^{3.7} \approx 13$.

Exponential equations and logarithmic equations are closely related. For every exponential equation there is a corresponding logarithmic equation, and for every logarithmic equation there is a corresponding exponential equation. Here are some examples.

Exponential Equation	Logarithmic Equation
$2^5 = 32$	$\log_2 32 = 5$
$3^2 = 9$	$\log_3 9 = 2$
$5^{-2} = \dfrac{1}{25}$	$\log_5 \dfrac{1}{25} = -2$
$7^0 = 1$	$\log_7 1 = 0$

The preceding discussion is summarized in the following definition.

INSTRUCTOR NOTE
Some students see the definition of logarithm as artificial. The following may help.

Defining a logarithm as the inverse of the exponential function is similar to defining square root as the inverse of square.

"If $49 = x^2$, what is x?" The answer is "the square root of 49."

Now consider the question, "If $8 = 2^x$, what is x?" The answer is "the logarithm, base 2, of 8."

This analogy can be extended to motivate the need for tables or calculators.

"If $19 = x^2$, what is x?"
"If $21 = 10^x$, what is x?"

Definition of Logarithm

If $x > 0$ and b is a positive constant, $b \neq 1$, then

$$y = \log_b x \quad \text{if and only if} \quad b^y = x.$$

The notation $\log_b x$ is read "the logarithm (or log) base b of x."
The function defined by $f(x) = \log_b x$ is a **logarithmic function** with base b.

QUESTION   *Which of the following is the logarithmic form of $4^3 = 64$?*

   a. $\log_4 3 = 64$ *b.* $\log_3 4 = 64$ *c.* $\log_4 64 = 3$

The equation $y = \log_b x$ is the logarithmic form of $b^y = x$ and the equation $b^y = x$ is the *exponential form* of $y = \log_b x$. These two forms state exactly the same relationship between x and y.

EXAMPLE 1 ■ **Write a Logarithmic Equation in Exponential Form and an Exponential Equation in Logarithmic Form**

a. Write $2 = \log_{10}(x + 5)$ in exponential form.

b. Write $2^{3x} = 64$ in logarithmic form.

Solution
Use the definition of logarithm: $y = \log_b x$ if and only if $b^y = x$.

a. $2 = \log_{10}(x + 5)$ if and only if $10^2 = x + 5$.

b. $2^{3x} = 64$ if and only if $\log_2 64 = 3x$.

CHECK YOUR PROGRESS 1
a. Write $\log_2(4x) = 10$ in exponential form.

b. Write $10^3 = 2x$ in logarithmic form.

Solution   See page S24.

The relationship between the exponential and logarithmic forms can be used to evaluate some logarithms. The solutions to these types of problems are based on the Equality of Exponents Property.

Equality of Exponents Property

If $b > 0$ and $b^x = b^y$, then $x = y$.

ANSWER   *c.* $\log_4 64 = 3$ *is equivalent to* $4^3 = 64$.

EXAMPLE 2 ■ **Evaluate Logarithmic Expressions**

Evaluate the logarithms.

a. $\log_8 64$ **b.** $\log_2\left(\dfrac{1}{8}\right)$

Solution

a. $\log_8 64 = x$ • Write an equation.

$\quad\quad 8^x = 64$ • Write the equation in its equivalent exponential form.

$\quad\quad 8^x = 8^2$ • Write 64 in exponential form using 8 as the base.

$\quad\quad\quad x = 2$ • Solve for x using the Equality of Exponents Property.

$\log_8 64 = 2$

b. $\log_2\left(\dfrac{1}{8}\right) = x$ • Write an equation.

$\quad\quad 2^x = \dfrac{1}{8}$ • Write the equation in its equivalent exponential form.

$\quad\quad 2^x = 2^{-3}$ • Write $\dfrac{1}{8}$ in exponential form using 2 as the base.

$\quad\quad\quad x = -3$ • Solve for x using the Equality of Exponents Property.

$\log_2\left(\dfrac{1}{8}\right) = -3$

CHECK YOUR PROGRESS 2   Evaluate the logarithms.

a. $\log_{10} 0.001$ **b.** $\log_5 125$

Solution   *See page S24.*

✔ **TAKE NOTE**

If you need to review material on negative exponents, see Lesson 6.1B on the CD that you received with this book.

EXAMPLE 3 ■ **Solve a Logarithmic Equation**

Solve: $\log_3 x = 2$

Solution

$\log_3 x = 2$

$\quad\quad 3^2 = x$ • Write the equation in its equivalent exponential form.

$\quad\quad\quad 9 = x$ • Simplify the exponential expression.

CHECK YOUR PROGRESS 3   Solve: $\log_2 x = 6$

Solution   *See page S24.*

Not all logarithms can be evaluated by rewriting the logarithm in its equivalent exponential form and using the Equality of Exponents Property. For instance, if we tried to evaluate $\log_{10} 18$, it would be necessary to solve the equivalent exponential equation $10^x = 18$. The difficulty here is trying to rewrite 18 in exponential form with 10 as a base.

Common and Natural Logarithms

Two of the most frequently used logarithmic functions are *common logarithms*, which have base 10, and *natural logarithms*, which have base e (the base of the natural exponential function).

> **Common and Natural Logarithms**
>
> The function defined by $f(x) = \log_{10} x$ is called the **common logarithmic function.** It is customarily written without the base as $f(x) = \log x$.
>
> The function defined by $f(x) = \log_e x$ is called the **natural logarithmic function.** It is customarily written as $f(x) = \ln x$.

Most scientific or graphing calculators have a `log` key for evaluating common logarithms and a `ln` key to evaluate natural logarithms. For instance,

$$\log 24 \approx 1.3802112 \qquad \text{and} \qquad \ln 81 \approx 4.3944492$$

EXAMPLE 4 ■ Solve Common and Natural Logarithmic Equations

Solve each of the following equations. Round to the nearest thousandth.

a. $\log x = -1.5$ **b.** $\ln x = 3$

Solution

a. $\log x = -1.5$
 $10^{-1.5} = x$ • Write the equation in its equivalent exponential form.
 $0.032 \approx x$ • Simplify the exponential expression.

b. $\ln x = 3$
 $e^3 = x$ • Write the equation in its equivalent exponential form.
 $20.086 \approx x$ • Simplify the exponential expression.

CHECK YOUR PROGRESS 4   Solve each of the following equations. Round to the nearest thousandth.

a. $\log x = -2.1$ **b.** $\ln x = 2$

Solution   *See page S24.*

Math Matters   Zipf's Law

George Zipf (1902–1950) was a lecturer in German and philology (the study of the change in a language over time) at Harvard University. Zipf's Law, as originally formulated, referred to the frequency of occurrence of words in a book, magazine article, or other written material, and its rank. The word used most frequently had rank 1, the next most frequent word had rank 2, and so on. Zipf hypothesized that if the x-axis were the logarithm of a word's rank and the y-axis were the logarithm of the frequency of the word, then the graph of rank versus frequency would lie on approximately a straight line.

(continued)

Zipf's law has been extended to demographics. For instance, let the population of a U.S. city be P and its rank in population be R. The graph of the points $(\log R, \log P)$ lie on approximately a straight line. The graph below shows Zipf's Law as applied to the populations and ranks of the 12 most populated states according to the 2000 Census.

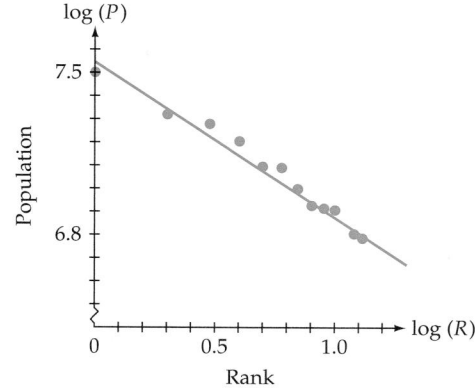

Recently, Zipf's Law has been applied to web site traffic, where web sites are ranked according to the number of hits they receive. Assumptions about web traffic are used by engineers and programmers who study ways to make the web more efficient.

Graphs of Logarithmic Functions

The graph of a logarithmic function can be drawn by first rewriting the function in its exponential form. This procedure is illustrated in Example 5.

EXAMPLE 5 ■ Graph a Logarithmic Function

Graph $f(x) = \log_3 x$.

Solution
To graph $f(x) = \log_3 x$, first write the equation in the form $y = \log_3 x$. Then write the equivalent exponential equation $x = 3^y$. Because this equation is solved for x, choose values of y and calculate the corresponding values of x, as shown in the table below.

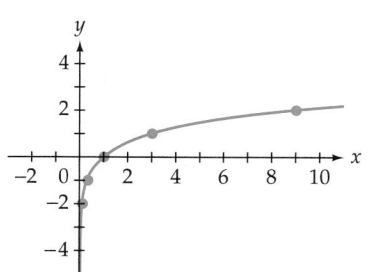

$x = 3^y$	$\frac{1}{9}$	$\frac{1}{3}$	1	3	9
y	-2	-1	0	1	2

Plot the ordered pairs and connect the points with a smooth curve, as shown at the left.

CHECK YOUR PROGRESS 5   Graph $f(x) = \log_5 x$.

Solution   *See page S24.*

The graphs of $y = \log x$ and $y = \ln x$ can be drawn on a graphing calculator by using the $\boxed{\texttt{log}}$ and $\boxed{\texttt{ln}}$ keys. The graphs are shown below for a TI-83 calculator.

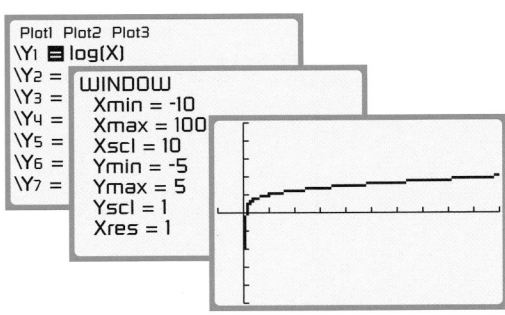

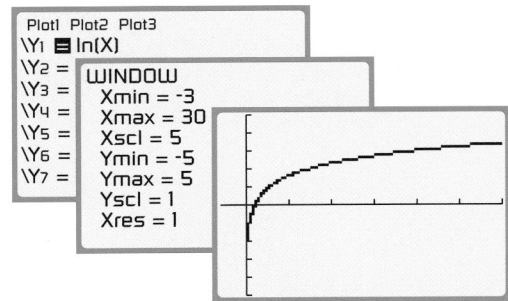

Applications of Logarithmic Functions

Many applications can be modeled by logarithmic functions.

EXAMPLE 6 ■ **Application of a Logarithmic Function**

During the years 1981 to 1999, the average time T of a major league baseball game tended to increase each year. If the year 1981 is represented by $x = 1$, then the function

$$T(x) = 149.57 + 7.63 \ln x$$

approximates the average time T, in minutes, of a major league baseball game for the years $x = 1$ to $x = 19$.

a. Use the function to determine the average time of a major league baseball game during the 1981 season and during the 1999 season.

b. By how much did the average time of a major league baseball game increase from 1981 to 1999?

Solution

a. The year 1981 is represented by $x = 1$ and the year 1999 by $x = 19$.

$$T(1) = 149.57 + 7.63 \ln(1) = 149.57$$

In 1981 the average time of a major league baseball game was about 149.57 minutes.

$$T(19) = 149.57 + 7.63 \ln(19) \approx 172.04$$

In 1999 the average time of a major league baseball game was about 172.04 minutes.

b. $T(19) - T(1) \approx 172.04 - 149.57 \approx 22.47$

From 1981 to 1999, the average time of a major league baseball game increased by about 22.47 minutes.

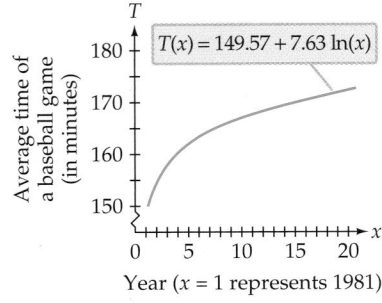

CHECK YOUR PROGRESS 6   The following function models the average typing speed S, in words per minute, of a student who has been typing for t months.

$$S(t) = 5 + 29 \ln(t + 1), \quad 0 \le t \le 9$$

a. Use the function to determine the student's average typing speed when the student first started to type and the student's average typing speed after 3 months.

b. By how much did the typing speed increase during the 3 months?

Solution   *See page S24.*

historical note

The Richter scale was created by the seismologist Charles Francis Richter (rĭk′tər) (1900–1985) in 1935.

Richter was born in Ohio. At the age of 16, he and his mother moved to Los Angeles, where he enrolled at the University of Southern California. He went on to study physics at Stanford University. Richter was a professor of seismology at the Seismological Laboratory at California Institute of Technology (Caltech) from 1936 until he retired in 1970. ■

Logarithmic functions are often used to convert very large or very small numbers into numbers that are easier to comprehend. For instance, the *Richter scale*, which measures the magnitude of an earthquake, uses a logarithmic function to scale the intensity of an earthquake's shock waves I into a number M, which for most earthquakes is in the range of 0 to 10. The intensity I of an earthquake is often given in terms of the constant I_0, where I_0 is the intensity of the smallest earthquake (called a **zero-level earthquake**) that can be measured on a seismograph near the earthquake's epicenter. The following formula is used to compute the Richter scale magnitude of an earthquake.

The Richter Scale Magnitude of an Earthquake

An earthquake with an intensity of I has a Richter scale magnitude of

$$M = \log\left(\frac{I}{I_0}\right)$$

where I_0 is the measure of the intensity of a zero-level earthquake.

EXAMPLE 7 ■ Find the Magnitude of an Earthquake

Find the Richter scale magnitude of the 1999 Joshua Tree, California earthquake, which had an intensity of $I = 12{,}589{,}254 I_0$. Round to the nearest tenth.

Solution

$$M = \log\left(\frac{I}{I_0}\right) = \log\left(\frac{12{,}589{,}254 I_0}{I_0}\right) = \log(12{,}589{,}254) \approx 7.1$$

The 1999 Joshua Tree earthquake had a Richter scale magnitude of 7.1.

✔ TAKE NOTE

Notice in Example 7 that we did not need to know the value of I_0 to determine the Richter scale magnitude of the quake.

CHECK YOUR PROGRESS 7   What is the Richter scale magnitude of an earthquake whose intensity is twice that of the Joshua Tree earthquake in Example 7?

Solution   *See page S24.*

If you know the Richter scale magnitude of an earthquake, then you can determine the intensity of the earthquake.

EXAMPLE 8 ■ Find the Intensity of an Earthquake

Find the intensity of the 1999 Taiwan earthquake, which measured 7.6 on the Richter scale. Round to the nearest thousand.

Solution

$$\log\left(\frac{I}{I_0}\right) = 7.6$$

$$\frac{I}{I_0} = 10^{7.6} \qquad \text{• Write in exponential form.}$$

$$I = 10^{7.6}I_0 \qquad \text{• Solve for } I.$$

$$I \approx 39{,}810{,}717 I_0$$

The 1999 Taiwan earthquake had an intensity that was approximately 39,811,000 times the intensity of a zero-level earthquake.

CHECK YOUR PROGRESS 8   On September 3, 2000, an earthquake measuring 5.2 on the Richter scale struck the Napa Valley, 50 miles north of San Francisco. Find the intensity of the quake. Round to the nearest thousand.

Solution   *See page S24.*

▼ **point of interest**

The pH scale was created by the Danish biochemist Søren Sørensen (sû′rn-sn) in 1909 to measure the acidity of water used in the brewing of beer. pH is an abbreviation for *pondus hydrogenii,* which translates as "potential hydrogen."

Logarithmic scales are also used in chemistry. Chemists use logarithms to determine the pH of a liquid, which is a measure of the liquid's **acidity** or **alkalinity.** (You may have tested the pH of a swimming pool or an aquarium.) Pure water, which is considered neutral, has a pH of 7.0. The pH scale ranges from 0 to 14, with 0 corresponding to the most acidic solutions and 14 to the most alkaline. Lemon juice has a pH of about 2, whereas household ammonia measures about 11.

Specifically, the acidity of a solution is a function of the hydronium-ion concentration of the solution. Because the hydronium-ion concentration of a solution can be very small (with values as low as 0.00000001), pH measures the acidity or alkalinity of a solution using a logarithmic scale.

✔ **TAKE NOTE**

One mole is equivalent to 6.022×10^{23} ions.

The pH of a Solution

The pH of a solution with a hydronium-ion concentration of H^+ moles per liter is given by

$$pH = -\log[H^+]$$

EXAMPLE 9 ■ Calculate the pH of a Liquid

Find the pH of each liquid.

a. Orange juice containing an H^+ concentration of 2.8×10^{-4} moles per liter

b. Milk containing an H^+ concentration of 3.97×10^{-7} moles per liter

c. A baking soda solution containing an H^+ concentration of 3.98×10^{-9} moles per liter

Solution

a. $pH = -\log[H^+]$
   $pH = -\log(2.8 \times 10^{-4}) \approx 3.6$
   The orange juice has a pH of about 3.6.

b. $pH = -\log[H^+]$
   $pH = -\log(3.97 \times 10^{-7}) \approx 6.4$
   The milk has a pH of about 6.4.

c. $pH = -\log[H^+]$
   $pH = -\log(3.98 \times 10^{-9}) \approx 8.4$
   The baking soda solution has a pH of about 8.4.

CHECK YOUR PROGRESS 9   Find the pH of each liquid. Round to the nearest tenth.

a. A cleaning solution containing bleach and an H^+ concentration of 2.41×10^{-13} moles per liter

b. A cola soft drink containing an H^+ concentration of 5.07×10^{-4} moles per liter

c. Rainwater containing an H^+ concentration of 6.31×10^{-5} moles per liter

Solution   See pages S24–S25.

The following figure illustrates the pH scale, along with the corresponding hydronium-ion concentrations. A solution with a pH less than 7 is an **acid,** and a solution with a pH greater than 7 is an **alkaline solution** or a **base.** Because the scale is logarithmic, a solution with a pH of 6 is 10 times more acidic than a solution with a pH of 5. From Example 9 we see that the orange juice and milk are acids, whereas the baking soda solution is a base.

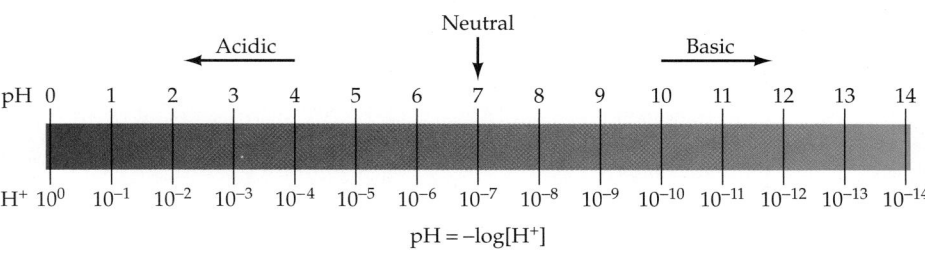

$$pH = -\log[H^+]$$

In Example 9, the hydronium-ion concentrations of orange juice and milk were given as 2.8×10^{-4} and 3.97×10^{-7}, respectively.

The figure above shows how the pH function scales small numbers on the H^+ axis into larger and more convenient numbers on the pH axis.

EXAMPLE 10 ■ Find the Hydronium-Ion Concentration of a Liquid

A sample of blood has a pH of 7.3. Find the hydronium-ion concentration of the blood.

Solution

$$pH = -\log[H^+]$$
$$7.3 = -\log[H^+] \qquad \text{• Substitute 7.3 for pH.}$$
$$-7.3 = \log[H^+] \qquad \text{• Multiply both sides by } -1.$$
$$10^{-7.3} = H^+ \qquad \text{• Change to exponential form.}$$
$$5.0 \times 10^{-8} \approx H^+ \qquad \text{• Evaluate } 10^{-7.3} \text{ and write the answer in scientific notation.}$$

The hydronium-ion concentration of the blood is about 5.0×10^{-8} moles per liter.

CHECK YOUR PROGRESS 10   The water in the Great Salt Lake in Utah has a pH of 10.0. Find the hydronium-ion concentration of the water.

Solution   See page S25.

Excursion

Benford's Law

▼ **point of interest**

Benford's Law has been used to identify fraudulent accountants. In most cases these accountants are unaware of Benford's Law, and have replaced valid numbers with numbers selected at random. Their numbers do not conform to Benford's Law. Hence, an audit is warranted.

The authors of this text know some interesting details about your finances. For instance, of the last 100 checks you have written, about 30% are for amounts that start with a 1. Also, you have written about three times as many checks for amounts that start with a 2 as you have for amounts that start with a 7.

We are sure of these results because of a mathematical formula known as *Benford's Law*. This law was first discovered by the mathematician Simon Newcomb in 1881, and then rediscovered by the physicist Frank Benford in 1938. Benford's Law states that the probability P that the first digit of a number selected from a wide range of numbers d is given by

$$P(d) = \log_{10}\left(1 + \frac{1}{d}\right)$$

1. Use Benford's Law to complete the table and bar graph shown below.

(continued)

d	$P(d) = \log_{10}(1 + 1/d)$
1	0.301
2	0.176
3	0.125
4	_____
5	_____
6	_____
7	_____
8	_____
9	_____

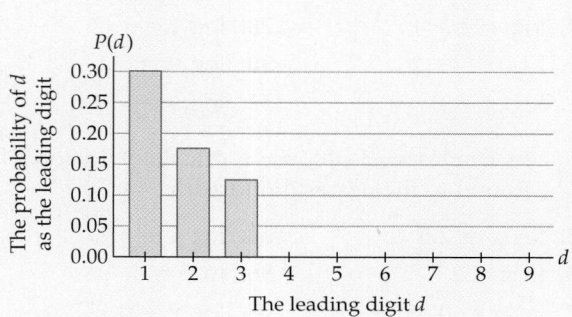

Benford's Law applies to most data with a wide range. For instance, it applies to

- the populations of the cities in the United States
- the numbers of dollars in the savings accounts at your local bank
- the numbers of miles driven during a month by each person in a state

2. Use Benford's Law to find the probability that, in a U.S. city selected at random, the number of telephones in that city will be a number starting with a 6.

3. Use Benford's Law to find how many times as many purchases you have made for dollar amounts that start with a 1 than for dollar amounts that start with a 9.

4. Explain why Benford's Law would not apply to the set of telephone numbers in a small city such as Le Mars, Iowa.

5. Explain why Benford's Law would not apply to the set of all the ages, in years, of the students at a local high school.

Exercise Set 6.6   (Suggested Assignment: 1–66, odds; 62)

Write the exponential equation in logarithmic form.

1. $7^2 = 49$

2. $10^3 = 1000$

3. $5^4 = 625$

4. $2^{-3} = \dfrac{1}{8}$

5. $10^{-4} = 0.0001$

6. $3^5 = 729$

7. $10^y = x$

8. $e^y = x$

Write the logarithmic equation in exponential form.

9. $\log_3 81 = 4$

10. $\log_2 16 = 4$

11. $\log_5 125 = 3$

12. $\log_4 64 = 3$

13. $\log_4 \dfrac{1}{16} = -2$

14. $\log_2 \dfrac{1}{16} = -4$

15. $\ln x = y$

16. $\log x = y$

In Exercises 17–24, evaluate the logarithm.

17. $\log_3 81$

18. $\log_7 49$

19. $\log 100$

20. $\log 0.001$

21. $\log_3 \dfrac{1}{9}$

22. $\log_7 \dfrac{1}{7}$

23. $\log_2 64$

24. $\log 0.01$

In Exercises 25–32, solve the equation for x.

25. $\log_3 x = 2$

26. $\log_5 x = 1$

27. $\log_7 x = -1$

28. $\log_8 x = -2$

29. $\log_3 x = -2$

30. $\log_5 x = 3$

31. $\log_4 x = 0$

32. $\log_4 x = -1$

In Exercises 33–40, use a calculator to solve for x. Round to the nearest hundredth.

33. $\log x = 2.5$

34. $\log x = 3.2$

35. $\ln x = 2$

36. $\ln x = 4$

37. $\log x = 0.35$

38. $\log x = 0.127$

39. $\ln x = \dfrac{8}{3}$

40. $\ln x = \dfrac{1}{2}$

In Exercises 41–46, graph the function.

41. $g(x) = \log_2 x$

42. $g(x) = \log_4 x$

43. $f(x) = \log_3(2x - 1)$

44. $f(x) = -\log_2 x$

45. $f(x) = \log_2(x - 1)$

46. $f(x) = \log_3(x - 2)$

The percent of light that will pass through a material is given by the equation $\log P = -kd$, where P is the percent of light passing through the material, k is a constant that depends on the material, and d is the thickness of the material in centimeters. Use this formula for Exercises 47 and 48.

47. The constant k for a piece of opaque glass that is 0.5 centimeters thick is 0.2. Find the percent of light that will pass through the glass. Round to the nearest percent.

48. The constant k for a piece of tinted glass is 0.5. How thick is a piece of this glass that allows 60% of the light incident to the glass to pass through it? Round to the nearest hundredth.

The number of decibels, D, of a sound can be given by the equation $D = 10(\log I + 16)$, where I is the power of the sound measured in watts. Use this formula for Exercises 49 and 50. Round to the nearest whole number.

49. Find the number of decibels of normal conversation. The power of the sound of normal conversation is approximately 3.2×10^{-10} watts.

50. The loudest sound made by an animal is made by the blue whale and can be heard over 500 miles away. The power of the sound is 630 watts. Find the number of decibels of the sound emitted by the blue whale.

For Exercises 51 and 52, use the equation $\mathrm{pH} = -\log(\mathrm{H}^+)$, where H^+ is the hydronium-ion concentration of a solution. Round to the nearest hundredth.

51. Find the pH of the digestive solution of the stomach, for which the hydronium-ion concentration is 0.045.

52. Find the pH of a morphine solution used to relieve pain, for which the hydronium-ion concentration is 3.2×10^{-10}.

For Exercises 53 to 57, use the Richter scale equation $M = \log \dfrac{I}{I_0}$, where M is the magnitude of an earthquake, I is the intensity of the shock waves, and I_0 is the measure of the intensity of a zero-level earthquake.

53. On July 14, 2000, an earthquake struck the Kodiak Island Region in Alaska. The earthquake had an intensity of $I = 6{,}309{,}573 I_0$. Find the Richter scale magnitude of the earthquake. Round to the nearest tenth.

54. The January 26, 2000 earthquake near Gujarat, India had an intensity of $I = 50{,}118{,}723 I_0$. Find the Richter scale magnitude of the earthquake. Round to the nearest tenth.

55. An earthquake in Japan on March 2, 1933 measured 8.9 on the Richter scale. Find the intensity of the earthquake. Round to the nearest whole number.

56. An earthquake that occurred in China in 1978 measured 8.2 on the Richter scale. Find the intensity of the earthquake in terms of I_0. Round to the nearest whole number.

57. How many times stronger is an earthquake whose magnitude is 8 than one whose magnitude is 6?

Astronomers use the distance modulus formula $M = 5 \log r - 5$, where M is the distance modulus and r is the distance of a star from Earth in parsecs. (One parsec is approximately 1.92×10^{13} miles, or approximately 20 trillion miles.) Use this formula for Exercises 58 and 59. Round to the nearest tenth.

58. The distance modulus of the star Betelgeuse is 5.89. How many parsecs from Earth is this star?

59. The distance modulus of Alpha Centauri is -1.11. How many parsecs from Earth is this star?

One model for the time it will take for the world's oil supply to be depleted is given by the equation $T = 14.29 \ln(0.00411r + 1)$, where r is the estimated oil reserves in billions of barrels and T is the time before that amount of oil is depleted. Use this formula for Exercises 60 and 61. Round to the nearest tenth.

60. How many barrels of oil are necessary to last 20 years?

61. How many barrels of oil are necessary to last 50 years?

Extensions
CRITICAL THINKING

62. As mentioned in this section, the main motivation for the development of logarithms was to aid astronomers and other scientists with arithmetic calculations. The idea was to allow scientists to multiply large numbers by adding the logarithms of the numbers. Dividing large numbers was accomplished by subtracting the logarithms of the numbers. Follow through the exercises below to see how this was accomplished. For simplicity, we will use small numbers (2 and 3) to illustrate the procedure.

    a. Write each of the equations $\log 2 = 0.30103$ and $\log 3 = 0.47712$ in exponential form.

    b. Replace 2 and 3 in $x = 2 \cdot 3$ with the exponential expressions from part a. (In this exercise, we know that x is 6. However, if the two numbers were very large, the value of x would not be obvious.)

    c. Simplify the exponential expression. Recall that to multiply two exponential expressions with the same base, *add* the exponents.

    d. If you completed part c correctly, you should have $x = 10^{0.77815.}$ Write this expression in logarithmic form.

    e. Using a calculator, verify that the solution of the equation you created in part d is 6. *Note:* When tables of logarithms were used, a scientist would have looked through the table to find 0.77815 and observed that it was the logarithm of 6.

63. Replace the denominator 2 and the numerator 3 in $x = \frac{3}{2}$ by the exponential expressions from part a of Exercise 62. Simplify the expression by *subtracting* the exponents. Write the answer in logarithmic form and verify that $x = 1.5$.

64. Write a few sentences explaining how adding the logarithms of two numbers can be used to find the product of the two numbers.

65. Write a few sentences explaining how subtracting the logarithms of two numbers can be used to find the quotient of the two numbers.

EXPLORATIONS

Seismologists generally determine the Richter scale magnitude of an earthquake by examining a *seismogram,* an example of which is shown below.

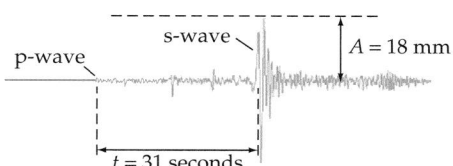

The magnitude of an earthquake cannot be determined just by examining the amplitude of a seismogram, because this amplitude decreases as the distance between the epicenter of the earthquake and the observation station increases. To account for the distance between the epicenter and the observation station, a seismologist examines a seismogram for both small waves, called *p-waves,* and larger waves, called *s-waves.* The Richter scale magnitude M of the earthquake is a function of both the amplitude A of the s-waves and the time t between the occurrence of the s-waves and the occurrence of the p-waves. In the 1950's, Charles Richter developed the following formula to determine the magnitude M of an earthquake from the data in a seismogram.

The Amplitude-Time-Difference Formula

The Richter scale magnitude of an earthquake is given by

$$M = \log A + 3 \log 8t - 2.92$$

where A is the amplitude, in millimeters, of the s-waves on a seismogram and t is the time, in seconds, between the s-waves and the p-waves.

66. Find the magnitude of the earthquake that produced the seismogram shown above.

67. Find the magnitude of the earthquake that produced the seismograph shown below.

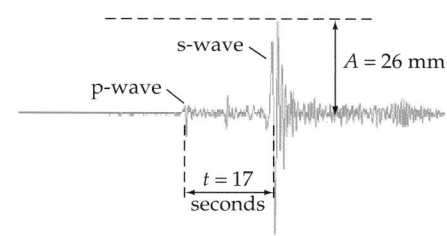

| CHAPTER 6 | **Summary** |

Key Terms

abscissa [p. 325]
acid [p. 393]
acidity and alkalinity [p. 392]
axis of symmetry [p. 361]
base [p. 393]
coefficient of determination [p. 355]
common logarithm [p. 388]
common logarithmic function [p. 388]
coordinate axes [p. 324]
coordinates of a point [p. 325]
correlation coefficient [p. 355]
dependent variable [p. 329]
domain [p. 328]
e [p. 376]
equation in two variables [p. 325]
evaluating a function [p. 329]
exponential decay function [p. 373]
exponential function [p. 373]
exponential growth function [p. 373]
first coordinate [p. 325]
function [p. 328]
functional notation [p. 329]
graph a point [p. 325]
graph of an equation in two variables [p. 326]
graph of an ordered pair [p. 325]
independent variable [p. 329]
linear function [p. 336]
logarithm [p. 386]
logarithmic function [p. 386]
maximum of a quadratic function [p. 364]
minimum of a quadratic function [p. 364]
natural exponential function [p. 377]
natural logarithm [p. 388]
natural logarithmic function [p. 388]
ordered pair [p. 325]
ordinate [p. 325]
origin [p. 324]
parabola [p. 360]
plot a point [p. 325]
quadrants [p. 324]
quadratic function [p. 360]
range [p. 328]
rectangular coordinate system [p. 324]
regression line [p. 354]
scatter diagram [p. 353]
second coordinate [p. 325]

slope [p. 340]
solution of an equation in two variables [p. 325]
value of a function [p. 329]
vertex [p. 361]
x-coordinate [p. 325]
x-intercept [pp. 337 and 362]
xy-plane [p. 324]
y-coordinate [p. 325]
y-intercept [p. 337]
zero-level earthquake [p. 391]

Essential Concepts

■ **Functional Notation**
The symbol $f(x)$ is the value of the function and represents the value of the dependent variable for a given value of the independent variable.

■ **Linear Function**
A linear function is one that can be written in the form $f(x) = mx + b$, where m is the coefficient of x and b is a constant. The slope of the graph of the function is m, the coefficient of x.

■ **y-Intercept of a Linear Function**
The y-intercept of the graph of $f(x) = mx + b$ is $(0, b)$.

■ **Slope of a Line**
Let (x_1, y_1) and (x_2, y_2) be two points on a line. Then the slope of the line through the two points is the ratio of the change in the y-coordinates to the change in the x-coordinates
$$m = \frac{\text{change in } y}{\text{change in } x} = \frac{y_2 - y_1}{x_2 - x_1}, x_1 \neq x_2$$

■ **Slope-Intercept Form of the Equation of a Line**
The graph of $f(x) = mx + b$ is a straight line with slope m and y-intercept $(0, b)$.

■ **Point–Slope Formula of a Straight Line**
Let (x_1, y_1) be a point on a line and let m be the slope of the line. Then the equation of the line can be found using the point–slope formula
$y - y_1 = m(x - x_1)$.

■ **Vertex of a Parabola**
Let $f(x) = ax^2 + bx + c, a \neq 0$, be the equation of a parabola. Then the coordinates of the vertex are
$$\left(-\frac{b}{2a}, f\left(-\frac{b}{2a} \right) \right).$$

■ **Equality of Exponents Property**
If $b > 0$ and $b^x = b^y$, then $x = y$.

■ **Richter Scale**
An earthquake with an intensity of I has a Richter scale magnitude of $M = \log\left(\dfrac{I}{I_0}\right)$, where I_0 is the measure of the intensity of a zero-level earthquake.

■ **The pH of a Solution**
The pH of a solution with a hydronium-ion concentration of H^+ moles per liter is given by $pH = -\log[H^+]$.

| CHAPTER 6 | **Review Exercises** |

---

1. Draw a line through all points with an x-coordinate of 4.

2. Draw a line through all points with a y-coordinate of -3.

3. Graph the ordered-pair solutions of $y = 2x^2$ when $x = -2, -1, 0, 1,$ and 2.

4. Graph the ordered-pair solutions of $y = x^2$ when $x = -2, -1, 0, 1,$ and 2.

5. Graph the ordered-pair solutions of $y = -2x + 1$ when $x = -2, -1, 0, 1,$ and 2.

6. Graph the ordered-pair solutions of $y = |x + 1|$ when $x = -5, -3, 0, 3,$ and 5.

In Exercises 7–16, graph the equation.

7. $y = -2x + 1$ **8.** $y = 3x + 2$

9. $f(x) = x^2 + 2$ **10.** $f(x) = x^2 - 3x + 1$

11. $y = |x + 4|$ **12.** $f(x) = 2|x| - 1$

13. $f(x) = 2^x - 3$ **14.** $f(x) = 3^{-x+2}$

15. $f(x) = e^{0.5x}$ **16.** $f(x) = \log_2(x - 2)$

In Exercises 17–23, evaluate the function for the given value.

17. $f(x) = 4x - 5; \; x = -2$

18. $g(x) = 2x^2 - x - 2; \; x = 3$

19. $s(t) = \dfrac{4}{3t - 5}; \; t = -1$

20. $R(s) = s^3 - 2s^2 + s - 3; \; s = -2$

21. $f(x) = 2^{x-3}; \; x = 5$

22. $g(x) = \left(\dfrac{2}{3}\right)^x; \; x = 2$

23. $T(r) = 2e^r + 1; \; r = 2$. Round to the nearest hundredth.

24. The volume of a sphere is a function of its radius and is given by $V(r) = \dfrac{4\pi r^3}{3}$, where r is the radius of the sphere.

   a. Find the volume of a sphere whose radius is 3 inches. Round to the nearest tenth.

   b. Find the volume of a sphere whose radius is 12 centimeters. Round to the nearest tenth.

25. The height h of a ball that is released 5 feet above the ground with an initial velocity of 96 feet per second is a function of the time t, in seconds, the ball is in the air and is given by $h(t) = -16t^2 + 96t + 5$.

   a. Find the height of the ball above the ground 2 seconds after it is released.

   b. Find the height of the ball above the ground 4 seconds after it is released.

26. The percent concentration P of sugar in a water solution depends on the amount of sugar that is added to the solution and is given by $P(x) = \dfrac{100x + 100}{x + 20}$, where x is the number of grams of sugar that are added.

   a. What is the original percent concentration of sugar?

   b. What is the percent concentration of sugar after 10 more grams of sugar are added? Round to the nearest tenth of a percent.

In Exercises 27–30, find the x- and y-intercepts of the graph of the function.

27. $f(x) = 2x + 10$ **28.** $f(x) = \dfrac{3}{4}x - 9$

29. $3x - 5y = 15$ **30.** $4x + 3y = 24$

31. The accountant for a small business uses the model $V(t) = 25{,}000 - 5000t$ to approximate the value, $V(t)$, of a small truck t years after its purchase. Find and discuss the meaning of the intercepts on the vertical and horizontal axes.

In Exercises 32–35, find the slope of the line containing the given points.

32. $(3, 2), (2, -3)$ **33.** $(-1, 4), (-3, -1)$

34. $(2, -5), (-4, -5)$ **35.** $(5, 2), (5, 7)$

36. The graph below shows how the number of worldwide automatic teller machine (ATM) transactions has been increasing. Find the slope of the line and write a sentence that explains the meaning of the slope in the context of the application.

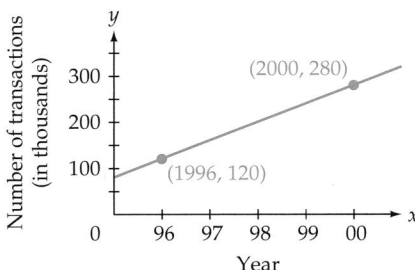

37. Graph the line that passes through the point $(3, -2)$ and has slope -2.

38. Graph the line that passes through the point $(-1, -3)$ and has slope $\frac{3}{4}$.

In Exercises 39–42, find the equation of the line that contains the given point and has the given slope.

39. $(-2, 3), m = 2$ **40.** $(1, -4), m = 1$

41. $(-3, 1), m = \frac{2}{3}$ **42.** $(4, 1), m = \frac{1}{4}$

43. Graph $f(x) = \frac{3}{2}x - 1$ using the slope and y-intercept.

44. A dentist's office is being recarpeted. The cost to install the new carpet is $100 plus $25 per square yard of carpeting.

  a. Determine a linear function for the cost to carpet the office.

  b. Use this function to determine the cost to carpet 32 square yards of floor space.

45. The manager of Valley Gas Mart has determined that 10,000 gallons of regular unleaded gasoline can be sold each week if the price is the same as that of Western QuickMart, a gas station across the street. If the manager increases the price $.02 above Western QuickMart's price, the manager will sell 500 less gallons per week. If the manager decreases the price $.02 below that of Western QuickMart, 500 more gallons of gasoline per week will be sold.

  a. Determine a linear function that will predict the number of gallons of gas per week that Valley Gas Mart can sell as a function of the price relative to that of Western QuickMart.

  b. Use the model to predict the number of gallons of gasoline Valley Gas Mart will sell if its price is $.03 below that of Western QuickMart.

46. The body mass index (BMI) of a person is a measure of the person's ideal body weight. The data in the table below show the BMI for different weights for a person 5′6″ tall.

BMI Data for Person 5′6″ Tall			
Weight (in pounds)	BMI	Weight (in pounds)	BMI
110	18	160	26
120	19	170	27
125	20	180	29
135	22	190	31
140	23	200	32
145	23	205	33
150	24	215	35

Source: Centers for Disease Control and Prevention

  a. Find the linear regression equation for this data.

  b. Use the regression equation to estimate the BMI for a person 5′6″ tall whose weight is 158 pounds. Round to the nearest whole number.

In Exercises 47–50, find the vertex of the graph of the function.

47. $y = x^2 + 2x + 4$ **48.** $y = -2x^2 - 6x + 1$

49. $f(x) = -3x^2 + 6x - 1$ **50.** $f(x) = x^2 + 5x - 1$

In Exercises 51–54, find the x-intercepts of the parabola given by the equation.

51. $y = x^2 + x - 20$ **52.** $y = x^2 + 2x - 1$

53. $f(x) = 2x^2 + 9x + 4$ **54.** $f(x) = x^2 + 4x + 6$

In Exercises 55–58, find the minimum or maximum value of the quadratic function. State whether the value is a maximum or a minimum.

55. $y = -x^2 + 4x + 1$ **56.** $y = 2x^2 + 6x - 3$

57. $f(x) = x^2 - 4x - 1$ **58.** $f(x) = -2x^2 + 3x - 1$

59. The height s, in feet, of a rock thrown upward at an initial speed of 80 feet per second from a cliff 25 feet above an ocean beach is given by the function $s(t) = -16t^2 + 80t + 25$, where t is the time in seconds. Find the maximum height above the beach that the rock will attain.

60. A manufacturer of rewritable CDs (CD-RW) estimates that the average daily cost C of producing a single CD-RW is given by $C(x) = 0.01x^2 - 40x + 50,000$, where x is the number of CD-RWs produced each day. Find the number of CD-RWs the company should produce in order to minimize the average daily cost.

61. A \$5000 certificate of deposit (CD) earns 6% annual interest compounded daily. What is the value of the investment after 15 years? Use the compound interest formula $A = P\left(1 + \frac{r}{n}\right)^{nt}$, where P is the amount deposited, A is the value of the money after t years, r is the annual interest rate as a decimal, and n is the number of compounding periods per year.

62. An isotope of technetium has a half-life of approximately 6 hours. If a patient is injected with 10 milligrams of this isotope, what will be the technetium level in the patient after 2 hours? Use the function $A = 10\left(\frac{1}{2}\right)^{t/6}$, where A is the technetium level, in milligrams, in the patient after t hours. Round to the nearest hundreth.

63. Iodine-131 has a half-life of approximately 8 days. If a patient is given an injection that contains 8 micrograms of iodine-131, what will be the iodine level in the patient after 10 days? Use the function $A = 8\left(\frac{1}{2}\right)^{t/8}$, where A is the amount of iodine-131, in micrograms, in the patient after t days. Round to the nearest hundredth.

64. A golf ball is dropped from a height of 6 feet. On each successive bounce, the ball rebounds to a height that is $\frac{2}{3}$ of the previous height.

a. Find an exponential model for the height of the ball after the nth bounce.

b. What is the height of the ball after the fifth bounce? Round to the nearest hundreth.

65. When a new movie is released, there is initially a surge in the number of people who go to see the movie. After 2 weeks, attendance generally begins to drop off. The data in the following table show the number of people attending a certain movie; $t = 0$ represents the number of patrons 2 weeks after the movie's initial release.

Number of weeks two weeks after release, t	0	1	2	3	4
Number of people (in thousands), N	250	162	110	65	46

a. Find an exponential regression equation for these data.

b. Use the equation to predict the number of people who will attend the movie 8 weeks after it has been released.

In Exercises 66–69, evaluate the logarithm.

66. $\log_3 243$ **67.** $\log_2 \dfrac{1}{16}$

68. $\log_4 \dfrac{1}{4}$ **69.** $\log_2 64$

In Exercises 70–73, solve for x. Round to the nearest ten-thousandth.

70. $\log_4 x = 3$ **71.** $\log_3 x = \dfrac{1}{3}$

72. $\ln x = 2.5$ **73.** $\log x = 2.4$

74. Use the distance modulus function $M = 5 \log r - 5$, where M is the distance modulus and r is the distance of a star from Earth in parsecs, to find the distance to a star whose distance modulus is 3.2. Round to the nearest tenth.

75. The number of decibels, D, of a sound can be given by the function $D = 10(\log I + 16)$, where I is the power of the sound measured in watts. The pain threshold for a sound for most humans is approximately 0.01 watt. Find the number of decibels of this sound.

In Exercises 1 and 2, evaluate the function for the given value of the independent variable.

1. $s(t) = -3t^2 + 4t - 1; t = -2$

2. $f(x) = 3^{x-4}; x = 2$

3. Evaluate: $\log_5 125$

4. Solve for x: $\log_6 x = 2$

In Exercises 5–8, graph the function.

5. $f(x) = 2x - 3$

6. $f(x) = x^2 + 2x - 3$

7. $f(x) = 2^x - 5$

8. $f(x) = \log_3(x - 1)$

9. Find the slope of the line that passes through $(3, -1)$ and $(-2, -4)$.

10. Find the equation of the line that passes through $(3, 5)$ and has slope $\frac{2}{3}$.

11. Find the vertex of the graph of $f(x) = x^2 + 6x - 1$.

12. Find the x-intercepts of the parabola given by the equation $y = x^2 + 2x - 8$.

13. Find the minimum or maximum value of the quadratic function $y = -x^2 - 3x + 10$ and state whether the value is a minimum or a maximum.

14. The distance d, in miles, a small plane is from its final destination is given by $d = 250 - 100t$, where t is the time, in hours, remaining for the flight. Find and discuss the meaning of the intercepts of the graph of the equation.

15. The height h, in feet, of a ball that is thrown straight up and released 4 feet above the ground is given by $h(t) = -16t^2 + 96t + 4$, where t is the time in seconds. Find the maximum height the ball attains.

16. A radioactive isotope has a half-life of 5 hours. If a chemist has a 10-gram sample of this isotope, what amount will remain after 8 hours? Use the function

$A = 10\left(\frac{1}{2}\right)^{t/5}$, where A is the amount of the isotope, in milligrams, remaining after t hours. Round to the nearest hundredth.

17. Two earthquakes struck Colombia, South America in the year 2000. One had a magnitude of 6.5 on the Richter scale and the second had a magnitude of 5 on the Richter scale. How many times greater was the intensity of the first earthquake than that of the second? Round to the nearest tenth.

18. The manager of an orange grove has determined that when there are 320 trees per acre, the average yield per tree is 260 pounds of oranges. If the number of trees is increased to 330 trees per acre, the average yield per tree decreases to 245 pounds. Find a linear model for the average yield per tree as a function of the number of trees per acre.

19. Giant pumpkin contests are popular at many state fairs. Suppose the data below show the weights, in pounds, of the winning pumpkins at recent fairs, where $x = 0$ represents 1997.

Year, x	0	1	2	3	4
Weight (in pounds), y	650	715	735	780	820

a. Find a linear regression equation for these data.

b. Use the regression equation to predict the winning weight in 2005. Round to the nearest pound.

20. The equation $\log P = -kd$ gives the relationship between the percent P, as a decimal, of light passing through a substance of thickness d. The value of k for a swimming pool is approximately 0.05. At what depth, in meters, will the percent of light be 75% of the light at the surface of the pool? Round to the nearest tenth.

As attorneys navigate through the legal process for their clients, they are often faced with strict deadlines imposed by the court. Documents that are vital to the case need to be filed within a certain amount of time (for example 10 days, 30 days, 60 days) of events within the case. If a deadline falls on a Saturday, Sunday, or holiday, then the document does not have to be filed until the following business day. By calculating on which day of the week the deadline will fall, the attorney can properly diary his or her calendar to ensure that the deadline is properly met. In the **Excursion exercises** on **page 414,** you will calculate the days of the week on which different events fall.

Mathematical Systems

Need help? For on-line student resources, such as section quizzes, visit this textbook's web site at **math.college.hmco.com/students.**

The Iran-Contra scandal surfaced in 1986, and in 1987 Oliver North testified before Congress regarding his involvement. One of the situations that came to light involved the transfer of $10 million to a Swiss bank account for the Contras. North had given the account number 368.430.22.1 to Secretary of State Elliott Abrams on a card typed by North's secretary, Fawn Hall. The number was reportedly incorrect, and the money was mistakenly transferred to the account of a Swiss businessman. The second and third digits of the correct account number, 386.430.22.1, had been transposed.

The transposing of two adjacent digits is one of the most common human errors made when copying a number or entering a number into a computer. Every day, millions of numbers are transmitted electronically. Credit card numbers travel across the Internet, money is transferred from one bank account to another, and goods are ordered by product number. To help prevent incorrect numbers from being transmitted, many numbers have an added digit called a *check digit.* Mathematical systems, such as *modular arithmetic,* have been used to design methods that detect errors using the check digit. Computers then usually can catch the error and alert the user before forwarding an incorrect number. Modular arithmetic and some of its applications—detecting errors in credit card numbers, product numbers (UPC or ISBN), and money order numbers—are discussed in Sections 7.1 and 7.2.

| **SECTION 7.1** | **Modular Arithmetic** |

▼ **point of interest**

The abbreviation A.M. comes from the Latin *ante* (before) *meridiem* (midday). The abbreviation P.M. comes from the Latin *post* (after) *meridiem* (midday).

Introduction to Modular Arithmetic

Many clocks have the familiar 12-hour design. We designate whether the time is before noon or after noon by using the abbreviations A.M. and P.M. A reference to 7:00 A.M. means 7 hours after 12:00 midnight; a reference to 7:00 P.M. means 7 hours after 12:00 noon. In both cases, once 12 is reached on the clock, we begin again with 1.

If we want to determine a time in the future or in the past, it is necessary to consider whether we have passed 12 o'clock. To determine the time 8 hours after 3 o'clock, we add 3 and 8. Because we did not pass 12 o'clock, the time is 11 o'clock (Figure 7.1a). However, to determine the time 8 hours after 9 o'clock, we must take into consideration that once we have passed 12 o'clock, we begin again with 1. Therefore, 8 hours after 9 o'clock is 5 o'clock, as shown in Figure 7.1b.

We will use the symbol $\oplus$ to denote that we are not adding in the usual sense. Using this notation,

$$3 \oplus 8 = 11 \quad \text{and} \quad 9 \oplus 8 = 5$$

on a 12-hour clock.

Figure 7.1a

We can also perform subtraction on a 12-hour clock. If the time now is 10 o'clock, then 7 hours ago the time was 3 o'clock, which is the difference between 10 and 7 ($10 - 7 = 3$). However, if the time now is 3 o'clock, then, using Figure 7.2, 7 hours ago it was 8 o'clock. If we use the symbol $\ominus$ to denote subtraction on a 12-hour clock, we can write

$$10 \ominus 7 = 3 \quad \text{and} \quad 3 \ominus 7 = 8$$

Figure 7.1b

EXAMPLE 1 ■ **Perform Clock Arithmetic**

Evaluate each of the following, where $\oplus$ and $\ominus$ indicate addition and subtraction, respectively, on a 12-hour clock.

a. $9 \oplus 8$ **b.** $7 \oplus 12$ **c.** $8 \ominus 11$ **d.** $2 \ominus 8$

Solution
Calculate using a 12-hour clock.

a. $9 \oplus 8 = 5$ **b.** $7 \oplus 12 = 7$ **c.** $8 \ominus 11 = 9$ **d.** $2 \ominus 8 = 6$

Figure 7.2

CHECK YOUR PROGRESS 1   Evaluate each of the following using a 12-hour clock.

a. $6 \oplus 10$ **b.** $5 \oplus 9$ **c.** $7 \ominus 11$ **d.** $5 \ominus 10$

Solution   *See page S25.*

Monday = 1	Friday = 5
Tuesday = 2	Saturday = 6
Wednesday = 3	Sunday = 7
Thursday = 4	

A similar example involves day-of-the-week arithmetic. If we associate each day of the week with a number, as shown at the left, then 6 days after Friday is Thursday; 16 days after Monday is Wednesday. Symbolically, we write

$$5 \boxplus 6 = 4 \quad \text{and} \quad 1 \boxplus 16 = 3$$

Note: We are using the $\boxplus$ symbol for days-of-the-week arithmetic to distinguish from the $\oplus$ symbol for clock arithmetic.

Another way to determine the day of the week is to note that when the sum $5 + 6 = 11$ is divided by 7, the number of days in a week, the remainder is 4, the number associated with Thursday. When $1 + 16 = 17$ is divided by 7, the remainder is 3, the number associated with Wednesday. This works because the days of the week repeat every 7 days.

The same method can be applied to 12-hour-clock arithmetic. From Example 1a, when $9 + 8 = 17$ is divided by 12, the number of hours on a 12-hour clock, the remainder is 5, the time 8 hours after 9 o'clock.

Situations such as these that repeat in cycles are represented mathematically by using **modular arithmetic** or **arithmetic modulo _n_.**

Modulo _n_

Two integers _a_ and _b_ are said to be **congruent modulo _n_,** where _n_ is a natural number called the **modulus,** if the remainder when _a_ is divided by _n_ equals the remainder when _b_ is divided by _n_. In this case, we write $a \equiv b \bmod n$. The statement $a \equiv b \bmod n$ is called a **congruence.**

For instance,

$29 \equiv 11 \bmod 3$ because $29 \div 3 = 9$ remainder 2 and $11 \div 3 = 3$ remainder 2. Both 29 and 11 have the same remainder when divided by 3.

$5 \not\equiv 37 \bmod 6$ because $5 \div 6 = 0$ remainder 5 and $37 \div 6 = 6$ remainder 1. The numbers 5 and 37 have different remainders when divided by 6.

QUESTION   *Is $33 \equiv 49 \bmod 4$?*

For a given number _n,_ there are infinitely many numbers that are congruent modulo _n._ For example, 2, 5, 8, 11, 14, … are all congruent to 2 modulo 3. Each number in the list has the same remainder, 2, when divided by 3.

The fact that there are infinitely many numbers congruent to a given modulus _n_ is used in personal digital assistants (PDAs), computers, calendar programs, and other electronic devices so that the day of the week at any future or past date can be calculated. For instance, if today is Tuesday, then 7 days from now, 14 days from now, 21 days from now, and so on with multiples of 7, will all be Tuesdays. Note that 7, 14, 21, … are all congruent to 0 modulo 7.

Now suppose today is Friday. To determine the day of the week 16 days from now, we observe that 14 days from now the day will be Friday, so 16 days from now the day will be Sunday. Note that the remainder when 16 is divided by 7 is 2, or, using modular notation, $16 \equiv 2 \bmod 7$. The 2 signifies 2 days after Friday, which is Sunday.

EXAMPLE 2 ■ Calculate a Day of the Week

July 4, 2005 is a Monday. What day of the week is July 4, 2010?

Solution

There are 5 years between the two dates. Each year has 365 days except 2008, which has one extra day because it is a leap year. So the total number of days between

INSTRUCTOR NOTE
You may wish to show students that $29 \equiv 11 \bmod 3$ also means that 29 can be arrived at by repeatedly adding 3, the modulus, to 11.

ANSWER   *Yes. $33 \div 4 = 8$ remainder 1 and $49 \div 4 = 12$ remainder 1. Both 33 and 49 have the same remainder when divided by 4.*

historical note

The Julian calendar, introduced by Julius Caesar in 46 B.C. and named after him, contained 12 months and 365 days. Every fourth year, an additional day was added to the year to compensate for the difference between an ordinary year and a solar year. However, a solar year actually consists of 365.2422 days, not 365.25 days as assumed by Caesar. The difference between these numbers is 0.0078 days.

Although this is a small number, after a few centuries had passed, the ordinary year and the solar year no longer matched. If something were not done to rectify the situation, after a period of time the summer season in the northern hemisphere would be in December and winter season would be in July.

To bring the seasons back in phase, Pope Gregory removed 10 days from October in 1582. The new calendar was called the Gregorian calendar and is the one we use today. This calendar did not win general acceptance until 1752. ∎

the two dates is $5 \cdot 365 + 1 = 1826$. Because $1826 \div 7 = 260$ remainder 6, $1826 \equiv 6 \bmod 7$. Any multiple of 7 days past a given day will be the same day of the week. So the day of the week 1826 days after July 4, 2005 will be the same as the day 6 days after July 4, 2005. Thus July 4, 2010 will be a Sunday.

CHECK YOUR PROGRESS 2 In 2004, Abraham Lincoln's birthday falls on Thursday, February 12. On what day of the week does Lincoln's birthday fall in 2013?

Solution See page S25.

Math Matters   A Leap-Year Formula

The calculation in Example 2 required that we consider whether the intervening years contained a leap year. There is a formula, based on modular arithmetic, that can be used to determine which years are leap years.

The calendar we use today is called the Gregorian calendar. This calendar differs from the Julian calendar (see the margin note at the left) in that leap years do not always occur every fourth year. Here is the rule: Let Y be the year. If $Y \equiv 0 \bmod 4$, then Y is a leap year unless $Y \equiv 0 \bmod 100$. In that case, Y is not a leap year unless $Y \equiv 0 \bmod 400$. Then Y is a leap year.

Using this rule, 2008 is a leap year because $2008 \equiv 0 \bmod 4$ and 2013 is not a leap year because $2013 \not\equiv 0 \bmod 4$. The year 1900 was *not* a leap year because $1900 \equiv 0 \bmod 100$ but $1900 \not\equiv 0 \bmod 400$. The year 2000 was a leap year because $2000 \equiv 0 \bmod 100$ and $2000 \equiv 0 \bmod 400$.

Arithmetic Operations Modulo *n*

Arithmetic modulo n, where n is a natural number, uses a variation of the standard rules of arithmetic we have used before. Perform the arithmetic operation and then divide by the modulus. The answer is the remainder. Thus the result of an arithmetic operation mod n is always a whole number less than n.

✔ TAKE NOTE

Remember that m mod n means the remainder when m is divided by n. For Example 3, we must find the remainder of the sum of $23 + 38$ when divided by 12.

EXAMPLE 3 ■ Addition Modulo *n*

Evaluate: $(23 + 38) \bmod 12$

Solution
Add $23 + 38$ to produce 61. Then divide by the modulus, 12. The answer is the remainder.

$$12\overline{)61} \quad\quad 23 + 38 = 61$$
$$\underline{60}$$
$$1$$

$$(23 + 38) \bmod 12 \equiv 1$$

The solution is 1.

CHECK YOUR PROGRESS 3   Evaluate: $(51 + 72) \bmod 3$

Solution   *See page S25.*

In modular arithmetic, adding the modulus to a number does not change the value of the number. For instance,

$13 \equiv 6 \bmod 7$		$10 \equiv 1 \bmod 3$	
$20 \equiv 6 \bmod 7$	• Add 7 to 13.	$13 \equiv 1 \bmod 3$	• Add 3 to 10.
$27 \equiv 6 \bmod 7$	• Add 7 to 20.	$16 \equiv 1 \bmod 3$	• Add 3 to 16.

To understand why the value does not change, consider $7 \equiv 0 \bmod 7$ and $3 \equiv 0 \bmod 3$. That is, in mod 7 arithmetic, 7 is equivalent to 0; in mod 3 arithmetic, 3 is equivalent to 0. Just as adding 0 to a number does not change the value of the number in regular arithmetic, in modular arithmetic adding the modulus to a number does not change the value of the number. This property of modular arithmetic is sometimes used in subtraction.

When subtracting two numbers modulo n, if the difference is positive, divide the difference by the modulus. The answer is the remainder. If the difference is negative, repeatedly add the modulus to the difference until you first reach a whole number (0 or a positive integer). The whole number is the answer.

EXAMPLE 4 ■ Subtraction Modulo n

Evaluate each of the following.

a. $(33 - 16) \bmod 6$ **b.** $(14 - 27) \bmod 5$

Solution

a. Subtract $33 - 16 = 17$. The result is positive. Divide the difference by the modulus, 6. The answer is the remainder.

$$
\begin{array}{r}
2 \\
6\overline{)17} \\
12 \\
\hline
5
\end{array}
$$

$(33 - 16) \bmod 6 \equiv 5$

b. Subtract $14 - 27 = -13$. The result is negative. Repeatedly add the modulus, 5, to -13 until a whole number is reached.

$$-13 + 5 = -8$$
$$-8 + 5 = -3$$
$$-3 + 5 = 2$$

$(14 - 27) \bmod 5 \equiv 2$

CHECK YOUR PROGRESS 4   Evaluate: $(21 - 43) \bmod 7$

Solution   *See page S25.*

The methods of adding and subtracting in modular arithmetic can be used for clock arithmetic and days-of-the-week arithmetic.

EXAMPLE 5 ▪ Calculating Times

Disregarding A.M. or P.M., if it is 5 o'clock now, what time was it 57 hours ago?

Solution
The time can be determined by calculating $(5 - 57) \bmod 12$. Because $5 - 57 = -52$, a negative number, repeatedly add the modulus, 12, to the difference until the first whole number is reached. By doing this, we get $(5 - 57) \bmod 12 \equiv 8$. Therefore, if it is 5 o'clock now, 57 hours ago it was 8 o'clock.

CHECK YOUR PROGRESS 5   If today is Tuesday, what day of the week will it be 93 days from now?

Solution   *See page S25.*

Problems involving multiplication can also be performed modulo *n*.

EXAMPLE 6 ▪ Multiplication Modulo *n*

Evaluate: $(15 \cdot 23) \bmod 11$

Solution
Find the product $15 \cdot 23$ and then divide by the modulus, 11. The answer is the remainder.

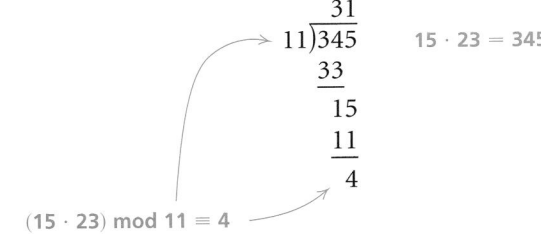

$$
\begin{array}{r}
31 \\
11\overline{)345} \\
33 \\
\hline
15 \\
11 \\
\hline
4
\end{array}
$$

$15 \cdot 23 = 345$

$(15 \cdot 23) \bmod 11 \equiv 4$

The answer is 4.

CHECK YOUR PROGRESS 6   Evaluate: $(33 \cdot 41) \bmod 17$

Solution   *See page S25.*

Solving Congruence Equations

Solving a congruence equation means finding all whole number values of the variable for which the congruence is true.

For example, to solve $3x + 5 \equiv 3 \bmod 4$, we search for whole number values of x for which the congruence is true.

$3(0) + 5 \not\equiv 3 \bmod 4$

$3(1) + 5 \not\equiv 3 \bmod 4$

$3(2) + 5 \equiv 3 \bmod 4$ 2 is a solution.

$3(3) + 5 \not\equiv 3 \bmod 4$

$3(4) + 5 \not\equiv 3 \bmod 4$

$3(5) + 5 \not\equiv 3 \bmod 4$

$3(6) + 5 \equiv 3 \bmod 4$ 6 is a solution.

If we continued trying values, we would find that 10 and 14 are also solutions. Note that the solutions 6, 10, and 14 are all congruent to 2 modulo 4. In general, once a solution is determined, additional solutions can be found by repeatedly adding the modulus to the original solution. Thus the solutions of $3x + 5 \equiv 3 \bmod 4$ are 2, 6, 10, 14, 18, ...

When solving a congruence equation, it is necessary to check only the whole numbers less than the modulus. For the congruence equation $3x + 5 \equiv 3 \bmod 4$, we needed to check only 0, 1, 2, and 3. Each time a solution is found, additional solutions can be found by repeatedly adding the modulus to it.

A congruence equation can have more than one solution among the whole numbers less than the modulus. The next example illustrates that you must check *all* whole numbers less than the modulus.

EXAMPLE 7 ■ **Solve a Congruence Equation**

Solve: $2x + 1 \equiv 3 \bmod 10$

Solution
Beginning with 0, substitute each whole number less than 10 into the congruence equation.

$x = 0$	$2(0) + 1 \not\equiv 3 \bmod 10$	Not a solution
$x = 1$	$2(1) + 1 \equiv 3 \bmod 10$	A solution
$x = 2$	$2(2) + 1 \not\equiv 3 \bmod 10$	Not a solution
$x = 3$	$2(3) + 1 \not\equiv 3 \bmod 10$	Not a solution
$x = 4$	$2(4) + 1 \not\equiv 3 \bmod 10$	Not a solution
$x = 5$	$2(5) + 1 \not\equiv 3 \bmod 10$	Not a solution
$x = 6$	$2(6) + 1 \equiv 3 \bmod 10$	A solution
$x = 7$	$2(7) + 1 \not\equiv 3 \bmod 10$	Not a solution
$x = 8$	$2(8) + 1 \not\equiv 3 \bmod 10$	Not a solution
$x = 9$	$2(9) + 1 \not\equiv 3 \bmod 10$	Not a solution

The solutions between 0 and 9 are 1 and 6; the remaining solutions are determined by repeatedly adding the modulus, 10, to these solutions. The solutions are 1, 6, 11, 16, 21, 26, ...

CHECK YOUR PROGRESS 7   Solve: $4x + 1 \equiv 5 \bmod 12$

Solution   *See page S25.*

Not all congruence equations have a solution. For instance, $5x + 1 \equiv 3 \bmod 5$ has no solution, as shown below.

$x = 0$	$5(0) + 1 \not\equiv 3 \bmod 5$	Not a solution
$x = 1$	$5(1) + 1 \not\equiv 3 \bmod 5$	Not a solution
$x = 2$	$5(2) + 1 \not\equiv 3 \bmod 5$	Not a solution
$x = 3$	$5(3) + 1 \not\equiv 3 \bmod 5$	Not a solution
$x = 4$	$5(4) + 1 \not\equiv 3 \bmod 5$	Not a solution

Because no whole number value of x less than the modulus is a solution, there is no solution.

Additive and Multiplicative Inverses in Modular Arithmetic

Recall that if the sum of two numbers is 0, then the numbers are *additive inverses* of each other. For instance, $8 + (-8) = 0$, so 8 is the additive inverse of -8, and -8 is the additive inverse of 8.

The same concept applies in modular arithmetic. For example, $(3 + 5) \equiv 0 \bmod 8$. Thus, in mod 8 arithmetic, 3 is the additive inverse of 5 and 5 is the additive inverse of 3. (Here we consider only those whole numbers smaller than the modulus.) Note that $3 + 5 = 8$; that is, the sum of a number and its additive inverse equals the modulus. Using this fact, we can easily find the additive inverse of a number for any modulus. For instance, in mod 11 arithmetic, the additive inverse of 5 is 6 because $5 + 6 = 11$.

EXAMPLE 8 ■ Find an Additive Inverse

Find the additive inverse of 7 in mod 16 arithmetic.

Solution
In mod 16 arithmetic, $7 + 9 = 16$, so the additive inverse of 7 is 9.

CHECK YOUR PROGRESS 8   Find the additive inverse of 6 in mod 12 arithmetic.

Solution   *See page S25.*

If the product of two numbers is 1, then the numbers are **multiplicative inverses** of each other. For instance, $2 \cdot \frac{1}{2} = 1$, so 2 is the multiplicative inverse of $\frac{1}{2}$ and $\frac{1}{2}$ is the multiplicative inverse of 2. The same concept applies to modular arithmetic (although the multiplicative inverses will always be natural numbers). For example, in mod 7 arithmetic, 5 is the multiplicative inverse of 3 (and 3 is the multiplicative inverse of 5) because $5 \cdot 3 \equiv 1 \bmod 7$. (Here we will only concern ourselves with numbers less than the modulus.) To find the multiplicative inverse of $a \bmod m$, solve the modular equation $ax \equiv 1 \bmod m$ for x.

TAKE NOTE

In mod m arithmetic, every number has an additive inverse but not necessarily a multiplicative inverse. For instance, in mod 12 arithmetic, 3 does not have a multiplicative inverse. You should verify this by trying to solve the equation $3x \equiv 1 \mod 12$ for x.

A similar situation occurs in standard arithmetic. The number 0 does not have a multiplicative inverse because there is no solution of the equation $0x = 1$.

EXAMPLE 9 ■ **Find a Multiplicative Inverse**

In mod 7 arithmetic, find the multiplicative inverse of 2.

Solution

To find the multiplicative inverse of 2, solve the equation $2x \equiv 1 \mod 7$ by trying different natural number values of x less than the modulus.

$$2x \equiv 1 \mod 7$$
$$2(1) \not\equiv 1 \mod 7 \quad \bullet \text{ Try } x = 1.$$
$$2(2) \not\equiv 1 \mod 7 \quad \bullet \text{ Try } x = 2.$$
$$2(3) \not\equiv 1 \mod 7 \quad \bullet \text{ Try } x = 3.$$
$$2(4) \equiv 1 \mod 7 \quad \bullet \text{ Try } x = 4.$$

In mod 7 arithmetic, the multiplicative inverse of 2 is 4.

CHECK YOUR PROGRESS 9   Find the multiplicative inverse of 5 in mod 11 arithmetic.

Solution   *See page S25.*

Excursion

Computing the Day of the Week

A function that is related to the modulo function is called the *floor function*. In the modulo function, we determine the remainder when one number is divided by another. In the floor function, we determine the quotient (and ignore the remainder) when one number is divided by another. The symbol for the floor function is $\lfloor \ \ \rfloor$. Here are some examples.

$$\left\lfloor \frac{2}{3} \right\rfloor = 0, \quad \left\lfloor \frac{10}{2} \right\rfloor = 5, \quad \left\lfloor \frac{17}{2} \right\rfloor = 8, \quad \text{and} \quad \left\lfloor \frac{2}{\sqrt{2}} \right\rfloor = 1$$

Using the floor function, we can write a formula that gives the day of the week for any date. The formula, known as the Zeller Congruence, is given by

$$x \equiv \left(\left\lfloor \frac{13m - 1}{5} \right\rfloor + \left\lfloor \frac{y}{4} \right\rfloor + \left\lfloor \frac{c}{4} \right\rfloor + d + y - 2c \right) \mod 7$$

where

d is the day of the month

m is the month using 1 for March, 2 for April, . . . , 10 for December; January and February are assigned the values 11 and 12 respectively

y is the last two digits of the year if the month is March through December; if the month is January or February, y is the last two digits of the year *minus 1*

c is the first two digits of the year

x is the day of the week (using 0 for Sunday, 1 for Monday, . . . , 6 for Saturday);

(continued)

For example, to determine the day of the week on July 4, 1776, we have $c = 17$, $y = 76$, $d = 4$, and $m = 5$. Using these values, we can calculate x.

$$x \equiv \left(\left\lfloor \frac{13(5) - 1}{5} \right\rfloor + \left\lfloor \frac{76}{4} \right\rfloor + \left\lfloor \frac{17}{4} \right\rfloor + 4 + 76 - 2(17) \right) \bmod 7$$

$$\equiv (12 + 19 + 4 + 4 + 76 - 34) \bmod 7$$

$$\equiv 81 \bmod 7$$

Solving $x \equiv 81 \bmod 7$ for x, we get $x = 4$. Therefore, July 4, 1776 was a Thursday.

Excursion Exercises

1. Determine the day of the week on which you were born.

2. Determine the day of the week on which Abraham Lincoln's birthday (February 12) will fall in 2010.

3. Determine the day of the week on which January 1, 2020 will fall.

4. Determine the day of the week on which Valentine's Day, February 14, 1950, fell.

Exercise Set 7.1   (Suggested Assignment: 1–77 odds; 82, 83, 85, 88, 89, 91)

In Exercises 1–16, evaluate each expression, where $\oplus$ and $\ominus$ indicate addition and subtraction, respectively, using a 12-hour clock.

1. $3 \oplus 5$
2. $6 \oplus 7$
3. $8 \oplus 4$
4. $5 \oplus 10$
5. $11 \oplus 3$
6. $8 \oplus 8$
7. $7 \oplus 9$
8. $11 \oplus 10$
9. $10 \ominus 6$
10. $2 \ominus 6$
11. $3 \ominus 8$
12. $5 \ominus 7$
13. $10 \ominus 11$
14. $5 \ominus 5$
15. $4 \ominus 9$
16. $1 \ominus 4$

In Exercises 17–24, evaluate each expression, where $\triangle$ and $\triangle$ indicate addition and subtraction, respectively, using military time. (Military time uses a 24-hour clock, where 2:00 A.M. is equivalent to 0200 hours and 10 P.M. is equivalent to 2200 hours.)

17. $1800 \triangle 0900$
18. $1600 \triangle 1200$
19. $0800 \triangle 2000$
20. $1300 \triangle 1300$
21. $1000 \triangle 1400$
22. $1800 \triangle 1900$
23. $0200 \triangle 0500$
24. $0600 \triangle 2200$

In Exercises 25–28, evaluate each expression, where $\boxplus$ and $\boxminus$ indicate addition and subtraction, respectively, using days-of-the-week arithmetic.

25. $6 \boxplus 4$
26. $3 \boxplus 5$
27. $2 \boxminus 3$
28. $3 \boxminus 6$

In Exercises 29–38, determine whether the congruence is true or false.

29. $5 \equiv 8 \bmod 3$
30. $11 \equiv 15 \bmod 4$
31. $5 \equiv 20 \bmod 4$
32. $7 \equiv 21 \bmod 3$
33. $21 \equiv 45 \bmod 6$
34. $18 \equiv 60 \bmod 7$
35. $88 \equiv 5 \bmod 9$
36. $72 \equiv 30 \bmod 5$
37. $100 \equiv 20 \bmod 8$
38. $25 \equiv 85 \bmod 12$

39. List five different positive integers that are congruent to 8 modulo 6.

40. List five different positive integers that are congruent to 10 modulo 4.

In Exercises 41–62, perform the modular arithmetic.

41. $(9 + 15) \bmod 7$
42. $(12 + 8) \bmod 5$
43. $(5 + 22) \bmod 8$
44. $(50 + 1) \bmod 15$

45. $(42 + 35)$ mod 3   **46.** $(28 + 31)$ mod 4

47. $(37 + 45)$ mod 12   **48.** $(62 + 21)$ mod 2

49. $(19 - 6)$ mod 5   **50.** $(25 - 10)$ mod 4

51. $(48 - 21)$ mod 6   **52.** $(60 - 32)$ mod 9

53. $(8 - 15)$ mod 12   **54.** $(3 - 12)$ mod 4

55. $(15 - 32)$ mod 7   **56.** $(24 - 41)$ mod 8

57. $(6 \cdot 8)$ mod 9   **58.** $(5 \cdot 12)$ mod 4

59. $(9 \cdot 15)$ mod 8   **60.** $(4 \cdot 22)$ mod 3

61. $(14 \cdot 18)$ mod 5   **62.** $(26 \cdot 11)$ mod 15

In Exercises 63–70, use modular arithmetic to determine each of the following.

63. Disregarding A.M. or P.M., if it is now 7 o'clock,
   a. what time will it be 59 hours from now?
   b. what time was it 62 hours ago?

64. Disregarding A.M. or P.M., if it is now 2 o'clock,
   a. what time will it be 40 hours from now?
   b. what time was it 34 hours ago?

65. If today is Friday,
   a. what day of the week will it be 25 days from now?
   b. what day of the week was it 32 days ago?

66. If today is Wednesday,
   a. what day of the week will it be 115 days from now?
   b. what day of the week was it 81 days ago?

67. In 2001, Halloween (October 31) fell on a Wednesday. On what day of the week will Halloween fall in the year 2011?

68. In 2002, April Fool's Day (April 1) fell on a Monday. On what day of the week will April Fool's Day fall in 2009?

69. Valentine's Day (February 14) fell on a Wednesday in 2001. On what day of the week will Valentine's Day fall in 2010?

70. Cinco de Mayo (May 5) fell on a Sunday in 2002. On what day of the week will Cinco de Mayo fall in 2018?

In Exercises 71–82, find all whole number solutions of the congruence equation.

71. $x \equiv 10$ mod 3   **72.** $x \equiv 12$ mod 5

73. $2x \equiv 12$ mod 5   **74.** $3x \equiv 8$ mod 11

75. $(2x + 1) \equiv 5$ mod 4   **76.** $(3x + 1) \equiv 4$ mod 9

77. $(2x + 3) \equiv 8$ mod 12 **78.** $(3x + 12) \equiv 7$ mod 10

79. $(2x + 2) \equiv 6$ mod 4   **80.** $(5x + 4) \equiv 2$ mod 8

81. $(4x + 6) \equiv 5$ mod 8   **82.** $(4x + 3) \equiv 3$ mod 4

In Exercises 83–88, find the additive inverse and the multiplicative inverse, if it exists, of the given number.

83. 4 in modulo 9 arithmetic

84. 4 in modulo 5 arithmetic

85. 7 in modulo 10 arithmetic

86. 11 in modulo 16 arithmetic

87. 3 in modulo 8 arithmetic

88. 6 in modulo 15 arithmetic

Modular division can be performed by considering the related multiplication problem. For instance, if $5 \div 7 = x$, then $x \cdot 7 = 5$. Similarly, the quotient $(5 \div 7)$ mod 8 is the solution to the congruence equation $x \cdot 7 \equiv 5$ mod 8, which is 3. In Exercises 89–94, find the given quotient.

89. $(2 \div 7)$ mod 8   **90.** $(4 \div 5)$ mod 8

91. $(6 \div 4)$ mod 9   **92.** $(2 \div 3)$ mod 5

93. $(5 \div 6)$ mod 7   **94.** $(3 \div 4)$ mod 7

Extensions

CRITICAL THINKING

95. Verify that the division $5 \div 8$ has no solution in modulo 8 arithmetic.

96. Verify that the division $4 \div 4$ has more than one solution in modulo 10 arithmetic.

97. Disregarding A.M. or P.M., if it is now 3:00, what time of day will it be in 3500 hours?

98. Using military time, if it is currently 1100 hours, determine what time of day it will be in 4250 hours.

99. There is only one whole number solution between 0 and 11 of the congruence equation $(x^2 + 3x + 7) \equiv 2$ mod 11. Find the solution.

100. There are only two whole number solutions between 0 and 27 of the congruence equation $(x^2 + 5x + 4) \equiv 1$ mod 27. Find the solutions.

EXPLORATIONS

101. Computers often need to generate lists of random numbers to simulate real world situations or even to run video games. However, this is not as simple as it sounds, so sometimes *pseudorandom numbers* are used instead. Pseudorandom numbers are not truly random, as they are determined by several initially chosen values, but they are still useful. One method of generating a list of pseudorandom numbers is called the *linear congruential method*. This method was developed by D. H. Lehmer in 1949. Start by choosing the first random natural number, x_0, also called the *seed*. Next choose natural number values for m, the modulus, a multiplier a, and an increment c. Each successive random number is determined from the preceding number by using the congruence equation

$$x_{n+1} \equiv (ax_n + c) \bmod m$$

For instance, if $m = 9$, $a = 5$, $c = 6$, and we chose $x_0 = 2$ as the seed, then

$$x_1 \equiv (5x_0 + 6) \bmod 9 \equiv (5 \cdot 2 + 6) \bmod 9$$
$$\equiv 16 \bmod 9 \equiv 7$$

and

$$x_2 \equiv (5x_1 + 6) \bmod 9 \equiv (5 \cdot 7 + 6) \bmod 9$$
$$\equiv 41 \bmod 9 \equiv 5$$

Each number generated will be a whole number between 0 and $m - 1$, inclusive.

a. To generate pseudorandom numbers from 0 to 10, we need to choose $m = 11$. For $a = 3$, $c = 4$, and $x_0 = 5$, determine the first five pseudorandom numbers that are generated.

b. Determine the next five pseudorandom numbers in part a. Can you observe a shortcoming of the linear congruential method? Explain.

c. There is a more complicated version of the linear congruential method that does not depend on knowing the previously generated number in the list:

$$x_n \equiv \left(a^n x_0 + c \cdot \frac{a^n - 1}{a - 1} \right) \bmod m$$

Use this formula to find x_3 and x_5 using the same initial values given in part a, and verify that you get the same pseudorandom numbers x_3 and x_5 as you did in part a.

d. Use the formula given in part c to compute the first eight pseudorandom numbers that are generated using the initial values $a = 13$, $c = 21$, $m = 80$, and $x_0 = 59$.

SECTION 7.2 | ## Applications of Modular Arithmetic

ISBN, UPC, and Credit Card Numbers

Every book that is cataloged in the Library of Congress must have an ISBN (International Standard Book Number). The ISBN for the third edition of the *American Heritage Dictionary* is 0-395-44895-6. The first number, 0, indicates that the book is written in English. The next three numbers, 395, indicate the publisher (Houghton Mifflin Company). The five numbers 44895 identify the book (*American Heritage Dictionary*) and the last digit is called a *check digit*. This digit is chosen so as to satisfy the following congruence.

0-395-44895-6

$$0(10) + 3(9) + 9(8) + 5(7) + 4(6) + 4(5) + 8(4) + 9(3) + 5(2) + x \equiv 0 \bmod 11$$
$$0 + 27 + 72 + 35 + 24 + 20 + 32 + 27 + 10 + x \equiv 0 \bmod 11$$
$$247 + x \equiv 0 \bmod 11$$

The digit 6 is selected as the check digit of the ISBN because $247 + 6 = 253$ and $253 \div 11 = 23$ remainder 0. Because the ISBN congruence equation has a modulus of 11, the value of x could be any number from 0 to 10. A check digit of 10 is coded as an X.

One purpose of the ISBN method of coding books is to ensure that orders placed for books are filled accurately. For instance, suppose a clerk sends an order for the *American Heritage Dictionary* and inadvertently enters the number 0-395-44985-6 (the 8 and the adjacent 9 have been transposed). Now

$$0(10) + 3(9) + 9(8) + 5(7) + 4(6) + 4(5) + 9(4) + 8(3) + 5(2) + 6 = 254$$

If we divide 254 by 11, the remainder is not zero. Thus $254 \not\equiv 0 \bmod 11$. Because the congruence is not true, we know there is an error in the order.

EXAMPLE 1 ■ Determine a Check Digit

Determine the ISBN check digit for the book *A Brief History of Time* by Stephen Hawking. The first nine digits of the ISBN are 0-553-05340-?.

Solution
Use the ISBN congruence equation.

$$0(10) + 5(9) + 5(8) + 3(7) + 0(6) + 5(5) + 3(4) + 4(3) + 0(2) + x \equiv 0 \bmod 11$$
$$155 + x \equiv 0 \bmod 11$$

To solve the congruence equation, try whole number values of x that are less than the modulus. Because $155 + 10 = 165 \equiv 0 \bmod 11$, the check digit is 10. Recall that a check digit of 10 is recorded as an X. Therefore, the ISBN is 0-553-05340-X.

CHECK YOUR PROGRESS 1   A purchase order for the book *Mathematica: A System for Doing Mathematics by Computer* includes the ISBN 0-201-15502-4. Determine if this is a valid ISBN.

Solution   See page S26.

✓ **TAKE NOTE**

As mentioned at the right, to solve $155 + x \equiv 0 \bmod 11$, choose whole number values of x that are less than the modulus.

$155 + x \equiv 0 \bmod 11$

$155 + 0 \not\equiv 0 \bmod 11$ • $x = 0$

$155 + 1 \not\equiv 0 \bmod 11$ • $x = 1$

$155 + 2 \not\equiv 0 \bmod 11$ • $x = 2$

⋮

$155 + 10 \equiv 0 \bmod 11$ • $x = 10$ results in a true congruence

QUESTION   *Could 1-805-53X97-X be a valid ISBN?*

Another coding scheme that is closely related to the ISBN is the UPC (Universal Product Code). This number is placed on many items and is particularly useful in grocery stores. A check-out clerk passes the product by a scanner, which reads the number from a bar code and records the price on the cash register. If the price of an item changes for a promotional sale, the price is updated in the computer, thereby relieving a clerk of having to reprice each item. In addition to pricing items, the UPC gives the store manager accurate information about inventory and the buying habits of the store's customers.

5 00041 10076 4

ANSWER   *No. X can only appear as the check digit.*

The UPC is a 12-digit number that satisfies a congruence equation. The last digit is the check digit. Here is the UPC for the DVD release of the film *The Matrix,* along with the UPC congruence equation.

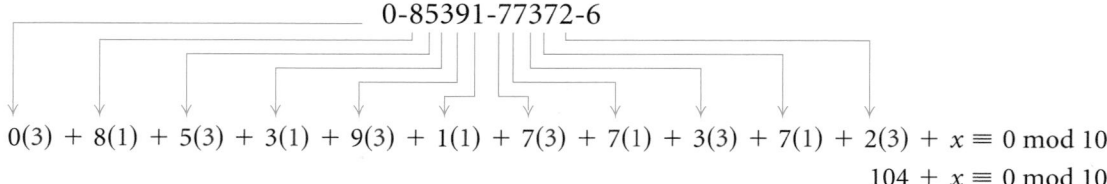

$$0(3) + 8(1) + 5(3) + 3(1) + 9(3) + 1(1) + 7(3) + 7(1) + 3(3) + 7(1) + 2(3) + x \equiv 0 \bmod 10$$
$$104 + x \equiv 0 \bmod 10$$

▼ **point of interest**

Bar codes were first patented in 1949. However, the first commercial computer did not arrive until 1951 (there were computers for military use prior to 1951), so bar codes were not extensively used. The UPC was developed in 1973.

In addition to the ISBN and UPC, there are other coding schemes, such as EAN for European Article Numbering and SKU for Stock Keeping Unit.

The digit 6 is selected as the last digit (the check digit) of the UPC because $104 + 6 = 110$ and $110 \div 10 = 11$ remainder 0. Because the UPC congruence equation has a modulus of 10, the value of x, the check digit, can be any whole number from 0 to 9.

The UPC method is not quite as accurate as the ISBN method when it comes to detecting transpositions of digits. Recall that in the ISBN method, when two digits are transposed, the congruence equation is no longer true. This does not always happen with the UPC method. For instance, suppose the 2 and the 7 preceding it in the UPC above are transposed. The new UPC is 0-85391-77327-6. Using this number and the UPC congruence equation, we have

$$0\text{-}85391\text{-}77327\text{-}6$$
$$0(3) + 8(1) + 5(3) + 3(1) + 9(3) + 1(1) + 7(3) + 7(1) +$$
$$3(3) + 2(1) + 7(3) + 6 = 120 \equiv 0 \bmod 10$$

Notice that the congruence is still true modulo 10 even though the UPC is not correct. This is the result of an unfortunate coincidence; if any other two digits were transposed, the result would have been a false congruence. It can be shown that undetected transposition errors of adjacent digits a and b occur only when $|a - b| = 5$. In this case, the two transposed digits are 2 and 7, and $|2 - 7| = |-5| = 5$.

EXAMPLE 2 ■ **Determine the Check Digit for a UPC**

A new product a company has developed has been assigned the UPC 0-21443-32912- ? . Determine the check digit for the UPC.

Solution
Use the UPC congruence equation.

$$0(3) + 2(1) + 1(3) + 4(1) + 4(3) + 3(1) + 3(3) +$$
$$2(1) + 9(3) + 1(1) + 2(3) + x \equiv 0 \bmod 10$$
$$69 + x \equiv 0 \bmod 10$$

INSTRUCTOR NOTE
Ask your students to find three products at home with a UPC and then verify the validity of each UPC.

To solve the congruence equation, try whole number values of x that are less than the modulus. Because $69 + 1 \equiv 0 \bmod 10$, the check digit is 1.

CHECK YOUR PROGRESS 2   Is 1-32342-65933-9 a valid UPC?

Solution   *See page S26.*

Companies that issue credit cards also use modular arithmetic to determine whether a credit card number is valid. This is especially necessary in e-commerce, where credit card information is frequently sent over the Internet. The primary coding method is based on the *Luhn algorithm*, which uses mod 10 arithmetic.

Credit card numbers are normally 13 to 16 digits long. The first one to four digits are used to identify the card issuer. The table below shows the identification prefixes used by four popular card issuers.

Card Issuer	Prefix	Number of digits
MasterCard	51 to 55	16
Visa	4	13 or 16
American Express	34 or 37	15
Discover	6011	16

The Luhn algorithm, used to determine whether a credit card number is valid, is calculated as follows. Beginning with the next-to-last digit (the last digit is the check digit) and reading from right to left, double every other digit. If a digit becomes a two-digit number after being doubled, treat the number as two individual digits. Now find the sum of the new list of digits; the final sum must equal 0 mod 10. The Luhn algorithm is demonstrated in the next example.

EXAMPLE 3 ■ Determine a Valid Credit Card Number

Determine whether 5234 8213 3410 1298 is a valid credit card number.

Solution
Highlight every other digit, beginning with the next-to-last digit and reading from right to left.

5 2 3 4 8 2 1 3 3 4 1 0 1 2 9 8

Next double each of the highlighted digits.

10 2 6 4 16 2 2 3 6 4 2 0 2 2 18 8

Finally, add all digits, treating two-digit numbers as two single digits.

$$(1 + 0) + 2 + 6 + 4 + (1 + 6) + 2 + 2 + 3 + 6 + 4 + 2 + 0 +$$
$$2 + 2 + (1 + 8) + 8 = 60$$

Because 60 ≡ 0 mod 10, this is a valid credit card number.

CHECK YOUR PROGRESS 3   Is 6011 0123 9145 2317 a valid credit card number?

Solution   *See page S26.*

Cryptology

Related to codes on books and grocery items are secret codes. These codes are used to send messages between people, companies, or nations. It is hoped that by devising a code that is difficult to break, the sender can prevent the communication from being read if it is intercepted by an unauthorized person. **Cryptology** is the study of making and breaking secret codes.

Before we discuss how messages are coded, we need to define a few terms. **Plaintext** is a message before it is coded. The line

SHE WALKS IN BEAUTY LIKE THE NIGHT

from Lord Byron's poem "Ode to a Nightingale" is in plaintext. **Ciphertext** is the message after it has been written in code. The line

ODA SWHGO EJ XAWQPU HEGA PDA JECDP

is the same line of the poem in ciphertext.

The method of changing from plaintext to ciphertext is called **encryption.** The line from the poem was encrypted by substituting each letter in plaintext with the letter that is 22 letters after that letter in the alphabet. (Continue from the beginning when the end of the alphabet is reached.) This is called a *cyclical coding scheme* because each letter of the alphabet is shifted the same number of positions. The original alphabet and the substitute alphabet are shown below.

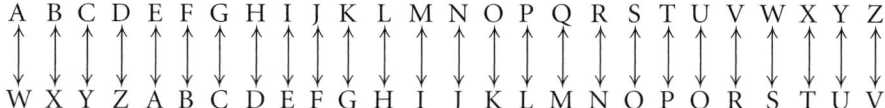

To **decrypt** a message means to take the ciphertext message and write it in plaintext. If a cryptologist thinks a message has been encrypted using a cyclical substitution code like the one shown above, the key to the code can be found by taking a word from the message (usually one of the longer words) and continuing the alphabet for each letter of the word. When a recognizable word appears, the key can be determined. This method is shown below using the ciphertext word XAWQPU.

X	A	W	Q	P	U
Y	B	X	R	Q	V
Z	C	Y	S	R	W
A	D	Z	T	S	X
B	E	A	U	T	Y

Shift four positions

Once a recognizable word has been found (BEAUTY), count the number of positions that the letters have been shifted (four, in this case). To decode the message, substitute the letter of the normal alphabet that comes four positions after the letter in the ciphertext.

Cyclical encrypting using the alphabet is related to modular arithmetic. We begin with the normal alphabet and associate each letter with a number as shown in Table 7.1.

Table 7.1 *Numerical Equivalents for the Letters of the Alphabet*

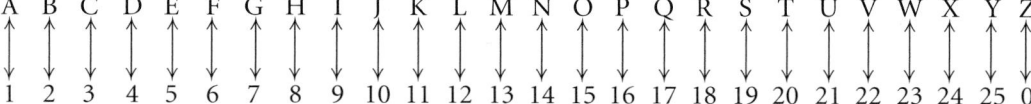

A	B	C	D	E	F	G	H	I	J	K	L	M	N	O	P	Q	R	S	T	U	V	W	X	Y	Z
1	2	3	4	5	6	7	8	9	10	11	12	13	14	15	16	17	18	19	20	21	22	23	24	25	0

If the encrypting code is to shift each letter of the plaintext message m positions, then the corresponding letter in the ciphertext message is given by $c \equiv (p + m) \bmod 26$, where p is the numerical equivalent of the plaintext letter and c is the numerical equivalent of the ciphertext letter. The letter Z is coded as 0 because $26 \equiv 0 \bmod 26$.

Each letter in Lord Byron's poem was shifted 22 positions ($m = 22$) to the right. To code the plaintext letter S in the word SHE, we use the congruence $c \equiv (p + m) \bmod 26$.

$c \equiv (p + m) \bmod 26$

$c \equiv (19 + 22) \bmod 26$ • $p = 19$ (S is the 19th letter.)

$c \equiv 41 \bmod 26$ $m = 22$, the number of positions the letter is shifted

$c = 15$

The 15th letter is O. Thus S is coded as O.

Once plaintext has been converted to ciphertext, there must be a method by which the person receiving the message can return the message to plaintext. For the cyclical code, the congruence is $p \equiv (c + n) \bmod 26$, where p and c are defined as before and $n = 26 - m$. The letter O in ciphertext is decoded below using the congruence $p \equiv (c + n) \bmod 26$.

$p \equiv (c + n) \bmod 26$

$p \equiv (15 + 4) \bmod 26$ • $c = 15$ (O is the 15th letter.)

$p \equiv 19 \bmod 26$ $n = 26 - m = 26 - 22 = 4$

$p = 19$

The 19th letter is S. Thus O is decoded as S.

Math Matters A Cipher of Caesar

Julius Caesar supposedly used a cyclical alphabetic encrypting code to communicate with his generals. The messages were encrypted using the congruence $c \equiv (p + 3) \bmod 26$ (in modern notation). Using this coding scheme, the message BEWARE THE IDES OF MARCH would be coded as EHZDUH WKH LGHV RI PDUFK.

EXAMPLE 4 ◼ **Write Messages Using Cyclical Coding**

Use the cyclical alphabetic encrypting code that shifts each letter 11 positions to code CATHERINE THE GREAT and decode TGLY ESP EPCCTMWP.

Solution

The encrypting congruence is $c \equiv (p + 11) \bmod 26$. Replace p by the numerical equivalent of each letter of plaintext and determine c. The results for CATHERINE are shown below.

C	$c \equiv (3 + 11) \bmod 26 \equiv 14 \bmod 26 \equiv 14$	Code C as N.
A	$c \equiv (1 + 11) \bmod 26 \equiv 12 \bmod 26 \equiv 12$	Code A as L.
T	$c \equiv (20 + 11) \bmod 26 \equiv 31 \bmod 26 \equiv 5$	Code T as E.
H	$c \equiv (8 + 11) \bmod 26 \equiv 19 \bmod 26 \equiv 19$	Code H as S.
E	$c \equiv (5 + 11) \bmod 26 \equiv 16 \bmod 26 \equiv 16$	Code E as P.
R	$c \equiv (18 + 11) \bmod 26 \equiv 29 \bmod 26 \equiv 3$	Code R as C.
I	$c \equiv (9 + 11) \bmod 26 \equiv 20 \bmod 26 \equiv 20$	Code I as T.
N	$c \equiv (14 + 11) \bmod 26 \equiv 25 \bmod 26 \equiv 25$	Code N as Y.
E	$c \equiv (5 + 11) \bmod 26 \equiv 16 \bmod 26 \equiv 16$	Code E as P.

Continuing, the plaintext would be coded as NLESPCTYP ESP RCPLE.

Because $m = 11$, $n = 26 - 11 = 15$. The ciphertext is decoded by using the congruence $p \equiv (c + 15) \bmod 26$. The results for TGLY are shown below.

T	$c \equiv (20 + 15) \bmod 26 \equiv 35 \bmod 26 \equiv 9$	Decode T as I.
G	$c \equiv (7 + 15) \bmod 26 \equiv 22 \bmod 26 \equiv 22$	Decode G as V.
L	$c \equiv (12 + 15) \bmod 26 \equiv 27 \bmod 26 \equiv 1$	Decode L as A.
Y	$c \equiv (25 + 15) \bmod 26 \equiv 40 \bmod 26 \equiv 14$	Decode Y as N.

Continuing, the ciphertext would be decoded as IVAN THE TERRIBLE.

CHECK YOUR PROGRESS 4 Use the cyclical alphabetic encrypting code that shifts each letter 17 positions to encode ALPINE SKIING and decode TIFJJ TFLEKIP JBZZEX.

Solution *See page S26.*

The practicality of a cyclical alphabetic coding scheme is limited because it is relatively easy for a cryptologist to determine the coding scheme. (Recall the method used for the line from Lord Byron's poem.) A coding scheme that is a little more difficult to break is based on the congruence $c \equiv (ap + m) \bmod 26$, where a and 26 do not have a common factor. (For instance, a cannot be 14 because 14 and 26 have a common factor of 2.) The reason that a and 26 cannot have a common factor is related to the procedure for determining the decoding congruence. However, we will leave that discussion to other math courses.

EXAMPLE 5 ◼ **Encode a Message**

Use the congruence $c \equiv (5p + 2) \bmod 26$ to encode the message LASER PRINTER.

Solution

The encrypting congruence is $c \equiv (5p + 2) \bmod 26$. Replace p by the numerical equivalent of each letter from Table 7.1 and determine c. The results for LASER are shown below.

L	$c \equiv (5 \cdot 12 + 2) \bmod 26 \equiv 62 \bmod 26 = 10$	Code L as J.
A	$c \equiv (5 \cdot 1 + 2) \bmod 26 \equiv 7 \bmod 26 = 7$	Code A as G.
S	$c \equiv (5 \cdot 19 + 2) \bmod 26 \equiv 97 \bmod 26 = 19$	Code S as S.
E	$c \equiv (5 \cdot 5 + 2) \bmod 26 \equiv 27 \bmod 26 = 1$	Code E as A.
R	$c \equiv (5 \cdot 18 + 2) \bmod 26 \equiv 92 \bmod 26 = 14$	Code R as N.

Continuing, the plaintext is coded in ciphertext as JGSAN DNUTXAN.

CHECK YOUR PROGRESS 5 Use the congruence $c \equiv (3p + 1) \bmod 26$ to encode the message COLOR MONITOR.

Solution *See page S27.*

Decoding a message that was encrypted using the congruence $c \equiv (ap + m) \bmod n$ requires solving the congruence for p. The method relies on multiplicative inverses, which were discussed in the last section.

Here we solve the congruence used in Example 5 for p.

$$c = 5p + 2$$
$$c - 2 = 5p \qquad \text{• Subtract 2 from each side of the equation.}$$
$$21(c - 2) = 21(5p) \qquad \text{• Multiply each side of the equation by the multiplicative inverse of 5. Because } 21 \cdot 5 \equiv 1 \bmod 26, \text{ multiply each side by 21.}$$

> **TAKE NOTE**
>
> $21(5p) \equiv p \bmod 26$ because $21 \cdot 5 = 105$ and $105 \div 26 = 4$ remainder 1.

$$[21(c - 2)] \bmod 26 \equiv p$$

Using this congruence equation, we can decode the ciphertext message JGSAN DNUTXAN. The method for decoding JGSAN is shown below.

J	$[21(10 - 2)] \bmod 26 \equiv 168 \bmod 26 \equiv 12$	Decode J as L.
G	$[21(7 - 2)] \bmod 26 \equiv 105 \bmod 26 \equiv 1$	Decode G as A.
S	$[21(19 - 2)] \bmod 26 \equiv 357 \bmod 26 \equiv 19$	Decode S as S.
A	$[21(1 - 2)] \bmod 26 \equiv (-21) \bmod 26 \equiv 5$	Decode A as E.
N	$[21(14 - 2)] \bmod 26 \equiv 252 \bmod 26 \equiv 18$	Decode N as R.

Note that to decode A it was necessary to determine $(-21) \bmod 26$. Recall that this requires adding the modulus until a whole number less than 26 results. Because $(-21) + 26 = 5$, we have $(-21) \bmod 26 \equiv 5$.

EXAMPLE 6 ▪ Decode a Message

Decode the message ACXUT CXRT, which was encrypted using the congruence $c \equiv (3p + 5) \bmod 26$.

Solution

Solve the congruence equation for p.

$$c = 3p + 5$$
$$c - 5 = 3p$$
$$9(c - 5) = 9(3p) \qquad \bullet \ 9(3) = 27 \text{ and } 27 \equiv 1 \bmod 26$$
$$[9(c - 5)] \bmod 26 \equiv p$$

The decoding congruence is $p \equiv [9(c - 5)] \bmod 26$.

Using this congruence, we will show the details for decoding ACXUT.

A	$[9(1 - 5)] \bmod 26 \equiv (-36) \bmod 26 \equiv 16$	Decode A as P.
C	$[9(3 - 5)] \bmod 26 \equiv (-18) \bmod 26 \equiv 8$	Decode C as H.
X	$[9(24 - 5)] \bmod 26 \equiv 171 \bmod 26 \equiv 15$	Decode X as O.
U	$[9(21 - 5)] \bmod 26 \equiv 144 \bmod 26 \equiv 14$	Decode U as N.
T	$[9(20 - 5)] \bmod 26 \equiv 135 \bmod 26 \equiv 5$	Decode T as E.

Continuing, we would decode the message as PHONE HOME.

CHECK YOUR PROGRESS 6 Decode the message IGHT OHGG, which was encrypted using the congruence $c \equiv (7p + 1) \bmod 26$.

Solution See page S27.

Math Matters   The Enigma Machine

During World War II, Nazi Germany encoded its transmitted messages using an Enigma machine, first patented in 1919. Each letter of the plaintext message was substituted with a different letter, but the machine changed the substitutions throughout the message so that the ciphertext appeared more random and would be harder to decipher if the message were intercepted. The Enigma machine accomplished its task by using three wheels of letters (chosen from a box of five) that rotated as letters were typed. In addition, electrical sockets corresponding to letters were connected in pairs with wires. The choice and order of the three wheels, their starting positions, and the arrangement of the electrical wires determined how the machine would encode a message. The receiver of the message needed to know the setup of the wheels and wires; with the receiver's Enigma machine configured identically, the ciphertext could be decoded. There are a staggering 150 quintillion number of different ways to configure the Enigma machine, making it very difficult to decode messages without knowing the setup that was used. A team of mathematicians and other codebreakers, led by Alan Turing, was assembled by the British government in an attempt to decode the German messages. The team was eventually successful, and their efforts helped change the course of World War II. One aspect of the Enigma machine that aided the codebreakers was the fact that the machine would never substitute a letter with itself.

PBS aired an excellent account of the breaking of the Enigma codes during World War II as part of its *Nova* series entitled *Decoding Nazi Secrets*. Information is available on the web at **http://www.pbs.org/wgbh/nova/decoding/**.

Excursion

▼ **point of interest**

Sophie Germain (zhĕr-män) (1776–1831) was born in Paris, France. Because enrollment in the university she wanted to attend was available only to men, Germain attended under the name Antoine-August Le Blanc. Eventually her ruse was discovered, but not before she came to the attention of Pierre Lagrange, one of the best mathematicians of the time. He encouraged her work and became a mentor to her.

A certain type of prime number named after Germain is called a Germain prime number. It is a prime p such that p and $2p + 1$ are both prime. For instance, 11 is a Germain prime because 11 and $2(11) + 1 = 23$ are both prime numbers. Germain primes are used in public key cryptography.

✔ **TAKE NOTE**

Although some calculators are capable of performing modular arithmetic, the numbers involved in these computations are so large that many calculators will give an error message. Mathematical software such as Maple and Mathematica can be used to obtain accurate results.

Public Key Cryptography

We end this section by discussing a relatively new method of coding that uses modular arithmetic. It is called *public key cryptography* and is based on the work of Whitfield Diffie and Martin E. Hellman that was published in 1976. Their patent based on this concept expired in 1997.

In public key cryptography, two *keys* are created, one public and one private. (A key is the basis on which the code is created. For instance, in Example 6, the key could be arithmetic mod 26.) The uniqueness of this system is that the public key is known to everyone; anyone having the public key can encode a message. However, only the people who have the *private* key can *decode* the message.

Here is how the method works. We will use very small numbers in our example. In practice, these numbers would be well over 100 digits!

Suppose Amanda and Brian each generate a private random number. Using a for Amanda and b for Brian, suppose $a = 21$ and $b = 19$. Amanda's number is known only to her; Brian's number is known only to him. Now Amanda and Brian decide on some prime number p and another number g, called the *generator*. These are public numbers. We will use $p = 23$ and $g = 13$.

Next Amanda and Brian calculate their public values and a private value based on p and g. Amanda's public value is $g^a \bmod p = 13^{21} \bmod 23$, which equals 16, and Brian's public value is $g^b \bmod p = 13^{19} \bmod 23$, which is 2. Amanda and Brian exchange these public values. Notice that the public values are different. However, using these public values, Amanda and Brian calculate the same private key. Amanda uses $(g^b)^a \bmod p$; Brian uses $(g^a)^b \bmod p$. By the rules of exponents $(g^b)^a = (g^a)^b$, so the private key is the same. Using this formula, the private key is $(13^{21})^{19} \bmod 23 \equiv 12$.

A refinement of the above method is called *RSA public key cryptography*. The initials RSA come from the first letters of the names of each of the mathematicians who devised the method: Ronald L. Rivest, Adi Shamir, and Leonard M. Adelman. This method is a refinement of regular public key cryptography in that it shows how to use the public and private keys to encrypt and decrypt messages.

To use RSA public key cryptography, begin by choosing two very large prime numbers p and q. Then choose a third odd number n (not necessarily a prime number) such that n and $(p - 1)(q - 1)$ have no common factors.

Now find m such that $mn \equiv 1 \bmod (p - 1)(q - 1)$. Recall that this means m is the multiplicative inverse of n in modular arithmetic. Your public key is the product pq (only the product, not the individual numbers that make up the product) and n. Your private key is m, the multiplicative inverse of $n \bmod (p - 1)(q - 1)$.

As an example, we will use $p = 59$, $q = 83$, and $n = 129$. From these values we can calculate $m = 1401$.

To encrypt a message, each letter of the alphabet is given a number; for instance, A = 11, B = 12, C = 13, ... Now replace each letter in the message with its number equivalent. Using our coding scheme, the word MATH would be 23 11 30 18. The letters are coded using the formula $a^n \bmod pq$, where a is the number equivalent of the letter. For the word MATH, we would send the message

$$23^{129} \bmod 4897 \equiv 3065 \qquad 11^{129} \bmod 4897 \equiv 2001$$

$$30^{129} \bmod 4897 \equiv 957 \qquad 18^{129} \bmod 4897 \equiv 2753$$

(continued)

The person receiving this message would decrypt it using the formula $c^m \bmod pq$, as shown below.

$$3065^{1401} \bmod 4897 \equiv 23 \qquad 2001^{1401} \bmod 4897 \equiv 11$$
$$957^{1401} \bmod 4897 \equiv 30 \qquad 2753^{1401} \bmod 4897 \equiv 18$$

The example we have shown is RSA in its simplest form. In practice, much larger prime numbers are used, and instead of coding one letter at a time, an entire block of letters is sent. For instance, we could send MA as $2311^{129} \bmod 4897 \equiv 3347$ and TH as $3018^{129} \bmod 4897 \equiv 625$.

Excursion Exercises

1. This Excursion will enable you to create your own secret coding scheme based on RSA public key cryptography. This exercise is intended to show an application of RSA public key cryptography and should not be used for any information that needs secure encryption. This exercise can be completed by visiting our web site at **math.college.hmco.com/students.**

2. Research the topic of *digital signatures* and write a report on what they are and how cryptography is used in conjunction with digital signatures.

Exercise Set 7.2   (Suggested Assignment: 1–61 every other odd; 42, 52, 62, 63)

In Exercises 1–6, determine whether the given number is a valid ISBN.

1. 0-281-44268-5
2. 1-56690-182-0
3. 0-446-31418-2
4. 0-614-35945-X
5. 0-684-85972-6
6. 0-231-10324-1

In Exercises 7–14, determine the correct check digit for each ISBN.

7. *Dune* by Frank Herbert, 0-441-17271-?
8. *The Hobbit* by J. R. R. Tolkien, 0-395-07122-?
9. *The Joy of Pi* by David Blatner, 0-802-71332-?
10. *The Mathematical Tourist* by Ivars Peterson, 0-716-72064-?
11. *War and Peace* by Leo Tolstoy, 0-140-44417-?
12. *The Grapes of Wrath* by John Steinbeck, 0-140-28162-?
13. *Relativity: The Special and the General Theory* by Albert Einstein, 0-517-88441-?
14. *The Shining* by Stephen King, 0-451-16091-?

In Exercises 15–22, determine the correct check digit for the UPC.

15. 0-51000-03139-? (Campbell's chicken vegetable soup)
16. 0-37000-46003-? (Jif peanut butter)
17. 0-37000-37717-? (Pringles potato chips)
18. 0-71142-00001-? (Arrowhead bottled water)
19. 0-25500-81121-? (Folgers coffee)
20. 3-8137-008238-? (Johnson's cotton swabs)
21. 7-17951-00461-? (*Fantasia* DVD)
22. 0-42280-00962-? (*Star Wars* soundtrack CD)

Some money orders have serial numbers that consist of a 10-digit number followed by a check digit. The check digit is chosen to be the sum of the first 10 digits mod 9. In Exercises 23–26, determine the check digit for the given money order serial number.

23. 0316615498-?
24. 5492877463-?
25. 1331497533-?
26. 3414793288-?

Many printed airline tickets contain a 10-digit document number followed by a check digit. The check digit is chosen to be the sum of the first 10 digits mod 7. In Exercises 27–30, determine whether the given number is a valid document number.

27. 1182649758 2   **28.** 1260429984 4

29. 2026178914 5   **30.** 2373453867 6

In Exercises 31–38, determine whether the given credit card number is a valid number.

31. 4417-5486-1785-6411 **32.** 5164-8295-1229-3674

33. 5591-4912-7644-1105 **34.** 6011-4988-1002-6487

35. 6011-0408-4977-3158 **36.** 4896-4198-8760-1970

37. 3715-548731-84466 **38.** 3401-714339-12041

In Exercises 39–44, encode the message by using a cyclical alphabetic encrypting code that shifts the message the stated number of positions.

39. 8 positions: THREE MUSKETEERS

40. 6 positions: FLY TONIGHT

41. 12 positions: IT'S A GIRL

42. 9 positions: MEET AT NOON

43. 3 positions: STICKS AND STONES

44. 15 positions: A STITCH IN TIME

In Exercises 45–48, use a cyclical alphabetic encrypting code that shifts the letters the stated number of positions to decode the encrypted message.

45. 18 positions: SYW GX WFDAYZLWFEWFL

46. 20 positions: CGUACHUNCIH LOFYM NBY QILFX

47. 15 positions: UGXTCS XC CTTS

48. 8 positions: VWJWLG QA XMZNMKB

In Exercises 49–52, use a cyclical alphabetic encrypting code to decode the encrypted message.

49. YVIBZM RDGG MJWDINJI

50. YBZAM HK YEBZAM

51. UDGIJCT RDDZXT

52. AOB HVS HCFDSRCSG

53. Julius Caesar supposedly used an encrypting code equivalent to the congruence $c \equiv (p + 3) \mod 26$. Use the congruence to encrypt the message "men willingly believe what they wish."

54. Julius Caesar supposedly used an encrypting code equivalent to the congruence $c \equiv (p + 3) \mod 26$. Use the congruence to decrypt the message WKHUH DUH QR DFFLGHQWV.

55. Use the encrypting congruence $c \equiv (3p + 2) \mod 26$ to code the message TOWER OF LONDON.

56. Use the encrypting congruence $c \equiv (5p + 3) \mod 26$ to code the message DAYLIGHT SAVINGS TIME.

57. Use the encrypting congruence $c \equiv (7p + 8) \mod 26$ to code the message PARALLEL LINES.

58. Use the encrypting congruence $c \equiv (3p + 11) \mod 26$ to code the message NONE SHALL PASS.

59. Decode the message LOFT JGMK LBS MNWMK that was encrypted using the congruence $c \equiv (3p + 4) \mod 26$.

60. Decode the message BTYW SCRBKN UCYN that was encrypted using the congruence $c \equiv (5p + 6) \mod 26$.

61. Decode the message SNUUHQ FM VFALHDZ that was encrypted using the congruence $c \equiv (5p + 9) \mod 26$.

62. Decode the message GBBZ OJQBWJ TBR GJHI that was encrypted using the congruence $c \equiv (7p + 1) \mod 26$.

Extensions

CRITICAL THINKING

63. Explain why the method used to determine the check digit of the money order serial numbers in Exercises 23–26 will not detect a transposition of digits among the first 10 digits.

64. Explain why the transposition of adjacent digits a and b in a UPC will go undetected only when $|a - b| = 5$.

EXPLORATIONS

65. When banks process an Electronic Funds Transfer (EFT), they assign a nine-digit routing number, where the ninth digit is a check digit. The check digit is chosen such that when the nine digits are multiplied in turn by the numbers 3, 7, 1, 3, 7, 1, 3, 7, and 1, the sum of the resulting products is 0 mod 10. For instance, 123456780 is a valid routing number because

$$1(3) + 2(7) + 3(1) + 4(3) + 5(7) + 6(1)$$
$$+ 7(3) + 8(7) + 0(1) = 150$$

and $150 \equiv 0 \mod 10$.

a. Compute the check digit for the routing number 72859372-?

b. Verify that 584926105 is a valid routing number.

c. If a bank clerk inadvertantly types the routing number in part b as 584962105, will the computer be able to detect the error?

d.  A bank employee accidentally transposed the 6 and the 1 in the routing number from part b and entered the number 584921605 into the computer system, but the computer did not detect an error. Explain why.

66. The Codabar system is a bar code method developed in 1972 that is commonly used in libraries.

Springfield Public Library

2 8888 00000 0001

Each bar code contains a 14-digit number, where the 14th digit is a check digit. The various check digit schemes discussed in this section are designed to detect errors, but they cannot catch every possible error. The Codabar method is more thorough and will detect any single mistyped digit as well as any adjacent digits that have been transposed, with the exception of 9−0 and 0−9. To determine the check digit, label the first 13 digits a_1 through a_{13}. The check digit is given by

$$(-2a_1 - a_2 - 2a_3 - a_4 - 2a_5 - a_6 - 2a_7 - a_8 \\ - 2a_9 - a_{10} - 2a_{11} - a_{12} - 2a_{13} - r) \bmod 10$$

where r is the number of digits from the digits a_1, a_3, a_5, a_7, a_9, a_{11}, and a_{13} that are greater than or equal to 5. For example, the bar code with the number 3194870254381-? has digits a_3, a_5, and a_9 that are greater than or equal to 5. Thus $r = 3$ and

$$-2(3) - 1 - 2(9) - 4 - 2(8) - 7 - 2(0) \\ - 2 - 2(5) - 4 - 2(3) - 8 - 2(1) - 3 = -87 \\ \equiv 3 \bmod 10$$

The check digit is 3.

a. Find the check digit for the bar code number 1982273895263-?.

b. Find the check digit for the bar code number 6328293856384-?.

Introduction to Group Theory

Introduction to Groups

An **algebraic system** is a set of elements along with one or more operations for combining the elements. The real numbers with the operations of addition and multiplication are an example of an algebraic system. Mathematicians call this particular algebraic system a *field*.

In the previous two sections of this chapter we discussed operations modulo n. Consider the set {0, 1, 2, 3, 4, 5} and addition modulo 6. The set of elements is {0, 1, 2, 3, 4, 5} and the operation is addition modulo 6. In this case there is only one operation. This is an example of an algebraic system called a *group*.

In addition to fields and groups, there are other types of algebraic systems. It is the properties of an algebraic system that distinguish it from another algebraic system. In this section, we will focus on groups.

> **Definition of a Group**
>
> A **group** is a set of elements, with one operation, that satisfies the following four properties.
>
> **1.** The set is closed with respect to the operation.
>
> **2.** The operation satisfies the associative property.
>
> **3.** There is an identity element.
>
> **4.** Each element has an inverse.

Note from the above definition that a group is an algebraic system with *one* operation and that the operation must have certain characteristics. The first of these characteristics is that the set is *closed* with respect to the operation. **Closure** means that if any two elements are combined using the operation, the result must be an element of the set.

For example, the set {0, 1, 2, 3, 4, 5} with addition modulo 6 as the operation is closed. If we add two numbers of this set, modulo 6, the result is always a number of the set. For instance, $(3 + 5) \mod 6 \equiv 2$ and $(1 + 3) \mod 6 \equiv 4$.

As another example, consider the whole numbers {0, 1, 2, 3, 4, ...} with multiplication as the operation. If we multiply two whole numbers, the result is a whole number. For instance, $5 \cdot 9 = 45$ and $12 \cdot 15 = 180$. So the set of whole numbers is closed using multiplication as the operation. However, the set of whole numbers is not closed with division as the operation. For example, even though $36 \div 9 = 4$ (a whole number), $6 \div 4 = 1.5$, which is not a whole number. Therefore, the set of whole numbers is not closed with respect to division.

QUESTION   ***a.*** *Is the set of whole numbers closed with respect to addition?*

   b. *Is the set of whole numbers closed with respect to subtraction?*

The second requirement of a group is that the operation must satisfy the associative property. Recall that the Associative Property of Addition states that $a + (b + c) = (a + b) + c$. Addition modulo 6 is an associative operation. For instance, if we use the symbol $\triangledown$ to represent addition modulo 6, then

$$2 \triangledown (5 \triangledown 3) = 2 \triangledown (8 \mod 6) = 2 \triangledown 2 = 4$$

and

$$(2 \triangledown 5) \triangledown 3 = (7 \mod 6) \triangledown 3 = 1 \triangledown 3 = 4$$

Thus $2 \triangledown (5 \triangledown 3) = (2 \triangledown 5) \triangledown 3$.

Although there are some operations that do not satisfy the associative property, we will assume that an operation used in this section is associative.

The third requirement of a group is that the set must contain an *identity element*. An **identity element** is an element that, when combined with a second element using the group's operation, always returns the second element. As an illustration, if zero is added to a number, there is no change: $6 + 0 = 6$ and $0 + 1.5 = 1.5$. The number 0 is called an *additive identity*. Similarly, if we multiply any number by 1, there is no change: $\frac{1}{2} \cdot 1 = \frac{1}{2}$ and $1 \cdot 5 = 5$. The number 1 is called a *multiplicative identity*. For the set {0, 1, 2, 3, 4, 5} with addition modulo 6 as the operation, the identity element is 0. As we will soon see, an identity element does not always have to be zero or one.

ANSWER   ***a.*** *Yes. The sum of any two whole numbers is a whole number.*   ***b.*** *No. The difference between any two whole numbers is not always a whole number. For instance, $5 - 8 = -3$, but -3 is not a whole number.*

The numbers 3 and −3 are called *additive inverses*. Adding these two numbers results in the additive identity: $3 + (−3) = 0$. The numbers $\frac{2}{3}$ and $\frac{3}{2}$ are called *reciprocals* or *multiplicative inverses*. Multiplying these two numbers results in the multiplicative identity: $\frac{2}{3} \cdot \frac{3}{2} = 1$. The last requirement of a group is that each element must have an *inverse*. This is a little more difficult to see for the set $\{0, 1, 2, 3, 4, 5\}$ with addition modulo 6 as the operation. However, using addition modulo 6, we have

$$(0 + 0) \bmod 6 \equiv 0 \qquad (2 + 4) \bmod 6 \equiv 0$$
$$(1 + 5) \bmod 6 \equiv 0 \qquad (3 + 3) \bmod 6 \equiv 0$$

The above equations show that 0 is its own inverse, 1 and 5 are inverses, 2 and 4 are inverses, and 3 is its own inverse. Therefore, every element has an inverse.

We have shown that the set $\{0, 1, 2, 3, 4, 5\}$ with addition modulo 6 as the operation is a group because it satisfies the four conditions stated in the definition of a group (see page 428).

EXAMPLE 1 ▧ **Verify the Properties of a Group**

Show that the integers with addition as the operation form a group.

Solution

We must show that the four properties of a group are satisfied.

1. The sum of two integers is always an integer. For instance,

$$−12 + 5 = −7 \qquad \text{and} \qquad −14 + (−21) = −35$$

   Therefore, the integers are closed with respect to addition.

2. The associative property of addition holds true for the integers.

3. The identity element is 0, which is an integer. Therefore, the integers have an identity element for addition.

4. Each element has an inverse. If a is an integer, then $−a$ is the inverse of a.

Because each of the four conditions of a group is satisfied, the integers with addition as the operation form a group.

CHECK YOUR PROGRESS 1 Do the integers with multiplication as the operation form a group?

Solution *See page S27.*

historical note

Niels Henrik Abel (1802–1829) was born near Stavenger, Norway. His mathematical prowess was recognized quite early in his schooling and he was encouraged to pursue a career in mathematics.

Abel became interested in trying to find a formula similar to the quadratic formula that could be used to solve all fifth-degree equations. In the process, he discovered that it was impossible to find such a formula. Part of Abel's proof investigated whether the equation $s_1 s_2 x = s_2 s_1 x$ was true; in other words, his conclusion depended on whether or not the commutative property held true. ∎

Recall that the **commutative property** for an operation states that the order in which two elements are combined does not affect the result. For each of the groups discussed thus far, the operation has satisfied the commutative property. For example, the group $\{0, 1, 2, 3, 4, 5\}$ with addition modulo 6 satisfies the commutative property, since for instance $2 + 5 = 5 + 2$. Groups in which the operation satisfies the commutative property are called **commutative groups** or **abelian groups,** after Niels Abel (see the note at the left). The type of group we will look at next is an example of a *nonabelian group*. A **nonabelian group** is a group whose operation does not satisfy the commutative property.

Symmetry Groups

The concept of a group is very general. The elements that make up a group do not have to be numbers and the operation does not have to be addition or multiplication. Also, a group does not have to satisfy the commutative property. We will now look at another group, called a *symmetry group,* an extension of which plays an important role in the study of atomic reactions. Symmetry groups are based on *regular polygons.* A **regular polygon** is a polygon all of whose sides have the same length and all of whose angles have the same measure.

Consider two equilateral triangles, one placed inside the other, with their vertices numbered clockwise from 1 to 3. The larger triangle is the *reference triangle.* If we pick up the smaller triangle, there are several different ways in which we can set it back in its place. Each possible positioning of the inner triangle will be an element of a group. For instance, we can pick up the triangle and replace it exactly as we found it. We will call this position *I*; it will represent no change in position.

Now pick up the smaller triangle, rotate it 120° clockwise, and set it down again on top of the reference triangle. The result is shown below.

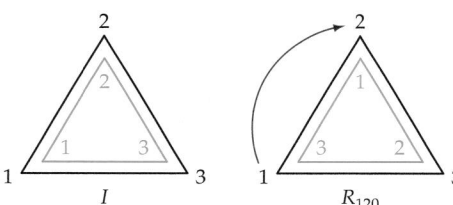

Notice that the vertex originally at vertex 1 of the reference triangle is now at vertex 2, 2 is now at 3, and 3 is now at 1. We call the rotation of the triangle 120° clockwise R_{120}.

Now return the smaller triangle to its original position, where the numbers on the vertices of the triangles coincide. Consider a 240° rotation of the smaller triangle. The result of this rotation is shown below.

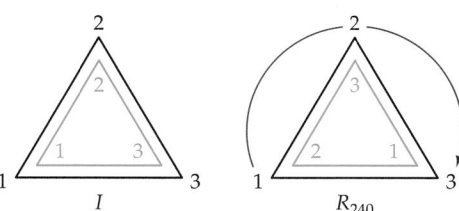

Notice that the vertex originally at vertex 1 of the reference triangle is now at vertex 3, 2 is now at 1, and 3 is now at 2. We call the rotation of the triangle 240° clockwise R_{240}.

If the original triangle were rotated 360°, there would be no apparent change. The vertex at 1 would return to 1, the vertex at 2 would return to 2, and the vertex at 3 would return to 3. Because this rotation does not produce a new arrangement of the vertices, we consider it the same as the element we named *I*.

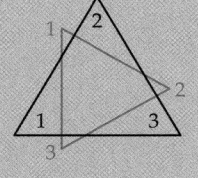

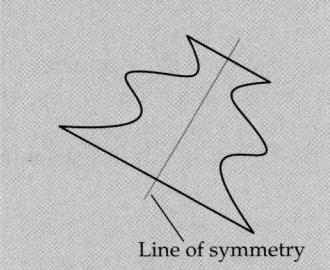
If we rotate the inner triangle *counterclockwise* 120°, the effect is the same as rotating it 240° clockwise. This rotation does not produce a different arrangement of the vertices. Similarly, a counterclockwise rotation of 240° is the same as a clockwise rotation of 120°.

In addition to rotating the triangle clockwise 120° and 240° as we did above, we could rotate the triangle about a line of symmetry that goes through a vertex before setting the triangle back down. Because there are three vertices, there are three possible results. These are shown below.

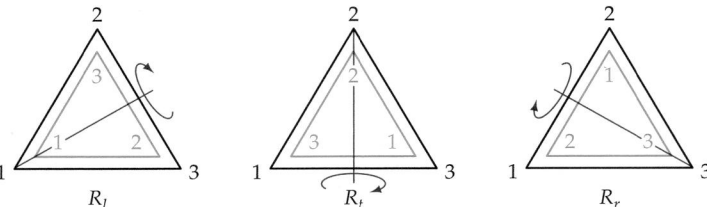

For rotation R_l, the bottom *left* vertex does not change, but vertices 2 and 3 are interchanged. If we rotate the triangle about the line of symmetry through the *top* vertex, rotation R_t, vertex 2 does not change but vertices 1 and 3 are interchanged. For rotation R_r, the bottom *right* vertex does not change but vertices 1 and 2 are interchanged.

The six positions of the vertices we have seen thus far are the only possibilities: I (the triangle without any rotation), R_{120}, R_{240}, R_l, R_t, and R_r. These positions are shown below (without the reference triangle).

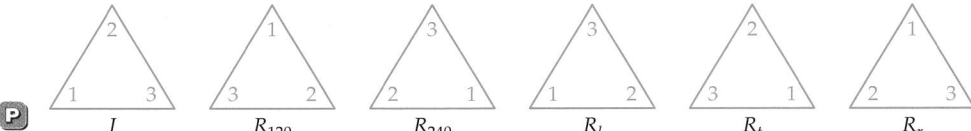

As we will see, these rotations are elements that form a group.

A group must have an operation, a method by which two elements of the group can be combined to form a third element that must also be a member of the group. (Recall that a group operation must be closed.) The operation we will use is called "followed by" and is symbolized by Δ. Next we show an example of how this operation works.

Consider $R_l \Delta R_{240}$. This means we rotate the original triangle, labeled as I, about the line of symmetry through vertex 1 "followed by" (without returning to the original position) a clockwise rotation of 240°. The result is one of the elements of the group, R_t. This operation is shown below.

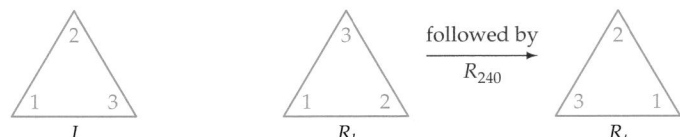

Now try reversing the operation and consider $R_{240} \Delta R_l$. This means we rotate the original triangle, I, clockwise 240° "followed by" (without returning to the original position) a rotation about the axis of symmetry through the bottom left vertex. Note that the result is an element of the group, namely R_r, as shown below.

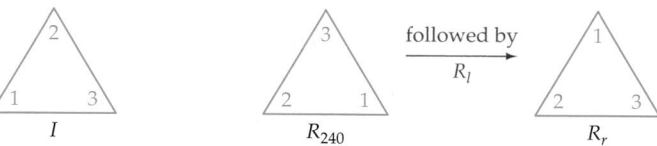

From these two examples, $R_l \Delta R_{240} = R_t$ and $R_{240} \Delta R_l = R_r$. Therefore, $R_l \Delta R_{240} \neq R_{240} \Delta_l$, which means the operation "followed by" is not commutative.

EXAMPLE 2 ■ Perform an Operation of a Symmetry Group

Find $R_l \Delta R_r$.

Solution

Rotate the original triangle, I, about the line of symmetry through the bottom left vertex, followed by a rotation about the line of symmetry through the bottom right vertex.

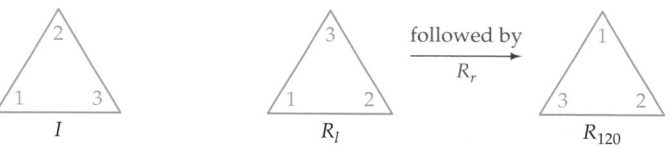

From the diagram, $R_l \Delta R_r = R_{120}$.

CHECK YOUR PROGRESS 2   Find $R_r \Delta R_{240}$.

Solution   See page S27.

We have stated that the elements R_{120}, R_{240}, R_l, R_t, R_r, and I, with the operation "followed by," form a group. However, we have not demonstrated this fact, so we will do so now.

As we have seen, the set is closed with respect to the operation Δ. To show that the operation Δ is associative, think about the meaning of $x\Delta(y\Delta z)$ and $(x\Delta y)\Delta z$, where x, y, and z are elements of the group. $x\Delta(y\Delta z)$ means x, followed by the result of y followed by z; $(x\Delta y)\Delta z$ means x followed by y, followed by z. For instance, $R_t\Delta(R_l\Delta R_r) = R_t\Delta R_{120} = R_l$ and $(R_t\Delta R_l)\Delta R_r = R_{120}\Delta R_r = R_l$. All the remaining combinations of elements can be verified similarly.

The identity element is I and can be thought of as "no rotation." To see that each element has an inverse, verify the following:

$$R_{120}\Delta R_{240} = I \qquad R_l\Delta R_l = I \qquad R_t\Delta R_t = I \qquad R_r\Delta R_r = I$$

Your work should show that every element has an inverse. (The inverse of I is I.) Thus the four conditions of a group are satisfied.

To determine the outcome of the "followed by" operation on two elements of the group, we drew triangles and then rotated them as directed by the type of rotation. From a mathematical point of view, it would be nice to have a symbolic (rather than geometric) way of determining the outcome. We can do this by creating a mathematical object that describes the geometric object.

Note that any rotation of the triangle changes the position of the vertices. For R_{120}, the vertex originally at position 1 moved to position 2, the vertex at 2 moved to 3, and the vertex at 3 moved to 1. We can represent this as

$$R_{120} = \begin{pmatrix} 1 & 2 & 3 \\ 2 & 3 & 1 \end{pmatrix}$$

Similarly, for R_t, the vertex originally at position 1 moved to position 3, the vertex at 2 remained at 2, and the vertex at 3 moved to 1. This can be represented as

$$R_t = \begin{pmatrix} 1 & 2 & 3 \\ 3 & 2 & 1 \end{pmatrix}$$

The remaining four elements can be represented as

$$I = \begin{pmatrix} 1 & 2 & 3 \\ 1 & 2 & 3 \end{pmatrix}, R_{240} = \begin{pmatrix} 1 & 2 & 3 \\ 3 & 1 & 2 \end{pmatrix}, R_l = \begin{pmatrix} 1 & 2 & 3 \\ 1 & 3 & 2 \end{pmatrix}, R_r = \begin{pmatrix} 1 & 2 & 3 \\ 2 & 1 & 3 \end{pmatrix}$$

EXAMPLE 3 ■ Symbolic Notation

Use symbolic notation to find $R_{120} \Delta R_t$.

Solution
To find the result of $R_{120} \Delta R_t$, we follow the movement of each vertex. The path vertex 2 follows is highlighted.

$$R_{120} \Delta R_t = \begin{pmatrix} 1 & 2 & 3 \\ 2 & 3 & 1 \end{pmatrix} \Delta \begin{pmatrix} 1 & 2 & 3 \\ 3 & 2 & 1 \end{pmatrix}$$

- $1 \rightarrow 2 \rightarrow 2$. Thus $1 \rightarrow 2$.
- $2 \rightarrow 3 \rightarrow 1$. Thus $2 \rightarrow 1$.
- $3 \rightarrow 1 \rightarrow 3$. Thus $3 \rightarrow 3$.

$$= \begin{pmatrix} 1 & 2 & 3 \\ 2 & 1 & 3 \end{pmatrix} = R_r$$

$R_{120} \Delta R_t = R_r$

CHECK YOUR PROGRESS 3   Use symbolic notation to find $R_r \Delta R_{240}$.

Solution   *See page S27.*

Math Matters   **A Group of Subatomic Particles**

In 1961, Murray Gell-Mann introduced what he called the *eightfold way* (after Buddha's Eightfold Path to enlightenment and bliss) to categorize subatomic particles into eight different families, where each family was based on a certain symmetry group. Because each family was a group, each family had to satisfy the closure requirement. For that to happen, Gell-Mann's theory indicated that certain particles that had not yet been discovered must exist. Scientists began looking for these particles. One such particle, discovered in 1964, is called omega-minus. Its existence was predicted purely on the basis of the known properties of groups.

Permutation Groups

The triangular symmetry group discussed previously is an example of a special kind of group called a *permutation group*. A **permutation** is a rearrangement of objects. For instance, if we start with the arrangement of objects [♦ ♥ ♣], then one permutation of these objects is the rearrangement [♥ ♣ ♦]. If we consider each permutation of these objects as an element of a set, then the set of all possible permutations forms a group. The elements of the group are not numbers or the objects themselves, but rather the different permutations of the objects that are possible.

If we start with the numbers 1 2 3, we can represent permutations of the numbers using the same symbolic method that we used for the triangular symmetry group. For example, the permutation that rearranges 1 2 3 to 2 3 1 can be written $\begin{pmatrix} 1 & 2 & 3 \\ 2 & 3 & 1 \end{pmatrix}$, which means that 1 is replaced by 2, 2 is replaced by 3, and 3 is replaced by 1.

There are only six distinct permutations of the numbers 1 2 3. We list and label them below. The identity element is named I.

$$I = \begin{pmatrix} 1 & 2 & 3 \\ 1 & 2 & 3 \end{pmatrix}, A = \begin{pmatrix} 1 & 2 & 3 \\ 2 & 3 & 1 \end{pmatrix}, B = \begin{pmatrix} 1 & 2 & 3 \\ 3 & 1 & 2 \end{pmatrix}$$

$$C = \begin{pmatrix} 1 & 2 & 3 \\ 1 & 3 & 2 \end{pmatrix}, D = \begin{pmatrix} 1 & 2 & 3 \\ 3 & 2 & 1 \end{pmatrix}, E = \begin{pmatrix} 1 & 2 & 3 \\ 2 & 1 & 3 \end{pmatrix}$$

The operation for the group is "followed by," which we will again denote by the symbol Δ. One can verify that the six elements along with this operation do indeed form a group.

EXAMPLE 4 ■ Perform an Operation in a Permutation Group

Find $B\Delta C$, where B and C are elements of the permutation group defined above.

Solution

$B\Delta C = \begin{pmatrix} 1 & 2 & 3 \\ 3 & 1 & 2 \end{pmatrix}\Delta\begin{pmatrix} 1 & 2 & 3 \\ 1 & 3 & 2 \end{pmatrix}$. In the first permutation, 1 is replaced by 3. In the second permutation, 3 is replaced by 2. Combining these two actions, 1 is ultimately replaced by 2. Similarly, in the first permutation 2 is replaced by 1, which remains as 1 in the second permutation. 3 is replaced by 2, which is in turn replaced by 3. Combining these actions, 3 remains as 3. The result of the operation is $\begin{pmatrix} 1 & 2 & 3 \\ 2 & 1 & 3 \end{pmatrix}$. Thus we see that $B\Delta C = E$.

CHECK YOUR PROGRESS 4 Compute $E\Delta B$, where E and B are elements of the permutation group defined previously.

Solution *See page S27.*

EXAMPLE 5 ■ Inverse Element of a Permutation Group

One of the requirements of a group is that each element must have an inverse. Find the inverse of the element B of the permutation group defined previously.

Solution

Because $B = \begin{pmatrix} 1 & 2 & 3 \\ 3 & 1 & 2 \end{pmatrix}$ replaces 1 with 3, 2 with 1, and 3 with 2, its inverse must reverse these replacements. Thus we need to replace 3 with 1, 1 with 2, and 2 with 3. The inverse is the element $\begin{pmatrix} 1 & 2 & 3 \\ 2 & 3 & 1 \end{pmatrix} = A$. To verify,

$$B \Delta A = \begin{pmatrix} 1 & 2 & 3 \\ 3 & 1 & 2 \end{pmatrix} \Delta \begin{pmatrix} 1 & 2 & 3 \\ 2 & 3 & 1 \end{pmatrix} = \begin{pmatrix} 1 & 2 & 3 \\ 1 & 2 & 3 \end{pmatrix} \text{ and}$$

$$A \Delta B = \begin{pmatrix} 1 & 2 & 3 \\ 2 & 3 & 1 \end{pmatrix} \Delta \begin{pmatrix} 1 & 2 & 3 \\ 3 & 1 & 2 \end{pmatrix} = \begin{pmatrix} 1 & 2 & 3 \\ 1 & 2 & 3 \end{pmatrix}.$$

CHECK YOUR PROGRESS 5 Find the inverse of the element D of the permutation group defined previously.

Solution See page S27.

> ✔ **TAKE NOTE**
>
> In a nonabelian group, the order in which the two elements are combined with the group operation can affect the result. For an element to be the inverse of another element, combining the elements must give the identity element regardless of the order used.

If we consider a group with more than three numbers or objects, the number of different permutations increases dramatically. The permutation group of the five numbers 1 2 3 4 5 has 120 elements, and there are over 40,000 permutations of the numbers 1 2 3 4 5 6 7 8. In the exercises, you are asked to examine the permutations of 1 2 3 4.

Excursion

Wallpaper Groups

The symmetry group we studied based on an equilateral triangle involved rotations of the triangle that retained the position of the triangle within a reference triangle. We will now look at symmetries of an infinitely large object.

Imagine that the pattern shown in Figure 7.3 on the following page continues forever in every direction. You can think of it as an infinitely large sheet of wallpaper. Overlaying this sheet is a transparent sheet with the same pattern.

Visualize shifting the transparent sheet of wallpaper upward until the printed cats realign in what appears to be the same position. This is called a **translation.** Any such translation, in which we can shift the paper in one direction until the pattern aligns with its original appearance, is an element of a group. There are several directions in which we can shift the wallpaper (up, down, and diagonally), and an infinite number of distances we can shift the wallpaper. So the group has an infinite number of elements. Several elements are shown in Figure 7.4. Not shifting the paper at all is also considered an element; we will call it *I*, because it is the identity element for the group. As with the symmetry groups, the operation is "followed by," denoted by Δ.

(continued)

> **INSTRUCTOR NOTE**
> To demonstrate the translations shown in Figure 7.4, make two transparencies of Figure 7.3. Overlay the transparencies, and then move one sheet in a direction until the wallpaper pattern coincides with the pattern on the second sheet.

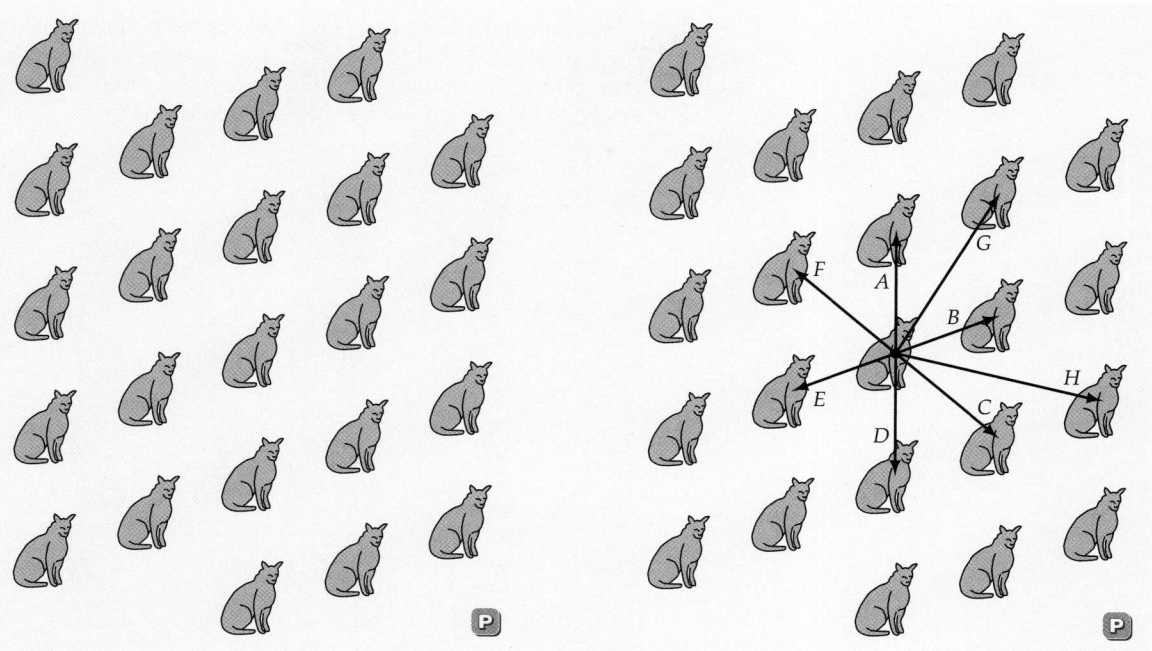

Figure 7.3

Figure 7.4

Some wallpaper patterns have rotational symmetries in addition to translation symmetries. The pattern in Figure 7.5 can be rotated 180° about the "✕" and the wallpaper will appear to be in the exact same position. Thus the wallpaper group derived from this pattern will include this rotation as an element. Excursion Exercise 4 asks you to identify other rotational elements. Note that the group also includes translational elements.

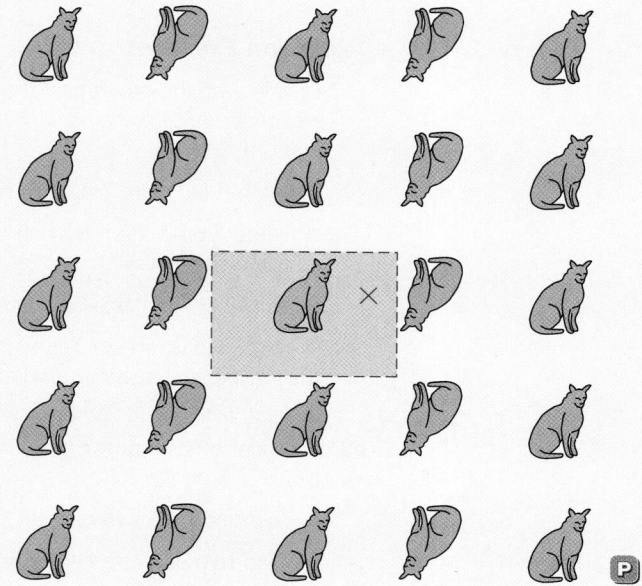

Figure 7.5

(continued)

We can also consider mirror reflections. The wallpaper shown below can be reflected across the dashed line and its appearance will be unchanged. (There are many other vertical axes that could also be used.) So, if we consider the group derived from symmetries of this wallpaper, one element will be the reflection of the entire sheet across this axis.

There is a fourth type of symmetry, called a *glide reflection,* that is a combination of a reflection and a shift. Mathematicians have determined that these are the only four types of symmetries possible, and that there are only 17 distinct wallpaper groups. Of course, the visual patterns you see on actual wallpaper can appear quite varied, but if we formed groups from their symmetries, each would have to be one of the 17 known groups. (If you wish to see all 17 patterns, an excellent resource is **http://www.clarku.edu/~djoyce/wallpaper/**.)

Excursion Exercises

1. Explain why the properties of a group are satisfied by the symmetries of the wallpaper pattern shown in Figure 7.3. (You may assume that the operation is associative.)

2. Verify that in Figure 7.4, $A\Delta B = G$.

3. In Figure 7.6 on the following page, determine where the black cat will be after applying the following translations illustrated in Figure 7.4.

   a. $F\Delta(D\Delta C)$ **b.** $(C\Delta D)\Delta G$ **c.** $(A\Delta H)\Delta(E\Delta D)$

4. In Figure 7.5, there are three other locations within the dashed box about which the wallpaper can be rotated and align to its original appearance. Find the rotational center points.

5. Consider the wallpaper group formed from symmetries of the pattern shown in Figure 7.7 on the following page.

   a. Identify three different elements of the group derived from translations.

   b. Find two different center points about which a 180° rotation is an element of the group.

(continued)

c. Find two different center points about which a 120° rotation is an element of the group.

d. Identify three distinct lines that could serve as axes of reflection for elements of the group.

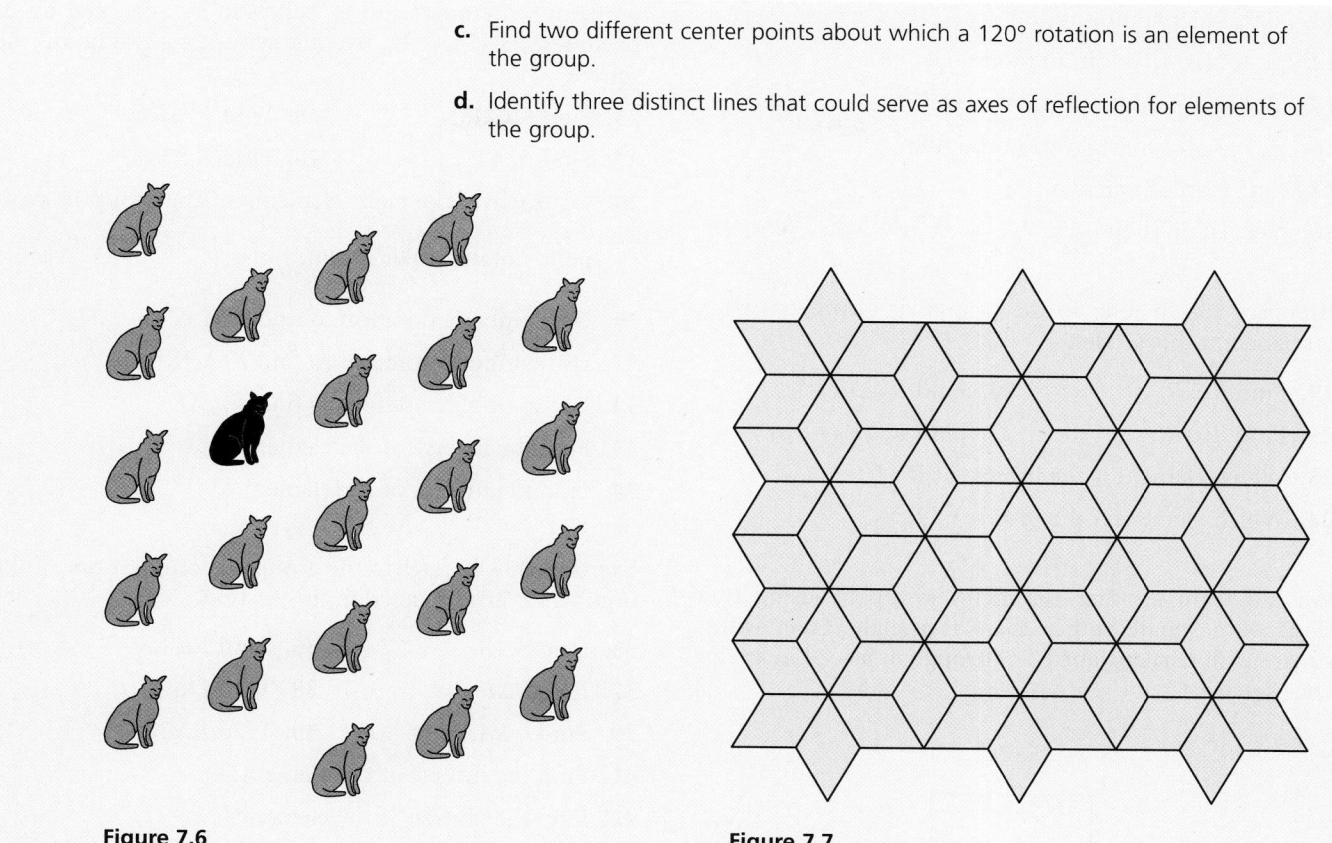

Figure 7.6   Figure 7.7

Exercise Set 7.3   (Suggested Assignment: 1–35 odds; 38, 41; 51–58)

1. Is the set $\{-1, 1\}$ closed with respect to the operation
  a. multiplication?
  b. addition?

2. Is the set of all even integers closed with respect to
  a. multiplication?
  b. addition?

3. Is the set of all odd integers closed with respect to
  a. multiplication?
  b. addition?

4. Is the set of all negative real numbers closed with respect to
  a. addition?   **b.** subtraction?
  c. multiplication?   **d.** division?

In Exercises 5–18, determine whether the set forms a group with respect to the given operation. (You may assume the operation is associative.) If the set does not form a group, determine which properties fail.

5. The even integers; addition

6. The even integers; multiplication

7. The real numbers; addition

8. The real numbers; division

9. The real numbers; multiplication

10. The real numbers *except* 0; multiplication

11. The rational numbers; addition
(Recall that a rational number is any number that can be expressed as a fraction.)

12. The positive rational numbers; multiplication

13. $\{0, 1, 2, 3\}$; addition modulo 4

14. $\{0, 1, 2, 3, 4\}$; addition modulo 6

15. $\{0, 1, 2, 3\}$; multiplication modulo 4

16. $\{1, 2, 3, 4\}$; multiplication modulo 5

17. $\{-1, 1\}$; multiplication

18. $\{-1, 1\}$; division

Exercises 19–24 refer to the triangular symmetry group discussed in this section.

19. Find $R_t \Delta R_{120}$.   **20.** Find $R_{240} \Delta R_r$.

21. Find $R_r \Delta R_t$.   **22.** Find $R_t \Delta R_l$.

23. Which element is the inverse of R_{240}?

24. Which element is the inverse of R_t?

We can form another symmetry group by using rotations of a square rather than a triangle. Start with a square with corners labeled 1 through 4, placed in a reference square.

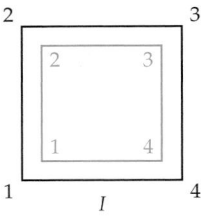

The current position of the inner square is labeled I. We can rotate the inner square clockwise 90° to give an element of the group, R_{90}. Similarly, we can rotate the square 180° or 270° to obtain the elements R_{180} and R_{270}. In addition, we can rotate the square about any of the lines of symmetry shown below.

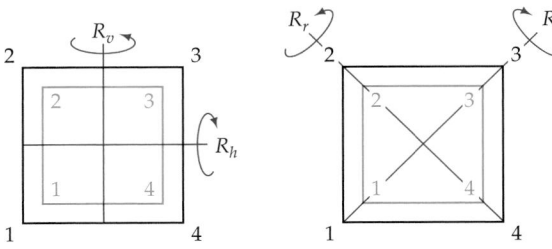

As before, the operation is "followed by," denoted by Δ. Exercises 25–34 refer to this symmetry group of the square.

25. Find $R_{90} \Delta R_v$.   **26.** Find $R_r \Delta R_{180}$.

27. Find $R_h \Delta R_{270}$.   **28.** Find $R_l \Delta R_h$.

29. List each of the eight elements of the group in symbolic notation. For instance, $I = \begin{pmatrix} 1 & 2 & 3 & 4 \\ 1 & 2 & 3 & 4 \end{pmatrix}$.

30. Use symbolic notation to find $R_v \Delta R_l$.

31. Use symbolic notation to find $R_h \Delta R_{90}$.

32. Use symbolic notation to find $R_{180} \Delta R_r$.

33. Find the inverse of the element R_{270}.

34. Find the inverse of the element R_v.

Exercises 35–42 refer to the group of permutations of the numbers 1 2 3 discussed in this section.

35. Find $A \Delta B$.   **36.** Find $B \Delta A$.

37. Find $B \Delta E$.   **38.** Find $D \Delta C$.

39. Find $C \Delta A$.   **40.** Find $B \Delta B$.

41. Find the inverse of the element A.

42. Find the inverse of the element E.

We can form a new permutation group by considering all permutations of the numbers 1 2 3 4. Exercises 43–50 refer to this group.

43. List all 24 elements of the group. Use symbolic notation, such as $I = \begin{pmatrix} 1 & 2 & 3 & 4 \\ 1 & 2 & 3 & 4 \end{pmatrix}$.

44. Find $\begin{pmatrix} 1 & 2 & 3 & 4 \\ 4 & 2 & 1 & 3 \end{pmatrix} \Delta \begin{pmatrix} 1 & 2 & 3 & 4 \\ 2 & 1 & 4 & 3 \end{pmatrix}$.

45. Find $\begin{pmatrix} 1 & 2 & 3 & 4 \\ 1 & 4 & 2 & 3 \end{pmatrix} \Delta \begin{pmatrix} 1 & 2 & 3 & 4 \\ 2 & 3 & 4 & 1 \end{pmatrix}$.

46. Find $\begin{pmatrix} 1 & 2 & 3 & 4 \\ 3 & 2 & 1 & 4 \end{pmatrix} \Delta \begin{pmatrix} 1 & 2 & 3 & 4 \\ 3 & 4 & 2 & 1 \end{pmatrix}$.

47. Find the inverse of $\begin{pmatrix} 1 & 2 & 3 & 4 \\ 2 & 4 & 1 & 3 \end{pmatrix}$.

48. Find the inverse of $\begin{pmatrix} 1 & 2 & 3 & 4 \\ 3 & 4 & 1 & 2 \end{pmatrix}$.

49. Find the inverse of $\begin{pmatrix} 1 & 2 & 3 & 4 \\ 4 & 3 & 2 & 1 \end{pmatrix}$.

50. Find the inverse of $\begin{pmatrix} 1 & 2 & 3 & 4 \\ 1 & 3 & 4 & 2 \end{pmatrix}$.

In Exercises 51–58, consider the set of elements $\{a, b, c, d\}$ with the operation ∇ as given in the table below.

∇	a	b	c	d
a	d	a	b	c
b	a	b	c	d
c	b	c	d	a
d	c	d	a	b

For example, to compute $c \nabla d$, find c in the left column and align it with d in the top row. The result is a.

51. Find $a \nabla a$.   **52.** Find $d \nabla c$.

53. Find $c \nabla b$.   **54.** Find $a \nabla d$.

55. Verify that the set with the operation ∇ is a group. (You may assume that the operation is associative.)

56. Which element of the group is the identity element?

57. Is the group commutative?

58. Identify the inverse of each element.

Extensions

CRITICAL THINKING

59. a. Verify that the integers $\{1, 2, 3, 4, 5, 6\}$ with the operation multiplication modulo 7 form a group.

   b. Verify that the integers $\{1, 2, 3, 4, 5\}$ with the operation multiplication modulo 6 do not form a group.

   c. For which values of n are the integers $\{1, 2, \ldots, n - 1\}$ a group under the operation multiplication modulo n?

60. The set of numbers $\{2^k\}$, where k is any integer, form a group with multiplication as the operation.

   a. Which element of the group is the identity element?

   b. Verify that the set is closed with respect to multiplication.

   c. Which element is the inverse of 2^8?

   d. Which element is the inverse of $\frac{1}{32}$?

   e. In general, what is the inverse of 2^k?

61. Consider the set $G = \{1, 2, 3, 4, 5\}$ and the operation $\oplus$ defined by $a \oplus b = a + b - 1 - r$, where $r = 5$ if $a + b \geq 7$ and $r = 0$ if $a + b \leq 6$. It can be shown that G forms a group with this operation.

   a. Make an operation table (similar to the one given for Exercises 51 to 58) under $\oplus$ for the group.

   b. Is the group commutative?

   c. Which element is the identity?

   d. Find the inverse of each element of the group.

EXPLORATIONS

62. The **quaternion group** is a famous noncommutative group; we will define the group as the set $Q = \{e, r, s, t, u, v, w, x\}$ and the operation ∇ that is defined by the table below.

∇	e	r	s	t	u	v	w	x
e	e	r	s	t	u	v	w	x
r	r	u	t	w	v	e	x	s
s	s	x	u	r	w	t	e	v
t	t	s	v	u	x	w	r	e
u	u	v	w	x	e	r	s	t
v	v	e	x	s	r	u	t	w
w	w	t	e	v	s	x	u	r
x	x	w	r	e	t	s	v	u

   a. Show that the group is noncommutative.

   b. Which element is the identity element?

   c. Find the inverse of every element in the group.

   d. Show that $r^4 = e$, where $r^4 = r \nabla r \nabla r \nabla r$.

   e. Show that the set $\{e, r, u, v\}$, which is a subset of Q, with the operation ∇ satisfies the properties of a group on its own. This is called a **subgroup** of the group.

   f. Find a subgroup that has only two elements.

| CHAPTER 7 | **Summary** |

Key Terms

additive inverse [p. 412]
algebraic system [p. 428]
check digit [p. 416]
ciphertext [p. 420]
clock arithmetic [p. 406]
closure [p. 429]
commutative group (abelian group) [p. 430]
congruence [p. 407]
cryptology [p. 420]
decryption [p. 420]
encryption [p. 420]
group [p. 428]
identity element [p. 429]
modulus [p. 407]
multiplicative inverse [p. 412]
nonabelian group [p. 430]
permutation [p. 435]
permutation group [p. 435]
plaintext [p. 420]
regular polygon [p. 431]
symmetry group [p. 431]

Essential Concepts

■ **Congruent Modulo n**
Two integers a and b are congruent modulo n, written $a \equiv b \bmod n$, if dividing a by n gives the same remainder as dividing b by n.

■ **Arithmetic Modulo n**
After performing the operation as usual, divide the result by n; the answer is the remainder.

■ **Solving Congruence Equations**
Individually check each whole number less than the modulus to see if it satisfies the congruence. Congruence equations may have more than one solution, or none at all.

■ **Finding Inverses in Modular Arithmetic**
To find the additive inverse of a number modulo n, subtract the number from n. To find the multiplicative inverse (if it exists) of a number a, solve the congruence equation $ax \equiv 1 \bmod n$ for x.

■ **ISBN Check Digit**
An ISBN is a 10-digit number $a_1 a_2 a_3 a_4 a_5 a_6 a_7 a_8 a_9 x$ in which the 10th digit x is a check digit. The check digit

must satisfy the congruence

$$a_1(10) + a_2(9) + a_3(8) + a_4(7) + a_5(6) + a_6(5)$$
$$+ a_7(4) + a_8(3) + a_9(2) + x \equiv 0 \bmod 11$$

The check digit has a value from 0 to 10; 10 is encoded as an X.

■ **UPC Check Digit**
A UPC is a 12-digit number $a_1 a_2 a_3 a_4 a_5 a_6 a_7 a_8 a_9 a_{10} a_{11} x$ in which the 12th digit x is a check digit. The check digit is a whole number from 0 to 9 that must satisfy the congruence

$$a_1(3) + a_2(1) + a_3(3) + a_4(1) + a_5(3) + a_6(1)$$
$$+ a_7(3) + a_8(1) + a_9(3) + a_{10}(1)$$
$$+ a_{11}(3) + x \equiv 0 \bmod 10$$

■ **Luhn Algorithm for Valid Credit Card Numbers**
The last digit of the credit card number is a check digit. Beginning with the next-to-last digit and reading from right to left, double every other digit. Treat any resulting two-digit number as two individual digits. Find the sum of the revised set of digits; the check digit is chosen such that the sum equals 0 mod 10.

■ **Cyclical Alphabetic Encryption**
A message can be encrypted by shifting each letter p of the plaintext message m positions through the alphabet. The encoded letter c satisfies the congruence $c \equiv (p + m) \bmod 26$, where p and c are the numerical equivalents of the plaintext and ciphertext letters, respectively. To decode the message, use the congruence $p \equiv (c + n) \bmod 26$, where $n = 26 - m$.

■ **The $c \equiv (ap + m) \bmod 26$ Encryption Scheme**
Choose a natural number a that does not have any common factors with 26 and a natural number m. A plaintext letter p is encoded to a ciphertext letter c using the congruence $c \equiv (ap + m) \bmod 26$. To decode the message, solve the congruence equation $c \equiv (ap + m) \bmod 26$ for p.

■ **Properties of a Group**
A group (a set of elements with one operation) must satisfy the following properties.

1. The set is closed with respect to the operation.
2. The operation satisfies the associative property.
3. There is an identity element.
4. Each element has an inverse.

■ **Symmetry Group of an Equilateral Triangle**
The group has six elements, which consist of the different possible rotations that return the triangle to its reference triangle:

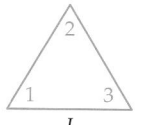 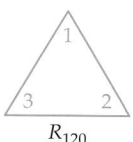 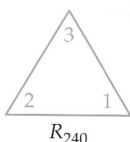 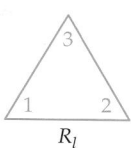 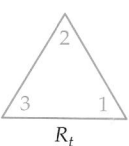 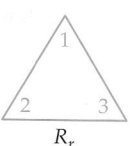

The group operation is "followed by," denoted by **Δ**. The operation is noncommutative.

■ **Symbolic Notation for the Symmetry Group of an Equilateral Triangle**
The six elements of the group are

$$I = \begin{pmatrix} 1 & 2 & 3 \\ 1 & 2 & 3 \end{pmatrix}, \qquad R_{120} = \begin{pmatrix} 1 & 2 & 3 \\ 2 & 3 & 1 \end{pmatrix},$$

$$R_{240} = \begin{pmatrix} 1 & 2 & 3 \\ 3 & 1 & 2 \end{pmatrix}, \qquad R_l = \begin{pmatrix} 1 & 2 & 3 \\ 1 & 3 & 2 \end{pmatrix},$$

$$R_t = \begin{pmatrix} 1 & 2 & 3 \\ 3 & 2 & 1 \end{pmatrix}, \qquad R_r = \begin{pmatrix} 1 & 2 & 3 \\ 2 & 1 & 3 \end{pmatrix}$$

where, for instance, $\begin{pmatrix} 1 & 2 & 3 \\ 2 & 3 & 1 \end{pmatrix}$ signifies that the vertex of the equilateral triangle originally at vertex 1 of the reference triangle moves to vertex 2, the vertex at 2 moves to 3, and the vertex at 3 moves to 1.

■ **Permutation Group of the Numbers 1, 2, 3**
The group consists of the six permutations of the numbers 1, 2, 3:

$$I = \begin{pmatrix} 1 & 2 & 3 \\ 1 & 2 & 3 \end{pmatrix}, \qquad A = \begin{pmatrix} 1 & 2 & 3 \\ 2 & 3 & 1 \end{pmatrix},$$

$$B = \begin{pmatrix} 1 & 2 & 3 \\ 3 & 1 & 2 \end{pmatrix}, \qquad C = \begin{pmatrix} 1 & 2 & 3 \\ 1 & 3 & 2 \end{pmatrix},$$

$$D = \begin{pmatrix} 1 & 2 & 3 \\ 3 & 2 & 1 \end{pmatrix}, \qquad E = \begin{pmatrix} 1 & 2 & 3 \\ 2 & 1 & 3 \end{pmatrix}$$

The group operation is "followed by," denoted by **Δ**.

CHAPTER 7   **Review Exercises**

In Exercises 1–8, evaluate each expression, where $\oplus$ and $\ominus$ indicate addition and subtraction, respectively, using a 12-hour clock.

1. $9 \oplus 5$ **2.** $8 \oplus 6$

3. $10 \oplus 7$ **4.** $7 \oplus 11$

5. $6 \ominus 9$ **6.** $2 \ominus 10$

7. $3 \ominus 4$ **8.** $7 \ominus 12$

In Exercises 9 and 10, evaluate each expression, where $\boxplus$ and $\boxminus$ indicate addition and subtraction, respectively, using days-of-the-week arithmetic.

9. $5 \boxplus 5$ **10.** $2 \boxminus 5$

In Exercises 11–14, determine whether the congruence is true or false.

11. $17 \equiv 2 \bmod 5$ **12.** $14 \equiv 24 \bmod 4$

13. $35 \equiv 53 \bmod 10$ **14.** $12 \equiv 36 \bmod 8$

In Exercises 15–22, perform the modular arithmetic.

15. $(8 + 12) \bmod 3$ **16.** $(15 + 7) \bmod 6$

17. $(42 - 10) \bmod 8$ **18.** $(19 - 8) \bmod 4$

19. $(7 \cdot 5) \bmod 9$ **20.** $(12 \cdot 9) \bmod 5$

21. $(15 \cdot 10) \bmod 11$ **22.** $(41 \cdot 13) \bmod 8$

In Exercises 23 and 24, use modular arithmetic to determine each of the following.

23. Disregarding A.M. or P.M., if it is now 5 o'clock,

  a. what time will it be 45 hours from now?

  b. what time was it 71 hours ago?

24. In 2002, April 15 (the day taxes are due in the U.S.) fell on a Monday. On what day of the week will April 15 fall in 2010?

In Exercises 25–28, find all whole number solutions of the congruence equation.

25. $x \equiv 7 \bmod 4$

26. $2x \equiv 5 \bmod 9$

27. $(2x + 1) \equiv 6 \bmod 5$

28. $(3x + 4) \equiv 5 \bmod 11$

In Exercises 29 and 30, find the additive inverse and the multiplicative inverse, if it exists, of the given number.

29. 5 in mod 7 arithmetic

30. 7 in mod 12 arithmetic

In Exercises 31 and 32, perform the modular division by solving a related multiplication problem.

31. $(2 \div 5) \bmod 7$

32. $(3 \div 4) \bmod 5$

In Exercises 33 and 34, determine the correct check digit for each ISBN.

33. *Oliver Twist* by Charles Dickens, 0-812-58003-?

34. *Interview with the Vampire* by Anne Rice, 0-394-49821-?

In Exercises 35 and 36, determine the correct check digit for the UPC.

35. 0-29000-07004-? (Planters Almonds)

36. 0-85391-89512-? (*Best in Show* DVD)

In Exercises 37–40, determine whether the given credit card number is a valid number.

37. 5126-6993-4231-2956

38. 5383-0118-3416-5931

39. 3412-408439-82594

40. 6011-5185-8295-8328

In Exercises 41 and 42, encode the message using a cyclical alphabetic encrypting code that shifts the message the stated number of positions.

41. 7 positions: MAY THE FORCE BE WITH YOU

42. 11 positions: CANCEL ALL PLANS

In Exercises 43 and 44, use a cyclical alphabetic encrypting code to decode the encrypted message.

43. PXXM UDLT CXVXAAXF

44. HVS ROM VOG OFFWJSR

45. Use the encrypting congruence $c \equiv (3p + 6) \bmod 26$ to encrypt the message END OF THE LINE.

46. Decode the message WEU LKGGMF NHM NMGN that was encrypted using the congruence $c \equiv (7p + 4) \bmod 26$.

In Exercises 47–50, determine whether the set is a group with respect to the given operation. (You may assume the operation is associative.) If the set is not a group, determine which properties fail.

47. All rational numbers *except* 0; multiplication

48. All multiples of 3; addition

49. All negative integers; multiplication

50. {1, 2, 3, 4, 5, 6, 7, 8, 9, 10}; multiplication modulo 11

Exercises 51 and 52 refer to the triangular symmetry group discussed in Section 7.3.

51. Find $R_t \Delta R_r$.

52. Find $R_{240} \Delta R_t$.

Exercises 53–56 refer to the square symmetry group discussed in Exercises 25–34 of Section 7.3.

53. Find $R_h \Delta R_{90}$.

54. Find $R_l \Delta R_v$.

55. Find the inverse of the element R_{180}.

56. Find the inverse of the element R_r.

In Exercises 57–61, a rectangle with corners labeled 1 through 4 is placed in a reference rectangle as shown below.

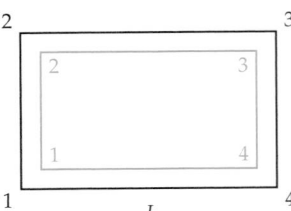

We can form a group in which each element is a rotation of the inner rectangle that allows us to reposition the inner rectangle within the reference rectangle, similar to the symmetry groups discussed in Section 7.3. We again use the "followed by" operation, Δ. (This group is called the *Klein 4-group*.)

57. Explain why this group has only four elements.

58. List each element in symbolic notation.

59. Is this group commutative?

60. Show that each element is its own inverse.

61. Of the three non-identity elements, show that the combination of any two by the operation Δ is always the third element.

Exercises 62 and 63 refer to the group of permutations of the numbers 1 2 3 discussed in Section 7.3.

62. Find $C\Delta E$.

63. Find $E\Delta D$.

In Exercises 64 and 65, consider the group of permutations of the numbers 1 2 3 4 5 with the operation Δ.

64. Find $\begin{pmatrix} 1 & 2 & 3 & 4 & 5 \\ 4 & 3 & 5 & 1 & 2 \end{pmatrix} \Delta \begin{pmatrix} 1 & 2 & 3 & 4 & 5 \\ 5 & 2 & 4 & 3 & 1 \end{pmatrix}$.

65. Find the inverse of $\begin{pmatrix} 1 & 2 & 3 & 4 & 5 \\ 3 & 5 & 1 & 4 & 2 \end{pmatrix}$.

CHAPTER 7	**Test**

1. Evaluate each expression, where $\oplus$ and $\ominus$ indicate addition and subtraction, respectively, using a 12-hour clock.

 a. $8 \oplus 7$ **b.** $2 \ominus 9$

2. Evaluate each expression, where $\triangle$ and $\triangle$ indicate addition and subtraction, respectively, using military time.

 a. $1500 \,\triangle\, 1100$ **b.** $0400 \,\triangle\, 1500$

3. Determine whether the congruence is true or false.

 a. $8 \equiv 20 \bmod 3$ **b.** $61 \equiv 38 \bmod 7$

In Exercises 4−6, perform the modular arithmetic.

4. $(25 + 9) \bmod 6$

5. $(31 - 11) \bmod 7$

6. $(5 \cdot 16) \bmod 12$

7. Disregarding A.M. or P.M., if it is now 3 o'clock, use modular arithmetic to determine

 a. what time it will be 27 hours from now.

 b. what time it was 58 hours ago.

In Exercises 8 and 9, find all whole number solutions of the congruence equation.

8. $x \equiv 5 \bmod 9$

9. $(2x + 3) \equiv 1 \bmod 4$

10. Find the additive inverse and the multiplicative inverse, if it exists, of 5 in modulo 9 arithmetic.

11. Determine the correct check digit for the ISBN 0-441-56959-? (*Neuromancer* by William Gibson).

12. Determine the correct check digit for the UPC 0-72878-27533-? (Herdez Salsa Verde).

13. Determine whether the credit card number 4232-8180-5736-4876 is a valid number.

14. Encrypt the plaintext message REPORT BACK using the cyclical alphabetic encrypting code that shifts letters 10 positions.

15. Decode the message UTSTG DPFM that was encrypted using the congruence $c \equiv (3p + 5) \bmod 26$.

16. a. Is the set of all odd positive integers closed with respect to multiplication? Explain.

 b. Is the set of all odd positive integers closed with respect to addition? Explain.

17. Determine whether the set of all odd integers with multiplication as the operation is a group. (You may assume the operation is associative.) If the set is not a group, explain which properties fail.

18. In the symmetry group of an equilateral triangle, determine the result of the operation.

 a. $R_{120}\Delta R_l$ **b.** $R_t\Delta R_{240}$

19. In the permutation group of the numbers 1, 2, 3, determine the result of the operation.

$$\begin{pmatrix} 1 & 2 & 3 \\ 3 & 1 & 2 \end{pmatrix} \Delta \begin{pmatrix} 1 & 2 & 3 \\ 3 & 2 & 1 \end{pmatrix}$$

20. In the permutation group of the numbers 1, 2, 3, 4 with the operation Δ, find the inverse element of $\begin{pmatrix} 1 & 2 & 3 & 4 \\ 3 & 1 & 4 & 2 \end{pmatrix}$.

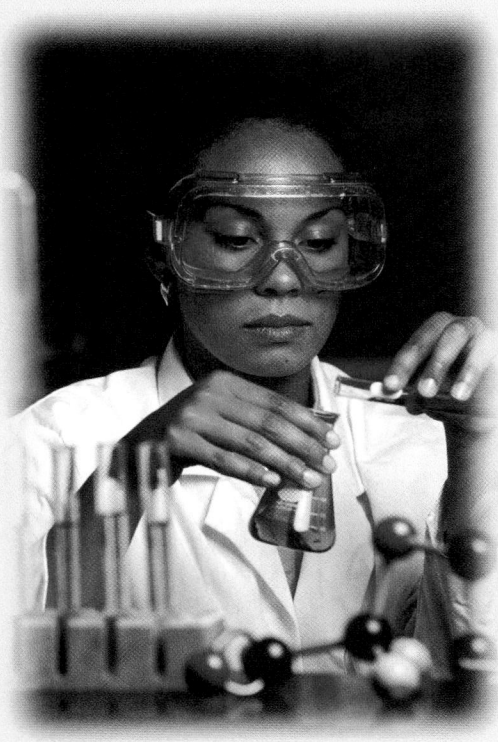

8

Geometry

It is difficult to use mathematical formulas to measure the volume of many three-dimensional objects because of their unusual shapes. The **Excursion exercises** on **pages 510–511** illustrate the use of water displacement as a means of measuring the volume of some of these objects.

8.1	**Basic Concepts of Euclidean Geometry**
8.2	**Perimeter and Area of Plane Figures**
8.3	**Similar Triangles**
8.4	**Volume and Surface Area**
8.5	**Non-Euclidean Geometry**
8.6	**Fractals**

Need help? For on-line student resources, such as section quizzes, visit this textbook's web site at **math.college.hmco.com/students.**

Take the time to look closely at each of the four rectangles shown below. Which rectangle seems to be the most pleasant to look at?

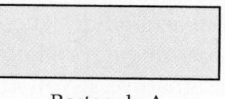
Rectangle A

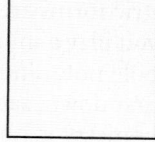
Rectangle B

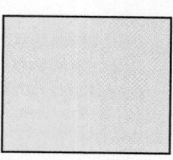
Rectangle C

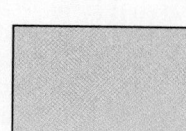
Rectangle D

The Parthenon

Record the responses of all the students in your class. Which rectangle was named most frequently as the most appealing?

The length of rectangle A is three times its width; the ratio of the length to the width is $3:1$. Rectangle B is a square, so its length is equal to its width; the ratio of the length to the width is $1:1$. The ratio of the length of rectangle C to its width is $1.2:1$.

The shape of rectangle D is referred to as the *golden rectangle*. In a **golden rectangle,** the ratio of the length to the width is approximately 1.6 to 1. The ratio is exactly

$$\frac{\text{length}}{\text{width}} = \frac{1 + \sqrt{5}}{2}$$

Did most students select rectangle D, the golden rectangle, as the most appealing rectangle?

The early Greeks considered the golden rectangle to be the most pleasing to the eye. Therefore, they used it extensively in their art and architecture. The Parthenon in Athens, Greece is an example of the use of the golden rectangle in Greek architecture. It was built around 435 B.C. as a temple to the goddess Athena.

LeCorbusier's United Nations building in New York City incorporates the golden rectangle. The building is L-shaped; the upright part is a golden rectangle.

Many artists have painted on canvases that have the dimensions of the golden rectangle. Albrecht Durer, Georges Seurut, Paul Signac, and Piet Mondrian are among them. But perhaps the most famous is Leonardo Da Vinci, who painted the Mona Lisa on a canvas having the shape of a golden rectangle.

Mona Lisa by Leonardo Da Vinci

The World Wide Web is a wonderful source of information on the golden rectangle. Simply enter "golden rectangle" in a search engine. It will respond with thousands of web sites where you can learn more about this rectangle's appearance in art, architecture, music, nature, and mathematics.

447

Lines and Angles

historical note

Geometry is one of the oldest branches of mathematics. Around 350 B.C., Euclid (yōō′klĭd) of Alexandria wrote *Elements,* which contained all of the known concepts of geometry. Euclid's contribution to geometry was to unify various concepts into a single deductive system that was based on a set of postulates. ■

The word *geometry* comes from the Greek words for *earth* and *measure.* In ancient Egypt, geometry was used by the Egyptians to measure land and to build structures such as the pyramids.

Today geometry is used in many fields, such as physics, medicine, and geology. Geometry is also used in applied fields such as mechanical drawing and astronomy. Geometric forms are used in art and design.

If you play a musical instrument, you know the meaning of the words measure, rest, whole note, and time signature. If you are a football fan, you have learned the terms first down, sack, punt, and touchback. Every field has its associated vocabulary. Geometry is no exception. We will begin by introducing two basic geometric concepts: point and line.

A **point** is symbolized by drawing a dot. A **line** is determined by two distinct points and extends indefinitely in both directions, as the arrows on the line at the right indicate. This line contains points A and B and is represented by $\overleftrightarrow{AB}$. A line can also be represented by a single letter, such as l.

A **ray** starts at a point and extends indefinitely in *one* direction. The point at which a ray starts is called the **endpoint** of the ray. The ray shown at the right is denoted $\overrightarrow{AB}$. Point A is the endpoint of the ray.

A **line segment** is part of a line and has two endpoints. The line segment shown at the right is denoted by $\overline{AB}$.

QUESTION *Classify each diagram as a line, a ray, or a line segment.*

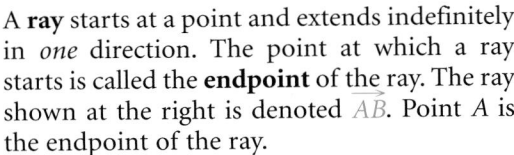

The distance between the endpoints of $\overline{AC}$ is denoted by AC. If B is a point on $\overline{AC}$, then AC (the distance from A to C) is the sum of AB (the distance from A to B) and BC (the distance from B to C).

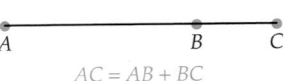

$AC = AB + BC$

ANSWER   *a. Ray*   *b. Line segment*   *c. Line*

Given $AB = 22$ cm and $BC = 13$ cm, find AC.

Make a drawing and write an equation to represent the distances between the points on the line segment.

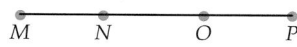

$$AB + BC = AC$$

Substitute the given distances for AB and BC into the equation. Solve for AC.

$$22 + 13 = AC$$
$$35 = AC$$
$$AC = 35 \text{ cm}$$

EXAMPLE 1 ▪ Find a Distance on a Line Segment

Given $MN = 14$ mm, $NO = 17$ mm, and $OP = 15$ mm on line segment MP, find MP.

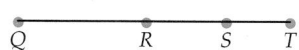

Solution

• **Make a drawing.**

$$MN + NO + OP = MP$$

• **Write an equation to represent the distances on the line segment.**

$$14 + 17 + 15 = MP$$

• **Replace *MN* by 14, *NO* by 17, and *OP* by 15.**

$$46 = MP$$

• **Solve for *MP*.**

$MP = 46$ mm

CHECK YOUR PROGRESS 1   Given $QR = 28$ cm, $RS = 16$ cm, and $ST = 10$ cm on $\overline{QT}$, find QT.

Solution   *See page S28.*

EXAMPLE 2 ▪ Use an Equation to Find a Distance on a Line Segment

Given $XY = 9$ m and YZ is twice XY, find XZ.

Solution

$$XZ = XY + YZ$$
$$XZ = XY + 2(XY)$$ • **YZ is twice XY.**
$$XZ = 9 + 2(9)$$ • **Replace XY by 9.**
$$XZ = 9 + 18$$ • **Solve for XZ.**
$$XZ = 27$$

$XZ = 27$ m

CHECK YOUR PROGRESS 2 Given $BC = 16$ ft and $AB = \frac{1}{4}(BC)$, find AC.

Solution *See page S28.*

In this section, we are discussing figures that lie in a plane. A **plane** is a flat surface and can be pictured as a tabletop or blackboard that extends forever in all direction. Figures that lie in a plane are called **plane figures.**

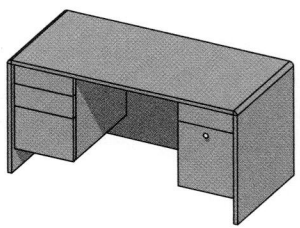

Lines in a plane can be intersecting or parallel. **Intersecting lines** cross at a point in the plane.

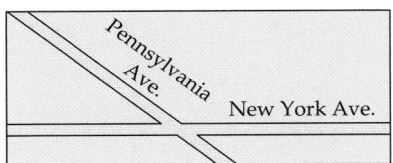

Parallel lines never intersect. The distance between them is always the same.

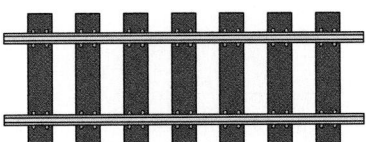

The symbol $\|$ means "is parallel to." In the figure at the right, $j \| k$ and $\overline{AB} \| \overline{CD}$. Note that j contains $\overline{AB}$ and k contains $\overline{CD}$. Parallel lines contain parallel line segments.

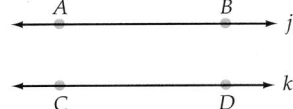

An **angle** is formed by two rays with the same endpoint. The **vertex** of the angle is the point at which the two rays meet. The rays are called the **sides** of the angle.

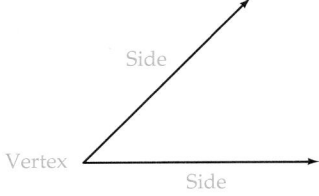

If A is a point on one ray of an angle, C is a point on the other ray, and B is the vertex, then the angle is called $\angle B$ or $\angle ABC$, where $\angle$ is the symbol for angle. Note that an angle can be named by the vertex, or by giving three points, where the second point listed is the vertex. $\angle ABC$ could also be called $\angle CBA$.

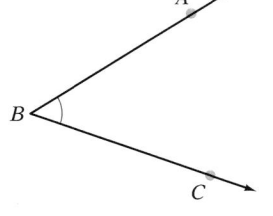

An angle can also be named by a variable written between the rays close to the vertex.

In the figure at the right, $\angle x$ and $\angle QRS$ are two different names for the same angle. $\angle y$ and $\angle SRT$ are two different names for the same angle. Note that in this figure, more than two rays meet at R. In this case, the vertex alone cannot be used to name $\angle QRT$.

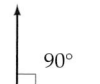

An angle is often measured in **degrees.** The symbol for degrees is a small raised circle, °. The angle formed by rotating a ray through a complete circle has a measure of 360°.

360°

A **protractor** is often used to measure an angle. Place the center of the protractor at the vertex of the angle with the edge of the protractor along a side of the angle. The angle shown in the figure below measures 58°.

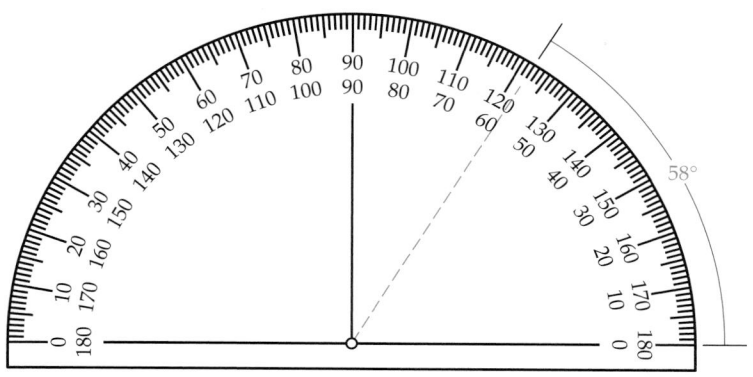

A 90° angle is called a **right angle.** The symbol ∟ represents a right angle.

90°

Perpendicular lines are intersecting lines that form right angles.

The symbol ⊥ means "is perpendicular to." In the figure at the right, $p \perp q$ and $\overline{AB} \perp \overline{CD}$. Note that line p contains $\overline{AB}$ and line q contains $\overline{CD}$. Perpendicular lines contain perpendicular line segments.

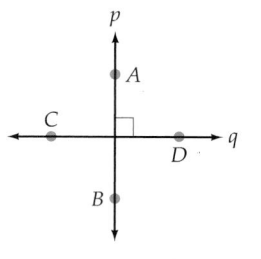

A **straight angle** is an angle whose measure is 180°. ∠*AOB* is a straight angle.

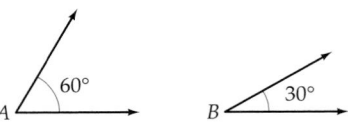

Straight angle

Complementary angles are two angles whose measures have the sum 90°.

∠*A* and ∠*B* at the right are complementary angles.

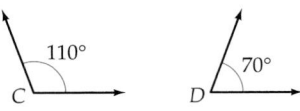

Supplementary angles are two angles whose measures have the sum 180°.

∠*C* and ∠*D* at the right are supplementary angles.

An **acute angle** is an angle whose measure is between 0° and 90°. ∠*D* above is an acute angle. An **obtuse angle** is an angle whose measure is between 90° and 180°. ∠*C* above is an obtuse angle.

The measure of ∠*C* is 110°. This is often written as $m\angle C = 110°$, where m is an abbreviation for "the measure of." This notation is used in Example 3.

EXAMPLE 3 ▪ Determine If Two Angles are Complementary

Are angles *E* and *F* complementary angles?

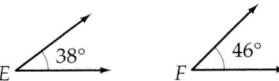

Solution

$$m\angle E + m\angle F = 38° + 46° = 84°$$

The sum of the measures of ∠*E* and ∠*F* is not 90°. Therefore, angles *E* and *F* are not complementary angles.

CHECK YOUR PROGRESS 3   Are angles *G* and *H* supplementary angles?

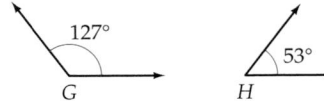

Solution   *See page S28.*

EXAMPLE 4 ▪ Find the Complement of an Angle

Find the complement of a 38° angle.

Solution
Complementary angles are two angles the sum of whose measures is 90°.
To find the complement, let x represent the complement of a 38° angle.
Write an equation and solve for x.

$$x + 38° = 90°$$
$$x = 52°$$

CHECK YOUR PROGRESS 4   Find the supplement of a 129° angle.

Solution   *See page S28.*

Adjacent angles are two angles that have a common vertex and a common side but have no interior points in common. In the figure at the right, $\angle DAC$ and $\angle CAB$ are adjacent angles.

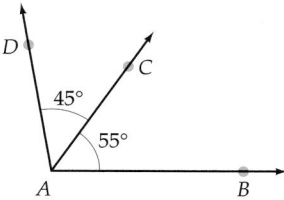

$$m\angle DAC = 45° \text{ and } m\angle CAB = 55°.$$
$$m\angle DAB = m\angle DAC + m\angle CAB$$
$$= 45° + 55° = 100°$$

In the figure at the right, $m\angle EDG = 80°$. The measure of $\angle FDG$ is three times the measure of $\angle EDF$. Find the measure of $\angle EDF$.

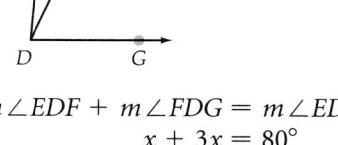

Let $x =$ the measure of $\angle EDF$.
Then $3x =$ the measure of $\angle FDG$.
Write an equation and solve for x.

$$m\angle EDF + m\angle FDG = m\angle EDG$$
$$x + 3x = 80°$$
$$4x = 80°$$
$$x = 20°$$

$$m\angle EDF = 20°$$

EXAMPLE 5 ▪ **Solve a Problem Involving Adjacent Angles**

Find the measure of $\angle x$.

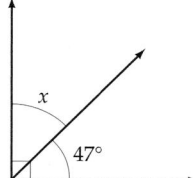

Solution
To find the measure of $\angle x$, write an equation using the fact that the sum of the measures of $\angle x$ and 47° is 90°. Solve for $m\angle x$.

$$m\angle x + 47° = 90°$$
$$m\angle x = 43°$$

CHECK YOUR PROGRESS 5 Find the measure of $\angle a$.

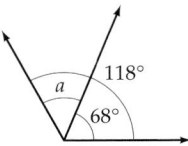

Solution *See page S28.*

Angles Formed by Intersecting Lines

Four angles are formed by the intersection of two lines. If the two lines are perpendicular, each of the four angles is a right angle.

If the two lines are not perpendicular, then two of the angles formed are acute angles and two of the angles formed are obtuse angles. The two acute angles are always opposite each other, and the two obtuse angles are always opposite each other.

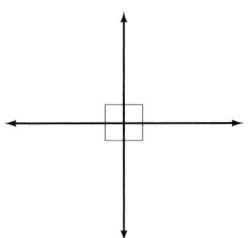

In the figure at the right, $\angle w$ and $\angle y$ are acute angles. $\angle x$ and $\angle z$ are obtuse angles.

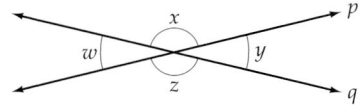

point of interest

Many cities in the New World, unlike those in Europe, were designed using rectangular street grids. Washington, D.C. was designed this way, except that diagonal avenues were added, primarily for the purpose of enabling troop movement in the event the city required defense. As an added precaution, monuments were constructed at major intersections so that attackers would not have a straight shot down a boulevard.

Two angles that are on opposite sides of the intersection of two lines are called **vertical angles.** Vertical angles have the same measure. $\angle w$ and $\angle y$ are vertical angles. $\angle x$ and $\angle z$ are vertical angles.

Vertical angles have the same measure.

$$m\angle w = m\angle y$$
$$m\angle x = m\angle z$$

Recall that two angles that share a common side are called adjacent angles. For the figure shown above, $\angle x$ and $\angle y$ are adjacent angles, as are $\angle y$ and $\angle z$, $\angle z$ and $\angle w$, and $\angle w$ and $\angle x$. Adjacent angles of intersecting lines are supplementary angles.

Adjacent angles of intersecting lines are supplementary angles.

$$m\angle x + m\angle y = 180°$$
$$m\angle y + m\angle z = 180°$$
$$m\angle z + m\angle w = 180°$$
$$m\angle w + m\angle x = 180°$$

EXAMPLE 6 ■ **Solve a Problem Involving Intersecting Lines**

In the diagram at the right, $m\angle b = 115°$. Find the measures of angles a, c, and d.

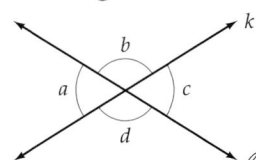

Solution

$m \angle a + m \angle b = 180°$ • $\angle a$ is supplementary to $\angle b$ because $\angle a$ and $\angle b$ are adjacent angles of intersecting lines.

$m \angle a + 115° = 180°$ • Replace $m \angle b$ with **115°**.

$m \angle a = 65°$ • Subtract 115° from each side of the equation.

$m \angle c = 65°$ • $m \angle c = m \angle a$ because $\angle c$ and $\angle a$ are vertical angles.

$m \angle d = 115°$ • $m \angle d = m \angle b$ because $\angle d$ and $\angle b$ are vertical angles.

$m \angle a = 65°$, $m \angle c = 65°$, and $m \angle d = 115°$.

CHECK YOUR PROGRESS 6

In the diagram at the right, $m \angle a = 35°$. Find the measures of angles *b*, *c*, and *d*.

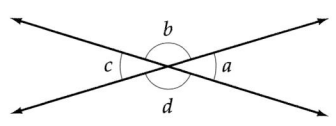

Solution   *See page S28.*

A line that intersects two other lines at different points is called a **transversal.**

If the lines cut by a transversal *t* are parallel lines and the transversal is perpendicular to the parallel lines, all eight angles formed are right angles.

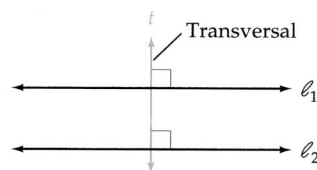

If the lines cut by a transversal *t* are parallel lines and the transversal is *not* perpendicular to the parallel lines, all four acute angles have the same measure and all four obtuse angles have the same measure. For the figure at the right:

$$m \angle b = m \angle d = m \angle x = m \angle z$$
$$m \angle a = m \angle c = m \angle w = m \angle y$$

If two lines in a plane are cut by a transversal, then any two non-adjacent angles that are on opposite sides of the transversal and between the parallel lines are **alternate interior angles.** In the figure above, $\angle c$ and $\angle w$ are alternate interior angles; $\angle d$ and $\angle x$ are alternate interior angles. Alternate interior angles formed by two parallel lines cut by a transversal have the same measure.

Two alternate interior angles formed by two parallel lines cut by a transversal have the same measure.

$$m \angle c = m \angle w$$
$$m \angle d = m \angle x$$

If two lines in a plane are cut by a transversal, then any two non-adjacent angles that are on opposite sides of the transversal and outside the parallel lines are **alternate exterior angles.** In the figure at the top of the following page,

Two alternate exterior angles formed by two parallel lines cut by a transversal have the same measure.

$$m \angle a = m \angle y$$
$$m \angle b = m \angle z$$

INSTRUCTOR NOTE
These concepts will be used in the discussion of similar triangles presented in Section 8.3.

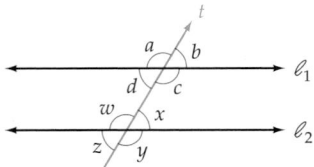

$\angle a$ and $\angle y$ are alternate exterior angles; $\angle b$ and $\angle z$ are alternate exterior angles. Alternate exterior angles formed by two parallel lines cut by a transversal have the same measure.

If two lines in a plane are cut by a transversal, then any two angles that are on the same side of the transversal and are both acute angles or both obtuse angles are **corresponding angles.** For the figure at the left, the following pairs of angles are corresponding angles: $\angle a$ and $\angle w$, $\angle d$ and $\angle z$, $\angle b$ and $\angle x$, $\angle c$ and $\angle y$. Corresponding angles formed by two parallel lines cut by a transversal have the same measure.

Two corresponding angles formed by two parallel lines cut by a transversal have the same measure.

$$m\angle a = m\angle w$$
$$m\angle d = m\angle z$$
$$m\angle b = m\angle x$$
$$m\angle c = m\angle y$$

QUESTION   *Which angles in the diagram at the left above have the same measure as angle a? Which angles have the same measure as angle b?*

EXAMPLE 7 ■ **Solve a Problem Involving Parallel Lines Cut by a Transversal**

In the diagram at the right, $\ell_1 \parallel \ell_2$ and $m\angle f = 58°$. Find the measures of $\angle a$, $\angle c$, and $\angle d$.

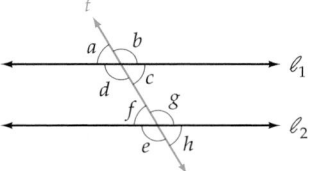

Solution

$m\angle a = m\angle f = 58°$ • $\angle a$ and $\angle f$ are corresponding angles.

$m\angle c = m\angle f = 58°$ • $\angle c$ and $\angle f$ are alternate interior angles.

$m\angle d + m\angle a = 180°$ • $\angle d$ is supplementary to $\angle a$.
$m\angle d + 58° = 180°$ • Replace $m\angle a$ with **58°**.
$m\angle d = 122°$ • Subtract 58° from each side of the equation.

$m\angle a = 58°$, $m\angle c = 58°$, and $m\angle d = 122°$.

ANSWER   *Angles c, w, and y have the same measure as angle a. Angles d, x, and z have the same measure as angle b.*

CHECK YOUR PROGRESS 7
In the diagram at the right, $\ell_1 \| \ell_2$ and $m\angle g = 124°$. Find the measures of $\angle b$, $\angle c$, and $\angle d$.

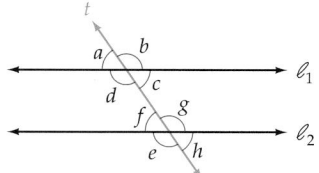

Solution   *See page S28.*

Math Matters   The Principle of Reflection

When a ray of light hits a flat surface, such as a mirror, the light is reflected back at the same angle at which it hit the surface. For example, in the diagram at the left, $m\angle x = m\angle y$.

This principle of reflection is in operation in a simple periscope. In a periscope light is reflected twice, with the result that light rays entering the periscope are parallel to the light rays at eye level.

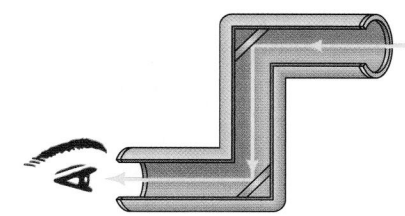

The same principle is in operation on a billiard table. Assuming it has no "side spin," a ball bouncing off the side of the table will bounce off at the same angle at which it hit the side.

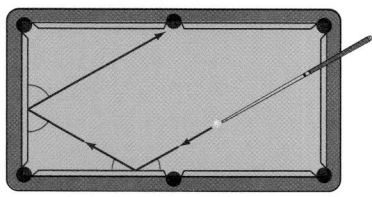

In the miniature golf shot illustrated below, $m\angle w = m\angle x$ and $m\angle y = m\angle z$.

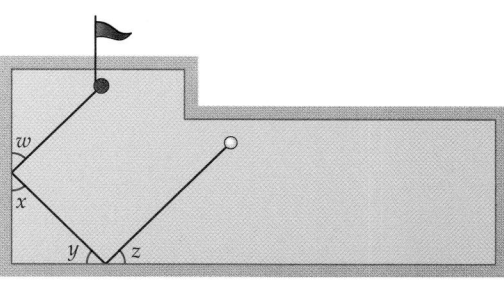

Angles of a Triangle

If the lines cut by a transversal are not parallel lines, the three lines will intersect at three points. In the figure at the right, the transversal t intersects lines p and q. The three lines intersect at points A, B, and C. These three points define three line segments, $\overline{AB}$, $\overline{BC}$, and $\overline{AC}$. The plane figure formed by these three line segments is called a **triangle.**

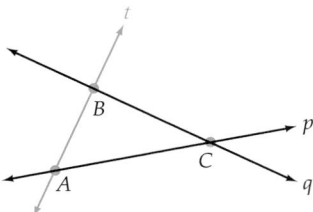

Each of the three points of intersection is the vertex of four angles. The angles within the region enclosed by the triangle are called **interior angles.** In the figure at the right, angles a, b, and c are interior angles. The sum of the measures of the interior angles of a triangle is 180°.

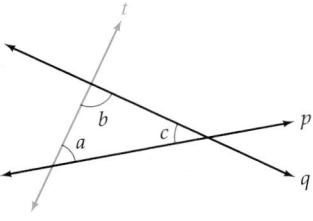

$$m\angle a + m\angle b + m\angle c = 180°$$

The Sum of the Measures of the Interior Angles of a Triangle

The sum of the measures of the interior angles of a triangle is 180°.

QUESTION   *Can the measures of the three interior angles of a triangle be 87°, 51°, and 43°?*

An **exterior angle of a triangle** is an angle which is adjacent to an interior angle of the triangle and is a supplement of the interior angle. In the figure at the right, angles m and n are exterior angles for angle a. The sum of the measures of an interior and an exterior angle is 180°.

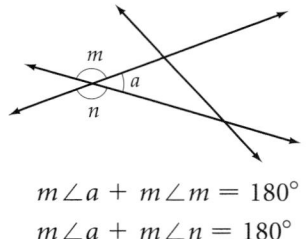

$$m\angle a + m\angle m = 180°$$
$$m\angle a + m\angle n = 180°$$

EXAMPLE 8 ■ Solve a Problem Involving the Angles of a Triangle

In the diagram at the right, $m\angle c = 40°$ and $m\angle e = 60°$. Find the measure of $\angle d$.

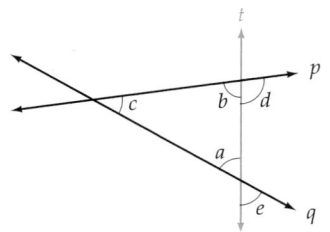

ANSWER   *No, because 87° + 51° + 43° = 181°, and the sum of the measures of the three interior angles of a triangle must be 180°.*

Solution

$m\angle a = m\angle e = 60°$ • $\angle a$ and $\angle e$ are vertical angles.

$m\angle c + m\angle a + m\angle b = 180°$ • The sum of the interior angles is **180°**.
$\quad\quad 40° + 60° + m\angle b = 180°$ • Replace $m\angle c$ with **40°** and $m\angle a$ with **60°**.
$\quad\quad\quad\quad\quad 100° + m\angle b = 180°$ • Add $40° + 60°$.
$\quad\quad\quad\quad\quad\quad\quad\quad m\angle b = 80°$ • Subtract 100° from each side of the equation.

$m\angle b + m\angle d = 180°$ • $\angle b$ and $\angle d$ are supplementary angles.
$\quad\quad 80° + m\angle d = 180°$ • Replace $m\angle b$ with **80°**.
$\quad\quad\quad\quad\quad m\angle d = 100°$ • Subtract 80° from each side of the equation.

$m\angle d = 100°$

CHECK YOUR PROGRESS 8
In the diagram at the right, $m\angle c = 35°$ and
$m\angle d = 105°$. Find the measure of $\angle e$.

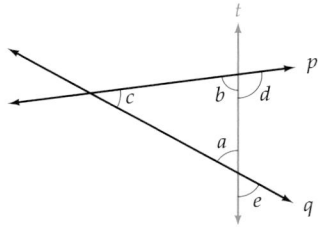

Solution   See page S28.

EXAMPLE 9 ■ Find the Measure of the Third Angle of a Triangle

Two angles of a triangle measure 43° and 86°. Find the measure of the third angle.

Solution
Use the fact that the sum of the measures of the interior angles of a triangle is 180°.
Write an equation using x to represent the measure of the third angle. Solve the
equation for x.

$x + 43° + 86° = 180°$
$\quad\quad x + 129° = 180°$ • Add $43° + 86°$.
$\quad\quad\quad\quad\quad\quad x = 51°$ • Subtract 129° from each side of the equation.

The measure of the third angle is 51°.

CHECK YOUR PROGRESS 9   One angle in a triangle is a right angle, and one
angle measures 27°. Find the measure of the third angle.

Solution   See page S28.

✔ **TAKE NOTE**

In this text, when we refer to the
angles of a triangle, we mean the
interior angles of the triangle unless specifically stated otherwise.

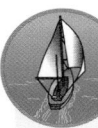

Excursion

Preparing a Circle Graph

On page 451, a protractor was used to measure angles. Preparing a circle graph requires the ability to use a protractor to draw angles.

To draw an angle of 142°, first draw a ray. Place a dot at the endpoint of the ray. This dot will be the vertex of the angle.

Place the straight bottom edge of the protractor on the ray as shown in the figure at the right. Make sure the center of the bottom edge of the protractor is located directly over the vertex point. Locate the position of the 142° mark. Place a dot next to the mark.

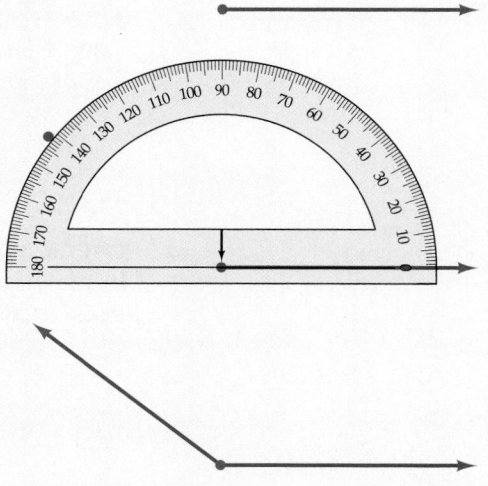

Remove the protractor and draw a ray from the vertex to the dot at the 142° mark.

Here is an example of how to prepare a circle graph.

In a presidential election in the United States, the popular vote is the number of people who vote for each candidate. The popular vote for the 2000 presidential election is shown below. The numbers have been rounded to the nearest hundred thousand.

Popular Vote in the 2000 Presidential Election	
Bush	50.5 million
Gore	51.0 million
Nader	2.9 million
Other	1.0 million

To draw a circle graph to illustrate the percent of the votes that each candidate received:

Find the total number of votes cast.

$$50.5 + 51.0 + 2.9 + 1.0 = 105.4$$

A total of 105.4 million votes were cast.

Find what percent of the total number of votes Bush, Gore, Nader, and all other candidates received.

Bush:

$$P \cdot B = A$$
$$P \cdot 105.4 = 50.5$$
$$\frac{P \cdot 105.4}{105.4} = \frac{50.5}{105.4}$$
$$P \approx 0.479$$
$$P \approx 47.9\%$$

Gore:

$$P \cdot B = A$$
$$P \cdot 105.4 = 51.0$$
$$\frac{P \cdot 105.4}{105.4} = \frac{51.0}{105.4}$$
$$P \approx 0.484$$
$$P \approx 48.4\%$$

(continued)

Nader:

$P \cdot B = A$

$P \cdot 105.4 = 2.9$

$\dfrac{P \cdot 105.4}{105.4} = \dfrac{2.9}{105.4}$

$P \approx 0.028$

$P \approx 2.8\%$

Other:

$P \cdot B = A$

$P \cdot 105.4 = 1.0$

$\dfrac{P \cdot 105.4}{105.4} = \dfrac{1.0}{105.4}$

$P \approx 0.009$

$P \approx 0.9\%$

Each percent represents that candidates' portion of the circle. Because a circle contains 360°, multiply each percent by 360° to find the angle measure for each sector. Round to the nearest degree.

Bush:

$0.479(360°) \approx 172°$

Nader:

$0.028(360°) \approx 10°$

Gore:

$0.484(360°) \approx 174°$

Other:

$0.009(360°) \approx 3°$

Draw a circle and use a protractor to draw the sectors representing the percent of the vote that each received. Label each sector and title the graph.

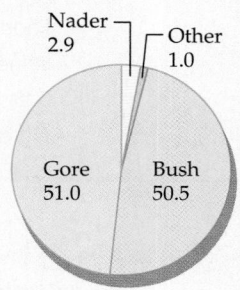

Nader 2.9

Other 1.0

Gore 51.0

Bush 50.5

Popular Vote (in millions) in the 2000 Presidential Election

Excursion Exercises

Prepare a circle graph for the data provided in each exercise.

1. Shown below are American adults' favorite pizza toppings. (*Source:* Market Facts for Bolla wines)

Pepperoni:	43%
Sausage:	19%
Mushrooms:	14%
Vegetables:	13%
Other:	7%
Onions:	4%

2. According to a Pathfinder Research Group survey, more than 94% of American adults have heard of the Three Stooges. The choices of a favorite among those who have one are:

Curly:	52%
Moe:	31%
Larry:	12%
Curly Joe:	3%
Shemp:	2%

(continued)

3. Only 18 million of the 67 million bicyclists in the United States own or have use of a helmet. The list below shows how often helmets are worn. (*Source:* U.S. Consumer Product Safety Commission)

Never or almost never:	13,680,000
Always or almost always:	2,412,000
Less than half the time:	1,080,000
More than half the time:	756,000
Unknown:	72,000

4. In a recent year, ten million tons of glass containers were produced. Given below is a list of the number of tons used by various industries. The category "Other" includes containers used for items such as drugs and toiletries. (*Source:* Salomon Bros.)

Beer:	4,600,000 tons
Food:	3,500,000 tons
Wine and Liquor:	900,000 tons
Soft Drinks:	500,000 tons
Other:	500,000 tons

Exercise Set 8.1

(Suggested Assignment: 9–65 odds; 68)

1. Provide three names for the angle below.

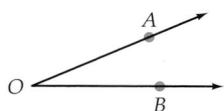

2. State the number of degrees in a full circle, a straight angle, and a right angle.

In Exercises 3–8, use a protractor to measure the angle to the nearest degree. State whether the angle is acute, obtuse, or right.

3.

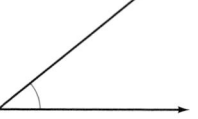

4.

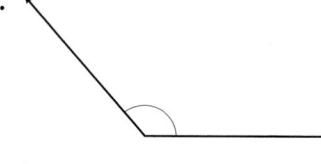

5.

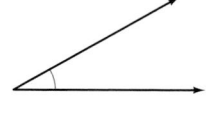

6.

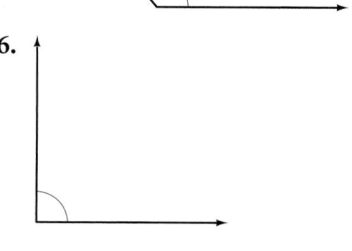

7.

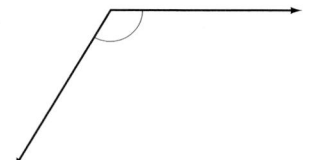

8.

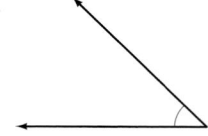

9. Are angles A and B complementary angles?

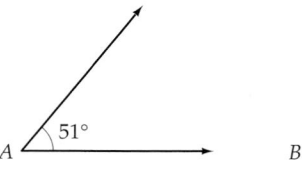

10. Are angles E and F complementary angles?

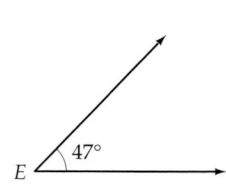

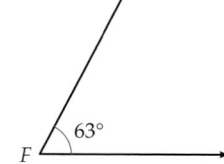

11. Are angles C and D supplementary angles?

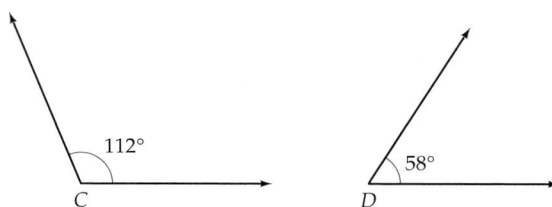

12. Are angles G and H supplementary angles?

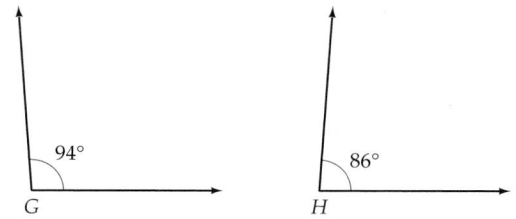

13. Find the complement of a 62° angle.

14. Find the complement of a 31° angle.

15. Find the supplement of a 162° angle.

16. Find the supplement of a 72° angle.

17. Given $AB = 12$ cm, $CD = 9$ cm, and $AD = 35$ cm, find the length of $\overline{BC}$.

18. Given $AB = 21$ mm, $BC = 14$ mm, and $AD = 54$ mm, find the length of $\overline{CD}$.

19. Given $QR = 7$ ft and RS is three times the length of $\overline{QR}$, find the length of $\overline{QS}$.

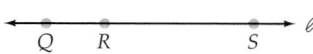

20. Given $QR = 15$ in. and RS is twice the length of $\overline{QR}$, find the length of $\overline{QS}$.

21. Given $EF = 20$ m and FG is $\frac{1}{2}$ the length of $\overline{EF}$, find the length of $\overline{EG}$.

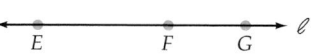

22. Given $EF = 18$ cm and FG is $\frac{1}{3}$ the length of $\overline{EF}$, find the length of $\overline{EG}$.

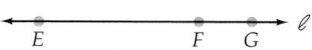

23. Given $m\angle LOM = 53°$ and $m\angle LON = 139°$, find the measure of $\angle MON$.

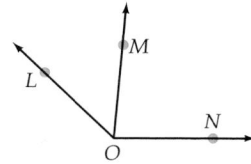

24. Given $m\angle MON = 38°$ and $m\angle LON = 85°$, find the measure of $\angle LOM$.

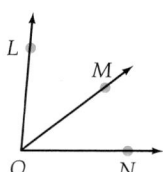

In Exercises 25 and 26, find the measure of $\angle x$.

25.

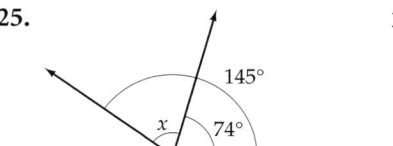

26.

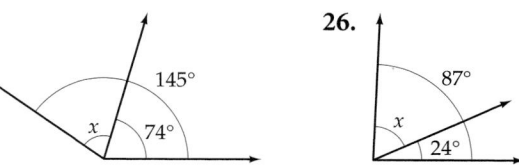

In Exercises 27–30, given that $\angle LON$ is a right angle, find the measure of $\angle x$.

27.

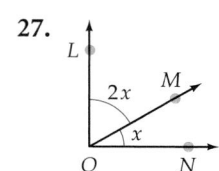

28.

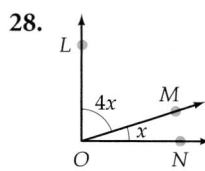

29.
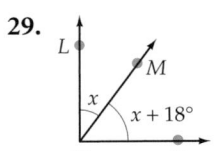

30.
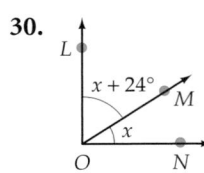

38.
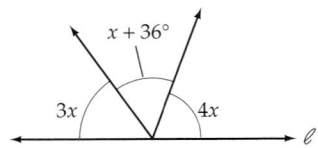

In Exercises 31–34, find the measure of ∠a.

31.
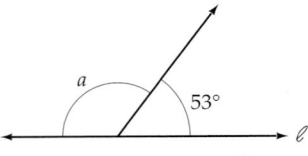

39.
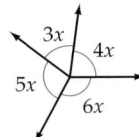

32.
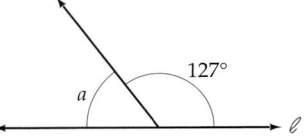

40.
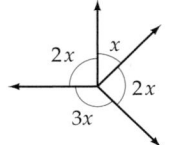

33.
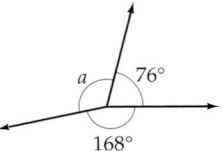

41. Given $m\angle a = 51°$, find the measure of ∠b.

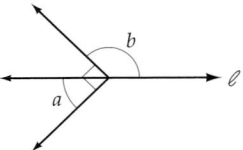

34.
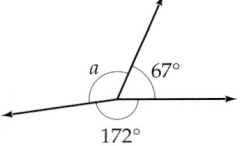

42. Given $m\angle a = 38°$, find the measure of ∠b.

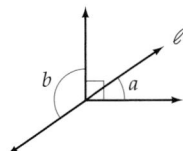

In Exercises 35–40, find the value of x.

35.
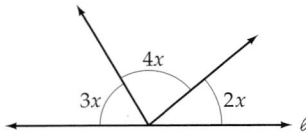

In Exercises 43 and 44, find the measure of ∠x.

43.
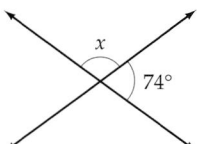

36.
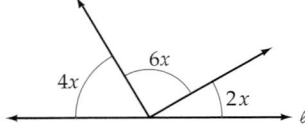

37.
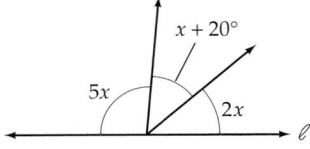

44.
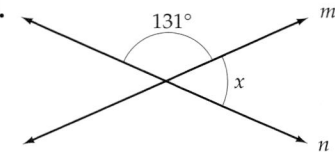

In Exercise 45 and 46, find the value of *x*.

45.

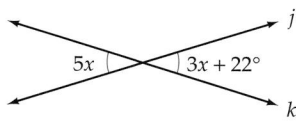

46.

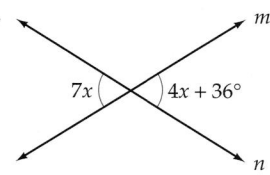

In Exercise 47–50, given that $\ell_1 \parallel \ell_2$, find the measures of angles *a* and *b*.

47.

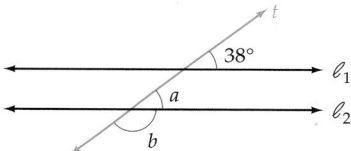

48.

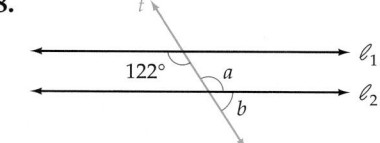

49.

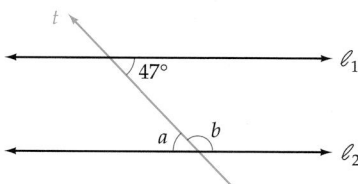

50.

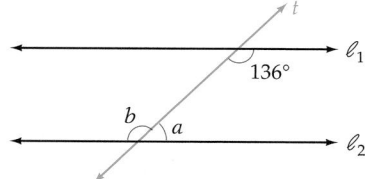

In Exercises 51–54, given that $\ell_1 \parallel \ell_2$, find *x*.

51.

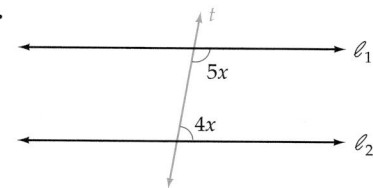

52.

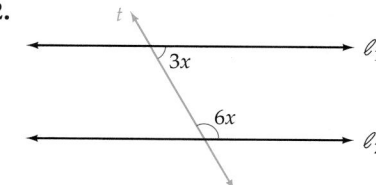

53.

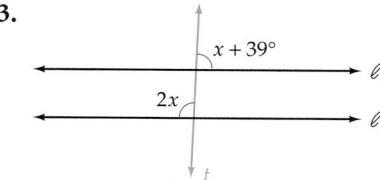

54.

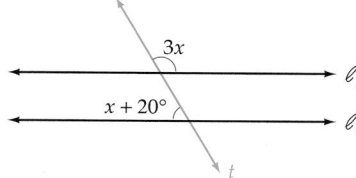

55. Given that $m \angle a = 95°$ and $m \angle b = 70°$, find the measures of angles *x* and *y*.

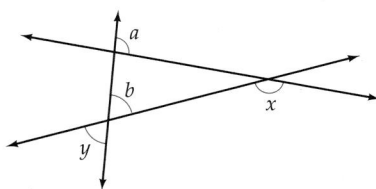

56. Given that $m \angle a = 35°$ and $m \angle b = 55°$, find the measures of angles *x* and *y*.

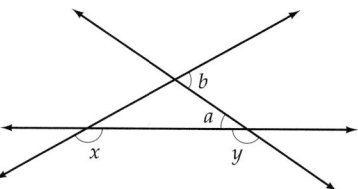

57. Given that $m \angle y = 45°$, find the measures of angles a and b.

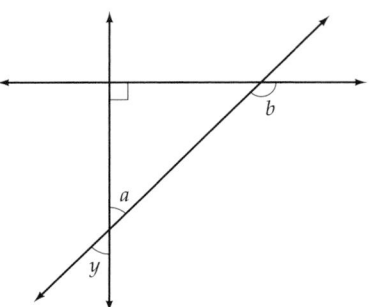

58. Given that $m \angle y = 130°$, find the measures of angles a and b.

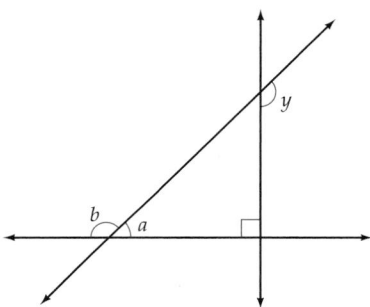

59. Given that $\overline{AO} \perp \overline{OB}$, express in terms of x the number of degrees in $\angle BOC$.

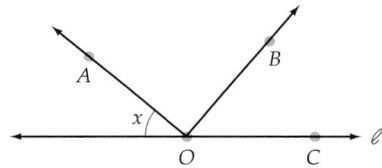

60. Given that $\overline{AO} \perp \overline{OB}$, express in terms of x the number of degrees in $\angle AOC$.

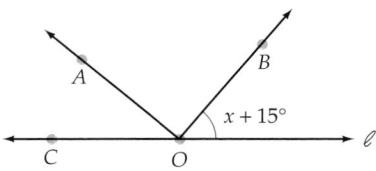

61. One angle in a triangle is a right angle, and one angle is equal to 30°. What is the measure of the third angle?

62. A triangle has a 45° angle and a right angle. Find the measure of the third angle.

63. Two angles of a triangle measure 42° and 103°. Find the measure of the third angle.

64. Two angles of a triangle measure 62° and 45°. Find the measure of the third angle.

65. A triangle has a 13° angle and a 65° angle. What is the measure of the third angle?

66. A triangle has a 105° angle and a 32° angle. What is the measure of the third angle?

67. Cut out a triangle and then tear off two of the angles, as shown below. Position the pieces you tore off so that angle a is adjacent to angle b and angle c is adjacent to angle b. Describe what you observe. What does this demonstrate?

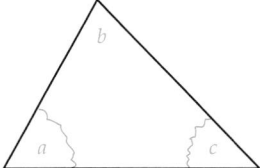

Extensions

CRITICAL THINKING

68. The road mileage between San Francisco, California and New York City is 3036 mi. The air distance between these two cities is 2571 mi. Why do the distances differ?

69. How many dimensions does a point have? a line? a line segment? a ray? an angle?

70. Which line segment is longer, $\overline{AB}$ or $\overline{CD}$?

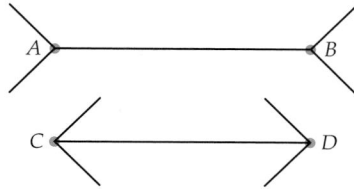

71. For the figure at the right, find the sum of the measures of angles x, y, and z.

72. For the figure at the right, explain why $m\angle a + m\angle b = m\angle x$. Write a rule that describes the relationship between the measure of an exterior angle of a triangle and the sum of the measures of its two opposite interior angles (the interior angles that are non-adjacent to the exterior angle). Use the rule to write an equation involving angles a, c, and z.

73. If $\overline{AB}$ and $\overline{CD}$ intersect at point O, and $m\angle AOC = m\angle BOC$, explain why $\overline{AB} \perp \overline{CD}$.

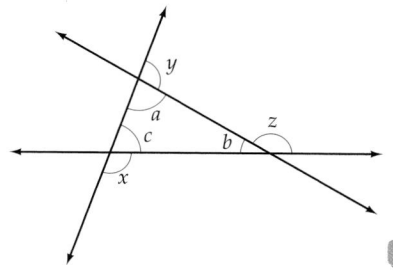

SECTION 8.2 | **Perimeter and Area of Plane Figures**

Perimeter of Plane Geometric Figures

A **polygon** is a closed figure determined by three or more line segments that lie in a plane. The line segments that form the polygon are called its **sides.** The figures below are examples of polygons.

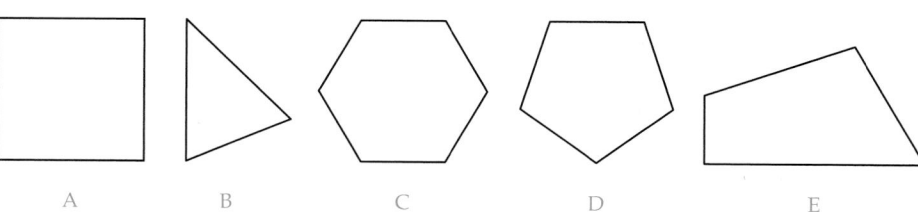

▼ **point of interest**

Although a polygon is described in terms of the number of its sides, the word actually comes from the Latin word *polygonum*, meaning "many *angles*."

A **regular polygon** is one in which each side has the same length and each angle has the same measure. The polygons in Figures A, C, and D above are regular polygons.

The name of a polygon is based on the number of its sides. The table below lists the names of polygons that have from 3 to 10 sides.

Number of Sides	Name of Polygon
3	Triangle
4	Quadrilateral
5	Pentagon
6	Hexagon
7	Heptagon
8	Octagon
9	Nonagon
10	Decagon

The Pentagon in Arlington, Virginia

Triangles and quadrilaterals are two of the most common types of polygons. Triangles are distinguished by the number of equal sides and also by the measures of their angles.

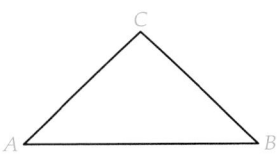

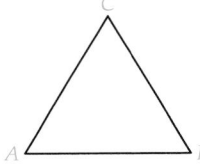

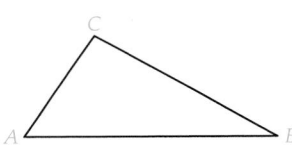

An **isosceles triangle** has exactly two sides of equal length. The angles opposite the equal sides are of equal measure.
$AC = BC$
$m \angle A = m \angle B$

The three sides of an **equilateral triangle** are of equal length. The three angles are of equal measure.
$AB = BC = AC$
$m \angle A = m \angle B$
$\qquad = m \angle C$

A **scalene triangle** has no two sides of equal length. No two angles are of equal measure.

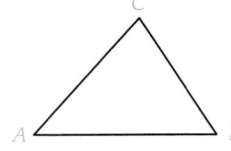

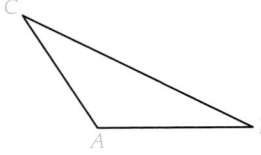

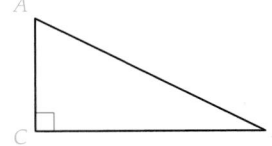

An **acute triangle** has three acute angles.

An **obtuse triangle** has one obtuse angle.

A **right triangle** has a right angle.

INSTRUCTOR NOTE
The diagram at the top of the following page shows the relationship among quadrilaterals. The description of each quadrilateral is within an example of that quadrilateral.

A **quadrilateral** is a four-sided polygon. Quadrilaterals are also distinguished by their sides and angles, as shown on the following page. Note that a rectangle, a square, and a rhombus are different forms of a parallelogram.

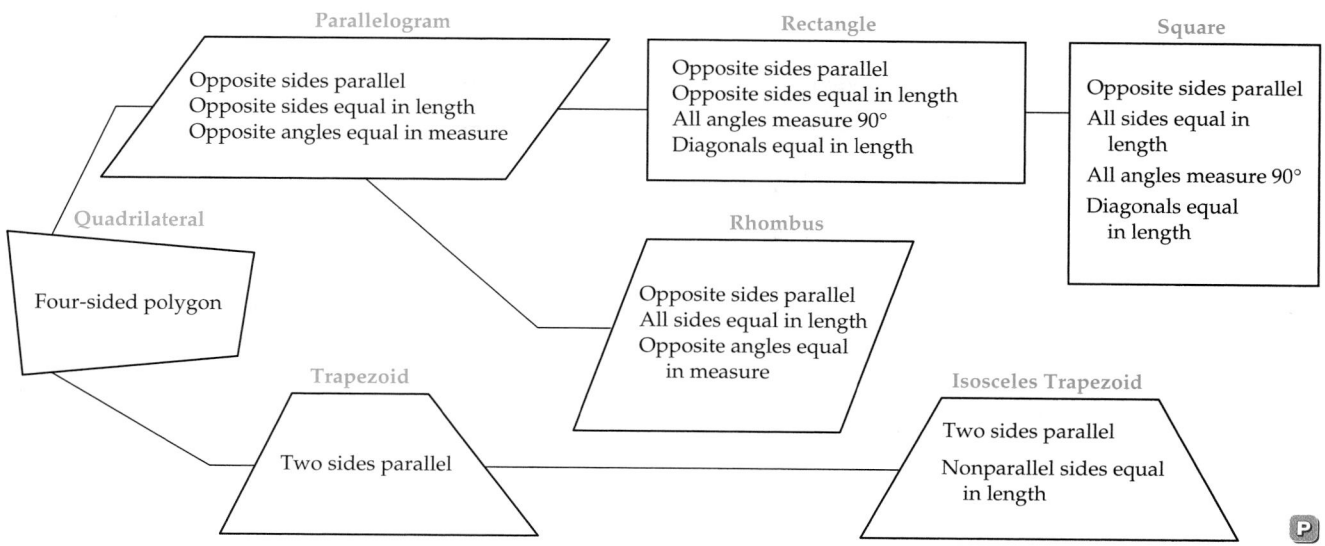

Parallelogram

Opposite sides parallel
Opposite sides equal in length
Opposite angles equal in measure

Rectangle

Opposite sides parallel
Opposite sides equal in length
All angles measure 90°
Diagonals equal in length

Square

Opposite sides parallel
All sides equal in
  length
All angles measure 90°
Diagonals equal
  in length

Quadrilateral

Four-sided polygon

Rhombus

Opposite sides parallel
All sides equal in length
Opposite angles equal
  in measure

Trapezoid

Two sides parallel

Isosceles Trapezoid

Two sides parallel
Nonparallel sides equal
  in length

QUESTION   ***a.*** *What distinguishes a rectangle from other parallelograms?*
   b. *What distinguishes a square from other rectangles?*

The **perimeter** of a plane geometric figure is a measure of the distance around the figure. Perimeter is used when buying fencing for a garden or determining how much baseboard is needed for a room.

The perimeter of a triangle is the sum of the lengths of the three sides.

Perimeter of a Triangle

Let a, b, and c be the lengths of the sides of a triangle. The perimeter, P, of the triangle is given by $P = a + b + c$.

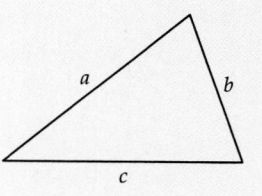

$$P = a + b + c$$

To find the perimeter of the triangle shown at the right, add the lengths of the three sides.

$P = 5 + 7 + 10 = 22$

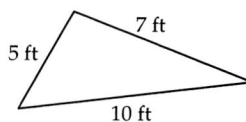

The perimeter is 22 ft.

EXAMPLE 1 ■ Find the Perimeter of a Triangle

You want to sew bias binding along the edges of a cloth flag that has sides that measure $1\frac{1}{4}$ ft, $3\frac{1}{2}$ ft, and $3\frac{3}{4}$ ft. Find the length of bias binding needed.

ANSWER   *a. In a rectangle, all angles measure 90°. **b.** In a square, all sides are equal in length.*

Solution

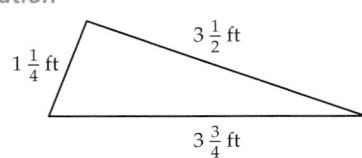

$P = a + b + c$

• Draw a diagram.

• Use the formula for the perimeter of a triangle.

$P = 1\dfrac{1}{4} + 3\dfrac{1}{2} + 3\dfrac{3}{4}$

• Replace a, b, and c with $1\frac{1}{4}$, $3\frac{1}{2}$, and $3\frac{3}{4}$. You can replace a with the length of any of the three sides, and then replace b and c with the lengths of the other two sides. The order does not matter. The result will be the same.

$P = 1\dfrac{1}{4} + 3\dfrac{2}{4} + 3\dfrac{3}{4}$

$P = 7\dfrac{6}{4}$

$P = 8\dfrac{1}{2}$

• $7\dfrac{6}{4} = 7\dfrac{3}{2} = 7 + \dfrac{3}{2} = 7 + 1\dfrac{1}{2} = \mathbf{8\dfrac{1}{2}}$

You need $8\frac{1}{2}$ ft of bias binding.

CHECK YOUR PROGRESS 1 A bicycle trail in the shape of a triangle has sides that measure $4\frac{3}{10}$ mi, $2\frac{1}{10}$ mi, and $6\frac{1}{2}$ mi. Find the total length of the bike trail.

Solution *See page S28.*

The perimeter of a quadrilateral is the sum of the lengths of its four sides.

A **rectangle** is a quadrilateral with all right angles and opposite sides of equal length. Usually the length, L, of a rectangle refers to the length of one of the longer sides of the rectangle and the width, W, refers to the length of one of the shorter sides. The perimeter can then be represented as $P = L + W + L + W$.

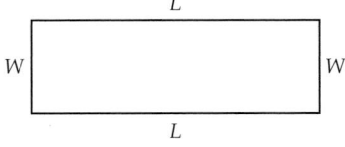

$P = L + W + L + W$

$P = 2L + 2W$

The formula for the perimeter of a rectangle is derived by combining like terms.

> **Perimeter of a Rectangle**
>
> Let L represent the length and W the width of a rectangle. The perimeter, P, of the rectangle is given by $P = 2L + 2W$.

EXAMPLE 2 ■ Find the Perimeter of a Rectangle

You want to trim a rectangular frame with a metal stripe. The frame measures 30 in. by 20 in. Find the length of metal stripe you will need to trim the frame.

Solution

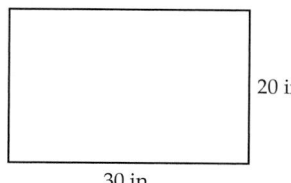

30 in.

• Draw a diagram.

$P = 2L + 2W$

• Use the formula for the perimeter of a rectangle.

$P = 2(30) + 2(20)$

• The length is **30** in. Substitute **30** for *L*. The width is **20** in. Substitute **20** for *W*.

$P = 60 + 40$

$P = 100$

You will need 100 in. of the metal stripe.

CHECK YOUR PROGRESS 2   Find the length of decorative molding needed to edge the top of the walls in a rectangular room that is 12 ft long and 8 ft wide.

Solution   *See page S28.*

A **square** is a rectangle in which each side has the same length. Letting *s* represent the length of each side of a square, the perimeter of the square can be represented $P = s + s + s + s$.

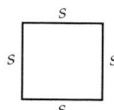

The formula for the perimeter of a square is derived by combining like terms.

$P = s + s + s + s$
$P = 4s$

Perimeter of a Square

Let *s* represent the length of a side of a square. The perimeter, *P*, of the square is given by $P = 4s$.

EXAMPLE 3 ■ Find the Perimeter of a Square

Find the length of fencing needed to surround a square corral that measures 60 ft on each side.

Solution

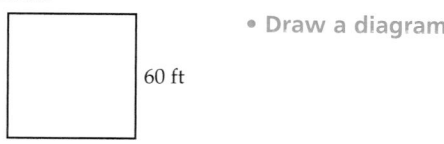

60 ft

• Draw a diagram.

$P = 4s$

• Use the formula for the perimeter of a square.

$P = 4(60)$

• The length of a side is **60** ft. Substitute **60** for *s*.

$P = 240$

240 ft of fencing is needed.

CHECK YOUR PROGRESS 3   A homeowner plans to fence in the area around the swimming pool in the back yard. The area to be fenced in is a square measuring 24 ft on each side. How many feet of fencing should the homeowner purchase?

Solution   See page S29.

Figure *ABCD* is a parallelogram. $\overline{BC}$ is the **base** of the parallelogram. Opposite sides of a parallelogram are equal in length, so $\overline{AD}$ is the same length as $\overline{BC}$, and $\overline{AB}$ is the same length as $\overline{CD}$.

Let *b* represent the length of the base and *s* the length of an adjacent side. Then the perimeter of a parallelogram can be represented as $P = b + s + b + s$.

The formula for the perimeter of a parallelogram is derived by combining like terms.

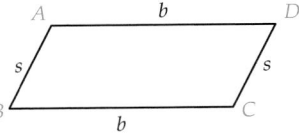

$$P = b + s + b + s$$

$$P = 2b + 2s$$

Perimeter of a Parallelogram

Let *b* represent the length of the base of a parallelogram and *s* the length of a side adjacent to the base. The perimeter, *P*, of the parallelogram is given by $P = 2b + 2s$.

EXAMPLE 4 ▪ **Find the Perimeter of a Parallelogram**

You plan to trim the edge of a kite with a strip of red fabric. The kite is in the shape of a parallelogram with a base measuring 40 in. and a side measuring 28 in. Find the length of the fabric needed to trim the kite.

Solution

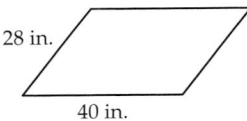

• Draw a diagram.

$P = 2b + 2s$

• Use the formula for the perimeter of a parallelogram.

$P = 2(40) + 2(28)$

• The base is **40** in. Substitute **40** for *b*. The length of a side is **28** in. Substitute **28** for *s*.

$P = 80 + 56$

• Simply using the Order of Operations Agreement.

$P = 136$

To trim the kite, 136 in. of fabric is needed.

CHECK YOUR PROGRESS 4   A flower bed is in the shape of a parallelogram that has a base of length 5 m and a side of length 7 m. Wooden planks are used to edge the garden. Find the length of wooden planks needed to surround the garden.

Solution   See page S29.

A **circle** is a plane figure in which all points are the same distance from point O, called the **center** of the circle.

A **diameter** of a circle is a line segment with endpoints on the circle and passing through the center. $\overline{AB}$ is a diameter of the circle at the right. The variable d is used to designate the length of a diameter of a circle.

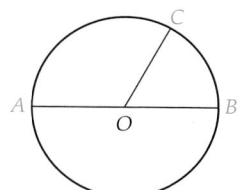

A **radius** of a circle is a line segment from the center of the circle to a point on the circle. $\overline{OC}$ is a radius of the circle at the right above. The variable r is used to designate the length of a radius of a circle.

The length of the diameter is twice the length of the radius.

$$d = 2r \quad \text{or} \quad r = \frac{1}{2}d$$

The distance around a circle is called the **circumference.** The formula for the circumference, C, of a circle is:

$$C = \pi d$$

Because $d = 2r$, the formula for the circumference can be written:

$$C = 2\pi r$$

> **Circumference of a Circle**
>
> The circumference, C, of a circle with diameter d and radius r is given by $C = \pi d$ or $C = 2\pi r$.

The formula for circumference uses the number π (pi), which is an irrational number. The value of π can be approximated by a fraction or by a decimal.

$$\pi \approx 3\frac{1}{7} \quad \text{or} \quad \pi \approx 3.14$$

The π key on a scientific calculator gives a closer approximation of π than 3.14. A scientific calculator is used in this section to find approximate values in calculations involving π.

Find the circumference of a circle with a diameter of 6 m.

The diameter of the circle is given. Use the circumference formula that involves the diameter. $d = 6$.

$$C = \pi d$$
$$C = \pi(6)$$

The exact circumference of the circle is 6π m.

$$C = 6\pi$$

An approximate measure can be found by using the π key on a calculator.

An approximate circumference is 18.85 m.

$$C \approx 18.85$$

EXAMPLE 5 ■ Find the Circumference of a Circle

Find the circumference of a circle with a radius of 15 cm. Round to the nearest hundredth.

Solution

$C = 2\pi r$ • The radius is given. Use the circumference formula that involves the radius.

$C = 2\pi(15)$ • Replace r with **15**.

$C = 30\pi$ • Multiply 2 times 15.

$C \approx 94.25$ • An approximation is asked for. Use the π key on a calculator.

The circumference of the circle is approximately 94.25 cm.

CHECK YOUR PROGRESS 5   Find the circumference of a circle with a diameter of 9 km. Give the exact measure.

Solution   *See page S29.*

EXAMPLE 6 ■ Application of Finding the Circumference of a Circle

A bicycle tire has a diameter of 24 in. How many feet does the bicycle travel when the wheel makes 8 revolutions? Round to the nearest hundredth.

24 in.

Solution

24 in. = 2 ft   • The diameter is given in inches, but the answer must be expressed in feet. Convert the diameter (24 in.) to feet. There are 12 in. in 1 ft. Divide 24 by 12.

$C = \pi d$ • The diameter is given. Use the circumference formula that involves the diameter.

$C = \pi(2)$ • Replace d with **2**.

$C = 2\pi$ • This is the distance traveled in 1 revolution.

$8C = 8(2\pi) = 16\pi \approx 50.27$ • Find the distance traveled in **8** revolutions.

The bicycle will travel about 50.27 ft when the wheel makes 8 revolutions.

CHECK YOUR PROGRESS 6   A tricycle tire has a diameter of 12 in. How many feet does the tricycle travel when the wheel makes 12 revolutions? Round to the nearest hundredth of a foot.

Solution   *See page S29.*

Area of Plane Geometric Figures

Area is the amount of surface in a region. Area can be used to describe, for example, the size of a rug, a parking lot, a farm, or a national park. Area is measured in square units.

A square that measures 1 inch on each side has an area of 1 square inch, written 1 in^2.

A square that measures 1 centimeter on each side has an area of 1 square centimeter, written 1 cm^2.

1 in^2

1 cm^2

Larger areas can be measured in square feet (ft^2), square meters (m^2), square miles (mi^2), acres (43,560 ft^2), or any other square unit.

QUESTION   **a.** *What is the area of a square that measures 1 yard on each side?*
b. *What is the area of a square that measures 1 kilometer on each side?*

The area of a geometric figure is the number of squares (each of area 1 square unit) that are necessary to cover the figure. In the figures below, two rectangles have been drawn and covered with squares. In the figure on the left, 12 squares, each of area 1 cm^2, were used to cover the rectangle. The area of the rectangle is 12 cm^2. In the figure on the right, six squares, each of area 1 in^2, were used to cover the rectangle. The area of the rectangle is 6 in^2.

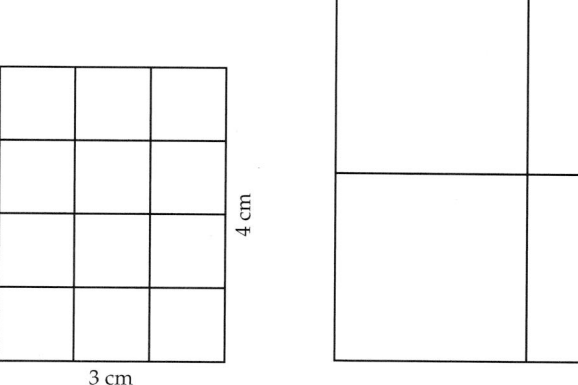

4 cm

2 in.

3 cm

3 in.

The area of the rectangle is 12 cm^2.

The area of the rectangle is 6 in^2.

Note from these figures that the area of a rectangle can be found by multiplying the length of the rectangle by its width.

ANSWER   *a. The area is 1 square yard, written 1 yd^2. **b.** The area is 1 square kilometer, written 1 km^2.*

> **Area of a Rectangle**
>
> Let L represent the length and W the width of a rectangle. The area, A, of the rectangle is given by $A = LW$.

QUESTION   *How many squares, each 1 inch on a side, are needed to cover a rectangle that has an area of 18 in²?*

EXAMPLE 7 ■ **Find the Area of a Rectangle**

How many square feet of sod are needed to cover a football field? A football field measures 360 ft by 160 ft.

Solution

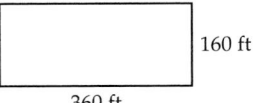

160 ft

360 ft

• Draw a diagram.

$A = LW$ • Use the formula for the area of a rectangle.

$A = 360(160)$ • The length is **360** ft. Substitute **360** for L. The width is **160** ft. Substitute **160** for W. Remember that LW means "L times W."

$A = 57,600$

57,600 ft² of sod is needed.   • Area is measured in square units.

CHECK YOUR PROGRESS 7   Find the amount of fabric needed to make a rectangular flag that measures 308 cm by 192 cm.

Solution   *See page S29.*

✔ **TAKE NOTE**

Recall that the rules of exponents state that when multiplying variables with like bases, we add the exponents.

A square is a rectangle in which all sides are the same length. Therefore, both the length and the width of a square can be represented by s, and $A = LW = s \cdot s$.

By the rules of exponents, $s \cdot s = s^2$.

s

$A = s \cdot s$
$A = s^2$

> **Area of a Square**
>
> Let s represent the length of a side of a square. The area, A, of the square is given by $A = s^2$.

ANSWER   *18 squares, each 1 inch on a side, are needed to cover the rectangle.*

EXAMPLE 8 ▪ Find the Area of a Square

A homeowner wants to carpet the family room. The floor is square and measures 6 m on each side. How much carpet should be purchased?

Solution

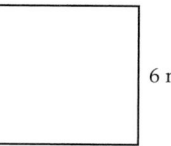

6 m

• Draw a diagram.

$A = s^2$

• Use the formula for the area of a square.

$A = 6^2$

• The length of a side is **6** m. Substitute **6** for *s*.

$A = 36$

36 m² of carpet should be purchased.

• Area is measured in square units.

CHECK YOUR PROGRESS 8   Find the area of the floor of a two-car garage that is in the shape of a square that measures 24 ft on a side.

Solution   See page S29.

Figure *ABCD* is a parallelogram. $\overline{BC}$ is the **base** of the parallelogram. $\overline{AE}$, perpendicular to the base, is the **height** of the parallelogram.

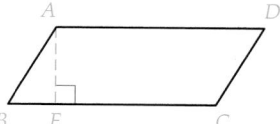

Any side of a parallelogram can be designated as the base. The corresponding height is found by drawing a line segment perpendicular to the base from the opposite side. In the figure at the right, $\overline{CD}$ is the base and $\overline{AE}$ is the height.

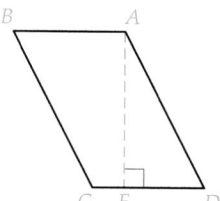

A rectangle can be formed from a parallelogram by cutting a right triangle from one end of the parallelogram and attaching it to the other end. The area of the resulting rectangle will equal the area of the original parallelogram.

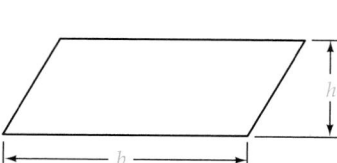

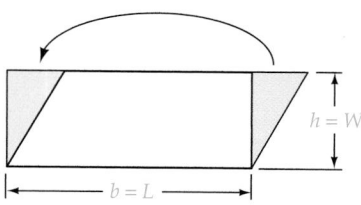

> **Area of a Parallelogram**
>
> Let b represent the length of the base and h the height of a parallelogram. The area, A, of the parallelogram is given by $A = bh$.

EXAMPLE 9 ▪ Find the Area of a Parallelogram

A solar panel is in the shape of a parallelogram that has a base of 2 ft and a height of 3 ft. Find the area of the solar panel.

Solution

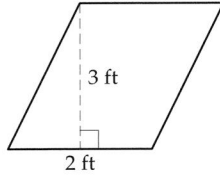

• Draw a diagram.

$A = bh$ • Use the formula for the area of a parallelogram.

$A = 2(3)$ • The base is **2** ft. Substitute **2** for *b*.
                  The height is **3** ft. Substitute **3** for *h*.
                  Remember that *bh* means "*b* times *h*."

$A = 6$

The area is 6 ft^2.   • Area is measured in square units.

CHECK YOUR PROGRESS 9   A fieldstone patio is in the shape of a parallelogram that has a base measuring 14 m and a height measuring 8 m. What is the area of the patio?

Solution   *See page S29.*

Figure ABC is a triangle. $\overline{AB}$ is the **base** of the triangle. $\overline{CD}$, perpendicular to the base, is the **height** of the triangle.

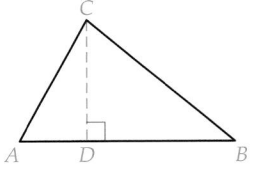

Any side of a triangle can be designated as the base. The corresponding height is found by drawing a line segment perpendicular to the base from the vertex opposite the base.

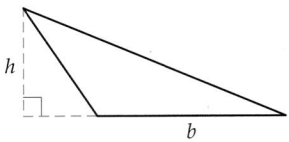

Consider the triangle with base b and height h shown at the right. By extending a line segment from C parallel to the base $\overline{AB}$ and equal in length to the base, a parallelogram is formed. The area of the parallelogram is bh and is twice the area of the triangle. Therefore, the area of the triangle is one half the area of the parallelogram, or $\frac{1}{2}bh$.

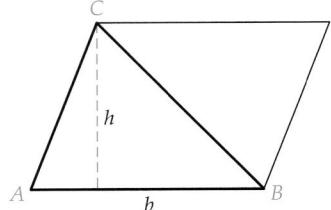

> **Area of a Triangle**
>
> Let b represent the length of the base and h the height of a triangle. The area, A, of the triangle is given by $A = \frac{1}{2}bh$.

EXAMPLE 10 ■ **Find the Area of a Triangle**

A riveter uses metal plates that are in the shape of a triangle with a base of 12 cm and a height of 6 cm. Find the area of one metal plate.

Solution

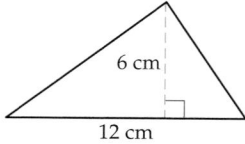

• Draw a diagram.

$A = \dfrac{1}{2}bh$ • Use the formula for the area of a triangle.

$A = \dfrac{1}{2}(12)(6)$ • The base is **12** cm. Substitute **12** for b. The height is **6** cm. Substitute **6** for h. Remember that bh means "b times h."

$A = 6(6)$

$A = 36$

The area is 36 cm². • Area is measured in square units.

CHECK YOUR PROGRESS 10   Find the amount of felt needed to make a banner that is in the shape of a triangle with a base of 18 in. and a height of 9 in.

Solution   See page S29.

✔ **TAKE NOTE**

The bases of a trapezoid are the parallel sides of the figure.

Figure $ABCD$ is a *trapezoid*. $\overline{AB}$, with length b_1, is one **base** of the trapezoid and $\overline{CD}$, with length b_2, is the other base. $\overline{AE}$, perpendicular to the two bases, is the **height.**

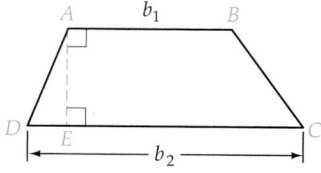

In the trapezoid at the right, the line segment $\overline{BD}$ divides the trapezoid into two triangles, ABD and BCD. In triangle ABD, b_1 is the base and h is the height. In triangle BCD, b_2 is the base and h is the height. The area of the trapezoid is the sum of the areas of the two triangles.

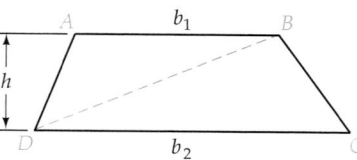

Area of trapezoid $ABCD$ = Area of triangle ABD + area of triangle BCD

$$= \frac{1}{2}b_1 h + \frac{1}{2}b_2 h = \frac{1}{2}h(b_1 + b_2)$$

Area of a Trapezoid

Let b_1 and b_2 represent the lengths of the bases and h the height of a trapezoid. The area, A, of the trapezoid is given by $A = \frac{1}{2}h(b_1 + b_2)$.

EXAMPLE 11 ■ Find the Area of a Trapezoid

A boat dock is built in the shape of a trapezoid with bases measuring 14 ft and 6 ft and a height of 7 ft. Find the area of the dock.

Solution

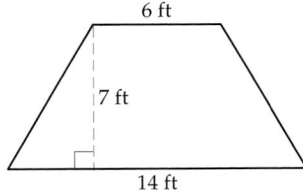

- Draw a diagram.

$$A = \frac{1}{2}h(b_1 + b_2)$$

- Use the formula for the area of a trapezoid.

$$A = \frac{1}{2} \cdot 7(14 + 6)$$

- The height is **7** ft. Substitute **7** for h. The bases measure **14** ft and **6** ft. Substitute **14** and **6** for b_1 and b_2.

$$A = \frac{1}{2} \cdot 7(20)$$

$$A = 70$$

The area is 70 ft².

- Area is measured in square units.

CHECK YOUR PROGRESS 11   Find the area of a patio that has the shape of a trapezoid with a height of 9 ft and bases measuring 12 ft and 20 ft.

Solution   *See page S29.*

The area of a circle is the product of π and the square of the radius.

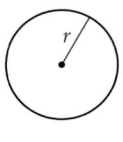

$$A = \pi r^2$$

The Area of a Circle

The area, A, of a circle with radius of length r is given by $A = \pi r^2$.

✔ **TAKE NOTE**

For your reference, all of the formulas for the perimeters and areas of the geometric figures presented in this section are listed in the Chapter Summary at the end of this chapter.

Find the area of a circle that has a radius of 6 cm.

Use the formula for the area of a circle. $r = 6$.

$$A = \pi r^2$$
$$A = \pi(6)^2$$
$$A = \pi(36)$$

The exact area of the circle is 36π cm².

$$A = 36\pi$$

An approximate measure can be found by using the π key on a calculator.

$$A \approx 113.10$$

The approximate area of the circle is 113.10 cm².

EXAMPLE 12 ■ **Find the Area of a Circle**

Find the area of a circle with a diameter of 10 m. Round to the nearest hundredth.

Solution

$$r = \frac{1}{2}d = \frac{1}{2}(10) = 5$$ • Find the radius of the circle.

$$A = \pi r^2$$ • Use the formula for the area of a circle.
$$A = \pi(5)^2$$ • Replace r with **5**.
$$A = \pi(25)$$ • Square 5.
$$A \approx 78.54$$ • An approximation is asked for. Use the π key on a calculator.

The area of the circle is approximately 78.54 m².

CHECK YOUR PROGRESS 12 Find the area of a circle with a diameter of 12 km. Give the exact measure.

Solution *See page S29.*

EXAMPLE 13 ■ **Find the Area of a Circle**

How large a cover is needed for a circular hot tub that is 8 ft in diameter? Round to the nearest tenth.

Solution

$$r = \frac{1}{2}d = \frac{1}{2}(8) = 4$$ • Find the radius of a circle with a diameter of 8 ft.

$$A = \pi r^2$$ • Use the formula for the area of a circle.
$$A = \pi(4)^2$$ • Replace r with **4**.
$$A = \pi(16)$$ • Square 4.
$$A \approx 50.3$$ • Use the π key on a calculator.

The cover for the hot tub must be 50.3 ft².

CHECK YOUR PROGRESS 13 How much material should be purchased to make a circular tablecloth that is to have a diameter of 4 ft? Round to the nearest hundredth.

Solution *See page S29.*

Math Matters   Mobius Bands

Cut out a long, narrow rectangular strip of paper. Shade one side of the paper.

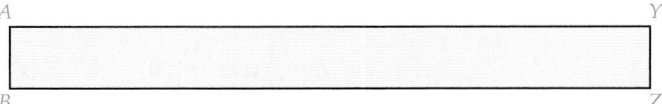

Give the strip of paper a half-twist.

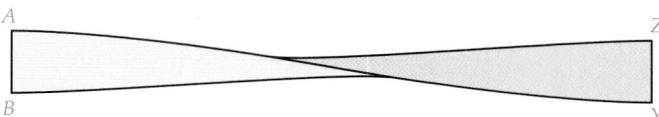

Put the ends together so that *Z* meets *Z* and *B* meets *Y*. Tape the ends together. The result is a *Mobius band.*

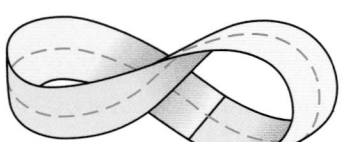
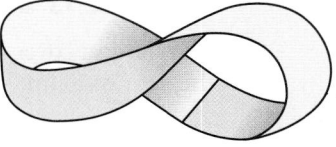

Make a Mobius band that is $1\frac{1}{2}$ in. wide. Use a pair of scissors to cut the Mobius band lengthwise down the middle, staying $\frac{3}{4}$ in. from the right-hand edge. Describe the result.

Make a Mobius band from plain white paper and then shade one side. Describe what remains unshaded on the Mobius band, and state the number of sides a Mobius band has.

Excursion

Slicing Polygons into Triangles[1]

Shown at the right is a triangle with three "slices" through it. Notice that the resulting pieces are six quadrilaterals and a triangle.

Excursion Exercises

1. Determine how you can slice a triangle so that all the resulting pieces are triangles. Use three slices. Each slice must cut all the way through the triangle.

2. Determine how you can slice each of the following polygons so that all the resulting pieces are triangles. Each slice must cut all the way through the figure. Record the number of slices required for each polygon. *Note:* There may be more than one solution for a figure.

   a. Quadrilateral   **b.** Pentagon   **c.** Hexagon   **d.** Heptagon

Using Patterns in Experimentation

3. Try to cut a pie into the greatest number of pieces with only five straight cuts of a knife. An illustration showing how five cuts can produce 13 pieces is shown below. The correct answer, however, yields more than thirteen pieces.

   A reasonable question is, "How do I know when I have the maximum number of pieces?" To determine the answer, we suggest that you start with one cut, then two cuts, then three cuts, and so on. Try to discover a pattern for the greatest number of pieces that each successive cut can produce.

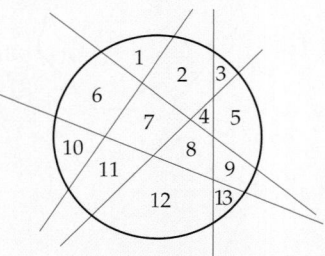

---

1. This activity is adapted from James Gary Propp's "The Slicing Game," *American Mathematical Monthly,* April, 1996.

Exercise Set 8.2   (Suggested Assignment: 17–79 odds)

1. Label the length of the rectangle *L* and the width of the rectangle *W*.

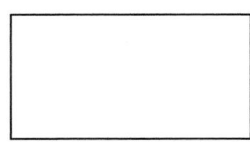

Figure A

2. Label the base of the parallelogram *b*, the side *s*, and the height *h*.

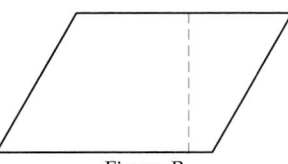

Figure B

3. What is wrong with each statement?
   a. The perimeter is 40 m².
   b. The area is 120 ft.

In Exercises 4–7, name each polygon.

4.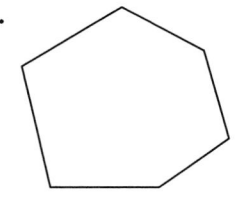

5.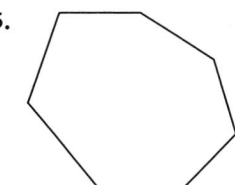

6.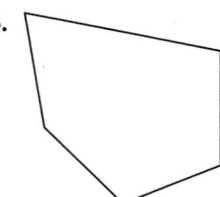

7.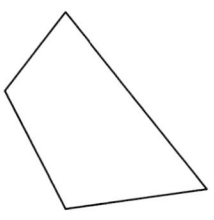

In Exercises 8–11, classify the triangle as isosceles, equilateral, or scalene.

8.

9.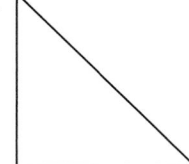

10.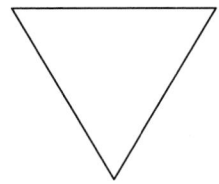

11.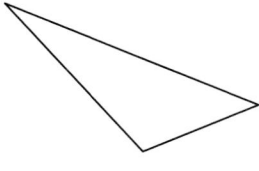

In Exercises 12–15, classify the triangle as acute, obtuse, or right.

12.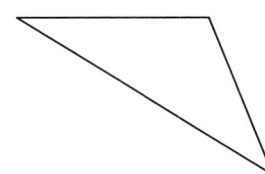

13.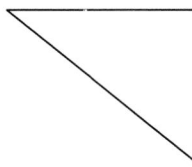

14.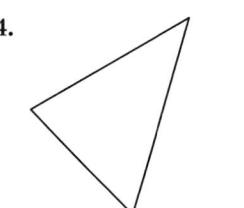

15.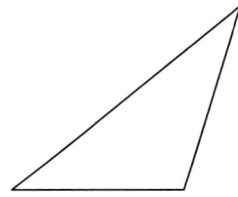

In Exercises 16–24, find (**a**) the perimeter and (**b**) the area of the figure.

16.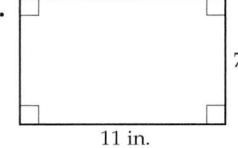
7 in.
11 in.

17.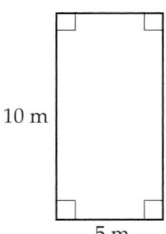
10 m
5 m

18.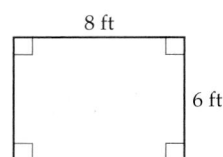
8 ft
6 ft

19.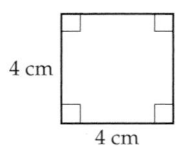
4 cm
4 cm

20.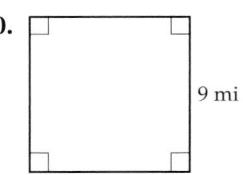
9 mi
9 mi

21.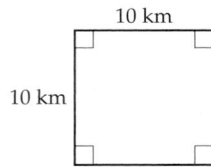
10 km
10 km

22.

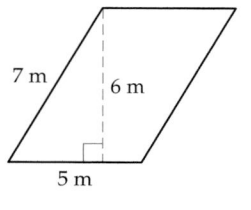

7 m   6 m

5 m

23.

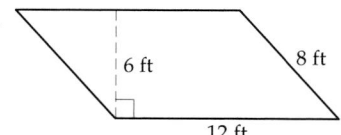

6 ft   8 ft

12 ft

24.

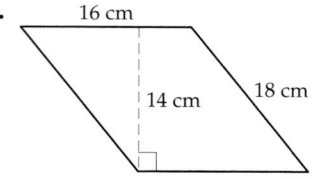

16 cm

14 cm   18 cm

In Exercises 25–30, find (**a**) the circumference and (**b**) the area of the figure. Give both exact values and approximations to the nearest hundredth.

25.

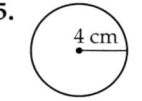

4 cm

26.

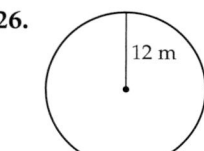

12 m

27.

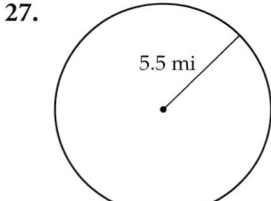

5.5 mi

28.

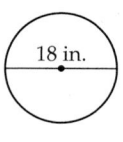

18 in.

29.

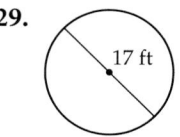

17 ft

30.

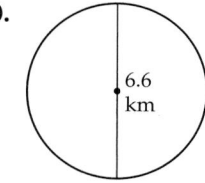

6.6 km

31. You need to fence in the triangular plot of land shown below. How many feet of fencing do you need?

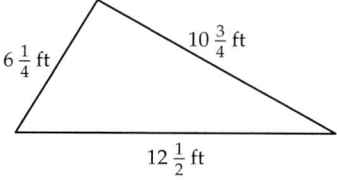

$6\frac{1}{4}$ ft $10\frac{3}{4}$ ft

$12\frac{1}{2}$ ft

32. A flower garden in the yard of a historical home is in the shape of a triangle, as shown below. The wooden beams lining the edge of the garden need to be replaced. Find the total length of wood beams that must be purchased in order to replace the old beams.

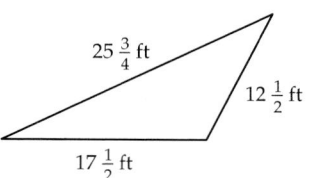

$25\frac{3}{4}$ ft $12\frac{1}{2}$ ft

$17\frac{1}{2}$ ft

33. The course of a yachting race is in the shape of a triangle with sides that measure $4\frac{3}{10}$ mi, $3\frac{7}{10}$ mi, and $2\frac{1}{2}$ mi. Find the total length of the course.

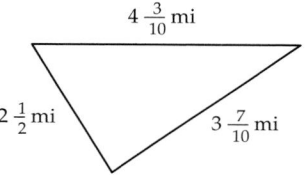

$4\frac{3}{10}$ mi

$2\frac{1}{2}$ mi $3\frac{7}{10}$ mi

34. An exercise course has stations set up along a path that is in the shape of a triangle with sides that measure $12\frac{1}{12}$ yd, $29\frac{1}{3}$ yd, and $26\frac{3}{4}$ yd. What is the entire length of the exercise course?

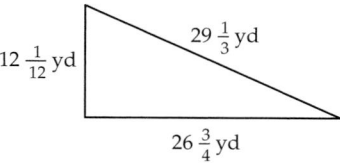

$12\frac{1}{12}$ yd $29\frac{1}{3}$ yd

$26\frac{3}{4}$ yd

35. How many feet of fencing should be purchased to enclose a rectangular garden that is 20 ft long and 14 ft wide?

36. How many meters of binding are required to bind the edge of a rectangular quilt that measures 4 m by 6 m?

37. Find the perimeter of a regular pentagon that measures 4 in. on one side.

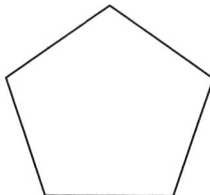

38. Wall-to-wall carpeting is installed in a room that is 15 ft long and 10 ft wide. The edges of the carpet are held down by tack strips. How many feet of tack-strip material are needed?

39. The length of a rectangular park is 62 yd. The width is 45 yd. How many yards of fencing are needed to surround the park?

40. What is the perimeter of a regular hexagon that measures 9 cm on each side?

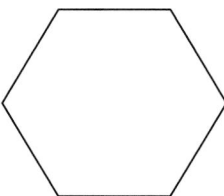

41. A cross-country course is in the shape of a parallelogram with a base of length 3 mi and a side of length 2 mi. What is the total length of the cross-country course?

42. A rectangular playground has a length of 160 ft and a width of 120 ft. Find the length of hedge that surrounds the playground.

43. Bias binding is to be sewn around the edge of a rectangular tablecloth measuring 68 in. by 42 in. If the bias binding comes in packages containing 15 ft of binding, how many packages of bias binding are needed for the tablecloth?

44. Find the area of a rectangular flower garden that measures 24 ft by 18 ft.

45. What is the area of a square patio that measures 12 m on each side?

46. Artificial turf is being used to cover a playing field. If the field is rectangular with a length of 110 yd and a width of 80 yd, how much artificial turf must be purchased to cover the field?

47. The perimeter of a square picture frame is 36 in. Find the length of each side of the frame.

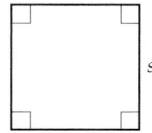

48. A square rug has a perimeter of 24 ft. Find the length of each edge of the rug.

49. The area of a rectangle is 400 in². If the length of the rectangle is 40 in., what is the width?

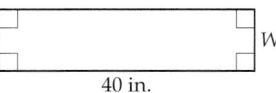

50. The width of a rectangle is 8 ft. If the area is 312 ft², what is the length of the rectangle?

51. The area of a parallelogram is 56 m². If the height of the parallelogram is 7 m, what is the length of the base?

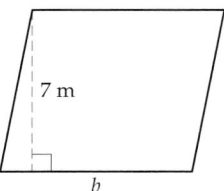

52. You want to rent a storage unit. You estimate that you will need 175 ft² of floor space. In the Yellow Pages, you see the ad shown below. You want to rent the smallest possible unit that will hold everything you want to store. Which of the six units pictured in the ad should you select?

53. A sail is in the shape of a triangle with a base of 12 m and a height of 16 m. How much canvas was needed to make the body of the sail?

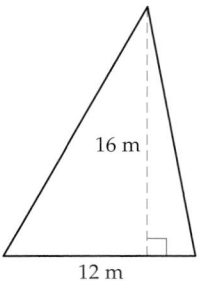

54. A vegetable garden is in the shape of a triangle with a base of 21 ft and a height of 13 ft. Find the area of the vegetable garden.

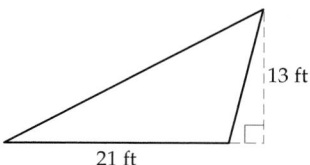

55. How much artificial turf should be purchased to cover an athletic field that is in the shape of a trapezoid with a height of 15 m and bases that measure 45 m and 36 m?

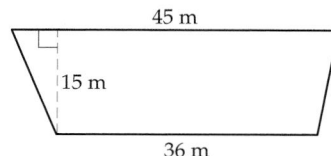

56. A township is in the shape of a trapezoid with a height of 10 km and bases measuring 9 km and 23 km. What is the land area of the township?

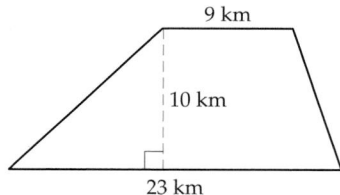

57. A city plans to plant grass seed in a public playground that has the shape of a triangle with a height of 24 m and a base of 20 m. Each bag of grass seed will seed 120 m². How many bags of seed should be purchased?

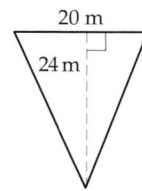

58. The family room of a home is on the third floor of the house. Because of the pitch of the roof, two walls of the room are in the shape of a trapezoid with bases that measure 16 ft and 24 ft and a height of 8 ft. You plan to wallpaper the two walls. One roll of wallpaper will cover 40 ft². How many rolls of wallpaper should you purchase?

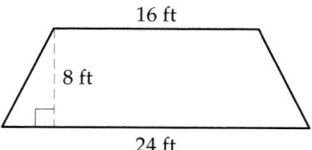

59. You plan to stain the wooden deck at the back of your house. The deck is in the shape of a trapezoid with bases that measure 10 ft and 12 ft and a height of 10 ft. A quart of stain will cover 55 ft². How many quarts of stain should you purchase?

60. A fabric wall hanging is in the shape of a triangle that has a base of 4 ft and a height of 3 ft. An additional 1 ft² of fabric is needed for hemming the material. How much fabric should be purchased to make the wall hanging?

61. You want to tile your kitchen floor. The floor measures 10 ft by 8 ft. How many square tiles that measure 2 ft along each side should you purchase for the job?

62. You are wallpapering two walls of a den, one measuring 10 ft by 8 ft and the other measuring 12 ft by 8 ft. The wallpaper costs $24 per roll, and each roll will cover 40 ft². What is the cost to wallpaper the two walls?

63. An urban renewal project involves reseeding a garden that is in the shape of a square, 80 ft on each side. Each bag of grass seed costs $8 and will seed 1500 ft². How much money should be budgeted for buying grass seed for the garden?

64. You want to install wall-to-wall carpeting in the family room. The floor plan is drawn below. If the cost of the carpet you would like to purchase is $19 per square yard, what is the cost of carpeting your family room? Assume there is no waste. *Hint:* 9 ft² = 1 yd²

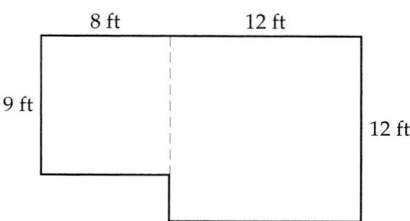

65. You want to paint the walls of your bedroom. Two walls measure 16 ft by 8 ft, and the other two walls measure 12 ft by 8 ft. The paint you wish to purchase costs $17 per gallon, and each gallon will cover 400 ft² of wall. Find the total amount you will spend on paint.

66. A walkway 2 m wide surrounds a rectangular plot of grass. The plot is 25 m long and 15 m wide. What is the area of the walkway?

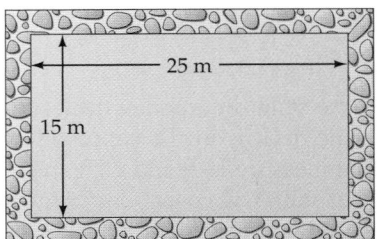

67. Pleated draperies for a window must be twice as wide as the width of the window. Draperies are being made for four windows, each 3 ft wide and 4 ft high. Because the drapes will fall slightly below the window sill and extra fabric is needed for hemming the drapes, 1 ft must be added to the height of the window. How much material must be purchased to make the drapes?

68. How many feet of fencing should be purchased to enclose a circular flower garden that has a diameter of 18 ft? Round to the nearest tenth.

69. Find the length of molding needed to put around a circular table that is 4.2 ft in diameter. Round to the nearest hundredth.

70. How much binding is needed to bind the edge of a circular rug that is 3 m in diameter? Round to the nearest hundredth.

71. Find the area of a circular flower garden that has a radius of 20 ft. Round to the nearest tenth.

72. A pulley system is diagrammed below. If pulley B has a diameter of 16 in. and is rotating at 240 revolutions per minute, how far does the belt travel each minute that the pulley system is in operation? Assume the belt does not slip as the pulley rotates. Round to the nearest inch.

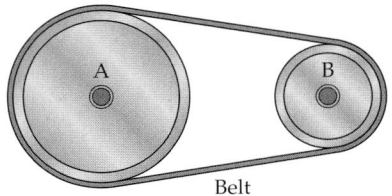

Belt

73. A bicycle tire has a diameter of 18 in. How many feet does the bicycle travel when the wheel makes 20 revolutions? Round to the nearest hundredth.

74. The front wheel of a tricycle has a diameter of 16 in. How many feet does the tricycle travel when the wheel makes 15 revolutions? Round to the nearest hundredth.

75. The lens located on an astronomical telescope has a diameter of 24 in. Find the area of the lens. Give the exact value.

76. An irrigation system waters a circular field that has a 50-foot radius. Find the area watered by the irrigation system. Give the exact value.

77. How much greater is the area of a pizza that has a radius of 10 in. than the area of a pizza that has a radius of 8 in.? Round to the nearest hundredth.

78. A restaurant serves a small pizza that has a radius of 6 in. The restaurant's large pizza has a radius that is twice the radius of the small pizza. How much larger is the area of the large pizza? Round to the nearest hundredth. Is the area of the large pizza more or less than twice the area of the small pizza?

79. There are two general types of satellite systems that orbit our Earth: geostationary Earth orbit (GEO) and non-geostationary, primarily low Earth orbit (LEO). Geostationary satellite systems orbit at a distance of 36,000 km above Earth. An orbit at this altitude allows the satellite to maintain a fixed position in relation to Earth. What is the distance traveled by a GEO satellite in one orbit around Earth? The radius of Earth at the equator is 6380 km. Round to the nearest kilometer.

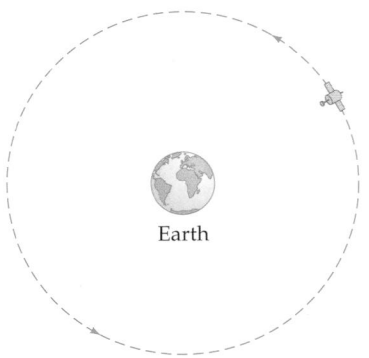

Earth

80. How much farther is it around the bases of a baseball diamond than around the bases of a softball dimond? *Hint:* Baseball and softball diamonds are squares.

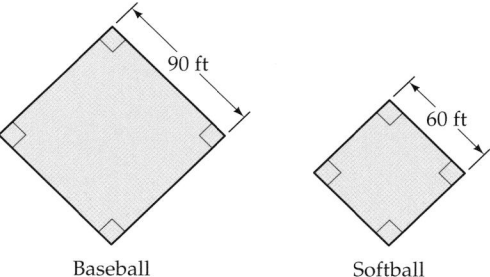

Baseball   Softball

81. Write an expression for the area of the shaded portion of the diagram. Leave the answer in terms of π and r.

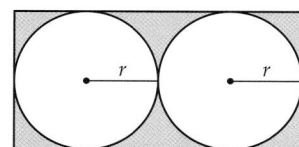

82. Write an expression for the area of the shaded portion of the diagram. Leave the answer in terms of π and r.

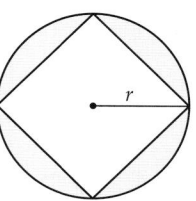

83. If both the length and width of a rectangle are doubled, how many times larger is the area of the resulting rectangle?

Extensions

CRITICAL THINKING

84. Determine whether the statement is always true, sometimes true, or never true.

   a. Two triangles that have the same perimeter have the same area.

   b. Two rectangles that have the same area have the same perimeter.

   c. If two squares have the same area, then the sides of the squares have the same length.

85. Find the dimensions of a rectangle that has the same area as the shaded region in the diagram below. Write the dimensions in terms of the variable a.

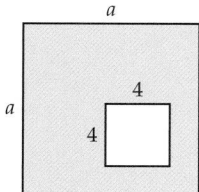

86. Find the dimensions of a rectangle that has the same area as the shaded region in the diagram below. Write the dimensions in terms of the variable x.

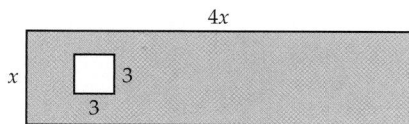

87. In the diagrams below, the length of one side of a square is 1 cm. Find the perimeter and the area of the eighth figure in the pattern.

a.

b.

c.

EXPLORATIONS

88. The perimeter of the square at the right is 4 units.

If two such squares are joined along one of the sides, the perimeter is 6 units. Note that it does not matter which sides are joined; the perimeter is still 6 units.

If three squares are joined, the perimeter of the resulting figure is 8 units for each possible placement of the squares.

Four squares can be joined in five different ways as shown. There are two possible perimeters: 10 units for A, B, C, and D, and 8 units for E.

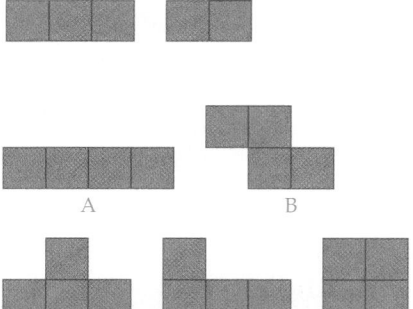

a. If five squares are joined, what is the maximum perimeter possible?

b. If five squares are joined, what is the minimum perimeter possible?

c. If six squares are joined, what is the maximum perimeter possible?

d. If six squares are joined, what is the minimum perimeter possible?

SECTION 8.3 | ## Similar Triangles

▼ **point of interest**

Model trains are similar to actual trains. They come in a variety of sizes that manufacturers have agreed upon, so that, for example, an engine made by manufacturer A is able to run on a track made by manufacturer B. Listed below are three model railroad sizes by name, along with the ratio of model size to actual size.

Name	Ratio
Z	1 : 220
N	1 : 160
HO	1 : 87

Z Scale

N Scale

HO Scale

HO's ratio of 1 : 87 means that in every dimension, a model railroad car is $\frac{1}{87}$ the size of the real railroad car.

Similar Triangles

Similar objects have the same shape but not necessarily the same size. A tennis ball is similar to a basketball. A model ship is similar to an actual ship.

Similar objects have corresponding parts; for example, the rudder on the model ship corresponds to the rudder on the actual ship. The relationship between the sizes of each of the corresponding parts can be written as a ratio, and each ratio will be the same. If the rudder on the model ship is $\frac{1}{100}$ the size of the rudder on the actual ship, then the model wheelhouse is $\frac{1}{100}$ of the size of the actual wheelhouse, the width of the model is $\frac{1}{100}$ the width of the actual ship, and so on.

The two triangles ABC and DEF shown at the right are similar. Side $\overline{AB}$ corresponds to side $\overline{DE}$, side $\overline{BC}$ corresponds to side $\overline{EF}$, and side $\overline{AC}$ corresponds to side $\overline{DF}$. The ratios of the lengths of corresponding sides are equal.

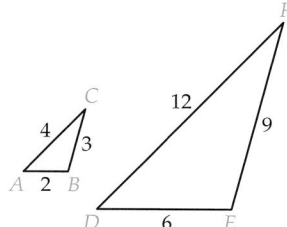

$$\frac{AB}{DE} = \frac{2}{6} = \frac{1}{3}, \quad \frac{BC}{EF} = \frac{3}{9} = \frac{1}{3}, \quad \text{and}$$

$$\frac{AC}{DF} = \frac{4}{12} = \frac{1}{3}$$

Because the ratios of corresponding sides are equal, several proportions can be formed.

$$\frac{AB}{DE} = \frac{BC}{EF}, \quad \frac{AB}{DE} = \frac{AC}{DF}, \quad \text{and} \quad \frac{BC}{EF} = \frac{AC}{DF}$$

The measures of corresponding angles in similar triangles are equal. Therefore,

$$m\angle A = m\angle D, \quad m\angle B = m\angle E, \quad \text{and}$$
$$m\angle C = m\angle F$$

Triangles ABC and DEF at the right are similar triangles. AH and DK are the heights of the triangles. The ratio of the heights of similar triangles equals the ratio of the lengths of corresponding sides.

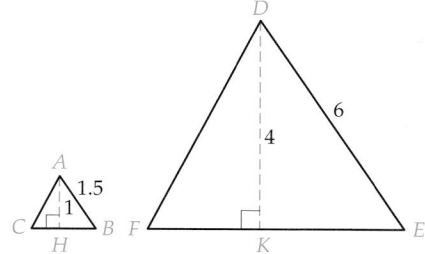

Ratio of corresponding sides $= \frac{1.5}{6} = \frac{1}{4}$

Ratio of heights $= \frac{1}{4}$

Properties of Similar Triangles

For similar triangles, the ratios of corresponding sides are equal. The ratio of corresponding heights is equal to the ratio of corresponding sides.

▼ **point of interest**

Congruent objects have the same shape *and* the same size. Congruent and similar triangles differ in that for congruent triangles, the measures of corresponding sides and angles must be equal; for similar triangles, the measures of corresponding angles must be equal, but corresponding sides are not necessarily the same length.

The two triangles at the right are similar triangles. Find the length of side $\overline{EF}$. Round to the nearest tenth.

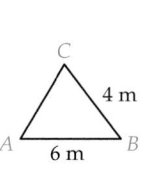 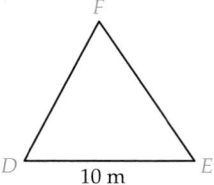

The triangles are similar, so the ratios of the lengths of corresponding sides are equal.

$$\frac{EF}{BC} = \frac{DE}{AB}$$

$$\frac{EF}{4} = \frac{10}{6}$$

$$6(EF) = 4(10)$$

$$6(EF) = 40$$

$$EF \approx 6.7$$

The length of side EF is approximately 6.7 m.

QUESTION *What are two other proportions that can be written for the similar triangles shown above?*

EXAMPLE 1 ■ **Use Similar Triangles to Find the Unknown Height of a Triangle**

Triangles ABC and DEF are similar. Find FG, the height of triangle DEF.

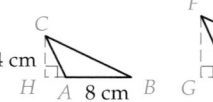 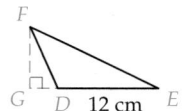

Solution

$$\frac{AB}{DE} = \frac{CH}{FG}$$ • For similar triangles, the ratio of corresponding sides equals the ratio of corresponding heights.

$$\frac{8}{12} = \frac{4}{FG}$$ • Replace *AB*, *DE*, and *CH* with their values.

$$8(FG) = 12(4)$$ • The cross products are equal.

$$8(FG) = 48$$

$$FG = 6$$ • Divide both sides of the equation by 8.

The height FG of triangle DEF is 6 cm.

CHECK YOUR PROGRESS 1
Triangles ABC and DEF are similar. Find FG, the height of triangle DEF.

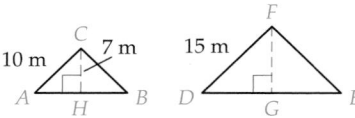

Solution *See page S29.*

INSTRUCTOR NOTE
The concept of similar triangles and the fact that the ratios of corresponding sides are equal are not obvious to students. Nonetheless, these concepts are essential in many practical applications.
  You might introduce the topic of similar triangles by using a magnifying glass. Illustrate that an object viewed under a magnifying lens appears larger, but its shape has not changed.

ANSWER *In addition to $\frac{EF}{BC} = \frac{DE}{AB}$, we can write the proportions $\frac{DE}{AB} = \frac{DF}{AC}$ and $\frac{EF}{BC} = \frac{DF}{AC}$. These three proportions can also be written using the reciprocal of each fraction: $\frac{BC}{EF} = \frac{AB}{DE}$, $\frac{AB}{DE} = \frac{AC}{DF}$, and $\frac{BC}{EF} = \frac{AC}{DF}$. Also, the right and left sides of each proportion can be interchanged.*

Triangles *ABC* and *DEF* are similar triangles. Find the area of triangle *ABC*.

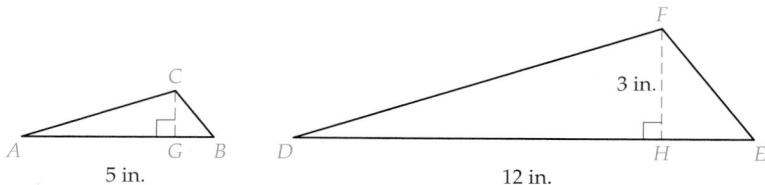

Solve a proportion to find the height of triangle *ABC*.

$$\frac{AB}{DE} = \frac{CG}{FH}$$

$$\frac{5}{12} = \frac{CG}{3}$$

$$12(CG) = 5(3)$$

$$12(CG) = 15$$

$$CG = 1.25$$

Use the formula for the area of a triangle.

$$A = \frac{1}{2}bh$$

The base is 5 in. The height is 1.25 in.

$$A = \frac{1}{2}(5)(1.25)$$

$$A = 3.125$$

The area of triangle *ABC* is 3.125 in².

If the three angles of one triangle are equal in measure to the three angles of another triangle, then the triangles are similar.

In triangle *ABC* at the right, line segment $\overline{DE}$ is drawn parallel to the base $\overline{AB}$. Because the measures of corresponding angles are equal, $m\angle x = m\angle r$ and $m\angle y = m\angle n$. We know that $m\angle C = m\angle C$. Thus the measures of the three angles of triangle *ABC* are equal, respectively, to the measures of the three angles of triangle *DEC*. Therefore, triangles *ABC* and *DEC* are similar triangles.

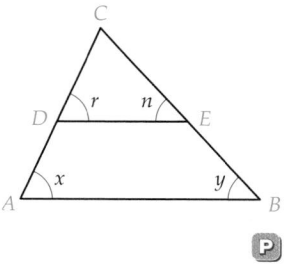

The sum of the measures of the three angles of a triangle is 180°. If two angles of one triangle are equal in measure to two angles of another triangle, then the third angles must be equal. Thus, we can say that if two angles of one triangle are equal in measure to two angles of another triangle, then the two triangles are similar.

In the figure at the right, $\overline{AB}$ intersects $\overline{CD}$ at point *O*. Angles *C* and *D* are right angles. Find the length of $\overline{DO}$.

First determine whether triangles *AOC* and *BOD* are similar.

$m\angle C = m\angle D$ because they are both right angles.

$m\angle x = m\angle y$ because vertical angles have the same measure.

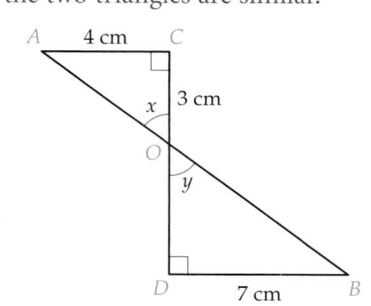

✔ **TAKE NOTE**

You can always create similar triangles by drawing a line segment inside the original triangle parallel to one side of the triangle. In the triangle below, $\overline{ST} \parallel \overline{QR}$ and triangle *PST* is similar to triangle *PQR*.

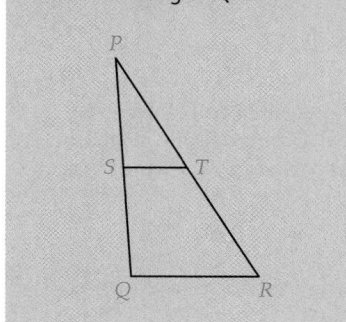

Because two angles of triangle *AOC* are equal in measure to two angles of triangle *BOD*, triangles *AOC* and *BOD* are similar.

Use a proportion to find the length of the unknown side.

$$\frac{AC}{BD} = \frac{CO}{DO}$$

$$\frac{4}{7} = \frac{3}{DO}$$

$$4(DO) = 7(3)$$

$$4(DO) = 21$$

$$DO = 5.25$$

The length of $\overline{DO}$ is 5.25 cm.

EXAMPLE 2 ■ Solve a Problem Involving Similar Triangles

In the figure at the right, $\overline{AB}$ is parallel to $\overline{DC}$, $\angle B$ and $\angle D$ are right angles, $AB = 12$ m, $DC = 4$ m, and $AC = 18$ m. Find the length of $\overline{CO}$.

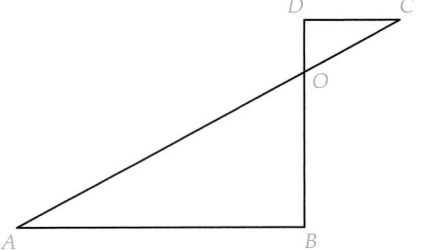

Solution

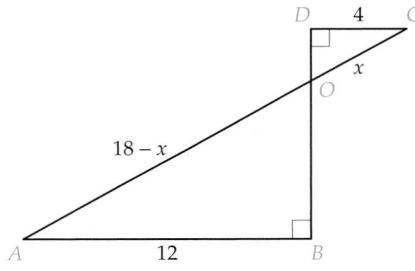

- Label the diagram using the given information. Let *x* represent *CO*. *AC* = *AO* + *CO*. Because *AC* = 18, *AO* = 18 − *x*.

$$\frac{DC}{AB} = \frac{CO}{AO}$$

$$\frac{4}{12} = \frac{x}{18 - x}$$

$$12x = 4(18 - x)$$

$$12x = 72 - 4x$$

$$16x = 72$$

$$x = 4.5$$

- Triangles *AOB* and *COD* are similar triangles. The ratios of corresponding sides are equal.

- Use the Distributive Property.

The length of $\overline{CO}$ is 4.5 m.

CHECK YOUR PROGRESS 2
In the figure at the right, $\overline{AB}$ is parallel to $\overline{DC}$, $\angle A$ and $\angle D$ are right angles, $AB = 10$ cm, $CD = 4$ cm, and $DO = 3$ cm. Find the area of triangle AOB.

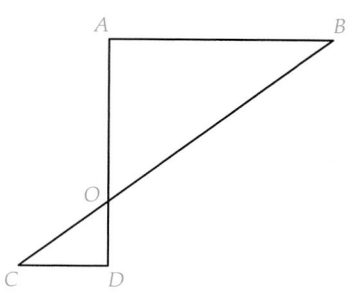

Solution   *See page S30.*

Math Matters   Similar Polygons

For similar triangles, the measures of corresponding angles are equal and the ratios of the lengths of corresponding sides are equal. The same is true for similar polygons: the measures of corresponding angles are equal and the lengths of corresponding sides are in proportion.

Quadrilaterals *ABCD* and *LMNO* are similar.

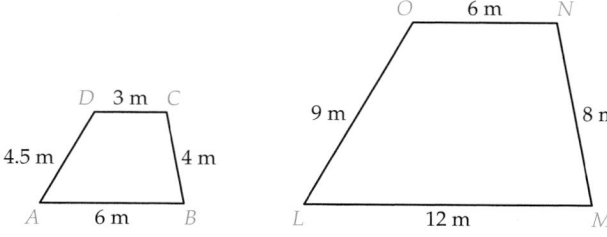

The ratio of the lengths of corresponding sides is: $\dfrac{AB}{LM} = \dfrac{6}{12} = \dfrac{1}{2}$

The ratio of the perimeter of *ABCD* to the perimeter of *LMNO* is:

$$\frac{\text{perimeter of } ABCD}{\text{perimeter of } LMNO} = \frac{17.5}{35} = \frac{1}{2}$$

Note that this ratio is the same as the ratio of corresponding sides. This is true for all similar polygons: If two polygons are similar, the ratio of their perimeters is equal to the ratio of the lengths of any pair of corresponding sides.

Excursion

Topology

In this section, we have been discussing similar figures; that is, figures with the same shape. The branch of geometry called *topology* is the study of even more basic properties of figures than their sizes and shapes. For example, look at the figures below. We could take a rubber band and stretch it into any one of these shapes.

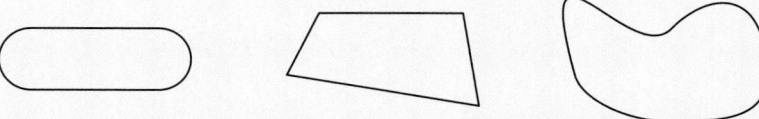

All three of these figures are different shapes, but each can be turned into one of the others by stretching the rubber band.

In topology, figures that can be stretched, molded, or bent into the same shape *without puncturing or cutting* belong to the same family. They are called **topologically equivalent.**

Rectangles, triangles, and circles are topologically equivalent.

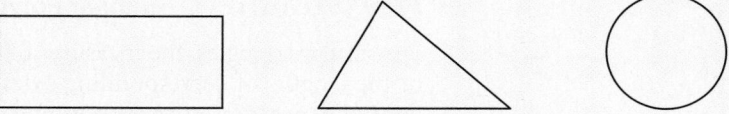

Line segments and wavy curves are topologically equivalent.

Note that the figures formed from a rubber band and a line segment are not topologically equivalent; to form a line segment from a rubber band, we would have to cut the rubber band.

In the following plane figures, the lines are joined where they cross. The figures are topologically equivalent. They are not topologically equivalent to any of the figures shown above.

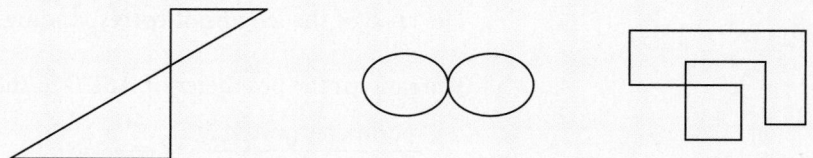

A **topologist** (a person who studies topology) is interested in identifying and describing different families of equivalent figures. Topology applies to solids as well as plane figures.

(continued)

For example, a topologist considers a brick, a potato, and a cue ball to be equivalent to each other. Think of using modeling clay to form each of these shapes.

Excursion Exercises

Which of the figures listed is not topologically equivalent to the others?

1. a. Parallelogram   **b.** Square   **c.** Ray   **d.** Trapezoid

2. a. Wedding ring   **b.** Doughnut   **c.** Fork   **d.** Sewing needle

3. a. A   **b.** D   **c.** O   **d.** P   **e.** T

Exercise Set 8.3   (Suggested Assignment: 5–27, odds)

In Exercises 1–4, find the ratio of the lengths of corresponding sides for the similar triangles.

1.

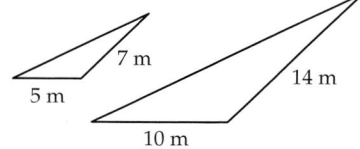

2.

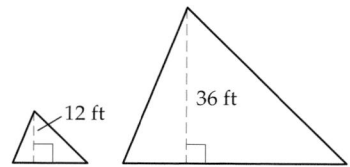

3.

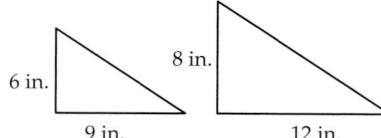

4.

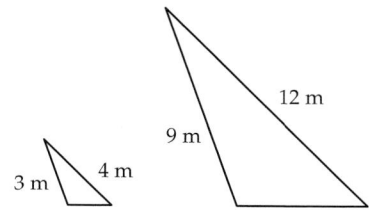

In Exercises 5–12, triangles ABC and DEF are similar triangles. Solve. Round to the nearest tenth.

5. Find side DE.

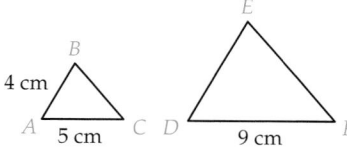

6. Find side DE.

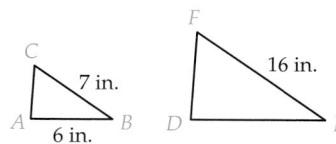

7. Find the height of triangle DEF.

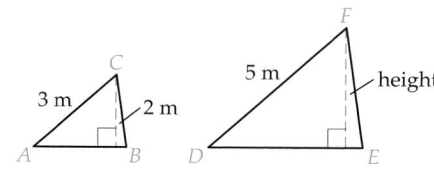

8. Find the height of triangle *ABC*.

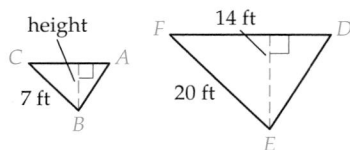

9. Find the perimeter of triangle *ABC*.

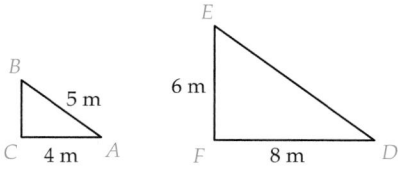

10. Find the perimeter of triangle *DEF*.

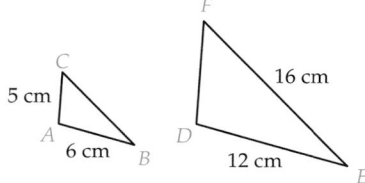

11. Find the perimeter of triangle *ABC*.

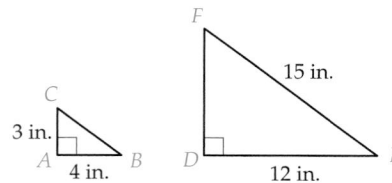

12. Find the area of triangle *DEF*.

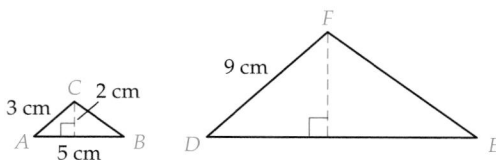

In Exercises 13 and 14, triangles *ABC* and *DEF* are similar triangles. Solve. Round to the nearest tenth.

13. Find the area of triangle *ABC*.

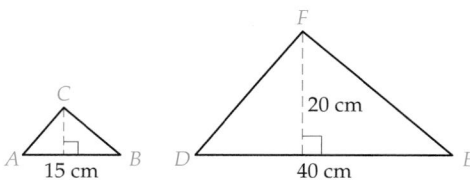

14. Find the area of triangle *DEF*.

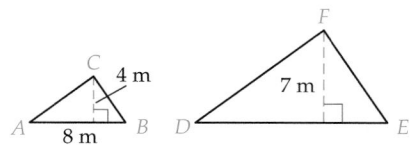

In Exercises 15–19, the given triangles are similar triangles. Use this fact to solve each exercise.

15. Find the height of the flagpole.

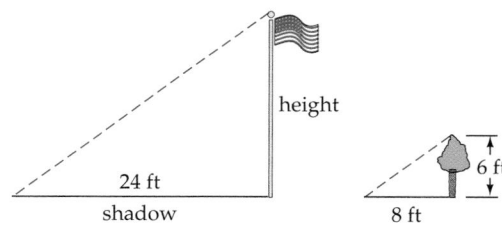

16. Find the height of the flagpole.

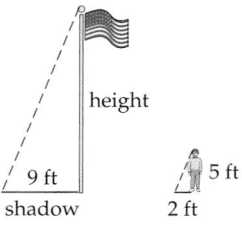

17. Find the height of the building.

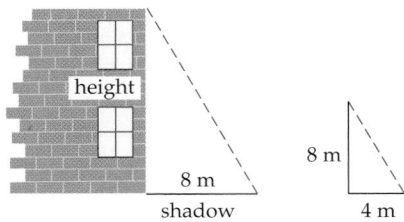

18. Find the height of the building.

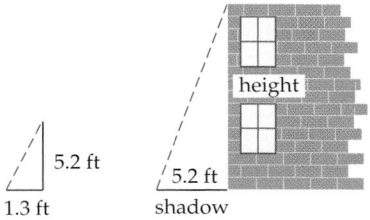

19. Find the height of the flagpole.

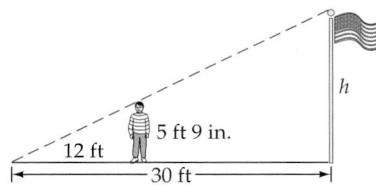

20. In the figure below, $\overline{BD} \parallel \overline{AE}$, BD measures 5 cm, AE measures 8 cm, and AC measures 10 cm. Find the length of $\overline{BC}$.

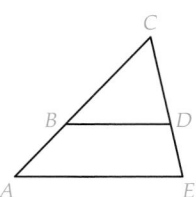

21. In the figure below, $\overline{AC} \parallel \overline{DE}$, BD measures 8 m, AD measures 12 m, and BE measures 6 m. Find the length of $\overline{BC}$.

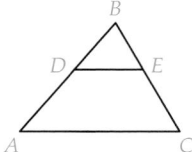

22. In the figure below, $\overline{DE} \parallel \overline{AC}$, DE measures 6 in., AC measures 10 in., and AB measures 15 in. Find the length of $\overline{DA}$.

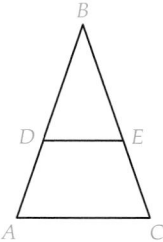

23. In the figure below, $\overline{AE} \parallel \overline{BD}$, AB measures 3 ft, ED measures 4 ft, and BC measures 3 ft. Find the length of $\overline{CE}$.

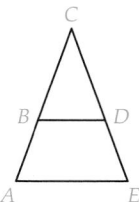

24. In the figure below, $\overline{MP}$ and $\overline{NQ}$ intersect at O, NO measures 25 ft, MO measures 20 ft, and PO measures 8 ft. Find the length of $\overline{QO}$.

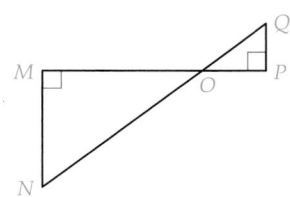

25. In the figure below, $\overline{MP}$ and $\overline{NQ}$ intersect at O, NO measures 24 cm, MN measures 10 cm, MP measures 39 cm, and QO measures 12 cm. Find the length of $\overline{OP}$.

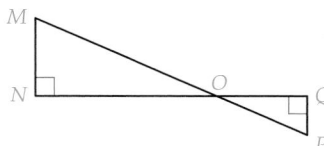

26. In the figure below, $\overline{MQ}$ and $\overline{NP}$ intersect at O, NO measures 12 m, MN measures 9 m, PQ measures 3 m, and MQ measures 20 m. Find the perimeter of triangle OPQ.

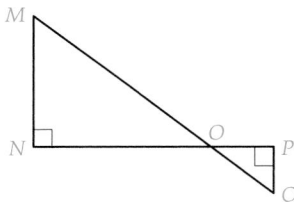

Surveyors use similar triangles to measure distances that cannot be measured directly. This is illustrated in Exercises 27 and 28.

27. The diagram below represents a river of width CD. Triangles AOB and DOC are similar. The distances AB, BO, and OC were measured and found to have the lengths given in the diagram. Find the width of the river.

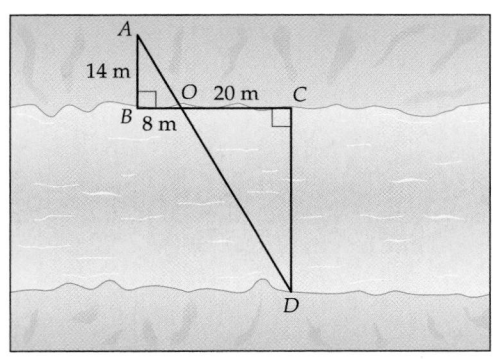

28. The diagram at the top of the next column shows how surveyors laid out similar triangles along the Winnepaugo River. Find the width, d, of the river.

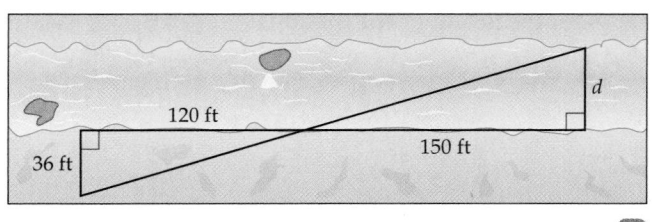

Extensions

CRITICAL THINKING

29. Determine whether the statement is always true, sometimes true, or never true.

a. If two angles of one triangle are equal to two angles of a second triangle, then the triangles are similar triangles.

b. Two isosceles triangles are similar triangles.

c. Two equilateral triangles are similar triangles.

d. If an acute angle of a right triangle is equal to an acute angle of another right triangle, then the triangles are similar triangles.

COOPERATIVE LEARNING

30. The height of a right triangle is drawn from the right angle perpendicular to the hypotenuse. (Recall that the hypotenuse of a right triangle is the side opposite the right angle.) Explain why the two smaller triangles formed are similar to the original triangle and similar to each other.

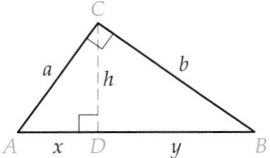

SECTION 8.4 | **Volume and Surface Area**

Volume

In Section 2 of this chapter, we developed the geometric concepts of perimeter and area. Perimeter and area refer to plane figures (figures that lie in a plane). We are now ready to introduce *volume* of geometric solids.

Geometric solids are figures in space. Figures in space include baseballs, ice cubes, milk cartons, and trucks.

Volume is a measure of the amount of space occupied by a geometric solid. Volume can be used to describe, for example, the amount of trash in a land fill, the amount of concrete poured for the foundation of a house, or the amount of water in a town's reservoir.

Five common geometric solids are rectangular solids, spheres, cylinders, cones, and pyramids.

A **rectangular solid** is one in which all six sides, called **faces**, are rectangles. The variable L is used to represent the length of a rectangular solid, W is used to represent its width, and H is used to represent its height. A shoe box is an example of a rectangular solid.

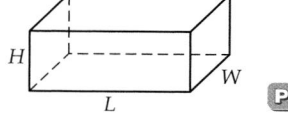

A **cube** is a special type of rectangular solid. Each of the six faces of a cube is a square. The variable s is used to represent the length of one side of a cube. A baby's block is an example of a cube.

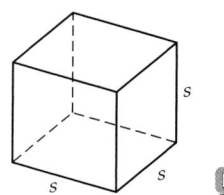

A cube that is 1 ft on each side has a volume of 1 cubic foot, which is written 1 ft³. A cube that measures 1 cm on each side has a volume of 1 cubic centimeter, written 1 cm³.

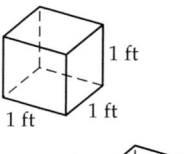

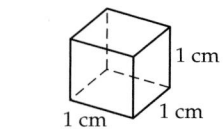

The volume of a solid is the number of cubes, each of volume 1 cubic unit, that are necessary to exactly fill the solid. The volume of the rectangular solid at the right is 24 cm³ because it will hold exactly 24 cubes, each 1 cm on a side. Note that the volume can be found by multiplying the length times the width times the height.

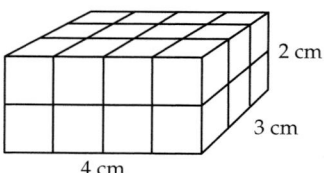

$$4 \cdot 3 \cdot 2 = 24$$

The volume of the solid is 24 cm³.

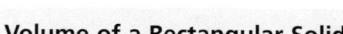

Volume of a Rectangular Solid

The volume, V, of a rectangular solid with length L, width W, and height H is given by $V = LWH$.

Volume of a Cube

The volume, V, of a cube with side of length s is given by $V = s^3$.

QUESTION   *Which of the following are rectangular solids: a juice box, a milk carton, a can of soup, a compact disk, the plastic container a compact disk is packaged in?*

A **sphere** is a solid in which all points are the same distance from a point O, called the **center** of the sphere. A **diameter** of a sphere is a line segment with endpoints on the sphere and passing through the center. A **radius** is a line segment from the center to a point on the sphere. $\overline{AB}$ is a diameter and $\overline{OC}$ is a radius of the sphere shown at the right. A basketball is an example of a sphere.

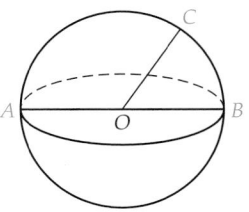

If we let d represent the length of a diameter and r represent the length of a radius, then $d = 2r$ or $r = \frac{1}{2}d$.

$$d = 2r \quad \text{or} \quad r = \frac{1}{2}d$$

Volume of a Sphere

The volume, V, of a sphere with radius of length r is given by $V = \frac{4}{3}\pi r^3$.

Find the volume of a rubber ball that has a diameter of 6 in.

First find the length of a radius of the sphere.

$$r = \frac{1}{2}d = \frac{1}{2}(6) = 3$$

Use the formula for the volume of a sphere.

$$V = \frac{4}{3}\pi r^3$$

Replace r with 3.

$$V = \frac{4}{3}\pi(3)^3$$

$$V = \frac{4}{3}\pi(27)$$

The exact volume of the sphere is 36π in^3.

$$V = 36\pi$$

An approximate measure can be found by using the π key on a calculator.

$$V \approx 113.10$$

The volume of the rubber ball is approximately 113.10 in^3.

ANSWER   *A juice box and the plastic container a compact disk is packaged in are rectangular solids.*

The most common cylinder, called a **right circular cylinder,** is one in which the bases are circles and are perpendicular to the height of the cylinder. The variable r is used to represent the length of the radius of a base of a cylinder, and h represents the height of the cylinder. In this text, only right circular cylinders are discussed. The hole inside a roll of paper towels is an example of a right circular cylinder.

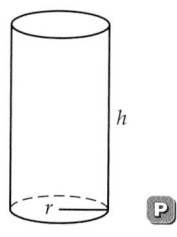

Volume of a Right Circular Cylinder

The volume, V, of a right circular cylinder is given by $V = \pi r^2 h$, where r is the radius of the base and h is the height of the cylinder.

A **right circular cone** is obtained when one base of a right circular cylinder is shrunk to a point, called the **vertex,** V. The variable r is used to represent the radius of the base of the cone, and h represents the height of the cone. The variable l is used to represent the **slant height,** which is the distance from a point on the circumference of the base to the vertex. In this text, only right circular cones are discussed. An ice cream cone is an example of a right circular cone.

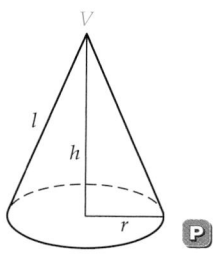

Volume of a Right Circular Cone

The volume, V, of a right circular cone is given by $V = \frac{1}{3}\pi r^2 h$, where r is the length of a radius of the circular base and h is the height of the cone.

The base of a **regular pyramid** is a regular polygon and the sides are isosceles triangles (two sides of the triangle are the same length). The height, h, is the distance from the vertex, V, to the base and is perpendicular to the base. The variable l is used to represent the **slant height,** which is the height of one of the isosceles triangles on the face of the pyramid. The regular square pyramid at the right has a square base. This is the only type of pyramid discussed in this text. Many Egyptian pyramids are regular square pyramids.

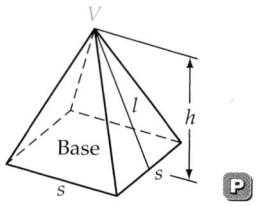

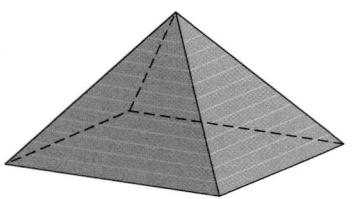

Volume of a Regular Square Pyramid

The volume, V, of a regular square pyramid is given by $V = \frac{1}{3}s^2 h$, where s is the length of a side of the base and h is the height of the pyramid.

QUESTION *Which of the following units could not be used to measure the volume of a regular square pyramid?*

a. ft^3 *b.* m^3 *c.* yd^2 *d.* cm^3 *e.* mi

EXAMPLE 1 ■ **Find the Volume of a Geometric Solid**

Find the volume of a cube that measures 1.5 m on a side.

Solution

$V = s^3$ • Use the formula for the volume of a cube.

$V = 1.5^3$ • Replace *s* with **1.5**.

$V = 3.375$

The volume of the cube is 3.375 m³.

CHECK YOUR PROGRESS 1 The length of a rectangular solid is 5 m, the width is 3.2 m, and the height is 4 m. Find the volume of the solid.

Solution See page S30.

EXAMPLE 2 ■ **Find the Volume of a Geometric Solid**

The radius of the base of a cone is 8 cm. The height of the cone is 12 cm. Find the volume of the cone. Round to the nearest hundredth.

Solution

$V = \dfrac{1}{3}\pi r^2 h$ • Use the formula for the volume of a cone.

$V = \dfrac{1}{3}\pi(8)^2(12)$ • Replace *r* with **8** and *h* with **12**.

$V = \dfrac{1}{3}\pi(64)(12)$

$V = 256\pi$

$V \approx 804.25$ • Use the π key on a calculator.

The volume of the cone is approximately 804.25 cm³.

CHECK YOUR PROGRESS 2 The length of a side of the base of a regular square pyramid is 15 m and the height of the pyramid is 25 m. Find the volume of the pyramid.

Solution See page S30.

ANSWER *Volume is measured in cubic units. Therefore, the volume of a regular square pyramid could be measured in ft^3, m^3, or cm^3, but not in yd^2 or mi.*

EXAMPLE 3 ■ Find the Volume of a Geometric Solid

An oil storage tank in the shape of a cylinder is 4 m high and has a diameter of 6 m. The oil tank is two-thirds full. Find the number of cubic meters of oil in the tank. Round to the nearest hundredth.

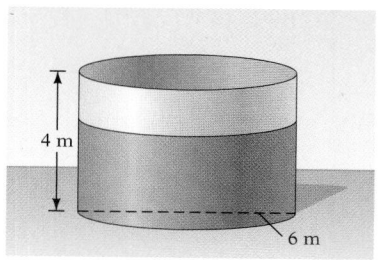

Solution

$$r = \frac{1}{2}d = \frac{1}{2}(6) = 3$$ • Find the radius of the base.

$$V = \pi r^2 h$$ • Use the formula for the volume of a cylinder.

$$V = \pi(3)^2(4)$$ • Replace *r* with **3** and *h* with **4**.

$$V = \pi(9)(4)$$

$$V = 36\pi$$

$$\frac{2}{3}(36\pi) = 24\pi$$ • Multiply the volume by $\frac{2}{3}$.

$$\approx 75.40$$ • Use the π key on a calculator.

There are approximately 75.40 m³ of oil in the storage tank.

CHECK YOUR PROGRESS 3   A silo in the shape of a cylinder is 16 ft in diameter and has a height of 30 ft. The silo is three-fourths full. Find the volume of the portion of the silo that is not being used for storage. Round to the nearest hundredth.

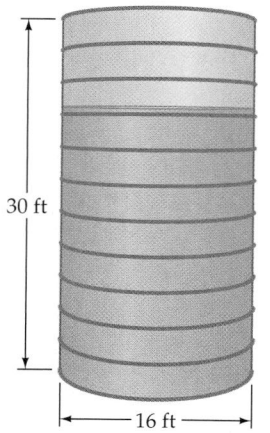

Solution   See page S30.

Surface Area

The **surface area** of a solid is the total area on the surface of the solid. Suppose you want to cover a geometric solid with wallpaper. The amount of wallpaper needed is equal to the surface area of the figure.

When a rectangular solid is cut open and flattened out, each face is a rectangle. The surface area, S, of the rectangular solid is the sum of the areas of the six rectangles:

$$S = LW + LH + WH + LW$$
$$+ WH + LH$$

which simplifies to

$$S = 2LW + 2LH + 2WH$$

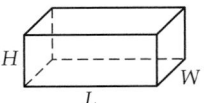

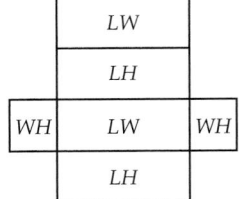

The surface area of a cube is the sum of the areas of the six faces of the cube. The area of each face is s^2. Therefore, the surface area S, of a cube is given by the formula $S = 6s^2$.

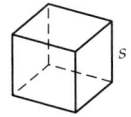

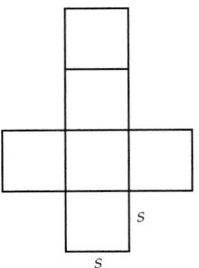

historical note

Pappus (păp′ŭs) of Alexandria (c. 290–350) was born in Egypt. We know the time period in which he lived because he wrote about his observation of the eclipse of the sun that took place in Alexandria on October 18, 320.

Pappas has been called the last of the great Greek geometers. His major work in geometry is *Synagoge*, or the *Mathematical Collection*. It consists of eight books. In Book V, Pappas proves that the sphere has a greater volume than any regular geometric solid with equal surface area. He also proves that if two regular solids have equal surface areas, the solid with the greater number of faces has the greater volume. ∎

When a cylinder is cut open and flattened out, the top and bottom of the cylinder are circles. The side of the cylinder flattens out to a rectangle. The length of the rectangle is the circumference of the base, which is $2\pi r$; the width is h, the height of the cylinder. Therefore, the area of the rectangle is $2\pi rh$. The surface area, S, of the cylinder is

$$S = \pi r^2 + 2\pi rh + \pi r^2$$

which simplifies to

$$S = 2\pi r^2 + 2\pi rh$$

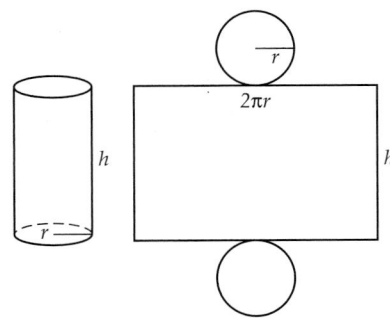

The surface area of a regular square pyramid is the area of the base plus the area of the four isosceles triangles. The length of a side of the square base is s; therefore, the area of the base is s^2. The slant height, l, is the height of each triangle, and s is the length of the base of each triangle. The surface area, S, of a regular square pyramid is

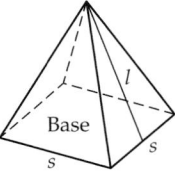

$$S = s^2 + 4\left(\frac{1}{2}sl\right)$$

which simplifies to

$$S = s^2 + 2sl$$

Formulas for the surface areas of geometric solids are given below.

Surface Areas of Geometric Solids

The surface area, S, of a **rectangular solid** with length L, width W, and height H is given by $S = 2LW + 2LH + 2WH$.

The surface area, S, of a **cube** with sides of length s is given by $S = 6s^2$.

The surface area, S, of a **sphere** with radius r is given by $S = 4\pi r^2$.

The surface area, S, of a **right circular cylinder** is given by $S = 2\pi r^2 + 2\pi rh$, where r is the radius of the base and h is the height.

The surface area, S, of a **right circular cone** is given by $S = \pi r^2 + \pi rl$, where r is the radius of the circular base and l is the slant height.

The surface area, S, of a **regular square pyramid** is given by $S = s^2 + 2sl$, where s is the length of a side of the base and l is the slant height.

QUESTION   *Which of the following units could not be used to measure the surface area of a rectangular solid?*

   a. *in^2*   ***b.*** *m^3*   ***c.*** *cm^2*   ***d.*** *ft^3*   ***e.*** *yd*

Find the surface area of a sphere with a diameter of 18 cm.

First find the radius of the sphere. $r = \dfrac{1}{2}d = \dfrac{1}{2}(18) = 9$

Use the formula for the surface area of a sphere.

$$S = 4\pi r^2$$
$$S = 4\pi(9)^2$$
$$S = 4\pi(81)$$
$$S = 324\pi$$

The exact surface area of the sphere is 324π cm².

An approximate measure can be found by using the π key on a calculator. $S \approx 1017.88$

The approximate surface area is 1017.88 cm².

ANSWER   *Surface area is measured in square units. Therefore, the surface area of a rectangular solid could be measured in in^2 or cm^2, but not in m^3, ft^3 or yd.*

EXAMPLE 4 ■ **Find the Surface Area of a Geometric Solid**

The diameter of the base of a cone is 5 m and the slant height is 4 m. Find the surface area of the cone. Round to the nearest hundredth.

Solution

$$r = \frac{1}{2}d = \frac{1}{2}(5) = 2.5$$ • Find the radius of the cone.

$$S = \pi r^2 + \pi r l$$ • Use the formula for the surface area of a cone.

$$S = \pi(2.5)^2 + \pi(2.5)(4)$$ • Replace *r* with **2.5** and *l* with **4**.

$$S = \pi(6.25) + \pi(2.5)(4)$$

$$S = 6.25\pi + 10\pi$$

$$S = 16.25\pi$$

$$S \approx 51.05$$

The surface area of the cone is approximately 51.05 m².

CHECK YOUR PROGRESS 4 The diameter of the base of a cylinder is 6 ft and the height is 8 ft. Find the surface area of the cylinder. Round to the nearest hundredth.

Solution *See page S30.*

EXAMPLE 5 ■ **Find the Surface Area of a Geometric Solid**

Find the area of a label used to cover a soup can that has a radius of 4 cm and a height of 12 cm. Round to the nearest hundredth.

Solution
The surface area of the side of a cylinder is given by $2\pi r h$.

$$\text{Area of the label} = 2\pi r h$$
$$= 2\pi(4)(12)$$
$$= 96\pi$$
$$\approx 301.59$$

The area of the label is approximately 301.59 cm².

CHECK YOUR PROGRESS 5 Which has a larger surface area, a cube with a side measuring 8 cm or a sphere with a diameter measuring 10 cm?

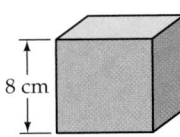
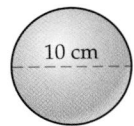

Solution *See page S30.*

Math Matters   Survival of the Fittest

The ratio of an animal's surface area to the volume of its body is a crucial factor in its survival. The more square units of skin for every cubic unit of volume, the more rapidly the animal loses body heat. Therefore, animals living in a warm climate benefit from a higher ratio of surface area to volume, whereas those living in a cool climate benefit from a lower ratio.

Excursion

Water Displacement

A recipe for peanut butter cookies calls for 1 cup of peanut butter. Peanut butter is difficult to measure. If you have ever used a measuring cup to measure peanut butter, you know that there tend to be pockets of air at the bottom of the cup. And trying to scrape all of the peanut butter out of the cup and into the mixing bowl is a challenge.

A more convenient method of measuring 1 cup of peanut butter is to fill a 2-cup measuring cup with 1 cup of water. Then add peanut butter to the water until the water reaches the 2-cup mark. (Make sure all the peanut butter is below the top of the water.) Drain off the water, and the one cup of peanut butter drops easily into the mixing bowl.

This method of measuring peanut butter works because when an object sinks below the surface of the water, the object displaces an amount of water that is equal to the volume of the object.

(continued)

A sphere with a diameter of 4 in. is placed in a rectangular tank of water that is 6 in. long and 5 in. wide. How much does the water level rise? Round to the nearest hundredth.

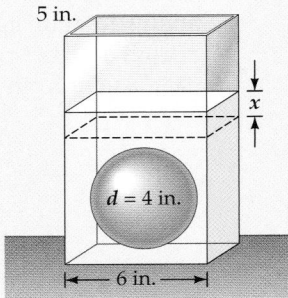

$$V = \frac{4}{3}\pi r^3$$ • Use the formula for the volume of a sphere.

$$V = \frac{4}{3}\pi(2^3) = \frac{32}{3}\pi$$ • $r = \frac{1}{2}d = \frac{1}{2}(4) = 2$

Let x represent the amount of the rise in water level. The volume of the sphere will equal the volume of the water displaced. As shown above, this volume is the rectangular solid with width 5 in., length 6 in., and height x in.

$$V = LWH$$ • Use the formula for the volume of a rectangular solid.

$$\frac{32}{3}\pi = (6)(5)x$$ • Substitute $\frac{32}{3}\pi$ for V, 6 for L, and 5 for W.

$$\frac{32}{90}\pi = x$$ • The exact height that the water will rise is $\frac{32}{90}\pi$.

$$1.12 \approx x$$ • Use a calculator to find an approximation.

The water will rise approximately 1.12 in.

Excursion Exercises

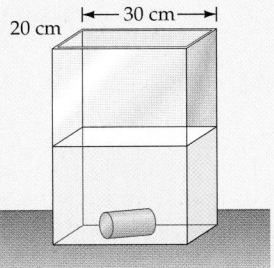

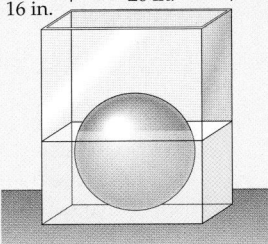

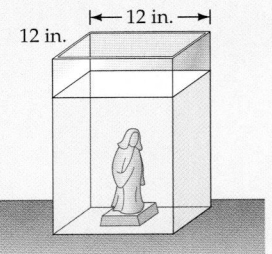

Figure 1　　　　　**Figure 2**　　　　　**Figure 3**

1. A cylinder with a 2-centimeter radius and a height of 10 cm is submerged in a tank of water that is 20 cm wide and 30 cm long (see Figure 1). How much does the water level rise? Round to the nearest hundredth.

(continued)

2. A sphere with a radius of 6 in. is placed in a rectangular tank of water that is 16 in. wide and 20 in. long (see Figure 2). The sphere displaces water until two-thirds of the sphere, with respect to its volume, is submerged. How much does the water level rise? Round to the nearest hundredth.

3. A chemist wants to know the density of a statue that weighs 15 lb. The statue is placed in a rectangular tank of water that is 12 in. long and 12 in. wide (see Figure 3). The water level rises 0.42 in. Find the density of the statue. Round to the nearest hundredth. *Hint:* Density = weight ÷ volume.

Exercise Set 8.4   (Suggested Assignment: 1–57, odds)

In Exercises 1–6, find the volume of the figure. For calculations involving π, give both the exact value and an approximation to the nearest hundredth.

1.
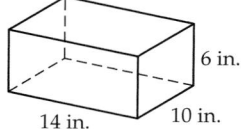
6 in.
14 in.
10 in.

2.
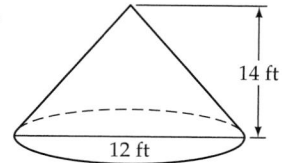
14 ft
12 ft

3.
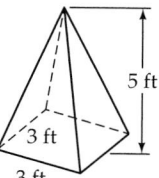
5 ft
3 ft
3 ft

4.
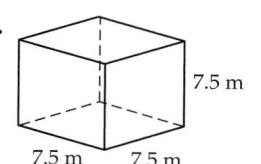
7.5 m
7.5 m
7.5 m

5.
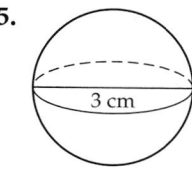
3 cm

6.
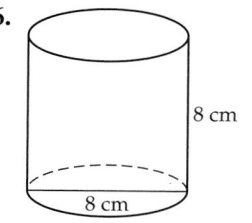
8 cm
8 cm

In Exercises 7–12, find the surface area of the figure. For calculations involving π, give both the exact value and an approximation to the nearest hundredth.

7.
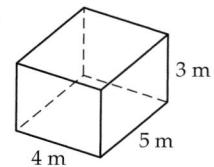
3 m
5 m
4 m

8.
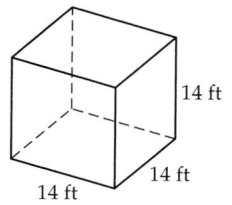
14 ft
14 ft
14 ft

9.
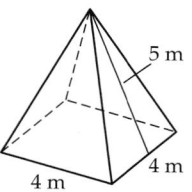
5 m
4 m
4 m

10.
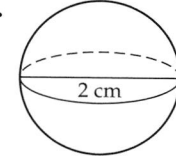
2 cm

11.
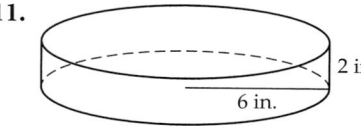
2 in.
6 in.

12.
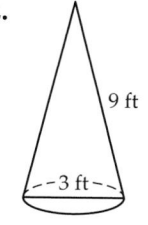
9 ft
3 ft

In Exercises 13–46, solve.

13. A rectangular solid has a length of 6.8 m, a width of 2.5 m, and a height of 2 m. Find the volume of a solid.

14. Find the volume of a rectangular solid that has a length of 4.5 ft, a width of 3 ft, and a height of 1.5 ft.

15. Find the volume of a cube whose side measures 2.5 in.

16. The length of a side of a cube is 7 cm. Find the volume of the cube.

17. The diameter of a sphere is 6 ft. Find the exact volume of the sphere.

18. Find the volume of a sphere that has a radius of 1.2 m.

19. The diameter of the base of a cylinder is 24 cm. The height of the cylinder is 18 cm. Find the volume of the cylinder. Round to the nearest hundredth.

20. The height of a cylinder is 7.2 m. The radius of the base is 4 m. Find the exact volume of the cylinder.

21. The radius of the base of a cone is 5 in. The height of the cone is 9 in. Find the exact volume of the cone.

22. The height of a cone is 15 cm. The diameter of the cone is 10 cm. Find the volume of the cone. Round to the nearest hundredth.

23. The length of a side of the base of a regular square pyramid is 6 in. and the height of the pyramid is 10 in. Find the volume of the pyramid.

24. The height of a regular square pyramid is 8 m and the length of a side of the base is 9 m. What is the volume of the pyramid?

25. The index finger of the Statue of Liberty is 8 ft long. The circumference at the second joint is 3.5 ft. Use the formula for the volume of a cylinder to approximate the volume of the index finger on the Statue of Liberty. Round to the nearest hundredth.

26. The height of a rectangular solid is 5 ft, the length is 8 ft, and the width is 4 ft. Find the surface area of the solid.

27. The width of a rectangular solid is 32 cm, the length is 60 cm, and the height is 14 cm. What is the surface area of the solid?

28. The side of a cube measures 3.4 m. Find the surface area of the cube.

29. Find the surface area of a cube with a side measuring 1.5 in.

30. Find the exact surface area of a sphere with a diameter of 15 cm.

31. The radius of a sphere is 2 in. Find the surface area of the sphere. Round to the nearest hundredth.

32. The radius of the base of a cylinder is 4 in. The height of the cylinder is 12 in. Find the surface area of the cylinder. Round to the nearest hundredth.

33. The diameter of the base of a cylinder is 1.8 m. The height of the cylinder is 0.7 m. Find the exact surface area of the cylinder.

34. The slant height of a cone is 2.5 ft. The radius of the base is 1.5 ft. Find the exact surface area of the cone. The formula for the surface area of a cone is given on page 507.

35. The diameter of the base of a cone is 21 in. The slant height is 16 in. What is the surface area of the cone? The formula for the surface area of a cone is given on page 507. Round to the nearest hundredth.

36. The length of a side of the base of a pyramid is 9 in., and the pyramid's slant height is 12 in. Find the surface area of the pyramid.

37. The slant height of a regular square pyramid is 18 m, and the length of a side of the base is 16 m. What is the surface area of the pyramid?

38. The volume of a freezer that is a rectangular solid with a length of 7 ft and a height of 3 ft is 52.5 ft^3. Find the width of the freezer.

39. The length of an aquarium is 18 in. and the width is 12 in. If the volume of the aquarium is 1836 in^3, what is the height of the aquarium?

40. The surface area of a rectangular solid is 108 cm^2. The height of the solid is 4 cm, and the length is 6 cm. Find the width of the rectangular solid.

41. The length of a rectangular solid is 12 ft and the width is 3 ft. If the surface area is 162 ft^2, find the height of the rectangular solid.

42. A can of paint will cover 300 ft^2 of surface. How many cans of paint should be purchased to paint a cylinder that has a height of 30 ft and a radius of 12 ft?

43. A hot air balloon is in the shape of a sphere. Approximately how much fabric was used to construct the balloon if its diameter is 32 ft? Round to the nearest whole number.

44. How much glass is needed to make a fish tank that is 12 in. long, 8 in. wide, and 9 in. high? The fish tank is open at the top.

45. Find the area of a label used to cover a cylindrical can of juice that has a diameter of 16.5 cm and a height of 17 cm. Round to the nearest hundredth.

46. The length of a side of the base of a regular square pyramid is 5 cm and the slant height of the pyramid is 8 cm. How much larger is the surface area of this pyramid than the surface area of a cone with a diameter of 5 cm and a slant height of 8 cm? Round to the nearest hundredth.

In Exercises 47–52, find the volume of the figure. Round to the nearest hundredth.

47.
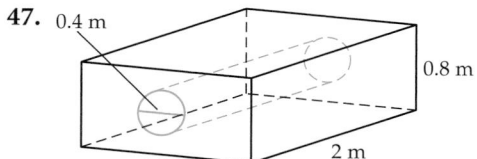

48.
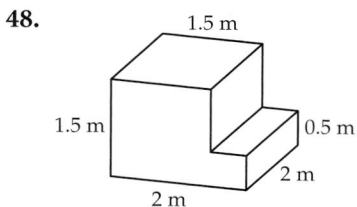

49.
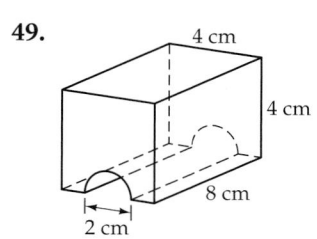

50.
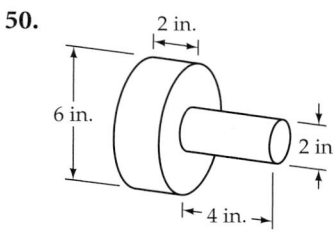

51.
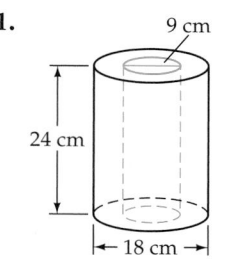

52.
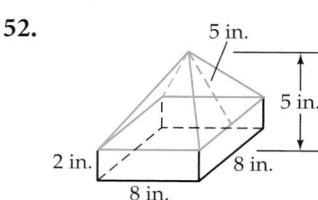

In Exercises 53–58, find the surface area of the figure. Round to the nearest hundredth.

53.
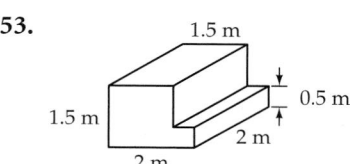

54.
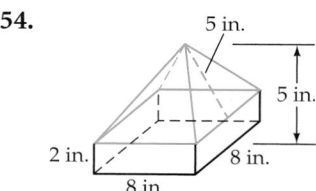

55.
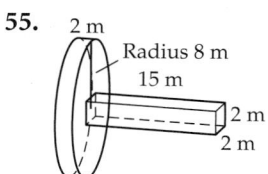

56.
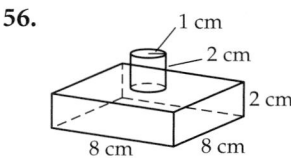

57.
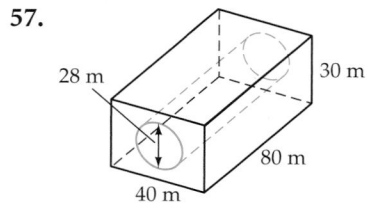

58.

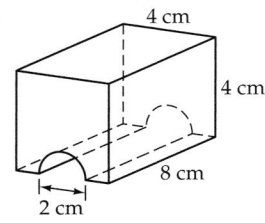

59. A truck is carrying an oil tank, as shown in the figure below. If the tank is half full, how many cubic feet of oil is the truck carrying? Round to the nearest hundredth.

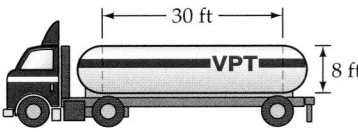

60. The concrete floor of a building is shown in the figure below. At a cost of $3.15 per cubic foot, find the cost of having the floor poured. Round to the nearest cent.

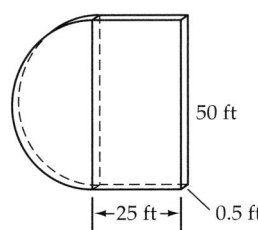

61. How many liters of water are needed to fill the swimming pool shown below? (1 m³ contains 1000 L.)

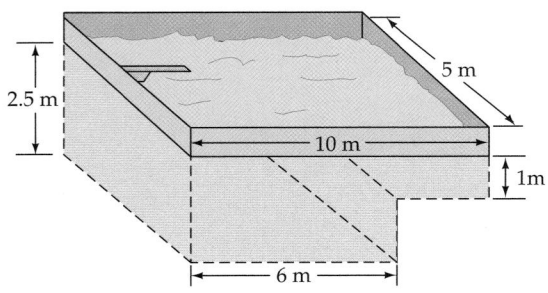

62. A can of paint will cover 250 ft² of surface. Find the number of cans of paint that should be purchased to

paint the exterior of the auditorium shown in the figure below.

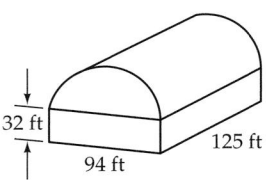

63. A piece of sheet metal is cut and formed into the shape shown below. Given that there are 0.24 g in 1 cm² of the metal, find the total number of grams of metal used. Round to the nearest hundredth.

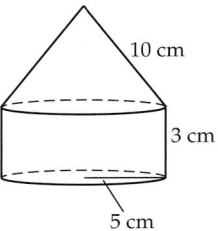

64. The walls of a room that is 25.5 ft long, 22 ft wide, and 8 ft high are being plastered. There are two doors in the room, each measuring 2.5 ft by 7 ft. Each of the six windows in the room measures 2.5 ft by 4 ft. At a cost of $.75 per square foot, find the cost of plastering the walls of the room.

65. A solid sphere of gold with a radius of 0.5 cm has a value of $180. Find the value of a sphere of the same type of gold with a radius of 1.5 cm.

66. A swimming pool is built in the shape of a rectangular solid. It holds 32,000 gal of water. If the length, width, and height of the pool are each doubled, how many gallons of water will be needed to fill the pool?

Extensions

CRITICAL THINKING

67. Half of a sphere is called a **hemisphere.** Derive formulas for the volume and surface area of a hemisphere.

68. a. Draw a two-dimensional figure that can be cut out and made into a right circular cone.

b. Draw a two-dimensional figure that can be cut out and made into a regular square pyramid.

69. A sphere fits inside a cylinder as shown below. The height of the cylinder equals the diameter of the sphere. Show that the surface area of the sphere equals the surface area of the side of the cylinder.

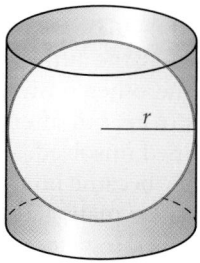

70. Determine whether the statement is always true, sometimes true, or never true.

a. The slant height of a regular square pyramid is longer than the height.

b. The slant height of a cone is shorter than the height.

c. The four triangular faces of a regular square pyramid are equilateral triangles.

71. a. What is the effect on the surface area of a rectangular solid of doubling the width and height?

b. What is the effect on the volume of a rectangular solid of doubling the length and width?

c. What is the effect on the volume of a cube of doubling the length of each side of the cube?

d. What is the effect on the surface area of a cylinder of doubling the radius and height?

EXPLORATIONS

72. Explain how you could cut through a cube so that the face of the resulting solid is

a. a square.

b. an equilateral triangle.

c. a trapezoid.

d. a hexagon.

SECTION 8.5 | **Non-Euclidean Geometry**

Euclidean Geometry versus Non-Euclidean Geometry

Learning to play a game by reading the rules can be frustrating. If the game has too many rules, you may decide that learning to play is not worth the effort. Most popular games are based on a few rules that are easy to learn but that still allow the game to develop into complex situations. The ancient Greek mathematician Euclid was immersed in the mathematical "game" of geometry. It was his desire to construct a substantial body of geometrical knowledge based on the *fewest* possible number of rules, which he called **postulates.** Euclid based his study of geometry on the following five postulates.

✔ **TAKE NOTE**

In addition to postulates, Euclid's geometry, referred to as *Euclidean geometry,* involves definitions and some *undefined terms.* For instance, the term *point* is an undefined term in Euclidean geometry. To define a point would require a description involving additional undefined terms. Because every term cannot be defined by simpler defined terms, Euclid chose to use *point* as an undefined term. In Euclidean geometry, the term *line* is also an undefined term.

Euclid's Postulates

P1: A line segment can be drawn from any point to any other point.

P2: A line segment can be extended continuously in a straight line.

P3: A circle can be drawn with any center and any radius.

P4: All right angles are equal to one another.

P5: *The Parallel Postulate*   Through a given point not on a given line, exactly one line can be drawn parallel to the given line.

Ⓟ

For many centuries, the truth of these postulates was felt to be self-evident. However, a few mathematicians suspected that the fifth postulate, known as the Parallel Postulate, could be deduced from the other postulates. Many individuals took on the task of proving the Parallel Postulate, but after more than 2000 years, the proof of the Parallel Postulate still remained unsolved.

Carl Friedrich Gauss (gaus′) (1777–1855) became interested in the problem of proving the Parallel Postulate as a teenager. After many failed attempts to establish the Parallel Postulate as a theorem, Gauss decided to take a different route. He came to the conclusion that the Parallel Postulate was an independent postulate and that it could be changed to produce a new type of geometry. This is analogous to changing one of the rules of a game to create a new game.

The diaries and letters written by Gauss show that he proposed the following alternative postulate.

Gauss's Alternative to the Parallel Postulate

Through a given point not on a given line, there are *at least two* lines parallel to the given line.

Thus Gauss was the first person to realize that non-Euclidean geometries could be created by merely changing or excluding the Parallel Postulate. It has been speculated that Gauss did not publish this result because he felt it would not be readily accepted by other mathematicians.

The idea of using an alternative postulate in place of the Parallel Postulate was first proposed *openly* by the Russian mathematician Nikolai Lobachevsky. In a lecture given in 1826, and in a series of monthly articles that appeared in the academic journal of the University of Kazan in 1829, Lobachevsky provided a detailed investigation into the problem of the Parallel Postulate. He proposed that a consistent new geometry could be developed by replacing the Parallel Postulate with the alternative postulate, which assumes that *more than one* parallel line can be drawn through a point not on a given line.

Lobachevsky described his geometry as *imaginary geometry,* because he could not comprehend a model of such a geometry. Independent of each other, Gauss and Lobachevsky had each conceived of the same idea; but whereas Gauss chose to keep the idea to himself, Lobachevsky professed the idea to the world. It is for this reason that we now refer to the geometry they each conceived as *Lobachevskian geometry.* Today Lobachevskian geometry is often called *hyperbolic geometry.*

The year 1826, in which Lobachevsky first lectured about a new non-Euclidean geometry, also marks the birth of the mathematician Bernhard Riemann. Although Riemann died of tuberculosis at age 39, he made major contributions in several areas of mathematics and physics, including the theory of functions of complex

historical note

Nikolai Lobachevsky (lō′bə-chĕf′skē) (1793–1856), was a noted Russian mathematician. Lobachevski's concept of a non-Euclidean geometry was so revolutionary that he is called the Copernicus of Geometry. ∎

historical note

Bernhard Riemann
(rē′mən) (1826–1866). "Riemann's achievement has taught mathematicians to disbelieve in *any* geometry, or any space, as a necessary mode of human perception."[2] ∎

variables, electrodynamics, and non-Euclidean geometry. Riemann was the first person to consider a geometry in which the Parallel Postulate was replaced with the following postulate.

Riemann's Alternative to the Parallel Postulate

Through a given point not on a given line, there exist *no* lines parallel to the given line.

Unlike the geometry developed by Lobachevsky, which was not based on a physical model, the non-Euclidean geometry of Riemann was closely associated with a sphere and the remarkable idea that because a line is an undefined term, a line on the surface of a sphere can be different from a line on a plane. Even though a line on a sphere can be different from a line on a plane, it seems reasonable that "spherical lines" should retain some of the properties of lines on a plane. For example, on a plane, the shortest distance between two points is measured along the line that connects the points. The line that connects the points is an example of what is called a *geodesic*.

Definition of a Geodesic

A **geodesic** is a curve C on a surface S such that for any two points on C, the portion of C between these points is the shortest path on S that joins these points.

On a sphere, the geodesic between two points is a *great circle* that connects the points.

Definition of a Great Circle

A **great circle** of a sphere is a circle on the surface of the sphere whose center is at the center of the sphere.

In *Riemannian geometry,* which is also called *spherical geometry* or *elliptical geometry,* great circles, which are the geodesics of a sphere, are thought of as lines. Figure 8.1 shows a sphere and two of its great circles. Because all great circles of a sphere intersect, a sphere provides us with a model of a geometry in which there are no parallel lines.

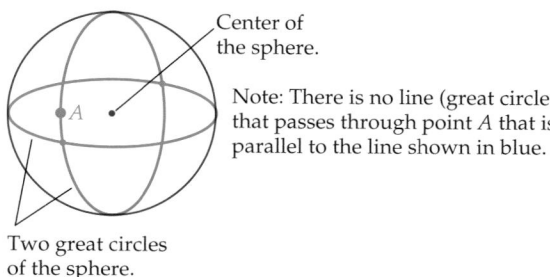

Center of the sphere.

Note: There is no line (great circle) that passes through point A that is parallel to the line shown in blue.

Two great circles of the sphere.

Figure 8.1 *A sphere and its great circles serve as a physical model for Riemannian geometry.*

2. Bell, E. T. *Men of Mathematics.* New York: Touchstone Books, Simon and Schuster, 1986.

In Riemannian geometry, a triangle may have as many as three right angles. Figure 8.2 illustrates a spherical triangle with one right angle, a spherical triangle with two right angles, and a spherical triangle with three right angles.

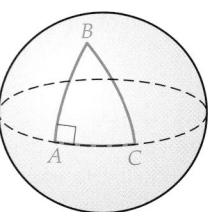 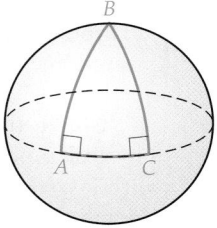 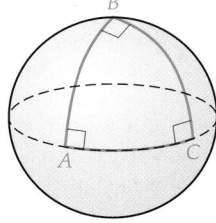

a. A spherical triangle with one right angle **b.** A spherical triangle with two right angles **c.** A spherical triangle with three right angles

Figure 8.2

In Euclidean geometry it is not possible to determine the area of a triangle from the measures of its angles alone.

In spherical geometry, however, any two triangles on the same sphere that are similar are also congruent. Therefore, in spherical geometry, the area of a triangle depends on the size of its angles and the radius of the sphere.

QUESTION *In Euclidean geometry, are two similar triangles necessarily congruent triangles?*

The Spherical Triangle Area Formula

The area S of the spherical triangle ABC on a sphere with radius r is given by

$$S = (m\angle A + m\angle B + m\angle C - 180°) \cdot \left(\frac{\pi}{180°}\right)r^2$$

where each angle is measured in degrees.

EXAMPLE 1 ■ **Find the Area of a Spherical Triangle**

Find the area of a spherical triangle with three right angles on a sphere with a radius of 1 ft. Find both the exact area and the approximate area rounded to the nearest hundredth.

ANSWER *No.*

TAKE NOTE

To check the result in Example 1, use the fact that the given triangle covers $\frac{1}{8}$ of the surface of the sphere. See Figure 8.2c. The total surface area of a sphere is $4\pi r^2$. In Example 1, $r = 1$ ft, so the sphere has a surface area of 4π ft^2. Thus the area of the spherical triangle in Example 1 should be

$$\left(\frac{1}{8}\right) \cdot (4\pi) = \frac{\pi}{2} \text{ ft}^2$$

Solution
Apply the spherical triangle area formula.

$$S = (m\angle A + m\angle B + m\angle C - 180°) \cdot \left(\frac{\pi}{180°}\right) r^2$$

$$= (90° + 90° + 90° - 180°) \cdot \left(\frac{\pi}{180°}\right)(1)^2$$

$$= (90°) \cdot \left(\frac{\pi}{180°}\right)$$

$$= \frac{\pi}{2} \text{ ft}^2 \quad \text{• Exact area}$$

$$\approx 1.57 \text{ ft}^2 \quad \text{• Approximate area}$$

CHECK YOUR PROGRESS 1 Find the area of the spherical triangle whose angles measure 200°, 90°, and 90° on a sphere with a radius of 6 in. Find both the exact area and the approximate area rounded to the nearest hundredth.

Solution *See page S30.*

Mathematicians have not been able to create a three-dimensional model that *perfectly* illustrates all aspects of hyperbolic geometry. However, an infinite saddle surface can be used to visualize some of the basic aspects of hyperbolic geometry. Figure 8.3 shows a portion of an infinite saddle surface.

INSTRUCTOR NOTE
Tell your students to visualize a large Pringle potato chip. This will help them comprehend the saddle surface shown to the right.

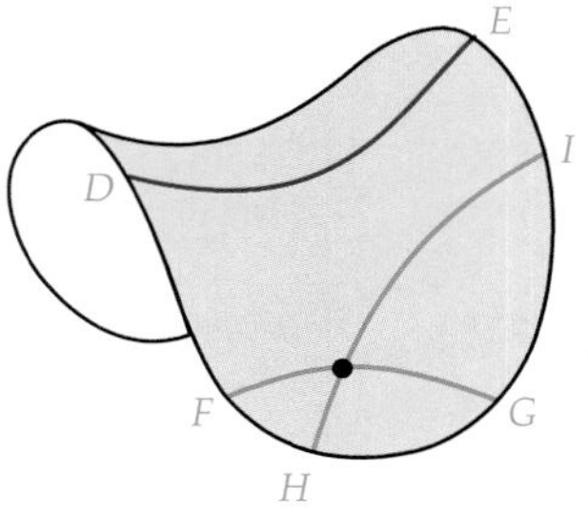

Figure 8.3 *A portion of an infinite saddle surface.*

A line (geodesic) can be drawn through any two points on the saddle surface. Most lines on the saddle surface have a concave curvature as shown by $\overleftrightarrow{DE}$, $\overleftrightarrow{FG}$, and $\overleftrightarrow{HI}$. Keep in mind that the saddle surface is an infinite surface. Figure 8.3 only shows a portion of the surface.

Parallel lines on an infinite saddle surface are defined as two lines that do not intersect. In Figure 8.3, $\overleftrightarrow{FG}$ and $\overleftrightarrow{HI}$ are *not* parallel because they intersect at a point. The lines $\overleftrightarrow{DE}$ and $\overleftrightarrow{FG}$ are parallel because they do not intersect. The lines $\overleftrightarrow{DE}$ and $\overleftrightarrow{HI}$ are also parallel lines. Figure 8.3 provides a geometric model of a hyperbolic geometry because for a given line, *more than one* parallel line exists through a point not on the given line.

TAKE NOTE

Several different physical models have been invented to illustrate hyperbolic geometry, but most of these models require advanced mathematics beyond the scope of this text.

Figure 8.4 shows a triangle drawn on a saddle surface. The triangle is referred to as a *hyperbolic triangle*. Due to the curvature of the sides of the hyperbolic triangle, the sum of the measures of the angles of the triangle is less than 180 degrees. This is true for all hyperbolic triangles.

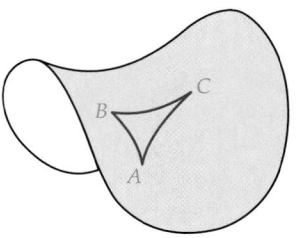

Figure 8.4 *A hyperbolic triangle*

The following chart compares and contrasts some of the properties of plane, hyperbolic, and spherical geometries.

Euclidean Geometry	Non-Euclidean Geometries	
Euclidean or Plane Geometry (circa 300 B.C.):	**Lobachevskian or Hyperbolic Geometry (1826):**	**Riemannian or Spherical Geometry (1855):**
Through a given point not on a given line, exactly one line can be drawn parallel to the given line.	*Through a given point not on a given line, there are at least two lines parallel to the given line.*	*Through a given point not on a given line, there exist no lines parallel to the given line.*
Geometry on a plane	Geometry on an infinite saddle surface	Geometry on a sphere
For any triangle *ABC*, $m\angle A + m\angle B + m\angle C = 180°$	For any triangle *ABC*, $m\angle A + m\angle B + m\angle C < 180°$	For any triangle *ABC*, $180° < m\angle A + m\angle B + m\angle C < 540°$
A triangle can have at most one right angle.	A triangle can have at most one right angle.	A triangle can have one, two, or three right angles.
The shortest path between two points is the line segment that connects the points.	The curves shown in the above figure illustrate some of the geodesics of an infinite saddle surface.	The shortest path between two points is the minor arc of a great circle that passes through the points.

Math Matters   Curved Space

In 1915, Albert Einstein proposed a revolutionary theory. This theory is now called the *general theory of relativity*. One of the major ideas of this theory is that space is curved, or "warped," by the mass of stars and planets. The greatest curvature occurs around those stars with the largest mass. Light rays in space do not travel in a straight path, but rather follow the geodesics of this curved space. Recall that the shortest path that joins two points on a given surface is on a geodesic of the surface.

It is interesting to consider the paths of light rays in space from a different perspective, in which the light rays travel along straight paths, in a space that is non-Euclidean. To better understand this concept, consider an airplane that flies from Los Angeles to London. We say that the airplane flies on a straight path between the two cities. The path is not really a straight path if we use Euclidean geometry as our frame of reference, but if we think in terms of Reimannian geometry, then the path *is* straight. We can use Euclidean geometry or a non-Euclidean geometry as our frame of reference. It does not matter which we choose, but it does change our concept about what is a straight path and what is a curved path.

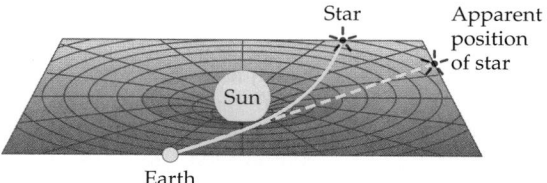

Light rays from a star follow the geodesics of space. Any light rays that pass near the sun are slightly bent. This causes some stars to appear to an observer on Earth to be in different positions than their actual positions.

Excursion

Finding Geodesics

Form groups of three or four students. Each group needs a roll of narrow tape or a ribbon and the two geometrical models shown at the right. The purpose of this Excursion is to use the tape (ribbon) to determine the geodesics of a surface.

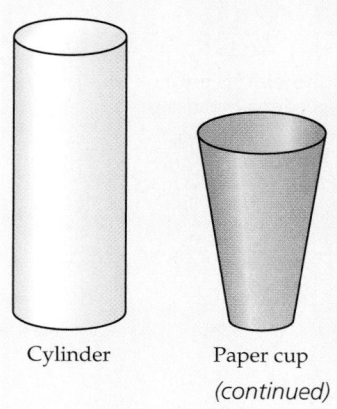

Cylinder   Paper cup

(continued)

The following three theorems can be used to determine the geodesics of a surface.

INSTRUCTOR NOTE
The experiments and the Excursion Exercises work best with ribbon or tape that is one-quarter to one-half inch in width.

Geodesic Theorems

Theorem 1   If a surface is smooth with no edges or holes, then the shortest path between any two points is on a geodesic of the surface.

Theorem 2   *The Tape Test*   If a piece of tape is placed so that it lies flat on a smooth surface, then the center line of the tape is on a geodesic of the surface.

Theorem 3   *Inverse of the Tape Test*   If a piece of tape does not lie flat on a smooth surface, then the center line of the tape is not on a geodesic of the surface.

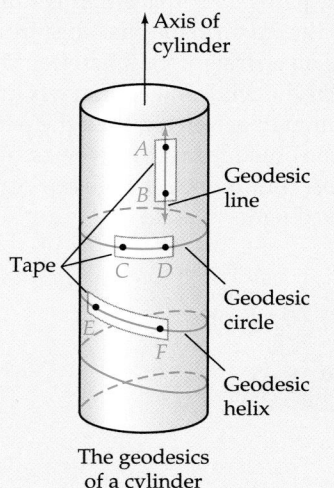

The geodesics of a cylinder

Figure 8.5   *The geodesics of a cylinder*

What are the geodesics of a cylinder? If two points A and B are as shown in Figure 8.5, then the vertical line segment between A and B is the shortest path between the points. A piece of tape can be placed so that it covers point A and point B and lies flat on the cylinder. Thus, by Theorem 2, line segment $\overline{AB}$ is on a geodesic of the cylinder.

If two points C and D are as shown in Figure 8.5, then the minor arc of a circle is the shortest path between the points. Once again we see that a piece of tape can be placed so that it covers points C and D and lies flat on the cylinder. Theorem 2 indicates that the arc $\overset{\frown}{CD}$ is on a geodesic of the cylinder.

To find a geodesic that passes through the two points E and F, start your tape at E and proceed slightly downward and to the right, toward point F. If your tape lies flat against the cylinder, you have found the geodesic for the two points. If your tape does not lie flat against the surface, then your path is not a geodesic and you need to experiment further. Eventually you will find the helix curve, shown in Figure 8.5 that allows the tape to lie flat on the cylinder.

Additional experiments with the tape and the cylinder should convince you that a geodesic of a cylinder is either (a) a line parallel to the axis of the cylinder, (b) a circle with center on the axis of the cylinder and diameter perpendicular to the axis, or (c) a circular helix curve that has a constant slope and a center on the axis of the cylinder.[3]

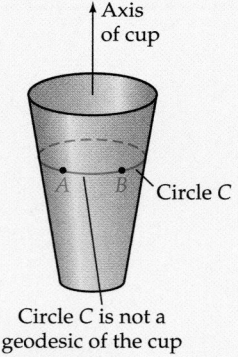

Circle C is not a geodesic of the cup

Figure 8.6

Excursion Exercises

1. a. Place two points A and B on a paper cup so that A and B are both at the same height, as in Figure 8.6. We know circle C that passes through A and B is *not* a geodesic of the cup because a piece of tape will not lie flat when placed directly on top of circle C. Experiment with a piece of tape to determine the *actual* geodesic that passes through A and B. Make a drawing that shows this geodesic and illustrate how it differs from circle C.

b. Use a cup similar to the one in Figure 8.6 and a piece of tape to determine two other types of geodesics of the cup. Make a drawing that shows each of these two additional types of geodesics.

(continued)

---

3. The tread of a bolt is an example of a circular helix.

2. The only geodesics of a sphere are great circles. Write a sentence that explains how you can use a piece of tape to show that circle *D* in Figure 8.7 is not a geodesic of the sphere.

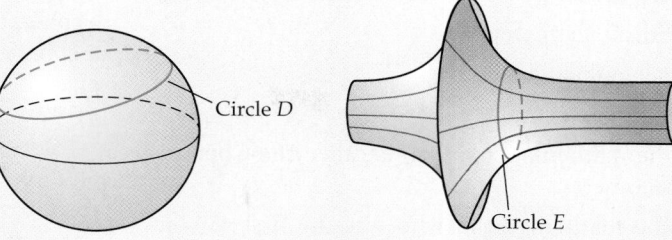

Figure 8.7 **Figure 8.8**

3. Write a sentence that explains how you know that circle *E* in Figure 8.8 is not a geodesic of the figure.

Excursion Exercises 4 to 6 require a world globe that shows the locations of major cities. We suggest that you use a thin ribbon, instead of tape, to determine the great circle routes in the following exercises, because tape may damage the globe.

4. A pilot flies a great circle route from Miami, Florida to Hong Kong. Which one of the following states will the plane fly over?

   a. California   **b.** Oregon   **c.** Washington   **d.** Alaska

5. A pilot flies a great circle route from Los Angeles to London. Which one of the following cities will the plane fly over?

   a. New York   **b.** Chicago   **c.** Godthaab, Greenland

   d. Vancouver, Canada

6. Washington, D.C. and Seoul, Korea both have a latitude of about 38°. How many miles (to the nearest 100 miles) will a pilot save by flying a great circle route between the cities as opposed to the route that follows the 38th parallel?

7. a. The surface at the left below is called a *hyperboloid of one sheet.* Explain how you can determine that the blue circle is a geodesic of the hyperboloid, but the red circles are not geodesics of the hyperboloid.

   b. The saddle surface at the right below is called a *hyperbolic paraboloid.* Explain how you can determine that the blue parabola is a geodesic of the hyperbolic paraboloid, but the red parabolas are not geodesics of the hyperbolic paraboloid.

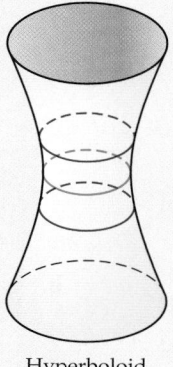

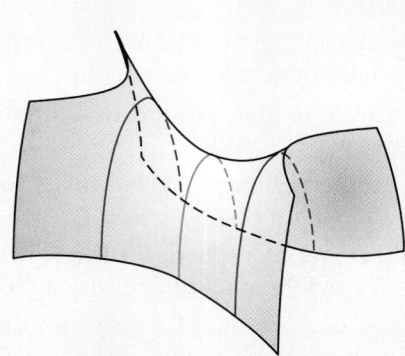

Hyperboloid
of one sheet

Hyperbolic paraboloid

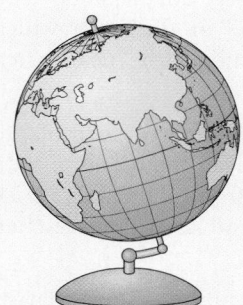

Exercise Set 8.5   (Suggested Assignment: 1–18)

1. State the parallel postulate for each of the following.

   a. Euclidean geometry

   b. Lobachevskian geometry

   c. Riemannian geometry

2. Name the mathematician who is called the Copernicus of Geometry.

3. Name the mathematician who was the first to consider a geometry in which Euclid's Parallel Postulate was replaced with "Through a given point not on a given line, there are *at least two* lines parallel to the given line."

4. Name the mathematician who was the first to create a non-Euclidean geometry using the postulate "Through a given point not on a given line, there exist no lines parallel to the given line."

5. What can be stated about the sum of the measures of the angles of a triangle in

   a. Euclidean geometry?

   b. Lobachevskian geometry?

   c. Riemannian geometry?

6. What is the maximum number of right angles a triangle can have in

   a. Euclidean geometry?

   b. Lobachevskian geometry?

   c. Riemannian geometry?

7. What name did Lobachevsky give to the geometry that he created?

8. Explain why great circles in Riemannian geometry are thought of as lines.

9. What is a geodesic?

10. In which geometry can two distinct lines be parallel to a third line but not parallel to each other?

11. What model was used in this text to illustrate hyperbolic geometry?

12. In which geometry do all perpendiculars to a given line intersect each other?

13. Find the exact area of a spherical triangle with angles of 150°, 120°, and 90° on a sphere with a radius of 1.

14. Find the area of a spherical triangle with three right angles on a sphere with a radius of 1980 mi. Round to the nearest ten thousand square miles.

A **spherical polygon** is a polygon on a sphere. Each side of a spherical polygon is an arc of a great circle. The area A of a spherical polygon of n sides on a sphere of radius r is given by

$$A = [\Sigma\theta - (n - 2) \cdot 180°] \cdot \left(\frac{\pi}{180°}\right) r^2$$

where $\Sigma\theta$ is the sum of the measures of the angles of the polygon in degrees.

15. Use the above spherical polygon area formula to determine the area of a spherical quadrilateral with angles of 90°, 90°, 100°, and 100° on the Earth, which has a radius of 1980 mi. Round to the nearest ten thousand square miles.

16. Use the above spherical polygon area formula to determine the area of a spherical quadrilateral with angles of 90°, 90°, 120°, and 120° on the Earth, which has a radius of 1980 mi. Round to the nearest ten thousand square miles.

17. The following flat map of Wyoming shows that its boundaries approximate a quadrilateral and the sum of the measures of its four angles is about 360°.

Wyoming

   a. What area does the spherical polygon area formula yield for the area of Wyoming with $\Sigma\theta = 360°$? Use 1980 mi as the radius of the Earth.

   b. Explain why the spherical polygon area formula and the data from part a do not yield a reasonable approximation of the actual area of Wyoming, which is 97,818 mi². *Hint:* It has nothing to do with mountains.

18. The following flat map shows the triangular route flown by a pilot. The angles of the triangle on this map measure about 78°, 63°, and 39°.

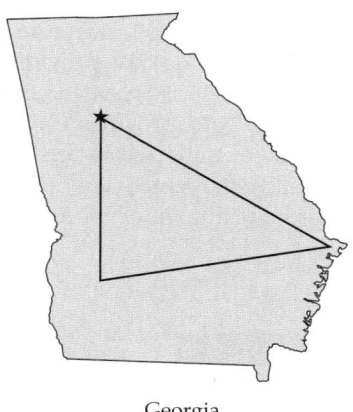

Georgia

Explain why the spherical triangle area formula cannot be used to find the actual area enclosed by the triangle shown on the map.

Extensions

CRITICAL THINKING

19. Consider a finite geometry with the five points A, B, C, D, and E. In this geometry a line is any two of the five points. For example, the two points A and B together form the line denoted by AB. Parallel lines are defined as two lines that do not share a common point.

a. How many lines are in this geometry?

b. How many of the lines are parallel to line AB?

20. Consider a finite geometry with the six points A, B, C, D, E, and F. In this geometry a line is any two of the six points. Parallel lines are defined as two lines that do not share a common point.

a. How many lines are in this geometry?

b. How many of the lines are parallel to line AB?

EXPLORATIONS

21. The mathematician Janos Bolyai (bōl′yoi) (1802–1860) proposed a geometry that excluded Euclid's Parallel Postulate. Concerning the geometry he created, he stated, "I have made such wonderful discoveries that I am myself lost in astonishment: Out of nothing I have created a strange new world." Use the Internet or a library to find information about Janos Bolyai and the mathematics he created. Write a report that summarizes the pertinent details concerning his work in the area of non-Euclidean geometry.

22. The mathematician Girolamo Saccheri (säk′kā-rē) (1667–1733) did important early work in non-Euclidean geometry. Use the Internet or a library to find information about Saccheri's work. Write a report that summarizes the pertinent details.

SECTION 8.6   Fractals

Fractals—Endlessly Repeated Geometric Figures

Have you ever used a computer program to enlarge a portion of a photograph? Sometimes the result is a satisfactory enlargement; however, if the photograph is enlarged too much, the image may become blurred. For example, the photograph in Figure 8.9 on the following page is shown at its original size. The image in Figure 8.10 is an enlarged portion of Figure 8.9. A computer monitor displays an image using small dots called *pixels*. If a computer image is enlarged using a program such as Adobe Photoshop™, the program must determine the color of each pixel in the enlargement. If the image file for the photograph cannot supply the needed color information for each pixel, then Photoshop determines the color of some pixels by *averaging* the numerical color values of neighboring pixels for which the image file has the color information. In the enlarged photograph in Figure 8.10, the image is somewhat blurred because some of the pixels do not provide accurate information about the original photograph. Figure 8.11 is an enlargement of a portion of Figure 8.10. The blurred image in Figure 8.11 provides even less detail than does Figure 8.10, because the colors of many of the pixels in this image were determined by averaging. If we continue to make enlargements of enlargements we will produce extremely blurred images that provide little information about the original photograph.

Figure 8.9

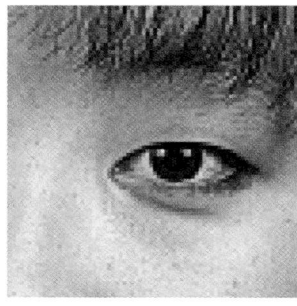

Figure 8.10

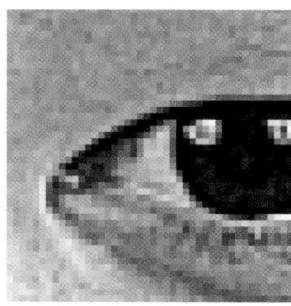

Figure 8.11

In the 1970s, the mathematician Benoit Mandelbrot discovered some remarkable methods that enable us to create geometric figures with a special property: if any portion of the figure is enlarged repeatedly, then additional details (not fewer details, as with the enlargement of a photograph) of the figure are displayed. Mandelbrot called these endlessly repeated geometric figures *fractals*. At the present time there is no universal agreement on the precise definition of a fractal, but the fractals that we will study in this lesson can be defined as follows. A **fractal** is a geometric figure in which a self-similar motif repeats itself on an ever-diminishing scale.

Fractals are generally constructed by using **iterative processes** in which the fractal is more closely approximated as a repeated cycle of procedures is performed. For example, a fractal known as the *Koch curve* is constructed as follows.

Construction of the Koch Curve

Step 0: Start with a line segment. This initial segment is shown as stage 0 in Figure 8.12. Stage 0 of a fractal is called the **initiator** of the fractal.

Step 1: On the middle third of the line segment, draw an equilateral triangle and remove its base. The resulting curve is stage 1 in Figure 8.12. Stage 1 of a fractal is called the **generator** of the fractal.

Step 2: Replace each initiator shape (line segment, in this example) with a *scaled version* of the generator to produce the next stage of the Koch curve. The width of the scaled version of the generator is the same as the width of the line segment it replaces. Continue to repeat this step ad infinitum to create additional stages of the Koch curve.

Three applications of step 2 produce stage 2, stage 3, and stage 4 of the Koch curve, as shown in Figure 8.12.

TAKE NOTE

The line segments at any stage of the Koch curve are $\frac{1}{3}$ the length of the line segments at the previous stage. To create the next stage of the Koch curve (after stage 1), replace each line segment with a scaled version of the generator.

At any stage after stage 2, the scaled version of the generator is $\frac{1}{3}$ the size of the preceding scaled generator.

historical note

The Koch curve was created by the mathematician Niels Fabian Helge von Koch (kôkh) (1870–1924). Von Koch attended Stockholm University and in 1911 became a professor of mathematics at Stockholm University. In addition to his early pioneering work with fractals, he wrote several papers on number theory. ∎

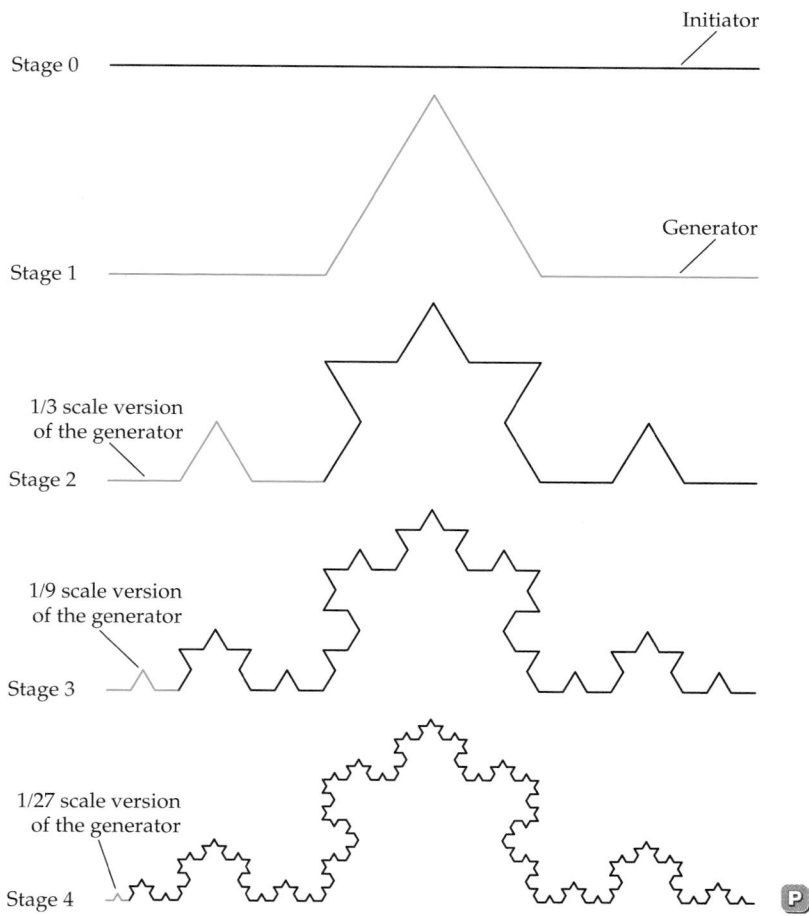

Figure 8.12 *The first five stages of the Koch curve*

None of the curves shown in Figure 8.12 is the Koch curve. The Koch curve is the curve that would be produced if step 2 in the above construction process were repeated ad infinitum. No one has ever seen the Koch curve, but we know that it is a very jagged curve in which the self-similar motif shown in Figure 8.12 repeats itself on an ever-diminishing scale.

The curves shown in Figure 8.13 are the first five stages of the *Koch snowflake*. Each stage of the Koch snowflake is composed of three congruent sides, each of which is a stage of the Koch curve.

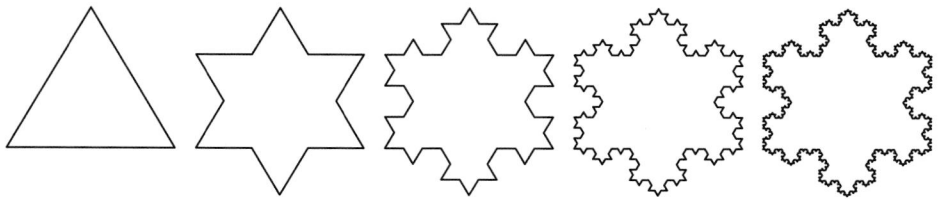

Figure 8.13 *The first five stages of the Koch snowflake*

EXAMPLE 1 ■ Draw Stages of a Fractal

Draw stage 2 and stage 3 of the *box curve*, which is defined by the following iterative process.

Step 0: Start with a line segment as the initiator. See stage 0 in Figure 8.14.

Step 1: On the middle third of the line segment, draw a square and remove its base. This produces the generator of the box curve. See stage 1 in Figure 8.14.

Step 2: Replace each initiator shape with a scaled version of the generator to produce the next stage.

Solution

Two applications of step 2 yield stage 2 and stage 3 of the box curve, as shown in Figure 8.14.

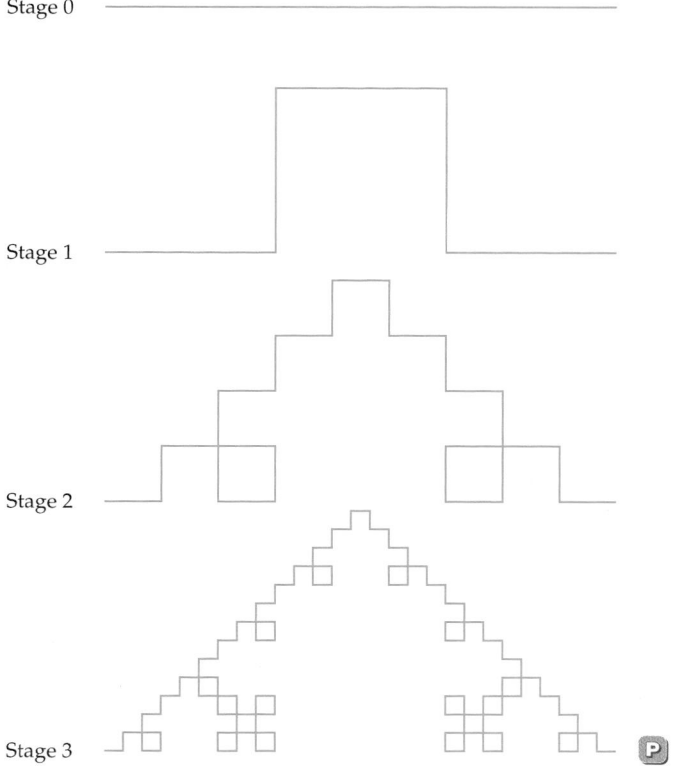

Figure 8.14 *The first four stages of the box curve*

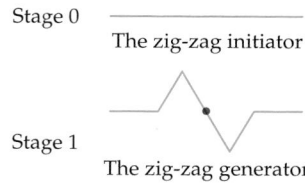

Stage 0 ——————————
The zig-zag initiator

Stage 1
The zig-zag generator

Figure 8.15

CHECK YOUR PROGRESS 1   Draw stage 2 of the *zig-zag curve*, which is defined by the following iterative process.

Step 0: Start with a line segment. See stage 0 of Figure 8.15.

Step 1: Remove the middle half of the line segment and draw a zig-zag, as shown in stage 1 of Figure 8.15. Each of the six line segments in the generator is a $\frac{1}{4}$-scale replica of the initiator.

Step 2: Replace each initiator shape with the scaled version of the generator to produce the next stage. Repeat this step to produce additional stages.

Solution   *See page S30.*

In each of the previous examples the initiator was a line segment. In Example 2, we use a triangle and its interior as the initiator.

Stage 0 (the initiator)

Stage 1 (the generator)

Figure 8.16

EXAMPLE 2 ■ Draw Stages of a Fractal

Draw stage 2 and stage 3 of the *Sierpinski gasket* (also known as the *Sierpinski triangle*), which is defined by the following iterative process.

Step 0: Start with an equilateral triangle and its interior. This is stage 0 of the Sierpinski gasket. See Figure 8.16.

Step 1: Form a new triangle by connecting the midpoints of the sides of the triangle. Remove this center triangle. The result is the three blue triangles shown in stage 1 in Figure 8.16.

Step 2: Replace each initiator (blue triangle) with a scaled version of the generator.

Solution

Two applications of step 2 of the above process produce stage 2 and stage 3 of the Sierpinski gasket, as shown in Figure 8.17.

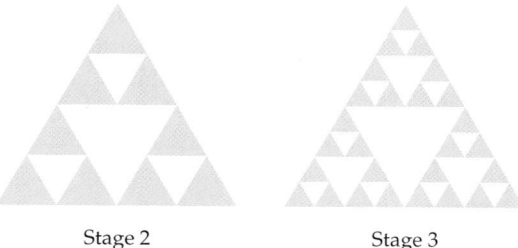

Stage 2   Stage 3

Figure 8.17

CHECK YOUR PROGRESS 2   Draw stage 2 of the *Sierpinski carpet,* which is defined by the following process.

Step 0: Start with a square and its interior. See stage 0 in Figure 8.18.

Step 1: Subdivide the square into nine smaller congruent squares and remove the center square. This yields stage 1 (the generator) shown in Figure 8.18.

Step 2: Replace each initiator (blue square) with a scaled version of the generator. Repeat this step to create additional stages of the Sierpinski carpet.

Stage 0

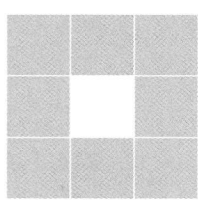
Stage 1

Figure 8.18

Solution   *See page S31.*

Math Matters   Benoit Mandelbrot (1924–   )

Benoit Mandelbrot is often called the Father of Fractal Geometry. He was not the first person to create a fractal, but he was the first person to discover how some of the ideas of earlier mathematicians such as Georg Cantor, Giuseppe Peano, Helge von Koch, Waclaw Sierpinski, and Gaston Julia could be united to form a new type of geometry. Mandelbrot also recognized that many fractals share characteristics with shapes and curves found in nature. For instance, the leaves of a fern, when compared with the whole fern, are almost identical in shape, only smaller. This self-similarity characteristic is evident (to some degree) in all fractals. The following quote by Mandelbrot is from his 1983 book, *The Fractal Geometry of Nature.*[4]

> Clouds are not spheres, mountains are not cones, coastlines are not circles, and bark is not smooth, nor does lightning travel in a straight line. More generally, I claim that many patterns of Nature are so irregular and fragmented, that, compared with Euclid—a term used in this work to denote all of standard geometry—Nature exhibits not simply a higher degree but an altogether different level of complexity.

4. Mandelbrot, Benoit B. *The Fractal Geometry of Nature.* New York: W. H. Freeman and Company, 1983, p. 1.

Strictly Self-Similar Fractals

All fractals show a self-similar motif on an ever-diminishing scale; however, some fractals are *strictly self-similar* fractals, according to the following definition.

> **Definition of a Strictly Self-Similar Fractal**
>
> A fractal is said to be **strictly self-similar** if any arbitrary portion of the fractal contains a replica of the entire fractal.

EXAMPLE 3 ■ Determine Whether a Fractal is Strictly Self-Similar

Determine whether the following fractals are strictly self-similar.

a. The Koch snowflake   **b.** The Koch curve

Solution

a. The Koch snowflake is a closed figure. Any portion of the Koch snowflake (like the portion circled in Figure 8.19) is not a closed figure. Thus the Koch snowflake is *not* a strictly self-similar fractal.

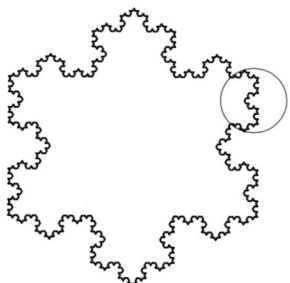

Figure 8.19 *The portion of the Koch snowflake shown in the circle is not a replica of the entire snowflake.*

b. Because any portion of the Koch curve replicates the entire fractal, the Koch curve is a strictly self-similar fractal. See Figure 8.20.

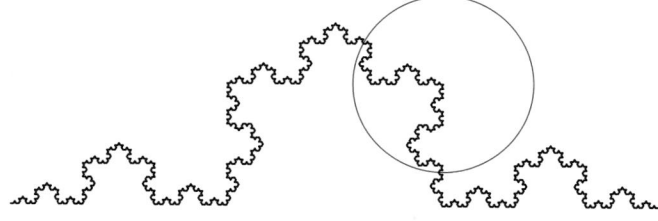

Figure 8.20 *Any portion of the Koch curve is a replica of the entire Koch curve.*

CHECK YOUR PROGRESS 3 Determine whether the following fractals are strictly self-similar.

a. The box curve   **b.** The Sierpinski gasket

Solution   See page S31.

Replacement Ratio and Scaling Ratio

Mathematicians like to assign numbers to fractals so that they can objectively compare fractals. Two numbers that are associated with many fractals are the *replacement ratio* and the *scaling ratio.*

Replacement Ratio and Scaling Ratio

- If the generator of a fractal consists of N replicas of the initiator, then the **replacement ratio** of the fractal is N.

- If the initiator of a fractal has linear dimensions that are r times the corresponding linear dimensions of its replicas in the generator, then the **scaling ratio** of the fractal is r.

EXAMPLE 4 ■ Find the Replacement Ratio and the Scaling Ratio of a Fractal

Find the replacement ratio and the scaling ratio of the

a. box curve.

b. Sierpinski gasket.

Solution

a. Figure 8.14 on page 528 shows that the generator of the box curve consists of five line segments and the initiator consists of only one line segment. Thus the replacement ratio of the box curve is $5:1$, or 5.

   The initiator of the box curve is a line segment that is 3 times longer than the replica line segments in the generator. Thus the scaling ratio of the box curve is $3:1$, or 3.

b. Figure 8.16 on page 529 shows that the generator of the Sierpinski gasket consists of three triangles and the initiator consists of only one triangle. Thus the replacement ratio of the Sierpinski gasket is $3:1$, or 3.

   The initiator triangle of the Sierpinski gasket has a width (height) that is 2 times the width (height) of the replica triangles in the generator. Thus the scaling ratio of the Sierpinski gasket is $2:1$, or 2.

CHECK YOUR PROGRESS 4   Find the replacement ratio and the scaling ratio of the

a. Koch curve.   **b.** zig-zag curve.

Solution   *See page S31.*

Similarity Dimension

INSTRUCTOR NOTE
In some texts on fractals, the scaling ratio is defined as the *reciprocal* of the scaling ratio *r* defined on the previous page. In these texts the similarity dimension *D* of a strictly self-similar fractal is given by

$$D = \frac{\log N}{\log\left(\dfrac{1}{r}\right)}$$

A number called the *similarity dimension* is used to quantify how densely a strictly self-similar fractal fills a region.

> The **similarity dimension** D of a strictly self-similar fractal is given by
>
> $$D = \frac{\log N}{\log r}$$
>
> where N is the replacement ratio of the fractal and r is the scaling ratio.

EXAMPLE 5 ■ Find the Similarity Dimension of a Fractal

Find the similarity dimension, to the nearest thousandth, of the

a. Koch curve.   **b.** Sierpinski gasket.

Solution

✔ **TAKE NOTE**

Because the Koch snowflake is not a strictly self-similar fractal, we cannot compute its similarity dimension.

a. Because the Koch curve is a strictly self-similar fractal, we can find its similarity dimension. Figure 8.12 on page 527 shows that stage 1 of the Koch curve consists of four line segments and stage 0 consists of only one line segment. Hence the replacement ratio is 4 : 1, or 4. The line segment in stage 0 is 3 times longer than each of the replica line segments in stage 1, so the scaling ratio is 3. Thus the Koch curve has a similarity dimension of

$$D = \frac{\log 4}{\log 3} \approx 1.262$$

b. In Example 4, we found that the Sierpinski gasket has a replacement ratio of 3 and a scaling ratio of 2. Thus the Sierpinski gasket has a similarity dimension of

$$D = \frac{\log 3}{\log 2} \approx 1.585$$

CHECK YOUR PROGRESS 5   Compute the similarity dimension, to the nearest thousandth, of the

a. box curve.   **b.** Sierpinski carpet.

Solution   *See page S31.*

The results of Example 5 show that the Sierpinski gasket has a larger similarity dimension than the Koch curve. This means that the Sierpinski gasket fills a flat two-dimensional surface more densely than does the Koch curve.

Computers are used to generate fractals such as those shown in Figure 8.21. These fractals were *not* rendered by using an initiator and a generator, but they were rendered using iterative procedures.

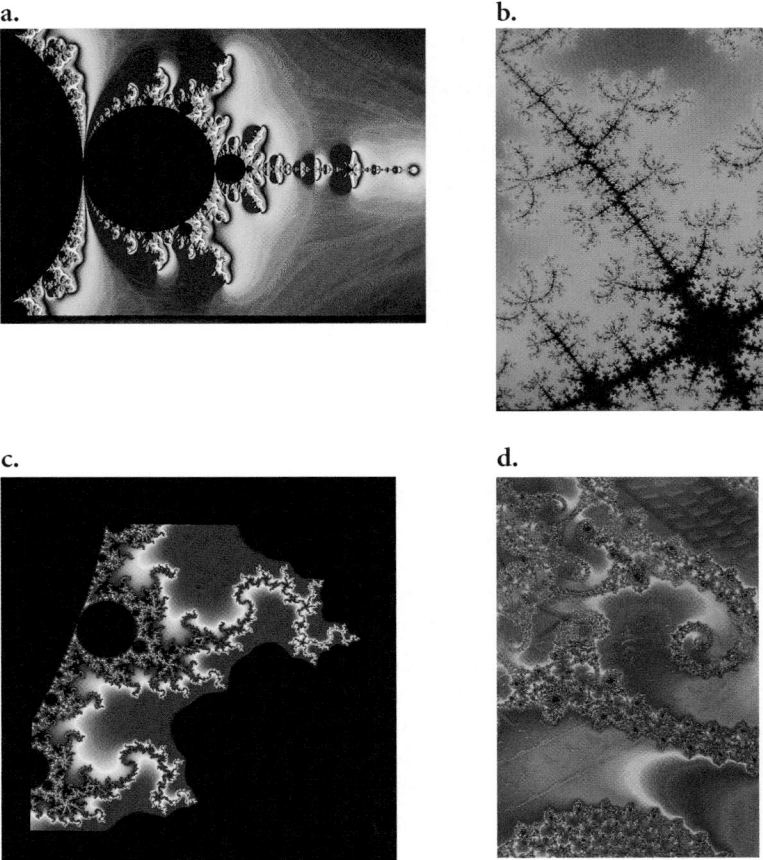

a.

b.

c.

d.

Figure 8.21 *Computer-generated fractals*

Fractals have other applications in addition to being used to produce intriguing images. For example, computer scientists have recently developed fractal image compression programs based on self-transformations of an image. An image compression program is a computer program that converts an image file to a smaller file that requires less computer memory. In some situations these fractal compression programs outperform standardized image compression programs such as JPEG (*jay-peg*), which was developed by the Joint Photographic Experts Group.

During the past few years, some cellular telephones have been manufactured with internal antennas that are fractal in design. Figure 8.22 shows a cellular telephone with an internal antenna in the shape of a stage on the Sierpinski carpet fractal. The antenna in Figure 8.23 is in the shape of a stage of the Koch curve.

Figure 8.22 *A Sierpinski carpet fractal antenna hidden inside a cellular telephone.*

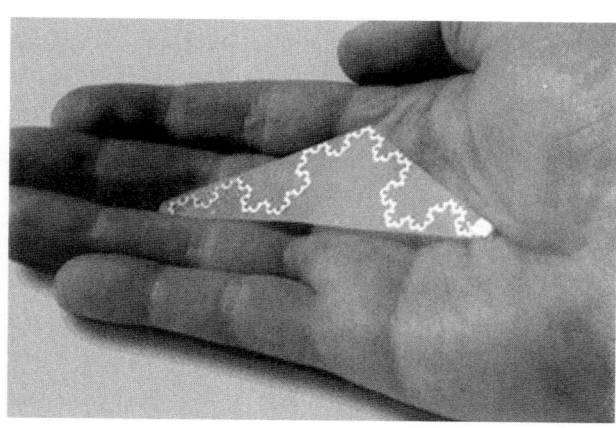

Figure 8.23 *A small Koch curve antenna designed for use in wireless communication devices.*

John Chenoweth, an engineer with T&W Antennas claimed that "fractal antennas are 25 percent more efficient than the rubbery 'stubby' found on most phones. In addition, they are cheaper to manufacture, operate on multiple bands—allowing, for example, a Global Positioning System receiver to be built into the phone—and can be tucked inside the phone body."[5]

Another quote from the article Wireless Communications, states "Just why these fractal antennas work so well was answered in part in the March issue of the journal *Fractals*. (Nathan) Cohen and his colleague Robert Hohlfeld proved mathematically that for an antenna to work equally well at all frequencies, it must satisfy two criteria. It must be symmetrical about a point. And it must be self-similar, having the same basic appearance at every scale—that is, it has to be fractal."[6]

Excursion

The Heighway Dragon Fractal

In this Excursion we illustrate two methods of constructing the stages of a fractal known as the *Heighway dragon*.

The Heighway Dragon via Paper Folding

The first few stages of the Heighway dragon fractal can be constructed by the repeated folding of a strip of paper. In the following discussion we use a 1-inch-wide strip of paper that is 14 inches in length as stage 0. To create stage 1 of the dragon fractal, just fold

(continued)

---

5. Musser, George. Technology and Business: Wireless Communications. *Scientific American,* July, 1999. **http://www.sciam.com/1999/0799issue/0799techbus3.html**
6. Ibid.

✓ TAKE NOTE

The easiest way to construct stage 4 is to make all four folds before you open the paper. All folds must be made in the same direction.

▼ point of interest

The Heighway dragon is sometimes called the "Jurassic Park fractal" because it was used in the book *Jurassic Park* to illustrate some of the surprising results that can be obtained by iterative processes.

INSTRUCTOR NOTE

Some textbooks and some Internet sites refer to the Heighway dragon fractal as the Harter-Heighway dragon.

the strip in half and open it so that the fold forms a right angle (see Figure 8.24). To create stage 2, fold the strip twice. The second fold should be in the same direction as the first fold. Open the paper so that each of the folds forms a right angle. Continue the iterative process of making an additional fold in the same direction as the first fold and then forming a right angle at each fold to produce additional stages. See Figure 8.24.

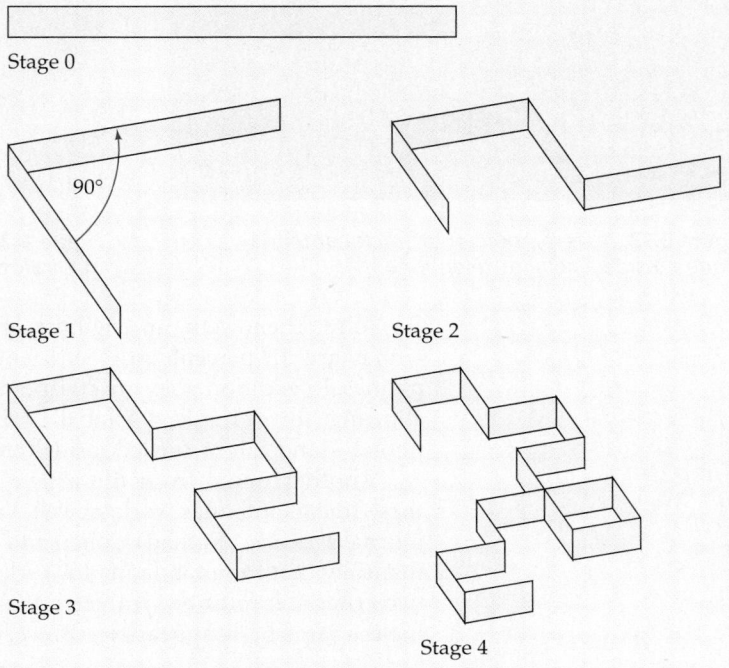

Figure 8.24 *The first five stages of the Heighway dragon via paper folding*

The Heighway Dragon via the Left-Right Rule

The *n*th stage of the Heighway dragon can also be created by the following drawing procedure.

Step 0: Draw a small vertical line segment. Label the bottom point of this segment as vertex $v = 0$ and label the top as vertex $v = 1$.

Step 1: Use the following left-right rule to determine whether to make a left turn or a right turn.

> **The Left-Right Rule**
>
> At vertex v, where v is an *odd* number, go
>
> - **right** if the remainder of v divided by 4 is 1.
> - **left** if the remainder of v divided by 4 is 3.
>
> At vertex 2, go to the right. At vertex v, where v is an *even* number greater than 2, go in the same direction in which you went at vertex $\frac{v}{2}$.

(continued)

Draw another line segment of the same length as the original segment. Label the endpoint of this segment with a number that is 1 larger than the number used for the preceding vertex.

Step 2: Continue to repeat step 1 until you reach the last vertex. The last vertex of an n-stage Heightway dragon is the vertex numbered 2^n.

Excursion Exercises

1. Use a strip of paper and the folding procedure explained above to create models of the first five stages (stage 0 through stage 4) of the Heighway dragon. Explain why it would be difficult to create the 10th stage of the Heighway dragon using the paper-folding procedure.

In Exercises 2 to 5, assume that your first move from vertex 0 is to the right.

2. Use the left-right rule to draw stage 2 of the Heighway dragon.

3. Use the left-right rule to determine the direction in which to turn at vertex 7 of the Heighway dragon.

4. Use the left-right rule to determine the direction in which to turn at vertex 50 of the Heighway dragon.

5. Use the left-right rule to determine the direction in which to turn at vertex 64 of the Heighway dragon.

Exercise Set 8.6   (Suggested Assignment: 1–23 odd)

In Exercises 1 and 2, use an iterative process to draw stage 2 and stage 3 of the fractal with the given initiator (stage 0) and the given generator (stage 1).

1. The Cantor point set

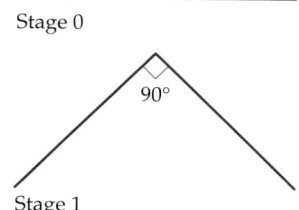

In Exercises 3–8, use an iterative process to draw stage 2 of the fractal with the given initiator (stage 0) and the given generator (stage 1).

3. The Sierpinski carpet, variation 1

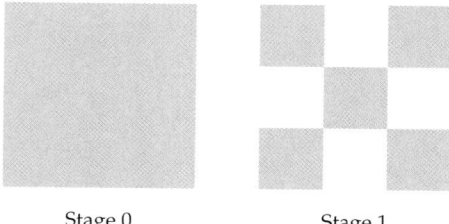

Stage 0   Stage 1

2. Lévy's curve

Stage 0

90°

Stage 1

4. The Sierpinski carpet, variation 2

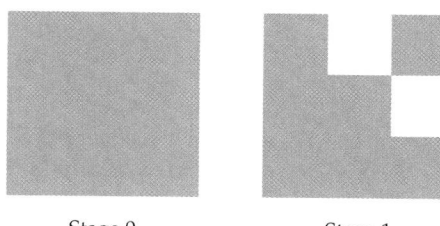

Stage 0   Stage 1

5. The river tree of Peano Cearo

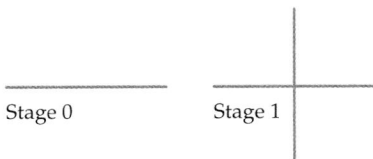

Stage 0   Stage 1

6. Minkowski's fractal

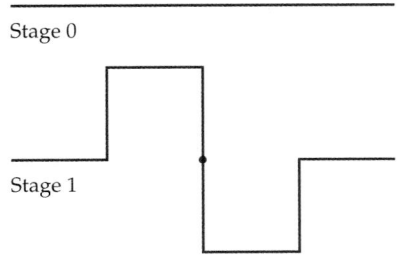

Stage 0

Stage 1

7. The square fractal

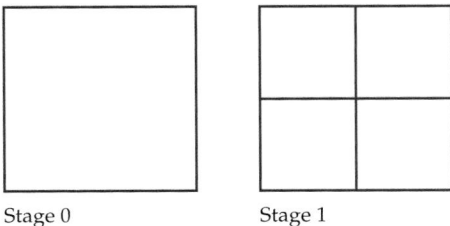

Stage 0   Stage 1

8. The cube fractal

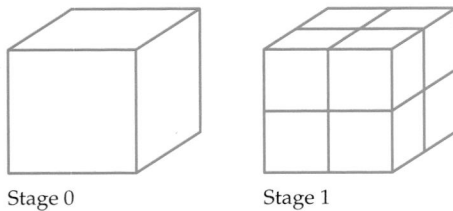

Stage 0   Stage 1

In Exercises 9 and 10, draw stage 3 and stage 4 of the fractal defined by the given iterative process.

9. The binary tree

Step 0: Start with a "⊤". This is stage 0 of the binary tree. The vertical line segment is the trunk of the tree and the horizontal line segment is the branch of the tree. The branch is half the length of the trunk.

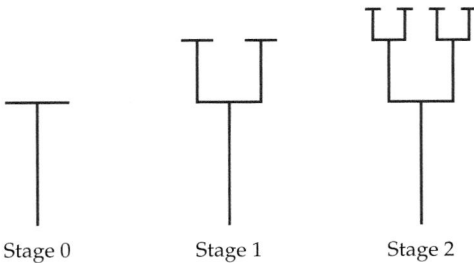

Stage 0   Stage 1   Stage 2

Step 1: At the ends of each branch, draw an upright ⊤ that is half the size of the ⊤ in the preceding stage.

Step 2: Continue to repeat step 1 to generate additional stages of the binary tree.

10. The I-fractal

Step 0: Start with the line segment shown as stage 0 below.

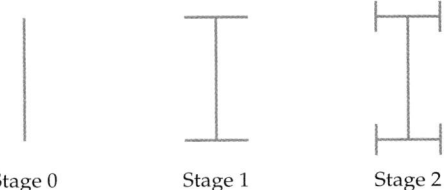

Stage 0   Stage 1   Stage 2

Step 1: At each end of the line segment, draw a crossbar that is half the length of the line segment it contacts. (This produces stage 1 of the I-fractal.)

Step 2: Use each crossbar from the preceding step as the connecting segment of a new "I." Attach new crossbars that are half the length of the connecting segment. Continue to repeat this step to generate additional stages of the I-fractal.

In Exercises 11–20, compute, if possible, the similarity dimension of the fractal. Round to the nearest thousandth.

11. The Cantor point set (See Exercise 1.)

12. Lévy's curve (See Exercise 2.)

13. The Sierpinski carpet, variation 1 (See Exercise 3.)

14. The Sierpinski carpet, variation 2 (See Exercise 4.)

15. The river tree of Peano Cearo (See Exercise 5.)

16. Minkowski's fractal (See Exercise 6.)

17. The square fractal (See Exercise 7.)

18. The cube fractal (See Exercise 8.)

19. The quadric Koch curve, defined by the following stages.

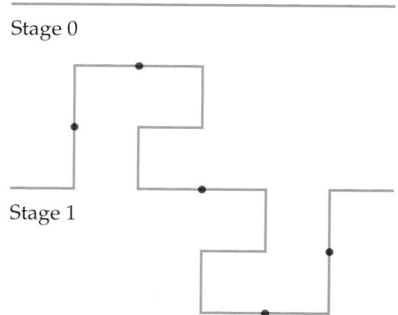

Stage 0

Stage 1

20. The Menger sponge, defined by the following stages.

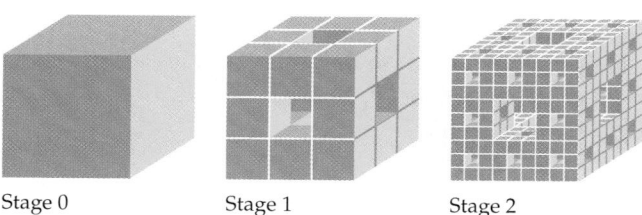

Stage 0   Stage 1   Stage 2

Extensions

CRITICAL THINKING

21. a. Rank, from largest to smallest, the similarity dimensions of the Sierpinski carpet; the Sierpinski carpet, variation 1 (see Exercise 3); and the Sierpinski carpet, variation 2 (see Exercise 4).

b. Which of the three fractals is the most dense?

22. The *Peano curve* is defined by the following stages.

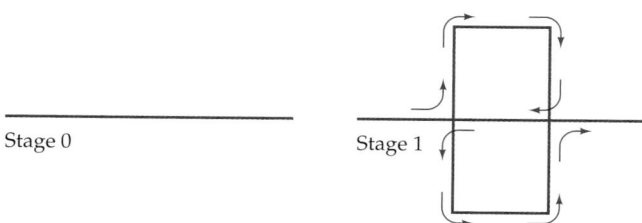

Stage 0

Stage 1

The arrows show the route used to trace the generator.

a. What is the similarity dimension of the Peano curve?

b. Explain why the Peano curve is referred to as a plane-filling curve.

23. Explain why the similarity dimension formula cannot be used to find the similarity dimension of the binary tree defined in Exercise 9.

24. Stage 0 and stage 1 of Lévy's curve and the Heighway dragon are identical, but the fractals start to differ at stage 2. Make two drawings that illustrate how they differ at stage 2.

EXPLORATIONS

25. Create a strictly self-similar fractal that is different from any of the fractals in this lesson.

a. Draw the first four stages of your fractal.

b. Compute the replacement ratio, the scaling ratio, and the similarity dimension of your fractal.

26. The following figure shows three stages of the *Sierpinski pyramid*. This fractal has been called a three-dimensional model of the Sierpinski gasket; however, its similarity dimension is 2, not 3. Build scale models of stage 0 and stage 1 of the Sierpinski pyramid. Use the models to demonstrate to your classmates why the Sierpinski pyramid fractal has a similarity dimension of 2.

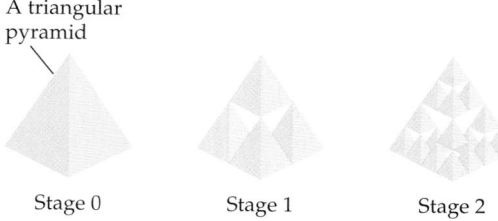

A triangular pyramid

Stage 0   Stage 1   Stage 2

The first three stages of the Sierpinski pyramid

Summary

Key Terms

acute angle [p. 452]
acute triangle [p. 468]
adjacent angles [p. 453]
alternate exterior angles [p. 455]
alternate interior angles [p. 455]
angle [p. 450]
area [p. 475]
box curve [p. 528]
center of a circle [p. 473]
circle [p. 473]
circumference [p. 473]
complementary angles [p. 452]
corresponding angles [p. 456]
cube [p. 501]
degree [p. 451]
diameter [p. 473]
diameter of a sphere [p. 502]
endpoint [p. 448]
equilateral triangle [p. 468]
Euclidean geometry [p. 516]
exterior angle of a triangle [p. 458]
face of a rectangular solid [p. 501]
fractal [p. 526]
generator [p. 526]
geodesic [p. 517]
great circle [p. 517]
hyperbolic geometry [p. 516]
hyperbolic triangle [p. 520]
imaginary geometry [p. 516]
initiator [p. 526]
interior angle of a triangle [p. 458]
intersecting lines [p. 450]
isosceles triangle [p. 468]
iterative process [p. 526]
Koch curve [pp. 526–527]
Koch snowflake [pp. 527–528]
line [p. 448]
line segment [p. 448]
Lobachevskian geometry [p. 516]
non-Euclidean geometry [p. 516]
obtuse angle [p. 468]
obtuse triangle [p. 452]
parallel lines [p. 450]
parallelogram [p. 469]

perimeter [p. 469]
perpendicular lines [p. 451]
plane [p. 450]
plane figure [p. 450]
point [p. 448]
polygon [p. 467]
postulate [p. 515]
protractor [p. 451]
quadrilateral [p. 468]
radius [p. 473]
radius of a sphere [p. 502]
ray [p. 448]
rectangle [p. 470]
rectangular solid [p. 501]
regular polygon [p. 467]
regular pyramid [p. 503]
replacement ratio [p. 532]
Riemmanian geometry [p. 517]
right angle [p. 451]
right circular cone [p. 503]
right circular cylinder [p. 503]
right triangle [p. 468]
scalene triangle [p. 468]
scaling ratio [p. 532]
Sierpinski carpet [p. 530]
Sierpinski gasket or Sierpinski triangle [p. 529]
similar objects [p. 491]
similar triangles [p. 491]
similarity dimension [p. 533]
slant height [p. 503]
sphere [p. 502]
spherical or elliptical geometry [p. 517]
spherical triangle [p. 518]
square [p. 471]
straight angle [p. 452]
strictly self-similar fractal [p. 531]
supplementary angles [p. 452]
surface area [p. 505]
transversal [p. 455]
trapezoid [p. 469]
triangle [p. 458]
undefined term [p. 516]
vertex [p. 450]
vertical angles [p. 454]
volume [p. 501]

Essential Concepts

- **Triangles**
  Sum of the measures of the interior angles = 180°
  Sum of the measures of an interior and a corresponding exterior angle = 180°

- **Perimeter Formulas**
Triangle:	$P = a + b + c$
Rectangle:	$P = 2L + 2W$
Square:	$P = 4s$
Parallelogram:	$P = 2b + 2s$
Circle:	$C = \pi d$ or $C = 2\pi r$

- **Area Formulas**
Triangle:	$A = \dfrac{1}{2}bh$
Rectangle:	$A = LW$
Square:	$A = s^2$
Circle:	$A = \pi r^2$
Parallelogram:	$A = bh$
Trapezoid:	$A = \dfrac{1}{2}h(b_1 + b_2)$

- **Volume Formulas**
Rectangular solid:	$V = LWH$
Cube:	$V = s^3$
Sphere:	$V = \dfrac{4}{3}\pi r^3$
Right circular cylinder:	$V = \pi r^2 h$
Right circular cone:	$V = \dfrac{1}{3}\pi r^2 h$
Regular pyramid:	$V = \dfrac{1}{3}s^2 h$

- **Surface Area Formulas**
Rectangular solid:	$S = 2LW + 2LH + 2WH$
Cube:	$S = 6s^2$
Sphere:	$S = 4\pi r^2$
Right circular cylinder:	$S = 2\pi r^2 + 2\pi rh$
Right circular cone:	$S = \pi r^2 + \pi rl$
Regular pyramid:	$S = s^2 + 2sl$

- **Similar Triangles**
  The ratios of corresponding sides are equal. The ratio of corresponding heights is equal to the ratio of corresponding sides.

- **Parallel Postulates**
  The Euclidean Parallel Postulate (*Plane Geometry*) Through a given point not on a given line, exactly one line can be drawn parallel to the given line.
  Gauss's Alternative to the Parallel Postulate (*Hyperbolic Geometry*) Through a given point not on a given line, there are *at least two* lines parallel to the given line.
  Riemann's Alternative to the Parallel Postulate (*Spherical Geometry*) Through a given point not on a given line, there exist *no* lines parallel to the given line.

- **The Spherical Triangle Area Formula**
  The area S of the spherical triangle ABC on a sphere with radius r is
  $$S = (m\angle A + m\angle B + m\angle C - 180°) \cdot \left(\frac{\pi}{180°}\right)r^2$$

- **Similarity Dimension of a Fractal**
  The similarity dimension D of a strictly self-similar fractal is $D = \dfrac{\log N}{\log r}$, where N is the replacement ratio of the fractal and r is the scaling ratio.

<table><tr><td>**CHAPTER 8**</td><td>## Review Exercises</td></tr></table>

1. Given that $m\angle a = 74°$ and $m\angle b = 52°$, find the measures of angles x and y.

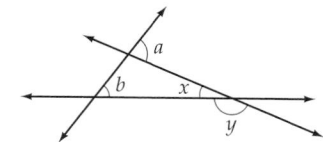

2. Triangles ABC and DEF are similar. Find the perimeter of triangle ABC.

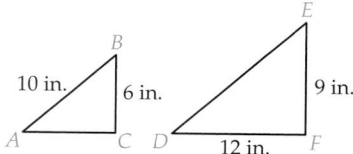

3. Find the volume of the figure.

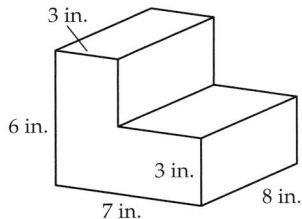

4. Find the measure of ∠x.

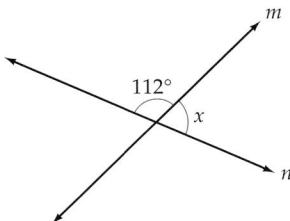

5. Find the surface area of the figure.

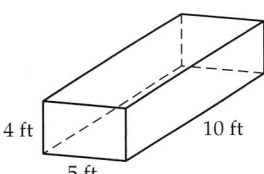

6. The length of a diameter of the base of a cylinder is 4 m, and the height of the cylinder is 8 m. Find the surface area of the cylinder.

7. Given $BC = 11$ cm and AB is three times the length of BC, find the length of AC.

8. Find x.

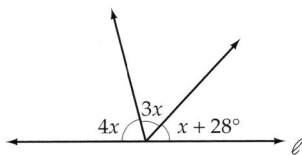

9. Find the area of a parallelogram that has a base of length 6 in. and a height of 4.5 in.

10. Find the volume of the figure.

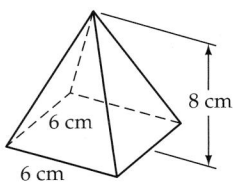

11. Find the circumference of a circle that has a diameter of 4.5 m. Round to the nearest hundredth.

12. Given that $l_1 \| l_2$, find the measures of angles a and b.

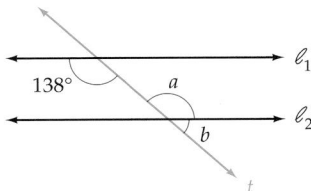

13. Find the supplement of a 32° angle.

14. Find the volume of a rectangular solid with a length of 6.5 ft, a width of 2 ft, and a height of 3 ft.

15. Two angles of a triangle measure 37° and 48°. Find the measure of the third angle.

16. The height of a triangle is 7 cm. The area of the triangle is 28 cm². Find the length of the base of the triangle.

17. Find the volume of a sphere that has a diameter of 12 mm. Give the exact value.

18. The perimeter of a square picture frame is 86 cm. Find the length of each side of the frame.

19. A can of paint will cover 200 ft² of surface. How many cans of paint should be purchased to paint a cylinder that has a height of 15 ft and a radius of 6 ft?

20. The length of a rectangular park is 56 yd. The width is 48 yd. How many yards of fencing are needed to surround the park?

21. What is the area of a square patio that measures 9.5 m on each side?

22. A walkway 2 m wide surrounds a rectangular plot of grass. The plot is 40 m long and 25 m wide. What is the area of the walkway?

23. Name the mathematician who was the first person to develop a non-Euclidean geometry, but chose not to publish his work.

24. Name the mathematician who called the geometry he developed "imaginary geometry."

25. What is another name for Riemannian geometry?

26. What is another name for Lobachevskian geometry?

27. Name a geometry in which the sum of the measures of the interior angles of a triangle is less than 180°.

28. Name a geometry in which there are no parallel lines.

In Exercises 29 and 30, determine the exact area of the spherical triangle.

29. Radius: 12 in.; angles: 90°, 150°, 90°

30. Radius: 5 ft; angles: 90°, 60°, 90°

In Exercises 31 and 32, determine the area of the spherical polygon.

31. Radius: 26 ft; angles: 90°, 90°, 95°, 95°

32. Radius: 1980 mi; angles: 90°, 90°, 110°, 110°

33. Draw stage 0, stage 1, and stage 2 of the Koch curve.

34. Draw stage 2 of the fractal with the following initiator and generator.

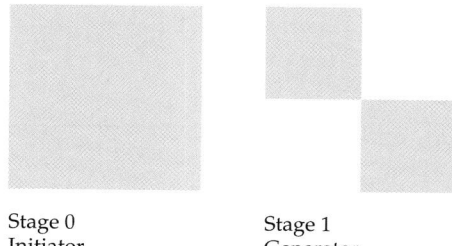

Stage 0
Initiator

Stage 1
Generator

35. Compute the similarity dimension of a strictly self-similar fractal with a replacement ratio of 5 and a scaling ratio of 4. Round to the nearest thousandth.

36. Determine the similarity dimension of the fractal defined in Exercise 34.

CHAPTER 8 **Test**

1. Find the volume of a cylinder with a height of 6 m and a radius of 3 m. Round to the nearest hundredth.

2. Find the perimeter of a rectangle that has a length of 2 m and a width of 1.4 m.

3. Find the complement of a 32° angle.

4. Find the area of a circle that has a diameter of 2 m. Round to the nearest hundredth.

5. In the figure below, lines l_1 and l_2 are parallel. Angle x measures 30°. Find the measure of angle y.

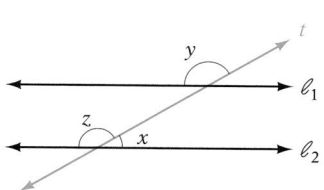

6. In the figure below, lines l_1 and l_2 are parallel. Angle x measures 45°. Find the measures of angles a and b.

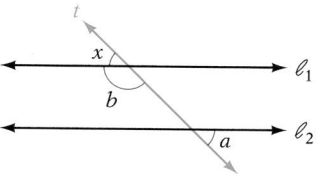

7. Find the area of a square that measures 2.25 ft on each side.

8. Find the volume of the figure.

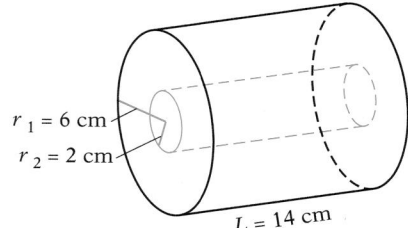

$r_1 = 6$ cm
$r_2 = 2$ cm
$L = 14$ cm

9. Triangles *ABC* and *DEF* are similar. Find side *BC*.

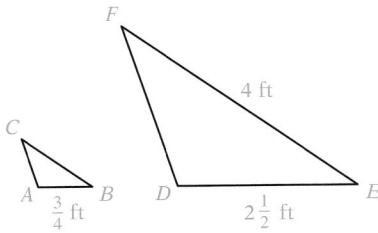

10. A right triangle has a 40° angle. Find the measures of the other two angles.

11. Find the measure of ∠*x*.

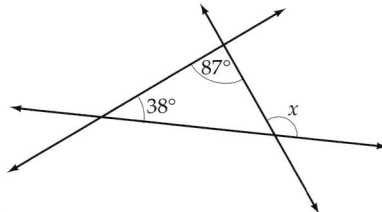

12. Find the area of the parallelogram shown below.

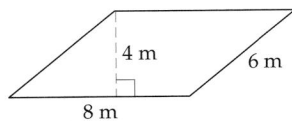

13. Find the width of the canal shown in the figure below.

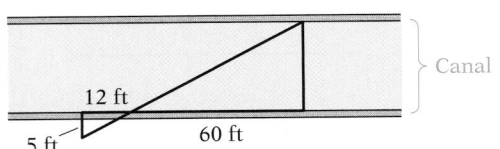

14. How much more area is in a pizza with radius 10 in. than in a pizza with radius 8 in.? Round to the nearest hundredth.

15. Find the cross-sectional area of a redwood tree that is 11 ft 6 in. in diameter. Round to the nearest hundredth of a square foot.

16. A toolbox is 1 ft 2 in. long, 9 in. wide, and 8 in. high. The sides and bottom of the toolbox are $\frac{1}{2}$ in. thick. The toolbox is open at the top. Find the volume of the interior of the toolbox in cubic inches.

17. a. State the Euclidean parallel postulate.

   b. State the parallel postulate used in Riemannian geometry.

18. What is the maximum number of right angles a triangle can have in

   a. Lobachevskian geometry?

   b. Riemannian geometry?

19. What is a great circle?

20. Find the area of a spherical triangle with a radius of 12 ft and angles of 90°, 100°, and 90°.

In Exercises 21 and 22, draw stage 2 of the fractal with the given initiator and generator.

21.

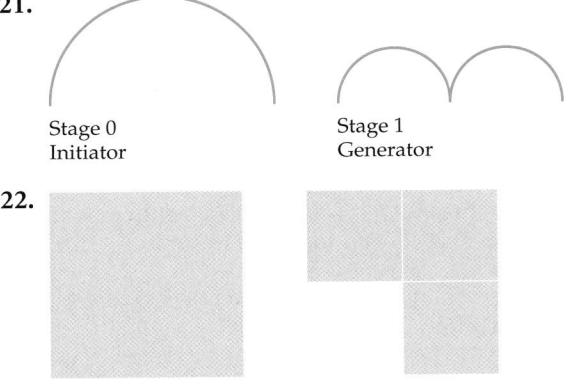

Stage 0
Initiator

Stage 1
Generator

22.

Stage 0
Initiator

Stage 1
Generator

23. Compute the replacement ratio, scale ratio, and similarity dimension of the fractal defined by the initiator and generator in Exercise 21.

24. Compute the replacement ratio, scale ratio, and similarity dimension of the fractal defined by the initiator and generator in Exercise 22.

CHAPTER

9

The Mathematics of Graphs

A lot of planning and preparation goes into creating roadways and intersections. Graphs can be used to model different traffic routes. These graphs are used to determine the placement and timing of traffic lights which create maximum traffic flow through an intersection. In the **Excursion exercises** on **page 606,** you will create graphs that represent the best possible traffic flow for different intersections.

9.1	**Traveling Roads and Visiting Cities**
9.2	**Efficient Routes**
9.3	**Planarity and Euler's Formula**
9.4	**Map Coloring and Graphs**

Need help? For on-line student resources, such as section quizzes, visit this textbook's web site at **math.college.hmco.com/students.**

In the early eighteenth century, the Pregel River in a city called Königsberg (located in modern-day Russia, now called Kaliningrad) surrounded an island before splitting in two. Seven bridges crossed the river and connected four different land areas, similar to the map drawn below.

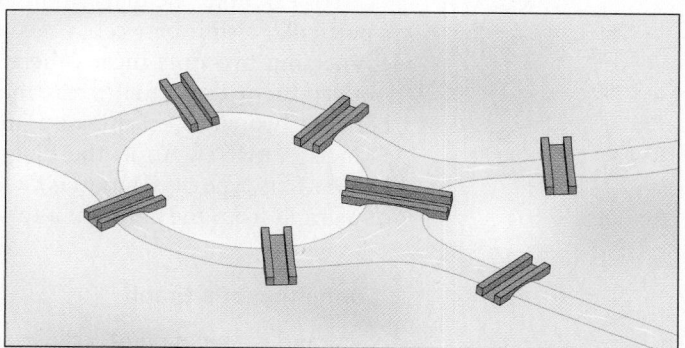

Many citizens of the time attempted to take a stroll that would lead them across each bridge and return them to the starting point without traversing the same bridge twice. None of them could do it, no matter where they chose to start. Try it for yourself with pencil and paper. You will see that it is not easy!

So is it impossible to take such a stroll? Or just difficult? In 1736 the Swiss mathematician Leonhard Euler (1707–1783) proved that it is, in fact, impossible to walk such a path. His analysis of the challenge laid the groundwork for a branch of mathematics known as *graph theory*. We will investigate how Euler approached the problem of the seven bridges of Königsberg in Section 9.1.

| **Traveling Roads and Visiting Cities**

Introduction to Graphs

The acts of traveling the streets of a city, routing data through nodes on the Internet, and flying between cities have a common link. The goal of each of these tasks is to go from one place to another along a specified path. Efficient strategies to accomplish these goals can be studied using a branch of mathematics called *graph theory.*

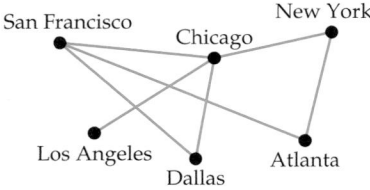

Figure 9.1

For example, the diagram in Figure 9.1 could represent the flights available on a particular airline between a selection of cities. Each dot represents a city, and a line connecting two dots means there is a flight between the two cities. The lines do not represent the actual paths the planes fly; they simply indicate a relationship between two cities (in this case, that a flight exists). If we wish to travel from San Francisco to New York, the diagram can help us examine the various possible routes. This type of diagram is called a *graph.* Note that this is a very different kind of a graph from the graph of a function that we discussed in Chapter 6.

> **Definition of a Graph**
>
> A **graph** is a set of points called **vertices** and line segments or arcs called **edges** that connect vertices.

Graphs can be used to represent many different scenarios. For instance, the three graphs in Figure 9.2 are the exact same graph as in Figure 9.1, but used in different contexts. In part (a), each vertex represents a baseball team, and an edge connecting two vertices might mean that the two teams played each other during the current season. Note that the placement of the vertices has nothing to do with geographical location; in fact, the vertices can be shown in any arrangement we choose. The important information is which vertices are connected by edges.

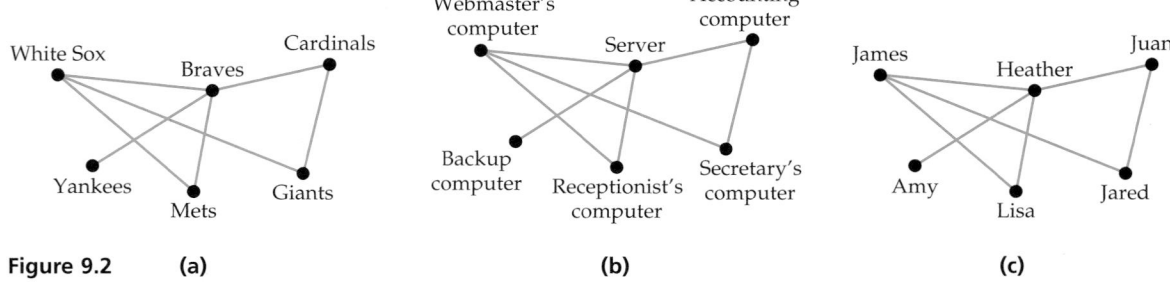

Figure 9.2   **(a)**   **(b)**   **(c)**

Figure 9.2(b) shows the computer network of a small business. Each vertex represents a computer, and the edges indicate which machines are directly connected to each other. The graph in Figure 9.2(c) could be used to indicate which students share a class together; each vertex represents a student, and an edge connecting two vertices means that those students share at least one class.

In general, graphs can contain vertices that are not connected to any edges, two or more edges that connect the same vertices (called **multiple edges**), or edges that loop back to the same vertex. We will usually deal with **connected graphs,** graphs in which any vertex can be reached from any other vertex by tracing along edges. (Essentially, the graph consists of only one "piece.") Several examples of graphs are shown below.

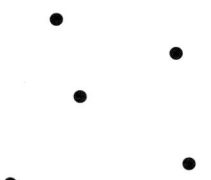

This graph has five vertices but no edges, and is referred to as a **null graph.** It is also an example of a *disconnected graph.*

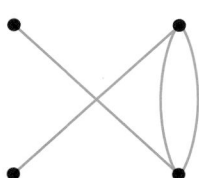

This is a connected graph that has a pair of *multiple edges.* Notice that two edges cross in the center, but there is no vertex there. Unless a dot is drawn, the edges are considered to pass over each other without touching.

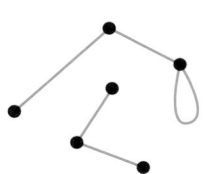

This graph is not connected; it consists of two different sections. It also contains a loop.

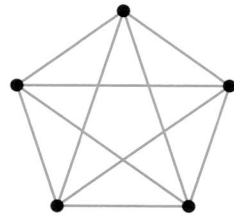

This is a connected graph in which every possible edge is drawn between vertices (without any multiple edges). Such a graph is called a **complete graph.**

Notice that it does not matter whether the edges are drawn straight or curved, and their lengths and positions are not important. Nor is the exact placement of the vertices important. The graph simply illustrates connections between vertices.

Consequently, the three graphs illustrated below are considered **equivalent graphs** because the edges form the same connections of vertices in each graph.

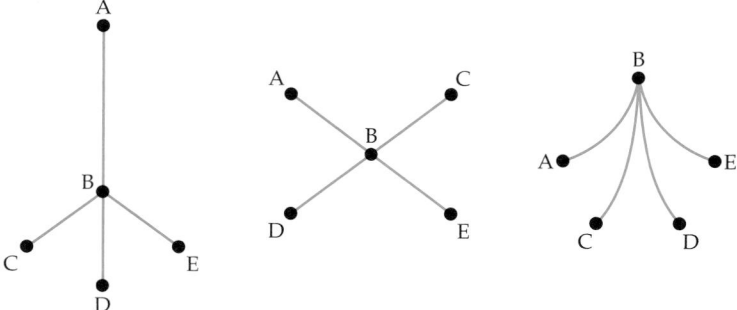

If you have difficulty seeing that these graphs are equivalent, use the labeled vertices to compare each graph. Notice that in each case, vertex B has an edge connecting it to each of the other four vertices, and no other edges exist.

EXAMPLE 1 ■ Equivalent Graphs

Determine whether the following two graphs are equivalent.

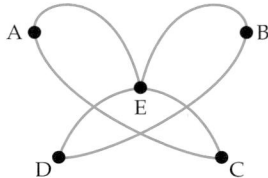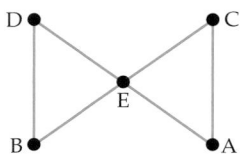

Solution

Despite the fact that the two graphs have different arrangements of vertices and edges, they are equivalent. To illustrate, we examine the edges of each graph. The first graph contains six edges; we can list them by indicating which two vertices they connect. The edges are AC, AE, BD, BE, CE, and DE. If we do the same for the second graph, we get the exact same six edges. Because the two graphs represent the same connections among the vertices, they are equivalent.

CHECK YOUR PROGRESS 1 Determine whether the following two graphs are equivalent.

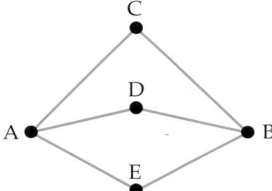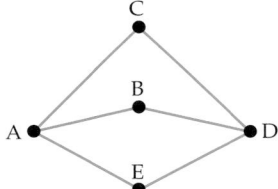

Solution See page S31.

MathMatters **Picture the World Wide Web as a Graph**

Graph theory is playing an ever-increasing role in the study of large and complex systems such as the World Wide Web. Internet Cartographer, available at **http://www.inventix.com,** is a software application that creates graphs from networks of web sites. Each vertex is a web site, and two vertices are joined by an edge if there is a link from one web site to the other. The software was used to generate the image on the following page. You can imagine how large the graph would be if all the billions of existing web pages were included. The Internet search engine Google uses these relationships to rank its search results. Web sites are ranked in

(continued)

part according to how many other sites link to them. In effect, the web sites represented by the vertices of the graph with the most connected edges would appear first in Google's search results.

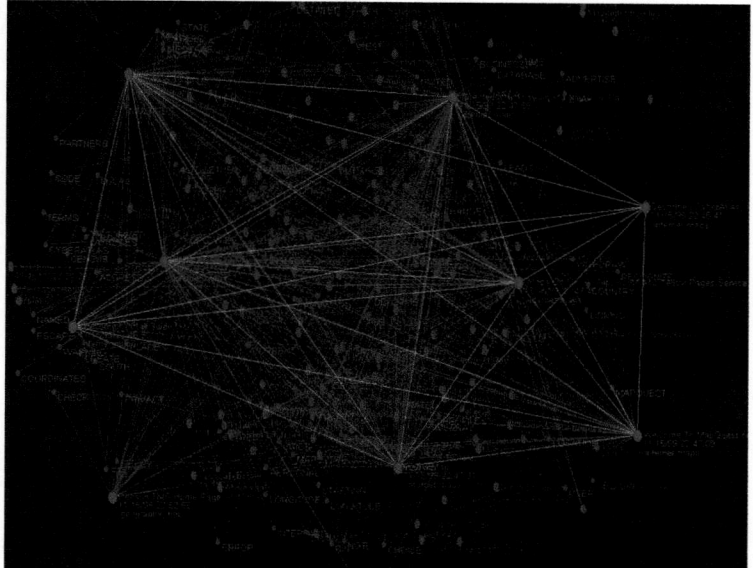

Another example of an extremely large graph is the graph that uses vertices to represent telephone numbers. An edge connects two vertices if one number called the other. James Abello of AT&T Shannon Laboratories analyzed a graph formed from one day's worth of calls. The graph had over 53 million vertices and 170 million edges! Interestingly, although the graph was not a connected graph (it contained 3.7 million separate components), over 80% of the vertices were part of one large connected component. Within this component, any telephone could be linked to any other through a chain of twenty or fewer calls.

Euler Circuits

To solve the Königsberg bridges problem presented on page 547, we can represent the arrangement of land areas and bridges with a graph. Let each land area be represented by a vertex, and then connect two vertices if there is a bridge spanning the corresponding land areas. Then the geographical situation shown in Figure 9.3 on the following page becomes the graph shown in Figure 9.4.

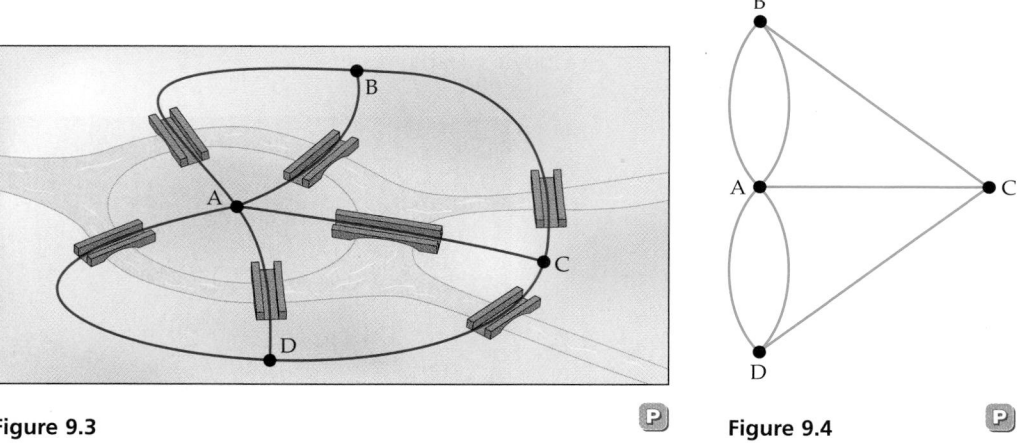

Figure 9.3 ⓟ   **Figure 9.4** ⓟ

historical note

Leonhard Euler
(oiʹlər)
(1707–1783) was
one of the most
prolific mathemati-
cians of all time.
He wrote hundreds
of papers in almost every area of
mathematics. In one of these
papers, published in 1736, Euler
proved that it was impossible to
traverse each of the bridges of
Königsberg exactly once and
return to the starting point.
   Although he did not present
his paper in the language of
graph theory, his arguments were
equivalent to our discussion in
this section. In fact, his proof was
more general than just a proof of
the Königsberg bridges problem,
and can be considered the first
paper in graph theory. ■

In terms of a graph, the original problem can be stated as follows: Can we start at any vertex, move through each edge once (but not more than once), and return to the starting vertex? Again, try it with pencil and paper. Every attempt seems to end in failure.

Before we can examine how Euler proved this task impossible, we need to establish some terminology. A **walk** in a graph can be thought of as a movement from one vertex to another by traversing edges. We can refer to our movement by vertex letters. For example, in the graph in Figure 9.4, one walk would be A–B–A–C. If a walk ends at the same vertex it started at, it is considered a **closed walk,** or **circuit.** A circuit that uses every edge, but never uses the same edge twice, is called an **Euler circuit.** So an Euler circuit is a walk that starts and ends at the same vertex and uses every edge of the graph exactly once. (The walk may cross through vertices more than once.) If we could find an Euler circuit in the graph in Figure 9.4, we would have a solution to the Königsberg bridges problem: a path that crosses each bridge exactly once and returns to the starting point.

Euler essentially proved that the graph in Figure 9.4 cannot have an Euler circuit. He accomplished this by examining the number of edges that met at each vertex. This is called the **degree** of a vertex. He made the observation that in order to complete the desired walk, every time you approached a vertex you would then need to leave that vertex. If you traveled through that vertex again, you would again need an approaching edge and a departing edge. Thus, for an Euler circuit to exist, the degree of every vertex would have to be an even number. Furthermore, he was able to show that any graph that has even degree at every vertex must have an Euler circuit. Consequently, such graphs are called **Eulerian.** If we now look at the graph in Figure 9.4, we can see that it is not Eulerian. No Euler circuit exists because not every vertex has even degree.

Eulerian Graph Theorem

A connected graph is Eulerian if and only if every vertex of the graph is of even degree.

✓ **TAKE NOTE**

For information on and examples of finding Eulerian circuits, search for "Fluery's Algorithm" on the World Wide Web.

The Eulerian Graph Theorem guarantees that when all vertices of a graph have even degree, an Euler circuit exists, but it does not tell us how to find one. Because the graphs we will examine here are relatively small, we will rely on trial and error to find Euler circuits. There is a systematic method, called *Fluery's Algorithm*, that can be used to find an Euler circuit in graphs with a large number of vertices.

EXAMPLE 2 ■ Find an Euler Circuit

Determine whether the graph shown below is Eulerian. If it is, find an Euler circuit. If it is not, explain how you know. The number beside each vertex indicates the degree of the vertex.

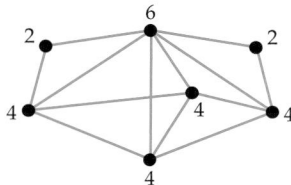

Solution

Each vertex has degree of either 2, 4, or 6, so by the Eulerian Graph Theorem, the graph is Eulerian. There are many possible Euler circuits in this graph. We do not have a formal method of locating one, so we just use trial and error. If we label the vertices as shown below, one Euler circuit is B–A–F–B–E–F–G–E–D–G–B–D–C–B.

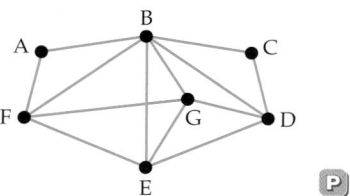

CHECK YOUR PROGRESS 2   Does the following graph have an Euler circuit? If so, find one. If not, explain why.

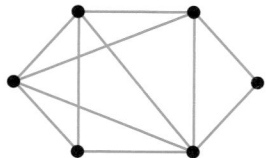

Solution   *See page S31.*

EXAMPLE 3 ■ **An Application of Euler Circuits**

The subway map below shows the tracks the subway trains traverse as well as the junctions where one can switch trains. Suppose an inspector needs to travel the full length of each track. Is it possible to plan a journey that traverses the tracks and returns to the starting point without traveling through any portion of a track more than once?

Solution

We can consider the subway map a graph, with a vertex at each junction. An edge represents a track that runs between two junctions. In order to find a travel route that does not traverse the same track twice, we need to find an Euler circuit in the graph. Notice, however, that the vertex representing the Civic Center junction has degree 3. Because a vertex has odd degree, the graph cannot be Eulerian, and it is impossible for the inspector not to travel at least some tracks twice.

CHECK YOUR PROGRESS 3 Suppose the city of Königsberg had the arrangement of islands and bridges pictured below instead of the arrangement we introduced previously. Would the citizens be able to complete a stroll across each bridge and return to their starting points without crossing the same bridge twice?

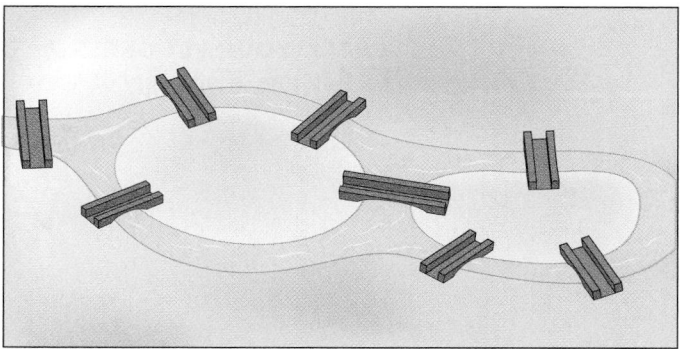

Solution *See page S31.*

Euler Walks

Perhaps the Königsberg bridge problem would have a solution if we did not need to return to the starting point. Give it a try; you will still find it difficult to find a solution.

If we do not need to return to our starting point, then what we are looking for in Figure 9.4 on page 552 is a walk (not necessarily a circuit) that uses every edge once and only once. We call such a walk an **Euler walk.** Euler showed that even with this relaxed condition, the bridge problem still was not solvable. The general result of his argument is given in the following theorem.

✔ **TAKE NOTE**

Note that an Euler *walk* does not require that we start and stop at the same vertex, whereas an Euler *circuit* does.

Euler Walk Theorem

A connected graph contains an Euler walk if and only if the graph has two vertices of odd degree with all other vertices of even degree. Furthermore, every Euler walk must start at one of the vertices of odd degree and end at the other.

To see why this theorem is true, notice that the only places at which an Euler walk differs from an Euler circuit are the start and end vertices. If we never return to the starting vertex, only one edge meets there and the degree of the vertex is 1. If we do return, we cannot stop there. So we depart again, giving the vertex a degree of 3. Similarly, any return trip means that an additional two edges meet at the vertex. Thus the degree of the start vertex must be odd. By similar reasoning, the ending vertex must also have odd degree. All other vertices, just as in the case of an Euler circuit, must have even degree.

EXAMPLE 4 ■ An Application of Euler Walks

A photographer would like to travel across all of the roads shown on the map below. The photographer will rent a car that need not be returned to the same city, so the trip can begin in any city. Is it possible for the photographer to design a trip that traverses all of the roads exactly once?

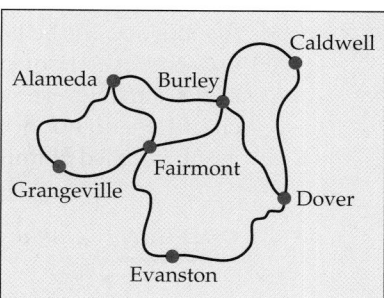

Solution

Looking at the map of roads as a graph, we see that a route that includes all of the roads but does not cover any road twice corresponds to an Euler walk of the graph. Notice that only two vertices have odd degree, the cities Alameda and Dover. Thus we know that an Euler walk exists, and so it is possible for the photographer to plan a route that travels each road once. Because (abbreviating the cities) A and D are vertices of odd degree, the photographer must start at one of these cities. With a little experimentation, we find that one Euler walk is A–B–C–D–B–F–A–G–F–E–D.

CHECK YOUR PROGRESS 4 A bicyclist wants to mountain bike through all the trails of a national park. A map of the park is shown below. Because the bicyclist will be dropped off in the morning by friends and picked up in the evening, she does not have a preference of where she begins and ends her ride. Is it possible for the cyclist to traverse all of the trails without repeating any portions of her trip?

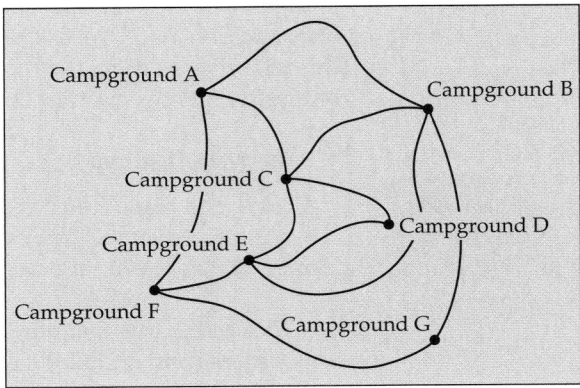

Solution   *See page S31.*

Hamiltonian Circuits

In Example 4, we tried to find a route that traversed every road once without repeating any roads. Suppose our priority is to visit cities rather than travel roads. We do not care if we use all the roads or not. Is there a route that visits each city once (without repeating any cities)?

You can verify that the route A–B–C–D–E–F–G visits each city in Example 4 just once. After visiting Grangeville, we could return to the starting city to complete the journey. In the language of graph theory, this route corresponds to a walk that uses every vertex of a (connected) graph, does not use any vertex twice, and returns to the starting vertex. Such a walk is called a **Hamiltonian circuit.** (Unlike an Euler circuit, we do not need to use every edge.) If a graph has a Hamiltonian circuit, the graph is called **Hamiltonian.**

QUESTION   *Can a graph be both Eulerian and Hamiltonian?*

Because we found the Hamiltonian Circuit A–B–C–D–E–F–G–A in the map of cities in Example 4, we know that the graph is Hamiltonian. Unfortunately, we do not have a straightforward criterion to guarantee that a graph will be Hamiltonian, but we do have the following theorem.

ANSWER   *Yes. For example, the graph shown here has an Euler circuit and a Hamiltonian circuit.*

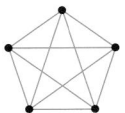

historical note

Paul A. M. Dirac (1902–1984) was born in England and earned an engineering degree in 1921 from Bristol University. He continued studies in mathematics and physics and was awarded a Ph.D. in 1926 for his thesis *Quantum Mechanics*. He became a professor of Mathematics at Cambridge, and in 1933 he shared the Nobel Prize in Physics with Erwin Schrödinger for his work in quantum mechanics. He moved to the United States in 1969 and became a professor of Physics at Florida State University shortly thereafter. ∎

Dirac's Theorem

Consider a connected graph with at least three vertices and no multiple edges. Let n be the number of vertices in the graph. If every vertex has degree of at least $n/2$, then the graph must be Hamiltonian.

We must be careful, however; if our graph does not meet the requirements of this theorem, it still might be Hamiltonian. Dirac's Theorem does not help us in this case.

EXAMPLE 5 ■ Apply Dirac's Theorem

The graph below shows the available flights of a popular airline. (An edge between two vertices in the graph means the airline has flights between the two corresponding cities.) Apply Dirac's Theorem to verify that the following graph is Hamiltonian. What does a Hamiltonian circuit represent in terms of flights?

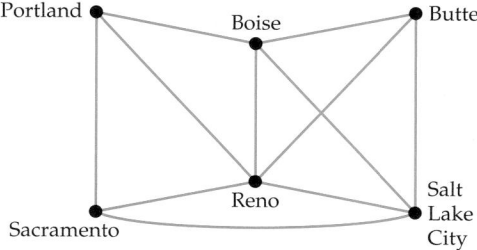

Solution

There are six vertices in the graph, so $n = 6$, and every vertex has degree of at least $n/2 = 3$. So by Dirac's Theorem, the graph is Hamiltonian. This means the graph contains a circuit that visits each vertex once and returns to the starting vertex without visiting any vertex twice. Here, a Hamiltonian circuit represents a sequence of flights that visits each city and returns to the starting city without visiting any city twice. Notice that Dirac's Theorem does not tell us how to find the Hamiltonian circuit; it just guarantees that one exists.

CHECK YOUR PROGRESS 5   A large law firm has offices in seven major cities. The firm has overnight document deliveries scheduled every day between certain offices. In the graph below, an edge between vertices indicates that there is delivery service between the corresponding offices. Use Dirac's Theorem to answer the question: Using the law firm's existing delivery service, is it possible to route a document to all the offices and return the document to its originating office without sending it through the same office twice?

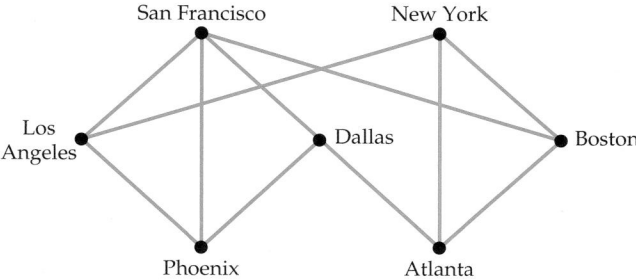

Solution   *See page S31.*

EXAMPLE 6 ■ An Application of Hamiltonian Circuits

The floor plan of an art gallery is pictured below. Draw a graph to represent the floor plan, where vertices correspond to rooms and edges correspond to doorways. Then use your graph to answer the following questions. Is it possible to take a walking tour through the gallery that visits every room and returns to the starting point without visiting any room twice? Is it possible to take a stroll that passes through every doorway without going through the same doorway twice? If so, does it matter whether we return to the starting point?

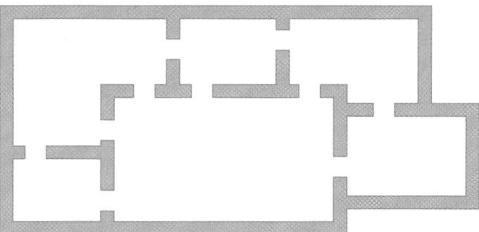

Solution

We can represent the floor plan by a graph if we let a vertex represent each room. Draw an edge between two vertices if there is a doorway between the two rooms, as shown in Figure 9.5.

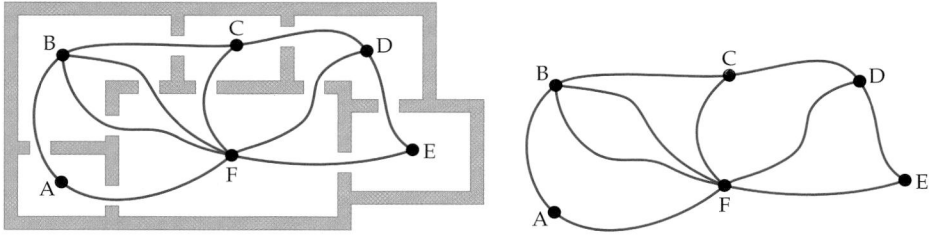

Figure 9.5   **Figure 9.6**

The graph in Figure 9.6 is equivalent to our floor plan. A walk through the gallery that visits each room just once and returns to the starting point corresponds to a visit to each vertex without visiting any vertex twice. In other words, we are looking for a Hamiltonian circuit. There are six vertices in the graph, and not all the vertices have degree of at least 3, so Dirac's Theorem does not guarantee that the graph is Hamiltonian. However, it is still possible that a Hamiltonian circuit exists. In fact, one is not too hard to find: A−B−C−D−E−F−A.

If we would like to tour the gallery and pass through every doorway once, we must find a walk on our graph that uses every edge once (and no more). Thus we are looking for an Euler walk. In the graph, two vertices have odd degree and the others have even degree. So we know that an Euler walk exists, but not an Euler circuit. Therefore, we cannot pass through each doorway once and only once if we want to return to the starting point, but we can do it if we end up somewhere else. Furthermore, we know we must start at a vertex of odd degree—either room C or room D. By trial and error, one such walk is C−B−F−B−A−F−E−D−C−F−D.

✔ **TAKE NOTE**

Recall that a Hamiltonian circuit visits each vertex once (and only once), whereas an Euler circuit uses each edge once (and only once).

CHECK YOUR PROGRESS 6   The floor plan of a warehouse is pictured below. Use a graph to represent the floor plan and answer the following questions. Is it possible to walk through the warehouse so that you pass through every doorway once but not twice? Does it matter whether you return to the starting point?

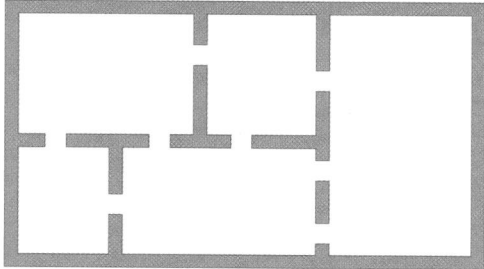

Solution   *See page S31.*

Excursion

Pen-Tracing Puzzles

You may have seen puzzles before like this one: Can you draw the following diagram without lifting your pencil from the paper, and without tracing over the same segment twice?

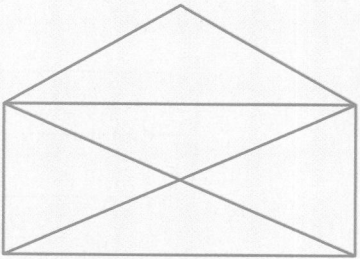

Before reading on, try it for yourself. By trial and error, you may discover a tracing that works. Even though there are several possible tracings, you may notice that only certain starting points seem to allow a complete tracing. How do we know which point to start from? How do we even know that a solution exists?

Puzzles like this, called "pen-tracing puzzles," are actually problems in graph theory. If we imagine a vertex placed wherever two lines meet or cross over each other, then we have a graph. Our task is to start at a vertex and find a path that traverses every edge of the graph, without repeating any edges. In other words, we need an Euler walk! (An Euler circuit would work as well.)

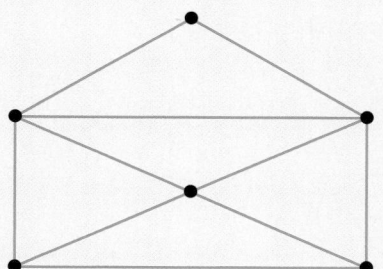

(continued)

As we learned in this section, a graph has an Euler walk only if two vertices are of odd degree and the remaining vertices are of even degree. Furthermore, the walk must start at one of the vertices of odd degree and end at the other. In the puzzle above, only two vertices are of odd degree—the two bottom corners. So we know that an Euler walk exists, and it must start from one of these two corners. Here is one possible solution.

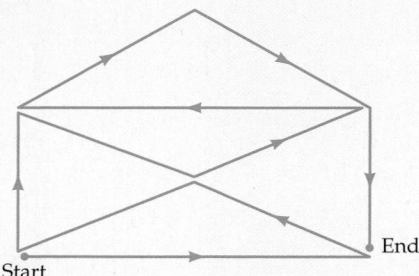

Of course, if every vertex is of even degree, we can solve the puzzle by finding an Euler circuit. If more than two vertices are of odd degree, then we know the puzzle cannot be solved.

Excursion Exercises

INSTRUCTOR NOTE
The exercises in this Excursion can be assigned as cooperative learning exercises for small groups of two or three students.

In Exercises 1–4, a pen-tracing puzzle is given. See if you can find a way to trace the shape without lifting your pen and without tracing over the same segment twice.

1.

2.

3.

4.

5. Explain why the following pen-tracing puzzle is impossible to solve.

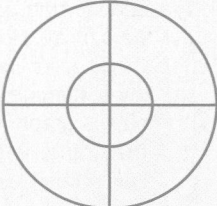

Exercise Set 9.1   (Suggested Assignment: 1–39, odd; 16, 30, 42)

In Exercises 1–4, (a) give the number of edges in the graph, (b) give the number of vertices in the graph, (c) determine the number of vertices that are of odd degree, (d) determine if the graph is connected, and (e) determine if the graph is a complete graph.

1.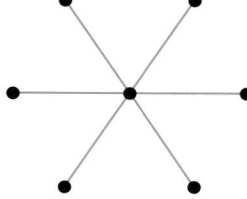

2.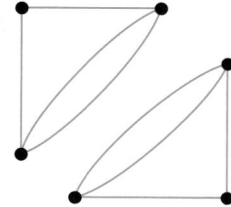

3.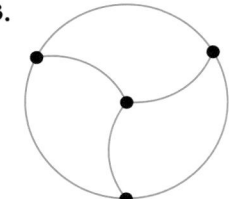

4.

5. An "X" in the table below indicates a direct train route between the corresponding cities. Draw a graph to represent this information, in which each vertex represents a city and an edge connects two vertices if there is a train route between the corresponding cities.

	Springfield	Riverside	Greenfield	Watertown	Midland	Newhope
Springfield	—		X		X	
Riverside		—		X	X	X
Greenfield	X		—	X	X	X
Watertown		X	X	—		
Midland	X	X	X		—	X
Newhope		X	X		X	—

6. The table below shows the nonstop flights offered by a small airline. Draw a graph to represent this information, where each vertex represents a city and an edge connects two vertices if there is a nonstop flight between the corresponding cities.

	Newport	Lancaster	Plymouth	Auburn	Dorset
Newport	—	no	yes	no	yes
Lancaster	no	—	yes	yes	no
Plymouth	yes	yes	—	yes	yes
Auburn	no	yes	yes	—	yes
Dorset	yes	no	yes	yes	—

In Exercises 7 and 8, a floor plan of a museum is given. Draw a graph to represent the floor plan, where each vertex represents a room and an edge connects two vertices if there is a doorway between the two rooms.

7.

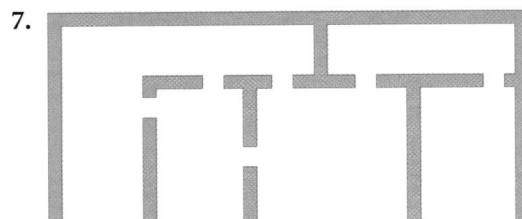

8.

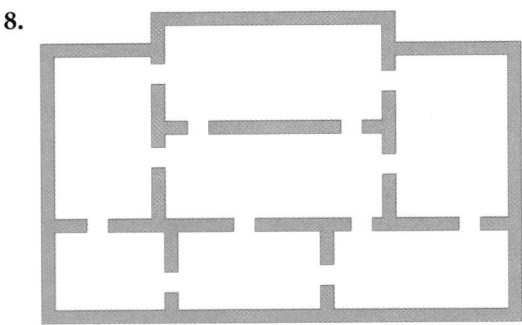

9. A group of friends is represented by the graph at the right. An edge connecting two names means that the two friends have spoken to each other in the last week.

    a. Have John and Stacy talked to each other in the last week?

    b. How many of the friends in this group has Steve talked to in the last week?

    c. Among this group of friends, who has talked to the most people in the last week?

    d. Why would it not make sense for this graph to contain a loop?

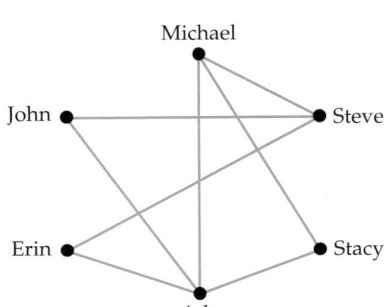

10. The local Little League baseball teams are represented by the graph at the right. An edge connecting two teams means that those teams have played a game against each other this season.

    a. Which team has played only one game this season?

    b. Which team has played the most games this season?

    c. Have any teams played each other twice this season?

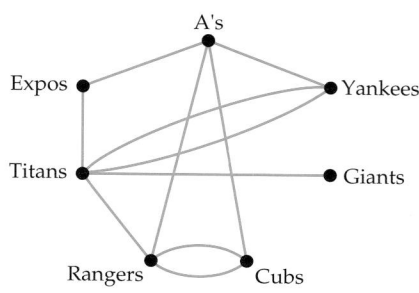

In Exercises 11–14, determine whether the two graphs are equivalent.

11.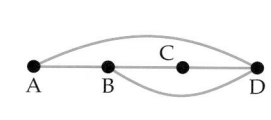

12.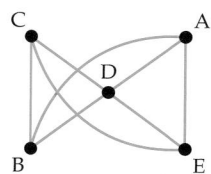

13.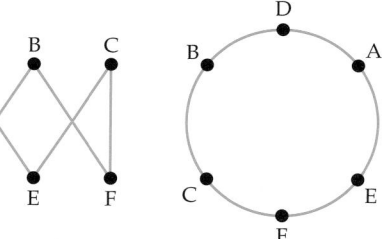

14.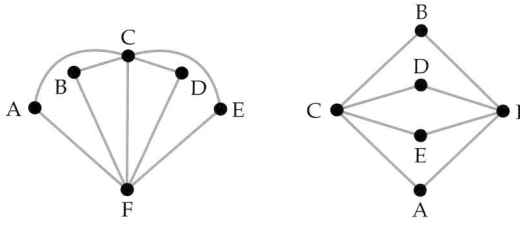

15. Explain why the following two graphs cannot be equivalent.

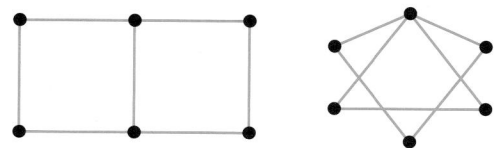

16. Label the vertices of the second graph so that it is equivalent to the first graph.

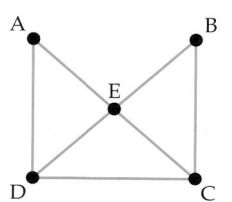 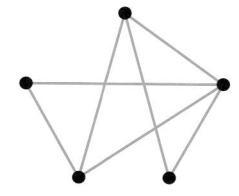

In Exercises 17–24, (a) determine if the graph is Eulerian. If it is, find an Euler circuit. If it is not, explain why. (b) If the graph does not have an Euler circuit, does it have an Euler walk? If so, find one. If not, explain why.

17. 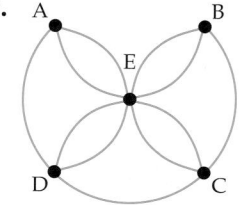   18.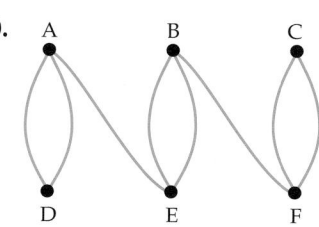

19. 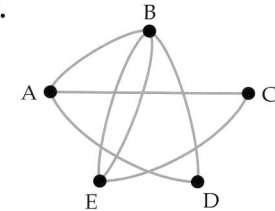   20.

21. 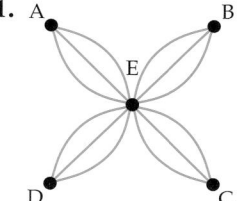   22.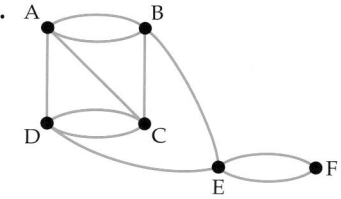

23.

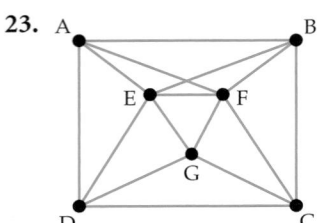

24.

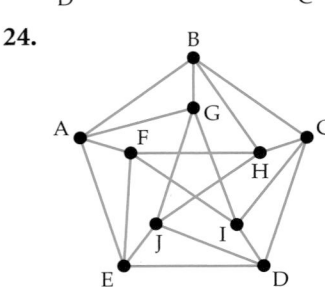

In Exercises 25 and 26, a map of a park is shown with bridges connecting islands in a river to the banks.

a. Represent the map as a graph. See Figures 9.3 and 9.4 on page 552.

b. Is it possible to take a walk that crosses each bridge once and return to the starting point without crossing any bridge twice? If not, can you do it if you do not end at the starting point? Explain how you know.

25.

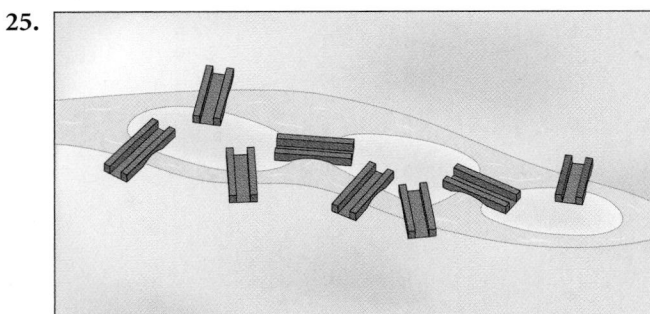

26.

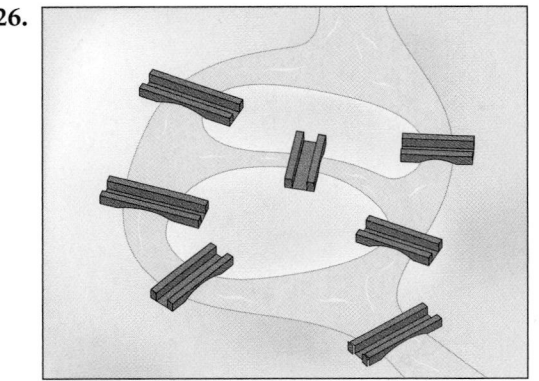

27. For the train routes given in Exercise 5, is it possible to travel along all of the train routes without traveling along any route twice? Explain how you reached your conclusion.

28. For the direct air flights given in Exercise 6, is it possible to start at one city and fly every route offered without repeating any flight if you return to the starting city? Explain how you reached your conclusion.

29. The diagram below shows the arrangement of a Habitrail cage for a pet hamster. (Plastic tubes connect different cages.) Is it possible for a hamster to travel through every tube without going through the same tube twice? If so, find a route for the hamster to follow. Can the hamster return to its starting point without repeating any tube passages?

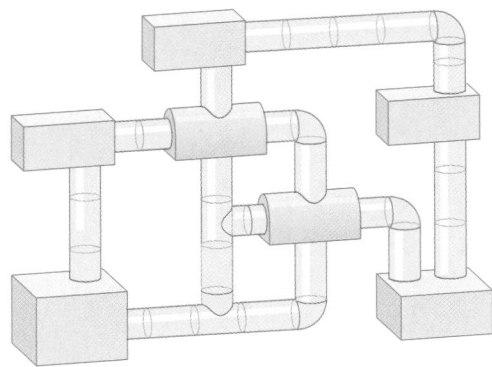

30. A subway map is shown below. Is it possible for a rider to travel the length of every subway route without repeating any segments? Justify your conclusion.

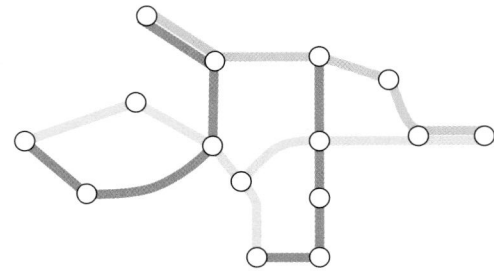

31. For the floor plan in Exercise 7, is it possible to walk through the museum and pass through each doorway without going through any doorway twice? Does it depend on whether you return to the room you started at? Justify your conclusion.

32. For the floor plan in Exercise 8, is it possible to walk through the museum and pass through each doorway without going through any doorway twice? Does it depend on whether you return to your starting point? Justify your conclusion.

In Exercises 33–36, use Dirac's Theorem to verify that the graph is Hamiltonian. Then find a Hamiltonian circuit.

33.

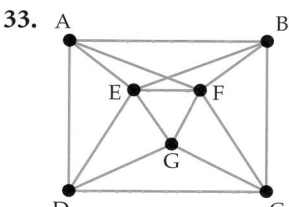

34.

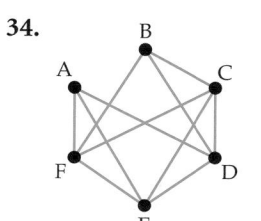

35.

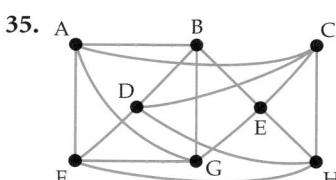

36.

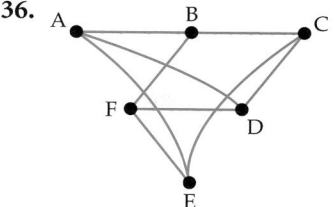

37. For the train routes given in Exercise 5, find a route that visits each city and returns to the starting city without visiting any city twice.

38. For the direct air flights given in Exercise 6, find a route that visits each city and returns to the starting city without visiting any city twice.

39. In Exercise 7, you were asked to draw a graph that represents a museum floor plan. Describe what a Hamiltonian circuit in the graph would correspond to in the museum.

40. Consider a subway map, like the one given in Exercise 30. If we draw a graph in which each vertex represents a train junction, and an edge between vertices means a train travels between those two junctions, what does a Hamiltonian circuit correspond to in regard to the subway?

Extensions

CRITICAL THINKING

41. A security officer patrolling a city neighborhood needs to drive every street each night. The officer has drawn a graph to represent the neighborhood in which the edges represent the streets and the vertices correspond to street intersections. Would the most efficient way to drive the streets correspond to an Euler circuit, a Hamiltonian circuit, or neither? (The officer must return to the starting location when finished.) Explain your answer.

42. A city engineer needs to inspect the traffic signs at each street intersection of a neighborhood. The engineer has drawn a graph to represent the neighborhood, where the edges represent the streets and the vertices correspond to street intersections. Would the most efficient route to drive correspond to an Euler circuit, a Hamiltonian circuit, or neither? (The engineer must return to the starting location when finished.) Explain your answer.

43. Is there an Euler circuit in the graph below? Is there an Euler walk? Is there a Hamiltonian circuit? Justify your answer. (You do not need to find any of the circuits or paths.)

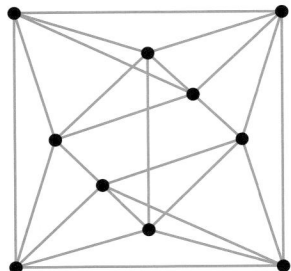

44. Is there an Euler circuit in the graph below? Is there an Euler walk? Is there a Hamiltonian circuit? Justify your answer. (You do not need to find any of the circuits or paths.)

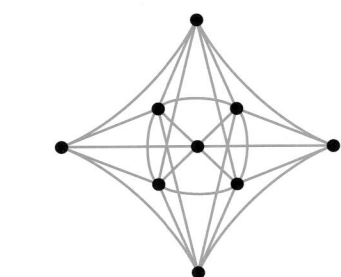

COOPERATIVE LEARNING

45. a. Draw a connected graph with six vertices that has no Euler circuits and no Hamiltonian circuits.

  b. Draw a graph with six vertices that has a Hamiltonian circuit but no Euler circuits.

  c. Draw a graph with five vertices that has an Euler circuit but no Hamiltonian circuits.

46. A map of South America is shown at the right.

  a. Draw a graph in which the vertices represent the 13 countries of South America, and two vertices are joined by an edge if the corresponding countries share a common border.

  b. Two friends are planning a driving tour of South America. They would like to drive across every border on the continent. Is it possible to plan such a route that never crosses the same border twice? What would the route correspond to on the graph?

  c. Find a route the friends can follow that will start and end in Venezuela and that crosses every border while repeating the fewest number of borders possible. *Hint:* On the graph, add multiple edges corresponding to border crossings that allow an Euler circuit.

SECTION 9.2 **Efficient Routes**

Weighted Graphs

In Section 9.1, we examined the problem of finding a path that enabled us to visit all of the cities represented on a graph, and we saw that in many cases there were a number of different paths we could use. What if we are concerned with the distances we must travel between cities? If there is more than one route that brings us through all of the cities, chances are some of the routes will involve a longer total distance than others. If we hope to minimize the number of miles we must travel in order to see the cities, we would be interested in finding the best route among all the possibilities.

We can represent this situation with a *weighted graph.* A **weighted graph** is a graph in which each edge is associated with a value, called a **weight.** The value can represent any quantity we desire. In the case of distances between cities, we can label each edge with the number of miles between the corresponding cities. (Note that the length of an edge does not necessarily correlate to its weight.) For instance, the graph in Figure 9.7 is the graph of airline flights shown in Figure 9.1, but with the weights of the edges added. We now know at a glance the distances between cities.

We previously looked for routes that visited each city and returned to the starting city; in other words, a Hamiltonian circuit. For each such circuit we find in a weighted graph, we can compute the total weight of the circuit by finding the sum of the weights of the edges we traverse. In this case, the sum gives us the total distance traveled along our route. We can then compare different routes and find the one that requires the least total distance. This is an example of a famous problem called the *traveling salesman problem.*

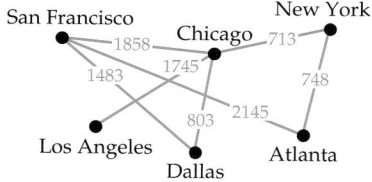

Figure 9.7

✔ **TAKE NOTE**

Remember that a Hamiltonian circuit in a graph is a closed walk that uses each vertex of the graph but does not use any vertex twice.

EXAMPLE 1 ■ Find Hamiltonian Circuits in a Weighted Graph

The table below lists the distances in miles between six popular cities that a particular airline flies to. Suppose a traveler would like to start in Chicago, visit the other five cities this airline flies to, and return to Chicago. Find three different routes that the traveler could follow, and find the total distance flown for each route.

	Chicago	New York	Washington, D.C.	Philadelphia	Atlanta	Dallas
Chicago	—	713	597	665	585	803
New York	713	—	no flights	no flights	748	1374
Washington, D.C.	597	no flights	—	no flights	544	1185
Philadelphia	665	no flights	no flights	—	670	1299
Atlanta	585	748	544	670	—	no flights
Dallas	803	1374	1185	1299	no flights	—

Solution

The various options will be simpler to analyze if we first organize the information in a graph. As in Figure 9.7, we can let each city be represented by a vertex. We will draw an edge between two vertices if there is a flight between the corresponding cities, and we will label each edge with a weight that represents the number of miles between the two cities.

✔ **TAKE NOTE**

Remember that the placement of the vertices is not important. There are many equivalent ways in which to draw the graph in Example 1.

A route that visits each city just once corresponds to a Hamiltonian circuit. Beginning at Chicago, one such circuit is Chicago–New York–Dallas–Philadelphia–Atlanta–Washington, D.C.–Chicago. By adding the weights of each edge in the circuit, we see that the total number of miles traveled is

$$713 + 1374 + 1299 + 670 + 544 + 597 = 5197$$

By trial and error, we can identify two additional routes. One is Chicago–Philadelphia–Dallas–Washington, D.C.–Atlanta–New York–Chicago. The total weight of the circuit is

$$665 + 1299 + 1185 + 544 + 748 + 713 = 5154$$

A third route is Chicago–Washington, D.C.–Dallas–New York–Atlanta–Philadelphia–Chicago. The total mileage is

$$597 + 1185 + 1374 + 748 + 670 + 665 = 5239$$

CHECK YOUR PROGRESS 1   A tourist visiting San Francisco is staying at a hotel near the Moscone Center. The tourist would like to visit five locations by bus tomorrow and then return to the hotel. The number of minutes spent traveling by bus between locations is given in the table on the following page. (N/A in the table indicates that no convenient bus route is available.) Find two different routes for the tourist to follow and compare the total travel times.

	Moscone Center	Civic Center	Union Square	Embarcardero Plaza	Fisherman's Wharf	Coit Tower
Moscone Center	—	18	6	22	N/A	N/A
Civic Center	18	—	14	N/A	33	N/A
Union Square	6	14	—	24	28	36
Embarcardero Plaza	22	N/A	24	—	N/A	18
Fisherman's Wharf	N/A	33	28	N/A	—	14
Coit Tower	N/A	N/A	36	18	14	—

Solution   *See page S32.*

Algorithms in Complete Graphs

In Example 1, the second route we found represented the smallest total distance out of the three options. Is there a way we can find the very best route to take? It turns out that this is no easy task. One method is to list every possible Hamiltonian circuit, compute the total weight of each one, and choose the smallest total weight. Unfortunately, the number of different possible circuits can be extremely large. For instance, the graph shown in Figure 9.8, with only six vertices, has 60 unique Hamiltonian circuits. If we have a graph with 12 vertices, and every vertex is connected to every other by an edge, there are almost 20 million different Hamiltonian circuits! Even by using computers, it can take too long to investigate each and every possibility.

Unfortunately, there is no known short cut for finding the best route in a weighted graph. We do, however, have two *algorithms* that we can use to find a pretty good solution to this type of problem. Both of these algorithms apply only to **complete graphs**—graphs in which every possible edge is drawn between vertices (without any multiple edges). For instance, the graph in Figure 9.8 is a complete graph with six vertices. Each algorithm can be applied to find a Hamiltonian circuit. The total weight of this circuit may not be the smallest possible, but it is often smaller than you would find by trial and error.

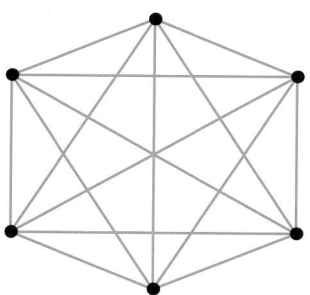

Figure 9.8

INSTRUCTOR NOTE
The number of Hamiltonian circuits in a complete graph with n vertices is $\dfrac{(n-1)!}{2}$.

✔ **TAKE NOTE**

An **algorithm** is a step-by-step procedure to accomplish a specific task.

The Greedy Algorithm

1. Choose a vertex to start at, then travel along the connected edge that has the smallest weight. (If two or more edges have the same weight, pick any one.)

2. After arriving at the next vertex, travel along the edge of smallest weight that connects to a vertex not yet visited. Continue this process until you have visited all vertices.

3. Return to the starting vertex.

EXAMPLE 2 ■ The Greedy Algorithm

Use the Greedy Algorithm to find a Hamiltonian circuit in the weighted graph shown in Figure 9.9. Start at vertex A.

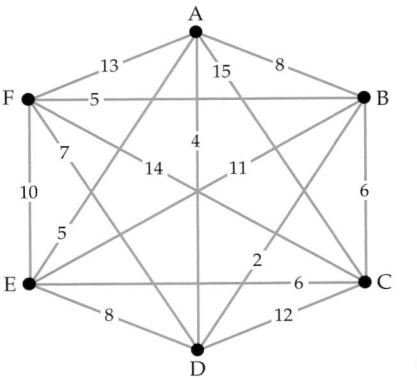

Figure 9.9

Solution

The edge at vertex A with the smallest weight is the edge to D, which has weight 4. From D, the edge with the smallest weight (besides the one we just traveled) has weight 2 and connects to B. The next edge, of weight 5, connects to vertex F. The edge from F of the least weight that has not yet been used is the edge to vertex D. However, we have already visited D, so we choose the edge with the next smallest weight, 10, to vertex E. The only remaining choice is the edge to vertex C, of weight 6, as all other vertices have been visited. Now we are at step 3 of the algorithm, so we return to starting vertex A by traveling along the edge of weight 15. Thus our Hamiltonian circuit is A−D−B−F−E−C−A, with a total weight of

$$4 + 2 + 5 + 10 + 6 + 15 = 42$$

CHECK YOUR PROGRESS 2 Use the Greedy Algorithm to find a Hamiltonian circuit starting at vertex A in the weighted graph shown below.

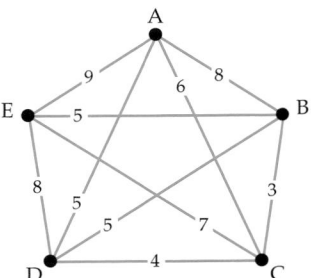

Solution *See page S32.*

The Greedy Algorithm gets its name because it has us choose the "cheapest" option at every chance we get. The next algorithm uses a different strategy to find an efficient route in a traveling salesman problem.

The Edge-Picking Algorithm

1. Mark the edge of smallest weight in the graph. (If two or more edges have the same weight, pick any one.)

2. Mark the edge of next smallest weight in the graph, as long as it does not complete a circuit and does not add a third marked edge to a single vertex.

3. Continue this process until you can no longer mark any edges. Then mark the final edge that completes the Hamiltonian circuit.

In Step 2 of the algorithm, we are instructed not to complete a circuit too early. Also, because a Hamiltonian circuit will always have exactly two edges at each vertex, we are warned not to mark an edge that would allow three edges to meet at one vertex. You can see the algorithm in action in the next example.

EXAMPLE 3 ■ **The Edge-Picking Algorithm**

Use the Edge-Picking Algorithm to find a Hamiltonian circuit in Figure 9.9.

Solution
The smallest weight appearing in the graph is 2, so we highlight the edge B–D of weight 2. The next smallest weight is 4, the weight of edge A–D.

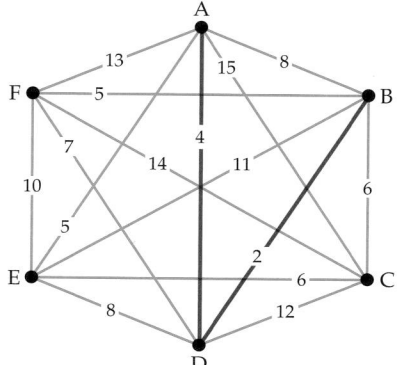

There are two edges of weight 5 (the next smallest weight), A–E and F–B, and we can highlight both of them. Next there are two edges of weight 6. However, we cannot use B–C because it would add a third marked edge to vertex B. We mark E–C.

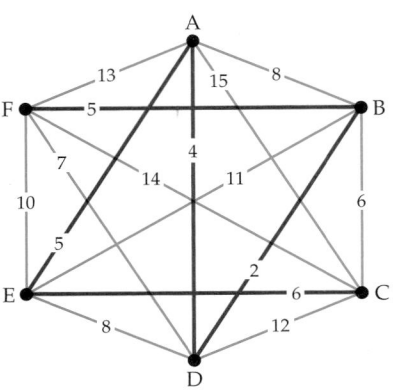

We are now at step 3 of the algorithm; any edge we mark will either complete a circuit or add a third edge to a vertex. So we mark the final edge to complete the circuit, F−C.

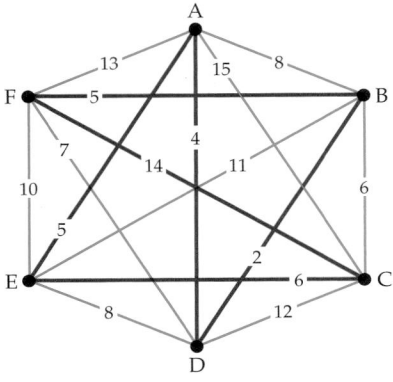

If we begin at vertex A, our Hamiltonian circuit is A−D−B−F−C−E−A. (In the reverse direction, an equivalent Hamiltonian circuit is A−E−C−F−B−D−A.) The total weight of the circuit is

$$4 + 2 + 5 + 14 + 6 + 5 = 36$$

CHECK YOUR PROGRESS 3   Use the Edge-Picking Algorithm to find a Hamiltonian circuit in the weighted graph in Check Your Progress 2.

Solution   *See page S32.*

Notice from Examples 2 and 3 that the two algorithms gave different Hamiltonian circuits, and in this case the Edge-Picking Algorithm gave the more efficient route. Is this the best route? We mentioned before that there is no known efficient method for finding the very best circuit. In fact, the Hamiltonian circuit A−D−F−B−C−E−A in Figure 9.9 has a total weight of 33, which is smaller than the weights of both routes given by the algorithms. So when we use the algorithms, we have no guarantee that we are finding the best route, but we can be sure that we are finding a reasonably good solution.

QUESTION   *Can the Greedy Algorithm or the Edge-Picking Algorithm be used to identify a route to visit the cities in Example 1?*

Math Matters   Computers and Traveling Salesman Problems

The traveling salesman problem is a long-standing mathematical problem, and has been analyzed since the 1920s by mathematicians, statisticians, and later, computer scientists. The algorithms given in this section can help us find good solutions, but they cannot guarantee that we will find the best solution.

At this point in time, the only way to find the optimal Hamiltonian circuit in a weighted graph is to find each and every possible circuit and compare the total weights of all circuits. Computers are well suited to such a task; however, as the number of vertices increases, the number of possible Hamiltonian circuits increases rapidly, and even computers are not fast enough to handle large graphs. There are so many possible circuits in a graph with a large number of vertices that finding them all could take hundreds or thousands of years, on even the fastest computers.

Computer scientists continue to improve their methods and produce more sophisticated algorithms. In 2001, four researchers (Applegate, Bixby, Chvatal, and Cook) were able to find the optimal route to visit 15,112 cities in Germany. They used a network of 110 computers for a combined computation time of 22.6 years. You can see their resulting circuit on the map below.

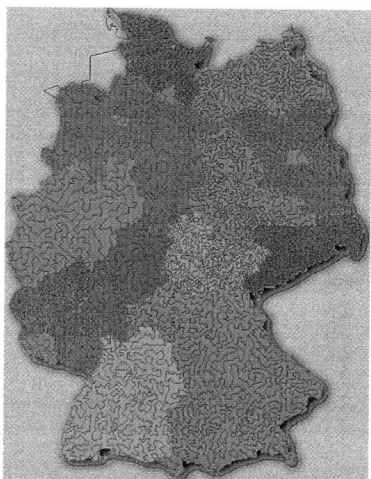

Applications of Weighted Graphs

In Example 1, we examined distances between cities. This is just one example of a weighted graph; the weight of an edge can be used to represent any quantity we like. For example, a traveler might be more interested in the cost of flights than the time

ANSWER   *No. Both algorithms apply only to complete graphs.*

or distance between cities. If we labeled each edge of the graph in Example 1 with the cost of traveling between the two cities, the total weight of a Hamiltonian circuit would be the total travel cost of the trip.

EXAMPLE 4 ■ An Application of the Greedy and Edge-Picking Algorithms

The cost of flying between various European cities is shown in the following table. Use both the Greedy Algorithm and the Edge-Picking Algorithm to find a low-cost route that visits each city just once and starts and ends in London. Which route is more economical?

	London, England	Berlin, Germany	Paris, France	Rome, Italy	Madrid, Spain	Vienna, Austria
London, England	—	$325	$160	$280	$250	$425
Berlin, Germany	$325	—	$415	$550	$675	$375
Paris, France	$160	$415	—	$495	$215	$545
Rome, Italy	$280	$550	$495	—	$380	$480
Madrid, Spain	$250	$675	$215	$380	—	$730
Vienna, Austria	$425	$375	$545	$480	$730	—

Solution

First we draw a weighted graph with vertices representing the cities and each edge labeled with the price of the flight between the corresponding cities.

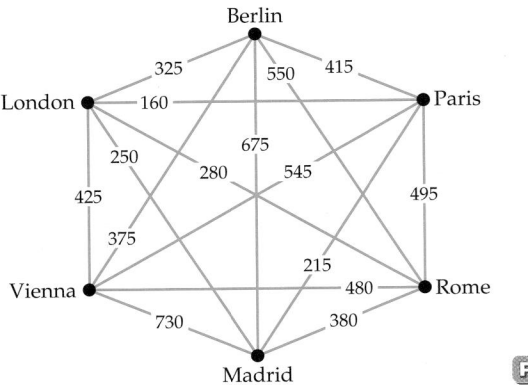

To use the Greedy Algorithm, start at London and travel along the edge with the smallest weight, 160, to Paris. The edge of smallest weight leaving Paris is the edge to Madrid. From Madrid, the edge of smallest weight (that we have not already traversed) is the edge to London, of weight 250. However, we cannot use this edge, as it would bring us to a city we have already seen. We can take the next-smallest-weight edge to Rome. We cannot yet return to London, so the next available edge is to Vienna, then to Berlin, and finally back to London.

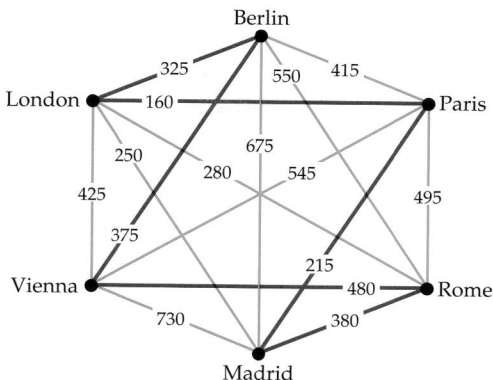

The total weight of the edges, and thus the total airfare for the trip, is

$$160 + 215 + 380 + 480 + 375 + 325 = \$1935$$

If we use the Edge-Picking Algorithm, the edges with the smallest weights that we can highlight are London–Paris and Madrid–Paris. The edge of next smallest weight is that of weight 250, but we cannot use this edge as it would complete a circuit. We can take the edge of next smallest weight, 280, from London to Rome. We cannot take the edge of next smallest weight, 325, because it would add a third edge to the London vertex, but we can take the edge Vienna–Berlin of weight 375. We must skip the edges of weights 380, 415, and 425, but can take the edge of weight 480, which is the Vienna–Rome edge. There are no more edges we can mark that will meet the requirements of the algorithm, so we mark the last edge to complete the circuit, Berlin–Madrid.

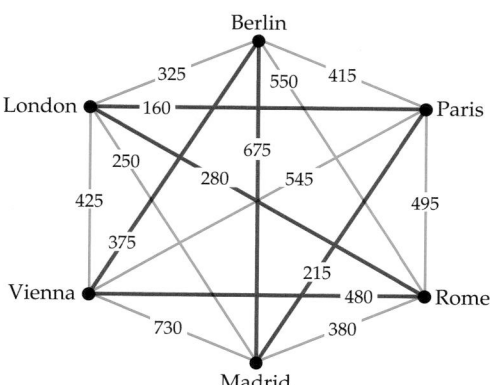

The resulting route is London–Paris–Madrid–Berlin–Vienna–Rome–London, for a total cost of

$$160 + 215 + 675 + 375 + 480 + 280 = \$2185$$

(We could also travel this route in the reverse order.)

CHECK YOUR PROGRESS 4   Susan needs to mail a package at the post office, pick up several items at the grocery store, return a rented video, and make a deposit at her bank. The estimated driving time, in minutes, between each of these locations is given in the table below.

	Home	Post office	Grocery store	Video rental store	Bank
Home	—	14	12	20	23
Post office	14	—	8	12	21
Grocery store	12	8	—	17	11
Video rental store	20	12	17	—	18
Bank	23	21	11	18	—

Use both of the algorithms from this section to design routes for Susan to follow that will help minimize her total driving time. Assume she must start from home and return home when her errands are done.

Solution   *See page S32.*

A wide variety of problems are actually traveling salesman problems in disguise, and can be analyzed using the algorithms from this section. An example follows.

EXAMPLE 5 ■ **An Application of the Edge-Picking Algorithm**

A toolmaker needs to use one machine to create four different tools. The toolmaker needs to make adjustments to the machine before starting each different tool. However, since the tools have parts in common, the amount of adjustment required depends on which tool the machine was previously used to create. The table below lists the estimated time (in minutes) to readjust the machine from one tool to another. The machine is currently configured for Tool A, and should be returned to that state when all the tools are finished.

	Tool A	Tool B	Tool C	Tool D
Tool A	—	25	6	32
Tool B	25	—	18	9
Tool C	6	18	—	15
Tool D	32	9	15	—

Use the Edge-Picking Algorithm to determine a sequence for the creation of tools.

Solution

Draw a weighted graph in which each vertex represents a tool configuration of the machine, and the weight of each edge is the number of minutes required to adjust the machine from one tool to another.

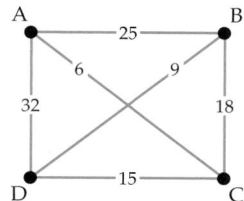

Using the Edge-Picking Algorithm, we first choose edge A–C, of weight 6. The next smallest weight is 9, or edge D–B, followed by edge D–C of weight 15. We end by completing the circuit with edge A–B. The final circuit, of total weight 25 + 9 + 15 + 6 = 55, is A–B–D–C–A. So the machine starts with tool A, is reconfigured for tool B, then for tool D, then for tool C, and finally returned to the settings used for tool A. (Notice that we can equivalently follow this sequence in the reverse order: from Tool A to C, to D, to B, and back to A.)

CHECK YOUR PROGRESS 5   Businesses often network their various computers. One option is to run wires from a central hub to each computer individually; another is to wire one computer to the next, and that one to the next, and so on until you return to the first computer. Thus the computers are all wired in a large loop. Suppose a company wishes to use the latter method, and the lengths of wire (in feet) required between computers are given in the table below.

	Computer A	Computer B	Computer C	Computer D	Computer E	Computer F	Computer G
Computer A	—	43	25	6	28	30	45
Computer B	43	—	26	40	37	22	25
Computer C	25	26	—	20	52	8	50
Computer D	6	40	20	—	30	24	45
Computer E	28	37	52	30	—	49	20
Computer F	30	22	8	24	49	—	41
Computer G	45	25	50	45	20	41	—

Use the Edge-Picking Algorithm to determine how the computers should be networked if the business wishes to use the least amount of wire possible.

Solution   *See page S32.*

Excursion

Extending the Greedy Algorithm

When we create a Hamiltonian circuit in a graph, it is a closed loop. We can start at any vertex, follow the path, and arrive back at the starting vertex. For instance, if we use the Greedy Algorithm to create a Hamiltonian circuit starting at vertex A, we are actually creating a circuit that could start at any vertex in the circuit.

If we use the Greedy Algorithm and start from different vertices, will we always get the same result? Try it on Figure 9.10 at the left. If we start at vertex A, we get the circuit A–C–E–B–D–A, with a total weight of 26 (see Figure 9.11). However, if we start at vertex B, we get B–E–D–C–A–B, with a total weight of 18 (see Figure 9.12). Even though we found the second circuit by starting at B, we could use the same circuit starting at A, namely A–B–E–D–C–A, or, equivalently, A–C–D–E–B–A. This circuit has a smaller total weight than the one we found by starting at A.

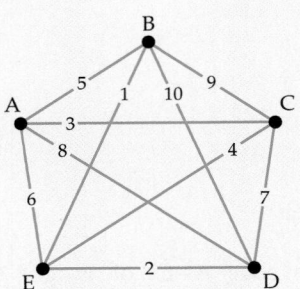

Figure 9.10

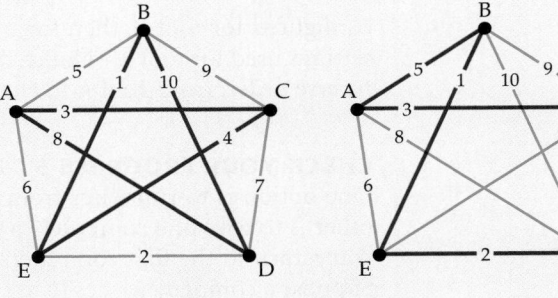

Figure 9.11   **Figure 9.12**

In other words, it may pay to try the Greedy Algorithm starting from different vertices, even when we know that we actually want to start from a particular vertex. To be thorough, we can extend the Greedy Algorithm by using it at each and every vertex, generating several different circuits. We then choose among these the circuit with the least total weight, and start at the vertex we want. The following exercises ask you to apply this approach for yourself.

Excursion Exercises

1. Continue investigating Hamiltonian circuits in Figure 9.10 by using the Greedy Algorithm starting at vertices C, D, and E. Then compare the various Hamiltonian circuits to identify the one with the smallest total weight.

2. Use the Greedy Algorithm and the weighted graph at the top of the following page to generate a Hamiltonian circuit starting from each vertex. Then compare the different circuits to find the one of smallest total weight.

(continued)

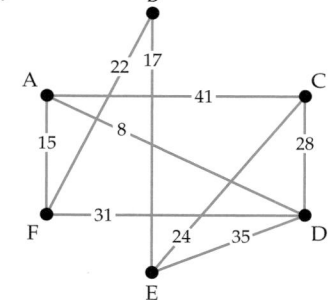

3. Use the Edge-Picking Algorithm to find a Hamiltonian circuit in the graph in Exercise 2. How does the weight of this circuit compare with the weights of the circuits found in Exercise 2?

Exercise Set 9.2

(Suggested Assignment: 1–23, odd; 8, 12, 24)

In Exercises 1–4, use trial and error to find two Hamiltonian circuits of different total weights, starting at vertex A in the weighted graph. Compute the total weight of each circuit.

1.

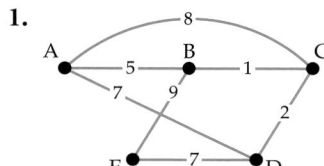

2.

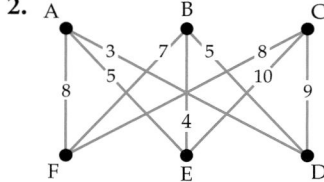

3.

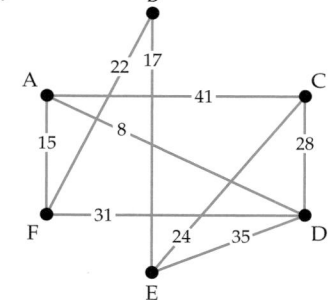

4.

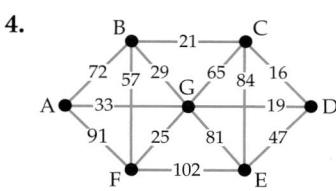

In Exercises 5–8, use the Greedy Algorithm to find a Hamiltonian circuit starting at vertex A in the weighted graph.

5.

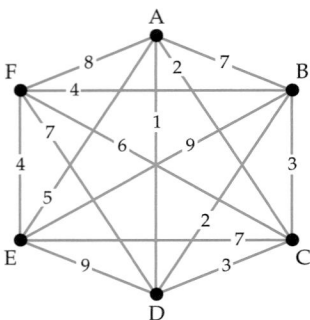

6.

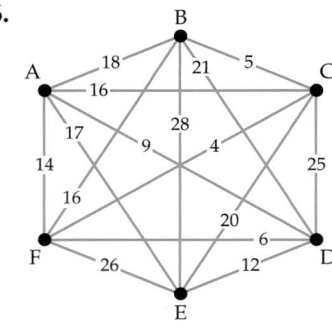

7.

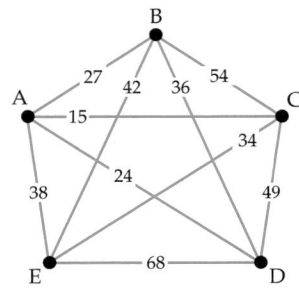

8.

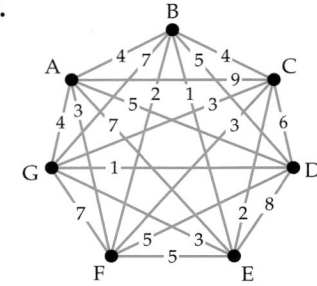

In Exercises 9–12, use the Edge-Picking Algorithm to find a Hamiltonian circuit in the indicated graph.

9. Graph in Exercise 5   **10.** Graph in Exercise 6

11. Graph in Exercise 7   **12.** Graph in Exercise 8

13. A company representative lives in Louisville, Kentucky and needs to visit offices in five different Indiana cities over the next few days. The representative wants to drive between cities and return to Louisville at the end of the trip. The estimated driving times, in hours, between cities are given in the table below. Represent the driving times by a weighted graph. Use the Greedy Algorithm to design an efficient route for the representative to follow.

	Louisville	Bloomington	Fort Wayne	Indianapolis	Lafayette	Evansville
Louisville	—	3.6	6.4	3.2	4.9	3.1
Bloomington	3.6	—	4.5	1.3	2.4	3.4
Fort Wayne	6.4	4.5	—	3.3	3.0	8.0
Indianapolis	3.2	1.3	3.3	—	1.5	4.6
Lafayette	4.9	2.4	3.0	1.5	—	5.0
Evansville	3.1	3.4	8.0	4.6	5.0	—

14. A tourist is staying in Toronto, Canada and would like to visit four other Canadian cities by train. The visitor wants to go from one city to the next and return to Toronto while minimizing the total travel distance. The distances between cities, in kilometers, are given in the table on the following page. Represent the distances between the cities using a weighted graph. Use the Greedy Algorithm to plan a route for the tourist.

	Toronto	Kingston	Niagara Falls	Ottawa	Windsor
Toronto	—	259	142	423	381
Kingston	259	—	397	174	623
Niagara Falls	142	397	—	562	402
Ottawa	423	174	562	—	787
Windsor	381	623	402	787	—

15. Use the Edge-Picking Algorithm to design a route for the company representative in Exercise 13.

16. Use the Edge-Picking Algorithm to design a route for the tourist in Exercise 14.

17. Nicole wants to tour Asia. She will start and end her journey in Tokyo and visit Hong Kong, Bangkok, Seoul, and Beijing. The airfares available to her between cities are given in the table. Use the Greedy Algorithm to find a low-cost route.

	Tokyo	Hong Kong	Bangkok	Seoul	Beijing
Tokyo	—	$845	$1275	$470	$880
Hong Kong	$845	—	$320	$515	$340
Bangkok	$1275	$320	—	$520	$365
Seoul	$470	$515	$520	—	$225
Beijing	$880	$340	$365	$225	—

18. The prices to travel between five cities in Colorado by bus are given in the table below. Represent the travel costs between cities using a weighted graph. Use the Greedy Algorithm to find a low-cost route that starts and ends in Boulder and visits each city.

	Boulder	Denver	Colorado Springs	Grand Junction	Durango
Boulder	—	$16	$25	$49	$74
Denver	$16	—	$22	$45	$72
Colorado Springs	$25	$22	—	$58	$59
Grand Junction	$49	$45	$58	—	$32
Durango	$74	$72	$59	$32	—

19. Use the Edge-Picking Algorithm to find a low-cost route for the traveler in Exercise 17.

20. Use the Edge-Picking Algorithm to find a low-cost bus route in Exercise 18.

21. Brian needs to visit the pet store, the shopping mall, the local farmers' market, and the pharmacy. His estimated driving times (in minutes) between the locations are given in the table below. Use the Greedy Algorithm and the Edge-Picking Algorithm to find two possible routes, starting and ending at home, that will help Brian minimize his total travel time.

	Home	Pet Store	Shopping mall	Farmers' market	Pharmacy
Home	—	18	27	15	8
Pet store	18	—	24	22	10
Shopping mall	27	24	—	20	32
Farmers' market	15	22	20	—	22
Pharmacy	8	10	32	22	—

22. A bike messenger downtown needs to deliver packages to five different buildings and return to the courier company. The estimated biking times (in minutes) between the buildings are given in the table below. Use the Greedy Algorithm and the Edge-Picking Algorithm to find two possible routes for the messenger to follow that will help minimize the total travel time.

	Courier company	Prudential building	Bank of America building	Imperial Bank building	GE Tower	Design Center
Courier company	—	10	8	15	12	17
Prudential building	10	—	10	6	9	8
Bank of America building	8	10	—	7	18	20
Imperial Bank building	15	6	7	—	22	16
GE Tower	12	9	18	22	—	5
Design Center	17	8	20	16	5	—

23. A research company has a large supercomputer that is used by different teams for a variety of computational tasks. In between each task, the software must be reconfigured. The time required depends on which tasks follow which, as some settings are shared by different tasks. The times (in minutes) required to reconfigure the machine from one task to another are given in the table on the following page. Use the Greedy Algorithm and the Edge-Picking Algorithm to find a time-efficient sequence in which to assign the tasks to the computer. The software configuration must start and end in the home state.

	Home state	Task A	Task B	Task C	Task D
Home state	—	35	15	40	27
Task A	35	—	30	18	25
Task B	15	30	—	35	16
Task C	40	18	35	—	32
Task D	27	25	16	32	—

24. A small office wishes to network its six computers in one large loop (see Check Your Progress 5). The lengths of cable, in meters, required between machines are given in the table below. Use the Edge-Picking Algorithm to find an efficient cable configuration in which to network the computers.

	Computer A	Computer B	Computer C	Computer D	Computer E	Computer F
Computer A	—	10	22	9	15	8
Computer B	10	—	12	14	16	5
Computer C	22	12	—	14	9	16
Computer D	9	14	14	—	7	15
Computer E	15	16	9	7	—	13
Computer F	8	5	16	15	13	—

Extensions

CRITICAL THINKING

25. Assign weights to the edges of the complete graph shown below so that the Edge-Picking Algorithm gives a circuit of lower total weight than the circuit given by the Greedy Algorithm. For the Greedy Algorithm, begin at vertex A.

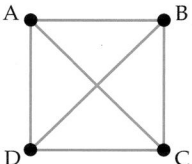

26. Assign weights to the edges of the complete graph in Exercise 25 so that the Greedy Algorithm gives a circuit of lower total weight than the circuit given by the Edge-Picking Algorithm. For the Greedy Algorithm, begin at vertex A.

27. Assign weights to the edges of the complete graph in Exercise 25 so that there is a circuit of lower total weight than the circuits given by both the Greedy Algorithm (beginning at vertex A) and the Edge-Picking Algorithm.

COOPERATIVE LEARNING

28. Form a team of five classmates and, using a map, identify a driving route that will visit each home, return to the starting home, and that you feel would be the smallest total driving distance. Next determine the driving distances between the homes of the classmates. Represent these distances in a weighted graph, and then use the Edge-Picking Algorithm to determine a driving route that would visit each home and return to the starting home. Did the algorithm find a more efficient route than your first route?

SECTION 9.3   # Planarity and Euler's Formula

Planarity

A puzzle that was posed some time ago goes something like this: Three utility companies each need to run pipes to three houses. Can they do so without crossing over each other's pipes at any point? It might be easier to think about the puzzle visually, as shown in Figure 9.13. Go ahead and try to draw pipes connecting each utility company to each house without letting any pipes cross over each other.

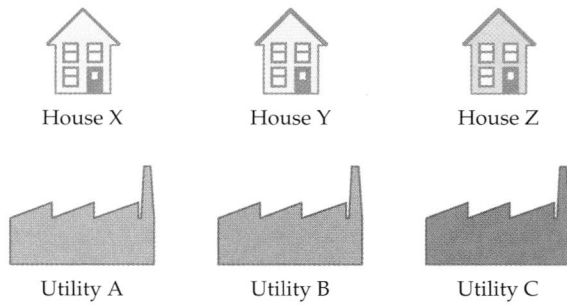

House X   House Y   House Z

Utility A   Utility B   Utility C

Figure 9.13

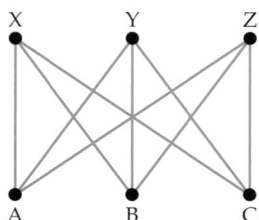

Figure 9.14 *The Utilities Graph*

One way to approach the puzzle is to express the situation in terms of a graph. Each of the houses and utility companies will be represented by a vertex, and we will draw an edge between two vertices if a pipe needs to run from one building to the other. If we were not worried about pipes crossing, we could easily draw a solution, as in Figure 9.14.

To solve the puzzle, we need to draw an equivalent graph in which no edges cross over each other. Such a graph is called a *planar graph*.

Definition of a Planar Graph

A **planar graph** is a graph that can be drawn so that no edges intersect each other (except at vertices).

✔ **TAKE NOTE**

A graph can be planar even if it
has edges that cross over each
other; for a graph to be planar,
we simply require that the graph
can be drawn in at least one
way in which the edges do
not intersect.

If the graph is drawn in such a way that no edges cross, we say that we have a
planar drawing of the graph.

EXAMPLE 1 ■ Identify a Planar Graph

Show that the graph below is planar.

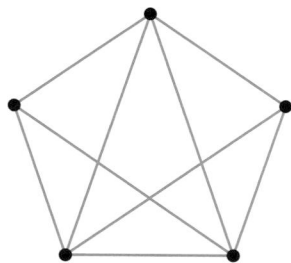

Solution

As the graph appears currently, several edges intersect each other. However, we can
redraw the graph in an equivalent form in which no edges touch except at vertices
by redrawing the two shaded edges shown below. To verify that the second graph is
equivalent to the first, we can label the vertices and check that the edges join the
same vertices in each graph. Because the given graph is equivalent to a graph whose
edges do not intersect, the graph is planar.

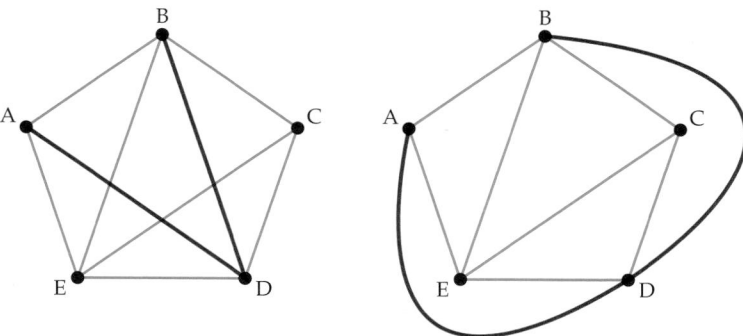

CHECK YOUR PROGRESS 1   Show that the following graph is planar.

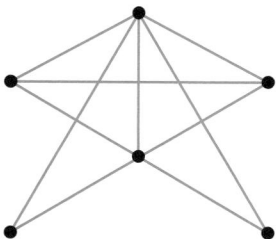

Solution   *See page S33.*

Returning to the puzzle of connecting the utilities to the houses (Figure 9.13), how can we determine whether the graph in Figure 9.14 is planar? If we attempt to redraw the graph, we will find it difficult to do without having edges cross over each other. It turns out that, in fact, the graph in Figure 9.14 is not planar. To see why this is true, notice that the graph is Hamiltonian, and one Hamiltonian circuit is A–X–B–Y–C–Z–A. If we redraw the graph so that this circuit is drawn in a loop (see Figure 9.15), then we need to add the edges A–Y, B–Z, and C–X. All three of these edges connect opposite vertices. We can draw only one of these edges inside the loop; otherwise two edges would cross. This means that the other two edges must be drawn outside the loop, but as you can see in Figure 9.16, those two edges would then have to cross. Thus the graph in Figure 9.14, which we will refer to as the *Utilities Graph,* is by definition not planar, and so the utilities puzzle is not solvable.

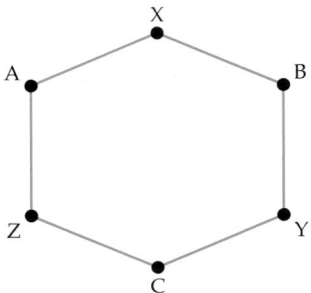

Figure 9.15

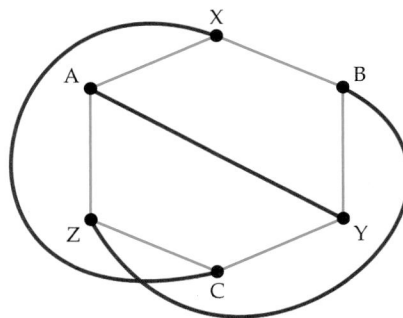

Figure 9.16

Generally speaking, it can be difficult to determine if a graph is planar, especially if the graph has a large number of edges. Of course, if you can find a planar drawing of a graph, then you know it is planar. If not, however, the graph may or may not be planar.

One strategy we can use to show that a graph is *not* planar is to find a portion of the graph that we know by itself is not planar. The Utilities Graph in Figure 9.14 is a common pattern to watch for. Another graph that is not planar is the complete graph with five vertices, denoted K_5, shown in Figure 9.17. (We can prove that K_5 is not planar by using a strategy similar to the strategy used for the Utilities Graph.)

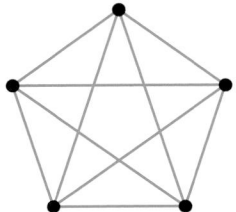

Figure 9.17 *The complete graph K_5*

What we are specifically looking for is a **subgraph**—a graph whose edges and vertices come from the given graph. Then the strategy is given by the following theorem.

> **Theorem**
>
> If a graph *G* has a subgraph that is not planar, then *G* is also not planar. In particular, if *G* contains the Utilities Graph or K_5 as a subgraph, *G* is not planar.

EXAMPLE 2 ■ A Non-Planar Graph

Show that the following graph is not planar.

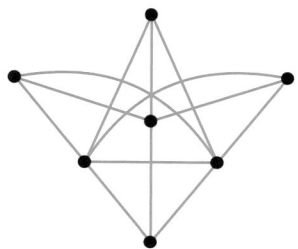

Solution

In the figure below, we have highlighted edges connecting the top six vertices. If we consider the highlighted edges and attached vertices as a subgraph, we can verify that the subgraph is the Utilities Graph. (The graph is slightly distorted compared with the version shown in Figure 9.14, but it is equivalent.) By the preceding theorem, we know that the graph is not planar.

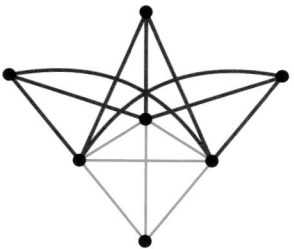

CHECK YOUR PROGRESS 2   Show that the following graph is not planar.

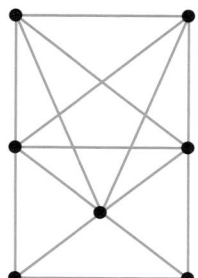

Solution   *See page S33.*

✔ TAKE NOTE

The figure below shows how the highlighted edges of the graph in Example 2 can be transformed into the Utilities Graph.

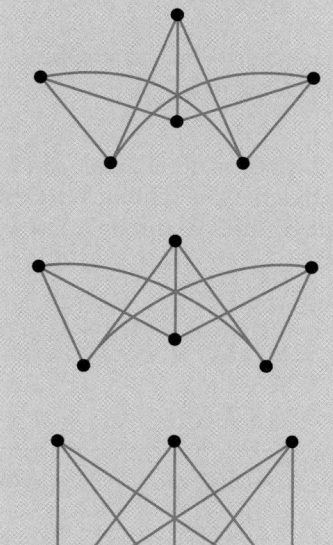

We can expand this strategy by considering *contractions* of a subgraph. A **contraction** of a graph is formed by "shrinking" an edge until the two vertices it connects come together and blend into one. If, in the process, the graph is left with any multiple edges, we merge them into one. The process is illustrated below.

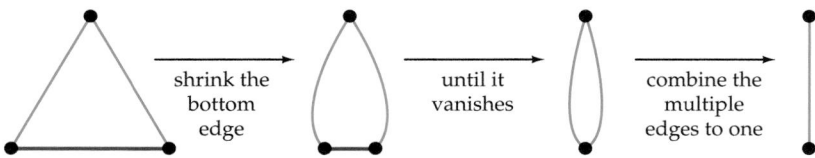

QUESTION   *Is the first graph below a contraction of the second graph?*

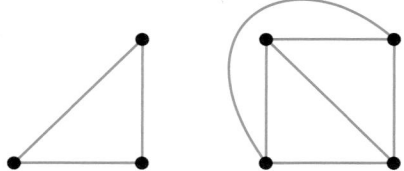

If we consider contractions, it turns out that the Utilities Graph and K_5 serve as building blocks for nonplanar graphs. In fact, it was proved in 1930 that any nonplanar graph will always have a subgraph that is the Utilities Graph or K_5, or a subgraph that can be contracted to the Utilities Graph or K_5. We can then expand our strategy, as given by the following theorem.

Nonplanar Graph Theorem

A graph is nonplanar if and only if it has the Utilities Graph or K_5 as a subgraph, or it has a subgraph that can be contracted to the Utilities Graph or K_5.

This gives us a definite test that will determine whether or not a graph can be drawn in such a way as to avoid crossing edges. (Note that the theorem includes the case in which the entire graph can be contracted to the Utilities Graph or K_5, as we can consider the graph a subgraph of itself.)

ANSWER   *Yes. Contract the edge colored red, then eliminate the multiple edges (and straighten out the curved edge).*

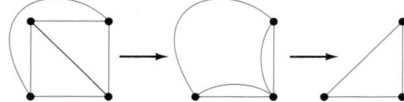

EXAMPLE 3 ■ A Non-Planar Graph

Show that the graph below is not planar.

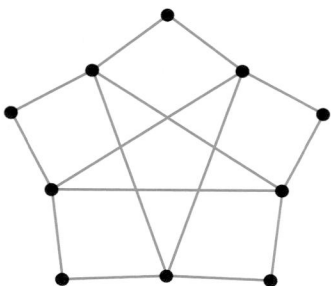

Solution

Notice that the graph looks similar to K_5. In fact, we can contract some edges and make the graph look like K_5. Choose a pair of adjacent outside edges and contract one of them, as shown in the figure below.

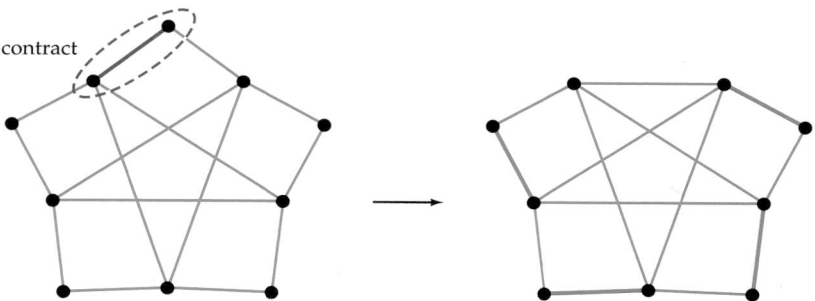

If we similarly contract the four edges colored green above, we arrive at the graph of K_5.

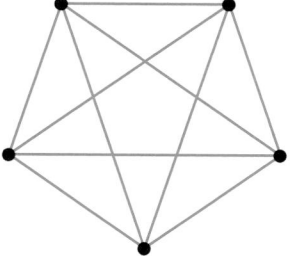

We were able to contract our graph to K_5, so by the Nonplanar Graph Theorem, the given graph is not planar.

CHECK YOUR PROGRESS 3 Show that the graph below is not planar.

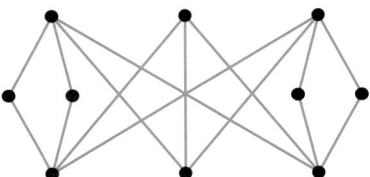

Solution   *See page S33.*

Math Matters   Planar Graphs on Circuit Boards

Trying to draw a graph as a planar drawing has many practical applications. For instance, the design of circuit boards used in computers and other electronic components depends on wires connecting different components without touching each other elsewhere. Circuit boards can be very complex, but in effect they require that collections of wires be arranged in a planar drawing of a planar graph. If this is impossible, special connections can be installed that allow one wire to "jump" over another without touching. Or, both sides of the board can be used, or sometimes more than one board is used, and the boards are then connected by wires. In effect, the graph is spread out among different surfaces.

Euler's Formula

Euler noticed a connection between various features of planar graphs. In addition to edges and vertices, he looked at *faces* of a graph. In a planar graph, the edges divide the graph into different regions called **faces**. The region surrounding the graph, or the exterior, is also considered a face, called the **infinite face**. (See the figure at the left.) The following relationship, called Euler's Formula, is always true.

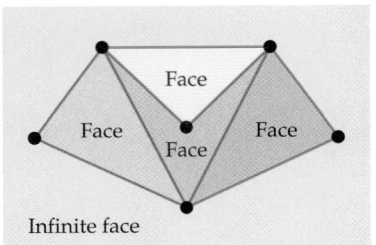

Euler's Formula

In a connected planar graph drawn with no intersecting edges, let v be the number of vertices, e the number of edges, and f the number of faces. Then $v + f = e + 2$.

EXAMPLE 4 ■ **Verify Euler's Formula in a Graph**

Count the number of edges, vertices, and faces in the planar graph below, and then verify Euler's Formula.

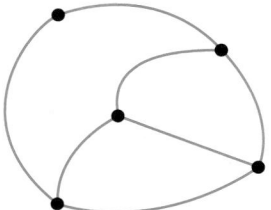

Solution

There are seven edges, five vertices, and four faces (counting the infinite face) in the graph. Thus $v + f = 5 + 4 = 9$ and $e + 2 = 7 + 2 = 9$, so $v + f = e + 2$, as Euler's Formula predicts.

CHECK YOUR PROGRESS 4   Verify Euler's Formula for the planar graph below.

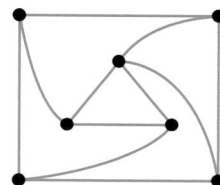

Solution   *See page S33.*

Excursion

The Five Regular Convex Polyhedra

A **regular polyhedron** is a three-dimensional object in which every face is identical. Specifically, each face is a regular polygon (meaning that each edge has the same length). In addition, the same number of faces must meet at each vertex, or corner, of the object. Here we will discuss convex polyhedra, which means that any line joining one vertex to another is entirely contained within the object. (In other words, there are no indentations.)

It was proven long ago that only five objects fit this description—the tetrahedron, the cube, the octahedron, the dodecahedron, and the icosahedron. One way mathematicians determined that there were only five such polyhedra was to visualize these three-dimensional objects in a two-dimensional way. We will use the cube to demonstrate.

(continued)

Imagine a standard cube, but with the edges made of wire and the faces empty. Now suspend the cube above a flat surface and place a light above the top face of the cube. If the light is shining downward, a shadow is created on the flat surface. This is called a *projection* of the cube.

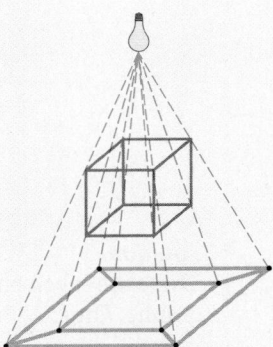

The shadow can be considered a graph. (Remember that we can redraw the graph if we wish.) Notice that each corner, or vertex, of the cube corresponds to a vertex of the graph, and each face of the cube corresponds to a face of the graph. The exception is the top face of the cube—its projection overlaps the entire graph. As a convention, we will identify the top face of the cube with the infinite face in the projected graph. Thus the three-dimensional cube is identified with the graph shown below.

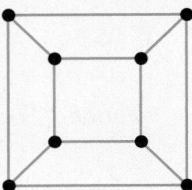

Because the same number of faces must meet at each vertex of a polyhedron, the same number of faces, and so the same number of edges, must meet at each vertex of the graph. We also know that each face of a polyhedron has the same shape. Although the corresponding faces of the graph are distorted, we must have the same number of edges around each face (including the infinite face). In addition, Euler's Formula must hold, because the projection is always a planar graph.

Thus we can investigate different polyhedra by looking at their features in two-dimensional graphs. It turns out that only five planar graphs satisfy these requirements, and the polyhedra that these graphs are the projections of are precisely the five pictured in Figure 9.18. The following exercises ask you to investigate some of the relationships between polyhedra and their projected graphs.

Excursion Exercises

1. The tetrahedron in Figure 9.18 consists of four faces, each of which is an equilateral triangle. Draw the graph that results from a projection of the tetrahedron.

2. The following graph is the projection of one of the polyhedra in Figure 9.18. Identify the polyhedron it represents by comparing features of the graph to features of the polyhedron.

tetrahedron

cube

octahedron

dodecahedron

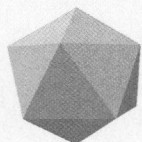

icosahedron

Figure 9.18

(continued)

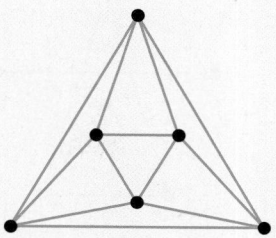

3. a. If we form a graph by a projection of the dodecahedron in Figure 9.18, what will the degree of each vertex be?

b. The dodecahedron has 12 faces, each of which has five edges. Use this information to determine the number of edges in the projected graph.

c. Use Euler's Formula to determine the number of vertices in the projected graph, and hence in the dodecahedron.

4. a. Give a reason why the graph below cannot be the projection of a *regular* convex polyhedron.

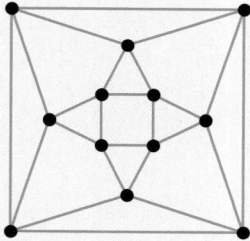

b. If the above graph is the projection of a three-dimensional convex polyhedron, how many faces does the polyhedron have?

c. Describe, in as much detail as you can, additional features of the polyhedron corresponding to the graph. If you are feeling adventurous, sketch a polyhedron that this graph could be a projection of!

Exercise Set 9.3   (Suggested Assignment: 1–23, odd; 14, 27, 28)

In Exercises 1–8, show that the graph is planar by finding a planar drawing.

1.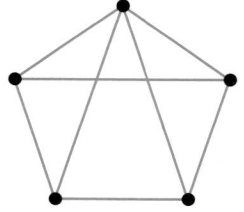

2.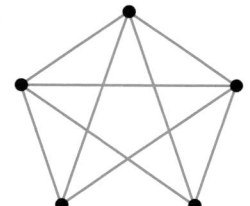

3.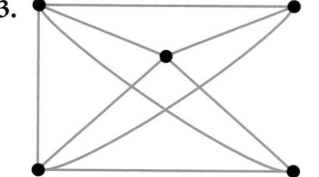

4.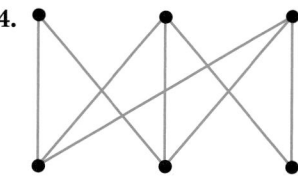

5.

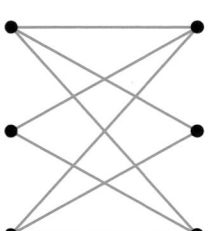

6.

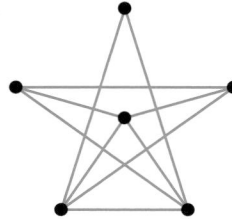

7.

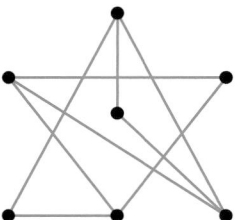

8.

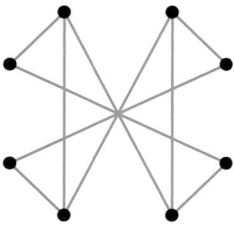

In Exercises 9–12, show that the graph is *not* planar.

9.

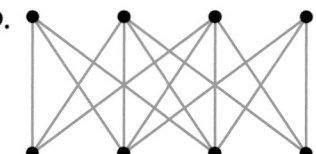

10.

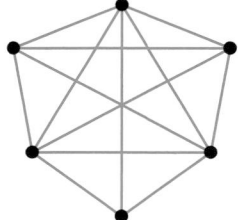

11.

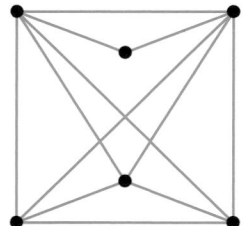

12.

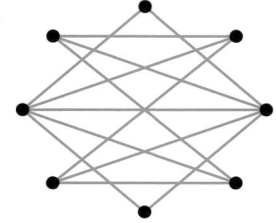

13. Show that the following graph contracts to K_5.

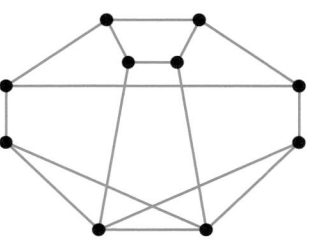

14. Show that the following graph contracts to the Utilities Graph.

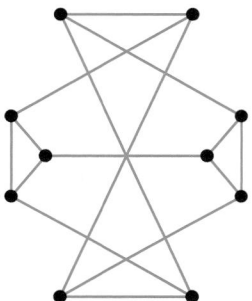

In Exercises 15 and 16, show that the graph is not planar by finding a subgraph whose contraction is the Utilities Graph or K_5.

15.

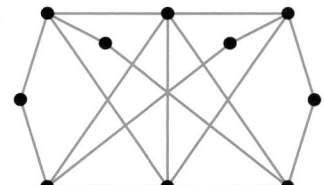

16.

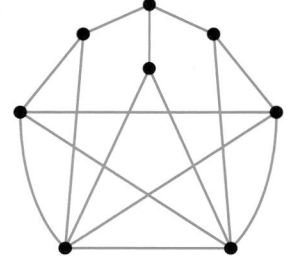

In Exercises 17–22, count the number of vertices, edges, and faces, and then verify Euler's Formula for the given graph.

17. **18.**

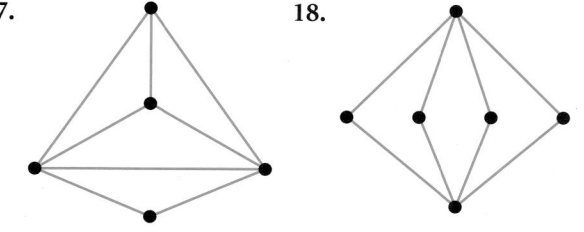

19. **20.**

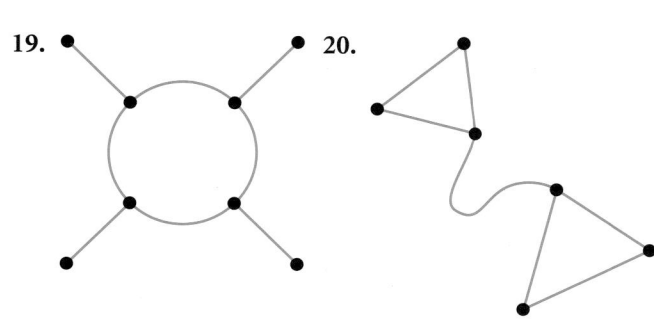

21. **22.**

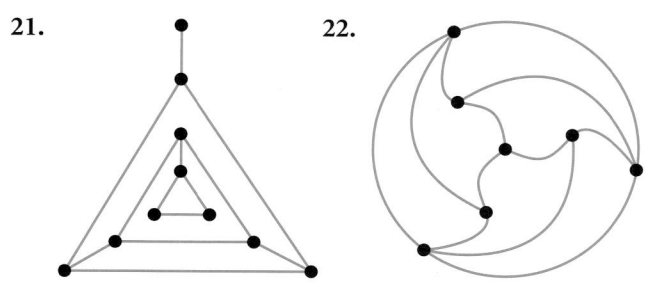

23. If a planar drawing of a graph has 15 edges and eight vertices, how many faces does the graph have?

24. If a planar drawing of a graph has 100 vertices and 50 faces, how many edges are in the graph?

Extensions

CRITICAL THINKING

25. In a planar drawing of a graph, each face is bordered by a circuit. Make a planar drawing of a graph that is equivalent to the graph shown, in which the faces are bordered by the same circuits as the given graph but where the face bordered by A−B−C−G−F−A is the infinite face.

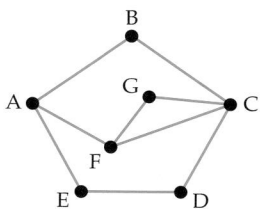

26. Make a planar drawing of a graph that is equivalent to the graph in Exercise 25, in which the faces are bordered by the same circuits as the graph in Exercise 25 but where the face bordered by C−F−G−C is the infinite face.

27. Sketch a planar graph (without multiple edges or loops) in which every vertex has degree 3.

28. Sketch a planar graph (without multiple edges or loops) in which every vertex has degree 4.

29. Explain why it is not possible to draw a planar graph that contains three faces and has the same number of vertices as edges.

30. If a planar drawing of a graph has twice as many edges as vertices, find a relationship between the number of faces and the number of vertices.

EXPLORATIONS

31. Every planar graph has what is called a *dual graph.* To form the dual graph, start with a planar drawing of the graph. With a different color pen, draw a dot in each face (including the infinite face). These will be the vertices of the dual graph. Now, for every edge in the original graph, draw a new edge that crosses over it and connects the two new vertices in the faces on each side of the original edge. The resulting dual graph is always itself a planar graph. (Note that it may contain multiple edges.) The procedure is illustrated below.

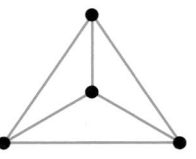

The original graph

Draw dots in each face, including the infinite face.

Connect adjacent faces with edges between the new vertices that cross each original edge.

The dual graph

a. Draw the dual graph of each of the planar graphs below.

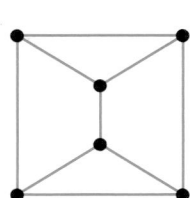 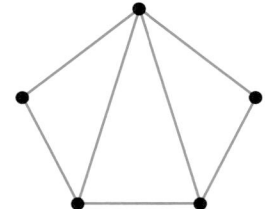 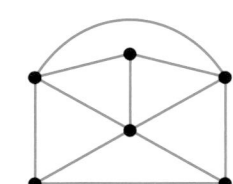

b. For each of the graphs in part a, count the number of faces, edges, and vertices in the original graph and compare your results with the numbers in the dual graph. What do you notice? Can you explain why this will always be true?

c. Start with your dual graph of the first graph in part a, and find *its* dual. What do you notice?

Figure 9.19

Coloring Maps

A map of South America is shown in Figure 9.19. Notice that each country is colored so that no two bordering countries are the same color. This is easy to accomplish with a large number of colors, but what is the least number of colors we would need to color the countries in such a way?

It was conjectured that four colors would always be enough, as no one had ever found a map that could not be colored as described with four or fewer colors. Many mathematicians attempted to prove that this would always be the case, but they were unsuccessful for many years. Finally, in 1977, the so-called *Four-Color Theorem* was proved.

EXAMPLE 1 ■ Coloring a Map

Color the map of South America in Figure 9.19 using only four colors such that no two neighboring countries are the same color.

Solution

We do not have a systematic way to go about coloring the countries, so we must use trial and error. One possible coloring is shown below.

CHECK YOUR PROGRESS 1   A map of some of the countries of western Europe is shown below. Color the map using only four colors such that no two neighboring countries are the same color.

Solution   *See page S33.*

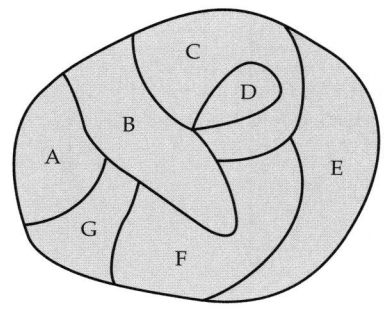

Figure 9.20

Maps Become Graphs

So what does coloring maps have to do with graph theory? In their attempts to prove the Four-Color Theorem, mathematicians first converted the map coloring question into a graph theory question. We will use a simple hypothetical map to illustrate.

Suppose the map in Figure 9.20 shows the countries, labeled as letters, of a continent. We will assume that no country is split into more than one piece, and countries that touch just at a corner point will not be considered neighbors. We can represent each country by a vertex, placed anywhere within the boundary of that country. We will then connect two vertices with an edge if the two corresponding countries are neighbors—that is, if they share a common boundary. The result is shown in Figure 9.21.

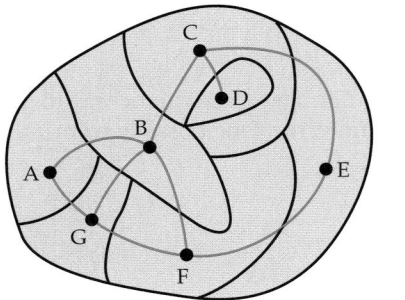

Figure 9.21

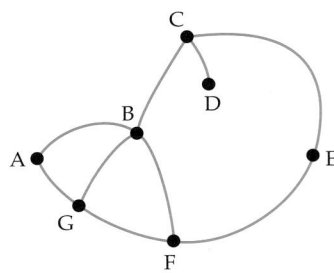

Figure 9.22

If we erase the boundaries of the countries, we are left with the graph in Figure 9.22. The resulting graph will always be a planar graph, because the edges simply connect neighboring countries. Our map coloring question then becomes: Can we give each vertex of the graph in Figure 9.22 a color such that no two vertices connected by an edge share the same color? How many different colors will be required? If this can be accomplished using four colors, for instance, we will say that the graph is **4-colorable.** In fact, the graph in Figure 9.22 is actually *3-colorable.* One possible coloring is given in Figure 9.23.

We can now formally state the Four-Color Theorem.

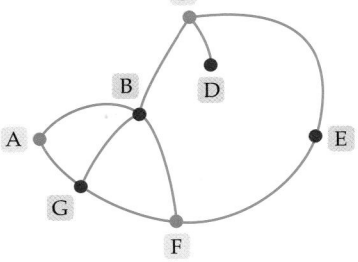

Figure 9.23

Four-Color Theorem

Every planar graph is 4-colorable.

QUESTION   *The graph shown at the right requires five colors if we wish to color it such that no edge joins two vertices of the same color. Does this contradict the Four-Color Theorem?*

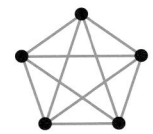

EXAMPLE 2 ■ Represent a Map as a Graph

The fictional map below shows the boundaries of countries on a rectangular continent. Represent the map as a graph, and then find a coloring of the graph using the least number of colors possible.

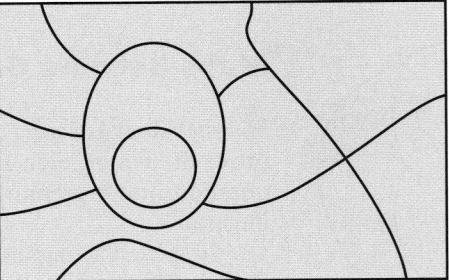

INSTRUCTOR NOTE
The following can be interesting to mention to students: The Four-Color Theorem does not hold true if we were to consider a map on a torus rather than on a plane or sphere. It has been proven that up to 7 colors may be required to color a map drawn on a torus. For a 2-holed torus, a maximum of 8 colors may be required. In general, the formula $\left\lfloor \frac{1}{2}\left(7 + \sqrt{48n + 1}\right) \right\rfloor$ gives the maximum number of colors required for an *n*-holed torus, where $\lfloor \ \rfloor$ denotes the *floor* function. The formula was discovered by Percy Heawood in 1890.

ANSWER   *No. This graph is not a planar graph, so the Four-Color Theorem does not apply.*

Solution

First draw a vertex in each country and then connect two vertices with an edge if the corresponding countries are neighbors. Now try to color the vertices of the resulting graph so that no edge connects two vertices of the same color. We know we will need at least two colors, so one strategy is to simply pick a starting vertex, give it a color, and then assign colors to the connected vertices one by one. Try to reuse the same colors, and only use a new color when there is no other option. For this graph we will need four colors. (The Four-Color Theorem guarantees that we will not need more than that.) To see why, notice that the one vertex colored green below connects to a ring of five vertices. Three different colors are required to color the five-vertex ring, and the green vertex connects to all these, so it must be colored a fourth color.

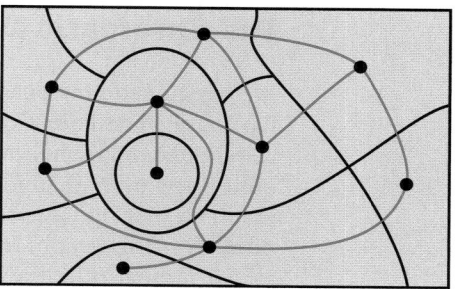

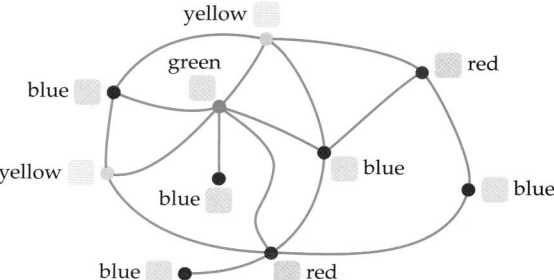

CHECK YOUR PROGRESS 2 Represent the fictional map of countries below as a graph, and then determine whether the graph is 2-colorable, 3-colorable, or 4-colorable by finding a suitable coloring of the graph.

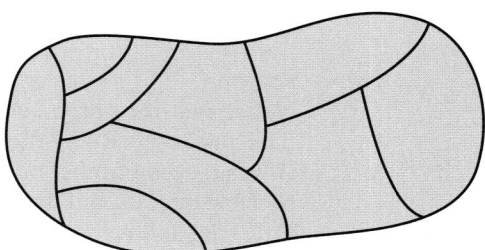

Solution   See page S33.

MathMatters   **Proving the Four-Color Theorem**

The Four-Color Theorem can be stated in a simple, short sentence, but proving it is anything but simple. The theorem was finally proved in 1976 by Wolfgang Haken and Kenneth Appel, two mathematicians at the University of Illinois. Mathematicians had long hunted for a short, elegant proof, but it turned out that the proof had to wait for the advent of computers to help sift through the many possible arrangements that can occur. The final proof used over 1000 hours of computer time (surely it would be less now!), and on paper came to several hundred pages that included some 10,000 diagrams.

The Chromatic Number of a Graph

We mentioned previously that representing a map as a graph always results in a planar graph. The Four-Color Theorem guarantees that we need only four colors to color a *planar* graph; however, if we wish to color a non-planar graph, we may need quite a few more than four colors. The smallest number of colors needed to color a graph so that no edge connects vertices of the same color is called the **chromatic number** of the graph. In general, there is no efficient method of finding the chromatic number of a graph, but we do have a theorem that can tell us if a graph is 2-colorable.

> **2-Colorable Graph Theorem**
>
> A graph is 2-colorable if and only if it has no circuits that consist of an odd number of vertices.

EXAMPLE 3 ■ Determine the Chromatic Number of a Graph

Find the chromatic number of the Utilities Graph from Section 9.3.

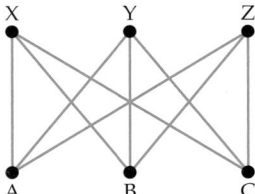

Solution
Notice that the graph contains circuits such as A−Y−C−Z−B−X−A with six vertices and A−Y−B−X−A with four vertices. Any circuit we find, in fact, seems to involve an even number of vertices. It is difficult to determine if we have looked at all possible circuits, but our observations suggest that the graph may be 2-colorable. A little trial and error confirms this if we simply color vertices A, B, and C one color and the remaining vertices another. Thus the Utilities Graph has a chromatic number of 2.

CHECK YOUR PROGRESS 3 Determine whether the following graph is 2-colorable.

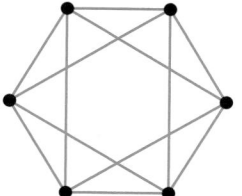

Solution   *See page S33.*

Applications of Graph Coloring

Determining the chromatic number of a graph and finding a corresponding coloring of the graph can solve a wide assortment of practical problems. One common application is in scheduling meetings or events. This is best shown by example.

EXAMPLE 4 ■ A Scheduling Application of Graph Coloring

Eight different school clubs want to schedule meetings on the last day of the semester. Some club members, however, belong to more than one of these clubs, so clubs that share members cannot meet at the same time. How many different time slots are required so that all members can attend all meetings? Clubs that have a member in common are indicated with an "X" in the table below.

	Ski club	Student government	Debate club	Honor society	Student newspaper	Community outreach	Campus democrats	Campus republicans
Ski club	—	X		X			X	X
Student government	X	—	X	X	X			
Debate club		X	—	X		X		X
Honor society	X	X	X	—	X	X		
Student newspaper		X		X	—	X	X	
Community outreach			X	X	X	—	X	X
Campus democrats	X				X	X	—	
Campus republicans	X		X			X		—

Solution

We can represent the given information by a graph. Each club can be represented by a vertex, and an edge will connect two vertices if the corresponding clubs have at least one common member.

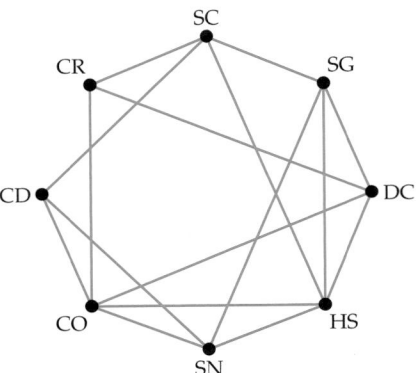

We now must choose time slots for the meetings so that two clubs that share members do not meet at the same time. In the graph, this means that two clubs that are connected by an edge cannot meet simultaneously. If we let a color correspond to a time slot, then we simply need to find a coloring of the graph that uses the least number of colors possible. The graph is not 2-colorable, because we can find circuits of odd length. However, by trial and error, we can find a 3-coloring. One example is shown below. So the chromatic number of the graph is 3, which means we need only three different time slots. Red vertices correspond to the clubs that will meet during the first time slot, green vertices to the clubs that will meet during the second time slot, and blue vertices to the clubs that will meet during the third time slot.

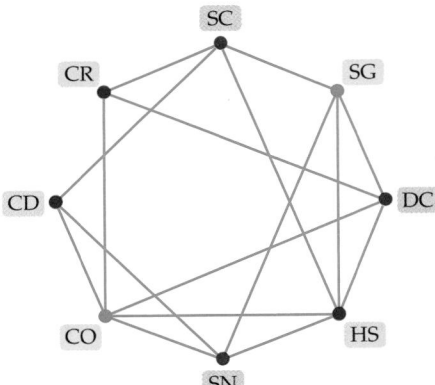

CHECK YOUR PROGRESS 4   Six friends are taking a film history course and, because they procrastinated, need to view several films the night before the final exam. They have rented a copy of each film on DVD, and they have a total of three DVD players in different dorm rooms. If each film is 2 hours long and they start watching at 8:00 P.M., how soon can they all be finished watching the required films? Can you make up a viewing schedule for the friends?

Film A needs to be viewed by Brian, Chris, and Damon.

Film B needs to be viewed by Allison and Fernando.

Film C needs to be viewed by Damon, Erin, and Fernando.

Film D needs to be viewed by Brian and Erin.

Film E needs to be viewed by Brian, Chris, and Erin.

Solution   *See page S33.*

EXAMPLE 5 ▪ A Scheduling Application of Graph Coloring

Five classes at an elementary school have arranged a tour at a zoo where the students get to feed the animals.

Classroom 1 wants to feed the elephants, giraffes, and hippos.

Classroom 2 wants to feed the monkeys, rhinos, and elephants.

Classroom 3 wants to feed the monkeys, deer, and sea lions.

Classroom 4 wants to feed the parrots, giraffes, and polar bears.

Classroom 5 wants to feed the sea lions, hippos, and polar bears.

If the zoo only allows animals to be fed once a day by one class of students, can the tour be accomplished in two days? If not, how many days will be required?

Solution

No animal is listed more than twice in the tour list, so you may be tempted to say that only two days will be required. However, to get a better picture of the problem, we can represent the situation with a graph. Use a vertex to represent each classroom, and connect two vertices with an edge if the corresponding classrooms want to feed the same animal. Then we can try to find a 2-coloring of the graph, where a different color represents a different day at the zoo.

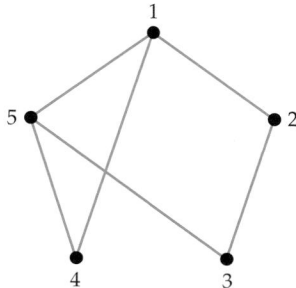

Notice that the graph contains a circuit, 1−4−5−1, comprised of three vertices. This circuit will require three colors, but the remaining vertices will not need more than that. So the chromatic number of the graph is 3; one possible coloring is given on the following page. Using this coloring, three days are required at the zoo. On the first day classrooms 2 and 5, represented by the blue vertices, will visit the zoo; on the second day classrooms 1 and 3, represented by the red vertices, will visit the zoo; and on the third day classroom 4, represented by the purple vertex, will visit the zoo.

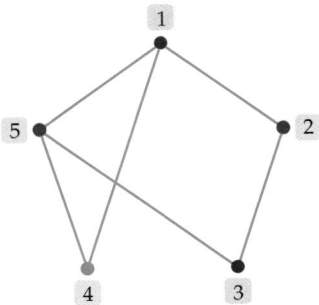

CHECK YOUR PROGRESS 5 Several delis in New York City have arranged deliveries to various buildings at lunchtime. The buildings' managements do not want more than one deli showing up at a particular building in one day, but the delis would like to deliver as often as possible. If they decide to agree on a delivery schedule, how many days will be required before each deli can return to the same building?

Deli A delivers to the Empire State Building, the Statue of Liberty, and Rockefeller Center.

Deli B delivers to the Chrysler Building, the Empire State Building, and the New York Stock Exchange.

Deli C delivers to the New York Stock Exchange, the American Stock Exchange, and the United Nations Building.

Deli D delivers to New York City Hall, the Chrysler Building, and Rockefeller Center.

Deli E delivers to Rockefeller Center, New York City Hall, and the United Nations Building.

Solution   *See page S34.*

Excursion

Modeling Traffic Lights with Graphs

Have you ever watched the cycles that traffic lights go through while you were waiting for a red light to turn green? Some intersections have lights that go through several stages, to allow all the different lanes of traffic to proceed safely.

Ideally, each stage of a traffic light cycle should allow as many lanes of traffic to proceed through the intersection as possible. We can design a traffic light cycle by modeling an intersection with a graph. Figure 9.24 shows a 3-way intersection where two

(continued)

2-way roads meet. Each direction of traffic has turn lanes, with left-turn lights. There are six different directions in which vehicles can travel, as indicated in the graph, and we have labeled each possibility with a letter.

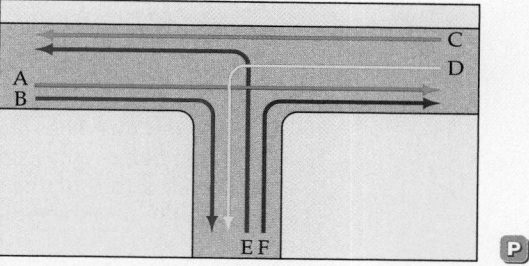

Figure 9.24

We can represent the traffic patterns with a graph; each vertex will represent one of the six possible traffic paths, and we will draw an edge between two vertices if the corresponding paths would allow vehicles to collide. The result is the graph shown in Figure 9.25. Because we do not want to allow vehicles to travel simultaneously along routes on which they could collide, any vertices connected by an edge can only allow traffic to move during different parts of the light cycle. We can represent each portion of the cycle by a color. Our job then is to color the graph using the fewest number of colors possible.

There is no 2-coloring of the graph because we have a circuit of length 3: A−D−E−A. We can, however, find a 3-coloring. One possibility is given in Figure 9.26.

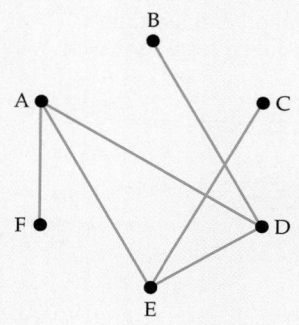

Figure 9.25

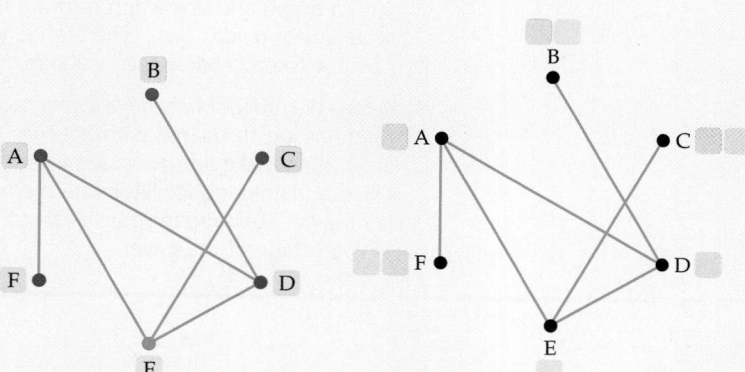

Figure 9.26   **Figure 9.27**

A 3-coloring of the graph means that the traffic lights at the intersection will have to go through a 3-stage cycle. One stage will allow the traffic routes corresponding to the red vertices to proceed, the next stage will let the paths corresponding to the blue vertices to proceed, and finally the third stage will start path E, colored green, moving.

While safety requires three stages for the lights, we can refine the design to allow more traffic to travel through the intersection. Notice that at the third stage, only one route, path E, is scheduled to be moving. However, there is no harm in allowing path B to move at the same time, since it is a right turn that doesn't conflict with route E. We could also allow path F to proceed at the same time. Adding these additional paths corresponds to adding colors to the graph in Figure 9.26. We do not want to use more than three colors, but we can add a second color to some of the vertices while maintaining

(continued)

the requirement that no edge can connect two vertices of the same color. The result is shown in Figure 9.27. Notice that the vertices in the triangular circuit A–D–E–A can only be assigned a single color, but the remaining vertices can handle two colors.

In summary, our design allows traffic paths A, B, and C to proceed during one stage of the cycle, paths C, D, and F during another, and paths B, E, and F during the third stage.

Excursion Exercises

1. A one-way road ends at a two-way street. The intersection and the different possible traffic routes are shown in the accompanying figure. The one-way road has a left-turn light. Represent the traffic routes with a graph and use graph coloring to determine the minimum number of stages required for a light cycle.

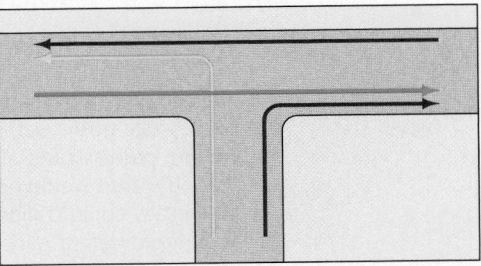

2. A one-way road intersects a two-way road in a four-way intersection. Each direction has turn lanes and left-turn lights. Represent the various traffic routes with a graph and use graph coloring to determine the minimum number of stages required for a light cycle. Then refine your design to allow as much traffic as possible to proceed at each stage of the cycle.

3. A two-way road intersects another two-way road in a four-way intersection. One road has left-turn lanes with left-turn lights, but on the other road cars are not allowed to make left turns. Represent the various traffic routes with a graph and use graph coloring to determine the minimum number of stages required for a light cycle. Then refine your design to allow as much traffic as possible to proceed at each stage of the cycle.

Exercise Set 9.4   (Suggested Assignment: 1–25, odd; 16, 18, 26, 29)

In Exercises 1–4, a fictional map of the countries of a continent is given. Use four colors to color the graph so that no two bordering countries share the same color.

1.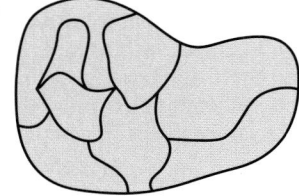

2.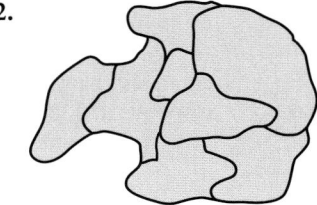

3.

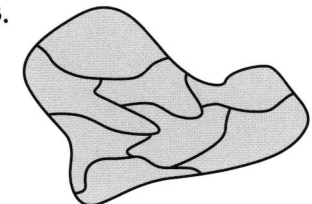

4.

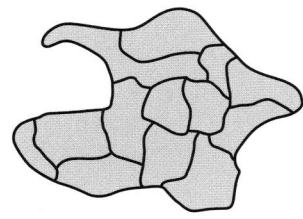

In Exercises 5–10, represent the map by a graph and then find a coloring of the graph that uses the fewest number of colors possible.

5. The map in Exercise 1

6. The map in Exercise 2

7. The map in Exercise 3

8. The map in Exercise 4

9. Western portion of the U.S.

10. Counties of New Hampshire

In Exercises 11–16, show that the graph is 2-colorable by finding a 2-coloring. If the graph is not 2-colorable, explain why.

11.

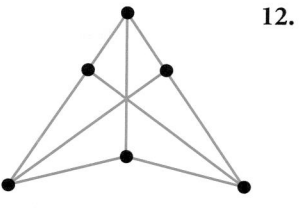

12.

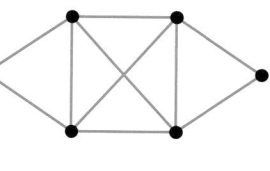

13.

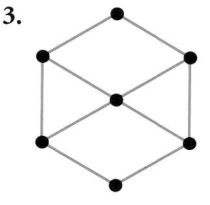

14.

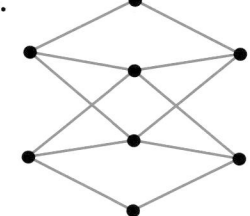

15.

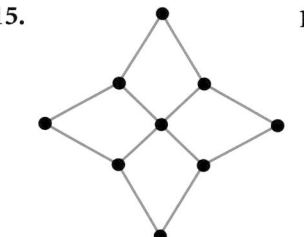

16.

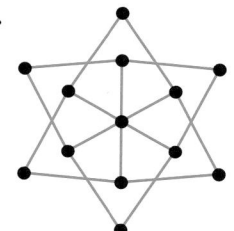

In Exercises 17–20, determine (by trial and error) the chromatic number of the graph.

17.

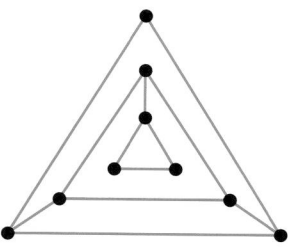

18.

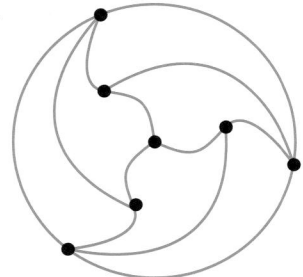

19.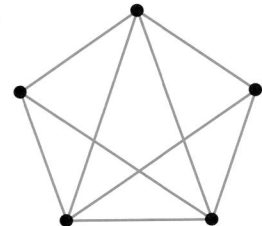

20.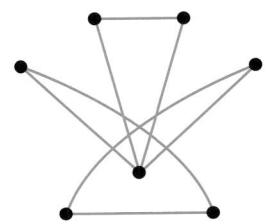

21. Six student clubs need to hold meetings on the same day, but some students belong to more than one club. In order to avoid members missing meetings, the meetings need to be scheduled during different time slots. An "X" in the table below indicates that the two corresponding clubs share at least one member. Use graph coloring to determine the least number of time slots necessary to ensure that all club members can attend all meetings.

	Student newspaper	Honor society	Biology assoc.	Gaming club	Debate team	Engineering club
Student newspaper	—		X		X	
Honor society		—	X		X	X
Biology assoc.	X	X	—	X		
Gaming club			X	—	X	X
Debate team	X	X		X	—	
Engineering club		X		X		—

22. Eight political committees must meet on the same day, but some members are on more than one committee. Thus any committees that have members in common cannot meet at the same time. An "X" in the table below indicates that the two corresponding committees share a member. Use graph coloring to determine the fewest number of meeting times required so that all members can attend the appropriate meetings.

	Appropriations	Budget	Finance	Judiciary	Education	Health	Foreign affairs	Housing
Appropriations	—		X			X	X	
Budget		—		X		X		
Finance	X		—	X			X	X
Judiciary		X	X	—		X		X
Education					—		X	X
Health	X	X		X		—		
Foreign affairs	X		X		X		—	
Housing			X	X	X			—

23. Six different groups of children would like to visit the zoo and feed different animals.

> Group 1 would like to feed the bears, dolphins, and gorillas.
>
> Group 2 would like to feed the bears, elephants, and hippos.
>
> Group 3 would like to feed the dolphins and elephants.
>
> Group 4 would like to feed the dolphins, zebras, and hippos.
>
> Group 5 would like to feed the bears and hippos.
>
> Group 6 would like to feed the gorillas, hippos, and zebras.

Use graph coloring to find the minimum number of days that are required so that all groups can feed the animals they would like to feed but no animals will be fed twice on the same day. Design a schedule to accomplish this goal.

24. Five different charity organizations send trucks on various routes to pick up donations residents leave on their doorsteps.

> Charity A covers Main St., First Ave., and State St.
>
> Charity B covers First Ave., Second Ave., and Third Ave.
>
> Charity C covers State St., City Dr., and Country Lane.
>
> Charity D covers City Dr., Second Ave., and Main St.
>
> Charity E covers Third Ave., Country Lane, and Fourth Ave.

Each charity has its truck travel down all three streets on its route on the same day, but no two charities wish to visit the same streets on the same day. Use graph coloring to design a schedule for the charities. Arrange their pickup routes so that no street is visited twice on the same day by different charities. The schedule should use the fewest number of days possible.

25. Students in a film class have volunteered to form groups and create several short films. The class has three digital video cameras that may be checked out for one day only, and it is expected that each group will need the entire day to finish shooting. All members of each group must participate in the film they volunteered for, so a student cannot work on more than one film on any given day.

> Film 1 will be made by Brian, Angela, and Kate.
>
> Film 2 will be made by Jessica, Vince, and Brian.
>
> Film 3 will be made by Corey, Brian, and Vince.
>
> Film 4 will be made by Ricardo, Sarah, and Lupe.
>
> Film 5 will be made by Sarah, Kate, and Jessica.
>
> Film 6 will be made by Angela, Corey, and Lupe.

Use graph coloring to design a schedule for loaning the cameras, using the fewest number of days possible, so that each group can shoot its film and all members can participate.

Extensions

CRITICAL THINKING

In Exercises 26–28, draw a map of a fictional continent with country boundaries corresponding to the graph given.

26.

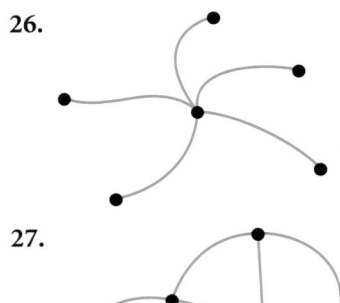

27.

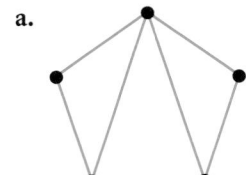

28.

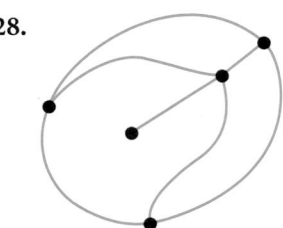

29. Draw a map of a fictional continent consisting of four countries, where the map cannot be colored with three or fewer colors without adjacent countries sharing a color.

30. If the chromatic number of a graph with five vertices is 1, what must the graph look like?

EXPLORATIONS

31. In this section, we colored vertices of graphs so that no edge connected two vertices of the same color. We can also consider coloring edges, rather than vertices, so that no vertex connects to more than one edge of the same color. In parts a to d, assign each edge in the graph a color so that no vertex connects to two or more edges of the same color. Use the fewest number of colors possible.

a.

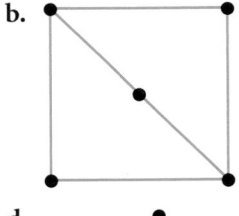

b.

c.

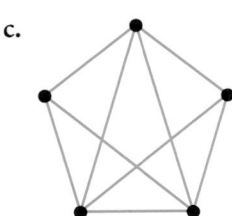

d.

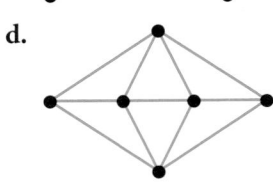

e. Explain why the number of colors required will always be at least the number of edges that meet at the vertex of highest degree in the graph.

32. Edge colorings, as explained in Exercise 31, can be used to solve scheduling problems. For instance, suppose five players are competing in a tennis tournament. Each player needs to play every other player in a match (but not more than once). Each player will participate in no more than one game per day, and two matches can occur at the same time when possible. How many days will be required for the tournament? Represent the tournament as a graph, where each vertex corresponds to a player and an edge joins two vertices if the corresponding players will compete against each other in a match. Next color the edges, where each different color corresponds to a different day of the tournament. Because one player will not be in more than one match per day, no two edges of the same color can meet at the same vertex. If we can find an edge coloring of the graph that uses the fewest number of colors possible, it will correspond to the fewest number of days required for the tournament. Sketch a graph that represents the tournament, find an edge coloring using the fewest number of colors possible, and use your graph to design a schedule of matches for the tournament that minimizes the number of days required.

CHAPTER 9 ## Summary

Key Terms

4-colorable [p. 598]
algorithm [p. 569]
chromatic number [p. 600]
circuit or closed walk [p. 552]
complete graph [p. 549]
connected graph [p. 549]
contraction [p. 588]
degree [p. 552]
edge [p. 548]
equivalent graphs [p. 549]
Euler circuit [p. 552]
Euler walk [p. 555]
Eulerian graph [p. 552]
face [p. 590]
graph [p. 548]
Hamiltonian circuit [p. 556]
Hamiltonian graph [p. 556]
infinite face [p. 590]
K_5 [p. 586]
multiple edges [p. 549]
null graph [p. 549]
planar drawing [p. 585]
planar graph [p. 584]
subgraph [p. 586]
traveling salesman problem [p. 567]
Utilities Graph [p. 586]
vertex [p. 548]
walk [p. 552]

weight [p. 567]
weighted graph [p. 567]

Essential Concepts

- **Eulerian Graph Theorem**
  A connected graph is *Eulerian* if and only if every vertex of the graph is of even degree.

- **Euler Walk Theorem**
  A connected graph contains an Euler walk if and only if the graph has two vertices of odd degree and the remaining vertices are of even degree. Furthermore, every Euler walk must start at one of the vertices of odd degree and end at the other.

- **Hamiltonian Graphs and Dirac's Theorem**
  Dirac's Theorem states that, in a connected graph with at least three vertices and with no multiple edges, if n is the number of vertices in the graph and every vertex has degree of at least $n/2$, then the graph must be Hamiltonian. If it is not the case that every vertex has degree of at least $n/2$, then the graph may or may not be Hamiltonian.

- **The Greedy Algorithm**
  One method of finding a Hamiltonian circuit in a complete weighted graph is given by the following procedure.

  1. Choose a vertex to start at, then travel along the connected edge of smallest weight. (If two

or more edges have the same weight, pick any one.)

2. After arriving at the next vertex, travel along the edge of smallest weight that connects to a vertex not yet visited. Continue this process until you have visited all vertices.

3. Return to the starting vertex.

The Greedy Algorithm attempts to give a circuit of minimal total weight, although it does not always succeed.

■ **The Edge-Picking Algorithm**
A second method of finding a Hamiltonian circuit in a complete weighted graph consists of the following steps.

1. Mark the edge of smallest weight in the graph. (If two or more edges have the same weight, pick any one.)

2. Mark the edge of next smallest weight in the graph, as long as it does not complete a circuit and does not add a third marked edge to a single vertex.

3. Continue this process until you can no longer mark any edges. Then mark the final edge that completes the Hamiltonian circuit.

■ **Nonplanar Graph Theorem**
A graph is nonplanar if and only if it has the Utilities Graph or K_5 as a subgraph, or it has a subgraph that can be contracted to the Utilities Graph or K_5.

■ **Euler's Formula**
In a connected planar graph drawn with no intersecting edges, let v be the number of vertices, e the number of edges, and f the number of faces. Then $v + f = e + 2$.

■ **The Four-Color Theorem**
Every planar graph is 4-colorable. (Note that fewer than four colors may be required; if the graph is not planar, more than four colors may be necessary.)

■ **Representing Maps as Graphs**
Draw a vertex in each region (country, state, etc.) of the map. Connect two vertices if the corresponding regions share a common border.

■ **2-Colorable Graph Theorem**
A graph is 2-colorable if and only if it has no circuits that consist of an odd number of vertices.

CHAPTER 9 **Review Exercises**

In Exercises 1 and 2, (a) give the number of edges in the graph, (b) find the number of vertices in the graph, (c) list the degree of each vertex, and (d) determine if the graph is connected.

1.
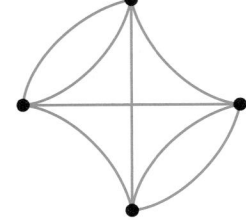

2.
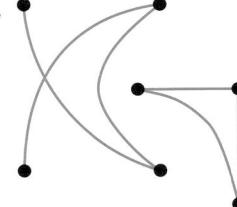

3. In the table on the following page, an "X" indicates teams from a junior soccer league that have played each other in the current season. Draw a graph to rep-

resent the games by using a vertex to represent each team. Connect two vertices with an edge if the corresponding teams played a game against each other this season.

	Mariners	Scorpions	Pumas	Stingrays	Vipers
Mariners	—	X		X	X
Scorpions	X	—	X		
Pumas		X	—	X	X
Stingrays	X		X	—	
Vipers	X		X		—

4. Each vertex in the graph at the right represents a freeway in the Los Angeles area. An edge connects two vertices if the corresponding freeways have interchanges allowing drivers to transfer from one freeway to the other.

   a. Can drivers transfer from the 105 freeway to the 10?

   b. Among the freeways represented, how many have interchanges with the 5 freeway?

   c. Which freeways have interchanges to all the other freeways in the graph?

   d. Of the freeways represented in the graph, which has the fewest number of interchanges to the other freeways?

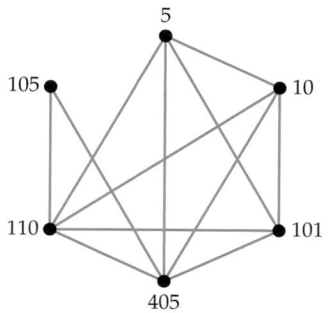

In Exercises 5 and 6, determine whether the two graphs are equivalent.

5.

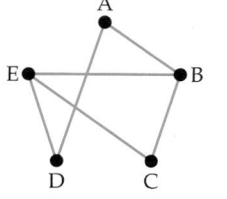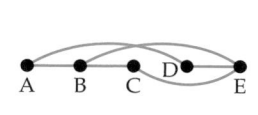

6.

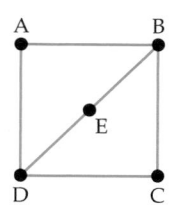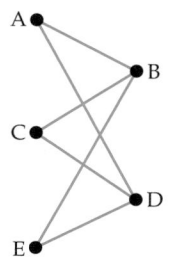

In Exercises 7–10, (a) find an Euler walk if possible, and (b) find an Euler circuit if possible.

7.

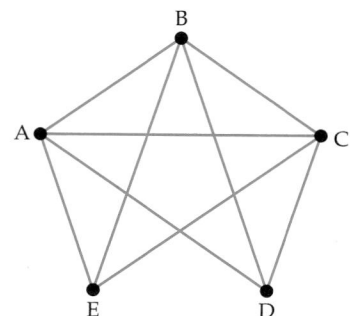

8.

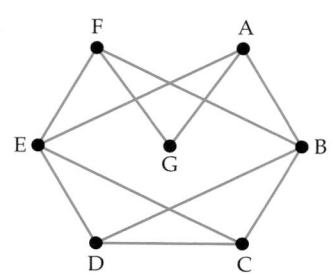

9.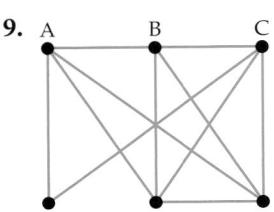

10.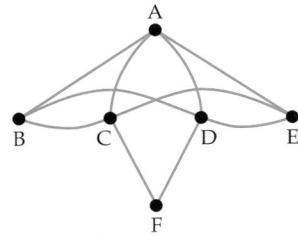

11. The figure shows an arrangement of bridges connecting land areas. Represent the map as a graph and then determine if it is possible to stroll across each bridge exactly once and return to the starting position.

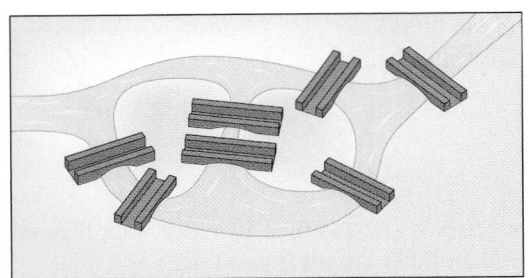

12. The floor plan of a sculpture gallery is shown. Is it possible to walk through each doorway exactly once? Is it possible to walk through each doorway exactly once and return to the starting point?

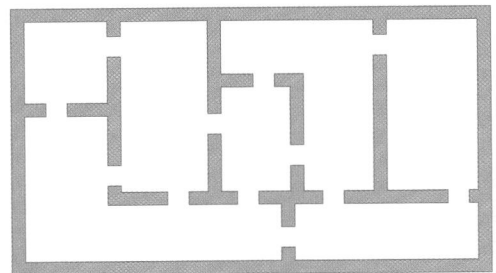

In Exercises 13 and 14, use Dirac's Theorem to verify that the graph is Hamiltonian, and then find a Hamiltonian circuit.

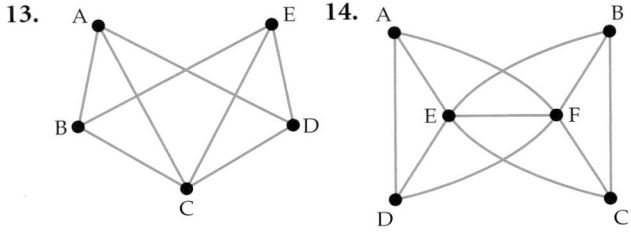

15. An "X" in the table indicates a direct flight offered by a small airline. Draw a graph that represents the direct flights, and use your graph to find a route that visits each city exactly once and returns to the starting city.

	Casper	Rapid City	Minneapolis	Des Moines	Topeka	Omaha	Boulder
Casper	—	X					X
Rapid City	X	—	X				
Minneapolis		X	—	X	X		X
Des Moines			X	—	X		
Topeka			X	X	—	X	X
Omaha					X	—	X
Boulder	X		X		X	X	—

16. For the direct flights given in Exercise 15, find a route that travels each flight exactly once and returns to the starting city. (You may visit cities more than once.)

In Exercises 17 and 18, use the Greedy Algorithm to find a Hamiltonian circuit starting at vertex A in the weighted graph.

17.

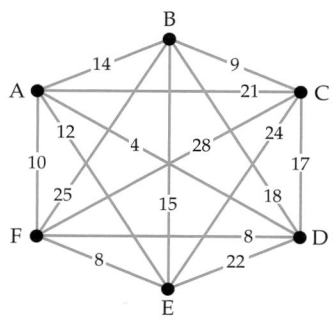

18.

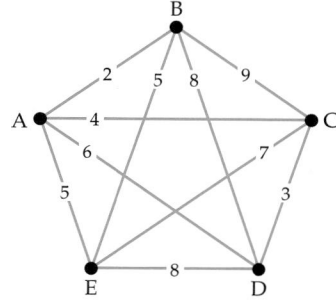

In Exercises 19 and 20, use the Edge-Picking Algorithm to find a Hamiltonian circuit starting at vertex A in the weighted graph.

19. The graph in Exercise 17

20. The graph in Exercise 18

21. The distances, in miles, between five different cities are given in the table. Sketch a weighted graph to represent the distances, and then use the Greedy Algorithm to design a route that starts in Memphis, visits each city, and returns to Memphis while attempting to minimize the total distance traveled.

	Memphis	Nashville	Atlanta	Birmingham	Jackson
Memphis	—	210	394	247	213
Nashville	210	—	244	189	418
Atlanta	394	244	—	148	383
Birmingham	247	189	148	—	239
Jackson	213	418	383	239	—

22. A small office needs to network five computers by connecting one computer to another and forming a large loop. The length of cable needed (in feet) between pairs of machines is given in the table. Use the Edge-Picking Algorithm to design a method to network the computers while attempting to use the least amount of cable possible.

	Computer A	Computer B	Computer C	Computer D	Computer E
Computer A	—	85	40	55	20
Computer B	85	—	35	40	18
Computer C	40	35	—	60	50
Computer D	55	40	60	—	30
Computer E	20	18	50	30	—

In Exercises 23 and 24, show that the graph is planar by finding a planar drawing.

23. 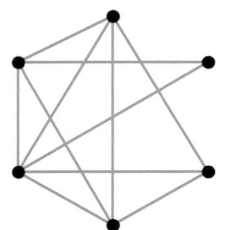 **24.**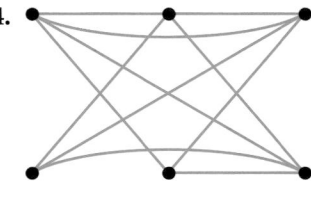

In Exercises 25 and 26, show that the graph is *not* planar.

25.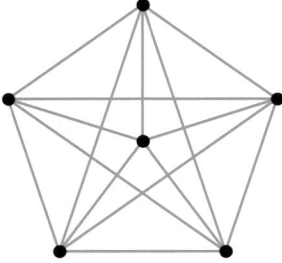

26.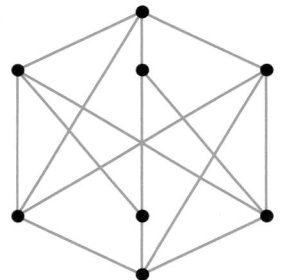

In Exercises 27 and 28, count the number of vertices, edges, and faces in the graph, and then verify Euler's Formula.

27.

28.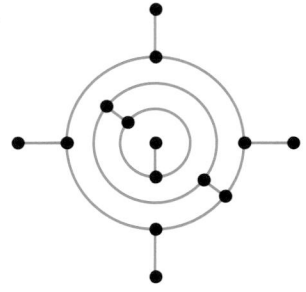

In Exercises 29 and 30, a fictional map is given showing the states of a country. Represent the map by a graph and find a coloring of the graph, using the fewest number of colors possible, such that no two neighboring countries share the same color.

29.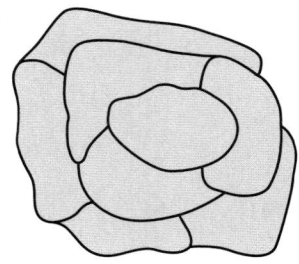

30.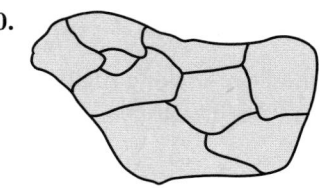

In Exercises 31 and 32, show that the graph is 2-colorable by finding a 2-coloring, or explain why the graph is not 2-colorable.

31.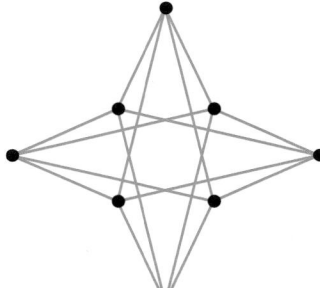

32.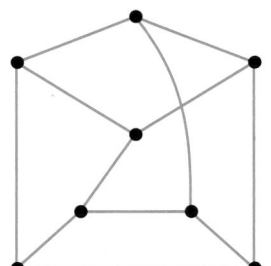

In Exercises 33 and 34, determine (by trial and error) the chromatic number of the graph.

33. 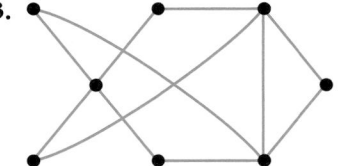 **34.**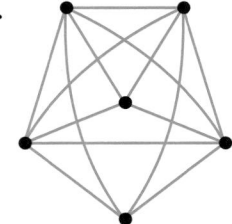

35. A company has scheduled a retreat at a hotel resort. It needs to hold meetings the first day there, and several conference rooms are available. Some employees must attend more than one meeting, however, so the meetings cannot all be scheduled at the same time. An "X" in the table below indicates that at least one employee must attend both of the corresponding meetings, and so those two meetings must be held at different times. Draw a graph in which each vertex represents a meeting, and an edge joins two vertices if the corresponding meetings require the attendance of the same employee. Then use graph coloring to design a meeting schedule that uses the fewest number of different time slots possible.

	Budget meeting	Marketing meeting	Executive meeting	Sales meeting	Research meeting	Planning meeting
Budget meeting	—	X	X		X	
Marketing meeting	X	—		X		
Executive meeting	X		—		X	X
Sales meeting		X		—		X
Research meeting	X		X		—	X
Planning meeting			X	X	X	—

CHAPTER 9   Test

1. Each vertex in the graph below represents a student. An edge connects two vertices if the corresponding students have at least one class in common during the current term.

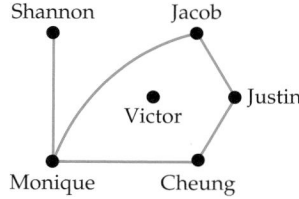

a. Is there a class that both Jacob and Cheung are enrolled in?

b. Who shares the most classes among the members of this group of students?

c. How many classes does Victor have in common with the other students?

d. Is this graph connected?

2. Determine whether the following two graphs are equivalent. Explain your reasoning.

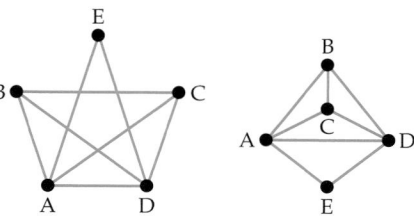

3. Answer the following questions for the graph shown below.

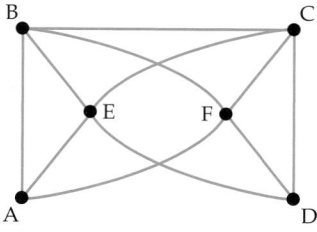

a. Is the graph Eulerian? If so, find an Euler circuit. If not, explain how you know.

b. Does the graph have an Euler walk? If so, find one. If not, explain how you know.

4. The illustration below depicts bridges connecting islands in a river to the banks. Can a person start at one location and take a stroll so that each bridge is crossed once, but no bridge is crossed twice? (The person does not need to return to the starting location.) Draw a graph to represent the land areas and bridges. Answer the question, using the graph to explain your reasoning.

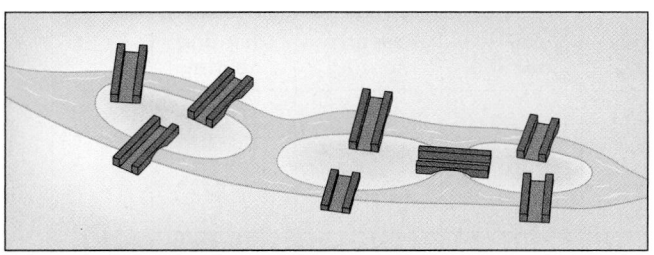

5. a. What does Dirac's Theorem state? Explain how it guarantees that the graph below is Hamiltonian.

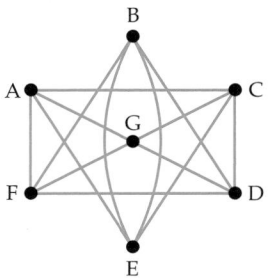

b. Find a Hamiltonian circuit in the graph.

6. The table below shows the cost of direct train travel between various cities.

   a. Draw a weighted graph that represents the train fares.

   b. Use the Edge-Picking Algorithm to find a route that begins in Angora, visits each city, and returns to Angora. What is the total cost of this route?

   c. Is it possible to travel along each train route and return to the starting city without traveling any routes more than once? Explain how you know.

	Angora	Bancroft	Chester	Davenport	Elmwood
Angora	—	$48	$52	$36	$90
Bancroft	$48	—	$32	$42	$98
Chester	$52	$32	—	$76	$84
Davenport	$36	$42	$76	—	$106
Elmwood	$90	$98	$84	$106	—

7. Use the Greedy Algorithm to find a Hamiltonian circuit beginning at vertex A in the weighted graph shown.

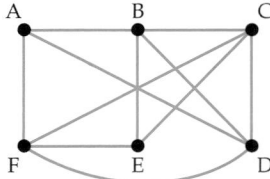

8. Sketch a planar drawing of the graph below.

9. Show that the graph below is not planar.

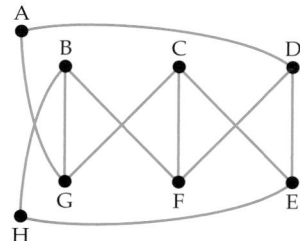

10. Answer the following questions for the graph shown below.

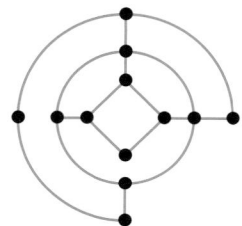

   a. How many vertices are of degree 3?

   b. How many faces does the graph consist of (including the infinite face)?

   c. Show that Euler's Formula is satisfied by this graph.

11. A fictional map of the countries of a continent is shown below.

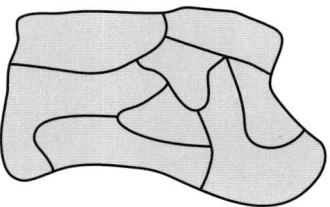

   a. Represent the map by a graph in which the vertices correspond to countries and the edges indicate which countries share a common border.

   b. What is the chromatic number of your graph?

   c. Use your work from part b to color the countries of the map with the fewest number of colors possible so that no two neighboring countries share the same color.

12. For the graph shown below, find a 2-coloring of the vertices or explain why a 2-coloring is impossible.

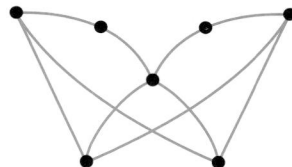

First time homebuyers are often amazed by how much there is to learn about buying real estate. They need to learn about down payments, fixed and adjustable rate mortgages, interest rates, fees, points, homeowner insurance, escrow accounts, and other related topics. Some of the financial considerations that arise when purchasing a home are presented in the **Excursion exercises** on **page 682.**

The Mathematics of Finance

Need help? For on-line student resources, such as section quizzes, visit this textbook's web site at **math.college.hmco.com/students.**

The table below shows the estimated annual expenses for a child born in 2000. The expenses are for the younger child in a two-child, two-parent home. From the table, we can see that the annual expense for a six-year-old child is $26,970, and that $4650 of that amount is spent on food. Use the table to answer the following questions. (Answers are provided on the next page.)

1. What is the estimated annual expense for a 10-year-old child? How much of that amount is spent on health care?

2. Which age group has the highest annual expense? What might be some explanations for this age group having the highest annual expense?

3. For which ages is the annual cost of child care and education highest? Provide an explanation for this.

Estimated Annual Expenses for a Child Born in 2000								
Child's Age	Total	Housing	Food	Transportation	Clothing	Health Care	Child Care and Education	Miscellaneous
0 – 2	26,220	9750	3180	3450	1320	1740	3930	2850
3 – 5	26,940	9660	3660	3390	1290	1680	4350	2910
6 – 8	26,970	9420	4650	3750	1440	1890	2790	3030
9 – 11	26,850	8760	5490	3990	1530	2070	1830	3120
12 – 14	29,070	9450	5520	4350	2670	2070	1350	3660
15 – 17	29,850	8130	6150	5490	2370	2190	2310	9940

Source: U.S. Department of Agriculture, Expenditures on Children by Families, 2000 Annual Report

This table can also be used to estimate expenses for an only child by multiplying the total for any age group by 1.24.

4. What are the estimated annual expenses for an only child aged 11? How much of that amount is spent on transportation?

5. Explain why the cost of raising an only child is greater than the cost of raising the second of two children.

6. What is the total estimated expense of raising an only child to the age of 18? And note that these data do not include the cost of college!

Given the cost of raising children, an understanding of the mathematics of finance is important for any parent. In this chapter, we will discuss the topics of simple interest, compound interest, consumer loans, and home ownership.

| **Simple Interest**

Simple Interest

historical note

The earliest loans date back to 3000 B.C., and interest on those loans may have extended over generations, not 4 or 5 years, as is the case for today's typical car loan. One of the first written records of an interest rate occurs in the Code of Hammurabi. Hammurabi ruled Babylon from 1795 to 1750 B.C. He is known for being the first ruler to write a set of laws that defined peoples' rights. In this Code, he allowed interest rates to be as high as $33\frac{1}{3}$%. ∎

When you deposit money in a bank—for example, in a savings account—you are permitting the bank to use your money. The bank may lend the deposited money to customers to buy cars or make renovations on their homes. The bank pays you for the privilege of using your money. The amount paid to you is called **interest.** If you are the one borrowing money from a bank, the amount you pay for the privilege of using that money is also called interest.

The amount deposited in a bank or borrowed from a bank is called the **principal.** The amount of interest paid is usually given as a percent of the principal. The percent used to determine the amount of interest is called the **interest rate.** If you deposit $1000 in a savings account paying 5% interest, $1000 is the principal and the interest rate is 5%.

Interest paid on the original principal is called **simple interest.** The formula used to calculate simple interest is given below.

> **Simple Interest Formula**
>
> The simple interest formula is
>
> $$I = Prt$$
>
> where I is the interest, P is the principal, r is the interest rate, and t is the time period.

In the simple interest formula, the time t is expressed in the same period as the rate. For example, if the rate is given as an annual interest rate, then the time is measured in years; if the rate is given as a monthly interest rate, then the time must be expressed in months.

Interest rates are most commonly expressed as annual interest rates. Therefore, unless stated otherwise, we will assume the interest rate is an annual interest rate.

Interest rates are generally given as percentages. Before performing calculations involving an interest rate, write the interest rate as a decimal.

INSTRUCTOR NOTE

Emphasize that the simple interest formula requires that the interest rate and the time have comparable units. If the *annual* interest rate is given, then the time must be in *years*. If a *monthly* interest rate is given (as on most credit cards), then the time must be in *months*.

ANSWERS TO QUESTIONS ON PAGE 621

1. *$26,850; $2070*

2. *15–17 years. For example, car insurance for teenage drivers.*

3. *3–5 years. For example, child care for preschoolers is an expense, whereas public school is free.*

4. *$33,294; $4947.60*

5. *For example, no hand-me-downs.*

6. *$617,148*

EXAMPLE 1 ▪ **Calculate Simple Interest**

Calculate the simple interest earned in 1 year on a deposit of $1000 if the interest rate is 5%.

Solution
Use the simple interest formula. Substitute the following values into the formula: $P = 1000$, $r = 5\% = 0.05$, and $t = 1$.

$$I = Prt$$
$$I = 1000(0.05)(1)$$
$$I = 50$$

The simple interest earned is $50.

CHECK YOUR PROGRESS 1 Calculate the simple interest earned in 1 year on a deposit of $500 if the interest rate is 4%.

Solution See page S34.

EXAMPLE 2 ■ Calculate Simple Interest

Calculate the simple interest due on a three-month loan of $2000 if the interest rate is 6.5%.

Solution

Use the simple interest formula. Substitute the values $P = 2000$ and $r = 6.5\% = 0.065$ into the formula. Because the interest rate is an annual rate, the time must be measured in years: $t = \frac{3 \text{ months}}{1 \text{ year}} = \frac{3 \text{ months}}{12 \text{ months}} = \frac{3}{12}$.

$$I = Prt$$
$$I = 2000(0.065)\left(\frac{3}{12}\right)$$
$$I = 32.5$$

The simple interest due is $32.50.

CHECK YOUR PROGRESS 2   Calculate the simple interest due on a four-month loan of $1500 if the interest rate is 5.25%.

Solution   *See page S34.*

EXAMPLE 3 ■ Calculate Simple Interest

Calculate the simple interest due on a two-month loan of $500 if the interest rate is 1.5% per month.

Solution

Use the simple interest formula. Substitute the values $P = 500$ and $r = 1.5\% = 0.015$ into the formula. Because the interest rate is *per month*, the time period of the loan is expressed as the number of months: $t = 2$.

$$I = Prt$$
$$I = 500(0.015)(2)$$
$$I = 15$$

The simple interest due is $15.

CHECK YOUR PROGRESS 3   Calculate the simple interest due on a five-month loan of $700 if the interest rate is 1.25% per month.

Solution   *See page S34.*

Remember that in the simple interest formula, time t is measured in the same period as the interest rate. Therefore, if the time period of a loan with an annual interest rate is given in days, it is necessary to convert the time period of the loan to a fractional part of a year. There are two methods for converting time from days to years: the exact method and the ordinary method. Using the exact method, the number of days of the loan is divided by 365, the number of days in a year.

Exact method: $t = \dfrac{\text{number of days}}{365}$

The ordinary method is based on there being an average of 30 days in a month and 12 months in a year ($30 \cdot 12 = 360$). Using this method, the number of days of the loan is divided by 360.

Ordinary method: $t = \dfrac{\text{number of days}}{360}$

The ordinary method is used by most businesses. Therefore, unless otherwise stated, the ordinary method will be used in this text.

EXAMPLE 4 ■ **Calculate Simple Interest**

historical note

One explanation for the origin of the dollar sign is that the symbol evolved from the Mexican or Spanish letters P and S, for pesos or piastres or pieces of eight. The S gradually came to be written over the P. The $ symbol was widely used before the adoption of the U.S. dollar in 1785. ■

Calculate the simple interest due on a 45-day loan of $3500 if the annual interest rate is 8%.

Solution
Use the simple interest formula. Substitute the following values into the formula: $P = 3500$, $r = 8\% = 0.08$, and $t = \frac{\text{number of days}}{360} = \frac{45}{360}$.

$$I = Prt$$

$$I = 3500(0.08)\left(\frac{45}{360}\right)$$

$$I = 35$$

The simple interest due is $35.

CHECK YOUR PROGRESS 4   Calculate the simple interest due on a 120-day loan of $7000 if the annual interest rate is 5.25%.

Solution   See page S34.

The simple interest formula can be used to find the interest rate on a loan when the interest, principal, and time period of the loan are known. An example is given below.

EXAMPLE 5 ■ **Calculate the Simple Interest Rate**

The simple interest charged on a six-month loan of $3000 is $150. Find the simple interest rate.

Solution
Use the simple interest formula. Solve the equation for r.

$$I = Prt$$

$$150 = 3000(r)\left(\frac{6}{12}\right)$$

$$150 = 1500r \qquad \bullet \ 3000\left(\frac{6}{12}\right) = \mathbf{1500}$$

$$0.10 = r \qquad \bullet \ \text{Divide each side of the equation by 1500.}$$

$$r = 10\% \qquad \bullet \ \text{Write the decimal as a percent.}$$

The simple interest rate on the loan is 10%.

CHECK YOUR PROGRESS 5   The simple interest charged on a six-month loan of $12,000 is $462. Find the simple interest rate.

Solution   *See page S35.*

Math Matters   Children Daydream about Being Rich

This chapter is about money, something everyone thinks about. But attitudes toward money differ. You might find it interesting to note that a survey of the world's six leading industrial nations revealed that about 62% of children aged 7 through 12 say that they think or daydream about "being rich." The breakdown by country is shown in the figure below.

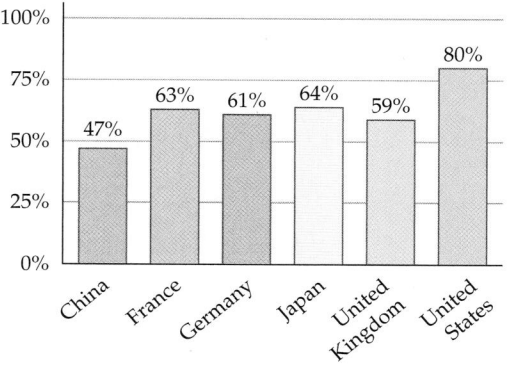

Percent of children aged 7–12 who say they think or daydream about "being rich"

Source: Roper Starch Worldwide for A. B. C. Research

---

point of interest

You may have heard the term *legal tender.* Section 102 of the Coinage Act of 1965 states, in part, that

All coins and currencies of the United States, regardless of when coins are issued, shall be legal tender for all debts, public and private, public charges, taxes, duties, and dues.

This means that when you offer U.S. currency to a creditor, you have made a valid and legal offer of payment of your debt.

Future Value and Maturity Value

When you borrow money, the total amount to be repaid to the lender is the sum of the principal and interest. This sum is calculated using the following future value or maturity value formula for simple interest.

Future Value or Maturity Value Formula for Simple Interest

The future or maturity value formula for simple interest is

$A = P + I$

where A is the amount after the interest, I, has been added to the principal, P.

This formula can be used for loans or investments. When used for a loan, A is the total amount to be repaid to the lender; this sum is called the **maturity value** of the loan. In Example 5, the simple interest charged on the loan of $3000 was $150. The maturity value of the loan is therefore $3150 ($3000 + $150).

For an investment, such as a deposit in a bank savings account, A is the total amount on deposit after the interest earned has been added to the principal. This sum is called the **future value** of the investment.

QUESTION   *Is the stated sum a maturity value or a future value?*

   a. *The sum of the principal and the interest on an investment*

   b. *The sum of the principal and the interest on a loan*

EXAMPLE 6 ▪ Calculate a Maturity Value

Calculate the maturity value of a simple interest, eight-month loan of $8000 if the interest rate is 9.75%.

Solution

Step 1: Find the interest. Use the simple interest formula. Substitute the values $P = 8000$, $r = 9.75\% = 0.0975$, and $t = \frac{8}{12}$ into the formula.

$$I = Prt$$

$$I = 8000(0.0975)\left(\frac{8}{12}\right)$$

$$I = 520$$

Step 2: Find the maturity value. Use the maturity value formula for simple interest. Substitute the values $P = 8000$ and $I = 520$ into the formula.

$$A = P + I$$

$$A = 8000 + 520$$

$$A = 8520$$

The maturity value of the loan is $8520.

CHECK YOUR PROGRESS 6   Calculate the maturity value of a simple interest, nine-month loan of $4000 if the interest rate is 8.75%.

Solution   *See page S35.*

Recall that the simple interest formula states that $I = Prt$. We can substitute Prt for I in the future or maturity value formula, as follows.

$$A = P + I$$

$$A = P + Prt$$

$$\mathbf{A = P(1 + rt)}$$ • Use the Distributive Property.

In the final equation, A is the future value of an investment or the maturity value of a loan, P is the principal, r is the interest rate, and t is the time period.

ANSWER   ***a.*** *Future value*   ***b.*** *Maturity value*

We used the formula $A = P + I$ in Example 6. The formula $A = P(1 + rt)$ is used in Examples 7 and 8. Note that two steps were required to find the solution in Example 6, but only one step is required in Examples 7 and 8.

EXAMPLE 7 ■ Calculate the Maturity Value Using $A = P(1 + rt)$

Calculate the maturity value of a simple interest, three-month loan of $3800. The interest rate is 6%.

Solution

Substitute the following values into the formula $A = P(1 + rt)$: $P = 3800$, $r = 6\% = 0.06$, and $t = \frac{3}{12}$.

$$A = P(1 + rt)$$
$$A = 3800\left[1 + 0.06\left(\frac{3}{12}\right)\right]$$
$$A = 3800(1 + 0.015)$$
$$A = 3800(1.015)$$
$$A = 3857$$

The maturity value of the loan is $3857.

CHECK YOUR PROGRESS 7   Calculate the maturity value of a simple interest, one-year loan of $6700. The interest rate is 8.9%.

Solution   *See page S35.*

EXAMPLE 8 ■ Calculate the Future Value Using $A = P(1 + rt)$

Find the future value after 1 year of $850 in an account earning 8.2% simple interest.

Solution

Because $t = 1$, $rt = r(1) = r$. Therefore, $1 + rt = 1 + r = 1 + 0.082 = 1.082$.

$$A = P(1 + rt)$$
$$A = 850(1.082)$$
$$A = 919.7$$

After 1 year, $919.70 is in the account.

CHECK YOUR PROGRESS 8   Find the future value after 1 year of $680 in an account earning 6.4% simple interest.

Solution   *See page S35.*

Recall that the formula $A = P + I$ states that A is the amount after the interest has been added to the principal. Subtracting P from each side of this equation yields the following formula.

$$I = A - P$$

This formula states that the amount of interest paid is equal to the total amount minus the principal. This formula is used in Example 9.

EXAMPLE 9 ■ Calculate the Simple Interest Rate

The maturity value of a three-month loan of $4000 is $4085. What is the simple interest rate?

Solution
First find the amount of interest paid. Subtract the principal from the maturity value.

$$I = A - P$$
$$I = 4085 - 4000$$
$$I = 85$$

Find the simple interest rate by solving the simple interest formula for r.

$$I = Prt$$
$$85 = 4000(r)\left(\frac{3}{12}\right)$$

$$85 = 1000r \qquad \bullet \ 4000\left(\frac{3}{12}\right) = 1000$$

$$0.085 = r \qquad \bullet \text{ Divide each side of the equation by 1000.}$$

$$r = 8.5\% \qquad \bullet \text{ Write the decimal as a percent.}$$

The simple interest rate on the loan is 8.5%.

CHECK YOUR PROGRESS 9   The maturity value of a four-month loan of $9000 is $9240. What is the simple interest rate?

Solution   See page S35.

Stocks and Bonds

Stocks, bonds, and mutual funds are investment vehicles, but they differ in nature.

   A **share of stock** in a corporation is a certificate that indicates partial ownership in the corporation. The owners of the certificates are called **stockholders** or **shareholders.** As owners, the stockholders share in the profits or losses of the corporation.

historical note

Have you heard the expression for making a fortune on Wall Street that goes something like, "buy low, sell high?" The phrase is adapted from a comment made by Hetty Green (1834–1916), who was also known as the Witch of Wall Street. She turned a $1 million inheritance into $100 million and became the richest woman in the world. In today's terms, that $100 million would be worth over $2 billion. When asked how she became so successful, she replied, "There is no secret in fortune making. All you have to do is buy cheap and sell dear, act with thrift and shrewdness and be persistent." ∎

▼ **point of interest**

Usually the dividends earned on shares of stock are sent to shareholders. However, in a **dividend reinvestment plan,** or DRIP, the dividends earned are applied toward the purchase of more shares.

▼ **point of interest**

Young adults aged 18 to 34 make up 19% of the stock investors in this country. According to a survey of investors in this age group, 93% plan to use investment money for retirement, 52% are saving for a child's college education, 23% report they will use proceeds to buy a home, and 19% have plans to start a business. (*Source:* Nasdaq Stock Market)

A **brokerage firm** is a dealer of stocks that acts as your agent when you want to buy or sell shares of stock. Most trading of stocks happens at a stock exchange. **Stock exchanges** are businesses whose purpose it is to bring together buyers and sellers of stock. Two of the major stock exchanges in the United States are the New York Stock Exchange and the American Stock Exchange. Each day these companies provide financial institutions, Internet web site hosts, newspapers, and other publications with data on the trading activity of all of the stocks traded on that exchange.

The **market value** of a stock is the price for which a stockholder is willing to sell a share of stock and a buyer is willing to purchase the stock. Shares are always sold to the highest bidder.

A **dividend** is a payment a corporation makes to its shareholders out of its earnings. It is usually expressed as a per-share amount. The **dividend yield,** which is used to compare companies' dividends, is the amount of the dividend divided by the stock price, and is expressed as a percent. Determining a dividend yield is similar to calculating the simple interest rate earned on an investment. You can think of the dividend as the interest earned, the stock price as the principal, and the yield as the interest rate.

EXAMPLE 10 ■ Calculate a Dividend Yield

A stock pays an annual dividend of $1.75 per share. The stock is trading at $70. Find the dividend yield.

Solution

$$I = Prt$$
$$1.75 = 70r(1)$$
• Let I = the annual dividend and P = the stock price. The time is 1 year.

$$1.75 = 70r$$
$$0.025 = r$$
• Divide each side of the equation by 70.

The dividend yield is 2.5%.

CHECK YOUR PROGRESS 10   A stock pays an annual dividend of $.82 per share. The stock is trading at $51.25. Find the dividend yield.

Solution   *See page S35.*

When a corporation issues stock, it is *selling* part of the company to the stockholders. When it issues a **bond,** the corporation is *borrowing* money from the bondholders; a **bondholder** lends money to a corporation. Corporations, the federal government, government agencies, states, and cities all issue bonds. These entities need money to operate—for example, to fund the federal deficit, repair roads, or build a new factory—so they borrow money from the public by issuing bonds.

Bonds are usually issued in units of $1000. The price paid for the bond is the **face value.** The issuer promises to repay the bondholder on a particular day, called the **maturity date,** at a given rate of interest, called the **coupon.**

Assume that a bond with a $1000 face value has a 5% coupon and a 10-year maturity date. The bondholder would collect interest payments of $50 in each of those 10 years. The payments are calculated using the simple interest formula, as shown below.

$$I = Prt$$
$$I = 1000(0.05)(1)$$
$$I = 50$$

At the end of the 10-year period, the bondholder receives from the issuer the $1000 face value of the bond.

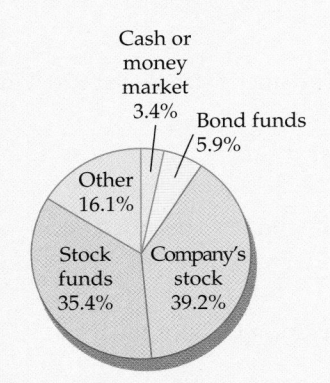
EXAMPLE 11 ▪ **Calculate Interest Payments on a Bond**

A bond with a $10,000 face value has a 3% coupon and a five-year maturity date. Calculate the total of the interest payments paid to the bondholder.

Solution
Use the simple interest formula to find the annual interest payment. Substitute the following values into the formula: $P = 10,000$, $r = 3\% = 0.03$, and $t = 1$.

$$I = Prt$$
$$I = 10,000(0.03)(1)$$
$$I = 300$$

Multiply the annual interest payment by the term of the bond.

$$300(5) = 1500$$

The total of the interest payments paid to the bondholder is $1500.

CHECK YOUR PROGRESS 11   A bond has a $15,000 face value, a four-year maturity, and a 3.5% coupon. What is the total of the interest payments paid to the bondholder?

Solution   *See page S35.*

A key difference between stocks and bonds is that stocks make no promises about dividends or returns, whereas the issuer of a bond guarantees to pay back the face value of the bond plus interest.

An **investment trust** is a company whose assets are stocks and bonds. The purpose of these companies is not to manufacture a product, but to purchase stocks and bonds with the hope that their value will increase. A **mutual fund** is an example of an investment trust. Owning shares of a mutual fund enables you to own stocks or bonds without having to choose which individual stocks or bonds to buy. The investment trust is responsible for managing the fund; that is, deciding when to buy and sell its holdings.

Math Matters   On-Line Brokerage Accounts

Where do Americans invest their money? You might correctly assume that the largest number of Americans invest in real estate, as every homeowner is considered to have an investment in real estate. But more and more Americans are investing their money in the stock market, and many of them are doing so by logging onto the World Wide Web. Consequently, the number of on-line brokerage accounts has been growing larger each year. The graph below shows the numbers of Internet brokerage accounts from 1996 to 2001.

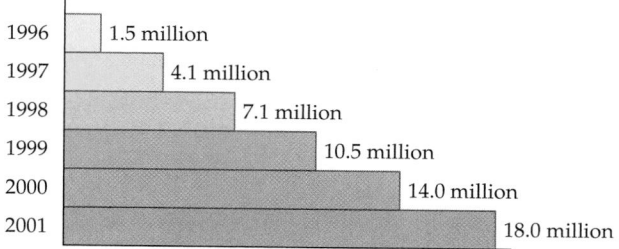

1996	1.5 million
1997	4.1 million
1998	7.1 million
1999	10.5 million
2000	14.0 million
2001	18.0 million

Growth in on-line investment accounts

Source: Gomez Advisers

Excursion

Treasury Bills

TAKE NOTE

You can obtain more information about the various Treasury securities at **http://www.publicdebt. treas.gov.** Using this site, investors can buy securities directly from the government, thereby avoiding the service fee charged by banks or brokerage firms.

The bonds issued by the United States government are called Treasuries. They are grouped into three categories.

U.S. Treasury bills have maturities of under 1 year.

U.S. Treasury notes have maturities ranging from 2 to 10 years.

U.S. Treasury bonds have maturities ranging from 10 to 30 years.

This Excursion will focus on Treasury bills.

The **face value** of a Treasury bill is the amount of money received on the maturity date of the bill. Treasury bills are sold on a **discount basis;** that is, the interest on the bill is computed and subtracted from the face value to determine its cost.

(continued)

Suppose a company invests in a $50,000 United States Treasury bill at 3.35% interest for 28 days. The bank through which the bill is purchased charges a service fee of $15. What is the cost of the Treasury bill?

To find the cost, first find the interest. Use the simple interest formula.

$$I = Prt$$

$$= 50{,}000(0.0335)\left(\frac{28}{360}\right)$$

$$\approx 130.278$$

The interest earned is $130.28.

Find the cost of the Treasury bill.

$$\text{Cost} = (\text{face value} - \text{interest}) + \text{service fee}$$

$$= (50{,}000 - 130.28) + 15$$

$$= 49{,}869.72 + 15$$

$$= 49{,}884.72$$

The cost of the Treasury bill is $49,884.72.

Excursion Exercises

1. The face value of a Treasury bill is $30,000. The interest rate is 2.32% and the bill matures in 182 days. The bank through which the bill is purchased charges a service fee of $15. What is the cost of the bill?

2. The face value of a 91-day Treasury bill is $20,000. The interest rate is 2.96%. The bank charges a service fee of $15. Calculate the cost of the bill.

3. A company invests in a 29-day, $60,000 United States Treasury bill at 2.28% interest. The bank charges a service fee of $35. Calculate the cost of the Treasury bill.

4. A $40,000 United States Treasury bill, purchased at 1.96% interest, matures in 92 days. The purchaser is charged a service fee of $20. What is the cost of the bill?

Exercise Set 10.1

(Suggested Assignment: 5–57, odds)

1. Explain how to convert a number of months to a fractional part of a year.

2. Explain how to convert a number of days to a fractional part of a year.

3. Explain what each variable in the simple interest formula represents.

In Exercises 4–17, calculate the simple interest earned. Round to the nearest cent.

4. $P = \$2000$, $r = 6\%$, $t = 1$ year
5. $P = \$8000$, $r = 7\%$, $t = 1$ year
6. $P = \$3000$, $r = 5.5\%$, $t = 6$ months
7. $P = \$7000$, $r = 6.5\%$, $t = 6$ months
8. $P = \$4200$, $r = 8.5\%$, $t = 3$ months
9. $P = \$9000$, $r = 6.75\%$, $t = 4$ months
10. $P = \$12{,}000$, $r = 7.8\%$, $t = 45$ days
11. $P = \$3000$, $r = 9.6\%$, $t = 21$ days
12. $P = \$4000$, $r = 8.4\%$, $t = 33$ days
13. $P = \$7000$, $r = 7.2\%$, $t = 114$ days
14. $P = \$800$, $r = 1.5\%$ monthly, $t = 3$ months
15. $P = \$2000$, $r = 1.25\%$ monthly, $t = 5$ months
16. $P = \$3500$, $r = 1.8\%$ monthly, $t = 4$ months
17. $P = \$1600$, $r = 1.75\%$ monthly, $t = 6$ months

In Exercises 18–23, use the formula $A = P(1 + rt)$ to calculate the maturity value of the simple interest loan.

18. $P = \$8500$, $r = 6.8\%$, $t = 6$ months
19. $P = \$15{,}000$, $r = 8.9\%$, $t = 6$ months
20. $P = \$4600$, $r = 9.75\%$, $t = 4$ months
21. $P = \$7200$, $r = 7.95\%$, $t = 4$ months
22. $P = \$13{,}000$, $r = 6.4\%$ monthly, $t = 3$ months
23. $P = \$2800$, $r = 9.2\%$, $t = 3$ months

In Exercises 24–29, calculate the simple interest rate.

24. $P = \$8000$, $I = \$500$, $t = 1$ year
25. $P = \$1600$, $I = \$120$, $t = 1$ year
26. $P = \$4000$, $I = \$190$, $t = 6$ months
27. $P = \$2000$, $I = \$105$, $t = 6$ months
28. $P = \$500$, $I = \$10.25$, $t = 3$ months
29. $P = \$1200$, $I = \$37.20$, $t = 4$ months

30. Calculate the simple interest earned in 1 year on a deposit of $1900 if the interest rate is 8%.
31. Calculate the simple interest earned in 1 year on a deposit of $2300 if the interest rate is 7%.
32. Calculate the simple interest due on a three-month loan of $1400 if the interest rate is 7.5%.

33. You deposit $1500 in an account earning 10.4% interest. Calculate the simple interest earned in 6 months.
34. Calculate the simple interest due on a two-month loan of $800 if the interest rate is 1.5% per month.
35. Calculate the simple interest due on a 45-day loan of $1600 if the interest rate is 9%.
36. Calculate the simple interest due on a 150-day loan of $4800 if the interest rate is 7.25%.
37. Calculate the maturity value of a simple interest, eight-month loan of $7000 if the interest rate is 8.7%.
38. Calculate the maturity value of a simple interest, 10-month loan of $6600 if the interest rate is 9.75%.
39. Calculate the maturity value of a simple interest, one-year loan of $5200. The interest rate is 10.2%.
40. You deposit $880 in an account paying 9.2% simple interest. Find the future value of the investment after 1 year.
41. You deposit $750 in an account paying 7.3% simple interest. Find the future value of the investment after 1 year.
42. The simple interest charged on a six-month loan of $6000 is $270. Find the simple interest rate.
43. The simple interest charged on a six-month loan of $18,000 is $918. Find the simple interest rate.
44. The maturity value of a four-month loan of $3000 is $3097. Find the simple interest rate.
45. Find the simple interest rate on a three-month loan of $5000 if the maturity value of the loan is $5125.
46. Your property tax bill is $1200. The county charges a penalty of 11% simple interest for late payments. How much do you owe if you pay the bill 2 months past the due date?
47. Your electric bill is $132. You are charged 12% simple interest for late payments. How much do you owe if you pay the bill 1 month past the due date?
48. If you withdraw part of your money from a certificate of deposit before the date of maturity, you must pay an interest penalty. Suppose you invested $5000 in a one-year certificate of deposit paying 8.5% interest. After 6 months, you decide to withdraw $2000. Your interest penalty is 3 months simple interest on the $2000. What interest penalty do you pay?
49. $10,000 is borrowed for 140 days at an 8% interest rate. Calculate the maturity value by the exact method and by the ordinary method. Which method yields the greater maturity value? Who benefits from using

the ordinary method rather than the exact method, the borrower or the lender?

50. Find the dividend yield for a stock that pays an annual dividend of $1.24 per share and has a current price of $49.375.

51. The Blackburn Computer Company has declared an annual dividend of $.50 per share. The stock is trading at $40 per share. Find the dividend yield.

52. A stock that pays an annual dividend of $.58 per share has a current price of $31.75. Find the dividend yield.

53. The Moreau Corporation is paying an annual dividend of $.65 per share. If the price of a share of the stock is $81.25, what is the dividend yield on the stock?

54. A bond with a face value of $6000 and a 4.2% coupon has a five-year maturity. Find the annual interest paid on the bond.

55. The face value on a bond is $15,000. It has a 10-year maturity and a 3.75% coupon. What is the annual interest paid to the bondholder?

56. A bond with an $8000 face value has a 3.5% coupon and a three-year maturity. What is the total of the interest payments paid to the bondholder?

57. A bond has a $12,000 face value, an eight-year maturity, and a 2.95% coupon. Find the total of the interest payments paid on the bond.

Extensions

CRITICAL THINKING

58. Interest has been described as a rental fee for money. Explain why this is an apt description of interest.

59. On July 31, at 4 P.M., you open a savings account that pays 5% interest, compounded daily, and you deposit $500 in the account. Your deposit is credited as of August 1. At the beginning of September, you receive a statement from the bank that shows that during the month of August, you received $2.15 in interest. The interest has been added to your account, bringing the total deposit to $502.15. At the beginning of October, you receive a statement from the bank that shows that during the month of September, you received $2.09 in interest on the $502.15 on deposit. Why did you receive less interest during the second month, when there was more money on deposit?

COOPERATIVE LEARNING

60. a. In the table below, the interest rate is an annual simple interest rate. Complete the table by calculating the simple interest due on the loan.

Loan Amount	Interest Rate	Period	Interest
$5000	6%	1 month	_____
$5000	6%	2 months	_____
$5000	6%	3 months	_____
$5000	6%	4 months	_____
$5000	6%	5 months	_____

Use the pattern of your answers in the table to find the simple interest due on a $5000 loan that has an annual simple interest rate of 6% for a period of:

b. 6 months   **c.** 7 months   **d.** 8 months   **e.** 9 months

Use your solutions to parts a through e to answer the following questions.

f. If you know the simple interest due on a one-month loan, explain how you can use that figure to calculate the simple interest due on a seven-month loan for the same principal and the same interest rate.

g. If the time period of a loan is doubled but the principal and interest rate remain the same, how many times larger is the simple interest due on the loan?

h. If the time period of a loan is tripled but the principal and interest rate remain the same, how many times larger is the simple interest due on the loan?

61. a. In the table below, the interest rate is an annual simple interest rate. Complete the table by calculating the simple interest due on the loan.

Loan Amount	Interest Rate	Period	Interest
$1000	6%	1 month	_____
$2000	6%	1 month	_____
$3000	6%	1 month	_____
$4000	6%	1 month	_____
$5000	6%	1 month	_____

Ⓟ

Use the pattern of your answers in the table to determine the simple interest due on a one-month loan that has an annual simple interest rate of 6% when the principal is:

b. $6000   **c.** $7000   **d.** $8000   **e.** $9000

Use your solutions to parts a through e to answer the following questions.

f. If you know the simple interest due on a $1000 loan, explain how you can use that figure to calculate the simple interest due on an $8000 loan for the same time period and the same interest rate.

g. If the principal of a loan is doubled but the time period and interest rate remain the same, how many times larger is the simple interest due on the loan?

h. If the principal of a loan is tripled but the time period and interest rate remain the same, how many times larger is the simple interest due on the loan?

Compound Interest

Compound Interest

Simple interest is generally used for loans of 1 year or less. For loans of more than 1 year, the interest paid on the money borrowed is called *compound interest*. **Compound interest** is interest calculated not only on the original principal, but also on any interest that has already been earned.

To illustrate compound interest, suppose you deposit $1000 in a savings account earning 5% interest, compounded annually (once a year).

During the first year, the interest earned is calculated as follows.

$$I = Prt$$
$$I = \$1000(0.05)(1) = \$50$$

At the end of the first year, the total amount in the account is

$$A = P + I$$
$$A = \$1000 + \$50 = \$1050$$

During the second year, the interest earned is calculated using the amount in the account at the end of the first year.

$$I = Prt$$
$$I = \$1050(0.05)(1) = \$52.50$$

Note that the interest earned during the second year ($52.50) is greater than the interest earned during the first year ($50). This is because the interest earned during the first year was added to the original principal, and the interest for the second year was calculated using this sum. If the account earned simple interest rather than compound interest, the interest earned each year would be the same ($50).

At the end of the second year, the total amount in the account is the sum of the amount in the account at the end of the first year and the interest earned during the second year.

$$A = P + I$$
$$A = \$1050 + 52.50 = \$1102.50$$

The interest earned during the third year is calculated using the amount in the account at the end of the second year ($1102.50).

$$I = Prt$$
$$I = \$1102.50(0.05)(1) = \$55.125 \approx \$55.13$$

The interest earned each year keeps increasing. This is the effect of compound interest.

In this example, the interest is compounded annually. However, compound interest can be compounded semiannually (twice a year), quarterly (four times a year), monthly, or daily. The frequency with which the interest is compounded is called the **compounding period.**

If, in the preceding example, interest is compounded quarterly rather than annually, then the first interest payment on the $1000 in the account occurs after 3 months

$\left(t = \frac{3}{12} = \frac{1}{4}; \text{3 months is one-quarter of a year}\right)$. That interest is then added to the account, and the interest earned for the second quarter is calculated using that sum.

End of 1st quarter: $I = Prt = \$1000(0.05)\left(\dfrac{3}{12}\right) = \$12.50$

$A = P + I = \$1000 + \$12.50 = \$1012.50$

End of 2nd quarter: $I = Prt = \$1012.50(0.05)\left(\dfrac{3}{12}\right) = \$12.65625 \approx \$12.66$

$A = P + I = \$1012.50 + \$12.66 = \$1025.16$

End of 3rd quarter: $I = Prt = \$1025.16(0.05)\left(\dfrac{3}{12}\right) = \$12.8145 \approx \$12.81$

$A = P + I = \$1025.16 + \$12.81 = \$1037.97$

End of 4th quarter: $I = Prt = \$1037.97(0.05)\left(\dfrac{3}{12}\right) = \$12.974625 \approx \$12.97$

$A = P + I = \$1037.97 + \$12.97 = \$1050.94$

The total amount in the account at the end of the first year is $1050.94.

When the interest is compounded quarterly, the account earns more interest ($50.94) than when the interest is compounded annually ($50). In general, an increase in the number of compounding periods results in an increase in the interest earned by an account.

In the example above, the formulas $I = Prt$ and $A = P + I$ were used to show the amount of interest added to the account each quarter. The formula $A = P(1 + rt)$ can be used to calculate A at the end of each quarter. For example, the amount in the account at the end of the first quarter is

$A = P(1 + rt)$

$A = 1000\left[1 + 0.05\left(\dfrac{3}{12}\right)\right]$

$A = 1000(1.0125)$

$A = 1012.50$

This amount, $1012.50, is the same as the amount calculated above using the formula $A = P + I$ to find the amount at the end of the first quarter.

The formula $A = P(1 + rt)$ is used in Example 1.

EXAMPLE 1 ■ Calculate Future Value

You deposit $500 in an account earning 6% interest, compounded semiannually. How much is in the account at the end of 1 year?

Solution

The interest is compounded every 6 months. Calculate the amount in the account after the first 6 months. $t = \frac{6}{12}$.

$A = P(1 + rt)$

$A = 500\left[1 + 0.06\left(\dfrac{6}{12}\right)\right]$

$A = 515$

Calculate the amount in the account after the second 6 months.

$$A = P(1 + rt)$$

$$A = 515\left[1 + 0.06\left(\frac{6}{12}\right)\right]$$

$$A = 530.45$$

The total amount in the account at the end of 1 year is $530.45.

CHECK YOUR PROGRESS 1   You deposit $2000 in an account earning 4% interest, compounded monthly. How much is in the account at the end of 6 months?

Solution   *See page S35.*

In calculations that involve compound interest, the sum of the principal and the interest that has been added to it is called the **compound amount.** In Example 1, the compound amount is $530.45.

The calculations necessary to determine compound interest and compound amounts can be simplified by using a formula. Consider an amount P deposited into an account paying an annual interest rate r, compounded annually.

The interest earned during the first year is

$$I = Prt$$
$$I = Pr(1) \qquad \bullet\ t = \mathbf{1}.$$
$$I = Pr$$

The compound amount A in the account after 1 year is the sum of the original principal and the interest earned during the first year:

$$A = P + I$$
$$A = P + Pr$$
$$A = P(1 + r) \qquad \bullet\ \text{Factor } P \text{ from each term.}$$

During the second year, the interest is calculated on the compound amount at the end of the first year, $P(1 + r)$.

$$I = Prt$$
$$I = P(1 + r)r(1) \qquad \bullet\ \text{Replace } P \text{ with } P(1 + r);\ t = \mathbf{1}.$$
$$I = P(1 + r)r$$

The compound amount A in the account after 2 years is the sum of the compound amount at the end of the first year and the interest earned during the second year:

$$A = P + I$$
$$A = P(1 + r) + P(1 + r)r \qquad \bullet\ \text{Replace } P \text{ with } \mathbf{P(1 + r)} \text{ and } I \text{ with } \mathbf{P(1 + r)r}.$$
$$A = 1[P(1 + r)] + [P(1 + r)]r$$
$$A = P(1 + r)(1 + r) \qquad \bullet\ \text{Factor } P(1 + r) \text{ from each term.}$$
$$A = P(1 + r)^2 \qquad \bullet\ \text{Write } (1 + r)(1 + r) \text{ as } (1 + r)^2.$$

During the third year, the interest is calculated on the compound amount at the end of the second year, $P(1 + r)^2$.

$$I = Prt$$
$$I = P(1 + r)^2 r(1) \qquad \text{• Replace } P \text{ with } P(1+r)^2;\ t = 1.$$
$$I = P(1 + r)^2 r$$

The compound amount A in the account after 3 years is the sum of the compound amount at the end of the second year and the interest earned during the third year:

$$A = P + I$$
$$A = P(1 + r)^2 + P(1 + r)^2 r \qquad \text{• Replace } P \text{ with } P(1+r)^2 \text{ and } I \text{ with } P(1+r)^2 r.$$

$$A = 1[P(1 + r)^2] + [P(1 + r)^2]r$$
$$A = P(1 + r)^2(1 + r) \qquad \text{• Factor } P(1+r)^2 \text{ from each term.}$$
$$A = P(1 + r)^3 \qquad \text{• Write } (1+r)^2(1+r) \text{ as } (1+r)^3.$$

Note that the compound amount at the end of each year is the previous year's compound amount multiplied by $(1 + r)$. The exponent on $(1 + r)$ is equal to the number of compounding periods. Generalizing from this, we can state that the compound amount after n years is $A = P(1 + r)^n$.

In deriving this equation, interest was compounded annually; therefore, $t = 1$. Applying a similar argument for more frequent compounding periods, we derive the following compound amount formula. This formula enables us to calculate the compound amount for any number of compounding periods per year.

Compound Amount Formula

The compound amount formula is

$$A = P(1 + i)^N$$

where A is the compound amount when P dollars are deposited at an interest rate of i per compounding period for N compounding periods.

Because i is the interest rate per compounding period and n is the number of compounding periods per year,

$$i = \frac{\text{annual interest rate}}{\text{number of compounding periods per year}} = \frac{r}{n}$$

$$N = (\text{number of compounding periods per year})(\text{number of years}) = nt$$

P

To illustrate how to calculate the values of i and N, consider an account earning 12% interest, compounded quarterly, for a period of 3 years.

The annual interest rate is 12%: $r = 12\%$.

When interest is compounded quarterly, there are four compounding periods per year: $n = 4$.

The time is 3 years: $t = 3$.

$$i = \frac{r}{n} = \frac{12\%}{4} = 3\%$$

$$N = nt = 4(3) = 12$$

The number of compounding periods in the 3 years is 12: $N = 12$. The account earns 3% interest per compounding period (per quarter): $i = 3\%$. Note that the interest rate per compounding period times the number of compounding periods per year equals the annual interest rate: $(3\%)(4) = 12\%$.

Recall that compound interest can be compounded annually (once a year), semiannually (twice a year), quarterly (four times a year), monthly, or daily. The possible values of n (the number of compounding periods per year) are recorded in the table below.

Values of n (number of compounding periods per year)

If interest is	then $n =$
compounded annually	1
compounded semiannually	2
compounded quarterly	4
compounded monthly	12
compounded daily	360

QUESTION   *What is the value of i when the interest rate is 6%, compounded monthly?*

Recall that the future value of an investment is the value of the investment after the original principal has been invested for a period of time. In other words, it is the principal plus the interest earned. Therefore, it is the compound amount A in the compound amount formula.

EXAMPLE 2 ■ **Calculate the Compound Amount**

Calculate the compound amount when $10,000 is deposited in an account earning 8% interest, compounded semiannually, for 4 years.

Solution
Use the compound amount formula. $r = 8\% = 0.08$, $n = 2$, $t = 4$,
$i = \dfrac{r}{n} = \dfrac{0.08}{2} = 0.04$, and $N = nt = 2(4) = 8$.

$$A = P(1 + i)^N$$
$$A = 10{,}000(1 + 0.04)^8$$
$$A = 10{,}000(1.04)^8$$
$$A \approx 10{,}000(1.368569)$$
$$A \approx 13{,}685.69$$

The compound amount after 4 years is approximately $13,685.69.

✔ **TAKE NOTE**

When using the compound amount formula, write the interest rate r as a decimal. Then calculate i.

ANSWER $i = \dfrac{r}{n} = \dfrac{6\%}{12} = 0.5\% = 0.005$

When using a scientific calculator to solve the compound amount formula for A, use the key-stroking sequence

$$r \div n + 1 = y^x \; N \times P =$$

When using a graphing calculator, use the sequence

$$P \; (\; 1 + r \div n \;) \; \wedge \; N$$

INSTRUCTOR NOTE
You might have students assume that they invest $2500 in an IRA today and that they don't add any additional money to the account before they retire. Have them find the value of the account in 40 or 45 years. Here are a few examples: For 40 years, compounded monthly, at a 7% annual interest rate, the value will be $40,778.53. For 45 years, compounded monthly, at a 8% annual interest rate, the value will be $90,409.00.

▼ **point of interest**

The majority of Americans say their children should start saving for retirement earlier than they did. Here is the breakdown, by age, of when parents started saving for retirement. (*Source: Lincoln Financial Group-Money Magazine Survey*)

Under 20: 9%
20–24: 19%
25–29: 18%
30–39: 21%
40–49: 11%
50 and over: 6%
Don't know: 16%

CHECK YOUR PROGRESS 2   Calculate the compound amount when $4000 is deposited in an account earning 6% interest, compounded monthly, for 2 years.

Solution   *See page S36.*

EXAMPLE 3 ■ **Calculate Future Value**

Calculate the future value of $5000 earning 9% interest, compounded daily, for 3 years.

Solution
Use the compound amount formula. $r = 9\% = 0.09$, $t = 3$. Interest is compounded daily. Use the ordinary method. $n = 360$, $i = \frac{r}{n} = \frac{0.09}{360} = 0.00025$, $N = nt = 360(3) = 1080$

$$A = P(1 + i)^N$$
$$A = 5000(1 + 0.00025)^{1080}$$
$$A = 5000(1.00025)^{1080}$$
$$A \approx 5000(1.3099202)$$
$$A \approx 6549.60$$

The future amount after 3 years is approximately $6549.60.

CHECK YOUR PROGRESS 3   Calculate the future value of $2500 earning 9% interest, compounded daily, for 4 years.

Solution   *See page S36.*

The formula $I = A - P$ was used in Section 10.1 to find the interest earned on an investment or the interest paid on a loan. This same formula is used for compound interest. It is used in Example 4 to find the interest earned on an investment.

EXAMPLE 4 ■ **Calculate Compound Interest**

How much interest is earned in 2 years on $4000 deposited in an account paying 6% interest, compounded quarterly?

Solution
Calculate the compound amount. Use the compound amount formula. $r = 6\% = 0.06$, $n = 4$, $t = 2$, $i = \frac{r}{n} = \frac{0.06}{4} = 0.015$, $N = nt = 4(2) = 8$

$$A = P(1 + i)^N$$
$$A = 4000(1 + 0.015)^8$$
$$A = 4000(1.015)^8$$
$$A \approx 4000(1.1264926)$$
$$A \approx 4505.97$$

Calculate the interest earned. Use the formula $I = A - P$.

$$I = A - P$$
$$I = 4505.97 - 4000$$
$$I = 505.97$$

The amount of interest earned is approximately $505.97.

CHECK YOUR PROGRESS 4   How much interest is earned in 6 years on $8000 deposited in an account paying 9% interest, compounded monthly?

Solution   *See page S36.*

An alternative to using the compound amount formula is to use a calculator that has finance functions built into it. The TI-83 graphing calculator is used in Example 5.

EXAMPLE 5 ■ Calculate Compound Interest

 Use the finance feature of a calculator to determine the compound amount when $2000 is deposited in an account earning an interest rate of 12%, compounded quarterly, for 10 years.

Solution

On a TI-83 calculator, press $\boxed{\text{2nd}}$ [Finance] to display the FINANCE CALC menu.

For the TI-83 Plus, press $\boxed{\text{APPS}}$ $\boxed{\text{ENTER}}$.

Press $\boxed{\text{ENTER}}$ to select 1: TVM Solver.

> After N =, enter 40.
>
> After I% =, enter 12.
>
> After PV =, enter −2000. (See the note below.)
>
> After PMT =, enter 0.
>
> After P/Y =, enter 4.
>
> After C/Y =, enter 4.

Use the up arrow key to place the cursor at FV =.

Press $\boxed{\text{ALPHA}}$ [Solve].

The solution is displayed to the right of FV =.

The compound amount is $6524.08.

Note: For most financial calculators and financial computer programs such as Excel, money that is paid out (such as the $2000 that is being deposited in this example) is entered as a negative number.

CHECK YOUR PROGRESS 5   Use the finance feature of a calculator to determine the compound amount when $3500 is deposited in an account earning an interest rate of 6%, compounded semiannually, for 5 years.

Solution   *See page S36.*

CALCULATOR NOTE

On the TI-83 calculator screen below, the variable N is the number of compounding periods, I% is the interest rate, PV is the present value of the money, PMT is the payment, P/Y is the number of payments per year, C/Y is the number of compounding periods per year, and FV is the future value of the money. TVM represents Time Value of money.

Present value is discussed a little later in this section. Enter the principal for the present value amount.

"Solve" is above the $\boxed{\text{ENTER}}$ key on the calculator. A typical screen is shown below.

```
N=40
I%=12
PV=-2000
PMT=0
■FV=6524.075584
P/Y=4
C/Y=4
PMT: END BEGIN
```

Math Matters   Federal Funds Rate

The Federal Reserve Board meets a number of times each year to decide, among other things, the federal funds rate. The **federal funds rate** is the interest rate banks charge one another for overnight loans. The Federal Reserve Board can decide to lower, raise, or leave unchanged the federal funds rate. Its decision has a significant effect on the economy. For example, if the board increases the rate, then banks have to pay more to borrow money. If banks must pay higher rates, they are going to charge their customers a higher rate to borrow money.

The table below provides a list of the Federal Reserve Board's changes to the federal funds rate during 2000 and 2001.

Date	Rate Change
February 2, 2000	Rate raised to 5.75%
March 21, 2000	Rate raised to 6.00%
May 16, 2000	Rate raised to 6.50%
January 3, 2001	Rate lowered to 6.00%
January 31, 2001	Rate lowered to 5.50%
March 20, 2001	Rate lowered to 5.00%
April 18, 2001	Rate lowered to 4.50%
May 15, 2001	Rate lowered to 4.00%
June 27, 2001	Rate lowered to 3.75%
August 21, 2001	Rate lowered to 3.50%
September 17, 2001	Rate lowered to 3.00%
October 2, 2001	Rate lowered to 2.50%
November 6, 2001	Rate lowered to 2.00%
December 11, 2001	Rate lowered to 1.75%

Changes in the federal funds rate during 2000 and 2001

Present Value

The **present value** of an investment is the original principal invested, or the value of the investment before it earns any interest. Therefore, it is the principal, P, in the compound amount formula. Present value is used to determine how much money must be invested today in order for an investment to have a specific value at a future date.

The formula for the present value of an investment is found by solving the compound amount formula for P.

$$A = P(1 + i)^N$$

$$\frac{A}{(1 + i)^N} = \frac{P(1 + i)^N}{(1 + i)^N} \qquad \text{• Divide each side by } (1 + i)^N$$

$$\frac{A}{(1 + i)^N} = P$$

> **Present Value Formula**
>
> The present value formula is
>
> $$P = \frac{A}{(1 + i)^N}$$
>
> where P is the original principal invested at an interest rate of i per compounding period for N compounding periods, and A is the compound amount.

The present value formula is used in Example 6.

EXAMPLE 6 ■ Calculate Present Value

How much money should be invested in an account that earns 8% interest, compounded quarterly, in order to have $30,000 in 5 years?

Solution
Use the present value formula. $r = 8\%$, $n = 4$, $t = 5$, $i = \frac{r}{n} = \frac{8\%}{4} = 2\% = 0.02$, $N = nt = 4(5) = 20$

$$P = \frac{A}{(1 + i)^N}$$

$$P = \frac{30,000}{(1 + 0.02)^{20}} = \frac{30,000}{1.02^{20}} \approx \frac{30,000}{1.485947}$$

$$P \approx 20,189.14$$

$20,189.14 should be invested in the account in order to have $30,000 in 5 years.

CALCULATOR NOTE

To calculate P in the present value formula using a calculator, you can first calculate $(1 + i)^N$. Next use the $\boxed{1/x}$ key or the $\boxed{x^{-1}}$ key. This will place the value of $(1 + i)^N$ in the denominator. Then multiply by the value of A.

CHECK YOUR PROGRESS 6   How much money should be invested in an account that earns 9% interest, compounded semiannually, in order to have $20,000 in 5 years?

Solution   *See page S36.*

EXAMPLE 7 ■ Calculate Present Value

Use the finance feature of a calculator to determine how much money should be invested in an account that earns 7% interest, compounded monthly, in order to have $50,000 in 10 years.

Solution
On a TI-83 calculator, press $\boxed{\text{2nd}}$ [FINANCE] to display the FINANCE CALC menu. For the TI-83 Plus, press $\boxed{\text{APPS}}$ $\boxed{\text{ENTER}}$.

Press $\boxed{\text{ENTER}}$ to select 1 TVM Solver.

     After N =, enter 120.
     After I% =, enter 7.
     After PMT =, enter 0.
     After FV =, enter 50000.
     After P/Y =, enter 12.
     After C/Y =, enter 12.

Use the up arrow key to place the cursor at PV =.

Press [ALPHA] [Solve].

The solution is displayed to the right of PV =.

$24,879.81 should be invested in the account in order to have $50,000 in 10 years.

CHECK YOUR PROGRESS 7　Use the finance feature of a calculator to determine how much money should be invested in an account that earns 6% interest, compounded daily, in order to have $25,000 in 15 years.

Solution　*See page S36.*

Inflation

We have discussed compound interest and its effect on the growth of an investment. After your money has been invested for a period of time in an account that pays interest, you will have more money than you originally deposited. But does that mean you will be able to buy more with the compound amount than you were able to buy with the original investment at the time you deposited the money? The answer is not necessarily, and the reason is because of the effect of inflation.

Suppose the price of a large-screen TV is $1500. You have enough money to purchase the TV, but decide to invest the $1500 in an account paying 6% interest, compounded monthly. After 1 year, the compound amount is $1592.52. But during that same year, the rate of inflation was 7%. The large-screen TV now costs:

$$\$1500 \text{ plus } 7\% \text{ of } \$1500 = \$1500 + 0.07(\$1500)$$
$$= \$1500 + \$105$$
$$= \$1605$$

Because $1592.52 < $1605, you have actually lost purchasing power. At the beginning of the year, you had enough money to buy the giant-screen TV; at the end of the year, the compound amount is not enough to pay for that same TV. Your money has actually lost value because it can buy less now than it could 1 year ago.

Inflation is an economic condition during which there are increases in the costs of goods and services. Inflation is expressed as a percent; for example, we speak of an annual inflation rate of 7%.

To calculate the effects of inflation, we use the same procedure we used to calculate compound amount. This process is illustrated in Example 8. Although inflation rates vary dramatically, in this section we will assume constant annual inflation rates, and we will use annual compounding in solving inflation problems. In other words, $n = 1$ for these exercises.

EXAMPLE 8 ■ **Calculate the Effect of Inflation on Salary**

Suppose your annual salary today is $35,000. You want to know what an equivalent salary will be in 20 years; that is, a salary that will have the same purchasing power. Assume a 6% inflation rate.

▼ **point of interest**

Inflation rates can change dramatically over time. As of December 2001, the annual inflation rate was 1.90%. The table below shows the five smallest and five largest inflation rates during the past 30 years. (*Source: Financial Trend Forecaster at* **www.fintrend.com**)

5 Smallest Inflation Rates	
Inflation Rate	Year
1.55%	1998
1.86%	1986
2.19%	1999
2.44%	1996
2.54%	1997

5 Largest Inflation Rates	
Inflation Rate	Year
13.50%	1980
11.35%	1979
11.04%	1974
10.32%	1981
9.13%	1975

Solution

Use the compound amount formula. $r = 6\% = 0.06$. The inflation rate is an annual rate, so $n = 1$. $t = 20$, $i = \frac{r}{n} = \frac{0.06}{1} = 0.06$, $N = nt = 1(20) = 20$

$$A = P(1 + i)^N$$
$$A = 35,000(1 + 0.06)^{20}$$
$$A = 35,000(1.06)^{20}$$
$$A \approx 35,000(3.20713547)$$
$$A \approx 112,249.74$$

Twenty years from now, you need to earn an annual salary of approximately $112,249.74 in order to have the same purchasing power.

CHECK YOUR PROGRESS 8   Assume that the average new car sticker price in 2003 is $28,000. Use an annual inflation rate of 5% to estimate the average new car sticker price in 2020.

Solution   *See page S36.*

The present value formula can be used to determine the effect of inflation on the future purchasing power of a given amount of money. Substitute the inflation rate for the interest rate in the present value formula. The compounding period is 1 year. Again we will assume a constant rate of inflation.

EXAMPLE 9 ■ **Calculate the Effect of Inflation on Future Purchasing Power**

Suppose you purchase an insurance policy in 2005 that will provide you with $250,000 when you retire in 2040. Assuming an annual inflation rate of 8%, what will be the purchasing power of the $250,000 in 2040?

Solution

Use the present value formula. $r = 8\%$ and $t = 35$. The inflation rate is an annual rate, so $n = 1$. $i = \frac{r}{n} = \frac{8\%}{1} = 8\% = 0.08$, $N = nt = 1(35) = 35$

$$P = \frac{A}{(1 + i)^N}$$
$$P = \frac{250,000}{(1 + 0.08)^{35}}$$
$$P \approx \frac{250,000}{14.785344}$$
$$P \approx 16,908.64$$

In 2040, the purchasing power of $250,000 will be approximately $16,908.64.

CHECK YOUR PROGRESS 9   Suppose you purchase an insurance policy in 2000 that will provide you with $500,000 when you retire in 40 years. Assuming an annual inflation rate of 7%, what will be the purchasing power of half a million dollars in 2040?

Solution   *See page S37.*

Math Matters   The Rule of 72

The **Rule of 72** states that the number of years for prices to double is approximately equal to 72 divided by the annual inflation rate.

$$\text{Years to double} = \frac{72}{\text{annual inflation rate}}$$

For example, at an annual inflation rate of 6%, prices will double in approximately 12 years.

$$\text{Years to double} = \frac{72}{\text{annual inflation rate}} = \frac{72}{6} = 12$$

Effective Interest Rate

When interest is compounded, the annual rate of interest is called the **nominal rate**. The **effective rate** is the simple interest rate that would yield the same amount of interest after 1 year. When a bank advertises a "7% annual interest rate compounded daily and yielding 7.25%," the nominal interest rate is 7% and the effective rate is 7.25%.

> **QUESTION**   *A bank offers a savings account that pays 2.75% annual interest, compounded daily and yielding 2.79%. What is the effective rate on this account? What is the nominal rate?*

▼ **point of interest**

As of July 31, 2000, of the $539,890,223,079 in U.S. currency in circulation through-out the world, $364,724,397,100 was in $100 bills. (*Source:* **www.moneyfactory.gov**)

Consider $100 deposited at 6%, compounded monthly, for 1 year.

The future value after 1 year is $106.17.

$A = P(1 + i)^N$
$A = 100(1 + 0.005)^{12}$
$A \approx 106.17$

The interest earned in 1 year is $6.17.

$I = A - P$
$I = 106.17 - 100$
$I = 6.17$

Now consider $100 deposited at an annual simple interest rate of 6.17%.

The interest earned in 1 year is $6.17.

$I = Prt$
$I = 100(0.0617)(1)$
$I = 6.17$

The interest earned on $100 is the same when it is deposited at 6% compounded monthly as when it is deposited at an annual simple interest rate of 6.17%. 6.17% is the effective annual rate of 6% compounded monthly.

ANSWER   *The effective rate is 2.79%. The nominal rate is 2.75%.*

In this example $100 was used as the principal. When we use $100 for *P* we multiply the interest rate by 100. Remember that the interest rate is written as a decimal in the equation $I = Prt$, and a decimal is written as a percent by multiplying by 100. Therefore, when $P = 100$, the interest earned on the investment ($6.17) is the same number as the effective annual rate (6.17%).

TAKE NOTE

To compare two investments or loan agreements, calculate the effective interest rate of each. When you are investing your money, you want the highest effective rate so that your money will earn more interest. If you are borrowing money, you want the lowest effective rate so that you will pay less interest on the loan.

EXAMPLE 10 ■ Calculate the Effective Interest Rate

A credit union offers a certificate of deposit at an annual interest rate of 3%, compounded monthly. Find the effective rate. Round to the nearest hundredth of a percent.

Solution

Use the compound amount formula to find the future value of $100 after 1 year.
$r = 3\% = 0.03$, $n = 12$, $t = 1$, $i = \frac{r}{n} = \frac{0.03}{12} = 0.0025$, $N = nt = 12(1) = 12$

$$A = P(1 + i)^N$$
$$A = 100(1 + 0.0025)^{12}$$
$$A = 100(1.0025)^{12}$$
$$A \approx 100(1.030415957)$$
$$A \approx 103.04$$

Find the interest earned on the $100.

$$I = A - P$$
$$I = 103.04 - 100$$
$$I = 3.04$$

The effective interest rate is 3.04%.

CHECK YOUR PROGRESS 10

A bank offers a certificate of deposit at an annual interest rate of 4%, compounded quarterly. Find the effective rate. Round to the nearest hundredth of a percent.

Solution See page S37.

To compare two investments or loan agreements, we could calculate the effective annual rate of each. However, a shorter method involves comparing the compound amounts of each. Because the value of $(1 + i)^N$ is the compound amount of $1, we can compare the value of $(1 + i)^N$ for each alternative.

EXAMPLE 11 ■ Compare Annual Yields

One bank advertises an interest rate of 5.5%, compounded quarterly, on a certificate of deposit. Another bank advertises an interest rate of 5.25%, compounded monthly. Which investment has the higher annual yield?

Solution
Calculate $(1 + i)^N$ for each investment.

$$i = \frac{r}{n} = \frac{0.055}{4} \qquad\qquad i = \frac{r}{n} = \frac{0.0525}{12}$$

$$N = nt = 4(1) = 4 \qquad\qquad N = nt = 12(1) = 12$$

$$(1 + i)^N = \left(1 + \frac{0.055}{4}\right)^4 \qquad (1 + i)^N = \left(1 + \frac{0.0525}{12}\right)^{12}$$

$$\approx 1.0561448 \qquad\qquad \approx 1.0537819$$

Compare the two compound amounts.

$$1.0561448 > 1.0537819$$

5.5% compounded quarterly has a higher annual yield than 5.25% compounded monthly.

CHECK YOUR PROGRESS 11   Which has the higher annual yield, 5% compounded quarterly or 5.25% compounded semiannually?

Solution   *See page S37.*

Math Matters   Saving for Retirement

The tables below show results of surveys in which workers were asked questions about saving for retirement.

How many workers fear they've fallen behind in saving				
	56 and older	47 – 55	37 – 46	25 and younger
On track to save enough for retirement	39%	30%	32%	33%
Ahead of schedule	6%	5%	5%	7%
A little behind schedule	21%	27%	24%	28%
A lot behind schedule	30%	37%	37%	30%

Source: 12th Annual Retirement Confidence Survey. Reprinted by permission of the Employee Benefits Research Institute.

Worker confidence in having enough money to live comfortably throughout their retirement years									
	1993	1994	1995	1996	1997	1998	1999	2000	2001
Very confident	18%	20%	21%	19%	24%	22%	22%	25%	22%
Somewhat confident	55%	45%	51%	41%	41%	45%	47%	47%	41%
Not too confident	19%	17%	19%	23%	19%	18%	21%	18%	18%
Not at all confident	6%	17%	8%	16%	15%	13%	9%	10%	17%

Source: 12th Annual Retirement Confidence Survey. Reprinted by permission of the Employee Benefits Research Institute.

Excursion

Consumer Price Index

An **index number** measures the change in a quantity, such as cost, over a period of time. One of the most widely used indexes is the Consumer Price Index (CPI). The CPI, which includes the selling prices of about 400 key consumer goods and services, indicates the relative change in the price of these items over time. It measures the effect of inflation on the cost of goods and services.

The main components of the Consumer Price Index, shown below, are the costs of housing, food and beverages, transportation, medical care, and clothing.

✓ **TAKE NOTE**

The category "Recreation" includes television sets, cable TV, pets and pet products, sports equipment, and admissions.

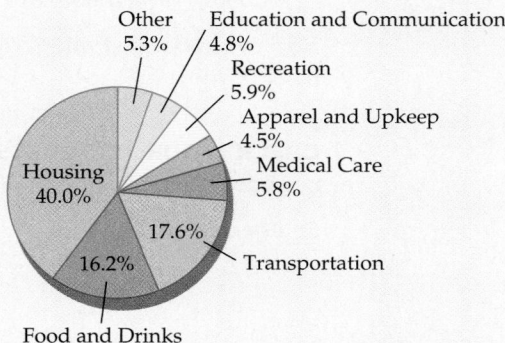

The Components of the Consumer Price Index

Source: U.S. Bureau of Labor Statistics

✓ **TAKE NOTE**

You can obtain current data on the Consumer Price Index by visiting the web site of the Bureau of Labor Statistics at **www.bls.gov.**

The CPI is a measure of the cost of living for consumers. The government publishes monthly and annual figures on the Consumer Price Index.

The Consumer Price Index has a base period, 1982–1984, from which to make comparisons. The CPI for the base period is 100. The CPI for November 2000 was 174.1. This means that $100 in the period 1982–1984 had the same purchasing power as $174.10 in November of 2000.

An index number is actually a percent written without a percent sign. The CPI of 174.1 for November 2000 means that the average cost of consumer goods at that time was 174.1% of their cost in the 1982–84 period.

The table below gives the CPI for various products in 2000.

Product	CPI
All items	172.2
Food and beverages	168.4
Housing	169.6
Apparel	129.6
Transportation	153.3
Medical care	260.8
Recreation	103.3
Edudation and communication	102.5

The Consumer Price Index, 2000

Source: U.S. Bureau of Labor Statistics

(continued)

Excursion Exercises

Solve the following.

1. The CPI for 1985 was 107.6. What percent of the base period prices were the consumer prices in 1985?

2. The CPI for 1986 was 109.6. The percent increase in consumer prices from 1986 to 1987 was 3.6%. Find the CPI for 1987. Round to the nearest tenth.

3. The CPI for 1988 was 118.3. The CPI for 1989 was 124. Find the percent increase in consumer prices for this time period.

4. Of the items listed in the table on the previous page, are there any items that cost at least twice as much in 2000 as they cost during the base years? If so, which ones?

5. Of the items listed in the table on the previous page, are there any items that cost more than one-and-one-half times as much in 2000 as they cost during the base years, but less than twice as much as they cost during the base years? If so, which ones?

6. If the cost of textbooks for one semester was $120 in the base years, how much did similar textbooks cost in 2000?

7. If a movie ticket cost $9 in 2000, what would a comparable movie ticket have cost during the base years?

8. The base year for the consumer confidence index is 1985. The consumer confidence index in November of 2001 was 82.2. Were consumers more confident in 1985 or in November of 2001?

Exercise Set 10.2   (Suggested Assignment: 27–89, odds)

 Use a calculator for Exercises 1–94.

In Exercises 1–8, calculate the compound amount. Use the compound amount formula and a calculator.

1. $P = \$1200$, $r = 7\%$ compounded semiannually, $t = 12$ years

2. $P = \$3500$, $r = 8\%$ compounded semiannually, $t = 14$ years

3. $P = \$500$, $r = 9\%$ compounded quarterly, $t = 6$ years

4. $P = \$7000$, $r = 11\%$ compounded quarterly, $t = 9$ years

5. $P = \$8500$, $r = 9\%$ compounded monthly, $t = 10$ years

6. $P = \$6400$, $r = 6\%$ compounded monthly, $t = 3$ years

7. $P = \$9600$, $r = 9\%$ compounded daily, $t = 3$ years

8. $P = \$1700$, $r = 9\%$ compounded daily, $t = 5$ years

In Exercises 9–14, calculate the compound amount. Use a calculator with a financial mode.

9. $P = \$1600$, $r = 8\%$ compounded quarterly, $t = 10$ years

10. $P = \$4200$, $r = 6\%$ compounded semiannually, $t = 8$ years

11. $P = \$3000$, $r = 12\%$ compounded monthly, $t = 5$ years

12. $P = \$9800$, $r = 10\%$ compounded quarterly, $t = 4$ years

13. $P = \$1700$, $r = 9\%$ compounded semiannually, $t = 3$ years

14. $P = \$8600$, $r = 11\%$ compounded semiannually, $t = 5$ years

In Exercises 15–20, calculate the future value.

15. $P = \$7500$, $r = 12\%$ compounded monthly, $t = 5$ years

16. $P = \$1800$, $r = 9.5\%$ compounded annually, $t = 10$ years

17. $P = \$4600$, $r = 10\%$ compounded semiannually, $t = 12$ years

18. $P = \$9000$, $r = 11\%$ compounded quarterly, $t = 3$ years

19. $P = \$22,000$, $r = 9\%$ compounded monthly, $t = 7$ years

20. $P = \$5200$, $r = 8.1\%$ compounded daily, $t = 9$ years

In Exercises 21–26, calculate the present value.

21. $A = \$25,000$, $r = 10\%$ compounded quarterly, $t = 12$ years

22. $A = \$20,000$, $r = 12\%$ compounded monthly, $t = 5$ years

23. $A = \$40,000$, $r = 7.5\%$ compounded annually, $t = 35$ years

24. $A = \$10,000$, $r = 11\%$ compounded semiannually, $t = 30$ years

25. $A = \$15,000$, $r = 8\%$ compounded quarterly, $t = 5$ years

26. $A = \$50,000$, $r = 18\%$ compounded monthly, $t = 5$ years

27. Calculate the compound amount when $8000 is deposited in an account earning 8% interest, compounded quarterly, for 5 years.

28. Calculate the compound amount when $3000 is deposited in an account earning 10% interest, compounded semiannually, for 3 years.

29. If you leave $2500 in an account earning 9% interest, compounded daily, how much money will be in the account after 4 years?

30. What is the compound amount when $1500 is deposited in an account earning an interest rate of 6%, compounded monthly, for 2 years?

31. What is the future value of $4000 earning 12% interest, compounded monthly, for 6 years?

32. Calculate the future value of $8000 earning 8% interest, compounded quarterly, for 10 years.

33. How much interest is earned in 3 years on $2000 deposited in an account paying 6% interest, compounded quarterly?

34. How much interest is earned in 5 years on $8500 deposited in an account paying 9% interest, compounded semiannually?

35. Calculate the amount of interest earned in 8 years on $15,000 deposited in an account paying 10% interest, compounded quarterly.

36. Calculate the amount of interest earned in 6 years on $20,000 deposited in an account paying 12% interest, compounded monthly.

37. How much money should be invested in an account that earns 6% interest, compounded monthly, in order to have $15,000 in 5 years?

38. How much money should be invested in an account that earns 7% interest, compounded quarterly, in order to have $10,000 in 5 years?

39. $1000 is deposited for 5 years in an account that earns 9% interest.

  a. Calculate the simple interest earned.

  b. Calculate the interest earned if interest is compounded daily.

  c. How much more interest is earned on the account when the interest is compounded daily?

40. $10,000 is deposited for 2 years in an account that earns 12% interest.

  a. Calculate the simple interest earned.

  b. Calculate the interest earned if interest is compounded daily.

  c. How much more interest is earned on the account when the interest is compounded daily?

41. $15,000 is deposited for 4 years in an account earning 8% interest.

  a. Calculate the future value of the investment if interest is compounded semiannually.

  b. Calculate the future value if interest is compounded quarterly.

  c. How much greater is the future value of the investment when the interest is compounded quarterly?

42. $25,000 is deposited for 3 years in an account earning 6% interest.

  a. Calculate the future value of the investment if interest is compounded annually.

  b. Calculate the future value if interest is compounded semiannually.

  c. How much greater is the future value of the investment when the interest is compounded semiannually?

43. $10,000 is deposited for 2 years in an account earning 8% interest.

  a. Calculate the interest earned if interest is compounded semiannually.

b. Calculate the interest earned if interest is compounded quarterly.

c. How much more interest is earned on the account when the interest is compounded quarterly?

44. $20,000 is deposited for 5 years in an account earning 6% interest.

a. Calculate the interest earned if interest is compounded annually.

b. Calculate the interest earned if interest is compounded semiannually.

c. How much more interest is earned on the account when the interest is compounded semiannually?

45. To help pay your college expenses, you borrow $7000 and agree to repay the loan at the end of 5 years at 8% interest, compounded quarterly.

a. What is the maturity value of the loan?

b. How much interest are you paying on the loan?

46. You borrow $6000 to help pay your college expenses. You agree to repay the loan at the end of 5 years at 10% interest, compounded quarterly.

a. What is the maturity value of the loan?

b. How much interest are you paying on the loan?

47. A couple plans to save for their child's college education. What principal must be deposited by the parents when their child is born in order to have $40,000 when the child reaches the age of 18? Assume the money earns 8% interest, compounded quarterly.

48. A couple plans to invest money for their child's college education. What principal must be deposited by the parents when their child turns 10 in order to have $30,000 when the child reaches the age of 18? Assume the money earns 8% interest, compounded quarterly.

49. You want to retire in 30 years with $1,000,000 in investments.

a. How much money would you have to invest today at 9% interest, compounded daily, in order to have $1,000,000 in 30 years?

b. How much will the $1,000,000 generate in interest each year if it is invested at 9% interest, compounded daily?

50. You want to retire in 40 years with $1,000,000 in investments.

a. How much money must you invest today at 8.1% interest, compounded daily, in order to have $1,000,000 in 40 years?

b. How much will the $1,000,000 generate in interest each year if it is invested at 8.1% interest, compounded daily?

51. You deposit $5000 in a two-year certificate of deposit (CD) earning 8.1% interest, compounded daily. At the end of the 2 years, you reinvest the compound amount in another two-year CD. The interest rate on the second CD is 7.2%, compounded daily. What is the compound amount when the second CD matures?

52. You deposit $7500 in a two-year certificate of deposit (CD) earning 9.9% interest, compounded daily. At the end of the 2 years, you reinvest the compound amount plus an additional $7500 in another two-year CD. The interest rate on the second CD is 10.8%, compounded daily. What is the compound amount when the second CD matures?

53. The average annual cost of renting a three-room apartment in Casper, Wyoming is $3180. Using an annual inflation rate of 7%, find the annual rent in 15 years.

54. The average annual cost of renting a three-room apartment in Corbin, Kentucky is $2400. Using an annual inflation rate of 7%, find the annual rent in 10 years.

55. Suppose your salary in 2005 is $40,000. Assuming an annual inflation rate of 7%, what salary do you need to earn in 2010 in order to have the same purchasing power?

56. Suppose your salary in 2005 is $50,000. Assuming an annual inflation rate of 6%, what salary do you need to earn in 2015 in order to have the same purchasing power?

57. In 2004 you purchase an insurance policy that will provide you with $125,000 when you retire in 2044. Assuming an annual inflation rate of 6%, what will be the purchasing power of the $125,000 in 2044?

58. You purchase an insurance policy in the year 2005 that will provide you with $250,000 when you retire in 25 years. Assuming an annual inflation rate of 8%, what will be the purchasing power of the quarter-of-a-million dollars in 2030?

59. A retired couple have a fixed income of $3500 per month. Assuming an annual inflation rate of 7%, what is the purchasing power of their monthly income in 5 years?

60. A retired couple have a fixed income of $46,000 per year. Assuming an annual inflation rate of 6%, what is the purchasing power of their annual income in 10 years?

In Exercises 61–68, calculate the effective annual rate. Round to the nearest hundredth of a percent.

61. 7.2% interest compounded quarterly

62. 8.4% interest compounded quarterly

63. 7.5% interest compounded monthly

64. 6.9% interest compounded monthly

65. 8.1% interest compounded daily

66. 6.3% interest compounded daily

67. 5.94% interest compounded monthly

68. 6.27% interest compounded monthly

In Exercises 69–76, you are given the 2000 price of an item. Use an inflation rate of 6% to calculate its price in 2005, 2010, and 2020. Round to the nearest cent.

69. Gasoline: $1.45 per gallon

70. Milk: $2.59 per gallon

71. Loaf of bread: $2.19

72. Sunday newspaper: $2.25

73. Ticket to a movie: $8

74. Paperback novel: $8.95

75. House: $175,000

76. Car: $24,000

In Exercises 77–82, calculate the purchasing power using an annual inflation rate of 7%. Round to the nearest cent.

77. $50,000 in 10 years

78. $25,000 in 8 years

79. $100,000 in 20 years

80. $30,000 in 15 years

81. $75,000 in 5 years

82. $20,000 in 25 years

83. a. Complete the table.

Nominal Rate	Effective Rate
4% annual compounding	_____
4% semiannual compounding	_____
4% quarterly compounding	_____
4% monthly compounding	_____
4% daily compounding	_____

b. As the number of compounding periods increases, does the effective rate increase or decrease?

84. Beth Chipman has money in a savings account that earns an annual interest rate of 4%, compounded monthly. What is the effective rate of interest on Beth's account? Round to the nearest hundredth of a percent.

85. Blake Hamilton has money in a savings account that earns an annual interest rate of 3%, compounded monthly. What is the effective rate of interest on Blake's savings? Round to the nearest hundredth of a percent.

86. One bank advertises an interest rate of 6.6%, compounded quarterly, on a certificate of deposit. Another bank advertises an interest rate of 6.25%, compounded monthly. Which investment has the higher annual yield?

87. Which has the higher annual yield, 6% compounded quarterly or 6.25% compounded semiannually?

88. Which investment has the higher annual yield, one earning 7.8% compounded monthly or one earning 7.5% compounded daily?

89. One bank advertises an interest rate of 5.8%, compounded quarterly, on a certificate of deposit. Another bank advertises an interest rate of 5.6%, compounded monthly. Which investment has a higher annual yield?

Extensions

CRITICAL THINKING

90. a. Using an 8% interest rate with interest compounded quarterly, calculate the future value in 2 years of $1000, $2000, and $4000.

b. Based on your answers to part a, what is the future value if the investment is doubled again, to $8000?

91. The future value of $2000 deposited in an account for 25 years with interest compounded semiannually is $22,934.80. Find the interest rate.

92. You want to retire with an interest income of $12,000 per month. How much principal must be invested at an interest rate of 8%, compounded monthly, to generate this amount of monthly income? Round to the nearest hundred thousand.

COOPERATIVE LEARNING

93. You want to buy a motorcycle costing $20,000. You have two options:

a. You can borrow $20,000 at an effective annual rate of 8% for 1 year.

b. You can save the money you would have made in loan payments during 1 year and purchase the motorcycle.

If you decide to save your money for 1 year, you will deposit the equivalent of 1 month's loan payment ($1800) into your savings account at the end of each month, and you will earn 5% interest on the account. (In determining the amount of interest earned each month, assume that each month is $\frac{1}{12}$ of 1 year.) You will pay 28% of the interest earned on the savings account in income taxes. Also, an annual inflation rate of 7% will have increased the price of the motorcycle by 7%. If you choose the option of saving your money for 1 year, how much money will you have left after you pay the income tax and purchase the motorcycle?

EXPLORATIONS

94. In our discussion of compound interest, we used annual, semiannual, monthly, quarterly, and daily compounding periods. When interest is compounded daily, it is compounded 360 times a year. If interest were compounded twice daily, it would be compounded $360(2) = 720$ times a year. If interest were compounded four times a day, it would be compounded $360(4) = 1440$ times a year. Remember that the more frequent the compounding period, the more interest earned on the account. Therefore, if interest is compounded more frequently than daily, an investment will earn even more interest than if interest is compounded daily.

Some banking institutions advertise **continuous compounding,** which means that the number of compounding periods per year gets very, very large. When compounding continuously, instead of using the compound amount formula $A = P(1 + i)^N$, the following formula is used.

$$A = Pe^{rt}$$

In this formula, A is the compounded amount when P dollars are deposited at an annual interest rate of r percent compounded continuously for t years. The number e is approximately equal to 2.7182818.

The number e is found in many real-world applications. It is an irrational number, so its decimal representation never terminates or repeats. Scientific calculators have an $\boxed{e^x}$ key for evaluating exponential expressions in which e is the base.

To calculate the compound amount when $10,000 is invested for 5 years at an interest rate of 10%, compounded continuously, use the formula for continuous compounding. Substitute the following values into the formula: $P = 10,000$, $r = 10\% = 0.10$, and $t = 5$.

$$A = Pe^{rt}$$
$$A = 10,000(2.7182818)^{0.10(5)}$$
$$A \approx 16,487.21$$

The compound amount is $16,487.21.

In the following exercises, calculate the compound interest when interest is compounded continuously.

a. $P = \$5000$, $r = 8\%$, $t = 6$ years

b. $P = \$8000$, $r = 7\%$, $t = 15$ years

c. $P = \$12,000$, $r = 9\%$, $t = 10$ years

d. $P = \$7000$, $r = 6\%$, $t = 8$ years

e. $P = \$3000$, $r = 7.5\%$, $t = 4$ years

f. $P = \$9000$, $r = 8.6\%$, $t = 5$ years

Solve the following exercises.

g. Calculate the compound amount when $2500 is deposited in an account earning 11% interest, compounded continuously, for 12 years.

h. What is the future value of $15,000 earning 9.5% interest, compounded continuously, for 7 years?

i. How much interest is earned in 9 years on $6000 deposited in an account paying 10% interest, compounded continuously?

j. $25,000 is deposited for 10 years in an account that earns 8% interest. Calculate the future value of the investment if interest is compounded quarterly and if interest is compounded continuously. How much greater is the future value of the investment when interest is compounded continuously?

SECTION 10.3 | # Credit Cards and Consumer Loans

Credit Cards

When a customer uses a credit card to make a purchase, the customer is actually receiving a loan. Therefore, there is frequently an added cost to the consumer who purchases on credit. This added cost may be in the form of an annual fee or interest charges on purchases. A **finance charge** is an amount paid in excess of the cash price; it is the cost to the customer for the use of credit.

Most credit card companies issue monthly bills. The due date on the bill is usually 1 month after the billing date (the date the bill is prepared and sent to the customer). If the bill is paid in full by the due date, the customer pays no finance charge. If the bill is not paid in full by the due date, a finance charge is added to the next bill.

Suppose a credit card billing date is the 10th day of each month. If a credit card purchase is made on April 15, then May 10 is the billing date (the 10th day of the month following April). The due date is June 10 (one month from the billing date). If the bill is paid in full before June 10, no finance charge is added. However, if the bill is not paid in full, interest charges on the outstanding balance will start to accrue (be added) on June 10, and any purchase made after June 10 will immediately start accruing interest.

The most common method of determining finance charges is the **average daily balance method.** Interest charges are based on the credit card's average daily balance, which is calculated by dividing the sum of the total amounts owed each day of the month by the number of days in the billing period.

▼ point of interest

The table below, based on information from CardWeb.com, Inc., shows the average credit card debt per household in the United States.

Year	Credit Card Debt per Household
1998	7188
1999	7564
2000	8123
2001*	8488

* Through June 30, 2001.

A 2000 survey by Nellie Mae, a provider of student loans, showed that college undergraduates had an average monthly credit card balance of $2748. This is more than twice the balance in 1993. The study also showed that approximately 10% of the students who graduate will have a balance exceeding $7000.

Average Daily Balance

$$\text{Average daily balance} = \frac{\text{sum of the total amounts owed each day of the month}}{\text{number of days in the billing period}}$$

An example of calculating the average daily balance follows.

Suppose an unpaid bill for $315 had a due date of April 10. A purchase of $28 was made on April 12, and $123 was charged on April 24. A payment of $50 was made on April 15. The next billing date is May 10. The interest on the average daily balance is 1.5% per month. Find the finance charge on the May 10 bill.

To find the finance charge, first prepare a table showing the unpaid balance for each purchase, the number of days the balance is owed, and the product of these numbers. A negative sign in the Payments or Purchases column of the table indicates that a payment was made on that date.

Date	Payments or Purchases	Balance Each Day	Number of Days Until Balance Changes	Unpaid Balance Times Number of Days
April 10 – 11		$315	2	$630
April 12 – 14	$28	$343	3	$1029
April 15 – 23	– $50	$293	9	$2637
April 24 – May 9	$123	$416	16	$6656
Total				$10,952

The sum of the total amounts owed each day of the month is $10,952.

Find the average daily balance.

$$\text{Average daily balance} = \frac{\text{sum of the total amounts owed each day of the month}}{\text{number of days in the billing period}}$$

$$= \frac{10,952}{30} \approx 365.07$$

Find the finance charge.

$$I = Prt$$
$$I = 365.07(0.015)(1)$$
$$I \approx 5.48$$

The finance charge on the May 10 bill is $5.48.

✓ TAKE NOTE

The program CreditCard.xls, which is on the CD that came with your textbook and can be accessed on the website math.college.hmco.com, calculates interest due on a credit card bill.

▼ point of interest

In 1950, Frank McNamara issued to 200 friends a card that could be used to pay for food at various restaurants in New York. The card, called a Diner's Card, spawned the credit card industry.

EXAMPLE 1 ▪ Calculate Interest on a Credit Card Bill

An unpaid bill for $620 had a due date of March 10. A purchase of $214 was made on March 15, and $67 was charged on March 30. A payment of $200 was made on March 22. The interest on the average daily balance is 1.5% per month. Find the finance charge on the April 10 bill.

Solution

First calculate the sum of the total amounts owed each day of the month.

Date	Payments or Purchases	Balance Each Day	Number of Days Until Balance Changes	Unpaid Balance times Number of Days
March 10 – 14		$620	5	$3100
March 15 – 21	$214	$834	7	$5838
March 22 – 29	– $200	$634	8	$5072
March 30 – April 9	$67	$701	11	$7711
Total				$21,721

The sum of the total amounts owed each day of the month is $21,721.

Find the average daily balance.

$$\text{Average daily balance} = \frac{\text{sum of the total amounts owed each day of the month}}{\text{number of days in the billing period}}$$

$$= \frac{21{,}721}{31} \approx \$700.68$$

Find the finance charge.

$$I = Prt$$
$$I = 700.68(0.015)(1)$$
$$I \approx 10.51$$

The finance charge on the April 10 bill is $10.51.

INSTRUCTOR NOTE
You might present a case in which the unpaid balance (say, $500) and the minimum payment due (perhaps $10) both remain constant. Use a charge of 1.5% per month on the unpaid balance. Illustrate the total amount paid if only the minimum payment is paid each month. Calculate the total amount paid in finance charges. Using the same figures, illustrate the difference in these totals when the customer pays more than the minimum payment each month.

CHECK YOUR PROGRESS 1   A bill for $1024 was due on July 1. Purchases of $315 were made on July 7, and $410 was charged on July 22. A payment of $400 was made on July 15. The interest on the average daily balance is 1.2% per month. Find the finance charge on the August 1 bill.

Solution   *See page S37.*

Math Matters   Credit Card Debt

The graph on the following page shows how long it would take you to pay off a credit card debt of $3000 and the amount of interest you would pay if you made the minimum monthly payment of 3% of the credit card balance each month.

If you have credit card debt and want to determine how long it will take you to pay off the debt, go to **http://www.cardweb.com.** There you will find a calculator
(continued)

that will calculate how long it will take to pay off the debt and the amount of interest you will pay. (*Source:* CardWeb.com)

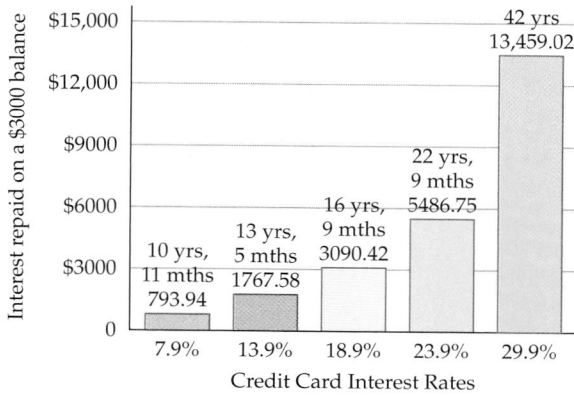

Source: **CardWeb.com,** reprinted by permission of Cardweb.com.

Annual Percentage Rate

Federal law, in the form of the Truth in Lending Act, requires that credit customers be made aware of the cost of credit. This law, passed by Congress in 1969, requires that a business issuing credit inform the consumer of all details of a credit transaction, including the true annual interest rate. The **true annual interest rate,** also called the **annual percentage rate (APR)** or **annual percentage yield (APY),** is the effective annual interest rate on which credit payments are based.

The idea behind the APR is that interest is owed only on the *unpaid balance* of the loan. For instance, suppose you decide to borrow $2400 from a bank that advertises a 10% *simple* interest rate. You want a six-month loan and agree to repay the loan in six equal monthly payments. The simple interest due on the loan is

$$I = Prt$$
$$I = \$2400(0.10)\left(\frac{6}{12}\right)$$
$$I = \$120$$

The total amount to be repaid to the bank is

$$A = P + I$$
$$A = \$2400 + \$120$$
$$A = \$2520$$

The amount of each monthly payment is

$$\text{Monthly payment} = \frac{2520}{6} = \$420$$

During the first month you owe $2400. The interest on that amount is

$$I = Prt$$

$$I = \$2400(0.10)\left(\frac{1}{12}\right)$$

$$I = \$20$$

At the end of the first month, of the $420 payment you make, $20 is the interest payment and $400 is applied to reducing the loan. Therefore, during the second month you owe $2400 − $400 = $2000.

During the second month you owe $2000. The interest on that amount is

$$I = Prt$$

$$I = \$2000(0.10)\left(\frac{1}{12}\right) \approx 16.667$$

$$I = \$16.67$$

At the end of the second month, of the $420 payment you make, $16.67 is the interest payment and $403.33 is applied to reducing the loan. Therefore, during the third month you owe $2000 − $403.33 = $1596.67.

The point of these calculations is to demonstrate that each month the amount you owe is decreasing, and not by a constant amount. From our calculations, the loan decreased by $400 the first month and by $403.33 the second month.

The Truth in Lending Act stipulates that the interest rate for a loan be calculated only on the amount owed at a particular time, not on the original amount borrowed. All loans must be stated according to this standard, thereby making it possible for a consumer to compare different loans.

We can use the following formula to estimate the annual percentage rate (APR) on a simple interest rate installment loan.

Approximate Annual Percentage Rate (APR) Formula

The annual percentage rate (APR) of a loan can be approximated by

$$\text{APR} \approx \frac{2nr}{n + 1}$$

where n is the number of payments and r is the simple interest rate.

For the loan described above, $n = 6$ and $r = 10\% = 0.10$.

$$\text{APR} \approx \frac{2nr}{n + 1}$$

$$\approx \frac{2(6)(0.10)}{6 + 1} = \frac{1.2}{7} \approx 0.171$$

The annual percentage rate on the loan is approximately 17.1%. Recall that the simple interest rate was 10%, much less than the actual rate. The Truth in Lending Act provides the consumer with a standard interest rate, APR, so that it is possible to compare loans. The 10% simple interest loan described above is equivalent to an APR loan of about 17%.

EXAMPLE 2 ■ Calculate a Finance Charge and an APR

You purchase a refrigerator for $675. You pay 20% down and agree to repay the balance in 12 equal monthly payments. The finance charge on the balance is 9% simple interest.

a. Find the finance charge.

b. Estimate the annual percentage rate. Round to the nearest tenth of a percent.

Solution

a. To find the finance charge, first calculate the down payment.

Down payment = percent down × purchase price
$$= 0.20 \times 675 = 135$$

Amount financed = purchase price − down payment
$$= 675 - 135 = 540$$

Calculate the interest owed on the loan.

Interest owed = finance rate × amount financed
$$= 0.09 \times 540 = 48.60$$

The finance charge is $48.60.

b. Use the APR formula to estimate the annual percentage rate.

$$\text{APR} \approx \frac{2nr}{n+1}$$

$$\approx \frac{2(12)(0.09)}{12+1} = \frac{2.16}{13} \approx 0.166$$

The annual percentage rate is approximately 16.6%.

CHECK YOUR PROGRESS 2 You purchase a washing machine and dryer for $750. You pay 20% down and agree to repay the balance in 12 equal monthly payments. The finance charge on the balance is 8% simple interest.

a. Find the finance charge.

b. Estimate the annual percentage rate. Round to the nearest tenth of a percent.

Solution See page S38.

Consumer Loans: Calculating Monthly Payments

The stated interest rate for most consumer loans, such as a car loan, is normally the annual percentage rate, APR, as required by the Truth in Lending Act. The payment amount for these loans is given by the following formula.

Payment Formula for an APR Loan

The payment for a loan based on APR is given by

$$PMT = A\left(\frac{i}{1 - (1 + i)^{-n}}\right)$$

where PMT is the payment, A is the loan amount, i is the interest rate per payment period, and n is the total number of payments.

It is important to note that in this formula i is the interest rate *per payment period*. For instance, if the annual interest rate is 9% and payments are made monthly, then

$$i = \frac{\text{annual interest rate}}{\text{number of payments per year}} = \frac{0.09}{12} = 0.0075$$

QUESTION   *For a four-year loan repaid on a monthly basis, what is the value of n in the formula above?*

The payment formula given above is used to calculate monthly payments on most consumer loans. In Example 3 we calculate the monthly payment for a new television and in Example 4 we calculate the monthly payment for a car loan.

EXAMPLE 3 ■ Calculate a Monthly Payment

Integrated Visual Technologies is offering anyone who purchases a television an annual interest rate of 9.5% for 4 years. If Andrea Smyer purchases a 50-inch, rear projection television for $5995 from Integrated Visual Technologies, find her monthly payment.

Solution

To calculate the monthly payment, you will need a calculator. The following keystrokes will work on most scientific calculators.

First calculate i and store the result.

$$i = \frac{\text{annual interest rate}}{\text{number of payments per year}} = \frac{0.095}{12}$$

Keystrokes: 0.095 $\boxed{\div}$ 12 $\approx$ 0.00791667 $\boxed{\text{STO}}$

Calculate the monthly payment. For a four-year loan, $n = 4(12) = 48$.

$$PMT = A\left(\frac{i}{1 - (1 + i)^{-n}}\right)$$
$$= 5995\left(\frac{0.095/12}{1 - (1 + 0.095/12)^{-48}}\right) \approx 150.61$$

Keystrokes: 5995 $\boxed{\times}$ $\boxed{\text{RCL}}$ $\boxed{=}$ $\boxed{\div}$ $\boxed{(}$ 1 $\boxed{-}$ $\boxed{(}$ 1 $\boxed{+}$ $\boxed{\text{RCL}}$ $\boxed{)}$ $\boxed{y^x}$ 48 $\boxed{+/-}$ $\boxed{)}$ $\boxed{=}$

The monthly payment is $150.61.

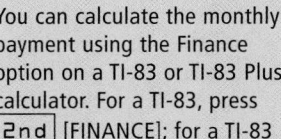

CALCULATOR NOTE

You can calculate the monthly payment using the Finance option on a TI-83 or TI-83 Plus calculator. For a TI-83, press $\boxed{\text{2nd}}$ [FINANCE]; for a TI-83 Plus, press $\boxed{\text{APPS}}$ $\boxed{\text{ENTER}}$ $\boxed{\text{ENTER}}$. Input the known values. Typically financial calculations use PV (present value) for the loan amount and FV (future value) for the amount owed at the end of the loan period, usually 0. P/Y = 12 and C/Y = 12 mean that payments and interest are calculated monthly (12 times a year). Now place the cursor on PMT and press $\boxed{\text{ALPHA}}$ [SOLVE].

```
N=48
I%=9.5
PV=5995
■PMT=-150.6132
FV=0
P/Y=12
C/Y=12
PMT: END BEGIN
```

The monthly payment is $150.61.

ANSWER $n = (\textit{number of years}) \times (\textit{number of payments per year}) = 4 \times 12 = 48$

CHECK YOUR PROGRESS 3   Carlos Menton purchases a new laptop computer from Knox Computer Solutions for $1499. If the sales tax is 4.25% of the purchase price and Carlos finances the total cost, including sales tax, for 3 years at an annual interest rate of 8.4%, find the monthly payment.

Solution   *See page S38.*

EXAMPLE 4 ■ Calculate a Car Payment

 A web page designer purchases a car for $18,395.

a. If the sales tax is 6.5% of the purchase price, find the amount of the sales tax.

b. If the car license fee is 1.2% of the purchase price, find the amount of the license fee.

c. If the designer makes a $2500 down payment, find the amount of the loan the designer needs.

d. Assuming the designer gets the loan in part c at an annual interest rate of 7.5% for 4 years, determine the monthly car payment.

Solution

a. Sales tax = 0.065(18,395) = 1195.675
The sales tax is $1195.68.

b. License fee = 0.012(18,395) = 220.74
The license fee is $220.74.

c. Loan amount = purchase price + sales tax + license fee − down payment
= 18,395 + 1195.68 + 220.74 − 2500
= 17,311.42
The loan amount is $17,311.42.

d. To calculate the monthly payment, you will need a calculator. The following keystrokes will work on most scientific calculators.
First calculate *i* and store the result.

$$i = \frac{\text{APR}}{12} = \frac{0.075}{12} = 0.00625$$

Keystrokes: 0.075 $\boxed{\div}$ 12 = 0.00625 $\boxed{\text{STO}}$

Calculate the monthly payment.

$$PMT = A\left(\frac{i}{1 - (1 + i)^{-n}}\right)$$
$$= 17,311.42\left(\frac{0.00625}{1 - (1 + 0.00625)^{-48}}\right) \approx 418.57$$

Keystrokes:
17311.42 $\boxed{\times}$ $\boxed{\text{RCL}}$ $\boxed{=}$ $\boxed{\div}$ $\boxed{(}$ 1 $\boxed{-}$ $\boxed{(}$ 1 $\boxed{+}$ $\boxed{\text{RCL}}$ $\boxed{)}$ $\boxed{y^x}$ 48 $\boxed{+/-}$ $\boxed{)}$ $\boxed{=}$

The monthly payment is $418.57.

CALCULATOR NOTE

A typical TI-83 screen for the calculation in Example 4 is shown below.

```
N=48
I%=7.5
PV=17311.42
■PMT=-418.5711
FV=0
P/Y=12
C/Y=12
PMT: END BEGIN
```

The monthly payment is $418.57.

CHECK YOUR PROGRESS 4   A school superintendent purchases a new station wagon for $26,788.

a. If the sales tax is 5.25% of the purchase price, find the amount of the sales tax.

b. The superintendent makes a $2500 down payment and the license fee is $145. Find the amount the superintendent must finance.

c. Assuming the superintendent gets the loan in part b at an annual interest rate of 8.1% for 5 years, determine the superintendent's monthly car payment.

Solution   *See page S38.*

Math Matters   Payday Loans

An ad reads

Get cash until payday! Loans of $100 or more available.

These ads refer to *payday* loans, which go by a variety of names, such as cash advance loans, check advance loans, post-dated check loans, or deferred deposit check loans. These types of loans are offered by finance companies and check-cashing companies.

Typically a borrower writes a personal check payable to the lender for the amount borrowed plus a *service fee*. The company gives the borrower the amount of the check minus the fee, which is normally a percent of the amount borrowed. The amount borrowed is usually repaid after payday, normally within a few weeks.

Under the Truth in Lending Act, the cost of a payday loan must be disclosed. Among other information, the borrower must receive, in writing, the APR for such a loan. To understand just how expensive these loans can be, suppose a borrower receives a loan for $100 for 2 weeks and pays a fee of $10. The APR for this loan can be calculated using a graphing or financial calculator. Typical screens for a graphing calculator are shown below.

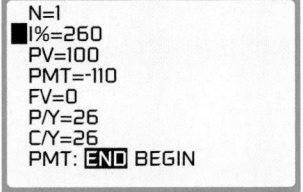

```
N=1
I%=0
PV=100
PMT=-110
FV=0
P/Y=26
C/Y=26
PMT: END BEGIN
```

N = 1 (number of payments)
I% is unknown.
PV = 100 (amount borrowed)
PMT = −110 (the payment)
FV = 0 (no money is owed after all the payments)
P/Y = 26 (There are 26 two-week periods in 1 year.)
C/Y = 26

```
N=1
■I%=260
PV=100
PMT=-110
FV=0
P/Y=26
C/Y=26
PMT: END BEGIN
```

Use the arrow key to highlight I%. Press ALPHA [SOLVE].

The annual interest rate is 260%.

To give you an idea of the enormity of a 260% APR, if the loan on the television set in Example 3, page 662, were based on a 260% interest rate, the monthly payment on the television set would be $1299.02!

The Federal Trade Commission (FTC) offers suggestions for consumers in need of credit. See **www.ftc.gov/bcp/conline/pubs/alerts/pdayalrt.htm.**

Consumer Loans: Calculating Loan Payoffs

Sometimes a consumer wants to pay off a loan before the end of the loan term. For instance, suppose you have a five-year car loan but would like to purchase a new car after owning your car for 4 years. Because there is still 1 year remaining on the loan, you must pay off the remaining loan before purchasing another car.

This is not as simple as just multiplying the monthly car payment by 12 to arrive at the payoff amount. The reason, as we mentioned earlier, is that each payment includes both interest and principal. By solving the Payment Formula for an APR Loan for A, the amount of the loan, we can calculate the payoff amount, which is just the remaining principal.

APR Loan Payoff Formula

The payoff amount for a loan based on APR is given by

$$A = PMT\left(\frac{1 - (1 + i)^{-n}}{i}\right)$$

where A is the loan payoff, PMT is the payment, i is the interest rate per payment period, and n is the number of *remaining* payments.

EXAMPLE 5 ■ Calculate a Payoff Amount

Allison Werke wants to pay off the loan on her jet ski that she has owned for 18 months. Allison's monthly payment is $284.67 on a two-year loan at an annual percentage rate of 8.7%. Find the payoff amount.

Solution

Because Allison has owned the jet ski for 18 months of a 24-month (two-year) loan, she has six payments remaining. Thus $n = 6$, the number of *remaining* payments. Here are the keystrokes to find the loan payoff.

Calculate i and store the result.

$$i = \frac{\text{APR}}{12} = \frac{0.087}{12} = 0.00725$$

Keystrokes: 0.087 ÷ 12 = 0.00725 STO

Use the APR Loan Payoff Formula.

$$A = PMT\left(\frac{1 - (1 + i)^{-n}}{i}\right)$$
$$= 284.67\left(\frac{1 - (1 + 0.00725)^{-6}}{0.00725}\right) \approx 1665.50$$

Keystrokes: 284.67 × (1 − (1 + RCL) y^x 6 +/−) ÷ RCL =

The loan payoff is $1665.50.

CHECK YOUR PROGRESS 5 Aaron Jefferson has a five-year car loan based on an annual percentage rate of 8.4%. The monthly payment is $592.57. After 3 years, Aaron decides to purchase a new car and must pay off his car loan. Find the payoff amount.

Solution *See page S38.*

Car Leases

Leasing a car may result in lower monthly car payments. The primary reason for this is that at the end of the lease term, you do not own the car. The car reverts back to the dealer, who can then sell it as a used car and realize the profit from the sale.

The value of the car at the end of the lease term is called the **residual value** of the car. The residual value of a car is frequently based on a percent of the manufacturer's suggested retail price (MSRP) and normally varies between 40% and 60% of the MSRP, depending on the type of lease.

For instance, suppose the MSRP of a car is $18,500 and the residual value is 45% of the MSRP. Then

$$\text{Residual value} = 0.45 \cdot 18,500$$
$$= 8325$$

The residual value is $8325. This is the amount the dealer thinks the car will be worth at the end of the lease period. The person leasing the car, the lessee, usually has the option of purchasing the car at that price at the end of the lease.

In addition to the residual value of the car, the monthly lease payment for a car takes into consideration *net capitalized cost, money factor, average monthly finance charge*, and *average monthly depreciation*. Each of these terms is defined below.

Net capitalized cost = negotiated price − down payment − trade-in value

$$\textbf{Money factor} = \frac{\text{annual interest rate as a percent}}{2400}$$

Average monthly finance charge
$$= (\text{net capitalized cost} + \text{residual value}) \times \text{money factor}$$

$$\textbf{Average monthly depreciation} = \frac{\text{net capitalized cost} - \text{residual value}}{\text{term of the lease in months}}$$

Using these definitions, we have the following formula for a monthly lease payment.

> **Monthly Lease Payment Formula**
>
> The monthly lease payment formula is given by $P = F + D$, where P is the monthly lease payment, F is the average monthly finance charge, and D is the average monthly depreciation of the car.

EXAMPLE 6 ■ Calculate a Monthly Lease Payment for a Car

The director of human resources for a company decides to lease a car for 30 months. Suppose the money factor is 0.0035, the negotiated price is $29,500, there is no trade-in, and the down payment is $5000. Find the monthly lease payment. Assume that the residual value is 55% of the MSRP of $33,400.

Solution

$$\text{Net capitalized cost} = \text{negotiated price} - \text{down payment} - \text{trade-in value}$$
$$= 29,500 - 5000 - 0 = 24,500$$

$$\text{Residual value} = 0.55(33,400) = 18,370$$

✓ TAKE NOTE

The money factor is sometimes written as the product of

$$\frac{\text{net capitalized cost} + \text{residual value}}{2}$$

and

$$\frac{\text{APR}}{12 \times 100}.$$

The division by 2 takes the average of the net capitalized cost and the residual value. The division by 12 converts the annual rate to a monthly rate. The division by 100 converts the percent to a decimal. The denominator 2400 in the money factor at the right is the product $2 \times 12 \times 100$.

Various methods are used to calculate monthly lease payments. The method we show here is typically used by car dealerships. In the exercises, we will give another method for calculating a lease payment.

TAKE NOTE

When a person purchases a car, any state sales tax must be paid at the time of the purchase. However, for a lease, the sales tax is added to the monthly lease payment.

Suppose, for instance, that in Example 6 the state sales tax is 6% of the lease payment. Then

Total lease payment
$$= 354.38 + 0.06(354.38)$$
$$\approx 375.64$$

Average monthly finance charge
$$= (\text{net capitalized cost} + \text{residual value}) \times \text{money factor}$$
$$= (24{,}500 + 18{,}370) \times 0.0035$$
$$\approx 150.05$$

Average monthly depreciation $= \dfrac{\text{net capitalized cost} - \text{residual value}}{\text{term of the lease in months}}$
$$= \dfrac{24{,}500 - 18{,}370}{30}$$
$$\approx 204.33$$

Monthly lease payment
$$= \text{average monthly finance charge} + \text{average monthly depreciation}$$
$$= 150.05 + 204.33$$
$$= 354.38$$

The monthly lease payment is $354.38.

CHECK YOUR PROGRESS 6 Find the monthly lease payment for a car for which the negotiated price is $31,900, the annual interest rate is 8%, the length of the lease is 5 years, and the residual value is 40% of the MSRP of $33,395.

Solution   *See page S38.*

Excursion

Leasing Versus Buying a Car

On November 1, 2001, the MSRP for a 2002 Ford Mustang GT Premium Coupe was $24,135. We will use this information to analyze the results of leasing versus buying the car. To ensure that we are making valid comparisons, we will assume:

The negotiated price is the MSRP and a down payment of $2500 is made.

The license fee is 1.1% of the negotiated price.

The state sales tax is 5.5% of the negotiated price or 5.5% of the monthly lease payment.

The annual interest rate for both a loan and a lease is 6% and the term of the loan or the lease is 60 months.

The residual value when leasing the car is 45% of the MSRP.

1. Determine the loan amount to purchase the car.

2. Determine the monthly car payment to purchase the car.

3. How much will you have paid for the car (excluding maintenance) over the five-year term of the loan?

(continued)

4. At the end of 5 years, you sell the car for 45% of the original MSRP (the residual lease value). Your net car ownership cost is the amount you paid over the 5 years minus the amount you realized from selling the car. What is your net ownership cost?

5. Determine the net capitalized cost for the car.

6. Determine the lease payment. Remember to include the sales tax that must be paid each month.

7. How much will you have paid to lease the car for the five-year term? Because the car reverts to the dealer after 5 years, the net ownership cost is the total of all the lease payments for the 5 years.

8. Which option, buying or leasing, results in the smaller net ownership cost?

9. List some advantages and disadvantages of buying or leasing a car.

Exercise Set 10.3   (Suggested Assignment: 5–35, odds)

In Exercises 1–4, calculate the finance charge for a credit card that has the given average daily balance and interest rate.

1. Average daily balance: $118.72; monthly interest rate: 1.25%

2. Average daily balance: $391.64; monthly interest rate: 1.75%

3. Average daily balance: $10,154.87; monthly interest rate: 1.5%

4. Average daily balance: $20,346.91; monthly interest rate: 1.25%

5. A credit card account had a $244 balance on March 5. A purchase of $152 was made on March 12, and a payment of $100 was made on March 28. Find the average daily balance if the billing date is April 5.

6. A credit card account has a $768 balance on April 1. A purchase of $316 was made on April 5, and a payment of $200 was made on April 18. Find the average daily balance if the new billing date is May 1.

7. A charge account had a balance of $944 on May 5. A purchase of $255 was made on May 17, and a payment of $150 was made on May 20. The interest on the average daily balance is 1.5% per month. Find the finance charge on the June 5 bill.

8. A charge account had a balance of $655 on June 1. A purchase of $98 was made on June 17, and a payment of $250 was made on June 15. The interest on the average daily balance is 1.2% per month. Find the finance charge on the July 1 bill.

9. On August 10, a credit card account had a balance of $345. A purchase of $56 was made on August 15, and $157 was charged on August 27. A payment of $75 was made on August 15. The interest on the average daily balance is 1.25% per month. Find the finance charge on the September 10 bill.

10. On May 1, a credit card account had a balance of $189. Purchases of $213 were made on May 5, and $102 was charged on May 21. A payment of $150 was made on May 25. The interest on the average daily balance is 1.5% per month. Find the finance charge on the June 1 bill.

In Exercises 11 and 12, you may want to use the spreadsheet program available at our web site at **http://college.hmco.com.** This spreadsheet automates the finance charge procedure shown in this section.

11. The activity date, company, and amount for a credit card bill are shown below. The due date of the bill is September 15. On August 15, there was an unpaid balance of $1236.43. Find the finance charge if the interest rate is 1.5% per month.

Activity Date	Company	Amount
August 15	Unpaid balance	1236.43
August 16	Vetenary clinic	125.00
August 17	Shell	23.56
August 18	Olive's restaurant	53.45
August 20	Seaside market	41.36
August 22	Monterey Hotel	223.65
August 25	Airline tickets	310.00
August 30	Bike 101	23.36
September 1	Trattoria Maria	36.45
September 9	Bookstore	21.39
September 12	Seaside Market	41.25
September 13	Credit card payment	−1345.00

12. The activity date, company, and amount for a credit card bill are shown below. The due date of the bill is July 10. On June 10, there was an unpaid balance of $987.81. Find the finance charge if the interest rate is 1.8% per month.

Activity Date	Company	Amount
June 10	Unpaid balance	987.81
June 11	Jan's Surf Shop	156.33
June 12	Albertson's	45.61
June 15	The Down Shoppe	59.84
June 16	NY Times Sales	18.54
June 20	Cardiff Delicatessen	23.09
June 22	The Olde Golf Mart	126.92
June 28	Lee's Hawaiian Restaurant	41.78
June 30	City Food Drive	100.00
July 2	Credit card payment	−1000.00
July 8	Safeway Stores	161.38

 Use a calculator for Exercises 13–42.

In Exercises 13–16, use the Approximate Annual Percentage Rate Formula.

13. Chuong Ngo borrows $2500 from a bank that advertises a 9% simple interest rate and repays the loan in three equal monthly payments. Estimate the APR. Round to the nearest tenth of a percent.

14. Charles Ferrara borrows $4000 from a bank that advertises an 8% simple interest rate. If he repays the loan in six equal monthly payments, estimate the APR. Round to the nearest tenth of a percent.

15. Kelly Ang buys a computer system for $2400 and makes a 15% down payment. If Kelly agrees to repay the balance in 24 equal monthly payments at an annual simple interest rate of 10%, estimate the APR for Kelly's loan.

16. Jill Richards purchases a stereo system for $1500. She makes a 20% down payment and agrees to repay the balance in 12 equal payments. If the finance charge on the balance is 7% simple interest, estimate the APR. Round to the nearest tenth of a percent.

17. Arrowood's Camera Store advertises a Canon Camedia 3.34-megapixel camera for $649.95, including taxes. If you finance the purchase of this camera for 1 year at an annual percentage rate of 6.9%, find the monthly payment.

18. Optics Mart offers a Meade ETX Astro telescope for $1249, including taxes. If you finance the purchase of this telescope for 2 years at an annual percentage rate of 7.2%, what is the monthly payment?

19. Alicia's Surf Shop offers its 9′1″-long WaveHanger surfboard for $649. The sales tax is 7.25% of the purchase price.
   a. What is the total cost, including sales tax?
   b. If you make a down payment of 25% of the total cost, find the down payment.
   c. Assuming you finance the remaining cost at an annual interest rate of 5.7% for six months, find the monthly payment.

20. A boat shop offers a Stratos 201 boat with a Johnson 225 engine for $18,250. The sales tax is 6.5% of the purchase price.
   a. What is the total cost, including sales tax?
   b. If you make a down payment of 20% of the total cost, find the down payment.
   c. Assuming you finance the remaining cost at an annual interest rate of 5.7% for three years, find the monthly payment.

21. After becoming a commercial pilot, Lorna Kao decides to purchase a Cessna 182 for $64,995. Assuming the sales tax is 5.5% of the purchase price, find each of the following.
   a. What is the total cost, including sales tax?
   b. If Lorna makes a down payment of 20% of the total cost, find the down payment.
   c. Assuming Lorna finances the remaining cost at an annual interest rate of 7.15% for 10 years, find the monthly payment.

22. Donald Savchenko purchased new living room furniture for $2488. Assuming the sales tax is 7.75% of the purchase price, find each of the following.
   a. What is the total cost, including sales tax?
   b. If Donald makes a down payment of 15% of the total cost, find the down payment.

   c. Assuming Donald finances the remaining cost at an annual interest rate of 8.16% for 2 years, find the monthly payment.

23. Luis Mahla purchases a Porsche Boxster for $42,600 and finances the entire amount at an annual interest rate of 5.7% for 5 years. Find the monthly payment. Assume the sales tax is 6% of the purchase price and the license fee is 1% of the purchase price.

24. Suppose you negotiate a selling price of $11,995 for a Ford Explorer. You make a down payment of 10% of the selling price and finance the remaining balance for 3 years at an annual interest rate of 7.5%. The sales tax is 7.5% of the selling price, and the license fee is 0.9% of the selling price. Find the monthly payment.

25. Margaret Hsi purchases a late model Corvette for $24,500. She makes a down payment of $3000 and finances the remaining amount for 4 years at an annual interest rate of 8.5%. The sales tax is 5.5% of the selling price and the license fee is $331. Find the monthly payment.

26. Chris Schmaltz purchases a Dodge Intrepid for $24,119. Chris makes a down payment of $5000 and finances the remaining amount for 5 years at an annual interest rate of 7.6%. The sales tax is 6.25% of the selling price, and the license fee is $429. Find the monthly payment.

27. Suppose you purchase a car for a total price of $15,445, including taxes and license fee, and finance that amount for 4 years at an annual interest rate of 8%.
   a. Find the monthly payment.
   b. What is the total amount of interest paid over the term of the loan?

28. Adele Paolo purchased a Chevrolet Blazer for a total price of $26,425, including taxes and license fee, and financed that amount for 5 years at an annual interest rate of 7.8%.
   a. Find the monthly payment.
   b. What is the total amount of interest paid over the term of the loan?

29. Angela Montery has a five-year car loan for a Jeep Wrangler at an annual interest rate of 6.3% and a monthly payment of $603.50. After 3 years, Angela decides to purchase a new car. What is the payoff on Angela's loan?

30. Suppose you have a four-year car loan at an annual interest rate of 7.2% and a monthly payment of $587.21. After $2\frac{1}{2}$ years, you decide to purchase a new car. What is the payoff on your loan?

31. Suppose you have a four-year car loan at an annual interest rate of 8.9% and a monthly payment of $303.52. After 3 years, you decide to purchase a new car. What is the payoff on your loan?

32. Ming Li has a three-year car loan for a Mercury Sable at an annual interest rate of 9.3% and a monthly payment of $453.68. After 1 year, Ming decides to purchase a new car. What is the payoff on his loan?

33. Suppose you decide to obtain a four-year lease for a car and negotiate a selling price of $28,990. The trade-in value of your old car is $3850. If you make a down payment of $2400, the money factor is 0.0027, and the residual value is $15,000, find each of the following.

   a. The net capitalized cost

   b. The average monthly finance charge

   c. The average monthly depreciation

   d. The monthly lease payment

34. Marcia Scripps obtains a five-year lease for a Ford F-10 pickup and negotiates a selling price of $31,115. The trade-in value of her old car is $2950. Assuming she makes a down payment of $3000, the money factor is 0.0035, and the residual value is $16,500, find each of the following.

   a. The net capitalized cost

   b. The average monthly finance charge

   c. The average monthly depreciation

   d. The monthly lease payment

35. Jorge Cruz obtains a three-year lease for a Dodge Stratus and negotiates a selling price of $22,100. The annual interest rate is 8.1%, the residual value is $15,000, and Jorge makes a down payment of $1000. Find each of the following.

   a. The net capitalized cost

   b. The money factor

   c. The average monthly finance charge

   d. The average monthly depreciation

   e. The monthly lease payment

36. Suppose you obtain a five-year lease for a Ferrari and negotiate a selling price of $165,000. The annual interest rate is 8.4%, the residual value is $85,000, and you make a down payment of $5000. Find each of the following.

   a. The net capitalized cost

   b. The money factor

   c. The average monthly finance charge

   d. The average monthly depreciation

   e. The monthly lease payment

Extensions

CRITICAL THINKING

37. Explain how the APR Loan Payoff Formula can be used to determine the selling price of a car when the monthly payment, the annual interest rate, and the term of the loan on the car are known. Using your process, determine the selling price of a car that is offered for $235 per month for 4 years if the annual interest rate is 7.2%.

COOPERATIVE LEARNING

38. You may have heard advertisements from car dealerships that say something like, "Bring in your car, paid for or not, and we'll take it as a trade-in for a new car." The advertisement does not go on to say that you have to pay off the remaining loan balance or that balance gets added to the price of the new car.

   a. Suppose you are making payments of $235.73 per month on a four-year car loan that has an annual interest rate of 8.4%. After making payments for 3 years, you decide to purchase a new car. What is the loan payoff?

   b. You negotiate a price, including taxes, of $18,234 for the new car. What is the actual amount you owe for the new car when the loan payoff is included?

   c. If you finance the amount in part b for 4 years at an annual interest rate of 8.4%, what is the new monthly payment?

39. The residual value of a car is based on "average" usage. To protect a car dealership from abnormal usage, most car leases stipulate an average number of miles driven annually, that the tires be serviceable when the car is returned, and (in many cases) that all the manufacturer's recommended services be performed over the course of the lease. Basically, the dealership wants a car that can be put on the lot and sold as a used car without much effort.

Suppose you decide to lease a car for 4 years. The net capitalized cost is $19,788, the residual value is 55% of the MSRP of $28,990, and the interest rate is 5.9%. In addition, you must pay a mileage penalty of $.20 for each mile over 48,000 miles that the car is driven during the 4 years of the lease.

a. Determine the monthly lease payment using the Monthly Lease Payment Formula given below. This formula is used by some financing agencies to calculate a monthly lease payment.

> **Monthly Lease Payment Formula**
>
> The monthly lease payment is given by $P = \dfrac{Ai(1 + i)^n - Vi}{(1 + i)^n - 1}$, where P is the monthly lease payment, A is the net capitalized cost, V is the residual value, i is the interest rate per payment period as a decimal, and n is the number of lease payments.

b. If the odometer reads 87 miles when you lease the car and you return the car with 61,432 miles, what mileage penalty will you have to pay?

c. If your car requires new tires at a cost of $635, what is the total cost, excluding maintenance, of the lease for the 4 years?

EXPLORATIONS

40. For most credit cards, no finance charge is added to the bill if the full amount owed is paid by the due date. However, if you do not pay the bill in full, the unpaid amount *and* all current charges are subject to a finance charge. This is a point that is missed by many credit card holders.

To illustrate, suppose you have a credit card bill of $500 and you make a payment of $499, $1 less than the amount owed. Your credit card activity is as shown in the statement below.

Activity Date	Company	Amount
October 10	Unpaid balance	1.00
October 11	Rick's Tires	455.69
October 12	Costa's Internet Appliances	128.54
October 15	The Belgian Lion Restaurant	64.31
October 16	Verizon Wireless	33.57
October 20	Milton's Cake and Pie Shoppe	22.33
October 22	Fleming's Perfumes	65.00
October 24	Union 76	27.63
October 26	Amber's Books	42.31
November 2	Lakewood Meadows	423.88
November 8	Von's Grocery	55.64

P

a. Find the finance charge if the interest rate is 1.8% per month. Assume the due date of the bill is November 10. On October 10, the unpaid balance was the $1 you did not pay. You may want to use the spreadsheet program mentioned above Exercise 11.

b. Now assume that instead of paying $499, you paid $200, which is $299 less than the amount paid in part a. Calculate the finance charge.

c. How much more interest did you pay in part b than in part a?

d. If you took the $299 difference in payment and deposited it into an account that earned 1.8% simple interest (the credit card rate) for 1 month, how much interest would you earn?

e. ✎ Explain why the answers to parts c and d are the same.

41. a. Use the formula in Exercise 39 to find the monthly lease payment, excluding sales tax, for a car for which the net capitalized cost is $23,488, the residual value is $12,500, the annual interest rate is 7%, and the term of the lease is 4 years.

b. Suppose you negotiate a net capitalized cost of $26,445 for a car and a residual value of $14,000. If the annual interest rate is 6.5% and the term of the lease is 5 years, find the monthly lease payment. The sales tax is 6.25% of the lease payment.

c. Use this formula to find the lease payment for Exercise 35. What is the difference between the payment calculated using this method and the payment calculated using the method in Exercise 35?

d. Use this formula to find the lease payment for Exercise 36. What is the difference between the payment calculated using this method and the payment calculated using the method in Exercise 36?

42. The APR Loan Payoff Formula can be used to determine how many months it would take to pay off a credit card debt if the minimum monthly payment is made each month. For instance, suppose you have a credit card bill of $620.50, the minimum payment is $13, and the interest rate is 18% per year. Using a graphing calculator, we can determine the number of months, n, it will take to pay off the debt. Enter the values shown on the calculator screen at the right. Move the cursor to N and press ALPHA [SOLVE]. It would take over 84 months (or approximately 7 years) to pay off the credit card debt, assuming you do not make additional purchases.

```
N=84.53746933
■I%=18
PV=620.5
PMT=-13
FV=0
P/Y=12
C/Y=12
PMT: END BEGIN
```

a. Find the number of months it would take to pay off a credit card debt of $1283.34 if the minimum payment is $27 and the annual interest rate is 19.6%. Round to the nearest month.

b. How much interest would be paid on the credit card debt in part a?

c. 🌐 If you have credit card debt, determine how many months it would take to pay off your debt by making the minimum monthly payments. How much interest would you pay? You may want to go to the web site at **http://www.cardweb.com,** which was mentioned in the Math Matters on page 658, to determine the answers.

| SECTION 10.4 | Home Ownership |

Initial Expenses

▼ **point of interest**

According to a recent survey by the Chicago Title and Trust Company, of all buyers of homes, approximately 45% were first-time buyers. The distribution of income levels for first-time buyers is given below.

Less than $30,000: 12.5%
$30,000–$49,999: 30.4%
$50,000–$69,000: 23.9%
$70,000–$89,999: 18.4%
$90,000 or more: 14.8%

When you purchase a home, you generally make a down payment and finance the remainder of the purchase price with a loan obtained through a bank or savings and loan association. The amount of the down payment can vary, but it is normally between 10% and 30% of the selling price. The **mortgage** is the amount that is borrowed to buy the real estate. The amount of the mortgage is the difference between the selling price and the down payment.

Mortgage = selling price − down payment

This formula is used to find the amount of the mortgage. For example, suppose you buy a $120,000 home with a down payment of 25%. First find the down payment by computing 25% of the purchase price.

Down payment = 25% of 120,000 = 0.25(120,000)
= 30,000

Then find the mortgage by subtracting the down payment from the selling price.

Mortgage = selling price − down payment
= 120,000 − 30,000
= 90,000

The mortgage is $90,000.

The down payment is generally the largest initial expense in purchasing a home, but there are other expenses associated with the purchase. These payments are due at the closing, when the sale of the house is finalized, and are called **closing costs.** The bank may charge fees for attorneys, credit reports, loan processing, and title searches. There may also be a **loan origination fee.** This fee is usually expressed in **points.** One point is equal to one percent of the mortgage.

Suppose you purchase a home and obtain a loan for $90,000. The bank charges a fee of 1.5 points. To find the charge for points, multiply the loan amount by 1.5%.

Points = 1.5% of 90,000 = 0.015(90,000)
= 1350

The charge for points is $1350.

✔ **TAKE NOTE**

1.5 points means 1.5%.
1.5% = 0.015

EXAMPLE 1 ■ Calculate Closing Costs

The purchase price of a home is $98,000. A down payment of 20% is made. The bank charges $450 in fees plus $2\frac{1}{2}$ points. Find the total of the down payment and the closing costs.

Solution
First find the down payment.

Down payment = 20% of 98,000 = 0.20(98,000)
= 19,600

The down payment is $19,600.

Next find the mortgage.

$$\text{Mortgage} = \text{selling price} - \text{down payment}$$
$$= 98{,}000 - 19{,}600$$
$$= 78{,}400$$

The mortgage is $78,400.

Next, calculate the charge for points.

$$\text{Points} = 2\frac{1}{2}\% \text{ of } 78{,}400 = 0.025(78{,}400)$$

$$= 1960$$

• $2\frac{1}{2}\% = 2.5\% = 0.025$

The charge for points is $1960.

Finally, find the sum of the down payment and the closing costs.

$$19{,}600 + 450 + 1960 = 22{,}010$$

The total of the down payment and the closing costs is $22,010.

CHECK YOUR PROGRESS 1   The purchase price of a home is $110,000. A down payment of 25% is made. The bank charges $375 in fees plus 1.75 points. Find the total of the down payment and the closing costs.

Solution   *See page S38.*

INSTRUCTOR NOTE
Many students will have heard of equity loans but do not understand what they are. You may want to discuss the following with them.
   Home equity is the difference between the fair market value of your home (the price at which it would probably sell if you were to put it on the market today) and the amount you still owe on the mortgage. For example, if the market value of your home is $125,000 and you still owe $75,000 on the mortgage, your home equity is $125,000 − $75,000 = $50,000.
   Many banks allow home owners to borrow on their equity. The bank charges fees to originate the loan and charges you interest, and you pay back the loan as you would any other loan, normally with monthly payments. In a home equity loan, your house serves as collateral, or security, for the loan.

Mortgages

When a bank agrees to provide you with a mortgage, you agree to pay off that loan in monthly payments. If you fail to make the payments, the bank has the right to **foreclose,** which means that the bank takes possession of the property and has the right to sell it.

There are many types of mortgages available to home buyers today, so the terms of mortgages differ considerably. Some mortgages are **adjustable rate mortgages (ARMs).** The interest rate charged on an ARM is adjusted periodically to more closely reflect current interest rates. The mortgage agreement specifies exactly how often and by how much the interest rate can change.

A **fixed rate mortgage,** or **conventional mortgage,** is one in which the interest rate charged on the loan remains the same throughout the life of the mortgage. For a fixed rate mortgage, the amount of the monthly payment also remains unchanged throughout the term of the loan.

The term of a mortgage can vary. Terms of 15, 20, 25, and 30 years are most common.

The monthly payment on a mortgage is the **mortgage payment.** The amount of the mortgage payment depends on the amount of the mortgage, the interest rate on the loan, and the term of the loan. This payment is calculated by using the Payment Formula for an APR Loan given in the last section. We will restate the formula here.

> **Mortgage Payment Formula**
>
> The mortgage payment for a mortgage is given by
>
> $$PMT = A\left(\frac{i}{1 - (1 + i)^{-n}}\right)$$
>
> where PMT is the mortgage payment, A is the amount of the mortgage, i is the interest rate per payment period, and n is the total number of payments.

EXAMPLE 2 ■ Calculate a Mortgage Payment

▼ point of interest

Home buyers rated the following characteristics as "extremely important" in their purchase decisions. (*Source:* American Lives, Inc. and Intercommunications, Inc. Republished with permission of *Wall Street Journal,* November 21, 1997; permission conveyed through Copyright Clearance Center, Inc.)

Natural, open space: 77%
Walking and biking paths: 74%
Gardens with native plants: 56%
Clustered retail stores: 55%
Wilderness area: 52%
Outdoor pool: 52%
Community recreation
   center: 52%
Interesting little parks: 50%

Suppose Allison Sommerset purchases a condominium and secures a loan of $134,000 for 30 years at an annual interest rate of 6.5%.

a. Find the monthly mortgage payment.
b. What is the total of the payments over the life of the loan?
c. Find the amount of interest paid on the loan over the 30 years.

Solution

a. First calculate i and store the result.

$$i = \frac{\text{annual interest rate}}{\text{number of payments per year}} = \frac{0.065}{12}$$

Keystrokes: $0.065 \boxed{\div} 12 \approx 0.00541667 \boxed{STO}$

Calculate the monthly payment. For a 30-year loan, $n = 30(12) = 360$.

$$PMT = A\left(\frac{i}{1 - (1 + i)^{-n}}\right)$$

$$= 134{,}000\left(\frac{0.065/12}{1 - (1 + 0.065/12)^{-360}}\right) \approx 846.97$$

Keystrokes:

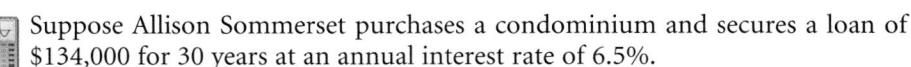

$134000 \boxed{\times} \boxed{RCL} \boxed{=} \boxed{\div} \boxed{(} 1 \boxed{-} \boxed{(} 1 \boxed{+} \boxed{RCL} \boxed{)} \boxed{y^x} 360 \boxed{+/-} \boxed{)} \boxed{=}$

The monthly mortgage payment is $846.97.

b. To determine the total of the payments, multiply the number of payments (360) by the monthly payment ($846.97).

$$846.97(360) = 304{,}909.20$$

The total of the payments over the life of the loan is $304,909.20.

c. To determine the amount of interest paid, subtract the mortgage from the total of the payments.

$$304{,}909.20 - 134{,}000 = 170{,}909.20$$

The amount of interest paid over the life of the loan is $170,909.20.

CHECK YOUR PROGRESS 2 Suppose Antonio Scarletti purchases a home and secures a loan of $123,000 for 25 years at an annual interest rate of 7%.

a. Find the monthly mortgage payment.
b. What is the total of the payments over the life of the loan?
c. Find the amount of interest paid on the loan over the 25 years.

Solution *See page S39.*

A portion of a mortgage payment pays the current interest owed on the loan, and the remaining portion of the mortgage payment is used to reduce the principal owed on the loan. This process of paying off the principal and the interest, which is similar to paying a car loan, is called **amortizing the loan**.

In Example 2, the mortgage payment on a $134,000 mortgage at 6.5% for 30 years was $846.97. The amount of the first payment that is interest and the amount that is applied to the principal can be calculated using the simple interest formula.

$$I = Prt$$

$$= 134{,}000(0.065)\left(\frac{1}{12}\right) \qquad \bullet\ P = 134{,}000\text{, the current loan amount;}$$
$$r = 0.065;\ t = \frac{1}{12}$$

$$\approx 725.83 \qquad \bullet\ \text{Round to the nearest cent.}$$

Of the $846.97 mortgage payment, $725.83 is an interest payment. The remainder is applied toward reducing the principal.

Principal reduction = $846.97 − $725.83 = $121.14

After the first month's mortgage payment, the balance on the loan (the amount that remains to be paid) is calculated by subtracting the principal paid on the mortgage from the mortgage.

Loan balance after first month = $134,000 − $121.14 = $133,878.86

The portion of the second mortgage payment that is applied to interest and the portion that is applied to the principal can be calculated in the same manner. In the calculation, the figure used for the principal, P, is the current balance on the loan, $133,878.86.

$$I = Prt$$

$$= 133{,}878.86(0.065)\left(\frac{1}{12}\right) \qquad \bullet\ P = 133{,}878.86\text{, the current loan amount;}$$
$$r = 0.065;\ t = \frac{1}{12}$$

$$\approx 725.18 \qquad \bullet\ \text{Round to the nearest cent.}$$

Principal reduction = $846.97 − $725.18 = $121.79

Of the second mortgage payment, $725.18 is an interest payment and $121.79 is a payment toward the principal.

Loan balance after second month = $133,878.86 − $121.79 = $133,757.07

The interest payment, principal payment, and balance on the loan can be calculated in this manner for all of the mortgage payments throughout the life of the loan—all 360 of them! Or, a computer can be programmed to make these calculations and print out the information. The printout is called an **amortization schedule.** It lists, for each mortgage payment, the payment number, the interest payment, the amount applied toward the principal, and the resulting balance to be paid.

Each month, the amount of the mortgage payment that is an interest payment decreases and the amount applied toward the principal increases. This is because you are paying interest on a decreasing balance each month. Mortgage payments early in the life of a mortgage are largely interest payments; mortgage payments late in the life of a mortgage are largely payments toward the principal.

The partial amortization schedule below shows the breakdown for the first 12 months of the loan in Example 2.

Amortization Schedule			
Loan Amount	$134,000.00		
Interest Rate	6.50%		
Term of Loan	30		
Monthly Payment	$846.97		
Month	Amount of Interest	Amount of Principal	New Loan Amount
1	$725.83	$121.14	$133,878.86
2	$725.18	$121.79	$133,757.07
3	$724.52	$122.45	$133,634.61
4	$723.85	$123.12	$133,511.50
5	$723.19	$123.78	$133,387.71
6	$722.52	$124.45	$133,263.26
7	$721.84	$125.13	$133,138.13
8	$721.16	$125.81	$133,012.32
9	$720.48	$126.49	$132,885.84
10	$719.80	$127.17	$132,758.66
11	$719.11	$127.86	$132,630.80
12	$718.42	$128.55	$132,502.25

QUESTION *Using the amortization schedule above, how much of the loan has been paid off after 1 year?*

EXAMPLE 3 ■ Calculate Principal and Interest for a Mortgage Payment

 You purchase a condominium for $98,750 and obtain a 30-year, fixed rate mortgage at 7.25%. After paying a down payment of 20%, how much of the second payment is interest and how much is applied toward the principal?

Solution
First find the down payment by multiplying the percent of the purchase price that is the down payment by the purchase price.

$0.20(98,750) = 19,750$

The down payment is $19,750.

Find the mortgage by subtracting the down payment from the purchase price.

$98,750 - 19,750 = 79,000$

The mortgage is $79,000.

ANSWER *After 1 year (12 months), the loan amount is $132,502.25. The original loan was $134,000. The amount that has been paid off is $134,000 - $132,502.25 = $1497.75.*

Calculate the mortgage payment.

$$i = \frac{\text{annual interest rate}}{\text{number of payments per year}} = \frac{0.0725}{12}$$

Keystrokes: 0.0725 $\boxed{\div}$ 12 $\approx$ 0.00604167 $\boxed{\text{STO}}$

Calculate the monthly payment. For a 30-year loan, $n = 30(12) = 360$.

$$PMT = A\left(\frac{i}{1 - (1 + i)^{-n}}\right)$$

$$= 79{,}000\left(\frac{0.0725/12}{1 - (1 + 0.0725/12)^{-360}}\right) \approx 538.92$$

Keystrokes:

79000 $\boxed{\times}$ $\boxed{\text{RCL}}$ $\boxed{=}$ $\boxed{\div}$ $\boxed{(}$ 1 $\boxed{-}$ $\boxed{(}$ 1 $\boxed{+}$ $\boxed{\text{RCL}}$ $\boxed{)}$ $\boxed{y^x}$ 360 $\boxed{+/-}$ $\boxed{)}$ $\boxed{=}$

The monthly payment is $538.92.

Find the amount of interest paid on the first mortgage payment by solving the simple interest formula for I.

$$I = Prt$$

$$= 79{,}000(0.0725)\left(\frac{1}{12}\right)$$

 • $P = 79{,}000$, the current loan amount; $r = 0.0725$; $t = \dfrac{1}{12}$

$$\approx 477.29$$

 • Round to the nearest cent.

Find the principal paid on the first mortgage payment by subtracting the interest paid from the monthly mortgage payment.

$$538.92 - 477.29 = 61.63$$

Calculate the balance on the loan after the first mortgage payment by subtracting the principal paid from the mortgage.

$$79{,}000 - 61.63 = 78{,}938.37$$

Find the amount of interest paid on the second mortgage payment.

$$I = Prt$$

$$= 78{,}938.37(0.0725)\left(\frac{1}{12}\right)$$

 • $P = 78{,}938.37$, the current loan amount; $r = 0.0725$; $t = \dfrac{1}{12}$

$$\approx 476.92$$

 • Round to the nearest cent.

The interest paid on the second payment was $476.92.

Find the principal paid on the second mortgage payment.

$$538.92 - 476.92 = 62.00$$

The principal paid on the second payment was $62.

CHECK YOUR PROGRESS 3   You purchase a home for $95,000. You obtain a 30-year conventional mortgage at 6.75% after paying a down payment of 25% of the purchase price. Of the first month's payment, how much is interest and how much is applied toward the principal?

Solution   *See page S39.*

When a home is sold before the term of the loan has expired, the homeowner must pay the lender the remaining balance on the loan. To calculate that balance, we can use the APR Loan Payoff Formula from the last section.

APR Loan Payoff Formula

The payoff amount for a mortgage is given by

$$A = PMT\left(\frac{1 - (1 + i)^{-n}}{i}\right)$$

where A is the loan payoff, PMT is the mortgage payment, i is the interest rate per payment period, and n is the number of *remaining* payments.

EXAMPLE 4 ■ Calculate a Mortgage Payoff

A homeowner has a monthly mortgage payment of $645.32 on a 30-year loan at an annual interest rate of 7.2%. After making payments for 5 years, the homeowner decides to sell the house. What is the payoff for the mortgage?

Solution

Use the APR Loan Payoff Formula. The homeowner has been making payments for 5 years, or 60 months. There are 360 months in a 30-year loan, so there are $360 - 60 = 300$ remaining payments.

$$A = PMT\left(\frac{1 - (1 + i)^{-n}}{i}\right)$$

$$= 645.32\left(\frac{1 - (1 + 0.006)^{-300}}{0.006}\right)$$

$$\approx 89{,}679.01$$

• $PMT = 645.32$; $i = \dfrac{0.072}{12} = \textbf{0.006}$; $n = \textbf{300}$, the number of remaining payments

Here are the keystrokes to compute the payoff on a scientific calculator. The same calculation using a graphing calculator is shown at the left.

Calculate i: 0.072 $\boxed{\div}$ 12 $\boxed{=}$ 0.006 $\boxed{\text{STO}}$

Calculate the payoff: 645.32 $\boxed{\times}$ $\boxed{(}$ $\boxed{(}$ 1 $\boxed{-}$ $\boxed{(}$ 1 $\boxed{+}$ $\boxed{\text{RCL}}$ $\boxed{)}$ $\boxed{y^x}$ 300 $\boxed{+/-}$ $\boxed{)}$ $\boxed{\div}$ $\boxed{\text{RCL}}$ $\boxed{=}$

The loan payoff is $89,679.01.

```
N=300
I%=7.2
■PV=89679.0079
PMT=-645.32
FV=0
P/Y=12
C/Y=12
PMT: END BEGIN
```

CHECK YOUR PROGRESS 4 Ava Rivera has a monthly mortgage payment of $423.41 on her condo. After making payments for 4 years, she decides to sell the condo. If she has a 25-year loan at an annual interest rate of 6.9%, what is the payoff for the mortgage?

Solution See page S39.

Math Matters   Biweekly and Two-Step Mortgages

A variation of the fixed rate mortgage is the *biweekly mortgage*. Borrowers make payments on a 30-year loan, but they pay half of a monthly payment every 2 weeks, which adds up to 26 half-payments a year, or 13 monthly payments. The extra monthly payment each year can result in the loan being paid off in about $17\frac{1}{2}$ years.

Another type of mortgage is the *two-step mortgage*. Its name is derived from the fact that the life of the loan has two stages. The first step is a low fixed rate for the first 7 years of the loan, and the second step is a different, and probably higher, fixed rate for the remaining 23 years of the loan. This loan is appealing to those homeowners who do not anticipate owning the home beyond the initial low-interest-rate period; they do not need to worry about the increased interest rate during the second step.

Ongoing Expenses

In addition to a monthly mortgage payment, there are other ongoing expenses associated with home ownership. Among these expenses are the costs of insurance, property tax, and utilities such as heat, electricity, and water.

Services such as schools, police and fire protection, road maintenance, and recreational services, which are provided by cities and counties, are financed by the revenue received from taxes levied on real property, or property taxes. Property tax is normally an annual expense that can be paid on a monthly, quarterly, semiannual, or annual basis.

In addition to property tax, homeowners must carry fire insurance. This insurance guarantees that the lender will be repaid in the event of a fire.

EXAMPLE 5 ■ Calculate a Total Monthly Payment

A homeowner has a monthly mortgage payment of $572.80 and an annual property tax bill of $1074. The annual cost of fire insurance is $600. Find the total monthly payment for the mortgage, property tax, and fire insurance.

Solution

Find the monthly property tax bill by dividing the annual property tax bill by 12.

$$1074 \div 12 = 89.50$$

The monthly property tax bill is $89.50.

Find the monthly fire insurance bill by dividing the annual fire insurance bill by 12.

$$600 \div 12 = 50$$

The monthly fire insurance bill is $50.

Find the sum of the mortgage payment, the monthly property tax bill, and the monthly fire insurance bill.

$$572.80 + 89.50 + 50.00 = 712.30$$

The monthly payment for the mortgage, property tax, and fire insurance is $712.30.

CHECK YOUR PROGRESS 5 A homeowner has a monthly mortgage payment of $497.63, an annual property tax bill of $777.60, and an annual fire insurance premium of $450. Find the total monthly payment for the mortgage, property tax bill, and fire insurance.

Solution   *See page S39.*

Excursion

Home Ownership Issues

There are a number of issues that a person must think about when purchasing a home. One such issue is the difference between the interest rate on which the loan payment is based and the APR. For instance, a bank may offer a loan at an annual interest rate of 6.5%, but then go on to say that the APR is 7.1%.

The discrepancy is a result of the Truth in Lending Act. This act requires that the APR be based on *all* loan fees. This includes points and other fees associated with the purchase. To calculate the APR, a computer or financial calculator is necessary.

Suppose you decide to purchase a home and you secure a 30-year, $95,000 loan at an annual interest rate of 6.5%.

1. Calculate the monthly payment for the loan.

2. If points are 1.5% of the loan amount, find the fee for points.

3. Add the fee for points to the loan amount. This is the modified mortgage on which the APR is calculated.

4. Using the result from Excursion Exercise 3 as the mortgage and the monthly payment from Excursion Exercise 1, determine the interest rate. (This is where the financial or graphing calculator is necessary. See Section 3 for details.) The result is the APR required by the Truth in Lending Act. For this example we have included only points. In most situations, other fees would be included as well.

Another issue to research when purchasing a home is that of points and mortgage interest rates. Usually paying higher points results in a lower mortgage interest rate. The question for the homebuyer is: Should I pay higher points for a lower mortgage interest rate, or pay lower points for a higher mortgage interest rate? The answer to that question depends on many factors, one of which is the amount of time the homeowner plans on staying in the home.

Consider two typical situations for a 30-year, $100,000 mortgage. Option 1 offers an annual mortgage interest rate of 7.25% and a loan origination fee of 1.5 points. Option 2 offers an annual interest rate of 7% and a loan origination fee of 2 points.

5. Calculate the monthly payments for Option 1 and Option 2.

6. Calculate the loan origination fees for Option 1 and Option 2.

7. What is the total amount paid, including points, after 2 years for each option?

8. What is the total amount paid, including points, after 3 years for each option?

9. Which option is more cost effective if you stay in the home for 2 years or less? Which option is more cost effective if you stay in the home for 3 years or more? Explain your answer.

Exercise Set 10.4

(Suggested Assignment: 1–35, odds)

1. You buy an $86,000 home with a down payment of 25%. Find the amount of the down payment and the mortgage amount.

2. Greg Walz purchases a home for $125,000 with a down payment of 10%. Find the amount of the down payment and the mortgage amount.

3. Clarrisa Madison purchases a home and secures a loan of $250,000. The bank charges a fee of 2.25 points. Find the charge for points.

4. Jerome Thurber purchases a home and secures a loan of $170,000. The bank charges a fee of $2\frac{3}{4}$ points. Find the charge for points.

5. The purchase price of a home is $109,000. A down payment of 30% is made. The bank charges $350 in fees plus 3 points. Find the total of the down payment and the closing costs.

6. The purchase price of a home is $81,000. A down payment of 20% is made. The bank charges $425 in fees plus 4 points. Find the total of the down payment and the closing costs.

7. The purchase price of a condominium is $121,500. A down payment of 25% is made. The bank charges $725 in fees plus $3\frac{1}{2}$ points. Find the total of the down payment and the closing costs.

8. The purchase price of a manufactured home is $159,000. A down payment of 20% is made. The bank charges $815 in fees plus 1.75 points. Find the total of the down payment and the closing costs.

9. Find the mortgage payment for a 25-year loan of $129,000 at an annual interest rate of 7.75%.

10. Find the mortgage payment for a 30-year loan of $245,000 at an annual interest rate of 6.5%.

11. Find the mortgage payment for a 15-year loan of $223,500 at an annual interest rate of 8.15%.

12. Find the mortgage payment for a 20-year loan of $149,900 at an annual interest rate of 8.5%.

13. Leigh King purchased a townhouse and obtained a 30-year loan of $152,000 at an annual interest rate of 7.75%.

   a. What is the mortgage payment?

   b. What is the total of the payments over the life of the loan?

c. Find the amount of interest paid on the mortgage loan over the 30 years.

14. Richard Miyashiro purchased a condominium and obtained a 25-year loan of $99,000 at an annual interest rate of 8.25%.

   a. What is the mortgage payment?

   b. What is the total of the payments over the life of the loan?

   c. Find the amount of interest paid on the mortgage loan over the 25 years.

15. Ira Patton purchased a home and obtained a 15-year loan of $219,990 at an annual interest rate of 8.7%. Find the amount of interest paid on the loan over the 15 years.

16. Leona Jefferson purchased a home and obtained a 30-year loan of $437,750 at an annual interest rate of 7.5%. Find the amount of interest paid on the loan over the 30 years.

17. Marcel Thiessen purchased a home for $208,500 and obtained a 15-year, fixed rate mortgage at 9% after paying a down payment of 10%. Of the first month's mortgage payment, how much is interest and how much is applied to the principal?

18. You purchase a condominium for $73,000. You obtain a 30-year, fixed rate mortgage loan at 12% after paying a down payment of 25%. Of the second month's mortgage payment, how much is interest and how much is applied to the principal?

19. You purchase a cottage for $85,000. You obtain a 20-year, fixed rate mortgage loan at 12.5% after paying a down payment of 30%. Of the second month's mortgage payment, how much is interest and how much is applied to the principal?

20. Fay Nguyen purchased a second home for $183,000 and obtained a 25-year, fixed rate mortgage loan at 9.25% after paying a down payment of 30%. Of the second month's mortgage payment, how much is interest and how much is applied to the principal?

21. After making payments of $456.55 for 6 years on your 30-year loan at 8.5%, you decide to sell your home. What is the loan payoff?

22. Christopher Chamberlain has a 25-year mortgage loan at an annual interest rate of 7.75%. After making

payments of $505.78 for $3\frac{1}{2}$ years, Christopher decides to sell his home. What is the loan payoff?

23. Iris Chung has a 15-year mortgage loan at an annual interest rate of 7.25%. After making payments of $672.39 for 4 years, Iris decides to sell her home. What is the loan payoff?

24. After making payments of $736.98 for 10 years on your 30-year loan at 6.75%, you decide to sell your home. What is the loan payoff?

25. A homeowner has a mortgage payment of $498.30, an annual property tax bill of $594, and an annual fire insurance premium of $300. Find the total monthly payment for the mortgage, property tax, and fire insurance.

26. Malcolm Rothschild has a mortgage payment of $876.73, an annual property tax bill of $1023, and an annual fire insurance premium of $780. Find the total monthly payment for the mortgage, property tax, and fire insurance.

27. Baka Onegin obtains a 25-year mortgage loan of $259,500 at an annual interest rate of 7.15%. Her annual property tax bill is $1320 and her annual fire insurance premium is $642. Find the total monthly payment for the mortgage, property tax, and fire insurance.

28. Suppose you obtain a 20-year mortgage loan of $198,000 at an annual interest rate of 8.4%. The annual property tax bill is $972 and the annual fire insurance premium is $486. Find the total monthly payment for the mortgage, property tax, and fire insurance.

29. Consider a mortgage loan of $150,000 at an annual interest rate of 8.125%.

   a. How much greater is the mortgage payment if the term is 15 years rather than 30 years?

   b. How much less is the amount of interest paid over the life of the 15-year loan than over the life of the 30-year loan?

30. Consider a mortgage loan of $89,990 at an annual interest rate of 7.875%.

   a. How much greater is the mortgage payment if the term is 15 years rather than 30 years?

   b. How much less is the amount of interest paid over the life of the 15-year loan than over the life of the 30-year loan?

31. The Mendez family is considering a mortgage loan of $349,500 at an annual interest rate of 6.75%.

   a. How much greater is their mortgage payment if the term is 20 years rather than 30 years?

   b. How much less is the amount of interest paid over the life of the 20-year loan than over the life of the 30-year loan?

32. Herbert Bloom is considering a mortgage loan of $322,495 at an annual interest rate of 7.5%.

   a. How much greater is his mortgage payment if the term is 20 years rather than 30 years?

   b. How much less is the amount of interest paid over the life of the 20-year loan than over the life of the 30-year loan?

33. A couple has saved $15,000 for a down payment on a home. Their bank requires a minimum down payment of 20%. What is the maximum price they can offer for a house in order to have enough money for the down payment?

34. You have saved $9000 for a down payment on a house. Your bank requires a minimum down payment of 15%. What is the maximum price you can offer for a home in order to have enough money for the down payment?

35. You have saved $19,700 to make a down payment and pay the closing costs on your future home. Your bank informs you that a 15% down payment is required and that the closing costs should be $380 plus 4 points. What is the maximum price you can offer for a home in order to have enough money for the down payment and the closing costs?

Extensions

CRITICAL THINKING

36. Suppose you have a 30-year mortgage loan for $119,500 at an annual interest rate of 8.25%. For which monthly payment does the amount of principal paid first exceed the amount of interest paid? For this exercise, you will need a spreadsheet program for producing amortization schedules. You can find one at our web site at **http://college.hmco.com.**

37. Does changing the amount of the loan in Exercise 36 change the number of the monthly payment for which the amount of principal paid first exceeds the amount of interest paid? For this exercise, you will need a spreadsheet program for producing amortization schedules. You can find one at our web site at **http://college.hmco.com.**

38. Does changing the interest rate of the loan in Exercise 36 change the number of the monthly payment for which the amount of principal paid first exceeds the amount of interest paid? For this exercise, you will need a spreadsheet program for producing amortization schedules. You can find one at our web site at **http://college.hmco.com.**

39. Suppose you are considering a mortgage loan for $150,000 at an annual interest rate of 8%. Explain why the monthly payment for a 15-year mortgage loan is not twice the monthly payment for a 30-year mortgage loan.

COOPERATIVE LEARNING

40. Suppose you buy a house for $208,750, make a down payment that is 30% of the purchase price, and secure a 30-year loan for the balance at an annual interest rate of 7.75%. The points on the loan are 1.5% and there are additional lender fees of $825.

a. How much is due at closing? Note that the down payment is due at closing.

b. After 5 years, you decide to sell your house. What is the loan payoff?

c. Because of inflation, you were able to sell your house for $248,000. Assuming the selling fees are 6% of the selling price, what are the proceeds of the sale after deducting selling fees? Do not include the interest paid on the mortgage. Remember to consider the loan payoff.

d. The percent return on an investment $= \dfrac{\text{proceeds from sale}}{\text{total closing costs}} \times 100$. Find the percent return on your investment. Round to the nearest percent.

41. Assume that you are a computer programmer for a company and that your gross pay is $36,000 per year. You are paid monthly and take home $2239.69 per month. You are considering purchasing a condominium and have looked at one you would like to buy. The purchase price is $67,000. The condominium development company will arrange financing for the purchase. A down payment of 20% of the selling price is required. The interest rate on the 30-year, fixed rate mortgage loan is 9%. The lender charges 2.5 points. The appraisal fee, title fee, and recording fee total $325. The property taxes on the condominium are currently $1152 per year.

The condominium management company charges a monthly maintenance fee for landscaping, trash removal, snow removal, maintenance on the buildings, management costs, and insurance on the property. The monthly fee is currently $60. This fee includes insurance on the buildings and land only, so you will need to purchase separate insurance coverage for your furniture and other personal possessions. You have talked to the insurance agent through whom you buy tenant homeowner's insurance (to cover the personal possessions in your apartment), and you were told that such coverage will cost you approximately $96 per year.

The condominium uses electricity for cooking, heating, and lighting. You have been informed that the unit will use approximately 8750 kilowatt-hours per year. The cost of electricity is currently $.07 per kilowatt-hour.

You have saved $18,000 in anticipation of buying a home, but you do not want your savings account balance to fall below $3000. (You want to have some funds to fall back on in case of an emergency.) You have $450 in your checking account, but you do not want your checking account balance to fall below $300.

a. Calculate the amount to be paid for the down payment and closing costs. Are you willing to take this much out of your accounts?

b. Determine the mortgage payment.

c. Calculate the total of the monthly payments related to ownership of the condominium (mortgage, maintenance fee, property tax, utilities, insurance).

d. Find the difference between your monthly take-home pay and the monthly expenses related to condominium ownership (part c). This figure indicates how much money is available to pay for all other expenses, including food, transportation, clothing, entertainment, dental bills, telephone bills, car bills, etc.

e. What percent of your monthly take-home pay are the condominium-related expenses (part c)? Round to the nearest tenth of a percent.

f. Do you think that, financially, you can handle the purchase of this condominium? Why?

CHAPTER 10   Summary

Key Terms

adjustable rate mortgage (ARM) [p. 675]
amortization schedule [p. 677]
amortize a loan [p. 677]
annual percentage rate (APR) [p. 659]
annual percentage yield (APY) [p. 659]
average daily balance [p. 656]
bond [p. 629]
bondholder [p. 629]
brokerage firm [p. 629]
closing costs [p. 674]
compound amount [p. 638]
compound interest [p. 636]
compounding period [p. 636]
conventional mortgage [p. 675]
coupon [p. 629]
dividend [p. 629]
dividend yield [p. 629]
effective interest rate [p. 647]
face value [p. 629]
finance charge [p. 656]
fixed rate mortgage [p. 675]
future value [p. 626]
inflation [p. 645]
interest [p. 622]
interest rate [p. 622]
investment trust [p. 630]
loan origination fee [p. 674]
market value [p. 629]

maturity date [p. 629]
maturity value [p. 625]
mortgage [p. 674]
mortgage payment [p. 675]
mutual fund [p. 630]
nominal interest rate [p. 647]
points [p. 674]
present value [p. 643]
principal [p. 622]
residual value [p. 666]
share of stock [p. 628]
simple interest [p. 622]
stock exchange [p. 629]
stockholders or shareholders [p. 628]

Essential Concepts

■ **Approximate Annual Percentage Rate (APR) Formula**

$$\text{APR} = \frac{2nr}{n + 1}$$

n is the number of payments and r is the stated rate of interest.

■ **APR Loan Payoff Formula**

$$A = PMT\left(\frac{1 - (1 + i)^{-n}}{i}\right)$$

A is the loan payoff, PMT is the payment, i is the interest rate per payment period, and n is the number of *remaining* payments.

- **Average Daily Balance Method**
Average daily balance
$$= \frac{\text{sum of the total amounts owed each day of the month}}{\text{number of days in the billing period}}$$

- **Compound Amount Formula**
$A = P(1 + i)^N$
A is the compound amount when P dollars are deposited at an interest rate i per compounding period for N compounding periods.
$$i = \frac{\text{annual interest rate}}{\text{number of compounding periods per year}} = \frac{r}{n}$$
N = (number of compounding periods per year)
$\times$ (number of years) $= nt$

- **Future Value/Maturity Value Formulas for Simple Interest**
$A = P + I$
$A = P(1 + rt)$

- **Monthly Lease Payment Formula**
$P = F + D$
P is the monthly lease payment, F is the average monthly finance charge, and D is the average monthly depreciation of the car, where:
Net capitalized cost
$= $ negotiated price $-$ down payment
$-$ trade-in value
$$\text{Money factor} = \frac{\text{annual interest rate as a percent}}{2400}$$
Average monthly finance charge
$= $ (net capitalized cost $+$ residual value)
$\times$ money factor

- Average monthly depreciation
$$= \frac{\text{net capitalized cost} - \text{residual value}}{\text{term of the lease in months}}$$

- **Mortgage Amount**
Mortgage $= $ selling price $-$ down payment

- **Payment Formula for an APR Loan**
$$PMT = A\left(\frac{i}{1 - (1 + i)^{-n}}\right)$$
PMT is the payment, A is the loan amount, i is the interest rate per payment period, and n is the total number of payments.

- **Present Value Formula**
$$P = \frac{A}{(1 + i)^N}$$
P is the original principal invested at an interest rate of i per compounding period for N compounding periods, and A is the compound amount.

- **Simple Interest Formula**
$I = Prt$
I is the interest, P is the principal, r is the interest rate, and t is the time period.

- **Time Period of a Loan: Converting from Days to Years**
Exact method: $t = \dfrac{\text{number of days}}{365}$
Ordinary method: $t = \dfrac{\text{number of days}}{360}$

CHAPTER 10   # Review Exercises

 Use a calculator for Exercises 1–31.

1. Calculate the simple interest due on a four-month loan of $2750 if the interest rate is 6.75%.

2. Find the simple interest due on an eight-month loan of $8500 if the interest rate is 1.15% per month.

3. What is the simple interest earned in 120 days on a deposit of $4000 if the interest rate is 6.75%?

4. Calculate the maturity value of a simple interest, 108-day loan of $7000 if the interest rate is 10.4%.

5. The simple interest charged on a three-month loan of $6800 is $127.50. Find the simple interest rate.

6. Calculate the compound amount when $3000 is deposited in an account earning 6.6% interest, compounded monthly, for 3 years.

7. What is the compound amount when $6400 is deposited in an account earning an interest rate of 6%, compounded quarterly, for 10 years?

8. Find the future value of $6000 earning 9% interest, compounded daily, for 3 years.

9. Calculate the amount of interest earned in 4 years on $600 deposited in an account paying 7.2% interest, compounded daily.

10. How much money should be invested in an account that earns 8% interest, compounded semiannually, in order to have $18,500 in 7 years?

11. To help pay your college expenses, you borrow $8000 and agree to repay the loan at the end of 5 years at 7% interest, compounded quarterly.

   a. What is the maturity value of the loan?

   b. How much interest are you paying on the loan?

12. A couple plans to save for their child's college education. What principal must be deposited by the parents when their child is born in order to have $80,000 when the child reaches the age of 18? Assume the money earns 8% interest, compounded quarterly.

13. A stock pays an annual dividend of $.66 per share. The stock is trading at $60. Find the dividend yield.

14. A bond with a $20,000 face value has a 4.5% coupon and a 10-year maturity. Calculate the total of the interest payments paid to the bondholder.

15. In 2001, the price of 1 pound of red delicious apples was $1.29. Use an annual inflation rate of 6% to calculate the price of 1 pound of red delicious apples in 2011. Round to the nearest cent.

16. You purchase a bond that will provide you with $75,000 in 8 years. Assuming an annual inflation rate of 7%, what will be the purchasing power of the $75,000 in 8 years?

17. Calculate the effective interest rate of 5.90% compounded monthly. Round to the nearest hundredth of a percent.

18. Which has the higher annual yield, 5.2% compounded quarterly or 5.4% compounded semiannually?

19. A credit card account had a $423.35 balance on March 11. A purchase of $145.50 was made on March 18, and a payment of $250 was made on March 29. Find the average daily balance if the billing date is April 11.

20. On September 10, a credit card account had a balance of $450. A purchase of $47 was made on September 20, and $157 was charged on September 25. A payment of $175 was made on September 28. The interest on the average daily balance is 1.25% per month. Find the finance charge on the October 10 bill.

21. Arlene McDonald borrows $1500 from a bank that advertises a 7.5% simple interest rate and repays the loan in six equal monthly payments.

   a. Find the monthly payment.

   b. Estimate the APR. Round to the nearest tenth of a percent.

22. Suppose you purchase a DVD player for $449, make a 10% down payment, and agree to repay the balance in 12 equal monthly payments. The finance charge on the balance is 7% simple interest.

   a. Find the monthly payment.

   b. Estimate the APR. Round to the nearest tenth.

23. Photo Experts offers a Nikon camera for $999, including taxes. If you finance the purchase of this camera for 2 years at an annual percentage rate of 8.5%, find the monthly payment.

24. Abeni Silver purchases a plasma high-definition television for $9499. The sales tax is 6.25% of the purchase price.

   a. What is the total cost, including sales tax?

   b. If Abeni makes a down payment of 20% of the total cost, find the down payment.

   c. Assuming Abeni finances the remaining cost at an annual interest rate of 8% for 3 years, find the monthly payment.

25. Suppose you decide to purchase a new car. You go to a credit union to get pre-approval for your loan. The credit union offers you an annual interest rate of 7.2% for 3 years. The purchase price of the car you select is $28,450, including taxes, and you make a 20% down payment. What is your monthly payment?

26. Dasan Houston obtains a $28,000, five-year car loan for a Windstar at an annual interest rate of 5.9%.

   a. Find the monthly payment.

   b. After 3 years, Dasan decides to purchase a new car. What is the payoff on his loan?

27. Nami Coffey obtains a five-year lease for a Chrysler 300M and negotiates a selling price of $32,450. Assuming she makes a down payment of $3000, the money factor is 0.004, and the residual value is $16,000, find each of the following.

   a. The net capitalized cost

   b. The average monthly finance charge

   c. The average monthly depreciation

   d. The monthly lease payment using a sales tax rate of 7.5%

28. The purchase price of a seaside cottage is $459,000. A down payment of 20% is made. The bank charges $815 in fees plus 1.75 points. Find the total of the down payment and the closing costs.

29. Suppose you purchase a condominium and obtain a 30-year loan of $127,900 at an annual interest rate of 6.75%.

a. What is the mortgage payment?

b. What is the total of the payments over the life of the loan?

c. Find the amount of interest paid on the mortgage loan over the 30 years.

30. Garth Santacruz purchased a condominium and obtained a 25-year loan of $189,000 at an annual interest rate of 7.5%.

a. What is the mortgage payment?

b. After making payments for 10 years, Garth decides to sell his home. What is the loan payoff?

31. Geneva Goldberg obtains a 15-year loan of $278,950 at an annual interest rate of 7%. Her annual property tax bill is $1134 and her annual fire insurance premium is $681. Find the total monthly payment for the mortgage, property tax, and fire insurance.

CHAPTER 10 **Test**

Use a calculator for Exercises 1–20.

1. Calculate the simple interest due on a three-month loan of $5250 if the interest rate is 8.25%.

2. Find the simple interest earned in 180 days on a deposit of $6000 if the interest rate is 6.75%.

3. Calculate the maturity value of a simple interest, 200-day loan of $8000 if the interest rate is 9.2%.

4. The simple interest charged on a two-month loan of $7600 is $114. Find the simple interest rate.

5. What is the compound amount when $4200 is deposited in an account earning an interest rate of 7%, compounded monthly, for 8 years?

6. Calculate the amount of interest earned in 3 years on $1500 deposited in an account paying 6.3% interest, compounded daily.

7. To help pay for a new truck, you borrow $10,500 and agree to repay the loan in 4 years at 9.5% interest, compounded monthly.

a. What is the maturity value of the loan?

b. How much interest are you paying on the loan?

8. A young couple wants to save money to buy a house. What principal must be deposited by the couple in order to have $30,000 in 5 years? Assume the money earns 6.25% interest, compounded daily.

9. A stock that has a market value of $40 pays an annual dividend of $.48 per share. Find the dividend yield.

10. Suppose you purchase a $5000 bond that has a 3.8% coupon and a 10-year maturity. Calculate the total of the interest payments you will receive.

11. In the fourth quarter of 1997, the cost per square foot of retail space in a shopping center in Fort Lauderdale was $102.63. Use an annual inflation rate of 7% to calculate the cost per square foot in the fourth quarter of 2017.

12. Calculate the effective interest rate of 6.25% compounded quarterly. Round to the nearest hundredth of a percent.

13. Which has the higher annual yield, 4.4% compounded monthly or 4.6% compounded semiannually?

14. On October 15, a credit card account had a balance of $515. A purchase of $75 was made on October 20, and a payment of $250 was made on October 28. The interest on the average daily balance is 1.8% per month. Find the finance charge on the November 15 bill.

15. Suppose you purchase a Joranda hand-held computer for $629, make a 15% down payment, and agree to repay the balance in 12 equal monthly payments. The finance charge on the balance is 9% simple interest.

a. Find the monthly payment.

b. Estimate the APR. Round to the nearest tenth.

16. Technology Pro offers a new computer for $1899, including taxes. If you finance the purchase of this computer for 3 years at an annual percentage rate of 9.25%, find your monthly payment.

17. Kalani Canfield purchases a high-speed color laser printer for $6575. The sales tax is 6.25% of the purchase price.

   a. What is the total cost, including sales tax?

   b. If Kalani makes a down payment of 20% of the total cost, find the down payment.

   c. Assuming Kalani finances the remaining cost at an annual interest rate of 7.8% for 3 years, find the monthly payment.

18. The purchase price of a house is $162,250. A down payment of 20% is made. The bank charges $815 in fees plus 3.25 points. Find the total of the down payment and the closing costs.

19. Bernard Mason purchased a house and obtained a 30-year loan of $236,000 at an annual interest rate of 6.75%.

   a. What is the mortgage payment?

   b. After making payments for 5 years, Bernard decides to sell his home. What is the loan payoff?

20. Zelda MacPherson obtains a 20-year loan of $312,000 at an annual interest rate of 7.25%. Her annual property tax bill is $1044 and her annual fire insurance premium is $516. Find the total monthly payment for the mortgage, property tax, and fire insurance.

Las Vegas is home to some of the world's most famous casinos. The casinos rely on games of chance, such as Keno and Chuck-a-luck, and favorable probability to generate large cash revenues, some of which support the city. In the **Excursion exercises** on **pages 737** and **757,** you will calculate the probabilities of some of the events that occur in these games of chance.

Combinatorics and Probability

 Need help? For on-line student resources, such as section quizzes, visit this textbook's web site at **math.college.hmco.com/students.**

In 1984, the people of the state of California passed a referendum authorizing a state lottery. One of the games of the lottery was called 6-49. To play this game a person selected six numbers from the numbers 1 through 49. If those six numbers matched the numbers chosen by the lottery commission, the person won the lottery's grand prize. It was possible for two or more people to select the same numbers. In that case, the grand prize was shared among the winning ticket holders. The number of possible lottery tickets a person could choose was 13,983,816.

If no one selected the correct numbers, the grand prize was increased for the next drawing. Because the chances of winning the grand prize were so small, frequently the grand prize was worth more than $20,000,000, and at one point it reached $65,000,000.

In 1989, the California lottery commission changed the 6-49 game to 6-53. Now, to win the grand prize, a player needed to select six numbers from 53. The number of possible lottery tickets was now 22,957,480. Because the number of possible winning tickets increased by approximately 9 million, there were longer periods between winning tickets. The result was that lottery prizes became larger, with one prize reaching 118.8 million dollars.

The California lottery was changed again in 1992, to the 6-51 game. The number of possible winning tickets was now 18,009,460. The problem with the 6-49 game was that jackpots were not high enough. With the 6-53 game, jackpots were not frequent enough. It was hoped that 6-51 would be just right.

However, the method of awarding the grand prize for the California lottery was changed again in 2000. In its current form, called Super Lotto Plus, the winner of the grand prize must pick five numbers from 1 to 47 and one MEGA number from 1 to 27. There are 41,416,353 ways this can be done. In 2002, one grand Super Lotto Plus prize exceeded $190 million.

Another lottery, called PowerBall, was formed when 21 states and the District of Columbia combined together to offer a lottery. To win this lottery, a player must choose five balls from balls numbered from 1 to 49 and one ball from balls numbered 1 to 42. Some PowerBall jackpots have exceeded $160 million.

This chapter focuses on the mathematics necessary to calculate the number of ways certain events can happen. This is part of the study of *combinatorics*. Once the number of ways an event can occur is calculated, it is possible to determine the *probability* of that event occurring. Probability is another topic of this chapter.

| # The Counting Principle

Counting by Making a List

Combinatorics is the study of counting the different outcomes of some task. For example, if a coin is flipped, the side facing upward will be a head or a tail. The outcomes can be listed as {H, T}. There are two possible outcomes.

If a regular six-sided die is rolled, the possible outcomes are ⚀, ⚁, ⚂, ⚃, ⚄, ⚅. The outcomes can also be listed as {1, 2, 3, 4, 5, 6}. There are six possible outcomes.

EXAMPLE 1 ■ Counting by Forming a List

List and then count the number of different outcomes that are possible when one letter from the word *Tennessee* is chosen.

Solution
The possible outcomes are {T, e, n, s}. There are four possible outcomes.

CHECK YOUR PROGRESS 1   List and then count the number of different outcomes that are possible when one letter is chosen from the word *Mississippi*.

Solution   See page S39.

In combinatorics, an **experiment** is an activity with an observable outcome. The set of all possible outcomes of an experiment is called the **sample space** of the experiment. Flipping a coin, rolling a die, and choosing a letter from the word *Tennessee* are experiments. The samples spaces are {H, T}, {1, 2, 3, 4, 5, 6}, and {T, e, n, s}, respectively.

An **event** is one or more of the possible outcomes of an experiment. Flipping a coin and having a head show on the upward face, rolling a 5 when a die is tossed, and choosing a T from one of the letters in the word *Tennessee* are all examples of events. An event is a *subset* of the sample space.

EXAMPLE 2 ■ Listing the Elements of an Event

One number is chosen from the sample space

$$S = \{1, 2, 3, 4, 5, 6, 7, 8, 9, 10, 11, 12, 13, 14, 15, 16, 17, 18, 19, 20\}$$

List the elements in the following events.

a. The number is even.
b. The number is divisible by 5.
c. The number is a prime number.

Solution
a. {2, 4, 6, 8, 10, 12, 14, 16, 18, 20}
b. {5, 10, 15, 20}
c. {2, 3, 5, 7, 11, 13, 17, 19}

CHECK YOUR PROGRESS 2 One digit is chosen from the digits 0 through 9. The sample space S is {0, 1, 2, 3, 4, 5, 6, 7, 8, 9}. List the elements in the following events.

a. The number is odd.

b. The number is divisible by 3.

c. The number is greater than 7.

Solution *See page S39.*

Counting by Making a Table

Each of the experiments given above illustrates a *single-stage experiment*. A **single-stage experiment** is an experiment for which there is a single outcome. Some experiments have two, three, or more stages. Such experiments are called **multi-stage experiments.** To count the number of outcomes of such an experiment, a systematic procedure is helpful. Using a table to record results is one such procedure.

Consider the two-stage experiment of rolling two dice, one red and one green. How many different outcomes are possible? To determine the number of outcomes, make a table with the different outcomes of rolling the red die across the top and the different outcomes of rolling the green die down the side.

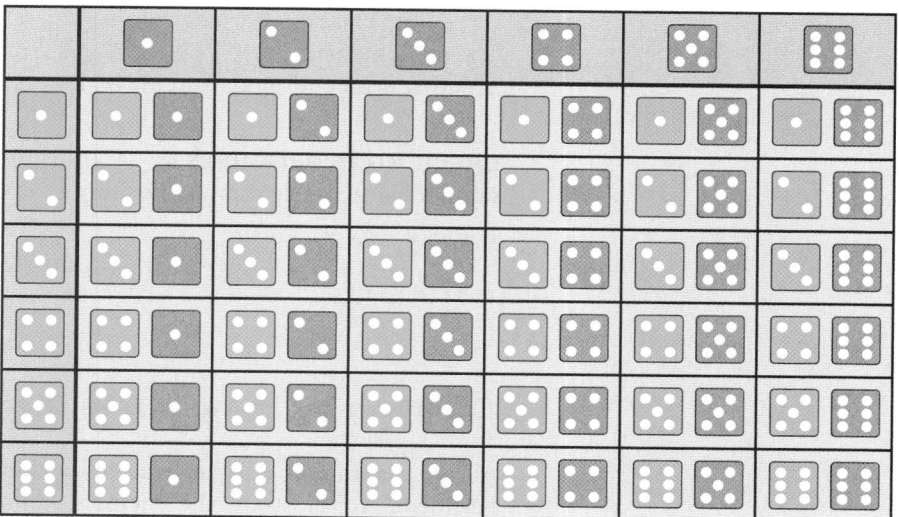

Outcomes of Rolling Two Dice

By counting the number of entries in the table, we see that there are 36 different outcomes of the experiment of rolling two dice. The sample space is

From the table, several different events can be discussed.

- The sum of the pips (dots) on the upward faces is 7. There are six outcomes of this event. They are $\{$⚀⚅, ⚁⚄, ⚂⚃, ⚃⚂, ⚄⚁, ⚅⚀$\}$.
- The sum of the pips on the upward faces is 11. There are two outcomes of this event. They are $\{$⚄⚅, ⚅⚄$\}$.
- The number of pips on the upward faces are equal. There are six outcomes of this event. They are $\{$⚀⚀, ⚁⚁, ⚂⚂, ⚃⚃, ⚄⚄, ⚅⚅$\}$.

EXAMPLE 3 ■ **Counting Using a Table**

Two-digit numbers are formed from the digits 1, 3, and 8. Find the sample space and determine the number of elements in the sample space.

Solution
Use a table to list all the different two-digit numbers that can be formed by using the digits 1, 3, and 8.

	1	**3**	**8**
1	11	13	18
3	31	33	38
8	81	83	88

The outcomes are {11, 13, 18, 31, 33, 38, 81, 83, 88}. There are nine two-digit numbers that can be formed from the digits 1, 3, and 8.

CHECK YOUR PROGRESS 3   A die is tossed and then a coin is flipped. Find the sample space and determine the number of elements in the sample space.

Solution   *See page S39.*

Counting by Using a Tree Diagram

A **tree diagram** is another way to organize the outcomes of a multi-stage experiment. To illustrate the method, consider a computer store that offers a three-component computer system. The system consists of a central processing unit (CPU), a hard drive, and a monitor. If there are two CPUs, three hard drives, and two monitors from which to choose, how many different computer systems are possible?

We can organize the information by letting C_1 and C_2 represent the two CPUs; H_1, H_2, and H_3 represent the three hard drives; and M_1 and M_2 represent the two monitors (see Figure 11.1).

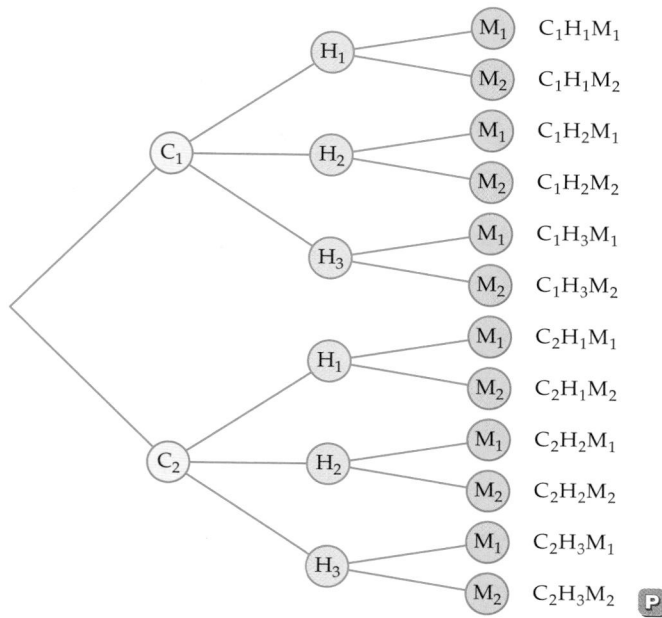

Figure 11.1

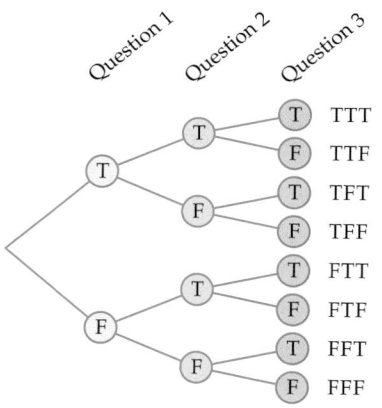

Figure 11.2

There are 12 possible computer systems.

EXAMPLE 4 ■ Counting Using a Tree Diagram

A true/false test consists of 10 questions. Draw a tree diagram to show the number of ways to answer the first three questions.

Solution
See the tree diagram in Figure 11.2. There are eight possible ways to answer the first three questions.

CHECK YOUR PROGRESS 4   Use a tree diagram to solve Example 3.

Solution   *See page S40.*

INSTRUCTOR NOTE
To convince students that explicitly listing all possible outcomes of an experiment can easily get out of hand, suggest that they draw a tree diagram for all 10 questions in Example 4. They should quickly see that it is not a reasonable request.

The Counting Principle

For each of the previous problems, the possible outcomes were listed and then counted to determine the number of different outcomes. However, it is not always possible or practical to list and count outcomes. For example, the number of different five-card poker hands that can be drawn from a standard deck of 52 playing cards is 2,598,960. Trying to create a list of these hands would be quite time consuming.

Consider again the problem of selecting a computer system having three components. By using a tree diagram, we listed the 12 possible computer systems. Another way to arrive at this result is to find the product of the numbers of options available for each component.

$$\begin{bmatrix} \text{number of} \\ \text{CPUs} \end{bmatrix} \times \begin{bmatrix} \text{number of} \\ \text{hard drives} \end{bmatrix} \times \begin{bmatrix} \text{number of} \\ \text{monitors} \end{bmatrix} = \begin{bmatrix} \text{number of} \\ \text{systems} \end{bmatrix}$$
$$2 \qquad\times\qquad 3 \qquad\times\qquad 2 \qquad=\qquad 12$$

For the example of tossing two dice, there were 36 possible outcomes. We can arrive at this result without listing the outcomes by finding the product of the number of possible outcomes of rolling the red die and the number of possible outcomes of rolling the green die.

$$\begin{bmatrix} \text{outcomes} \\ \text{of red die} \end{bmatrix} \times \begin{bmatrix} \text{outcomes} \\ \text{of green die} \end{bmatrix} = \begin{bmatrix} \text{number of} \\ \text{outcomes} \end{bmatrix}$$
$$6 \qquad\times\qquad 6 \qquad=\qquad 36$$

This method of counting all of the outcomes of a multi-stage experiment without listing them is called the **counting principle.**

Counting Principle

Let E be a multi-stage experiment. If $n_1, n_2, n_3, \ldots, n_k$ are the number of possible outcomes of each of the k stages of E, then there are $n_1 \cdot n_2 \cdot n_3 \cdot \cdots \cdot n_k$ possible outcomes for E.

EXAMPLE 5 ■ Counting by Using the Counting Principle

In horse racing, a *trifecta* consists of choosing the exact order of the first three horses across the finish line. If there are eight horses in a race, how many trifectas are possible, assuming there are no ties?

Solution
Any one of the eight horses can be first, so $n_1 = 8$. Because a horse cannot finish both first and second, there are seven horses that can finish second; thus $n_2 = 7$. Similarly, there are six horses that can finish third; $n_3 = 6$. By the counting principle, there are $8 \cdot 7 \cdot 6 = 336$ possible trifectas.

CHECK YOUR PROGRESS 5   Nine runners are entered in a 100-meter dash for which a gold, silver, and bronze medal will be awarded for first, second, and third place finishes, respectively. In how many possible ways can the medals be awarded? (Assume there are no ties.)

Solution   See page S40.

Math Matters   **Coding Characters for Computers**

The American Standard Code for Information Interchange (ASCII) is a code used to interchange information between computers. Having a uniform standard of coding information on which all computer manufacturers agree enables data to be transferred from one computer to another.

Each upper- and lower-case letter, numeral, and punctuation mark, as well as some other special symbols (for example, &, %, and #) has a representation as an eight-digit binary number. Here are some examples.

$$A = 0100\ 0001 \quad B = 0100\ 0010 \quad a = 0110\ 0001 \quad \& = 0010\ 0110$$

Using the counting principle, we can count the number of possible characters that can be represented by this scheme. Each of the eight digits in the binary number can be a 0 or a 1. Thus there are two choices for the first digit, two choices for the second digit, and so on until the eighth digit, for which there are two choices. By the counting principle, there are $2 \cdot 2 \cdot 2 \cdot 2 \cdot 2 \cdot 2 \cdot 2 \cdot 2 = 2^8 = 256$ possible characters that can be represented using the ASCII method.

This may seem like enough, and it certainly is for the English language. However, if we want to allow computers from different countries to communicate, it is necessary to extend this coding method. Currently, new standards of coding are being discussed that would allow at least 64,000 different characters to have a unique code.

Counting With and Without Replacement

Consider an experiment in which balls colored red, blue, and green are placed in a box. A person reaches into the box and repeatedly pulls out a colored ball, keeping note of the color picked. The sequence of colors that can result depends on whether or not the balls are returned to the box after each pick. This is referred to as performing the experiment *with replacement* or *without replacement*.

QUESTION   *Does a multi-stage experiment performed with replacement generally have more or fewer possible outcomes than the same experiment performed without replacement?*

Consider the following two situations.

1. How many four-digit numbers can be formed from the digits 1 through 9 if no digit can be repeated?

2. How many four-digit numbers can be formed from the digits 1 through 9 if a digit can be used repeatedly?

ANSWER   *More. Each stage of an experiment (after the first) performed without replacement will have fewer possible outcomes than the preceding stage. Performed with replacement, each stage of the experiment has the same number of outcomes.*

In the first case, there are nine choices for the first digit ($n_1 = 9$). Because a digit cannot be repeated, the first digit chosen cannot be used again. Thus there are only eight choices for the second digit ($n_2 = 8$). Because neither of the first two digits can be used as the third digit, there are only seven choices for the third digit ($n_3 = 7$). Similarly, there are six choices for the fourth digit ($n_4 = 6$). By the counting principle there are $9 \cdot 8 \cdot 7 \cdot 6 = 3024$ four-digit numbers in which no digit is repeated.

In the second case, there are nine choices for the first digit ($n_1 = 9$). Because a digit can be used repeatedly, the first digit chosen can be used again. Thus there are nine choices for the second digit ($n_2 = 9$), and, similarly, there are nine choices for the third and fourth digits ($n_3 = 9$, $n_4 = 9$). By the counting principle, there are $9 \cdot 9 \cdot 9 \cdot 9 = 6561$ four-digit numbers when digits can be used repeatedly.

The set of four-digit numbers created without replacement includes numbers such as 3867, 7941, and 9128. For these numbers, no digit is repeated. However, numbers such as 6465, 9911, and 2222, each of which contains at least one repeated digit, can be created only with replacement.

EXAMPLE 6 ■ Counting With and Without Replacement

From the letters a, b, c, d, and e, how many four-letter groups can be formed if

a. a letter can be used more than once?

b. each letter can be used exactly once?

Solution

a. Because each letter can be repeated, there are $5 \cdot 5 \cdot 5 \cdot 5 = 625$ possible four-letter groups.

b. Because each letter can be used only once, there are $5 \cdot 4 \cdot 3 \cdot 2 = 120$ four-letter groups in which no letter is repeated.

CHECK YOUR PROGRESS 6   In how many ways can a mail carrier place three letters in five mailboxes if

a. each box may receive more than one letter?

b. each mailbox may receive no more than one letter?

Solution   *See page S40.*

Excursion

Decision Trees

Decision trees are tree diagrams that are used to solve problems that involve many choices. To illustrate, we will consider a particular puzzle. Suppose we are given eight coins, one of which is counterfeit and slightly heavier than the other seven. Using a balance scale, we must find the counterfeit coin.

Designate the coins as c_1, c_2, c_3, c_4, c_5, c_6, c_7, and c_8. One way to determine the counterfeit coin is to weigh coins in pairs. This method is illustrated by the decision tree

(continued)

in Figure 11.3. In this case, it would take from one to four weighings to determine the counterfeit coin.

A second method is to divide the coins into two groups of four coins each, place each group on a pan of the balance scale, and take the coins from the side that goes down. Now divide these four coins into two groups of two coins each and weigh them on the balance scale. Again keep the coins from the heavier side. Weighing one of these coins against the other will reveal the counterfeit coin. The decision tree for this procedure is shown in Figure 11.4. Using this method, the counterfeit coin will be found in three weighings.

A third method is even more efficient. Divide the coins into three groups. Two of the groups contain three coins and the third group contains two coins. Place each of the three-coin groups on the balance scale. If they balance, the counterfeit coin is in the third group. Placing a coin from the third group on each of the balance pans will determine the counterfeit coin. If the three-coin groups do not balance, then weigh two of the coins in the pan that goes down. If these balance, the third coin is the counterfeit. If not, the counterfeit is the coin on the pan that goes down. This method requires only two weighings and is shown in the decision tree in Figure 11.5.

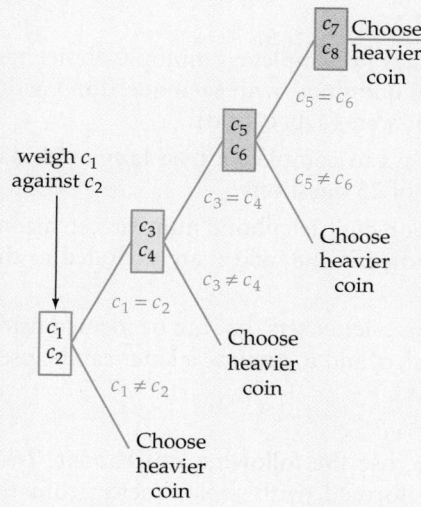

Figure 11.3

Using this method, it may take up to four weighings to determine which is the heaviest coin.

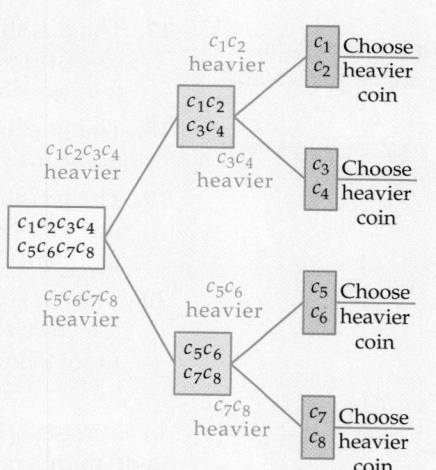

Figure 11.4

Using this method, it will always take three weighings to determine which is the heaviest coin.

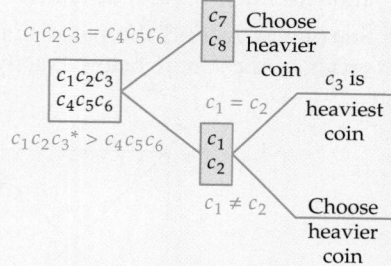

*We are assuming that $c_1c_2c_3$ is heavier than $c_4c_5c_6$. If $c_4c_5c_6$ is heavier, use those coins in the final weighing.

Figure 11.5

Using this method, it will take only two weighings to determine which is the heaviest coin.

INSTRUCTOR NOTE
There is a much more difficult version of this puzzle in which it is not known whether the counterfeit coin is heavier or lighter than the other coins. See:
http://mathforum.org/library/drmath/view/57947.html.

Excursion Exercises

For each of the following problems, draw a decision tree and determine the minimum number of weighings necessary to identify the counterfeit coin.

1. In a stack of 12 identical-looking coins, one is counterfeit and is lighter than the remaining 11 coins. Using a balance scale, identify the counterfeit coin in as few weighings as possible.

2. In a stack of 13 identical-looking coins, one is counterfeit and is heavier than the remaining 12 coins. Using a balance scale, identify the counterfeit coin in as few weighings as possible.

Exercise Set 11.1   (Suggested Assignment: 1–33 odds; 14, 16, 28, 34)

In Exercises 1–14, list the elements of the sample space defined by each experiment.

1. Select an even single-digit whole number.

2. Select an odd single-digit whole number.

3. Select one day from the days of the week.

4. Select one month from the months of the year.

5. Toss a coin twice.

6. Toss a coin three times.

7. Roll a single die and then toss a coin.

8. Toss a coin and then choose a digit from the digits 1 through 4.

9. Choose a complete dinner from a dinner menu that allows a customer to choose from two salads, three entrees, and two desserts.

10. Choose a car during a new car promotion that allows a buyer to choose from three body styles, two radios, and two interior color schemes.

11. Starting at A, find the paths that pass through each vertex of the square below exactly once.

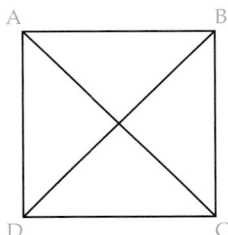

12. Starting at A, find the number of paths that pass through each vertex of the pentagon below exactly once.

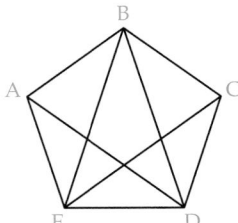

13. Three letters addressed to A, B, and C are randomly placed in three envelopes addressed to A, B, and C.

14. The first of three characters in a customer identification code is chosen from the first letters of North, South, East, and West. The second character is chosen from the digits 3 and 4 (for 2003 or 2004) and the third character is chosen from the first letters of Yes and No (to identify previous customers). Find the sample space of the beginning three characters of the identification codes for this company.

In Exercises 15–20, use the counting principle to determine the number of elements in the sample space.

15. Two digits are selected without replacement from the digits 1, 2, 3, and 4.

16. Two digits are selected with replacement from the digits 1, 2, 3, and 4.

17. The possible ways to complete a multiple-choice test consisting of 20 questions, with each question having four possible answers (a, b, c, or d)

18. The possible ways to complete a true-false examination consisting of 25 questions

19. The possible four-digit telephone number extensions that can be formed if 0, 8, and 9 are excluded as the first digit

20. The possible three-letter sets that can be formed using the vowels a, e, i, o, and u. Assume a letter can be used more than once.

In Exercises 21–26, use the following experiment. Two-digit numbers are formed, with replacement, from the digits 0 through 9.

21. How many two-digit numbers are possible?

22. How many two-digit even numbers are possible?

23. How many numbers are divisible by 5?

24. How many numbers are divisible by 3?

25. How many numbers are greater than 37?

26. How many numbers are less than 59?

In Exercises 27–30, use the following experiment. Four cards labeled A, B, C, and D are randomly placed in four boxes labeled A, B, C, and D. Each box receives exactly one card.

27. In how many ways can the cards be placed in the boxes?

28. Count the number of elements in the event that no box contains a card with the same letter as the box.

29. Count the number of elements in the event that *at least* one card is placed in the box with the corresponding letter.

30. If you add the answer for Exercise 28 and the answer for Exercise 29, is the sum the answer for Exercise 27? Why or why not?

In Exercises 31–34, use the following experiment. A state lottery game consists of choosing one card from each of the four suits in a standard deck of playing cards. (There are 13 cards in each suit.)

31. Count the number of elements in the sample space.

32. Count the number of elements in the event that an ace, a king, a queen, and a jack are chosen.

33. Count the number of ways in which four aces can be chosen.

34. Count the number of ways in which four cards, each of a different face value, can be chosen.

35. Write a lesson that you could use to explain the meanings of the words *experiment, sample space,* and *event* as they apply to combinatorics.

36. Explain how a tree diagram is used to count the number of ways an experiment can be performed.

Extensions

CRITICAL THINKING

37. A main component of any computer programming language is its method of repeating a series of computations. Each language has its own syntax for performing those "loops." In one language, BASIC, the structure is similar to the display below.

```
FOR I = 1 TO 10          ——— Starting with I = 1
    FOR J = 1 TO 15      ——— Starting with J = 1
        SUM = I + J
    NEXT J              ←—— Increase J by 1 until J exceeds 15
NEXT I              ←—— Increase I by 1 until I exceeds 10
END
```

The program repeats each loop until the index variables, I and J in this case, exceed a certain value. After this program is executed, how many times will the instruction SUM = I + J have been executed? What is the final value of SUM?

38. One way in which a software engineer can write a program to sort a list of numbers from smallest to largest is to use a *bubble sort.* The following code segment, written in BASIC, will perform a bubble sort of n numbers. We have left out the details of how the numbers are sorted in the list.

```
FOR J = 1 TO n − 1
    FOR K = J + 1 TO n
        <Sorting takes place here.>
    NEXT K
NEXT J
END
```

a. If the list contains 10 numbers ($n = 10$), how many times will the program loop before reaching the END statement?

b. Suppose there are 20 numbers in the list. How many times will the program loop before reaching the END statement?

EXPLORATIONS

39. Use a tree diagram to display the relationships among the junior- and senior-level courses you must take for your major and their prerequisites.

40. Review the rules of the game checkers, and make a tree diagram that shows all of the first two moves that are possible by one player of a checker game. Assume that no moves are blocked by the opponent's checkers. (*Hint:* it may help to number the squares of the checkerboard.)

| **SECTION 11.2** | **Permutations and Combinations** |

Factorial

Suppose four different colored squares are arranged in a row. One possibility is shown below.

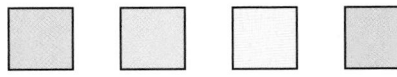

How many different ways are there to order the colors? There are four choices for the first square, three choices for the second square, two choices for the third square, and one choice for the fourth square. By the counting principle, there are $4 \cdot 3 \cdot 2 \cdot 1 = 24$ different arrangements of the four squares. Note from this example that the number of arrangements equals the product of the natural numbers n through 1, where n is the number of objects. This product is called a *factorial*.

> **Definition of *n* Factorial**
>
> ***n* factorial** is the product of the natural numbers n through 1 and is symbolized by ***n*!**.
>
> $$n! = n \cdot (n - 1) \cdot (n - 2) \cdot \cdots \cdot 3 \cdot 2 \cdot 1$$

Here are some examples.

$$5! = 5 \cdot 4 \cdot 3 \cdot 2 \cdot 1 = 120$$
$$8! = 8 \cdot 7 \cdot 6 \cdot 5 \cdot 4 \cdot 3 \cdot 2 \cdot 1 = 40,320$$
$$1! = 1$$

On some occasions it will be necessary to use 0! (zero factorial). Because it is impossible to define zero factorial in terms of a product of natural numbers, a standard definition is used.

> **Definition of Zero Factorial**
>
> $$0! = 1$$

CALCULATOR NOTE

Most scientific calculators can compute factorials. The numbers get very large, however. A calculator will normally give these large results in scientific notation. Many calculators cannot compute the factorial of numbers greater than 69. In fact, 70! has more than 100 digits.

A factorial can be written in terms of smaller factorials. This is useful when calculating large factorials. For example,

$$10! = 10 \cdot 9!$$
$$10! = 10 \cdot 9 \cdot 8!$$
$$10! = 10 \cdot 9 \cdot 8 \cdot 7!$$

EXAMPLE 1 ■ Simplify Factorials

Evaluate: **a.** $5! - 3!$ **b.** $\dfrac{9!}{6!}$

Solution

a. $5! - 3! = (5 \cdot 4 \cdot 3 \cdot 2 \cdot 1) - (3 \cdot 2 \cdot 1) = 120 - 6 = 114$

b. $\dfrac{9!}{6!} = \dfrac{9 \cdot 8 \cdot 7 \cdot \cancel{6!}}{\cancel{6!}} = 9 \cdot 8 \cdot 7 = 504$

CHECK YOUR PROGRESS 1   Evaluate: **a.** $7! + 4!$ **b.** $\dfrac{8!}{4!}$

Solution   *See page S40.*

Permutations

Determining the number of possible arrangements of distinct objects in a definite order, as we did with the squares earlier, is one application of the counting principle. Each arrangement of this type is called a *permutation*.

> **Definition of Permutation**
>
> A **permutation** is an arrangement of objects in a definite order.

For example, abc and cba are two different permutations of the letters a, b, and c. As a second example, 122 and 212 are two different permutations of the digits 1 and 2.

The counting principle is used to count the number of different permutations of any set of objects. We will begin our discussion with *distinct* objects. For instance, the objects a, b, c, d are distinct, whereas the objects ♥ ♥ ♠ ♦ are not all distinct.

Consider the two objects ❑ ▼. Counting without replacement, there are two ways to choose the first object; $n_1 = 2$. There is one way to choose the second object; $n_2 = 1$. By the counting principle, there are $2 \cdot 1 = 2! = 2$ permutations of the two objects. These are shown below.

Permutation 1   ❑ ▼

Permutation 2   ▼ ❑

For the three objects ○ ■ ❑, there are (counting without replacement) three choices for the first object, two choices for the second object, and one choice for the third object. By the counting principle, there are $3 \cdot 2 \cdot 1 = 3! = 6$ permutations of the three objects.

Permutation 1 ○ ■ ❑
Permutation 2 ○ ❑ ■
Permutation 3 ❑ ○ ■
Permutation 4 ❑ ■ ○
Permutation 5 ■ ○ ❑
Permutation 6 ■ ❑ ○

For the four objects ❑ ▼ ○ ■, there are, without replacement, four choices for the first object, three choices for the second object, two choices for the third object, and one choice for the fourth object. There are $4 \cdot 3 \cdot 2 \cdot 1 = 4! = 24$ permutations of the four objects.

The arrangements of objects that we have discussed include arranging *all* of the objects. In many cases, we may use only *some* of the objects. For example, consider the five objects ❑ ◗ ✳ ○ ▼. When only 3 objects are selected from the five without replacement, there are five choices for the first object, four choices for the second object, and three choices for the third object. By the counting principle, there are $5 \cdot 4 \cdot 3 = 60$ possible permutations.

The following formula can be used to determine the number of permutations of n distinct objects, of which k are selected.

Permutation Formula for Distinct Objects

The number of permutations $P(n, k)$ of n distinct objects selected k at a time is

$$P(n, k) = \frac{n!}{(n - k)!}$$

EXAMPLE 2 ■ Counting Permutations

A university tennis team consists of six players who are ranked from 1 through 6. If a tennis coach has 10 players from which to choose, how many different tennis teams can the coach select?

Solution

Because the players on the tennis team are ranked from 1 through 6, a team with player A in position 1 is different from a team with player A in position 2. Therefore, the number of different teams is the number of permutations of 10 players selected six at a time.

$$P(10, 6) = \frac{10!}{(10 - 6)!} = \frac{10!}{4!} = \frac{10 \cdot 9 \cdot 8 \cdot 7 \cdot 6 \cdot 5 \cdot 4!}{4!}$$
$$= 10 \cdot 9 \cdot 8 \cdot 7 \cdot 6 \cdot 5 = 151{,}200$$

There are 151,200 possible tennis teams.

CHECK YOUR PROGRESS 2   A college golf team consists of five players who are ranked from 1 through 5. If a golf coach has eight players from which to choose, how many different golf teams can the coach select?

Solution   See page S40.

EXAMPLE 3 ■ Counting Permutations

 In 2001, the Kentucky Derby had 17 horses entered in the race. How many different finishes of first, second, third, and fourth place were possible?

Solution

Because the order in which the horses finish the race is important, the number of possible finishes of first, second, third, and fourth place is $P(17, 4)$.

$$P(17, 4) = \frac{17!}{(17 - 4)!} = \frac{17!}{13!} = \frac{17 \cdot 16 \cdot 15 \cdot 14 \cdot 13!}{13!}$$
$$= 17 \cdot 16 \cdot 15 \cdot 14 = 57{,}120$$

There were 57,120 possible finishes of first, second, third, and fourth place.

CHECK YOUR PROGRESS 3 A 10-K marathon has 20 people entered. In how many different ways can the first, second, and third place prizes be awarded?

Solution *See page S40.*

A Standard Deck of
Playing Cards

Math Matters How Many Shuffles?

A standard deck of playing cards consists of 52 different cards divided into four suits: spades (♠), hearts (♥), diamonds (♦), and clubs (♣). Each shuffle of the deck results in a new arrangement of the cards. Another way of stating this is to say that each shuffle results in a new permutation of the cards. There are $P(52, 52) = 52! \approx 8 \times 10^{67}$ (that's 8 with 67 zeros after it) possible arrangements.

Suppose a deck has each of the four suits arranged in order from two through ace. How many shuffles are necessary to achieve a randomly ordered deck in which any card is equally likely to occur in any position in the deck?

Two mathematicians, Dave Bayer of Columbia University and Persi Diaconis of Harvard University, have shown that seven shuffles are enough. Their proof has many applications to complicated counting problems. One problem in particular is that of analyzing speech patterns. Solving this problem is critical to enabling computers to interpret human speech.

Applying Several Counting Techniques

The permutation formula is derived from the counting principle. This formula is just a convenient way to express the number of ways the items in an ordered list can be arranged. Some counting problems require using both the permutation formula and the counting principle.

EXAMPLE 4 ■ Counting Using Several Methods

Five women and four men are to be seated in a row of nine chairs. How many different seating arrangements are possible if

a. there are no restrictions on the seating arrangements?

b. the women sit together and the men sit together?

Solution

Because seating arrangements have a definite order, they are permutations.

a. If there are no restrictions on the seating arrangements, then the number of seating arrangements is $P(9, 9)$.

$$P(9, 9) = \frac{9!}{(9 - 9)!} = \frac{9!}{0!} = 9! = 362{,}880$$

There are 362,880 seating arrangements.

b. This is a multi-stage experiment, so both the permutation formula and the counting principle will be used. There are 5! ways to arrange the women and 4! ways to arrange the men. We must also consider that either the women or the men could be seated at the beginning of the row. There are two ways to do this. By the counting principle, there are $2 \cdot 5! \cdot 4!$ ways to seat the women together and the men together.

$$2 \cdot 5! \cdot 4! = 5760$$

There are 5760 arrangements in which women sit together and men sit together.

CHECK YOUR PROGRESS 4   There are seven tutors, three juniors and four seniors, who must be assigned to the seven hours that a math center is open each day. If each tutor works one hour per day, how many different tutoring schedules are possible if

a. there are no restrictions?

b. the juniors tutor during the first three hours and the seniors tutor during the last four hours?

Solution   See page S40.

Permutations of Indistinguishable Objects

Up to this point we have been counting the number of permutations of *distinct* objects. We now look at the situation of arranging objects when some of them are identical. In the case of identical or indistinguishable objects, a modification of the permutation formula is necessary. The general idea for modifying the formula is to count the number of permutations as if all of the objects were distinct, and then remove the permutations that are not different in appearance.

Consider the permutations of the letters *bbbcc*. We first assume the letters are all different—for example, $b_1 b_2 b_3 c_1 c_2$. Using the permutation formula, there are $5! = 120$ permutations. Now we need to remove repeated permutations. Consider for a moment the letter b in the permutation $bbbc_1c_2$. If the b's are written as b_1, b_2, and b_3 (so that they are distinguishable), then

$$b_1b_2b_3c_1c_2 \quad b_1b_3b_2c_1c_2 \quad b_2b_1b_3c_1c_2 \quad b_2b_3b_1c_1c_2 \quad b_3b_2b_1c_1c_2 \quad b_3b_1b_2c_1c_2$$

are all distinct permutations that end with c_1c_2. There are six of these permutations. Note that $3! = 6$, where 3 is the number of b's.

However, because the b's are not distinct, each of these permutations of b's should have been counted only once. Thus there are six times too many permutations of b's for each arrangement of c_1 and c_2.

A similar argument applies to the c's. If the c's are identical, then c_1c_2bbb and c_2c_1bbb are the same permutation. For each arrangement of b's, there are two arrangements of c's that yield identical permutations. Note that there are two c's and that $2! = 2$, the number of identical permutations. Thus there are two times too many permutations of c's for each arrangement of b's.

Combining the results above, the number of permutations of $bbbcc$ is

$$\frac{5!}{3! \cdot 2!} = \frac{5 \cdot 4 \cdot 3 \cdot 2 \cdot 1}{(3 \cdot 2 \cdot 1) \cdot (2 \cdot 1)} = 10.$$

There are 10 distinct permutations of $bbbcc$.

Permutations of Objects, Some of Which are Identical

The number of distinguishable permutations of n objects of r different types, where k_1 identical objects are of one type, k_2 of another, and so on, is given by

$$\frac{n!}{k_1! \cdot k_2! \cdots \cdot k_r!}$$

where $k_1 + k_2 + \cdots + k_r = n$.

EXAMPLE 5 ■ Permutations of Identical Objects

If seven identical dice are rolled, find the number of ways two 4's, one 5, and four 6's can appear on the upward faces.

Solution
We are looking for the number of permutations of the digits 4456666. With $n = 7$ (number of dice), $k_1 = 2$ (number of 4's), $k_2 = 1$ (number of 5's), and $k_3 = 4$ (number of 6's), we have

$$\frac{7!}{2! \cdot 1! \cdot 4!} = 105$$

There are 105 permutations of the dice.

CHECK YOUR PROGRESS 5   Eight coins—three pennies, two nickels, and three dimes—are placed in a single stack. How many different stacks are possible if

a. there are no restrictions on the placement of the coins?
b. the dimes must stay together?

Solution   See page S40.

Combinations

INSTRUCTOR NOTE
Students often confuse permutations with combinations.
Emphasize that if the particular
order of objects makes a
difference, then we are counting
permutations. You can also point
out that there are more
permutations of objects than
combinations.

For some arrangements of objects, the order of the arrangement is important. These are permutations. If a telephone extension is 2537, then the digits must be dialed in exactly that order. On the other hand, if you were to receive a one-dollar bill, a five-dollar bill, and a ten-dollar bill, you would have $16 regardless of the order in which you received the bills. A **combination** is a collection of objects for which the order is not important. The three-letter sequences acb and bca are *different* permutations but the *same* combination.

> *QUESTION*
> ———————
> *From a group of 45 applicants, five identical scholarships will be awarded. Is the number of ways in which the scholarships can be awarded determined by permutations or combinations?*

The formula for finding the number of combinations is derived in much the same manner as the formula for finding the number of permutations of identical objects. Consider the problem of finding the number of combinations possible when choosing three letters from the letters a, b, c, d, and e, without replacement. For each choice of three letters, there are 3! permutations. For example, choosing the letters a, d, and e gives the following six permutations.

ade   aed   dea   dae   ead   eda

Because there are six permutations and each permutation is the *same* combination, the number of permutations is six times the number of combinations. This is true each time three letters are selected. Therefore, to find the number of combinations of five objects chosen three at a time, divide the number of permutations by $3! = 6$. The number of combinations of five objects chosen 3 at a time is

$$\frac{P(5, 3)}{3!} = \frac{5!}{3! \cdot (5-3)!} = \frac{5!}{3! \cdot 2!} = \frac{5 \cdot 4 \cdot 3!}{3! \cdot 2!} = \frac{5 \cdot 4}{2 \cdot 1} = 10$$

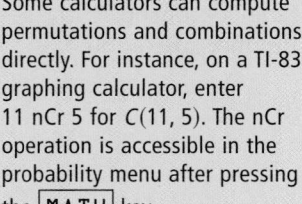

CALCULATOR NOTE

Some calculators can compute
permutations and combinations
directly. For instance, on a TI-83
graphing calculator, enter
11 nCr 5 for $C(11, 5)$. The nCr
operation is accessible in the
probability menu after pressing
the MATH key.

> **Combination Formula**
>
> The number of combinations of n objects chosen k at a time is
>
> $$C(n, k) = \frac{P(n, k)}{k!} = \frac{n!}{k! \cdot (n-k)!}$$

In Section 11.1, we stated that there were 2,598,960 possible five-card poker hands. This number was calculated using the combination formula. Because the five-card hand ace of hearts, king of diamonds, queen of clubs, jack of spades, ten of hearts is exactly the same as the five-card hand king of diamonds, jack of spades, queen of clubs, ten of hearts, ace of hearts, the order of the cards is not important and therefore the number of hands is a combination. The number of different five-

> *ANSWER*
> ————————
> *Combinations. The order in which the scholarship winners are chosen is not important.*

card poker hands is the combination of 52 cards chosen five at a time, which is given by $C(52, 5)$.

$$C(52, 5) = \frac{52!}{5! \cdot (52 - 5)!} = \frac{52!}{5! \cdot 47!} = \frac{52 \cdot 51 \cdot 50 \cdot 49 \cdot 48 \cdot 47!}{5! \cdot 47!}$$

$$= \frac{52 \cdot 51 \cdot 50 \cdot 49 \cdot 48}{5 \cdot 4 \cdot 3 \cdot 2 \cdot 1} = 2{,}598{,}960$$

EXAMPLE 6 ■ Counting Using the Combination Formula

A basketball team consists of 11 players. In how many different ways can a coach choose the five starting players, assuming the position of a player is not considered?

Solution

This is a combination problem, because the order in which the coach chooses the players is not important. The five starting players P_1, P_2, P_3, P_4, P_5 are the same as the five starting players P_3, P_5, P_1, P_2, P_4.

$$C(11, 5) = \frac{11!}{5! \cdot (11 - 5)!} = \frac{11!}{5! \cdot 6!} = \frac{11 \cdot 10 \cdot 9 \cdot 8 \cdot 7 \cdot 6!}{5! \cdot 6!}$$

$$= \frac{11 \cdot 10 \cdot 9 \cdot 8 \cdot 7}{5 \cdot 4 \cdot 3 \cdot 2 \cdot 1} = 462$$

There are 462 possible five-player starting teams.

CHECK YOUR PROGRESS 6   A softball team consists of 16 players. In how many ways can a coach choose the 9 starting players? (Assume the position of a player is not considered.)

Solution   See page S40.

Not considering the position of a player in Example 6 is an important assumption. If the positions center, left forward, right forward, left guard, and right guard are considered, then the problem is no longer a combination problem. If the coach chooses a team by position, then P_1 at center and P_2 at right guard would be a different team than P_1 at right guard and P_2 at center. In this case, because the position of a player is important, the number of teams is the permutations of 11 players chosen 5 at a time: $P(11, 5) = 55{,}440$.

EXAMPLE 7 ■ Counting Using the Combination Formula and the Counting Principle

A committee of five is chosen from five mathematicians and six economists. How many different committees are possible if the committee must include two mathematicians and three economists?

Solution

Because a committee of professors A, B, C, D, and E is exactly the same as a committee of professors B, D, E, A, and C, choosing a committee is an example of choosing a combination. There are five mathematicians from whom two are chosen, which is equivalent to $C(5, 2)$ combinations. There are six economists from whom three are chosen, which is equivalent to $C(6, 3)$ combinations. Therefore, by the counting principle, there are $C(5, 2) \cdot C(6, 3)$ ways to choose two mathematicians and three economists.

$$C(5, 2) \cdot C(6, 3) = \frac{5!}{2! \cdot 3!} \cdot \frac{6!}{3! \cdot 3!} = 10 \cdot 20 = 200$$

There are 200 possible committees consisting of two mathematicians and three economists.

CHECK YOUR PROGRESS 7 An IRS auditor randomly chooses five tax returns to audit from a stack of 10 tax returns, four of which are from corporations and six of which are from individuals. In how many different ways can the auditor choose the tax returns if the auditor wants to include three corporate and two individual returns?

Solution *See page S40.*

Math Matters   Buying Every Possible Lottery Ticket

A lottery prize in Pennsylvania reached $65 million. A resident of the state suggested that it might be worth buying a ticket for every possible combination of numbers. To win the $65 million, a person needed to correctly select six of 50 numbers. Each ticket costs one dollar. Because the order of the numbers drawn is not important, the number of different possible tickets is $C(50, 6) = 15{,}890{,}700$. Thus it would cost the resident $15,890,700 to purchase tickets for every possible combination of numbers.

It might seem that an approximately $16 million investment to win $65 million is reasonable. Unfortunately, when prize levels reach such lofty heights, many more people play the lottery. This increases the chances that more than one person will select the winning combination of numbers. In fact, there were eight people with the winning numbers and each received approximately $8 million. Now the $16 million investment does not look very appealing.

Some of the problems in this section require a basic knowledge of what comprises a standard deck of playing cards. In a standard deck, there are four suits: spades, hearts, diamonds, and clubs. Each suit has 13 cards: 2 through 10, jack, queen, king, and ace. See the Math Matters on page 707.

EXAMPLE 8 ■ Counting Problems with Cards

From a standard deck of playing cards, five cards are chosen. How many five-card combinations contain

a. two kings and three queens?

b. five hearts?

c. five cards of the same suit?

Solution

a. There are $C(4, 2)$ ways of choosing two kings from four kings and $C(4, 3)$ ways of choosing three queens from four queens. By the counting principle, there are $C(4, 2) \cdot C(4, 3)$ ways of choosing two kings and three queens.

$$C(4, 2) \cdot C(4, 3) = \frac{4!}{2! \cdot 2!} \cdot \frac{4!}{3! \cdot 1!} = 6 \cdot 4 = 24$$

There are 24 ways of choosing two kings and three queens.

b. There are $C(13, 5)$ ways of choosing five hearts from 13 hearts.

$$C(13, 5) = \frac{13!}{5! \cdot 8!} = 1287$$

There are 1287 ways to choose five hearts from 13 hearts.

c. From part b, there would also be 1287 ways of choosing five spades from 13 spades, five clubs from 13 clubs, or five diamonds from 13 diamonds. Because there are four suits from which to choose and $C(13, 5)$ ways of choosing five cards from a suit, by the counting principle there are $4 \cdot C(13, 5)$ ways to choose five cards of the same suit.

$$4 \cdot C(13, 5) = 4 \cdot 1287 = 5148$$

There are 5148 ways of choosing five cards of the same suit from a standard deck of playing cards.

CHECK YOUR PROGRESS 8   From a standard deck of playing cards, five cards are chosen. How many five-card combinations contain four cards of the same suit?

Solution   *See page S40.*

Five cards of the same suit is called a *flush* in poker.

Excursion

Choosing Numbers in Keno

A popular gambling game called Keno, first introduced in China over 2000 years ago, is played in many casinos. In Keno, there are 80 balls numbered from 1 to 80. The casino randomly chooses 20 balls from the 80 balls. These are "lucky balls" because if a gambler also chooses some of the numbers on these balls, there is a possibility of winning money. The amount that is won depends on the number of lucky numbers the gambler has selected. The number of ways in which a casino can choose 20 balls from 80 is

$$C(80, 20) = \frac{80!}{20! \cdot 60!} \approx 3,535,000,000,000,000,000$$

Once the casino chooses the 20 lucky balls, the remaining 60 balls are unlucky for the gambler. A gambler who chooses five numbers will have from zero to five lucky numbers.

(continued)

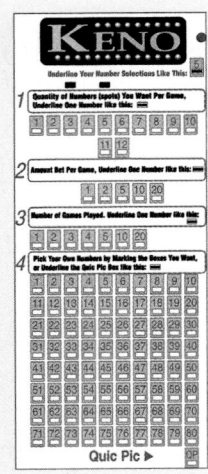

A Keno card used to mark the numbers chosen.

Let's consider the case in which two of the five numbers chosen by the gambler are lucky numbers. Because five numbers were chosen, there must be three unlucky numbers among the five numbers. The number of ways of choosing two lucky numbers from 20 lucky numbers is $C(20, 2)$. The number of ways of choosing three unlucky numbers from 60 unlucky numbers is $C(60, 3)$. By the counting principle, there are $C(20, 2) \cdot C(60, 3) = 190 \cdot 34{,}220 = 6{,}501{,}800$ ways to choose two lucky and three unlucky numbers.

Excursion Exercises

For each of the following exercises, assume that a gambler playing Keno has randomly chosen four numbers.

1. In how many ways can the gambler choose no lucky numbers?
2. In how many ways can the gambler choose exactly one lucky number?
3. In how many ways can the gambler choose exactly two lucky numbers?
4. In how many ways can the gambler choose exactly three lucky numbers?
5. In how many ways can the gambler choose four lucky numbers?

Exercise Set 11.2

(Suggested Assignment: 1–43 every other odd; 45–69 odds; 48, 56, 58)

In Exercises 1–36, evaluate each expression.

1. $8!$
2. $5!$
3. $9! - 5!$
4. $(9 - 5)!$
5. $(8 - 3)!$
6. $8! - 3!$
7. If $7! = 5040$, find $8!$.
8. If $9! = 362{,}880$, find $10!$.
9. If $6! = 720$, find $5!$.
10. If $11! = 39{,}916{,}800$, find $10!$.
11. If $P(n, 2) = 42$, find n.
12. If $P(n, 4) = 360$, find n.
13. $P(8, 5)$
14. $P(7, 2)$
15. $P(9, 7)$
16. $P(10, 5)$
17. $P(8, 0)$
18. $P(7, 0)$
19. $P(8, 8)$
20. $P(10, 10)$
21. $P(8, 2) \cdot P(5, 3)$
22. $\dfrac{P(10, 4)}{P(8, 4)}$
23. $\dfrac{P(6, 0)}{P(6, 6)}$
24. $\dfrac{P(6, 3) \cdot P(5, 2)}{P(4, 3)}$
25. $C(9, 2)$
26. $C(8, 6)$
27. $C(12, 0)$
28. $C(11, 11)$
29. $C(6, 2) \cdot C(7, 3)$
30. $C(8, 5) \cdot C(9, 4)$

31. $\dfrac{C(10, 4) \cdot C(5, 2)}{C(15, 6)}$
32. $\dfrac{C(4, 3) \cdot C(5, 2)}{C(9, 5)}$
33. $\dfrac{C(9, 7) \cdot C(5, 3)}{C(14, 10)}$
34. $3! \cdot C(8, 5)$
35. $4! \cdot C(10, 3)$
36. $5! \cdot C(18, 0)$

In Exercises 37–42, how many combinations are possible? Assume the items are distinct.

37. Seven items chosen five at a time
38. Eight items chosen three at a time
39. 12 items chosen seven at a time
40. Nine items chosen two at a time
41. 11 items chosen 11 at a time
42. 10 items chosen 10 at a time

43. Is it possible to calculate $C(7, 9)$? Think of your answer in terms of seven items chosen nine at a time.
44. Is it possible to calculate $C(n, k)$ where $k > n$? See Exercise 43 for some help.
45. A student downloaded five music files to a portable MP3 player. In how many different orders can the songs be played?

46. The board of directors of a corporation must select a president, a secretary, and a treasurer. In how many possible ways can this be accomplished if there are 20 members on the board of directors?

47. A committee of 16 students must select a president, a vice-president, a secretary, and a treasurer. In how many possible ways can this be accomplished?

48. A gold, a silver, and a bronze medal are awarded in an Olympic event. In how many possible ways can the medals be awarded for a 200-meter sprint in which there are nine runners?

49. A student must read three of seven books for an English class. How many different selections can the student make?

50. The personnel director of a company must select four finalists from a group of 10 candidates for a job opening. How many different groups of four can the director select as finalists?

51. A quality control inspector receives a shipment of 15 DVD players. How many different groups of three players can the inspector choose to test?

52. There are 15 CDs from which a librarian can choose to add to the library's CD collection. How many different sets of 10 CDs can the librarian choose to purchase?

53. Twelve horses are entered in a race. Prizes will be awarded to the owners of the horses finishing first, second, third, and fourth. In how many possible ways can the owners be awarded prizes?

54. The Coast Guard uses signal flags as a method of communicating between ships. If four different flags are available and the order in which the flags are raised is important, how many different signals are possible?

55. How many games are necessary in a softball league consisting of eight teams if each team must play each of the other teams once?

56. A math quiz is generated by randomly choosing five questions from a test bank consisting of 50 questions. How many different quizzes are possible?

57. A football league consists of six teams. How many games must be scheduled if each team must play each other team twice?

58. A committee of six people is chosen from eight women and eight men. How many different committees are possible that consist of three women and three men?

59. Ten identical coins are tossed. How many possible arrangements of the coins include five heads and five tails?

60. Twelve identical coins are tossed. How many possible arrangements of the coins include eight heads and four tails?

61. Seven points are drawn in a plane, no three of which are on the same straight line. How many different lines can be drawn through the seven points?

62. A pentagon is a five-sided plane figure. A diagonal is a line segment connecting any two nonadjacent vertices. How many diagonals are possible?

63. A hexagon is a six-sided plane figure. A diagonal is a line segment connecting any two nonadjacent vertices. How many diagonals are possible?

64. Seven distinct points are drawn on a circle. How many different triangles can be drawn in which each vertex of the triangle is at one of the seven points?

65. Eighteen people decide to play softball. In how many ways can the 18 people be divided into two teams of nine people?

66. Fifteen people decide to form a bowling league. In how many ways can the 15 people be divided into three teams of five people each?

67. The "get-out-the-vote" committee of a political action group has 12 members. In how many ways can the 12 members be divided into groups of two people?

68. A restaurant offers a special pizza with any five toppings. If the restaurant has 12 toppings from which to choose, how many different special pizzas are possible?

69. How many different letter arrangements are possible using all the letters of the word *committee*?

70. How many different letter arrangements are possible using all the letters of the word *calculus*?

71. One shift at a manufacturing plant requires the possible operation of four machines. If eight people are qualified to operate any of the four machines, how many different shifts are possible?

72. A car-racing consortium has three race cars entered in an event. In how many ways can five drivers be assigned to the three cars?

Exercises 73–78 refer to a standard deck of playing cards. Assume five cards are randomly chosen from the deck.

73. How many hands contain four aces?

74. How many hands contain two aces and two kings?

75. How many hands contain exactly three jacks?

76. How many hands contain exactly three jacks and two queens?

77. How many hands contain exactly two 7's?

78. How many hands contain exactly two 7's and two 8's?

79. 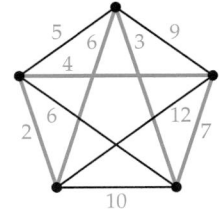 Write a few sentences that explain the difference between a permutation and a combination of distinct objects. Give examples of each.

80. Francis Bacon, a contemporary of William Shakespeare, invented a cipher (a secret code) based on permutations of the letters a and b. Write an essay on Bacon's method and the intended use of his scheme.

Extensions

CRITICAL THINKING

81. If the expression $(x + y)^5$ is expanded, the result is initially 32 terms. The terms consist of every possible product of the form $a_1a_2a_3a_4a_5$ where each a_i is either x or y. After combining like terms, the expansion simplifies to 6 terms with variable parts x^5, x^4y, x^3y^2, x^2y^3, xy^4, and y^5. Use the formula for the arrangements of nondistinct objects to find the coefficient of each term and then write the expansion of $(x + y)^5$.

82. If the expression $(x + y)^6$ is expanded, the result is initially 64 terms. The terms consist of every possible product of the form $a_1a_2a_3a_4a_5a_6$ where each a_i is either x or y. After combining like terms, the expansion simplifies to 7 terms with variable parts x^6, x^5y, x^4y^2, x^3y^3, x^2y^4, xy^5, and y^6. Use the formula for the arrangements of nondistinct objects to find the coefficient of each term and then write the expansion of $(x + y)^6$.

EXPLORATIONS

83. In Chapter 9, we examined the problem of determining certain paths through a network (or graph). One particular path, called a *Hamiltonian circuit*, visited each vertex (dot) of a graph exactly once and returned to the starting vertex. A Hamiltonian circuit is shown as the shaded path in the network at the right. The number associated with each line (edge) of the graph indicates the distance between two vertices. (The indicated distance does not match the physical distance in the drawing.) In most applications, the object is to find the *shortest* path. The shaded path shown here is the shortest path that visits each vertex exactly once and returns to the starting vertex.

One method of searching for the shortest Hamiltonian circuit is by trial and error. For a network with a small number of vertices, this procedure will work fine. As the number of vertices increases, the likelihood of finding a trial-and-error solution becomes extremely remote.

The counting principle can be used to find the number of possible paths through a network in which every pair of vertices possible is connected by a line. For the complete network with five vertices shown in the figure at the

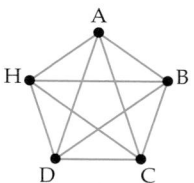

right, beginning at the home vertex labeled H, there are four choices for the next vertex to visit, three choices for the next, two choices for the next, and finally one choice that returns to H. By the counting principle there are $4 \cdot 3 \cdot 2 \cdot 1 = 24$ possible circuits. Traveling circuit HABCDH is the same as traveling the circuit in the reverse order, HDCBAH. Thus there are $\frac{24}{2} = 12$ possible circuits. In general, there are $\frac{(n-1)!}{2}$ possible Hamiltonian circuits through a network of n vertices, where $n \geq 3$.

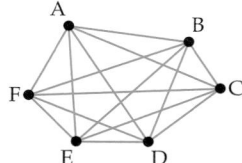

a. A network has eight vertices and each vertex is connected to the other vertices. How many Hamiltonian circuits are possible?

b. For the network at the right, find the number of possible Hamiltonian circuits.

c. Suppose a network has 20 vertices (with every pair of vertices connected by a line) and a computer is available that can analyze one million paths per second. How many years would it take this computer to find all the possible Hamiltonian circuits of this network? (Assume a year is 365 days.)

d. Suppose a network has 40 vertices (with every pair of vertices connected by a line) and a computer is available that can analyze one trillion paths per second. How many years would it take this computer to find all the possible Hamiltonian circuits?

84. The PowerBall multi-state lottery is played by choosing five numbers between 1 and 49 (regular numbers) and one number between 1 and 42 (bonus number). A player wins a prize under any of the following conditions:

   i) Only the bonus number is chosen correctly.

   ii) The bonus number and one regular number are chosen correctly.

   iii) The bonus number and two regular numbers are chosen correctly.

   iv) The bonus number and three regular numbers are chosen correctly.

   v) Three regular numbers are chosen correctly.

   vi) Four regular numbers are chosen correctly.

   vii) The bonus number and four regular numbers are chosen correctly.

   viii) Five regular numbers are chosen correctly.

   ix) The bonus number and five regular numbers are chosen correctly. This combination results in winning the jackpot.

In how many different ways can a person who plays this game win a prize?

SECTION 11.3 **Probability and Odds**

Introduction to Probability

In California, the likelihood of selecting the winning lottery numbers in the Super Lotto Plus game is approximately 1 chance in 41,000,000. In contrast, the likelihood of being struck by lightning is about 1 chance in 1,000,000. Comparing the likelihood of winning the Super Lotto Plus to the likelihood of being struck by lightning, you are 41 times more likely to be struck by lightning than to pick the winning California lottery numbers.

Which is more likely?

The likelihood of the occurrence of a particular event is described by a number between 0 and 1. (You can think of this as a percentage between 0% and 100%, inclusive.) This number is called the **probability** of the event. An event that is not very likely has a probability close to 0; an event that is very likely has a probability close to 1 (100%). For instance, the probability of being struck by lightning is close to 0. However, if you randomly choose a basketball player from the National Basketball Association, the probability that the player is over 6 feet tall is very likely and is therefore close to 1.

Because any event has from a 0% to 100% chance of occurring, probabilities are always between 0 and 1, inclusive. If an event *must* occur, its probability is 1. If an event *cannot* occur, its probability is 0.

Probabilities can be calculated by considering experiments. Recall from Section 11.1 that experiments are activities with observable outcomes. Here are some examples of experiments.

- Flip a coin and observe the outcome as a head or a tail.
- Select a company and observe its annual profit.
- Record the time a person spends at the checkout line in a supermarket.

The sample space of an experiment is the set of all possible outcomes of the experiment. For example, consider tossing a coin three times and observing the outcome as a head or a tail. Using H for head and T for tail, the sample space is

$$S = \{HHH, HHT, HTH, HTT, THH, THT, TTH, TTT\}$$

Note that the sample space consists of *every* possible outcome of tossing three coins.

EXAMPLE 1 ■ Find a Sample Space

A single die is rolled once. What is the sample space for this experiment?

Solution
The sample space is the set of possible outcomes of the experiment.

$$S = \left\{ \boxed{\cdot}, \boxed{\because}, \boxed{\therefore}, \boxed{::}, \boxed{:\cdot:}, \boxed{:::} \right\}$$

CHECK YOUR PROGRESS 1 A coin is tossed twice. What is the sample space for this experiment?

Solution *See page S41.*

Formally, an event is a subset of a sample space. Using the sample space of Example 1, here are some possible events:

- There are an even number of pips (dots) facing up. The event is

$$E_1 = \left\{ \boxed{\because}, \boxed{::}, \boxed{:::} \right\}.$$

- The number of pips facing up is greater than 4. The event is $E_2 = \left\{ \boxed{:\cdot:}, \boxed{:::} \right\}.$

■ The number of pips facing up is less than 20. The event is

$$E_3 = \left\{ \boxed{\cdot}, \boxed{\because}, \boxed{\therefore}, \boxed{::}, \boxed{\because\cdot}, \boxed{:::} \right\}.$$

Because the number of pips facing up is always less than 20, this event will always occur. The event and the sample space are the same.

■ The number of pips facing up is greater than 15. The event is $E_4 = \varnothing$, the empty set. This is an impossible event; the number of pips facing up cannot be greater than 15.

Outcomes of some experiments are **equally likely**, which means that the chance of any one outcome is just as likely as the chance of another. For instance, if four balls of the same size but different colors—red, blue, green, and white—are placed in a box and a blindfolded person chooses one ball, the chance of choosing a green ball is the same as the chance of choosing any other color ball.

In the case of equally likely outcomes, the probability of an event is based on the number of elements in the event and the number of elements in the sample space. We will use $n(E)$ to denote the number of elements in the event E and $n(S)$ to denote the number of elements in the sample space S.

> ✔ **TAKE NOTE**
>
> As an example of an experiment whose outcomes are not equally likely, consider tossing a thumbtack and recording whether it lands with the point up or on its side. There are only two possible outcomes for this experiment, but the outcomes are not equally likely.

Probability of an Event

For an experiment with sample space S of *equally likely outcomes*, the probability $P(E)$ of an event E is given by

$$P(E) = \frac{n(E)}{n(S)} = \frac{\text{number of elements in } E}{\text{total number of elements in sample space } S}$$

Because each outcome of rolling a fair die is equally likely, the probability of the events E_1 through E_4 described above can be determined from the formula for the probability of an event.

$$P(E_1) = \frac{3}{6} \quad \xleftarrow{\hspace{1em}} \text{Number of elements in } E_1$$
$$ \quad \xleftarrow{\hspace{1em}} \text{Number of elements in the sample space}$$
$$= \frac{1}{2}$$

The probability of rolling an even number of pips on a single roll of one die is $\frac{1}{2}$ (or 50%).

$$P(E_2) = \frac{2}{6} \quad \xleftarrow{\hspace{1em}} \text{Number of elements in } E_2$$
$$ \quad \xleftarrow{\hspace{1em}} \text{Number of elements in the sample space}$$
$$= \frac{1}{3}$$

The probability of rolling a number greater than 4 on a single roll of one die is $\frac{1}{3}$.

$$P(E_3) = \frac{6}{6} \quad \xleftarrow{\hspace{1em}} \text{Number of elements in } E_3$$
$$ \quad \xleftarrow{\hspace{1em}} \text{Number of elements in the sample space}$$
$$= 1$$

The probability of rolling a number less than 20 on a single roll of one die is 1 (or 100%). Recall that the probability of any event that is certain to occur is 1.

$$P(E_4) = \frac{0}{6} \quad \xleftarrow{\text{\quad}} \text{Number of elements in } E_4$$

$$\xleftarrow{\text{\quad}} \text{Number of elements in the sample space}$$

$$= 0$$

The probability of rolling a number greater than 15 on a single roll of one die is 0. (It is not possible to roll any number greater than 15.)

EXAMPLE 2 ■ Probability of Equally Likely Outcomes

A fair coin—one for which it is equally likely that heads or tails will result from a single toss—is tossed three times. What is the probability that two heads and one tail are tossed?

Solution

Determine the number of elements in the sample space. The sample space must include every possible toss of a head or a tail (in order) in three tosses of the coin.

$$S = \{HHH, HHT, HTH, HTT, THH, THT, TTH, TTT\}$$

The elements in the event are $E = \{HHT, HTH, THH\}$.

$$P(E) = \frac{n(E)}{n(S)} = \frac{3}{8}$$

The probability is $\frac{3}{8}$.

CHECK YOUR PROGRESS 2   If a fair die is rolled once, what is the probability that an odd number will show on the upward face?

Solution   *See page S41.*

✓ **TAKE NOTE**

In Example 2, we calculated the probability as $\frac{3}{8}$. However, we could have expressed the probability as a decimal, 0.375, or as a percent, 37.5%. A probability can always be expressed as a fraction, a decimal, or a percent.

QUESTION   *Is it possible that the probability of some event could be 1.23?*

EXAMPLE 3 ■ Calculating Probabilities with Dice

Two fair dice are tossed, one after the other. What is the probability that the sum of the pips on the upward faces of the two dice equals 8?

ANSWER   *No. All probabilities must be between 0 and 1, inclusive.*

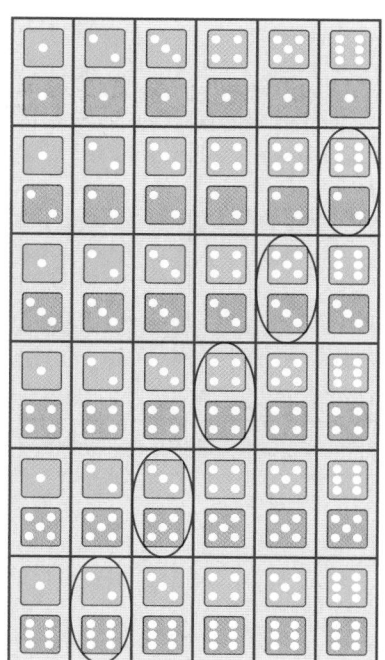

Figure 11.6

Outcomes of the roll of two dice

Point up	15
Side	85
Total	100

Solution

The dice must be considered as distinct, so there are 36 possible outcomes. (and are considered different outcomes.) Therefore, $n(S) = 36$. The sample space is shown at the left. (See also Section 11.1.) Let E represent the event that the sum of the pips on the upward faces is 8. These outcomes are circled in Figure 11.6. By counting the number of circled pairs, $n(E) = 5$.

$$P(E) = \frac{n(E)}{n(S)} = \frac{5}{36}$$

The probability that the sum of the pips is 8 is $\frac{5}{36}$.

CHECK YOUR PROGRESS 3   Two fair dice are tossed. What is the probability that the sum of the pips on the upward faces of the two dice equals 7?

Solution   *See page S41.*

Empirical Probability

Probabilities such as those calculated in the preceding examples are sometimes referred to as *theoretical* probabilities. In Example 2, we assume that, in theory, we have a perfectly balanced coin and we calculate the probability based on the fact that each event is equally likely. Similarly, we assume the dice in Example 3 are equally likely to land with any of the six faces upward.

When a probability is based on data gathered from an experiment, it is called an **experimental probability** or an **empirical probability.** For instance, if we tossed a thumbtack 100 times and recorded the number of times it landed "point up," the results might be as shown in the table at the left. From this experiment, the empirical probability of "point up" is

$$P(\text{point up}) = \frac{15}{100} = 0.15$$

Empirical Probability of an Event

If an experiment is performed repeatedly and the occurrence of the event E is observed, the probability $P(E)$ of the event E is given by

$$P(E) = \frac{\text{number of times event } E \text{ occurred}}{\text{number of times the experiment was performed}}$$

EXAMPLE 4 ■ Calculate an Empirical Probability

A survey of the Registrar of Voters office in a certain city showed the following information on the ages and party affiliations of registered voters. If one voter is chosen from this survey, what is the probability that the voter is a Republican?

Age	Republican	Democrat	Independent	Other	Total
18–28	205	432	98	112	847
29–38	311	301	109	83	804
39–49	250	251	150	122	773
50+	272	283	142	107	804
Total	1038	1267	499	424	3228

Solution

Let R be the event that a Republican is selected. Then

$$P(R) = \frac{1038}{3228} \quad \longleftarrow \text{ Number of Republicans in the survey}$$
$$\phantom{P(R) = \frac{1038}{3228}} \longleftarrow \text{ Total number of people surveyed}$$
$$\approx 0.32$$

The probability that the selected person is a Republican is approximately 0.32.

CHECK YOUR PROGRESS 4   Using the data from Example 4, what is the probability that a randomly selected person is between the ages of 39 and 49?

Solution   *See page S41.*

historical note

Gregor Mendel
(mĕn′dl)
(1822–1884) was
an Augustinian
monk and teacher.
His study of the
transmission of
traits from one generation to
another was actually started to
confirm the prevailing theory of
the day that environment
influenced the traits of a plant.
However, his research seemed to
suggest that plants have certain
"determiners" that dictate the
characteristics of the next
generation. Thus began the
study of heredity. It took over
30 years for Mendel's conclusions
to be accepted in the scientific
community. ■

Application to Genetics

The Human Genome Project is a 13-year project designed to completely map the genetic make-up of Homo sapiens. Researchers hope to use this information to treat and prevent certain hereditary diseases.

The concept behind this project began with Gregor Mendel and his work on flower color and how it was transmitted from generation to generation. From his studies, Mendel concluded that flower color seems to be predictable in future generations by making certain assumptions about a plant's color "determiner." He concluded that red was a *dominant* determiner of color and that white was a *recessive* determiner. Today, geneticists talk about the *gene* for flower color and the *allele* of the gene. A gene consists of two dominant alleles (two red), a dominant and a recessive allele (red and white), or two recessive alleles (two white). Because red is the dominant allele, a flower will be white only if no dominant allele is present.

Later work by Reginald Punnett (1875–1967) showed how to determine the probability of certain flower colors by using a **Punnett square.** Using a capital letter for a dominant allele (R for red) and the corresponding lower-case letter for recessive allele (r for white), Punnett arranged the alleles of the parents in a square. A parent could be RR, Rr, or rr.

Suppose that the genotype (genetic composition) of one parent is rr and the genotype of the other is Rr. The first parent is represented in the left column of the square and the second parent is represented in the top row. The genotypes of the offspring are shown in the body of the table and are the result of combining one allele from each parent.

Parents	R	r
r	Rr	rr
r	Rr	rr

Because each of the genotypes of the offspring are equally likely, the probability that a flower is red is $\frac{1}{2}$ (two Rr genotypes of the four possible genotypes) and the probability that a flower is white is $\frac{1}{2}$ (two rr genotypes of the four possible genotypes).

EXAMPLE 5 ■ Probability Using a Punnett Square

A child will have cystic fibrosis if the child inherits the recessive gene from both parents. Using F for the normal allele and f for the mutant recessive allele, suppose a parent who is Ff (said to be a *carrier*) and a parent who is FF (does not have the mutant allele) decide to have a child.

a. What is the probability that the child will have cystic fibrosis? (To have the disease, the child must be ff.)

b. What is the probability that the child will be a carrier?

Solution

Make a Punnett square.

Parents	F	F
F	FF	FF
f	Ff	Ff

a. To have the disease, the child must be ff. From the table, there is no combination of the alleles that will produce ff. Therefore, the child cannot have the disease, and the probability is 0.

b. To be a carrier, exactly one allele must be f. From the table, there are two cases out of four of a genotype with one f. The probability that the child will be a carrier is $\frac{2}{4} = \frac{1}{2}$.

CHECK YOUR PROGRESS 5   For a certain type of hamster, the color cinnamon, C, is dominant and the color white, c, is recessive. If both parents are Cc, what is the probability that an offspring will be white?

Solution   See page S41.

✔ TAKE NOTE

The probability that a parent will pass a certain genetic characteristic on to a child is one-half. However, the probability that a parent has a certain genetic characteristic is not one-half. For instance, the probability of passing on the allele for cystic fibrosis by a parent who has the mutant allele is 0.5. However, the probability that a person randomly selected from the population has the mutant allele is less than 0.025. We will discuss this further in Section 11.5.

Calculating Odds

For the Super Bowl game in 2002, some Las Vegas casinos put the *odds in favor* of the St. Louis Rams beating the New England Patriots at 5 to 1. These odds indicate that bettors thought that if six games were played, the Rams would win five and the Patriots would win one.

When the racehorse Cigar went for his record-setting 17th consecutive victory, the *odds against* Cigar winning were given as 1 to 2. In this case, bettors thought it was two times more likely that Cigar would lose than that he would win. Thus it was estimated that in three races, Cigar would lose twice and win once.

A **favorable outcome** of an experiment is one that satisfies some event. For instance, in the case of the 2002 Super Bowl, bettors were assuming that in six games there would be five favorable outcomes (the Rams would win five times). The opposite event, the event of the Patriots winning, is an **unfavorable outcome.** (To many gamblers' surprise, the Rams lost.) Odds are frequently expressed in terms of favorable and unfavorable outcomes.

INSTRUCTOR NOTE
Make sure that students understand that a fraction such as $\frac{1}{4}$ used to express odds is being used in a very different way when the same fraction $\frac{1}{4}$ is used to express a probability.

Odds of an Event

Let E be an event in a sample space of equally likely outcomes. Then

$$\textbf{Odds in favor of } E = \frac{\text{number of favorable outcomes}}{\text{number of unfavorable outcomes}}$$

$$\textbf{Odds against } E = \frac{\text{number of unfavorable outcomes}}{\text{number of favorable outcomes}}$$

When the odds of an event are written in fractional form, the fraction bar is read as the word *to*. Thus odds of $\frac{3}{2}$ are read as "3 to 2." We can also write odds of $\frac{3}{2}$ as $3:2$. This form is also read as "3 to 2."

EXAMPLE 6 ■ Calculate Odds

If a pair of fair dice are rolled once, what are the odds in favor of rolling a sum of 7?

Solution
Let E be the event of rolling a sum of 7. From Figure 11.6, the six favorable outcomes are $E = \{$ ⚃⚂, ⚄⚁, ⚅⚀, ⚀⚅, ⚁⚄, ⚂⚃ $\}$. The unfavorable outcomes are the remaining 30 possibilities. (Because there are 36 possible outcomes when tossing two dice and six of them are favorable, then $36 - 6 = 30$ are unfavorable.)

$$\text{Odds in favor of } E = \frac{\text{number of favorable outcomes}}{\text{number of unfavorable outcomes}} = \frac{6}{30} = \frac{1}{5}$$

The odds in favor of rolling a sum of 7 are 1 to 5.

CHECK YOUR PROGRESS 6   If three red, four white, and five blue balls are placed in a box and one ball is randomly selected from the box, what are the odds against the ball being blue?

Solution   *See page S41.*

✓ **TAKE NOTE**

Recall that if *A* is a set formed from the elements of a universal set *U*, then the complement of *A* is the set of elements in *U* but not in *A*. See Section 2.2 for a further discussion.

Odds express the likelihood of an event and are therefore related to probability. When the odds of an event are known, the probability of the event can be determined. Conversely, when the probability of an event is known, the odds of the event can be determined.

The Relationship Between Odds and Probability

1. Suppose *E* is an event in a sample space and that the *odds in favor* of *E* are $\frac{a}{b}$. Then $P(E) = \frac{a}{a + b}$.

2. Suppose *E* is an event in a sample space. Then the *odds in favor* of *E* are $\frac{P(E)}{1 - P(E)}$.

EXAMPLE 7 ■ Determine Probability from Odds

In 2000, the horse Fusaichi Pegasus won the 126th Kentucky Derby. It was the first time the favorite had won the Derby since 1979, when Spectacular Bid won the race. The odds against Fusaichi Pegasus winning the race were 3 to 1. What was the probability of Fusaichi Pegasus winning the race?

Solution

Because the odds against Fusaichi Pegasus winning the race were 3 to 1, the odds in favor of Fusaichi Pegasus winning the race were 1 to 3. Write the odds as a fraction: $\frac{1}{3}$. Now use the formula for calculating the probability of an event when the odds in favor are known.

$$P(E) = \frac{a}{a + b}$$

$$P(E) = \frac{1}{1 + 3} = \frac{1}{4} \qquad \bullet \ a = 1, \ b = 3$$

The probability of Fusaichi Pegasus winning the race was $\frac{1}{4}$ (or 25%).

CHECK YOUR PROGRESS 7   The probability of tossing five coins with a result of three heads and two tails is $\frac{5}{16}$. What are the odds in favor of three heads and two tails?

Solution   *See page S41.*

Math Matters   The Birth of Probability Theory

Pierre Fermat and Blaise Pascal are generally considered to be the founders of the study of probability. For many historians, the starting point of the formal discussion of probability is contained in a letter from Pascal to Fermat written in July of 1654.

> The Chevalier de Mere said to me that he found a falsehood in the theory of numbers for the following reason. If one wants to throw a six with a single die, there is an advantage in undertaking to do it in four throws, as the odds are 671 to 625. If one throws two sixes with a pair of dice, there is a disadvantage in having only 24 throws. However, 24 to 36 (which is the number of cases for two dice) is 4 to 6 (which is the number of cases on one die). This is the great "scandal" which makes him proclaim loftily that the theorems are not constant and Arithmetic is self-contradictory.

> Basically (although it certainly isn't obvious from this letter), the Chevalier was claiming that there is better than a 50% chance of tossing a 6 in four rolls of a single die. By his reasoning, he concluded that there should be a better than 50% chance of rolling a pair of sixes in 24 tosses of two dice. However, the Chevalier tested his theory and found that 25 tosses were needed. From his tests, he concluded a "falsehood in the theory of numbers." In this letter, Pascal was mocking the inability of the Chevalier to correctly determine the probabilities involved.

Excursion

The Value of Pi by Simulation[1]

A **simulation** is an activity designed to approximate a given situation. For instance, pilots fly a simulator to practice maneuvers that would be dangerous to try in an airplane. Chemists use computer programs to simulate chemical processes. This Excursion uses a simulation to approximate the value of π.

This Excursion will work better if four or five people work together. Get 25 toothpicks, and then tape several sheets of blank paper together to form a large rectangular shape. Use a large ruler to draw parallel lines on the paper such that the distance between the lines is the length of one toothpick. Then drop all of the toothpicks from approximately knee height onto the paper. (Make sure none of the toothpicks lands off the paper.) Record the number of toothpicks that cross any of the parallel lines. Repeat this process ten times for a total of 250 toothpicks dropped.

Excursion Exercises

1. Using the recorded data, calculate the empirical probability that a dropped toothpick will cross a line.

2. The theoretical probability that a toothpick will cross a line is $\frac{2}{\pi}$. Use a calculator to find the value of $\frac{2}{\pi}$, rounded to four decimal places.

(continued)

---

1. See our web site at **math.college.hmco.com/students** for a computer program that can be used to simulate the dropping of the toothpicks. This program will enable you to perform the experiment many thousands of times.

3. The experimental value you calculated in Exercise 1 should be close to the theoretical value you calculated in Exercise 2 (at least it should be if you dropped a toothpick around 1000 times). Calculate the percent error between the experimental value and the theoretical value. (The percent error can be determined by finding the difference between the two values and dividing it by the theoretical value.)

4. Explain how performing this experiment thousands of times could give you an approximate value for π.

5. Using your data, what approximate value for π do you calculate?

6. This experiment is based on a famous eighteenth century problem called the *Buffon Needle problem.* Look up "Buffon Needle problem" in the library or on the Internet and write a short essay about this problem and how it applies to this Excursion.

Exercise Set 11.3

(Suggested Assignment: 1–85 every other odd; 36, 62, 83, 96)

In Exercises 1–6, list the elements of the sample space for each experiment.

1. A coin is flipped three times.

2. An even number between 1 and 11 is selected at random.

3. One day in the first two weeks of November is selected.

4. A current U.S. coin is selected from a piggybank.

5. A state is selected from the states in the U.S. whose name begins with the letter A.

6. A month is selected from the months that have exactly 30 days.

In Exercises 7–15, assume that it is equally likely for a child to be born a boy or a girl, and that the Lin family is planning on having three children.

7. List the elements of the sample space for the genders of the three children.

8. List the elements of the event that the Lins have two boys and one girl.

9. List the elements of the event that the Lins have at least two girls.

10. List the elements of the event that the Lins have no girls.

11. List the elements of the event that the Lins have at least one girl.

12. Compute the probability that the Lins will have two boys and one girl.

13. Compute the probability that the Lins will have at least two girls.

14. Compute the probability that the Lins will have no girls.

15. Compute the probability that the Lins will have at least one girl.

In Exercises 16–22, a coin is tossed four times. Assuming that the coin is equally likely to land on heads or tails, compute the probability of each event occurring.

16. Two heads and two tails

17. One head and three tails

18. All tails

19. All four coin tosses are identical

20. At least three heads

21. At least two tails

22. At least one head

In Exercises 23–26, a dodecahedral die (one with 12 sides numbered from 1 to 12) is tossed once. Find each of the following probabilities.

23. The number on the upward face is 12.

24. The number on the upward face is not 10.

25. The number on the upward face is divisible by 4.

26. The number on the upward face is less than 5 or greater than 9.

In Exercises 27–36, two regular six-sided dice are tossed. Compute the probability that the sum of the pips on the upward faces of the two dice is each of the following. (See Figure 11.6 for the sample space of this experiment.)

27. 6   **28.** 11

29. 2   **30.** 12

31. 1   **32.** 14

33. At least 10 **34.** At most 5

35. An even number **36.** An odd number

37. If two dice are rolled, compute the probability of rolling doubles (both dice show the same number of pips).

38. If two dice are rolled, compute the probability of *not* rolling doubles.

In Exercises 39–44, a card is selected at random from a standard deck of playing cards.

39. Compute the probability that the card is red.

40. Compute the probability that the card is a spade.

41. Compute the probability that the card is a 9.

42. Compute the probability that the card is a face card (jack, queen, or king).

43. Compute the probability that the card is between 5 and 9.

44. Compute the probability that the card is between 3 and 6.

In Exercises 45–50, use the data given in Example 4 to compute the probability that a randomly chosen voter from the survey will satisfy the following.

45. The voter is a Democrat.

46. The voter is not a Republican.

47. The voter is 50 years old or older.

48. The voter is under 39 years old.

49. The voter is between 39 and 49, and is registered as an Independent.

50. The voter is under 29 and is registered as a Democrat.

In Exercises 51–54, a survey asked 850 respondents about their highest levels of completed education. The results are given in the table below.

Education completed	Number of respondents
No high school diploma	52
High school diploma	234
Associate's degree or two years of college	274
Bachelor's degree	187
Master's degree	67
Ph.D. or professional degree	36

If a respondent from the survey is selected at random, compute the probability of each of the following.

51. The respondent did not complete high school.

52. The respondent has an associate's degree or two years of college (but not more).

53. The respondent has a Ph.D. or professional degree.

54. The respondent has a degree beyond a bachelor's degree.

A random survey asked respondents about their current annual salaries. The results are given in the table below. Use the table for Exercises 55–58.

Salary range	Number of respondents
Below $18,000	24
$18,000–$27,999	41
$28,000–$35,999	52
$36,000–$45,999	58
$46,000–$59,999	43
$60,000–$79,999	39
$80,000–$99,999	22
$100,000 or more	14

If a respondent from the survey is selected at random, compute the probability of the following. Round to the nearest hundredth.

55. The respondent earns from $36,000 to $45,999 annually.

56. The respondent earns from $60,000 to $79,999 per year.

57. The respondent earns at least $80,000 per year.

58. The respondent earns less than $36,000 annually.

59. During a recent year in Alaska, 5238 boys and 4984 girls were born. If a baby is selected at random, what is the probability that the baby is a girl?

60. For the Fall 2000 semester, the entering freshman class at the University of Oregon consisted of 1265 male and 1515 female students. If a freshman student is selected randomly, what is the probability that the student is male?

61. The following Punnett square for flower color shows two parents of genotype Rr, where R corresponds to

the dominant red flower allele and r represents the recessive white flower allele. (See Example 5.)

Parents	R	r
R	RR	Rr
r	Rr	rr

What is the probability that the offspring of these parents will have white flowers?

62. One parent plant with red flowers has genotype RR and the other with white flowers has genotype rr, where R is the dominant allele for a red flower and r is the recessive allele for a white flower. Compute the probability of one of the offspring having white flowers. Hint: Draw a Punnett square.

63. The eye color of mice is determined by a dominant allele E, corresponding to black eyes, and a recessive allele e, corresponding to red eyes. If two mice, one of genotype EE and the other of genotype ee, have offspring, compute the probability of one of the offspring having red eyes. Hint: Draw a Punnett square.

64. The height of a certain plant is determined by a dominant allele T corresponding to tall plants, and a recessive allele t corresponding to short (or dwarf) plants. If both parent plants have genotype Tt, compute the probability that the offspring plants will be tall. Hint: Draw a Punnett square.

65. Explain the difference between the probability of an event and the odds of the same event.

66. Give an example of an event that has probability 0 and one that has probability 1.

In Exercises 67–72, the odds in favor of an event occurring are given. Compute the probability of the event occurring.

67. 1 to 2

68. 1 to 4

69. 3 to 7

70. 3 to 5

71. 8 to 5

72. 11 to 9

In Exercises 73–78, the probability of an event occurring is given. Find the odds in favor of the event occurring.

73. 0.2

74. 0.6

75. 0.375

76. 0.28

77. 0.55

78. 0.81

79. If a pair of fair dice are rolled once, what are the odds in favor of rolling a sum of 9?

80. If a single fair die is rolled, what are the odds in favor of rolling an even number?

81. If a card is randomly pulled from a standard deck of playing cards, what are the odds in favor of pulling a heart?

82. A coin is tossed four times. What are the odds against the coin showing heads all four times?

83. A bookmaker has placed 8 to 3 odds *against* a particular football team winning its next game. What is the probability, in the bookmaker's view, of the team winning?

84. A contest is advertising that the odds against winning first prize are 100 to 1. What is the probability of winning?

85. A snack-size bag of M&Ms candies contains 12 red candies, 12 blue, 7 green, 13 brown, 3 orange, and 10 yellow. If a candy is randomly picked from the bag, compute

a. the odds of getting a green M&M.

b. the probability of getting a green M&M.

86. A snack-size bag of Skittles candies contains 10 red candies, 15 blue, 9 green, 8 purple, 15 orange, and 13 yellow. If a candy is randomly picked from the bag, compute

a. the odds of picking a purple Skittle.

b. the probability of picking a purple Skittle.

Extensions

CRITICAL THINKING

87. If four cards labeled A, B, C, and D are randomly placed in four boxes also labeled A, B, C, and D, one to each box, find the probability that no card will be in a box with the same letter.

88. Determine the probability that if 10 coins are tossed, five heads and five tails will result.

89. In a family of three children, all of whom are girls, a family member new to probability reasons that the probability that each child would be a girl is 0.5. Therefore, the probability that the family would have three girls is $0.5 + 0.5 + 0.5 = 1.5$. Explain why this reasoning is not valid.

In Exercises 90 and 91, a hand of five cards is dealt from a standard deck of playing cards. You may want to review the material on combinations before doing these exercises.

90. Find the probability that the hand will contain all four aces.

91. Find the probability that the hand will contain three jacks and two queens.

EXPLORATIONS

Exercises 92 to 97 use the casino game roulette. Roulette is played by spinning a wheel with 38 numbered slots. The numbers 1 through 36 appear on the wheel, half of them colored black and half colored red. Two slots, numbered 0 and 00, are colored green. A ball is placed on the spinning wheel and allowed to come to rest in one of the slots. Bets are placed on where the ball will land.

92. You can place a bet that the ball will stop in a black slot. If you win, the casino will pay you $1 for each dollar you bet. What is the probability of winning this bet?

93. You can bet that the ball will land on an odd number. If you win, the casino will pay you $1 for each dollar you bet. What is the probability of winning this bet?

94. You can bet that the ball will land on any number from 1 to 12. If you win, the casino will pay you $2 for each dollar you bet. What is the probability of winning this bet?

95. You can bet that the ball will land on any particular number. If you win, the casino will pay you $35 for each dollar you bet. What is the probability of winning this bet?

96. You can bet that the ball will land on one of 0 or 00. If you win, the casino will pay you $17 for each dollar you bet. What is the probability of winning this bet?

97. You can bet that the ball will land on certain groups of six numbers (such as 1−6). If you win, the casino will pay you $5 for each dollar you bet. What is the probability of winning this bet?

SECTION 11.4 | **Addition and Complement Rules**

The Addition Rule for Probabilities

Suppose you draw a single card from a standard deck of playing cards. The sample space S consists of the 52 cards of the deck. Therefore, $n(S) = 52$. Now consider the events

E_1 = a four is drawn = $\{\spadesuit 4, \heartsuit 4, \diamondsuit 4, \clubsuit 4\}$

E_2 = a spade is drawn

$\quad = \{\spadesuit A, \spadesuit 2, \spadesuit 3, \spadesuit 4, \spadesuit 5, \spadesuit 6, \spadesuit 7, \spadesuit 8, \spadesuit 9, \spadesuit 10, \spadesuit J, \spadesuit Q, \spadesuit K\}$

It is possible, on one draw, to satisfy the conditions of both events: the $\spadesuit 4$ could be drawn. This card is an element of both E_1 and E_2.

Now consider the events

E_3 = a five is drawn = $\{\spadesuit 5, \heartsuit 5, \diamondsuit 5, \clubsuit 5\}$

E_4 = a king is drawn = $\{\spadesuit K, \heartsuit K, \diamondsuit K, \clubsuit K\}$

In this case, it is not possible to draw one card that satisfies the conditions of both events. There are no elements common to both sets. Two events that cannot both occur at the same time are called **mutually exclusive events.** The events E_3 and E_4 are mutually exclusive events, whereas E_1 and E_2 are not.

> **Mutually Exclusive Events**
>
> Two events A and B are mutually exclusive if they cannot occur at the same time. That is, A and B are mutually exclusive when $A \cap B = \varnothing$.

QUESTION *A die is rolled once. Let E be the event that an even number is rolled and let O be the event that an odd number is rolled. Are the events E and O mutually exclusive?*

The probability of either of two mutually exclusive events occurring can be determined by adding the probabilities of the individual events.

ANSWER *Yes. It is not possible to roll an even number and an odd number on a single roll of the die.*

Probability of Mutually Exclusive Events

If A and B are two mutually exclusive events, then the probability of A or B occurring is

$$P(A \cup B) = P(A) + P(B)$$

EXAMPLE 1 ■ Probability of Mutually Exclusive Events

Suppose a single card is drawn from a standard deck of playing cards. Find the probability of drawing a five or a king.

Solution

Let $A = \{\spadesuit 5, \heartsuit 5, \diamondsuit 5, \clubsuit 5\}$ and $B = \{\spadesuit K, \heartsuit K, \diamondsuit K, \clubsuit K\}$. There are 52 cards in a standard deck of playing cards; thus $n(S) = 52$. Because the events are mutually exclusive, we can use the formula for the probability of mutually exclusive events.

$$P(A \text{ or } B) = P(A) + P(B) \qquad \bullet \text{ Formula for the probability of mutually exclusive events}$$

$$= \frac{1}{13} + \frac{1}{13} = \frac{2}{13} \qquad \bullet P(A) = \frac{4}{52} = \frac{1}{13}, P(B) = \frac{4}{52} = \frac{1}{13}$$

The probability of drawing a five or a king is $\frac{2}{13}$.

CHECK YOUR PROGRESS 1 Two fair dice are tossed once. What is the probability of rolling a 7 or an 11? For the sample space for this experiment, see page 721.

Solution *See page S41.*

Consider the experiment of rolling two dice. Let A be the event of rolling a sum of 8 and let B be the event of rolling a double (the same number on both dice).

$$A = \left\{ \boxed{} \boxed{}, \boxed{} \boxed{}, \boxed{} \boxed{}, \boxed{} \boxed{}, \boxed{} \boxed{} \right\}$$
$$B = \left\{ \boxed{} \boxed{}, \boxed{} \boxed{}, \boxed{} \boxed{}, \boxed{} \boxed{}, \boxed{} \boxed{}, \boxed{} \boxed{} \right\}$$

These events are *not* mutually exclusive because it is possible to satisfy the conditions of each event on one toss of the dice—a $\boxed{}\,\boxed{}$ could be rolled. Therefore, $P(A \cup B)$, the probability of a sum of 8 or a double, cannot be calculated using the formula for the probability of mutually exclusive events. However, a modification of that formula can be used.

✔ TAKE NOTE

The $P(A \cap B)$ term in the Addition Rule for Probabilities is subtracted to compensate for the overcounting of the first two terms of the formula. If two events are mutually exclusive, then $A \cap B = \varnothing$. Therefore, $n(A \cap B) = 0$ and $P(A \cap B) = \frac{n(A \cap B)}{n(S)} = 0$. For mutually exclusive events, the Addition Rule for Probabilities is the same as the formula for the probability of mutually exclusive events.

Addition Rule for Probabilities

Let A and B be two events in a sample space S. Then

$$P(A \cup B) = P(A) + P(B) - P(A \cap B)$$

In other words, the Addition Rule states that

$$P(A \text{ or } B) = P(A) + P(B) - P(A \text{ and } B)$$

Using this formula with

$$A = \left\{ \boxed{} \boxed{}, \boxed{} \boxed{}, \boxed{} \boxed{}, \boxed{} \boxed{}, \boxed{} \boxed{} \right\}$$
$$B = \left\{ \boxed{} \boxed{}, \boxed{} \boxed{}, \boxed{} \boxed{}, \boxed{} \boxed{}, \boxed{} \boxed{}, \boxed{} \boxed{} \right\}$$
$$A \cap B = \left\{ \boxed{} \boxed{} \right\}$$

the probability of A or B can be calculated.

$$P(A \cup B) = P(A) + P(B) - P(A \cap B)$$
$$= \frac{5}{36} + \frac{6}{36} - \frac{1}{36} \qquad \bullet \ P(A) = \frac{5}{36}, P(B) = \frac{6}{36}, P(A \cap B) = \frac{1}{36}$$
$$= \frac{10}{36} = \frac{5}{18}$$

On a single roll of two dice, the probability of rolling a sum of 8 or a double is $\frac{5}{18}$.

	F	**No F**	**Total**
V	21	198	219
No V	76	195	271
Total	97	393	490

V: Vaccinated
F: Contracted the flu

EXAMPLE 2 ■ Use the Addition Rule for Probabilities

The table at the left shows data from an experiment conducted to test the effectiveness of a flu vaccine. If one person is selected from this population, what is the probability that the person was vaccinated or contracted the flu?

Solution

Let $V = \{$people who were vaccinated$\}$ and $F = \{$people who contracted the flu$\}$. These events are not mutually exclusive because there are 21 people who were vaccinated and who contracted the flu. The sample space S consists of the 490 people who participated in the experiment. From the table, $n(V) = 219$, $n(F) = 97$, and $n(V \text{ and } F) = 21$.

$$P(V \text{ or } F) = P(V) + P(F) - P(V \text{ and } F)$$
$$= \frac{219}{490} + \frac{97}{490} - \frac{21}{490}$$
$$= \frac{295}{490} \approx 0.602$$

The probability of selecting a person who was vaccinated or who contracted the flu is approximately 60.2%.

CHECK YOUR PROGRESS 2   The data in the table below show the starting salaries of college graduates with selected degrees. If one person is chosen from this population, what is the probability that the person has a degree in business or has a starting salary between \$20,000 and \$24,999?

	Degree			
Salary (in \$)	**Engineering**	**Business**	**Chemistry**	**Psychology**
Less than 20,000	0	4	1	12
20,000–24,999	4	16	3	16
25,000–29,999	7	21	5	15
30,000–34,999	12	35	5	7
35,000 or more	12	22	4	5

Solution   *See page S41.*

The Complement of an Event

Consider the experiment of tossing a single die once. The sample space is

$$S = \left\{ \boxed{\cdot}, \boxed{\because}, \boxed{\therefore}, \boxed{::}, \boxed{:\because}, \boxed{:::} \right\}$$

Now consider the event $E = \left\{ \boxed{\because} \right\}$, that is, the event of tossing a $\boxed{\because}$. The probability of E is

$$P(E) = \frac{1}{6} \quad \begin{array}{l} \longleftarrow \text{ Number of elements in } E \\ \longleftarrow \text{ Number of elements in the sample space} \end{array}$$

The **complement** of E, symbolized by E^c, is the "opposite" event of E. The complement includes all those outcomes of S that are not in E and excludes the outcomes in E. For the event E above, E^c is the event of not tossing a $\boxed{\because}$. Thus

$$E^c = \left\{ \boxed{\cdot}, \boxed{\therefore}, \boxed{::}, \boxed{:\because}, \boxed{:::} \right\}$$

Notice that because E and E^c are opposite events, they are mutually exclusive, and their union is the entire sample space S. Thus $P(E) + P(E^c) = P(S)$. But $P(S) = 1$, so $P(E^c) = 1 - P(E)$.

> **Probability of the Complement of an Event**
>
> If E is an event and E^c is the complement of the event, then
>
> $$P(E^c) = 1 - P(E)$$

Continuing our example, the probability of not tossing a $\boxed{\because}$ is given by

$$P\left(\text{not a } \boxed{\because}\right) = 1 - P\left(\boxed{\because}\right)$$

$$= 1 - \frac{1}{6} = \frac{5}{6}$$

You can also verify the probability of E^c directly:

$$P\left(\text{not a } \boxed{\because}\right) = \frac{5}{6} \quad \begin{array}{l} \longleftarrow \text{ Number of elements in } E^c \\ \longleftarrow \text{ Number of elements in the sample space} \end{array}$$

INSTRUCTOR NOTE
It may be helpful to have students practice formulating a probability in terms of the complement of an event. Ask the students how they might use the complement to determine the probability of the following situations.
- not rolling doubles in the game of Monopoly
- a randomly selected student on campus is enrolled in at least two classes
- a randomly selected student in the class has received at least one traffic violation

EXAMPLE 3 ■ Find a Probability by the Complement Rule

The probability of tossing a sum of 11 on the toss of two dice is $\frac{1}{18}$. What is the probability of not tossing a sum of 11 on the toss of two dice?

Solution
Use the formula for the probability of the complement of an event. Let $E = \{\text{toss a sum of } 11\}$. Then $E^c = \{\text{toss a sum that is not } 11\}$.

$$P(E^c) = 1 - P(E)$$

$$= 1 - \frac{1}{18} = \frac{17}{18} \qquad \bullet \ P(E) = \frac{1}{18}$$

The probability of not tossing a sum of 11 is $\frac{17}{18}$.

CHECK YOUR PROGRESS 3 The probability that a person has type A blood is 34%. What is the probability that a person does not have type A blood?

Solution See page S41.

TAKE NOTE

The phrase "at least one" means one or more. Tossing a coin three times and asking the probability of getting at least one head means to calculate the probability of getting one, two, or three heads.

Suppose we toss a coin three times and want to calculate the probability of having heads occur *at least once*. We could list all the possibilities of tossing three coins, as shown below, and then find the ones that contain at least one head.

{HHH, HHT, HTH, HTT, THH, THT, TTH, TTT}

at least one head

The probability of at least one head is $\frac{7}{8}$.

Another way to calculate this result is to use the formula for the probability of the complement of an event. Let $E = \{$at least one head$\}$. From the list above, note that E contains every outcome except TTT (no heads). Thus $E^c = \{$TTT$\}$ and we have

$$P(E) = 1 - P(E^c)$$
$$= 1 - \frac{1}{8} = \frac{7}{8} \qquad \bullet \ P(E^c) = \frac{n(E^c)}{n(S)} = \frac{1}{8}$$

This is the same result that we calculated above. As we will see, sometimes working with a complement is much less work than proceeding directly.

In many cases, the principles of counting that were discussed in Sections 11.1 and 11.2 are part of the process of calculating a probability.

EXAMPLE 4 ■ Find a Probability by the Complement Rule

A die is tossed four times. What is the probability that a ⚃ will show on the upward face at least once?

Solution

Let $E = \{$at least one 6$\}$. Then $E^c = \{$no 6's$\}$. To calculate the number of elements in the sample space (all possible outcomes of tossing a die four times) and the number of items in E^c, we will use the counting principle.

Because on each toss of the die there are six possible outcomes,

$$n(S) = 6 \cdot 6 \cdot 6 \cdot 6 = 1296$$

On each toss of the die there are five numbers that are not 6's. Therefore,

$$n(E^c) = 5 \cdot 5 \cdot 5 \cdot 5 = 625$$
$$P(E) = 1 - P(E^c)$$
$$= 1 - \frac{625}{1296} = \frac{671}{1296}$$
$$\approx 0.518$$

When a die is tossed four times, the probability that a ⚃ will show on the upward face at least once is approximately 0.518.

CHECK YOUR PROGRESS 4   A pair of dice are rolled three times. What is the probability that a sum of 7 will occur at least once?

Solution   *See page S41.*

Math Matters The Monte Hall Problem

A famous probability puzzle began with the game show *Let's Make a Deal,* of which Monte Hall was the host, and goes something like the following. (See also the Chapter 1 opener on page 3.) Suppose you appear on the show and are shown three closed curtains. Behind one of the curtains is the grand prize; behind the other two curtains are less desirable prizes (like a goat!). If you select the door hiding the grand prize, you win that prize. The probability of randomly choosing the grand prize, of course, is 1/3. After you choose a curtain, the show's host (who knows where the grand prize is) does not immediately open it to show you what you have won. Instead, he opens one of the other two curtains and reveals a goat. Obviously you are relieved that you did not choose that particular curtain, but he then asks if you would like to switch your choice to the third curtain. Should you stay with your original choice, or switch? Most people would say at first that it makes no difference. However, computer simulations that play the game over and over have shown that you *should* switch. In fact, you double your chances of winning the grand prize if you give up your first choice! See Exercise 56 on page 740 for a mathematical investigation, and then try a simulation of your own with Exercises 58 and 59.

Combinatoric Formulas and Probability

We end this section with another example of using counting principles in the calculation of a probability. In this case we will use the combination formula $C(n, r) = \frac{n!}{r!(n-r)!}$, which gives the number of ways r objects can be chosen from n objects.

EXAMPLE 5 ■ Find a Probability Using the Combination Formula

Every manufacturing process has the potential to produce a defective article. Suppose a manufacturing process for tableware produces 40 dinner plates, of which three are defective. (For instance, there is a paint flaw.) If five plates are randomly selected from the 40, what is the probability that at least one is defective?

Solution
Let $E = \{$at least one plate is defective$\}$. It is easier to work with the complement event, $E^c = \{$no plates are defective$\}$. To calculate the number of elements in the sample space (all possible outcomes of choosing five plates from 40), use the combination formula with $n = 40$ (the number of plates) and $r = 5$ (the number of plates that are chosen). Then

$$n(S) = C(40, 5) = \frac{40!}{5!(40-5)!} \qquad \bullet\, n = 40, r = 5$$

$$= \frac{40!}{5!\,35!} = 658{,}008$$

To find the number of outcomes that contain no defective plates, all of the plates chosen must come from the 37 nondefective plates. Therefore, we must calculate the number of ways five objects can be chosen from 37. Thus $n = 37$ (the number of nondefective plates) and $r = 5$ (the number of plates chosen).

$$n(E^c) = C(37, 5) = \frac{37!}{5!\,(37 - 5)!} \qquad \bullet\ n = 37, r = 5$$

$$= \frac{37!}{5!\,32!} = 435{,}897$$

$$P(E) = 1 - P(E^c)$$

$$= 1 - \frac{435{,}897}{658{,}008}$$

$$\approx 0.338$$

The probability is approximately 0.338, or 33.8%.

CHECK YOUR PROGRESS 5 The winner of a contest will be blindfolded and then allowed to reach into a hat containing 31 $1 bills and four $100 bills. The winner can remove four bills from the hat and keep the money. Find the probability that the winner will pull out at least one $100 bill.

Solution *See page S42.*

Excursion

Keno Revisited

In Section 11.2 (see pages 713–714), we looked at the popular casino game Keno, in which a player chooses numbers from 1 to 80 and hopes that the casino will draw balls with the same lucky numbers.

A player can choose only one number, or as many as 20. The casino will then pick 20 numbered balls from the 80 possible; if enough of the player's numbers match the lucky numbers the casino chooses, the player wins money. The amount won varies according to how many numbers were chosen and how many match.

Excursion Exercises

1. A gambler playing Keno has randomly chosen five numbers. What is the probability that the gambler will match at least one lucky number?

2. If five numbers are chosen, compute the probability of matching fewer than five lucky numbers.

3. If the Keno player chooses 15 numbers and bets $1, and matches 13 of the lucky numbers, the gambler will be paid $12,000. What is the probability of this occurring?

(continued)

4. If the Keno player chooses 15 numbers and matches five or six of the lucky numbers, the gambler gets the bet back but is not paid any extra. What is the probability of this occurring?

5. Some casinos will let you choose up to 20 numbers. In this case, if you don't match any of the lucky numbers, the casino pays you! Although this may seem unusual, it is actually more difficult not to match any of the lucky numbers than it is to match a few of them. Compute the probability of not matching any of the lucky numbers at all, and compare it to the probability of matching five lucky numbers.

6. If 20 numbers are chosen, find the probability of matching *at least one* lucky number.

Exercise Set 11.4

(Suggested Assignment: 3–43 odds; 34, 47–48, 52)

1. What are mutually exclusive events? How do you calculate the probability of mutually exclusive events?

2. Give an example of two mutually exclusive events and an example of two events that are not mutually exclusive.

In Exercises 3–6, first verify that the compound event consists of two mutually exclusive events, and then compute the probability of the compound event occurring.

3. A single card is drawn from a standard deck of playing cards. Find the probability of drawing a 4 or an ace.

4. A single card is drawn from a standard deck. Find the probability of drawing a heart or a club.

5. Two dice are rolled. Find the probability of rolling a 2 or a 10.

6. Two dice are rolled. Find the probability of rolling a 7 or an 8.

7. If $P(A) = 0.2$, $P(B) = 0.5$, and $P(A \text{ and } B) = 0.1$, find $P(A \text{ or } B)$.

8. If $P(A) = 0.6$, $P(B) = 0.4$, and $P(A \cap B) = 0.2$, find $P(A \cup B)$.

9. If $P(A) = 0.3$, $P(B) = 0.8$, and $P(A \cup B) = 0.9$, find $P(A \cap B)$.

10. If $P(A) = 0.7$, $P(A \cap B) = 0.4$, and $P(A \cup B) = 0.8$, find $P(B)$.

In Exercises 11–14, suppose you ask a friend to randomly choose an integer between 1 and 10, inclusive.

11. What is the probability that the number will be more than 6 or odd?

12. What is the probability that the number will be less than 5 or even?

13. What is the probability that the number will be even or prime?

14. What is the probability that the number will be prime or greater than 7?

In Exercises 15–20, two dice are rolled. Determine the probability of each of the following. ("Doubles" means both dice show the same number)

15. Rolling a 6 or doubles

16. Rolling a 7 or doubles

17. Rolling an even number or doubles

18. Rolling a number greater than 7 or doubles

19. Rolling an odd number or a number less than 6

20. Rolling an even number or a number greater than 9

In Exercises 21–26, a single card is drawn from a standard deck. Find the probability of each of the following events.

21. Drawing an 8 or a spade

22. Drawing an ace or a red card

23. Drawing a jack or a face card

24. Drawing a red card or a face card
25. Drawing a diamond or a black card
26. Drawing a spade or a red card

In Exercises 27–30, use the data in the table below, which shows the employment status of individuals in a particular town by age group.

Age	Full-time	Part-time	Unemployed
0–17	24	164	371
18–25	185	203	148
26–34	348	67	27
35–49	581	179	104
50+	443	162	173

27. If a person is randomly chosen from the town's population, what is the probability that the person is aged 26–34 or is employed part-time?
28. If a person is randomly chosen from the town's population, what is the probability that the person is at least 50 years old or unemployed?
29. If a person is randomly chosen from the town's population, what is the probability that the person is under 18 or employed part-time?
30. If a person is randomly chosen from the town's population, what is the probability that the person is 18 or older or employed full-time?
31. If the probability of winning a particular contest is 0.04, what is the probability of not winning the contest?
32. Suppose the probability that it will rain tomorrow is 0.38. What is the probability that it will not rain tomorrow?
33. If there is a 1 in 2500 chance of an individual being involved in a car accident, find the probability of not being involved in a car accident.
34. The odds in favor of a candidate winning an election are given as 3 to 5. What is the probability that the candidate will lose the election?

In Exercises 35–42, use the formula for the probability of the complement of an event.

35. Two dice are tossed. What is the probability of not tossing a 7?
36. Two dice are tossed. What is the probability of not getting doubles?
37. Two dice are tossed. What is the probability of getting a sum of at least 4?
38. Two dice are tossed. What is the probability of getting a sum of at most 11?
39. A single card is drawn from a deck. What is the probability of not drawing an ace?
40. A single card is drawn from a deck. What is the probability of not drawing a face card?
41. A coin is flipped four times. What is the probability of getting at least one tail?
42. A coin is flipped four times. What is the probability of getting at least two heads?
43. A single die is rolled three times. What is the probability that a 1 will show on the upward face at least once?
44. A single die is rolled four times. Find the probability that a 5 will be rolled at least once.
45. A pair of dice are rolled three times. What is the probability that a sum of 8 on the two dice will occur at least once?
46. A pair of dice are rolled four times. Compute the probability that a sum of 11 on the two dice will occur at least once.
47. A magician shuffles a standard deck of playing cards and allows an audience member to pull out a card, look at it, and replace it in the deck. Two additional people do the same. Find the probability that of the three cards drawn, at least one is a face card.
48. As in the preceding exercise, four audience members pull a card from a standard deck one-by-one and replace it. What is the probability that at least one ace is drawn?
49. If a person draws three cards from a standard deck (without replacing them), what is the probability that at least one of the cards is a face card?
50. If a person draws four cards from a standard deck (without replacing them), what is the probability that at least one of the cards is an ace?
51. A video rental store purchases 30 copies of a new movie. Unbenownst to the video store, four of the tapes are defective. If the store rents 12 of these videos the first day, what is the probability that at least one of the 12 renters will get a defective video?
52. An electronics store currently has 28 new DVD players in stock, of which five are defective. If a customer

buys three DVD players, what is the probability that at least one of them will be defective?

53. Three employees of a restaurant each contributed one business card for a random drawing to win a prize. Forty-two business cards were received in all, and three cards will be drawn for prizes. Determine the probability that at least one of the restaurant employees will win a prize.

54. A bag contains 44 U.S. quarters and six Canadian quarters. (The coins are identical in size.) If five quarters are randomly picked from the bag, what is the probability of getting at least one Canadian quarter?

Extensions
CRITICAL THINKING

55. In the game blackjack, a player is dealt two cards from a standard deck of playing cards. The player has a blackjack if one card is an ace and the other card is a 10, a jack, a queen, or a king. In some casinos, blackjack is played with more than one standard deck of playing cards. Does using more than one deck of cards change the probability of getting a blackjack?

56. The *Monte Hall problem* is described in the Math Matters on page 736. The question arises, "If the contestant changes his or her original choice of curtain, what is the probability of choosing the curtain hiding the grand prize?" To answer this question, complete the following.

  a. What is the probability that the contestant will choose the grand prize on the first try?

  b. What is the probability that the prize is not behind the curtain chosen by the contestant?

  c. What is the probability that the person will choose the grand prize by switching?

57. A planned community has 300 homes, each with an automatic garage door opener. The door opener has eight switches that a homeowner can set to 0 or 1. For example, a door opener code might be 01101001. Assuming all the homes in the community are sold, what is the probability that at least two homeowners will set their switches to the same code, and will therefore be able to open each other's garage doors?

COOPERATIVE LEARNING

58. When someone is first presented with the Monte Hall problem in Exercise 56, there is a tendency for that person to think that switching his or her choice of curtain does not make any difference. You can actually simulate the Monte Hall game using playing cards and show that one is twice as likely to win by switching from the first choice. To do this you will need at least two people and three cards, say the ace of spades to represent the grand prize, and the two of hearts and the two of diamonds, which represent the other two prizes. Shuffle the cards and place them face down on the table. Choose one of the cards. The other player picks up the remaining two cards, removes one of the cards that is not the ace (it is possible that both cards will not be the ace), and places the other card face down on the table. Now you may either stay with your original selection or change to the remaining card. Perform this experiment 30 times staying with your original selection and thirty times where you change to the other card. Keep a record of how many times staying with your original selection resulted in selecting the ace and how many times switching cards resulted in choosing the ace. On the basis of your results, is staying or switching the better strategy?

59. The benefit of switching doors in the Monte Hall problem is even more dramatic if there are more hidden prizes from which to choose. Instead of using three cards as in Exercise 58, use five cards, the four 2's and the ace of spades. Shuffle them well, place them face down on a table, and choose one card. Another person then picks up the remaining four cards, removes three 2's, and places the remaining card face down on the table. Now you may either stay with your original selection or switch to the remaining card. Perform this experiment 30 times staying with your original selection and 30 where you change to the other card. Keep a record of how many times staying with your original selection resulted in selecting the ace and how many times changing resulted in choosing the ace. On the basis of your results, is staying or switching the better strategy? About how many more times did switching result in choosing the ace than did staying with your original choice?

EXPLORATIONS

60. In five-card stud poker, a hand containing five cards of the same suit from a standard deck is called a flush. In this Exploration, you will compute the probability of getting a flush.

  a. How many five-card poker hands are possible?

  b. How many five-card poker hands are possible that contain all spades?

  c. What is the probability of getting a five-card poker hand containing all spades?

  d. What is the probability of getting a five-card poker hand containing all hearts? all diamonds? all clubs?

  e. Are the events of getting either five spades, five hearts, five diamonds, or five clubs mutually exclusive?

  f. What is the probability of getting a flush in five-card stud poker?

SECTION 11.5 | **Conditional Probability**

Conditional Probability

	F	**No F**	**Total**
V	21	198	219
No V	76	195	271
Total	97	393	490

V: Vaccinated
F: Contracted the flu

	F	**No F**	**Total**
V	21	198	219

In the preceding section, we discussed the effectiveness of a flu vaccination in preventing the onset of the flu. The table from that discussion is shown again at the left. From the table, we can calculate the probability that one person randomly selected from this population will have the flu.

$$P(F) = \frac{n(F)}{n(S)} = \frac{97}{490} \approx 0.198$$ • $n(F) = 97$, $n(S) = 490$
(S denotes the sample space)

Now consider a slightly different situation. We could ask, "What is the probability that a person randomly chosen from this population will contract the flu *given* that the person received the flu vaccination?"

In this case we know that the person received a flu vaccination and we want to determine the probability that the person will contract the flu. Therefore, the only part of the table that is of concern to us is the top row. In this case, we have

$$P(F \text{ given } V) = \frac{21}{219} \approx 0.096$$

Thus the probability that an individual will contract the flu given that the individual has been vaccinated is about 0.096.

The probability of an event B occurring given that we know some other event A has already occurred is called a **conditional probability,** and is denoted $P(B \mid A)$.

> **Conditional Probability Formula**
>
> Let A and B be two events in a sample space S. Then the conditional probability of B given that A has occurred is
>
> $$P(B \mid A) = \frac{P(A \cap B)}{P(A)}$$
>
> The symbol $P(B \mid A)$ is read "the probability of B given A."

To see how this formula applies to the flu data above, let

$S = \{$all people participating in the test$\}$
$F = \{$people who contracted the flu$\}$
$V = \{$people who were vaccinated$\}$

Then $F \cap V = \{$people who contracted the flu *and* were vaccinated$\}$

$$P(F \mid V) = \frac{P(F \cap V)}{P(V)} = \frac{\frac{21}{490}}{\frac{219}{490}}$$

• $P(F \cap V) = \dfrac{n(F \cap V)}{n(S)} = \dfrac{21}{490}$

• $P(V) = \dfrac{n(V)}{n(S)} = \dfrac{219}{490}$

$$= \frac{21}{219} \approx 0.096$$

The probability that one person selected from this population contracted the flu given that the person received the vaccination, $P(F \mid V)$, is about 0.096. Our answer agrees with the calculation we performed directly from the table, but the Conditional Probability Formula enables us to find conditional probabilities even when we cannot compute them directly.

QUESTION *In the preceding example, what is the interpretation of $P(V \mid F)$?*

EXAMPLE 1 ■ Determine a Conditional Probability

The data in the table below show the results of a survey used to determine the number of adults who received financial help from their parents for certain purchases.

Age	College tuition	Buy a car	Buy a house	Total
18–28	405	253	261	919
29–39	389	219	392	1000
40–49	291	146	245	682
50–59	150	71	112	333
60+	62	15	98	175
Total	1297	704	1108	3109

If one person is selected from this survey, what is the probability that the person received financial help for purchasing a home, given that the person is between the ages of 29 and 39?

ANSWER *$P(V \mid F)$ is the probability that a person selected from this population has been vaccinated, given that the person contracted the flu.*

Solution

Let $B = \{$adults who received financial help for a home purchase$\}$ and $A = \{$adults between 29 and 39$\}$. From the table, $n(A \cap B) = 392$, $N(A) = 1000$, and $n(S) = 3109$. Using the Conditional Probability Formula, we have

$$P(B \mid A) = \frac{P(A \cap B)}{P(A)} = \frac{\dfrac{392}{3109}}{\dfrac{1000}{3109}}$$

- $P(A \cap B) = \dfrac{n(A \cap B)}{n(S)} = \dfrac{392}{3109}$

- $P(A) = \dfrac{n(A)}{n(S)} = \dfrac{1000}{3109}$

$$= \frac{392}{1000} = 0.392$$

The probability that a person received financial help for purchasing a home, given that the person is between the ages of 29 and 39, is 0.392.

CHECK YOUR PROGRESS 1   Two dice are tossed, one after the other. What is the probability that the result is a sum of 6, given that the first die is not a 3?

Solution   *See page S42.*

Math Matters   **Reversing Conditional Probabilities**

Bayes Theorem, named after Thomas Bayes (1702–1761), gives a method of computing conditional probabilities by knowing the reverse conditional probabilities. Consider the following problem, known as Bertrand's Box Paradox. Three boxes each contain two coins. One box has two gold coins, another has two silver coins, and the third box has one gold coin and one silver coin. A box is chosen at random, and a coin is pulled out (without looking at the other coin). If the coin that is taken out is gold, what is the probability that the other coin in the same box is also gold? Many people initially guess that the probability is $\frac{1}{2}$. To check, let B_1 be the event that the first box (with two gold coins) is chosen, B_2 the event that the second box (with two silver coins) is chosen, and B_3 the event that the third box (with one silver and one gold) is chosen. In addition, let G represent the event that a gold coin is pulled from a box, and S the event that a silver coin is chosen. The conditional probability of pulling a gold coin from a box, given that we know which box was chosen, is simple to compute. What we want to find, however, is the reverse: the conditional probability that we have chosen the first box, given that a gold coin was chosen, $P(B_1 \mid G)$. Bayes Theorem gives us a way to find this probability. In this case, the theorem states

$$P(B_1 \mid G) = \frac{P(G \mid B_1)P(B_1)}{P(G \mid B_1)P(B_1) + P(G \mid B_2)P(B_2) + P(G \mid B_3)P(B_3)}$$

$$= \frac{1 \cdot \frac{1}{3}}{1 \cdot \frac{1}{3} + 0 \cdot \frac{1}{3} + \frac{1}{2} \cdot \frac{1}{3}} = \frac{2}{3}$$

Thus, there is actually a $\frac{2}{3}$ chance that if a gold coin is pulled from a box, the other coin is also gold.

Product Rule for Probabilities

Suppose two cards are drawn, without replacement, from a standard deck of playing cards. Let A be the event that an ace is drawn on the first draw and B the event that an ace is drawn on the second draw. Then the probability that an ace is drawn on the first *and* second draw is $P(A \text{ and } B) = P(A \cap B)$. To find this probability, we can solve the Conditional Probability Formula for $P(A \cap B)$.

 **TAKE NOTE**

Recall that $A \cap B$ contains the elements that are common to both *A and B*.

$$\frac{P(A \cap B)}{P(A)} = P(B \mid A)$$

$$P(A) \cdot \frac{P(A \cap B)}{P(A)} = P(A) \cdot P(B \mid A) \qquad \text{• Multiply each side by } P(A).$$

$$P(A \cap B) = P(A) \cdot P(B \mid A)$$

This is called the Product Rule for Probabilities.

> ### Product Rule for Probabilities
>
> If A and B are two events from the sample space S, then
>
> $$P(A \text{ and } B) = P(A \cap B) = P(A) \cdot P(B \mid A).$$

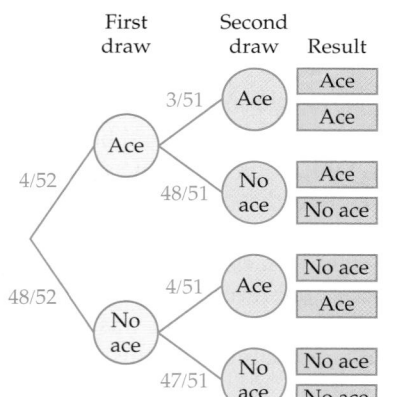

For the problem of drawing an ace from a standard deck of playing cards on the first and second draws, $P(A \text{ and } B) = P(A \cap B)$ is the product of $P(A)$, the probability that the first drawn card is an ace, and $P(B \mid A)$, the probability of an ace on the second draw *given* that the first card drawn was an ace.

The tree diagram at the left shows the possible outcomes of drawing two cards from a deck without replacement. On the first draw, there are four aces in the deck of 52 cards. Therefore, $P(A) = \frac{4}{52} = \frac{1}{13}$. On the second draw, there are only 51 cards remaining and only three aces (an ace was drawn on the first draw). Therefore, $P(B \mid A) = \frac{3}{51} = \frac{1}{17}$. Putting these calculations together, we have

$$P(A \cap B) = P(A) \cdot P(B \mid A)$$

$$= \frac{1}{13} \cdot \frac{1}{17} = \frac{1}{221}$$

The probability of drawing an ace on the first and second draws is $\frac{1}{221}$.

The Product Rule for Probabilities can be extended to more than two events. The probability that a certain sequence of events will occur in succession is the product of the probabilities of each of the events *given* that the preceding events have occurred.

> ### Probability of Successive Events
>
> The probability of two or more events occurring in succession is the product of the conditional probabilities of each of the events.

EXAMPLE 2 ■ Find the Probability of Successive Events

A box contains four red, three white, and five green balls. Suppose three balls are randomly selected from the box in succession, without replacement.

a. What is the probability that first a red, then a white, and then a green ball are selected?

b. What is the probability that two white balls followed by one green ball are selected?

Solution

a. Let $A = \{$a red ball is selected first$\}$, $B = \{$a white ball is selected second$\}$, and $C = \{$a green ball is selected third$\}$. Then

$$P(A \text{ followed by } B \text{ followed by } C) = P(A) \cdot P(B|A) \cdot P(C|A \text{ and } B)$$
$$= \frac{4}{12} \cdot \frac{3}{11} \cdot \frac{5}{10}$$
$$= \frac{1}{22}$$

The probability of choosing a red, then a white, then a green ball is $\frac{1}{22}$.

b. Let $A = \{$a white ball is selected first$\}$, $B = \{$a white ball is selected second$\}$, and $C = \{$a green ball is selected third$\}$. Then

$$P(A \text{ followed by } B \text{ followed by } C) = P(A) \cdot P(B|A) \cdot P(C|A \text{ and } B)$$
$$= \frac{3}{12} \cdot \frac{2}{11} \cdot \frac{5}{10}$$
$$= \frac{1}{44}$$

The probability of choosing two white balls followed by one green ball is $\frac{1}{44}$.

CHECK YOUR PROGRESS 2 A standard deck of playing cards is shuffled and three cards are dealt. Find the probability that the cards dealt are a spade followed by a heart followed by another spade.

Solution See page S42.

> ✔ **TAKE NOTE**
>
> In part a, there are originally 12 balls in the box. After a red ball is selected, there are only 11 balls remaining, of which three are white. Thus $P(B|A) = \frac{3}{11}$. After a red ball and a white ball are selected, there are 10 balls left, of which five are green. Thus $P(C|A \text{ and } B) = \frac{5}{10}$.
>
> In part b, we have a similar situation. However, after a white ball is selected, there are 11 balls remaining, of which only two are white. Therefore, $P(B|A) = \frac{2}{11}$.

Independent Events

Earlier in this section we considered the probability of drawing two aces in a row from a standard deck of playing cards. Because the cards were drawn without replacement, the probability of an ace on the second draw *depended* on the result of the first draw.

Now consider the case of tossing a coin twice. The outcome of the first coin toss has no effect on the outcome of the second toss. So the probability of the coin flipping to a head or a tail on the second toss is not affected by the result of the first toss. When the outcome of a first event does not affect the outcome of a second event, the events are called *independent*.

> **Independent Events**
>
> If A and B are two events in a sample space and $P(B \mid A) = P(B)$, then A and B are called **independent events.**

For a mathematical verification, consider tossing a coin twice. We can compute the probability that the second toss comes up heads, given that the first coin toss came up heads. If A is the event of a head on the first toss, then $A = \{HH, HT\}$. Let B be the event of a head on the second toss. Then $B = \{HH, TH\}$. The sample space is $S = \{HH, HT, TH, TT\}$. The conditional probability $P(B \mid A)$ (the probability of a head on the second toss given a head on the first toss) is

$$P(B \mid A) = \frac{P(A \cap B)}{P(A)} = \frac{\dfrac{1}{4}}{\dfrac{1}{2}} = \frac{1}{2}$$

- $P(A \cap B) = \dfrac{n(A \cap B)}{n(S)} = \dfrac{1}{4}$
- $P(A) = \dfrac{n(A)}{n(S)} = \dfrac{2}{4} = \dfrac{1}{2}$

Thus $P(B \mid A) = \frac{1}{2}$. Note, however, that $P(B) = \frac{n(B)}{n(S)} = \frac{2}{4} = \frac{1}{2}$. Therefore, in the case of tossing a coin twice, the probability of the second event does not depend on the outcome of the first event, and we have $P(B \mid A) = P(B)$.

In general, this result enables us to simplify the product rule when two events are independent; the probability of two independent events occurring in succession is simply the product of the probabilities of each of the individual events.

✓ **TAKE NOTE**

The Product Rule for Independent Events can be extended to more than two events. If E_1, E_2, E_3, and E_4 are independent events, then the probability that all four events will occur is $P(E_1) \cdot P(E_2) \cdot P(E_3) \cdot P(E_4)$.

> **Product Rule for Independent Events**
>
> If A and B are two independent events from the sample space S, then $P(A \text{ and } B) = P(A \cap B) = P(A) \cdot P(B)$.

✓ **TAKE NOTE**

See page 721 for all the possible outcomes of the roll of two dice.

EXAMPLE 3 ■ **Probability of Independent Events**

A pair of dice are tossed twice. What is the probability that the first roll is a sum of 7 and the second roll is a sum of 11?

Solution
The rolls of a pair of dice are independent; the probability of a sum of 11 on the second roll does not depend on the outcome of the first roll. Let $A = \{\text{sum of 7 on the first roll}\}$ and $B = \{\text{sum of 11 on the second roll}\}$. Then

$$P(A \cap B) = P(A) \cdot P(B) = \frac{6}{36} \cdot \frac{2}{36} = \frac{1}{108}$$

CHECK YOUR PROGRESS 3   A coin is tossed three times. What is the probability that heads appears on all three tosses?

Solution *See page S42.*

Applications of Conditional Probability

Conditional probability is used in many real-world situations, such as to determine the efficacy of a drug test, to verify the accuracy of genetic testing, and to analyze evidence in legal proceedings.

EXAMPLE 4 ■ Drug Testing and Conditional Probability

Suppose that a company claims it has a test that is 95% effective in determining whether an athlete is using a steroid. That is, if an athlete is using a steroid, the test will be positive 95% of the time. In the case of a negative result, the company says its test is 97% accurate. That is, even if an athlete is not using steroids, it is possible that the test will be positive in 3% of the cases. Such an occurrence is called a **false positive.** Suppose this test is given to a group of athletes in which 10% of the athletes are using steroids. What is the probability that a randomly chosen athlete actually uses steroids, given that the athlete's test is positive?

Solution

Let S be the event that an athlete uses steroids and let T be the event that the test is positive. Then the probability we wish to determine is $P(S \mid T)$. Using the conditional probability formula, we have

$$P(S \mid T) = \frac{P(S \text{ and } T)}{P(T)}$$

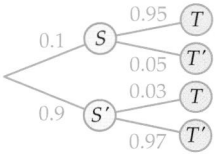

A tree diagram, shown at the left, can be used to calculate this probability. A positive test result can occur in two ways: either an athlete using steroids correctly tests positive, or an athlete not using steroids incorrectly tests positive. The probability of a positive test result, $P(T)$, corresponds to an athlete following path ST or path $S'T$ in the tree diagram. (S' symbolizes no steroid use.) $P(S \cap T)$, the probability of using steroids and getting a positive test result, is path ST. Thus,

$$P(S \mid T) = \frac{P(S \text{ and } T)}{P(T)}$$

$$= \frac{0.1 \cdot 0.95}{0.1 \cdot 0.95 + 0.9 \cdot 0.03} \approx 0.779$$

Given that an athlete tests positive, the probability that the athlete actually uses steroids is approximately 77.9%.

This may be lower than you would expect, especially considering that the manufacturer claims that its test is 95% accurate. In fact, the prevalence of false-positive test results can be much more dramatic in cases of conditions that are present in a small percentage of the population. (See Exercise 61.)

▼ **point of interest**

The resulting calculation in Example 4 is actually an illustration of Bayes Theorem described in the Math Matters on page 743.

CHECK YOUR PROGRESS 4   A pharmaceutical company has a test that is 95% effective in determining whether a person has a certain genetic defect. It is possible, however, that the test may give a false positive result in 4% of cases. Suppose this particular genetic defect occurs in 2% of the population. Given that a person tests positive, what is the probability that the person actually has the defect?

Solution See page S42.

Excursion

Sharing Birthdays

Have you ever been introduced to someone at a party or other social gathering and discovered that you shared the same birthday? It seems like an amazing coincidence when it occurs. In fact, how rare is this?

As an example, suppose four people have gathered for a dinner party. We can determine the probability that at least two of the guests have the same birthday. (For simplicity, we will ignore the February 29th birthday from leap years.) Let E be the event that at least two people share a birthday. It is easier to look at the complement E^c, the event that no one shares the same birthday.

If we start with one of the guests, then the second guest cannot share the same birthday, so that person has 364 possible dates for his or her birthday from a total of 365. Thus the conditional probability that the second guest has a different birthday, given that we know the first person's birthday, is $\frac{364}{365}$. Similarly, the third person has 363 possible birthday dates that do not coincide with those of the first two guests. So the conditional probability that the third person does not share a birthday with either of the first two guests, given that we know the birthdays of the first two people, is $\frac{363}{365}$. The probability of the fourth guest having a distinct birthday is, similarly, $\frac{362}{365}$. We can use the Product Rule for Probabilities to find the probability that all of these conditions are met; that is, none of the four guests share a birthday.

$$P(E^c) = \frac{364}{365} \cdot \frac{363}{365} \cdot \frac{362}{365} \approx 0.984$$

INSTRUCTOR NOTE
As suggested, you can introduce this Excursion by first asking the class how likely they think it is that two of the students in the class would have the same birthday. Then ask each student when his or her birthday is and determine if any of the students share a birthday. (If you have a large number of students you can proceed month by month.) If two or more students do have a common birthday, it can serve as evidence that perhaps it doesn't require as many people present as one might think to have a good chance of a shared birthday.

Then $P(E) = 1 - P(E^c) \approx 0.016$, so there is about a 1.6% chance that in a group of four people, two or more will have the same birthday.

It would require 366 people gathered together to *guarantee* that two people in the group will have the same birthday. But how many people would be required to guarantee that the chance that at least two of them share the same birthday is at least 50/50? Make a guess before you proceed through the exercises. The results may surprise you!

Excursion Exercises

1. If eight people are present at a meeting, find the probability that at least two share a common birthday.

2. Compute the probability that at least two people among a group of 15 have the same birthday.

3. If 23 people are in attendance at a party, what is the probability that at least two share a birthday?

4. In a group of 40 people, what would you estimate to be the probability that at least two people share a birthday? If you have the patience, compute the probability to check your guess.

Exercise Set 11.5   (Suggested Assignment: 3–43 odds; 46, 54, 57, 60–62)

1. What is a conditional probability?

2. Explain the difference between independent events and dependent events.

In Exercises 3–6, compute the conditional probabilities $P(A \mid B)$ and $P(B \mid A)$.

3. $P(A) = 0.7$, $P(B) = 0.4$, $P(A \cap B) = 0.25$
4. $P(A) = 0.45$, $P(B) = 0.8$, $P(A \cap B) = 0.3$
5. $P(A) = 0.61$, $P(B) = 0.18$, $P(A \cap B) = 0.07$
6. $P(A) = 0.2$, $P(B) = 0.5$, $P(A \cap B) = 0.2$

In Exercises 7–10, use the data in the table below, which shows the employment status of individuals in a particular town by age group.

	Full-time	Part-time	Unemployed
0–17	24	164	371
18–25	185	203	148
26–34	348	67	27
35–49	581	179	104
50+	443	162	173

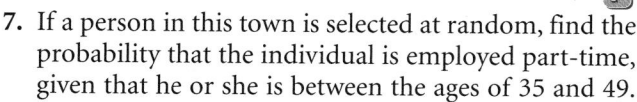

7. If a person in this town is selected at random, find the probability that the individual is employed part-time, given that he or she is between the ages of 35 and 49.

8. If a person in the town is randomly selected, what is the probability that the individual is unemployed, given that he or she is over 50 years old?

9. A person from the town is randomly selected; what is the probability that the individual is employed full-time, given that he or she is between 18 and 49 years of age?

10. A person from the town is randomly selected; what is the probability that the individual is employed part-time, given that he or she is at least 35 years old?

In Exercises 11–14, use the data in the following table, which shows the results of a survey of 2000 gamers about their favorite home video game systems, organized by age group. If a survey participant is selected at random, determine the probability of each of the following.

	Sony Play-station 2	Microsoft Xbox	Nintendo GameCube	Sega Dreamcast
0–12	63	84	55	51
13–18	105	139	92	113
19–24	248	217	83	169
25+	191	166	88	136

11. The participant prefers the Playstation 2 system.

12. The participant prefers the Microsoft Xbox, given that the person is between the ages of 13 and 18.

13. The participant prefers Nintendo GameCube, given that the person is between the ages of 13 and 24.

14. The participant is under 12 years of age, given that the person prefers the Dreamcast machine.

15. A pair of dice are tossed. Find the probability that the sum on the two dice is 8, given that the sum is even.

16. A pair of dice are tossed. Find the probability that the sum on the two dice is 12, given that doubles are rolled.

17. A pair of dice are tossed. What is the probability that doubles are rolled, given that the sum on the two dice is less than 7?

18. A pair of dice are tossed. What is the probability that the sum on the two dice is 8, given that the sum is more than 6?

19. What is the probability of drawing two cards in succession (without replacement) from a standard deck and having them both be face cards?

20. Two cards are drawn from a standard deck without replacement. Find the probability that both cards are hearts.

21. Two cards are drawn from a standard deck without replacement. What is the probability that the first card is a spade and the second card is red?

22. Two cards are drawn from a standard deck without replacement. What is the probability that the first card is a king and the second card is not?

In Exercises 23–26, a snack-size bag of M&Ms candies is opened. Inside, there are 12 red candies, 12 blue, 7 green, 13 brown, 3 orange, and 10 yellow. Three candies are pulled from the bag in succession, without replacement.

23. Determine the probability that the first candy drawn is blue, the second is red, and the third is green.

24. Determine the probability that the first candy drawn is brown, the second is orange, and the third is yellow.

25. What is the probability that the first two candies drawn are green and the third is red?

26. What is the probability that the first candy drawn is orange, the second is blue, and the third is orange?

In Exercises 27–30, three cards are dealt from a shuffled standard deck of playing cards.

27. Find the probability that the first card dealt is red, the second is black, and the third is red.

28. Find the probability that the first two cards dealt are clubs and the third is a spade.

29. What is the probability that the three cards dealt are, in order, an ace, a face card, and an 8? (A face card is a jack, queen, or king.)

30. What is the probability that the three cards dealt are, in order, a red card, a club, and another red card?

In Exercises 31–34, the probability that a student enrolled at a local high school will be absent on a particular day is 0.04, assuming that the student was in attendance the previous school day. However, if a student is absent, the probability that he or she will be absent again the following day is 0.11. For each exercise, assume that the student was in attendance the previous day.

31. What is the probability that a student will be absent three days in a row?

32. What is the probability that a student will be absent two days in a row, but then show up on the third day?

33. Find the probability that a student will be absent, attend the next day, but then will be absent again the third day.

34. Find the probability that a student will be absent four days in a row.

In Exercises 35–38, determine whether the events are independent.

35. A single die is rolled and then rolled a second time.

36. Numbered balls are pulled from a bin one-by-one to determine the winning lottery numbers.

37. Numbers are written on slips of paper in a hat; one person pulls out a slip of paper without replacing it, then a second person pulls out a slip of paper.

38. In order to determine who goes first in a game, one person picks a number between 1 and 10, then a second person picks a number between 1 and 10.

In Exercises 39–44, a pair of dice are tossed twice.

39. Find the probability that both rolls give a sum of 8.

40. Find the probability that the first roll is a sum of 6 and the second roll is a sum of 12.

41. Find the probability that the first roll is a total of at least 10 and the second roll is a total of at least 11.

42. Find the probability that both rolls result in doubles.

43. Find the probability that both rolls give even sums.

44. Find the probability that both rolls give at most a sum of 4.

45. A fair coin is tossed four times in succession. Find the probability of getting two heads followed by two tails.

46. In the game of Monopoly, a player is sent to jail if he or she rolls doubles with a pair of dice three times in a row. What is the probability of rolling doubles three times in succession?

47. Find the probability of tossing a pair of dice three times in succession and getting a sum of at least 10 on all three tosses.

48. Find the probability of tossing a pair of dice three times in succession and getting a sum of at most 3 on all three tosses.

In Exercises 49–54, a card is drawn from a standard deck and replaced. After the deck is shuffled, another card is pulled.

49. What is the probability that both cards pulled are aces?

50. What is the probability that both cards pulled are face cards?

51. What is the probability that the first card drawn is a spade and the second card is a diamond?

52. What is the probability that the first card drawn is an ace and the second card is not an ace?

53. Find the probability that the first card drawn is a heart and the second card is a spade.

54. Find the probability that the first card drawn is a face card and the second card is black.

55. A standard deck of playing cards is shuffled and three people each choose a card. Find the probability that the first two cards chosen are diamonds and the third card is black if

  a. the cards are chosen *with* replacement.

  b. the cards are chosen *without* replacement.

56. A standard deck of playing cards is shuffled and three people each choose a card. Find the probability that all three cards are face cards if

  a. the cards are chosen *with* replacement.

  b. the cards are chosen *without* replacement.

57. A bag contains five red marbles, four green marbles, and eight blue marbles. Find the probability of pulling two red marbles followed by a green marble if the marbles are pulled from the bag

  a. with replacement.

  b. without replacement.

58. A box contains three medium t-shirts, five large t-shirts, and four extra-large t-shirts. If someone randomly chooses three t-shirts from the box, find the probability that the first t-shirt is large, the second is medium, and the third is large if the shirts are chosen

  a. with replacement.

  b. without replacement.

59. A company that performs drug testing guarantees that its test determines a positive result with 97% accuracy. However, the test also gives 6% false positives. If 5% of those being tested actually have the drug present in their bloodstream, find the probability that a person testing positive has actually been using drugs.

60. A test for a genetic disorder can detect the disorder with 94% accuracy. However, the test will incorrectly report positive results for 3% of those without the disorder. If 12% of the population has the disorder, find the probability that a person testing positive actually has the genetic disorder.

61. A pharmaceutical company has developed a test for a scarce disease that is present in 0.5% of the population. The test is 98% accurate in determining a positive result, and the chance of a false positive is 4%. What is the probability that someone who tests positive actually has the disease?

62. When used together, the ELISA and Western Blot tests for HIV are more than 99% accurate in determining a positive result. The chance of a false positive is between 1 and 5 for every 100,000 tests. If we assume a 99% accuracy rate for correctly identifying positive results and 5/100,000 false positives, find the probability that someone who tests positive has HIV. (The Centers for Disease Control and Prevention estimate that about 0.3% of U.S. residents are infected with HIV.)

Extensions
CRITICAL THINKING

63. Suppose you are standing at a street corner and flip a coin to decide whether you will go north or south from your current position. When you reach the next intersection, you repeat the procedure. (This problem is a simplified version of what is called a *random walk* problem. Problems of this type are important in economics, physics, chemistry, biology, and other disciplines.)

  a. After performing this experiment three times, what is the probability that you will be three blocks north of your original position?

  b. After performing this experiment four times, what is the probability that you will be two blocks north of your original position?

  c. After performing this experiment four times, what is the probability that you will be back at your original position?

COOPERATIVE LEARNING

64. From a standard deck of playing cards, choose four red cards and four black cards. Deal the eight cards, face up, in two rows of four cards each. A good event is that each column contains a red and a black card. (The column can be red/black or black/red.) A bad event is any other situation. Although this problem is stated in terms of cards, it has a very practical application. If several proteins in a cell break (say, from radiation therapy) and then reattach, it is possible that the new protein is harmful to the cell.

a. Perform this experiment 50 times and keep a record of the number of good events and bad events.

b. Use your data to predict the probability of a good event.

c. Repeat parts a and b again.

d. Use your data from the complete 100 trials to predict the probability of a good event.

e. Calculate the theoretical probability of a good event.

EXPLORATIONS

65. This is another explanation, using conditional probability, of the Monte Hall problem discussed in Exercise 56 of the last section. Let the curtain chosen by the contestant be labeled A and the other two curtains be labeled B and C. In the following exercises, we will use A to represent the event that the grand prize is behind curtain A (and similarly for curtains B and C), and $\overline{A}$ to represent the event that Monte Hall opens curtain A (and similarly for curtains B and C).

a. What is the probability that Monte Hall opens curtain B given that the grand prize is behind curtain A? This is $P(\overline{B}|A)$.

b. What is the probability that Monte Hall opens curtain B given that the grand prize is behind curtain B? This is $P(\overline{B}|B)$.

c. What is the probability that Monte Hall opens curtain B given that the grand prize is behind curtain C? This is $P(\overline{B}|C)$.

d. The probability that the grand prize is behind curtain A given that curtain B is opened (you have not switched choices) is given by Bayes Theorem in the following form.

$$P(A|\overline{B}) = \frac{P(\overline{B}|A)P(A)}{P(\overline{B}|A)P(A) + P(\overline{B}|B)P(B) + P(\overline{B}|C)P(C)}$$

What is the probability?

e. Find the probability of choosing the grand prize if you switch curtains. That is, find

$$P(C|\overline{B}) = \frac{P(\overline{B}|C)P(C)}{P(\overline{B}|A)P(A) + P(\overline{B}|B)P(B) + P(\overline{B}|C)P(C)}$$

f. Is switching the better strategy?

SECTION 11.6 | # Expectation

Expectation

Suppose a barrel contains a large number of balls, half of which have the number 1000 painted on them and the other half of which have the number 500 painted on them. As the grand prize winner of a contest, you get to reach into the barrel (blindfolded, of course) and select 10 balls. Your prize is the sum of the numbers on the balls in cash.

If you are very lucky, all of the balls will have 1000 painted on them and you will win $10,000. If you are very unlucky, all of the balls will have 500 painted on them and you will win $5000. Most likely, however, approximately one-half of the balls will have 1000 painted on them and one-half will have 500 painted on them. The amount of your winnings in this case will be 5(1000) + 5(500), or $7500. Because 10 balls were drawn, your amount of winnings per ball is $\frac{\$7500}{10} = \$750$.

The number $750 is called the *expected value* or the *expectation* of the game. You cannot win $750 on one draw, but if given the opportunity to draw many times, you will win, on average, $750 per ball.

<hr>

QUESTION *Can an expectation be negative?*

Another way we can calculate expectation is to use probabilities. For the game above, one-half of the balls have the number 1000 painted on them and one-half have the number 500 painted on them. Therefore, $P(1000) = \frac{1}{2}$ and $P(500) = \frac{1}{2}$. The expectation is calculated as follows.

Expectation
= (probability of winning $1000) $\cdot$ $1000 + (probability of winning $500) $\cdot$ $500
= $P(1000) \cdot \$1000 + P(500) \cdot \$500$
= $\frac{1}{2} \cdot \$1000 + \frac{1}{2} \cdot \$500 = \$750$

The general result for experiments with numerical outcomes follows.

INSTRUCTOR NOTE
You can point out to students that expectation is like a weighted average. We are finding the average value of the outcomes, but each outcome is weighted by the probability of it occurring.

Expectation

Let $S_1, S_2, S_3, \ldots, S_n$ be the possible outcomes of an experiment and let $P(S_1), P(S_2), P(S_3), \ldots, P(S_n)$ be the probabilities of those outcomes. Then the **expectation** of the experiment is

$$P(S_1) \cdot S_1 + P(S_2) \cdot S_2 + P(S_3) \cdot S_3 + \cdots + P(S_n) \cdot S_n$$

That is, to find the expectation of an experiment, multiply the probability of each outcome of the experiment by the outcome and then add the results.

EXAMPLE 1 ■ Expectation in Gambling

One of the wagers in roulette is to place a bet on one of the numbers from 0 to 36 or on 00. If that number comes up, the player wins 35 times the amount bet (and keeps the original bet). Suppose a player bets $1 on a number. What is the player's expectation?

<hr>

ANSWER *Yes. For instance, if the expected value of a gambling game is negative, it simply means that, in the long run, a person will lose that amount of money, on average, on each play.*

✔ **TAKE NOTE**

In Example 1, suppose the player bets $5 instead of $1. The payoff is then $175 (35 · 5) if the player wins and −$5 if the player loses. The expectation is $\frac{1}{38}(175) + \frac{37}{38}(-5) = -\frac{5}{19}$. Note that the bet is 5 times greater and the expectation, $-\frac{5}{19}$, is 5 times the expectation when $1 is bet. Thus a player who makes $5 bets can expect to lose 5 times as much money as a player who makes $1 bets.

Solution

Let S_1 be the event that the player's number comes up and the player wins $35. Because there are 38 numbers from which to choose, $P(S_1) = \frac{1}{38}$. Let S_2 be the event that the player's number does not come up and the player therefore loses $1. Then $P(S_2) = 1 - \frac{1}{38} = \frac{37}{38}$.

$$\text{Expectation} = P(S_1) \cdot S_1 + P(S_2) \cdot S_2$$
$$= \frac{1}{38}(35) + \frac{37}{38}(-1) = -\frac{1}{19}$$
$$\approx -0.053$$

• The amount the player can win is entered as a positive number. The amount that can be lost is entered as a negative number.

The player's expectation is approximately −$.05. This means that, on average, the player will lose about $.05 every time this bet is made.

CHECK YOUR PROGRESS 1 In roulette it is possible to place a wager that one of the numbers between 1 and 12 (inclusive) will come up. If it does, the player wins twice the amount bet. Suppose a player bets $5 that a number between 1 and 12 will come up. What is the player's expectation?

Solution See page S42.

In Example 1, the fact that the player is losing approximately 5 cents on each dollar bet means that the casino's expectation is positive 5 cents; it is earning (on average) 5 cents for every dollar spent making that particular bet at the roulette wheel. An individual player may get lucky, but over time the casino can plan on a predictable profit.

historical note

Daniel Bernoulli (bər-noo´lē) (1700–1782) was the son of Jean Bernoulli I and the nephew of Jacques Bernoulli. For a time he was a professor in St. Petersburg, Russia, where he collaborated with Leonhard Euler. There he wrote a paper on probability and expectation in which he discussed the game described at the right, now known as the St. Petersburg Paradox. ∎

Math Matters   **A Bargain at Any Price?**

The Swiss mathematician Daniel Bernoulli discussed the following game in a paper he published in 1738. A fair coin is repeatedly tossed until the coin comes up tails. Let $n =$ the number of total coin flips. When the coin comes up tails, you are paid 2^n dollars. Thus, if the first flip is tails, you are paid $2. If the coin comes up heads five times in a row and the sixth flip comes up tails, you are paid $2^6 = $64. How much would you pay to play such a game? $5? $20? The game becomes interesting when we compute the expected value:

$$\text{Expectation}$$
$$= P(\text{T}) \cdot 2^1 + P(\text{HT}) \cdot 2^2 + P(\text{HHT}) \cdot 2^3 + P(\text{HHHT}) \cdot 2^4 + \cdots$$
$$= \frac{1}{2} \cdot 2 + \frac{1}{4} \cdot 4 + \frac{1}{8} \cdot 8 + \frac{1}{16} \cdot 16 + \cdots$$
$$= 1 + 1 + 1 + 1 + \cdots$$

There is no maximum number of times the coin can be flipped, and the expectation is infinite! Theoretically, it would be worthwhile to play no matter how high the fee is.

Expectation in Insurance

When an insurance company sells a life insurance policy, the premium (the cost to purchase the policy) is based on many factors, but one of the most important is the probability that the insured person will outlive the term of the policy. Such probabilities are found in *mortality tables,* which give the probability that a person of a certain age will live one more year. For the life insurance company, it is very much like gambling. The company wants to know its expectation on a policy—that is, how much it will pay out, on average, for each policy it writes. Here is an example.

historical note

Edmond Halley (hăl′ē) (1656–1742), of Halley's comet fame, also created one of the first mortality tables for the city of Breslau, Germany, which he published in 1693. This was one of the first attempts to relate mortality and age in a population. ■

EXAMPLE 2 ■ Expectation in Insurance

According to mortality tables published in the National Vital Statistics Report, the probability that a 27-year-old will die within one year is approximately 0.001. Suppose that the premium for a one-year, $25,000 life insurance policy for a 27-year-old is $32. What is the insurance company's expectation for this policy?

Solution

Let S_1 be the event that the person dies within one year. Then $P(S_1) = 0.001$ and the company must pay out $25,000. (Because the company charged $32 for the policy, the company's actual loss is $24,968.) Let S_2 be the event that the policy holder does not die during the year of the policy. Then $P(S_2) = 0.999$ and the company keeps the premium of $32. The expectation is

$$\text{Expectation} = P(S_1) \cdot S_1 + P(S_2) \cdot S_2$$
$$= 0.001(-24{,}968) + 0.999(32)$$
$$= 7.00$$

• The amount the company pays out is entered as a negative number. The amount the company receives is entered as a positive number.

The company's expectation is $7.00, so the company earns, on average, $7 for each policy sold.

CHECK YOUR PROGRESS 2   The probability that a 51-year-old will die within one year is approximately 0.005. Suppose that the premium for a one-year, $10,000 life insurance policy for a 51-year-old is $45. What is the insurance company's expectation for this policy?

Solution   See page S42.

historical note

Pierre Simon Laplace (lə-pläs′) (1749–1827) made many important contributions to mathematics and astronomy. In 1812, he published his *Théorie Analytique des Probabilitiés,* in which he extended the theories of probability beyond analyzing games of chance and applied them to many practical and scientific problems. In his book he states, "The most important questions of life are indeed, for the most part, really only problems of probability. ■

Expectation is also used when a company bids on a project. The company must try to predict the costs and amount of work involved to give a bid that allows it to make a profit from completing the project. At the same time, if the bid is too high, the client may reject the offer. Because it is impossible to predict in advance the exact requirements of the job, probabilities can be used to analyze the likelihood of making a profit, as the next example demonstrates.

EXAMPLE 3 ■ Expected Company Profits

Suppose a software company bids on a project to update the database program for an accounting firm. The software company assesses its potential profit as shown in the following table.

Profit/Loss	Probability
$75,000	0.10
$50,000	0.25
$20,000	0.50
−$10,000	0.10
−$25,000	0.05

What is the profit expectation for the company?

Solution

The company's expected profit is

Expectation
$$= 0.10(75,000) + 0.25(50,000) + 0.50(20,000) + 0.10(-10,000) + 0.05(-25,000)$$
$$= 7500 + 12,500 + 10,000 - 1000 - 1250 = 27,750$$

The company's expected profit is $27,750.

Profit/Loss	Probability
$500,000	0.05
$250,000	0.30
$150,000	0.35
−$100,000	0.20
−$350,000	0.10

CHECK YOUR PROGRESS 3 A road construction company bids on a project to build a new freeway. The company estimates its potential profit as shown in the table at the left. What is the profit expectation for the company?

Solution See page S43.

Excursion

Chuck-a-luck

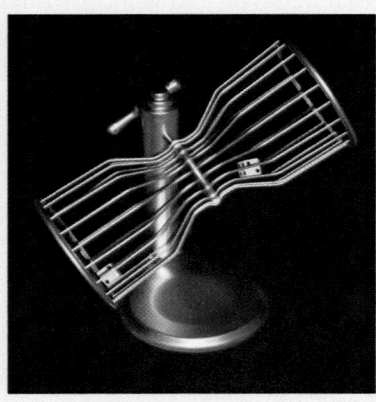

Chuck-a-luck is a game of chance in which three dice in a cage are tumbled. Bets can be placed on the values that the three dice will show. The table below shows the different bets and the amount won if $1 is wagered.

Numbers bet	Bet on a number from 1 through 6	
	if one die matches	Pays $1
	if two dice match	Pays $2
	if all three dice match	Pays $10
Field bet	Bet that the sum of the numbers showing on all three dice will be 5, 6, 7, 8, 13, 14, 15, or 16	Pays $1
Over 10	Bet that the sum of the numbers showing on all three dice will be more than 10	Pays $1
Under 11	Bet that the sum of the numbers showing on all three dice will be less than 11	Pays $1

(continued)

Excursion Exercises

1. Compute the probability that if a bet is placed on the number 5, all three dice will show a 5.

2. What is the probability that *exactly* one die will show a 5?

3. What is the probability that two (but not three) dice will show a 5?

4. Find the probability that none of the dice will show a 5.

5. What is the expectation for a $1 bet placed on the number 5?

6. Determine the expectation for wagering $1 on the Over 10 bet.

7. Determine the expectation for wagering $1 on the Under 11 bet.

8. Which bet is most favorable for the player? Which is most favorable for the casino? (The expectation for a $1 Field bet is −4 cents.)

Exercise Set 11.6   (Suggested Assignment: 1–19 odds; 4, 22)

1. The outcomes of an experiment and the probability of each outcome are given in the table below. Compute the expectation for this experiment.

Outcome	Probability
30	0.15
40	0.2
50	0.4
60	0.05
70	0.2

2. The outcomes of an experiment and the probability of each outcome are given in the table below. Compute the expectation for this experiment.

Outcome	Probability
5	0.4
6	0.3
7	0.1
8	0.08
9	0.07
10	0.05

3. One of the wagers in the game of roulette is to place a bet that the ball will land on a black number.

(Eighteen of the numbers are black, 18 are red, and two are green.) If the ball lands on a black number, the player wins the amount of his bet. If a player bets $1, find the player's expectation.

4. One of the wagers in roulette is to bet that the ball will stop on a number that is a multiple of 3. (Both 0 and 00 are not included.) If the ball stops on such a number, the player wins double the amount bet. If a player bets $1, compute the player's expectation.

Many casinos have a game called the Big Six Money Wheel, in which a large wheel with various dollar amounts is spun. Players may bet on one or more denominations; if the wheel stops on that denomination, the player wins that amount for each dollar bet. The wheel has 54 slots; the number of slots marked with each denomination is given in the following table. Exercises 5 to 8 use this game.

Denomination	Number of slots
$40	2
$20	2
$10	4
$5	8
$2	15
$1	23

5. If a player bets $1 on the $40 denomination, find the player's expectation.

6. If a player bets $1 on the $20 denomination, find the player's expectation.

7. If a player bets $1 on the $5 denomination, find the player's expectation.

8. If a player bets $1 on the $2 denomination, find the player's expectation.

9. The probability that a 44-year-old female in the U.S. will die within one year is approximately 0.002. If an insurance company sells a one-year, $25,000 life insurance policy to such a person for $75, what is the company's expectation?

10. The probability that a 54-year-old male in the U.S. will die within one year is approximately 0.006. If an insurance company sells a one-year, $20,000 life insurance policy to such a person for $155, what is the company's expectation?

11. The probability that an 80-year-old male in the U.S. will die within one year is approximately 0.073. If an insurance company sells a one-year, $10,000 life insurance policy to such a person for $495, what is the company's expectation?

12. The probability that an 80-year-old female in the U.S. will die within one year is approximately 0.051. If an insurance company sells a one-year, $15,000 life insurance policy to such a person for $860, what is the company's expectation?

13. The probability that a 65-year-old male in the U.S. will die within one year is about 0.02. An insurance company is preparing to sell a 65-year-old male a one-year, $30,000 life insurance policy. How much should it charge for its premium in order to have a positive expectation for the policy?

14. The probability that a 65-year-old female in the U.S. will die within one year is about 0.013. An insurance company is preparing to sell a 65-year-old female a one-year, $75,000 life insurance policy. How much should it charge for its premium in order to have a positive expectation for the policy?

15. A construction company has been hired to build a custom home. The builder estimates the probabilities of potential profit (or loss) as shown in the table below. What is the profit expectation for the company?

Profit/Loss	Probability
$100,000	0.10
$60,000	0.40
$30,000	0.25
0	0.15
−$20,000	0.08
−$40,000	0.02

16. A professional painter has been hired to paint a commercial building for $18,000. From this fee, the painter must buy supplies and pay employees. The painter estimates the potential profit as shown in

the table below. What is the profit expectation for the painter?

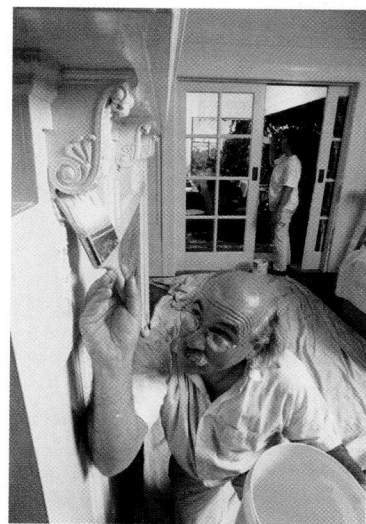

Profit/Loss	Probability
$10,000	0.15
$8,000	0.35
$5,000	0.2
$3,000	0.2
$1,000	0.1

17. A consultant has been hired to redesign a company's production facility. The consultant estimates the probabilities of her potential profit as shown in the table below. What is her profit expectation?

Profit/Loss	Probability
$40,000	0.05
$30,000	0.2
$20,000	0.5
$10,000	0.2
$5,000	0.05

18. A real estate company has purchased an office building with the intention of renting office space to small businesses. The company estimates the probabilities of potential profit (or loss) as shown in the table below. What is the profit expectation for the company?

Profit/Loss	Probability
$700,000	0.15
$400,000	0.25
$200,000	0.25
$50,000	0.20
−$100,000	0.10
−$250,000	0.05

19. On October 3, 2001, the jackpot in Florida's state lottery was $10 million. The probability of winning the jackpot (by choosing the correct six numbers) was 1 in 22,957,480. If you matched five of six numbers (a probability of 1 in 81,410), you would have won $8378.50. If you had matched four of six numbers (probability of 1 in 1416) you would have won $90.50, and if you had matched three numbers (probability of 1 in 71) you would have won $5.50. Assuming the jackpot was not split among multiple winners, find the expectation for buying a $1 lottery ticket.

20. A little league team has sold 500 raffle tickets at $2 each. One first prize of $300 will be awarded, along with three second prizes of $100. What is the expectation on the purchase of one ticket?

Extensions
CRITICAL THINKING

21. If a pair of regular dice are tossed once, use the expectation formula to determine the expected sum of the numbers on the upward faces of the two dice.

22. Consider rolling a pair of unusual dice, for which the faces have the number of pips indicated.

Die 1: {1, 2, 3, 4, 5, 6}
Die 2: {0, 0, 0, 6, 6, 6}

   a. List the sample space for the experiment.
   b. Compute the probability of each possible sum of the upward faces on the dice.
   c. What is the expected value of the sum of the numbers on the upward faces of the two dice?

23. Two dice, one labeled 1, 2, 2, 3, 3, 4 and the other labeled 1, 3, 4, 5, 6, 8, are rolled once. Use the formula for expectation to determine the expected sum of the numbers on the upward faces of the two dice. Dice such as these are called *Sicherman dice.*

24. Suppose you purchase a ticket for a prize and your expectation is −$1. What is the meaning of this expectation?

COOPERATIVE LEARNING

25. Suppose you are offered one of two pairs of dice, a red pair or a green pair, that are labeled as follows.

Red die 1: 0, 0, 4, 4, 4, 4

Red die 2: 2, 3, 3, 9, 10, 11

Green die 1: 3, 3, 3, 3, 3, 3

Green die 2: 0, 1, 7, 8, 8, 8

After you choose, your friend will receive the other pair. Which pair should you choose if you are going to play a game in which each of you rolls your dice and the player with the higher sum wins? Dice such as these are part of a set of four pairs of dice called *Efron's dice*. Explain why you chose the dice you did.

26. The PowerBall lottery commission chooses five white balls from a drum containing 49 balls marked with the numbers 1 through 49, and one red ball from a separate drum containing 42 balls. The following table shows the odds of winning a certain prize if the numbers you choose match those chosen by the lottery commission.

Match	Prize	Odds
⚪⚪⚪⚪⚪ + ⚫	Jackpot	1:80,089,128
⚪⚪⚪⚪⚪	$100,000	1:1,953,393
⚪⚪⚪⚪ + ⚫	$5,000	1:364,042
⚪⚪⚪⚪	$100	1:8,879
⚪⚪⚪ + ⚫	$100	1:8,466
⚪⚪⚪	$7	1:207
⚪⚪ + ⚫	$7	1:605
⚪ + ⚫	$4	1:118
⚪	$3	1:74

Overall odds: 1:35. Above odds based on $1 play.

Source: http://www.powerball.com/pbprizesNodds.shtm

a. Assuming that the jackpot for a certain drawing is $20,000,000, what is your expectation if you purchase one ticket for $1?

b. What is unusual about the two prizes for $7?

CHAPTER 11   **Summary**

Key Terms

Addition Rule [p. 732]
combination [p. 710]
combinatorics [p. 694]
complement of an event [p. 734]
conditional probability [p. 741]
counting with replacement [p. 699]
counting without replacement [p. 699]
counting principle [p. 698]
empirical probability (experimental probability) [p. 721]
equally likely outcomes [p. 719]
event [p. 694]
expectation [p. 753]
experiment [p. 694]
false positive [p. 747]
favorable outcome [p. 724]
independent events [p. 746]
multi-stage experiment [p. 695]
mutually exclusive events [p. 731]
n factorial, *n*! [p. 704]

odds [p. 724]
outcome [p. 694]
permutation [p. 705]
probability [p. 718]
Product Rule [p. 744]
Punett square [p. 722]
sample space [p. 694]
single-stage experiment [p. 695]
theoretical probability [p. 721]
tree diagram [p. 696]
unfavorable outcome [p. 724]

Essential Concepts

■ **Counting Principle**

Let *E* be a multi-stage experiment. If $n_1, n_2, n_3, \ldots, n_k$ are the number of possible outcomes of each of the *k* stages of *E*, then there are $n_1 \cdot n_2 \cdot n_3 \cdot \cdots \cdot n_k$ possible outcomes for *E*.

■ **Permutation Formula for Distinct Objects**
The number of permutations $P(n, k)$ of n distinct objects selected k at a time is

$$P(n, k) = \frac{n!}{(n - k)!}$$

■ **Permutations of Objects, Some of Which Are Identical**
The number of permutations of n objects of r different types, where k_1 identical objects are of one type, k_2 of another, and so on, is given by

$$\frac{n!}{k_1! \cdot k_2! \cdot \cdots \cdot k_r!}$$

where $k_1 + k_2 + \cdots + k_r = n$.

■ **Combination Formula**
The number of combinations of n objects chosen k at a time is

$$C(n, k) = \frac{n!}{k! \, (n - k)!}$$

■ **Probability of an Event**
For an experiment with sample space S of equally likely outcomes, the probability $P(E)$ of an event E is given by

$$P(E) = \frac{n(E)}{n(S)}$$

where $n(E)$ is the number of elements in the event and $n(S)$ is the number of elements in the sample space.

■ **Empirical Probability of an Event**
If an experiment is performed repeatedly and the occurrence of the event E is observed, the probability $P(E)$ of the event E is given by

$$P(E) = \frac{\text{number of times event } E \text{ occurred}}{\text{number of times the experiment was performed}}$$

■ **Odds of an Event**
Let E be an event in a sample space of equally likely events. Then

$$\text{Odds in favor of } E = \frac{\text{number of favorable outcomes}}{\text{number of unfavorable outcomes}}$$

$$\text{Odds against } E = \frac{\text{number of unfavorable outcomes}}{\text{number of favorable outcomes}}$$

■ **Converting Odds to Probability**
Suppose E is an event in a sample space and that the *odds in favor* of E are $\frac{a}{b}$. Then $P(E) = \frac{a}{a + b}$.

■ **Converting Probability to Odds**
Suppose E is an event in a sample space. Then the *odds in favor* of E are $\frac{P(E)}{1 - P(E)}$.

■ **Probability of Mutually Exclusive Events**
Two events A and B are mutually exclusive if they cannot occur at the same time. That is, A and B are mutually exclusive when $A \cap B = \varnothing$. In this case, the probability of A or B occurring is

$$P(A \cup B) = P(A) + P(B)$$

■ **Addition Rule for Probabilities**
Let A and B be two events in a sample space S. Then

$$P(A \cup B) = P(A) + P(B) - P(A \cap B)$$

■ **Probability of the Complement of an Event**
If E is an event and E^c is the complement of the event, then $P(E^c) = 1 - P(E)$.

■ **Conditional Probability Formula**
Let A and B be two events in a sample space S. Then the conditional probability of B given that A has occurred is

$$P(B \,|\, A) = \frac{P(A \cap B)}{P(A)}$$

■ **Product Rule for Probabilities**
If A and B are two events from the sample space S, then

$$P(A \text{ and } B) = P(A \cap B) = P(A) \cdot P(B \,|\, A)$$

■ **Probability of Successive Events**
The probability of two or more events occurring in succession is the product of the conditional probabilities of each of the events.

■ **Product Rule for Independent Events**
If A and B are two events in a sample space and $P(B \,|\, A) = P(B)$, then A and B are called independent events. In this case,

$$P(A \text{ and } B) = P(A \cap B) = P(A) \cdot P(B)$$

■ **Expectation**
Let $S_1, S_2, S_3, \ldots, S_n$ be the possible numerical outcomes of an experiment and let $P(S_1)$, $P(S_2)$, $P(S_3), \ldots, P(S_n)$ be the probabilities of those outcomes. Then the expectation of the experiment is

$$P(S_1) \cdot S_1 + P(S_2) \cdot S_2 + P(S_3) \cdot S_3 + \cdots + P(S_n) \cdot S_n$$

| CHAPTER 11 | **Review Exercises** |

In Exercises 1–4, list the elements of the sample space for the given experiment.

1. Two-digit numbers are formed, with replacement, from the digits 1, 2, and 3.

2. Two-digit numbers are formed, without replacement, from the digits 2, 6, and 8.

3. Use a tree diagram to list the number of outcomes that result from tossing four coins.

4. Use a table to list the number of two-character codes that can be formed from one of the digits 7, 8, or 9 followed by one of the letters A or B.

5. An athletic shoe store sells jogging shoes in three styles that come in four colors. Each color comes in six sizes. How many distinct shoes are available?

6. The combination for a lock to a bicycle chain contains four numbers chosen from the numbers 0 through 9. How many different combinations are possible?

7. Five characters in a registration code for the Adobe Systems FrameMaker program are formed by alternating a letter with a single nonzero digit: for instance, A3W4C. How many five-character sequences are possible?

8. A biquinary code is a code that consists of two distinct binary digits (a binary digit is a zero or one) followed by five binary digits for which there are no restrictions. How many biquinary codes are possible?

In Exercises 9–14, evaluate each expression.

9. $7!$

10. $8! - 4!$

11. $\dfrac{9!}{2!\,3!\,4!}$

12. $P(10, 6)$

13. $P(8, 3)$

14. $\dfrac{C(6, 2) \cdot C(8, 3)}{C(14, 5)}$

15. In how many different ways can seven people arrange themselves in a line to receive service from a bank teller?

16. A matching test has seven definitions that are to be paired with seven words. Assuming each word corresponds to exactly one definition, how many different matches are possible by random matching?

17. A matching test has seven definitions to be matched with five words. Assuming each word corresponds to exactly one definition, how many different matches are possible by random matching?

18. How many distinct arrangements are possible using the letters of the word *letter*?

19. Twelve identical coins are tossed. How many distinct arrangements are possible consisting of four heads and eight tails?

20. Three positions are open at a manufacturing plant: the day shift, the swing shift, and the night shift. In how many different ways can five people be assigned to the three shifts?

21. A professor assigns 25 homework problems, of which 10 will be graded. How many different sets of 10 problems can the professor choose to grade?

22. A stockbroker recommends 11 stocks to a client. If the client will invest in three of the stocks, how many different three-stock portfolios can be selected?

23. A quality control inspector receives a shipment of 15 computer monitors, of which three are defective. If the inspector randomly chooses five monitors, how many different sets can be formed that consist of three nondefective monitors and two defective monitors?

24. How many ways can nine people be seated in nine chairs if two of the people refuse to sit next to each other?

25. How many five-card poker hands consist of four of a kind (four aces, four kings, four queens, and so on)?

26. If it is equally likely that a child will be born a boy or a girl, compute the probability that a family of four children will have one boy and three girls.

27. If a coin is tossed three times, what is the probability of getting one head and two tails?

28. A large company currently employs 5739 men and 7290 women. If an employee is selected at random, what is the probability that the employee is a woman?

In Exercises 29 and 30, use the table below, which shows the number of students at a university who are currently in each class level.

Class level	Number of students
freshman	642
sophomore	549
junior	483
senior	445
graduate student	376

29. If a student is selected at random, what is the probability that the student is an upper-division undergraduate student (junior or senior)?

30. If a student is selected at random, what is the probability that the student is not a graduate student?

In Exercises 31–36, a pair of dice are tossed.

31. Find the probability that the sum of the pips on the two upward faces is 9.

32. Find the probability that that sum on the two dice is not 11.

33. Find the probability that the sum on the two dice is at least 10.

34. Find the probability that the sum on the two dice is an even number or a number less than 5.

35. What is the probability that the sum on the two dice is 9, given that the sum is odd?

36. What is the probability that the sum on the two dice is 8, given that doubles were rolled?

In Exercises 37–40, a single card is selected from a standard deck of playing cards.

37. What is the probability that the card is a heart or a black card?

38. What is the probability that the card is a heart or a jack?

39. What is the probability that the card is not a 3?

40. What is the probability that the card is red, given that it is not a club?

41. If a pair of dice are rolled, what are the odds in favor of getting a sum of 6?

42. If one card is drawn from a standard deck of playing cards, what are the odds that the card is a heart?

43. If the odds against an event occurring are 4 to 5, compute the probability of the event occurring.

44. The hair length of a particular rodent is determined by a dominant allele H, corresponding to long hair, and a recessive allele h, corresponding to short hair. Draw a Punnett square for parents of genotypes Hh and hh, and compute the probability that the offspring of the parents will have short hair.

45. Two cards are drawn from a standard deck of playing cards. The probability that exactly one card is an ace is 0.145. The probability that exactly one card is a face card (jack, queen, or king) is 0.362 and the probability that a selection of two cards will contain an ace or a face card is 0.489. Find the probability that the two cards are an ace and a face card.

46. A recent survey asked 1000 people whether they liked cheese-flavored corn chips (642 people), jalapeno-flavored chips (487 people), or both (302 people). If one person is chosen from this survey, what is the probability that the person does not like either of the two flavors?

In Exercises 47–51, a box contains 24 different colored chips that are identical in size. Five are black, four are red, eight are white, and seven are yellow.

47. If a chip is selected at random, what is the probability that the chip will be yellow or white?

48. If a chip is selected at random, what are the odds in favor of getting a red chip?

49. If a chip is selected at random, find the probability that the chip is yellow given that it is not white.

50. If five chips are randomly chosen, without replacement, what is the probability that none of them is red?

51. If three chips are chosen without replacement, find the probability that the first one is yellow, the second is white, and the third is yellow.

In Exercises 52–56, use the table below, which shows the number of voters in a city who voted for or against a proposition (or abstained from voting) according to their party affiliations.

	For	Against	Abstained
Democrat	8452	2527	894
Republican	2593	5370	1041
Independent	1225	712	686

52. If a voter is chosen at random, compute the probability that the person voted against the proposition.

53. If a voter is chosen at random, compute the probability that the person is a Democrat or an Independent.

54. If a voter is randomly chosen, what is the probability that the person abstained from voting on the proposition and is not a Republican?

55. A voter is randomly selected. What is the probability that the individual voted for the proposition, given that the voter is a registered Independent?

56. A voter is randomly selected. What is the probability that the individual is registered as a Democrat, given that the person voted against the proposition?

57. A single die is rolled three times in succession. What is the probability that each roll gives a 6?

58. A single die is rolled five times in succession. Find the probability that a 6 will be rolled at least once.

59. A single die is rolled five times in succession. Find the probability that exactly two of the rolls give a 6.

60. A person draws a card from a standard deck and replaces it; she then does this three more times. What is the probability that she drew a spade at least once?

61. A veterinarian uses a test to determine whether or not a dog has a disease that affects 7% of the dog population. The test correctly gives a positive result for 98% of dogs that have the disease, but gives false positives for 4% of dogs that do not have the disease. If a dog tests positive, what is the probability that the dog has the disease?

62. Suppose that in your area in the wintertime, if it rains one day there is a 65% chance that it will rain the next day. If it does not rain on a given day, there is only a 15% chance that it will rain the following day. What is the probability that, if it didn't rain today, it will rain the next two days but not the following two?

63. About 1.2% of AA batteries produced by a particular manufacturer are defective. If a consumer buys a box of 12 of these batteries, what is the probability that at least one battery is defective?

64. Suppose it costs $4 to play a game in which a single die is rolled and you win the amount of dollars that the die shows. What is the expectation for the game?

65. I will flip two coins. If both coins come up tails, I will pay you $5. If one shows heads and one shows tails,

you will pay me $2. If both coins come up heads, we will call it a draw. What is your expectation for this game?

66. For a fundraiser, an elementary school is selling 800 raffle tickets for $1 each. From these, five tickets will be drawn. One of the winners gets $200 and the others each get $75. If you buy one raffle ticket, what is your expectation?

67. If a pair of dice are rolled 65 times, on how many rolls can we expect to get a total of 4?

68. The probability that a 45-year-old in the U.S. will die within one year is approximately 0.003. If an insurance company sells a one-year, $40,000 life insurance policy to a 45-year-old for $320, what is the company's expectation?

69. The probability that a 74-year-old in the U.S. will die within one year is approximately 0.035. If an insurance company sells a one-year, $25,000 life insurance policy to a 74-year-old for $795, what is the company's expectation?

70. A construction company has bid on a building renovation project. The company estimates the probabilities of potential profit (or loss) as shown in the table below. What is the profit expectation for the company?

Profit/loss	Probability
$25,000	0.20
$15,000	0.25
$10,000	0.20
$5,000	0.15
$0	0.10
−$5,000	0.10

1. A certain driver's license number begins with one of the letters A, D, G, or K and is followed by one of the digits 2, 3, or 4. List the elements in the sample space of the first two digits of this driver's license number.

2. A computer system can be configured using one of three processors of different speeds, one of four disk drives of different sizes, one of three monitors, and one of two graphics cards. How many different computer systems are possible?

3. In a very simple computer chip, 10 wires go from one transistor to a second transistor. For the computer to function, instructions must be sent between these two transistors. In how many ways can four different instructions be sent between the two transistors if no two instructions can be sent along the same wire?

4. A matching test asks students to match 10 words with 15 definitions. If each word can be paired with only one definition, how many different pairs are possible?

5. Twenty computers are attached through a network to four printers. If any computer can be connected to any printer, how many different networks are possible?

6. A coin and a regular six-sided die are tossed together once. What is the probability that the coin shows a head or the die has a 5 on the upward face?

7. The probability that a certain brace A in a structure will break when a 1000-pound force is applied is 0.4. The probability that a brace B in the same structure will fail when the same force is applied is 0.5, and the probability that both braces will fail when the 1000-pound force is applied is 0.2. Are the events of A breaking and B breaking independent? Explain your answer.

8. What is the probability of drawing two cards in succession (without replacement) from a standard deck of playing cards and having them both be hearts?

9. Four cards are drawn from a standard deck of playing cards without replacement. What is the probability that none of them are 9's?

10. Three coins are tossed once. What are the odds in favor of the coins showing all heads?

11. A new medical test can determine whether or not a human has a disease that affects 5% of the population. The test correctly gives a positive result for 99% of people who have the disease, but gives a false positive for 3% of people who do not have the disease. If a person tests positive, what is the probability that the person has the disease?

12. The table below shows the number of men and the number of women who responded either positively or negatively to a new commercial. If one person is chosen from this group, find the probability that the person is a woman, given that the person responded negatively.

	Positive	Negative	Total
Men	684	736	1420
Women	753	642	1395
Total	1437	1378	2815

13. Straight or curly hair for a hamster is determined by a dominant allele S, corresponding to straight hair, and a recessive allele s, which gives curly hair. If one parent is of genotype Ss and the other is of genotype ss, compute the probability that the offspring of the parents will have curly hair.

14. A software company is preparing a bid to create a new inventory program for an auto parts company. The software company estimates the probabilities of potential profit (or loss) as shown in the table below. What is the profit expectation for the company?

Profit/loss	Probability
$75,000	0.18
$50,000	0.36
$25,000	0.31
0	0.08
−$10,000	0.05
−$20,000	0.02

Scientists have discovered that the speed of an animal can be closely estimated from the length of the stride of the animal. Scientists have used information to predict that some dinosaurs could attain speeds up to 32 miles per hour. In the **Excursion exercises** on **page 831** you will use the stride length of an animal to estimate its speed.

Statistics

Need help? For on-line student resources, such as section quizzes, visit this textbook's web site at **math.college.hmco.com/students.**

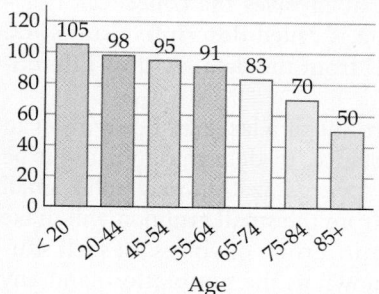

Men for Every 100 Women

Age

Source: U.S. Census Bureau, 2000

Every 10 years the U.S. Census Bureau collects data on the population of the U.S. It then issues *statistical* reports that indicate changes and trends in the U.S. population. For instance, according to the 2000 Census, there were approximately 105 men for every 100 women in the category of people under the age of 20. However, in the category of people 85 and older, there were only 50 men for every 100 women. See the graph at the left.

Here are some other statistics:

- The *mean* (average) commuting time to work is approximately 24 minutes. The mean is a statistical measure discussed in this chapter.

- Approximately 2% of the households in the U.S. have annual incomes in excess of $200,000. About 16% have annual incomes of $15,000 or less.

- In 1910, the mean annual family income in the United States was $687 a year. In 2000, the mean annual family income was $64,000. However, the *median* (another type of average) family income in 2000 was approximately $50,000. The difference between the mean and the median is discussed in this chapter.

In addition to the U.S. Census Bureau, there are a number of other organizations and groups that keep statistics. Here are some statistics from **www.puzzlegrid.com.**

- More education means longer life. Research shows that college graduates live longer than people who did not complete high school.

- The first Rolls-Royce sold for $600 in 1906. Today, the cheapest model goes for close to $200,000.

- More people are killed by donkeys annually than are killed in plane crashes.

- Eighty percent of all people hit by lightning are men.

- Chinese Mandarin is the most commonly spoken language in the world. Spanish follows in second place, English is third, and Bengali is fourth.

- In the famous Parker Brothers game Monopoly, the space on which a player has the greatest statistical chance of landing is Illinois Avenue. This is followed by the B&O Railroad, Free Parking, Tennessee Avenue, New York Avenue, and the Reading Railroad.

- In 1900, the life expectancy in the U.S. was 46.7 years for males and 48.9 years for females. Today, the average American male life expectancy is around 74 years. For American women the life expectancy is approximately 79 years. (*Source:* National Institute of Health)

- According to the American Medical Association, in 1970, approximately 7.6% of all physicians were women. In 2000, about 24% of all physicians were women. (*Source:* American Medical Society)

Measures of Central Tendency

The Arithmetic Mean

Statistics involves the collection, organization, summarization, presentation, and interpretation of data. The branch of statistics that involves the collection, organization, summarization, and presentation of data is called **descriptive statistics.** The branch that interprets and draws conclusions from the data is called **inferential statistics.**

Statisticians often collect data from small portions of a large group in order to determine information about the group. For instance, to determine who will be elected as the next president of the United States, an organization may poll a small group of voters. From the information it obtains from the small group, it will make conjectures about the voting preferences of the entire group of voters. In such situations the entire group under consideration is known as the **population,** and any subset of the population is called a **sample.**

Because of practical restraints such as time and money, it is common to apply descriptive statistical procedures to a sample of a population and then to make use of inferential statistics to deduce conclusions about the population. Obviously, some samples are more representative of the population than others.

One of the most basic statistical concepts involves finding *measures of central tendency* of a set of numerical data. Here is a scenario in which it would be helpful to find numerical values that locate, in some sense, the *center* of a set of data. Elle is a senior at a university. In a few months she plans to graduate and start a career as a graphic artist. A sample of five graphic artists from her class shows that they have received job offers with the following yearly salaries.

$41,000 $39,500 $34,000 $32,500 $30,500

Before Elle interviews for a job, she wishes to determine an *average* of these five salaries. This average should be a "central" number around which the salaries cluster. We will consider three types of averages, known as the *arithmetic mean,* the *median,* and the *mode.* Each of these averages is a **measure of central tendency** for numerical data.

The *arithmetic mean* is the most commonly used measure of central tendency. The arithmetic mean of a set of numbers is often referred to as simply the *mean.* To find the mean for a set of data, find the sum of the data values and divide by the number of data values. For instance, to find the mean of the five salaries listed above, divide the sum of the salaries by 5.

$$\text{Mean} = \frac{\$41{,}000 + \$39{,}500 + \$34{,}000 + \$32{,}500 + \$30{,}500}{5}$$

$$= \frac{\$177{,}500}{5} = \$35{,}500$$

In statistics it is often necessary to find the sum of a set of numbers. The traditional symbol used to indicate a summation is the Greek letter *sigma,* Σ. Thus the notation Σx denotes the sum of all the numbers in a given set. The use of summation notation enables us to define the mean as follows.

Mean

The **mean** of n numbers is the sum of the numbers divided by n.

$$\text{mean} = \frac{\Sigma x}{n}$$

It is traditional to denote the mean of a *sample* by $\bar{x}$ (which is read as "x bar") and to denote the mean of a *population* by the Greek letter μ (lower case mu).

EXAMPLE 1 ■ Find a Mean

Six friends in a biology class received test grades of

92,   84,   65,   76,   88,   and   90

Find the mean of these test scores.

Solution

$$\bar{x} = \frac{\Sigma x}{n} = \frac{92 + 84 + 65 + 76 + 88 + 90}{6} = \frac{495}{6} = 82.5$$

The mean of these test scores is 82.5.

CHECK YOUR PROGRESS 1   Four separate blood tests revealed that a patient had total blood cholesterol levels of

245,   235,   220,   and   210

Find the mean of the blood cholesterol levels.

Solution   See page S43.

From a physical perspective, numerical data can be represented by weights on a seesaw. The mean of the data is represented by the balance point of the seesaw. For instance, the mean of 1, 3, 5, 5, 5, and 8 is $4\frac{1}{2}$. If equal weights are placed at the locations 1, 3, and 8 and three weights are placed at 5, then the seesaw will balance at $4\frac{1}{2}$.

A Physical Interpretation of the Mean

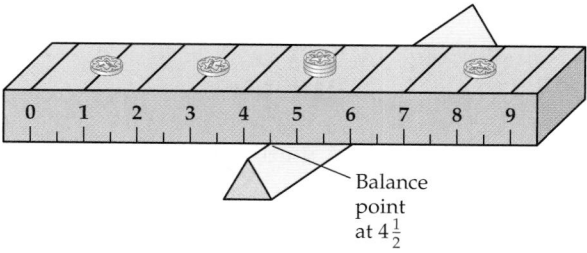

Balance point at $4\frac{1}{2}$

The Median

Another type of average is the *median*. Essentially, the median is the *middle number* or the *mean of the two middle numbers* in a list of numbers that have been arranged in numerical order from smallest to largest or from largest to smallest. Any list of numbers that is arranged in numerical order from smallest to largest or from largest to smallest is a **ranked list.**

▼ **point of interest**

The average price of the homes in a neighborhood is often stated in terms of the median price of the homes that have been sold over a given time period. The median price, rather than the mean, is used because it is easy to calculate and is less sensitive to extreme prices. The median price of a home can vary dramatically, even for areas that are relatively close. For instance, in December of 2001, the median price of a home in North Santa Barbara County was $225,760, whereas in Santa Barbara, South Coast, the median price was $552,730. (*Source:* California Association of Realtors)

> **Median**
>
> The **median** of a *ranked list* of n numbers is:
> - the middle number if n is odd.
> - the mean of the two middle numbers if n is even.

EXAMPLE 2 ■ **Find a Median**

Find the median for the data in the following lists.

a. 4, 8, 1, 14, 9, 21, 12   **b.** 46, 23, 92, 89, 77, 108

Solution

a. The list 4, 8, 1, 14, 9, 21, 12 contains seven numbers. The median of a list with an odd number of numbers is found by ranking the numbers and finding the middle number. Ranking the numbers from smallest to largest gives 1, 4, 8, 9, 12, 14, 21. The middle number is 9. Thus 9 is the median.

b. The list 46, 23, 92, 89, 77, 108 contains six numbers. The median of a list of data with an even number of numbers is found by ranking the numbers and computing the mean of the two middle numbers. Ranking the numbers from smallest to largest gives 23, 46, 77, 89, 92, 108. The two middle numbers are 77 and 89. The mean of 77 and 89 is 83. Thus 83 is the median of the data.

CHECK YOUR PROGRESS 2   Find the median for the data in the following lists.

a. 14, 27, 3, 82, 64, 34, 8, 51   **b.** 21.3, 37.4, 11.6, 82.5, 17.2

Solution   *See page S43.*

QUESTION   *The median of the ranked list 3, 4, 7, 11, 17, 29, 37 is 11. If the maximum value 37 is increased to 55, what effect will this have on the median?*

The Mode

A third type of average is the *mode.*

> **Mode**
>
> The **mode** of a list of numbers is the number that occurs most frequently.

ANSWER   *The median will remain the same because 11 will still be the middle number in the ranked list.*

Some lists of numbers do not have a mode. For instance, in the list 1, 6, 8, 10, 32, 15, 49, each number occurs exactly once. Because no number occurs more often than the other numbers, there is no mode.

A list of numerical data can have more than one mode. For instance, in the list 4, 2, 6, 2, 7, 9, 2, 4, 9, 8, 9, 7, the number 2 occurs three times and the number 9 occurs three times. Each of the other numbers occurs less than three times. Thus 2 and 9 are both modes for the data.

EXAMPLE 3 ■ **Find a Mode**

Find the mode for the data in the following lists.

a. 18, 15, 21, 16, 15, 14, 15, 21   **b.** 2, 5, 8, 9, 11, 4, 7, 23

Solution

a. In the list 18, 15, 21, 16, 15, 14, 15, 21, the number 15 occurs more often than the other numbers. Thus 15 is the mode.

b. Each number in the list 2, 5, 8, 9, 11, 4, 7, 23 occurs only once. Because no number occurs more often than the others, there is no mode.

CHECK YOUR PROGRESS 3   Find the mode for the data in the following lists.

a. 3, 3, 3, 3, 3, 4, 4, 5, 5, 5, 8   **b.** 12, 34, 12, 71, 48, 93, 71

Solution   *See page S43.*

The mean, the median, and the mode are all averages; however, they are generally not equal and they have different properties. The following summary illustrates some of the properties of each type of average.

Comparative Properties of the Mean, the Median, and the Mode

The *mean* of a set of data:

■ is the most sensitive of the averages. A change in any of the numbers changes the mean.

■ can be different from each of the numbers in the set.

■ can be changed drastically by changing an extreme value.

The *median* of a set of data:

■ is usually not changed by changing an extreme value.

■ is generally easy to compute.

The *mode* of a set of data:

■ may not exist, and when it does exist it may not be unique.

■ is one of the numbers in the set, provided a mode exists.

■ is generally not changed by changing an extreme value.

■ is generally easy to compute.

▼ **point of interest**

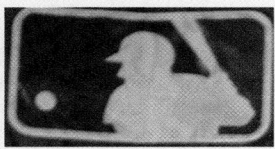

For some data, the mean and the median can differ by a large amount. For instance, during the 1994 baseball strike, the baseball team owners reported that the average (mean) player's salary was $1.2 million. This was true; however, the median salary was $500,000. (*Source: A Mathematician Reads the Newspaper*, 1995, by John Allen Paulos)

In the following example, we compare the mean, the median, and the mode for the salaries of five employees of a small company.

Salaries: $370,000 $60,000 $32,000 $16,000 $16,000

The sum of the five salaries is $494,000. Hence the mean is

$$\frac{\$494,000}{5} = \$98,800$$

The median is the middle number, $32,000. Because the $16,000 salary occurs the most, the mode is $16,000. The data contain one extreme value that is much larger than the other values. This extreme value makes the mean considerably larger than the median. Most of the employees of this company would probably agree that the median of $32,000 better represents the average of the salaries than does either the mean or the mode.

Math Matters   **Average Rate for a Round Trip**

Suppose you average 60 miles per hour on a one-way trip of 60 miles. On the return trip you average 30 miles per hour. You might be tempted to think that the average of 60 miles per hour and 30 miles per hour, which is 45 miles per hour, is the average rate for the entire trip. However, this is not the case. Because you were traveling more slowly on the return trip, the return trip took longer than the time spent going to your destination. More time was spent traveling at the slower speed. Thus the average rate for the round trip is less than the average (mean) of 60 miles per hour and 30 miles per hour.

To find the actual average rate for the round trip, use the formula

$$\text{Average rate} = \frac{\text{total distance}}{\text{total time}}$$

The total round-trip distance is 120 miles. The time spent going to your destination was 1 hour and the time spent on the return trip was 2 hours. The total time for the round trip was 3 hours. Thus

$$\text{Average rate} = \frac{\text{total distance}}{\text{total time}} = \frac{120}{3} = 40 \text{ miles per hour}$$

The Weighted Mean

A value called the *weighted mean* is often used when some data values are more important than others. For instance, many professors determine a student's course grade from the student's tests and the final examination. Consider the situation in which a professor counts the final examination score as two test scores. To find the weighted mean of the student's scores, the professor first assigns a weight to each score. In this case the professor could assign each of the test scores a weight of 1 and the final exam score a weight of 2. A student with test scores of 65, 70, and 75 and a final examination score of 90 has a weighted mean of

$$\frac{(65 \times 1) + (70 \times 1) + (75 \times 1) + (90 \times 2)}{5} = \frac{390}{5} = 78$$

Note that the numerator of the above weighted mean is the sum of the products of each test score and its corresponding weight. The number 5 in the denominator is the sum of all the weights. The above procedure can be generalized as follows.

The Weighted Mean

The **weighted mean** of the n numbers $x_1, x_2, x_3, \ldots, x_n$ with the respective assigned weights $w_1, w_2, w_3, \ldots, w_n$ is

$$\text{Weighted mean} = \frac{\Sigma(x \cdot w)}{\Sigma w}$$

where $\Sigma(x \cdot w)$ is the sum of the products formed by multiplying each number by its assigned weight, and Σw is the sum of all the weights.

Ⓟ

---

point of interest

Grade point averages (GPA) can be determined by using the weighted mean formula. See Exercises 21 and 22 page 778.

EXAMPLE 4 ■ Find a Weighted Mean

An instructor determines a student's weighted mean from quizzes, tests, and a project. Each test counts as four quizzes and the project counts as eight quizzes. Larry has quiz scores of 70 and 55. His test scores are 90, 72, and 68. His project score is 85. Find Larry's weighted mean for the course.

Solution

If we assign each quiz score a weight of 1, then each test score will have a weight of 4, and the project will have a weight of 8. The sum of all the weights is 22.

Weighted mean

$$= \frac{(70 \times 1) + (55 \times 1) + (90 \times 4) + (72 \times 4) + (68 \times 4) + (85 \times 8)}{22}$$

$$= \frac{1725}{22} \approx 78.4$$

Larry's weighted mean is approximately 78.4.

CHECK YOUR PROGRESS 4   Find Larry's weighted mean in Example 4 if a test score counts as two quiz scores and the project counts as six quiz scores.

Solution   See page S43.

Data that have not been organized or manipulated in any manner are called **raw data.** A large collection of raw data may not provide much pertinent information that can be readily observed. A **frequency distribution,** which is a table that lists observed events and the frequency of occurrence of each observed event, is often used to organize raw data. For instance, consider the following table, which lists the number of cable television connections for each of 40 homes in a subdivision.

Table 12.1 *Numbers of Cable Television Connections per Household*

2	0	3	1	2	1	0	4
2	1	1	7	2	0	1	1
0	2	2	1	3	2	2	1
1	4	2	5	2	3	1	2
2	1	2	1	5	0	2	5

The frequency distribution in Table 12.2 below was constructed using the data from Table 12.1. The first column of the frequency distribution consists of the numbers 0, 1, 2, 3, 4, 5, 6, and 7. The corresponding frequency of occurrence, f, of each of the numbers in the first column is listed in the second column.

Table 12.2 *A Frequency Distribution for Table 12.1*

Observed event Number of cable television connections, x	Frequency Number of households, f, with x cable television connections
0	5
1	12
2	14
3	3
4	2
5	3
6	0
7	1
	40 total

This row indicates that there are 14 households with two cable television connections.

The formula for a weighted mean can be used to find the mean of the data in a frequency distribution. The only change is that the weights w_1, w_2, w_3, ..., w_n are replaced with the frequencies $f_1, f_2, f_3, \ldots, f_n$. This procedure is illustrated in the next example.

> **EXAMPLE 5 ■ Find the Mean of Data Displayed in a Frequency Distribution**
>
> Find the mean of the data in Table 12.2.
>
> *Solution*
> The numbers in the right-hand column of Table 12.2 are the frequencies f for the numbers in the first column. The sum of all the frequencies is 40.
>
> Mean
> $$= \frac{\Sigma(x \cdot f)}{\Sigma f}$$
> $$= \frac{(0 \cdot 5) + (1 \cdot 12) + (2 \cdot 14) + (3 \cdot 3) + (4 \cdot 2) + (5 \cdot 3) + (6 \cdot 0) + (7 \cdot 1)}{40}$$
> $$= \frac{79}{40}$$
> $$= 1.975$$
>
> The mean number of cable connections per household for the homes in the subdivision is 1.975.

CHECK YOUR PROGRESS 5   A housing division consists of 45 homes. The following frequency distribution shows the number of homes in the subdivision that are two-bedroom homes, the number that are three-bedroom homes, the number that are four-bedroom homes, and the number that are five-bedroom homes. Find the mean number of bedrooms for the 45 homes.

Observed event Number of bedrooms, *x*	*Frequency* Number of homes with *x* bedrooms
2	5
3	25
4	10
5	5
	45 total

Solution   *See page S43.*

Excursion

Linear Interpolation and Animation

Linear interpolation is a method used to find a particular number between two given numbers. For instance, if a table lists the two entries 0.3156 and 0.8248, then the value exactly halfway between the numbers is the mean of the numbers, which is 0.5702. To find the number that is 0.2 of the way from 0.3156 to 0.8248, compute 0.2 times the difference between the numbers and, because the first number is smaller than the second number, add this result to the smaller number.

$$0.8248 - 0.3156 = 0.5092 \quad \longleftarrow \text{Difference between the table entries}$$
$$0.2 \cdot (0.5092) = 0.10184 \quad \longleftarrow \text{0.2 of the above difference}$$
$$0.3156 + 0.10184 = 0.41744 \quad \longleftarrow \text{Interpolated result, which is 0.2 of the way between the two table entries}$$

The above linear interpolation process can be used to find an intermediate number that is any specified fraction of the difference between two given numbers.

Excursion Exercises

1. Use linear interpolation to find the number that is 0.7 of the way from 1.856 to 1.972.

2. Use linear interpolation to find the number that is 0.3 of the way from 0.8765 to 0.8652. Note that because 0.8765 is larger than 0.8652, three-tenths of the difference between 0.8765 and 0.8652 must be subtracted from 0.8765 to find the desired number.

3. A calculator shows that $\sqrt{2} \approx 1.414$ and $\sqrt{3} \approx 1.732$. Use linear interpolation to estimate $\sqrt{2.4}$. *Hint:* Find the number that is 0.4 of the difference between 1.414 and 1.732 and add this number to the smaller number, 1.414. Round your estimate to the nearest thousandth.

4. We know that $2^1 = 2$ and $2^2 = 4$. Use linear interpolation to estimate $2^{1.2}$.

5. At the present time a football player weighs 325 pounds. There are 90 days until the player needs to report to spring training at a weight of 290 pounds. The player wants to lose weight at a constant rate. That is, the player wants to lose the same amount of weight each day of the 90 days. What weight, to the nearest tenth of a pound, should the player attain in 25 days?

Graphic artists use computer drawing programs, such as Adobe Illustrator, to draw the intermediate frames of an animation. For instance, in the following figure, the artist drew the small green apple on the left and the large ripe apple on the right. The drawing program used interpolation procedures to draw the five apples between the two apples drawn by the artist.

(continued)

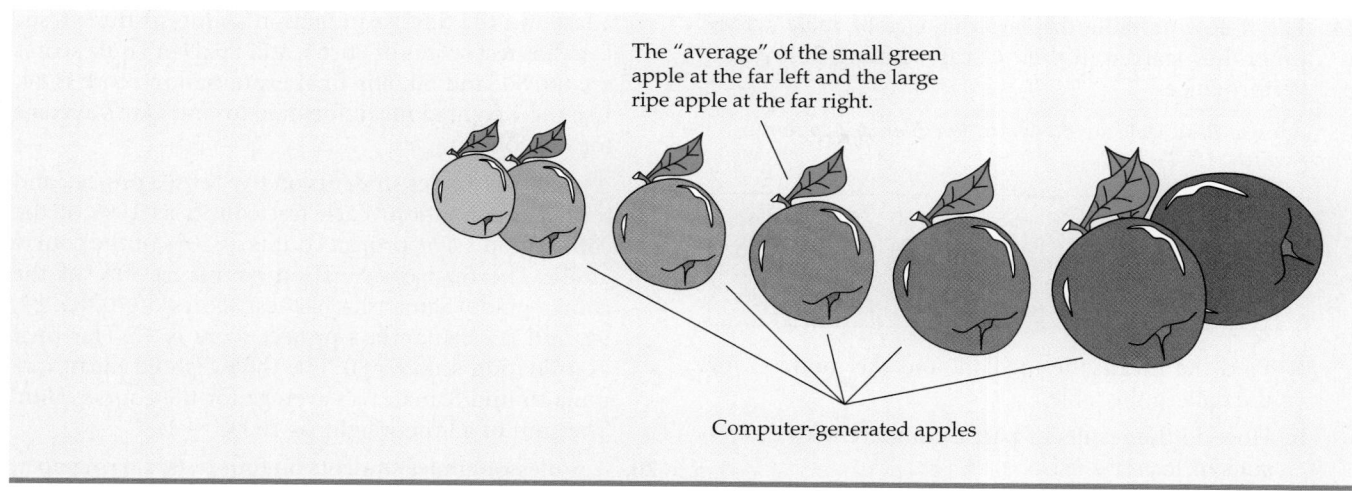

The "average" of the small green apple at the far left and the large ripe apple at the far right.

Computer-generated apples

Exercise Set 12.1   (Suggested Assignment: 1–37 odds)

In Exercises 1–10, find the mean, the median, and the mode(s), if any, for the given data. Round noninteger means to the nearest tenth.

1. 2, 7, 5, 7, 14
2. 8, 3, 3, 17, 9, 22, 19
3. 11, 8, 2, 5, 17, 39, 52, 42
4. 101, 88, 74, 60, 12, 94, 74, 85
5. 2.1, 4.6, 8.2, 3.4, 5.6, 8.0, 9.4, 12.2, 56.1, 78.2
6. 5, 5, 5, 5, 5, 5, 5, 5, 5, 5, 5, 5
7. 255, 178, 192, 145, 202, 188, 178, 201
8. 118, 105, 110, 118, 134, 155, 166, 166, 118
9. −12, −8, −5, −5, −3, 0, 4, 9, 21
10. −8.5, −2.2, 4.1, 4.1, 6.4, 8.3, 9.7

11. **a.** If one number in a set of data is changed, will this necessarily change the mean of the set? Explain.

b. If one number in a set of data is changed, will this necessarily change the median of the set? Explain.

12. If a set of data has a mode, then *must* the mode be one of the numbers in the set? Explain.

13. The following table displays the ages of female actors when they starred in their Oscar-winning Best Actor performances.

Ages of Best Female Actor Award Recipients, Academy Awards, 1971–2002

35	34	34	26	37	42	39	35	31	41	33
31	74	33	49	38	61	22	41	26	80	44
29	33	35	44	49	39	34	25	32	33	

Find the mean, the median, and the mode(s) for the data in the table.

14. The following table displays the ages of male actors when they starred in their Oscar-winning Best Actor performances.

Ages of Best Male Actor Award Recipients, Academy Awards, 1971–2002

62	43	40	48	48	56	38	60	30	40	42
37	76	38	52	44	35	61	43	51	31	42
54	53	36	37	31	45	60	40	35	47	

a. Find the mean, the median, and the mode(s) for the data in the table.

b. How do the results of part a compare with the results of Exercise 13?

15. Dental schools provide urban statistics to their students.

a. Use the following data to decide which of the two cities you would pick to set up your practice in.

Cloverdale: Population, 18,250;

median price of a home, $167,000;

dentists, 12; median age, 49;

mean number of patients, 1294.5

Barnbridge: Population, 27,840;

median price of a home, $204,400;

dentists, 17.5; median age, 53;

mean number of patients, 1148.7

b. Explain how you made your decision.

16. A salesperson records the following daily expenditures during a 10-day trip.

$185.34, $234.55, $211.86, $147.65, $205.60
$216.74, $1345.75, $184.16, $320.45, $88.12

In your opinion, does the mean or the median of the expenditures best represent the salesperson's average daily expenditure? Explain your reasoning.

17. A professor grades students on three tests, four quizzes, and a final examination. Each test counts as two quizzes and the final examination counts as two tests. Sara has test scores of 85, 75, and 90. Sara's quiz scores are 96, 88, 60, and 76. Her final examination score is 85. Use the weighted mean formula to find Sara's average for the course.

18. A professor grades students on four tests, four quizzes, and a final examination. Each quiz counts as one-half

a test and the final examination counts as three tests. Dan has test scores of 80, 65, and 86. Dan's quiz scores are 86, 80, and 60. His final examination score is 84. Use the weighted mean formula to find Dan's average for the course.

19. A professor grades students on five tests, a project, and a final examination. Each test counts as 10% of the course grade. The project counts as 20% of the course grade. The final examination counts as 30% of the course grade. Samantha has test scores of 70, 65, 82, 94, and 85. Samantha's project score is 92. Her final examination score is 80. Use the weighted mean formula to find Samantha's average for the course. *Hint:* The sum of all the weights is 100% = 1.

20. A professor grades students on four tests, a term paper, and a final examination. Each test counts as 15% of the course grade. The term paper counts as 20% of the course grade. The final examination counts as 20% of the course grade. Alan has test scores of 80, 78, 92, and 84. Alan received an 84 on his term paper. His final examination score was 88. Use the weighted mean formula to find Alan's average for the course.

Many colleges use the four-point grading system: A = 4, B = 3, C = 2, D = 1, and F = 0. In Exercises 21 and 22, use the weighted mean formula to compute the grade point average for each student. *Hint:* The units represent the "weights" of the letter grades that the student received in a course.

21. *Dillon's Grades, Fall Semester*

Course	Course grade	Course units
English	B	4
History	A	3
Chemistry	D	3
Algebra	C	4

22. *Janet's Grades, Spring Semester*

Course	Course grade	Course units
Biology	A	4
Statistics	B	3
Business Law	C	3
Psychology	F	2
CAD	B	2

In Exercises 23–26, find the mean, the median, and all modes for the data in the given frequency distribution.

23. *Points Scored by Lynn*

Points scored in a basketball game	Frequency
2	6
4	5
5	6
9	3
10	1
14	2
19	1

24. *Mystic Pizza Company*

Hourly pay rates for employees	Frequency
$8.00	14
$11.50	9
$14.00	8
$16.00	5
$19.00	2
$22.50	1
$35.00	1

25. *Quiz Scores*

Scores on a biology quiz	Frequency
2	1
4	2
6	7
7	12
8	10
9	4
10	3

26. *Ages of Science Fair Contestants*

Age	Frequency
7	3
8	4
9	6
10	15
11	11
12	7
13	1

Another measure of central tendency for a set of data is called the *midrange*. The **midrange** is defined as the value that is halfway between the minimum data value and the maximum data value. That is,

$$\text{Midrange} = \frac{\text{minimum value} + \text{maximum value}}{2}$$

The midrange is often stated as the *average* of a set of data in situations in which there are a large amount of data and the data are constantly changing. Many weather reports state the average daily temperature of a city as the midrange of the temperatures achieved during that day. For instance, if the minimum daily temperature of a city was 60° and the maximum daily temperature was 90°, then the midrange of the temperatures is $\frac{60° + 90°}{2} = 75°$.

27. Find the midrange of the following daily temperatures, which were recorded at three-hour intervals.

$$52°, 65°, 71°, 74°, 76°, 75°, 68°, 57°, 54°$$

28. Find the midrange of the following daily temperatures, which were recorded at three-hour intervals.

$$-6°, 4°, 14°, 21°, 25°, 26°, 18°, 12°, 2°$$

29. During a 24-hour period on January 23–24, 1916, the temperature in Browning, Montana decreased from a high of 44°F to a low of −56°F. Find the midrange of the temperatures during this 24-hour period. (*Source: Time Almanac 2002*, page 609)

30. During a two-minute period on January 22, 1943, the temperature in Spearfish, South Dakota increased from a low of −4°F to a high of 45°F. Find the midrange of the temperatures during this two-minute period. (*Source: Time Almanac 2002*, page 609)

31. After six biology tests, Ruben has a mean score of 78. What score does Ruben need on the next test to raise his average (mean) to 80?

32. After four algebra tests, Alisa has a mean score of 82. One more 100-point test is to be given in this class. All of the test scores are of equal importance. Is it possible for Alisa to raise her average (mean) to 90? Explain.

33. For the first half of a baseball season, a player had 92 hits out of 274 times at bat. The player's batting average was $\frac{92}{274} \approx 0.336$. During the second half of the season, the player had 60 hits out of 282 times at bat. The player's batting average was $\frac{60}{282} \approx 0.213$.

  a. What is the average (mean) of 0.336 and 0.213?

  b. What is the player's batting average for the complete season?

  c. Does the answer in part a equal the average in part b?

34. Mark averaged 60 miles per hour during the 30-mile trip to college. Because of heavy traffic he was able to average only 40 miles per hour during the return trip. What was Mark's average speed for the round trip?

Extensions

CRITICAL THINKING

35. The mean of 12 numbers is 48. Removing one of the numbers causes the mean to decrease to 45. What number was removed?

36. Find eight numbers such that the mean, the median, and the mode of the numbers are all 45, and no more than two of the numbers are the same.

37. The average rate for a trip is given by

$$\text{Average rate} = \frac{\text{total distance}}{\text{total time}}$$

If a person travels to a destination at an average rate of r_1 miles per hour and returns over the same route to the original starting point at an average rate of r_2 miles per hour, then show that the average rate for the round trip is

$$r = \frac{2r_1 r_2}{r_1 + r_2}$$

38. Pick six numbers and compute the mean and the median of the numbers.

  a. Now add 12 to each of your original numbers and compute the mean and the median for this new set of numbers.

  b. How does the mean of the new set of data compare with the mean of the original set of data?

  c. How does the median of the new set of data compare with the median of the original set of data?

COOPERATIVE LEARNING

Consider the data in the following table.

Summary of Yards Gained in Two Football Games

	Game 1	Game 2	Combined statistics for both games
Warren	12 yards on four carries Average: 3 yards/carry	78 yards on 16 carries Average: 4.875 yards/carry	90 yards on 20 carries Average: 4.5 yards/carry
Barry	120 yards on 30 carries Average: 4 yards/carry	100 yards on 20 carries Average: 5 yards/carry	220 yards on 50 carries Average: 4.4 yards/carry

- In the first game Barry has the best average.
- In the second game Barry has the best average.
- If the statistics for the games are combined, Warren has the best average.

You may be surprised by the above results. After all, how can it be that Barry has the best average in game 1 and game 2, but he does not have the best average for both games? In statistics, an example such as this is known as a **Simpson's paradox.**

Form groups of three or four students to work Exercises 39 to 41.

39. Consider the following data.

Batting Statistics for Two Baseball Players

	First month	Second month	Both months
Dawn	2 hits; 5 at-bats Average: ?	19 hits; 49 at-bats Average: ?	? hits; ? at-bats Average: ?
Joanne	29 hits; 73 at-bats Average: ?	31 hits; 80 at-bats Average: ?	? hits; ? at-bats Average: ?

Is this an example of a Simpson's paradox? Explain.

40. Consider the following data.

Test Scores for Two Students

	English	History	English and history combined
Wendy	84, 65, 72, 91, 99, 84 Average: ?	66, 84, 75, 77, 94, 96, 81 Average: ?	Average: ?
Sarah	90, 74 Average: ?	68, 78, 98, 76, 68, 92, 88, 86 Average: ?	Average: ?

Is this an example of a Simpson's paradox? Explain.

EXPLORATIONS

41. Create your own example of a Simpson's paradox.

SECTION 12.2 **Measures of Dispersion**

Table 12.3 *Test Scores*

Alan	Tara
55	80
80	76
97	77
80	83
68	84
100	80
Mean: 80	Mean: 80
Median: 80	Median: 80
Mode: 80	Mode: 80

The Range

In the preceding section we introduced the mean, the median, and the mode. Each of these statistics is a type of average that is designed to measure central tendencies of the data from which it was derived. Some characteristics of a set of data may not be evident from an examination of averages. For instance, consider the test scores for Alan and Tara, as shown in Table 12.3.

The mean, the median, and the mode of Alan's test scores and Tara's test scores are identical; however, an inspection of the test scores shows that Alan's scores are widely scattered, whereas all of Tara's scores are within a few units of the mean. This example shows that average values do not reflect the *spread* or *dispersion* of data. To measure the spread or dispersion of data, we must introduce statistical values known as the *range* and the *standard deviation*.

Robert Wadlow

> **Range**
>
> The **range** of a set of data values is the difference between the largest data value and the smallest data value.

EXAMPLE 1 ■ Find a Range

Find the range of Alan's test scores in Table 12.3.

Solution
Alan's largest test score is 100 and his smallest test score is 55. The range of Alan's test score is $100 - 55 = 45$.

CHECK YOUR PROGRESS 1   Find the range of Tara's test scores in Table 12.3.

Solution   See page S43.

MathMatters   **A World Record Range**

The tallest man for whom there is irrefutable evidence was Robert Pershing Wadlow. On June 27, 1940, Wadlow was 8 feet 11.1 inches tall. The shortest man for whom there is reliable evidence is Gul Mohammad. On July 19, 1990, he was 22.5 inches tall. (*Source:* Guinness World Records 2001) The range of the heights of these men is $107.1 - 22.5 = 84.6$ inches.

The Standard Deviation

220-Yard Dash (times in seconds)

Race	Sprinter 1	Sprinter 2
1	23.8	24.1
2	24.0	24.2
3	24.1	24.1
4	24.4	24.2
5	23.9	24.1
6	24.5	25.8
Range	0.7	1.7

The range of a set of data is easy to compute, but it can be deceiving. The range is a measure that depends only on the two most extreme values, and as such it is very sensitive. For instance, the table at the left shows the times for two sprinters in six track meets. The range of times for the first sprinter is 0.7 second, and the range for the second sprinter is 1.7 seconds. If you consider only range values, then you might conclude that the first sprinter's times are more consistent than those of the second sprinter. However, a closer examination shows that if you exclude the time of 25.8 seconds by the second sprinter in the sixth race, then the second sprinter has a range of 0.1 second. On this basis one could argue that the second sprinter has a more consistent performance record.

The next measure of dispersion that we will consider is called the *standard deviation*. It is less sensitive to a change in an extreme value than is the range. The standard deviation of a set of numerical data makes use of the individual amount that each data value deviates from the mean. These deviations, represented by $(x - \bar{x})$, are positive when the data value x is greater than the mean $\bar{x}$, and are negative when x is less than the mean $\bar{x}$. The sum of all the deviations $(x - \bar{x})$ is 0 for all sets of data. For instance, consider the sample data 2, 6, 11, 12, 14. For these data, $\bar{x} = 9$. The individual deviation of each data value from the mean is shown in the table on the following page. Note that the sum of the deviations is 0.

Because the sum of all the deviations of the data values from the mean is *always* 0, we cannot use the sum of the deviations as a measure of dispersion for a set of

Deviations from the Mean

x	$x - \bar{x}$
2	$2 - 9 = -7$
6	$6 - 9 = -3$
11	$11 - 9 = 2$
12	$12 - 9 = 3$
14	$14 - 9 = 5$

Sum of the deviations ⟶ 0

✓ **TAKE NOTE**

You may question why a denominator of $n - 1$ is used instead of n when we compute a sample standard deviation. The reason is because a sample standard deviation is often used to estimate the population standard deviation and it can be shown mathematically that the use of $n - 1$ tends to yield better estimates.

data. What is needed is a procedure that can be applied to the deviations so that the sum of the numbers that are derived by adjusting the deviations is not always 0. The procedure that we will make use of *squares* each of the deviations $(x - \bar{x})$ to make each of them nonnegative. The sum of the squares of the deviations is then divided by a constant that depends on the number of data values. We then compute the square root of this result. The following definitions show that the formula for calculating the standard deviation of a population differs slightly from the formula used to calculate the standard deviation of a sample.

> **Standard Deviations for Samples and Populations**
>
> If $x_1, x_2, x_3, \ldots, x_n$ is a *population* of n numbers with a mean of μ, then the **standard deviation** of the population is $\sigma = \sqrt{\dfrac{\Sigma(x - \mu)^2}{n}}$ (1).
>
> If $x_1, x_2, x_3, \ldots, x_n$ is a *sample* of n numbers with a mean of $\bar{x}$, then the **standard deviation** of the sample is $s = \sqrt{\dfrac{\Sigma(x - \bar{x})^2}{n - 1}}$ (2).

Most statistical applications involve a sample rather than a population, which is the complete set of data values. Sample standard deviations are designated by the lower-case letter s. In those cases in which we *do* work with a population, we designate the standard deviation of the population by σ, which is the lower-case Greek letter sigma. To calculate the standard deviation of n numbers, it is helpful to use the following procedure.

> **Procedure for Computing a Standard Deviation**
>
> 1. Determine the mean of the n numbers.
> 2. For each number, calculate the deviation (difference) between the number and the mean of the numbers.
> 3. Calculate the square of each of the deviations and find the sum of these squared deviations.
> 4. If the data is a *population,* then divide the sum by n. If the data is a *sample,* then divide the sum by $n - 1$.
> 5. Find the square root of the quotient in Step 4.

EXAMPLE 2 ■ Find the Standard Deviation

The following numbers were obtained by sampling a population.

   2, 4, 7, 12, 15

Find the standard deviation of the sample.

Solution
Step 1: The mean of the numbers is

$$\bar{x} = \frac{2 + 4 + 7 + 12 + 15}{5} = \frac{40}{5} = 8$$

Step 2: For each number, calculate the deviation between the number and the mean.

x	$x - \overline{x}$
2	$2 - 8 = -6$
4	$4 - 8 = -4$
7	$7 - 8 = -1$
12	$12 - 8 = 4$
15	$15 - 8 = 7$

Step 3: Calculate the square of each of the deviations in Step 2, and find the sum of these squared deviations.

x	$x - \overline{x}$	$(x - \overline{x})^2$
2	$2 - 8 = -6$	$(-6)^2 = 36$
4	$4 - 8 = -4$	$(-4)^2 = 16$
7	$7 - 8 = -1$	$(-1)^2 = 1$
12	$12 - 8 = 4$	$4^2 = 16$
15	$15 - 8 = 7$	$7^2 = \underline{49}$
		118 ← ——— The sum of the squared deviations

Step 4: Because we have a sample of $n = 5$ values, divide the sum 118 by $n - 1$, which is 4.

$$\frac{118}{4} = 29.5$$

Step 5: The standard deviation of the sample is $s = \sqrt{29.5}$. To the nearest hundredth, the standard deviation is $s = 5.43$.

CHECK YOUR PROGRESS 2 A student has the following quiz scores: 5, 8, 16, 17, 18, 20. Find the standard deviation for this population of quiz scores.

Solution See page S43.

In the next example we use standard deviations to determine which company produces batteries that are most consistent with regard to their life expectancy.

EXAMPLE 3 ■ Use Standard Deviations

A consumers group has tested a sample of eight size D batteries from each of three companies. The results of the tests are shown in the following table. According to these tests, which company produces batteries for which the values representing hours of constant use have the least standard deviation?

Company	Hours of Constant Use per Battery
EverSoBright	6.2, 6.4, 7.1, 5.9, 8.3, 5.3, 7.5, 9.3
Dependable	6.8, 6.2, 7.2, 5.9, 7.0, 7.4, 7.3, 8.2
Beacon	6.1, 6.6, 7.3, 5.7, 7.1, 7.6, 7.1, 8.5

INSTRUCTOR NOTE

You may wish to show your students the formulas

$$\sigma = \sqrt{\frac{\Sigma(x^2)}{n} - \bar{x}^2} \quad \text{and}$$

$$s = \sqrt{\frac{n(\Sigma x^2) - (\Sigma x)^2}{n(n-1)}}$$

These formulas provide a more efficient method of computing, by hand, the standard deviation of $x_1, x_2, \ldots, x_n$ for large values of n.

If your students have access to graphing calculators, the procedure shown in Example 4 will enable them to quickly and effortlessly find the standard deviation of a set of data.

Solution

The mean for each sample of batteries is 7 hours.

The batteries from EverSoBright have a standard deviation of

$$s_1 = \sqrt{\frac{(6.2 - 7)^2 + (6.4 - 7)^2 + \cdots + (9.3 - 7)^2}{7}}$$

$$= \sqrt{\frac{12.34}{7}} \approx 1.328 \text{ hours}$$

The batteries from Dependable have a standard deviation of

$$s_2 = \sqrt{\frac{(6.8 - 7)^2 + (6.2 - 7)^2 + \cdots + (8.2 - 7)^2}{7}}$$

$$= \sqrt{\frac{3.62}{7}} \approx 0.719 \text{ hours}$$

The batteries from Beacon have a standard deviation of

$$s_3 = \sqrt{\frac{(6.1 - 7)^2 + (6.6 - 7)^2 + \cdots + (8.5 - 7)^2}{7}}$$

$$= \sqrt{\frac{5.38}{7}} \approx 0.877 \text{ hours}$$

The batteries from Dependable have the *least* standard deviation. According to these results, the Dependable company produces the most consistent batteries with regard to life expectancy under constant use.

CHECK YOUR PROGRESS 3   A consumer testing agency has tested the strengths of three brands of $\frac{1}{8}$-inch rope. The results of the tests are shown in the following table. According to the sample test results, which company produces $\frac{1}{8}$-inch rope for which the breaking point has the least standard deviation?

Company	Breaking Point of $\frac{1}{8}$-inch Rope, in Pounds
Trustworthy	122, 141, 151, 114, 108, 149, 125
Brand X	128, 127, 148, 164, 97, 109, 137
NeverSnap	112, 121, 138, 131, 134, 139, 135

Solution   *See page S44.*

Many calculators have built-in features for calculating the mean and standard deviation of a set of numbers. The next example illustrates these features on a TI-83 graphing calculator.

EXAMPLE 4 ■ Use a Calculator to Find the Mean and Standard Deviation

Use a graphing calculator to find the mean and standard deviation of the times in the following table. Because the table contains all the winning times for this race (up to the year 2000), the data set is a population.

Olympic Women's 400-Meter Dash Results, in Seconds, 1964–2000

| 52.0 | 52.0 | 51.08 | 49.29 | 48.88 | 48.83 | 48.65 | 48.83 | 48.25 | 49.11 |

Solution

On a TI-83 calculator, press $\boxed{\text{STAT}}$ $\boxed{\text{ENTER}}$ and then enter the above times into list **[L1]**. See the calculator display below. Press $\boxed{\text{STAT}}$ $\boxed{\blacktriangleright}$ $\boxed{\text{ENTER}}$ $\boxed{\text{ENTER}}$. The calculator displays the mean and standard deviations shown below. Because we are working with a population, we are interested in the population standard deviation. From the calculator screen, $\bar{x} \approx 49.692$ and $\sigma x \approx 1.356$ seconds.

TI-83 Display of List 1

L1	L2	L3	1
52	- - - - - -	- - - - - -	
52			
51.08			
49.29			
48.88			
48.83			
48.65			
L1(1) = 52			

TI-83 Display of $\bar{x}$, s and σ

1-Var Stats
$\bar{x}$=49.692 ← Mean
Σx=496.92
Σx^2=24711.3398
Sx=1.429497192 ← Sample standard deviation
σx=1.356140111 ← Population standard deviation
↓n=10

CHECK YOUR PROGRESS 4   Use a calculator to find the mean and the population standard deviation of the race times in the following table.

Olympic Men's 400-Meter Dash Results, in Seconds, 1896–2000

54.2	49.4	49.2	53.2	50.0	48.2	49.6	47.6	47.8
46.2	46.5	46.2	45.9	46.7	44.9	45.1	43.8	44.66
44.26	44.60	44.27	43.87	43.50	43.49	43.84		

Solution   *See page S44.*

The Variance

A statistic known as the *variance* is also used as a measure of dispersion. The **variance** for a given set of data is the square of the standard deviation of the data. The following chart shows the mathematical notations that are used to denote standard deviations and variances.

Notations for Standard Deviation and Variance

σ is the standard deviation of a population.

σ^2 is the variance of a population.

s is the standard deviation of a sample.

s^2 is the variance of a sample.

EXAMPLE 5 ■ Find the Variance

Find the variance for the sample given in Example 2.

Solution

In Example 2, we found $s = \sqrt{29.5}$. Variance is the square of the standard deviation. Thus the variance is $s^2 = \left(\sqrt{29.5}\right)^2 = 29.5$.

CHECK YOUR PROGRESS 5   Find the variance for the population given in Check Your Progress 2.

Solution   *See page S44.*

QUESTION   *Can the variance of a data set be smaller than the standard deviation of the data set?*

Although the variance of a set of data is an important measure of dispersion, it does have a disadvantage that is not shared by the standard deviation: the variance does not have the same unit of measure as the original data. For instance, if a set of data consists of times measured in hours, then the variance of the data will be measured in *square* hours. The standard deviation of this data set is the square root of the variance, and as such it is measured in hours, which is a more intuitive unit of measure.

ANSWER   *Yes. The variance is smaller than the standard deviation whenever the standard deviation is less than 1.*

Excursion

A Geometric View of Variance and Standard Deviation[1]

The following geometric explanation of the variance and standard deviation of a set of data is designed to provide you with a deeper understanding of these important concepts.

Consider the data $x_1, x_2, \ldots, x_n$, which are arranged in ascending order. The average, or mean, of these data is

$$\mu = \frac{\Sigma x_i}{n}$$

and the variance is

$$s^2 = \frac{\Sigma(x_i - \mu)^2}{n}$$

In the last formula, each term $(x_i - \mu)^2$ can be pictured as the area of a square whose sides are of length $|x_i - \mu|$, the distance between the ith data value and the mean. We will refer to these squares as *tiles,* denoting by T_i the area of the tile associated with the data value x_i. Thus $\sigma^2 = \frac{\Sigma T_i}{n}$, which means that the variance may be thought of as the *area of the averaged-sized tile* and the standard deviation σ as the length of a side of this averaged-sized tile. By drawing the tiles associated with a data set, as shown below, you can visually estimate an averaged-size tile and thus you can roughly approximate the variance and standard deviation.

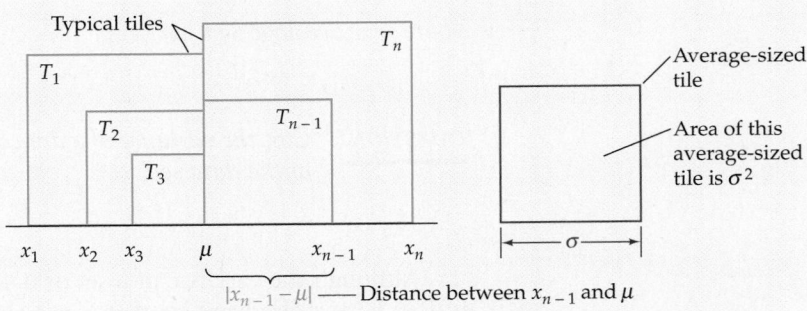

A typical data set, with its associated tiles and average-sized tile

These geometric representations of variance and standard deviation enable us to see visually how these values are used as measures of the dispersion of a set of data. If all of the data are bunched up near the mean, it is clear that the average-sized tile will be small and, consequently, so will its side length, which represents the standard deviation. But if even a small portion of the data lies far from the mean, the average-sized tile may be rather large, and thus its side length will also be large.

(continued)

1. Adapted with permission from "Chebyshev's Theorem: A Geometric Approach," *The College Mathematics Journal,* Vol. 26, No. 2, March 1995. Article by Pat Touhey, College Misericordia, Dallas, PA 18612.

Excursion Exercises

1. This exercise makes use of the geometric procedure explained on the previous page to calculate the variance and standard deviation of the population 2, 5, 7, 11, 15. The following figure shows the given set of data labeled on a number line, along with its mean, which is 8.

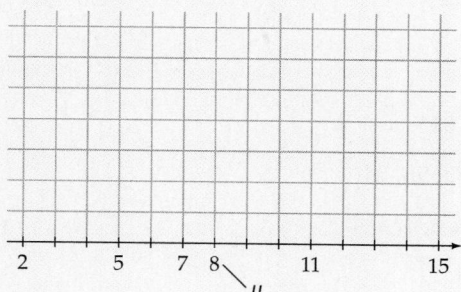
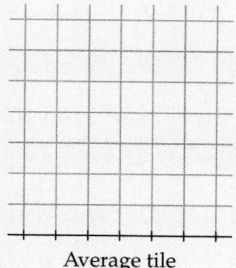

Average tile

a. Draw the tile associated with each of the five data values 2, 5, 7, 11, and 15.

b. Label each tile with its area.

c. Find the sum of the areas of all the tiles.

d. Find the average (mean) of the areas of all five tiles.

e. To the right of the above number line, draw a tile whose area is the average found in part d.

f. What is the variance of the data? What geometric figure represents the variance?

g. What is the standard deviation of the data? What geometric figure represents the standard deviation?

2. **a. to g.** Repeat all of the steps described in Excursion Exercise 1 for the data set

6, 8, 9, 11, 16

h. Which of the data sets in these two Excursion Exercises has the larger mean? Which data set has the larger standard deviation?

Exercise Set 12.2   (Suggested Assignment: 1–25 odds)

1. During a 12-hour period on December 24, 1924, the temperature in Fairfield, Montana dropped from a high of 63°F to a low of −21°F. What was the range of the temperatures during this period? (*Source: Time Almanac 2002*, page 609)

2. During a two-hour period on January 12, 1911, the temperature in Rapid City, South Dakota dropped from a high of 49°F to a low of −13°F. What was the range of the temperatures during this period? (*Source: Time Almanac 2002*, page 609)

In Exercises 3–12, find the range, the standard deviation, and the variance for the given *samples*. Round noninteger results to the nearest tenth.

3. 1, 2, 5, 7, 8, 19, 22

4. 3, 4, 7, 11, 12, 12, 15, 16

5. 2.1, 3.0, 1.9, 1.5, 4.8

6. 5.2, 11.7, 19.1, 3.7, 8.2, 16.3

7. 48, 91, 87, 93, 59, 68, 92, 100, 81

8. 93, 67, 49, 55, 92, 87, 77, 66, 73, 96, 54

9. 4, 4, 4, 4, 4, 4, 4, 4, 4, 4, 4, 4, 4, 4, 4, 4

10. 8, 6, 8, 6, 8, 6, 8, 6, 8, 6, 8, 6, 8

11. −8, −5, −12, −1, 4, 7, 11

12. −23, −17, −19, −5, −4, −11, −31

13. A mountain climber plans to buy some rope to use as a lifeline. Which of the following would be the better choice? Explain why you think your choice is the better choice.

Rope A: Mean breaking strength: 500 pounds; standard deviation of 300 pounds

Rope B: Mean breaking strength: 400 pounds; standard deviation of 40 pounds

14. Two longtime golfers are trying to break a course record of 66. Golfer A has an average of 75 with a standard deviation of 3. Golfer B has an average of 77 with a standard deviation of 5. Which golfer do you think has the better chance of breaking the course record in the next golf season? Explain.

15. Which would you expect to have the larger standard deviation: the resting pulse rates of the students in a history class, or the pulse rates of the same students after a 5-minute break period? Explain.

16. Which would you expect to have the larger standard deviation: the SAT scores of 30 students in a physics class at a large university, or the SAT scores of 30 students chosen at random from a community college? Explain.

17. The following table lists the winning and losing scores for all of the Super Bowl games up to the year 2001.

Super Bowl Results, 1967–2001

35–10	24–7	27–10	42–10	49–26
33–14	16–6	26–21	20–16	27–17
16–7	21–17	27–17	55–10	35–21
23–7	32–14	38–9	20–19	31–24
16–13	27–10	38–16	37–24	34–19
24–3	35–31	46–10	52–17	23–16
14–7	31–19	39–20	30–13	34–7

a. Find the mean and the *population* standard deviation of the winning scores. Round each result to the nearest tenth.

b. Find the mean and the *population* standard deviation of the losing scores. Round each result to the nearest tenth.

c. Which of the two data sets has the larger mean? Which of the two data sets has the larger standard deviation?

18. The following tables list the ages of female and male actors when they starred in their Oscar-winning Best Actor performances.

Ages of Best Female Actor Award Recipients, Academy Awards, 1971–2001

35	34	34	26	37	42	39	35	31	41	33
31	74	33	49	38	61	22	41	26	80	44
29	33	35	44	49	39	34	25	32		

*Ages of Best Male Actor Award Recipients,
Academy Awards, 1971–2001*

62	43	40	48	48	56	38	60	30	40	42
37	76	38	52	44	35	61	43	51	31	42
54	53	36	37	31	45	60	40	35		

a. Find the mean and the *sample* standard deviation of the ages of the female recipients. Round each result to the nearest tenth.

b. Find the mean and the *sample* standard deviation of the ages of the male recipients. Round each result to the nearest tenth.

c. Which of the two data sets has the larger mean? Which of the two data sets has the larger standard deviation?

19. The following tables list the numbers of home runs hit by the leaders in the National and American Leagues from 1971 to 2001.

Home Run Leaders, 1971–2001

National League										
48	40	44	36	38	38	52	40	48	48	
31	37	40	36	37	37	49	39	47	40	
38	35	46	43	40	47	49	70	65	50	73

American League										
33	37	32	32	36	32	39	46	45	41	
22	39	39	43	40	40	49	42	36	51	
44	43	46	40	50	52	56	56	48	47	52

a. Find the mean and the *population* standard deviation of the number of home runs hit by the leaders in the National League. Round each result to the nearest tenth.

b. Find the mean and the *population* standard deviation of the number of home runs hit by the leaders in the American League. Round each result to the nearest tenth.

c. Which of the two data sets has the larger mean? Which of the two data sets has the larger standard deviation?

20. The following table lists the winning times for the men's and women's Ironman Triathlon World Championships, held in Kailua-Kona, Hawaii.

*Ironman Triathlon World Championships
(Winning times rounded to the nearest minute)*

Men	(1978 – 2000)		Women	(1979 – 2000)	
11:47	8:29	8:20	12:55	9:35	9:17
11:16	8:34	8:21	11:21	9:01	9:07
9:25	8:31	8:04	12:01	9:01	9:32
9:38	8:09	8:33	10:54	9:14	9:24
9:08	8:28	8:24	10:44	9:08	9:13
9:06	8:19	8:17	10:25	8:55	9:26
8:54	8:09	8:21	10:25	8:58	
8:51	8:08		9:49	9:20	

a. Find the mean and the *population* standard deviation of the winning times of the female athletes. *Note:* Convert each time to hours. For instance, a time of 12:55 (12 hours 55 minutes) is equal to $12 + \frac{55}{60} = 12.91\overline{6}$ hours. Round each result to the nearest tenth.

b. Find the mean and the *population* standard deviation of the winning times of the male athletes. Round each result to the nearest tenth.

c. Which of the two data sets has the larger mean? Which of the two data sets has the larger standard deviation?

21. The table on the following page lists the U.S. presidents and their ages at inauguration. President Cleveland has two entries because he served two nonconsecutive terms.

Washington	57	J. Adams	61
Jefferson	57	Madison	57
Monroe	58	J. Q. Adams	57
Jackson	61	Van Buren	54
W. H. Harrison	68	Tyler	51
Polk	49	Taylor	64
Fillmore	50	Pierce	48
Buchanan	65	Lincoln	52
A. Johnson	56	Grant	46
Hayes	54	Garfield	49
Arthur	50	Cleveland	47, 55
B. Harrison	55	McKinley	54
T. Roosevelt	42	Taft	51
Wilson	56	Harding	55
Coolidge	51	Hoover	54
F. D. Roosevelt	51	Truman	60
Eisenhower	62	Kennedy	43
L. B. Johnson	55	Nixon	56
Ford	61	Carter	52
Reagan	69	G. H. W. Bush	64
Clinton	46	G. W. Bush	54

Source: Time Almanac 2002, page 61

Find the mean and the *population* standard deviation of the ages. Round each result to the nearest tenth.

22. The following table lists the deceased U.S. presidents as of March 2002, and their ages at death.

Washington	67	J. Adams	90
Jefferson	83	Madison	85
Monroe	73	J. Q. Adams	80
Jackson	78	Van Buren	79
W. H. Harrison	68	Tyler	71
Polk	53	Taylor	65
Fillmore	74	Pierce	64
Buchanan	77	Lincoln	56
A. Johnson	66	Grant	63
Hayes	70	Garfield	49
Arthur	56	Cleveland	71
B. Harrison	67	McKinley	58
T. Roosevelt	60	Taft	72
Wilson	67	Harding	57
Coolidge	60	Hoover	90
F. D. Roosevelt	63	Truman	88
Eisenhower	78	Kennedy	46
L. B. Johnson	64	Nixon	81

Source: Time Almanac 2002, page 61

Find the mean and the *population* standard deviation of the ages. Round each result to the nearest tenth.

Extensions
CRITICAL THINKING

23. Pick five numbers and compute the *population* standard deviation of the numbers.

a. Add a nonzero constant c to each of your original numbers and compute the standard deviation of this new population.

b. Use the results of part a and inductive reasoning to state what happens to the standard deviation of a population when a nonzero constant c is added to each data item.

24. Pick six numbers and compute the *population* standard deviation of the numbers.

a. Double each of your original numbers and compute the standard deviation of this new population.

b. Use the results of part a and inductive reasoning to state what happens to the standard deviation of a population when each data item is multiplied by a positive constant k.

25. a. All of the numbers in a sample are the same number. What is the standard deviation of the sample?

b. If the standard deviation of a sample is 0, must all of the numbers in the sample be the same number?

c. If two samples both have the same standard deviation, are the samples necessarily identical?

26. Under what condition would the variance of a sample be equal to the standard deviation of the sample?

EXPLORATIONS

27. a. ▣ Use a calculator to compare the standard deviation of the population 1, 2, 3, 4, 5 and the value $\sqrt{\dfrac{5^2 - 1}{12}}$.

b. ▣ Use a calculator to compare the standard deviation of the population 1, 2, 3, ..., 10 and the value $\sqrt{\dfrac{10^2 - 1}{12}}$.

c. ▣ Use a calculator to compare the standard deviation of the population 1, 2, 3, ..., 15 and the value $\sqrt{\dfrac{15^2 - 1}{12}}$.

d. Make a conjecture about the standard deviation of the population 1, 2, 3, ..., n and the value $\sqrt{\dfrac{n^2 - 1}{12}}$.

28. Find, without using a calculator, the standard deviation of the population

$$3001, 3002, 3003, 3004, \ldots, 3010$$

Hint: Use your answer to part b of Exercise 23 and your conjecture from part d of Exercise 27.

SECTION 12.3 | **Measures of Relative Position**

z-Scores

When you take a course in college, it is natural to wonder how you will do compared to the other students. Will you finish in the top 10% or will you be closer to the middle? One statistic that is often used to measure the position of a data value with respect to other values is known as the *z-score* or the *standard score*.

> **z-Score**
>
> The **z-score** for a given data value x is the number of standard deviations that x is above or below the mean of the data. The following formulas show how to calculate the z-score for a data value x in a population and in a sample.
>
> Population: $z_x = \dfrac{x - \mu}{\sigma}$ Sample: $z_x = \dfrac{x - \bar{x}}{s}$

QUESTION   *Must the z-score for a data value be a positive number?*

In the next example, we use a student's z-scores for two tests to determine how well the student did on each test in comparison to the other students.

ANSWER   *No. The z-score for a data value x is positive if x is greater than the mean, it is 0 if x is equal to the mean, and it is negative if x is less than the mean.*

EXAMPLE 1 ■ Compare z-Scores

Raul has taken two tests in his chemistry class. He scored 72 on the first test, for which the mean of all scores was 65 and the standard deviation was 8. He received a 60 on a second test, for which the mean of all scores was 45 and the standard deviation was 12. In comparison to the other students, did Raul do better on the first test or the second test?

Solution
Find the z-score for each test.

$$z_{72} = \frac{72 - 65}{8} = 0.875 \qquad z_{60} = \frac{60 - 45}{12} = 1.25$$

Raul scored 0.875 standard deviation above the mean on the first test and 1.25 standard deviations above the mean on the second test. These z-scores indicate that in comparison to his classmates, Raul scored better on the second test than he did on the first test.

CHECK YOUR PROGRESS 1 Cheryl has taken two quizzes in her history class. She scored 15 on the first quiz, for which the mean of all scores was 12 and the standard deviation was 2.4. Her score on the second quiz, for which the mean of all scores was 11 and the standard deviation was 2.0, was 14. In comparison to her classmates, did Cheryl do better on the first quiz or the second quiz?

Solution See page S44.

The z-score equation $z_x = \dfrac{x - \bar{x}}{s}$ involves four variables. If the values of any three of the four variables are known, you can solve for the unknown variable. This procedure is illustrated in the next example.

EXAMPLE 2 ■ Use z-Scores

A consumers group tested a sample of 100 light bulbs. It found that the mean life expectancy of the bulbs was 842 hours, with a standard deviation of 90. One particular light bulb from the DuraBright Company had a z-score of 1.2. What was the life span of this light bulb?

Solution
Substitute the given values into the z-score equation and solve for *x*.

$$z_x = \frac{x - \bar{x}}{s}$$
$$1.2 = \frac{x - 842}{90}$$
$$108 = x - 842$$
$$950 = x$$

The light bulb had a life span of 950 hours.

CHECK YOUR PROGRESS 2 Roland received a score of 70 on a test for which the mean score was 65.5. Roland has learned that the z-score for his test is 0.6. What is the standard deviation for this set of test scores?

Solution See page S44.

Percentiles

Most standardized examinations provide scores in terms of *percentiles,* which are defined as follows.

> **pth Percentile**
>
> A value x is called the **pth percentile** of a data set provided $p\%$ of the data values are less than x.

EXAMPLE 3 ■ Use Percentiles

In 1999, the median yearly salary for orthopedic surgeons was $293,525. (*Source:* American Medical Group Association, Medical Group Compensation and Productivity Survey, 2000) If the 90th percentile for the 1999 yearly salary of orthopedic surgeons was $360,000, find the percentage of surgeons whose 1999 yearly salary was

a. more than $293,525.

b. less than $360,000.

c. between $293,525 and $360,000.

Solution

a. The median is by definition the 50th percentile. Therefore, 50% of the surgeons earned more than $293,525.

b. Because $360,000 is the 90th percentile, we know that 90% of the surgeons earned less than $360,000.

c. From parts a and b, we know that 90% − 50% = 40% of the surgeons earned between $293,525 and $360,000.

CHECK YOUR PROGRESS 3   In 1999, the median yearly salary for pediatricians was $139,307. (*Source:* American Medical Group Association, Medical Group Compensation and Productivity Survey, 2000) If the 20th percentile for the 1999 yearly salary of pediatricians was $95,000, find the percentage of pediatricians whose 1999 yearly salary was

a. more than $139,307.

b. more than $95,000.

c. between $95,000 and $139,307.

Solution   See page S44.

The following formula can be used to find the percentile that corresponds to a particular data value in a set of data.

> **Percentile for a Given Data Value**
>
> Given a set of data and a data value x,
>
> $$\text{Percentile of score } x = \frac{\text{number of data values less than } x}{\text{total number of data values}} \cdot 100$$

EXAMPLE 4 ■ **Find a Percentile**

On a reading examination given to 900 students, Elaine's score of 602 was higher than the scores of 576 of the students who took the examination. What is the percentile for Elaine's score?

Solution

$$\text{Percentile} = \frac{\text{number of data values less than 602}}{\text{total number of data values}} \cdot 100$$

$$= \frac{576}{900} \cdot 100$$

$$= 64$$

Elaine's score of 602 places her at the 64th percentile.

CHECK YOUR PROGRESS 4   On an examination given to 8600 students, Hal's score of 405 was higher than the scores of 3952 of the students who took the examination. What is the percentile for Hal's score?

Solution   *See page S44.*

Math Matters   Standardized Tests and Percentiles

Standardized tests, such as the Scholastic Assessment Test (SAT), are designed to measure all students by a *single* standard. The SAT is used by many colleges as part of their admissions criteria. The SAT I is a three-hour examination that measures verbal and mathematical reasoning skills. Scores on each portion of the test range from 200 to 800 points. SAT scores are generally reported in points and percentiles. Sometimes students are confused by the percentile score. For instance, if a student scores 650 points on the mathematics portion of the SAT and is told that this score is in the 85th percentile, this *does not* indicate that the student answered 85% of the questions correctly. An 85th percentile score means that the student scored higher than 85% of the students who took the test. Consequently, the student scored lower than 15% (100% − 85%) of the students who took the test.

Quartiles

The three numbers Q_1, Q_2, and Q_3 that partition a data set into four (approximately) equal portions are called the **quartiles** of the data. For instance, for the data set below, the values $Q_1 = 11$, $Q_2 = 29$, and $Q_3 = 104$ are the quartiles of the data.

2, 5, 5, 8, 11, 12, 19, 22, 23, 29, 31, 45, 83, 91, 104, 159, 181, 312, 354

Q_1 Q_2 Q_3

The quartile Q_1 is called the *first quartile*. The quartile Q_2 is called the *second quartile*. It is the median of the data. The quartile Q_3 is called the *third quartile*. The following method of finding quartiles makes use of medians.

> **The Median Procedure for Finding Quartiles**
>
> **1.** Rank the data.
>
> **2.** Find the median of the data. This is the second quartile, Q_2.
>
> **3.** The first quartile, Q_1, is the median of the data values smaller than Q_2. The third quartile, Q_3, is the median of the data values larger than Q_2.

EXAMPLE 5 ■ Use Medians to Find the Quartiles of a Data Set

The following table lists the calories per 100 milliliters of 25 popular beers. Find the quartiles for the data.

Calories, per 100 milliliters, of Selected Beers

43	37	42	40	53	62	36	32	50	49
26	53	73	48	45	39	45	48	40	56
41	36	58	42	39					

Solution

Step 1: Rank the data as shown in the following table.

1) 26	**2)** 32	**3)** 36	**4)** 36	**5)** 37	**6)** 39	**7)** 39	**8)** 40	**9)** 40
10) 41	**11)** 42	**12)** 42	**13)** 43	**14)** 45	**15)** 45	**16)** 48	**17)** 48	**18)** 49
19) 50	**20)** 53	**21)** 53	**22)** 56	**23)** 58	**24)** 62	**25)** 73		

Step 2: The median of these 25 data values has a rank of 13. Thus the median is 43. The second quartile Q_2 is the median of the data, so $Q_2 = 43$.

Step 3: There are 12 data values less than the median and 12 data values greater than the median. The first quartile is the median of the data values less than the median. Thus Q_1 is the average of the data values with ranks of 6 and 7.

$$Q_1 = \frac{39 + 39}{2} = 39$$

The third quartile is the median of the data values greater than the median. Thus Q_3 is the average of the data values with ranks of 19 and 20.

$$Q_3 = \frac{50 + 53}{2} = 51.5$$

CHECK YOUR PROGRESS 5   The following table lists the weights, in ounces, of 15 avocados in a random sample. Find the quartiles for the data.

Weights, in ounces, of Avocadoes

12.4	10.8	14.2	7.5	10.2	11.4	12.6	12.8	13.1	15.6
9.8	11.4	12.2	16.4	14.5					

Solution   *See page S44.*

Box-and-Whisker Plots

A **box-and-whisker plot** (sometimes called a **box plot**) is often used to provide a visual summary of a set of data. A box-and-whisker plot shows the median, the first and third quartiles, and the minimum and maximum values of a data set. See the figure below.

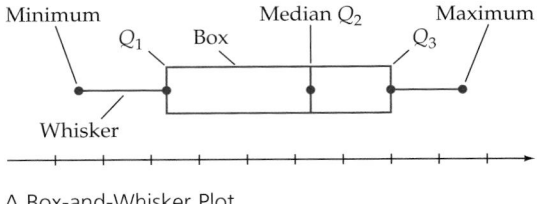

A Box-and-Whisker Plot

Construction of a Box-and-Whisker Plot

1. Draw a horizontal scale that extends from the minimum data value to the maximum data value.

2. Above the scale, draw a rectangle (box) with its left side at Q_1 and its right side at Q_3.

3. Draw a vertical line segment across the rectangle at the median, Q_2.

4. Draw a horizontal line segment, called a **whisker,** that extends from Q_1 to the minimum and another whisker that extends from Q_3 to the maximum.

EXAMPLE 6 ■ Construct a Box-and-Whisker Plot

Construct a box-and-whisker plot for the data set in Example 5.

Solution
For the data set in Example 5, we determined that $Q_1 = 39$, $Q_2 = 43$, and $Q_3 = 51.5$. The minimum data value for the data set is 26 and the maximum data value is 73. Thus the box-and-whisker plot is as shown below.

Figure 12.1 *A box-and-whisker plot of the data in Example 5*

CHECK YOUR PROGRESS 6   Construct a box-and-whisker plot for the following set of test scores.

The Number of Rooms Occupied in a Resort during an 18-day Period

86	77	58	45	94	96	83	76	75
65	68	72	78	85	87	92	55	61

Solution   *See page S44.*

Box plots have become popular because they are easy to construct and they illustrate several important features of a data set in a simple diagram. Note from Figure 12.1 that we can easily estimate

- the quartiles of the data.
- the range of the data.
- the position of the middle half of the data as shown by the length of the box.

Some graphing calculators, such as the TI-83, can be used to produce box-and-whisker plots. For instance, on a TI-83, you enter the data into a list as shown in Figure 12.2. The $\boxed{\texttt{WINDOW}}$ menu is used to enter appropriate boundaries that contain all the data. Use the key sequence $\boxed{\texttt{2nd}}$ $\boxed{\texttt{[STAT PLOT]}}$ $\boxed{\texttt{ENTER}}$ and choose from the $\texttt{Type}$ menu the box-and-whisker plot icon. The $\boxed{\texttt{GRAPH}}$ key is then used to display the box-and-whisker plot. After the calculator displays a box-and-whisker plot, the $\boxed{\texttt{TRACE}}$ key and the $\boxed{\blacktriangleright}$ key enable you to view Q_1, Q_2, Q_3, and the minimum and maximum of your data set.

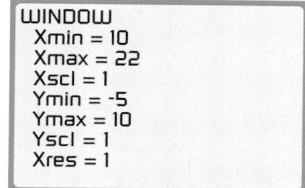

TAKE NOTE

The following data was used to produce the box plot shown to the right.

21.2, 20.5, 17.0, 16.8, 16.8, 16.5, 16.2, 14.0, 13.7, 13.3, 13.1, 13.0, 12.4, 12.1, 12.0

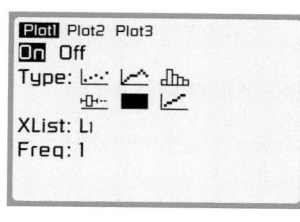

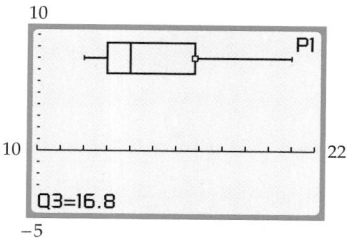

Figure 12.2 *TI-83 Screen Displays*

Excursion

Stem-and-Leaf Diagrams

The relative position of each data value in a small set of data can be graphically displayed by using a *stem-and-leaf diagram*. For instance, consider the following history test scores:

65, 72, 96, 86, 43, 61, 75, 86, 49, 68, 98, 74, 84, 78, 85, 75, 86, 73

In the stem-and-leaf diagram on the following page, we have organized the history test scores by placing all of the scores that are in the 40s in the top row, the scores that are

(continued)

A Stem-and-Leaf Diagram of a Set of History Test Scores

Stems	Leaves
4	3 9
5	
6	1 5 8
7	2 3 4 5 5 8
8	4 5 6 6 6
9	6 8

Legend: 8│6 represents 86

in the 50s in the second row, the scores that are in the 60s in the third row, and so on. The tens digits of the scores have been placed to the left of the vertical line. In this diagram they are referred to as **stems.** The ones digits of the test scores have been placed in the proper row to the right of the vertical line. In this diagram they are the **leaves.** It is now easy to make observations about the distribution of the scores. Only two of the scores are in the 90s. Six of the scores are in the 70s, and none of the scores are in the 50s. The lowest score is 43 and the highest is 98.

Steps in the Construction of a Stem-and-Leaf Diagram

1. Determine the stems and list them in a column from smallest to largest.

2. List the remaining digit of each stem as a leaf to the right of the stem.

3. Include a *legend* that explains the meaning of the stems and the leaves. Include a *title* for the diagram.

The choice of how many leading digits to use as the stem will depend on the particular data set. For instance, consider the following data set, in which a travel agent has recorded the amount spent by customers for a cruise.

Amount Spent for a Cruise, Summer of 2003

$3600	$4700	$7200	$2100	$5700	$4400	$9400
$6200	$5900	$2100	$4100	$5200	$7300	$6200
$3800	$4900	$5400	$5400	$3100	$3100	$4500
$4500	$2900	$3700	$3700	$4800	$4800	$2400

One method of choosing the stems is to let each thousands digit be a stem and each hundreds digit be a leaf. If the stems and leaves are assigned in this manner, then the notation 2│1, with a stem of 2 and a leaf of 1, represents a cost of $2100, and 5│4 represents a cost of $5400. A stem-and-leaf diagram can now be constructed by writing all of the stems in a column from smallest to largest to the left of a vertical line, and writing the corresponding leaves to the right of the line.

Amount Spent for a Cruise

Stems	Leaves
2	1 1 4 9
3	1 1 6 7 7 8
4	1 4 5 5 7 8 8 9
5	2 4 4 7 9
6	2 2
7	2 3
8	
9	4

Legend:
7│3 represents $7300

(continued)

Sometimes two sets of data can be compared by using a **back-to-back stem-and-leaf diagram,** in which common stems are listed in the middle column of the diagram. Leaves from one data set are displayed to the right of the stems, and leaves from the other data set are displayed to the left. For instance, the back-to-back stem-and-leaf diagram below shows the test scores for two classes that took the same test. It is easy to see that the 8 A.M. class did better on the test because it had more scores in the 80s and 90s and fewer scores in the 40s, 50s, and 60s. The number of scores in the 70s was the same for both classes.

Biology Test Scores

8 A.M. class		10 A.M. class
2	4	5 8
7	5	6 7 9 9
5 8	6	2 3 4 8
1 2 3 3 3 7 8	7	1 3 3 5 5 6 8
4 4 5 5 6 8 8 9	8	2 3 6 6 6
2 4 5 5 8	9	4 5

Legend: 3 | 7   Legend: 8 | 2
represents 73   represents 82

Excursion Exercises

1. The following table lists the ages of customers who purchased a cruise. Construct a stem-and-leaf diagram for the data.

Ages of Customers Who Purchased a Cruise

32	45	66	21	62	68	72
61	55	23	38	44	77	64
46	50	33	35	42	45	51
51	28	40	41	52	52	33

2. Construct a back-to-back stem-and-leaf diagram for the winning and losing scores given in Exercise 17 of Section 12.2 (page 790). What information is revealed by your diagram?

3. Construct a back-to-back stem-and-leaf diagram for the data given in the two tables in Exercise 18 of Section 12.2 (pages 790–791). What information is revealed by your diagram?

4. Construct a back-to-back stem-and-leaf diagram for the data given in the two tables in Exercise 19 of Section 12.2 (page 791). What information is revealed by your diagram?

Exercise Set 12.3   (Suggested Assignment: 1–23 odds)

In Exercises 1–4, round each z-score to the nearest hundredth.

1. A data set has a mean of $\bar{x} = 75$ and a standard deviation of 11.5. Find the z-score for each of the following.

    a. $x = 85$ **b.** $x = 95$

    c. $x = 50$ **d.** $x = 75$

2. A data set has a mean of $\bar{x} = 212$ and a standard deviation of 40. Find the z-score for each of the following.

    a. $x = 200$ **b.** $x = 224$

    c. $x = 300$ **d.** $x = 100$

3. A data set has a mean of $\bar{x} = 6.8$ and a standard deviation of 1.9. Find the z-score for each of the following.

    a. $x = 6.2$ **b.** $x = 7.2$

    c. $x = 9.0$ **d.** $x = 5.0$

4. A data set has a mean of $\bar{x} = 4010$ and a standard deviation of 115. Find the z-score for each of the following.

    a. $x = 3840$ **b.** $x = 4200$

    c. $x = 4300$ **d.** $x = 4030$

5. A blood pressure test was given to 450 women ages 20 to 36. It showed that their mean systolic blood pressure was 119.4 mm Hg, with a standard deviation of 13.2 mm Hg.

    a. Determine the z-score, to the nearest hundredth, for a woman who had a systolic blood pressure reading of 110.5 mm Hg.

    b. The z-score for one woman was 2.15. What was her systolic blood pressure reading?

6. A random sample of 1000 oranges showed that the mean amount of juice per orange was 7.4 fluid ounces, with a standard deviation of 1.1 fluid ounces.

    a. Determine the z-score, to the nearest hundredth, of an orange that produced 6.6 fluid ounces of juice.

    b. The z-score for one orange was 3.15. How much juice was produced by this orange? Round to the nearest tenth of a fluid ounce.

7. A test involving 380 men ages 20 to 24 found that their blood cholesterol levels had a mean of 182 mg/dl and a standard deviation of 44.2 mg/dl.

    a. Determine the z-score, to the nearest hundredth, for one of the men who had a blood cholesterol level of 214 mg/dl.

    b. The z-score for one man was −1.58. What was his blood cholesterol level?

8. A random sample of 80 tires showed that the mean mileage per tire was 41,700 miles, with a standard deviation of 4300 miles.

    a. Determine the z-score, to the nearest hundredth, for a tire that provided 46,300 miles of wear.

    b. The z-score for one tire was −2.44. What mileage did this tire provide? Round your result to the nearest hundred miles.

9. Which of the following three test scores is the highest relative score?

    a. A score of 65 on a test with a mean of 72 and a standard deviation of 8.2

    b. A score of 102 on a test with a mean of 130 and a standard deviation of 18.5

    c. A score of 605 on a test with a mean of 720 and a standard deviation of 116.4

10. Which of the following fitness scores is the highest relative score?

    a. A score of 42 on a test with a mean of 31 and a standard deviation of 6.5

    b. A score of 1140 on a test with a mean of 1080 and a standard deviation of 68.2

    c. A score of 4710 on a test with a mean of 3960 and a standard deviation of 560.4

11. On a reading test, Shaylen's score of 455 was higher than the scores of 4256 of the 7210 students who took the test. Find the percentile, rounded to the nearest percent, for Shaylen's score.

12. On a placement examination, Rick scored lower than 1210 of the 12,860 students who took the exam. Find the percentile, rounded to the nearest percent, for Rick's score.

13. Kevin scored at the 65th percentile on a test given to 9840 students. How many students scored lower than Kevin?

14. Rene scored at the 84th percentile on a test given to 12,600 students. How many students scored higher than Rene?

15. In 1999, the median four-person family income was $59,981. (*Source: Time Almanac 2002,* page 635.) If the 88th percentile for the 1999 median four-person family income was $73,400, find the percentage of four-person families whose 1999 income was

   a. more than $59,981.

   b. more than $73,400.

   c. between $59,981 and $73,400.

16. In 1992, the median prison sentence for robbery was 6.0 years. (*Source:* F.B.I.) If the first quartile for prison sentences for robbery was 3.6 years, then find the percentage of prison sentences for robbery that were

   a. more than 3.6 years.

   b. less than 6.0 years.

   c. between 3.6 years and 6.0 years.

17. A survey was given to 18 students. One question asked about the one-way distance the student had to travel to attend college. The results, in miles, are shown in the following table. Use the median procedure for finding quartiles to find the first, second, and third quartiles for the data.

Distance Traveled to Attend College								
12	18	4	5	26	41	1	8	10
10	3	28	32	10	85	7	5	15

18. The following table shows the number of prescriptions a doctor wrote each day for a 36-day period. Use the median procedure for finding quartiles to find the first, second, and third quartiles for the data.

Number of Prescriptions Written per Day					
8	12	14	10	9	16
7	14	10	7	11	16
11	12	8	14	13	10
9	14	15	12	10	8
10	14	8	7	12	15
14	10	9	15	10	12

19. The following table shows the earned run averages (ERAs) for the baseball teams in the American League and the National League during the 2001 season. Draw box-and-whisker plots for the teams' ERAs in each league. Place one plot above the other so that you can compare the data. Write a few sentences that explain any differences you found by using the box-and-whisker plots.

Team Pitching Statistics, 2001 Regular Season Earned Run Averages (ERAs)

American League		National League	
Seattle	3.54	Atlanta	3.59
Oakland	3.59	Arizona	3.87
NY Yankees	4.02	St. Louis	3.93
Boston	4.15	Chicago Cubs	4.03
Anaheim	4.20	NY Mets	4.07
Toronto	4.28	Philadelphia	4.15
Minnesota	4.51	San Francisco	4.18
Chicago Sox	4.55	Los Angeles	4.25
Cleveland	4.64	Florida	4.32
Baltimore	4.67	Houston	4.37
Kansas City	4.87	San Diego	4.52
Tampa Bay	4.94	Milwaukee	4.64
Detroit	5.01	Montreal	4.68
Texas	5.71	Cincinnati	4.77
		Pittsburgh	5.05
		Colorado	5.29

Source: ESPN Baseball
http://sports.espn.go.com/mlb/statistics

20. Draw a box-and-whisker plot for the data in each of the following tables. Place one plot above the other so that you can compare the data. Write a few sentences that explain any differences you found by using the box-and-whisker plots.

Points Scored by the Denver Broncos in the 2001 Season									
31	38	13	20	21	10	31	28	26	10
26	10	20	23	23	10				

Points Scored by the Jacksonville Jaguars in the 2001 Season									
21	13	14	15	10	17	24	30	7	21
21	14	15	33	26	13				

21. The table below shows the percentage of males and females who smoked for selected years from 1965 to 1998. Draw box-and-whisker plots for the male percentages and the female percentages. Place one plot above the other so that you can compare the data. Write a few sentences that explain any differences you found.

Smoking Prevalence Among U.S. Adults, 1965–1998 (as a percent of the population 18 years of age and older)

Year	Males	Females
1965	51.2%	33.7%
1974	42.8%	32.2%
1983	34.8%	29.4%
1985	32.2%	27.9%
1990	28.0%	22.9%
1992	28.1%	24.6%
1993	27.3%	22.6%
1994	27.6%	23.1%
1995	26.5%	22.7%
1997	27.1%	22.2%
1998	25.9%	22.1%

Source: Centers for Disease Control and Prevention, *Time Almanac 2002*, page 562

22. The following table shows the median sales prices of existing single-family homes in the United States in the four regions of the country for the years 1990 through 1996. Prices have been rounded to the nearest thousand.

a. Draw a box-and-whisker plot of the data for each of the four regions. Write a few sentences that explain any differences you found.

b. Use the Internet to find the median price of a single-family home in your region of the United States for the current year. How does this median price compare with the 1996 median price shown in the following table?

Median Prices of Houses Sold in the United States

Year	Northeast	Midwest	South	West
1990	141	74	86	140
1991	142	78	89	147
1992	140	82	92	144
1993	140	85	95	143
1994	139	88	96	147
1995	137	94	98	147
1996	141	100	103	152

Source: National Association of Realtors

Extensions

CRITICAL THINKING

23. a. The population 3, 4, 9, 14, and 20 has a mean of 10 and a standard deviation of 6.356. The z-scores for each of the five data values are $z_3 \approx -1.101$, $z_4 \approx -0.944$, $z_9 \approx -0.157$, $z_{14} \approx 0.629$, and $z_{20} \approx 1.573$. Find the mean and the standard deviation of these z-scores.

b. The population 2, 6, 12, 17, 22, and 25 has a mean of 14 and a standard deviation of 8.226. The z-scores for each of the six data values are $z_2 \approx -1.459$, $z_6 \approx -0.973$, $z_{12} \approx -0.243$, $z_{17} \approx 0.365$, $z_{22} \approx 0.973$, and $z_{25} \approx 1.337$. Find the mean and the standard deviation of these z-scores.

c. Use the results of part a and part b to make a conjecture about the mean and standard deviation of the z-scores for any set of data.

24. For each of the following, determine whether the statement is true or false.

a. For any given set of data, the median of the data equals the mean of Q_1 and Q_3.

b. For any given set of data, $Q_3 - Q_2 = Q_2 - Q_1$.

c. A z-score for a given data value x in a set of data can be a negative number.

d. If a student answers 75% of the questions on a test correctly, then the student's score on the test will place the student at the 75th percentile.

25. Some data sets include values so large or so small that they differ significantly from the rest of the data. Such data values are referred to as **outliers.** An outlier may be the result of an error, such as an incorrect measurement or a recording error, or it may be a legitimate data value. Consult a statistics text to find the mathematical formula that is used to determine whether a given data value in a set of data is an outlier. Use the formula to find all outliers for the data set in Exercise 17 page 803.

SECTION 12.4   Normal Distributions

Frequency Distributions and Histograms

Large sets of data are often displayed using a *grouped frequency distribution* or a *histogram.* For instance, consider the following situation. An Internet service provider (ISP) has installed new computers. To estimate the new download times its subscribers will experience, the ISP surveyed 1000 of its subscribers to determine the time required for each subscriber to download a particular file from the Internet site **music.net.** The results of that survey are summarized in the table below.

Table 12.4 *A grouped frequency distribution with 12 classes*

Download time (in seconds)	Number of subscribers
0–5	6
5–10	17
10–15	43
15–20	92
20–25	151
25–30	192
30–35	190
35–40	149
40–45	90
45–50	45
50–55	15
55–60	10

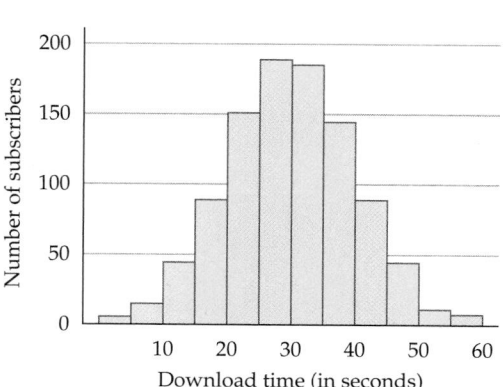

Figure 12.3 *A histogram for the frequency distribution in Table 12.4*

Table 12.4 is called **a grouped frequency distribution.** It shows how often (frequently) certain events occurred. Each interval, 0–5, 5–10, and so on, is called a **class.** This distribution has 12 classes. For the 10–15 class, 10 is the **lower class**

boundary and 15 is the **upper class boundary.** Any data value that lies on a common boundary is assigned to the higher class. The *graph* of a frequency distribution is called a **histogram.** A histogram provides a pictorial view of how the data are distributed. In Figure 12.3 on the previous page, the height of each bar of the histogram indicates how many subscribers experienced the download times shown by the class on the base of the bar.

Examine the distribution in Table 12.5. It shows the *percent* of subscribers that are in each class, as opposed to the frequency distribution in Table 12.4 on the previous page, which shows the *number* of customers in each class. The type of frequency distribution that lists the *percent* of data in each class is called a **relative frequency distribution.** The **relative frequency histogram** in Figure 12.4 was drawn by using the data in the relative frequency distribution. It shows the *percent* of subscribers along its vertical axis.

Table 12.5 *A relative frequency distribution*

Download time (in seconds)	Percent of subscribers
0–5	0.6
5–10	1.7
10–15	4.3
15–20	9.2
20–25	15.1
25–30	19.2
30–35	19.0
35–40	14.9
40–45	9.0
45–50	4.5
50–55	1.5
55–60	1.0

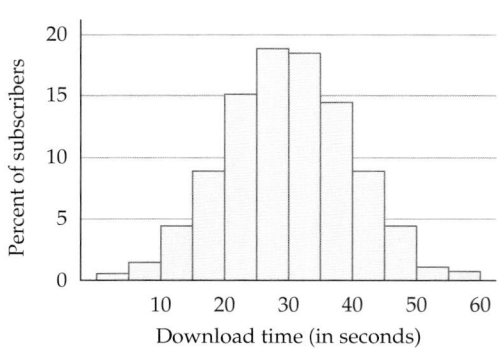

Figure 12.4 *A relative frequency histogram*

One advantage of using a relative frequency distribution instead of a grouped frequency distribution is that there is a direct correspondence between the percent values of the relative frequency distribution and probabilities. For instance, in the relative frequency distribution in Table 12.5, the percent of the data that lies between 35 and 40 seconds is 14.9%. Thus, if a subscriber is chosen at random, the probability that the subscriber will require at least 35 seconds but less than 40 seconds to download the music file is 0.149.

EXAMPLE 1 ■ Use a Relative Frequency Distribution

Use the relative frequency distribution in Table 12.5 to determine the

a. *percent* of subscribers who required at least 25 seconds to download the file.

b. *probability* that a subscriber chosen at random will require at least 5 but less than 20 seconds to download the file.

Solution

a. The percent of data in all the classes with a lower boundary of 25 seconds or more is the sum of the percent for all of the classes highlighted in red in the distribution below. Thus the percent of subscribers who required at least 25 seconds to download the file is 69.1%. See Table 12.6.

Table 12.6

Download time (in seconds)	Percent of subscribers	
0–5	0.6	
5–10	1.7	
10–15	4.3	Sum is 15.2%
15–20	9.2	
20–25	15.1	
25–30	19.2	
30–35	19.0	
35–40	14.9	
40–45	9.0	Sum is 69.1%
45–50	4.5	
50–55	1.5	
55–60	1.0	

b. The percent of data in all the classes with a lower boundary of at least 5 seconds and an upper boundary of 20 seconds or less is the sum of the percents in all of the classes highlighted in blue in the distribution above. Thus the percent of subscribers who required at least 5 but less than 20 seconds to download the file is 15.2%. The probability that a subscriber chosen at random will require at least 5 but less than 20 seconds to download the file is 0.152. See Table 12.6.

CHECK YOUR PROGRESS 1 Use the relative frequency distribution in Table 12.5 to determine the

a. *percent* of subscribers who required less than 25 seconds to download the file.

b. *probability* that a subscriber chosen at random will require at least 10 seconds but less than 30 seconds to download the file.

Solution *See page S45.*

There is a geometric analogy between percent of data and probabilities and the relative frequency histogram for the data. For instance, the percent of data described in part a of Example 1 corresponds to the *area* represented by the red bars in the histogram in Figure 12.5. The percent of data described in part b of Example 1 corresponds to the area represented by the blue bars in the histogram in Figure 12.6.

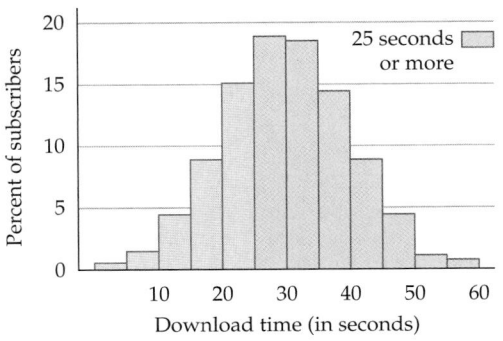

Figure 12.5

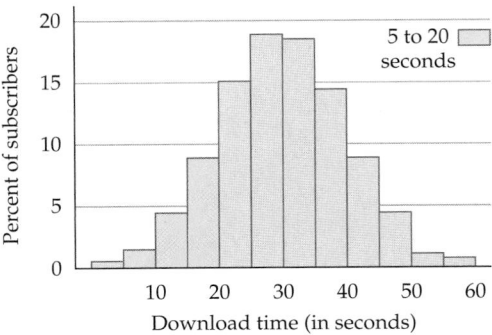

Figure 12.6

Normal Distributions and The Empirical Rule

A histogram of a set of data provides us with a tool that can indicate patterns or trends in the distribution of the data. The terms *uniform, bimodal, symmetrical, skewed* and *normal* are used to describe the distribution of sets of data.

A **uniform distribution,** shown in the figure below, occurs when all of the observed events occur with the same frequency. The graph of a uniform distribution remains at the same height over the range of the data. Some random processes produce distributions that are uniform or nearly uniform. For example, if the spinner below is used to generate numbers, then in the *long run* each of the numbers 1, 2, 3, ..., 8 will be generated with approximately the same frequency.

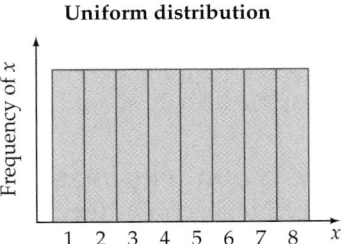

Uniform distribution

Random number generator

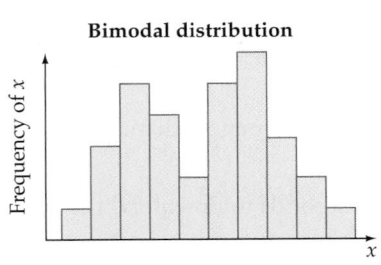

Bimodal distribution

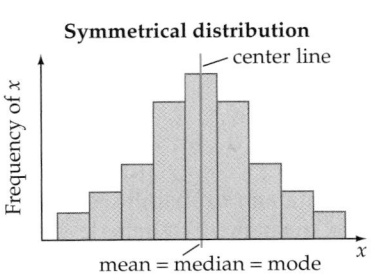

Symmetrical distribution

A **bimodal distribution,** shown in the histogram at the left, is produced when two *nonadjacent* classes occur more frequently than any of the other classes. Bimodal distributions are often produced by a sample that contains data from two very different populations.

A **symmetrical distribution,** shown at the left, is symmetrical about a vertical center line. If you fold a symmetrical distribution along the center line, the right side of the distribution will match the left side. The following sets of data are examples of distributions that are nearly symmetrical: the weights of all male students, the heights of all teenage females, the prices of a gallon of regular gasoline in a large city, the mileages for a particular type of automobile tire, and the amounts of soda dispensed by a vending machine per day. In a symmetrical distribution, the mean, the median, and the mode are all equal and are located at the center of the distribution.

Skewed distributions, shown in the figures below, can be identified by the fact that their distributions have a longer *tail* on one side of the distribution and a shorter tail on the other side. A distribution is skewed to the *left* if it has a longer tail on the left and is skewed to the *right* if it has a longer tail on the right. In a distribution that is skewed to the left, the mean is less than the median, which is less than the mode. In a distribution that is skewed to the right, the mode is less than the median, which is less than the mean.

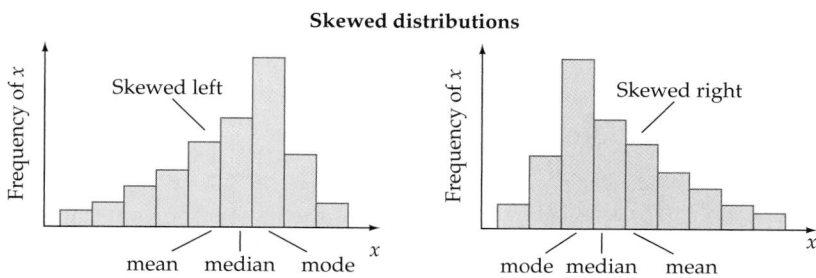

Skewed distributions

Many examinations yield test scores that have skewed distributions. For instance, if a test designed for students in the sixth grade is given to students in a ninth-grade class, most of the scores will be high. The distribution of the test scores will be skewed to the left, as shown above.

Discrete data are separated from each other by an increment, or "space." For instance, only whole numbers are used to record the number of points that a basketball player scores in a game. The possible points that the player can score, which we will represent by s, is restricted to 0, 1, 2, 3, 4, ... The variable s is a **discrete variable.** Different scores are separated from each other by at least 1. A variable that is based on counting procedures is a discrete variable. Histograms are generally used to show the distributions of discrete variables.

Continuous data can take on the values of all real numbers in some interval. For example, the possible times that it takes to drive to the grocery store are continuous data. The times are not restricted to natural numbers such as 4 minutes or 5 minutes. In fact, the time may be any part of a minute, or even of a second, if we care to measure that precisely. A variable such as the time t, that is based on *measuring* with smaller and smaller units, is called a **continuous variable.** Continuous curves, rather than histograms, are used to show the distributions of continuous variables.

Distributions of Continuous Variables

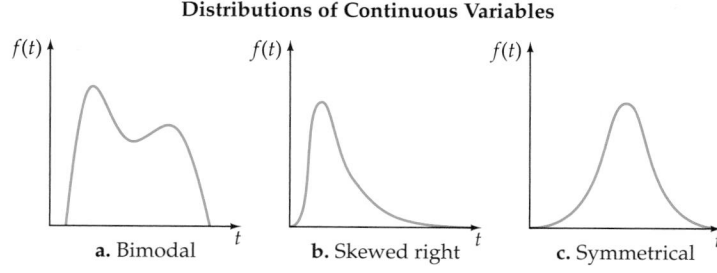

In some cases we use a continuous curve to display the distribution of a set of discrete data. For instance, in those cases in which we have a large set of data and very small class intervals, the shape of the top of the histogram approaches a smooth curve. See the two figures below. When graphing the distributions of very large sets of data with very small class intervals, it is common practice to replace the histogram with a smooth continuous curve.

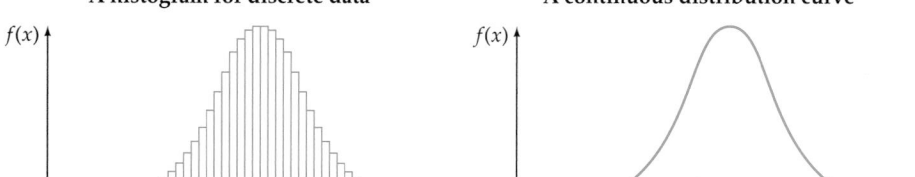

A histogram for discrete data   A continuous distribution curve

One of the most important statistical distributions is known as a *normal distribution*. The precise mathematical definition of a normal distribution is given by the equation in the Take Note at the left. However, for many applied problems, it is sufficient to know that all normal distributions have the following properties.

Properties of a Normal Distribution

The distribution of data in a normal distribution has a bell shape that is symmetrical about a vertical line through its center. The mean, the median, and the mode of a normal distribution are all equal and they are located at the center of the distribution.

A normal distribution

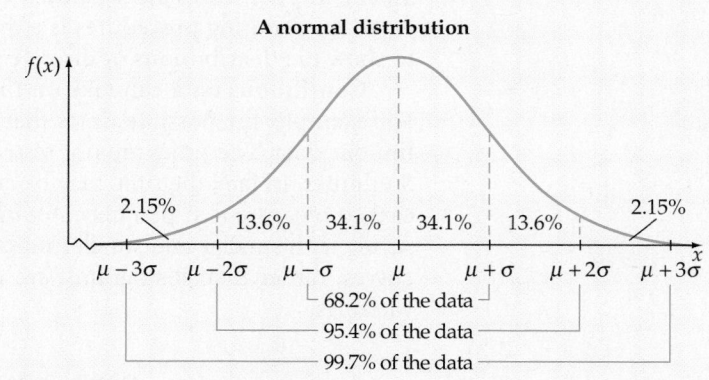

The Empirical Rule: In a normal distribution, about

68.2% of the data lies within 1 standard deviation of the mean.
95.4% of the data lies within 2 standard deviations of the mean.
99.7% of the data lies within 3 standard deviations of the mean.

The Empirical Rule can be used to solve many applied problems.

EXAMPLE 2 ■ Use the Empirical Rule to Solve an Application

A survey of 1000 U.S. gas stations found that the price charged for a gallon of regular gas could be closely approximated by a normal distribution with a mean of $1.50 and a standard deviation of $0.20. How many of the stations charge

a. between $1.10 and $1.90 for a gallon of regular gas?

b. less than $1.70 for a gallon of regular gas?

c. more than $1.90 for a gallon of regular gas?

Solution

a. The $1.10 per gallon price is 2 standard deviations below the mean. The $1.90 price is 2 standard deviations above the mean. In a normal distribution, 95.4% of all data lies within 2 standard deviations of the mean. See Figure 12.7. Therefore, approximately

$$(95.4\%)(1000) = (0.954)(1000) = 954$$

of the stations charge between $1.10 and $1.90 for a gallon of regular gas.

b. The $1.70 price is 1 standard deviation above the mean. See Figure 12.8. In a normal distribution, 34.1% of all data lies between the mean and 1 standard deviation above the mean. Thus, approximately

$$(34.1\%)(1000) = (0.341)(1000) = 341$$

of the stations charge between $1.50 and $1.70 for a gallon of regular gasoline. Half of the stations charge less than the mean. Therefore, about 341 + 500 = 841 of the stations charge less than $1.70 for a gallon of regular gas.

c. The $1.90 price is 2 standard deviations above the mean. In a normal distribution, 95.4% of all data is within 2 standard deviations of the mean. This means that the other 4.6% of the data will lie either above 2 standard deviations of the mean or below 2 standard deviations of the mean. We are interested only in the data that are more than 2 standard deviations above the mean, which is $\frac{1}{2}$ of 4.6%, or 2.3%, of the data. See Figure 12.9. Thus about $(2.3\%)(1000) = (0.023)(1000) = 23$ of the stations charge more than $1.90 for a gallon of regular gas.

CHECK YOUR PROGRESS 2 A vegetable distributor knows that during the month of August, the weights of its tomatoes are normally distributed with a mean of 0.61 pound and a standard deviation of 0.15 pound.

a. What percent of the tomatoes weigh less than 0.76 pound?

b. In a shipment of 6000 tomatoes, how many tomatoes can be expected to weigh more than 0.31 pound?

c. In a shipment of 4500 tomatoes, how many tomatoes can be expected to weigh from 0.31 pound to 0.91 pound?

Solution See page S45.

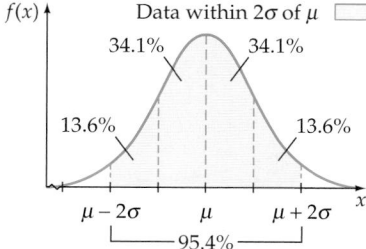

Figure 12.7

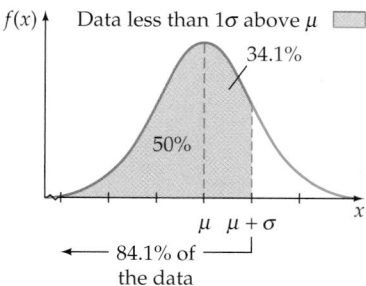

Figure 12.8

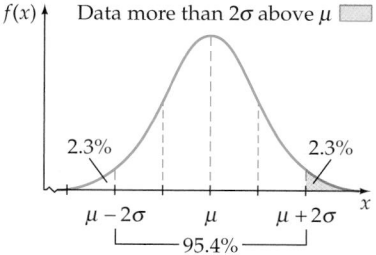

Figure 12.9

QUESTION   *Can the Empirical Rule be applied to all data sets?*

The Standard Normal Distribution

It is often helpful to convert the values of a continuous variable x to z-scores, as we did in the previous section by using the z-score formulas:

$$z_x = \frac{x - \bar{x}}{s} \quad \text{or} \quad z_x = \frac{x - \mu}{\sigma}$$

If the original distribution of x values is a normal distribution, then the corresponding distribution of z-scores will also be a normal distribution. This normal distribution of z-scores is called the *standard normal distribution*. See Figure 12.10. It has a mean of 0 and a standard deviation of 1, and it was first used by the French mathematician Abraham De Moivre (1667–1754).

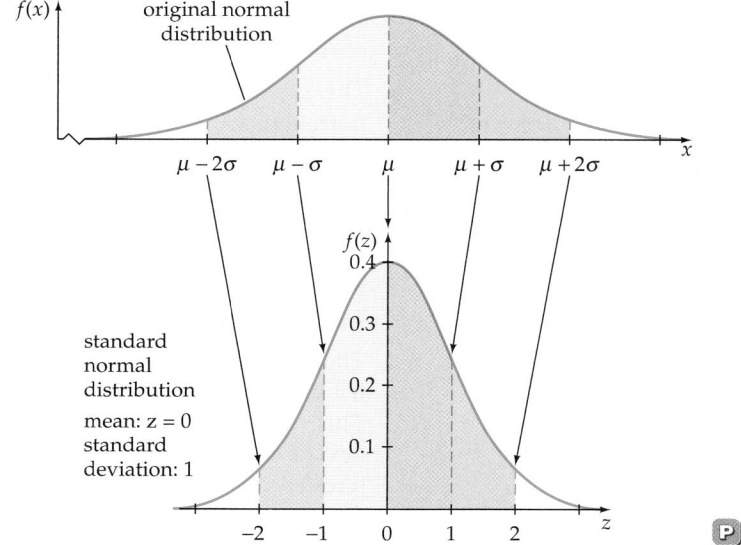

Figure 12.10 *Conversion of a normal distribution to the standard normal distribution*

The Standard Normal Distribution

The **standard normal distribution** is the normal distribution for the continuous variable z that has a mean of 0 and a standard deviation of 1.

   Tables and calculators are often used to determine the area of a portion of the standard normal distribution. For example, Table 12.7 gives the approximate areas of the standard normal distribution between the mean 0 and z standard deviations from the mean. Table 12.7 indicates that the area A of the standard normal distribution from the mean 0 up to $z = 1.34$ is 0.410 square unit.

ANSWER   *No. The Empirical Rule can only be applied to normal distributions.*

Table 12.7 *Area Under the Standard Normal Curve*

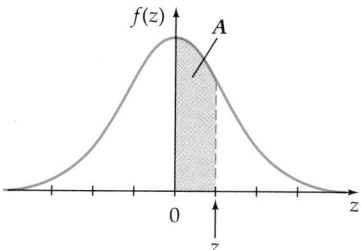

Figure 12.11 *A is the area of the shaded region*

z	A	z	A	z	A	z	A	z	A	z	A
.00	.000	.56	.212	1.12	.369	1.68	.454	2.24	.487	2.80	.497
.01	.004	.57	.216	1.13	.371	1.69	.454	2.25	.488	2.81	.498
.02	.008	.58	.219	1.14	.373	1.70	.455	2.26	.488	2.82	.498
.03	.012	.59	.222	1.15	.375	1.71	.456	2.27	.488	2.83	.498
.04	.016	.60	.226	1.16	.377	1.72	.457	2.28	.489	2.84	.498
.05	.020	.61	.229	1.17	.379	1.73	.458	2.29	.489	2.85	.498
.06	.024	.62	.232	1.18	.381	1.74	.459	2.30	.489	2.86	.498
.07	.028	.63	.236	1.19	.383	1.75	.460	2.31	.490	2.87	.498
.08	.032	.64	.239	1.20	.385	1.76	.461	2.32	.490	2.88	.498
.09	.036	.65	.242	1.21	.387	1.77	.462	2.33	.490	2.89	.498
.10	.040	.66	.245	1.22	.389	1.78	.462	2.34	.490	2.90	.498
.11	.044	.67	.249	1.23	.391	1.79	.463	2.35	.491	2.91	.498
.12	.048	.68	.252	1.24	.393	1.80	.464	2.36	.491	2.92	.498
.13	.052	.69	.255	1.25	.394	1.81	.465	2.37	.491	2.93	.498
.14	.056	.70	.258	1.26	.396	1.82	.466	2.38	.491	2.94	.498
.15	.060	.71	.261	1.27	.398	1.83	.466	2.39	.492	2.95	.498
.16	.064	.72	.264	1.28	.400	1.84	.467	2.40	.492	2.96	.498
.17	.067	.73	.267	1.29	.401	1.85	.468	2.41	.492	2.97	.499
.18	.071	.74	.270	1.30	.403	1.86	.469	2.42	.492	2.98	.499
.19	.075	.75	.273	1.31	.405	1.87	.469	2.43	.492	2.99	.499
.20	.079	.76	.276	1.32	.407	1.88	.470	2.44	.493	3.00	.499
.21	.083	.77	.279	1.33	.408	1.89	.471	2.45	.493	3.01	.499
.22	.087	.78	.282	1.34	.410	1.90	.471	2.46	.493	3.02	.499
.23	.091	.79	.285	1.35	.411	1.91	.472	2.47	.493	3.03	.499
.24	.095	.80	.288	1.36	.413	1.92	.473	2.48	.493	3.04	.499
.25	.099	.81	.291	1.37	.415	1.93	.473	2.49	.494	3.05	.499
.26	.103	.82	.294	1.38	.416	1.94	.474	2.50	.494	3.06	.499
.27	.106	.83	.297	1.39	.418	1.95	.474	2.51	.494	3.07	.499
.28	.110	.84	.300	1.40	.419	1.96	.475	2.52	.494	3.08	.499
.29	.114	.85	.302	1.41	.421	1.97	.476	2.53	.494	3.09	.499
.30	.118	.86	.305	1.42	.422	1.98	.476	2.54	.494	3.10	.499
.31	.122	.87	.308	1.43	.424	1.99	.477	2.55	.495	3.11	.499
.32	.126	.88	.311	1.44	.425	2.00	.477	2.56	.495	3.12	.499
.33	.129	.89	.313	1.45	.426	2.01	.478	2.57	.495	3.13	.499
.34	.133	.90	.316	1.46	.428	2.02	.478	2.58	.495	3.14	.499
.35	.137	.91	.319	1.47	.429	2.03	.479	2.59	.495	3.15	.499
.36	.141	.92	.321	1.48	.431	2.04	.479	2.60	.495	3.16	.499
.37	.144	.93	.324	1.49	.432	2.05	.480	2.61	.495	3.17	.499
.38	.148	.94	.326	1.50	.433	2.06	.480	2.62	.496	3.18	.499
.39	.152	.95	.329	1.51	.434	2.07	.481	2.63	.496	3.19	.499
.40	.155	.96	.331	1.52	.436	2.08	.481	2.64	.496	3.20	.499
.41	.159	.97	.334	1.53	.437	2.09	.482	2.65	.496	3.21	.499
.42	.163	.98	.336	1.54	.438	2.10	.482	2.66	.496	3.22	.499
.43	.166	.99	.339	1.55	.439	2.11	.483	2.67	.496	3.23	.499
.44	.170	1.00	.341	1.56	.441	2.12	.483	2.68	.496	3.24	.499
.45	.174	1.01	.344	1.57	.442	2.13	.483	2.69	.496	3.25	.499
.46	.177	1.02	.346	1.58	.443	2.14	.484	2.70	.497	3.26	.499
.47	.181	1.03	.348	1.59	.444	2.15	.484	2.71	.497	3.27	.499
.48	.184	1.04	.351	1.60	.445	2.16	.485	2.72	.497	3.28	.499
.49	.188	1.05	.353	1.61	.446	2.17	.485	2.73	.497	3.29	.499
.50	.191	1.06	.355	1.62	.447	2.18	.485	2.74	.497	3.30	.500
.51	.195	1.07	.358	1.63	.448	2.19	.486	2.75	.497	3.31	.500
.52	.198	1.08	.360	1.64	.449	2.20	.486	2.76	.497	3.32	.500
.53	.202	1.09	.362	1.65	.451	2.21	.486	2.77	.497	3.33	.500
.54	.205	1.10	.364	1.66	.452	2.22	.487	2.78	.497		
.55	.209	1.11	.367	1.67	.453	2.23	.487	2.79	.497		

Because the standard normal distribution is symmetrical about the mean of 0, we can also use Table 12.7 to find the area of a region that is located to the left of the mean. This process is explained in Example 3.

EXAMPLE 3 ■ Use Symmetry to Determine an Area

Find the area of the standard normal distribution between $z = -1.44$ and $z = 0$.

Solution

Because the standard normal distribution is symmetrical about the center line $z = 0$, the area of the standard normal distribution between $z = -1.44$ and $z = 0$ is equal to the area between $z = 0$ and $z = 1.44$. See Figure 12.12. The entry in Table 12.7 associated with $z = 1.44$ is 0.425. Thus the area of the standard normal distribution between $z = -1.44$ and $z = 0$ is 0.425 square unit.

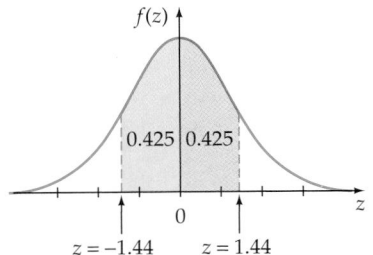

Figure 12.12 *Symmetrical region*

CHECK YOUR PROGRESS 3   Find the area of the standard normal distribution between $z = -0.67$ and $z = 0$.

Solution   *See page S45.*

In Figure 12.13, the region to the right of $z = 0.82$ is called a *tail region*. A **tail region** is a region of the standard normal distribution to the right of a positive z-value or to the left of a negative z-value. To find the area of a tail region, we subtract the entry in Table 12.7 from 0.500. This procedure is illustrated in the next example.

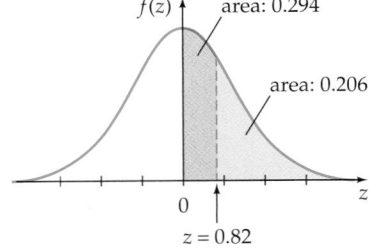

Figure 12.13 *Area of a tail region*

EXAMPLE 4 ■ Find the Area of a Tail Region

Find the area of the standard normal distribution to the right of $z = 0.82$.

Solution

Table 12.7 indicates that the area from $z = 0$ to $z = 0.82$ is 0.294 square unit. The area to the right of $z = 0$ is 0.500 square unit. Thus the area to the right of $z = 0.82$ is $0.500 - 0.294 = 0.206$ square unit. See Figure 12.13.

CHECK YOUR PROGRESS 4   Find the area of the standard normal distribution to the left of $z = -1.47$.

Solution   *See page S45.*

The Standard Normal Distribution, Areas, Percentages, and Probabilities

In the standard normal distribution, the *area* of the distribution from $z = a$ to $z = b$ represents

■ the *percentage* of z-values that lie in the interval from a to b.

■ the *probability* that z lies in the interval from a to b.

Because the area of a portion of the standard normal distribution can be interpreted as a percentage of the data or as a probability that the variable lies in an interval, we can use the standard normal distribution to solve many application problems.

EXAMPLE 5 ■ Solve an Application

A soda machine dispenses soda into 14-ounce cups. Tests show that the actual amount of soda dispensed is normally distributed, with a mean of 12 ounces and a standard deviation of 0.8 ounce.

a. What percent of cups will receive less than 11 ounces of soda?

b. What percent of cups will receive between 10.8 ounces and 12.2 ounces of soda?

c. If a cup is chosen at random, what is the probability that the machine will overflow the cup?

Solution

a. The z-score for 11 ounces is

$$z_{11} = \frac{11 - 12}{0.8} = -1.25$$

Table 12.7 indicates that 0.394 (39.4%) of the data in a normal distribution is between $z = 0$ and $z = 1.25$. Because the data are normally distributed, 39.4% of the data is also between $z = 0$ and $z = -1.25$. The percent of data to the left of $z = -1.25$ is 50% − 39.4% = 10.6%. See Figure 12.14. Thus 10.6% of the cups filled by the soda machine will receive less than 11 ounces of soda.

b. The z-score for 12.2 ounces is

$$z_{12.2} = \frac{12.2 - 12}{0.8} = 0.25$$

Table 12.7 indicates that 0.099 (9.9%) of the data in a normal distribution is between $z = 0$ and $z = 0.25$.

The z-score for 10.8 ounces is

$$z_{10.8} = \frac{10.8 - 12}{0.8} = -1.5$$

Table 12.7 indicates that 0.433 (43.3%) of the data in a normal distribution is between $z = 0$ and $z = 1.5$. Because the data are normally distributed, 43.3% of the data is also between $z = 0$ and $z = -1.5$. See Figure 12.15. Thus the percent of the cups that the vending machine will fill with between 10.8 ounces and 12.2 ounces of soda is

43.3% + 9.9% = 53.2%

c. A cup will overflow if it receives more than 14 ounces of soda. The z-score for 14 ounces is

$$z_{14} = \frac{14 - 12}{0.8} = 2.5$$

Table 12.7 indicates that 0.494 (49.4%) of the data in the standard normal distribution is between $z = 0$ and $z = 2.5$. The percent of data to the right of $z = 2.5$ is determined by subtracting 49.4% from 50%. See Figure 12.16. Thus 0.6% of the time the machine produces an overflow, and the probability that a cup chosen at random will overflow is 0.006.

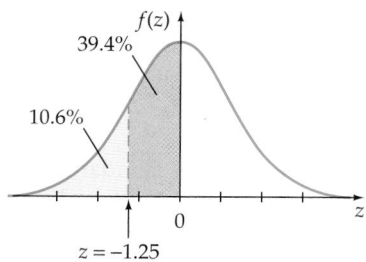

Figure 12.14 *Portion of data to the left of* $z = -1.25$

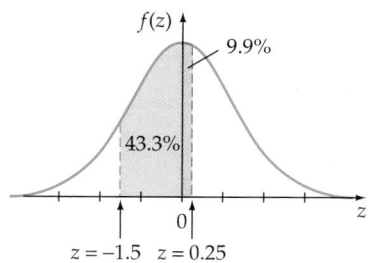

Figure 12.15 *Portion of data between two z-scores*

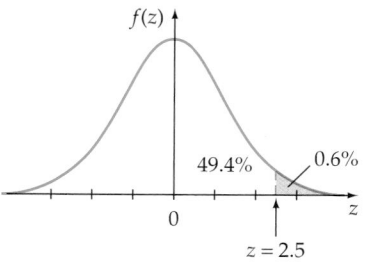

Figure 12.16 *Portion of data to the right of* $z = 2.5$

CHECK YOUR PROGRESS 5   A study of the careers of professional football players shows that the lengths of their careers are nearly normally distributed, with a mean of 6.1 years and a standard deviation of 1.8 years.

a. What percent of professional football players have a career of more than 9 years?

b. If a professional football player is chosen at random, what is the probability that the player will have a career of between 3 and 4 years?

Solution   See page S45.

MathMatters   **Find the Area of a Portion of the Standard Normal Distribution by Using a Calculator**

Some calculators can be used to find the area of a portion of the standard normal distribution. For instance, the TI-83 screen displays below both indicate that the area of the standard normal distribution from a lower bound of $z = 0$ to an upper bound of $z = 1.34$ is about 0.409877 square unit. This is a more accurate value than the entry given in Table 12.7, which is 0.410.

Select the `normalcdf(` function from the DISTR menu. Enter your lower bound, followed by your upper bound. Press ENTER.

The ShadeNorm instruction in the DISTR, DRAW menu draws the standard normal distribution and shades the area between your lower bound and your upper bound.

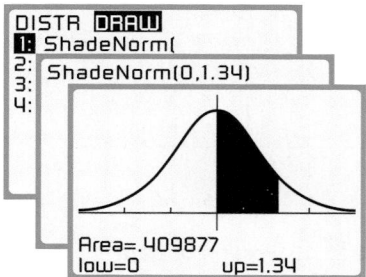

Figure 12.17 *TI-83 screen displays*

Excursion

Cut-Off Scores

There are many applications for which it is necessary to find a **cut-off score,** which is a score that separates data into two groups such that the data in one group satisfy a certain requirement and the data in the other group do not satisfy the requirement. If the data are normally distributed, then we can find a cut-off score by the method shown in the following example.

(continued)

EXAMPLE

The OnTheGo company manufactures laptop computers. A study indicates that the life spans of their computers are normally distributed, with a mean of 4.0 years and a standard deviation of 1.2 years. How long should the company warrant its computers if the company wishes less than 4% of its computers to fail during the warranty period?

Solution

Figure 12.18 shows a standard normal distribution with 4% of the data to the left of some unknown z-score and 46% of the data to the right of the z-score but to the left of the mean of 0. Using Table 12.7, we find that the z-score associated with an area of $A = 0.46$ is 1.75. Our unknown z-score is to the left of 0, so it must be negative. Thus $z_x = -1.75$. If we let x represent the time in years that a computer is in use, then x is related to the z-scores by the formula

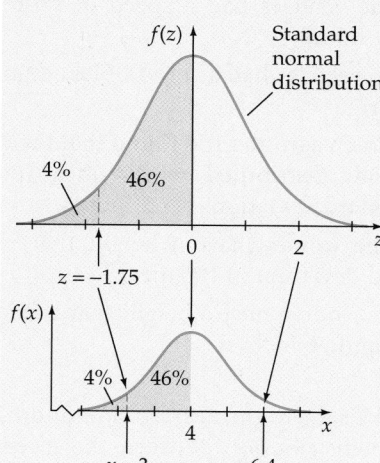

$$z_x = \frac{x - \mu}{s}$$

Figure 12.18 *Finding a cut-off score*

Solving for x with $\mu = 4.0$, $s = 1.2$, and $z = -1.75$ gives us

$$-1.75 = \frac{x - 4.0}{1.2}$$

$$(-1.75)(1.2) = x - 4.0$$

$$x = 4.0 - 2.1$$

$$x = 1.9 \qquad \text{The cut-off score}$$

Hence the company can provide a 1.9-year warranty and expect less than 4% of its computers to fail during the warranty period.

Excursion Exercises

1. A professor finds that the grades in a large class are normally distributed. The mean of the grades is 64 and the standard deviation is 10. If the professor decides to give an A grade to the students in the top 9% of the class, what is the cut-off score for an A?

2. The results of a statewide examination of the reading skills of sixth-grade students were normally distributed, with a mean score of 104 and a standard deviation of 16. The students in the top 10% are to receive an award, and those in the bottom 14% will be required to take a special reading class.

   a. What score does a student need in order to receive an award?

   b. What is the cut-off score that will be used to determine whether a student will be required to take the special reading class?

3. A secondary school system finds that the 440-yard dash times of its female students are normally distributed, with an average time of 72 seconds and a standard deviation of 5.5 seconds. What time does a runner need in order to be in the 9% of the runners with the best times? Round to the nearest hundredth of a second.

Exercise Set 12.4   (Suggested Assignment: 1–53 every other odd)

1. The 1999 median income for family practice physicians was \$141,560. (*Source:* American Medical Group Association, Medical Group Compensation and Productivity Survey, © 2000). The distribution of the physicians' incomes is skewed to the right. Is the mean of these incomes greater or less than \$141,560?

2. At a university, 500 law students took an examination. One student completed the exam in 24 minutes; however, the mode for the times is 50 minutes. The distribution of the times the students took to complete the exam is skewed to the left. Is the mean of these times greater or less than 50 minutes?

In Exercises 3–8, use the Empirical Rule to answer each question.

3. In a normal distribution, what percent of the data lies
   a. within 2 standard deviations of the mean?
   b. above 1 standard deviation of the mean?
   c. between 1 standard deviation below the mean and 2 standard deviations above the mean?

4. In a normal distribution, what percent of the data lies
   a. within 3 standard deviations of the mean?
   b. below 2 standard deviations of the mean?
   c. between 2 standard deviations below the mean and 3 standard deviations above the mean?

5. During 1 week an overnight delivery company found that the weights of its parcels were normally distributed, with a mean of 24 ounces and a standard deviation of 6 ounces.
   a. What percent of the parcels weighed between 12 ounces and 30 ounces?
   b. What percent of the parcels weighed more than 42 ounces?

6. A baseball franchise finds that the attendance at its home games is normally distributed, with a mean of 16,000 and a standard deviation of 4000.
   a. What percent of the home games have an attendance between 12,000 and 20,000 people?
   b. What percent of the home games have an attendance of less than 8000?

7. A highway study of 8000 vehicles that passed by a checkpoint found that their speeds were normally dis-

tributed, with a mean of 61 miles per hour and a standard deviation of 7 miles per hour.
   a. How many of the vehicles had a speed of more than 68 miles per hour?
   b. How many of the vehicles had a speed of less than 40 miles per hour?

8. A survey of 1000 women ages 20 to 30 found that their heights were normally distributed, with a mean of 65 inches and a standard deviation of 2.5 inches.
   a. How many of the women have a height that is within 1 standard deviation of the mean?
   b. How many of the women have a height that is between 60 inches and 70 inches?

In Exercises 9–16, find the area, to the nearest thousandth, of the standard normal distribution between the given z-scores.

9. $z = 0$ and $z = 1.5$
10. $z = 0$ and $z = 1.9$
11. $z = 0$ and $z = -1.85$
12. $z = 0$ and $z = -2.3$
13. $z = 1$ and $z = 1.9$
14. $z = 0.7$ and $z = 1.92$
15. $z = -1.47$ and $z = 1.64$
16. $z = -0.44$ and $z = 1.82$

In Exercises 17–24, find the area, to the nearest thousandth, of the indicated region of the standard normal distribution.

17. The region where $z > 1.3$
18. The region where $z > 1.92$
19. The region where $z < -2.22$
20. The region where $z < -0.38$
21. The region where $z > -1.45$
22. The region where $z < 1.82$
23. The region where $z < 2.71$
24. The region where $z < 1.92$

In Exercises 25–30, find the z-score, to the nearest hundredth, that satisfies the given condition.

25. 0.200 square unit of the standard normal distribution is to the right of z.
26. 0.227 square unit of the standard normal distribution is to the right of z.

27. 0.184 square unit of the standard normal distribution is to the left of z.

28. 0.330 square unit of the standard normal distribution is to the left of z.

29. 0.363 square unit of the standard normal distribution is to the right of z.

30. 0.440 square unit of the standard normal distribution is to the left of z.

In Exercises 31–40, use Table 12.7 to answer each question. Note: Round z-scores to the nearest hundredth and then find the required A values using Table 12.7 on page 813.

31. A population is normally distributed with a mean of 44.8 and a standard deviation of 12.4.

   a. What percent of the data is greater than 51.0?
   b. What percent of the data is between 47.9 and 63.4?

32. A population is normally distributed with a mean of 6.8 and a standard deviation of 1.2.

   a. What percent of the data is less than 7.2?
   b. What percent of the data is between 7.1 and 9.5?

33. A population is normally distributed with a mean of 580 and a standard deviation of 160.

   a. What percent of the data is less than 404?
   b. What percent of the data is between 460 and 612?

34. A population is normally distributed with a mean of 3010 and a standard deviation of 640.

   a. What percent of the data is greater than 2818?
   b. What percent of the data is between 2562 and 4162?

35. The weights of all the boxes of corn flakes filled by a machine are normally distributed, with an average weight of 14.5 ounces and a standard deviation of 0.4 ounce. What percent of the boxes will

   a. weigh less than 14 ounces?
   b. weigh between 13.5 ounces and 15.5 ounces?

36. A telephone company has found that the lengths of its long distance telephone calls are normally distributed, with a mean of 225 seconds and a standard deviation of 55 seconds. What percent of its long distance calls are

   a. longer than 340 seconds?
   b. between 200 and 300 seconds?

37. The breaking point of a particular type of rope is normally distributed, with a mean of 350 pounds and a standard deviation of 24 pounds. What is the proba-

bility that a piece of this rope chosen at random will have a breaking point of

   a. less than 320 pounds?
   b. between 340 and 370 pounds?

38. The mileage for WearEver tires is normally distributed, with a mean of 48,000 miles and a standard deviation of 7400 miles. What is the probability that the WearEver tires you purchase will provide a mileage of

   a. more than 60,000 miles?
   b. between 40,000 and 50,000 miles?

39. The amount of time customers spend waiting in line at a bank is normally distributed, with a mean of 2.5 minutes and a standard deviation of 0.75 minute. Find the probability that the time a customer spends waiting is

   a. less than 3 minutes.
   b. less than 1 minute.

40. A psychologist finds that the intelligence quotients of a group of patients are normally distributed, with a mean of 102 and a standard deviation of 16. Find the percent of the patients with IQs

   a. above 114.
   b. between 90 and 118.

41. Consider the data set of the heights of all babies born in the United States during a particular year. Do you think this data set is nearly normally distributed? Explain.

42. Consider the data set of the weights of all Valencia oranges grown in California during a particular year. Do you think this data set is nearly normally distributed? Explain.

Extensions

CRITICAL THINKING

In Exercises 43–52, determine whether the given statement is true or false.

43. The standard normal distribution has a mean of 0.

44. Every normal distribution is a bell-shaped distribution.

45. In a normal distribution, the mean, the median, and the mode of the distribution are all located at the center of the distribution.

46. If a distribution is symmetrical about a vertical line down its center, then it is a normal distribution.

47. The mean of a normal distribution is always larger than the standard deviation of the distribution.

48. The standard deviation of a normal distribution can be 0.

49. If a data value x from a normal distribution is positive, then its z-score must also be positive.

50. All normal distributions have a mean of 0.

51. Let x be the number of people who attended a baseball game today. The variable x is a discrete variable.

52. The time of day, d, in the lobby of a bank is measured with a digital clock. The variable d is a continuous variable.

53. **a.** How does the area of the standard normal distribution for $0 \leq z < 1$ compare with the area of the standard normal distribution for $0 \leq z \leq 1$?

   b. Explain the reasoning you used to answer part a.

54. **a.** Make a sketch of two normal distributions that have the same standard deviation but different means.

   b. Make a sketch of two normal distributions that have the same mean but different standard deviations.

55. Determine the two z-scores that bound the middle 60% of the data in a normal distribution.

56. Determine the approximate z-scores for the first quartile and the third quartile of the standard normal distribution.

EXPLORATIONS

57. The mathematician Pafnuty Chebyshev (cha-bĭ´shôf) (1821–1894) is well known for a theorem that concerns the distribution of data in any data set. This theorem is known as Chebyshev's Theorem. Consult a statistics text or the Internet to find information on Chebyshev's Theorem. Write a statement of Chebyshev's Theorem and use the theorem to find the minimum percent of data in any data set that must be within 2 standard deviations of the mean.

SECTION 12.5 | **Linear Regression and Correlation**

Linear Regression

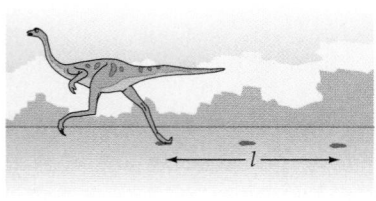

In many applications, scientists try to determine whether two variables are related. If they are related, the scientists then try to find an equation that can be used to *model* the relationship. For instance, the zoology professor R. McNeill Alexander wanted to determine whether the *stride length* of a dinosaur, as shown by its fossilized footprints, could be used to estimate the speed of the dinosaur. Stride length for an animal is defined as the distance l from a particular point on a footprint to that same point on the next footprint of the same foot. (See the figure at the left.) Because no dinosaurs were available, he and fellow scientist A. S. Jayes carried out experiments with many types of animals, including adult men, dogs, camels, ostriches, and elephants. The results of these experiments tended to support the idea that the speed s of an animal is related to the animal's stride length l. To better understand this relationship, examine the data in Table 12.8, which is similar to, but less extensive than, the data collected by Alexander and Jayes.

Table 12.8 *Speed for Selected Stride Lengths*

a. Adult men

Stride length (meters)	2.5	3.0	3.3	3.5	3.8	4.0	4.2	4.5
Speed (meters per second)	3.4	4.9	5.5	6.6	7.0	7.7	8.3	8.7

b. Dogs

Stride length (meters)	1.5	1.7	2.0	2.4	2.7	3.0	3.2	3.5
Speed (meters per second)	3.7	4.4	4.8	7.1	7.7	9.1	8.8	9.9

c. Camels

Stride length (meters)	2.5	3.0	3.2	3.4	3.5	3.8	4.0	4.2
Speed (meters per second)	2.3	3.9	4.4	5.0	5.5	6.2	7.1	7.6

A graph of the ordered pairs in Table 12.8 is shown in Figure 12.19. In this graph, which is called a **scatter diagram** or **scatter plot,** the horizontal axis represents the stride lengths in meters and the vertical axis represents the average speeds in meters per second. The scatter diagram seems to indicate that for each of the three species, a larger stride length generally produces a faster speed. Also notice that for each species, a straight line can be drawn such that all of the points for that species lie on or very close to the line. Thus the relationship between speed and stride length appears to be a linear relationship.

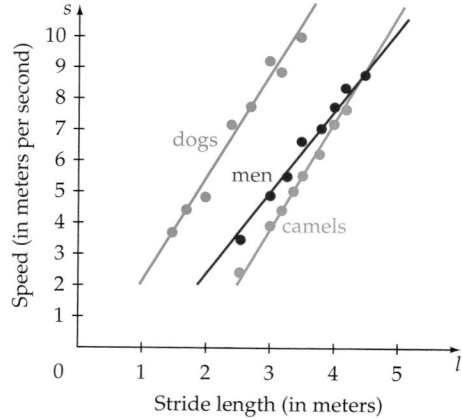

Figure 12.19 *Scatter diagram for Table 12.8*

After a relationship between paired data, which are referred to as **bivariate data,** has been discovered, a scientist tries to model the relationship with an equation. One method of determining a linear relationship for bivarite data is called **linear regression.** To see how linear regression is carried out, let us concentrate on the bivariate data for the dogs, which is shown by the green points in Figures 12.19

and 12.20. There are many lines that can be drawn such that the data points lie close to the line; however, scientists are generally interested in the line called the *line of best fit* or the *least-squares regression line*.

Definition of the Least-Squares Regression Line

The **least-squares regression line** for a set of data is the line that minimizes the sum of the squares of the vertical deviations from each data point to the line.

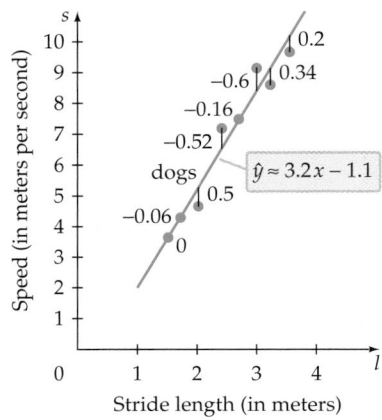

Figure 12.20 *Vertical deviations*

The least-squares regression line is also called the **least-squares line.** The approximate equation of the least-squares line for the bivariate data for the dogs is $\hat{y} = 3.2x - 1.1$. Figure 12.20 shows the graph of these data and the graph of $\hat{y} = 3.2x - 1.1$. In Figure 12.20, the vertical deviations from the ordered pairs to the graph of $\hat{y} = 3.2x - 1.1$ are 0, −0.06, 0.5, −0.52, −0.16, −0.6, 0.34 and 0.2.

It is traditional to use the symbol $\hat{y}$ (pronounced *y*-hat) in place of *y* in the equation of a least-squares line. This also helps us differentiate the line's *y*-values from the *y*-values of the given ordered pairs.

The next theorem can be used to determine the equation of the least-squares line for a given set of ordered pairs.

The Formula for the Least-Squares Line

The equation of the least-squares line for the *n* ordered pairs

$$(x_1, y_1), (x_2, y_2), (x_3, y_3), \ldots, (x_n, y_n)$$

is $\hat{y} = ax + b$, where

$$a = \frac{n\Sigma xy - (\Sigma x)(\Sigma y)}{n\Sigma x^2 - (\Sigma x)^2} \quad \text{and} \quad b = \bar{y} - a\bar{x}$$

P

In the above theorem, Σx represents the sum of all the *x* values, Σy represents the sum of all the *y* values, and Σxy represents the sum of the *n* products $x_1 y_1, x_2 y_2, \ldots, x_n y_n$. The notation $\bar{x}$ represents the mean of the *x* values, and $\bar{y}$ represents the mean of the *y* values. The following example illustrates a procedure that can be used to calculate efficiently the sums needed to find the equation of the least-squares line for a given set of data.

EXAMPLE 1 ■ **Find the Equation of a Least-Squares Line**

Find the equation of the least-squares line for the ordered pairs in Table 12.8a.

Solution
The ordered pairs are

$$(2.5, 3.4), (3.0, 4.9), (3.3, 5.5), (3.5, 6.6), (3.8, 7.0), (4.0, 7.7), (4.2, 8.3), (4.5, 8.7)$$

The number of ordered pairs is $n = 8$. Organize the data in four columns, as shown in Table 12.9 on the following page. Then find the sum of each column.

Table 12.9

x	y	x^2	xy
2.5	3.4	6.25	8.50
3.0	4.9	9.00	14.70
3.3	5.5	10.89	18.15
3.5	6.6	12.25	23.10
3.8	7.0	14.44	26.60
4.0	7.7	16.00	30.80
4.2	8.3	17.64	34.86
4.5	8.7	20.25	39.15
$\Sigma x = 28.8$	$\Sigma y = 52.1$	$\Sigma x^2 = 106.72$	$\Sigma xy = 195.86$

Find the slope a.

$$a = \frac{n\Sigma xy - (\Sigma x)(\Sigma y)}{n\Sigma x^2 - (\Sigma x)^2}$$

$$= \frac{(8)(195.86) - (28.8)(52.1)}{(8)(106.72) - (28.8)^2}$$

$$\approx 2.7303$$

Find $\bar{x}$ and $\bar{y}$.

$$\bar{x} = \frac{\Sigma x}{n} = \frac{28.8}{8} = 3.6 \qquad \bar{y} = \frac{\Sigma y}{n} = \frac{52.1}{8} = 6.5125$$

Find the y-intercept b.

$$b = \bar{y} - a\bar{x}$$

$$\approx 6.5125 - (2.7303)(3.6)$$

$$= -3.31658$$

If a and b are each rounded to the nearest tenth, to reflect the accuracy of the original data, then we have as our equation of the least-squares line:

$$\hat{y} = ax + b$$

$$\hat{y} \approx 2.7x - 3.3$$

See Figure 12.21.

CHECK YOUR PROGRESS 1 Find the equation of the least-squares line for the stride length and speed of camels given in Table 12.8c.

Solution *See page S45.*

Once the equation of the least-squares line is found, it can be used to make predictions. This procedure is illustrated in the next example.

✔ **TAKE NOTE**

It can be proved that for any set of ordered pairs, the graph of the ordered pair $(\bar{x}, \bar{y})$ is a point on the least-squares line for the ordered pairs. This can serve as a check. If you have calculated the least-squares line for a set of ordered pairs and you find that $(\bar{x}, \bar{y})$ is not a point on your least-squares line, then you know that you have made an error.

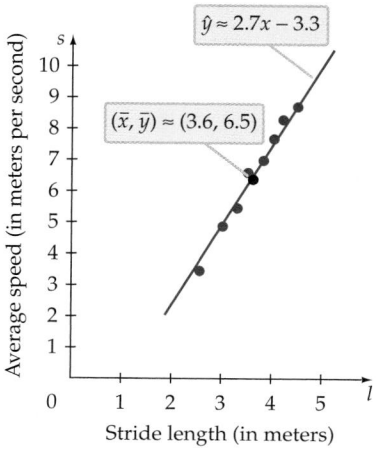

Figure 12.21 *Least-squares line for speed versus stride length in adult men*

EXAMPLE 2 ■ **Use a Least-Squares Line to Make a Prediction**

Use the equation of the least-squares line from Example 1 to predict the average speed of an adult man for each of the following stride lengths. Round your results to the nearest tenth of a meter per second (mps).

a. 2.8 meters   **b.** 4.8 meters

Solution

a. In Example 1, we found the equation of the least-squares line to be $\hat{y} = 2.7x - 3.3$. Substituting 2.8 for x gives

$$\hat{y} = 2.7(2.8) - 3.3 = 4.26$$

Rounding 4.26 to the nearest tenth produces 4.3. Thus 4.3 meters per second is the predicted average speed for an adult man with a stride length of 2.8 meters.

b. In Example 1, we found the equation of the least-squares line to be $\hat{y} = 2.7x - 3.3$. Substituting 4.8 for x gives

$$\hat{y} = 2.7(4.8) - 3.3 = 9.66$$

Rounding 9.66 to the nearest tenth produces 9.7. Thus 9.7 meters per second is the predicted average speed for an adult man with a stride length of 4.8 meters.

CHECK YOUR PROGRESS 2   Use the equation of the least-squares line from Check Your Progress 1 to predict the average speed of a camel for each of the following stride lengths. Round your results to the nearest tenth of a meter per second.

a. 2.7 meters   **b.** 4.5 meters

Solution   *See page S46.*

✔ **TAKE NOTE**

Sometimes values predicted by extrapolation are not reasonable. For instance, if we wish to predict the speed of a man with a stride length of $x = 20$ meters, the least-squares equation $\hat{y} = 2.7x - 3.3$ gives us a speed of 50.7 meters per second. Because the maximum stride length of adult men is considerably less than 20 meters, we should not trust this prediction.

The procedure in Example 2a made use of an equation to determine a point between given data points. This procedure is referred to as **interpolation.** In Example 2b, an equation was used to determine a point to the right of the given data points. The process of using an equation to determine a point to the right or left of given data points is referred to as **extrapolation.** See Figure 12.22.

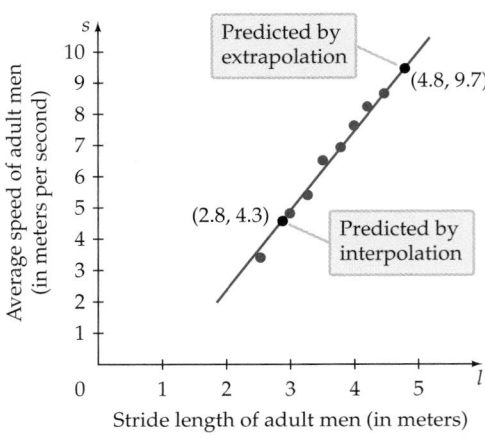

Figure 12.22 *Interpolation and extrapolation*

Linear Correlation Coefficient

To determine the strength of a linear relationship between two variables, statisticians use a statistic called the *linear correlation coefficient*, which is denoted by the variable r and is defined as follows.

Linear Correlation Coefficient

For the n ordered pairs (x_1, y_1), (x_2, y_2), (x_3, y_3), $\ldots$, (x_n, y_n), the **linear correlation coefficient** r is given by

$$r = \frac{n(\Sigma xy) - (\Sigma x)(\Sigma y)}{\sqrt{n(\Sigma x^2) - (\Sigma x)^2} \cdot \sqrt{n(\Sigma y^2) - (\Sigma y)^2}}$$

---

historical note

Karl Pearson (pîr'sən) spent most of his career as a mathematics professor at University College, London. Some of his major contributions concerned the development of statistical procedures such as regression analysis and correlation. He was particularly interested in applying these statistical concepts to the study of heredity. The term *standard deviation* was invented by Pearson, and because of his work in the area of correlation, the formal name given to the linear correlation coefficient is the *Pearson product moment coefficient of correlation.* Pearson was a co-founder of the statistical journal *Biometrika.* ■

If the linear correlation coefficient r is positive, the relationship between the variables has a **positive correlation.** In this case, if one variable increases, the other variable also tends to increase. If r is negative, the linear relationship between the variables has a **negative correlation.** In this case, if one variable increases, the other variable tends to decrease. Figure 12.23 shows some scatter diagrams along with the type of linear correlation that exists between the x and y variables. If r is positive, then the closer r is to 1 the stronger the linear relationship between the variables. If r is negative, then the closer r is to -1 the stronger the linear relationship between the variables.

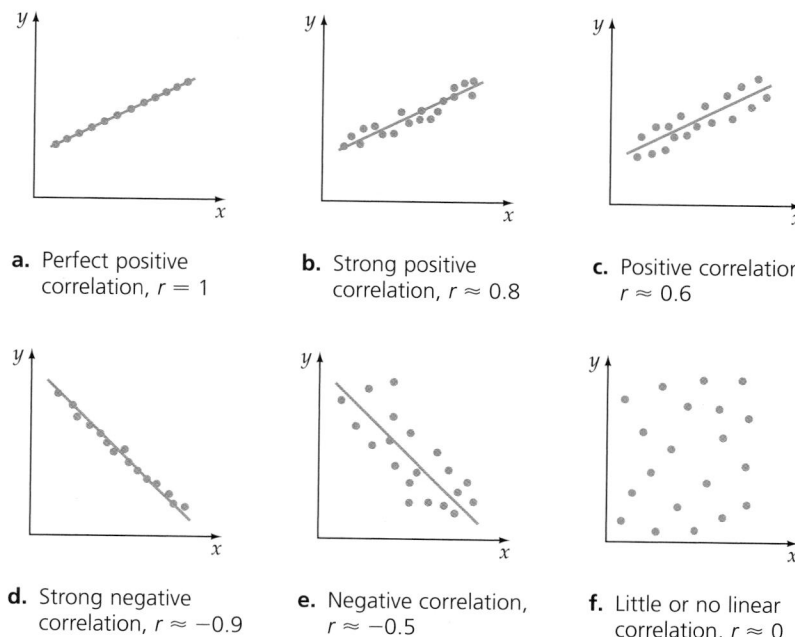

a. Perfect positive correlation, $r = 1$

b. Strong positive correlation, $r \approx 0.8$

c. Positive correlation, $r \approx 0.6$

d. Strong negative correlation, $r \approx -0.9$

e. Negative correlation, $r \approx -0.5$

f. Little or no linear correlation, $r \approx 0$

Figure 12.23 *Linear correlation*

EXAMPLE 3 ■ Find a Linear Correlation Coefficient

Find the linear correlation coefficient for stride length versus speed of an adult man. Use the data in Table 12.8a. Round your result to the nearest hundredth.

Solution

The ordered pairs are

$$(2.5, 3.4), (3.0, 4.9), (3.3, 5.5), (3.5, 6.6), (3.8, 7.0), (4.0, 7.7), (4.2, 8.3), (4.5, 8.7)$$

The number of ordered pairs is $n = 8$. In Table 12.9 on page 823 we found:

$$\Sigma x = 28.8 \qquad \Sigma y = 52.1 \qquad \Sigma x^2 = 106.72 \qquad \Sigma xy = 195.86$$

The only additional value that is needed is

$$\Sigma y^2 = 3.4^2 + 4.9^2 + 5.5^2 + 6.6^2 + 7.0^2 + 7.7^2 + 8.3^2 + 8.7^2$$
$$= 362.25$$

Substituting the above values into the equation for the linear correlation coefficient gives us

$$r = \frac{n(\Sigma xy) - (\Sigma x)(\Sigma y)}{\sqrt{n(\Sigma x^2) - (\Sigma x)^2} \cdot \sqrt{n(\Sigma y^2) - (\Sigma y)^2}}$$
$$= \frac{8(195.86) - (28.8)(52.1)}{\sqrt{8(106.72) - (28.8)^2} \cdot \sqrt{8(362.25) - (52.1)^2}}$$
$$\approx 0.993715$$

To the nearest hundredth, the linear correlation coefficient is 0.99.

CHECK YOUR PROGRESS 3   Find the linear correlation coefficient for stride length versus speed of a camel as given in Table 12.8c. Round your result to the nearest hundredth.

Solution   *See page S46.*

QUESTION   *What is the significance of the fact that the linear correlation coefficient is positive in Example 3?*

The linear correlation coefficient indicates the strength of a linear relationship between two variables; however, it does *not* indicate the presence of a *cause-and-effect relationship*. For instance, the data in Table 12.10 show the hours per week that a student spent playing pool and the student's weekly algebra test scores for those same weeks.

ANSWER   *It indicates a positive correlation between a man's stride length and his speed. That is, as a man's stride length increases, his speed also increases.*

Table 12.10 *Algebra Test Scores vs. Hours Spent Playing Pool*

Hours per week spent playing pool	4	5	7	8	10
Weekly algebra test score	52	60	72	79	83

The linear correlation coefficient for the ordered pairs in the table is $r \approx 0.98$. Thus there is a strong positive linear relationship between the student's algebra test scores and the time the student spent playing pool. This does not mean that the higher algebra test scores were caused by the increased time spent playing pool. The fact that the student's test scores increased with the increase in the time spent playing pool could be due to many other factors or it could just be a coincidence.

In your work with applications that involve the linear correlation coefficient r, it is important to remember the following properties of r.

Properties of the Linear Correlation Coefficient

1. The linear correlation coefficient r is always a real number between 1 and −1, inclusive. In the case in which

- all of the ordered pairs lie on a line with positive slope, r is 1.
- all of the ordered pairs lie on a line with negative slope, r is −1.

2. For any set of ordered pairs, the linear correlation coefficient r and the slope of the least-squares line both have the same sign.

3. Interchanging the variables in the ordered pairs does not change the value of r. Thus the value of r for the ordered pairs (x_1, y_1), (x_2, y_2), ..., (x_n, y_n) is the same as the value of r for the ordered pairs (y_1, x_1), (y_2, x_2), ..., (y_n, x_n).

4. The value of r does not depend on the units used. You can change the units of a variable from, for example, feet to inches and the value of r will remain the same.

MathMatters   **Use a Calculator to Find the Equation of the Least-Squares Line and the Linear Correlation Coefficient**

Calculators can be used to estimate the slope and y-intercept of the least-squares line for bivariate data. Many calculators will also estimate the linear correlation coefficient. **A TI-83 calculator displays the linear correlation coefficient only if you have used the** `DiagnosticOn` **command, which is found in the** `CATALOG` menu. Press $\boxed{\text{2nd}}$ `[CATALOG]`, scroll down to the `DiagnosticOn` command, and press $\boxed{\text{ENTER}}$. Now enter the first components of the ordered pairs into list `L1` and the second components into list `L2`, as shown in Figure 12.24 on the following page. The key sequence $\boxed{\text{STAT}}$ $\boxed{\blacktriangleright}$ $\boxed{4}$ $\boxed{\text{VARS}}$ $\boxed{\blacktriangleright}$ $\boxed{\text{ENTER}}$ $\boxed{\text{ENTER}}$ $\boxed{\text{ENTER}}$ stores the equation for the least-squares line in `Y1` and produces the `LinReg` display in which `a` is the slope of the least-squares line, `b` is the y-intercept of the least-squares line, and `r` is the linear correlation coefficient.

(continued)

1. Enter the data.

2. LinReg(ax + b) display

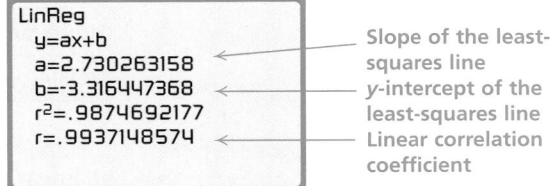

Slope of the least-squares line

y-intercept of the least-squares line

Linear correlation coefficient

3. The equation of the least-squares line is stored in the Y= menu.

```
Plot1 Plot2 Plot3
\Y1 ▤ 2.7302631578947X+-3
.3164473684209
\Y2 =
\Y3 =
\Y4 =
\Y5 =
\Y6 =
```

Figure 12.24 *TI-83 Screen Displays*

To display a scatter diagram of the ordered pairs and a graph of the least-squares line, use the WINDOW menu to enter appropriate values for Xmin, Xmax, Ymin, and Ymax. Use the key sequence [2nd] [STAT PLOT] to display the STAT PLOT menu. Select the scatter diagram icon and enter L1 to the right of Xlist: and L2 to the right of Ylist:. See Figure 12.25. Press the [GRAPH] key to display the scatter diagram of the data and the least-squares line.

1. Enter window settings.

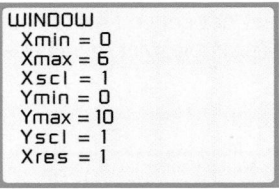

2. Use the STAT PLOT menu to choose settings.

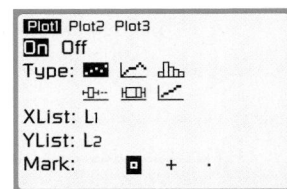

3. Press GRAPH to diplay scatter diagram and least-squares line.

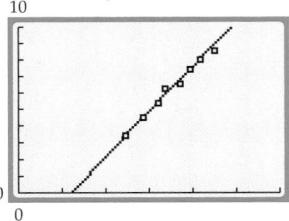

Figure 12.25 *TI-83 Screen Displays*

Excursion

An Application of Linear Regression

At this point, the work by Alexander and Jayes presented in this lesson cannot be used to estimate the speed of a dinosaur because for each animal species, we calculated a different least-squares regression line. Also, no dinosaurs are available to provide data for

(continued)

stride lengths and speeds. Motivated by a strong desire to find a mathematical model that could be used to estimate the speed of any animal from its stride length, Alexander came up with the idea of using *relative stride lengths*. A **relative stride length** is the number obtained by dividing the stride length of an animal by the animal's leg length. That is,

$$\text{Relative stride length} = \frac{\text{stride length}}{\text{leg length}} \qquad \text{(I)}$$

Thus a person with a leg length of 0.9 meter (distance from the hip to the ground) who runs steadily with a stride length of 4.5 meters has a relative stride length of (4.5 meters) ÷ (0.9 meters) = 5. Note that a relative stride length is a dimensionless quantity.

Because Alexander found it helpful to convert stride length to a dimensionless quantity (relative stride length), it was somewhat natural for him to also convert speed to a dimensionless quantity. His definition of *dimensionless speed* is as follows.

$$\text{Dimensionless speed} = \frac{\text{speed}}{\sqrt{\text{leg length} \times g}} \qquad \text{(II)}$$

where g is the gravitational acceleration constant of 9.8 meters per second per second. At this point you may feel that things are getting a bit complicated and that you weren't really all that interested in the speed of a dinosaur anyway. However, once Alexander and Jayes converted stride lengths to *relative* stride lengths and speeds to *dimensionless* speeds, they discovered that many graphs of their data, even for different species, were nearly linear! To illustrate this concept, examine Table 12.11, in which the ordered pairs were formed by converting each ordered pair of Table 12.8 (page 821) from the form (stride length, speed) to the form (relative stride length, dimensionless speed). The conversions were calculated by using leg lengths of 0.8 meter for the adult men, 0.5 meter for the dogs, and 1.2 meters for the camels.

Table 12.11 *Dimensionless Speed for Relative Stride Lengths*

a. Adult men

Relative stride length (x)	3.1	3.8	4.1	4.4	4.8	5.0	5.3	5.6
Dimensionless speed (y)	1.2	1.8	2.0	2.4	2.5	2.8	3.0	3.1

b. Dogs

Relative stride length (x)	3.0	3.4	4.0	4.8	5.4	6.0	6.4	7.0
Dimensionless speed (y)	1.7	2.0	2.2	3.2	3.5	4.1	4.0	4.5

c. Camels

Relative stride length (x)	2.1	2.5	2.7	2.8	2.9	3.2	3.3	3.5
Dimensionless speed (y)	0.7	1.1	1.3	1.5	1.6	1.8	2.1	2.2

(continued)

A scatter diagram of the data in Table 12.11 is shown in Figure 12.26. The scatter diagram shows a strong linear correlation. (You didn't expect a perfect linear correlation, did you? After all, we are working with camels, dogs, and adult men.) Although we have only considered three species, Alexander and Jayes were able to show a strong linear correlation for several species.

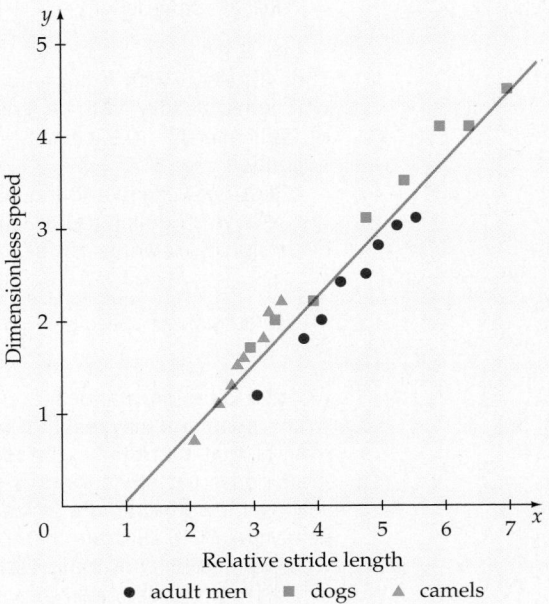

Figure 12.26

Finally it is time to estimate the speed of a dinosaur. Consider a large theropod with a leg length of 2.5 meters. If the theropod's fossilized footprints show a stride length of 5 meters, then its relative stride length is $\frac{5 \text{ m}}{2.5 \text{ m}} = 2$. The least-squares regression line in Figure 12.26 shows that a relative stride length of 2 has a dimensionless speed of about 0.8. If we use 2.5 meters for the leg length and 0.8 for dimensionless speed and solve equation (II) (page 829) for speed, we get

$$\text{Speed} = (\text{dimensionless speed})\sqrt{\text{leg length} \times 9.8}$$
$$= (0.8)\sqrt{2.5 \times 9.8}$$
$$\approx 4.0 \text{ meters per second}$$

For more information about estimating the speeds of dinosaurs, consult the following article by Alexander and Jayes. "A dynamic similarity hypothesis for the gaits of quadrupedal mammals." *Journal of Zoology* 201:135–152, 1983.

Excursion Exercises

1. **a.** Use a calculator to find the equation of the least-squares regression line for *all* of the data in Table 12.11 on the previous page.

   b. Find the linear correlation coefficient for the least-squares line in part a.

(continued)

sauropod

pachycephalosaur

2. The photograph at the right shows a set of sauropod tracks and a set of tracks made by a carnivore. These tracks were discovered by Roland Bird in 1938 in the Paluxy River bed, near the town of Glen Rose, Texas. Measurements of the sauropod tracks indicate an average stride length of about 4.0 meters. Assume that the sauropod that made the tracks had a leg length of 3.0 meters. Use the equation of the least-squares regression line from Excursion Exercise 1 to estimate the speed of the sauropod that produced the tracks.

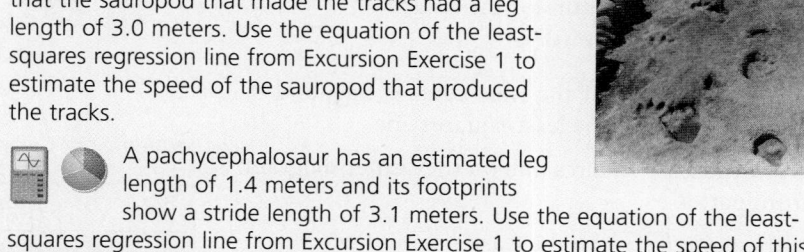

3. A pachycephalosaur has an estimated leg length of 1.4 meters and its footprints show a stride length of 3.1 meters. Use the equation of the least-squares regression line from Excursion Exercise 1 to estimate the speed of this pachycephalosaur.

Exercise Set 12.5   (Suggested Assignment: 1–21 odds)

1. Which of the scatter diagrams below suggests the

  a. strongest positive linear correlation between the x and y variables?

  b. strongest negative linear correlation between the x and y variables?

2. Which of the scatter diagrams below suggests

  a. a near perfect positive linear correlation between the x and y variables?

  b. little or no linear correlation between the x and y variables?

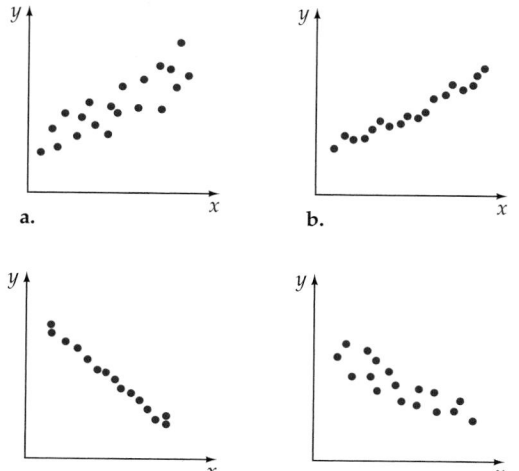

a.

b.

c.

d.

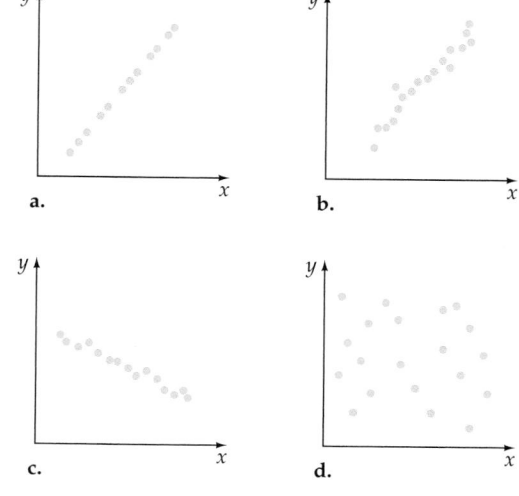

a.

b.

c.

d.

3. Given the bivariate data:

x	1	2	3	5	6
y	7	5	3	2	1

  a. Draw a scatter diagram for the data.

  b. Find n, Σx, Σy, Σx^2, $(\Sigma x)^2$, and Σxy.

  c. Find a, the slope of the least-squares line, and b, the y-intercept of the least-squares line.

  d. Draw the least-squares line on the scatter diagram from part a.

  e. Is the point $(\bar{x}, \bar{y})$ on the least-squares line?

  f. Use the equation of the least-squares line to predict the value of y when $x = 3.4$.

  g. Find, to the nearest hundredth, the linear correlation coefficient.

4. Given the bivariate data:

x	3	4	5	6	7
y	2	3	3	5	5

  a. Draw a scatter diagram for the data.

  b. Find n, Σx, Σy, Σx^2, $(\Sigma x)^2$, and Σxy.

  c. Find a, the slope of the least-squares line, and b, the y-intercept of the least-squares line.

  d. Draw the least-squares line on the scatter diagram from part a.

  e. Is the point $(\bar{x}, \bar{y})$ on the least-squares line?

  f. Use the equation of the least-squares line to predict the value of y when $x = 7.3$.

  g. Find, to the nearest hundredth, the linear correlation coefficient.

In Exercises 5–10, find the equation of the least-squares regression line and the linear correlation coefficient for the given data. Round the constants, a, b, and r, to the nearest hundredth.

5. {(2, 6), (3, 6), (4, 8), (6, 11), (8, 18)}

6. {(2, −3), (3, −4), (4, −9), (5, −10), (7, −12)}

7. {(−3, 11.8), (−1, 9.5), (0, 8.6), (2, 8.7), (5, 5.4)}

8. {(−7, −11.7), (−5, −9.8), (−3, −8.1), (1, −5.9), (2, −5.7)}

9. {(1, 4.1), (2, 6.0), (4, 8.2), (6, 11.5), (8, 16.2)}

10. {(2, 5), (3, 7), (4, 8), (6, 11), (8, 18), (9, 21)}

11. 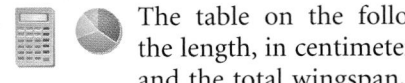 The following table shows the number of students per computer in U.S. public schools for recent years.

Students per Computer in U.S. Public Schools

Year	'94–5	'95–6	'96–7	'97–8	'98–9	'99–00
Students per computer	10.5	10	7.8	6.1	5.7	5.4

Source: Quality Education Data, Inc., Denver, CO, Technology in Public Schools, 1983–2000. Data extracted from *The World Almanac and Book of Facts 2002*, page 233.

  a. Find the equation of the least-squares regression line for the data, with 0 representing the school year 1994–95, 1 representing the year 1995–96, and so on.

  b. Use the equation from part a to predict the number of students per computer for the 2001–02 school year.

  c. Find the linear correlation coefficient for the data. Round to the nearest hundredth.

  d. What is the significance of the fact that the linear correlation coefficient is negative for these data?

12. 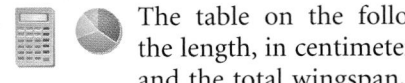 The table on the following page shows the length, in centimeters, of the humerus and the total wingspan, in centimeters, of several pterosaurs, which are extinct flying reptiles. (*Source:* Southwest Educational Development Laboratory)

Pterosaur Data

Humerus	Wingspan	Humerus	Wingspan
24	600	20	500
32	750	27	570
22	430	15	300
17	370	15	310
13	270	9	240
4.4	68	4.4	55
3.2	53	2.9	50
1.5	24		

a. Find the equation of the least-squares regression line for the data.

b. Use the equation from part a to determine the projected wingspan of the *plerosaur quetzalcoatlus northropi* if its humerus is 54 centimeters.

13. The following table shows the approximate number of sport utility vehicle sales in the U.S. for recent years.

Sport Utility Vehicle Sales in the U.S.

Year	1995	1996	1997	1998	1999	2000
Sales of SUVs, in millions	1.7	2.1	2.4	2.8	2.9	3.0

Source: Ward's Communications. Data extracted from *The World Almanac and Book of Facts 2002*, page 226.

a. Find the equation of the least-squares regression line for the data, with 0 representing the year 1995, 1 representing the year 1996, and so on.

b. Use the equation from part a to predict the number of SUV sales for the year 2002.

14. The data in the following table are based on a study by R. A. Fisher of various flowers of the iris family. The width, in centimeters, and length, in centimeters, of the petals for selected flowers are displayed in the following table.

Iris Petal Data

Width	Length	Width	Length
2	14	24	56
23	51	10	36
20	52	19	51
13	45	16	47
17	45	14	47
16	31	17	45
14	47	16	31

a. Find the equation of the least-squares regression line for the data.

b. Use the equation from part a to estimate the length of an iris petal that has a width of 18 centimeters.

15. The following table shows the approximate number of cellular telephone subscriptions in the U.S. for recent years.

U.S. Cellular Telephone Subscriptions

Year	1995	1996	1997	1998	1999	2000
Subscriptions, in millions	34	44	55	69	86	109

Source: CTIA Semiannual Wireless Survey. Data extracted from *The World Almanac and Book of Facts 2002*, page 637.

a. Find the linear correlation coefficient for the data.

b. On the basis of the value of the linear correlation coefficient, would you conclude, at the $|r| > 0.9$ level, that the data can be reasonably modeled by a linear equation? Explain.

16. The average remaining lifetimes for men in the United States are given in the following table. (*Source:* National Institute of Health)

Average Remaining Lifetimes for Men

Age	Years	Age	Years
0	73.6	65	15.9
15	59.4	75	9.9
35	40.8		

Use the linear correlation coefficient to determine whether there is a strong correlation, at the $|r| > 0.9$ level, between a man's age and the average remaining lifetime for that man.

17. The following table lists the median incomes for women and men for selected years from 1975 to 1998. The incomes are in thousands of dollars.

Median Income per Year
(in thousands of dollars; based on 1998 dollars)

	1975	1980	1985	1990	1995	1998
Women (x)	9.8	9.7	10.9	12.6	13.0	14.4
Men (y)	25.7	24.8	24.7	25.3	24.1	26.5

Source: Data extracted from *Time Almanac 2001 with Information Please*, page 635.

a. Find the linear correlation coefficient for the data.

b. Based on your answer to part a, would you say that there is a strong linear relationship, at the $|r| > 0.9$ level, between the median incomes of the women and the median incomes of the men? Explain.

18. The average remaining lifetimes for women in the United States are given in the following table. (*Source:* National Institute of Health)

Average Remaining Lifetimes
for Women

Age	Years	Age	Years
0	79.4	65	19.2
15	65.1	75	12.1
35	45.7		

a. Find the equation of the least-squares regression line for the data.

b. Use the equation from part a to estimate the remaining lifetime of a woman of age 25.

19. A tourist remembers the data given in the following table, which shows equivalent temperatures on the Celsius temperature scale and the Fahrenheit temperature scale.

Celsius ($x°$)	−40	0	100
Fahrenheit ($y°$)	−40	32	212

a. Find the linear correlation coefficient for the data.

b. What is the significance of the value found in part a?

c. Find the equation of the least-squares regression line.

d. Use the equation of the least-squares line from part c to predict the Fahrenheit temperature that corresponds to a Celsius temperature of 35°.

e. Is the procedure in part d an example of interpolation or extrapolation?

20. An aerobic exercise instructor remembers the data given in the following table, which shows the recommended maximum exercise heart rates for individuals of the given ages.

Age (x years)	20	40	60
Maximum heart rate (y beats per minute)	170	153	136

a. Find the linear correlation coefficient for the data.

b. What is the significance of the value found in part a?

c. Find the equation of the least-squares regression line.

d. Use the equation of the least-squares line from part c to predict the maximum exercise heart rate for person who is 72.

e. Is the procedure in part d an example of interpolation or extrapolation?

Extensions

CRITICAL THINKING

21. The following table shows the Environmental Protection Agency (EPA) fuel efficiency estimates for city and highway driving for 10 selected luxury cars. (*Source:* **www.money.com**, May 26, 2000)

EPA Miles-per-Gallon Estimates for City and Highway Driving for Selected Luxury Cars

Car	City mpg	Highway mpg
Acura RL	18	24
Audi A8, 4.2L	17	24
BMW 528i	21	29
Cadillac Deville	17	27
Infiniti Q45	18	23
Jaguar XJ8	17	24
Lexus LS400	18	25
Lincoln Continental	17	25
Mercedes S500	16	23
Saab	18	24

Is there a strong linear relationship, at the $|r| > 0.9$ level, between city miles per gallon and highway miles per gallon for these cars? Explain.

22. The following table shows the approximate numbers of bachelor's degrees conferred on women and men for selected years from 1959 to 2002. Data for 2001–2002 are projected. (*Source:* National Center for Educational Statistics. Extracted from the *World Almanac and Book of Facts 2002*, page 237.)

Bachelor's Degrees Conferred (in hundreds of thousands)

Year	59–60	69–70	79–80	89–90	01–02
Women	1.4	3.3	4.6	5.6	6.7
Men	2.5	4.6	4.8	5.0	5.1

In parts a and b, use $x = 0$ to represent 1959–1960, $x = 10$ to represent 1969–1970, and so on.

   a. Find the equation of the least-squares regression line for the number of bachelor's degrees conferred on women as a function of the year.

   b. Find the equation of the least-squares regression line for the number of bachelor's degrees conferred on men as a function of the year.

   c. Is there a strong linear relationship, at the $r > 0.9$ level, between the number of bachelor's degrees conferred on women in a given year and the number of bachelor's degrees conferred on men in the same year? Explain.

EXPLORATIONS

23. Another linear model that can be used to model data is called the *median-median line*. Use a statistics text or the Internet to read about the median-median line.

   a. Find the equation of the median-median line for the data given in Exercise 13.

   b. Explain the type of situation in which it would be better to model data using the median-median line than it would be to model the data using the least-squares line.

24. Search for bivariate data (in a magazine, a newspaper, an almanac, or on the Internet) that can be closely modeled by a linear equation.

   a. Draw a scatter diagram of the data.

   b. Find the equation of the least-squares line and the linear correlation coefficient for the data.

   c. Graph the least-squares line on the scatter diagram in part a.

   d. Use the equation of the least-squares line to predict a range value for a specific domain value.

CHAPTER 12 **Summary**

Key Terms

average [p. 768]
bimodal distribution [p. 808]
bivariate data [p. 821]
box-and-whisker plot [p. 798]

class [p. 805]
continuous data [p. 809]
continuous variable [p. 809]
descriptive statistics [p. 768]

Essential Concepts

■ The *mean* of n numbers is the sum of the numbers divided by n. The mean of a sample is denoted by $\bar{x}$ and the mean of a population is denoted by μ.

■ The *median* of a *ranked list* of n numbers is the middle number if n is odd and the mean of the two middle numbers if n is even.

■ The *mode* of a list of numbers is the number that occurs most frequently.

■ The *weighted mean* of the n numbers $x_1, x_2, x_3, \ldots, x_n$ with the respective assigned weights $w_1, w_2, w_3, \ldots, w_n$ is $\frac{\Sigma(x \cdot w)}{\Sigma w}$, where $\Sigma(x \cdot w)$ is the sum of the products formed by multiplying each number by its assigned weight, and Σw is the sum of all the weights.

■ The *range* of a set of data values is the difference between the largest data value and the smallest data value.

■ The *standard deviation* of the *sample* $x_1, x_2, x_3, \ldots, x_n$ with a mean of $\bar{x}$ is $s = \sqrt{\dfrac{\Sigma(x - \bar{x})^2}{n - 1}}$.

■ The *standard deviation* of the *population* $x_1, x_2, x_3, \ldots, x_n$ with a mean of μ is
$$\sigma = \sqrt{\frac{\Sigma(x - \mu)^2}{n}}.$$

■ The *z-score* for a given data value x is the number of standard deviations that x lies above or below the mean of the data.

Population: $z_x = \dfrac{x - \mu}{\sigma}$ Sample: $z_x = \dfrac{x - \bar{x}}{s}$

■ A value x is called the *pth percentile* of a data set provided $p\%$ of the data is less than x.

■ **The Empirical Rule**
In a normal distribution, about
   68.2% of the data lies within 1 standard deviation of the mean.
   95.4% of the data lies within 2 standard deviations of the mean.
   99.7% of the data lies within 3 standard deviations of the mean.

■ The *standard normal distribution* is the normal distribution for the continuous variable z that has a mean of 0 and a standard deviation of 1.

■ In the standard normal distribution, the *area* of the distribution from $z = a$ to $z = b$ represents the percentage of z-values that lie in the interval from a to b and the probability that z lies in the interval from a to b.

■ The *least-squares regression line* is the line that minimizes the sum of the squares of the vertical deviations from each data point to the line.

■ The equation of the *least-squares line* for the n ordered pairs $(x_1, y_1), (x_2, y_2), (x_3, y_3), \ldots, (x_n, y_n)$ is $\hat{y} = ax + b$, where $a = \dfrac{n\Sigma xy - (\Sigma x)(\Sigma y)}{n\Sigma x^2 - (\Sigma x)^2}$ and $b = \bar{y} - a\bar{x}$.

■ For the n ordered pairs $(x_1, y_1), (x_2, y_2), (x_3, y_3), \ldots, (x_n, y_n)$, the *linear correlation coefficient* r is given by
$$r = \frac{n(\Sigma xy) - (\Sigma x)(\Sigma y)}{\sqrt{n(\Sigma x^2) - (\Sigma x)^2} \cdot \sqrt{n(\Sigma y^2) - (\Sigma y)^2}}.$$

Review Exercises

1. Find the mean, the median, and the mode for the following data. Round noninteger values to the nearest tenth.

$$12, 17, 14, 12, 8, 19, 21$$

2. A set of data has a mean of 16, a median of 15, and a mode of 14. Which of these numbers must be a value in the data set?

3. Write a set of data with five data values for which the mean, median, and mode are all 55.

4. State whether the mean, the median, or the mode is being used.

   a. In 2002, there were as many people aged 25 and younger in the world as there were people aged 25 and older.

   b. The majority of full-time students carry a load of 15 credit hours per semester.

   c. The average annual return on an investment is 6.5%.

5. The lengths of cantilever bridges in the United States are shown below. Find the mean, the median, the mode, and the range of the data.

   Bridge Length (in feet)
   Baton Rouge (Louisiana), 1235
   Commodore John Barry (Pennsylvania), 1644
   Greater New Orleans (Louisiana), 1576
   Longview (Washington), 1200
   Patapsco River (Maryland), 1200
   Queensboro (New York), 1182
   Tappan Zee (New York), 1212
   Transbay Bridge (California), 1400

6. Cleone traveled 45 miles to her sister's house in 1 hour. The return trip took 1.5 hours. What was Cleone's average rate for the entire trip?

7. In a 4.0 grading system, each letter grade has the following numerical value.

A = 4.00	B− = 2.67	D+ = 1.33
A− = 3.67	C+ = 2.33	D = 1.00
B+ = 3.33	C = 2.00	D− = 0.67
B = 3.00	C− = 1.67	F = 0.00

   Use the weighted mean formula to find the grade point average for a student with the following grades.

Course	Credits	Grade
Mathematics	3	A
English	3	C+
Computer	2	B−
Biology	4	B
Art	1	A

8. A teacher finds that the test scores of a group of 40 students have a mean of 72 and a standard deviation of 8.

   a. If Ann has a test score of 82, what is Ann's z-score?

   b. What is Ann's percentile score?

9. An airline recorded the times it took for a ground crew to unload the baggage from an airplane. The recorded times, in minutes, were 12, 18, 20, 14, and 16. Find the *sample* standard deviation and the variance of these times. Round your results to the nearest hundredth.

10. The following table gives the average annual admission prices to U.S. movie theaters for the years 1991 to 2000.

    Average Annual Admission Price

1991	$4.21	1996	$4.42
1992	$4.15	1997	$4.59
1993	$4.14	1998	$4.69
1994	$4.08	1999	$5.08
1995	$4.35	2000	$5.39

    Source: NATO average ticket price based on Ernst & Young survey; MPAA Worldwide Market Research

    Find the mean, the median, and the standard deviation for this *sample* of admission prices.

11. One student received test scores of 85, 92, 86, and 89. A second student received scores of 90, 97, 91, and 94 (exactly five points more on each test).

    a. What is the relationship between the means of the two students' test scores?

    b. What is the relationship between the standard deviations of the two students' test scores?

12. A *population* data set has a mean of 81 and a standard deviation of 5.2. Find the *z*-scores for each of the following. Round to the nearest hundredth.

a. $x = 72$

b. $x = 84$

13. The cholesterol levels for 10 adults are shown below. Draw a box-and-whiskers plot of the data.

Cholesterol Levels

310	185	254	221	170
214	172	208	164	182

14. The following histogram shows the distribution of the test scores for a history test.

a. How many students scored at least 84 on the test?

b. How many students took the test?

c. What is the uniform class width?

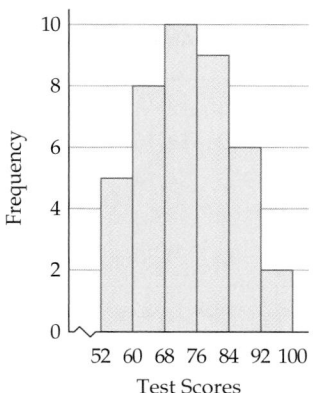

15. Use the following relative frequency distribution to determine the

a. *percent* of the states that paid an average teacher salary of at least $45,000.

b. *probability* that a state selected at random paid an average teacher salary of at least $30,000 but less than $39,000.

Average Salaries of Public School Teachers, 1998–1999

Average Salary, s	Number of States	Relative Frequency
$27,000 ≤ s < $30,000	3	6%
$30,000 ≤ s < $33,000	7	14%
$33,000 ≤ s < $36,000	12	24%
$36,000 ≤ s < $39,000	9	18%
$39,000 ≤ s < $42,000	6	12%
$42,000 ≤ s < $45,000	3	6%
$45,000 ≤ s < $48,000	5	10%
$48,000 ≤ s < $51,000	3	6%
$51,000 ≤ s < $54,000	2	4%

Source: NEA Average Salaries of Public School Teachers, 1998–99.

16. The following histogram appeared in the newspaper *USA TODAY*.

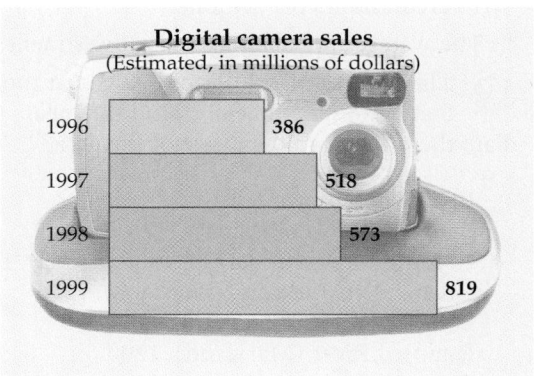

Source: http://www.usa-today.com/snapshot/money/msnap023.htm.
© Copyright 2002 *USA Today*, a division of Gannet Co.

a. Determine the linear correlation coefficient for the number of camera sales as a function of the year. Represent the year 1996 by $x = 0$, the year 1997 by $x = 1$, and so on.

b. If you computed the linear correlation coefficient in part a by using $x = 1996$ to represent the year 1996, $x = 1997$ to represent the year 1997, and so on, what effect would this have on your result?

c. On the basis of the value of the linear correlation coefficient, would you conclude that the data can be reasonably modeled by a linear equation? *Note:* Assume that data can be reasonably modeled by a linear equation provided $|r| > 0.9$.

17. The following histogram appeared in the newspaper *USA TODAY*.

Teens feel most pressure to...

Fit in socially — 29%
Get into college — 32%
Get good grades — 44%

Source: © Copyright 1997 *USA Today*, a division of Gannet Co. Inc.

Many statistics texts suggest that a histogram should have between 5 and 15 classes. The above histogram has only three classes. What reason might a graphic artist give for using such a small number of classes?

18. A professor gave a final examination to 110 students. Eighteen students had examination scores that were more than one standard deviation above the mean. With this information, can you conclude that 18 of the students had examination scores that were less than one standard deviation below the mean? Explain.

19. The amount of time customers spend waiting in line at the ticket counter of an amusement park is normally distributed, with a mean of 6.5 minutes and a standard deviation of 1 minute. Find the probability that the time a customer will spend waiting is:

a. less than 8 minutes.   **b.** less than 6 minutes.

20. The weights of all the sacks of dog food filled by a machine are normally distributed, with an average weight of 50 pounds and a standard deviation of 0.5 pound. What percent of the sacks will

a. weigh less than 49.6 pounds?

b. weigh between 49 and 51 pounds?

21. A telephone company finds that the life spans of its telephones are normally distributed, with a mean of 6.5 years and a standard deviation of 0.5 year.

a. What percent of its telephones will last at least 7.25 years?

b. What percent of its telephones will last between 5.8 years and 6.8 years?

c. What percent of its telephones will last less than 6.9 years?

22. Find the equation of the least-squares regression line and the linear correlation coefficient for the following data.

$$\{(2, 7), (3, 9), (4, 14), (7, 15)\}$$

23. Given the bivariate data

x	10	12	14	15	16
y	8	7	5	4	1

a. Draw a scatter diagram for the data.

b. Find n, Σx, Σy, Σx^2, $(\Sigma x)^2$, and Σxy.

c. Find a, the slope of the least-squares regression line, and b, the y-intercept of the least-squares line.

d. Draw the least-squares line on the scatter diagram from part a.

e. Is the point $(\bar{x}, \bar{y})$ on the least-squares line?

f. Use the equation of the least-squares line to predict the value of y for $x = 8$.

g. Find the linear correlation coefficient.

24. A student has recorded the data in the following table, which shows the distance a spring stretches in inches for a given weight in pounds.

Weight, x	80	100	110	150	170
Distance, y	6.2	7.4	8.3	11.1	12.7

a. Find the linear correlation coefficient.

b. Find the equation of the least-squares line.

c. Use the equation of the least-squares line from part b to predict the distance a weight of 195 pounds will stretch the spring.

25. A test of an Internet service provider showed the following download times (in seconds) for files of various sizes (in kilobytes).

Download Times			
Size	Time	Size	Time
10.5	0.20	110	2.01
12.9	0.24	156	2.68
15	0.27	163	2.87
20	0.36	175	3.10
60	1.09	200	3.64
75	1.42	250	4.61

a. Find the equation of the least-squares line for these data.

b. On the basis of the value of the linear correlation coefficient, is a linear model of these data a reasonable model?

c. Use the equation of the least-squares line from part a to predict the expected download time of a file that is 100 kilobytes in size.

26. The U.S. Consumer Price Index (CPI) is a measure of the average change in prices over time. In the following table, the CPI for each year is based on a cost of $100 in 1967. For example, the CPI for the year 1985 is 322.2. This means that in 1985 it took $322.20 to purchase the same goods that cost $100 in the year 1967.

Consumer Price Index, 1970–2001 (1967 has a CPI of 100)

Year	1970	1975	1980	1985	1990	1995	2001*
CPI	116.3	161.2	248.8	322.2	391.4	456.5	529.1

*Average for first half of 2001
Source: Bureau of Labor Statistics, U.S. Dept. of Labor. Data taken from *The World Almanac and Book of Facts 2002,* page 103.

a. Find the equation of the least-squares regression line and the linear correlation coefficient for the data in the table. Let the year 1970 be represented by $x = 0$, 1975 by $x = 5$, and so on.

b. Use the equation of the least-squares regression line to predict the CPI for the year 2003.

CHAPTER 12 Test

1. Find the mean, the median, and the mode for the following data. Round noninteger values to the nearest tenth.

$$3, 7, 11, 12, 7, 9, 15$$

2. A professor grades students on three tests, two quizzes, and a final examination. Each quiz counts as one-half a test and the final examination counts as two tests. Pam has test scores of 90, 75, and 84. Pam's quiz scores are 86 and 50. Her final examination score is 88. Use the weighted mean formula to find Pam's average for the course. Round to the nearest tenth.

3. Find the range, the standard deviation, and the variance for the following sample data.

$$7, 11, 12, 15, 31, 22$$

4. A *sample* data set has a mean of $\bar{x} = 65$ and a standard deviation of 10.2. Find the z-scores for each of the following. Round to the nearest hundredth.

a. $x = 77$ **b.** $x = 60$

5. Draw a box-and-whisker plot for the following data.

Points Scored by the Green Bay Packers in the 2001 Regular Season

28	37	28	10	31	13	21	20
20	29	28	17	20	30	24	34

6. Use the following frequency distribution to estimate what *percent* of the movie attendees were

a. at least 40 years of age.

b. at least 21 but less than 40 years of age.

Movie Attendance by Age Group, 1999

Age Group	Percent of Total Yearly Admissions
12–15	11%
16–20	20%
21–24	10%
25–29	12%
30–39	18%
40–49	14%
50–59	7%
60+	8%

Source: MPA Worldwide Market Research

7. During 1 month, an overnight delivery company found that the weights of its parcels were normally distributed, with a mean of 34 ounces and a standard deviation of 10 ounces. Use the Empirical Rule to determine

a. what percent of the parcels weighed between 34 ounces and 54 ounces.

b. what percent of the parcels weighed less than 24 ounces.

8. The weights of all the boxes of cake mix filled by a machine are normally distributed, with an average weight of 18.0 ounces and a standard deviation of 0.8 ounce. What percent of the boxes will

a. weigh less than 17 ounces?

b. weigh between 18.4 and 19.0 ounces?

9. Find the equation of the least-squares regression line and the linear correlation coefficient for the following data. Round constants to the nearest hundredth.

$$\{(2, 5), (3, 7), (4, 11), (7, 12)\}$$

10. The following table shows the percent of water and the number of calories in various canned soups to which 100 grams of water are added.

% Water	Calories
93.2	28
92.3	26
91.9	39
89.5	56
89.6	56
90.5	36
91.9	32
91.7	32

Percent Water in Soups

a. Find the equation of the least-squares regression line for the data. Round constants to the nearest hundredth.

b. Use the equation in part a to find the expected number of calories in a soup that is 89% water. Round to the nearest whole number.

Apportionment and Voting

The two houses of Congress are the Senate and House of Representatives, which convene in the Capitol building. There are 50 senators, two from each state in the United States. The 435 representatives in the House of Representatives are apportioned among the states according to their populations. Different formulas, including the Jefferson and Huntington-Hill methods, have been used to determine the number of representatives from each state. The **Excursion exercises on page 858** will demonstrate how these different methods affect the result.

13.1	Introduction to Apportionment
13.2	Introduction to Voting
13.3	Weighted Voting Systems

Need help? For on-line student resources, such as section quizzes, visit this textbook's web site at **math.college.hmco.com/students.**

George Washington

In this chapter, we discuss two of the most fundamental principles of a democracy: the right to vote and the value of that vote. There are really two issues here. The right to vote does not necessarily guarantee that each vote cast has the same influence on the outcome of an election. For instance, in a dictatorship, people may vote, but those votes can be overruled by the dictator.

The U.S. Constitution, Article I, Section 2, states in part

The House of Representatives shall be composed of members chosen every second year by the people of the several states, and the electors in each state shall have the qualifications requisite for electors of the most numerous branch of the state legislature.... Representatives and direct taxes shall be apportioned among the several states which may be included within this union, according to their respective numbers.... The actual Enumeration shall be made within three years after the first meeting of the Congress of the United States, and within every subsequent term of ten years, in such manner as they shall by law direct. The number of Representatives shall not exceed one for every thirty thousand, but each state shall have at least one Representative;...

This article of the Constitution requires that "Representatives...be *apportioned* (our italics) among the several states...according to their respective numbers...". That is, the number of representatives each state sends to Congress should be based on its population. Because populations change over time, this article also requires that the number of people within a state be counted "within every subsequent term of ten years." This is why we have a census every 10 years.

The way in which representatives are *apportioned* has been a contentious issue since the founding of the United States. The first presidential veto was issued by George Washington in 1792 because he did not approve of the way the House of Representatives had decided to apportion the number of representatives for each state. Under the plan passed by Congress (the Hamilton plan), Washington's home state of Virginia would receive 17 representatives. However, a competing method of apportionment (the Jefferson plan) would give Virginia 18 representatives. Congress could not override the veto and eventually voted to accept the Jefferson plan. Ever since that first veto, the issue of how to apportion the House of Representatives has been revisited many times.

Introduction to Apportionment

historical note

The original Article I, Section 2 of the U.S. Constitution included wording as to how to count citizens. According to this article, the numbers "shall be determined by adding to the whole number of free persons, including those bound to service for a term of years, and excluding Indians not taxed, three fifths of all other Persons." The "three fifths of all other Persons" meant that a slave was counted as only $\frac{3}{5}$ of a person. This article was modified by Section 1 of the 14th Amendment to the Constitution, which, in part, states that "All persons born or naturalized in the United States, and subject to the jurisdiction thereof, are citizens of the United States and of the state wherein they reside...." ∎

The mathematical investigation into **apportionment,** which is a method of dividing a whole into various parts, has its roots in the U.S. Constitution. (See the chapter opener on page 843.) Since 1790, when the House of Representatives first attempted to apportion itself, various methods have been used to decide how many voters would be represented by each member of the House. The two competing plans in 1790 were put forward by Alexander Hamilton and Thomas Jefferson.

To illustrate how the Hamilton and Jefferson plans were used to calculate the number of representatives each state should have, we will consider the fictitious country of Andromeda, with a population of 20,000 and five states. The population of each state is given in the table at the right.

Andromeda's constitution calls for 25 representatives to be chosen from these states. The number of representatives is to be apportioned according to the states' respective populations.

Andromeda

State	Population
Apus	11,123
Libra	879
Draco	3518
Cephus	1563
Orion	2917
Total	20,000

The Hamilton Plan

Under the Hamilton plan, the total population of the country (20,000) is divided by the number of representatives (25). This gives the number of citizens represented by each representative. The number is called the *standard divisor.*

Standard Divisor

$$\text{Standard divisor} = \frac{\text{total population}}{\text{number of people to apportion}}$$

For Andromeda, we have

$$\text{Standard divisor} = \frac{\text{total population}}{\text{number of people to apportion}} = \frac{20,000}{25} = 800$$

QUESTION *What is the meaning of the number 800 calculated above?*

Now divide the population of each state by the standard divisor and round the quotient *down* to a whole number. For example, both 15.1 and 15.9 would be rounded to 15. Each whole number quotient is called a *standard quota.*

✓ **TAKE NOTE**

Today apportionment is applied to situations other than a population of people. For instance, population could refer to the number of math classes offered at a college or the number of fire stations in a city. Nonetheless, the definition of standard divisor is still given as if people were involved.

ANSWER *It is the number of citizens represented by each representative.*

Standard Quota

The **standard quota** is the whole number part of the quotient of a population divided by the standard divisor.

State	Population	Quotient	Standard quota
Apus	11,123	$\frac{11{,}123}{800} \approx 13.904$	13
Libra	879	$\frac{879}{800} \approx 1.099$	1
Draco	3518	$\frac{3518}{800} \approx 4.398$	4
Cephus	1563	$\frac{1563}{800} \approx 1.954$	1
Orion	2917	$\frac{2917}{800} \approx 3.646$	3
		Total	22

From the calculations in the above table, the total number of representatives is 22, not 25 as required by Andromeda's constitution. When this happens, the Hamilton plan calls for revisiting the calculation of the quotients and assigning an additional representative to the state with the largest decimal remainder. This process is continued until the number of representatives equals the number required by the constitution. For Andromeda, we have

TAKE NOTE

Additional representatives are assigned according to the largest decimal remainders. Because the sum of the standard quotas came to only 22 representatives, we must add three more representatives. The states with the three highest decimal remainders are Cephus (1.954), Apus (13.904), and Orion (3.646). Thus each of these states gets an additional representative.

State	Population	Quotient	Standard quota	Number of representatives
Apus	11,123	$\frac{11{,}123}{800} \approx 13.904$	13	14
Libra	879	$\frac{879}{800} \approx 1.099$	1	1
Draco	3518	$\frac{3518}{800} \approx 4.398$	4	4
Cephus	1563	$\frac{1563}{800} \approx 1.954$	1	2
Orion	2917	$\frac{2917}{800} \approx 3.646$	3	4
		Total	22	25

The Jefferson Plan

As we saw with the Hamilton plan, dividing by the standard divisor and then rounding down does not always yield the correct number of representatives. In the example above, we were three representatives short. The Jefferson plan attempts to overcome this difficulty by using a *modified standard divisor*. This number is chosen, by trial and error, so that the sum of the standard quotas is the total number of representatives. For the following calculations, we used 740 as the modified standard divisor.

State	Population	Quotient	Standard quota	Number of representatives
Apus	11,123	$\frac{11{,}123}{740} \approx 15.031$	15	15
Libra	879	$\frac{879}{740} \approx 1.188$	1	1
Draco	3518	$\frac{3518}{740} \approx 4.754$	4	4
Cephus	1563	$\frac{1563}{740} \approx 2.112$	2	2
Orion	2917	$\frac{2917}{740} \approx 3.942$	3	3
		Total	25	25

The table below shows how the results of the Hamilton and Jefferson apportionment methods differ. Note that each method assigns a different number of representatives to certain states.

State	Population	Hamilton plan	Jefferson plan
Apus	11,123	14	15
Libra	879	1	1
Draco	3518	4	4
Cephus	1563	2	2
Orion	2917	4	3
	Total	25	25

Although we have applied apportionment to allocating representatives to a congress, there are other applications of apportionment. For instance, nurses can be assigned to hospitals according to the number of patients requiring care; police officers can be assigned to precincts based on the number of reported crimes; math classes can be scheduled based on student demand for those classes. These are only some of the ways apportionment can be used.

Ruben County

City	Population
Cardiff	7020
Solana	2430
Vista	1540
Pauma	3720
Pacific	5290

EXAMPLE 1 ■ **Apportioning Board Members Using the Hamilton Method**

Suppose the 18 members on the board of the Ruben County environmental agency are selected according to the populations of the five cities in the county, as shown in the table at the left.

a. Use the Hamilton method to determine the number of board members each city should have.

b. Use the Jefferson method to determine the number of board members each city should have.

Solution

a. First find the total population of the counties.

$$7020 + 2430 + 1540 + 3720 + 5290 = 20{,}000$$

Now calculate the standard divisor.

$$\text{Standard divisor} = \frac{\text{population of county}}{\text{number of board members}} = \frac{20{,}000}{18} \approx 1111.11$$

Use the standard divisor to find the standard quota for each city.

City	Population	Quotient	Standard quota	Number of board members
Cardiff	7020	$\frac{7020}{1111.11} \approx 6.318$	6	6
Solana	2430	$\frac{2430}{1111.11} \approx 2.187$	2	2
Vista	1540	$\frac{1540}{1111.11} \approx 1.386$	1	2
Pauma	3720	$\frac{3720}{1111.11} \approx 3.348$	3	3
Pacific	5290	$\frac{5290}{1111.11} \approx 4.761$	4	5
		Total	16	18

The sum of the standard quotas is 16, so we must add two more members. The two cities with the largest decimal remainders are Pacific and Vista. Each of these two cities gets one additional board member. Thus the composition of the environmental board is Cardiff: 6, Solana: 2, Vista: 2, Pauma: 3, Pacific: 5.

CALCULATOR NOTE

Using lists and the iPart function (which returns only the whole number part of a number) of a TI-83 calculator can be helpful when trying to find a modified standard divisor. Press STAT ENTER to display the list editor. If a list is present in L1, use the up arrow key to high-light L1, then press CLEAR ENTER. Enter the populations of each city in L1.

Press 2nd [QUIT]. To divide each number in L1 by a modified divisor (we are using 950 for this example), enter the following.

MATH ▶ 3 2nd [L1]

÷ 950)

STO 2nd [L2] ENTER

```
iPart(L1/950)→L2
            {7 2 1 3 5}
```

The standard quota is shown on the screen. The sum of these numbers is 18, the desired number of representatives.

b. To use the Jefferson method, we must find a *modified* standard divisor that is less than the standard divisor we calculated in part a. We must do this by trial and error. For instance, if we choose 925 as the modified standard divisor, we have the following result.

City	Population	Quotient	Standard quota	Number of board members
Cardiff	7020	$\frac{7020}{925} \approx 7.589$	7	7
Solana	2430	$\frac{2430}{925} \approx 2.627$	2	2
Vista	1540	$\frac{1540}{925} \approx 1.665$	1	1
Pauma	3720	$\frac{3720}{925} \approx 4.022$	4	4
Pacific	5290	$\frac{5290}{925} \approx 5.719$	5	5
		Total	19	19

This result yields too many board members. Thus we must increase the modified standard divisor. By experimenting with different divisors, we find that 950 is a possible modified standard divisor. Using 950 as the standard divisor gives the results shown in the table below.

City	Population	Quotient	Standard quota	Number of board members
Cardiff	7020	$\frac{7020}{950} \approx 7.389$	7	7
Solana	2430	$\frac{2430}{950} \approx 2.558$	2	2
Vista	1540	$\frac{1540}{950} \approx 1.621$	1	1
Pauma	3720	$\frac{3720}{950} \approx 3.916$	3	3
Pacific	5290	$\frac{5290}{950} \approx 5.568$	5	5
		Total	18	18

Thus the composition of the environmental board using the Jefferson method is Cardiff: 7, Solana: 2, Vista: 1, Pauma: 3, Pacific: 5.

European Countries

Country	Population
France	59,500,000
Germany	82,300,000
Italy	57,900,000
Spain	39,500,000
Belgium	10,300,000

Source: Adapted from
http://www.planetalk.org

CHECK YOUR PROGRESS 1   Suppose the 20 members of a committee from five European countries are selected according to the populations of the five countries, as shown in the table at the left.

a. Use the Hamilton method to determine the number of representatives each country should have.

b. Use the Jefferson method to determine the number of representatives each country should have.

Solution   See page S46.

Suppose that the environmental agency in Example 1 decides to add one more member to the board even though the population of each city remains the same. The total number of members is now 19 and we must determine how the members of the board will be apportioned.

The standard divisor is now $\frac{20,000}{19} \approx 1052.63$. Using Hamilton's method, the calculations necessary to apportion the board members are shown below.

City	Population	Quotient	Standard quota	Number of board members
Cardiff	7020	$\frac{7020}{1052.63} \approx 6.669$	6	7
Solana	2430	$\frac{2430}{1052.63} \approx 2.309$	2	2
Vista	1540	$\frac{1540}{1052.63} \approx 1.463$	1	1
Pauma	3720	$\frac{3720}{1052.63} \approx 3.534$	3	4
Pacific	5290	$\frac{5290}{1052.63} \approx 5.026$	5	5
		Total	17	19

The table below summarizes the number of board members each city would have if the board consisted of 18 members (Example 1) or 19 members.

City	Hamilton apportionment with 18 board members	Hamilton apportionment with 19 board members
Cardiff	6	7
Solana	2	2
Vista	2	1
Pauma	3	4
Pacific	5	5
Total	18	19

Notice that although one more board member was added, Vista lost a board member, even though the populations of the cities did not change. This is called the *Alabama paradox* and has a negative effect on fairness. In the interest of fairness, an apportionment method should not exhibit the Alabama paradox. (See the Math Matters below for other paradoxes.)

Math Matters   Apportionment Paradoxes

The Alabama paradox, although it was not given that name until later, was first noticed after the 1870 census. At the time, the House of Representatives had 270 seats. However, when the number of representatives in the House was increased to 280 seats, Rhode Island lost a representative.

After the 1880 census, C. W. Seaton, the chief clerk of the U.S. Census Office, calculated the number of representatives each state would have if the number were set at some number between 275 and 300. He noticed that when the number of representatives was increased from 299 to 300, Alabama lost a representative.

There are other paradoxes that involve apportionment methods. Two of them are the *population paradox* and the *new states paradox*. It is possible for the population of one state to be increasing faster than that of another state and for the state to still lose a representative. This is an example of the **population paradox.**

In 1907, when Oklahoma was added to the Union, the size of the House was increased by five representatives to account for Oklahoma's population. However, when the complete apportionment of the Congress was recalculated, New York lost a seat and Maine gained a seat. This is an example of the **new states paradox.**

Fairness in Apportionment

To decide which plan—the Hamilton or the Jefferson—is better, we might try to determine which plan is fairer. Of course, *fair* can be quite a subjective term, so we will try to state conditions by which an apportionment plan is judged fair. One criterion of fairness for an apportionment plan is that it should satisfy the *quota rule.*

> **Quota Rule**
>
> The number of representatives apportioned to a state is the standard quota or one more than the standard quota.

We can show that the Jefferson plan does not satisfy the quota rule by calculating the standard quota of Apus (see page 846).

$$\text{Standard quota} = \frac{\text{population of Apus}}{\text{standard divisor}} = \frac{11{,}123}{800} \approx 13$$

The standard quota of Apus is 13. However, the Jefferson plan assigns 15 representatives to that state, two more than its standard quota. Therefore, the Jefferson method violates the quota rule.

As we have seen, the choice of apportionment method affects the number of representatives a state will have. Given that fact, mathematicians and others have tried to work out an apportionment method that is fair. The difficulty lies in trying to define what is fair.

Another measure of fairness is *average constituency*. This is the population of a state divided by the number of representatives from the state and then rounded to the nearest whole number.

Average Constituency

$$\text{Average constituency} = \frac{\text{population of a state}}{\text{number of representatives from the state}}$$

Consider the two states Hampton and Shasta in the table below.

State	Population	Representatives	Average constituency
Hampton	16,000	10	$\frac{16,000}{10} = 1600$
Shasta	8340	5	$\frac{8340}{5} = 1668$

TAKE NOTE

The idea of average constituency is an essential aspect of our democracy. To understand this, suppose state A has an average constituency of 1000 and state B has an average constituency of 10,000. When a bill is voted on in the House of Representatives, each vote has equal weight. However, a vote from a representative from state A would represent 1000 people, but a vote from a representative from state B would represent 10,000 people. Consequently, in this situation, we do not have "equal representation."

Because the average constituencies are approximately equal, it seems natural to say that both states are equally represented. See the Take Note to the left.

QUESTION *Although the average constituencies of Hampton and Shasta are approximately equal, which state has the more favorable representation?*

Now suppose that one representative will be added to one of the states. Which state is more deserving of the new representative? In other words, to be fair, which state should receive the new representative?

The changes in the average constituency are shown below.

State	Average constituency (old)	Average constituency (new)
Hampton	$\frac{16,000}{10} = 1600$	$\frac{16,000}{11} \approx 1455$
Shasta	$\frac{8340}{5} = 1668$	$\frac{8340}{6} = 1390$

From the table, there are two possibilities for adding one representative. If Hampton receives the representative, its average constituency will be 1455 and Shasta's will remain at 1668. The difference in the average constituencies is $1668 - 1455 = 213$. This difference is called the *absolute unfairness of the apportionment.*

ANSWER *Because Hampton's average constituency is smaller than Shasta's, Hampton has the more favorable representation.*

> **Absolute Unfairness of an Apportionment**
>
> The **absolute unfairness of an apportionment** is the absolute value of the difference between the average constituency of state A and the average constituency of state B.
>
> |Average constituency of A − average constituency of B|

If Shasta receives the representative, its average constituency will be 1390 and Hampton's will remain at 1600. The absolute unfairness of apportionment is $1600 - 1390 = 210$. This is summarized below.

	Hampton's average constituency	Shasta's average constituency	Absolute unfairness of apportionment
Hampton receives the new representative	1455	1668	213
Shasta receives the new representative	1600	1390	210

Because the smaller absolute unfairness of apportionment occurs if Shasta receives the new representative, it might seem that Shasta should receive the representative. However, this is not necessarily true.

To understand this concept, let's consider a somewhat different situation. Suppose an investor makes two investments, one of $10,000 and another of $20,000. One year later, the first investment is worth $11,000 and the second investment is worth $21,500. This is shown in the table below.

	Original investment	One year later	Increase
Investment A	$10,000	$11,000	$1000
Investment B	$20,000	$21,500	$1500

Although there is a larger increase in investment B, the increase per dollar of the original investment is $\frac{1500}{20,000} = 0.075$. On the other hand, the increase per dollar of investment A is $\frac{1000}{10,000} = 0.10$. Another way of saying this is that each $1 of investment A produced a return of 10 cents (0.10), whereas each $1 of investment B produced a return of 7.5 cents (0.075). Therefore, even though the increase in investment A was less than the increase in investment B, investment A was more productive.

A similar process is used when deciding which state should receive another representative. Rather than look at the difference in the absolute unfairness in apportionment, we determine the *relative unfairness* of adding the representative.

> **Relative Unfairness of an Apportionment**
>
> The **relative unfairness of an apportionment** is the quotient of the absolute unfairness of apportionment and the average constituency of the state receiving the new representative.
>
> $$\frac{\text{Absolute unfairness of the apportionment}}{\text{Average constituency of the state receiving the new representative}}$$

EXAMPLE 2 ■ **Determine the Relative Unfairness of an Apportionment**

Determine the relative unfairness of an apportionment that gives a new representative to Hampton rather than Shasta.

Solution

Using the table for Hampton and Shasta shown on the previous page, we have

Relative unfairness of the apportionment

$$= \frac{\text{absolute unfairness of the apportionment}}{\text{average constituency of Hampton with a new representative}}$$

$$= \frac{213}{1455} \approx 0.146$$

The relative unfairness of the apportionment is approximately 0.146.

CHECK YOUR PROGRESS 2 Determine the relative unfairness of an apportionment that gives a new representative to Shasta rather than Hampton.

Solution See page S47.

The relative unfairness of an apportionment is used in the following way.

Apportionment Principle

When adding a new representative to a state, the representative is assigned to the state so as to give the smallest relative unfairness of apportionment.

From Example 2, the relative unfairness of adding a representative to Hampton is approximately 0.146. From Check Your Progress 2, the relative unfairness of adding a representative to Shasta is approximately 0.151. Because the smaller relative unfairness results from adding the representative to Hampton, that state should receive the new representative.

Although we have focused on assigning representatives to states, the apportionment principle can be used in many other situations.

EXAMPLE 3 ■ **Use the Apportionment Principle**

The table below shows the number of paramedics and the annual number of paramedic calls for two cities. If a new paramedic is hired, use the apportionment principle to determine to which city the paramedic should be assigned.

	Paramedics	Annual paramedic calls
Tahoe	125	17,526
Erie	143	22,461

Solution

Calculate the relative unfairness of the apportionment that assigns the paramedic to Tahoe and the relative unfairness of the apportionment that assigns the paramedic to Erie. In this case, average constituency is the annual paramedic calls divided by the number of paramedics.

	Tahoe's annual paramedic calls per paramedic	Erie's annual paramedic calls per paramedic	Absolute unfairness of apportionment
Tahoe receives a new paramedic	$\dfrac{17{,}526}{125+1} \approx 139$	$\dfrac{22{,}461}{143} \approx 157$	$157 - 139 = 18$
Erie receives a new paramedic	$\dfrac{17{,}526}{125} \approx 140$	$\dfrac{22{,}461}{143+1} \approx 156$	$156 - 140 = 16$

If Tahoe receives the new paramedic, the relative unfairness of the apportionment is

Relative unfairness of the apportionment

$$= \frac{\text{absolute unfairness of the apportionment}}{\text{Tahoe's average constituency with a new paramedic}}$$

$$= \frac{18}{139} \approx 0.129$$

If Erie receives the new paramedic, the relative unfairness of the apportionment is

Relative unfairness of the apportionment

$$= \frac{\text{absolute unfairness of the apportionment}}{\text{Erie's average constituency with a new paramedic}}$$

$$= \frac{16}{156} \approx 0.103$$

Because the smaller relative unfairness results from adding the paramedic to Erie, that city should receive the paramedic.

CHECK YOUR PROGRESS 3   The table below shows the number of first and second grade teachers in a school district and the number of students in each of those grades. If a new teacher is hired, use the apportionment principle to determine to which grade the teacher should be assigned.

	Number of teachers	Number of students
First grade	512	12,317
Second grade	551	15,439

Solution   *See page S47.*

historical note
———

According to the U.S. Bureau of the Census, methods of apportioning the House of Representatives have changed over time.

1790–1830: Jefferson method

1840: Webster method (See the paragraph following Exercise 20, page 862.)

1850–1900: Hamilton method

1910, 1930: Webster method

Note that 1920 is missing. In direct violation of the U.S. Constitution, the House of Representatives failed to reapportion the House in 1920.

1940–2000: Method of equal proportions or the Huntington-Hill method

All apportionment plans enacted by the House have created some controversy. For instance, the constitutionality of the Huntington-Hill method was challenged by Montana in 1992 because it lost a seat to Washington after the 1990 census. For more information on this subject, see **http://www.census.gov/ population/www/censusdata/ apportionment/history.html.** ∎

Huntington-Hill Apportionment Method

As we mentioned earlier, the members of the House of Representatives are apportioned among the states every 10 years. The present method used by the House is based on the apportionment principle and is called the *method of equal proportions* or the *Huntington-Hill method*. This method has been used since 1940.

The Huntington-Hill method is implemented by calculating what is called a *Huntington-Hill number*. This number is derived from the apportionment principle. Let

P_A = population of state A a = number of representatives from state A

P_B = population of state B b = number of representatives from state B

$\dfrac{P_A}{a+1}$ = average constituency of A when it receives a new representative

$\dfrac{P_B}{b}$ = average constituency of B without a new representative

Then the relative unfairness of apportionment by giving A the new member is

$$\frac{\text{Absolute unfairness of apportionment}}{\text{Average constituency of } A}$$

$$= \frac{\text{average constituency of } B - \text{average constituency of } A}{\text{average constituency of } A}$$

$$= \frac{\dfrac{P_B}{b} - \dfrac{P_A}{a+1}}{\dfrac{P_A}{a+1}}$$

According to the apportionment principle, state A should receive the next representative instead of state B if the relative unfairness to A is less than the relative unfairness to B. That is,

Relative unfairness to A $<$ relative unfairness to B

$$\frac{\dfrac{P_B}{b} - \dfrac{P_A}{a+1}}{\dfrac{P_A}{a+1}} < \frac{\dfrac{P_A}{a} - \dfrac{P_B}{b+1}}{\dfrac{P_B}{b+1}}$$

• This is the symbolic form for relative unfairness.

Simplifying this inequality gives $\dfrac{(P_B)^2}{b(b+1)} < \dfrac{(P_A)^2}{a(a+1)}$.

✔ **TAKE NOTE**

Note that after simplifying, the inequality $\dfrac{(P_A)^2}{a(a+1)}$ is the *larger* quantity. This fact is important later in the discussion.

> **Huntington-Hill Number**
>
> The value of $\dfrac{(P_A)^2}{a(a+1)}$, where P_A is the population of state A and a is the current number of representatives from state A, is called the **Huntington-Hill number** for state A.

When the Huntington-Hill method is used to apportion representatives between two states, the state with the greater Huntington-Hill number receives the next representative. This method can be extended to more than two states.

Huntington-Hill Apportionment Principle

When there is a choice of adding one representative to a number of states, the representative should be added to the state with the greatest Huntington-Hill number.

EXAMPLE 4 ■ **Use the Huntington-Hill Apportionment Principle**

The table below shows the numbers of lifeguards that are assigned to three different beaches and the numbers of rescues made by lifeguards at those beaches. Use the Huntington-Hill apportionment principle to determine to which beach a new lifeguard should be assigned.

Beach	Number of lifeguards	Number of rescues
Mellon	37	1227
Donovan	51	1473
Ferris	24	889

Solution

Calculate the Huntington-Hill number for each of the beaches. In this case, the population is the number of rescues and the number of representatives is the number of lifeguards.

Mellon:
$$\frac{1227^2}{37(37 + 1)} \approx 1071$$

Donovan:
$$\frac{1473^2}{51(51 + 1)} \approx 818$$

Ferris:
$$\frac{889^2}{24(24 + 1)} \approx 1317$$

Ferris has the greatest Huntington-Hill number. Thus, according to the Huntington-Hill Apportionment Principle, the new lifeguard should be assigned to Ferris.

CHECK YOUR PROGRESS 4   A university has a president's council that is composed of students from each of the undergraduate classes. If a new student representative is added to the council, use the Huntington-Hill apportionment principle to determine which class the new student council member should represent.

Class	Number of representatives	Number of students
First year	12	2015
Second year	10	1755
Third year	9	1430
Fourth year	8	1309

Solution   *See page S48.*

Now that we have looked at various apportionment methods, it seems reasonable to ask which is the best method. Unfortunately, all apportionment methods have some flaws. This was proved by Michael Balinski and H. Peyton Young.

> **Balinski-Young Impossibility Theorem**
>
> Any apportionment method will either violate the quota rule or will produce paradoxes such as the Alabama paradox.

Although there is no perfect apportionment method, Balinski and Young went on to present a strong case that the Webster method (preceding Exercise 20 on page 862) is the system that most closely satisfies the goal of one person, one vote. However, political expediency sometimes overrules mathematical proof. Some historians have suggested that although the Huntington-Hill apportionment method was better than some of the previous methods, President Franklin Roosevelt chose this method in 1941 because it alloted one more seat to Arkansas and one less to Michigan. This essentially meant that the House of Representatives would have one more seat for the Democrats, Roosevelt's party.

Excursion

Apportioning the 1790 House of Representatives

historical note

According to the U.S. Constitution, each state must have at least one representative to the House of Representatives. The remaining representatives (there are 435 in all) are then assigned to the states using the Huntington-Hill apportionment principle. ■

 The first apportionment of the House of Representatives, using the 1790 census, is given in the table below. This apportionment was calculated by using the Jefferson method. (See our web site at **college.hmco.com** for an Excel spreadsheet that will help with the computations.)

Apportionment Using the Jefferson Method

State	Population	Number of representatives
Connecticut	236,841	7
Delaware	55,540	1
Georgia	70,835	2
Maryland	278,514	8
Massachusetts	475,327	14
Kentucky	68,705	2
New Hampshire	141,822	4
Vermont	85,533	2
New York	331,589	10
New Jersey	179,570	5
Pennsylvania	432,879	13
North Carolina	353,523	10
South Carolina	206,236	6
Virginia	630,560	19
Rhode Island	68,446	2

Source: **http://www.uwm.edu/~margo/apport/datasets.htm**   *(continued)*

Excursion Exercises

1. Verify this apportionment using the Jefferson method. You will have to experiment with various modified divisors until you reach the given representation. See the Calculator Note on page 848.

2. Find the apportionment that would have resulted if the Hamilton method had been used.

3. Give each state one representative. Use the Huntington-Hill method with $n = 1$ to determine the state that receives the next representative. The Calculator Note in this section will help. With the populations stored in L1, enter `2nd` L1 `x²` $\div$ 2 `STO` L2 `ENTER`. Now scroll through L2 to find the largest number.

4. Find the apportionment that would have resulted if the Huntington-Hill method (the one used for the 2000 census) had been used in 1790. See our web site at **math.college.hmco.com** for a spreadsheet that will help with the calculations.

Exercise Set 13.1
(Suggested Assignment: 1–7, odd; 8, 9, 10, 11–25, odd)

1. Explain how to calculate the standard divisor of an apportionment for a total population p with n items to apportion.

2. The U.S. House of Representatives currently has 435 members to represent the 281,424,177 citizens of the U.S. as determined by the 2000 census.

   a. Calculate the standard divisor for the apportionment of these representatives and explain the meaning of this standard divisor in the context of this exercise.

   b. According to the 2000 census, the population of Delaware was 785,068. Delaware currently has only one representative in the House of Representatives. Is Delaware currently overrepresented or underrepresented in the House of Representatives? Explain.

   c. According to the 2000 census, the population of Vermont was 609,890. Vermont currently has only one representative in the House of Representatives. Is Vermont currently overrepresented or underrepresented in the House of Representatives? Explain.

3. In the Hamilton apportionment method, explain how to calculate the standard quota for a particular state (group).

4. What is the quota rule?

5. There are a total of 25 teacher aides that are to be apportioned among seven classes at a new elementary school. The enrollment in each of the seven classes is shown in the following table.

Class	Number of students
Kindergarten	38
First grade	39
Second grade	35
Third grade	27
Fourth grade	21
Fifth grade	31
Sixth grade	33
Total	224

   a. Determine the standard divisor. What is the meaning of the standard divisor in the context of this exercise?

b. Use the Hamilton method to determine the number of teacher aides to be apportioned to each class.

c. Use the Jefferson method to determine the number of teacher aides to be apportioned to each class. Is this apportionment in violation of the quota rule?

d. How do the apportionment results produced using the Jefferson method compare with the results produced using the Hamilton method?

6. The following table shows the enrollment for each of the four divisions of a college. The four divisions are liberal arts, business, humanities, and science. There are 180 new computers that are to be apportioned among the divisions based on the enrollments.

Division	Enrollment
Liberal arts	3455
Business	5780
Humanities	1896
Science	4678
Total	15,809

a. What is the standard divisor for an apportionment of the computers? What is the meaning of the standard divisor in the context of this exercise?

b. Use the Hamilton method to determine the number of computers to be apportioned to each division.

c. If the computers are to be apportioned using the Jefferson method, explain why neither 86 nor 87 can be used as a modified standard divisor. Explain why 86.5 can be used as a modified standard divisor.

d. Explain why the modified standard divisor used in the Jefferson method cannot be larger than the standard divisor.

e. Use the Jefferson method to determine the number of computers to be apportioned to each division. Is this apportionment in violation of the quota rule?

f. How do the apportionment results produced using the Jefferson method compare with the results produced using the Hamilton method?

7. A hospital district consists of six hospitals. The district administrators have decided that 48 new nurses should be apportioned based on the number of beds in each of the hospitals. The following table shows the number of beds in each hospital.

Hospital	Number of beds
Sharp	242
Palomar	356
Tri-City	308
Del Raye	190
Rancho Verde	275
Bel Aire	410
Total	1781

a. Determine the standard divisor. What is the meaning of the standard divisor in the context of this exercise?

b. Use the Hamilton method to determine the number of nurses to be apportioned to each hospital.

c. Use the Jefferson method to determine the number of nurses to be apportioned to each hospital.

d. How do the apportionment results produced using the Jefferson method compare with the results produced using the Hamilton method?

8. What is the Alabama paradox?

9.  What is the population paradox?

10. What is the new states paradox?

11. What is the Balinski-Young Impossibility Theorem?

12. Consider the apportionment of 27 projectors for a school district with four campus locations labeled *A*, *B*, *C*, and *D*. The following table shows the apportionment of the projectors using the Hamilton method.

Campus	A	B	C	D
Enrollment	840	1936	310	2744
Apportionment of 27 projectors	4	9	1	13

a. If the number of projectors to be apportioned increases from 27 to 28, what will be the apportionment if the Hamilton method is used? Did the Alabama paradox occur? Explain.

b. If the number of projectors to be apportioned using the Hamilton method increases from 28 to 29, will the Alabama paradox occur? Explain.

13. A company operates four resorts. The CEO of the company decides to use the Hamilton method to apportion 115 new flat-screen digital television sets to the resorts based on the number of guest rooms at each resort.

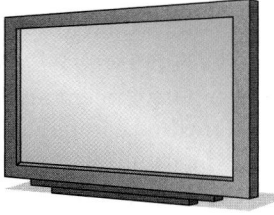

Resort	A	B	C	D
Number of guest rooms	23	256	182	301
Apportionment of 115 televisions	4	39	27	45

a. If the number of television sets to be apportioned by the Hamilton method increases from 115 to 116, will the Alabama paradox occur?

b. If the number of television sets to be apportioned by the Hamilton method increases from 116 to 117, will the Alabama paradox occur?

c. If the number of television sets to be apportioned by the Hamilton method increases from 117 to 118, will the Alabama paradox occur?

14. A college apportions 40 security personnel among three education centers according to their enrollments. The following table shows the present enrollments at each of the centers.

Center	A	B	C
Enrollment	356	1054	2590

a. Use the Hamilton method to apportion the security personnel.

b. After a semester the centers have the following enrollments.

Center	A	B	C
Enrollment	370	1079	2600

Center *A* has an increased enrollment of 14 students, which is an increase of $\frac{14}{356} \approx 0.039 = 3.9\%$. Center *B* has an increased enrollment of 25 students, which is an increase of $\frac{25}{1054} \approx 0.024 = 2.4\%$.

If the security personnel are reapportioned using the Hamilton method, will the population paradox occur? Explain.

15. Scientific Research Corporation has offices in Boston and Chicago. The number of employees at each office

is shown in the following table. There are 22 vice presidents to be apportioned between the offices.

Office	Boston	Chicago
Employees	151	1210

a. Use the Hamilton method to find each office's apportionment of vice presidents.

b. The corporation opens an additional office in San Francisco with 135 employees and decides to have a total of 24 vice presidents. If the vice presidents are reapportioned using the Hamilton method, will the new states paradox occur? Explain.

16. The science division of a college consists of the three departments mathematics, physics, and chemistry. The number of students enrolled in each department is shown in the following table. There are 19 clerical assistants to be apportioned among the departments.

Department	Math	Physics	Chemistry
Student enrollment	4325	520	1165

a. Use the Hamilton method to find each department's apportionment of clerical assistants.

b. The division opens a new computer science department with a student enrollment of 495. The division decides to have a total of 20 clerical assistants. If the clerical assistants are reapportioned using the Hamilton method, will the new states paradox occur? Explain.

17. The following table shows the number of fifth and sixth grade teachers in a school district and the number of students in each of those grades. The number of teachers for each of the grade levels was determined by using the Huntington-Hill apportionment method.

	Number of teachers	Number of students
Fifth grade	19	604
Sixth grade	21	698

The district has decided to hire a new teacher for either the fifth or sixth grade.

a. Use the apportionment principle to determine to which grade the new teacher should be assigned.

b. Use the Huntington-Hill apportionment principle to determine to which grade the new teacher should be assigned. How does this result compare with the result in part a?

18. The following table shows the number of social workers and the number of cases (case load) handled by the social workers for two offices. The number of social workers for each office was determined by using the Huntington-Hill apportionment method.

	Number of social workers	Case load
Hill Street office	20	584
Valley office	24	712

A new social worker is to be hired for one of the offices.

a. Use the apportionment principle to determine to which office the social worker should be assigned.

b. Use the Huntington-Hill apportionment principle to determine to which office the new social worker should be assigned. How does this result compare with the result in part a?

c. The results of part b indicate that the new social worker should be assigned to the Valley office. At this moment the Hill Street office has 20 social workers and the Valley office has 25 social workers. Use the Huntington-Hill apportionment principle to determine to which office the *next* new social worker should be assigned. Assume the case loads remain the same.

19. The table below shows the number of computers that are assigned to four different schools and the number of students in those schools. Use the Huntington-Hill

apportionment principle to determine to which school a new computer should be assigned.

School	Number of computers	Number of students
Rose	26	625
Lincoln	22	532
Midway	26	620
Valley	31	754

20. Currently, the U.S. House of Representatives has 435 members who have been apportioned by the Huntington-Hill apportionment method. If the number of representatives were to be increased to 436, then, according to the 2000 census figures, Utah would be given the new representative. How must Utah's 2000 census Huntington-Hill number compare with the 2000 census Huntington-Hill numbers for the other 49 states? Explain.

The *Webster method of apportionment* is similar to the Jefferson method except that quotas are rounded up when the decimal remainder is 0.5 or greater and down when the decimal remainder is less than 0.5. This method of rounding is referred to as rounding **to the nearest integer.** For instance, using the Jefferson method, a quotient of 15.91 would be rounded to 15; using the Webster method, it would be rounded to 16. A quotient of 15.49 would be rounded to 15 in both methods.

To use the Webster method you must still experiment to find a *modified standard divisor* for which the sum of the quotas rounded to the nearest integer equals the number of items to be apportioned. Webster's method is similar to Jefferson's method; however, Webster's method is generally more difficult to apply because the modified divisor may be less than, equal to, or more than the standard divisor.

A calculator can be very helpful in testing a possible modified standard divisor *md* when applying the Webster method of apportionment. For instance, on a TI-83 calculator, first store the populations in L1. Then enter

$$\text{iPart(L1/}md\text{+.5)}\rightarrow\text{L2}$$

where *md* is the modified division. Then press the ENTER key. Now scroll through L2 to view the quotas rounded to the *nearest integer*. If the sum of these quotas equals the total number of items to be apportioned, then you have the Webster apportionment. If the sum of the quotas in L2 is less than the total number of items to be apportioned, try a smaller modified standard divisor. If the sum of the

quotas in L2 is greater than the total number of items to be apportioned, try a larger modified standard divisor.

21. Use the Webster method to apportion the computers in Exercise 6, page 859. How do the apportionment results produced using the Webster method compare with the results produced using the
  a. Hamilton method?
  b. Jefferson method?

22. The table below shows the populations of five European countries. A committee of 20 people from these countries is to be formed using the Webster method of apportionment.

Country	Population
France	59,000,000
Germany	82,000,000
Italy	57,000,000
Spain	39,000,000
Belgium	10,000,000
Total	247,000,000

Source: Adapted from Time Almanac 2000 with Information Please

  a. Explain why 12,600,000 *cannot* be used as a modified standard divisor.
  b. Explain why 12,700,000 *cannot* be used as a modified standard divisor.
  c. Explain why 12,650,000 *can* be used as a modified standard divisor.
  d. Use the Webster apportionment method to determine the apportionment of the 20 committee members.

23. Which of the following apportionment methods can violate the quota rule?
  ■ Hamilton method
  ■ Jefferson method
  ■ Webster method
  ■ Huntington-Hill method

24. According to Michael Balinski and H. Peyton Young, which of the apportionment methods most closely satisfies the goal of one person, one vote?

25. What method is presently used to apportion the members of the U.S. House of Representatives?

Extensions

CRITICAL THINKING

26. According to the 2000 census, what is the population of your state? How many representatives does your state have in the U.S. House of Representatives? Is your state underrepresented or overrepresented in the House of Representatives? Explain. $\left(\textit{Hint:}\ \text{The result of Exercise 2a on page 858 shows that the ratio of representatives to citizens is about}\ \frac{1}{646,952}.\right)$ How does your state's current number of representatives compare with the number of representatives it had after the 1990 census?

27. Create an apportionment problem in which the Hamilton, Jefferson, and Webster methods produce the same apportionment.

EXPLORATIONS

28. John Quincy Adams, the sixth president of the United States, proposed an apportionment method. Research this method, known as the Adams method of apportionment. Describe how this method works. Also indicate whether it satisfies the quota rule and if it is susceptible to any paradoxes.

29. In the Huntington-Hill method of apportionment, each state is first given one representative and then additional representatives are assigned, one at a time, to the state currently having the highest Huntington-Hill number. This way of implementing the Huntington-Hill apportionment method is time consuming. Another process for implementing the Huntington-Hill apportionment method consists of using a modified divisor and a special rounding procedure that involves the *geometric mean* of two consecutive integers. Research this process for implementing the Huntington-Hill apportionment method. Apply this process to apportion 22 new security vehicles to each of the following schools, based on their student populations.

School	Number of students	Number of security vehicles
Del Mar	5230	?
Wheatly	12,375	?
West	8568	?
Mountain View	14,245	?

It can be shown that the results of the rounding procedure used in the Huntington-Hill method described in this exercise differ only slightly from the results of the rounding procedure used in the Webster apportionment method. Thus both methods often produce the same apportionment. Verify that for this exercise, the Webster method produces the same apportionment as the Huntington-Hill method.

30. Simplify $\dfrac{\dfrac{P_B}{b} - \dfrac{P_A}{a+1}}{\dfrac{P_A}{a+1}} < \dfrac{\dfrac{P_A}{a} - \dfrac{P_B}{b+1}}{\dfrac{P_B}{b+1}}$. Assume that $\dfrac{P_B}{b} - \dfrac{P_A}{a+1}$ and $\dfrac{P_A}{a} - \dfrac{P_B}{b+1}$ are positive numbers.

SECTION 13.2 **Introduction to Voting**

 TAKE NOTE

When an issue requires a **majority** vote, it means that over 50% of the people voting must vote for the issue. This is not the same as a **plurality**, in which the person or issue with the most votes wins.

Plurality Method of Voting

One of the most revered privileges of those of us who live in a democracy is the right to vote for our representatives. Sometimes, however, we are puzzled by the fact that the best candidate did not get elected. Unfortunately, because of the way our *plurality* voting system works, it is possible to elect someone or pass a proposition that has less than *majority* support. As we proceed through this section, we will look at the problems with plurality voting and alternatives to this system. We start with a definition.

> **The Plurality Method of Voting**
>
> Each voter votes for one candidate, and the candidate with the most votes wins. The winning candidate does not have to have a majority of the votes.

EXAMPLE 1 ■ Determine the Winner Using Plurality Voting

Fifty people were asked to rank their preferences of five varieties of chocolate candy, using 1 for their favorite and 5 for their least favorite. This type of ranking of choices is called a **preference schedule.** The results are shown in the table below.

	Rankings					
Caramel center	5	4	4	4	2	4
Vanilla center	1	5	5	5	5	5
Almond center	2	3	2	1	3	3
Toffee center	4	1	1	3	4	2
Solid chocolate	3	2	3	2	1	1
Number of voters:	17	11	9	8	3	2

According to the table (see the column in blue), 3 voters ranked solid chocolate first, caramel centers second, almond centers third, toffee centers fourth, and vanilla centers fifth. According to this table, which variety of candy would win the taste test using the plurality voting system?

Solution
To answer the question, we will make a table showing the number of first-place votes for each candy.

	First-place votes
Caramel center	0
Vanilla center	17
Almond center	8
Toffee center	11 + 9 = 20
Solid chocolate	3 + 2 = 5

Because toffee centers received 20 first-place votes, this type of candy wins the plurality taste test.

CHECK YOUR PROGRESS 1   According to the table in Example 1, which variety of candy would win second place using the plurality voting system?

Solution   *See page S48.*

Example 1 can be used to show the difference between plurality and majority. There were 20 first-place votes for toffee-centered chocolate, so it wins the taste test. However, toffee-centered chocolate was the first choice of only 40% $\left(\frac{20}{50} = 40\%\right)$ of the people voting. Thus less than half of the people voted for toffee-centered chocolate as number one, so it did not receive a majority vote.

Math Matters   Gubernatorial and Presidential Elections

In 1998, in a three-party race, plurality voting resulted in the election of former wrestler Jesse Ventura as governor of Minnesota, despite the fact that over 60% of the state's voters did not vote for him. In fact, he won the governor's race with only 37% of the voters choosing him. Ventura won not because the majority of voters chose him, but because of the plurality voting method.

There are many situations that can be cited to show that plurality voting can lead to unusual results. When plurality voting is mixed with other voting methods, even the plurality winner may not win. In the 2000 presidential election, George Bush received 47.87% of the vote and Al Gore received 48.38% of the vote. Nonetheless, George Bush was elected president, even though he was not the plurality choice. This occurred because there are two votings for president: the vote of the people and the vote of the Electoral College. The Electoral College actually elects the president.

Borda Count Method of Voting

historical note

The issue of whether plurality voting methods are fair has been around for over 200 years. Jean C. Borda (1733–1799) was a member of the French Academy of Sciences when he first started thinking about the way in which people were elected to the Academy. He was concerned that the plurality method of voting might not result in the best candidate being elected. The Borda Count method was born out of these concerns. It was the first attempt to mathematically quantify voting systems. ∎

The problem with plurality voting is that alternate choices are not considered. For instance, the result of the Minnesota governor's contest may have been quite different if voters had been asked, "Choose the candidate you prefer, but if that candidate does not receive a *majority* of the votes, which candidate would be your second choice?"

To see why this might be a reasonable alternative to plurality voting, consider the following situation. Thirty-six senators are considering an educational funding measure. Because the senate leadership wants an educational funding measure to pass, the leadership first determined that the senators preferred measure A for $50 million over measure B for $30 million. However, because of an unexpected dip in state revenues, measure A was removed from consideration and a new measure C, for $15 million, was proposed. The senate leadership determined that senators favored measure B over measure C. In summary, we have

A majority of senators favor measure A over measure B.

A majority of senators favor measure B over measure C.

From these results, it seems reasonable to think that a majority of senators would prefer Measure A over Measure C. However, when the senators were asked about their preferences between the two measures, Measure C was preferred over Measure A. To understand how this could happen, consider the preference schedule for the senators shown in the table below.

	Rankings		
Measure A: $50 million	1	2	3
Measure B: $30 million	2	3	1
Measure C: $15 million	3	1	2
Number of senators:	15	12	9

Notice that 15 senators prefer Measure A over Measure C, but $12 + 9 = 21$ senators, a majority of the 36 senators, prefer Measure C over Measure A. This means that if all three measures were on the ballot, A would come in first, B would come in second, and C would come in third. However, if just A and C were on the ballot, C would win over A. This paradoxical result was first discussed by Jean C. Borda in 1770.

In an attempt to remove such paradoxical results from voting, Borda proposed that voters rank their choices by giving each choice a certain number of points.

✔ **TAKE NOTE**

Paradoxes occur in voting only when there are three or more candidates or issues on a ballot. If there are only two candidates in a race, then the candidate receiving the majority of the votes cast is the winner. In a two-candidate race, the majority and the plurality are the same.

The Borda Count Method of Voting

If there are n candidates or issues in an election, each voter ranks the candidates by giving n points to the voter's first choice, $n - 1$ points to the voter's second choice, and so on, with the voter's least favorite choice receiving 1 point. The candidate who receives the most total points is the winner.

Applying the Borda Count method to the education measures, a measure receiving a first-place vote receives 3 points. (There are three different measures.) Each measure receiving a second-place vote receives 2 points, and each measure receiving a third-place vote receives 1 point. This is summarized in the table below.

▼ **point of interest**

One way to see the difference between a plurality voting system (sometimes called a "winner-take-all" method) and the Borda Count method is to consider grades earned in school. Suppose a student is going to be selected for a scholarship based on grades. If one student has 10 A's and 20 F's and another student has five A's and 25 B's, it would seem that the second student should receive the scholarship. However, if the scholarship is awarded by the plurality of A's, the first student will get the scholarship. The Borda Count method is closely related to the method used to calculate grade-point average (GPA).

Points per vote

Measure A:	15 first-place votes:	$15 \cdot 3 = 45$
	12 second-place votes:	$12 \cdot 2 = 24$
	9 third-place votes:	$9 \cdot 1 = 9$
	Total:	78

Measure B:	9 first-place votes:	$9 \cdot 3 = 27$
	15 second-place votes:	$15 \cdot 2 = 30$
	12 third-place votes:	$12 \cdot 1 = 12$
	Total:	69

Measure C:	12 first-place votes:	$12 \cdot 3 = 36$
	9 second-place votes:	$9 \cdot 2 = 18$
	15 third-place votes:	$15 \cdot 1 = 15$
	Total:	69

Using the Borda Count method, measure A is the clear winner. Notice also that measure A is the plurality winner, although it does not always happen that the Borda Count method and the plurality method yield the same winner.

EXAMPLE 2 ■ Use the Borda Count Method

The members of a club are going to elect a president from four nominees using the Borda Count method. If the 100 members of the club mark their ballots as shown in the table below, who will be elected president?

	Rankings					
Avalon	2	2	2	2	3	2
Branson	1	4	4	4	2	1
Columbus	3	3	1	3	1	3
Dunkirk	4	1	3	1	4	4
Number of voters:	30	24	18	12	10	6

Solution

Using the Borda Count method, each first-place vote receives 4 points, each second-place vote receives 3 points, each third-place vote receives 2 points, and each last-place vote receives 1 point. The summary for each candidate is shown below.

Avalon: 0 first-place votes $0 \cdot 4 = 0$
           90 second-place votes $90 \cdot 3 = 270$
           10 third-place votes $10 \cdot 2 = 20$
           0 fourth-place votes $0 \cdot 1 = 0$
                                     Total   290

Branson: 36 first-place votes $36 \cdot 4 = 144$
           10 second-place votes $10 \cdot 3 = 30$
           0 third-place votes $0 \cdot 2 = 0$
           54 fourth-place votes $54 \cdot 1 = 54$
                                     Total   228

Columbus: 28 first-place votes $28 \cdot 4 = 112$
           0 second-place votes $0 \cdot 3 = 0$
           72 third-place votes $72 \cdot 2 = 144$
           0 fourth-place votes $0 \cdot 1 = 0$
                                     Total   256

Dunkirk: 36 first-place votes $36 \cdot 4 = 144$
           0 second-place votes $0 \cdot 3 = 0$
           18 third-place votes $18 \cdot 2 = 36$
           46 fourth-place votes $46 \cdot 1 = 46$
                                     Total   226

Because Avalon has the largest total score, Avalon is elected president.

✔ TAKE NOTE

Notice in Example 2 that Avalon was the winner even though that candidate did not receive any first-place votes. The Borda Count method was devised to allow voters to say, "If my first choice does not win, then consider my second choice."

CHECK YOUR PROGRESS 2 The preference schedule given earlier for the 50 people who were asked to rank their preferences of five varieties of chocolate candy is shown again below.

	Rankings					
Caramel center	5	4	4	4	2	4
Vanilla center	1	5	5	5	5	5
Almond center	2	3	2	1	3	3
Toffee center	4	1	1	3	4	2
Solid chocolate	3	2	3	2	1	1
Number of voters:	17	11	9	8	3	2

Determine the taste test favorite using the Borda Count method.

Solution   See page S48.

Plurality with Elimination

✓ **TAKE NOTE**

When the second round of voting occurs, the two ballots that listed Bremerton as the first choice must be adjusted. The second choice on those ballots becomes the first, the third choice becomes the second, and the fourth choice becomes the third. The order of preference does not change. Similar adjustments must be made to the 12 ballots that listed Bremerton as the second choice. Because that choice is no longer available, Apple Valley becomes the second choice and Del Mar becomes the third choice. This also applies to the 11 ballots that listed Bremerton as the third choice. The fourth choice of those ballots, Del Mar, becomes the third choice.

A variation of the plurality method of voting is called *plurality with elimination.* Like the Borda Count method, the method of plurality with elimination considers a voter's alternate choices.

Suppose that 30 members of a regional planning board must decide where to build a new airport. The airport consultants to the regional board have recommended four different sites. The preference schedule for the board members is shown in the table below.

	Rankings			
Apple Valley	3	1	2	3
Bremerton	2	3	3	1
Cochella	1	2	4	2
Del Mar	4	4	1	4
Number of ballots:	12	11	5	2

Using the plurality with elimination method, the board members first eliminate the site with the fewest number of first-place votes. In this case, Bremerton is eliminated because it received only two first-place votes. Now a vote is retaken using the following important assumption: *Voters do not change their preferences from round to round.* This means that after Bremerton is deleted, the twelve people in the first column would adjust their preference so that Apple Valley becomes their second choice, Cochella remains their first choice, and Del Mar becomes their third choice.

▼ **point of interest**

Variations of plurality with elimination are used in states such as Alaska and New Mexico, in cities such as Ann Arbor and New York, and in countries such as Australia and Ireland.

A variation of this method is also used to select the Academy Award nominees.

For the eleven voters in the second column, Apple Valley remains their first choice, Cochella remains their second choice, and Del Mar becomes their third choice. Similar adjustments are made by the remaining voters. The new preference schedule is

	Rankings			
Apple Valley	2	1	2	2
Cochella	1	2	3	1
Del Mar	3	3	1	3
Number of ballots:	12	11	5	2

The board members now repeat the process and eliminate the site with the fewest first-place votes. In this case it is Del Mar. The new adjusted preference schedule is

	Rankings			
Apple Valley	2	1	1	2
Cochella	1	2	2	1
Number of ballots:	12	11	5	2

From this table, Apple Valley has 16 first-place votes and Cochella has 14 first-place votes. Therefore, Apple Valley is the selected site for the new airport.

EXAMPLE 3 ■ Use the Plurality with Elimination Voting Method

A university wants to add a new sport to its existing program. To help ensure that the new sport will have student support, the students of the university are asked to rank the four sports under consideration. The results are shown in the table below.

	Rankings					
Lacrosse	3	2	3	1	1	2
Squash	2	1	4	2	3	1
Rowing	4	3	2	4	4	4
Golf	1	4	1	3	2	3
Number of ballots:	326	297	287	250	214	197

Use the plurality with elimination method to determine which of these sports should be added to the university's program.

✔ **TAKE NOTE**

Remember to shift the preferences at each stage of the elimination method. The 297 students who chose rowing as their third choice and golf as their fourth choice now have golf as their third choice. Check each preference schedule and update it as necessary.

Solution

Because rowing received no first-place votes, it is eliminated from consideration. The new preference schedule is shown below.

	Rankings					
Lacrosse	3	2	2	1	1	2
Squash	2	1	3	2	3	1
Golf	1	3	1	3	2	3
Number of ballots:	326	297	287	250	214	197

From this table, lacrosse has the fewest first-place votes, so it is eliminated. The new preference schedule is shown below.

	Rankings					
Squash	2	1	2	1	2	1
Golf	1	2	1	2	1	2
Number of ballots:	326	297	287	250	214	197

From this table, squash received $297 + 250 + 197 = 744$ first-place votes and golf received $326 + 287 + 214 = 827$ first-place votes. Therefore, golf is added to the sports program.

CHECK YOUR PROGRESS 3 A service club is going to sponsor a dinner to raise money for a charity. The club has decided to serve Italian, Mexican, Thai, Chinese, or Indian food. The members of the club were surveyed to determine their preferences. The results are shown in the table below.

	Rankings				
Italian	2	5	1	4	3
Mexican	1	4	5	2	1
Thai	3	1	4	5	2
Chinese	4	2	3	1	4
Indian	5	3	2	3	5
Number of ballots:	33	30	25	20	18

Use the plurality with elimination method to determine the food preference of the club members.

Solution *See page S49.*

Pairwise Comparison Voting Method

historical note

Maria Nicholas Caritat (kä-re-tä) (1743–1794) the Marquis de Condorcet, was, like Borda, a member of the French Academy of Sciences. Around 1780, he showed that the Borda Count method also had flaws and proposed what is now called the *Condorcet criterion.*

In addition to his work on the theory of voting, Condorcet contributed to the writing of a French constitution in 1793. He was an advocate of equal rights for women and the abolishment of slavery. Condorcet was also one of the first mathematicians to try, although without much success, to use mathematics to discover principles in the social sciences. ∎

The *pairwise comparison* method of voting is sometimes referred to as the "head-to-head" method. In this method, each candidate is compared one-on-one with each of the other candidates. A candidate receives 1 point for a win, 0.5 points for a tie, and 0 points for a loss. The candidate with the greatest number of points wins the election.

A voting method that elects the candidate who wins all head-to-head matchups is said to satisfy the Condorcet criterion.

> **Condorcet criterion**
>
> A candidate who wins all possible head-to-head matchups should win an election when all candidates appear on the ballot.

This is one of the *fairness criteria* that a voting method should exhibit. We will discuss other fairness criteria later in this section.

EXAMPLE 4 ■ Use the Pairwise Comparison Voting Method

There are four proposals for the name of a new football stadium at a college: Panther Stadium, after the team mascot; Sanchez Stadium, after a large university contributor; Mosher Stadium, after a famous alumnus known for humanitarian work; and Fritz Stadium, after the college's most winning football coach. The preference schedule cast by alumni and students is shown below.

	Rankings				
Panther Stadium	2	3	1	2	4
Sanchez Stadium	1	4	2	4	3
Mosher Stadium	3	1	4	3	2
Fritz Stadium	4	2	3	1	1
Number of ballots:	752	678	599	512	487

Use the pairwise comparison voting method to determine the name of the stadium.

Solution

We will create a table to keep track of each of the head-to-head comparisons. Before we begin, note that a matchup between, say, Panther and Sanchez is the same as the matchup between Sanchez and Panther. Therefore, we will shade the duplicate cells and the cells between the same candidates. This is shown below.

versus	Panther	Sanchez	Mosher	Fritz
Panther				
Sanchez				
Mosher				
Fritz				

✔ **TAKE NOTE**

Although we have shown the totals for both Panther over Sanchez and Sanchez over Panther, it is only necessary to do one of the matchups. You can then just subtract that number from the total number of ballots cast, which in this case is 3028.

Panther over Sanchez: 1789

Sanchez over Panther:
$3028 - 1789 = 1239$

We could have used this calculation method for all of the other matchups. For instance, in the final matchup of Mosher versus Fritz, we have

Mosher over Fritz: 1430

Fritz over Mosher:
$3028 - 1430 = 1598$

To complete the table, we will place the name of the winner in the cell of the head-to-head match. For instance, for the Panther–Sanchez matchup,

✓ Panther was favored over Sanchez on $678 + 599 + 512 = 1789$ ballots.
   Sanchez was favored over Panther on $752 + 487 = 1239$ ballots.

The winner of this matchup is Panther, so that name is placed in the Panther versus Sanchez cell. Do this for each of the matchups.

✓ Panther was favored over Mosher on $752 + 599 + 512 = 1863$ ballots.
   Mosher was favored over Panther on $678 + 487 = 1165$ ballots.

   Panther was favored over Fritz on $752 + 599 = 1351$ ballots.
✓ Fritz was favored over Panther on $678 + 512 + 487 = 1677$ ballots.

   Sanchez was favored over Mosher on $752 + 599 = 1351$ ballots.
✓ Mosher was favored over Sanchez on $678 + 512 + 487 = 1677$ ballots.

   Sanchez was favored over Fritz on $752 + 599 = 1351$ ballots.
✓ Fritz was favored over Sanchez on $678 + 512 + 487 = 1677$ ballots.

   Mosher was favored over Fritz on $752 + 678 = 1430$ ballots.
✓ Fritz was favored over Mosher on $599 + 512 + 487 = 1598$ ballots.

versus	Panther	Sanchez	Mosher	Fritz
Panther		Panther	Panther	Fritz
Sanchez			Mosher	Fritz
Mosher				Fritz
Fritz				

From the above table, Fritz has three wins, Panther has two wins, and Mosher has one win. Therefore, Fritz Stadium is the winning name.

CHECK YOUR PROGRESS 4   One hundred restaurant critics were asked to rank their favorite restaurants from a list of four. The preference schedule for the critics is shown in the table below.

	Rankings				
Sanborn's Fine Dining	3	1	4	3	1
The Apple Inn	4	3	3	2	4
May's Steak House	2	2	1	1	3
Tory's Seafood	1	4	2	4	2
Number of ballots:	31	25	18	15	11

Use the pairwise voting method to determine the critics' favorite restaurant.

Solution   *See page S49.*

Fairness of Voting Methods and Arrow's Theorem

Kenneth Arrow

Now that we have examined various voting options, we will stop to ask which of these options is the *fairest*. To answer that question, we must first determine what we mean by fair.

In 1948, Kenneth J. Arrow was trying to develop material for his Ph.D. dissertation. As he studied, it occurred to him that he might be able to apply the principles of order relations to problems in social choice or voting. (An example of an order relation for real numbers is "less than.") His investigation led him to outline various criteria for a fair voting system. A paraphrasing of four fairness criteria is given below.

Fairness Criteria

1. *Majority criterion:* The candidate who receives a majority of the first-place votes is the winner.

2. *Monotonicity criterion:* If candidate *A* wins an election, then candidate *A* will also win the election if the only change in the voters' preferences is that supporters of a different candidate change their votes to support candidate *A*.

3. *Condorcet criterion:* A candidate who wins all possible head-to-head matchups should win an election when all candidates appear on the ballot.

4. *Independence of irrelevant alternatives:* If a candidate wins an election, the winner should remain the winner in any recount in which losing candidates withdraw from the race.

There are other criteria, such as the *dictator criterion,* which we will discuss in the next section. However, what Kenneth Arrow was able to prove is that no matter what kind of voting system we devise, it is impossible for it to satisfy the fairness criteria.

Arrow's Impossibility Theorem

There is no voting method involving three or more choices that satisfies the fairness criteria.

By Arrow's Impossibility Theorem, none of the voting methods we have discussed are fair. Not only that, we cannot construct a fair voting system for three or more candidates. We will now give some examples of each of the methods we have discussed and show which of the fairness criteria are not satisfied.

EXAMPLE 5 ■ Show that the Borda Count Method Violates the Majority Criterion

Suppose the preference schedule for three candidates, Alpha, Beta, and Gamma, is given by the table below.

	Rankings		
Alpha	1	3	3
Beta	2	1	2
Gamma	3	2	1
Number of ballots:	55	50	3

Show that using the Borda Count method violates the majority criterion.

Solution

The calculations for Borda's method are shown below.

Alpha

55 first-place votes	$55 \cdot 3 = 165$
0 second-place votes	$0 \cdot 2 = 0$
53 third-place votes	$53 \cdot 1 = 53$
	Total 218

Beta

50 first-place votes	$50 \cdot 3 = 150$
58 second-place votes	$58 \cdot 2 = 116$
0 third-place votes	$0 \cdot 1 = 0$
	Total 266

Gamma

3 first-place votes	$3 \cdot 3 = 9$
50 second-place votes	$50 \cdot 2 = 100$
55 third-place votes	$55 \cdot 1 = 55$
	Total 164

From these calculations, Beta should win the election. However, Alpha has the majority (more than 50%) of the first-place votes. This result violates the majority criterion.

CHECK YOUR PROGRESS 5   Using the table in Example 5, show that the Borda Count method violates the Condorcet criterion.

Solution   *See page S49.*

QUESTION   *Does the pairwise comparison voting method satisfy the Condorcet criterion?*

EXAMPLE 6 ■ **Show that Plurality with Elimination Violates the Monotonicity Criterion**

Suppose the preference schedule for three candidates, Alpha, Beta, and Gamma, is given by the table below.

	Rankings			
Alpha	2	3	1	1
Beta	3	1	2	3
Gamma	1	2	3	2
Number of ballots:	25	20	16	10

ANSWER   *Yes. The pairwise comparison voting method elects the person who wins all head-to-head matchups.*

a. Show that, using plurality with elimination voting, Gamma wins the election.

b. Suppose that the 10 people who voted for Alpha first and Gamma second changed their votes such that they all voted for Alpha second and Gamma first. Show that, using plurality with elimination voting, Beta will now be elected.

c. Explain why this result violates the montonicity criterion.

Solution

a. Beta received the fewest first-place votes, so Beta is eliminated. The new preference schedule is

	Rankings			
Alpha	2	2	1	1
Gamma	1	1	2	2
Number of ballots:	25	20	16	10

From this schedule, Gamma has 45 first-place votes and Alpha has 26 first-place votes, so Gamma is the winner.

b. If the 10 people who voted for Alpha first and Gamma second changed their votes such that they all voted for Alpha second and Gamma first, the preference schedule would be

	Rankings			
Alpha	2	3	1	2
Beta	3	1	2	3
Gamma	1	2	3	1
Number of ballots:	25	20	16	10

From this schedule, Alpha has the fewest first-place votes and is eliminated. The new preference schedule is

	Rankings			
Beta	2	1	1	2
Gamma	1	2	2	1
Number of ballots:	25	20	16	10

From this schedule, Gamma has 35 first-place votes and Beta has 36 first-place votes, so Beta is the winner.

c. This result violates the monotonicity criterion because Gamma, who won the first election, loses the second election even though Gamma received a larger number of first-place votes.

CHECK YOUR PROGRESS 6 The table below shows the preferences for three new car colors.

	Rankings		
Radiant silver	1	3	3
Electric red	2	2	1
Lightning blue	3	1	2
Number of votes:	30	27	2

Show that the Borda Count method violates the independence of irrelevant alternatives criterion.

Solution *See page S49.*

Excursion

Variations of the Borda Count Method

Sixty people were asked to select their preferences among plain ice tea, lemon-flavored ice tea, and raspberry-flavored ice tea. The preference schedule is shown in the table below.

	Rankings		
Plain ice tea	1	3	3
Lemon ice tea	2	2	1
Raspberry ice tea	3	1	2
Number of ballots:	25	20	15

1. Using the Borda method of voting, which flavor of ice tea is preferred by this group? Which is second? Which is third?

2. Instead of using the normal Borda method, suppose the Borda method used in Exercise 1 of this Excursion assigned 1 point for first, 0 points for second, and −1 point for third place. Does this alter the preferences you found in Exercise 1?

3. Suppose the Borda method used in Exercise 1 of this Excursion assigned 10 points for first, 5 points for second, and 0 points for third place. Does this alter the preferences you found in Exercise 1?

(continued)

4. Suppose the Borda method used in Exercise 1 of this Excursion assigned 20 points for first, 5 points for second, and 0 points for third place. Does this alter the preferences you found in Exercise 1?

5. Suppose the Borda method used in Exercise 1 of this Excursion assigned 25 points for first, 5 points for second, and 0 points for third place. Does this alter the preferences you found in Exercise 1?

6. Can the assignment of points for first, second, and third place change the preference order when the Borda method of voting is used?

7. Suppose the assignment of points for first, second, and third place for the Borda method of voting are consecutive integers. Can the value of the starting integer change the outcome of the preferences?

Exercise Set 13.2   (Suggested Assignment: 1–39, odd; 40, 44)

1. What is the difference between a majority and a plurality? Is it possible to have one without the other?

2. Explain why the plurality voting system may not be the best system to use in some situations.

3. Explain how the Borda Count method of voting works.

4. Explain how the plurality with elimination voting method works.

5. Explain how the pairwise comparison voting method works.

6. What does the Condorcet criterion say?

7. Is there a "best" voting method? Is one method more fair than the others?

8. Explain why, if only two candidates are running, the plurality and Borda Count methods will determine the same winner.

9. In the 2000 presidential election, the following votes were cast: (*Source:* **http://www.radix.net**)

Pat Buchanan	0.45 million
George W. Bush	49.82 million
Al Gore	50.16 million
Ralph Nader	2.78 million

If the plurality method were used, who would win the election? Does the winner have a majority?

10. Sixteen people were asked to rank three breakfast cereals in order of preference. Their responses are given below.

Corn Flakes	3	1	1	2	3	3	2	2	1	3	1	3	1	2	1	2
Raisin Bran	1	3	2	3	1	2	1	1	2	1	2	2	3	1	3	3
Mini Wheat	2	2	3	1	2	1	3	3	3	2	3	1	2	3	2	1

If the plurality method of voting is used, which cereal is the group's first preference?

11. A kindergarten class was surveyed to determine the childrens' favorite cartoon characters among Mickey Mouse, Bugs Bunny, and Scooby Doo. The students ranked the characters in order of preference; the results are shown in the preference schedule below.

	Rankings					
Mickey Mouse	1	1	2	2	3	3
Bugs Bunny	2	3	1	3	1	2
Scooby Doo	3	2	3	1	2	1
Number of students:	6	4	6	5	6	8

   a. How many students are in the class?

   b. How many votes are required for a majority?

   c. Using plurality voting, which character is the childrens' favorite?

12. A 15-person committee is having lunch catered for a meeting. Three caterers, each specializing in a different cuisine, are available. In order to choose a caterer for the group, each member is asked to rank the cuisine options in order of preference. The results are given in the preference schedule below.

	Rankings				
Italian	1	1	2	3	3
Mexican	2	3	1	1	2
Japanese	3	2	3	2	1
Number of votes:	2	4	1	5	3

Using plurality voting, which caterer should be chosen?

13. Fifty consumers were surveyed about their movie watching habits. They were asked to rank the likelihood that they would participate in each listed activity. The results are summarized in the table below.

	Rankings				
Go to a theater	2	3	1	2	1
Rent a video or DVD	3	1	3	1	2
Watch pay-per-view	1	2	2	3	3
Number of votes:	8	13	15	7	7

Using the Borda Count method of voting, which activity is the most popular choice among this group of consumers?

14. Use the Borda Count method of voting to determine the preferred breakfast cereal in Exercise 10.

15. Use the Borda Count method of voting to determine the childrens' favorite cartoon character in Exercise 11.

16. Use the Borda Count method of voting to determine which caterer the committee should hire in Exercise 12.

17. A senior high school class held an election for class president. Instead of just voting for one candidate, the students were asked to rank all four candidates in order of preference. The results are shown below.

	Rankings					
Raymond Lee	2	3	1	3	4	2
Suzanne Brewer	4	1	3	4	1	3
Elaine Garcia	1	2	2	2	3	4
Michael Turley	3	4	4	1	2	1
Number of votes:	36	53	41	27	31	45

Using the Borda Count method, which student should be class president?

18. A journalist reviewing various cellular phone services surveyed 200 customers and asked each one to rank four service providers in order of preference. The group's results are shown below.

	Rankings				
Verizon	3	4	2	3	4
Sprint PCS	1	1	4	4	3
Cingular	2	2	1	2	1
Nextel	4	3	3	1	2
Number of votes:	18	38	42	63	39

Using the Borda Count method, which provider is the favorite of these customers?

19. A Little League baseball team must choose the colors for its uniforms. The coach offered four different choices, and the players ranked them in order of preference, as shown in the table below.

	Rankings			
Red and white	2	3	3	2
Green and yellow	4	1	4	1
Red and blue	3	4	2	4
Blue and white	1	2	1	3
Number of votes:	4	2	5	4

Using the plurality with elimination method, what color should the uniforms be?

20. A number of college students were asked to rank four radio stations in order of preference. The responses are given in the table below.

	Rankings				
WNNX	3	1	1	2	4
WKLS	1	3	4	1	2
WWVV	4	2	2	3	1
WSTR	2	4	3	4	3
Number of votes:	57	72	38	61	15

Use plurality with elimination to determine the students' favorite radio station among the four.

21. Use plurality with elimination to choose the class president in Exercise 17.

22. Use plurality with elimination to determine the preferred cellular phone service in Exercise 18.

23. A campus club has money left over in its budget and must spend it before the school year ends. The members arrived at five different possibilities, and each member ranked them in order of preference. The results are shown in the table below.

	Rankings				
Establish a scholarship	1	2	3	3	4
Pay for several members to travel to a convention	2	1	2	1	5
Buy new computers for the club	3	3	1	4	1
Throw an end-of-year party	4	5	5	2	2
Donate to charity	5	4	4	5	3
Number of votes:	8	5	12	9	7

a. Using the plurality voting system, how should the club spend the money?

b. Use the plurality with elimination method to determine how the money should be spent.

c. Using the Borda Count method of voting, how should the money be spent?

d. ✎ In your opinion, which of the previous three methods seems most appropriate in this situation? Why?

24. A company is planning its annual summer retreat and has asked its employees to rank five different choices of recreation in order of preference. The results are given in the table below.

	Rankings				
Picnic in a park	1	2	1	3	4
Water skiing at a lake	3	1	2	4	3
Amusement park	2	5	5	1	2
Riding horses at a ranch	5	4	3	5	1
Dinner cruise	4	3	4	2	5
Number of votes:	10	18	6	28	16

a. Using the plurality voting system, what activity should be planned for the retreat?

b. Use the plurality with elimination method to determine which activity should be chosen.

c. Using the Borda Count method of voting, which activity should be planned?

25. Fans of the *Star Wars* movies have been debating on a web site regarding which of the films is the best. To see what the overall opinion is, visitors to the web site can rank the four films in order of preference. The results are shown in the preference schedule below.

	Rankings			
Star Wars	1	2	1	3
The Empire Strikes Back	4	4	2	1
Return of the Jedi	2	1	3	2
The Phantom Menace	3	3	4	4
Number of votes:	429	1137	384	582

Using pairwise comparison, which film is the favorite of the visitors to the web site who voted?

26. The Nelson family is trying to decide where to hold a family reunion. They have asked all their family members to rank four choices in order of preference. The results are shown in the preference schedule below.

	Rankings				
Grand Canyon	3	1	2	3	1
Yosemite	1	2	3	4	4
Bryce Canyon	4	4	1	2	2
Yellowstone	2	3	4	1	3
Number of votes:	7	3	12	8	13

Use the pairwise comparison method to determine the best choice for the reunion.

27. A new college needs to pick a mascot for its football team. The students were allowed to rank four choices in order of preference; the results are tallied below.

	Rankings				
Bulldog	3	4	4	1	3
Panther	2	1	2	4	2
Hornet	4	2	1	2	4
Bobcat	1	3	3	3	1
Number of votes:	638	924	525	390	673

Using the pairwise comparison method of voting, which mascot should be chosen?

28. Five candidates are running for president of a charity organization. Interested persons were asked to rank the candidates in order of preference. The results are given below.

	Rankings				
P. Gibson	5	1	2	1	2
E. Yung	2	4	5	5	3
R. Allenbaugh	3	2	1	3	5
T. Meckley	4	3	4	4	1
G. DeWitte	1	5	3	2	4
Number of votes:	16	9	14	9	4

Use the pairwise comparison method to determine the president of the organization.

29. Use the pairwise comparison method to choose the colors for the Little League uniforms in Exercise 19.

30. Use the pairwise comparison method to determine the favorite radio station in Exercise 20.

31. Does the winner in Exercise 11 satisfy the Condorcet criterion?

32. Does the winner in Exercise 12 satisfy the Condorcet criterion?

33. Does the winner in Exercise 23c satisfy the Condorcet criterion?

34. Does the winner in Exercise 24c satisfy the Condorcet criterion?

35. Does the winner in Exercise 17 satisfy the majority criterion?

36. Does the winner in Exercise 20 satisfy the majority criterion?

37. Three candidates are running for mayor. A vote was taken in which the candidates were ranked in order of preference. The results are shown in the preference schedule below.

	Rankings		
John Lorenz	1	3	3
Marcia Beasley	3	1	2
Stephen Hyde	2	2	1
Number of votes:	2691	2416	237

a. Use the Borda Count method to determine the winner of the election.

b. Verify that the majority criterion has been violated.

c. Identify a candidate who wins all head-to-head comparisons.

d. Explain why the Condorcet criterion has been violated.

e. If Marcia Beasley drops out of the race for mayor (and voter preferences remain the same), determine the winner of the election again, using the Borda Count method.

f. Explain why the independence of irrelevant alternatives criterion has been violated.

38. Three films have been selected as finalists in a national student film competition. Seventeen judges have viewed each of the films and ranked them in order of preference. The results are given in the preference schedule below.

	Rankings			
Film A	1	3	2	1
Film B	2	1	3	3
Film C	3	2	1	2
Number of votes:	4	6	5	2

a. Using the plurality with elimination method, which film should win the competition?

b. Suppose the first vote is declared invalid and a revote is taken. All of the judges' preferences remain the same except for the votes represented by the last column of the table. The judges who cast these votes both decide to switch their first place vote to Film C, so their preference now is C first, then A, and then B. Which film now wins using the plurality with elimination method?

c. Has the monotonicity criterion been violated?

Extensions

CRITICAL THINKING

39. A campus club needs to elect four officers: a president, a vice president, a secretary, and a treasurer. The club has five volunteers. Rather than vote individually for each position, the club members will rank the candidates in order of preference. The votes will then be tallied using the Borda Count method. The candidate receiving the highest number of points will be president, the candidate receiving the next highest number of points is vice president, the candidate receiving the next highest number of points is secretary, and the candidate receiving the next highest number of points will be treasurer. For the preference schedule shown below, determine who wins each position in the club.

	Rankings				
Cynthia	4	2	5	2	3
Andrew	2	3	1	4	5
Jen	5	1	2	3	2
Hector	1	5	4	1	4
Medin	3	4	3	5	1
Number of votes:	22	10	16	6	27

40. Use the plurality with elimination method of voting to determine the four officers in Exercise 39. Is your result the same as the result you arrived at using the Borda Count method?

41. The members of a scholarship committee have ranked four finalists competing for a scholarship in order of preference. The results are shown in the preference schedule below.

	Rankings			
Francis Chandler	3	4	4	1
Michael Huck	1	2	3	4
David Chang	2	3	1	2
Stephanie Owen	4	1	2	3
Number of votes:	9	5	7	4

If you are one of the voting members and you want David Chang to win the scholarship, which voting method would you suggest that the committee use?

COOPERATIVE LEARNING

42. Suppose you and three friends, David, Sara, and Cliff, are trying to decide on a pizza restaurant. You like Pizza Hut best, Round Table pizza is acceptable to you, and you definitely do not want to get pizza from Domino's. Domino's is David's favorite, and he also likes Round Table, but he won't eat pizza from Pizza Hut. Sara says she will only eat Round Table pizza. Cliff prefers Domino's, but he will also eat Round Table pizza. He doesn't like Pizza Hut.

a. Given the preferences of the four friends, which pizza restaurant would be the best choice?

b. If you use the plurality system of voting to determine which pizza restaurant to go to, which restaurant wins? Does this seem like the best choice for the group?

c. If you use the pairwise comparison method of voting, which pizza restaurant wins?

d. Does one of the four voting methods discussed in this section give the same winner as your choice in part a? Does the method match your reasoning in making your choice?

EXPLORATIONS

43. Another method of voting is to assign a "weight," or score, to each candidate rather than ranking the candidates in order. All candidates must receive a score, and two or more candidates can receive the same score from a voter. A score of 5 represents the strongest endorsement of a candidate. The scores range down to 1, which corresponds to complete disapproval of a candidate. A score of 3 represents indifference. The candidate with the most total points wins the election. The results of a sample election are given in the table.

	Rankings							
Candidate A	2	1	2	5	4	2	4	5
Candidate B	5	3	5	3	2	5	3	2
Candidate C	4	5	4	1	4	3	2	1
Candidate D	1	3	2	4	5	1	3	2
Number of votes:	26	42	19	33	24	8	24	33

a. Find the winner of the election.

b. If plurality were used (assuming that a person's vote would go to the candidate that he or she gave the highest score to), verify that a different winner would result.

44. *Approval voting* is a system in which voters may vote for more than one candidate. Each vote counts equally, and the candidate with the most total votes wins the election. Many feel that this is a better system for large elections than simple plurality because it considers a voter's second choices and is a stronger measure of overall voter support for each candidate. Some organizations use approval voting to elect their officers. The United Nations uses this method to elect the secretary-general.

a. Suppose a math class is going to show a film involving mathematics or mathematicians on the last day of class. The options are *Stand and Deliver, Good Will Hunting, A Beautiful Mind, Pi,* and *Contact*. The students vote using approval voting. The results are as follows.

   8 students vote for all five films.

   8 students vote for *Good Will Hunting, A Beautiful Mind,* and *Contact*.

   8 students vote for *Stand and Deliver, Good Will Hunting,* and *Contact*.

   8 students vote for *A Beautiful Mind* and *Pi*.

   8 students vote for *Stand and Deliver* and *Pi*.

   8 students vote for *Good Will Hunting* and *Contact*.

   1 student votes for *Pi*.

   Which film will be chosen for the last day of class screening?

b. Use approval voting in Exercise 42 to determine the pizza restaurant the group of friends should choose. Does your result agree with your answer to part a of Exercise 42?

SECTION 13.3 | ## Weighted Voting Systems

Biased Voting Systems

A **weighted voting system** is one in which some voters have more weight on the outcome of an election. Examples of weighted voting systems are fairly common. A few examples are the stockholders of a company, the Electoral College, the United Nations Security Council, and the European Union.

> ## Math Matters   The Electoral College
>
> As mentioned in the Historical Note on the following page, the Electoral College elects the president of the United States. The number of electors representing each state is equal to the sum of the number of senators (2) and the number of members in the House of Representatives for that state. The original intent of the framers of the Constitution was to protect the smaller states. We can verify this by computing the number of people represented by each elector. In the 2000 election, each Vermont elector represented about 188,000 people; each California elector represented about
>
> *(continued)*

historical note

The U.S. Constitution, Article 2, Section 1 states that the members of the Electoral College elect the president of the United States. The original article directed members of the College to vote for two people. However, it did not stipulate that one name was for president and the other name was for vice president. The article goes on to state that the person with the greatest number of votes becomes president and the one with the next highest number of votes becomes vice president. In 1800, Thomas Jefferson and Aaron Burr received exactly the same number of votes even though they were running on a Jefferson for president, Burr for vice president ticket. Thus the House of Representatives was asked to select the president. It took 36 different votes by the House before Jefferson was elected president. In 1804, the 12th Amendment to the Constitution was ratified to prevent a recurrence of the 1800 election problems. ■

551,000 people. To see how this gives a state with a smaller population more *power* (a word we will discuss in more detail later in this section), note that three electoral votes from Vermont represent approximately the same size population as does one electoral vote from California. Each vote does not represent the same number of people.

Another peculiarity related to the Electoral College system is that it is very sensitive to small vote swings. For instance, in the 2000 election, if an additional 0.01% of the voters in Florida had cast their votes for Al Gore instead of George Bush, Gore would have won the presidential election.

Consider a small company with a total of 100 shares of stock and three stockholders, *A*, *B*, and *C*. Suppose that *A* owns 45 shares of the stock (which means *A* has 45 votes), *B* owns 45 shares, and *C* owns 10 shares. If a vote of 51 or greater is required to approve any measure before the owners, then a measure cannot be passed without two of the three owners voting for the measure. Even though *C* has only 10 shares, *C* has the same voting power as *A* and *B*.

Now suppose that a new stockholder is brought into the company and the shares of the company are redistributed so that *A* has 27 shares, *B* has 26 shares, *C* has 25 shares, and *D* has 22 shares. Note, in this case, that any two of *A*, *B*, or *C* can pass a measure, but *D* paired with any of the other shareholders cannot pass a measure. *D* has virtually no power even though *D* has only three shares less than *C*.

The number of votes that are required to pass a measure is called a **quota.** For the two stockholder examples above, the quota was 51. The **weight** of a voter is the number of votes controlled by the voter. In the case of the company whose stock was split *A*−27 shares, *B*−26 shares, *C*−25 shares, and *D*−22 shares, the weight of *A* is 27, the weight of *B* is 26, the weight of *C* is 25, and the weight of *D* is 22. Rather than write out in sentence form the quota and weight of each voter, we use the notation

Quota ——— Weights

$$\{51: 27, 26, 25, 22\}$$

• The four numbers after the colon indicate that there are a total of four voters in this system.

This notation is very convenient. We state its more general form in the definition below.

> ### Weighted Voting System
>
> A weighted voting system of *n* voters is written $\{q: w_1, w_2, \ldots, w_n\}$, where *q* is the quota and w_1 through w_n represent the weights of each of the *n* voters.

Using this notation, we can describe various voting systems.

- **One person, one vote:** For instance, $\{5: 1, 1, 1, 1, 1, 1, 1, 1, 1\}$. In this system, each person has one vote and five votes, a majority, are required to pass a measure.

- **Dictatorship:** For instance, $\{20: 21, 6, 5, 4, 3\}$. In this system, the person with 21 votes can pass any measure. Even if the remaining four people get together, their votes do not total the quota of 20.

■ **Null system:** For instance, {28: 6, 3, 5, 2}. If all the members of this system vote for a measure, the total number of votes is 16, which is less than the quota. Therefore, no measure can be passed.

■ **Veto power system:** For instance, {21: 6, 5, 4, 3, 2, 1}. In this case, the sum of all the votes is 21, the quota. Therefore, if any one voter does not vote for the measure, it will fail. Each voter is said to have **veto power.** In this case, this means that even the voter with one vote can veto a measure (cause the measure not to pass). If at least one voter in a voting system has veto power, the system is a veto power system.

Math Matters   UN Security Council: An Application of Inequalities

UN Security Council Chamber

The United Nations Security Council consists of five permanent members (United States, China, France, Great Britain, and Russia) and 10 members that are elected by the General Assembly for a two-year term. In 2000, the 10 nonpermanent members were Argentina, Bangladesh, Canada, Jamaica, Malaysia, Mali, Namibia, Netherlands, Tunisia, and Ukraine.

For a resolution to pass the Security Council,

1. Nine countries must vote for the resolution; and

2. If one of the five permanent members votes against the resolution, it fails.

This situation can be described using inequalities. Let x be the weight of the vote of one permanent member of the Council and let q be the quota for a vote to pass. Then, by condition 1, $q \leq 5x + 4$. (The weights of the five votes of the permanent members plus the single votes of four nonpermanent members must be greater than or equal to the quota.)

By condition 2, we have $4x + 10 < q$. (If one of the permanent members opposes the resolution, it fails even if all of the nonpermanent members vote for it.)

Combining the inequalities from condition 1 and condition 2, we have

$$4x + 10 < 5x + 4$$
$$10 < x + 4 \qquad \bullet \text{ Subtract } 4x \text{ from each side.}$$
$$6 < x \qquad \bullet \text{ Subtract 4 from each side.}$$

The smallest whole number greater than 6 is 7. Therefore, the weight of each permanent member is 7. Substituting 7 into $q \leq 5x + 4$ and $4x + 10 < q$, we find that $q = 39$. Thus the weighted voting system of the Security Council is given by {39: 7, 7, 7, 7, 7, 1, 1, 1, 1, 1, 1, 1, 1, 1, 1, 1}.

QUESTION   *Is the UN Security Council voting system a veto power system?*

In a weighted voting system, a **coalition** is a set of voters each of whom votes the same way, either for or against a resolution. A **winning coalition** is a set of voters the sum of whose votes is greater than or equal to the quota. A **losing coali-**

ANSWER   *Yes. If any of the permanent members votes against a resolution, the resolution cannot pass.*

tion is a set of voters the sum of whose votes is less than the quota. A voter who leaves a winning coalition and thereby turns it into a losing coalition is called a **critical voter.**

As shown in the next theorem, for large numbers of voters, there are many possible coalitions.

Number of Possible Coalitions of _n_ Voters

The number of possible coalitions of _n_ voters is $2^n - 1$.

As an example, if all electors of each state to the Electoral College cast their ballots for one candidate, then there are $2^{51} - 1 \approx 2.25 \times 10^{15}$ possible coalitions (the District of Columbia is included). The number of _winning_ coalitions is far less. For instance, any coalition of 10 or fewer states cannot be a winning coalition because the largest 10 states do not have enough electoral votes to elect the president. As we proceed through this section, we will not attempt to list all the coalitions, only the winning coalitions.

EXAMPLE 1 ■ Determine Winning Coalitions in a Weighted Voting System

Suppose that the four owners of a company, Ang, Bonhomme, Carmel, and Diaz, own, respectively, 500 shares, 375 shares, 225 shares, and 400 shares. The weighted voting system for this company is {751: 500, 375, 225, 400}.

a. Determine the winning coalitions.

b. For each winning coalition, determine the critical voters.

Solution

a. A winning coalition must represent at least 751 votes. We will list these coalitions in the table below, in which we use _A_ for Ang, _B_ for Bonhomme, _C_ for Carmel, and _D_ for Diaz.

Winning coalition	Number of votes
{A, B}	875
{A, D}	900
{B, D}	775
{A, B, C}	1100
{A, B, D}	1275
{A, C, D}	1125
{B, C, D}	1000
{A, B, C, D}	1500

b. A voter who leaves a winning coalition and thereby creates a losing coalition is a critical voter. For instance, for the winning coalition {A, B, C}, if A leaves, the

TAKE NOTE

The number of coalitions of _n_ voters is the number of subsets that can be formed from _n_ voters. From Chapter 2, this is 2^n. Because a coalition must contain at least one voter, the empty set is not a possible coalition. Therefore, the number of coalitions is $2^n - 1$.

TAKE NOTE

The coalition {A, C} is not a winning coalition because the total number of votes for that coalition is 725, which is less than 751.

number of remaining votes is 600, which is not enough to pass a resolution. If B leaves, the number of remaining votes is 725—again, not enough to pass a resolution. Therefore, A and B are critical voters for the coalition {A, B, C} and C is not a critical voter. The table below shows the critical voters for each winning coalition.

Winning coalition	Number of votes	Critical voters
{A, B}	875	A, B
{A, D}	900	A, D
{B, D}	775	B, D
{A, B, C}	1100	A, B
{A, B, D}	1275	None
{A, C, D}	1125	A, D
{B, C, D}	1000	B, D
{A, B, C, D}	1500	None

CHECK YOUR PROGRESS 1   Many countries must govern by forming coalitions from among many political parties. Suppose a country has five political parties named A, B, C, D, and E. The numbers of votes, respectively, for each party are 22, 18, 17, 10, and 5.

a.   Determine the winning coalitions if 37 votes are required to pass a resolution.

b.   For each winning coalition, determine the critical voters.

Solution   *See page S50.*

QUESTION   *Is the voting system in Example 1 a dictatorship? What is the total number of possible coalitions in Example 1?*

Banzhaf Power Index

There are a number of measures of the *power* of a voter. For instance, as we saw from the Electoral College example, some electors represent fewer people and therefore their votes may have more power. As an extreme case, suppose two electors, A and B, each represent 10 people and a third elector, C, represents 1000 people. If a measure passes when two of the three electors vote for the measure, then A and B voting together could pass a resolution even though they represent only 20 people.

ANSWER   *No. There is no one shareholder who has 751 or more shares of stock. The number of possible coalitions is $2^4 - 1 = 15$.*

Another measure of power, called the *Banzhaf power index*, was derived by John F. Banzhaf III in 1965. The purpose of this index is to determine the power of a voter in a weighted voting system.

Banzhaf Power Index

The **Banzhaf power index** of a voter v, symbolized by $BPI(v)$, is given by

$$BPI(v) = \frac{\text{number of times voter } v \text{ is a critical voter}}{\text{number of times any voter is a critical voter}}$$

Consider four people A, B, C, and D and the one-person, one-vote system given by $\{3: 1, 1, 1, 1\}$.

Winning coalition	Number of votes	Critical voters
{A, B, C}	3	A, B, C
{A, B, D}	3	A, B, D
{A, C, D}	3	A, C, D
{B, C, D}	3	B, C, D
{A, B, C, D}	4	None

To find $BPI(A)$, we look under the critical voters column and find that A is a critical voter three times. The number of times any voter is a critical voter, the denominator of the Banzhaf power index, is 12. (A is a critical voter three times, B is a critical voter three times, C is a critical voter three times, and D is a critical voter three times. The sum is $3 + 3 + 3 + 3 = 12$.) Thus

$$BPI(A) = \frac{3}{12} = 0.25$$

Similarly, we can calculate the Banzhaf power index for each of the other voters.

$$BPI(B) = \frac{3}{12} = 0.25 \qquad BPI(C) = \frac{3}{12} = 0.25 \qquad BPI(D) = \frac{3}{12} = 0.25$$

In this case, each voter has the same power. This is expected in a voting system in which each voter has one vote.

Now suppose that three people A, B, and C belong to a dictatorship given by $\{3: 3, 1, 1\}$.

Winning coalition	Number of votes	Critical voters
{A}	3	A
{A, B}	4	A
{A, C}	4	A
{A, B, C}	5	A

The sum of the critical voters in all winning coalitions is 4. To find $BPI(A)$, we look under the critical voters column and find that A is a critical voter four times. Thus

$$BPI(A) = \frac{4}{4} = 1 \qquad BPI(B) = \frac{0}{4} = 0 \qquad BPI(C) = \frac{0}{4} = 0$$

Thus A has all the power. This is expected in a dictatorship.

EXAMPLE 2 ■ Compute the BPI for a Weighted Voting System

Suppose the stock in a company is held by five people, A, B, C, D, and E. The voting system for this company is {625: 350, 300, 250, 200, 150}. Determine the Banzhaf power index of A and E.

Solution
Determine all of the winning coalitions and the critical voters in each coalition.

Winning coalition	Critical voters	Winning coalition	Critical voters
{A, B}	A, B	{B, C, E}	B, C, E
{A, B, C}	A, B	{B, D, E}	B, D, E
{A, B, D}	A, B	{A, B, C, D}	None
{A, B, E}	A, B	{A, B, C, E}	None
{A, C, D}	A, C, D	{A, B, D, E}	None
{A, C, E}	A, C, E	{A, C, D, E}	A
{A, D, E}	A, D, E	{B, C, D, E}	B
{B, C, D}	B, C, D	{A, B, C, D, E}	None

The number of times all voters are critical is 28. To find $BPI(A)$, we look under the critical voters columns and find that A is a critical voter eight times. Thus

$$BPI(A) = \frac{8}{28} \approx 0.29$$

To find $BPI(E)$, we look under the critical voters columns and find that E is a critical voter four times. Thus

$$BPI(E) = \frac{4}{28} \approx 0.14$$

CHECK YOUR PROGRESS 2 Suppose that a government is composed of four political parties, A, B, C, and D. The voting system for this government is {25: 18, 16, 10, 6}. Determine the Banzhaf power index of A and D.

Solution *See page S50.*

In many cities, the only time the mayor votes on a resolution is when there is a tie vote by the members of the city council. This is also true of the United States Senate. The vice president only votes when there is a tie vote by the senators.

In Example 3, we will calculate the Banzhaf power index for a voting system in which one voter votes only to break a tie.

EXAMPLE 3 ■ Use the BPI to Determine a Voter's Power

Suppose a city council consists of four members, A, B, C, and D, and a mayor M. The mayor votes only when there is a tie vote among the members of the council. In all cases, a resolution receiving three or more votes passes. Show that the Banzhaf power index for the mayor is the same as the Banzhaf power index for each city council member.

Solution

We first list all of the winning coalitions that do not include the mayor. To this list, we add the winning coalitions in which the mayor votes to break a tie.

Winning coalition (without mayor)	Critical voters	Winning coalition (mayor voting)	Critical voters
{A, B, C}	A, B, C	{A, B, M}	A, B, M
{A, B, D}	A, B, D	{A, C, M}	A, C, M
{A, C, D}	A, C, D	{A, D, M}	A, D, M
{B, C, D}	B, C, D	{B, C, M}	B, C, M
{A, B, C, D}	None	{B, D, M}	B, D, M
		{C, D, M}	C, D, M

By examining the table, we see that A, B, C, D, and M each occur in exactly six winning coalitions. The total number of critical voters in all winning coalitions is 30. Therefore, each member of the council and the mayor have the same Banzhaf power index, which is $\frac{6}{30} = 0.2$.

CHECK YOUR PROGRESS 3   The European Economic Community (EEC) was founded in 1958 and originally consisted of Belgium, France, Germany, Italy, Luxembourg, and the Netherlands. The weighted voting system was {12: 2, 4, 4, 4, 1, 2}. Find the Banzhaf power index for each country.

Solution   *See page S51.*

Solution   See page S51.

▼ **point of interest**

In 2002, Austria, Belgium, Denmark, Finland, France, Germany, United Kingdom, Greece, Ireland, Italy, Luxembourg, the Netherlands, Portugal, Spain, and Sweden were full members of the organization known as the European Union (EU), which at one time was referred to as the Common Market or the European Economic Community (EEC) (see Check Your Progress 3). Of these members, all except Denmark, United Kingdom, and Sweden use a common currency called the *euro*.

By working Check Your Progress 3, you will find that the Banzhaf power index for the original EEC gave little power to Belgium and the Netherlands and no power to Luxembourg. However, there was an implicit understanding among the countries that a resolution would not pass unless all countries voted for it, thereby effectively giving each country veto power.

Excursion

Blocking Coalitions and the Banzhaf Power Index

The four members A, B, C, and D of an organization adopted the weighted voting system {6: 4, 3, 2, 1}. The table below shows the winning coalitions.

Winning coalition	Number of Votes	Critical voters
{A, B}	7	A, B
{A, C}	6	A, C
{A, B, C}	9	A
{A, B, D}	8	A, B
{A, C, D}	7	A, C
{B, C, D}	6	B, C, D
{A, B, C, D}	10	None

Using the Banzhaf power index, we have $BPI(A) = \frac{5}{12}$.

A **blocking coalition** is a group of voters who can prevent passage of a resolution. In this case, a critical voter is one who leaves a blocking coalition, thereby producing a coalition that is no longer capable of preventing the passage of a resolution. For the voting system above, we have

Blocking coalition	Number of votes	Number of remaining votes	Critical voters
{A, B}	7	3	A, B
{A, C}	6	4	A, C
{A, D}	5	5	A, D
{B, C}	5	5	B, C
{A, B, C}	9	1	None
{A, B, D}	8	2	A
{A, C, D}	7	3	A, C
{B, C, D}	6	4	B

If we count the number of times A is a critical voter in a winning or blocking coalition, we find what is called the *Banzhaf index*. In this case, the Banzhaf index is 10 and we write $BI(A) = 10$. Using both the winning and blocking coalition tables, we find that

(continued)

$BI(B) = 6$, $BI(C) = 6$, and $BI(D) = 2$. This information can be used to create an alternative definition of the Banzhaf power index.

Banzhaf Power Index—Alternative Definition

$$BPI(A) = \frac{BI(A)}{\text{sum of all Banzhaf indices for the voting system}}$$

Applying this definition to the voting system given above, we have

$$BPI(A) = \frac{BI(A)}{BI(A) + BI(B) + BI(C) + BI(D)} = \frac{10}{10 + 6 + 6 + 2} = \frac{10}{24} = \frac{5}{12}$$

Excursion Exercises

1. Using the data in Example 1 on page 889, list all blocking coalitions.

2. For the data in Example 1 on page 889, calculate the Banzhaf power indices for A, B, C, and D using the alternative definition.

3. Using the data in Check Your Progress 3 on page 893, list all blocking coalitions.

4. For the data in Check Your Progress 3 on page 893, calculate the Banzhaf power indices for Belgium and Luxembourg using the alternative definition.

5. Create a voting system with three members that is a dictatorship. Calculate the Banzhaf power index of each voter for this system using the alternative definition.

6. Create a voting system with four members in which one member has veto power. Calculate the Banzhaf power index for this system using the alternative definition.

7. Create a voting system with five members that meets the one-person, one-vote rule. Calculate the Banzhaf power index for this system using the alternative definition.

Exercise Set 13.3

(Suggested Assignment: 1–25, odd; 28, 29, 33)

In each of the following exercises, the weighted voting systems for the voters A, B, C, ... are given in the form $\{q: w_1, w_2, w_3, w_4, \ldots, w_n\}$. The weight of voter A is w_1, the weight of voter B is w_2, the weight of voter C is w_3, and so on.

1. A weighted voting system is given by $\{6: 4, 3, 2, 1\}$.
   a. What is the quota?
   b. How many voters are in this system?
   c. What is the weight of voter B?
   d. What is the weight of the coalition $\{A, C\}$?
   e. Is $\{A, D\}$ a winning coalition?
   f. Which voters are critical voters in the coalition $\{A, C, D\}$?

   g. How many coalitions can be formed?
   h. How many coalitions consist of exactly two voters?

2. A weighted voting system is given by $\{16: 8, 7, 4, 2, 1\}$.
   a. What is the quota?
   b. How many voters are in this system?
   c. What is the weight of voter C?
   d. What is the weight of the coalition $\{B, C\}$?
   e. Is $\{B, C, D, E\}$ a winning coalition?
   f. Which voters are critical voters in the coalition $\{A, B, D\}$?
   g. How many coalitions can be formed?
   h. How many coalitions consist of exactly three voters?

In Exercises 3–12, calculate, if possible, the Banzhaf power index for each voter.

3. $\{6: 4, 3, 2\}$

4. $\{10: 7, 6, 4\}$

5. $\{10: 7, 3, 2, 1\}$

6. $\{14: 7, 5, 1, 1\}$

7. $\{19: 14, 12, 4, 3, 1\}$

8. $\{3: 1, 1, 1, 1\}$

9. $\{18: 18, 7, 3, 3, 1, 1\}$

10. $\{14: 6, 6, 4, 3, 1\}$

11. $\{80: 50, 40, 30, 25, 5\}$

12. $\{85: 55, 40, 25, 5\}$

13. Which, if any, of the voting systems in Exercises 3 to 12 is

a. a dictatorship?

b. a veto power system? *Note:* A voting system is a veto power system if any of the voters have veto power.

c. a null system?

d. a one-person, one-vote system?

14. Explain why it is impossible to calculate the Banzhaf power index for any voter in the null system $\{8: 3, 2, 1, 1\}$.

15. A music department consists of a band director and a music teacher. Decisions on motions are made by voting. If both members vote in favor of a motion, it passes. If both members vote against a motion, it fails. In the event of a tie vote, the principal of the school votes to break the tie. For this voting scheme, determine the Banzhaf power index for each department member and for the principal. *Hint:* See Example 3, page 893.

16. Four voters A, B, C, and D make decisions by using the voting scheme $\{4: 3, 1, 1, 1\}$, except when there is a tie. In the event of a tie, a fifth voter E casts a vote to break the tie. For this voting scheme, determine the Banzhaf power index for each voter, including voter E. *Hint:* See Example 3, page 893.

17. In a criminal trial, each of the 12 jurors has one vote and all of the jurors must agree to reach a verdict. Otherwise the judge will declare a mistrial.

a. Write the weighted voting system, in the form $\{q: w_1, w_2, w_3, w_4, \ldots, w_{12}\}$, used by these jurors.

b. Is this weighted voting system a one-person, one-vote system?

c. Is this weighted voting system a veto power system?

d. Explain an easy way to determine the Banzhaf power index for each voter.

18. In California civil court cases, each of the 12 jurors has one vote and at least nine of the jury members must agree to reach a verdict.

a. Write the weighted voting system, in the form $\{q: w_1, w_2, w_3, w_4, \ldots, w_{12}\}$, used by these jurors.

b. Is this weighted voting system a one-person, one-vote system?

c. Is this weighted voting system a veto power system?

d. Explain an easy way to determine the Banzhaf power index for each voter.

A voter who has a weight that is greater than or equal to the quota is called a **dictator.** In a weighted voting system, the dictator has all the power. A voter who is never a critical voter has no power and is referred to as a **dummy.** This term is not meant to be a comment on the voter's intellectual powers. It just indicates that the voter has no ability to influence an election.

In Exercises 19–22, identify any dictator and all dummies for each weighted voting system.

19. $\{16: 16, 5, 4, 2, 1\}$

20. $\{15: 7, 5, 3, 2\}$

21. $\{19: 12, 6, 3, 1\}$

22. $\{45: 40, 6, 2, 1\}$

23. At the beginning of each football season, the coaching staff at Vista High School must vote to decide which players to select for the team. They use the weighted voting system $\{4: 3, 2, 1\}$. In this voting system, the head coach A has a weight of 3, the assistant coach B has a weight of 2, and the junior varsity coach C has a weight of 1.

a. Compute the Banzhaf power index for each of the coaches.

b. Explain why it seems reasonable that the assistant coach and the junior varsity coach have the same Banzhaf power index in this voting system.

24. The head coach in Exercise 23 has decided that next year the coaching staff should use the weighted voting system {5: 4, 3, 1}. The head coach is still voter A, the assistant coach is still voter B, and the junior varsity coach is still voter C.

a. Compute the Banzhaf power index for each coach under this new system.

b. How do the Banzhaf power indices for this new voting system compare with the Banzhaf power indices in Exercise 23? Did the head coach gain any power, according to the Banzhaf power indices, with this new voting system?

25. Consider the weighted voting system {60: 4, 56, 58}.

a. Compute the Banzhaf power index for each voter in this system.

b. Voter B has a weight of 56 compared to only 4 for voter A, yet the results of part a show that voter A and voter B both have the same Banzhaf power index. Explain why it seems reasonable, in this voting system, to assign voters A and B the same Banzhaf power index.

Extensions
CRITICAL THINKING

26. It can be proved that for any counting number constant c, the weighted voting systems $\{q: w_1, w_2, w_3, w_4, \ldots, w_n\}$ and $\{cq: cw_1, cw_2, cw_3, cw_4, \ldots, cw_n\}$ both have the same Banzhaf power index distribution. Verify that this theorem is true for the weighted voting system {14: 8, 7, 6, 3} and the constant $c = 2$.

27. Consider the weighted voting system $\{q: 8, 3, 3, 2\}$, with q an integer and $9 \leq q \leq 16$.

a. For what values of q is there a dummy?

b. For what values of q do all voters have the same power?

c. If a voter is a dummy for a given quota, must the voter be a dummy for all larger quotas?

28. Consider the weighted voting system {17: 7, 7, 7, 2}.

a. Explain why voter D is a dummy in this system.

b. Explain an *easy* way to compute the Banzhaf power indices for this system.

29. In a weighted voting system, two voters have the same weight. Must they also have the same Banzhaf power index?

30. In a weighted voting system, Voter A has a larger weight than voter B. Must the Banzhaf power index for voter A be larger than the Banzhaf power index for voter B?

31. Consider the one-person, one-vote strict majority system {2: 1, 1, 1} and the weighted voting system {5: 3, 3, 3}. Explain why the systems are essentially equivalent.

32. Consider the voting system {7: 3, 2, 2, 2, 2}. A student has determined that the Banzhaf power index for voter A is $\frac{5}{13}$.

a. Explain an *easy* way to calculate the Banzhaf power index for each of the other voters.

b. If the given voting system were changed to {8: 3, 2, 2, 2, 2}, which voters, if any, would lose power according to the Banzhaf power index?

EXPLORATIONS

33. Another index that is used to measure the power of voters in a weighted voting system is the *Shapley-Shubik power index*. Write a short report on the Shapley-Shubik power index. Explain how to calculate the Shapley-Shubik power index for each voter in the weighted voting system {6: 4, 3, 2}. How do these Shapley-Shubik power indices compare with the Banzhaf power indices for this voting system? (See your results from Exercise 3.) Explain under what circumstances you would choose to use the Shapley-Shubik power index over the Banzhaf power index.

34. The United Nations Security Council consists of five permanent members and 10 members that are elected for a two-year term. (See the *Math Matters* on page 888.) The weighted voting system of the Security Council is given by

$$\{39: 7, 7, 7, 7, 7, 1, 1, 1, 1, 1, 1, 1, 1, 1, 1\}$$

In this system, each permanent member has a voting weight of 7 and each elected member has a voting weight of 1.

a. Use the spreadsheet program "BPI" at the web site **www.math.temple.edu/~cow/bpi.html** to compute the Banzhaf power index distribution for the members of the Security Council.

b. According to the Banzhaf power index distribution from part a, each permanent member has about how many times more power than an elected member?

c. Some people think that the voting system used by the United Nations Security Council gives too much power to the permanent members. A weighted voting system that would reduce the power of the permanent members is given by

$$\{30: 5, 5, 5, 5, 5, 1, 1, 1, 1, 1, 1, 1, 1, 1, 1\}$$

In this new system, each permanent member has a voting weight of 5 and each elected member has a voting weight of 1. Use the spreadsheet pro-

gram "BPI" on the Aufmann web site at **www.math.temple.edu/~cow/bpi.html** to compute the Banzhaf power indices for the members of the Security Council under this vot-ing system.

d. According to the Banzhaf power indices from part c, each permanent member has about how many times more power than an elected member under this new voting system?

35. The European Community 2000 consists of 15 countries. Issues are decided by the weighted voting system

$$\{62: 10, 10, 10, 10, 8, 5, 5, 5, 5, 4, 4, 3, 3, 3, 2\}$$

The following table shows the weights for each country. (*Source:* **http://europa.eu.int/inst/en/cl.htm**)

Germany, France, Italy, United Kingdom	10
Spain	8
Belgium, Greece, the Netherlands, Portugal	5
Austria, Sweden	4
Italy, Denmark, Finland	3
Luxembourg	2

a. Use the spreadsheet program "BPI" on the web site at **www.math.temple.edu/~cow/bpi.html** to compute the Banzhaf power indices for the countries in the European Community 2000.

b. According to the Banzhaf power indices from part a, how many times more voting power does Italy have than Luxembourg?

CHAPTER 13 **Summary**

Key Terms

Essential Concepts

■ **Standard Divisor**

$$\text{Standard divisor} = \frac{\text{total population}}{\text{number of items to apportion}}$$

■ **Standard Quota**

The *standard quota* is the whole number part of the quotient of a population and the standard divisor.

■ **Quota Rule**

The number of representatives apportioned to a state is the standard quota or one more than the standard quota.

■ **Huntington-Hill Number**

The value of $\dfrac{(P_A)^2}{a(a+1)}$, where P_A is the population of state A and a is the current number of representatives from state A, is called the Huntington-Hill number for state A.

■ **Balinski-Young Impossibility Theorem**

Any apportionment method will either violate the quota rule or will produce paradoxes such as the Alabama paradox.

■ **Plurality Method of Voting**

Each voter votes for one candidate, and the candidate with the most votes wins.

■ **Borda Count Method of Voting**

With n candidates in an election, each voter ranks the candidates by giving n points to the voter's first choice, $n - 1$ points to the voter's second choice, and so on. The candidate who receives the most total points is the winner.

■ **Plurality with Elimination Method of Voting**

Eliminate the candidate with the fewest number of first-place votes. Retake a vote, keeping the same ranking preferences, and eliminate the candidate with the fewest number of first-place votes. Continue until only one candidate remains.

■ **Pairwise Comparison Method of Voting**

Compare each candidate head-to-head with each other candidate. Award 1 point for a win, 0.5 point for a tie, and 0 points for a loss. The candidate with the greatest number of points wins the election.

■ **Majority Criterion**

The candidate who receives a majority of the first-place votes is the winner.

■ **Monotonicity Criterion**

If candidate A wins the election, then candidate A will also win the election if the only change in the voters' preferences is that supporters of a different candidate change their votes to support candidate A.

■ **Condorcet Criterion**

A candidate who wins all possible head-to-head matchups should win an election when all candidates appear on the ballot.

■ **Independence of Irrelevant Alternatives Criterion**

If a candidate wins an election, the winner should remain the winner in any recount in which losing candidates withdraw from the race.

■ **Arrow's Impossibility Theorem**

There is no voting method for an election with three or more candidates that satisfies all four fairness criteria.

■ **Banzhaf Power Index**

The Banzhaf power index of a voter v, symbolized by $BPI(v)$, is given by

$$BPI(v) = \frac{\text{number of times voter } v \text{ is a critical voter}}{\text{number of times any voter is a critical voter}}$$

1. The following table shows the enrollments for each of the four divisions of a college. There are 50 new overhead projectors that are to be apportioned among the divisions based on the enrollments.

Division	Enrollment
Health	1280
Business	3425
Engineering	1968
Science	2936
Total	9609

   a. Use the Hamilton method to determine the number of projectors to be apportioned to each division.

   b. Use the Jefferson method to determine the number of projectors to be apportioned to each division.

   c. Use the Webster method to determine the number of projectors to be apportioned to each division.

2. The following table shows the numbers of ticket agents at five airports for a small airline company. The company has hired 35 new security employees who are to be apportioned among the airports based on the number of ticket agents at each airport.

Airport	Number of ticket agents
Newark	28
Cleveland	19
Chicago	34
Philadelphia	13
Detroit	16
Total	110

   a. Use the Hamilton method to apportion the new security employees among the airports.

   b. Use the Jefferson method to apportion the new security employees among the airports.

   c. Use the Webster method to apportion the new security employees among the airports.

3. A company has four offices. The president of the company uses the Hamilton method to apportion 66 new computer printers among the offices based on the number of employees at each office.

Office	A	B	C	D
Number of employees	19	195	308	402
Apportionment of 66 printers	1	14	22	29

   a. If the number of printers to be apportioned by the Hamilton method increases from 66 to 67, will the Alabama paradox occur? Explain.

   b. If the number of printers to be apportioned by the Hamilton method increases from 67 to 68, will the Alabama paradox occur? Explain.

4. Consider the apportionment of 27 automobiles to the sales departments of a business with five regional centers labeled A, B, C, D, and E. The following table shows the Hamilton apportionment of the automobiles based on the number of sales personnel at each center.

Center	A	B	C	D	E
Number of sales personnel	31	108	70	329	49
Apportionment of 27 automobiles	2	5	3	15	2

   a. If the number of automobiles to be apportioned increases from 27 to 28, what will be the apportionment if the Hamilton method is used? Does the Alabama paradox occur? Explain.

   b. If the number of automobiles to be apportioned using the Hamilton method increases from 28 to 29, will the Alabama paradox occur? Explain.

5. MusicGalore.net has offices in Los Angeles and Newark. The number of employees at each office is

shown in the following table. There are 11 new computer file servers to be apportioned between the offices according to the number of employees.

Office	Los Angeles	Newark
Employees	1430	235

a. Use the Hamilton method to find each office's apportionment of file servers.

b. The corporation opens an additional office in Kansas City with 111 employees and decides to have a total of 12 file servers. If the file servers are reapportioned using the Hamilton method, will the new states paradox occur? Explain.

6. A city apportions 34 building inspectors among three regions according to their populations. The following table shows the present population of each region.

Region	A	B	C
Population	14,566	3321	29,988

a. Use the Hamilton method to apportion the inspectors.

b. After a year the regions have the following populations.

Region	A	B	C
Population	15,008	3424	30,109

Region *A* has an increase in population of 442, which is an increase of 3.03%. Region *B* has an increase in population of 103, which is an increase of 3.10%. Region *C* has an increase in population of 121, which is an increase of 0.40%. If the inspectors are reapportioned using the Hamilton method, will the population paradox occur? Explain.

7. Is the Hamilton apportionment method susceptible to the population paradox?

8. Is the Jefferson apportionment method susceptible to the new states paradox?

9. The Huntington-Hill apportionment method has been used to apportion 86 security guards among each of three corporate office buildings according to the number of employees at each building. See the following table.

Building	Number of security guards	Number of employees
A	25	414
B	43	705
C	18	293

The corporation has decided to hire a new security guard.

a. Use the Huntington-Hill apportionment principle to determine to which building the new security guard should be assigned.

b. If another security guard is hired, bringing the total number of guards to 88, to which building should this guard be assigned?

10. Three high school students are running for Homecoming Queen. Students at the school were allowed to rank the candidates in order of preference. The results are shown in the preference schedule below.

	Rankings		
Cynthia L.	3	2	1
Hannah A.	1	3	3
Shannon M.	2	1	2
Number of votes:	112	97	11

a. Use the Borda Count method to find the winner.

b. Find a candidate who wins all head-to-head comparisons.

c. Explain why the Condorcet criterion has been violated.

d. Who wins using the plurality voting system?

e. Explain why the majority criterion has been violated.

11. In Exercise 10, suppose Cynthia L. withdraws from the Homecoming Queen election.

a. Assuming voter preferences between the remaining two candidates remain the same, who will be crowned Homecoming Queen using the Borda Count method?

b. Explain why the independence of irrelevant alternatives criterion has been violated.

12. A scholarship committee must choose a winner from three finalists, Jean, Margaret, and Terry. Each member of the committee ranked the three finalists and Margaret was selected winner using the plurality with elimination method. This vote was later declared invalid and a new vote was taken. All members voted using the same rankings except one, who changed her first choice from Terry to Margaret. This time Jean won the scholarship using the same voting method. Which fairness criterion was violated, and why?

13. A weighted voting system for the voters A, B, C, and D is given by {18: 12, 7, 6, 1}. The weight of voter A is 12, the weight of voter B is 7, the weight of voter C is 6, and the weight of voter D is 1.

a. What is the quota?

b. What is the weight of the coalition {A, C}?

c. Is {A, C} a winning coalition?

d. Which voters are critical voters in the coalition {A, C, D}?

e. How many coalitions can be formed?

f. How many coalitions consist of exactly two voters?

14. A weighted voting system for the voters A, B, C, D, and E is given by {35: 29, 11, 8, 4, 2}. The weight of voter A is 29, the weight of voter B is 11, the weight of voter C is 8, the weight of voter D is 4, and the weight of voter E is 2.

a. What is the quota?

b. What is the weight of the coalition {A, D, E}?

c. Is {A, D, E} a winning coalition?

d. Which voters are critical voters in the coalition {A, C, D, E}?

e. How many coalitions can be formed?

f. How many coalitions consist of exactly two voters?

15. Calculate the Banzhaf power indices for the voters A, B, and C in the weighted voting system {9: 6, 5, 3}.

16. Calculate the Banzhaf power indices for the voters A, B, C, D, and E in the one-person, one-vote system {3: 1, 1, 1, 1, 1}.

17. Calculate the Banzhaf power indices for the voters A, B, C, and D in the weighted voting system {31: 19, 15, 12, 10}.

18. Calculate the Banzhaf power indices for the voters A, B, C, D, and E in the weighted voting system {35: 29, 11, 8, 4, 2}.

In Exercises 19 and 20, identify any dictator and all dummies for each weighted voting system.

19. Voters A, B, C, D, and E: {15: 15, 10, 2, 1, 1}

20. Voters A, B, C, and D: {28: 19, 6, 4, 2}

21. Four voters A, B, C, and D make decisions by using the weighted voting system {5: 4, 2, 1, 1}. In the event of a tie, a fifth voter E casts a vote to break the tie. For this voting scheme, determine the Banzhaf power index for each voter, including voter E.

22. Four finalists are competing in an essay contest. Judges have read and ranked each essay in order of preference. The results are shown in the preference schedule below.

	Rankings			
Crystal Kelley	3	2	2	1
Manuel Ortega	1	3	4	3
Peter Nisbet	2	4	1	2
Sue Toyama	4	1	3	4
Number of votes:	8	5	4	6

a. Using the plurality voting system, who is the winner of the essay contest?

b. Does this winner have a majority?

c. Use the Borda Count method of voting to determine the winner of the essay contest.

23. A campus ski cub is trying to decide where to hold its winter break ski trip. The members of the club were surveyed and asked to rank five choices in order of preference. Their responses are tallied in the table below.

	Rankings				
Aspen	1	1	3	2	3
Copper Mountain	5	4	2	4	4
Powderhorn	3	2	5	1	5
Telluride	4	5	4	5	2
Vail	2	3	1	3	1
Number of votes:	14	8	11	18	12

a. Use the plurality method of voting to determine which resort the club should choose.

b. Use the Borda Count method to choose the ski resort the club should visit.

24. Four students are running for the Activities Director position on campus. Students were asked to rank the four candidates in order of preference. The results are shown in the table below.

	Rankings			
G. Reynolds	2	3	1	3
L. Hernandez	1	4	4	2
A. Kim	3	1	2	1
J. Schneider	4	2	3	4
Number of votes:	132	214	93	119

Use the plurality with elimination method to determine the winner of the election.

25. A group of consumers were surveyed about their favorite candy bars. Each participant was asked to rank four candy bars in order of preference. The results are given in the table below.

	Rankings				
Nestle Crunch	1	4	4	2	3
Snickers	2	1	2	4	1
Milky Way	3	2	1	1	4
Twix	4	3	3	3	2
Number of votes:	15	38	27	16	22

Use the plurality with elimination method to determine the group's favorite candy bar.

26. Use the pairwise comparison method of voting to choose the winner of the election in Exercise 24.

27. Use the pairwise comparison method of voting to choose the group's favorite candy bar in Exercise 25.

---

1. The following table shows the number of employees for each of the four divisions of a corporation. There are 85 new computers that are to be apportioned among the divisions based on the number of employees in each division.

Division	Number of employees
Sales	1008
Advertising	234
Service	625
Manufacturing	3114
Total	4981

a. Use the Hamilton method to determine the number of computers to be apportioned to each division.

b. Use the Jefferson method to determine the number of computers to be apportioned to each division.

Does this particular apportionment violate the quota rule?

2. The following table shows the number of counselors and the number of students for each of two high schools. The current number of counselors for each school was determined using the Huntington-Hill apportionment method.

School	Number of counselors	Number of students
Cedar Falls	9	2646
Lake View	7	1984

A new counselor is to be hired for one of the schools.

a. Calculate the Huntington-Hill number for each of the schools.

b. Use the Huntington-Hill apportionment principle to determine to which school the new counselor should be assigned.

3. A weighted voting system for the voters A, B, C, D, and E is given by {33: 21, 14, 12, 7, 6}.

   a. What is the quota?

   b. What is the weight of the coalition {B, C}?

   c. Is {B, C} a winning coalition?

   d. Which voters are critical voters in the coalition {A, C, D}?

   e. How many coalitions can be formed?

   f. How many coalitions consist of exactly two voters?

4. One hundred consumers ranked three brands of bottled water in order of preference. The results are shown in the preference schedule below.

	Rankings				
Arrowhead	2	3	1	3	2
Evian	3	2	2	1	1
Aquafina	1	1	3	2	3
Number of votes:	22	17	31	11	19

   a. Using the plurality system of voting, which brand of bottled water is the preferred brand?

   b. Does the winner have a majority?

   c. Use the Borda Count method of voting to determine the favored brand of water.

5. A company with offices across the country will hold its annual executive meeting in one of four locations. All of the executives were asked to rank the locations in order of preference. The results are shown in the table below.

	Rankings			
New York	3	1	2	2
Dallas	2	3	1	4
Los Angeles	4	2	4	1
Atlanta	1	4	3	3
Number of votes:	19	24	7	35

   Use the pairwise comparison method to determine which location should be chosen.

6. A professor is preparing an extra review session the Monday before final exams and she wants as many students as possible to be able to attend. She asks all of the students to rank different times of day in order of preference. The results are given below.

	Rankings				
Morning	4	1	4	2	3
Noon	3	2	2	3	1
Afternoon	1	3	1	4	2
Evening	2	4	3	1	4
Number of votes:	12	16	9	5	13

   a. Using the plurality with elimination method, for what time of day should the professor schedule the review session?

   b. Use the Borda Count method to determine the best time of day for the review.

7. A committee must vote on which of three budget proposals to adopt. Each member of the committee has ranked the proposals in order of preference. The results are shown below.

	Rankings		
Proposal A	1	3	3
Proposal B	2	2	1
Proposal C	3	1	2
Number of votes:	40	9	39

   a. Using the plurality system of voting, which proposal should be adopted?

   b. If Proposal C is found to be invalid and is eliminated, which proposal wins using the plurality system?

   c. Explain why the independence of irrelevant alternatives criterion has been violated.

   d. Verify that Proposal B wins all head-to-head comparisons.

   e. Explain why the Condorcet criterion has been violated.

8. The four staff members A, B, C, and D of a college drama department use the weighted voting system {8: 5, 4, 3, 2} to make casting decisions for a play they

are producing. In this voting system voter A has a weight of 5, voter B has a weight of 4, voter C has a weight of 3, and voter D has a weight of 2. Compute the Banzhaf power index for each member of the department. Round each result to the nearest hundredth.

9. Three voters A, B, and C make decisions by using the weighted voting system {6: 5, 4, 1}. In the case of a tie, a fourth voter D casts a single vote to break the tie. For this voting scheme, determine the Banzhaf power index for each voter, including voter D.

10. Identify any dictator and all dummies in the following weighted voting systems. Assume the voters are listed as A, B, C, D.

    a. {36: 24, 7, 5, 2}   **b.** {20: 21, 15, 3, 1}

| APPENDIX | ## The Metric System of Measurement |

point of interest

To learn more about these tests, go to **www.ed.gov.** Use the search feature and enter TIMSS.

International trade, or trade between nations, is a vital and growing segment of business in the world today. The opening of McDonald's restaurants around the globe is testimony to the expansion of international business.

The United States, as a nation, is dependent on world trade. And world trade is dependent on internationally standardized units of measurement: the metric system. The third International Mathematics and Science Study (TIMSS) compared the performances of half a million students from 41 countries at five different grade levels on tests of their mathematics and science knowledge. One area of mathematics in which the U.S. average was below the international average was measurement, due in large part to the fact that units cited in the questions were metric units. Because the United States has not yet converted to the metric system, its citizens are less familiar with it.

In this Appendix we will present the metric system of measurement and explain how to convert between different units.

The basic unit of *length,* or distance, in the metric system is the **meter** (m). One meter is approximately the distance from a doorknob to the floor. All units of length in the metric system are derived from the meter. Prefixes added to the basic unit denote the length of the unit. For example, the prefix *centi-* means "one hundredth;" therefore, 1 centimeter is 1 one hundredth of a meter (0.01 m).

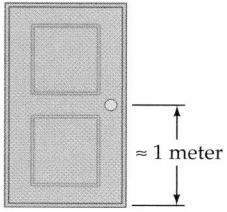

≈ 1 meter

kilo- = 1 000	1 kilometer (km) = 1 000 meters (m)
hecto- = 100	1 hectometer (hm) = 100 m
deca- = 10	1 decameter (dam) = 10 m
	1 meter (m) = 1 m
deci- = 0.1	1 decimeter (dm) = 0.1 m
centi- = 0.01	1 centimeter (cm) = 0.01 m
milli- = 0.001	1 millimeter (mm) = 0.001 m

point of interest

Originally the meter (spelled *metre* in some countries) was defined as $\frac{1}{10,000,000}$ of the distance from the equator to the North Pole. Modern scientists have redefined the meter as 1,650,753.73 wavelengths of the orange-red light given off by the element krypton.

Notice that in this list 1000 is written as 1 000, with a space between the 1 and the zeros. When writing numbers using metric units, each group of three numbers is separated by a space instead of a comma. A space is also used after each group of three numbers to the right of a decimal. For example, 31,245.2976 is written 31 245.297 6 in metric notation.

QUESTION   *Which unit in the metric system is one thousandth of a meter?*

Mass and weight are closely related. *Weight* is a measure of how strongly gravity is pulling on an object. Therefore, an object's weight is less in space than on Earth's surface. However, the amount of material in the object, its *mass,* remains the same. On the surface of Earth, the terms *mass* and *weight* can be used interchangeably.

ANSWER   *The millimeter is one thousandth of a meter.*

The basic unit of mass in the metric system is the **gram** (g). If a box 1 centimeter long on each side is filled with pure water, the mass of that water is 1 gram.

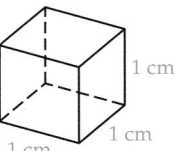

1 gram = the mass of water in a box that is 1 centimeter long on each side

The units of mass in the metric system have the same prefixes as the units of length.

1 kilogram (kg)   = 1 000 grams (g)
1 hectogram (hg) = 100 g
1 decagram (dag) = 10 g
1 gram (g)   = 1 g
1 decigram (dg) = 0.1 g
1 centigram (cg) = 0.01 g
1 milligram (mg) = 0.001 g

The gram is a small unit of mass. A paperclip weighs about 1 gram. In many applications, the kilogram (1 000 grams) is a more useful unit of mass. This textbook weighs about 1 kilogram.

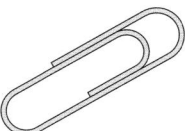

Weight ≈ 1 gram

QUESTION   *Which unit in the metric system is equal to 1000 grams?*

Liquid substances are measured in units of *capacity*.

The basic unit of capacity in the metric system is the **liter** (L). One liter is defined as the capacity of a box that is 10 centimeters long on each side.

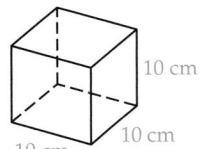

1 liter = the capacity of a box that is 10 centimeters long on each side

The units of capacity in the metric system have the same prefixes as the units of length.

1 kiloliter (kl)   = 1 000 liters (L)
1 hectoliter (hl) = 100 L
1 decaliter (dal) = 10 L
1 liter (L)   = 1 L
1 deciliter (dl)   = 0.1 L
1 centiliter (cl)   = 0.01 L
1 milliliter (ml) = 0.001 L

ANSWER   *The kilogram is equal to 1000 grams.*

▼ **point of interest**

The definition of 1 inch has been changed as a consequence of the wide acceptance of the metric system. One inch is now exactly 25.4 millimeters.

Converting between units in the metric system involves moving the decimal point to the right or to the left. Listing the units in order from largest to smallest will indicate how many places to move the decimal point and in which direction.

To convert 3 800 centimeters to meters, write the units of length in order from largest to smallest.

km   hm   dam   m   dm   cm   mm
2 positions

- Converting from cm to m requires moving 2 positions to the left.

3 800 cm = 38.00 m
2 places

- Move the decimal point the same number of places and in the same direction.

Convert 27 kilograms to grams.

kg   hg   dag   g   dg   cg   mg
3 positions

- Write the units of mass in order from largest to smallest.
- Converting kg to g requires moving 3 positions to the right.

27 kg = 27 000 g
3 places

- Move the decimal point the same number of places and in the same direction.

EXAMPLE 1 ■ **Convert Units in the Metric System of Measurement**

Convert.

a. 4.08 meters to centimeters   **b.** 5.93 grams to milligrams

c. 82 milliliters to liters   **d.** 9 kiloliters to liters

Solution

a. Write the units of length from largest to smallest.

km   hm   dam   ⓜ   dm   ⓒⓜ   mm

Converting m to cm requires moving 2 positions to the right.

4.08 m = 408 cm

b. Write the units of mass from largest to smallest.

kg   hg   dag   ⓖ   dg   cg   ⓜⓖ

Converting g to mg requires moving 3 positions to the right.

5.93 g = 5 930 mg

c. Write the units of capacity from largest to smallest.

kl   hl   dal   Ⓛ   dl   cl   ⓜⓛ

Converting ml to L requires moving 3 positions to the left.

82 ml = 0.082 L

d. Write the units of capacity from largest to smallest.

ⓚⓛ   hl   dal   Ⓛ   dl   cl   ml

Converting kl to L requires moving 3 positions to the right.

9 kl = 9 000 L

CHECK YOUR PROGRESS 1 Convert.

a. 1 295 meters to kilometers   **b.** 7 543 grams to kilograms
c. 6.3 liters to milliliters   **d.** 2 kiloliters to liters

Solution *See page S51.*

Other prefixes in the metric system are becoming more common as a result of technological advances in the computer industry. For example:

tera- = 1 000 000 000 000
giga- = 1 000 000 000
mega- = 1 000 000

micro- = 0.000 001
nano- = 0.000 000 001
pico- = 0.000 000 000 001

A **bit** is the smallest unit of code that computers can read; it is a binary digit, either a 0 or a 1. Usually bits are grouped into bytes of 8 bits. Each byte stands for a letter, number, or any other symbol we might use in communicating information. For example, the letter W can be represented by 01010111. The amount of memory in a computer hard drive is measured in terabytes, gigabytes, and megabytes. The speed of a computer used to be measured in microseconds and then nanoseconds, but now the speed is measured in picoseconds.

Here are a few more examples of how these prefixes are used.

The mass of Earth gains 40 Gg (gigagrams) each year from captured meteorites and cosmic dust.

The average distance from Earth to the moon is 384.4 Mm (megameters) and the average distance from Earth to the sun is 149.5 Gm (gigameters).

The wavelength of yellow light is 590 nm (nanometers).

The diameter of a hydrogen atom is about 70 pm (picometers).

There are additional prefixes in the metric system, representing both larger and smaller units. We may hear them more and more often as computer chips hold more and more information, as computers get faster and faster, and as we learn more and more about objects in our universe and beyond that are great distances away.

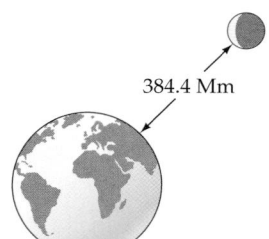

384.4 Mm

Exercises

(Suggested Assignment: 1–69, odds)

1. In the metric system, what is the basic unit of length? of liquid measure? of weight?

2. a. Explain how to convert meters to centimeters.

   b. Explain how to convert milliliters to liters.

In Exercises 3 to 26, name the most convenient unit in the metric system that can be used to measure each of the following.

3. The distance from New York to London
4. The weight of a truck
5. A person's waist
6. The amount of coffee in a mug
7. The weight of a thumbtack
8. The amount of water in a swimming pool
9. The distance a baseball player hits a baseball
10. A person's hat size
11. The amount of fat in a slice of cheddar cheese
12. A person's weight
13. The maple syrup served with pancakes
14. The amount of water in a water cooler
15. The amount of vitamin C in a vitamin tablet
16. A serving of cereal
17. The width of a hair
18. A person's height
19. The amount of medication in an aspirin
20. The weight of a lawnmower
21. The weight of a slice of bread
22. The contents of a bottle of salad dressing
23. The amount of water a family uses monthly
24. The newspapers collected at a recycling center
25. The amount of liquid in a bowl of soup
26. The distance to the bank

27. **a.** Complete the table.

Metric system prefix	Symbol	Magnitude	Means multiply the basic unit by:
tera-	T	10^{12}	1 000 000 000 000
giga-	G	?	1 000 000 000
mega-	M	10^6	?
kilo-	?	?	1 000
hecto-	h	?	100
deca-	da	10^1	?
deci-	d	$\dfrac{1}{10}$	?
centi-	?	$\dfrac{1}{10^2}$	?
milli-	?	?	0.001
micro-	u (mu)	$\dfrac{1}{10^6}$	?
nano-	n	$\dfrac{1}{10^9}$	?
pico-	p	?	0.000 000 000 001

b. How can the magnitude column in the table above be used to determine how many places to move the decimal point when converting to the basic unit in the metric system?

In Exercises 28 to 57, convert the given measure.

28. 42 cm = _____ mm
29. 91 cm = _____ mm
30. 360 g = _____ kg
31. 1 856 g = _____ kg
32. 5 194 ml = _____ L

33. 7 285 ml = _____ L

34. 2 m = _____ mm

35. 8 m = _____ mm

36. 217 mg = _____ g

37. 34 mg = _____ g

38. 4.52 L = _____ ml

39. 0.029 7 L = _____ ml

40. 8 406 m = _____ km

41. 7 530 m = _____ km

42. 2.4 g = _____ mg

43. 9.2 kg = _____ g

44. 6.18 kl = _____ L

45. 0.036 kl = _____ L

46. 9.612 km = _____ m

47. 2.35 km = _____ m

48. 0.24 g = _____ mg

49. 0.083 g = _____ mg

50. 298 cm = _____ m

51. 71.6 cm = _____ m

52. 2 431 L = _____ kl

53. 6 302 L = _____ kl

54. 0.66 m = _____ cm

55. 4.58 m = _____ cm

56. 243 mm = _____ cm

57. 92 mm = _____ cm

58. a. One of the events in the summer Olympics is the 50 000-meter walk. How many kilometers do the entrants in this event walk?

b. One of the events in the winter Olympic games is the 10 000-meter speed skating event. How many kilometers do the entrants in this event skate?

59. A carat is a unit of weight equal to 200 milligrams. Find the weight in grams of a 10-carat precious stone.

60. How many pieces of material, each 75 centimeters long, can be cut from a bolt of fabric that is 6 meters long?

61. An athletic club uses 800 milliliters of chlorine each day for its swimming pool. How many liters of chlorine are used in a month of 30 days?

62. Each of the four shelves in a bookcase measures 175 centimeters. Find the cost of the shelves when the price of lumber is $15.75 per meter.

63. The printed label from a container of milk is shown below. To the nearest whole number, how many 230-milliliter servings are in the container?

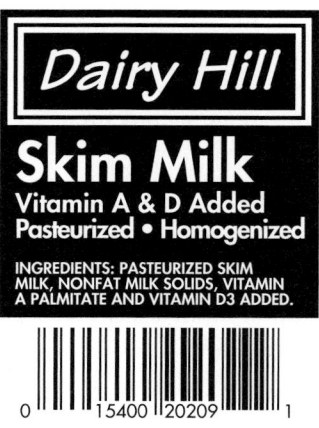

1 GAL. (3.78 L)

64. A 1.19-kilogram container of Quaker Oats contains 30 servings. Find the number of grams in one serving of the oatmeal. Round to the nearest gram.

65. A patient is advised to supplement her diet with 2 grams of calcium per day. The calcium tablets she purchases contain 500 milligrams of calcium per tablet. How many tablets per day should the patient take?

66. A laboratory assistant is in charge of ordering acid for three chemistry classes of 30 students each. Each student requires 80 milliliters of acid. How many liters of acid should be ordered? The assistant must order by the whole liter.

67. A case of 12 one-liter bottles of apple juice costs $19.80. A case of 24 cans, each can containing 340 milliliters of apple juice, costs $14.50. Which case of apple juice costs less per milliliter?

68. A column assembly is being constructed in a building. The components are shown in the diagram below. What height column must be cut?

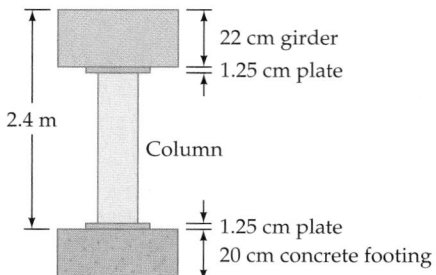

69. The distance between Earth and the sun is 150 000 000 kilometers. Light travels 300 000 000 meters in 1 second. How many seconds does it take for light to reach Earth from the sun?

70. Explain why is it advantageous to have internationally standardized units of measure?

Extensions

CRITICAL THINKING

71. A health food store buys nuts in 10-kilogram containers and repackages the nuts for resale. The store packages the nuts in 200-gram bags costing $.04 each, and sells them for $1.89 per bag. Find the profit on a 10-kilogram container of nuts costing $50.

72. For $99.50, a cosmetician buys 5 liters of moisturizer and repackages it in 125-milliliter jars. Each jar costs the cosmetician $.35. Each jar of moisturizer is sold for $5.95. Find the profit on the 5 liters of moisturizer.

73. A service station operator bought 85 kiloliters of gasoline for $19,250. The gasoline was sold for $.329 per liter. Find the profit on the 85 kiloliters of gasoline.

COOPERATIVE LEARNING

74. Form two debating teams. One team should argue in favor of changing to the metric system in the United States, and the other should argue against it.

CHAPTER 1

SECTION 1.1

CHECK YOUR PROGRESS 1, *page 4*

a. Each successive number is 5 larger than the preceding number. Thus we predict that the next number in the list is 5 larger than 25, which is 30.

b. The first two numbers differ by 3. The second and third numbers differ by 5. It appears that the difference between any two numbers is always 2 more than the preceding difference. Thus we predict that the next number will be 11 more than 26, which is 37.

CHECK YOUR PROGRESS 2, *page 5*

If the original number is 2, then $\dfrac{2 \times 9 + 15}{3} - 5 = 6$, which is

three times the original number.

If the original number is 7, then $\dfrac{7 \times 9 + 15}{3} - 5 = 21$, which is

three times the original number.

If the original number is -12, then $\dfrac{-12 \times 9 + 15}{3} - 5 = -36$,

which is three times the original number.

It appears, by inductive reasoning, that the procedure produces a number that is three times the original number.

CHECK YOUR PROGRESS 3, *page 6*

a. It appears that when the velocity of a tsunami is doubled, its height is quadrupled.

b. It appears that when the velocity of a tsunami is doubled, its height is quadrupled. A tsunami with a velocity of 30 feet per second will have a height that is four times that of a tsunami with a speed of 15 feet per second. We predict a height of $4 \times 25 = 100$ feet for a tsunami with a velocity of 30 feet per second.

CHECK YOUR PROGRESS 4, *page 7*

a. Let $x = 0$. Then $\dfrac{x}{x} \neq 1$, because division by 0 is undefined.

b. Let $x = 1$. Then $\dfrac{x + 3}{3} = \dfrac{1 + 3}{3} = \dfrac{4}{3}$, whereas

$x + 1 = 1 + 1 = 2$.

c. Let $x = 3$. Then $\sqrt{x^2 + 16} = \sqrt{3^2 + 16} = \sqrt{25} = 5$, whereas $x + 4 = 3 + 4 = 7$.

CHECK YOUR PROGRESS 5, *page 8*

Let n represent the original number.
Multiply the number by 6: $6n$
Add 10 to the product: $6n + 10$

Divide the sum by 2: $\dfrac{6n + 10}{2} = 3n + 5$

Subtract 5: $3n + 5 - 5 = 3n$

The procedure always produces a number that is three times the original number.

CHECK YOUR PROGRESS 6, *page 10*

a. The conclusion is a specific case of a general assumption, so the argument is an example of deductive reasoning.

b. The argument reaches a conclusion based on specific examples, so the argument is an example of inductive reasoning.

SECTION 1.2

CHECK YOUR PROGRESS 1, *page 16*

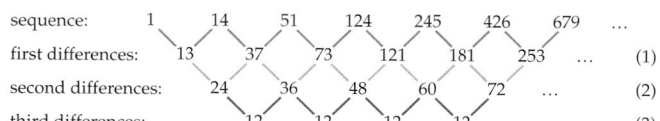

sequence:	1		14		51		124		245		426		679	...	
first differences:		13		37		73		121		181		253		...	(1)
second differences:			24		36		48		60		72		...		(2)
third differences:				12		12		12		12		...			(3)

Using the method of extending the difference table, we predict that 679 is the next term in the sequence.

CHECK YOUR PROGRESS 2, *page 18*

a. Each figure after the first figure consists of a square region and a "tail." The number of tiles in the square region is n^2 and the number of tiles in the tail is $n - 1$. Thus the nth term formula for the number of tiles in the nth figure is $a_n = n^2 + n - 1$.

b. Let $n = 10$. Then $n^2 + n - 1 = (10)^2 + (10) - 1 = 109$.

c. $n^2 + n - 1 = 419$
$n^2 + n - 420 = 0$
$(n + 21)(n - 20) = 0$

$n + 21 = 0$ or $n - 20 = 0$
$n = -21$ $n = 20$

We consider only the positive result. The 20th figure will consist of 419 tiles.

CHECK YOUR PROGRESS 3, *page 20*

$F_9 = F_8 + F_7$
$= 21 + 13$
$= 34$

CHECK YOUR PROGRESS 4, *page 21*

a. The inequality $2F_n > F_{n+1}$ is true for $n = 3, 4, 5, 6, \ldots, 10$. Thus, by inductive reasoning, we conjecture that the statement is a true statement.

b. The equality $2F_n + 3 = F_{n+2}$ is not true for $n = 4$, because $2F_4 + 3 = 2(3) + 3 = 9$ and $F_{4+2} = F_6 = 8$. Thus the statement is a false statement.

SECTION 1.3

CHECK YOUR PROGRESS 1, *page 31*

Understand the Problem   In order to go past Starbucks, Allison must walk along Third Avenue from Boardwalk to Park Avenue.

Devise a Plan   Label each intersection that Allison can pass through with the number of routes to that intersection. If she can reach an intersection from two different routes, then the number of routes to that intersection is the sum of the numbers of routes to the two adjacent intersections.

Carry Out the Plan   The following figure shows the number of routes to each of the intersections that Allison could pass through. Thus there are a total of nine routes that Allison can take if she wishes to walk directly from point A to point B and pass by Starbucks.

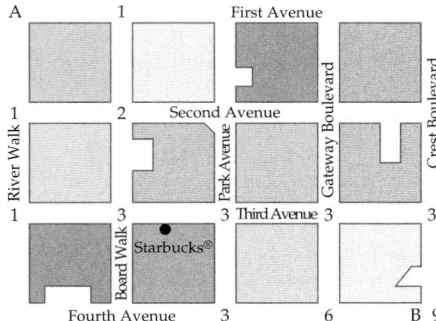

Review the Solution   The total of nine routes seems reasonable. We know from Example 1 that if Allison can take any route, the total number of routes is 35. Requiring Allison to go past Starbucks eliminates several routes.

CHECK YOUR PROGRESS 2, *page 31*

Understand the Problem   There are several ways to answer the questions so that two answers are "false" and three answers are "true." One way is TTTFF and another is FFTTT.

Devise a Plan   Make an organized list. Try the strategy of listing a T unless doing so will produce too many T's or a duplicate of one of the previous orders in your list.

Carry Out the Plan   (Start with 3 T's in a row.)

TTTFF	(1)
TTFTF	(2)
TTFFT	(3)
TFTTF	(4)
TFTFT	(5)
TFFTT	(6)
FTTTF	(7)
FTTFT	(8)
FTFTT	(9)
FFTTT	(10)

Review the Solution   Each entry in the list has two F's and three T's. Since the list is complete and has no duplications, we know that there are 10 ways for a student to mark two questions with "false" and the other three with "true."

CHECK YOUR PROGRESS 3, *page 32*

Understand the Problem   There are six people, and each person shakes hands with each of the other people.

Devise a Plan   Each person will shake hands with five other people (a person won't shake his or her own hand; that would be silly). Since there are six people, we could multiply 6 times 5 to get the total number of handshakes. However, this procedure would count each handshake exactly twice, so we must divide this product by 2 for the actual answer.

Carry Out the Plan   6 times 5 is 30. 30 divided by 2 is 15.

Review the Solution   Denote the people by the letters A, B, C, D, E, and F. Make an organized list. Remember that AB and BA represent the same people shaking hands, so do not list both AB and BA.

AB	BC	CD	DE	EF
AC	BD	CE	DF	
AD	BE	CF		
AE	BF			
AF				

The method of making an organized list verifies that if six people shake hands with each other there will be a total of 15 handshakes.

CHECK YOUR PROGRESS 4, *page 33*

Understand the Problem   We need to find the ones digit of 4^{200}.

Devise a Plan   Compute a few powers of 4 to see if there are any patterns. $4^1 = 4$, $4^2 = 16$, $4^3 = 64$, and $4^4 = 256$. It appears that the last digit (ones digit) of 4^{200} must be either a 4 or a 6.

Carry Out the Plan   If the exponent n is an even number, then 4^n has a ones digit of 6. If the exponent n is an odd number, then 4^n has a ones digit of 4. Because 200 is an even number, we conjecture that 4^{200} has a ones digit of 6.

Review the Solution   You could try to check the answer by using a calculator, but you would find that 4^{200} is too large to be displayed. Thus we need to rely on the patterns we have observed to conclude that 6 is indeed the ones digit of 4^{200}.

CHECK YOUR PROGRESS 5, *page 34*

Understand the Problem   We are asked to find the possible numbers that Melody could have started with.

Devise a Plan   Work backward from 18 and do the inverse of each operation that Melody performed.

Carry Out the Plan   To get 18, Melody subtracted 30 from a number, so that number was $18 + 30 = 48$. To get 48, she divided a number by 3, so that number was $48 \times 3 = 144$. To get 144, she squared a number. She could have squared either 12 or -12 to produce 144. If the number she squared was 12, then she must have doubled 6 to get 12. If the number she squared was -12, then the number she doubled was -6.

Review the Solution   We can check by starting with 6 or -6. If we do exactly as Melody did, we end up with 18. The operation that prevents us from knowing with 100% certainty which number she started with is the squaring operation. We have no way of knowing whether the number she squared was a positive number or a negative number.

CHECK YOUR PROGRESS 6, *page 35*

Understand the Problem   We need to find Diophantus's age when he died.

Devise a Plan   Read the hint and then look for clues that will help you make an educated guess. You know from the given information that Diophantus's age must be divisible by 6, 12, 7, and 2. Find a number divisible by all of these numbers and check to see if it is a possible solution to the problem.

Carry Out the Plan   All multiples of 12 are divisible by 6 and 2, but the smallest multiple of 12 that is divisible by 7 is $12 \times 7 = 84$. Thus we conjecture that Diophantus's age when he died was $x = 84$ years. If $x = 84$, then $\frac{1}{6}x = 14$, $\frac{1}{12}x = 7$, $\frac{1}{7}x = 12$, and $\frac{1}{2}x = 42$. Then $\frac{1}{6}x + \frac{1}{12}x + \frac{1}{7}x + 5 + \frac{1}{2}x + 4 = 14 + 7 + 12 + 5 + 42 + 4 = 84$. It seems that 84 years is a correct solution to the problem.

Review the Solution   After 84, the next multiple of 12 that is divisible by 7 is 168. The number 168 also satisfies all the conditions of the problem, but it is unlikely that Diophantus died at the age of 168 years or at any age older than 168 years. Hence the only reasonable solution is 84 years.

CHECK YOUR PROGRESS 7, *page 37*

Understand the Problem   We need to determine two U.S. coins that have a total value of 35¢, given that one of the coins is not a quarter.

Devise a Plan   Experiment with different coins to try to produce 35¢. After a few attempts, you should conclude that one of the coins must be a quarter. Consider that the problem may be a *trick problem*.

Carry Out the Plan   A total of 35¢ can be produced by using a dime and a quarter. One of the coins is a quarter, but it is also true that *one of the coins, the dime, is not a quarter.*

Review the Solution   A dime and a quarter satisfy all the conditions of the problem. No other combination of coins satisfies the conditions of the problem. Thus the only solution is a dime and a quarter.

CHAPTER 2

SECTION 2.1

CHECK YOUR PROGRESS 1, *page 51*

a. $\{0, 1, 2, 3\}$

b. $\{12, 13, 14, 15, 16, 17, 18, 19\}$

c. $\{-4, -3, -2, -1\}$

CHECK YOUR PROGRESS 2, *page 52*

a. False

b. True

c. True

d. True

CHECK YOUR PROGRESS 3, *page 52*

a. $\{x \mid x \in I \text{ and } x < 9\}$

b. $\{x \mid x \in N \text{ and } x > 4\}$

CHECK YOUR PROGRESS 4, *page 53*

a. $n(C) = 5$

b. $n(D) = 1$

c. $n(E) = 0$

CHECK YOUR PROGRESS 5, *page 54*

a. The sets are not equal but they both contain six elements. Thus the sets are equivalent.

b. The sets are not equal but they both contain 16 elements. Thus the sets are equivalent.

SECTION 2.2

CHECK YOUR PROGRESS 1, *page 60*

a. $M = \{0, 4, 6, 17\}$. The set of elements in $U = \{0, 2, 3, 4, 6, 7, 17\}$ but not in M is $M' = \{2, 3, 7\}$.

b. $P = \{2, 4, 6\}$. The set of elements in $U = \{0, 2, 3, 4, 6, 7, 17\}$ but not in P is $P' = \{0, 3, 7, 17\}$.

CHECK YOUR PROGRESS 2, *page 61*

a. False. The number 3 is an element of the first set but not an element of the second set. Therefore, the first set is not a subset of the second set.

b. True. The set of counting numbers is the same set as the set of natural numbers, and every set is a subset of itself.

c. True. The empty set is a subset of every set.

d. True. Each element of the first set is an integer.

CHECK YOUR PROGRESS 3, *page 62*

a. Yes, because every natural number is a whole number, and the whole numbers include 0, which is not a natural number.

b. The first set is not a proper subset of the second set because the sets are equal.

CHECK YOUR PROGRESS 4, *page 63*

Subsets with zero elements: { }

Subsets with one element: $\{a\}, \{b\}, \{c\}, \{d\}, \{e\}$

Subsets with two elements: $\{a, b\}, \{a, c\}, \{a, d\}, \{a, e\}, \{b, c\}, \{b, d\},$ $\{b, e\}, \{c, d\}, \{c, e\}, \{d, e\}$

Subsets with three elements: $\{a, b, c\}, \{a, b, d\}, \{a, b, e\}, \{a, c, d\},$ $\{a, c, e\}, \{a, d, e\}, \{b, c, d\}, \{b, c, e\}, \{b, d, e\}, \{c, d, e\}$

Subsets with four elements: $\{a, b, c, d\}, \{a, b, c, e\}, \{a, b, d, e\},$ $\{a, c, d, e\}, \{b, c, d, e\}$

Subsets with five elements: $\{a, b, c, d, e\}$

CHECK YOUR PROGRESS 5, *page 63*

a. $2^3 = 8$

b. $2^{15} = 32{,}768$

c. $2^{10} = 1024$

SECTION 2.3

CHECK YOUR PROGRESS 1, *page 69*

a. $D \cap E = \{0, 3, 8, 9\} \cap \{3, 4, 8, 9, 11\}$
$= \{3, 8, 9\}$

b. $D \cap F = \{0, 3, 8, 9\} \cap \{0, 2, 6, 8\}$
$= \{0, 8\}$

CHECK YOUR PROGRESS 2, *page 70*

a. $D \cup E = \{0, 4, 8, 9\} \cup \{1, 5, 4, 7\}$
$= \{0, 1, 4, 5, 7, 8, 9\}$

b. $D \cup F = \{0, 4, 8, 9\} \cup \{2, 6, 8\}$
$= \{0, 2, 4, 6, 8, 9\}$

CHECK YOUR PROGRESS 3, *page 71*

a. The set $D \cap (E' \cup F)$ can be described as "the set of all elements that are in D, and in F or not E."

b. The set $L' \cup M$ can be described as "the set of all elements that are in M or are not in L."

CHECK YOUR PROGRESS 4, *page 71*

The following Venn diagrams show that $(A \cap B)'$ is equal to $A' \cup B'$.

The white region represents $(A \cap B)$.
The shaded region represents $(A \cap B)'$.

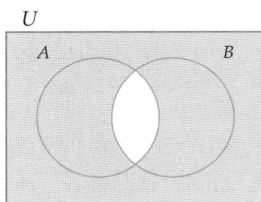

The diagonal patterned region below represents B'.
The grey shaded region below represents A'.
$A' \cup B'$ is the union of the grey shaded region and the diagonal patterned region.

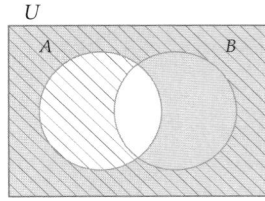

CHECK YOUR PROGRESS 5, *page 73*
The following Venn diagrams show that
$A \cup (B \cap C) = (A \cup B) \cap (A \cup C)$.

The grey shaded region represents $A \cup (B \cap C)$.

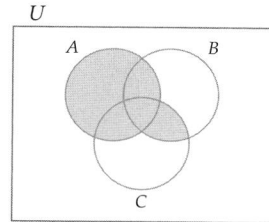

The grey shaded region below represents $A \cup B$.
The diagonal patterned region below represents $A \cup C$.
The intersection of the grey shaded region and the diagonal patterned region represents $(A \cup B) \cap (A \cup C)$.

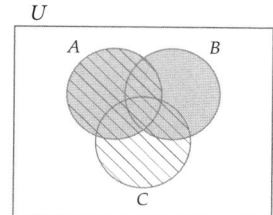

CHECK YOUR PROGRESS 6, *page 74*

a. Because Alex is in blood group A, not in blood group B, and is Rh+, his blood type is A+.

b. Roberto is in both blood group A and blood group B. Roberto is not Rh+. Thus Roberto's blood type is AB−.

CHECK YOUR PROGRESS 7, *page 75*

a. Alex's blood type is A+. The blood transfusion table shows that a person with blood type A+ can safely receive type A− blood.

b. The blood transfusion table shows that a person with type AB+ blood can safely receive each of the eight different types of blood. Thus a person with AB+ blood is classified as a universal recipient.

SECTION 2.4

CHECK YOUR PROGRESS 1, *page 82*
The intersection of the two sets includes the 85 students who like both volleyball and basketball.

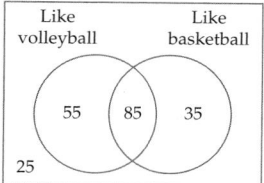

a. Because 140 students like volleyball and 85 like both sports, there must be $140 - 85 = 55$ students who like only volleyball.

b. Because 120 students like basketball and 85 like both sports, there must be $120 - 85 = 35$ students who like only basketball.

c. The Venn diagram shows that the number of students who like only volleyball plus the number who like only basketball plus the number who like both sports is $55 + 35 + 85 = 175$. Thus of the 200 students surveyed, only $200 - 175 = 25$ do not like either of the sports.

CHECK YOUR PROGRESS 2, *page 83*
The intersection of the three sets includes the 15 people who like all three activities.

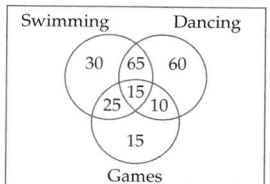

a. There are 25 people who like dancing and games. This includes the 15 people who like all three activities. Thus there must be another $25 - 15 = 10$ people who like only dancing and games. There are 40 people who like swimming and games. Thus there must be another $40 - 15 = 25$ people who like only swimming and games. There are 80 people who like swimming and dancing. Thus there must be another $80 - 15 = 65$ people who like only swimming and dancing. Hence $10 + 25 + 65 = 100$ people who like exactly two of the three activities.

b. There are 135 people who like swimming. We have determined that 15 people like all three activities, 25 like only swimming and games, and 65 like only swimming and dancing. This means that $135 - (15 + 25 + 65) = 30$ people like only swimming.

c. There are a total of 240 passengers surveyed. The Venn diagram shows that $15 + 25 + 10 + 15 + 30 + 65 + 60 = 220$ passengers like at least one of the activities. Thus $240 - 220 = 20$ passengers like none of the activities.

CHECK YOUR PROGRESS 3, *page 85*
Let $B = \{$the set of students who play basketball$\}$.
Let $S = \{$the set of students who play soccer$\}$.

$$n(B \cup S) = n(B) + n(S) - n(B \cap S)$$
$$= 80 + 60 - 24$$
$$= 116$$

Using the inclusion-exclusion principle, we see that 116 students play either basketball or soccer.

CHECK YOUR PROGRESS 4, *page 85*

$$n(A \cup B) = n(A) + n(B) - n(A \cap B)$$
$$852 = 785 + 162 - n(A \cap B)$$
$$852 = 947 - n(A \cap B)$$
$$n(A \cap B) = 947 - 852$$
$$n(A \cap B) = 95$$

CHECK YOUR PROGRESS 5, *page 86*

$$p(A \cup \text{Rh}+) = p(A) + p(\text{Rh}+) - p(A \cap \text{Rh}+)$$
$$91\% = 44\% + 84\% - p(A \cap \text{Rh}+)$$
$$91\% - (44\% + 84\%) = -p(A \cap \text{Rh}+)$$
$$-37\% = -p(A \cap \text{Rh}+)$$
$$37\% = p(A \cap \text{Rh}+)$$

Thus about 37% of the U.S. population has the A antigen and is Rh+.

CHECK YOUR PROGRESS 6, *page 87*

a. The table shows that 410 children surveyed use Yahoo as a search engine. Thus $n(Y \cap C) = 410$.

b. The set $L \cap M'$ is the set of surveyed Lycos users who are women or children. The number in this set is $325 + 40 = 365$. Thus $n(L \cap M') = 365$.

c. The set $A \cap M$ represents the set of surveyed Alta Vista users who are men. The table shows that this set includes 440 people. The set $A \cap W$ represents the set of surveyed Alta Vista users who are women. The table shows that this set includes 390 people. Thus $n((A \cap M) \cup (A \cap W)) = 440 + 390 = 830$.

SECTION 2.5

CHECK YOUR PROGRESS 1, *page 93*

Write the sets so that one is aligned below the other. Draw arrows to show how you wish to pair the elements of each set. One possible method is shown in the following figure.

$$N = \{1, 2, 3, 4, \ldots, \quad n, \quad \ldots\}$$
$$D = \{1, 3, 5, 7, \ldots, 2n - 1, \ldots\}$$

In the preceding correspondence, each natural number $n \in N$ is paired with the odd number $(2n - 1) \in D$. The *general correspondence* $n \leftrightarrow (2n - 1)$ enables us to determine exactly which element of D will be paired with any given element of N, and vice versa. For instance, under this correspondence, $8 \in N$ is paired with the odd number $2 \cdot 8 - 1 = 15$ and $21 \in D$ is paired with the natural number $\frac{21 + 1}{2} = 11$. The general correspondence $n \leftrightarrow (2n - 1)$ establishes a one-to-one correspondence between the sets.

CHECK YOUR PROGRESS 2, *page 94*

One proper subset of V is $P = \{41, 42, 43, 44, \ldots, 41 + n\ldots\}$, which was produced by deleting 40 from set V. To establish a one-to-one correspondence between V and P, consider the following diagram.

$$V = \{40, 41, 42, 43, \ldots, 40 + n, \ldots\}$$
$$P = \{41, 42, 43, 44, \ldots, 41 + n, \ldots\}$$

In the above correspondence, each element of the form $40 + n$ from set V is paired with an element of the form $41 + n$ from set P. The general correspondence $(40 + n) \leftrightarrow (41 + n)$ establishes a one-to-one correspondence between V and P. Because V can be placed in a one-to-one correspondence with a proper subset of itself, V is an infinite set.

CHECK YOUR PROGRESS 3, *page 95*

The following figure shows that we can establish a one-to-one correspondence between M and the set of natural numbers N by pairing $\frac{1}{n + 1}$ of set M with n of set N.

$$M = \left\{\frac{1}{2}, \frac{1}{3}, \frac{1}{4}, \frac{1}{5}, \ldots, \frac{1}{n + 1}, \ldots\right\}$$
$$N = \{1, 2, 3, 4, \ldots, n, \ldots\}$$

Thus the cardinality of M must be the same as the cardinality of N, which is $\aleph_0$.

CHAPTER 3

SECTION 3.1

CHECK YOUR PROGRESS 1, *page 111*

a. The sentence "Open the door" is a command. It is not a statement.

b. The word *large* is not a precise term. It is not possible to determine whether the sentence "7055 is a large number" is true or false and thus the sentence is not a statement.

c. The sentence $4 + 5 = 8$ is a false statement.

d. At this time we do not know whether the given sentence is true or false, but we know that the sentence is either true or false and that it is not both true and false. Thus the sentence is a statement.

e. The sentence $x > 3$ is a statement because for any given value of x, the inequality $x > 3$ is true or false, but not both.

CHECK YOUR PROGRESS 2, *page 113*

a. 1001 is not divisible by 7.

b. 5 is not an even number.

c. That fire engine is red.

CHECK YOUR PROGRESS 3, *page 114*

a. $\sim p \wedge r$

b. $\sim s \wedge \sim r$

c. $r \leftrightarrow q$

d. $p \rightarrow \sim r$

CHECK YOUR PROGRESS 4, *page 114*

a. The game will be shown on CBS or the game will be shown on ESPN.

b. If the game will not be shown on ESPN, then the game will be shown on CBS.

c. The game will be shown on CBS if and only if the game will not be shown on ESPN.

CHECK YOUR PROGRESS 5, *page 115*

a. True. A conjunction is true provided both components are true.

b. True. A disjunction is true provided at least one component is true.

c. False. If both components of a disjunction are false, then the disjunction is false.

CHECK YOUR PROGRESS 6, *page 116*

a. Some bears are not brown.

b. Some math classes are fun.

c. All vegetables are green.

SECTION 3.2

CHECK YOUR PROGRESS 1, *page 122*

a.

p	q	$\sim p$	$\sim q$	$p \wedge \sim q$	$\sim p \vee q$	$(p \wedge \sim q) \vee (\sim p \vee q)$	
T	T	F	F	F	T	T	Row 1
T	F	F	T	T	F	T	Row 2
F	T	T	F	F	T	T	Row 3
F	F	T	T	F	T	T	Row 4
		1	2	3	4	5	

b. p is true and q is false in row 2 of the above truth table. The truth value of $(p \wedge \sim q) \vee (\sim p \vee q)$ in row 2 is T (true).

CHECK YOUR PROGRESS 2, *page 123*

a.

p	q	r	$\sim p$	$\sim r$	$\sim p \wedge r$	$q \wedge \sim r$	$(\sim p \wedge r) \vee (q \wedge \sim r)$	
T	T	T	F	F	F	F	F	Row 1
T	T	F	F	T	F	T	T	Row 2
T	F	T	F	F	F	F	F	Row 3
T	F	F	F	T	F	F	F	Row 4
F	T	T	T	F	T	F	T	Row 5
F	T	F	T	T	F	T	T	Row 6
F	F	T	T	F	T	F	T	Row 7
F	F	F	T	T	F	F	F	Row 8
			1	2	3	4	5	

b. p is false, q is true, and r is false in row 6 of the above truth table. The truth value of $(\sim p \wedge r) \vee (q \wedge \sim r)$ in row 6 is T (true).

CHECK YOUR PROGRESS 3, *page 124*

p	q	~p	∨	(p	∧	q)
T	T	F	T	T	T	T
T	F	F	F	T	F	F
F	T	T	T	F	F	T
F	F	T	T	F	F	F

1	5	2	4	3

CHECK YOUR PROGRESS 4, *page 125*

p	q	p	∨	(p	∧	~q)
T	T	T	T	T	F	F
T	F	T	T	T	T	T
F	T	F	F	F	F	F
F	F	F	F	F	F	T

1	5	2	4	3

The above truth table shows that $p \equiv p \vee (p \wedge \sim q)$.

CHECK YOUR PROGRESS 5, *page 126*

Let d represent "I am going to the dance." Let g represent "I am going to the game." The original sentence in symbolic form is $\sim(d \wedge g)$. Applying one of De Morgan's laws, we find that $\sim(d \wedge g) \equiv \sim d \vee \sim g$. Thus an equivalent form of "It is not true that I am going to the dance and I am going to the game" is "I am not going to the dance or I am not going to the game."

CHECK YOUR PROGRESS 6, *page 126*

The following truth table shows that $p \wedge (\sim p \wedge q)$ is always false. Thus $p \wedge (\sim p \wedge q)$ is a self-contradiction.

p	q	p	∧	(~p	∧	q)
T	T	T	F	F	F	T
T	F	T	F	F	F	F
F	T	F	F	T	T	T
F	F	F	F	T	F	F

1	5	2	4	3

SECTION 3.3

CHECK YOUR PROGRESS 1, *page 131*

a. *Antecedent:* I study for at least 6 hours
   Consequent: I will get an A on the test

b. *Antecedent:* I get the job
   Consequent: I will buy a new car

c. *Antecedent:* you can dream it
   Consequent: you can do it

CHECK YOUR PROGRESS 2, *page 133*

a. Because the antecedent is true and the consequent is false, the statement is a false statement.

b. Because the antecedent is false, the statement is a true statement.

c. Because the consequent is true, the statement is a true statement.

CHECK YOUR PROGRESS 3, *page 133*

p	q	[p	∧	(p	→	q)]	→	q
T	T	T	T	T	T	T	T	T
T	F	T	F	T	F	F	T	F
F	T	F	F	F	T	T	T	T
F	F	F	F	F	T	F	T	F

1	6	2	5	3	7	4

CHECK YOUR PROGRESS 4, *page 134*

a. I will move to Georgia or I will live in Houston.

b. The number is not divisible by 2 or the number is even.

CHECK YOUR PROGRESS 5, *page 135*

a. I finished the report and I did not go to the concert.

b. The square of n is 25 and n is not 5 or -5.

CHECK YOUR PROGRESS 6, *page 135*

a. Let $x = 6.5$. Then the first component of the biconditional is false and the second component of the biconditional is true. Thus the given biconditional statement is false.

b. Both components of the biconditional are true for $n > 2$ and both components are false for $n \leq 2$. Because both components have the same truth value for any real number n, the given biconditional is true.

SECTION 3.4

CHECK YOUR PROGRESS 1, *page 140*

a. If a geometric figure is a square, then it is a rectangle.

b. If I am older than 30, then I am at least 21.

CHECK YOUR PROGRESS 2, *page 141*

Converse: If we are not going to have a quiz tomorrow, then we will have a quiz today.

Inverse: If we don't have a quiz today, then we will have a quiz tomorrow.

Contrapositive: If we have a quiz tomorrow, then we will not have a quiz today.

CHECK YOUR PROGRESS 3, *page 142*

a. The second statement is the inverse of the first statement. Thus the statements are not equivalent. This can also be demonstrated by the fact that the first statement is true for $c = 0$ and the second statement is false for $c = 0$.

b. The second statement is the contrapositive of the first statement. Thus the statements are equivalent.

CHECK YOUR PROGRESS 4, *page 142*

a. *Contrapositive:* If x is an odd integer, then $3 + x$ is an even integer. The contrapositive is true and so the original statement is also true.

b. *Contrapositive:* If two triangles are congruent triangles, then the two triangles are similar triangles. The contrapositive is true and so the original statement is also true.

c. *Contrapositive:* If tomorrow is Thursday, then today is Wednesday. The contrapositive is true and so the original statement is also true.

SECTION 3.5

CHECK YOUR PROGRESS 1, *page 147*

Let p represent the statement "She got on the plane." Let r represent the statement "She will regret it." Then the symbolic form of the argument is

$$\sim p \rightarrow r$$
$$\underline{\sim r}$$
$$\therefore p$$

CHECK YOUR PROGRESS 2, *page 149*

Let r represent the statement "The stock market rises." Let f represent the statement "The bond market will fall." Then the symbolic form of the argument is

$$r \rightarrow f$$
$$\underline{\sim f}$$
$$\therefore \sim r$$

The truth table for this argument is as follows.

		First premise	Second premise	Conclusion	
r	f	$r \rightarrow f$	$\sim f$	$\sim r$	
T	T	T	F	F	Row 1
T	F	F	T	F	Row 2
F	T	T	F	T	Row 3
F	F	T	T	T	Row 4

Row 4 is the only row in which all the premises are true, so it is the only row that we examine. Because the conclusion is true in row 4, the argument is valid.

CHECK YOUR PROGRESS 3, *page 150*

Let a represent the statement "I arrive before 8 A.M." Let f represent the statement "I will make the flight." Let p represent the statement "I will give the presentation." Then the symbolic form of the argument is

$$a \rightarrow f$$
$$\underline{f \rightarrow p}$$
$$\therefore a \rightarrow p$$

The truth table for this argument is as follows.

a	f	p	First premise $a \rightarrow f$	Second premise $f \rightarrow p$	Conclusion $a \rightarrow p$	
T	T	T	T	T	T	Row 1
T	T	F	T	F	F	Row 2
T	F	T	F	T	T	Row 3
T	F	F	F	T	F	Row 4
F	T	T	T	T	T	Row 5
F	T	F	T	F	T	Row 6
F	F	T	T	T	T	Row 7
F	F	F	T	T	T	Row 8

The only rows in which all the premises are true are rows 1, 5, 7, and 8. In each of these rows the conclusion is also true. Thus the argument is a valid argument.

CHECK YOUR PROGRESS 4, *page 151*
Let f represent "I go to Florida for spring break." Let $\sim s$ represent "I will not study." Then the symbolic form of the argument is

$$f \rightarrow \sim s$$
$$\underline{\sim f}$$
$$\therefore s$$

This argument has the form of the fallacy of the inverse. Thus the argument is invalid.

CHECK YOUR PROGRESS 5, *page 152*
Let r represent "I read a math book." Let f represent "I start to fall asleep." Let d represent "I drink a soda." Let e represent "I eat a candy bar." Then the symbolic form of the argument is

$$r \rightarrow f$$
$$f \rightarrow d$$
$$\underline{d \rightarrow e}$$
$$\therefore r \rightarrow e$$

The argument has the form of the extended law of syllogism. Thus the argument is valid.

CHECK YOUR PROGRESS 6, *page 153*
We are given the following premises:

$$\sim m \lor t$$
$$t \rightarrow \sim d$$
$$e \lor g$$
$$\underline{e \rightarrow d}$$
$$\therefore ?$$

The first premise can be written as $m \rightarrow t$, the third premise can be written as $\sim e \rightarrow g$, and the fourth premise can be written as

$\sim d \rightarrow \sim e$. Thus the argument can be expressed in the following equivalent form:

$$m \rightarrow t$$
$$t \rightarrow \sim d$$
$$\sim e \rightarrow g$$
$$\underline{\sim d \rightarrow \sim e}$$
$$\therefore ?$$

If we switch the order of the third and fourth premises, then we have the following equivalent form.

$$m \rightarrow t$$
$$t \rightarrow \sim d$$
$$\sim d \rightarrow \sim e$$
$$\underline{\sim e \rightarrow g}$$
$$\therefore ?$$

An application of the extended law of syllogism produces $m \rightarrow g$ as a valid conclusion for the argument. *Note:* Although $m \rightarrow \sim e$ is also a valid conclusion for the argument, we do not list it as our answer because it can be obtained without using all of the given premises.

SECTION 3.6

CHECK YOUR PROGRESS 1, *page 159*
The following Euler diagram shows that the argument is valid.

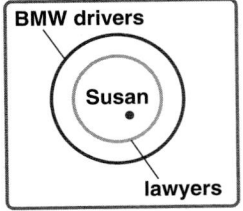

CHECK YOUR PROGRESS 2, *page 160*

From the given premises we can conclude that 7 may or may not be a prime number. Thus the argument is invalid.

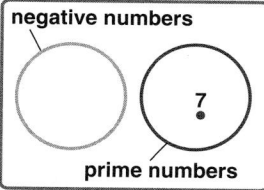 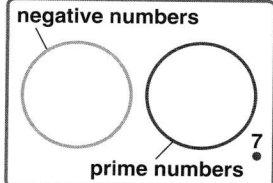

CHECK YOUR PROGRESS 3, *page 161*

From the given premises we can construct two possible Euler diagrams.

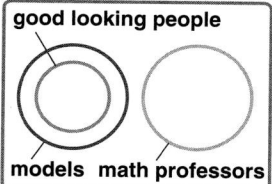 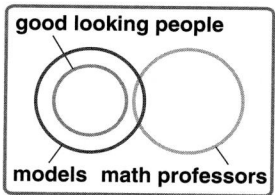

From the rightmost Euler diagram we can determine that the argument is invalid.

CHECK YOUR PROGRESS 4, *page 162*

The following Euler diagram illustrates that all squares are quadrilaterals, so the argument is a valid argument.

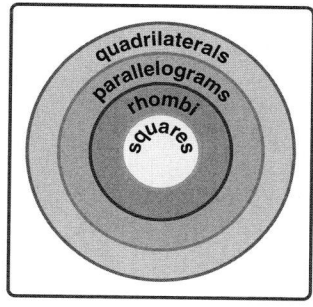

CHECK YOUR PROGRESS 5, *page 163*

The following Euler diagrams illustrate two possible cases. In both cases we see that all white rabbits like tomatoes.

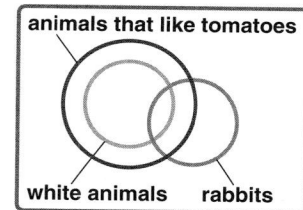 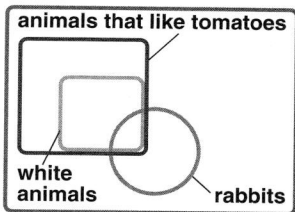

CHAPTER 4

SECTION 4.1

CHECK YOUR PROGRESS 1, *page 175*

CHECK YOUR PROGRESS 2, *page 175*

$1,000,000 + (3 \times 100,000) + 10,000 + (4 \times 1000)$
$+ (3 \times 100) + (2 \times 10) + 6 = 1,314,326$

CHECK YOUR PROGRESS 3, *page 176*

$$\begin{array}{r} 23,341 \\ + 10,562 \end{array}$$

Replace 10 heel bones with one scroll to produce:

which is
33,903.

CHECK YOUR PROGRESS 4, *page 176*

$$\begin{array}{r} 61,432 \\ - 45,121 \end{array}$$

Replace one pointing finger with 10 lotus flowers to produce:

which is
16,311.

CHECK YOUR PROGRESS 5, *page 178*

$MCDXLV = M + (CD) + (XL) + V$
$= 1000 + 400 + 40 + 5 = 1445$

CHECK YOUR PROGRESS 6, *page 178*

$473 = 400 + 70 + 3 = CD + LXX + III = CDLXXIII$

CHECK YOUR PROGRESS 7, *page 179*

a. $\overline{\text{VII}}\text{CCLIV} = \overline{\text{VII}} + \text{CCLIV} = 7000 + 254 = 7254$

b. $8070 = 8000 + 70 = \overline{\text{VIII}} + \text{LXX} = \overline{\text{VIII}}\text{LXX}$

SECTION 4.2

CHECK YOUR PROGRESS 1, *page 183*

$17{,}325 = 10{,}000 + 7000 + 300 + 20 + 5$
$= (1 \times 10{,}000) + (7 \times 1000) + (3 \times 100) + (2 \times 10) + 5$
$= (1 \times 10^4) + (7 \times 10^3) + (3 \times 10^2) + (2 \times 10^1)$
$+ (5 \times 10^0)$

CHECK YOUR PROGRESS 2, *page 184*

$(5 \times 10^4) + (9 \times 10^3) + (2 \times 10^2) + (7 \times 10^1) + (4 \times 10^0)$
$= 5 \times 10{,}000 + 9 \times 1000 + 2 \times 100 + 7 \times 10 + 4 \times 1$
$= 50{,}000 + 9000 + 200 + 70 + 4$
$= 59{,}274$

CHECK YOUR PROGRESS 3, *page 184*

$\begin{array}{r} 152 = (1 \times 100) + (5 \times 10) + 2 \\ + \; 234 = (2 \times 100) + (3 \times 10) + 4 \\ \hline (3 \times 100) + (8 \times 10) + 6 = 386 \end{array}$

CHECK YOUR PROGRESS 4, *page 185*

$\begin{array}{r} 147 = (1 \times 100) + (4 \times 10) + \; 7 \\ + \; 329 = (3 \times 100) + (2 \times 10) + \; 9 \\ \hline (4 \times 100) + (6 \times 10) + 16 \end{array}$

Replace 16 with $(1 \times 10) + 6$

$= (4 \times 100) + (6 \times 10) + (1 \times 10) + 6$
$= (4 \times 100) + (7 \times 10) + 6$
$= 476$

CHECK YOUR PROGRESS 5, *page 185*

$\begin{array}{r} 382 = (3 \times 100) + (8 \times 10) + 2 \\ - \; 157 = (1 \times 100) + (5 \times 10) + 7 \\ \hline \end{array}$

Because $7 > 2$, it is necessary to borrow by rewriting (8×10) as $(7 \times 10) + 10$.

$\begin{array}{r} 382 = (3 \times 100) + (7 \times 10) + 12 \\ - \; 157 = (1 \times 100) + (5 \times 10) + \; 7 \\ \hline (2 \times 100) + (2 \times 10) + \; 5 = 225 \end{array}$

CHECK YOUR PROGRESS 6, *page 186*

≪𝌆 𝌆𝌆𝌆𝌆𝌆 ≪≪𝌆𝌆𝌆𝌆

$= (21 \times 3600) + (5 \times 60) + (34 \times 1)$
$= 75{,}600 + 300 + 34 = 75{,}934$

CHECK YOUR PROGRESS 7, *page 187*

$$\begin{array}{r} 3 \\ 3600\overline{)12578} \\ 10800 \\ \hline 1778 \end{array} \qquad \begin{array}{r} 29 \\ 60\overline{)1778} \\ 120 \\ \hline 578 \\ 540 \\ \hline 38 \end{array}$$

Thus $12{,}578 = (3 \times 3600) + (29 \times 60) + 38 =$

𝌆𝌆𝌆 ≪𝌆𝌆𝌆𝌆𝌆𝌆𝌆𝌆𝌆 ≪≪≪𝌆𝌆𝌆𝌆𝌆𝌆𝌆𝌆

CHECK YOUR PROGRESS 8, *page 189*

a. $(16 \times 360) + (0 \times 20) + (1 \times 1) = 5761$

b. $(9 \times 7200) + (1 \times 360) + (10 \times 20) + (4 \times 1) = 65{,}364$

CHECK YOUR PROGRESS 9, *page 189*

$$\begin{array}{r} 1 \\ 7200\overline{)11480} \\ 7200 \\ \hline 4280 \end{array} \qquad \begin{array}{r} 11 \\ 360\overline{)4280} \\ 360 \\ \hline 680 \\ 360 \\ \hline 320 \end{array} \qquad \begin{array}{r} 16 \\ 20\overline{)320} \\ 20 \\ \hline 120 \\ 120 \\ \hline 0 \end{array}$$

Thus $11{,}480 = 1 \times 7200 + 11 \times 360 + 16 \times 20 + 0 \times 1$.
In Mayan numerals this is

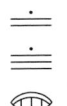

SECTION 4.3

CHECK YOUR PROGRESS 1, *page 193*

$3156_{\text{seven}} = (3 \times 7^3) + (1 \times 7^2) + (5 \times 7^1) + (6 \times 7^0)$
$= (3 \times 343) + (1 \times 49) + (5 \times 7) + (6 \times 1)$
$= 1029 + 49 + 35 + 6$
$= 1119$

CHECK YOUR PROGRESS 2, *page 194*

$111000101_{\text{two}} = (1 \times 2^8) + (1 \times 2^7) + (1 \times 2^6) + (0 \times 2^5)$
$+ (0 \times 2^4) + (0 \times 2^3) + (1 \times 2^2)$
$+ (0 \times 2^1) + (1 \times 2^0)$
$= (1 \times 256) + (1 \times 128) + (1 \times 64) + (0 \times 32)$
$+ (0 \times 16) + (0 \times 8) + (1 \times 4)$
$+ (0 \times 2) + (1 \times 1)$
$= 256 + 128 + 64 + 0 + 0 + 0 + 4 + 0 + 1$
$= 453$

CHECK YOUR PROGRESS 3, *page 194*

$A5B_{\text{twelve}} = (10 \times 12^2) + (5 \times 12^1) + (11 \times 12^0)$
$= 1440 + 60 + 11$
$= 1511$

CHECK YOUR PROGRESS 4, *page 195*

$C24F_{\text{sixteen}} = (12 \times 16^3) + (2 \times 16^2) + (4 \times 16^1) + (15 \times 16^0)$
$= 49{,}152 + 512 + 64 + 15$
$= 49{,}743$

CHECK YOUR PROGRESS 5, *page 196*

a.
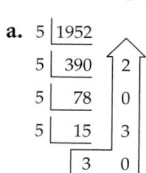

5 | 1952
5 | 390 2
5 | 78   0
5 | 15   3
    3   0

$$1952 = 30302_{\text{five}}$$

b.

12 | 1952
12 | 162 8
12 | 13   6
      1   1

$$1952 = 1168_{\text{twelve}}$$

CHECK YOUR PROGRESS 6, *page 197*

6	3	2	1	0$_{\text{eight}}$
‖	‖	‖	‖	‖
110	011	010	001	000$_{\text{two}}$

$$63210_{\text{eight}} = 110011010001000_{\text{two}}$$

CHECK YOUR PROGRESS 7, *page 197*

111	010	011	100$_{\text{two}}$
‖	‖	‖	‖
7	2	3	4$_{\text{eight}}$

$$111010011100_{\text{two}} = 7234_{\text{eight}}$$

CHECK YOUR PROGRESS 8, *page 198*

C	5	A$_{\text{sixteen}}$
‖	‖	‖
1100	0101	1010$_{\text{two}}$

$$C5A_{\text{sixteen}} = 110001011010_{\text{two}}$$

CHECK YOUR PROGRESS 9, *page 198*

— Insert a zero to make a group of four.

0101	0001	1101	0010$_{\text{two}}$
‖	‖	‖	‖
5	1	D	2$_{\text{sixteen}}$

$$101000111010010_{\text{two}} = 51D2_{\text{sixteen}}$$

CHECK YOUR PROGRESS 10, *page 199*

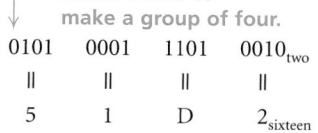

dabble	dabble	double	double	dabble	double
3	7	14	28	57	114

| 1 | 1 | 1 | 0 | 0 | 1 | 0$_{\text{two}}$ |

$$1110010_{\text{two}} = 114$$

SECTION 4.4

CHECK YOUR PROGRESS 1, *page 204*

```
  1 1   1
    1 1 0 0 1 two
+     1 1 0 1 two
  1 0 0 1 1 0 two
```

CHECK YOUR PROGRESS 2, *page 205*

```
   1 1
     4 2 five
+    2 3 five
   1 2 0 five
```

CHECK YOUR PROGRESS 3, *page 205*

```
   1 2
     3 5 seven
     4 6 seven
+    2 4 seven
   1 4 1 seven
```

CHECK YOUR PROGRESS 4, *page 206*

```
   1 1
   A C 4 sixteen
+    6 E 8 sixteen
   1 1 A C sixteen
```

CHECK YOUR PROGRESS 5, *page 207*

```
 2 + 1                  10                    16
  З 6 5 nine           2 6 5 nine           2 6̸ 5 nine
- 1 8 3 nine         - 1 8 3 nine         - 1 8 3 nine
                                            1 7 2 nine
```

Because $8_{\text{nine}} > 6_{\text{nine}}$, it is necessary to borrow from the 3 in the first column.

Borrow 1 nine from the first column and add $9 = 10_{\text{nine}}$ to the 6 in the middle column.

$$16_{\text{nine}} - 8_{\text{nine}} = 15 - 8$$
$$= 7$$
$$= 7_{\text{nine}}$$

CHECK YOUR PROGRESS 6, *page 208*

```
   7  10
   8̸   3   A twelve
-  4   6   7 twelve
   3   9   3 twelve
```

- $A_{\text{twelve}} - 7_{\text{twelve}} = 10 - 7 = 3 = 3_{\text{twelve}}$
- $10_{\text{twelve}} + 3_{\text{twelve}} = 13_{\text{twelve}} = 15$
  $15 - 6 = 9 = 9_{\text{twelve}}$
- $7_{\text{twelve}} - 4_{\text{twelve}} = 3_{\text{twelve}}$

CHECK YOUR PROGRESS 7, *page 208*

```
          1
      2   1   3 four
×             2 four
  1   0   3   2 four
```

- $2_{\text{four}} \times 3_{\text{four}} = 12_{\text{four}}$
- $2_{\text{four}} \times 1_{\text{four}} + 1_{\text{four}} = 3_{\text{four}}$
- $2_{\text{four}} \times 2_{\text{four}} = 10_{\text{four}}$

CHECK YOUR PROGRESS 8, *page 210*

$$
\begin{array}{r}
2 \\
3 \quad 4_{\text{eight}} \\
\times \quad 2 \quad 5_{\text{eight}} \\
\hline
2 \quad 1 \quad 4_{\text{eight}}
\end{array}
$$

- $5_{\text{eight}} \times 4_{\text{eight}} = 20 = 24_{\text{eight}}$
- $5_{\text{eight}} \times 3_{\text{eight}} + 2_{\text{eight}} = 15 + 2 = 17 = 21_{\text{eight}}$

$$
\begin{array}{r}
1 \\
3 \quad 4_{\text{eight}} \\
\times \quad 2 \quad 5_{\text{eight}} \\
\hline
2 \quad 1 \quad 4_{\text{eight}} \\
7 \quad 0_{\text{eight}} \\
\hline
1 \quad 1 \quad 1 \quad 4_{\text{eight}}
\end{array}
$$

- $2_{\text{eight}} \times 4_{\text{eight}} = 8 = 10_{\text{eight}}$
- $2_{\text{eight}} \times 3_{\text{eight}} + 1_{\text{eight}} = 6 + 1 = 7 = 7_{\text{eight}}$

CHECK YOUR PROGRESS 9, *page 211*

First list a few multiples of 3_{five}.

$3_{\text{five}} \times 0_{\text{five}} = 0_{\text{five}}$

$3_{\text{five}} \times 1_{\text{five}} = 3_{\text{five}}$

$3_{\text{five}} \times 2_{\text{five}} = 11_{\text{five}}$

$3_{\text{five}} \times 3_{\text{five}} = 14_{\text{five}}$

$3_{\text{five}} \times 4_{\text{five}} = 22_{\text{five}}$

$$
\begin{array}{r}
1 \\
3_{\text{five}})\overline{3\ 2\ 4}_{\text{five}} \\
\underline{3} \\
2
\end{array}
\qquad
\begin{array}{r}
1\ 0 \\
3_{\text{five}})\overline{3\ 2\ 4}_{\text{five}} \\
\underline{3} \\
2 \\
0 \\
\hline
2\ 4
\end{array}
\qquad
\begin{array}{r}
1\ 0\ 4 \\
3_{\text{five}})\overline{3\ 2\ 4}_{\text{five}} \\
\underline{3} \\
2 \\
0 \\
\hline
2\ 4 \\
\underline{2\ 2} \\
2
\end{array}
$$

Thus $324_{\text{five}} \div 3_{\text{five}} = 104_{\text{five}}$ with a remainder of 2_{five}.

CHECK YOUR PROGRESS 10, *page 211*

The divisor is 10_{two}. The multiples of the divisor are $10_{\text{two}} \times 0_{\text{two}} = 0_{\text{two}}$ and $10_{\text{two}} \times 1_{\text{two}} = 10_{\text{two}}$.

$$
\begin{array}{r}
1\ 1\ 1\ 0\ 0\ 1_{\text{two}} \\
10_{\text{two}})\overline{1\ 1\ 1\ 0\ 0\ 1\ 1}_{\text{two}} \\
\underline{1\ 0} \\
1\ 1 \\
\underline{1\ 0} \\
1\ 0 \\
\underline{1\ 0} \\
0\ 0 \\
\underline{0} \\
0\ 1 \\
\underline{0} \\
1\ 1 \\
\underline{1\ 0} \\
1
\end{array}
$$

Thus $1110011_{\text{two}} \div 10_{\text{two}} = 111001_{\text{two}}$ with a remainder of 1_{two}.

SECTION 4.5

CHECK YOUR PROGRESS 1, *page 215*

a. Divide 9 by 1, 2, 3, . . . , 9 to determine that the only natural number divisors of 9 are 1, 3, and 9.

b. Divide 11 by 1, 2, 3, . . . , 11 to determine that the only natural number divisors of 11 are 1 and 11.

c. Divide 24 by 1, 2, 3, . . . , 24 to determine that the only natural number divisors of 24 are 1, 2, 3, 4, 6, 8, 12, and 24.

CHECK YOUR PROGRESS 2, *page 216*

a. The only divisors of 47 are 1 and 47. Thus 47 is a prime number.

b. 171 is divisible by 3, 9, 19, and 57. Thus 171 is a composite number.

CHECK YOUR PROGRESS 3, *page 218*

a. The sum of the digits of 341,565 is 24; therefore, 341,565 is divisible by 3.

b. The number 341,565 is not divisible by 4 because the number formed by last two digits, 65, is not divisible by 4.

c. The number 341,565 is not divisible by 10 because it does not end in 0.

d. The sum of the digits with even place-value powers is 14. The sum of the digits with odd place-value powers is 10. The difference of these sums is 4. Thus 341,565 is not divisible by 11.

CHECK YOUR PROGRESS 4, *page 219*

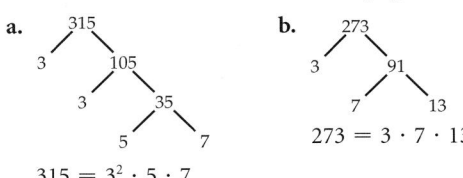

a.

$315 = 3^2 \cdot 5 \cdot 7$

b.

$273 = 3 \cdot 7 \cdot 13$

SECTION 4.6

CHECK YOUR PROGRESS 1, *page 226*

a. The proper factors of 24 are 1, 2, 3, 4, 6, 8, and 12. The sum of these proper factors is 36. Because 24 is less than the sum of its proper factors, 24 is an abundant number.

b. The proper factors of 28 are 1, 2, 4, 7, and 14. The sum of these proper factors is 28. Because 28 equals the sum of its proper factors, 28 is a perfect number.

c. The proper factors of 35 are 1, 5, and 7. The sum of these proper factors is 13. Because 35 is larger than the sum of its proper factors, 35 is a deficient number.

CHECK YOUR PROGRESS 2, *page 226*

$2^7 - 1 = 127$, which is a prime number.

CHECK YOUR PROGRESS 3, *page 227*

The exponent $n = 61$ is a prime number and we are given that $2^{61} - 1$ is a prime number, so the perfect number we seek is $2^{60}(2^{61} - 1)$.

CHECK YOUR PROGRESS 4, *page 230*

First consider $2^{2976221}$. The base b is 2. The exponent x is 2,976,221.

$$
\begin{aligned}
(x \log b) + 1 &= (2{,}976{,}221 \log 2) + 1 \\
&\approx 895{,}931.8 + 1 \\
&= 895{,}932.8
\end{aligned}
$$

The greatest integer of 895,932.8 is 895,932. Thus $2^{2976221}$ has 895,932 digits. The number $2^{2976221}$ is not a power of 10, so the Mersenne number $2^{2976221} - 1$ also has 895,932 digits.

CHECK YOUR PROGRESS 5, *page 231*

Substituting 9 for x, 11 for y, and 4 for n in $x^n + y^n = z^n$ yields

$$9^4 + 11^4 = z^4$$
$$6561 + 14{,}641 = z^4$$
$$21{,}202 = z^4$$

The real solution of $z^4 = 21{,}202$ is $\sqrt[4]{21{,}202} \approx 12.066858$, which is not a natural number. Thus $x = 9$, $y = 11$, and $n = 4$ do not satisfy the equation $x^n + y^n = z^n$ where z is a natural number.

CHAPTER 5

SECTION 5.1

CHECK YOUR PROGRESS 1, *page 243*

a.
$$
\begin{aligned}
c - 6 &= -13 \\
c - 6 + 6 &= -13 + 6 \\
c &= -7
\end{aligned}
$$

The solution is -7.

b.
$$
\begin{aligned}
4 &= -8z \\
\frac{4}{-8} &= \frac{-8z}{-8} \\
-\frac{1}{2} &= z
\end{aligned}
$$

The solution is $-\frac{1}{2}$.

c.
$$
\begin{aligned}
22 + m &= -9 \\
22 - 22 + m &= -9 - 22 \\
m &= -31
\end{aligned}
$$

The solution is -31.

d.
$$
\begin{aligned}
5x &= 0 \\
\frac{5x}{5} &= \frac{0}{5} \\
x &= 0
\end{aligned}
$$

The solution is 0.

CHECK YOUR PROGRESS 2, *page 245*

a.
$$
\begin{aligned}
4x + 3 &= 7x + 9 \\
4x - 7x + 3 &= 7x - 7x + 9 \\
-3x + 3 &= 9 \\
-3x + 3 - 3 &= 9 - 3 \\
-3x &= 6 \\
\frac{-3x}{-3} &= \frac{6}{-3} \\
x &= -2
\end{aligned}
$$

The solution is -2.

b.
$$
\begin{aligned}
7 - (5x - 8) &= 4x + 3 \\
7 - 5x + 8 &= 4x + 3 \\
15 - 5x &= 4x + 3 \\
15 - 5x - 4x &= 4x - 4x + 3 \\
15 - 9x &= 3 \\
15 - 15 - 9x &= 3 - 15 \\
-9x &= -12 \\
\frac{-9x}{-9} &= \frac{-12}{-9} \\
x &= \frac{4}{3}
\end{aligned}
$$

The solution is $\frac{4}{3}$.

c.
$$
\begin{aligned}
\frac{3x - 1}{4} + \frac{1}{3} &= \frac{7}{3} \\
12\left(\frac{3x - 1}{4} + \frac{1}{3}\right) &= 12\left(\frac{7}{3}\right) \\
12 \cdot \frac{3x - 1}{4} + 12 \cdot \frac{1}{3} &= 12 \cdot \frac{7}{3} \\
9x - 3 + 4 &= 28 \\
9x + 1 &= 28 \\
9x + 1 - 1 &= 28 - 1 \\
9x &= 27 \\
\frac{9x}{9} &= \frac{27}{9} \\
x &= 3
\end{aligned}
$$

The solution is 3.

CHECK YOUR PROGRESS 3, *page 246*

a. $P = 0.05Y - 95$
$$
\begin{aligned}
P &= 0.05(1990) - 95 \\
P &= 99.5 - 95 \\
P &= 4.5
\end{aligned}
$$

The amount of garbage was about 4.5 pounds per day.

b.
$$P = 0.05Y - 95$$
$$5.25 = 0.05Y - 95$$
$$5.25 + 95 = 0.05Y - 95 + 95$$
$$100.25 = 0.05Y$$
$$\frac{100.25}{0.05} = \frac{0.05Y}{0.05}$$
$$2005 = Y$$

The year will be 2005.

CHECK YOUR PROGRESS 4, *page 247*

$$\boxed{\begin{array}{c}\text{\$7.50 for the first}\\ \text{25 words +\$.25 for}\\ \text{each word over 25}\end{array}} = \boxed{\text{\$14}}$$

Let w = the number of words in the ad.

$$7.50 + 0.25(w - 25) = 14$$
$$7.50 + 0.25w - 6.25 = 14$$
$$1.25 + 0.25w = 14$$
$$1.25 - 1.25 + 0.25w = 14 - 1.25$$
$$0.25w = 12.75$$
$$\frac{0.25w}{0.25} = \frac{12.75}{0.25}$$
$$w = 51$$

You can place 51 words in the ad.

CHECK YOUR PROGRESS 5, *page 248*

$$\boxed{\begin{array}{c}\text{The 1990 population of}\\ \text{Vermont plus an annual}\\ \text{increase times } n\end{array}} = \boxed{\begin{array}{c}\text{The 1990 population of}\\ \text{North Dakota minus an}\\ \text{annual decrease times } n\end{array}}$$

$$562{,}576 + 5116n = 638{,}800 - 1370n$$
$$562{,}576 + 5116n + 1370n = 638{,}800 - 1370n + 1370n$$
$$562{,}576 + 6486n = 638{,}800$$
$$562{,}576 - 562{,}576 + 6486n = 638{,}800 - 562{,}576$$
$$6486n = 76{,}224$$
$$\frac{6486n}{6486} = \frac{76{,}224}{6486}$$
$$n \approx 12$$

$$1990 + 12 = 2002$$

The populations would be the same in 2002.

CHECK YOUR PROGRESS 6, *page 250*

a.
$$s = \frac{A + L}{2}$$
$$2 \cdot s = 2 \cdot \frac{A + L}{2}$$
$$2s = A + L$$
$$2s - A = A - A + L$$
$$2s - A = L$$

b.
$$L = a(1 + ct)$$
$$\frac{L}{a} = \frac{a(1 + ct)}{a}$$
$$\frac{L}{a} = 1 + ct$$
$$\frac{L}{a} - 1 = 1 - 1 + ct$$
$$\frac{L}{a} - 1 = ct$$
$$\frac{\frac{L}{a} - 1}{t} = \frac{ct}{t}$$
$$\frac{\frac{L}{a} - 1}{t} = c$$
$$\frac{L}{at} - \frac{1}{t} = c$$

SECTION 5.2

CHECK YOUR PROGRESS 1, *page 258*

$$4.92 \div 1.5 = 3.28$$

$$\frac{\$4.92}{1.5 \text{ pounds}} = \frac{\$3.28}{1 \text{ pound}} = \$3.28/\text{pound}$$

The hamburger costs $3.28 per pound.

CHECK YOUR PROGRESS 2, *page 259*

Find the increase per hour.

$$\frac{\$5.15}{1 \text{ hour}} - \frac{\$4.75}{1 \text{ hour}} = \frac{\$.40}{1 \text{ hour}}$$

Multiply by the number of hours per week.

$$\frac{\$.40}{1 \text{ hour}} \cdot \frac{35 \text{ hours}}{1 \text{ week}} = \$14.00/\text{week}$$

The increase was $14 per week.

CHECK YOUR PROGRESS 3, *page 259*

$$\frac{\$2.99}{32 \text{ ounces}} \approx \frac{\$.093}{1 \text{ ounce}} \qquad \frac{\$3.99}{48 \text{ ounces}} \approx \frac{\$.083}{1 \text{ ounce}}$$

$$\$.093 > \$.083$$

The more economical purchase is 48 ounces of detergent for $3.99.

CHECK YOUR PROGRESS 4, *page 261*

a. $20,000(1.602) = 32,040$
32,040 Canadian dollars would be needed to pay for an order costing $20,000.

b. $25,000(1.676) = 41,900$
41,900 Swiss francs would be exchanged for $25,000.

CHECK YOUR PROGRESS 5, *page 262*

a. $\dfrac{24 \text{ hours}}{1 \text{ day}} \cdot 7 \text{ days} = (24 \text{ hours})(7) = 168 \text{ hours}$

$\dfrac{120 \text{ hours}}{1 \text{ week}} = \dfrac{120 \text{ hours}}{168 \text{ hours}} = \dfrac{120}{168} = \dfrac{5}{7}$

The ratio is $\frac{5}{7}$.

b. $\dfrac{60 \text{ hours}}{168 - 60} = \dfrac{60 \text{ hours}}{108} = \dfrac{60}{108} = \dfrac{5}{9}$

The ratio is 5 to 9.

CHECK YOUR PROGRESS 6, *page 263*

$5507 + 6221 = 11{,}728$

$\dfrac{11{,}728}{777} \approx \dfrac{15.09}{1} \approx \dfrac{15}{1}$

The ratio is 15 to 1.

CHECK YOUR PROGRESS 7, *page 265*

$\dfrac{42}{x} = \dfrac{5}{8}$

$42 \cdot 8 = x \cdot 5$

$336 = 5x$

$\dfrac{336}{5} = \dfrac{5x}{5}$

$67.2 = x$

The solution is 67.2.

CHECK YOUR PROGRESS 8, *page 266*

$\dfrac{15 \text{ kilometers}}{2 \text{ centimeters}} = \dfrac{x \text{ kilometers}}{7 \text{ centimeters}}$

$\dfrac{15}{2} = \dfrac{x}{7}$

$15 \cdot 7 = 2 \cdot x$

$105 = 2x$

$\dfrac{105}{2} = \dfrac{2x}{2}$

$52.5 = x$

The distance between the two cities is 52.5 kilometers.

CHECK YOUR PROGRESS 9, *page 267*

$\dfrac{7}{5} = \dfrac{\$28{,}000}{x \text{ dollars}}$

$\dfrac{7}{5} = \dfrac{28{,}000}{x}$

$7 \cdot x = 5 \cdot 28{,}000$

$7x = 140{,}000$

$\dfrac{7x}{7} = \dfrac{140{,}000}{7}$

$x = 20{,}000$

The other partner receives $20,000.

CHECK YOUR PROGRESS 10, *page 268*

$\dfrac{10.1 \text{ deaths}}{1{,}000{,}000 \text{ people}} = \dfrac{d \text{ deaths}}{4{,}000{,}000 \text{ people}}$

$10.1(4{,}000{,}000) = 1{,}000{,}000 \cdot d$

$40{,}400{,}000 = 1{,}000{,}000 d$

$\dfrac{40{,}400{,}000}{1{,}000{,}000} = \dfrac{1{,}000{,}000 d}{1{,}000{,}000}$

$40.4 = d$

Approximately 40 people aged 5 to 34 die from asthma each year in New York City.

SECTION 5.3

CHECK YOUR PROGRESS 1, *page 278*

a. $74\% = 0.74$

b. $152\% = 1.52$

c. $8.3\% = 0.083$

d. $0.6\% = 0.006$

CHECK YOUR PROGRESS 2, *page 278*

a. $0.3 = 30\%$

b. $1.65 = 165\%$

c. $0.072 = 7.2\%$

d. $0.004 = 0.4\%$

CHECK YOUR PROGRESS 3, *page 279*

a. $8\% = 8\left(\dfrac{1}{100}\right) = \dfrac{8}{100} = \dfrac{2}{25}$

b. $180\% = 180\left(\dfrac{1}{100}\right) = \dfrac{180}{100} = 1\dfrac{80}{100} = 1\dfrac{4}{5}$

c. $2.5\% = 2.5\left(\dfrac{1}{100}\right) = \dfrac{2.5}{100} = \dfrac{25}{1000} = \dfrac{1}{40}$

d. $66\dfrac{2}{3}\% = \dfrac{200}{3}\% = \dfrac{200}{3}\left(\dfrac{1}{100}\right) = \dfrac{2}{3}$

CHECK YOUR PROGRESS 4, *page 280*

a. $\dfrac{1}{4} = 0.25 = 25\%$

b. $\dfrac{3}{8} = 0.375 = 37.5\%$

c. $\dfrac{5}{6} = 0.83\overline{3} = 83.\overline{3}\%$

d. $1\dfrac{2}{3} = 1.66\overline{6} = 166.\overline{6}\%$

CHECK YOUR PROGRESS 5, *page 281*

$$\frac{\text{Percent}}{100} = \frac{\text{amount}}{\text{base}}$$

$$\frac{70}{100} = \frac{22{,}400}{B}$$

$$70 \cdot B = 100(22{,}400)$$

$$70B = 2{,}240{,}000$$

$$\frac{70B}{70} = \frac{2{,}240{,}000}{70}$$

$$B = 32{,}000$$

The Blazer cost $32,000 when it was new.

CHECK YOUR PROGRESS 6, *page 282*

$$\frac{\text{Percent}}{100} = \frac{\text{amount}}{\text{base}}$$

$$\frac{p}{100} = \frac{416{,}000}{1{,}300{,}000}$$

$$p \cdot 1{,}300{,}000 = 100(416{,}000)$$

$$1{,}300{,}000p = 41{,}600{,}000$$

$$\frac{1{,}300{,}000p}{1{,}300{,}000} = \frac{41{,}600{,}000}{1{,}300{,}000}$$

$$p = 32$$

32% of the enlisted people are over the age of 30.

CHECK YOUR PROGRESS 7, *page 283*

$$\frac{\text{Percent}}{100} = \frac{\text{amount}}{\text{base}}$$

$$\frac{3.5}{100} = \frac{A}{32{,}500}$$

$$3.5(32{,}500) = 100(A)$$

$$113{,}750 = 100A$$

$$\frac{113{,}750}{100} = \frac{100A}{100}$$

$$1137.5 = A$$

The customer would receive a rebate of $1137.50.

CHECK YOUR PROGRESS 8, *page 283*

$$PB = A$$

$$0.05(32{,}685) = A$$

$$1634.25 = A$$

The teacher contributes $1634.25.

CHECK YOUR PROGRESS 9, *page 284*

$$PB = A$$

$$0.06B = 14{,}370$$

$$\frac{0.06B}{0.06} = \frac{14{,}370}{0.06}$$

$$B = 239{,}500$$

The selling price of the home was $239,500.

CHECK YOUR PROGRESS 10, *page 284*

$$PB = A$$

$$P \cdot 90 = 63$$

$$\frac{P \cdot 90}{90} = \frac{63}{90}$$

$$P = 0.7$$

$$P = 70\%$$

You answered 70% of the questions correctly.

CHECK YOUR PROGRESS 11, *page 285*

$$PB = A$$

$$0.90(21{,}262) = A$$

$$19{,}135.80 = A$$

$$21{,}262 - 19{,}135.80 = 2126.20$$

The difference between the cost of the remodeling and the increase in value of your home is $2126.20.

CHECK YOUR PROGRESS 12, *page 287*

$$5.67 - 1.82 = 3.85$$

$$PB = A$$

$$P \cdot 1.82 = 3.85$$

$$\frac{P \cdot 1.82}{1.82} = \frac{3.85}{1.82}$$

$$P \approx 2.115$$

The percent increase in the federal debt from 1985 to 2000 was 211.5%.

CHECK YOUR PROGRESS 13, *page 289*

$$\frac{\text{Percent}}{100} = \frac{\text{amount}}{\text{base}}$$

$$\frac{40}{100} = \frac{A}{23{,}985}$$

$$40(23{,}985) = 100(A)$$

$$959{,}400 = 100A$$

$$\frac{959{,}400}{100} = \frac{100A}{100}$$

$$9594 = A$$

$$23{,}985 - 9594 = 14{,}391$$

The resale value of a Ford Mustang in 2002 was expected to be $14,391.

SECTION 5.4

CHECK YOUR PROGRESS 1, *page 301*

$$2s^2 = 6 - 4s$$

$$2s^2 + 4s = 6 - 4s + 4s$$

$$2s^2 + 4s = 6$$

$$2s^2 + 4s - 6 = 6 - 6$$

$$2s^2 + 4s - 6 = 0$$

CHECK YOUR PROGRESS 2, *page 302*

$$(n + 5)(2n - 3) = 0$$

$$
\begin{array}{ll}
n + 5 = 0 & 2n - 3 = 0 \\
n = -5 & 2n = 3 \\
 & n = \dfrac{3}{2}
\end{array}
$$

Check:

$(n + 5)(2n - 3) = 0$		$(n + 5)(2n - 3) = 0$	
$(-5 + 5)[2(-5) - 3]$	0	$\left(\dfrac{3}{2} + 5\right)\left(2 \cdot \dfrac{3}{2} - 3\right)$	0
$0(-13)$	0	$\dfrac{13}{2}(3 - 3)$	0
	$0 = 0$		$0 = 0$

The solutions are -5 and $\frac{3}{2}$.

CHECK YOUR PROGRESS 3, *page 303*

$$2x^2 = x + 1$$

$$2x^2 - x = x - x + 1$$

$$2x^2 - x = 1$$

$$2x^2 - x - 1 = 1 - 1$$

$$2x^2 - x - 1 = 0$$

$$(2x + 1)(x - 1) = 0$$

$$
\begin{array}{ll}
2x + 1 = 0 & x - 1 = 0 \\
2x = -1 & x = 1 \\
x = -\dfrac{1}{2} &
\end{array}
$$

Check:

$2x^2 = x + 1$		$2x^2 = x + 1$	
$2\left(-\dfrac{1}{2}\right)^2$	$-\dfrac{1}{2} + 1$	$2(1)^2$	$1 + 1$
$2\left(\dfrac{1}{4}\right)$	$\dfrac{1}{2}$	$2(1)$	2
$\dfrac{1}{2} = \dfrac{1}{2}$		$2 = 2$	

The solutions are $-\frac{1}{2}$ and 1.

CHECK YOUR PROGRESS 4, *page 304*

$$2x^2 = 8x - 5$$

$$2x^2 - 8x + 5 = 0$$

$$a = 2,\ b = -8,\ c = 5$$

$$x = \frac{-b \pm \sqrt{b^2 - 4ac}}{2a}$$

$$x = \frac{-(-8) \pm \sqrt{(-8)^2 - 4(2)(5)}}{2(2)} = \frac{8 \pm \sqrt{64 - 40}}{4}$$

$$= \frac{8 \pm \sqrt{24}}{4} = \frac{8 \pm 2\sqrt{6}}{4} = \frac{2(4 \pm \sqrt{6})}{2(2)} = \frac{4 \pm \sqrt{6}}{2}$$

The exact solutions are $\dfrac{4 + \sqrt{6}}{2}$ and $\dfrac{4 - \sqrt{6}}{2}$.

$$\frac{4 + \sqrt{6}}{2} \approx 3.225 \qquad \frac{4 - \sqrt{6}}{2} \approx 0.775$$

To the nearest thousandth, the solutions are 3.225 and 0.775.

CHECK YOUR PROGRESS 5, *page 305*

$$z^2 = -6 - 2z$$

$$z^2 + 2z + 6 = 0$$

$$a = 1,\ b = 2,\ c = 6$$

$$z = \frac{-b \pm \sqrt{b^2 - 4ac}}{2a}$$

$$z = \frac{-(2) \pm \sqrt{(2)^2 - 4(1)(6)}}{2(1)}$$

$$= \frac{-2 \pm \sqrt{4 - 24}}{2} = \frac{-2 \pm \sqrt{-20}}{2}$$

$\sqrt{-20}$ is not a real number.

The equation has no real number solutions.

CHECK YOUR PROGRESS 6, *page 306*

$$h = 64t - 16t^2$$
$$0 = 64t - 16t^2$$
$$16t^2 - 64t = 0$$
$$16t(t - 4) = 0$$

$$16t = 0 \qquad t - 4 = 0$$
$$t = 0 \qquad t = 4$$

The object will be on the ground at 0 seconds and after 4 seconds.

CHECK YOUR PROGRESS 7, *page 307*

$$h = -16t^2 + 32t + 6.5$$
$$10 = -16t^2 + 32t + 6.5$$
$$16t^2 - 32t + 3.5 = 0$$

$$a = 16, b = -32, c = 3.5$$
$$t = \frac{-b \pm \sqrt{b^2 - 4ac}}{2a}$$

$$t = \frac{-(-32) \pm \sqrt{(-32)^2 - 4(16)(3.5)}}{2(16)} = \frac{32 \pm \sqrt{800}}{32}$$

$$t = \frac{32 + \sqrt{800}}{32} \approx 1.88 \qquad t = \frac{32 - \sqrt{800}}{32} \approx 0.12$$

The solution $t \approx 0.12$ second is not reasonable. The ball hits the basket 1.88 seconds after the ball is released.

CHAPTER 6

SECTION 6.1

CHECK YOUR PROGRESS 1, *page 326*

x	−2x + 3 = y	(x, y)
−2	−2(−2) + 3 = 7	(−2, 7)
−1	−2(−1) + 3 = 5	(−1, 5)
0	−2(0) + 3 = 3	(0, 3)
1	−2(1) + 3 = 1	(1, 1)
2	−2(2) + 3 = −1	(2, −1)
3	−2(3) + 3 = −3	(3, −3)

CHECK YOUR PROGRESS 2, *page 327*

x	−x² + 1 = y	(x, y)
−3	−(−3)² + 1 = −8	(−3, −8)
−2	−(−2)² + 1 = −3	(−2, −3)
−1	−(−1)² + 1 = 0	(−1, 0)
0	−(0)² + 1 = 1	(0, 1)
1	−(1)² + 1 = 0	(1, 0)
2	−(2)² + 1 = −3	(2, −3)
3	−(3)² + 1 = −8	(3, −8)

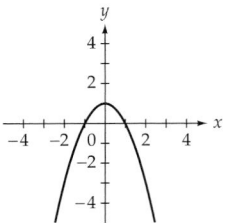

CHECK YOUR PROGRESS 3, *page 330*

$$f(z) = z^2 - z$$
$$f(-3) = (-3)^2 - (-3)$$
$$= 12$$

The value of the function is 12 when $z = -3$.

CHECK YOUR PROGRESS 4, *page 330*

$$N(s) = \frac{s^2 - 3s}{2}$$

$$N(12) = \frac{(12)^2 - 3(12)}{2}$$

$$= \frac{144 - 36}{2}$$

$$= 54$$

A polygon with 12 sides has 54 diagonals.

CHECK YOUR PROGRESS 5, *page 332*

x	$f(x) = 2 - \dfrac{3}{4}x$	(x, y)
-3	$f(-3) = 2 - \dfrac{3}{4}(-3) = 4\dfrac{1}{4}$	$\left(-3, 4\dfrac{1}{4}\right)$
-2	$f(-2) = 2 - \dfrac{3}{4}(-2) = 3\dfrac{1}{2}$	$\left(-2, 3\dfrac{1}{2}\right)$
-1	$f(-1) = 2 - \dfrac{3}{4}(-1) = 2\dfrac{3}{4}$	$\left(-1, 2\dfrac{3}{4}\right)$
0	$f(0) = 2 - \dfrac{3}{4}(0) = 2$	$(0, 2)$
1	$f(1) = 2 - \dfrac{3}{4}(1) = 1\dfrac{1}{4}$	$\left(1, 1\dfrac{1}{4}\right)$
2	$f(2) = 2 - \dfrac{3}{4}(2) = \dfrac{1}{2}$	$\left(2, \dfrac{1}{2}\right)$
3	$f(3) = 2 - \dfrac{3}{4}(3) = -\dfrac{1}{4}$	$\left(3, -\dfrac{1}{4}\right)$

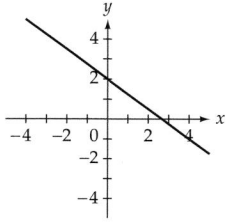

CHECK YOUR PROGRESS 1, *page 338*

$$f(x) = \frac{1}{2}x + 3 \qquad\qquad f(x) = \frac{1}{2}x + 3$$

$$0 = \frac{1}{2}x + 3 \qquad\qquad f(0) = \frac{1}{2}(0) + 3$$

$$-3 = \frac{1}{2}x \qquad\qquad\qquad = 3$$

$$-6 = x$$

The x-intercept is $(-6, 0)$.   The y-intercept is $(0, 3)$.

CHECK YOUR PROGRESS 2, *page 339*

$$g(t) = -20t + 8000$$
$$g(0) = -20(0) + 8000 = 8000$$

The intercept on the vertical axis is $(0, 8000)$. This means that the plane is at an altitude of 8000 feet when it begins its descent.

$$g(t) = -20t + 8000$$
$$0 = -20t + 8000$$
$$-8000 = -20t$$
$$400 = t$$

The intercept on the horizontal axis is $(400, 0)$. This means that the plane reaches the ground 400 seconds after beginning its descent.

CHECK YOUR PROGRESS 3, *page 342*

a. $(x_1, y_1) = (-6, 5), (x_2, y_2) = (4, -5)$

$$m = \frac{y_2 - y_1}{x_2 - x_1} = \frac{-5 - 5}{4 - (-6)} = \frac{-10}{10} = -1$$

The slope is -1.

b. $(x_1, y_1) = (-5, 0), (x_2, y_2) = (-5, 7)$

$$m = \frac{y_2 - y_1}{x_2 - x_1} = \frac{7 - 0}{-5 - (-5)} = \frac{7}{0}$$

The slope is undefined.

c. $(x_1, y_1) = (-7, -2), (x_2, y_2) = (8, 8)$

$$m = \frac{y_2 - y_1}{x_2 - x_1} = \frac{8 - (-2)}{8 - (-7)} = \frac{10}{15} = \frac{2}{3}$$

The slope is $\frac{2}{3}$.

d. $(x_1, y_1) = (-6, 7), (x_2, y_2) = (1, 7)$

$$m = \frac{y_2 - y_1}{x_2 - x_1} = \frac{7 - 7}{1 - (-6)} = \frac{0}{7} = 0$$

The slope is 0.

CHECK YOUR PROGRESS 4, *page 343*

For the linear function $d(t) = 50t$, the slope is the coefficient of t. Therefore, the slope is 50. This means that a homing pigeon can fly 50 miles for each 1 hour of flight time.

CHECK YOUR PROGRESS 5, *page 344*

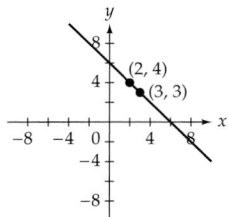

CHECK YOUR PROGRESS 6, *page 345*

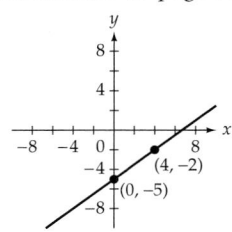

SECTION 6.3

CHECK YOUR PROGRESS 1, *page 351*

$f(a) = ma + b$

$f(a) = -3.5a + 100$

The linear function is $f(a) = -3.5a + 100$, where $f(a)$ is the boiling point of water at an altitude of a kilometers above sea level.

CHECK YOUR PROGRESS 2, *page 352*

$y - y_1 = m(x - x_1)$

$y - 2 = -\dfrac{1}{2}[x - (-2)]$

$y - 2 = -\dfrac{1}{2}x - 1$

$y = -\dfrac{1}{2}x + 1$

CHECK YOUR PROGRESS 3, *page 353*

$C - C_1 = m(t - t_1)$

$C - 191 = 3.8(t - 50)$

$C - 191 = 3.8t - 190$

$C = 3.8t + 1$

A linear function that models the number of calories burned is $C(t) = 3.8t + 1$.

CHECK YOUR PROGRESS 4, *page 354*

$m = \dfrac{y_2 - y_1}{x_2 - x_1} = \dfrac{1 - 3}{4 - (-2)} = \dfrac{-2}{6} = -\dfrac{1}{3}$

$y - y_1 = m(x - x_1)$

$y - 3 = -\dfrac{1}{3}[x - (-2)]$

$y - 3 = -\dfrac{1}{3}x - \dfrac{2}{3}$

$y = -\dfrac{1}{3}x + \dfrac{7}{3}$

CHECK YOUR PROGRESS 5, *page 355*

The regression equation is $y = 5.63x - 252.86$.

The estimated weight of a woman swimmer who is 63 inches tall is approximately 102 pounds.

SECTION 6.4

CHECK YOUR PROGRESS 1, *page 361*

$a = 1, b = 0; -\dfrac{b}{2a} = -\dfrac{0}{2(1)} = 0$

$y = x^2 - 2$

$y = (0)^2 - 2$

$y = -2$

The vertex is $(0, -2)$.

CHECK YOUR PROGRESS 2, *page 363*

a. $y = 2x^2 - 5x + 2$

$0 = 2x^2 - 5x + 2$

$0 = (2x - 1)(x - 2)$

$2x - 1 = 0 \qquad x - 2 = 0$

$x = \dfrac{1}{2} \qquad\qquad x = 2$

The x-intercepts are $\left(\dfrac{1}{2}, 0\right)$ and $(2, 0)$.

b. $y = x^2 + 4x + 4$

$0 = x^2 + 4x + 4$

$0 = (x + 2)(x + 2)$

$x + 2 = 0 \qquad x + 2 = 0$

$x = -2 \qquad\qquad x = -2$

The x-intercept is $(-2, 0)$.

CHECK YOUR PROGRESS 3, *page 364*

$a = 2, b = -3; -\dfrac{b}{2a} = -\dfrac{-3}{2(2)} = \dfrac{3}{4}$

$f(x) = 2x^2 - 3x + 1$

$f\left(\dfrac{3}{4}\right) = 2\left(\dfrac{3}{4}\right)^2 - 3\left(\dfrac{3}{4}\right) + 1$

$f\left(\dfrac{3}{4}\right) = -\dfrac{1}{8}$

The vertex is $\left(\dfrac{3}{4}, -\dfrac{1}{8}\right)$. The minimum value of the function is $-\dfrac{1}{8}$, the y-coordinate of the vertex.

CHECK YOUR PROGRESS 4, *page 365*

$a = -16, b = 64; -\dfrac{b}{2a} = -\dfrac{64}{2(-16)} = 2$

The ball reaches its maximum height in 2 seconds.

$s(t) = -16t^2 + 64t + 4$

$s(2) = -16(2)^2 + 64(2) + 4$

$s(2) = 68$

The maximum height of the ball is 68 feet.

CHECK YOUR PROGRESS 5, *page 366*

Perimeter: $w + l + w + l = 44$

$2w + 2l = 44$

$w + l = 22$

$l = -w + 22$

Area: $A = lw$

$= (-w + 22)w$

$A = -w^2 + 22w$

$w = -\dfrac{b}{2a} = -\dfrac{22}{2(-1)} = 11$

The width is 11 feet.

$l = -w + 22$

$l = -(11) + 22 = 11$

The length is 11 feet.

The dimensions of the rectangle with maximum area are 11 feet by 11 feet.

SECTION 6.5

CHECK YOUR PROGRESS 1, *page 374*

$g(x) = \left(\dfrac{1}{2}\right)^x$

$g(3) = \left(\dfrac{1}{2}\right)^3 = \dfrac{1}{8}$

$g(-1) = \left(\dfrac{1}{2}\right)^{-1} = \dfrac{1}{\frac{1}{2}} = 2$

$g(\sqrt{3}) = \left(\dfrac{1}{2}\right)^{\sqrt{3}} \approx \left(\dfrac{1}{2}\right)^{1.732} \approx 0.301$

CHECK YOUR PROGRESS 2, *page 376*

Because the base $\dfrac{3}{2}$ is greater than 1, f is an exponential growth function.

x	$f(x) = \left(\dfrac{3}{2}\right)^x$	(x, y)
-3	$\left(\dfrac{3}{2}\right)^{-3} = \dfrac{8}{27}$	$\left(-3, \dfrac{8}{27}\right)$
-2	$\left(\dfrac{3}{2}\right)^{-2} = \dfrac{4}{9}$	$\left(-2, \dfrac{4}{9}\right)$
-1	$\left(\dfrac{3}{2}\right)^{-1} = \dfrac{2}{3}$	$\left(-1, \dfrac{2}{3}\right)$
2	$\left(\dfrac{3}{2}\right)^{2} = \dfrac{9}{4}$	$\left(2, \dfrac{9}{4}\right)$
3	$\left(\dfrac{3}{2}\right)^{3} = \dfrac{27}{8}$	$\left(3, \dfrac{27}{8}\right)$

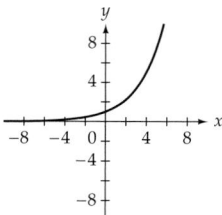

CHECK YOUR PROGRESS 3, *page 377*

x	-2	-1	0	1	2
$f(x) = e^{-x} + 2$	9.4	4.7	3	2.4	2.1

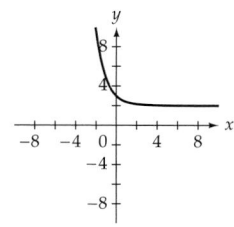

CHECK YOUR PROGRESS 4, *page 378*

$$N(t) = 1.5\left(\tfrac{1}{2}\right)^{t/193.7}$$

$$N(24) = 1.5\left(\tfrac{1}{2}\right)^{24/193.7}$$

$$\approx 1.5(0.9177) \approx 1.3766$$

After 24 hours, there are approximately 1.3766 grams of the isotope in the body.

CHECK YOUR PROGRESS 5, *page 379*

$$A(t) = 200e^{-0.014t}$$

$$A(45) = 200e^{-0.014(45)}$$

$$\approx 107$$

After 45 minutes, there are approximately 107 milligrams of aspirin in the patient's bloodstream.

CHECK YOUR PROGRESS 6, *pages 380–381*

The regression equation is $y = 10.1468(0.8910)^x$.

The atmospheric pressure at an altitude of 24 kilometers is approximately 0.6 newton per square centimeter.

SECTION 6.6

CHECK YOUR PROGRESS 1, *page 386*

a. $2^{10} = 4x$

b. $\log_{10} 2x = 3$

CHECK YOUR PROGRESS 2, *page 387*

a. $\log_{10} 0.001 = x$ **b.** $\log_5 125 = x$

$\quad 10^x = 0.001$ $5^x = 125$

$\quad 10^x = 10^{-3}$ $5^x = 5^3$

$\qquad x = -3$ $x = 3$

$\log_{10} 0.001 = -3$ $\log_5 125 = 3$

CHECK YOUR PROGRESS 3, *page 387*

$\log_2 x = 6$

$\quad 2^6 = x$

$\quad 64 = x$

CHECK YOUR PROGRESS 4, *page 388*

a. $\log x = -2.1$ **b.** $\ln x = 2$

$\quad 10^{-2.1} = x$ $e^2 = x$

$\quad 0.008 \approx x$ $7.389 \approx x$

CHECK YOUR PROGRESS 5, *page 389*

$y = \log_5 x$

$5^y = x$

$x = 5^y$	$\frac{1}{25}$	$\frac{1}{5}$	1	5	25
y	-2	-1	0	1	2

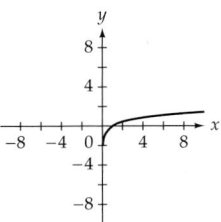

CHECK YOUR PROGRESS 6, *page 391*

a. $S(0) = 5 + 29\ln(0 + 1) = 5$

The average typing speed when the student first started to type was 5 words per minute.

$S(3) = 5 + 29\ln(3 + 1) \approx 45$

The average typing speed after 3 months was about 45 words per minute.

b. $S(3) - S(0) = 45 - 5 = 40$

The typing speed increased by 40 words per minute during the 3 months.

CHECK YOUR PROGRESS 7, *page 391*

$I = 2 \cdot (12{,}589{,}254 I_0) = 25{,}178{,}508 I_0$

$$M = \log\left(\frac{I}{I_0}\right) = \log\left(\frac{25{,}178{,}508 I_0}{I_0}\right) = \log(25{,}178{,}508) \approx 7.4$$

The Richter scale magnitude of an earthquake whose intensity is twice that of the Joshua Tree earthquake is 7.4.

CHECK YOUR PROGRESS 8, *page 392*

$$\log\left(\frac{I}{I_0}\right) = 5.2$$

$$\frac{I}{I_0} = 10^{5.2}$$

$$I = 10^{5.2} I_0$$

$$I \approx 158{,}489 I_0$$

The September 3, 2000 Napa Valley earthquake had an intensity that was approximately 158,000 times the intensity of a zero-level earthquake.

CHECK YOUR PROGRESS 9, *page 393*

a. $\text{pH} = -\log[\text{H}^+] = -\log(2.41 \times 10^{-13}) \approx 12.6$

The bleach has a pH of 12.6.

b. $\text{pH} = -\log[\text{H}^+] = -\log(5.07 \times 10^{-4}) \approx 3.3$

The cola soft drink has a pH of 3.3.

c. $pH = -\log[H^+] = -\log(6.31 \times 10^{-5}) \approx 4.2$
The rainwater has a pH of 4.2.

CHECK YOUR PROGRESS 10, *page 394*

$$pH = -\log[H^+]$$
$$10.0 = -\log[H^+]$$

$$-10.0 = \log[H^+]$$
$$10^{-10.0} = H^+$$
$$1.0 \times 10^{-10} = H^+$$

The hydronium-ion concentration of the water in the Great Salt Lake in Utah is 1.0×10^{-10} moles per liter.

CHAPTER 7

SECTION 7.1

CHECK YOUR PROGRESS 1, *page 406*

a. $6 \oplus 10 = 4$

b. $5 \oplus 9 = 2$

c. $7 \ominus 11 = 8$

d. $5 \ominus 10 = 7$

CHECK YOUR PROGRESS 2, *page 408*

The years 2004, 2008, and 2012 are leap years, so there are 3 years between the two dates with 366 days and 6 years with 365 days. The total number of days between the dates is $3 \cdot 366 + 6 \cdot 365 = 3288$. $3288 \div 7 = 469$ remainder 5, so $3288 \equiv 5 \bmod 7$. The day of the week 3288 days after Thursday, February 12, 2004 will be the same as the day 5 days later, a Tuesday.

CHECK YOUR PROGRESS 3, *page 409*

$51 + 72 = 123$, and $123 \div 3 = 41$ remainder 0, so $(51 + 72) \bmod 3 \equiv 0$.

CHECK YOUR PROGRESS 4, *page 409*

$21 - 43 = -22$, a negative number. Repeatedly add the modulus 7 to the difference until a whole number is reached.

$$-22 + 7 = -15$$
$$-15 + 7 = -8$$
$$-8 + 7 = -1$$
$$-1 + 7 = 6$$

$(21 - 43) \bmod 7 \equiv 6$.

CHECK YOUR PROGRESS 5, *page 410*

Tuesday corresponds to 2 (see the chart on page 406), so the day of the week 93 days from now is represented by $(2 + 93) \bmod 7$. Because $95 \div 7 = 13$ remainder 4, $(2 + 93) \bmod 7 \equiv 4$, which corresponds to Thursday.

CHECK YOUR PROGRESS 6, *page 410*

$33 \cdot 41 = 1353$ and $1353 \div 17 = 79$ remainder 10, so $(33 \cdot 41) \bmod 17 \equiv 10$.

CHECK YOUR PROGRESS 7, *page 412*

Substitute each whole number from 0 to 11 into the congruence.

$4(0) + 1 \not\equiv 5 \bmod 12$	Not a solution
$4(1) + 1 \equiv 5 \bmod 12$	1 is a solution.
$4(2) + 1 \not\equiv 5 \bmod 12$	Not a solution
$4(3) + 1 \not\equiv 5 \bmod 12$	Not a solution
$4(4) + 1 \equiv 5 \bmod 12$	4 is a solution.
$4(5) + 1 \not\equiv 5 \bmod 12$	Not a solution
$4(6) + 1 \not\equiv 5 \bmod 12$	Not a solution
$4(7) + 1 \equiv 5 \bmod 12$	7 is a solution.
$4(8) + 1 \not\equiv 5 \bmod 12$	Not a solution
$4(9) + 1 \not\equiv 5 \bmod 12$	Not a solution
$4(10) + 1 \equiv 5 \bmod 12$	10 is a solution.
$4(11) + 1 \not\equiv 5 \bmod 12$	Not a solution

The solutions from 0 to 11 are 1, 4, 7, and 10. The remaining solutions are obtained by repeatedly adding the modulus 12 to these solutions. So the solutions are 1, 4, 7, 10, 13, 16, 19, 22, ...

CHECK YOUR PROGRESS 8, *page 412*

In mod 12 arithmetic, $6 + 6 = 12$, so the additive inverse of 6 is 6.

CHECK YOUR PROGRESS 9, *page 413*

Solve the congruence equation $5x \equiv 1 \bmod 11$ by substituting whole number values of x less than the modulus.

$$5(1) \not\equiv 1 \bmod 11$$
$$5(2) \not\equiv 1 \bmod 11$$
$$5(3) \not\equiv 1 \bmod 11$$
$$5(4) \not\equiv 1 \bmod 11$$
$$5(5) \not\equiv 1 \bmod 11$$
$$5(6) \not\equiv 1 \bmod 11$$
$$5(7) \not\equiv 1 \bmod 11$$
$$5(8) \not\equiv 1 \bmod 11$$
$$5(9) \equiv 1 \bmod 11$$

In mod 11 arithmetic, the multiplicative inverse of 5 is 9.

CHECK YOUR PROGRESS 1, *page 417*
Check the ISBN congruence equation.

$$0(10) + 2(9) + 0(8) + 1(7) + 1(6) + 5(5) + 5(4) + 0(3) + 2(2) + 4 \equiv ? \bmod 11$$
$$84 \equiv 7 \bmod 11$$

Because $84 \not\equiv 0 \bmod 11$, the ISBN is invalid.

CHECK YOUR PROGRESS 2, *page 418*
Check the UPC congruence equation.

$$1(3) + 3(1) + 2(3) + 3(1) + 4(3) + 2(1) + 6(3) + 5(1) + 9(3) + 3(1) + 3(3) + 9 \equiv ? \bmod 10$$
$$100 \equiv 0 \bmod 10$$

Because $100 \equiv 0 \bmod 10$, the UPC is valid.

CHECK YOUR PROGRESS 3, *page 419*
Highlight every other digit, reading from right to left:

6 0 1 1 0 1 2 3 9 1 4 5 2 3 1 7

Double the highlighted digits:

12 0 2 1 0 1 4 3 18 1 8 5 4 3 2 7

Add all the digits, treating two-digit numbers as two single digits:

$$(1 + 2) + 0 + 2 + 1 + 0 + 1 + 4 + 3 + (1 + 8) + 1 + 8 + 5 + 4 + 3 + 2 + 7 = 53$$

Because $53 \not\equiv 0 \bmod 10$, this is not a valid credit card number.

CHECK YOUR PROGRESS 4, *page 422*
The encrypting congruence is $c \equiv (p + 17) \bmod 26$.

A	$c \equiv (1 + 17) \bmod 26 \equiv 18 \bmod 26 \equiv 18$	Code A as R.
L	$c \equiv (12 + 17) \bmod 26 \equiv 29 \bmod 26 \equiv 3$	Code L as C.
P	$c \equiv (16 + 17) \bmod 26 \equiv 33 \bmod 26 \equiv 7$	Code P as G.
I	$c \equiv (9 + 17) \bmod 26 \equiv 26 \bmod 26 \equiv 0$	Code I as Z.
N	$c \equiv (14 + 17) \bmod 26 \equiv 31 \bmod 26 \equiv 5$	Code N as E.
E	$c \equiv (5 + 17) \bmod 26 \equiv 22 \bmod 26 \equiv 22$	Code E as V.
S	$c \equiv (19 + 17) \bmod 26 \equiv 36 \bmod 26 \equiv 10$	Code S as J.
K	$c \equiv (11 + 17) \bmod 26 \equiv 28 \bmod 26 \equiv 2$	Code K as B.
G	$c \equiv (7 + 17) \bmod 26 \equiv 24 \bmod 26 \equiv 24$	Code G as X.

Thus the plaintext ALPINE SKIING is coded as RCGZEV JBZZEX.

To decode, because $m = 17$, $n = 26 - 17 = 9$, and the decoding congruence is $p \equiv (c + 9) \bmod 26$.

T	$c \equiv (20 + 9) \bmod 26 \equiv 29 \bmod 26 \equiv 3$	Decode T as C.
I	$c \equiv (9 + 9) \bmod 26 \equiv 18 \bmod 26 \equiv 18$	Decode I as R.
F	$c \equiv (6 + 9) \bmod 26 \equiv 15 \bmod 26 \equiv 15$	Decode F as O.
J	$c \equiv (10 + 9) \bmod 26 \equiv 19 \bmod 26 \equiv 19$	Decode J as S.

Continuing, the ciphertext TIFJJ TFLEKIP JBZZEX decodes as CROSS COUNTRY SKIING.

CHECK YOUR PROGRESS 5, *page 423*

The encrypting congruence is $c \equiv (3p + 1)$ mod 26.

C $c \equiv (3 \cdot 3 + 1)$ mod 26 $\equiv 10$ mod 26 $\equiv 10$ Code C as J.

O $c \equiv (3 \cdot 15 + 1)$ mod 26 $\equiv 46$ mod 26 $\equiv 20$ Code O as T.

L $c \equiv (3 \cdot 12 + 1)$ mod 26 $\equiv 37$ mod 26 $\equiv 11$ Code L as K.

R $c \equiv (3 \cdot 18 + 1)$ mod 26 $\equiv 55$ mod 26 $\equiv 3$ Code R as C.

Continuing, the plaintext COLOR MONITOR is coded as JTKTC NTQBITC.

CHECK YOUR PROGRESS 6, *page 424*

Solve the congruence equation $c \equiv (7p + 1)$ mod 26 for p.

$$c = 7p + 1$$
$$c - 1 = 7p \qquad \text{• Subtract 1 from each side of the equation.}$$
$$15(c - 1) = 15(7p) \qquad \text{• Multiply each side of the equation by the multiplicative inverse of 7.}$$

         Because $7 \cdot 15 \equiv 1$ mod 26, multiply each side by 15.

$$[15(c - 1)] \text{ mod } 26 \equiv p$$

The decoding congruence is $p \equiv [15(c - 1)]$ mod 26.

I $p \equiv [15(9 - 1)]$ mod 26 $\equiv 120$ mod 26 $\equiv 16$ Decode I as P.

G $p \equiv [15(7 - 1)]$ mod 26 $\equiv 90$ mod 26 $\equiv 12$ Decode G as L.

H $p \equiv [15(8 - 1)]$ mod 26 $\equiv 105$ mod 26 $\equiv 1$ Decode H as A.

T $p \equiv [15(20 - 1)]$ mod 26 $\equiv 285$ mod 26 $\equiv 25$ Decode T as Y.

Continuing, the ciphertext IGHT OHGG decodes as PLAY BALL.

SECTION 7.3

CHECK YOUR PROGRESS 1, *page 430*

Check to see if the four properties of a group are satisfied.

1. The product of two integers is always an integer, so the integers are closed with respect to multiplication.

2. The associative property of multiplication is true for integers.

3. The integers have an identity element for multiplication, namely 1.

4. Not every integer has a multiplicative inverse that is also an integer. For instance, $\frac{1}{2}$ is the multiplicative inverse of 2, but $\frac{1}{2}$ is not an integer. There is no integer that can be multiplied by 2 that gives the identity element 1.

Because property 4 is not satisfied, the integers with multiplication do not form a group.

CHECK YOUR PROGRESS 2, *page 433*

Rotate the original triangle, I, about the line of symmetry through the bottom right vertex, followed by a clockwise rotation of 240°.

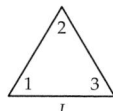 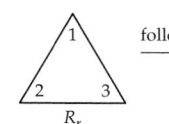 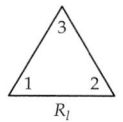

Therefore, $R_r \Delta R_{240} = R_l$.

CHECK YOUR PROGRESS 3, *page 434*

$$R_r \Delta R_{240} = \begin{pmatrix} 1 & 2 & 3 \\ 2 & 1 & 3 \end{pmatrix} \Delta \begin{pmatrix} 1 & 2 & 3 \\ 3 & 1 & 2 \end{pmatrix}$$

• $1 \rightarrow 2 \rightarrow 1$. Thus $1 \rightarrow 1$.
• $2 \rightarrow 1 \rightarrow 3$. Thus $2 \rightarrow 3$.
• $3 \rightarrow 3 \rightarrow 2$. Thus $3 \rightarrow 2$.

$$= \begin{pmatrix} 1 & 2 & 3 \\ 1 & 3 & 2 \end{pmatrix} = R_l$$

CHECK YOUR PROGRESS 4, *page 435*

$$E\Delta B = \begin{pmatrix} 1 & 2 & 3 \\ 2 & 1 & 3 \end{pmatrix} \Delta \begin{pmatrix} 1 & 2 & 3 \\ 3 & 1 & 2 \end{pmatrix}.$$ 1 is replaced by 2, which is then replaced by 1 in the second permutation. So 1 remains as 1. 2 is replaced by 1, which is then replaced by 3, so ultimately, 2 is replaced by 3. Finally, 3 remains as 3 in the first permutation but is replaced by 2 in the second, so ultimately 3 is replaced by 2.

The result is $\begin{pmatrix} 1 & 2 & 3 \\ 1 & 3 & 2 \end{pmatrix}$, which is C. Thus $E\Delta B = C$.

CHECK YOUR PROGRESS 5, *page 436*

$D = \begin{pmatrix} 1 & 2 & 3 \\ 3 & 2 & 1 \end{pmatrix}$ replaces 1 with 3, 2 with 2, and 3 with 1.

Reversing these, we need to replace 3 with 1, leave 2 alone, and replace 1 with 3. This is the element $\begin{pmatrix} 1 & 2 & 3 \\ 3 & 2 & 1 \end{pmatrix}$, which is D again.

So D is its own inverse.

CHAPTER 8

SECTION 8.1

CHECK YOUR PROGRESS 1, *page 449*

$$QR + RS + ST = QT$$
$$28 + 16 + 10 = QT$$
$$54 = QT$$

$QT = 54$ cm

CHECK YOUR PROGRESS 2, *page 450*

$$AB + BC = AC$$
$$\frac{1}{4}(BC) + BC = AC$$
$$\frac{1}{4}(16) + 16 = AC$$
$$4 + 16 = AC$$
$$20 = AC$$

$AC = 20$ ft

CHECK YOUR PROGRESS 3, *page 452*

$\angle G + \angle H = 127° + 53° = 180°$

The sum of the measures of $\angle G$ and $\angle H$ is 180°. Angles G and H are supplementary angles.

CHECK YOUR PROGRESS 4, *page 453*

Supplementary angles are two angles the sum of whose measures is 180°. To find the supplement, let x represent the supplement of a 129° angle.

$$x + 129 = 180$$
$$x = 51$$

The supplement of a 129° angle is a 51° angle.

CHECK YOUR PROGRESS 5, *page 454*

$$m\angle a + 68° = 118°$$
$$m\angle a = 50°$$

The measure of $\angle a$ is 50°.

CHECK YOUR PROGRESS 6, *page 455*

$$m\angle b + m\angle a = 180°$$
$$m\angle b + 35° = 180°$$
$$m\angle b = 145°$$

$$m\angle c = m\angle a = 35°$$

$$m\angle d = m\angle b = 145°$$

$m\angle b = 145°$, $m\angle c = 35°$, and $m\angle d = 145°$.

CHECK YOUR PROGRESS 7, *page 457*

$m\angle b = m\angle g = 124°$

$m\angle d = m\angle g = 124°$

$$m\angle c + m\angle b = 180°$$
$$m\angle c + 124° = 180°$$
$$m\angle c = 56°$$

$m\angle b = 124°$, $m\angle c = 56°$, and $m\angle d = 124°$.

CHECK YOUR PROGRESS 8, *page 459*

$$m\angle b + m\angle d = 180°$$
$$m\angle b + 105° = 180°$$
$$m\angle a + 110° = 180°$$
$$m\angle b = 75°$$

$$m\angle a + m\angle b + m\angle c = 180°$$
$$m\angle a + 75° + 35° = 180°$$
$$m\angle a + 110° = 180°$$
$$m\angle a = 70°$$

$m\angle e = m\angle a = 70°$

CHECK YOUR PROGRESS 9, *page 459*

Let x represent the measure of the third angle.

$$x + 90° + 27° = 180°$$
$$x + 117° = 180°$$
$$x = 63°$$

The measure of the third angle is 63°.

SECTION 8.2

CHECK YOUR PROGRESS 1, *page 470*

$$P = a + b + c$$
$$P = 4\frac{3}{10} + 2\frac{1}{10} + 6\frac{1}{2}$$
$$P = 4\frac{3}{10} + 2\frac{1}{10} + 6\frac{5}{10}$$
$$P = 12\frac{9}{10}$$

The total length of the bike trail is $12\frac{9}{10}$ mi.

CHECK YOUR PROGRESS 2, *page 471*

$$P = 2L + 2W$$
$$P = 2(12) + 2(8)$$
$$P = 24 + 16$$
$$P = 40$$

You will need 40 ft of molding to edge the top of the walls.

CHECK YOUR PROGRESS 3, *page 472*
$P = 4s$
$P = 4(24)$
$P = 96$

The homeowner should purchase 96 ft of fencing.

CHECK YOUR PROGRESS 4, *page 472*
$P = 2b + 2s$
$P = 2(5) + 2(7)$
$P = 10 + 14$
$P = 24$

24 m of plank are needed to surround the garden.

CHECK YOUR PROGRESS 5, *page 474*
$C = \pi d$
$C = 9\pi$

The circumference of the circle is 9π km.

CHECK YOUR PROGRESS 6, *page 474*
12 in. = 1 ft

$C = \pi d$
$C = \pi(1)$
$C = \pi$

$12C = 12\pi \approx 37.70$

The tricycle travels approximately 37.70 ft when the wheel makes 12 revolutions.

CHECK YOUR PROGRESS 7, *page 476*
$A = LW$
$A = 308(192)$
$A = 59{,}136$

59,136 cm² of fabric are needed.

CHECK YOUR PROGRESS 8, *page 477*
$A = s^2$
$A = 24^2$
$A = 576$

The area of the floor is 576 ft².

CHECK YOUR PROGRESS 9, *page 478*
$A = bh$
$A = 14(8)$
$A = 112$

The area of the patio is 112 m².

CHECK YOUR PROGRESS 10, *page 479*
$A = \dfrac{1}{2}bh$

$A = \dfrac{1}{2}(18)(9)$

$A = 9(9)$
$A = 81$

81 in² of felt are needed.

CHECK YOUR PROGRESS 11, *page 480*
$A = \dfrac{1}{2}h(b_1 + b_2)$

$A = \dfrac{1}{2} \cdot 9(12 + 20)$

$A = \dfrac{1}{2} \cdot 9(32)$

$A = \dfrac{9}{2} \cdot (32)$

$A = 144$

The area of the patio is 144 ft².

CHECK YOUR PROGRESS 12, *page 481*
$r = \dfrac{1}{2}d = \dfrac{1}{2}(12) = 6$

$A = \pi r^2$
$A = \pi(6)^2$
$A = 36\pi$

The area of the circle is 36π km².

CHECK YOUR PROGRESS 13, *page 482*
$r = \dfrac{1}{2}d = \dfrac{1}{2}(4) = 2$

$A = \pi r^2$
$A = \pi(2)^2$
$A = \pi(4)$
$A \approx 12.57$

Approximately 12.57 ft² of material should be purchased.

SECTION 8.3

CHECK YOUR PROGRESS 1, *page 492*
$$\dfrac{AC}{DF} = \dfrac{CH}{FG}$$
$$\dfrac{10}{15} = \dfrac{7}{FG}$$
$10(FG) = 7(15)$
$10(FG) = 105$
$\quad FG = 10.5$

The height *FG* of triangle *DEF* is 10.5 m.

CHECK YOUR PROGRESS 2, *page 495*

$$\frac{AO}{DO} = \frac{AB}{CD}$$

$$\frac{AO}{3} = \frac{10}{4}$$

$4(AO) = 10(3)$
$4(AO) = 30$
$AO = 7.5$

$$A = \frac{1}{2}bh$$

$$A = \frac{1}{2}(10)(7.5)$$

$A = 5(7.5)$
$A = 37.5$

The area of triangle *AOB* is 37.5 cm².

SECTION 8.4

CHECK YOUR PROGRESS 1, *page 504*
$V = LWH$
$V = 5(3.2)(4)$
$V = 64$

The volume of the solid is 64 m³.

CHECK YOUR PROGRESS 2, *page 504*

$$V = \frac{1}{3}s^2h$$

$$V = \frac{1}{3}(15)^2(25)$$

$$V = \frac{1}{3}(225)(25)$$

$V = 1875$

The volume of the pyramid is 1875 m³.

CHECK YOUR PROGRESS 3, *page 505*

$$r = \frac{1}{2}d = \frac{1}{2}(16) = 8$$

$V = \pi r^2 h$
$V = \pi(8)^2(30)$
$V = \pi(64)(30)$
$V = 1920\pi$

$$\frac{1}{4}(1920\pi) = 480\pi$$

$$\approx 1507.96$$

There are approximately 1507.96 ft³ not being used for storage.

CHECK YOUR PROGRESS 4, *page 508*

$$r = \frac{1}{2}d = \frac{1}{2}(6) = 3$$

$S = 2\pi r^2 + 2\pi rh$
$S = 2\pi(3)^2 + 2\pi(3)(8)$
$S = 2\pi(9) + 2\pi(3)(8)$
$S = 18\pi + 48\pi$
$S = 66\pi$
$S \approx 207.35$

The surface area of the cylinder is approximately 207.35 ft².

CHECK YOUR PROGRESS 5, *page 508*
Surface area of the cube $= 6s^2$
$ = 6(8)^2$
$ = 6(64)$
$ = 384$ cm²

Surface area of the sphere $= 4\pi r^2$
$ = 4\pi(5)^2$
$ = 4\pi(25)$
$ \approx 314.16$ cm²

The cube has a larger surface area.

SECTION 8.5

CHECK YOUR PROGRESS 1, *page 519*

$$S = (m\angle A + m\angle B + m\angle C - 180°) \cdot \left(\frac{\pi}{180°}\right)r^2$$

$$= (200° + 90° + 90° - 180°) \cdot \left(\frac{\pi}{180°}\right)(6)^2$$

$$= (200°) \cdot \left(\frac{\pi}{180°}\right) \cdot (36)$$

$$= 40\pi \quad \bullet \text{ Exact area}$$

$$\approx 125.66 \quad \bullet \text{ Approximate area}$$

SECTION 8.6

CHECK YOUR PROGRESS 1, *page 529*
Replace each line segment with a scaled version of the generator. As you move from left to right, your first zig should be to the left.

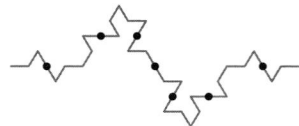

Stage 2 of the zig-zag curve

CHECK YOUR PROGRESS 2, *page 530*
Replace each square with a scaled version of the generator.

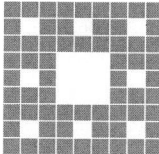

Stage 2 of the
Sierpinski carpet

CHECK YOUR PROGRESS 3, *page 532*

a. Any portion of the box curve replicates the entire fractal, so the box curve is a strictly self-similar fractal.

b. Any portion of the Sierpinski gasket replicates the entire fractal, so the Sierpinski gasket is a strictly self-similar fractal.

CHECK YOUR PROGRESS 4, *page 533*

a. The generator of the Koch curve consists of four line segments and the initiator consists of only one line segment. Thus the

replacement ratio of the Koch curve is 4:1, or 4. The initiator of the Koch curve is a line segment that is 3 times longer than the replica line segments in the generator. Thus the scaling ratio of the Koch curve is 3:1, or 3.

b. The generator of the zig-zag curve consists of six line segments and the initiator consists of only one line segment. Thus the replacement ratio of the zig-zag curve is 6:1, or 6. The initiator of the zig-zag curve is a line segment that is 4 times longer than the replica line segments in the generator. Thus the scaling ratio of the zig-zag curve is 4:1, or 4.

CHECK YOUR PROGRESS 5, *page 533*

a. In Example 4 we determined that the replacement ratio of the box curve is 5 and the scaling ratio of the box curve is 3. Thus the similarity dimension of the box curve is $D = \frac{\log 5}{\log 3} \approx 1.465$.

b. The replacement ratio of the Sierpinski carpet is 8 and the scaling ratio of the Sierpinski carpet is 3. Thus the similarity dimension of the Sierpinski carpet is $D = \frac{\log 8}{\log 3} \approx 1.893$.

CHAPTER 9

SECTION 9.1

CHECK YOUR PROGRESS 1, *page 550*
Because the second graph has edge AB and the first graph does not, the two graphs are not equivalent.

CHECK YOUR PROGRESS 2, *page 553*
One vertex in the graph is of degree 3, and another is of degree 5. Because not all vertices are of even degree, the graph is not Eulerian.

CHECK YOUR PROGRESS 3, *page 554*
Represent the land areas and bridges with a graph, as we did for the Königsberg bridges earlier in the section. The vertices of the resulting graph, shown at the right below, all have even degree. Thus we know that the graph has an Euler circuit. An Euler circuit corresponds to a stroll that crosses each bridge and returns to the starting point without crossing any bridge twice.

CHECK YOUR PROGRESS 4, *page 556*
Consider the campground map as a graph. A route through all the trails that does not repeat any trails corresponds to an Euler walk. Because only two vertices (A and F) are of odd degree, we know that an Euler walk exists. Furthermore, the walk must begin at A and end at F or begin at F and end at A. By trial and error, one Euler walk is A–B–C–D–E–B–G–F–E–C–A–F.

CHECK YOUR PROGRESS 5, *page 557*
The graph has seven vertices, so $n = 7$ and $n/2 = 3.5$. Several vertices are of degree less than $n/2$, so Dirac's Theorem does not apply. Still, a routing for the document may be possible. By trial and error, one such route is Los Angeles–New York–Boston–Atlanta–Dallas–Phoenix–San Francisco–Los Angeles.

CHECK YOUR PROGRESS 6, *page 559*
Represent the floor plan with a graph, as in Example 6.

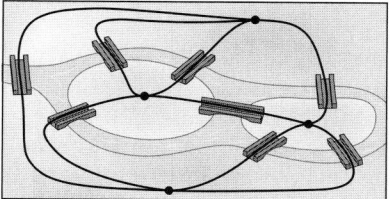 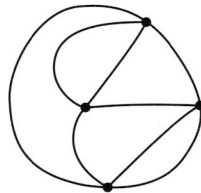

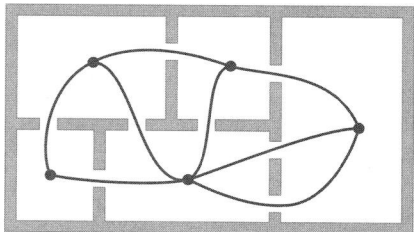

A stroll passing through each doorway just once corresponds to an Euler circuit or walk. Because four vertices are of odd degree, no Euler circuit or walk exists, so it is not possible to take such a stroll.

SECTION 9.2

CHECK YOUR PROGRESS 1, *page 568*

Draw a graph in which the vertices represent locations and the edges indicate available bus routes between locations. Each edge should be given a weight corresponding to the number of minutes for the bus ride.

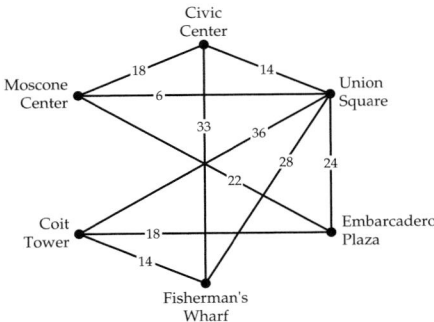

A route that visits each location and returns to the Moscone Center corresponds to a Hamiltonian circuit. Using the graph, one such route is Moscone Center–Civic Center–Union Square–Fisherman's Wharf–Coit Tower–Embarcadero Plaza–Moscone Center, with a total weight of
18 + 14 + 28 + 14 + 18 + 22 = 114. Another route is Moscone Center–Union Square–Embarcadero Plaza–Coit Tower–Fisherman's Wharf–Civic Center–Moscone Center, with a total weight of 6 + 24 + 18 + 14 + 33 + 18 = 113.

CHECK YOUR PROGRESS 2, *page 570*

Starting at vertex A, the edge of smallest weight is the edge to D, with weight 5. From D, take the edge of weight 4 to C, then the edge of weight 3 to B. From B, the edge of least weight to a vertex not yet visited is the edge to vertex E (with weight 5). This is the last vertex, so we return to A along the edge of weight 9. Thus the Hamiltonian circuit is A–D–C–B–E–A, with a total weight of 26.

CHECK YOUR PROGRESS 3, *page 572*

The smallest weight appearing in the graph is 3, so we mark edge BC. The next smallest weight is 4, on edge CD. Three edges have weight 5, but we cannot mark edge BD, as it would complete a circuit. We can, however, mark edge AD. The next valid edge of smallest weight is BE, also of weight 5. No more edges can be marked without completing a circuit or adding a third edge to a vertex, so we mark the final edge, AE, to complete the Hamiltonian Circuit. In this case, the Edge-Picking Algorithm generated the same circuit as the Greedy Algorithm did in Check Your Progress 2.

CHECK YOUR PROGRESS 4, *page 576*

Represent the time between locations with a weighted graph.

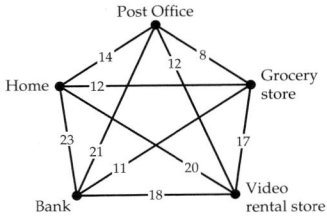

Starting at the home vertex and using the Greedy Algorithm, we first use the edge to the grocery store (of weight 12) followed by the edge of weight 8 to the post office and then the edge of weight 12 to the video rental store. The edge of next smallest weight is to the grocery store, but that vertex has already been visited, so we take the edge to the bank, with weight 18. All vertices have now been visited, so we select the last edge, of weight 23, to return home. The total weight is 73, corresponding to a total driving time of 73 minutes.

For the Edge-Picking Algorithm, we first select the edge of weight 8, followed by the edge of weight 11. Two edges have weight 12, but one adds a third edge to the grocery store vertex, so we must choose the edge from the post office to the video rental store. The next smallest weight is 14, but that edge would add a third edge to a vertex, as would the edge of weight 17. The edge of weight 18 would complete a circuit too early, so the next edge we can select is that of weight 20, the edge from home to the video rental store. The final step is to select the edge from home to the bank to complete the circuit. The resulting route is home–video rental store–post office–grocery store–bank–home (we could travel the same route in the reverse order) with a total travel time of 74 minutes.

CHECK YOUR PROGRESS 5, *page 577*

Represent the computer network by a graph in which the weights of the edges indicate the distances between computers.

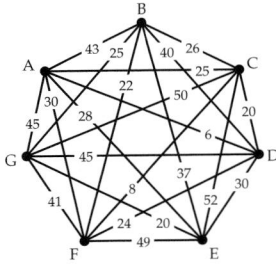

The edges with the smallest weights, which can all be chosen, are those of weights 6, 8, 20, 20, and 22. The edge of next smallest weight, 24, cannot be selected. There are two edges of weight 25; edge AC would add a third edge to vertex C, but edge BG can be chosen. All that remains is to complete the circuit with edge AE. The computers should be networked in this order: A, D, C, F, B, G, E, and back to A.

CHECK YOUR PROGRESS 1, *page 585*
First redraw the highlighted edge as shown below.

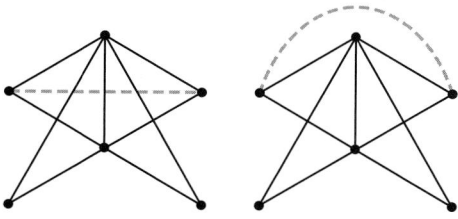

Now redraw the two lower vertices and the edges that meet there, as shown below.

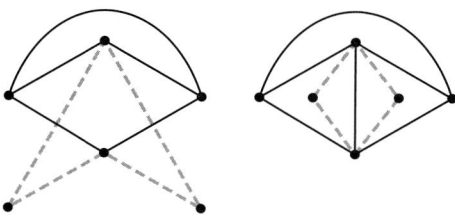

The result is a graph with no intersecting edges.

CHECK YOUR PROGRESS 2, *page 587*
The highlighted edges in the graph, considered as a subgraph, form the graph K_5. (It is upside-down and slightly distorted compared with the version shown in Figure 9.17.)

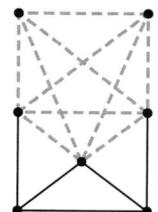

CHECK YOUR PROGRESS 3, *page 590*
The graph looks similar to the Utilities Graph. Contract edges as shown below, and combine the resulting multiple edges.

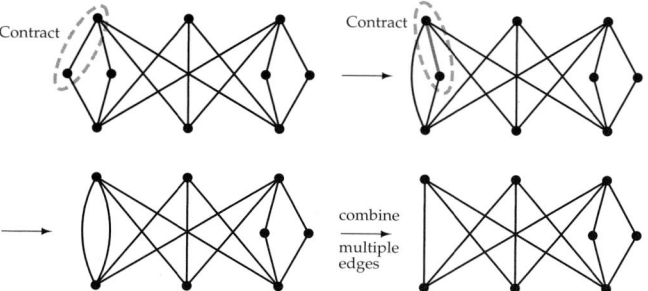

If we do the same on the right side of the graph, we are left with the Utilities Graph.

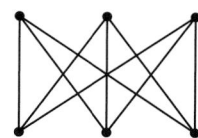

Therefore, the graph is not planar.

CHECK YOUR PROGRESS 4, *page 591*
There are 11 edges in the graph, seven vertices, and six faces (including the infinite face). Then $v + f = 7 + 6 = 13$ and $e + 2 = 11 + 2 = 13$, so $v + f = e + 2$.

CHECK YOUR PROGRESS 1, *page 597*
One possible coloring is given below.

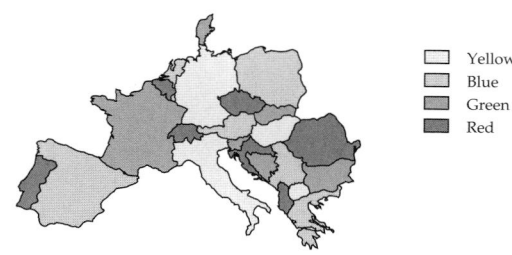

□ Yellow
□ Blue
□ Green
■ Red

CHECK YOUR PROGRESS 2, *page 599*
Draw a graph on the map as in Example 2. More than two colors are required to color the resulting graph, but, by experimenting, the graph can be colored with 3 colors. Thus the graph is 3-colorable.

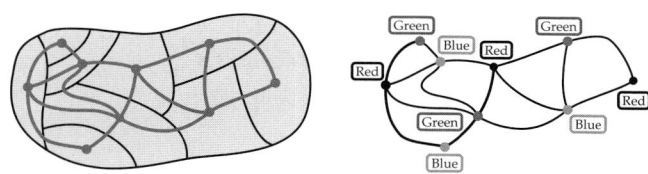

CHECK YOUR PROGRESS 3, *page 601*
There are several locations in the graph at which three edges form a triangle. Because a triangle is a circuit with an odd number of vertices, the graph is not 2-colorable.

CHECK YOUR PROGRESS 4, *page 602*
Draw a graph in which each vertex corresponds to a film and an edge joins two vertices if one person needs to view both of the corresponding films. We can use colors to represent the different times at which the films can be viewed. No two vertices connected

by an edge can share the same color, because that would mean one person would have to watch two films at the same time.

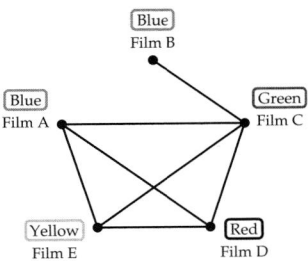

It is not possible to color the vertices with only three colors; one possible 4-coloring is shown. This means that four different time slots will be required to view the films, and the soonest all the friends can finish watching them is 4:00 A.M. A schedule can be set using the coloring in the graph. From 8 to 10, the films labeled blue, film A and film B, can be shown in two different rooms. The remaining films are represented by unique colors so will require their own viewing times. Film C can be shown from 10 to 12, film D from 12 to 2, and film E from 2 to 4.

CHECK YOUR PROGRESS 5, *page 604*

Draw a graph in which each vertex represents a deli, and an edge connects two vertices if the corresponding delis deliver to a common building. Try to color the vertices using the least number of colors possible; each color can correspond to a day of the week that the delis can deliver.

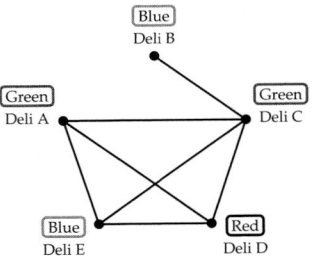

As shown, a 3-coloring is possible (but a 2-coloring is not). Therefore three different delivery days will be necessary—Delis A and C deliver on one day, B and E on another day, and D on a third day.

CHAPTER 10

SECTION 10.1

CHECK YOUR PROGRESS 1, *page 622*

$P = 500$, $r = 4\% = 0.04$, $t = 1$

$I = Prt$
$I = 500(0.04)(1)$
$I = 20$

The simple interest earned is $20.

CHECK YOUR PROGRESS 2, *page 623*

$P = 1500$, $r = 5.25\% = 0.0525$

$$t = \frac{4 \text{ months}}{1 \text{ year}} = \frac{4 \text{ months}}{12 \text{ months}} = \frac{4}{12}$$

$I = Prt$

$$I = 1500(0.0525)\left(\frac{4}{12}\right)$$

$I = 26.25$

The simple interest due is $26.25.

CHECK YOUR PROGRESS 3, *page 623*

$P = 700$, $r = 1.25\% = 0.0125$, $t = 5$

$I = Prt$
$I = 700(0.0125)(5)$
$I = 43.75$

The simple interest due is $43.75.

CHECK YOUR PROGRESS 4, *page 624*

$P = 7000$, $r = 5.25\% = 0.0525$

$$t = \frac{\text{number of days}}{360} = \frac{120}{360}$$

$I = Prt$

$$I = 7000(0.0525)\left(\frac{120}{360}\right)$$

$I = 122.5$

The simple interest due is $122.50.

CHECK YOUR PROGRESS 5, *page 625*

$I = Prt$

$462 = 12{,}000(r)\left(\dfrac{6}{12}\right)$

$462 = 6000r$

$0.077 = r$

$r = 7.7\%$

The simple interest rate on the loan is 7.7%.

CHECK YOUR PROGRESS 6, *page 626*

$P = 4000, r = 8.75\% = 0.0875, t = \dfrac{9}{12}$

$I = Prt$

$I = 4000(0.0875)\left(\dfrac{9}{12}\right)$

$I = 262.50$

$A = P + I$

$A = 4000 + 262.50$

$A = 4262.50$

The maturity value of the loan is $4262.50.

CHECK YOUR PROGRESS 7, *page 627*

$P = 6700, r = 8.9\% = 0.089, t = 1$

$A = P(1 + rt)$

$A = 6700[1 + 0.089(1)]$

$A = 6700(1 + 0.089)$

$A = 6700(1.089)$

$A = 7296.30$

The maturity value of the loan is $7296.30.

CHECK YOUR PROGRESS 8, *page 627*

$P = 680, r = 6.4\% = 0.064, t = 1$

$A = P(1 + rt)$

$A = 680[1 + 0.064(1)]$

$A = 680(1 + 0.064)$

$A = 680(1.064)$

$A = 723.52$

After 1 year, $723.52 is in the account.

CHECK YOUR PROGRESS 9, *page 628*

$I = A - P$

$I = 9240 - 9000$

$I = 240$

$I = Prt$

$240 = 9000(r)\left(\dfrac{4}{12}\right)$

$240 = 3000r$

$0.08 = r$

$r = 8.0\%$

The simple interest rate on the loan is 8.0%.

CHECK YOUR PROGRESS 10, *page 629*

$I = Prt$

$0.82 = 51.25(r)(1)$

$0.82 = 51.25(r)$

$0.016 = r$

The dividend yield is 1.6%.

CHECK YOUR PROGRESS 11, *page 630*

$P = 15{,}000, r = 3.5\% = 0.035, t = 1$

$I = Prt$

$I = 15{,}000(0.035)(1)$

$I = 525$

$525(4) = 2100$

The total of the interest payments paid to the bondholder is $2100.

SECTION 10.2

CHECK YOUR PROGRESS 1, *page 638*

$A = P(1 + rt)$

$A = 2000\left[1 + 0.04\left(\dfrac{1}{12}\right)\right]$

$A \approx 2006.67$

$A = P(1 + rt)$

$A \approx 2006.67\left[1 + 0.04\left(\dfrac{1}{12}\right)\right]$

$A \approx 2013.36$

$A = P(1 + rt)$

$A = 2013.36\left[1 + 0.04\left(\dfrac{1}{12}\right)\right]$

$A \approx 2020.07$

$A = P(1 + rt)$

$A = 2020.07\left[1 + 0.04\left(\dfrac{1}{12}\right)\right]$

$A \approx 2026.80$

$$A = P(1 + rt)$$
$$A = 2026.80\left[1 + 0.04\left(\frac{1}{12}\right)\right]$$
$$A \approx 2033.56$$

$$A = P(1 + rt)$$
$$A = 2033.56\left[1 + 0.04\left(\frac{1}{12}\right)\right]$$
$$A \approx 2040.34$$

The total amount in the account at the end of 6 months is $2040.34.

CHECK YOUR PROGRESS 2, *page 641*

$$r = 6\% = 0.06, \ n = 12, \ t = 2, \ i = \frac{r}{n} = \frac{0.06}{12} = 0.005,$$

$$N = nt = 12(2) = 24$$

$$A = P(1 + i)^N$$
$$A = 4000(1 + 0.005)^{24}$$
$$A = 4000(1.005)^{24}$$
$$A \approx 4000(1.127160)$$
$$A \approx 4508.64$$

The compound amount after 2 years is approximately $4508.64.

CHECK YOUR PROGRESS 3, *page 641*

$$r = 9\% = 0.09, \ n = 360, \ t = 4, \ i = \frac{r}{n} = \frac{0.09}{360} = 0.00025,$$

$$N = nt = 360(4) = 1440$$

$$A = P(1 + i)^N$$
$$A = 2500(1 + 0.00025)^{1440}$$
$$A = 2500(1.00025)^{1440}$$
$$A \approx 2500(1.4332649)$$
$$A \approx 3583.16$$

The future amount after 4 years is approximately $3583.16.

CHECK YOUR PROGRESS 4, *page 642*

$$r = 9\% = 0.09, \ n = 12, \ t = 6, \ i = \frac{r}{n} = \frac{0.09}{12} = 0.0075,$$

$$N = nt = 12(6) = 72$$

$$A = P(1 + i)^N$$
$$A = 8000(1 + 0.0075)^{72}$$
$$A = 8000(1.0075)^{72}$$
$$A \approx 8000(1.7125527)$$
$$A \approx 13{,}700.42$$

$$I = A - P$$
$$I = 13{,}700.42 - 8000$$
$$I = 5700.42$$

The amount of interest earned is approximately $5700.42.

CHECK YOUR PROGRESS 5, *page 642*

Press 2nd [Finance] to display the FINANCE CALC menu.

Press ENTER to select 1: TVM Solver.

After N =, enter 10.

After I% =, enter 6.

After PV =, enter −3500.

After PMT =, enter 0.

After P/Y =, enter 2.

After C/Y =, enter 2.

Use the up arrow key to place the cursor at FV =.

Press ALPHA [Solve].

The solution is displayed to the right of FV =.

The compound amount is $4703.71.

CHECK YOUR PROGRESS 6, *page 644*

$$r = 9\%, \ n = 2, \ t = 5, \ i = \frac{r}{n} = \frac{9\%}{2} = 4.5\% = 0.045,$$

$$N = nt = 2(5) = 10$$

$$P = \frac{A}{(1 + i)^N}$$

$$P = \frac{20{,}000}{(1 + 0.045)^{10}}$$

$$P \approx \frac{20{,}000}{1.552970}$$

$$P \approx 12{,}878.55$$

$12,878.55 should be invested in the account.

CHECK YOUR PROGRESS 7, *page 645*

Press 2nd [Finance] to display the FINANCE CALC menu.

Press ENTER to select 1: TVM Solver.

After N =, enter 5400 (15 × 360).

After I% =, enter 6.

After PMT =, enter 0.

After FV =, enter 25000.

After P/Y =, enter 360.

After C/Y =, enter 360.

Use the up arrow key to place the cursor at PV =.

Press ALPHA [Solve].

The solution is displayed to the right of PV =.

$10,165.00 should be invested in the account.

CHECK YOUR PROGRESS 8, *page 646*

$$r = 5\% = 0.05, \ n = 1, \ t = 17, \ i = \frac{r}{n} = \frac{0.05}{1} = 0.05,$$

$$N = nt = 1(17) = 17$$

$A = P(1 + i)^N$

$A = 28{,}000(1 + 0.05)^{17}$

$A = 28{,}000(1.05)^{17}$

$A \approx 28{,}000(2.2920183)$

$A \approx 64{,}176.51$

The average new car sticker price in 2020 will be approximately $64,176.51.

CHECK YOUR PROGRESS 9, *page 647*

$r = 7\%, \ n = 1, \ t = 40, \ i = \dfrac{r}{n} = \dfrac{7\%}{1} = 7\% = 0.07,$

$N = nt = 1(40) = 40$

$P = \dfrac{A}{(1 + i)^N}$

$P = \dfrac{500{,}000}{(1 + 0.07)^{40}}$

$P \approx \dfrac{500{,}000}{14.9744578}$

$P \approx 33{,}390.19$

In 2040, the purchasing power of $500,000 will be approximately $33,390.19.

CHECK YOUR PROGRESS 10, *page 648*

$r = 4\% = 0.04, \ n = 4, \ t = 1, \ i = \dfrac{r}{n} = \dfrac{0.04}{4} = 0.01,$

$N = nt = 4(1) = 4$

$A = P(1 + i)^N$

$A = 100(1 + 0.01)^4$

$A = 100(1.01)^4$

$A = 100(1.040604)$

$A \approx 104.06$

$I = A - P$

$I = 104.06 - 100$

$I = 4.06$

The effective interest rate is 4.06%.

CHECK YOUR PROGRESS 11, *page 649*

$i = \dfrac{r}{n} = \dfrac{0.05}{4}$

$N = nt = 4(1) = 4$

$(1 + i)^N = \left(1 + \dfrac{0.05}{4}\right)^4$

≈ 1.050945

$i = \dfrac{r}{n} = \dfrac{0.0525}{2}$

$N = nt = 2(1) = 2$

$(1 + i)^N = \left(1 + \dfrac{0.0525}{2}\right)^2$

≈ 1.053189

5.25% compounded semiannually has a higher annual yield than 5% compounded quarterly.

SECTION 10.3

CHECK YOUR PROGRESS 1, *page 658*

Date	Payments or Purchases	Balance Each Day	Number of Days Until Balance Changes	Unpaid Balance Times Number of Days
July 1–6		$1024	6	$6144
July 7–14	$315	$1339	8	$10,712
July 15–21	−$400	$939	7	$6573
July 22–31	$410	$1349	10	$13,490
Total				$36,919

Average daily balance $= \dfrac{\text{sum of the total amounts owed each day of the month}}{\text{number of days in the billing period}}$

$= \dfrac{36{,}919}{31} \approx \$1190.94$

$I = Prt$

$I = 1190.94(0.012)(1)$

$I \approx 14.29$

The finance charge on the August 1 bill is $14.29.

CHECK YOUR PROGRESS 2, *page 661*

a. Down payment = Percent down × purchase price
$$= 0.20 \times 750 = 150$$

Amount financed = purchase price − down payment
$$= 750 - 150 = 600$$

Interest owed = finance rate × amount financed
$$= 0.08 \times 600 = 48$$

The finance charge is $48.

b. APR $\approx \dfrac{2nr}{n+1}$

$\approx \dfrac{2(12)(0.08)}{12+1} \approx \dfrac{1.92}{13} \approx 0.148$

The annual percentage rate is approximately 14.8%.

CHECK YOUR PROGRESS 3, *page 663*

Sales tax amount = sales tax rate × purchase price
$$= 0.0425 \times 1499 \approx 63.71$$

Amount financed = purchase price + sales tax amount
$$= 1499 + 63.71 = 1562.71$$

$i = \dfrac{\text{annual interest rate}}{\text{number of payments per year}} = \dfrac{0.084}{12} = 0.007$

$n = 3(12) = 36$

$PMT = A\left(\dfrac{i}{1-(1+i)^{-n}}\right)$

$PMT = 1562.71\left(\dfrac{0.007}{1-(1+0.007)^{-36}}\right)$

$PMT \approx 49.26$

The monthly payment is $49.26.

CHECK YOUR PROGRESS 4, *page 664*

a. Sales tax = 0.0525(26,788) = 1406.37

b. Loan amount
= purchase price + sales tax + license fee − down payment
$$= 26{,}788 + 1406.37 + 145 - 2500$$
$$= 25{,}839.37$$

c. $i = \dfrac{APR}{12} = \dfrac{0.081}{12} = 0.00675$

$PMT = A\left(\dfrac{i}{1-(1+i)^{-n}}\right)$

$PMT = 25{,}839.37\left(\dfrac{0.00675}{1-(1+0.00675)^{-60}}\right)$

$PMT \approx 525.17$

The monthly payment is $525.17.

CHECK YOUR PROGRESS 5, *page 665*

$i = \dfrac{APR}{12} = \dfrac{0.084}{12} = 0.007$

$A = PMT\left(\dfrac{1-(1+i)^{-n}}{i}\right)$

$A = 592.57\left(\dfrac{1-(1+0.007)^{-24}}{0.007}\right)$

$A \approx 13{,}049.34$

The loan payoff is $13,049.34.

CHECK YOUR PROGRESS 6, *page 667*

Residual value = 0.40(33,395) = 13,358

Money factor = $\dfrac{\text{annual interest rate as a percent}}{2400} = \dfrac{8}{2400}$

≈ 0.00333

Average monthly finance charge
= (net capitalized cost + residual value) × money factor
$$= (31{,}900 + 13{,}358) \times 0.00333$$
$$\approx 150.71$$

Average monthly depreciation
$= \dfrac{\text{net capitalized cost − residual value}}{\text{term of the lease in months}}$

$= \dfrac{31{,}900 - 13{,}358}{60}$

≈ 309.03

Monthly lease payment
= average monthly finance charge + average monthly depreciation
$$= 150.71 + 309.03$$
$$= 459.74$$

The monthly lease payment is $459.74.

SECTION 10.4

CHECK YOUR PROGRESS 1, *page 675*

Down payment = 25% of 110,000 = 0.25(110,000)
$$= 27{,}500$$

Mortgage = selling price − down payment
$$= 110{,}000 - 27{,}500$$
$$= 82{,}500$$

Points = 1.75% of 82,500 = 0.0175(82,500)
$$= 1443.75$$

Total = 27,500 + 375 + 1443.75 = 29,318.75

The total of the down payment and the closing costs is $29,318.75.

CHECK YOUR PROGRESS 2, *page 676*

a. $i = \dfrac{0.07}{12} \approx 0.005833$

$n = 25(12) = 300$

$PMT = A\left(\dfrac{i}{1 - (1 + i)^{-n}}\right)$

$PMT \approx 123{,}000\left(\dfrac{0.005833}{1 - (1 + 0.005833)^{-300}}\right)$

$PMT \approx 869.31$

The monthly payment is $869.31.

b. Total $= 869.31(300) = 260{,}793$
The total of the payments over the life of the loan is $260,793.

c. Interest $= 260{,}793 - 123{,}000 = 137{,}793$
The amount of interest paid over the life of the loan is $137,793.

CHECK YOUR PROGRESS 3, *page 679*

Down payment $= 0.25(95{,}000) = 23{,}750$
Mortgage $= 95{,}000 - 23{,}750 = 71{,}250$

$i = \dfrac{0.0675}{12} = 0.005625$

$n = 30(12) = 360$

$PMT = A\left(\dfrac{i}{1 - (1 + i)^{-n}}\right)$

$PMT = 71{,}250\left(\dfrac{0.005625}{1 - (1 + 0.005625)^{-360}}\right)$

$PMT \approx 462.13$
The monthly payment is $462.13.

$I = Prt$

$= 71{,}250(0.0675)\left(\dfrac{1}{12}\right)$

$= 400.78$

The interest paid on the first payment is $400.78.

Principal $= 462.13 - 400.78 = 61.35$
The principal paid on the first payment is $61.35.

CHECK YOUR PROGRESS 4, *page 680*

$i = \dfrac{0.069}{12} = 0.00575$

$n = 25(12) - 4(12) = 300 - 48 = 252$

$A = PMT\left(\dfrac{1 - (1 + i)^{-n}}{i}\right)$

$A = 423.41\left(\dfrac{1 - (1 + 0.00575)^{-252}}{0.00575}\right)$

$A \approx 56{,}274.40$

The mortgage payoff is $56,274.40.

CHECK YOUR PROGRESS 5, *page 682*

Monthly property tax $= 777.60 \div 12 = 64.80$
Monthly fire insurance $= 450 \div 12 = 37.50$
Total monthly payment $= 497.63 + 64.80 + 37.50 = 599.93$

The total monthly payment for mortgage, property tax, and fire insurance is $599.93.

CHAPTER 11

SECTION 11.1

CHECK YOUR PROGRESS 1, *page 694*

The possible outcomes are {M, i, s, p}. There are four possible outcomes.

CHECK YOUR PROGRESS 2, *page 695*

a. $\{1, 3, 5, 7, 9\}$

b. $\{0, 3, 6, 9\}$

c. $\{8, 9\}$

CHECK YOUR PROGRESS 3, *page 696*

	H	T
1	1H	1T
2	2H	2T
3	3H	3T
4	4H	4T
5	5H	5T
6	6H	6T

The sample space has 12 elements:
{1H, 1T, 2H, 2T, 3H, 3T, 4H, 4T, 5H, 5T, 6H, 6T}

CHECK YOUR PROGRESS 4, *page 697*

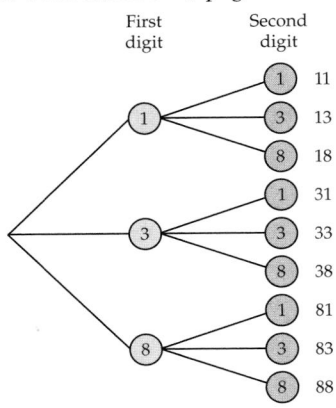

First digit   Second digit

1 — 1   11
1 — 3   13
1 — 8   18
3 — 1   31
3 — 3   33
3 — 8   38
8 — 1   81
8 — 3   83
8 — 8   88

CHECK YOUR PROGRESS 5, *page 698*
Any of the nine runners could win the gold medal, so $n_1 = 9$. That leaves $n_2 = 8$ runners that could win silver, and $n_3 = 7$ possibilities for bronze. By the counting principle, there are $9 \cdot 8 \cdot 7 = 504$ possible ways the medals can be awarded.

CHECK YOUR PROGRESS 6, *page 700*

a. Because a mailbox can accept more than one letter and each letter has five possible destinations, there are $5 \cdot 5 \cdot 5 = 125$ ways to place the letters.

b. Once a mailbox has a letter, it cannot be used, so there are $5 \cdot 4 \cdot 3 = 60$ different ways to place the letters.

SECTION 11.2

CHECK YOUR PROGRESS 1, *page 705*

a. $7! + 4! = (7 \cdot 6 \cdot 5 \cdot 4 \cdot 3 \cdot 2 \cdot 1) + (4 \cdot 3 \cdot 2 \cdot 1)$
$= 5040 + 24 = 5064$

b. $\dfrac{8!}{4!} = \dfrac{8 \cdot 7 \cdot 6 \cdot 5 \cdot \cancel{4!}}{\cancel{4!}} = 8 \cdot 7 \cdot 6 \cdot 5 = 1680$

CHECK YOUR PROGRESS 2, *page 706*
Because the players are ranked, the number of different golf teams possible is the number of permutations of eight players selected five at a time.

$$P(8, 5) = \frac{8!}{(8 - 5)!} = \frac{8!}{3!} = \frac{8 \cdot 7 \cdot 6 \cdot 5 \cdot 4 \cdot 3!}{3!}$$
$$= 8 \cdot 7 \cdot 6 \cdot 5 \cdot 4 = 6720$$

There are 6720 possible golf teams.

CHECK YOUR PROGRESS 3, *page 707*
The order in which the runners finish is important, so the number of ways to place first, second, and third is

$$P(20, 3) = \frac{20!}{(20 - 3)!} = \frac{20!}{17!} = \frac{20 \cdot 19 \cdot 18 \cdot 17!}{17!} = 6840$$

There are 6840 different ways to award the first, second, and third place prizes.

CHECK YOUR PROGRESS 4, *page 708*

a. With no restrictions, there are seven tutors available for seven hours, so the number of schedules is

$$P(7, 7) = \frac{7!}{(7 - 7)!} = \frac{7!}{0!} = 7! = 5040$$

b. This is a multi-stage experiment; there are 3! ways to schedule the juniors and 4! ways to schedule the seniors. By the counting principle, the number of different tutoring schedules is $3! \cdot 4! = 6 \cdot 24 = 144$.

CHECK YOUR PROGRESS 5, *page 709*

a. With $n = 8$ coins and $k_1 = 3$ (number of pennies), $k_2 = 2$ (number of nickels), and $k_3 = 3$ (number of dimes), the number of different possible stacks is

$$\frac{8!}{3! \cdot 2! \cdot 3!} = \frac{8 \cdot 7 \cdot 6 \cdot 5 \cdot 4 \cdot 3!}{3! \cdot 2! \cdot 3!} = \frac{8 \cdot 7 \cdot 6 \cdot 5 \cdot 4}{2 \cdot 6} = 560$$

b. Not including the dimes, there are $\frac{5!}{3! \cdot 2!} = 10$ ways to stack the pennies and nickels. The dimes are identical, so there is only one way to arrange the dimes together, but there are six different locations in the stack of pennies and nickels into which the dimes could be placed. By the counting principle, the total number of ways in which the stack of coins can be arranged if the dimes are together is $10 \cdot 6 = 60$.

CHECK YOUR PROGRESS 6, *page 711*
The order in which the players are chosen is not important, so the number of ways to choose 9 players from 16 is

$$C(16, 9) = \frac{16!}{9! \cdot (16 - 9)!} = \frac{16!}{9! \cdot 7!}$$
$$= \frac{16 \cdot 15 \cdot 14 \cdot 13 \cdot 12 \cdot 11 \cdot 10 \cdot 9!}{9! \cdot 7!}$$
$$= \frac{16 \cdot 15 \cdot 14 \cdot 13 \cdot 12 \cdot 11 \cdot 10}{7 \cdot 6 \cdot 5 \cdot 4 \cdot 3 \cdot 2 \cdot 1} = 11{,}440$$

CHECK YOUR PROGRESS 7, *page 712*
There are $C(4, 3)$ ways for the auditor to choose three corporate tax returns and $C(6, 2)$ ways to choose two individual tax returns. By the counting principle, the total number of ways in which the auditor can choose the returns is

$$C(4, 3) \cdot C(6, 2) = \frac{4!}{3! \cdot 1!} \cdot \frac{6!}{2! \cdot 4!} = 4 \cdot 15 = 60$$

CHECK YOUR PROGRESS 8, *page 713*
For any one suit, there are $C(13, 4)$ ways of choosing four cards. That leaves $52 - 13 = 39$ cards of other suits from which to choose the fifth card. In addition, there are four different suits we could start with. By the counting principle, the number of five-card combinations containing four cards of the same suit is

$$4 \cdot C(13, 4) \cdot 39 = 4 \cdot \frac{13!}{4! \cdot 9!} \cdot 39 = 4 \cdot 715 \cdot 39 = 111{,}540$$

SECTION 11.3

CHECK YOUR PROGRESS 1, *page 718*

$S = \{HH, HT, TH, TT\}$

CHECK YOUR PROGRESS 2, *page 720*

The sample space for rolling a single die is $S = \{1, 2, 3, 4, 5, 6\}$. The elements in the event that an odd number will be rolled are $E = \{1, 3, 5\}$. Then

$$P(E) = \frac{n(E)}{n(S)} = \frac{3}{6} = \frac{1}{2}$$

The probability that an odd number will be rolled is $\frac{1}{2}$.

CHECK YOUR PROGRESS 3, *page 721*

The sample space is shown in Figure 11.6 on page 721. Let E be the event that the sum of the pips on the upward faces is 7; the elements of this event are

$E = \{$  $\}$.

Then the probability of rolling a 7 is

$$P(E) = \frac{n(E)}{n(S)} = \frac{6}{36} = \frac{1}{6}$$

CHECK YOUR PROGRESS 4, *page 722*

Let E be the event that a person between the ages of 39 and 49 is selected. Then

$$P(E) = \frac{773}{3228} \approx 0.24$$

CHECK YOUR PROGRESS 5, *page 723*

Make a Punnett square.

Parents	C	c
C	CC	Cc
c	Cc	cc

To be white, the child must be cc. From the table, only one of the four possible genotypes is cc, so the probability that an offspring will be white is $\frac{1}{4}$.

CHECK YOUR PROGRESS 6, *page 725*

Let E be the event of selecting a blue ball. Because there are five blue balls, there are five favorable outcomes, leaving seven unfavorable outcomes.

$$\text{Odds against } E = \frac{\text{number of unfavorable outcomes}}{\text{number of favorable outcomes}} = \frac{7}{5}$$

The odds against selecting a blue ball from the box are 7 to 5.

CHECK YOUR PROGRESS 7, *page 725*

Let E be the event of a result of three heads and two tails after tossing five coins. Then the odds in favor of E are

$$\frac{P(E)}{1 - P(E)} = \frac{\frac{5}{16}}{1 - \frac{5}{16}} = \frac{\frac{5}{16}}{\frac{11}{16}} = \frac{5}{11}$$

SECTION 11.4

CHECK YOUR PROGRESS 1, *page 732*

Let A be the event of rolling a 7, and let B be the event of rolling an 11. From the sample space on page 721, $P(A) = \frac{6}{36} = \frac{1}{6}$ and $P(B) = \frac{2}{36} = \frac{1}{18}$. Because A and B are mutually exclusive events,

$$P(A \text{ or } B) = P(A) + P(B) = \frac{1}{6} + \frac{1}{18} = \frac{4}{18} = \frac{2}{9}$$

The probability of rolling a 7 or an 11 is $\frac{2}{9}$.

CHECK YOUR PROGRESS 2, *page 733*

Let $A = \{$people with a degree in business$\}$ and $B = \{$people with a starting salary between \$20,000 and \$24,999$\}$. Then, from the table, $n(A) = 4 + 16 + 21 + 35 + 22 = 98$, $n(B) = 4 + 16 + 3 + 16 = 39$, and $n(A \text{ and } B) = 16$. The total number of people represented in the table is 206.

$$P(A \text{ or } B) = P(A) + P(B) - P(A \text{ and } B)$$
$$= \frac{98}{206} + \frac{39}{206} - \frac{16}{206} = \frac{121}{206} \approx 0.587$$

The probability of choosing a person who has a degree in business or a starting salary between \$20,000 and \$24,999 is about 58.7%.

CHECK YOUR PROGRESS 3, *page 734*

If E is the event that a person has type A blood, then E^C is the event that the person does not have type A blood, and

$$P(E^c) = 1 - P(E) = 1 - 0.34 = 0.66$$

The probability that a person does not have type A blood is 66%.

CHECK YOUR PROGRESS 4, *page 735*

Let $E = \{$at least one roll of sum 7$\}$; then $E^c = \{$no sum of 7 is rolled$\}$. Using the table on page 721, there are 36 possibilities for each toss of the dice. Thus, $n(S) = 36 \cdot 36 \cdot 36 = 46,656$. For each roll of the dice, there are 30 numbers that do not total 7, so $n(E^c) = 30 \cdot 30 \cdot 30 = 27,000$. Then

$$P(E) = 1 - P(E^c) = 1 - \frac{27,000}{46,656} = \frac{19,656}{46,656} \approx 0.421$$

There is about a 42.1% chance of rolling a sum of 7 at least once.

CHECK YOUR PROGRESS 5, *page 737*

Let $E = \{$at least one \$100 bill$\}$; then $E^c = \{$no \$100 bills$\}$. The number of elements in the sample space is the number of ways to choose four bills from 35:

$$n(S) = C(35, 4) = \frac{35!}{4!\,(35 - 4)!} = \frac{35!}{4!\,31!} = 52,360$$

To count the number of ways not to choose any \$100 bills, we need to compute the number of ways to choose four \$1 bills from the 31 \$1 bills available.

$$n(E^c) = C(31, 4) = \frac{31!}{4!\,(31 - 4)!} = \frac{31!}{4!\,27!} = 31,465$$

$$P(E) = 1 - P(E^c) = 1 - \frac{n(E^c)}{n(S)}$$

$$= 1 - \frac{31,465}{52,360} = \frac{20,895}{52,360} \approx 0.399$$

The probability of pulling out at least one \$100 bill is about 39.9%.

SECTION 11.5

CHECK YOUR PROGRESS 1, *page 743*

Let $B = \{$the sum is 6$\}$ and $A = \{$the first die is not a 3$\}$. From the table on page 721, there are four possible rolls of the dice for which the first die is not a 3 and the sum is 6. So $P(A \cap B) = \frac{4}{36} = \frac{1}{9}$. There are 30 possibilities for which the first die is not a 3, so $P(A) = \frac{30}{36} = \frac{5}{6}$. Then

$$P(B\,|\,A) = \frac{P(A \cap B)}{P(A)} = \frac{\frac{1}{9}}{\frac{5}{6}} = \frac{2}{15}$$

The probability of rolling a 6 given that the first die is not a 3 is $\frac{2}{15}$.

CHECK YOUR PROGRESS 2, *page 745*

Let $A = \{$a spade is dealt first$\}$, $B = \{$a heart is dealt second$\}$, and $C = \{$a spade is dealt third$\}$. Then

$$P(A \text{ and } B \text{ and } C) = P(A) \cdot P(B\,|\,A) \cdot P(C\,|\,A \text{ and } B)$$

$$= \frac{13}{52} \cdot \frac{13}{51} \cdot \frac{12}{50} = \frac{13}{850}$$

The probability is $\frac{13}{850}$, or about 0.015.

CHECK YOUR PROGRESS 3, *page 746*

Each coin toss is independent of the others, because the probability of getting heads on any toss is not affected by the results of the other coin tosses. Let $E_1 = \{$heads on the first toss$\}$, $E_2 = \{$heads on the second toss$\}$, and $E_3 = \{$heads on the third toss$\}$. The events are independent and the probability of flipping heads is $\frac{1}{2}$, so

$$P(E_1 \cap E_2 \cap E_3) = P(E_1) \cdot P(E_2) \cdot P(E_3) = \frac{1}{2} \cdot \frac{1}{2} \cdot \frac{1}{2} = \frac{1}{8}$$

CHECK YOUR PROGRESS 4, *page 747*

Let D be the event that a person has the genetic defect and let T be the event that the test for the defect is positive. We are asked for $P(D\,|\,T)$, which can be calculated by

$$P(D\,|\,T) = \frac{P(D \text{ and } T)}{P(T)}$$

A tree diagram will help us compute the needed probabilities.

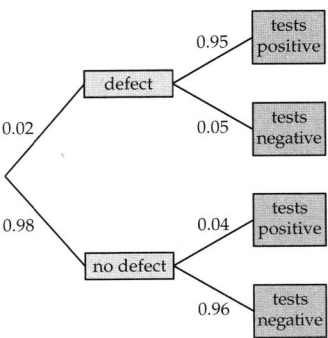

From the diagram, $P(D \text{ and } T) = P(D) \cdot P(T\,|\,D) = (0.02)(0.95)$. To compute $P(T)$ we need to combine two branches from the diagram, one corresponding to a correct positive test result when the person has the defect, and one corresponding to a false positive result when the person does not have the defect:
$P(T) = (0.02)(0.95) + (0.98)(0.04)$. Then

$$P(D\,|\,T) = \frac{P(D \text{ and } T)}{P(T)} = \frac{(0.02)(0.95)}{(0.02)(0.95) + (0.98)(0.04)} \approx 0.326$$

There is only a 32.6% chance that a person who tests positive actually has the defect.

SECTION 11.6

CHECK YOUR PROGRESS 1, *page 754*

Let S_1 be the event that the roulette ball lands on a number from 1 to 12, in which case the player wins \$10. There are 38 possible numbers, so $P(S_1) = \frac{12}{38}$. Let S_2 be the event that a number from 1 to 12 does not come up, in which case the player loses \$5. Then $P(S_2) = 1 - \frac{12}{38} = \frac{26}{38}$.

$$\text{Expectation} = P(S_1) \cdot S_1 + P(S_2) \cdot S_2$$

$$= \frac{12}{38}(10) + \frac{26}{38}(-5) = -\frac{5}{19} \approx -0.263$$

The player's expectation is about \$$-0.263$.

CHECK YOUR PROGRESS 2, *page 755*

Let S_1 be the event that the person dies within one year. Then $P(S_1) = 0.005$ and the company must pay out \$10,000. (Since the company charged \$45 for the policy, the company's actual loss is

$9955.) Let S_2 be the event that the policy holder does not die during the year of the policy. Then $P(S_2) = 0.995$ and the company keeps the premium of $45. The expectation is

$$\text{Expectation} = P(S_1) \cdot S_1 + P(S_2) \cdot S_2$$
$$= 0.005(-9955) + 0.995(45)$$
$$= -5$$

The company's expectation is a loss of $5.00.

CHAPTER 12

SECTION 12.1

CHECK YOUR PROGRESS 1, *page 769*

$$\bar{x} = \frac{\Sigma x}{n} = \frac{245 + 235 + 220 + 210}{4} = \frac{910}{4} = 227.5$$

The mean of the patient's blood cholesterol levels is 227.5.

CHECK YOUR PROGRESS 2, *page 770*

a. The list 14, 27, 3, 82, 64, 34, 8, 51 contains eight numbers. The median of a list of data with an even number of numbers is found by ranking the numbers and computing the mean of the two middle numbers. Ranking the numbers from smallest to largest gives 3, 8, 14, 27, 34, 51, 64, 82. The two middle numbers are 27 and 34. The mean of 27 and 34 is 30.5. Thus 30.5 is the median of the data.

b. The list 21.3, 37.4, 11.6, 82.5, 17.2 contains five numbers. The median of a list of data with an odd number of numbers is found by ranking the numbers and finding the middle number. Ranking the numbers from smallest to largest gives 11.6, 17.2, 21.3, 37.4, 82.5. The middle number is 21.3. Thus 21.3 is the median.

CHECK YOUR PROGRESS 3, *page 771*

a. In the list 3, 3, 3, 3, 3, 4, 4, 5, 5, 5, 8, the number 3 occurs more often than the other numbers. Thus 3 is the mode.

b. In the list 12, 34, 12, 71, 48, 93, 71, the numbers 12 and 71 both occur twice and the other numbers all occur only once. Thus 12 and 71 are both modes for the data.

CHECK YOUR PROGRESS 4, *page 773*

Weighted mean

$$= \frac{(70 \times 1) + (55 \times 1) + (90 \times 2) + (72 \times 2) + (68 \times 2) + (85 \times 6)}{14}$$
$$= \frac{1095}{14} \approx 78.2$$

Larry's weighted mean is approximately 78.2.

CHECK YOUR PROGRESS 3, *page 756*

$$\text{Expectation} = 0.05(500,000) + 0.30(250,000) + 0.35(150,000)$$
$$+ 0.20(-100,000) + 0.10(-350,000)$$
$$= 97,500$$

The company's profit expectation is $97,500.

CHECK YOUR PROGRESS 5, *page 775*

$$\text{Mean} = \frac{\Sigma(x \cdot f)}{\Sigma f}$$
$$= \frac{(2 \cdot 5) + (3 \cdot 25) + (4 \cdot 10) + (5 \cdot 5)}{45}$$
$$= \frac{150}{45}$$
$$= 3\frac{1}{3}$$

The mean number of bedrooms per household for the homes in the subdivision is $3\frac{1}{3}$.

SECTION 12.2

CHECK YOUR PROGRESS 1, *page 782*

Tara's largest test score is 84 and her smallest test score is 76. The range of Tara's test scores is $84 - 76 = 8$.

CHECK YOUR PROGRESS 2, *page 784*

$$\mu = \frac{5 + 8 + 16 + 17 + 18 + 20}{6} = \frac{84}{6} = 14$$

x	$x - \mu$	$(x - \mu)^2$
5	$5 - 14 = -9$	$(-9)^2 = 81$
8	$8 - 14 = -6$	$(-6)^2 = 36$
16	$16 - 14 = 2$	$2^2 = 4$
17	$17 - 14 = 3$	$3^2 = 9$
18	$18 - 14 = 4$	$4^2 = 16$
20	$20 - 14 = 6$	$6^2 = 36$
		Sum: 182

$$\sigma = \sqrt{\frac{\Sigma(x - \mu)^2}{n}} = \sqrt{\frac{182}{6}} \approx \sqrt{30.33} \approx 5.51$$

The standard deviation for this population is approximately 5.51.

CHECK YOUR PROGRESS 3, *page 785*

The rope from Trustworthy has a breaking point standard deviation of

$$s_1 = \sqrt{\frac{(122 - 130)^2 + (141 - 130)^2 + \cdots + (125 - 130)^2}{6}}$$

$$= \sqrt{\frac{1752}{6}} \approx 17.1 \text{ pounds}$$

The rope from Brand X has a breaking point standard deviation of

$$s_2 = \sqrt{\frac{(128 - 130)^2 + (127 - 130)^2 + \cdots + (137 - 130)^2}{6}}$$

$$= \sqrt{\frac{3072}{6}} \approx 22.6 \text{ pounds}$$

The rope from NeverSnap has a breaking point standard deviation of

$$s_3 = \sqrt{\frac{(112 - 130)^2 + (121 - 130)^2 + \cdots + (135 - 130)^2}{6}}$$

$$= \sqrt{\frac{592}{6}} \approx 9.9 \text{ pounds}$$

The rope from NeverSnap has the lowest breaking point standard deviation.

CHECK YOUR PROGRESS 4, *page 786*

The mean is 46.6796.

The population standard deviation is approximately 2.885.

L1	L2	L3	1
54.200	-----	-----	
49.400			
49.200			
53.200			
50.000			
48.200			
49.600			
L1(1) = 54.2			

```
1-Var Stats
x̄=46.6796          ← Mean
Σx=1166.99
Σx²=54682.7787
Sx=2.944997849     ← Sample standard deviation
σx=2.88549681      ← Population standard
↓n=25                deviation
```

CHECK YOUR PROGRESS 5, *page 787*

In Check Your Progress 2 we found $\sigma \approx \sqrt{30.33}$. Variance is the square of the standard deviation. Thus the variance is $\sigma^2 \approx \left(\sqrt{30.33}\right)^2 = 30.33$.

SECTION 12.3

CHECK YOUR PROGRESS 1, *page 794*

$$z_{15} = \frac{15 - 12}{2.4} = 1.25 \qquad z_{14} = \frac{14 - 11}{2.0} = 1.5$$

These z-scores indicate that in comparison to her classmates, Cheryl did better on the second quiz than she did on the first quiz.

CHECK YOUR PROGRESS 2, *page 794*

$$z_x = \frac{x - \bar{x}}{\sigma}$$

$$0.6 = \frac{70 - 65.5}{\sigma}$$

$$\sigma = \frac{4.5}{0.6} = 7.5$$

The standard deviation for this set of test scores is 7.5.

CHECK YOUR PROGRESS 3, *page 795*

a. The median is by definition the 50th percentile. Therefore, 50% of the pediatricians earned more than $139,307.

b. Because $95,000 is the 20th percentile, we know that 80% of the pediatricians earned more than $95,000.

c. 50% − 20% = 30% of the pediatricians earned between $95,000 and $139,307.

CHECK YOUR PROGRESS 4, *page 796*

$$\text{Percentile} = \frac{\text{number of data values less than 405}}{\text{total number of data values}} \cdot 100$$

$$= \frac{3952}{8600} \cdot 100$$

$$\approx 46$$

Hal's score of 405 places him at the 46th percentile.

CHECK YOUR PROGRESS 5, *page 797*

Rank the data.

7.5 9.8 10.2 10.8 11.4 11.4 12.2 12.4 12.6 12.8 13.1 14.2 14.5 15.6 16.4

The median of these 15 data values has a rank of 8. Thus the median is 12.4. The second quartile, Q_2, is the median of the data, so $Q_2 = 12.4$.

The first quartile is the median of the seven values less than Q_2. Thus Q_1 has a rank of 4, so $Q_1 = 10.8$.

The third quartile is the median of the values greater than Q_2. Thus Q_3 has a rank of 12, so $Q_3 = 14.2$.

CHECK YOUR PROGRESS 6, *page 798*

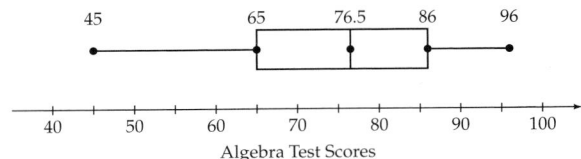

Algebra Test Scores

SECTION 12.4

CHECK YOUR PROGRESS 1, *page 807*

a. The percent of data in all classes with an upper bound of 25 seconds or less is the sum of the percents for the first five classes in Table 12.5. Thus the percent of subscribers who required less than 25 seconds to download the file is 30.9%.

b. The percent of data in all the classes with a lower bound of at least 10 seconds and an upper bound of 30 seconds or less is the sum of the percent in the third through sixth classes in Table 12.5. Thus the percent of subscribers who required from 10 to 30 seconds to download the file is 47.8%. The probability that a subscriber chosen at random will require from 10 to 30 seconds to download the file is 0.478.

CHECK YOUR PROGRESS 2, *page 811*

a. 0.76 pound is 1 standard deviation above the mean of 0.61 pound. In a normal distribution, 34.1% of all data lies between the mean and 1 standard deviation above the mean, and 50% of all data lies below the mean. Thus 34.1% + 50% = 84.1% of the tomatoes weigh less than 0.76 pound.

b. 0.31 pound is 2 standard deviations below the mean of 0.61 pound. In a normal distribution, 47.7% of all data lies between the mean and 2 standard deviations below the mean, and 50% of all data lies above the mean. This gives a total of 47.7% + 50% = 97.7% of the tomatoes that weigh more than 0.31 pound. Therefore

$$(97.7\%)(6000) = (0.977)(6000) = 5862$$

of the tomatoes can be expected to weigh more than 0.31 pound.

c. 0.31 pound is 2 standard deviations below the mean of 0.61 pound and 0.91 pound is 2 standard deviations above the mean of 0.61 pound. In a normal distribution, 95.4% of all data lies within 2 standard deviations of the mean. Therefore

$$(95.4\%)(4500) = (0.954)(4500) = 4293$$

of the tomatoes can be expected to weigh from 0.31 pound to 0.91 pound.

CHECK YOUR PROGRESS 3, *page 814*

The area of the standard normal distribution between $z = -0.67$ and $z = 0$ is equal to the area between $z = 0$ and $z = 0.67$. The entry in Table 12.7 associated with $z = 0.67$ is 0.249. Thus the area of the standard normal distribution between $z = -0.67$ and $z = 0$ is 0.249 square unit.

CHECK YOUR PROGRESS 4, *page 814*

Table 12.7 indicates that the area from $z = 0$ to $z = -1.47$ is 0.429 square unit. The area to the left of $z = 0$ is 0.500 square unit. Thus the area to the left of $z = -1.47$ is $0.500 - 0.429 = 0.071$ square unit.

CHECK YOUR PROGRESS 5, *page 816*

Round z-scores to the nearest hundredth so you can use Table 12.7.

a. $z_9 = \dfrac{9 - 6.1}{1.8} \approx 1.61$

Table 12.7 indicates that 0.446 (44.6%) of the data in the standard normal distribution is between $z = 0$ and $z = 1.61$. The percent of the data to the right of $z = 1.61$ is $50\% - 44.6\% = 5.4\%$.

Approximately 5.4% of professional football players have careers of more than 9 years.

b. $z_3 = \dfrac{3 - 6.1}{1.8} \approx -1.72 \qquad z_4 = \dfrac{4 - 6.1}{1.8} \approx -1.17$

From Table 12.7:

$A_{1.72} = 0.457 \qquad A_{1.17} = 0.379$

$0.457 - 0.379 = 0.078$

The probability that a professional football player chosen at random will have a career of between 3 and 4 years is about 0.078.

SECTION 12.5

CHECK YOUR PROGRESS 1, *page 823*

x	y	x²	xy
2.5	2.3	6.25	5.75
3.0	3.9	9.00	11.70
3.2	4.4	10.24	14.08
3.4	5.0	11.56	17.00
3.5	5.5	12.25	19.25
3.8	6.2	14.44	23.56
4.0	7.1	16.00	28.40
4.2	7.6	17.64	31.92
$\Sigma x = 27.6$	$\Sigma y = 42.0$	$\Sigma x^2 = 97.38$	$\Sigma xy = 151.66$

$n = 8$

$$a = \frac{n\Sigma xy - (\Sigma x)(\Sigma y)}{n\Sigma x^2 - (\Sigma x)^2}$$
$$= \frac{(8)(151.66) - (27.6)(42.0)}{(8)(97.38) - (27.6)^2}$$
$$\approx 3.1296$$

$\bar{x} = \dfrac{\Sigma x}{n} = \dfrac{27.6}{8} = 3.45 \qquad \bar{y} = \dfrac{\Sigma y}{n} = \dfrac{42.0}{8} = 5.25$

$b = \bar{y} - a\bar{x}$
$= 5.25 - (3.1296)(3.45)$
$= -5.54712$

$\hat{y} = ax + b$
$\hat{y} \approx 3.1x - 5.5$

CHECK YOUR PROGRESS 2, *page 824*

a. $\hat{y} = 3.1x - 5.5$

$\hat{y} = 3.1(2.7) - 5.5 \approx 2.9$

The predicted average speed of a camel with a stride length of 2.7 meters is approximately 2.9 meters per second.

b. $\hat{y} = 3.1x - 5.5$

$\hat{y} = 3.1(4.5) - 5.5 \approx 8.5$

The predicted average speed of a camel with a stride length of 4.5 meters is approximately 8.5 meters per second.

CHECK YOUR PROGRESS 3, *page 826*

$n = 8$ $\Sigma x = 27.6$ $\Sigma y = 42.0$ $\Sigma x^2 = 97.38$ $\Sigma xy = 151.66$

$\Sigma y^2 = 2.3^2 + 3.9^2 + 4.4^2 + 5.0^2 + 5.5^2 + 6.2^2 + 7.1^2 + 7.6^2$
$\quad = 241.72$

$$r = \frac{n(\Sigma xy) - (\Sigma x)(\Sigma y)}{\sqrt{n(\Sigma x^2) - (\Sigma x)^2} \cdot \sqrt{n(\Sigma y^2) - (\Sigma y)^2}}$$
$$= \frac{8(151.66) - (27.6)(42.0)}{\sqrt{8(97.38) - (27.6)^2} \cdot \sqrt{8(241.72) - (42.0)^2}}$$
$$\approx 0.998498$$

The linear correlation coefficient, rounded to the nearest hundredth, is 1.00.

CHAPTER 13

SECTION 13.1

CHECK YOUR PROGRESS 1, *page 849*

a. The standard divisor is the sum of all the populations (249,500,000) divided by the number of representatives (20).

$$\text{Standard divisor} = \frac{249,500,000}{20} = 12,475,000$$

Country	Population	Quotient	Standard quota	Number of representatives
France	59,500,000	$\frac{59,500,000}{12,475,000} \approx 4.770$	4	5
Germany	82,300,000	$\frac{82,300,000}{12,475,000} \approx 6.597$	6	6
Italy	57,900,000	$\frac{57,900,000}{12,475,000} \approx 4.641$	4	5
Spain	39,500,000	$\frac{39,500,000}{12,475,000} \approx 3.166$	3	3
Belgium	10,300,000	$\frac{10,300,000}{12,475,000} \approx 0.826$	0	1
		Total	17	20

Because the sum of the standard quotas is 17 and not 20, we add one representative to each of the three countries with the largest decimal remainders. These are Belgium, France, and Italy.

b. To use the Jefferson method, we must find a modified divisor such that the sum of the standard quotas is 20. This modified divisor is found by trial and error, but is always less than or equal to the standard divisor. We are using 11,000,000 for the modified standard divisor.

Country	Population	Quotient	Standard quota	Number of representatives
France	59,500,000	$\dfrac{59{,}500{,}000}{11{,}000{,}000} \approx 5.409$	5	5
Germany	82,300,000	$\dfrac{82{,}300{,}000}{11{,}000{,}000} \approx 7.482$	7	7
Italy	57,900,000	$\dfrac{57{,}900{,}000}{11{,}000{,}000} \approx 5.264$	5	5
Spain	39,500,000	$\dfrac{39{,}500{,}000}{11{,}000{,}000} \approx 3.591$	3	3
Belgium	10,300,000	$\dfrac{10{,}300{,}000}{11{,}000{,}000} \approx 0.936$	0	0
		Total	20	20

CHECK YOUR PROGRESS 2, *page 853*

$$\begin{aligned} \text{Relative unfairness of} \\ \text{the apportionment} \end{aligned} = \frac{\text{absolute unfairness of the apportionment}}{\text{average constituency of Shasta with a new representative}}$$

$$= \frac{210}{1390} \approx 0.151$$

The relative unfairness of the apportionment is approximately 0.151.

CHECK YOUR PROGRESS 3, *page 854*

Calculate the relative unfairness of the apportionment that assigns the teacher to the first grade and the relative unfairness of the apportionment that assigns the teacher to the second grade. In this case, the average constituency is the number of students divided by the number of teachers.

	First grade number of students per teacher	Second grade number of students per teacher	Absolute unfairness of apportionment
First grade receives teacher	$\dfrac{12{,}317}{512 + 1} \approx 24$	$\dfrac{15{,}439}{551} \approx 28$	$28 - 24 = 4$
Second grade receives teacher	$\dfrac{12{,}317}{512} \approx 24$	$\dfrac{15{,}439}{551 + 1} \approx 28$	$28 - 24 = 4$

If the first grade receives the new teacher, then the relative unfairness of the apportionment is

$$\begin{aligned} \text{Relative unfairness of} \\ \text{the apportionment} \end{aligned} = \frac{\text{absolute unfairness of the apportionment}}{\text{first grade's average constituency with a new teacher}}$$

$$= \frac{4}{24} \approx 0.167$$

If the second grade receives the new teacher, then the relative unfairness of the apportionment is

$$\text{Relative unfairness of the apportionment} = \frac{\text{absolute unfairness of the apportionment}}{\text{second grade's average constituency with a new teacher}}$$

$$= \frac{4}{28} \approx 0.143$$

Because the smaller relative unfairness results from adding the teacher to the second grade, that class should receive the new teacher.

CHECK YOUR PROGRESS 4, *page 856*

Calculate the Huntington-Hill number for each of the classes. In this case, the population is the number of students.

First year:

$$\frac{2015^2}{12(12 + 1)} \approx 26,027$$

Second year:

$$\frac{1755^2}{10(10 + 1)} \approx 28,000$$

Third year:

$$\frac{1430^2}{9(9 + 1)} \approx 22,721$$

Fourth year:

$$\frac{1309^2}{8(8 + 1)} \approx 23,798$$

Because the second year class has the greatest Huntington-Hill number, the new representative should represent the second year class.

SECTION 13.2

CHECK YOUR PROGRESS 1, *page 864*

To answer the question, we will make a table showing the number of second-place votes for each candy.

	Second-place votes
Caramel center	3
Vanilla center	0
Almond center	17 + 9 = 26
Toffee center	2
Solid chocolate	11 + 8 = 19

The largest number of second-place votes (26) were for almond centers.

CHECK YOUR PROGRESS 2, *page 868*

Using the Borda Count method, each first-place vote receives 5 points, each second-place vote receives 4 points, each third-place vote receives 3 points, and each fourth-place vote receives 2 points, and each last-place vote receives 1 point. The summary for each flavor is shown below.

Caramel:

0 first-place votes	0 · 5 = 0
3 second-place votes	3 · 4 = 12
0 third-place votes	0 · 3 = 0
30 fourth-place votes	30 · 2 = 60
17 fifth-place votes	17 · 1 = 17
Total	89

Vanilla:

17 first-place votes	17 · 5 = 85
0 second-place votes	0 · 4 = 0
0 third-place votes	0 · 3 = 0
0 fourth-place votes	0 · 2 = 0
33 fifth-place votes	33 · 1 = 33
Total	118

Almond:

8 first-place votes	8 · 5 = 40
26 second-place votes	26 · 4 = 104
16 third-place votes	16 · 3 = 48
0 fourth-place votes	0 · 2 = 0
0 fifth-place votes	0 · 1 = 0
Total	192

Toffee:

20 first-place votes	20 · 5 = 100
2 second-place votes	2 · 4 = 8
8 third-place votes	8 · 3 = 24
20 fourth-place votes	20 · 2 = 40
0 fifth-place votes	0 · 1 = 0
Total	172

Chocolate:

5 first-place votes	5 · 5 = 25
19 second-place votes	19 · 4 = 76
26 third-place votes	26 · 3 = 78
0 fourth-place votes	0 · 2 = 0
0 fifth-place votes	0 · 1 = 0
Total	179

Using the Borda Count method, almond centers is the first choice.

CHECK YOUR PROGRESS 3, *page 870*

	Rankings				
Italian	2	5	1	4	3
Mexican	1	4	5	2	1
Thai	3	1	4	5	2
Chinese	4	2	3	1	4
Indian	5	3	2	3	5
Number of ballots:	33	30	25	20	18

Indian food received no first place votes, so it is eliminated.

	Rankings				
Italian	2	4	1	3	3
Mexican	1	3	4	2	1
Thai	3	1	3	4	2
Chinese	4	2	2	1	4
Number of ballots:	33	30	25	20	18

In this ranking, Chinese food received the fewest first-place votes, so it is eliminated.

	Rankings				
Italian	2	3	1	2	3
Mexican	1	2	3	1	1
Thai	3	1	2	3	2
Number of ballots:	33	30	25	20	18

In this ranking, Italian food received the fewest first-place votes, so it is eliminated.

	Rankings				
Mexican	1	2	2	1	1
Thai	2	1	1	2	2
Number of ballots:	33	30	25	20	18

In this ranking, Thai food received the fewest first-place votes, so it is eliminated. The preference for the banquet food is Mexican.

CHECK YOUR PROGRESS 4, *page 872*

Do a head-to-head comparison for each of the restaurants and enter the winner in the table below. For instance, in the Sanborn's versus Apple Inn comparison, Sanborn's was favored by $31 + 25 + 11 = 67$ critics. In the Apple Inn versus Sanborn's comparison, Apple Inn was favored by $18 + 15 = 33$ critics. Therefore, Sanborn's wins this head-to-head match. The completed table is shown below.

versus	Sanborn's	Apple Inn	May's	Tory's
Sanborn's		Sanborn's	May's	Sanborn's
Apple Inn			May's	Tory's
May's				May's
Tory's				

From the table, May's has three wins, so it is the critics' choice.

CHECK YOUR PROGRESS 5, *page 874*

Do a head-to-head comparison for each of the candidates and enter the winner in the table below.

versus	Alpha	Beta	Gamma
Alpha		Alpha	Alpha
Beta			Beta
Gamma			

From this table, Alpha is the winner. However, using the Borda Count method (See Example 5), Beta is the winner. Thus the Borda Count method violates the Condorcet criterion.

CHECK YOUR PROGRESS 6, *page 876*

	Rankings		
Radiant silver	1	3	3
Electric red	2	2	1
Lightning blue	3	1	2
Number of votes:	30	27	2

Using the Borda Count method, we have

Silver:

30 first-place votes	$30 \cdot 3 = 90$
0 second-place votes	$0 \cdot 2 = 0$
29 third-place votes	$29 \cdot 1 = 29$
	Total 119

Red:

2 first-place votes	$2 \cdot 3 = 6$
57 second-place votes	$57 \cdot 2 = 114$
0 third-place votes	$0 \cdot 1 = 0$
	Total 120

Blue:

27 first-place votes	$27 \cdot 3 = 81$
2 second-place votes	$2 \cdot 2 = 4$
30 third-place votes	$30 \cdot 1 = 30$
	Total 115

Using this method, electric red is the preferred color. Now suppose we eliminate the third-place choice (lightning blue). This gives the following table.

	Rankings		
Radiant silver	1	2	2
Electric red	2	1	1
Number of votes:	30	27	2

Recalculating the results, we have

Silver:

30 first-place votes	$30 \cdot 2 = 60$
29 second-place votes	$29 \cdot 1 = 29$
	Total 89

Red:

29 first-place votes	$29 \cdot 2 = 58$
30 second-place votes	$30 \cdot 1 = 30$
	Total 88

Now radiant silver is the preferred color. By deleting an alternative, the result of the voting changed. This violates the irrelevant alternatives criterion.

SECTION 13.3

CHECK YOUR PROGRESS 1, *page 890*

a. and b.

Winning coalition	Number of votes	Critical voters
{A, B}	40	A, B
{A, C}	39	A, C
{A, B, C}	57	A
{A, B, D}	50	A, B
{A, B, E}	45	A, B
{A, C, D}	49	A, C
{A, C, E}	44	A, C
{A, D, E}	37	A, D, E
{B, C, D}	45	B, C, D
{B, C, E}	40	B, C, E
{A, B, C, D}	67	None
{A, B, C, E}	62	None
{A, B, D, E}	55	A
{A, C, D, E}	54	A
{B, C, D, E}	50	B, C
{A, B, C, D, E}	72	None

CHECK YOUR PROGRESS 2, *page 892*

Winning coalition	Number of votes	Critical voters
{A, B}	34	A, B
{A, C}	28	A, C
{B, C}	26	B, C
{A, B, C}	44	None
{A, B, D}	40	A, B
{A, C, D}	34	A, C
{B, C, D}	32	B, C
{A, B, C, D}	50	None

The number of times all voters are critical is 12.

$$BPI(A) = \frac{4}{12} = \frac{1}{3}$$

$$BPI(D) = \frac{0}{12} = 0$$

CHECK YOUR PROGRESS 3, *page 893*

The countries are represented as follows: B, Belgium; F, France; G, Germany; I, Italy; L, Luxembourg; N, Netherlands.

Winning coalition	Number of votes	Critical voters
{F, G, I}	12	F, G, I
{B, F, G, I}	14	F, G, I
{B, F, G, I, L}	15	F, G, I
{B, F, G, I, N}	16	None
{B, F, G, I, L, N}	17	None
{B, F, G, N}	12	B, F, G, N
{B, F, I, N}	12	B, F, I, N
{B, G, I, N}	12	B, G, I, N
{B, G, I, N, L}	13	B, G, I, N
{B, F, I, N, L}	13	B, F, I, N
{B, F, G, N, L}	13	B, F, G, N
{F, G, I, L}	13	F, G, I
{F, G, I, L, N}	15	F, G, I
{F, G, I, N}	14	F, G, I

The number of times all votes are critical is 42. The BPIs of the nations are:

$$BPI(B) = \frac{6}{42} = \frac{1}{7}$$

$$BPI(F) = \frac{10}{42} = \frac{5}{28}$$

$$BPI(G) = \frac{10}{42} = \frac{5}{28}$$

$$BPI(I) = \frac{10}{42} = \frac{5}{28}$$

$$BPI(L) = \frac{0}{42} = 0$$

$$BPI(W) = \frac{6}{42} = \frac{1}{7}$$

APPENDIX

APPENDIX

CHECK YOUR PROGRESS 1, *page 910*

a. 1 295 m = 1.295 km

b. 7 543 g = 7.543 kg

c. 6.3 L = 6 300 ml

d. 2 kl = 2 000 L

CHAPTER 1

EXCURSION EXERCISES, SECTION 1.1   *page 11*

1. a. The second player can guarantee a win.   **b.** Answers will vary.   **2. a.** The first player can guarantee a win.
b. Answers will vary.   **3.** yes

EXERCISE SET 1.1   *page 12*

1. 28   **2.** 41   **3.** 45   **4.** 216   **5.** 64   **6.** 35   **7.** $\dfrac{15}{17}$ **8.** $\dfrac{7}{8}$ **9.** -13 **10.** 51

11. correct   **12.** correct   **13.** correct   **14.** correct   **15.** incorrect   **16.** correct   **17.** no effect
18. no effect   **19.** 150 inches   **20.** 216 inches   **21.** The distance is quadrupled.   **22.** The distance is increased
by a factor of 9.   **23.** 0.5 second   **24.** 13.5 inches   **25.** inductive   **26.** deductive   **27.** deductive
28. deductive   **29.** deductive   **30.** inductive   **31.** inductive   **32.** inductive
In Exercises 33–40, only one possible answer is given. Your answers may vary from the given answers.

33. $x = \dfrac{1}{2}$ **34.** $x = -3$ **35.** $x = \dfrac{1}{2}$ **36.** $x = -2$ and $y = 5$ **37.** $x = -3$ **38.** $x = 1$

39. Consider 1 and 3. $1 + 3$ is even, but $1 \cdot 3$ is odd.   **40.** Consider 2 and 3. $2 \cdot 3$ is even, but 3 is not even.
41. It does not work for 121.   **42.** 101 is the same with its digits reversed, but 101 is not a multiple of 11.
43. n **44.** n **45.** N, because the first letter of Nine is N.

$6n + 8$
$\dfrac{6n + 8}{2} = 3n + 4$
$3n + 4 - 2n = n + 4$
$n + 4 - 4 = n$

$n + 4$
$3(n + 4) = 3n + 12$
$3n + 12 - 7 = 3n + 5$
$3n + 5 - 3n = 5$

46. A 6 preceded by a backward 6. That is 06.   **47.** d   **48. a.** 1010 is a multiple of 101 because $10 \times 101 = 1010$, but
11×1010 equals 11,110, and the digits of 11,110 are not all the same.   **b.** For $n = 11$, $n^2 - n + 11 = (11)^2 - 11 + 11 = 11^2$,
which is not a prime number.   **49.** Galileo discovered that the time of descent for a ball that rolls on an inverted cycloidial path
from A to B was less than the time of descent for a ball that rolls on a straight incline from A to B.   **50. a.** Answers will vary.
b. Most students will find, by inductive reasoning, that the best strategy for winning the grand prize is by switching.

EXCURSION EXERCISES, SECTION 1.2   *page 23*

1.

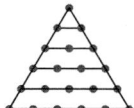

The sixth triangular
number is 21.

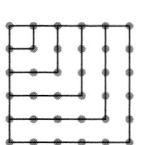

The sixth square
number is 36.

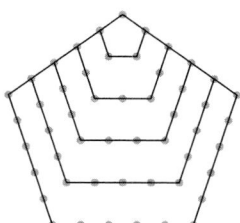

The sixth pentagonal
number is 51.

2. a.

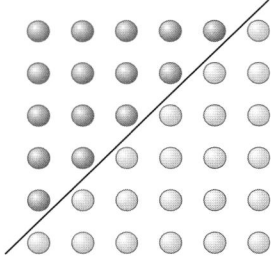

The fifth triangular number is 15. The sixth triangular number is 21. $15 + 21 = 36$, which is the sixth square number.

b. $1275 + 1326 = 2601 = 51^2$

c. $\dfrac{n(n + 1)}{2} + \dfrac{(n + 1)(n + 2)}{2} = \dfrac{2n^2 + 4n + 2}{2} = n^2 + 2n + 1 = (n + 1)^2$

3.

The fourth hexagonal number is 28.

EXERCISE SET 1.2   *page 23*

1. 97   **2.** 40   **3.** 329   **4.** 280   **5.** 159   **6.** 305   **7.** $\dfrac{3}{2}, 5, \dfrac{21}{2}, 18, \dfrac{55}{2}$ **8.** $\dfrac{1}{2}, \dfrac{2}{3}, \dfrac{3}{4}, \dfrac{4}{5}, \dfrac{5}{6}$

9. 2, 14, 36, 68, 110   **10.** 1, 12, 45, 112, 225   **11.** $a_n = n^2 + n - 1$ **12.** $a_n = 3n + 2$ **13.** $a_n = 2n$

14. $a_n = (n + 2)^2 - 1 = n^2 + 4n + 3$ **15. a.** There are 56 cannonballs in the sixth pyramid and 84 cannonballs in the seventh pyramid.   **b.** The eighth pyramid has eight levels of cannonballs. The number of cannonballs in the nth level is given by the nth number in the sequence 1, 3, 6, 10, 15, 21, 28, 36. Thus the total number of cannonballs in the eighth pyramid is $1 + 3 + 6 + 10 + 15 + 21 + 28 + 36 = 120$.   **16.** $tet_{10} = 220$ **17. a.** Five cuts produce six pieces and six cuts produce 7 pieces.   **b.** $a_n = n + 1$ **18. a.** 29   **b.** The nth pizza-slicing number is one more than the nth triangular number.

19. a. 26   **b.** 7   **20. a.** valid   **b.** not valid   **c.** valid   **d.** valid   **21.** $a_3 = 7, a_4 = 9, a_5 = 11$

22. $a_3 = -1, a_4 = 2, a_5 = -3$ **23.** $F_{20} = 6765, F_{30} = 832,040, F_{40} = 102,334,155$ **24.** $F_{16} = 987, F_{21} = 10,946,$

$F_{32} = 2,178,309$ **25.** n^2 **26. a.** $a_n = 2\left[\dfrac{n(n - 1)(n - 2)(n - 3)(n - 4)}{4 \cdot 3 \cdot 2 \cdot 1}\right] + 2n$

b. $a_n = 4\left[\dfrac{n(n - 1)(n - 2)(n - 3)(n - 4)}{4 \cdot 3 \cdot 2 \cdot 1}\right] + 2n$ **27. a.** 38.8 AU   **b.** $38.8 - 30.6 = 8.2$ AU. The prediction is not close compared to the results obtained for the inner planets.   **28.** Bode's tenth number is 77.2, which is not close to Pluto's average distance of 39.52 AU. However, 39.52 AU is relatively close to Bode's ninth number, 38.8.   **29. a.** 154 AU

b. No. The bodies in the Kuiper Belt are much closer to the sun than the distance of 154 AU predicted by Bode's 11th number.

30. a. 0.40 unit and 0.77 unit   **b.** The results in part a are relatively close to Bode's first two numbers, 0.4 and 0.7.

c. It is interesting that the results compare so closely with Bode's numbers, but we cannot say, based on this one example, that Bode's rule provides an accurate model for the placement of planets in other solar systems.

31. a. 1   **b.** 3   **c.** 7   **d.** 15   **e.** 31   **f.** $2^n - 1$ **g.** about 5.85×10^{11} years

32. $\begin{aligned} F_{n+1} &= F_n + F_{n-1} \\ &= (F_{n-1} + F_{n-2}) + F_{n-1} \\ &= 2F_{n-1} + F_{n-2} \\ &= 2F_{n-1} + 2F_{n-2} - F_{n-2} \\ &= 2(F_{n-1} + F_{n-2}) - F_{n-2} \\ &= 2F_n - F_{n-2} \end{aligned}$

EXCURSION EXERCISES, SECTION 1.3   *page 37*

1. 1, 4, 6, 4, 1   **2.** 1, 8, 28, 56, 70   **3.** The figure is symmetrical about a vertical line from A to K. Since J is the same distance to the left of the line AK as L is to the right of AK, the same number of routes lead from A to J as lead from A to L.   **4.** More routes lead to the center bin than to any of the other bins.

EXERCISE SET 1.3   *page 38*

1. 195   **2.** 12.5 feet and 19 feet   **3.** 91   **4.** 9   **5.** $25   **6.** 132   **7.** 18   **8. a.** 4
b. 3   **c.** 2   **9.** $2^{12} = 4096$ **10.** 8   **11. a.** B: 1, C: 9, D: 36, E: 84, F: 126, G: 126, H: 84   **b.** Region F is the fifth region from the left and region G is the fifth region from the right. For any path from A to F, the ball makes a number of left or right turns. If the turns are reversed, right turns instead of left turns and left turns instead of right turns, then the ball ends up in region G. Thus for each path from A to F, there is exactly one symmetrical path, about a central vertical line, from A to G. Also, from each path from A to G, there is exactly one symmetrical path from A to F. Thus there are the same numbers of paths to both regions.   **12. a.** row 0, sum is 1; row 1, sum is 2; row 2, sum is 4; row 3, sum is 8; row 4, sum is 16; row 5, sum is 32. It appears that the sum for the *n*th row is 2^n. Thus we predict that the numbers in the ninth row have a sum of $2^9 = 512$.   **b.** They are the same.   **c.** They appear in the third diagonals.   **13.** 28   **14.** 276   **15.** 21 ducks, 14 pigs   **16.** 40 seconds   **17.** 12   **18.** 20
19. 6   **20.** 8   **21.** 7   **22.** 8   **23. a.** 80,200   **b.** 151,525   **c.** 1892   **24.** One method is to apply the procedure by Gauss to find the sum of the numbers from 1 to 64 and then add 65 to this total.   **25.** 1221   **26.** No. $11^5 = 161,051$, which is not a palindromic number.   **27.** 121, 484, and 676   **28.** 1331 is the only four-digit palindromic number that is the cube of a natural number.   **29.** $1\frac{1}{2}$ inches   **30.** Yes, it is possible, if you "think outside the box."

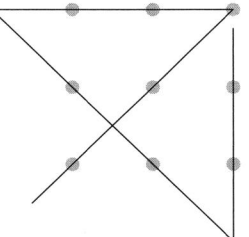

31. Only dead people are buried in cemeteries.   **32.** 22   **33.** the 11th day   **34.** 1 minute 45 seconds
35. a. Let A, B, C, and D represent the four people of weights 80, 100, 150, and 170 pounds, respectively. A and B make the first trip across, then A comes back alone. C crosses the river and B comes back alone. C is now on the opposite bank. Repeat this procedure to get D to the opposite bank. Then one more trip will get both A and B to the opposite bank.   **b.** The minimum number of crossings is nine.
36. 91   **37.** No. If his wife is a widow, then he is dead.   **38.** $9^9 = 387,420,489$ **39.** The faster runner will also win the second race.   **40.** No. The conditions imposed on the boundary lines require that the piece of land straddle the equator, and the U.S. does not own any land mass south of the equator.   **41.** Fill the 5-gallon jug. Pour from it into the 3-gallon jug until the 3-gallon jug is full. Now empty the 3-gallon jug. You now have 2 gallons in the 5-gallon jug. Pour these 2 gallons into the 3-gallon jug. Fill the 5-gallon jug again and pour from it into the 3-gallon jug. At this point there are already 2 gallons in the 3-gallon jug, so only 1 more gallon will fit. Thus, when the 3-gallon jug is full, the 5-gallon jug will contain exactly 4 gallons.   **42. a.** Place four coins on the left balance pan and the other four on the right balance pan. The pan that is the highest contains the fake coin. Take the four coins from the highest pan and use the balance scale to compare the weight of two of these coins to the weight of the other two. The pan that is the highest contains the fake coin. Take the two coins from the highest pan and use the balance scale to compare the weight of one of these coins to the weight of the other. The pan that is the highest contains the fake coin. This procedure enables you to determine the fake coin in three weighings.   **b.** Place three of the coins on one of the balance pans and another three coins on the other. If the pans balance, then the fake coin is one of the two remaining coins. You can use the balance scale to determine which of the remaining coins is the fake coin because it will be lighter than the other coin. If the three coins on the left pan do not balance with the three coins on the right pan, then the fake coin must be one of the three coins on the highest pan. Pick any two coins from these three and place one on each balance pan. If these two coins do not balance, then the one that is the highest is the fake. If the coins balance, then the third coin (the one that you did not place on the balance pan) is the fake. In any case, this procedure enables you to determine the fake coin in two weighings.
43. a. 5:40 P.M. EST   **b.** 8:00 P.M. CST   **44. a.** Understand the problem. Devise a plan. Carry out the plan. Review the solution.   **b.** Answers will vary.   **45.** 8   **46. a.** $3^{18} = 387,420,489$ times larger   **b.** 4^{240} times larger
47. a. People born in 1980 will be 45 in 2025. ($2025 = 45^2$)   **b.** 2070, because people born in 2070 will be 46 in 2116.

$(2116 = 46^2)$ **48.** Adding 83 is the same as adding 100 and subtracting 17. Thus, after you add 83, you will have a number that has 1 as the hundreds digit. The number formed by the tens digit and the units digit will be 17 less than your original number. After you add the hundreds digit, 1, to the other two digits of this new number, you will have a number that is 16 less than your original number. If you subtract this number from your original number you must get 16.   **49.** 612   **50.** It is not possible. The conditions of the problem require that each of the 31 dominoes cover one red square and one black square, but the checkerboard has 30 red squares and 32 black squares.   **51.** Answers will vary.   **52.** $M = 1, S = 9, E = 5, N = 6, D = 7, O = 0, R = 8, Y = 2$

53. a. The Collatz problem (the $3n + 1$ problem): Start with any counting number $n > 1$. Now generate a sequence of numbers using the following rules.

- If n is even, divide n by 2.
- If n is odd, multiply n by 3 and add 1.

Repeat the above procedure on the new number you have just generated. Keep applying the above procedure until you obtain the number 1. Collatz conjectured that the procedure would always generate a sequence of numbers that would eventually reach the number 1, regardless of the starting number n. Thus far no one has been able to prove that Collatz's conjecture is true or to show that it is false. The sequences generated are sometimes called "hailstone" sequences because the numbers in the sequences tend to bounce up and down, much like a hailstone in a storm.

b. Here are the Collatz sequences for the counting numbers 2 through 10.

    2: 2, 1
    3: 3, 10, 5, 16, 8, 4, 2, 1
    4: 4, 2, 1
    5: 5, 16, 8, 4, 2, 1
    6: 6, 3, 10, 5, 16, 8, 4, 2, 1
    7: 7, 22, 11, 34, 17, 52, 26, 13, 40, 20, 10, 5, 16, 8, 4, 2, 1
    8: 8, 4, 2, 1
    9: 9, 28, 14, 7, 22, 11, 34, 17, 52, 26, 13, 40, 20, 10, 5, 16, 8, 4, 2, 1
   10: 10, 5, 16, 8, 4, 2, 1

54. Student reports will vary, but should include information on Erdos's love of mathematics and his passion for working on unsolved problems. Erdos was one of the great mathematicians of the twentieth century, and one of the most prolific. Erdos loved difficult problems that were easy to state and understand without learning a lot of definitions. Much of his work was in the area of combinatorics and number theory. Concerning the Collatz problem Erdos stated, "Mathematics is not yet ready for such problems."

CHAPTER 1 REVIEW EXERCISES   *page 43*

1. deductive [Sec 1.1]   **2.** inductive [Sec. 1.1]   **3.** inductive [Sec. 1.1]   **4.** deductive [Sec. 1.1]
5. $x = 0$ provides a counterexample because $0^4 = 0$ and 0 is not greater than 0. [Sec. 1.1]   **6.** $x = 4$ provides a counterexample because $\dfrac{(4)^3 + 5(4) + 6}{6} = 15$, which is not an even number. [Sec. 1.1]   **7.** $x = 1$ provides a counterexample because $[(1) + 4]^2 = 25$, but $(1)^2 + 16 = 17$. [Sec. 1.1]   **8.** Let $a = 1$ and $b = 1$. Then $(a + b)^3 = (1 + 1)^3 = 2^3 = 8$. However, $a^3 + b^3 = 1^3 + 1^3 = 2$. [Sec. 1.1]   **9. a.** 112   **b.** 479 [Sec. 1.2]   **10. a.** -72 **b.** -768 [Sec. 1.2]
11. $a_1 = 1, a_2 = 12, a_3 = 31, a_4 = 58, a_5 = 93, a_{20} = 1578$ [Sec. 1.2]   **12.** $a_1 = 3, a_2 = -6, a_3 = -39, a_4 = -108, a_5 = -225$ $a_{25} = -31{,}125$ [Sec. 1.2]   **13.** $a_n = 3n$ [Sec. 1.2]   **14.** $a_n = n^2 + 3n + 4$ [Sec. 1.2]   **15.** 320 feet by 1600 feet [Sec. 1.3]   **16.** $3^{15} = 14{,}348{,}907$ [Sec. 1.3]   **17. a.** 50,000 miles   **b.** 60,000 miles [Sec. 1.3]
18. On the first trip the rancher takes the rabbit across the river. The rancher returns alone. The rancher takes the dog across the river and returns with the rabbit. The rancher next takes the carrots across the river and returns alone. On the final trip the rancher takes the rabbit across the river. [Sec. 1.3]   **19.** $300 [Sec. 1.3]   **20.** 0 feet [Sec. 1.3]   **21.** 7 [Sec. 1.3]
22. 105 [Sec. 1.3]   **23.** Answers will vary. [Sec. 1.3]   **24.** Answers will vary. [Sec. 1.3]   **25. a.** $2^{10} = 1024$
b. $2^{30} = 1{,}073{,}741{,}824$ [Sec. 1.2]   **26.** A represents 1, B represents 9, and D represents 0. [Sec. 1.3]   **27.** 5 [Sec. 1.3]
28. 10 [Sec. 1.3]   **29.** 1 [Sec. 1.3]   **30.** 3 [Sec. 1.3]
31. n

    $4n$

    $4n + 12$

    $\dfrac{4n + 12}{2} = 2n + 6$

    $2n + 6 - 6 = 2n$ [Sec. 1.1]

32. 9 [Sec. 1.3]

33. Every nickel is worth 5 cents. Thus 2004 nickels are worth $2004 \times 5 = 10{,}020$ cents, or \$100.20. [Sec. 1.1] **34.** numbers to indicate the address of his house [Sec. 1.1/1.3] **35.** 28 [Sec. 1.3] **36.** Jennifer [Sec. 1.1/1.3] **37.** 5005 [Sec. 1.3] **38.** There are no narcissistic numbers. [Sec. 1.3] **39. a.** 10 **b.** yes [Sec. 1.2] **40. a.** 22 **b.** No. $9^9 = 387{,}420{,}489$. Thus $9^{(9^9)}$ is the product of 387,420,489 nines. At one multiplication per second this computation would take about 12.3 years. [Sec. 1.1/1.3]

CHAPTER 1 TEST *page 45*

1. deductive [Sec. 1.1] **2.** inductive [Sec. 1.1] **3.** inductive [Sec. 1.1] **4.** deductive [Sec. 1.1]
5. 384 [Sec. 1.2] **6.** 1, 1, 2, 3, 5, 8, 13, 21, 34, 55 [Sec. 1.2] **7.** $a_n = 4n$ [Sec. 1.2] **8.** $a_n = 3n + 1$ [Sec. 1.2]
9. $0, 1, -3, 6, -10$, and -5460 [Sec. 1.2] **10.** 131, 212, 343 [Sec. 1.2] **11.** Understand the problem. Devise a plan. Carry out the plan. Review the solution. [Sec. 1.3] **12.** 6 [Sec. 1.3] **13.** 15 [Sec. 1.3] **14.** 3 [Sec. 1.3]
15. \$672 [Sec. 1.3] **16.** 126 [Sec. 1.3] **17.** 36 [Sec. 1.3] **18.** 11 [Sec. 1.3] **19.** 525 [Sec. 1.3]
20. $x = 4$ provides a counterexample because division by zero is undefined. [Sec. 1.1]

CHAPTER 2

EXCURSION EXERCISES, SECTION 2.1 *page 55*

1. a. Erica **b.** Larry **c.** Answers will vary. **2. a.** 0 **b.** 0.75 **c.** 1 **d.** 30 **e.** (40, 0.5)
3. a. 0 **b.** 0.5 **c.** 0 **d.** (3.5, 0.5) and (4.5, 0.5) **4. a.** 0.5 **b.** 1 **c.** (40, 0.5) and (60, 0.5)
5. Answers will vary.

EXERCISE SET 2.1 *page 57*

1. $\{-5, -4, -3, -2, -1\}$ **2.** $\{0, 1, 2, 3, 4, 5, 6, 7, 8, 9, 10, 11\}$ **3.** $\{7\}$ **4.** $\{-5\}$ **5.** $\varnothing$ **6.** $\{0, 1, 2, 3, 4\}$
7. {Bashful, Dopey, Doc, Grumpy, Happy, Sleepy, Sneezy} **8.** {Pacific Ocean, Atlantic Ocean, Indian Ocean, Arctic Ocean}
9. {Ford, Carter, Reagan, G. H. W. Bush, Clinton, G. W. Bush} **10.** {April, June, September, November} **11.** True
12. True **13.** False; $b \in \{a, b, c\}$, but $\{b\} \subset \{a, b, c\}$. **14.** True **15.** False; $\{0\}$ has one element, whereas $\varnothing$ has no elements. **16.** False; the word "large" is a relative term. **17.** False; the word "good" is subjective. **18.** True
19. False; 0 is an element of the first set, but not of the second set. **20.** False; the empty set has no elements.
In Exercises 21–32, only one possible answer is given. Your answers may vary from the given answers.
21. $\{x \mid x \in N \text{ and } x < 13\}$ **22.** $\{x \mid x \text{ is a multiple of 5 that ends with a 5, and } x \text{ is between 40 and 80}\}$
23. $\{x \mid x \text{ is a multiple of 5 and } 4 < x < 16\}$ **24.** $\{x \mid x \text{ is a positive square number less than or equal to 81}\}$
25. $\{x \mid x \text{ is the name of a month that has 31 days}\}$ **26.** $\{x \mid x \text{ is a state with a name that has exactly four letters}\}$
27. $\{x \mid x \text{ is the name of a U.S. state that begins with the letter A}\}$ **28.** $\{x \mid x \text{ is a country that shares a boundary with the United States}\}$
29. $\{x \mid x \text{ is the name of a season of the year}\}$ **30.** $\{x \mid x \in N \text{ and } 1900 \leq x \leq 1999\}$
31. $\{x \mid x \text{ is the name of an actor who played Harry, Ron, or Hermione in the film } Harry Potter and the Sorcerer's Stone\}$
32. $\{x \mid x \text{ is the elected president or vice president of the United States for the 2001–2005 term}\}$ **33.** 11 **34.** 8
35. 0 **36.** 50 **37.** 4 **38.** 13 **39.** 16 **40.** 32 **41.** 121 **42.** 101 **43.** Neither
44. Neither **45.** Both **46.** Neither **47.** Equivalent **48.** Equivalent **49.** Both
50. Equivalent **51.** Neither **52.** Neither **53.** Not well defined **54.** Well defined **55.** Not well defined **56.** Well defined **57.** Well defined **58.** Well defined **59.** Well defined **60.** Well defined
61. Not well defined **62.** Not well defined **63.** Not well defined **64.** Not well defined **65.** $A = B$
66. $A = B$ **67.** $A \neq B$ **68.** $A = B$ **69.** Answers will vary; however, the set of all real numbers between 0 and 1 is one example of a set that cannot be written using the roster method. **70.** A cardinal number indicates how many elements are in a set. For example in a race between Sue, Joanne, and Mary the number 3, that indicates the number of contestants, is a cardinal number. Ordinal numbers are used to indicate the order or rank of elements. For instance, if Sue finished 2nd in the race then 2nd is an ordinal number. Nominal numbers are used to name elements or to identify elements. For example, if Mary's jersey number was 7, then this number 7 is a nominal number.

EXCURSION EXERCISES, SECTION 2.2   *page 65*

1. Yes, because the membership value of each element of J is less than or equal to its membership value in set K.   **2.** Yes, because the graph of the fuzzy set *ADOLESCENT* is always below or at the same height as the graph of the fuzzy set *YOUNG*.

3. $G' = \{(A, 0), (B, 0.3), (C, 0.6), (D, 0.9), (F, 1)\}$ **4.** $C' = \{(\text{Ferrari}, 0.1), (\text{Ford Mustang}, 0.4), (\text{Dodge Neon}, 0.5), (\text{Hummer}, 0.3)\}$

5.

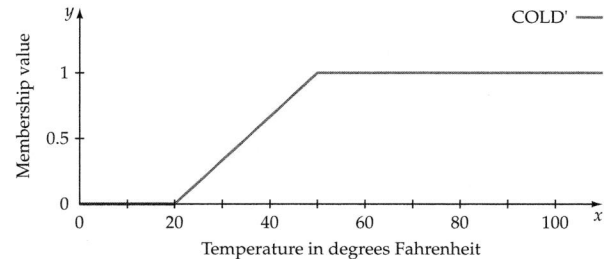

EXERCISE SET 2.2   *page 66*

1. $\{0, 1, 3, 5, 8\}$ **2.** $\{0, 1, 2, 4, 5, 7, 8\}$ **3.** $U = \{0, 1, 2, 3, 4, 5, 6, 7, 8\}$ **4.** $\{0, 1, 2, 3\}$ **5.** $\{0, 7, 8\}$
6. $\{6, 7, 8\}$ **7.** $\{0, 2, 4, 6, 8\}$ **8.** $\{0, 1, 3, 5, 7\}$ **9.** $\subseteq$ **10.** $\not\subseteq$ **11.** $\not\subseteq$ **12.** $\subseteq$ **13.** $\subseteq$
14. $\not\subseteq$ **15.** $\subseteq$ **16.** $\not\subseteq$ **17.** $\subseteq$ **18.** $\not\subseteq$ **19.** True   **20.** False   **21.** True   **22.** False
23. True   **24.** False   **25.** False   **26.** False   **27.** True   **28.** False   **29.** True   **30.** False
31. False   **32.** True   **33.** False   **34.** True   **35.** False   **36.** False   **37.** 18 hours
38. 136 years   **39.** $\varnothing, \{\alpha\}, \{\beta\}, \{\alpha, \beta\}$ **40.** $\varnothing, \{\alpha\}, \{\beta\}, \{\Gamma\}, \{\Delta\}, \{\alpha, \beta\}, \{\alpha, \Gamma\}, \{\alpha, \Delta\}, \{\beta, \Gamma\}, \{\beta, \Delta\}, \{\Gamma, \Delta\}, \{\alpha, \beta, \Gamma\},$
$\{\alpha, \beta, \Delta\}, \{\alpha, \Gamma, \Delta\}, \{\beta, \Gamma, \Delta\}, \{\alpha, \beta, \Gamma, \Delta\}$ **41.** $\varnothing, \{I\}, \{II\}, \{III\}, \{I, II\}, \{I, III\}, \{II, III\}, \{I, II, III\}$ **42.** $\varnothing$ **43.** 4
44. 8   **45.** 128   **46.** 64   **47.** 2048   **48.** 67,108,864   **49.** 1   **50.** 512   **51. a.** 15
b. 10   **c.** Two different sets of coins can have the same value.   **52. a.** $2^{18} = 262,144$ **b.** Answers will vary.
$2^{19} = 524,288$
$2^{20} = 1,048,576$

2^{33} for the TI-83 calculator.   **53. a.** 6   **b.** 2   **c.** 4   **d.** 1   **54. a.** 256   **b.** 11
55. a. 1024   **b.** 12   **56. a.** 4096   **b.** 14   **57. a.** $\{1, 2, 3\}$ has only three elements, namely 1, 2, and 3. Because
$\{2\}$ is not equal to 1, 2, or 3, $\{2\} \notin \{1, 2, 3\}$.   **b.** 1 is not a set, so it cannot be a subset.   **c.** The given set has the elements 1
and $\{1\}$. Because $1 \neq \{1\}$, there are exactly two elements in $\{1, \{1\}\}$.   **58. a.** 10   **b.** 8   **c.** Yes. The empty set has
exactly one subset.   **59. a.** $\{A, B, C\}, \{A, B, D\}, \{A, B, E\}, \{A, C, D\}, \{A, C, E\}, \{A, D, E\}, \{B, C, D\}, \{B, C, E\}, \{B, D, E\}, \{C, D, E\},$
$\{A, B, C, D\}, \{A, B, C, E\}, \{A, B, D, E\}, \{A, C, D, E\}, \{B, C, D, E\}, \{A, B, C, D, E\}$ **b.** $\{A\}, \{B\}, \{C\}, \{D\}, \{E\}, \{A, B\}, \{A, C\}, \{A, D\},$
$\{A, E\}, \{B, C\}, \{B, D\}, \{B, E\}, \{C, D\}, \{C, E\}, \{D, E\}$ **60. a.** 1, 5, 10, 10, 5, 1   **b.** Row 6: 1, 6, 15, 20, 15, 6, 1. The 20 subsets of
$\{a, b, c, d, e, f\}$ that have exactly three elements are: $\{a, b, c\}, \{a, b, d\}, \{a, b, e\}, \{a, b, f\}, \{a, c, d\}, \{a, c, e\}, \{a, c, f\}, \{a, d, e\}, \{a, d, f\}, \{a, e, f\},$
$\{b, c, d\}, \{b, c, e\}, \{b, c, f\}, \{b, d, e\}, \{b, d, f\}, \{b, e, f\}, \{c, d, e\}, \{c, d, f\}, \{c, e, f\}, \{d, e, f\}$

EXCURSION EXERCISES, SECTION 2.3   *page 76*

1. $M \cup J = \{(A, 1), (B, 0.8), (C, 0.6), (D, 0.5), (F, 0)\}$ **2.** $M \cap J = \{(A, 1), (B, 0.75), (C, 0.5), (D, 0.1), (F, 0)\}$
3. $E \cup J' = \{(A, 1), (B, 0.2), (C, 0.4), (D, 0.9), (F, 1)\}$ **4.** $J \cap L' = \{(A, 0), (B, 0), (C, 0), (D, 0), (F, 0)\}$
5. $J \cap (M' \cup L') = \{(A, 0), (B, 0.25), (C, 0.5), (D, 0.1), (F, 0)\}$ **6.**

7. Yes

EXERCISE SET 2.3 *page 78*

1. {1, 2, 4, 5, 6, 8}   **2.** {2}   **3.** {4, 6}   **4.** {2, 5, 8}   **5.** {3, 7}   **6.** {2, 3, 4, 6, 7}
7. U = {1, 2, 3, 4, 5, 6, 7, 8}   **8.** {2}   **9.** ∅   **10.** A' = {1, 3, 5, 7, 8}   **11.** B = {1, 2, 5, 8}   **12.** ∅
13. U = {1, 2, 3, 4, 5, 6, 7, 8}   **14.** {1, 3, 4, 5, 6, 7, 8}   **15.** {2, 5, 8}   **16.** U = {1, 2, 3, 4, 5, 6, 7, 8}
17. {1, 3, 4, 6, 7}   **18.** ∅   **19.** {2, 5, 8}   **20.** {1, 3, 4, 5, 6, 7, 8}
In Exercises 21–28, one possible answer is given. Your answers may vary from the given answers.
21. The set of all elements that are not in L or are in T.   **22.** The set of all elements that are in K and are not in J.   **23.** The
set of all elements that are in A, or are in C but not in B.   **24.** The set of all elements that are in either A or B, and are not in C.
25. The set of all elements that are in T, and are also in J or not K.   **26.** The set of all elements that are in both A and B, or in C.
27. The set of all elements that are in both W and V, or are in both W and Z.   **28.** The set of all elements that are in D, but are not
in either E or F.   **29.**

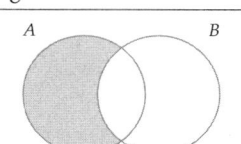   **30.** 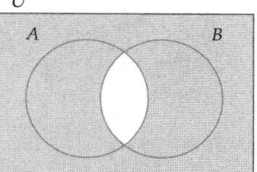   **31.**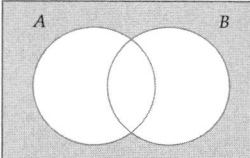

32. 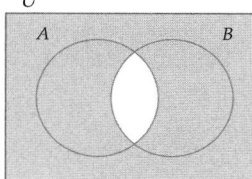   **33.** 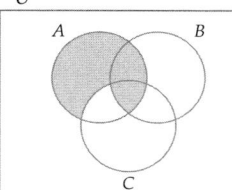   **34.**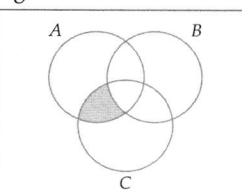

35. 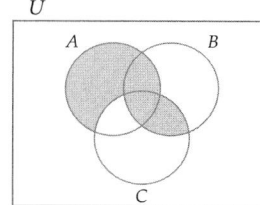   **36.** 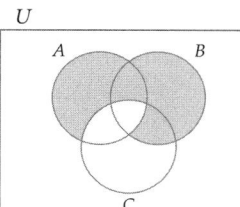   **37.** Not equal   **38.** Not equal   **39.** Equal

40. Equal   **41.** Not equal   **42.** Not equal   **43.** Not equal   **44.** Equal   **45.** Equal
46. Not equal   **47.** Yellow   **48.** Magenta   **49.** Cyan   **50.** Blue   **51.** Red   **52.** Green
In Exercises 53–62, one possible answer is given. Your answers may vary from the given answers.   **53.** $A \cap B'$
54. $(A \cup B) \cap (A \cap B)'$ **55.** $(A \cup B)'$ **56.** $A \cap (B \cup C)$ **57.** $B \cup C$ **58.** $C \cup (A \cap B)$
59. $C \cap (A \cup B)'$ **60.** $(A \cup B)'$ **61.** $(A \cup B)' \cup (A \cap B \cap C)$ **62.** $C \cup (A \cap B')$
63. 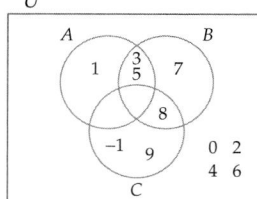   **64.** 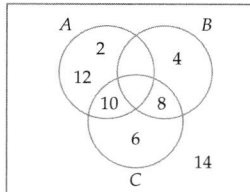   **65.**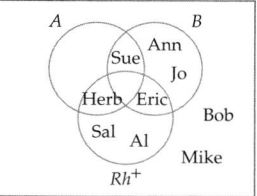

66. 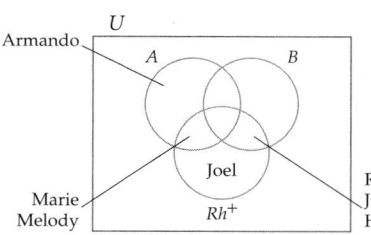   **69.** {3, 9}   **70.** {4, 6}   **71.** {2, 8}   **72.** {1, 5, 7}

73. $\{3, 9\}$ **74.** $\{1, 5, 7\}$ **75.** Responses will vary. **76.**

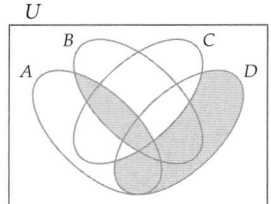

77.

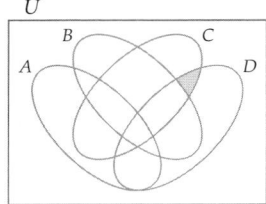

78.

A Venn diagram for five sets.

A Venn diagram for six sets.

Source: **http://www.combinatorics.org/Surveys/ds5/VennWhatEJC.html**

EXCURSION EXERCISES, SECTION 2.4 *page 88*

1. a. {Ryan, Susan}, {Ryan, Trevor}, {Susan, Trevor}, {Ryan, Susan, Trevor} **b.** $\varnothing$, {Ryan}, {Susan}, {Trevor}
2. {M, N}, {M, P}, {M, N, P} **3.** {M, N}, {M, P}

EXERCISE SET 2.4 *page 89*

1. 7 **2.** 7 **3.** 8 **4.** 9 **5.** 8 **6.** 2 **7.** 12 **8.** 0 **9.** $n(A \cup B) = 7$;
$n(A) + n(B) - n(A \cap B) = 4 + 5 - 2 = 7$ **10.** $n(A \cup C) = 6$; $n(A) + n(C) - n(A \cap C) = 4 + 3 - 1 = 6$ **11.** 113
12. 970 **13.** 1060 **14.** 155 **15.** **16.**

17. a. 180 **b.** 200 **18. a.** 440 **b.** 700 **19. a.** 15% **b.** 13% **20. a.** 995 **b.** 575
c. 1110 **21. a.** 450 **b.** 140 **c.** 130 **22. a.** 6% **b.** 38% **c.** 7% **23. a.** 72
b. 47 **c.** 25 **d.** 0 **24. a.** 35 **b.** 16 **c.** 16 **d.** 0 **25. a.** 200 **b.** 271
c. 16 **26.** d

EXCURSION EXERCISES, SECTION 2.5 *page 100*

1. Let $A = \{0\}$. Then
$$n(N) + n(A) = n(N \cup A)$$
$$\aleph_0 + 1 = n(W)$$
$$\aleph_0 + 1 = \aleph_0$$

2. Let W be the set of whole numbers and let B be the set of negative integers. Then
$$n(W) + n(B) = n(W \cup B)$$
$$\aleph_0 + \aleph_0 = n(I)$$
$$\aleph_0 + \aleph_0 = \aleph_0$$

3. Let $C = \{1, 2, 3, 4, 5, 6\}$. Then
$$n(N) - n(C) = n(N \cap C')$$
$$\aleph_0 - 6 = n(\{7, 8, 9, 10, \ldots\})$$
$$\aleph_0 - 6 = \aleph_0$$

4. a. Let E be the set of even natural numbers and let
D be the set of odd natural numbers. Then,
$$n(N) - n(E) = n(N \cap E')$$
$$\aleph_0 - \aleph_0 = n(D)$$
$$\aleph_0 - \aleph_0 = \aleph_0$$
Now let W be the set of whole numbers. Then
$$n(W) - n(N) = n(W \cap N')$$
$$\aleph_0 - \aleph_0 = n(\{0\})$$
$$\aleph_0 - \aleph_0 = 1$$

b. The subtraction of transfinite numbers is an undefined operation
because of inconsistent results such as those demonstrated in Excursion
Exercise 4a.

EXERCISE SET 2.5   *page 101*

1. a. One possible one-to-one correspondence
between V and M is given by:
$$V = \{a, e, i\}$$

$$M = \{3, 6, 9\}$$

b. 6   **2.** Pair n of N with $5n$ of F.

3. Pair $(2n - 1)$ of D with $(3n)$ of M.   **4.** 4   **5.** $\aleph_0$ **6.** $\aleph_0$ **7.** c **8.** c **9.** c **10.** 16

11. Equivalent   **12.** Not equivalent   **13.** Equivalent   **14.** Not equivalent   **15.** Let
$S = \{10, 20, 30, \ldots, 10n, \ldots\}$. Then S is a proper subset of A. A rule for a one-to-one correspondence between A and S is $(5n) \leftrightarrow (10n)$.
Because A can be placed in a one-to-one correspondence with a proper subset of itself, A is an infinite set.   **16.** Let
$F = \{15, 19, 23, 27, \ldots, 4n + 11, \ldots\}$. Then F is a proper subset of B. A rule for a one-to-one correspondence between B and F is
$(4n + 7) \leftrightarrow (4n + 11)$. Because B can be placed in a one-to-one correspondence with a proper subset of itself, B is an infinite set.

17. Let $R = \left\{ \dfrac{3}{4}, \dfrac{5}{6}, \dfrac{7}{8}, \ldots, \dfrac{2n + 1}{2n + 2}, \ldots \right\}$. Then R is a proper subset of C. A rule for a one-to-one correspondence between C and R is

$\left(\dfrac{2n - 1}{2n} \right) \leftrightarrow \left(\dfrac{2n + 1}{2n + 2} \right)$. Because C can be placed in a one-to-one correspondence with a proper subset of itself, C is an infinite set.

18. Let $H = \left\{ \dfrac{1}{3}, \dfrac{1}{4}, \dfrac{1}{5}, \dfrac{1}{6}, \ldots, \dfrac{1}{n + 2}, \ldots \right\}$. Then H is a proper subset of D. A rule for a one-to-one correspondence between D and H is

$\left(\dfrac{1}{n + 1} \right) \leftrightarrow \left(\dfrac{1}{n + 2} \right)$. Because D can be placed in a one-to-one correspondence with a proper subset of itself, D is an infinite set.

In Exercises 19–26, let $N = \{1, 2, 3, 4, \ldots, n, \ldots\}$. Then a one-to-one correspondence between the given sets and the set of natural
numbers N is given by the following general correspondences.

19. $(n + 49) \leftrightarrow n$ **20.** $(-5n + 15) \leftrightarrow n$ **21.** $\left(\dfrac{1}{3^{n-1}} \right) \leftrightarrow n$ **22.** $(-6n - 6) \leftrightarrow n$ **23.** $(10^n) \leftrightarrow n$

24. $\left(\dfrac{1}{2^n} \right) \leftrightarrow n$ **25.** $(n^3) \leftrightarrow n$ **26.** $(10^{-n}) \leftrightarrow n$

27. a. For any natural number n, the two natural numbers preceding $3n$ are not multiples of 3. Pair these two numbers, $3n - 2$ and
$3n - 1$, with the multiples of 3 given by $6n - 3$ and $6n$, respectively. Using the two general correspondences $(6n - 3) \leftrightarrow (3n - 2)$ and
$(6n) \leftrightarrow (3n - 1)$ (as shown below), we can establish a one-to-one correspondence between the multiples of 3 (set M) and the set K of all
natural numbers that are not multiples of 3.
$$M = \{3, 6, 9, 12, 15, 18, \ldots, 6n - 3, \quad 6n, \quad \ldots\}$$

$$K = \{1, 2, 4, 5, \quad 7, \quad 8, \quad \ldots, 3n - 2, 3n - 1, \ldots\}$$
The following answers in parts b and c were produced by using the correspondences established in part a.   **b.** 302   **c.** 1800

28. a. In the following figure, the line from E that passes through $\overline{AB}$ and $\overline{CD}$ illustrates a method of establishing a one-to-one
correspondence between the sets $\{x \mid 0 \le x \le 1\}$ and $\{x \mid 0 \le x \le 5\}$.

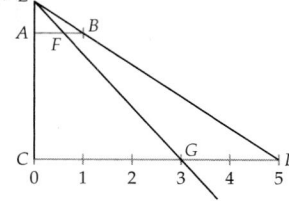

b. In the following figure, the line from E that passes through $\overline{AB}$ and $\overline{CD}$ illustrates a method of establishing a one-to-one correspondence between the sets $\{x \mid 2 \le x \le 5\}$ and $\{x \mid 1 \le x \le 8\}$.

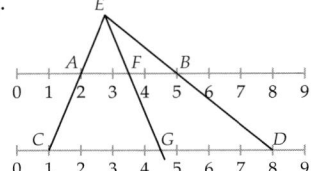

29. The set of real numbers x such that $0 < x < 1$ is equivalent to the set of all real numbers. **30.** Place a point P in the interior of the circle. Any ray from point P pairs exactly one point on the circle with exactly one point on the square, and vice versa. The one-to-one correspondence shown by these rays verifies that the set of points on the circle and the set of points on the square are equivalent. **31.** In the Hilbert Hotel there is always room for one more guest, even when the hotel is full. For example, if every room is occupied, a new guest can be accommodated by having each of the current guests move to the room with the next higher natural number. This will allow the new guest to occupy room 1. Even if a bus with $\aleph_0$ new guests arrives, the manager of the hotel can make room for the new guests by having each of the current guests move to the room that has a number twice as large as the guest's current room number. Now the new guests can be assigned to empty rooms in the following manner. The first new arrival will get room 1, the second will get room 3, and, in general, the nth new arrival will get room $2n - 1$. **32.** In 1938, Kurt Gödel showed that the Continuum Hypothesis cannot be disproved using the axioms of set theory. In 1963, Paul Cohen showed that the Continuum Hypothesis cannot be proved using the axioms of set theory. Thus the Continuum Hypothesis is independent of the axioms of set theory and cannot be proved or disproved.

CHAPTER 2 REVIEW EXERCISES *page 103*

1. $\{0, 1, 2, 3, 4, 5, 6, 7\}$ [Sec. 2.1] **2.** $\{-8, 8\}$ [Sec. 2.1] **3.** $\{1, 2, 3, 4\}$ [Sec. 2.1] **4.** $\{1, 2, 3, 4, 5, 6\}$ [Sec. 2.1]
5. $\{x \mid x \in I \text{ and } x > -6\}$ [Sec. 2.1] **6.** $\{x \mid x \text{ is the name of a month with exactly 30 days}\}$ [Sec. 2.1]
7. $\{x \mid x \text{ is the name of a U.S. state that begins with the letter K}\}$ [Sec. 2.1] **8.** $\{x^3 \mid x = 1, 2, 3, 4, \text{ or } 5\}$ [Sec. 2.1]
9. False [Sec. 2.1] **10.** True [Sec. 2.1] **11.** True [Sec. 2.1] **12.** False [Sec. 2.1]
13. $\{6, 10\}$ [Sec. 2.3] **14.** $\{2, 6, 10, 16, 18\}$ [Sec. 2.3] **15.** $C = \{14, 16\}$ [Sec. 2.3]
16. $\{2, 6, 8, 10, 12, 16, 18\}$ [Sec. 2.3] **17.** $\{2, 6, 10, 16\}$ [Sec. 2.3] **18.** $\{8, 12\}$ [Sec. 2.3]
19. $\{6, 8, 10, 12, 14, 16, 18\}$ [Sec. 2.3] **20.** $\{8, 12\}$ [Sec. 2.3] **21.** Proper subset [Sec. 2.2]
22. Proper subset [Sec. 2.2] **23.** Not a proper subset [Sec. 2.2] **24.** Not a proper subset [Sec. 2.2]
25. $\varnothing, \{I\}, \{II\}, \{I, II\}$ [Sec. 2.2] **26.** $\varnothing, \{s\}, \{u\}, \{n\}, \{s, u\}, \{s, n\}, \{u, n\}, \{s, u, n\}$ [Sec. 2.2]
27. $\varnothing$, {penny}, {nickel}, {dime}, {quarter}, {penny, nickel}, {penny, dime}, {penny, quarter}, {nickel, dime},{nickel, quarter}, {dime, quarter}, {penny, nickel, dime}, {penny, nickel, quarter}, {penny, dime, quarter}, {nickel, dime, quarter}, {penny, nickel, dime, quarter} [Sec. 2.2] **28.** $\varnothing$, {A}, {B}, {C}, {D}, {E}, {A, B}, {A, C}, {A, D}, {A, E}, {B, C}, {B, D}, {B, E}, {C, D}, {C, E}, {D, E}, {A, B, C}, {A, B, D}, {A, B, E}, {A, C, D}, {A, C, E}, {A, D, E}, {B, C, D}, {B, C, E}, {B, D, E}, {C, D, E}, {A, B, C, D}, {A, B, C, E}, {A, B, D, E}, {A, C, D, E}, {B, C, D, E}, {A, B, C, D, E} [Sec. 2.2] **29.** $2^4 = 16$ [Sec. 2.2]
30. $2^{26} = 67{,}108{,}864$ [Sec. 2.2] **31.** $2^{15} = 32{,}768$ [Sec. 2.2] **32.** $2^7 = 128$ [Sec. 2.2]
33.

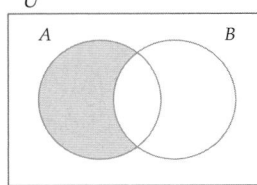

[Sec. 2.3]

34.

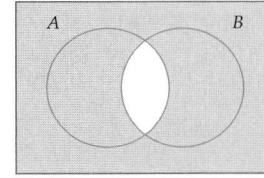

[Sec. 2.3]

35.

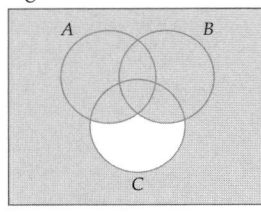

[Sec. 2.3]

36.

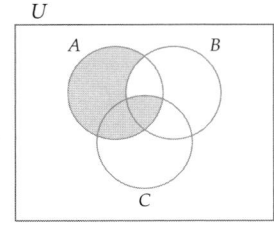

[Sec. 2.3]

37. Equal [Sec. 2.3]

38. Not equal [Sec. 2.3] **39.** Not equal [Sec. 2.3] **40.** Not equal [Sec. 2.3]

41. $(A \cup B)' \cap C$ or $C \cap (A' \cap B')$ [Sec. 2.3]

42. $(A \cap B) \cup (B \cap C')$ [Sec. 2.3]

43.
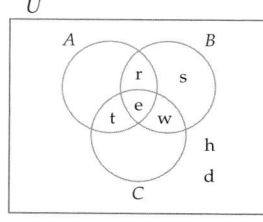
[Sec. 2.3]

44.
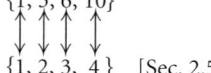
[Sec. 2.3]

45. 391 [Sec. 2.4]

46. a. 42 **b.** 31 **c.** 20 **d.** 142 [Sec. 2.4]

47. One possible one-to-one correspondence between $\{1, 3, 6, 10\}$ and $\{1, 2, 3, 4\}$ is given by
$$\{1, 3, 6, 10\}$$
$$\updownarrow\ \updownarrow\ \updownarrow\ \updownarrow$$
$$\{1, 2, 3, 4\}$$ [Sec. 2.5]

48. $\{x \mid x > 10 \text{ and } x \in N\} = \{11, 12, 13, 14, \ldots, n + 10, \ldots\}$
Thus a one-to-one correspondence between the sets is given by
$$\{11, 12, 13, 14, \ldots, n + 10, \ldots\}$$
$$\updownarrow\ \ \updownarrow\ \ \updownarrow\ \ \updownarrow\ \ \ \ \ \ \updownarrow$$
$$\{2,\ \ 4,\ \ 6,\ \ 8,\ \ \ldots,\ \ 2n,\ \ \ldots\}$$ [Sec. 2.5]

49. One possible one-to-one correspondence between the sets is given by
$$\{3,\ \ 6,\ \ 9,\ \ \ldots, 3n, \ldots\}$$
$$\updownarrow\ \ \updownarrow\ \ \updownarrow\ \ \ \ \updownarrow$$
$$\{10, 100, 1000, \ldots, 10^n, \ldots\}$$ [Sec. 2.5]

50. In the following figure, the line from E that passes through $\overline{AB}$ and $\overline{CD}$ illustrates a method of establishing a one-to-one correspondence between the sets $\{x \mid 0 \le x \le 1\}$ and $\{x \mid 0 \le x \le 4\}$.

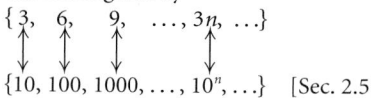
[Sec. 2.5]

51. A proper subset of A is
$S = \{10, 14, 18, \ldots, 4n + 6, \ldots\}$. A one-to-one correspondence between A and S is given by
$$A = \{6,\ 10, 14, 18,\ \ldots, 4n + 2, \ldots\}$$
$$\updownarrow\ \ \updownarrow\ \ \updownarrow\ \ \updownarrow\ \ \ \ \ \updownarrow$$
$$S = \{10, 14, 18, 22, \ldots, 4n + 6, \ldots\}$$
Because A can be placed in a one-to-one correspondence with a proper subset of itself, A is an infinite set. [Sec. 2.5]

52. A proper subset of B is $T = \left\{\dfrac{1}{2}, \dfrac{1}{4}, \dfrac{1}{8}, \dfrac{1}{16}, \ldots, \dfrac{1}{2^n}, \ldots\right\}$. A one-to-one correspondence between B and T is given by
$$B = \left\{1, \dfrac{1}{2}, \dfrac{1}{4}, \dfrac{1}{8},\ \ldots, \dfrac{1}{2^{n-1}}, \ldots\right\}$$
$$\updownarrow\ \ \updownarrow\ \ \updownarrow\ \ \updownarrow\ \ \ \ \ \updownarrow$$
$$T = \left\{\dfrac{1}{2}, \dfrac{1}{4}, \dfrac{1}{8}, \dfrac{1}{16}, \ldots, \dfrac{1}{2^n}, \ldots\right\}$$
Because B can be placed in a one-to-one correspondence with a proper subset of itself, B is an infinite set. [Sec. 2.5]

53. 5 [Sec. 2.1] **54.** 10 [Sec. 2.1] **55.** 2 [Sec. 2.1] **56.** 5 [Sec. 2.1] **57.** $\aleph_0$ [Sec. 2.5]
58. $\aleph_0$ [Sec. 2.5] **59.** c [Sec. 2.5] **60.** c [Sec. 2.5] **61.** $\aleph_0$ [Sec. 2.5] **62.** $\aleph_0$ [Sec. 2.5]
63. $\aleph_0$ [Sec. 2.5] **64.** c [Sec. 2.5] **65.** c [Sec. 2.5] **66.** c [Sec. 2.5] **67.** $\aleph_0$ [Sec. 2.5]
68. c [Sec. 2.5]

CHAPTER 2 TEST *page 105*

1. $\{2, 3, 5, 7, 8, 9, 10\}$ [Sec. 2.1/2.3] **2.** $\{2, 9, 10\}$ [Sec. 2.1/2.3] **3.** $\{1, 2, 4, 5, 6, 7, 9, 10\}$ [Sec. 2.1/2.3]
4. $\{2, 9, 10\}$ [Sec. 2.1/2.3] **5.** $\{1, 2, 3, 4, 6, 9, 10\}$ [Sec. 2.1/2.3] **6.** $\{5, 7, 8\}$ [Sec. 2.1/2.3]
7. $\{x \mid x \in W \text{ and } x < 7\}$ [Sec. 2.1/2.3] **8.** $\{x \mid x \in I \text{ and } -3 \le x \le 2\}$ [Sec. 2.1/2.3] **9. a.** 4
b. $\aleph_0$ [Sec. 2.1/2.5] **10. a.** $\aleph_0$ **b.** c [Sec. 2.5] **11. a.** Equivalent **b.** Equivalent [Sec. 2.5]
12. a. Neither **b.** Equivalent [Sec. 2.5]
13. $\varnothing, \{a\}, \{b\}, \{c\}, \{d\}$ **14.** $2^{21} = 2{,}097{,}152$ [Sec. 2.2] **15. a.** False **b.** True **c.** False
$\{a, b\}, \{a, c\}, \{a, d\}, \{b, c\}, \{b, d\}, \{c, d\}$
$\{a, b, c\}, \{a, b, d\}, \{a, c, d\}, \{b, c, d\}$
$\{a, b, c, d\}$ [Sec. 2.2]

d. True   [Sec. 2.1/2.2/2.5]   **16. a.** 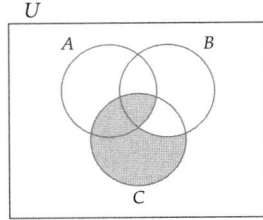   **b.** 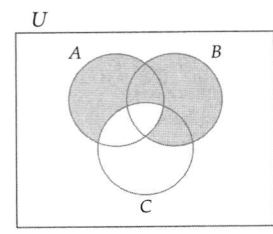   [Sec. 2.3]

17. $B \cup (A \cap C)$ [Sec. 2.3]   **18.** 541   [Sec. 2.4]

19. $\{5, 10, 15, 20, 25, \ldots, 5n, \ldots\}$
$\{0, 1, 2, 3, 4, \ldots, n-1, \ldots\}$
$(5n) \leftrightarrow (n-1)$ [Sec. 2.5]

20. $\{3, 6, 9, 12, \ldots, 3n, \ldots\}$
$\{6, 12, 18, 24, \ldots, 6n, \ldots\}$
$(3n) \leftrightarrow (6n)$ [Sec. 2.5]

CHAPTER 3

EXCURSION EXERCISES, SECTION 3.1 *page 118*

1. $S \wedge (\sim R \vee \sim Q \vee P)$ **2.** $\sim S \wedge (\sim Q \vee P) \wedge R$ **3.** $(\sim P \vee Q) \wedge (\sim R \vee S)$ **4.** $[(R \vee P) \wedge Q] \wedge (R \vee \sim P)$

5. $[\sim R \vee (Q \wedge S)] \vee (R \wedge \sim S)$ **6.** $(P \vee \sim Q) \vee (\sim R \vee S)$ **7.** 2, 5, 6   **8.** 3, 4, 5, 6

9. 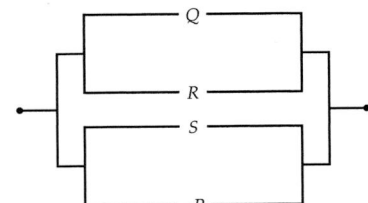   **10.** 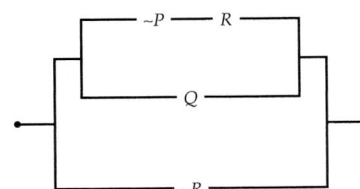   **11.**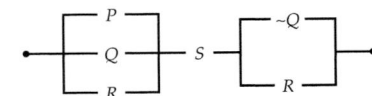

12.

13.

14.

EXERCISE SET 3.1 *page 118*

1. Statement   **2.** Statement   **3.** Statement   **4.** Statement   **5.** Not a statement   **6.** Not a statement
7. Statement   **8.** Statement   **9.** Not a statement   **10.** Statement   **11.** One component is "The principal will attend the class on Tuesday." The other component is "The principal will attend the class on Wednesday."   **12.** One component is "5 is an odd number." The other component is "6 is an even number."   **13.** One component is "A triangle is an acute triangle." The other component is "It has three acute angles."   **14.** One component is "Some birds can swim." The other component is "Some fish can fly."   **15.** One component is "I ordered a salad." The other component is "I ordered a cola."   **16.** One component is "This is Saturday." The other component is "Tomorrow is Sunday."   **17.** One component is $5 + 2 > 6$. The other component is $5 + 2 = 6$.   **18.** One component is $9 - 1 < 8$. The other component is $9 - 1 = 8$.   **19.** The Giants did not lose the game.
20. The lunch was not served at noon.   **21.** The game went into overtime.   **22.** The game was shown on ABC.
23. $w \rightarrow t$; conditional   **24.** $\sim t$; negation   **25.** $s \rightarrow r$; conditional   **26.** $p \wedge s$; conjunction
27. $l \leftrightarrow a$; biconditional   **28.** $e \leftrightarrow t$; biconditional   **29.** $d \rightarrow f$; conditional   **30.** $p \rightarrow t$; conditional
31. $m \vee c$; disjunction   **32.** $p \rightarrow s$; conditional   **33.** The tour goes to Italy and the tour does not go to Spain.

34. We will go to Venice or we will go to Florence.   **35.** If we go to Venice, then we will not go to Florence.   **36.** If the tour goes to Italy, then we will go to Venice.   **37.** We will go to Florence if and only if we do not go to Venice.   **38.** The hotel fees are not included and the meals are not included.   **39.** All cats have claws.   **40.** All dogs are friendly.   **41.** Some classic movies were not first produced in black and white.   **42.** Some people did not enjoy the dinner.   **43.** Some of the numbers were even numbers.   **44.** No student received an A.   **45.** Some irrational numbers can be written as terminating decimals.   **46.** Some cameras do not use film.   **47.** Some cars do not run on gasoline.   **48.** Some of the students took my advice.   **49.** Some items are not on sale.   **50.** Some of the telephone lines are busy.   **51.** True   **52.** True   **53.** True   **54.** True   **55.** True   **56.** True   **57.** True   **58.** True   **59.** True   **60.** False   **61.** True   **62.** True   **63.** True   **64.** False   **65.** $p \rightarrow q$, where p represents "music be the food of love" and q represents "play on."   **66.** $e \rightarrow f$, where e represents "you aren't fired with enthusiasm" and f represents "you will be fired with enthusiasm."   **67.** $p \rightarrow q$, where p represents "you do not learn from history" and q represents "you are condemned to repeat it."   **68.** $e \rightarrow p$, where e represents "man does not extend the circle of his compassion to all living things" and p represents "man will not himself find peace."   **69.** $p \rightarrow q$, where p represents "people concentrated on the really important things in life" and q represents "there'd be a shortage of fishing poles."   **70.** $k \rightarrow l$, where k represents "you're killed" and l represents "you've lost a very important part of your life."   **71.** $p \leftrightarrow q$, where p represents "an angle is a right angle" and q represents "its measure is 90°."   **72.** $i \rightarrow r$, where i represents "an angle is inscribed in a semicircle" and r represents "the angle is a right angle."   **73.** $p \rightarrow q$, where p represents "two sides of a triangle are equal in length" and q represents "the angles opposite those sides are congruent."   **74.** $t \rightarrow s$, where t represents "this sum is the sum of the measures of the three angles of a triangle" and s represents "this sum equals 180°."   **75.** $p \rightarrow q$, where p represents "it is a square" and q represents "it is a rectangle."   **76.** $p \rightarrow s$, where p represents "the corresponding sides of two triangles are proportional" and s represents "the triangles are similar."   **77.** Raymond Smullyan was born in 1919 in Far Rockaway, New York. Smullyan dropped out of high school because he found it boring and wanted to pursue his own mathematical interests. He attended several universities and eventually received a degree in mathematics from the University of Chicago. After earning a Ph.D. at Princeton in 1959, he taught mathematics and logic at Dartmouth, Princeton, and the City University of New York. In 1981 he became a member of the faculty at Indiana University. Smullyan has written several popular puzzle books and many books on mathematical logic.

EXCURSION EXERCISES, SECTION 3.2   *page 128*

1. The network is closed if P and Q are both closed. Otherwise the network is open.

P	Q	P	$\wedge$	$(\sim P \vee Q)$
1	1	1	1	1
1	0	1	0	0
0	1	0	0	1
0	0	0	0	1

2. The network is never closed.

P	Q	[$\sim P$	$\wedge$	$(\sim Q \vee P)$]	$\wedge$	Q
1	1	0	0	1	0	1
1	0	0	0	1	0	0
0	1	1	0	0	0	1
0	0	1	1	1	0	0

3. The network is closed if Q is closed or if both P and R are open.

P	Q	R	$(\sim P \vee Q)$	$\wedge$	$(\sim R \vee Q)$
1	1	1	1	1	1
1	1	0	1	1	1
1	0	1	0	0	0
1	0	0	0	0	1
0	1	1	1	1	1
0	1	0	1	1	1
0	0	1	1	0	0
0	0	0	1	1	1

4. The network is closed if R and Q are both closed.

P	Q	R	[$(R \vee P)$	$\wedge$	Q]	$\wedge$	$(R \vee \sim P)$
1	1	1	1	1	1	1	1
1	1	0	1	0	1	0	0
1	0	1	1	0	0	0	1
1	0	0	1	0	0	0	0
0	1	1	1	1	1	1	1
0	1	0	0	0	1	0	1
0	0	1	1	0	0	0	1
0	0	0	0	0	0	0	1

5. The network is closed except when *P* is open and *R* is closed.

P	Q	R	[~R	∨	(Q∧P)]	∨	(R∧P)
1	1	1	0	1	1	1	1
1	1	0	1	1	1	1	0
1	0	1	0	0	0	1	1
1	0	0	1	1	0	1	0
0	1	1	0	0	0	0	0
0	1	0	1	1	0	1	0
0	0	1	0	0	0	0	0
0	0	0	1	1	0	1	0

6. The network is always closed.

P	Q	R	(P∨~Q)	∨	(~R∨~P)
1	1	1	1	1	0
1	1	0	1	1	1
1	0	1	1	1	0
1	0	0	1	1	1
0	1	1	0	1	1
0	1	0	0	1	1
0	0	1	1	1	1
0	0	0	1	1	1

EXERCISE SET 3.2 *page 129*

1. True **2.** False **3.** False **4.** True **5.** False **6.** True **7.** False **8.** True
9. a. If *p* is false, then $p \land (q \lor r)$ must be a false statement. **b.** For a conjunctive statement to be true, it is necessary that all components of the statement be true. Because it is given that one of the components (*p*) is false, $p \land (q \lor r)$ must be a false statement.
10. a. It is a true statement. **b.** A disjunction is a true statement if at least one of its component statements is a true statement.

p	q	11.	12.	13.	14.	15.	16.
T	T	T	F	F	T	F	F
T	F	F	T	T	T	T	F
F	T	T	T	F	T	T	T
F	F	T	T	F	T	T	F

p	q	r	17.	18.	19.	20.	21.	22.	23.	24.
T	T	T	F	T	T	T	T	F	T	T
T	T	F	F	T	T	F	F	T	T	F
T	F	T	F	F	T	F	F	F	T	T
T	F	F	F	T	T	F	F	F	T	F
F	T	T	F	T	F	T	T	F	T	T
F	T	F	F	T	F	F	F	F	T	F
F	F	T	F	T	T	T	T	F	F	T
F	F	F	T	T	T	F	F	F	T	F

See the *Solution Manual* for the solutions to Exercises 25–30. **31.** It did not rain and it did not snow. **32.** It is not true that I passed the test or completed the course. **33.** She did not visit either France or Italy. **34.** I did not buy a new car or I did not move to Florida. **35.** She did not get a promotion and she did not receive a raise. **36.** The students did not cut classes and did not take part in the demonstration. **37.** Tautology **38.** Tautology **39.** Tautology **40.** Tautology **41.** Tautology **42.** Tautology **43.** Self-contradiction **44.** Self-contradiction **45.** Self-contradiction **46.** Self-contradiction **47.** Not a self-contradiction **48.** Self-contradiction **49.** The symbol ≤ means "less than or equal to." **50. a.** The given statement is true because 5 < 7. **b.** The given statement is true because 7 = 7. **51.** $2^5 = 32$ **52.** There is no whole number *n* for which $2^n = 100$. **53.** F F F T T T F T F F F T F F F F

54. F F F F T F F F T F T F T F T F **55. a.** $(p \wedge q \wedge r) \vee (p \wedge \sim q \wedge r) \vee (\sim p \wedge \sim q \wedge r)$ **b.** The disjunctive normal form is a valuable concept because it provides an easy mechanical method for naming any proposition defined by a truth table.
56. $(\sim p \vee \sim q \vee r) \wedge (\sim p \vee q \vee r) \wedge (p \vee \sim q \vee \sim r) \wedge (\sim p \vee q \vee \sim r) \wedge (\sim p \vee \sim q \vee \sim r)$

EXCURSION EXERCISES, SECTION 3.3 *page 137*

1. a. 1011 **b.** 0001 **c.** 10101011 **2.**

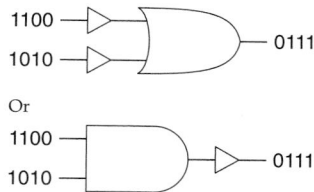

EXERCISE SET 3.3 *page 138*

1. *Antecedent:* I had the money
Consequent: I would buy the painting
2. *Antecedent:* Shelly goes on the trip
Consequent: she will not be able to take part in the graduation ceremony
3. *Antecedent:* they had a guard dog
Consequent: no one would trespass on their property
4. *Antecedent:* I don't get to school before 7:30
Consequent: I won't be able to find a parking place
5. *Antecedent:* I change my major
Consequent: I must reapply for admission
6. *Antecedent:* your blood type is O−
Consequent: you are classified as a universal donor
7. True **8.** False **9.** True **10.** True **11.** True **12.** True **13.** False **14.** True

p	q	15.	16.	17.	18.
T	T	T	T	T	T
T	F	T	T	T	T
F	T	T	T	T	T
F	F	T	T	T	T

p	q	r	19.	20.	21.	22.	23.	24.
T	T	T	T	T	T	T	T	T
T	T	F	T	T	F	T	T	T
T	F	T	T	T	T	T	T	T
T	F	F	T	T	T	T	T	F
F	T	T	T	F	T	T	T	T
F	T	F	T	T	F	T	T	T
F	F	T	T	T	T	T	T	T
F	F	F	T	T	T	T	T	T

25. She cannot sing or she would be perfect for the part. **26.** He will get frustrated or he will be able to complete the job.
27. Either x is not an irrational number or x is not a terminating decimal. **28.** Mr. Hyde does not have a brain or he would be dangerous. **29.** The fog must lift or our flight will be cancelled. **30.** The Yankees won't win the pennant or else Carol would be happy. **31.** They offered me the contract and I didn't accept. **32.** I painted the house and I did not get the money. **33.** Pigs have wings and they still can't fly. **34.** We had a telescope and we could not see the comet.
35. She traveled to Italy and she didn't visit her relatives. **36.** Paul could play better defense and he could not be a professional basketball player. **37.** False **38.** True **39.** True **40.** False **41.** False **42.** True
43. True **44.** False **45.** True **46.** False **47.** $p \rightarrow v$ **48.** $\sim t \rightarrow v$ **49.** $t \rightarrow \sim v$
50. $\sim v \rightarrow (\sim t \wedge p)$ **51.** $(\sim t \wedge p) \rightarrow v$ **52.** $p \rightarrow (t \wedge v)$ **53.** Not equivalent **54.** Not equivalent
55. Not equivalent **56.** Equivalent **57.** Equivalent **58.** Equivalent **59.** If a number is a rational number, then it is a real number. **60.** If a number is a whole number, then it is an integer. **61.** If a number is a repeating decimal, then it is a rational number. **62.** If a number is a multiple of 5, then it ends with a 0 or with a 5. **63.** If an animal is a Sauropod, then it is herbivorous. **64.** If a painting was painted by Vincent van Gogh, then the painting is valuable.
65. Student demonstration. **66.** Answers will vary.

EXCURSION EXERCISES, SECTION 3.4 *page 144*

1. a.

p	q	p		(q | q)
T	T	T	T	F
T	F	T	F	T
F	T	F	T	F
F	F	F	T	T

b. $p \mid (q \mid q) \equiv p \rightarrow q$

2.

p	q	(p | q)		(p | q)
T	T	F	T	F
T	F	T	F	T
F	T	T	F	T
F	F	T	F	T

b. $(p \mid q) \mid (p \mid q) \equiv p \wedge q$ **3. a.** 1000   **b.** AND gate   **4.**

Input streams   Output stream

1100

1010

1110

EXERCISE SET 3.4 *page 144*

1. If we take the aerobics class, then we will be in good shape for the ski trip.   **2.** If we get a dog, then we will install a fence around the back yard.   **3.** If the number is an odd prime number, then it is greater than 2.   **4.** If the length of the hypotenuse is twice the length of the shorter leg, then the triangle is a 30-60-90 triangle.   **5.** If he has the talent to play a keyboard, then he can join the band.   **6.** If the dinosaur is a theropod, then it is carnivorous.   **7.** If I was able to prepare for the test, then I had the textbook.   **8.** If Education 147 is offered in the spring semester, then I will be able to receive my credential.
9. If you ran the Boston marathon, then you are in excellent shape.   **10.** If it is an ankylosaur, then it is quadrupedal.
11. a. If I quit this job, then I am rich.   **b.** If I were not rich, then I would not quit this job.   **c.** If I would not quit this job, then I would not be rich.   **12. a.** If we were able to take the class, then we had a car.   **b.** If we did not have a car, then we were not able to take the class.   **c.** If we were not able to take the class, then we did not have a car.   **13. a.** If we are not able to attend the party, then she did not return soon.   **b.** If she returns soon, then we will be able to attend the party.   **c.** If we are able to attend the party, then she returned soon.   **14. a.** If I can do the same comedy routine that I did for the banquet, then I will be in the talent show.   **b.** If I am not in the talent show, then I was not allowed to do the same comedy routine that I did for the banquet.   **c.** If I cannot do the same comedy routine that I did for the banquet, then I will not be in the talent show.
15. a. If a figure is a quadrilateral, then it is a parallelogram.   **b.** If a figure is not a parallelogram, then it is not a quadrilateral.   **c.** If a figure is not a quadrilateral, then it is not a parallelogram.   **16. a.** If you need to move to Denver, then you got the promotion.   **b.** If you did not get the promotion, then you will not need to move to Denver.   **c.** If you do not need to move to Denver, then you did not get the promotion.   **17. a.** If I am able to get current information about astronomy, then I have access to the Internet.   **b.** If I do not have access to the Internet, then I will not be able to get current information about astronomy.
c. If I am not able to get current information about astronomy, then I don't have access to the Internet.   **18. a.** If you need four-wheel drive, then you are going on a trip to Death Valley.   **b.** If you are not going on a trip to Death Valley, then you will not need four-wheel drive.   **c.** If you do not need four-wheel drive, then you are not going on a trip to Death Valley.   **19. a.** If we don't have enough money for dinner, then we took a taxi.   **b.** If we did not take a taxi, then we will have enough money for dinner.
c. If we have enough money for dinner, then we did not take a taxi.   **20. a.** If your age is at least 35, then you are the president of the United States.   **b.** If you are not the president of the United States, then your age is less than 35.   **c.** If your age is less than 35, then you are not the president of the United States.   **21. a.** If she can extend her vacation for at least two days, then she will visit Kauai.   **b.** If she does not visit Kauai, then she could not extend her vacation for at least two days.   **c.** If she cannot extend her vacation for at least two days, then she will not visit Kauai.   **22. a.** If the acute angles of a triangle are complementary, then the triangle is a right triangle.   **b.** If a triangle is not a right triangle, then the acute angles are not complementary.
c. If the acute angles of a triangle are not complementary, then the triangle is not a right triangle.   **23. a.** If two lines are parallel, then the two lines are perpendicular to a given line.   **b.** If two lines are not perpendicular to a given line, then the two lines are not parallel.   **c.** If two lines are not parallel, then the two lines are not both perpendicular to a given line.   **24. a.** If $x = 7$, then $x + 5 = 12$.   **b.** If $x + 5 \neq 12$, then $x \neq 7$.   **c.** If $x \neq 7$, then $x + 5 \neq 12$.   **25.** Not equivalent   **26.** Not equivalent   **27.** Equivalent   **28.** Not equivalent   **29.** Not equivalent   **30.** Equivalent   **31.** If $x = 7$, then $3x - 7 \neq 11$. Original statement is true.   **32.** If $5x + 7 = 22$, then $x = 3$. Original statement is true.   **33.** If $|a| = 3$, then $a = 3$. Original statement is false.   **34.** If a is not divisible by 3 or b is not divisible by 3, then $a + b$ is not divisible by 3. The

original statement is false.   **35.** If $a + b = 25$, then $\sqrt{a + b} = 5$. Original statement is true.   **36.** If x is an odd integer, then x^2 is an odd integer. The original statement is true.   **37.** $p \rightarrow q$ **38.** $p \rightarrow q$ **39. a. and b.** Answers will vary.
40. a. and b. Answers will vary.   **41.** If you can dream it, then you can do it.   **42.** If I had a dime, I would spend it.
43. If I were a dancer, then I would not be a singer.   **44.** Pigs do not have wings or pigs could fly. This answer can also be written as "If pigs have wings, then pigs can fly."   **45.** A conditional statement and its contrapositive are equivalent. They always have the same truth values.   **46.** Yes. The conditional $p \rightarrow q$ is false only when p is true (T) and q is false (F). Under these conditions the converse $q \rightarrow p$ would be of the form F $\rightarrow$ T, which is true.   **47.** The Hatter is telling the truth.   **48.** Answers will vary.

EXCURSION EXERCISES, SECTION 3.5   *page 156*

1.–4. Answers will vary.   **5.** The step in which each side is divided by $(a - b)$ is not valid because $a - b = 0$, and division by zero is an undefined operation.

EXERCISE SET 3.5   *page 156*

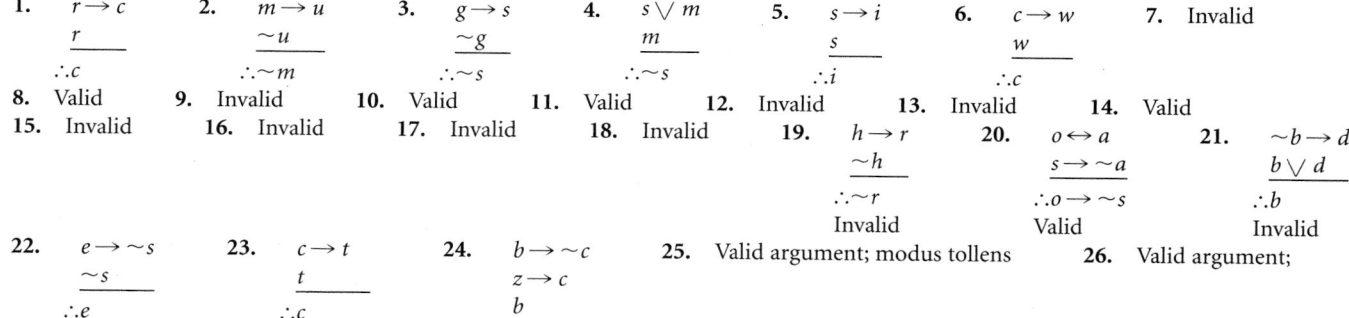

1.
$r \rightarrow c$
r
$\therefore c$

2.
$m \rightarrow u$
$\sim u$
$\therefore \sim m$

3.
$g \rightarrow s$
$\sim g$
$\therefore \sim s$

4.
$s \vee m$
m
$\therefore \sim s$

5.
$s \rightarrow i$
s
$\therefore i$

6.
$c \rightarrow w$
w
$\therefore c$

7. Invalid

8. Valid   **9.** Invalid   **10.** Valid   **11.** Valid   **12.** Invalid   **13.** Invalid   **14.** Valid
15. Invalid   **16.** Invalid   **17.** Invalid   **18.** Invalid   **19.** $h \rightarrow r$ **20.** $o \leftrightarrow a$ **21.** $\sim b \rightarrow d$
$\sim h$
$\therefore \sim r$
Invalid
$s \rightarrow \sim a$
$\therefore o \rightarrow \sim s$
Valid
$b \vee d$
$\therefore b$
Invalid

22. $e \rightarrow \sim s$
$\sim s$
$\therefore e$
Invalid

23. $c \rightarrow t$
t
$\therefore c$
Invalid

24. $b \rightarrow \sim c$
$z \rightarrow c$
b
$\therefore \sim z$
Valid

25. Valid argument; modus tollens   **26.** Valid argument;
disjunctive syllogism   **27.** Invalid argument; fallacy of the inverse   **28.** Invalid argument; fallacy of the converse
29. Valid argument; law of syllogism   **30.** Invalid argument; fallacy of the inverse   **31.** Valid argument; modus ponens
32. Invalid argument; fallacy of the inverse   **33.** Valid argument; modus tollens   **34.** Valid argument; modus tollens
See the *Solution Manual* for the solutions to Exercises 35–40.   **41.** q **42.** s **43.** it is not a theropod   **44.** you
will make monthly payments   **45.** Valid   **46.** Answers will vary.

EXCURSION EXERCISES, SECTION 3.6   *page 164*

1. $T = 1, S = 5, O = 0$ **2.** $U = 8, S = 5, A = 1, L = 0$ **3.** $C = 8, O = 1, A = 6, L = 0, S = 2, I = 9$
4. $A = 3, T = 6, E = 7, S = 1, W = 4, O = 2, U = 0, H = 8$

EXERCISE SET 3.6   *page 165*

1. Valid   **2.** Invalid   **3.** Valid   **4.** Invalid   **5.** Valid   **6.** Invalid   **7.** Valid
8. Invalid   **9.** Invalid   **10.** Valid   **11.** Invalid   **12.** Invalid   **13.** Invalid   **14.** Valid
15. Valid   **16.** Invalid   **17.** Valid   **18.** Invalid   **19.** Invalid   **20.** Invalid   **21.** All Reuben
sandwiches need mustard.   **22.** Boomer is not a cat.   **23.** 1001 ends with a 5.   **24.** If it isn't broke, then I don't
get paid.   **25.** Some horses are grey.   **26.** If we like to ski, then we will buy a condo.   **27.**

1	2
2	1
3	4
8	4

28. The following diagram shows a bilateral diagram as devised by Charles Dodgson. Note that the diagram has four regions, much like a Venn diagram for two sets, but in the bilateral diagram each region is a square and all of the regions are the same size. The upper-left region marked xy corresponds to the $X \cap Y$ region of a Venn diagram. The fact that the regions in a bilateral diagram are all

large square regions allowed Dodgson to place red and grey counters in the squares or on the boundaries between the squares. A red (dark) counter indicated the existence of that type of element. A grey (light) counter was used to indicate that a particular type of element did not exist. For instance, the bilateral diagram on the right below indicates the existence of an *xy* element and the nonexistence of an *xy'* element. These bilateral diagrams, along with the red and grey counters, allowed Dodgson to solve many logic problems in a visual manner. He also devised a trilateral diagram that could be used to solve logic problems involving three sets of elements. *Source:*
http://durendal.org/lcsl052.html

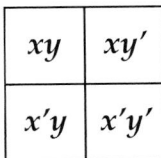   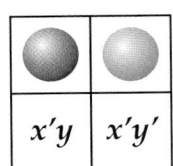

A bilateral diagram

CHAPTER 3 REVIEW EXERCISES   *page 167*

1. Not a statement [Sec. 3.1]   **2.** Statement [Sec. 3.1]   **3.** Statement [Sec. 3.1]   **4.** Statement [Sec. 3.1]
5. Not a statement [Sec. 3.1]   **6.** Statement [Sec. 3.1]   **7.** $m \wedge b$; conjunction [Sec. 3.1]
8. $d \rightarrow e$; conditional [Sec. 3.1]   **9.** $g \leftrightarrow d$; biconditional [Sec. 3.1]   **10.** $t \rightarrow s$; conditional [Sec. 3.1]
11. No dogs bite. [Sec. 3.1]   **12.** Some desserts at the Cove restaurant are not good. [Sec. 3.1]   **13.** Some winners do not receive a prize. [Sec. 3.1]   **14.** All cameras use film. [Sec. 3.1]   **15.** Some of the students received an A. [Sec. 3.1]
16. Nobody enjoyed the story. [Sec. 3.1]   **17.** True [Sec. 3.1]   **18.** True [Sec. 3.1]   **19.** True [Sec. 3.1]
20. True [Sec. 3.1]   **21.** True [Sec. 3.1]   **22.** True [Sec. 3.1]   **23.** False [Sec. 3.2]
24. False [Sec. 3.2/3.3]   **25.** True [Sec. 3.2]   **26.** True [Sec. 3.2/3.3]   **27.** True [Sec. 3.2/3.3]
28. False [Sec. 3.2.3.3]

p	q	**29.** [Sec. 3.2/3.3]	**30.** [Sec. 3.2/3.3]	**31.** [Sec. 3.2/3.3]	**32.** [Sec. 3.2/3.3]
T	T	T	F	F	T
T	F	T	F	F	T
F	T	T	T	F	F
F	F	F	F	F	T

p	q	r	**33.** [Sec. 3.2/3.3]	**34.** [Sec. 3.2/3.3]	**35.** [Sec. 3.2/3.3]	**36.** [Sec. 3.2/3.3]
T	T	T	T	F	F	T
T	T	F	T	T	F	T
T	F	T	T	T	T	T
T	F	F	F	T	T	T
F	T	T	T	F	F	F
F	T	F	T	F	F	T
F	F	T	T	T	F	T
F	F	F	T	T	F	T

37. Bob passed the English proficiency test or he did not register for a speech course. [Sec. 3.2]   **38.** It is not true that Ellen went to work this morning or she took her medication. [Sec. 3.2]   **39.** It is not the case that Wendy will not go to the store this afternoon and she will be able to fix her fettuccine al pesto recipe. [Sec. 3.2]   **40.** It is not the case that Gina did not enjoy the movie or she enjoyed the party. [Sec. 3.2/3.3]   See the *Solution Manual* for solutions to Exercises 41–44.

45. Self-contradiction [Sec. 3.2]   **46.** Tautology [Sec. 3.2/3.3]   **47.** Tautology [Sec. 3.2/3.3]
48. Tautology [Sec. 3.2/3.3]   **49.** *Antecedent:* he has talent   **50.** *Antecedent:* I had a credential
Consequent: he will succeed [Sec. 3.3]   *Consequent:* I could get the job [Sec. 3.3]
51. *Antecedent:* I join the fitness club   **52.** *Antecedent:* I will attend
Consequent: I will follow the exercise program [Sec. 3.3]   *Consequent:* it is free [Sec. 3.3]
53. She is not tall or she would be on the volleyball team. [Sec. 3.3]   **54.** He cannot stay awake or he would finish
the report. [Sec. 3.3]   **55.** Rob is ill or he would start. [Sec. 3.3]   **56.** Sharon will not be promoted or she closes
the deal. [Sec. 3.3]   **57.** I get my paycheck and I do not purchase a ticket. [Sec. 3.3]   **58.** The tomatoes will get big and
you did not provide plenty of water [Sec. 3.3]   **59.** You entered Cleggmore University and you did not have a high score on the
SAT exam. [Sec. 3.3]   **60.** Ryan enrolled at a university and he did not enroll at Yale. [Sec. 3.3]   **61.** False [Sec. 3.3]
62. True [Sec. 3.3]   **63.** True [Sec. 3.3]   **64.** False [Sec. 3.3]   **65.** False [Sec. 3.3]
66. False [Sec. 3.3]   **67.** If a real number has a nonrepeating, nonterminating decimal form, then the real number is
irrational. [Sec. 3.4]   **68.** If you are a politician, then you are well known. [Sec. 3.4]   **69.** If I can sell my condominium,
then I can buy the house. [Sec. 3.4]   **70.** If a number is divisible by 9, then the number is divisible by 3. [Sec. 3.4]
71. a. *Converse:* If $x > 3$, then $x + 4 > 7$.   **b.** *Inverse:* If $x + 4 \leq 7$, then $x \leq 3$.   **c.** *Contrapositive:* If $x \leq 3$, then
$x + 4 \leq 7$. [Sec. 3.4]   **72. a.** *Converse:* If the recipe can be prepared in less than 20 minutes, then the recipe is in this book.
b. *Inverse:* If the recipe is not in this book, then the recipe cannot be prepared in less than 20 minutes.   **c.** *Contrapositive:* If the
recipe cannot be prepared in less than 20 minutes, then the recipe is not in this book. [Sec. 3.4]   **73. a.** *Converse:* If $(a + b)$ is
divisible by 3, then a and b are both divisible by 3.   **b.** *Inverse:* If a and b are not both divisible by 3, then $(a + b)$ is not
divisible by 3.   **c.** *Contrapositive:* If $(a + b)$ is not divisible by 3, then a and b are not both divisible by 3. [Sec. 3.4]
74. a. *Converse:* If they come, then you built it.   **b.** *Inverse:* If you do not build it, then they will not come.
c. *Contrapositive:* If they do not come, then you did not build it. [Sec. 3.4]   **75. a.** *Converse:* If it has exactly two
parallel sides, then it is a trapezoid.   **b.** *Inverse:* If it is not a trapezoid, then it does not have exactly two parallel sides.
c. *Contrapositive:* If it does not have exactly two parallel sides, then it is not a trapezoid. [Sec. 3.4]   **76. a.** *Converse:* If they
returned, then they liked it.   **b.** *Inverse:* If they do not like it, then they will not return.   **c.** *Contrapositive:* If they do not
return, then they did not like it. [Sec. 3.4]   **77.** $q \rightarrow p$, the converse of the original statement [Sec. 3.4]   **78.** $p \rightarrow q$, the
original statement [Sec. 3.4]   **79.** If x is an odd prime number, then $x > 2$. [Sec. 3.4]   **80.** If the senator attends the
meeting, then she will vote on the motion. [Sec. 3.4]   **81.** If their manager contacts me, then I will purchase some of their
products. [Sec. 3.4]   **82.** If I can rollerblade, then Ginny can rollerblade. [Sec. 3.4]   **83.** Valid [Sec. 3.5]
84. Valid [Sec. 3.5]   **85.** Invalid [Sec. 3.5]   **86.** Valid [Sec. 3.5]   **87.** Valid argument; disjunctive
syllogism [Sec. 3.5]   **88.** Valid argument; law of syllogism [Sec. 3.5]   **89.** Invalid argument; fallacy of the
inverse [Sec. 3.5]   **90.** Valid argument; disjunctive syllogism [Sec. 3.5]   **91.** Valid argument; modus tollens [Sec. 3.5]
92. Invalid argument; fallacy of the inverse [Sec. 3.5]   **93.** Valid [Sec. 3.6]   **94.** Invalid [Sec. 3.6]
95. Invalid [Sec. 3.6]   **96.** Valid [Sec. 3.6]

CHAPTER 3 TEST   *page 170*

1. a. Not a statement   **b.** Statement [Sec. 3.1]   **2. a.** All trees are green.   **b.** Some of the kids had seen the
movie [Sec. 3.1]   **3. a.** False   **b.** True [Sec. 3.1]   **4. a.** False   **b.** True [Sec. 3.2/3.3]

p	q	5. [Sec. 3.2/3.3]
T	T	T
T	F	T
F	T	T
F	F	T

p	q	r	6. [Sec. 3.2/3.3]
T	T	T	F
T	T	F	T
T	F	T	F
T	F	F	F
F	T	T	F
F	T	F	T
F	F	T	T
F	F	F	F

7. It is not the case that Elle ate breakfast or took a lunch break. [Sec. 3.2] **8.** A tautology is a statement that is always true. [Sec. 3.2] **9.** $\sim p \vee q$ [Sec. 3.3] **10. a.** False **b.** False [Sec. 3.2/3.3] **11. a.** *Converse:* If $x > 4$, then $x + 7 > 11$. **b.** *Inverse:* If $x + 7 \leq 11$, then $x \leq 4$. **c.** *Contrapositive:* If $x \leq 4$, then $x + 7 \leq 11$. [Sec. 3.4]

12. $p \rightarrow q$
$\underline{\quad p \quad\quad}$
$\therefore q$ [Sec. 3.5]

13. $p \rightarrow q$
$\underline{q \rightarrow r}$
$\therefore p \rightarrow r$ [Sec. 3.5]

14. Valid [Sec. 3.5] **15.** Invalid [Sec. 3.5]

16. Invalid argument; the argument is a fallacy of the inverse. [Sec. 3.5] **17.** Valid argument; the argument is a disjunctive syllogism. [Sec. 3.5] **18.** Invalid argument, as shown by an Euler diagram. [Sec. 3.6] **19.** Invalid argument, as shown by an Euler diagram. [Sec. 3.6] **20.** Invalid argument; the argument is a fallacy of the converse. [Sec. 3.5]

CHAPTER 4

EXCURSION EXERCISES, SECTION 4.1 *page 180*

1.

2. **3.** **4. a.** 896 **b.** 2465 **5. a.** 4 **b.** 7

6. Answers will vary.

EXERCISE SET 4.1 *page 181*

1. **2.** **3.** **4.**

5. **6.** **7.**

8. **9.** 845 **10.** 287 **11.** 1232 **12.** 232,333 **13.** 65,769

14. 2,436,723 **15.** 5,122,406 **16.** 2,601,400 **17.** 94 **18.** 125 **19.** 666 **20.** 747
21. 32 **22.** 71 **23.** 56 **24.** 362 **25.** 650 **26.** 1110 **27.** 1409 **28.** 1702
29. 9044 **30.** 7517 **31.** 11,461 **32.** 4221 **33.** CLVII **34.** CCXXXI **35.** DCCLXXXVII
36. MCCCXLIII **37.** DCLXXXIII **38.** CMLIX **39.** $\overline{\text{VI}}$DCCCXCVIII **40.** $\overline{\text{IV}}$CCCLVII **41.** 504
42. 228 **43.** 203 **44.** 297 **45.** 595 **46.** 1118 **47.** 2484 **48.** 15,480
49. a. and b. Answers will vary. **50.** MMMDCCCLXXXVIII = 3888 **51.** The Ionic Greek numeration system gained popularity in Greco-Roman times, starting around 100 B.C. It consisted of 24 Greek letters and three letters taken from the Phoenician alphabet. The first nine letters were used to represent the counting numbers from 1 to 9. The next nine letters were used to represent the multiples of 10 from 10 to 90. The third set of nine letters were used to represent the multiples of 100 from 100 to 900. For example, α (alpha) represented 1, λ (lambda) represented 30, and ϕ (phi) represented 500. In the Ionic Greek numeration system, the value of a numeral was the sum of each of the individual numerals that comprised it. For instance, 531 was represented by $\alpha\lambda\phi$ (500 + 30 + 1). A prime symbol was used to multiply the value of a numeral by 1000. One of the advantages of the Ionic Greek system was that large numbers could be written using very few symbols. A disadvantage of the Ionic Greek system was that it was difficult to perform arithmetic using Ionic numerals.

52. The following reasons for using IIII in place of IV on clock faces have been suggested.
- The numeral for 4 is directly across from the Roman numeral for 8, which is VIII. For the sake of symmetry the numeral for 4 is written as IIII, which is also composed of four symbols.
- The early Roman numeral system, utilized before 100 A.D., used IIII instead of IV for the numeral 4. Thus many clock makers also employed the use of IIII for the numeral 4.
- It is reported that Emperor Charles V (1500–1588) admonished a clock maker for using IV instead of IIII on the face of a clock. Clock makers began to use IIII in place of IV to avoid the wrath of the emperor.

53. The method of false position is used in Problems 24 to 29 of the Rhind papyrus. Ahmes, the scribe who wrote the Rhind papyrus, illustrates the method of false position in the solution of Problem 24. Problem 24 asks the question, "a quantity added to one-quarter of the quantity is 15, what is the quantity?" In modern notation the problem is $x + \frac{x}{4} = 15$. To solve the problem, make a guess of, say, $x = 4$. Then $x + \frac{x}{4} = 4 + \frac{4}{4} = 5$. This result is only one-third the desired result, so triple your guess to produce the correct solution $x = 12$. This method of making a guess, checking it, and adjusting the original guess became known as the method of false position.

EXCURSION EXERCISES, SECTION 4.2 *page 191*

1. 373　　**2.** 797　　**3.** 83,748　　**4.** 166,346　　**5.** 2,281,546　　**6.** 46,648,870

7. The nines complement and the end-around carry procedure produce the correct answer to a subtraction problem because applying these processes **is equivalent to adding and subtracting the same number from a given difference.**

> *Example:*　$641 - 235 = 641 + 999 - 235 - 999$ ● Add 999 and subtract 999.
> $= 641 + (999 - 235) - 999$ ● $999 - 235$ is 764, which can also be obtained by taking the nines complement of 235.
> $= (641 + 764) - 999$
> $= 1405 - 1000 + 1$ ● From $641 + 764 = 1405$, subtract 999 by subtracting 1000 and adding 1. This is the end-around carry procedure.
> $= 406$

Adding the nines complement of 235 to 641 produces a result that is 999 larger than $641 - 235$. Applying the end-around carry procedure decreases the previous sum by 999. Thus adding the nines complement and performing the end-around carry nullify each other and produce the correct difference.

EXERCISE SET 4.2 *page 191*

1. $(4 \times 10^1) + (8 \times 10^0)$　　**2.** $(9 \times 10^1) + (3 \times 10^0)$　　**3.** $(4 \times 10^2) + (2 \times 10^1) + (0 \times 10^0)$
4. $(5 \times 10^2) + (0 \times 10^1) + (1 \times 10^0)$　　**5.** $(6 \times 10^3) + (8 \times 10^2) + (0 \times 10^1) + (3 \times 10^0)$
6. $(9 \times 10^3) + (0 \times 10^2) + (4 \times 10^1) + (5 \times 10^0)$　　**7.** $(1 \times 10^4) + (0 \times 10^3) + (2 \times 10^2) + (0 \times 10^1) + (8 \times 10^0)$
8. $(6 \times 10^4) + (7 \times 10^3) + (4 \times 10^2) + (8 \times 10^1) + (2 \times 10^0)$　　**9.** 456　　**10.** 763　　**11.** 5076　　**12.** 3128
13. 35,407　　**14.** 230,675　　**15.** 683,040　　**16.** 53,007,902　　**17.** 76　　**18.** 98　　**19.** 395
20. 813　　**21.** 27　　**22.** 36　　**23.** 3363　　**24.** 21,225　　**25.** 23　　**26.** 45　　**27.** 97
28. 746　　**29.** 72,133　　**30.** 76,272　　**31.** 2,171,466　　**32.** 4,575,694　　**33.** 𒌋𒌋𒌋𒌋𒐕𒐕

34. 𒌋𒌋𒌋𒌋𒌋𒐕𒐕𒐕𒐕𒐕𒐕𒐕　　**35.** 𒐕𒐕 𒐕𒐕𒐕𒐕𒐕𒐕𒐕𒐕　　**36.** 𒐕𒐕𒐕𒐕𒐕𒐕𒐕𒐕𒐕 ▲

37. 𒐕 𒌋𒌋𒌋𒐕𒐕𒐕𒐕 𒌋𒌋𒌋𒐕𒐕𒐕𒐕𒐕𒐕𒐕𒐕𒐕　　**38.** 𒐕𒐕 𒌋 𒌋𒌋𒐕　　**39.** 𒐕𒐕 𒌋𒌋𒌋𒌋𒌋𒐕𒐕𒐕𒐕𒐕 𒌋𒌋𒐕𒐕𒐕𒐕

40. 𒐕𒐕𒐕 𒌋𒌋𒌋𒐕 𒌋𒌋𒐕𒐕𒐕𒐕𒐕𒐕𒐕　　**41.** 194　　**42.** 302　　**43.** 1803　　**44.** 3100　　**45.** 14,492

46. 2420　　**47.** 36,103　　**48.** 57,706　　**49.** 　　**50.** 　　**51.** 　　**52.**

53. 　　**54.** 　　**55.** 　　**56.** 　　**57. a. and b.** Answers will vary.　　**58.** If the space between

a first and second pair of wedge-shaped numerals is too small, then the numeral will appear as four wedges, which is the Babylonian numeral for 4.　　**59. a.** 7　　**b.** 20　　**c.** 35　　**d.** UUZU　　**e.** UZZTZ　　**f.** UTUZZ

1. The cards are in numeric order from smallest, $00001_{two} = 1$, to largest, $11111_{two} = 31$.
   - The first implementation of the sorting procedure places all odd numerals behind the even numerals. Thus the base two numerals of the form xxxx1 are behind the base two numerals of the form xxxx0, where x represents either a 0 or a 1.
   - The second implementation of the sorting procedure places the base two numerals of the form xxx1x behind the base two numerals of the form xxx0x. Thus, from front to back, the base two numerals are now in the order xxx00, xxx01, xxx10, xxx11.
   - The third implementation of the sorting procedure places the base two numerals of the form xx1xx behind the base two numerals of the form xx0xx. Thus the order of the cards from front to back is now given by the base two numerals xx000, xx001, xx010, xx011, xx100, xx101, xx110, xx111.
   - The fourth implementation of the sorting procedure places the base two numerals of the form x1xxx behind the base two numerals of the form x0xxx. Thus the order of the cards from front to back is now given by the base two numerals x0000, x0001, x0010, x0011, x0100, x0101, x0110, x0111, x1000, x1001, x1010, x1011, x1100, x1101, x1110, x1111.
   - The fifth implementation of the sorting procedure places the base two numerals of the form 1xxxx behind the base two numerals of the form 0xxxx. Thus the order of the cards from front to back is now given by the base two numerals 00001, 00010, 00011, 00100, 00101, 00110, 00111, 01000, 01001, 01010, 01011, 01100, 01101, 01110, 01111, 10000, 10001, 10010, 10011, 10100, 10101, 10110, 10111, 11000, 11001, 11010, 11011, 11100, 11101, 11110, 11111. In terms of base ten numerals, the cards are now arranged in the order 1, 2, 3, 4, ..., 31.

2. 10; 14   3. A base three number system requires three numerals, such as 0, 1, and 2. A notch in a card can be used to represent a 0 and a hole in a card can be used to represent a 1, as in base two, but there is no convenient method that can be used to represent the numeral 2.

1. 73   2. 82   3. 61   4. 379   5. 718   6. 477   7. 485   8. 305   9. 181
10. 58   11. 2032_{five} 12. 552_{eight} 13. 12540_{six} 14. 133220_{four} 15. 22886_{nine} 16. 105226_{seven}
17. 111111011100_{two} 18. 21210010_{three} 19. $1B7_{twelve}$ 20. $18A_{sixteen}$ 21. 27   22. 45   23. 100
24. 1960   25. 41   26. 116   27. 1338   28. 1028   29. 26_{eight} 30. 212_{five} 31. 23033_{four}
32. 1143_{six} 33. 24_{five} 34. 543_{six} 35. 2446_{nine} 36. 611_{eight} 37. 126_{eight} 38. 123_{six}
39. $7C_{sixteen}$ 40. 149_{twelve} 41. 11101010_{two} 42. 10100100_{two} 43. 312_{eight} 44. $EE5_{sixteen}$
45. $151_{sixteen}$ 46. 101110111010001_{two} 47. 1011111011110011_{two} 48. $1101010011110111000_{two}$
49. Answers will vary.   50. a. true   b. true   c. true   51. 54   52. 702
53.   54.   55.   56.   57. 256   58. 383

59. a. and b. Answers will vary.   60. a. and b. Answers will vary.
61. The following chart shows the American Standard Code for Information Interchange (ASCII) character set. Many modern computers make use of ASCII to represent internally the characters that can be typed on a computer keyboard. The chart for coding ASCII characters uses the hexadecimal numeration system. For instance, the capital letter A is shown in row 40 and column 1. Thus the ASCII code for the capital letter A is $41_{sixteen}$, which is 65 in base ten and 01000001 as an 8-bit binary numeral. The lower-case letter m is shown in row 60 and column D. The ASCII code for a lower-case m is $6D_{sixteen}$, which is 109 in base ten and 01101101 as an 8-bit binary numeral. (*Source:* **http://www.aelius.com/products/asciitable.phtml**)

	0	1	2	3	4	5	6	7	8	9	A	B	C	D	E	F
0	NUL	SOH	STX	ETX	EOT	ENQ	ACK	BEL	BS	HT	LF	VT	FF	CR	SO	SI
1	DLE	DC1	DC2	DC3	DC4	NAK	SYN	ETB	CAN	EM	SUB	ESC	FS	GS	RS	US
2	SPC	!	"	#	$	%	&	'	(	)	*	+	,	−	.	/
3	0	1	2	3	4	5	6	7	8	9	:	;	<	=	>	?
4	@	A	B	C	D	E	F	G	H	I	J	K	L	M	N	O
5	P	Q	R	S	T	U	V	W	X	Y	Z	[	\	]	^	_
6	`	a	b	c	d	e	f	g	h	i	j	k	l	m	n	o
7	p	q	r	s	t	u	v	w	x	y	z	{	|	}	~	DEL

62. All Postnet codes consist of tall bars and short bars that can be read by a machine. Every Postnet code for a zip code + 4 starts and ends with a tall bar. Each digit in a zip code +4 is represented by five bars, of which two are tall bars and three are short bars. The following table shows how each digit is represented. (*Source:* **http://www.cedar.buffalo.edu/Adserv/postcode.html**)

0	‖ııı	5	ıʰıʰı
1	ııı‖	6	ıʰıı
2	ıʰıʰı	7	‖ıı
3	ııʰʰı	8	ʰııʰı
4	ıʰıı	9	ʰıʰıı

Postnet Code

For every digit except 0, each bar position has a weight, or place value, assigned to it. From left to right, the weights are 7, 4, 2, 1, and 0. A tall bar can be thought of as a 1 and a short bar as a 0. For example, the digit 3 is

$$(0 \times 7) + (0 \times 4) + (1 \times 2) + (1 \times 1) + (0 \times 0) = 2 + 1 = 3$$

Postnet code also utilizes a "checksum" digit at the end. This checksum digit is chosen such that the sum of all the digits will be a multiple of 10. If a machine makes a mistake in reading one of the digits, the checksum will alert the operator to that fact.

EXCURSION EXERCISES, SECTION 4.4 *page 213*

1. 101_{two} **2.** 1001_{two} **3.** 11101101_{two} **4.** 110110010010_{two} **5.** 1110011100_{two} **6.** 1101110001110_{two}

EXERCISE SET 4.4 *page 213*

1. 332_{five} **2.** 1201_{four} **3.** 6562_{seven} **4.** 10000_{two} **5.** 1001000_{two} **6.** 100110001_{two} **7.** 1271_{twelve}
8. $BCF_{sixteen}$ **9.** $D036_{sixteen}$ **10.** 253_{six} **11.** 6542_{eight} **12.** 5256_{nine} **13.** 1111_{two} **14.** 100100_{two}
15. 1111001_{two} **16.** $33E_{sixteen}$ **17.** $411A_{twelve}$ **18.** $B639_{twelve}$ **19.** 1201_{three} **20.** 32133_{five}
21. 45234_{eight} **22.** 1010001_{two} **23.** 1010100_{two} **24.** 100000100_{two} **25.** 14207_{eight} **26.** 104130_{six}
27. 321222_{four} **28.** 31403_{twelve} **29.** $3A61_{sixteen}$ **30.** $109C8_{sixteen}$ **31.** Quotient 33_{four}; remainder 0_{four}
32. Quotient 130_{eight}; remainder 2_{eight} **33.** Quotient 1223_{six}; remainder 1_{six} **34.** Quotient 1101_{two}; remainder 1_{two}
35. Quotient 1110_{two}; remainder 0_{two} **36.** Quotient 10110_{two}; remainder 11_{two} **37.** Quotient $A8_{twelve}$; remainder 3_{twelve}
38. Quotient $12B_{sixteen}$; remainder $5_{sixteen}$ **39.** Quotient 14_{five}; remainder 11_{five} **40.** Base six **41.** Base eight
42. Base nine **43. a.** 629 **b.** $384 = 110000000_{two}$; $245 = 11110101_{two}$ **c.** 1001110101_{two} **d.** 629
e. Same **44. a.** 139 **b.** $457 = 111001001_{two}$; $318 = 100111110_{two}$ **c.** 10001011_{two} **d.** 139 **e.** Same
45. a. 6422 **b.** $247 = 11110111_{two}$; $26 = 11010_{two}$ **c.** 1100100010110_{two} **d.** 6422 **e.** Same
46. If you borrow in a base eight numeration system, you must borrow a power of 8. In the subtraction a group of 10, which is not a power of 8, was borrowed to produce the incorrect result 625_{eight}. **47.** Base seven **48.** No. 12 is an even number regardless of the base in which it is written. The numeral for 12 in base seven is 15_{seven}, which ends with a 5, but this does not make 12 an odd number. **49.** In a base one numeration system, 0 would be the only numeral and the place values would be $1^0, 1^1, 1^2, 1^3, \ldots$, each of which equals 1. Thus 0 is the only number you could write using a base one numeration system. **50.** N = 1, A = 3, T = 2, and O = 0 **51.** M = 1, A = 4, S = 3, and O = 0 **52. a.** 8; −59; 18 **b.** $1413_{negative\ five}$ **c.** $11101_{negative\ three}$
d. $292_{negative\ ten}$

EXCURSION EXERCISES, SECTION 4.5 *page 222*

1. The first number in the list is divisible by 2, the second is divisible by 3, the third is divisible by 4, and so on, and the last number is divisible by 1,000,001. Thus each number in the list is a composite number.
2. One solution is provided by 13! + 2, 13! + 3, 13! + 4, ..., 13! + 13.
6,227,020,802; 6,227,020,803; 6,227,020,804; 6,227,020,805; 6,227,020,806; 6,227,020,807; 6,227,020,808; 6,227,020,809; 6,227,020,810; 6,227,020,811; 6,227,020,812; 6,227,020,813.
3. a. 21! + 2, 21! + 3, 21! + 4, ..., 21! + 21 **b.** 500,001! + 2, 500,001! + 3, 500,001! + 4, ..., 500,001! + 500,001
c. 7,000,000,001! + 2, 7,000,000,001! + 3, 7,000,000,001! + 4, ..., 7,000,000,001! + 7,000,000,001

EXERCISE SET 4.5 *page 222*

1. 1, 2, 4, 5, 10, 20　　**2.** 1, 2, 4, 8, 16, 32　　**3.** 1, 5, 13, 65　　**4.** 1, 3, 5, 15, 25, 75　　**5.** 1, 41　　**6.** 1, 79
7. 1, 2, 5, 10, 11, 22, 55, 110　　**8.** 1, 2, 3, 5, 6, 10, 15, 25, 30, 50, 75, 150　　**9.** Composite　　**10.** Prime　　**11.** Prime
12. Composite　　**13.** Prime　　**14.** Composite　　**15.** Prime　　**16.** Composite　　**17.** 2, 3, 5, 6, and 10
18. 2　　**19.** 3　　**20.** 2, 3, 4, 6, and 8　　**21.** 2, 3, 4, 6, and 8　　**22.** 3 and 5　　**23.** 2, 5, and 10　　**24.** 3
25. $2 \cdot 3^2$　　**26.** $2^4 \cdot 3$　　**27.** $2^3 \cdot 3 \cdot 5$　　**28.** $2^2 \cdot 5 \cdot 19$　　**29.** $5^2 \cdot 17$　　**30.** 5^4　　**31.** 2^{10}
32. $2 \cdot 3 \cdot 5 \cdot 47$　　**33.** $2^3 \cdot 3 \cdot 263$　　**34.** $5 \cdot 631$　　**35.** $2 \cdot 3^2 \cdot 1013$　　**36.** $5 \cdot 53 \cdot 73$　　**37.** 2, 3, 5, 7, 11,
13, 17, 19, 23, 29, 31, 37, 41, 43, 47, 53, 59, 61, 67, 71, 73, 79, 83, 89, 97, 101, 103, 107, 109, 113, 127, 131, 137, 139, 149, 151, 157, 163, 167,
173, 179, 181, 191, 193, 197, 199　　**38. a.** 15　　**b.** 10　　**c.** 10　　**d.** 11　　**39.** 3 and 5, 5 and 7, 11 and 13, 17
and 19, 29 and 31, 41 and 43, 59 and 61, 71 and 73, 101 and 103, 107 and 109, 137 and 139, 149 and 151, 179 and 181, 191 and 193, 197
and 199　　**40.** One of *every* three consecutive odd natural numbers is divisible by 3. If *n* is a natural number greater than 3, one of
the numbers *n*, *n* + 2, or *n* + 4 is divisible by 3 and hence at least one of the numbers *n*, *n* + 2, or *n* + 4 is not a prime number. Thus 3,
5, and 7 are the only prime triplets.　　In Exercise 41, parts a to d, only one possible sum is given.　　**41. a.** 24 = 5 + 19
b. 50 = 3 + 47　　**c.** 144 = 5 + 139　　**d.** 210 = 11 + 199　　**42.** Conjecture: Every perfect square has an odd number
of distinct natural number factors.　　**43.** Yes　　**44.** Yes　　**45.** Yes　　**46.** No　　**47.** No　　**48.** No
49. Yes　　**50.** Yes　　**51.** Yes　　**52.** Yes　　**53.** Yes　　**54.** Yes　　**55.** No　　**56.** No　　**57.** Yes
58. Yes　　**59.** To determine whether a given number is divisible by 17, multiply the ones digit of the given number by 5. Find the
difference between this result and the number formed by omitting the ones digit from the given number. Keep repeating this procedure
until you obtain a small final difference. If the final difference is divisible by 17, then the given number is divisible by 17. If the final
difference is not divisible by 17, then the given number is not divisible by 17.　　**60.** To determine whether a given number is divisible
by 19, multiply the ones digit of the given number by 2. Find the sum of this result and the number formed by omitting the ones digit
from the given number. Keep repeating this procedure until you obtain a small final sum. If the final sum is divisible by 19, then the given
number is divisible by 19. If the final sum is not divisible by 19, then the given number is not divisible by 19.　　**61.** 12　　**62.** 12
63. 18　　**64.** 24　　**65. a.** *N* is a product of all the prime numbers. Thus *N* − 1, which must have at least one prime factor,
must share at least one prime factor with the number *N*.　　**b.** The distributive property of multiplication over addition (subtraction)
66. a. If a number consists of *k* 1's, where *k* is a prime number, then the number is a prime number. This statement is not true because
111 consists of exactly three 1's and 111 is not a prime number.　　**b.** The theorem stated in Exercise 66 applies only to *prime numbers*
of the form 111...1.　　**67. a.** Definition of a divisor　　**b.** *aj* > *aj*　　**c.** Because *aj* = *n*, where *a* and *j* are both natural
numbers, both *a* and *j* are by definition divisors of *n*.　　**68.** RSA is a powerful method of encryption. It allows one to communicate
securely through an insecure channel such as the Internet. The RSA encryption algorithm is a fairly simple mathematical procedure that
involves large prime numbers. Computers can easily implement the encryption algorithm. One important feature of the algorithm is that
anyone (the public) can use the algorithm to encrypt a message, but only those people with the "private key" can decrypt a message. The
RSA algorithm is considered very secure because in order to crack a coded message, one must factor a very large number. By present
techniques this is extremely time-consuming, even with modern high-speed computers.

EXCURSION EXERCISES, SECTION 4.6 *page 233*

1. The sum of the proper factors of 200 is 265. Thus, 200 is an abundant number.　　**2.** The sum of the proper factors of 262 is 134.
Thus, 262 is a deficient number.　　**3.** The sum of the proper factors of 325 is 109. Thus, 325 is a deficient number.
4. The sum of the proper factors of 496 is 496. Thus, 496 is a perfect number.　　**5.** The only divisors of a prime number are 1 and
the number itself. Thus the only proper factor of a prime number is 1. Hence every prime number is deficient.　　**6.** Because we
cannot find a multiple of 6 greater than 6 that is *not* abundant, we conjecture, using inductive reasoning, that every multiple of 6 greater
than 6 is an abundant number.

EXERCISE SET 4.6 *page 234*

1. Abundant　　**2.** Deficient　　**3.** Deficient　　**4.** Deficient　　**5.** Deficient　　**6.** Abundant
7. Abundant　　**8.** Deficient　　**9.** Deficient　　**10.** Deficient　　**11.** Deficient　　**12.** Deficient
13. Prime　　**14.** Prime　　**15.** Prime　　**16.** Prime　　**17.** $2^{126}(2^{127} - 1)$　　**18.** $2^{520}(2^{521} - 1)$　　**19.** 6
20. 39,751　　**21.** 420,921　　**22.** 909,526　　**23.** 12　　**24.** 130,100　　**25.** 757,263　　**26.** 1,791,864
27. $9^5 + 15^5 = 818{,}424$. Because $15^5 < 818{,}424$ and $16^5 > 818{,}424$, we know there is no natural number *z* such that $z^5 = 9^5 + 15^5$.
28. $7^6 + 19^6 = 47{,}163{,}530$. Because $19^6 < 47{,}163{,}530$ and $20^6 > 47{,}163{,}530$, we know there is no natural number *z* such that
$z^6 = 7^6 + 19^6$.　　**29.** For any natural number *n*, the number 231^n has a ones digit of 1. For any natural number *n*, the number 455^n

has a ones digit of 5. Thus, for any natural number n, the number $231^n + 455^n$ has a ones digit of 6. For any natural number n, the number 1347^n has a ones digit of either 7, 9, 3, or 1. Therefore, it is not possible for $231^n + 455^n$ to be equal to 1347^n for any natural number n. **30.** For any natural number n, the number 4078^n has a ones digit of 8, 4, 2, or 6. For any natural number n, the number 3433^n has a ones digit of 3, 9, 7, or 1. Thus, for any natural number n, the number $4078^n + 3433^n$ has an *odd* ones digit. For any natural number n, the number $12,046^n$ has a ones digit of 6. Therefore, it is not possible for $4078^n + 3433^n$ to equal $12,046^n$ for any natural number n. **31. a.** $12^7 - 12 = 35,831,796$, which is divisible by 7. **b.** $8^{11} - 8 = 8,589,934,584$, which is divisible by 11. **32. a.** No **b.** Yes **33.** 945 **34.** The first five Fermat numbers formed using $m = 0, 1, 2, 3,$ and 4 are all prime numbers. In 1732, Euler discovered that the sixth Fermat number $4,294,967,297$, formed using $m = 5$, is not a prime number because it is divisible by 641.

CHAPTER 4 REVIEW EXERCISES   *page 235*

1. ⟨numeral symbols⟩ [Sec. 4.1] **2.** ⟨numeral symbols⟩ [Sec. 4.1]
3. 223,013 [Sec. 4.1] **4.** 221,354 [Sec. 4.1] **5.** 349 [Sec. 4.1] **6.** 774 [Sec. 4.1] **7.** 9640 [Sec. 4.1]
8. 92,444 [Sec. 4.1] **9.** DLXVII [Sec. 4.1] **10.** DCCCXXIII [Sec. 4.1] **11.** MMCDLXXXIX. [Sec. 4.1]
12. MCCCXXXV [Sec. 4.1] **13.** $(4 \times 10^2) + (3 \times 10^1) + (2 \times 10^0)$ [Sec. 4.2]
14. $(4 \times 10^5) + (5 \times 10^4) + (6 \times 10^3) + (3 \times 10^2) + (2 \times 10^1) + (7 \times 10^0)$ [Sec. 4.2] **15.** 5,038,204 [Sec. 4.2]
16. 387,960 [Sec. 4.2] **17.** 801 [Sec. 4.2] **18.** 1603 [Sec. 4.2] **19.** 76,441 [Sec. 4.2]
20. 87,393 [Sec. 4.2] **21.** ⟨numeral symbols⟩ [Sec. 4.2] **22.** ⟨numeral symbols⟩ [Sec. 4.2]
23. ⟨numeral symbols⟩ [Sec. 4.2] **24.** ⟨numeral symbols⟩ [Sec. 4.2]
25. 194 [Sec. 4.2] **26.** 267 [Sec. 4.2] **27.** 2178 [Sec. 4.2] **28.** 6580 [Sec. 4.2]
29. ⟨numeral symbol⟩ [Sec. 4.2] **30.** ⟨numeral symbol⟩ [Sec. 4.2] **31.** ⟨numeral symbol⟩ [Sec. 4.2] **32.** ⟨numeral symbol⟩ [Sec. 4.2] **33.** 29 [Sec. 4.3]
34. 146 [Sec. 4.3] **35.** 227 [Sec. 4.3] **36.** 286 [Sec. 4.3] **37.** 1153_{six} [Sec. 4.3]
38. 640_{eight} [Sec. 4.3] **39.** 458_{nine} [Sec. 4.3] **40.** $B62_{\text{twelve}}$ [Sec. 4.3] **41.** 34_{eight} [Sec. 4.3]
42. 124_{eight} [Sec. 4.3] **43.** $38D_{\text{sixteen}}$ [Sec. 4.3] **44.** 754_{sixteen} [Sec. 4.3] **45.** 10101_{two} [Sec. 4.3]
46. 1100111010_{two} [Sec. 4.3] **47.** 1001010_{two} [Sec. 4.3] **48.** $110001110010_{\text{two}}$ [Sec. 4.3]
49. 423_{six} [Sec. 4.4] **50.** 1240_{eight} [Sec. 4.4] **51.** 536_{nine} [Sec. 4.4] **52.** 1113_{four} [Sec. 4.4]
53. 16412_{eight} [Sec. 4.4] **54.** 324203_{five} [Sec. 4.4] **55.** Quotient 11100_{two}; remainder 1_{two} [Sec. 4.4]
56. Quotient 21_{four}; remainder 3_{four} [Sec. 4.4] **57.** $3^2 \cdot 5$ [Sec. 4.5] **58.** $2 \cdot 3^3$ [Sec. 4.5]
59. $3^2 \cdot 17$ [Sec. 4.5] **60.** $3 \cdot 5 \cdot 19$ [Sec. 4.5] **61.** Composite [Sec. 4.5] **62.** Composite [Sec. 4.5]
63. Composite [Sec. 4.5] **64.** Composite [Sec. 4.5] **65.** Perfect [Sec. 4.6] **66.** Deficient [Sec. 4.6]
67. Abundant [Sec. 4.6] **68.** Abundant [Sec. 4.6] **69.** $2^{60}(2^{61} - 1)$ [Sec. 4.6] **70.** $2^{1278}(2^{1279} - 1)$ [Sec. 4.6]
71. 368 [Sec. 4.1] **72.** 513 [Sec. 4.1] **73.** 1162 [Sec. 4.1] **74.** 3003 [Sec. 4.1] **75.** 410 [Sec. 4.3]
76. 277 [Sec. 4.3] **77.** 1041 [Sec. 4.3] **78.** 1616 [Sec. 4.3] **79.** A base ten number is divisible by 3 provided the sum of the digits of the number is divisible by 3. [Sec. 4.5] **80.** A number is divisible by 6 provided the number is divisible by 2 and by 3. [Sec. 4.5] **81.** Every composite number can be written as a unique product of prime numbers (disregarding the order of the factors). [Sec. 4.5] **82.** Zero [Sec. 4.6] **83.** 39,751 [Sec. 4.6] **84.** 895,932 [Sec. 4.6]

CHAPTER 4 TEST   *page 237*

1. ⟨numeral symbols⟩ [Sec. 4.1] **2.** 4263 [Sec. 4.1] **3.** 1447 [Sec. 4.1] **4.** MMDCIX [Sec. 4.1]
5. $(6 \times 10^4) + (7 \times 10^3) + (4 \times 10^2) + (8 \times 10^1) + (5 \times 10^0)$ [Sec. 4.2] **6.** 530,284 [Sec. 4.2] **7.** 37,274 [Sec. 4.2]
8. ⟨numeral symbols⟩ [Sec. 4.2] **9.** 1305 [Sec. 4.2] **10.** ⟨numeral symbol⟩ [Sec. 4.2] **11.** 854 [Sec. 4.3]
12. a. 4144_{eight} **b.** $12B0_{\text{twelve}}$ [Sec. 4.3] **13.** $100101110111_{\text{two}}$ [Sec. 4.3] **14.** $AB7_{\text{sixteen}}$ [Sec. 4.3]
15. 112_{five} [Sec. 4.4] **16.** 313_{eight} [Sec. 4.4] **17.** 11100110_{two} [Sec. 4.4]
18. Quotient 61_{seven}; remainder 3_{seven} [Sec. 4.4] **19.** $2 \cdot 5 \cdot 23$ [Sec. 4.5] **20.** Composite [Sec. 4.5]
21. a. No **b.** Yes **c.** No [Sec. 4.5] **22. a.** Yes **b.** No **c.** No [Sec. 4.5]
23. Abundant [Sec. 4.6] **24.** $2^{16}(2^{17} - 1)$ [Sec. 4.6]

CHAPTER 5

EXCURSION EXERCISES, SECTION 5.1 *page 251*

1. 21.3; low risk **2.** 29.8, moderate risk **3.** 179 pounds **4.** 135 pounds **5.** 30 pounds
6. 69 pounds **7. a.** 7.9 points **b.** 58 pounds **8. a.** 5.1 points **b.** 33 pounds

EXERCISE SET 5.1 *page 252*

1. An equation expresses the equality of two mathematical expressions. An equation contains an equals sign. An expression does not.
2. The solution is 8. Explanations will vary. **3.** Substitute the solution back into the original equation and confirm the equality.
4. -12 **5.** 12 **6.** -1 **7.** 22 **8.** -50 **9.** -8 **10.** 12 **11.** -20 **12.** -4

13. 8 **14.** -20 **15.** -1 **16.** 8 **17.** 1 **18.** $-\dfrac{2}{5}$ **19.** $-\dfrac{1}{3}$ **20.** $\dfrac{3}{4}$ **21.** $\dfrac{2}{3}$

22. 3 **23.** -2 **24.** 3 **25.** 4 **26.** 2 **27.** 2 **28.** -1 **29.** 2 **30.** $\dfrac{1}{2}$ **31.** $\dfrac{1}{4}$

32. 5 **33.** -2 **34.** $-\dfrac{1}{9}$ **35.** 4 **36.** 6 **37.** 8 **38.** 2 **39.** 1 **40.** 8

41. -32 **42.** $14,450.87 **43.** $16,859.34 **44.** 30 inches **45.** 1350 inches **46.** 11,250 inches
47. 60 feet **48.** 80 feet **49.** 1952 **50.** 1987 **51.** 168 feet **52.** 136 feet
53. 18.6 degrees Celsius **54.** 24.3 degrees Celsius **55.** 175 **56.** 168.75 **57.** $22,000
58. 300 megahertz **59. a.** $22,063 **b.** $24,578 **60. a.** $40 million **b.** $47.5 million **61.** $117.75
62. 5 hours **63. a.** 163,000 kilograms **b.** 1930 kilograms **64.** 150 research assistants **65.** $12.50
66. $2000 and $3000 **67.** 200 vertical pixels **68.** more than 11 ounces but not over 12 ounces **69.** more than

16 minutes but not over 17 minutes **70.** $h = \dfrac{2A}{b}$ **71.** $b = P - a - c$ **72.** $t = \dfrac{d}{r}$ **73.** $R = \dfrac{E}{I}$

74. $R = \dfrac{PV}{nT}$ **75.** $r = \dfrac{I}{Pt}$ **76.** $W = \dfrac{P - 2L}{2}$ **77.** $C = \dfrac{5}{9}(F - 32)$ **78.** $C = R - P$ **79.** $t = \dfrac{A - P}{Pr}$

80. $V_0 = \dfrac{S + 16t^2}{t}$ **81.** $f = \dfrac{T + gm}{m}$ **82.** $R = Pn + C$ **83.** $S = C - Rt$ **84.** $h = \dfrac{3V}{\pi r^2}$

85. $b_2 = \dfrac{2A}{h} - b_1$ **86.** $d = \dfrac{a_n - a_1}{n - 1}$ **87.** $h = \dfrac{S}{2\pi r} - r$ **88.** $y = 2x - 4$ **89.** $y = 2 - \dfrac{4}{3}x$

90. $x = \dfrac{-by - c}{a}$ **91.** $x = \dfrac{y - y_1}{m} + x_1$ **92.** No solution **93.** 0 **94.** Every real number is a solution.

95. Every real number is a solution. **96. a.** 0 **b.** -20 **97.** $x = \dfrac{d - b}{a - c}$. $a \neq c$ or the denominator equals zero and

the expression is undefined. **98.** Here is one possibility: Row 1–1, 2, 4, 3; Row 2–4, 3, 1, 2; Row 3–3, 4, 2, 1; Row 4–2, 1, 3, 4
99. a. $4125 **b.** 4325 passenger cars **c.** $T = 225m + 0.03x$, where T is the total cost, m is the number of months, and x
is the number of copies **d.** $T = 750m + 30,000$, where T is the total number of miles driven **e.** $C = 2.50 + 1.75(h - 1)$,
where C is the parking charge **f.** $C = 19.95d + 0.25(m - 100)$, where C is the total cost, d is the number of days, and m is the
number of miles driven **g.** Answers will vary.

EXCURSION EXERCISES, SECTION 5.2 *page 269*

1. 1.71 **2.** 4.33 **3.** 1999; 0.82 **4.** 2.77 **5.** Answers will vary.

EXERCISE SET 5.2 *page 270*

1. Examples will vary. **2.** Unit rates are used so that different rates can be easily compared. **3.** The purpose is to allow
currency from one country to be converted into the currency of another country. **4.** Here are a few examples: Investors talk of
price-earnings ratios; accountants use the current ratio, which is the ratio of current assets to current liabilities; metallurgists use ratios to
make various grades of steel. **5.** Explanations will vary. **6.** It means that the product of the means in a proportion is equal

to the product of the extremes.	**7.** The cross-products method is a shortcut for multiplying each side of the proportion by the least common multiple of the denominators.	**8.** 48.5 miles per hour	**9.** 23 miles per gallon	**10.** 68 words per minute
11. 12.5 meters per second	**12.** $26 per share	**13.** 400 square feet per gallon	**14.** $14 per hour
15. 213,600 gallons per minute	**16. a.** 336 feet per minute	**b.** 89 seconds	**17.** A 24-ounce jar of mayonnaise for $2.09	**18.** 18 ounces for $2.89	**19.** $16.50 per hour	**20. a.** 9.2, 9, 10.4, 10.1, 9.6, 9.7, 9.5, 10.7, 11.8, 7.3, 9.7, 6.5
b. Barry Bonds; Mark McGwire	**c.** Explanations may vary.	**21. a.** Australia	**b.** 819 more people per square mile
22. a. 2, 2.1, 2.2, 2.4, 2.5, 2.7, 3, 3.3, 3.5, 3.7, 29.4	**b.** 12.25 times greater	**23.** 73,576 krona	**24.** 2,165,850 rupees
25. 24,035 pounds	**26.** 2,904,000 yen	**27.** For each state, the ratio is 3.125 to 1.	**28. a.** No
b. 297,108 deaths	**29.** 13:1 or 13 to 1. There is one faculty member for every 13 students at Syracuse University.
30. Georgetown University	**31.** Boston College	**32.** Syracuse University and Villanova University	**33.** No
34. 4.5	**35.** 17.14	**36.** 42.86	**37.** 25.6	**38.** 216	**39.** 20.83	**40.** 4.31	**41.** 2.22
42. 4.35	**43.** 13.71	**44.** 10.97	**45.** 39.6	**46.** 1.15	**47.** 0.52	**48.** 38.73	**49.** 6.74
50. 29 pounds	**51.** $45,000	**52.** 50 pounds	**53.** 5.5 milligrams	**54.** 329 miles	**55.** 24 feet; 15 feet
56. 16 miles	**57.** 160,000 people	**58.** 3.5 gallons	**59.** 198,000 miles	**60.** $1.25 million
61. 11.25 grams. Explanations will vary.	**62. a.** There were 196 murders during the year, or 47 murders for every 100,000 people in the city.	**b.** The population of each city was needed.	**63. a.** True	**b.** True	**c.** True	**d.** False
64.	$\dfrac{a}{b} = \dfrac{c}{d}$	**65. a.–c.** Explanations will vary.	**66. a.** To make comparisons easier	**b.** $\dfrac{18.3}{109.5} = \dfrac{45}{168}$; no

$$\dfrac{a}{b} + 1 = \dfrac{c}{d} + 1$$

$$\dfrac{a}{b} + \dfrac{b}{b} = \dfrac{c}{d} + \dfrac{d}{d}$$

$$\dfrac{a+b}{b} = \dfrac{c+d}{d}$$

c. Answers will vary.	**67.** Answers will vary.

EXCURSION EXERCISES, SECTION 5.3 *page 291*

1. $15,010.51	**2.** $5330.75	**3.** $10,333; balance due of $3209	**4.** $16,856.25; a tax refund of $1563.75
5. No. The taxpayer pays 30.5% on the amount earned over a given figure.	**6.** $17,250
7. $4057.50 = 15% of $27,050; $14,645 = (15% of $27,050) + [27.5% of ($65,550 − $27,050)]
8. $6780 = 15% of $45,200; $24,393.75 = (15% of $45,200) + [27.5% of ($109,250 − $45,200)]

EXERCISE SET 5.3 *page 291*

1. Answers will vary.	**2.** 100% = 1. Multiplying a number by 1 does not change the value of the number.	**3.** 3
4. One ratio is $\dfrac{\text{percent}}{100}$, in which the percent is not written as a decimal. The other ratio is $\dfrac{\text{amount}}{\text{base}}$.	**5.** Employee B's salary is now the highest because Employee B had the highest initial salary, and the percent raises were the same for all employees.

6. Employee B's salary is now the highest because Employee B had the greatest percent raise.	**7.** 0.5; 50%	**8.** $\dfrac{3}{4}$; 75%

9. $\dfrac{2}{5}$; 0.4	**10.** 0.375; 37.5%	**11.** $\dfrac{7}{10}$; 70%	**12.** $\dfrac{9}{16}$; 0.5625	**13.** 0.55; 55%	**14.** $\dfrac{13}{25}$; 52%

15. $\dfrac{5}{32}$; 0.15625	**16.** 0.18; 18%	**17. a.** 73 fans	**b.** More fans approved.
c. 7%; 100% − 73% − 20% = 7%	**18.** 25.2 hours	**19.** $26.6 billion	**20.** 30%	**21.** 23.7%
22. 160 students	**23. a.** $1381	**b.** $1628	**24. a.** $14,600	**b.** Food: 27.5%; veterinary: 26.9%; grooming, etc.: 20.3%; training: 8.4%; flea and tick treatment: 7.3%; other: 9.6%	**c.** The percents would be lower.
d. Explanations will vary.	**25. a.** 48.2 hours	**b.** 33.6 hours	**c.** 3.4 hours	**26. a.** 1990: 9.2%; 1996: 9.8%
b. 1,141,022 state inmates; more; the number of federal inmates is growing at a more rapid rate	**c.** Explanations will vary.
27. a. 1998 to 2000	**b.** 2004 to 2006	**c.** More slowly	**28. a.** Mississippi	**b.** 2,697,000 people, 0.00456%
c. 1,828,000 people, 0.00290%	**d.** 870,000 people, 0.00345%	**e.** 0.00327%	**f.** The rate is 1000 times the percent.

29. a.   Arizona: 118,153; California: 625,041; Colorado: 165,038; Maryland: 142,718; Massachusetts: 246,833; Minnesota: 229,543; New Hampshire: 51,714; New Jersey: 229,543; Vermont: 17,274; Virginia: 185,900   **b.**   New Hampshire; California   **c.**   More than half   **d.**   Massachusetts: 7.5%; New Jersey: 5.5%; Virginia: 5.2%   **e.**   5.3%   **f.**   The rate is 10 times the percent.
30.   260 eggs per person per year   **31. a.**   900%   **b.**   60%   **c.**   1500%   **d.**   Explanations will vary.
32. a.   102.5%   **b.**   122.2%   **c.**   350%   **d.**   It is 4.5 times larger. Convert the percent to a decimal and add 1.
33. a.   Teacher aides and educational assistants   **b.**   Personal and home care aides   **c.**   158,100 people   **d.**   183,600 people   **e.**   Guards   **f.**   The percent increases are based on different original employment figures.   **g.**   Answers will vary.   **34. a.**   31,820 more people   **b.**   598,650 voters   **c.**   2,097,600 people   **35.**   Less than   **36.**   16.79%
37.   51.7 months   **38. a.**   Ages 5–17   **b.**   65 and older   **c.**   55 and older   **d.**   Under 5   **e.**   Ages 18–24
f.   Under 25; 25 and older   **g.**   62.8%; yes; 6.7%   **h.**   7.4%   **i.**   27.8%   **j.**   52.0%; 46.4%
k.   The percent of the population in the work force is decreasing while the percent of the population of retirement age is increasing.
l.   Answers will vary.   **m.**   Answers will vary.   **39. a.**   11,000,000 TV households; 58,100,000 TV households
b.   16,600,000 TV households; 59,100,000 TV households   **c.**   2.5 people   **d.**   Answers will vary.

EXCURSION EXERCISES, SECTION 5.4   *page 309*

1. -2 and 5 **2.** 4 **3.** -3 and $\frac{4}{3}$ **4.** -3 and $\frac{1}{3}$ **5.** $3 + 2\sqrt{3}$ and $3 - 2\sqrt{3}$ **6.** $1 + \sqrt{6}$ and $1 - \sqrt{6}$

7. $\frac{1 + \sqrt{2}}{2}$ and $\frac{1 - \sqrt{2}}{2}$ **8.** $\frac{1 + \sqrt{5}}{2}$ and $\frac{1 - \sqrt{5}}{2}$ **9.** $x^2 - 5x - 6 = 0$ **10.** $x^2 + 9x + 20 = 0$
11. $2x^2 - 7x + 3 = 0$ **12.** $4x^2 - 5x - 6 = 0$ **13.** $8x^2 + 10x - 3 = 0$ **14.** $9x^2 - 4 = 0$
15. $x^2 - 4x + 2 = 0$ **16.** $x^2 - 2x - 2 = 0$

EXERCISE SET 5.4   *page 310*

1.   The Principle of Zero Products states that if the product of two factors is zero, then one of the two factors equals zero. A second-degree equation must be written in standard form so that the variable expression is equal to zero. Then, when the variable expression is factored, the Principle of Zero Products can be used to set each factor equal to zero.   **2.**   If $a = 0$, then there is no x^2 term in $ax^2 + bx + c = 0$, and it is the x^2 term that makes an equation quadratic.   **3.**   Answers will vary.   **4.**   Answers will vary.

5. -2 and 5 **6.** -6 and 1 **7.** $\frac{1 + \sqrt{5}}{2}$ and $\frac{1 - \sqrt{5}}{2}$; -0.618 and 1.618 **8.** $\frac{1 + \sqrt{13}}{2}$ and $\frac{1 - \sqrt{13}}{2}$; -1.303 and 2.303 **9.** $3 + \sqrt{13}$ and $3 - \sqrt{13}$; -0.606 and 6.606 **10.** $-2 + \sqrt{6}$ and $-2 - \sqrt{6}$; -4.449 and 0.449

11. 0 and 2 **12.** 0 and -5 **13.** $\frac{1 + \sqrt{17}}{2}$ and $\frac{1 - \sqrt{17}}{2}$; -1.562 and 2.562 **14.**   No real number solutions

15. $\frac{2 + \sqrt{14}}{2}$ and $\frac{2 - \sqrt{14}}{2}$; -0.871 and 2.871 **16.** $\frac{3 + 2\sqrt{3}}{3}$ and $\frac{3 - 2\sqrt{3}}{3}$; -0.155 and 2.155 **17.** $2 + \sqrt{11}$ and $2 - \sqrt{11}$; -1.317 and 5.317 **18.** $-3 + \sqrt{10}$ and $-3 - \sqrt{10}$; -6.162 and 0.162 **19.** $-\frac{3}{2}$ and 6 **20.** $-\frac{2}{3}$ and 2

21.   No real number solutions   **22.**   No real number solutions   **23.** -4 and $\frac{1}{4}$ **24.** $-\frac{3}{4}$ and $\frac{1}{2}$ **25.** $-\frac{1}{2}$ and $\frac{4}{3}$

26. $-\frac{1}{3}$ and $\frac{3}{2}$ **27.** $1 + \sqrt{5}$ and $1 - \sqrt{5}$; -1.236 and 3.236 **28.** $7 + \sqrt{30}$ and $7 - \sqrt{30}$; 1.523 and 12.477
29.   0.75 second and 3 seconds   **30.**   10 meters   **31.** $T = 0.5(1)^2 + 0.5(1) = 1$; $T = 0.5(2)^2 + 0.5(2) = 3$; $T = 0.5(3)^2 + 0.5(3) = 6$; $T = 0.5(4)^2 + 0.5(4) = 10$; 10 rows   **32.**   After 1.73 seconds and after 5.77 seconds
33. a.   240 feet   **b.**   30 miles per hour   **34. a.**   6.29 feet   **b.**   Yes; 1 foot   **c.**   No
35.   5.51 seconds   **36.**   1.58 seconds   **37.**   244.10 feet   **38.**   6 meters   **39.**   No
40.   151.64 kilometers per hour   **41.**   1.74 seconds and 10.76 seconds   **42.**   No   **43.**   62 cents
44. $ax^2 + bx = 0$ **45.**   If the discriminant is not a perfect square, the radical expression in the quadratic formula
$x(ax + b) = 0$
$x = 0$ $ax + b = 0$
$ax = -b$
$x = -\dfrac{b}{a}$
will not simplify to a whole number.   **46.**   7.507 feet and 0.493 feet   **47.** $-12a$ and $-4a$ **48.** $3b$ and $5b$

49. $-\dfrac{b}{2}$ and $-b$ **50.** c and $\dfrac{c}{3}$ **51.** $-y$ and $\dfrac{3}{2}y$ **52.** $2y$ **53.** $b^2 - 4ac = b^2 - 4(1)(-1) = b^2 + 4$. If $b^2 - 4ac \geq 0$, then the equation has real number solutions. Because $b^2 \geq 0$ for all real numbers b, $b^2 + 4 \geq 4$ for all real numbers b. Therefore, the equation has real number solutions regardless of the value of b. **54.** $b^2 - 4ac = b^2 - 4(2)(-2) = b^2 + 16$. If $b^2 - 4ac \geq 0$, then the equation has real number solutions. Because $b^2 \geq 0$ for all real numbers b, $b^2 + 16 \geq 16$ for all real numbers b. Therefore, the equation has real number solutions regardless of the value of b.

CHAPTER 5 REVIEW EXERCISES *page 314*

1. 4 [Sec. 5.1] **2.** $\dfrac{1}{8}$ [Sec. 5.1] **3.** -2 [Sec. 5.1] **4.** $\dfrac{10}{3}$ [Sec. 5.2] **5.** No real number solutions [Sec. 5.4] **6.** -5 and 6 [Sec. 5.4] **7.** $2 + \sqrt{3}$ and $2 - \sqrt{3}$ [Sec. 5.4] **8.** $\dfrac{1 + \sqrt{13}}{2}$ and $\dfrac{1 - \sqrt{13}}{2}$ [Sec. 5.4] **9.** $y = -\dfrac{4}{3}x + 4$ [Sec. 5.1] **10.** $t = \dfrac{f - v}{a}$ [Sec. 5.1] **11.** 2450 feet [Sec. 5.1] **12.** 3 seconds [Sec. 5.1] **13.** 60°C [Sec. 5.1] **14.** 39 minutes [Sec. 5.1] **15.** 28.4 miles per gallon [Sec. 5.2] **16.** $\dfrac{1}{3}$ [Sec. 5.2] **17.** 6883 complaints [Sec. 5.1] **18. a.** New York, Chicago, Philadelphia, Los Angeles, Houston **b.** 21,596 more people per square mile [Sec. 5.2] **19. a.** 17:1, 17 to 1. There are 17 students for every one faculty member at the university. **b.** Grand Canyon University, DeVry Institute of Technology/Phoenix **c.** Prescott College and the University of Arizona [Sec. 5.2] **20.** Department A: $10,500; Department B: $24,500 [Sec. 5.2] **21.** 7.5 tablespoons [Sec. 5.2] **22. a.** No **b.** 17:8 **c.** $1,147.5 billion **d.** $388.125 billion [Sec. 5.2] **23. a.** 51.0% **b.** Less than [Sec. 5.3] **24.** 735,000 people [Sec. 5.3] **25.** 69.0% [Sec. 5.3] **26. a.** 83.$\overline{3}$% **b.** 100% **c.** 50% [Sec. 5.3] **27.** 80% [Sec. 5.3] **28. a.** Ages 9–10 **b.** Ages 15–16 **c.** 46.5%; less than **d.** 7230 boys **e.** 549,400 young people [Sec. 5.3] **29.** 1 second and 5 seconds [Sec. 5.4] **30.** 0.5 second and 1.5 seconds [Sec. 5.4]

CHAPTER 5 TEST *page 318*

1. 14 [Sec. 5.1] **2.** 10 [Sec. 5.1] **3.** $\dfrac{21}{4}$ [Sec. 5.2] **4.** 3 and 9 [Sec. 5.4] **5.** $\dfrac{2 + \sqrt{7}}{3}$ and $\dfrac{2 - \sqrt{7}}{3}$ [Sec. 5.4] **6.** $y = \dfrac{1}{2}x - \dfrac{15}{2}$ [Sec. 5.1] **7.** $F = \dfrac{9}{5}C + 32$ [Sec. 5.1] **8.** 2.5 minutes [Sec. 5.1] **9.** 10 days [Sec. 5.1] **10.** 54.8 miles per hour [Sec. 5.2] **11.** 843 acres [Sec. 5.1] **12. a.** 2.727, 2.905, 2.777, 2.808, 2.901, 2.904 **b.** Ty Cobb, Rogers Hornsby, Joe Jackson, Tris Speaker, Ted Williams, Billy Hamilton [Sec. 5.2] **13.** $\dfrac{4}{7}$ [Sec. 5.2] **14.** $112,500 and $67,500 [Sec. 5.2] **15.** 2.75 pounds [Sec. 5.2] **16. a.** Miami-Dade **b.** 9420 violent crimes [Sec. 5.2] **17.** 14.4% [Sec. 5.3] **18. a.** 31.8% **b.** Between 1994 and 1995 **c.** 80.9% [Sec. 5.3] **19. a.** 20% **b.** 1.6 million working farms **c.** Answers will vary. [Sec. 5.3] **20.** 0.2 second and 1.6 seconds [Sec. 5.4]

CHAPTER 6

EXCURSION EXERCISES, SECTION 6.1 *page 333*

1. a. 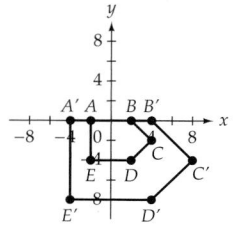 **b.** 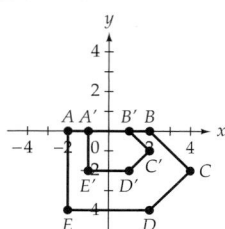 **2.** Drawings will vary. The angles of the

dilated figure have the same measures as the corresponding angles of the original figure.   **3.**   Drawings will vary.
4.   Drawings will vary.   **5.**   The center of dilation is the center of the edge of the paper along the width.

EXERCISE SET 6.1   *page 334*

1.
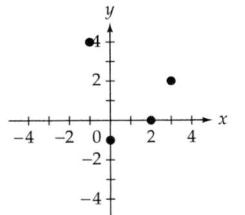

2.
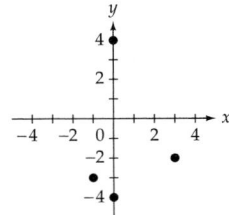

3.
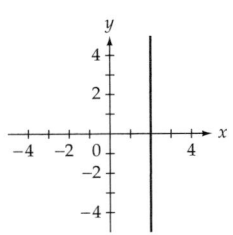

4.
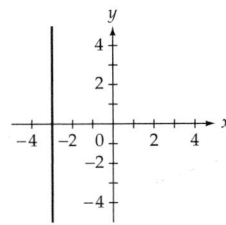

5.
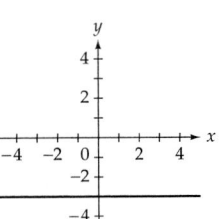

6.
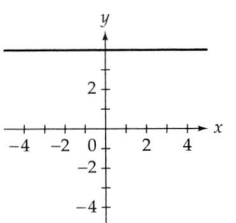

7.
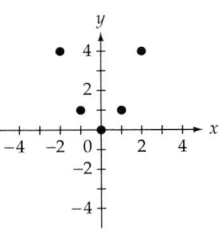

8.
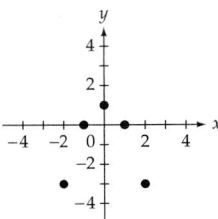

9.
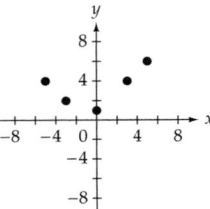

10.
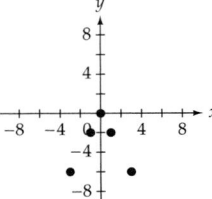

11.
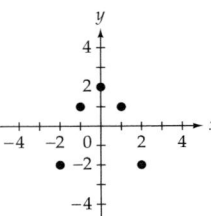

12.
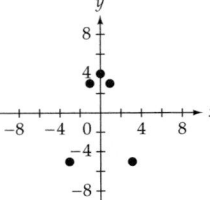

13.
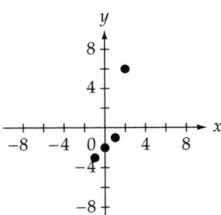

14.
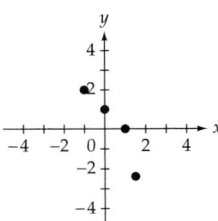

15.
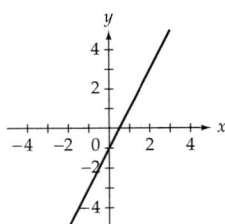

16.
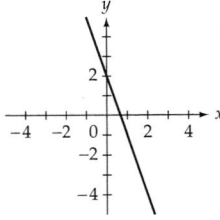

17.
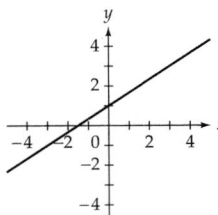

18.
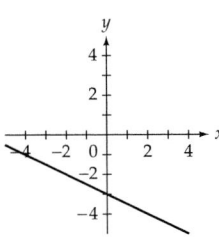

19.
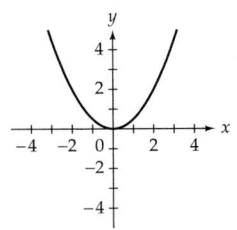

20.
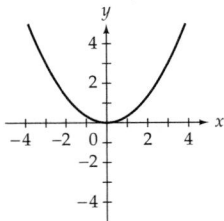

21.
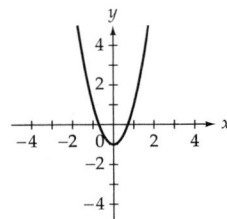

22.
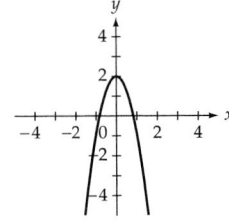

23.
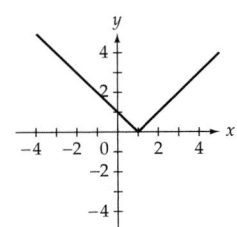

24.
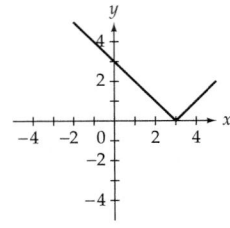

25. 3   **26.** 13   **27.** 3   **28.** 14   **29.** 10   **30.** 15   **31.** 0   **32.** $\dfrac{4}{3}$ **33. a.** 16 meters

b. 20 feet   **34. a.** 28.3 square inches   **b.** 452.4 square centimeters   **35. a.** 100 feet   **b.** 68 feet
36. a. 6.7 miles   **b.** 8.9 miles   **37. a.** 1087 feet per second   **b.** 1136 feet per second   **38. a.** 30 games
b. 45 games   **39. a.** 10%   **b.** 40%   **40. a.** 1.92 seconds   **b.** 1.0 seconds

41.   **42.**   **43.**   **44.**

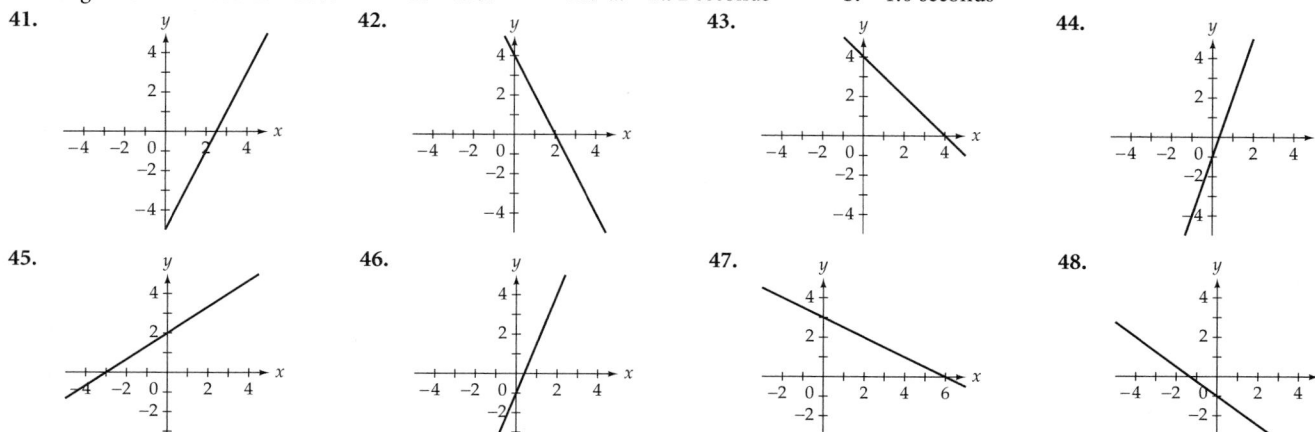

45.   **46.**   **47.**   **48.**

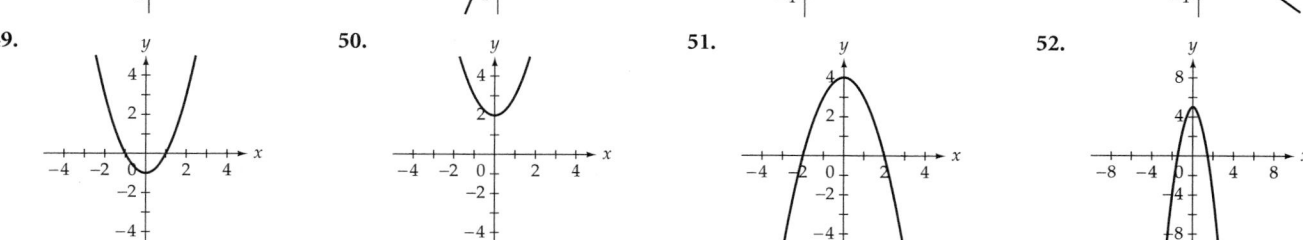

49.   **50.**   **51.**   **52.**

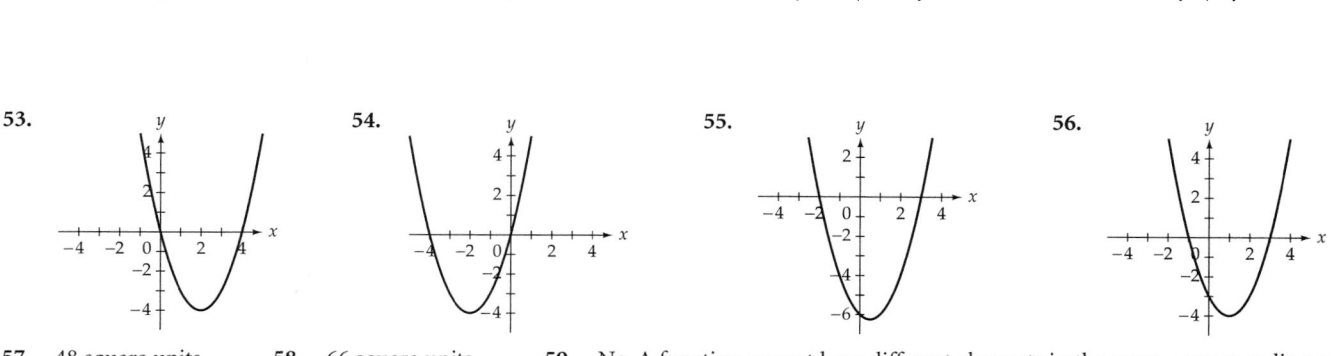

53.   **54.**   **55.**   **56.**

57. 48 square units   **58.** 66 square units   **59.** No. A function cannot have different elements in the range corresponding to
one element in the domain.   **60.** No. Explanations will vary. For example, if $f(x) = x^2$, $f(2) = 4$ and $f(-2) = 4$, but $2 \neq -2$.
61. 2   **62.** $a = 3$ or -3 **63.** 17   **64.** 77   **65.** $x = 0$ **66.** 2   **67. a.** $M(10, 8) = 10$;
$M(-3, -1) = -1$; $M(12, -13) = 12$; $M(-11, 15) = 15$ **b.** Answers will vary.   **c.** The value of the function is equal to
the larger of the two input values.   **d.** It is a good name for the function because the value of the function is equal to the
maximum of x and y.   **e.** Answers will vary. For example, $M(x, y) = \dfrac{x + y}{2} - \dfrac{|x - y|}{2}$.

EXCURSION EXERCISES, SECTION 6.2   *page 345*

1. 0 to 2 hours   **2.** 0 to 2 hours   **3.** 2 to 6 hours   **4.** 2 to 6 hours   **5.** 120 miles   **6.** 6 hours
7. 60 miles per hour   **8.** −30 miles per hour   **9.** 0   **10.** 0 miles per hour; no   **11.** 2 to 3 hours

EXERCISE SET 6.2 *page 346*

1. $(2, 0), (0, -6)$ **2.** $(-4, 0), (0, 8)$ **3.** $(6, 0), (0, -4)$ **4.** $(8, 0), (0, 6)$ **5.** $(-4, 0), (0, -4)$

6. $(2, 0), (0, 1)$ **7.** $(4, 0), (0, 3)$ **8.** $(2, 0), (0, -5)$ **9.** $\left(\frac{9}{2}, 0\right), (0, -3)$ **10.** $(2, 0), \left(0, \frac{8}{3}\right)$

11. $(2, 0), (0, 3)$ **12.** $(3, 0), (0, -2)$ **13.** $(1, 0), (0, -2)$ **14.** $(-4, 0), (0, 3)$ **15.** $\left(\frac{30}{7}, 0\right)$; At $\frac{30}{7}$ °C the cricket

stops chirping.   **16.** The intercept on the horizontal axis, $(12, 0)$, means that after flying for 12 hours, the remaining distance is 0 miles. The intercept on the vertical axis, $(0, 6000)$, means that before the trip begins, the remaining distance is 6000 miles.
17. The intercept on the vertical axis is $(0, -15)$. This means that the temperature of the object is $-15°F$ before it is removed from the freezer. The intercept on the horizontal axis is $(5, 0)$. This means that it takes 5 minutes for the temperature of the object to reach $0°F$.
18. The intercept on the horizontal axis, $(40, 0)$, means that 40 months after withdrawals begin, there is $0 remaining in the account. The intercept on the vertical axis, $(0, 100,000)$, means that before withdrawals begin, there is $100,000 in the account.   **19.** -1

20. $-\frac{2}{3}$ **21.** $\frac{1}{3}$ **22.** -3 **23.** $-\frac{2}{3}$ **24.** -2 **25.** $-\frac{3}{4}$ **26.** $\frac{3}{2}$ **27.** Undefined

28. Undefined   **29.** $\frac{7}{5}$ **30.** $\frac{3}{5}$ **31.** 0 **32.** 0 **33.** $-\frac{1}{2}$ **34.** Undefined   **35.** Undefined

36. -2 **37.** The slope is 40, which means the motorist was traveling at 40 miles per hour.   **38.** The slope is -5000, which means the value of the house decreases $5000 each year.   **39.** The slope is 0.28, which means the tax rate for an income range of $22,101 to $54,500 is 28%.   **40.** The slope is 0.007, which means the payment per dollar of mortgage is $.007.
41. The slope is approximately 385.5, which means the runner traveled at a rate of 385.5 meters per minute.   **42.** The slope is approximately 368.5, which means the average speed of the runner was 368.5 meters per minute.   **43.**

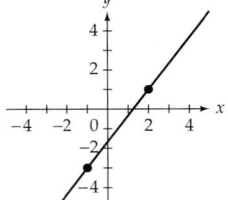

44.

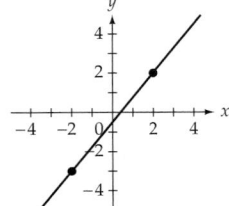

45.

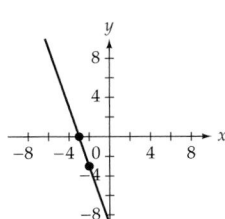

46.

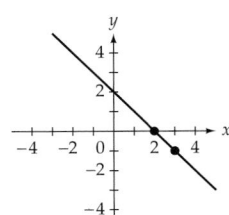

47.

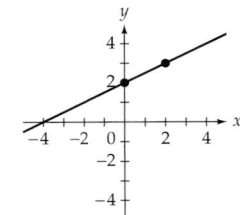

48.

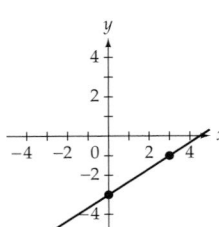

49.

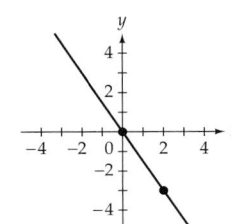

50.

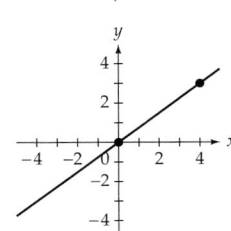

51.

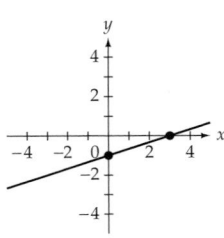

52.

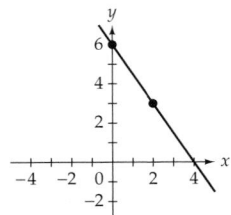

53. Line A represents the distance traveled by Lois in x hours, line B represents the distance traveled

by Tanya in x hours, and line C represents the distance between them in x hours.   **54. a.** Line A represents can 1
b. approximately 5 millimiters   **55.** No   **56.** 168 inches   **57.** 7   **58.** -4 **59.** 2   **60.** -7
61. It rotates the line counterclockwise.   **62.** It rotates the line clockwise.   **63.** It raises the line on the rectangular
coordinate system.   **64.** It lowers the line on the rectangular coordinate system.   **65.** No. A vertical line through $(4, 0)$ does
not have a y-intercept.   **66.** Yes   **67. a.** $0.91\overline{6}$ **b.** $0.6\overline{3}$ **c.** New design. $0.\overline{6}$ is closer to $0.6\overline{3}$ than to $0.91\overline{6}$.

d. Designs will vary.   **e.** Answers will vary.   **68. a.** $m_1 = \dfrac{BC}{AC} = \dfrac{BC}{1} = BC$ **b.** $-m_2 = \dfrac{CD}{AC} = \dfrac{CD}{1} = CD$

c. Lines l_1 and l_2 are perpendicular. Thus $\angle BAD = 90°$. Therefore,
$$90° = \angle x + \angle v$$
$$90° = \angle x + \angle y$$
$$\angle x + \angle v = \angle x + \angle y$$
$$\angle v = \angle y$$
In a similar manner, $\angle u = \angle x$. Thus the corresponding angles of the two triangles are equal and the triangles are similar.
d. Because $\triangle ACB$ is similar to $\triangle ACD,$ the ratios of corresponding sides are equal. Therefore,
$$\frac{BC}{AC} = \frac{AC}{CD}$$
$$\frac{m_1}{1} = \frac{1}{-m_2}$$
e. $\dfrac{m_1}{1} = \dfrac{1}{-m_2}$
$m_1 m_2 = -1$ Multiply each side by m_2.

EXCURSION EXERCISES, SECTION 6.3   *page 356*

1. $C = 28n + 13{,}500$ **2.** $R = 45n$ **3.** Approximately 794 cartridges   **4.** Approximately 4465 cartridges
5. Approximately $18.65 per hour   **6.** Approximately 489 additional cartridges   **7.** $C = 28n + 18{,}300; R = 45n;$
approximately 1076 cartridges; approximately $17.73 per hour

EXERCISE SET 6.3   *page 356*

1. $y = 2x + 5$ **2.** $y = \dfrac{1}{2}x + 2$ **3.** $y = -3x + 4$ **4.** $y = \dfrac{1}{2}x$ **5.** $y = -\dfrac{2}{3}x + 7$ **6.** $y = -x - 3$

7. $y = -3$ **8.** $y = -2x + 3$ **9.** $y = x + 2$ **10.** $y = -2x - 3$ **11.** $y = -\dfrac{3}{2}x + 3$ **12.** $y = x - 1$

13. $y = \dfrac{1}{2}x - 1$ **14.** $y = -4$ **15.** $y = -\dfrac{5}{2}x$ **16.** $y = x - 1$

17. $R(x) = -\dfrac{3}{5}x + 545; R(100) \approx 485$ rooms   **18.** $y = 85x + 30{,}000; \$183{,}000$ **19.** $D(t) = 415t; D(4.5) = 1867.5$ miles
20. $y = 0.04x + 1000; \$4400$ **21.** $N(x) = -5x + 110{,}000; N(12{,}500) = 47{,}500$ trucks   **22.** $y = -400x + 48{,}000;$
18,000 calculators   **23. a.** $y = 0.56x + 41.71$ **b.** 89   **24. a.** $y = 0.10x + 7.34; \$8.34$
25. a. $y = 42.50x + 2613.76$ **b.** 3209 thousand students   **26. a.** $y = -0.14x + 31.20$ **b.** 22 miles per gallon
27. a. $y = -1.35x + 106.98$ **b.** 46°F   **28. a.** $y = 0.32x + 0.40$ **b.** 3.6 meters per second   **29.** Answers
will vary. For example, $(0, 3), (1, 2),$ and $(3, 0).$ **30.** 0   **31.** 7   **32.** 1   **33.** No. The three points do not lie on a
straight line.   **34.** $y = -x + 6.$ The slope of the line between any two of the points is $-1.$ **35.** 3°. The car is climbing.

36. a. $(0.75, -1)$ **b.** $1\dfrac{1}{3}$ miles   **c.** $y = -\dfrac{4}{3}x$ **37. a.** The parametric equations giving the plane's position after
t minutes are $x = 9000t$ and $y = 100t + 5000.$ **b.** 5500 feet   **c.** Approximately 5133 feet

---

EXCURSION EXERCISES, SECTION 6.4   *page 366*

1. $(0, 0.625)$ **2.** 659.75 inches   **3.** Answers will vary.   **4.** Answers will vary. The light bulb should be placed at the
focus so that all light hitting the paraboloid will be bounced off to form a strong beam of light from the flashlight.

EXERCISE SET 6.4   *page 368*

1. $(0, -2)$ **2.** $(0, 2)$ **3.** $(0, -1)$ **4.** $(0, 3)$ **5.** $(0, 2)$ **6.** $(0, 0)$ **7.** $(0, -1)$ **8.** $(1, -1)$

9. $\left(\dfrac{1}{2}, -\dfrac{9}{4}\right)$ **10.** $\left(\dfrac{3}{2}, -\dfrac{1}{4}\right)$ **11.** $\left(\dfrac{1}{4}, -\dfrac{41}{8}\right)$ **12.** $\left(\dfrac{1}{4}, -\dfrac{25}{8}\right)$ **13.** $(0, 0), (2, 0)$ **14.** $(0, 0), (-2, 0)$

15. $(-2, 0), \left(-\dfrac{3}{4}, 0\right)$ **16.** $(3, 0), (-3, 0)$ **17.** $\left(-1 + \sqrt{2}, 0\right), \left(-1 - \sqrt{2}, 0\right)$

18. $\left(-2 + \sqrt{7}, 0\right), \left(-2 - \sqrt{7}, 0\right)$ **19.** None   **20.** $\left(-1 + \sqrt{2}, 0\right), \left(-1 - \sqrt{2}, 0\right)$ **21.** $\left(2 + \sqrt{5}, 0\right), \left(2 - \sqrt{5}, 0\right)$

22. None   **23.** $\left(-\dfrac{1}{2}, 0\right), (3, 0)$ **24.** $\left(\sqrt{2}, 0\right), \left(-\sqrt{2}, 0\right)$ **25.** Minimum: 2   **26.** Minimum: -2

27. Maximum: -3 **28.** Minimum: $-\dfrac{11}{4}$ **29.** Minimum: -3.25 **30.** Minimum: $-\dfrac{9}{8}$ **31.** Maximum: $\dfrac{9}{4}$

32. Maximum: $-\dfrac{2}{3}$ **33.** c   **34.** 114 feet   **35.** 150 feet   **36.** $250   **37.** 100 lenses   **38.** 5 days

39. 24.36 feet   **40. a.** 48 years   **b.** $148,000   **41.** Yes   **42.** 1.5 feet   **43.** 141.6 feet
44. 20 miles per hour   **45. a.** 41 miles per hour   **b.** 33.658 miles per gallon   **46.** Length: 100 feet; width: 50 feet
47. 7   **48.** -8 **49.** 4   **50.** -8 and 8   **51.** $f(2) = 2^2 + 2 - 6 = 4 + 2 - 6 = 0$ **52.** $-4, 1$
53. $-1, 5$ **54.** $-4.35, 0.95$ **55.** $-0.32, 0.59$ **56.** $-0.38, 1.69$ **57.** $-1.77, 1.06$ **58.** $-2, 1, 3$
59. $-4, -1, 3$ **60.** $-4, -2, 1, 3$

EXCURSION EXERCISES, SECTION 6.5   *page 381*

1. a. 1, 2, 4, 8, 16, 32, 64   **b.** 3   **c.** 7   **d.** 15   **e.** 31   **f.** 63   **g.** 127   **2.** Answers will vary.
For example, $f(n) = 2^{n-1} + (2^{n-1} - 1)$ or $f(n) = 2(2^{n-1}) - 1$.   **3.** Approximately 18,000,000,000,000,000,000 grains of wheat
4. Approximately 144 trillion kilograms   **5.** 228 years

EXERCISE SET 6.5   *page 382*

1. a. 9   **b.** 1   **c.** $\dfrac{1}{9}$ **2. a.** $\dfrac{1}{8}$ **b.** 1   **c.** 4   **3. a.** 16   **b.** 4   **c.** $\dfrac{1}{4}$ **4. a.** $\dfrac{1}{729}$

b. $\dfrac{1}{27}$ **c.** $\dfrac{1}{9}$ **5. a.** 1   **b.** $\dfrac{1}{8}$ **c.** 16   **6. a.** 3   **b.** $\dfrac{1}{27}$ **c.** 729   **7. a.** 7.3891

b. 0.3679   **c.** 1.2840   **8. a.** 0.0183   **b.** 0.2636   **c.** 54.5982   **9. a.** 54.5982   **b.** 1
c. 0.1353   **10. a.** 4.4817   **b.** 0.1353   **c.** 0.7788   **11. a.** 16   **b.** 16   **c.** 1.4768

12. a. $\dfrac{1}{512}$ **b.** $\dfrac{1}{2}$ **c.** $\dfrac{1}{16}$ **13. a.** 0.1353   **b.** 0.1353   **c.** 0.0111   **14. a.** 8.3891   **b.** 1.0025
c. 55.5982   **15.**   **16.**   **17.**

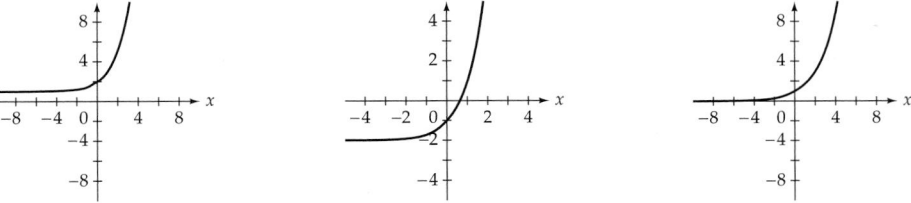

18.   **19.**   **20.**   **21.**

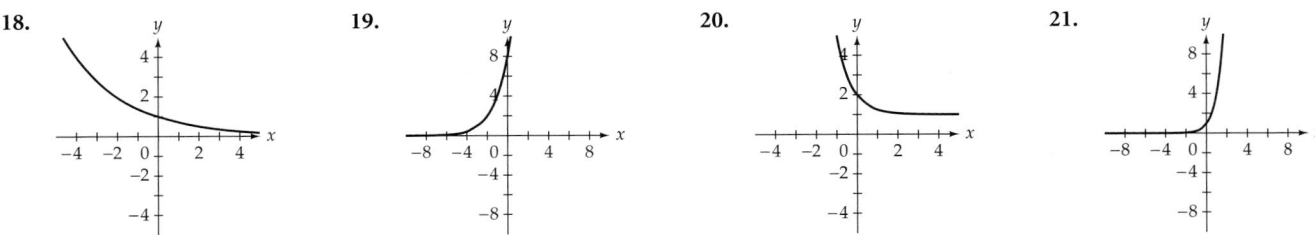

22.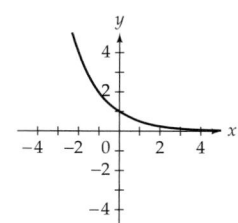

23.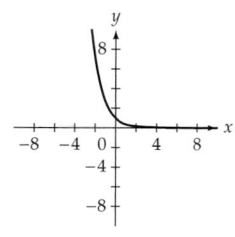

24.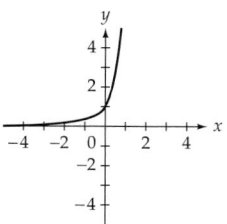

25. $11,202.50

26. $27,180.96   **27.** $3210.06   **28.** 21.2 milligrams   **29.** 5.2 micrograms   **30.** 68%

31. $F(n) = 440(2^{n/12})$ **32. a.** 0.098 newton per square centimeter   **b.** 10.13 newtons per square centimeter

c. Decreases   **33.** $y = 8000(1.1066^t)$; 552,200,000 automobiles   **34.** $y = 3200(0.919^t)$; 20 pandas

35. $y = 100(0.99^t)$ **36. a.** $y = 119,311.81(1.18^t)$ **b.** 1,428,610 ATMs   **37. a.** $y = 4.959(1.063^x)$

b. 12.4 milliliters   **38. a.** $y = 2.78(1.06^t)$ **b.** 11.9 cubic feet per minute   **39. a.** 5.9 billion people

b. 7.2 billion people   **c.** The maximum population that Earth can support is 70.0 billion people.

40. a. $P(t) = \dfrac{500,000}{500 + 500e^{-0.025t}}$ **b.** 593 wolves   **41. a.** 0.0075   **b.** $P = \dfrac{Ar(1 + r)^n}{(1 + r)^n - 1}$ **c.** $107.33

d. $193.33   **e.** There is no residual value on the car when it is purchased.   **f.** $3999.92; $6000.08   **g.** $10,000.14;

approximately $0   **h.** The $6000 remaining on the leased car is its residual value.

EXCURSION EXERCISES, SECTION 6.6   *page 394*

1.

d	$P(d) = \log_{10}(1 + 1/d)$
1	0.301
2	0.176
3	0.125
4	0.097
5	0.079
6	0.067
7	0.058
8	0.051
9	0.046

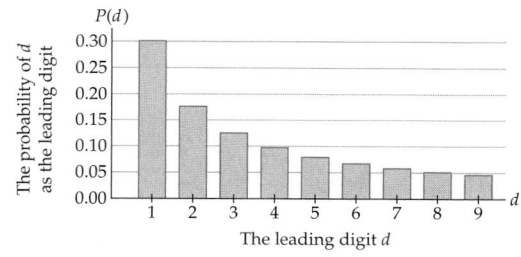

2. 6.7%   **3.** About 6.5 times   **4.** In a small city, not all four-digit numbers are used. Benford's Law applies only to data

with a wide range.   **5.** There is not a wide range of values.

EXERCISE SET 6.6   *page 395*

1. $\log_7 49 = 2$ **2.** $\log_{10} 1000 = 3$ **3.** $\log_5 625 = 4$ **4.** $\log_2 \dfrac{1}{8} = -3$ **5.** $\log_{10} 0.0001 = -4$

6. $\log_3 729 = 5$ **7.** $\log_{10} x = y$ **8.** $\ln x = y$ **9.** $3^4 = 81$ **10.** $2^4 = 16$ **11.** $5^3 = 125$

12. $4^3 = 64$ **13.** $4^{-2} = \dfrac{1}{16}$ **14.** $2^{-4} = \dfrac{1}{16}$ **15.** $e^y = x$ **16.** $10^y = x$ **17.** 4

18. 2   **19.** 2   **20.** -3 **21.** -2 **22.** -1 **23.** 6   **24.** -2 **25.** 9   **26.** 5

27. $\dfrac{1}{7}$ **28.** $\dfrac{1}{64}$ **29.** $\dfrac{1}{9}$ **30.** 125   **31.** 1   **32.** $\dfrac{1}{4}$ **33.** 316.23   **34.** 1584.89

35. 7.39   **36.** 54.60   **37.** 2.24   **38.** 1.34   **39.** 14.39   **40.** 1.65

41.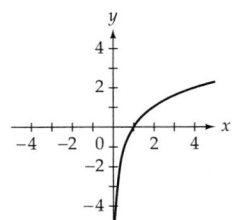

42.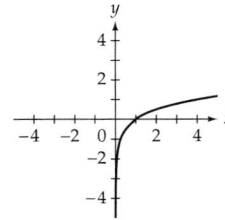

43.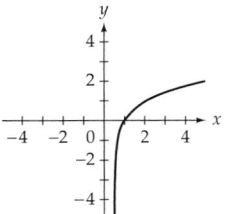

44.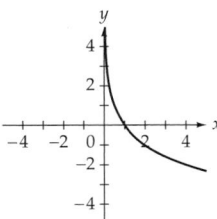

45.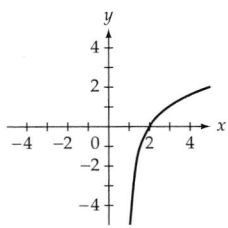

46.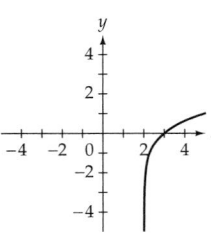

47. 79%   **48.** 0.44 centimeter   **49.** 65 decibels

50. 188 decibels   **51.** 1.35   **52.** 9.49   **53.** 6.8   **54.** 7.7   **55.** $794,328,235I_0$ **56.** $158,489,319I_0$

57. 100 times stronger   **58.** 150.7 parsecs   **59.** 6.0 parsecs   **60.** 742.9 billion barrels

61. 7805.5 billion barrels   **62. a.** $10^{0.30103} = 2; 10^{0.47712} = 3$ **b.** $x = 10^{0.30103} \cdot 10^{0.47712}$ **c.** $x = 10^{0.77815}$

d. $\log x = 0.77815$ **e.** $10^{0.77815} \approx 6$ **63.** $x = \dfrac{10^{0.47712}}{10^{0.30103}} = 10^{0.17609} \approx 1.5$ **64.** Answers will vary.

65. Answers will vary.   **66.** Answers will vary.   **67.** 4.9

CHAPTER 6 REVIEW EXERCISES *page 399*

1.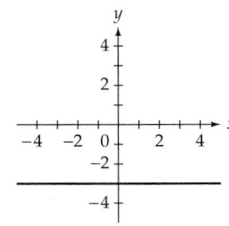

[Sec. 6.1]

2.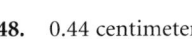

[Sec. 6.1]

3.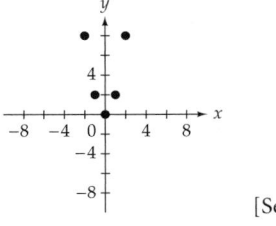

[Sec. 6.1]

4.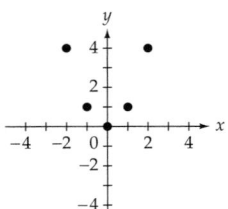

[Sec. 6.1]

5.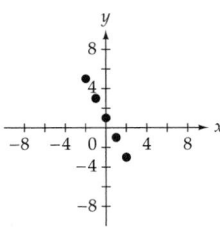

[Sec. 6.1]

6.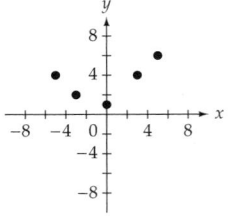

[Sec. 6.1]

7.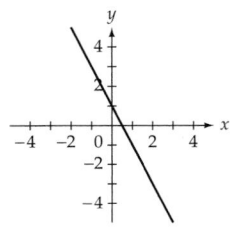

[Sec. 6.1]

8.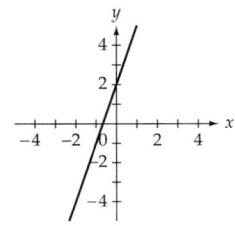

[Sec. 6.1]

9.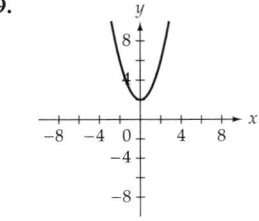

[Sec. 6.1]

10.

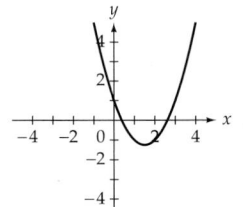

[Sec. 6.1]

11.

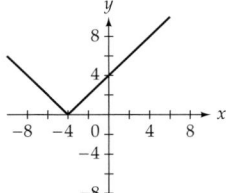

[Sec. 6.1]

12.

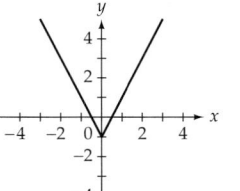

[Sec. 6.1]

13.

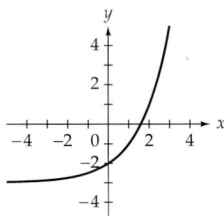

[Sec. 6.5]

14.

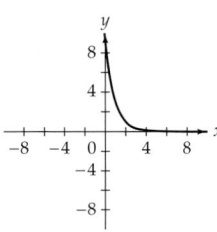

[Sec. 6.5]

15.

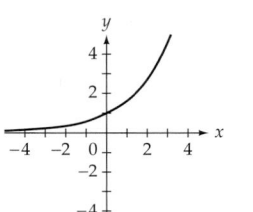

[Sec. 6.6]

16.

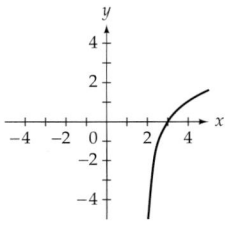

[Sec. 6.6]

17. -13 [Sec. 6.1]   **18.** 13 [Sec. 6.1]   **19.** $-\dfrac{1}{2}$ [Sec. 6.1]

20. -21 [Sec. 6.1]   **21.** 4 [Sec. 6.5]   **22.** $\dfrac{4}{9}$ [Sec. 6.5]   **23.** 15.78 [Sec. 6.5]   **24. a.** 113.1 cubic inches

b. 7238.2 cubic centimeters [Sec. 6.1]   **25. a.** 133 feet   **b.** 133 feet [Sec. 6.1]   **26. a.** 5%

b. 36.7% [Sec. 6.1]   **27.** $(-5, 0), (0, 10)$ [Sec. 6.2]   **28.** $(12, 0), (0, -9)$ [Sec. 6.2]   **29.** $(5, 0), (0, -3)$ [Sec. 6.2]

30. $(6, 0), (0, 8)$ [Sec. 6.2]   **31.** The intercept on the vertical axis is (0, 25,000). This means that the value of the truck was $25,000 when it was new. The intercept on the horizontal axis is (5, 0). This means that after 5 years the truck will be worth $0. [Sec. 6.2]

32. 5 [Sec. 6.2]   **33.** $\dfrac{5}{2}$ [Sec. 6.2]   **34.** 0 [Sec. 6.2]   **35.** Undefined [Sec. 6.2]

36. The slope is 40, which means that the number of ATM transactions increases by 40,000 annually. [Sec. 6.2]

37.

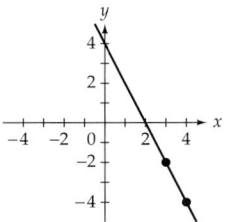

[Sec. 6.2]

38.

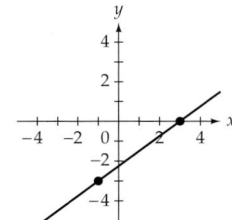

[Sec. 6.2]

39. $y = 2x + 7$ [Sec. 6.3]

40. $y = x - 5$ [Sec. 6.3]   **41.** $y = \dfrac{2}{3}x + 3$ [Sec. 6.3]   **42.** $y = \dfrac{1}{4}x$ [Sec. 6.3]

43.

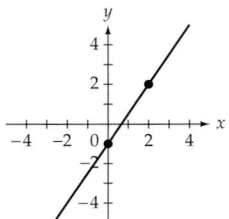

[Sec. 6.2]

44. a. $f(x) = 25x + 100$ **b.** $900 [Sec. 6.3]

45. a. $A(p) = -25,000p + 10,000$ **b.** 10,750 gallons [Sec. 6.3]   **46. a.** $y = 0.16204x - 0.12674$

b. 25 [Sec. 6.3]   **47.** $(-1, 3)$ [Sec. 6.4]   **48.** $\left(-\dfrac{3}{2}, \dfrac{11}{2}\right)$ [Sec. 6.4]   **49.** $(1, 2)$ [Sec. 6.4]

50. $(-2.5, -7.25)$ [Sec. 6.4]   **51.** $(4, 0), (-5, 0)$ [Sec. 6.4]   **52.** $\left(-1 + \sqrt{2}, 0\right), \left(-1 - \sqrt{2}, 0\right)$ [Sec. 6.4]

53. $\left(-\dfrac{1}{2}, 0\right), (-4, 0)$ [Sec. 6.4]   **54.** None [Sec. 6.4]   **55.** 5, maximum [Sec. 6.4]

56. $-\dfrac{15}{2}$, minimum [Sec. 6.4]   **57.** -5, minimum [Sec. 6.4]   **58.** $\dfrac{1}{8}$, maximum [Sec. 6.4]

59. 125 feet [Sec. 6.4]   **60.** 2000 CD-RWs [Sec. 6.4]   **61.** $12,297.11 [Sec. 6.5]

62. 7.94 milligrams [Sec. 6.5]   **63.** 3.36 micrograms [Sec. 6.5]   **64. a.** $H(n) = 6\left(\dfrac{2}{3}\right)^n$ **b.** 0.79 foot [Sec. 6.5]

65. a. $N = 250.2056(0.6506)^t$ **b.** 8 thousand people [Sec. 6.5]   **66.** 5 [Sec. 6.6]   **67.** -4 [Sec. 6.6]
68. -1 [Sec. 6.6]   **69.** 6 [Sec. 6.6]   **70.** 64 [Sec. 6.6]   **71.** 1.4422 [Sec. 6.6]   **72.** 12.1825 [Sec. 6.6]
73. 251.1886 [Sec. 6.6]   **74.** 43.7 parsecs [Sec. 6.6]   **75.** 140 decibels [Sec. 6.6]

CHAPTER 6 TEST   *page 402*

1. -21 [Sec. 6.1]   **2.** $\dfrac{1}{9}$ [Sec. 6.5]   **3.** 3 [Sec. 6.6]   **4.** 36 [Sec. 6.6]

5.
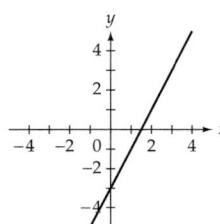
[Sec. 6.1]

6.
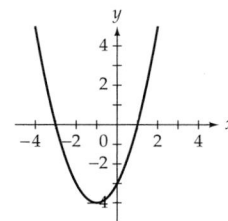
[Sec. 6.1]

7.
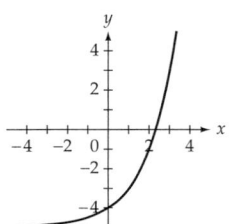
[Sec. 6.5]

8.
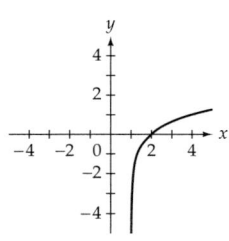
[Sec. 6.6]

9. $\dfrac{3}{5}$ [Sec. 6.2]   **10.** $y = \dfrac{2}{3}x + 3$ [Sec. 6.3]   **11.** $(-3, -10)$ [Sec. 6.4]

12. $(-4, 0), (2, 0)$ [Sec. 6.4]   **13.** $\dfrac{49}{4}$, maximum [Sec. 6.4]   **14.** The vertical intercept is $(0, 250)$. This means that the plane starts 250 miles from its destination. The horizontal intercept is $(2.5, 0)$. This means that it takes the plane 2.5 hours to reach its destination. [Sec. 6.2]   **15.** 148 feet [Sec. 6.4]   **16.** 3.30 grams [Sec. 6.5]   **17.** 31.6 times greater [Sec. 6.6]
18. $y = -1.5x + 740$ [Sec. 6.3]   **19. a.** $y = 40.5x + 659$ **b.** 983 pounds [Sec. 6.3]   **20.** 2.5 meters [Sec. 6.6]

CHAPTER 7

EXCURSION EXERCISES, SECTION 7.1   *page 414*

1. Answers will vary.   **2.** Friday   **3.** Wednesday   **4.** Tuesday

EXERCISE SET 7.1   *page 414*

1. 8   **2.** 1   **3.** 12   **4.** 3   **5.** 2   **6.** 4   **7.** 4   **8.** 9   **9.** 4   **10.** 8   **11.** 7
12. 10   **13.** 11   **14.** 12   **15.** 7   **16.** 9   **17.** 0300   **18.** 0400   **19.** 0400   **20.** 0200
21. 2000   **22.** 2300   **23.** 2100   **24.** 0800   **25.** 3   **26.** 1   **27.** 6   **28.** 4   **29.** True
30. True   **31.** False   **32.** False   **33.** True   **34.** True   **35.** False   **36.** False   **37.** True
38. True   **39.** Possible answers are 2, 8, 14, 20, 26, 32, 38, ...   **40.** Possible answers are 2, 6, 10, 14, 18, 22, 26, 30, ...
41. 3   **42.** 0   **43.** 3   **44.** 6   **45.** 2   **46.** 3   **47.** 10   **48.** 1   **49.** 3   **50.** 3
51. 3   **52.** 1   **53.** 5   **54.** 3   **55.** 4   **56.** 7   **57.** 3   **58.** 0   **59.** 7   **60.** 1
61. 2   **62.** 1   **63. a.** 6 o'clock   **b.** 5 o'clock   **64. a.** 6 o'clock   **b.** 4 o'clock   **65. a.** Tuesday
b. Monday   **66. a.** Saturday   **b.** Saturday   **67.** Monday   **68.** Wednesday   **69.** Sunday
70. Saturday   **71.** 1, 4, 7, 10, 13, 16, . . .   **72.** 2, 7, 12, 17, 22, 27, . . .   **73.** 1, 6, 11, 16, 21, 26, . . .
74. 10, 21, 32, 43, 54, 65, . . .   **75.** 0, 2, 4, 6, 8, 10, 12, . . .   **76.** 1, 4, 7, 10, 13, 16, . . .   **77.** No solutions
78. 5, 15, 25, 35, 45, 55, . . .   **79.** 0, 2, 4, 6, 8, 10, 12, . . .   **80.** 6, 14, 22, 30, 38, 46, . . .   **81.** No solutions
82. All whole numbers   **83.** 5, 7   **84.** 1, 4   **85.** 3, 3   **86.** 5, 3   **87.** 5, 3   **88.** 9, none
89. 6   **90.** 4   **91.** 6   **92.** 4   **93.** 2   **94.** 6   **97.** 11:00   **98.** 1300   **99.** 4
100. 3, 19   **101. a.** 8, 6, 0, 4, 5   **b.** 8, 6, 0, 4, 5; the pattern of numbers repeats   **c.** $x_3 = 0, x_5 = 5$
d. 68, 25, 26, 39, 48, 5, 6, 19

EXERCISE SET 7.2   *page 426*

1. No   **2.** Yes   **3.** No   **4.** No   **5.** Yes   **6.** No   **7.** 7   **8.** 4   **9.** 7   **10.** 7
11. 3   **12.** 2   **13.** 0   **14.** 6   **15.** 6   **16.** 9   **17.** 7   **18.** 2   **19.** 5   **20.** 5
21. 1   **22.** 5   **23.** 7   **24.** 1   **25.** 3   **26.** 4   **27.** Yes   **28.** No   **29.** Yes
30. Yes   **31.** Yes   **32.** Yes   **33.** No   **34.** No   **35.** No   **36.** Yes   **37.** Yes   **38.** No
39. BPZMM UCASMBMMZA   **40.** LRE ZUTOMNZ   **41.** UF'E M SUDX   **42.** VNNC JC WXXW
43. VWLFNV DQG VWRQHV   **44.** P HIXIRW XC IXBT   **45.** AGE OF ENLIGHTENMENT
46. IMAGINATION RULES THE WORLD   **47.** FRIEND IN NEED   **48.** NOBODY IS PERFECT
49. DANGER WILL ROBINSON   **50.** FIGHT OR FLIGHT   **51.** FORTUNE COOKIE
52. MAN THE TORPEDOES   **53.** PHQ ZLOOLQJOB EHOLHYH ZKDW WKHB ZLVK   **54.** THERE ARE
NO ACCIDENTS   **55.** JUSQD UT LURNUR   **56.** WHXKVLQY THIVULT YVPB   **57.** PODONNQN NSBQK
58. ADAZ PINUU GNPP   **59.** TURN BACK THE CLOCK   **60.** THIS MORTAL COIL
61. BARREL OF MONKEYS   **62.** LOOK BEFORE YOU LEAP   **63.** Because the check digit is simply the sum of the first
10 digits mod 9, the same digits in a different order will give the same sum and hence the same check digit.   **64.** Every other digit is
multiplied by 3 before the sum mod 10 is computed. Two adjacent digits a and b could be transposed and still give the same sum mod 10 if
$3a + b \equiv a + 3b$ mod 10, which is satisfied when $a \equiv b$ mod 5, or, equivalently, $|a - b| = 5$.   **65. a.** 1   **c.** Yes
d. The digits 6 and 1 are multiplied by 1 and 3, so the situation is identical to that in Exercise 64.   **66. a.** 5   **b.** 6

EXCURSION EXERCISES, SECTION 7.3   *page 438*

1. Any translation followed by another is itself a translation, so the set of translations is closed. Translations are associative,
the identity element is I, and any translation can be reversed, so each element has an inverse.

3. a.–c.   **4.**   **5. a.**

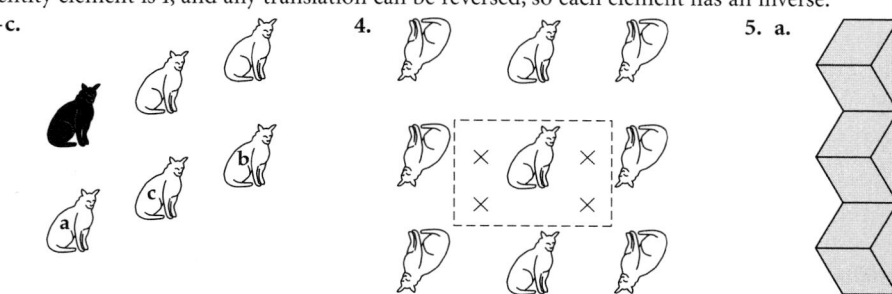
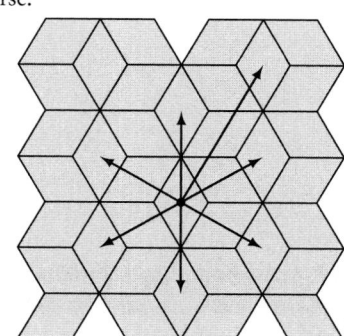

b.
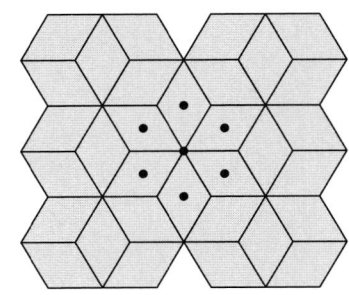

c.
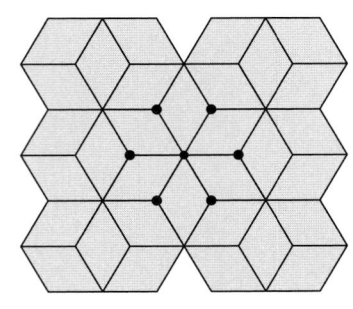

d.
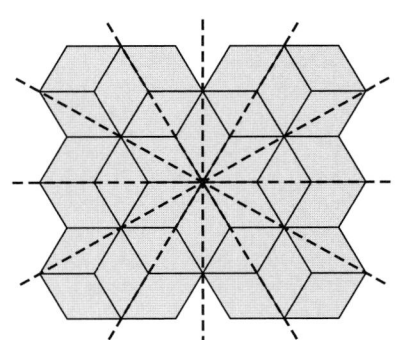

EXERCISE SET 7.3 *page 439*

1. a. Yes   **b.** No   **2. a.** Yes   **b.** Yes   **3. a.** Yes   **b.** No   **4. a.** Yes   **b.** No
c. No   **d.** No   **5.** Yes   **6.** No; properties 3 and 4 fail.   **7.** Yes   **8.** No; properties 1, 2, and 4 fail.
9. No; property 4 fails.   **10.** Yes   **11.** Yes   **12.** Yes   **13.** Yes   **14.** No; properties 1 and 4 fail.
15. No; property 4 fails.   **16.** Yes   **17.** Yes   **18.** Yes   **19.** R_l **20.** R_t **21.** R_{120}
22. R_{120} **23.** R_{120} **24.** R_t **25.** R_r **26.** R_l **27.** R_l **28.** R_{90}

29. $I = \begin{pmatrix} 1 & 2 & 3 & 4 \\ 1 & 2 & 3 & 4 \end{pmatrix}$, $R_{90} = \begin{pmatrix} 1 & 2 & 3 & 4 \\ 2 & 3 & 4 & 1 \end{pmatrix}$, $R_{180} = \begin{pmatrix} 1 & 2 & 3 & 4 \\ 3 & 4 & 1 & 2 \end{pmatrix}$, $R_{270} = \begin{pmatrix} 1 & 2 & 3 & 4 \\ 4 & 1 & 2 & 3 \end{pmatrix}$, $R_v = \begin{pmatrix} 1 & 2 & 3 & 4 \\ 4 & 3 & 2 & 1 \end{pmatrix}$,

$R_h = \begin{pmatrix} 1 & 2 & 3 & 4 \\ 2 & 1 & 4 & 3 \end{pmatrix}$, $R_r = \begin{pmatrix} 1 & 2 & 3 & 4 \\ 3 & 2 & 1 & 4 \end{pmatrix}$, $R_l = \begin{pmatrix} 1 & 2 & 3 & 4 \\ 1 & 4 & 3 & 2 \end{pmatrix}$ **30.** R_{90} **31.** R_r **32.** R_l **33.** R_{90}

34. R_v **35.** I **36.** I **37.** D **38.** A **39.** E **40.** A **41.** B **42.** E

43. $\begin{pmatrix} 1 & 2 & 3 & 4 \\ 1 & 2 & 3 & 4 \end{pmatrix}, \begin{pmatrix} 1 & 2 & 3 & 4 \\ 1 & 2 & 4 & 3 \end{pmatrix}, \begin{pmatrix} 1 & 2 & 3 & 4 \\ 1 & 3 & 2 & 4 \end{pmatrix}, \begin{pmatrix} 1 & 2 & 3 & 4 \\ 1 & 3 & 4 & 2 \end{pmatrix}, \begin{pmatrix} 1 & 2 & 3 & 4 \\ 1 & 4 & 2 & 3 \end{pmatrix}, \begin{pmatrix} 1 & 2 & 3 & 4 \\ 1 & 4 & 3 & 2 \end{pmatrix}, \begin{pmatrix} 1 & 2 & 3 & 4 \\ 2 & 1 & 3 & 4 \end{pmatrix},$

$\begin{pmatrix} 1 & 2 & 3 & 4 \\ 2 & 1 & 4 & 3 \end{pmatrix}, \begin{pmatrix} 1 & 2 & 3 & 4 \\ 2 & 3 & 1 & 4 \end{pmatrix}, \begin{pmatrix} 1 & 2 & 3 & 4 \\ 2 & 3 & 4 & 1 \end{pmatrix}, \begin{pmatrix} 1 & 2 & 3 & 4 \\ 2 & 4 & 1 & 3 \end{pmatrix}, \begin{pmatrix} 1 & 2 & 3 & 4 \\ 2 & 4 & 3 & 1 \end{pmatrix}, \begin{pmatrix} 1 & 2 & 3 & 4 \\ 3 & 1 & 2 & 4 \end{pmatrix}, \begin{pmatrix} 1 & 2 & 3 & 4 \\ 3 & 1 & 4 & 2 \end{pmatrix},$

$\begin{pmatrix} 1 & 2 & 3 & 4 \\ 3 & 2 & 1 & 4 \end{pmatrix}, \begin{pmatrix} 1 & 2 & 3 & 4 \\ 3 & 2 & 4 & 1 \end{pmatrix}, \begin{pmatrix} 1 & 2 & 3 & 4 \\ 3 & 4 & 1 & 2 \end{pmatrix}, \begin{pmatrix} 1 & 2 & 3 & 4 \\ 3 & 4 & 2 & 1 \end{pmatrix}, \begin{pmatrix} 1 & 2 & 3 & 4 \\ 4 & 1 & 2 & 3 \end{pmatrix}, \begin{pmatrix} 1 & 2 & 3 & 4 \\ 4 & 1 & 3 & 2 \end{pmatrix}, \begin{pmatrix} 1 & 2 & 3 & 4 \\ 4 & 2 & 1 & 3 \end{pmatrix},$

$\begin{pmatrix} 1 & 2 & 3 & 4 \\ 4 & 2 & 3 & 1 \end{pmatrix}, \begin{pmatrix} 1 & 2 & 3 & 4 \\ 4 & 3 & 1 & 2 \end{pmatrix}, \begin{pmatrix} 1 & 2 & 3 & 4 \\ 4 & 3 & 2 & 1 \end{pmatrix}$ **44.** $\begin{pmatrix} 1 & 2 & 3 & 4 \\ 3 & 1 & 2 & 4 \end{pmatrix}$ **45.** $\begin{pmatrix} 1 & 2 & 3 & 4 \\ 2 & 1 & 3 & 4 \end{pmatrix}$ **46.** $\begin{pmatrix} 1 & 2 & 3 & 4 \\ 2 & 4 & 3 & 1 \end{pmatrix}$

47. $\begin{pmatrix} 1 & 2 & 3 & 4 \\ 3 & 1 & 4 & 2 \end{pmatrix}$ **48.** $\begin{pmatrix} 1 & 2 & 3 & 4 \\ 3 & 4 & 1 & 2 \end{pmatrix}$ **49.** $\begin{pmatrix} 1 & 2 & 3 & 4 \\ 4 & 3 & 2 & 1 \end{pmatrix}$ **50.** $\begin{pmatrix} 1 & 2 & 3 & 4 \\ 1 & 4 & 2 & 3 \end{pmatrix}$ **51.** d **52.** a

53. c **54.** c **55.** Answers will vary.   **56.** b **57.** Yes   **58.** a and c are inverses of each other;
b and d are their own inverses.   **59. a. and b.** Answers will vary.   **c.** Values of n that are prime   **60. a.** 2^0
b. Answers will vary.   **c.** 2^{-8} **d.** 2^5 **e.** 2^{-k} **61. a.**   **b.** Yes

$\oplus$	1	2	3	4	5
1	1	2	3	4	5
2	2	3	4	5	1
3	3	4	5	1	2
4	4	5	1	2	3
5	5	1	2	3	4

c. 1   **d.** 1 is its own inverse; 2 and 5 are inverses; 3 and 4 are inverses.   **62. a.** $r \nabla s \neq s \nabla r$ **b.** e **c.** e and u
are their own inverses; r and v, s and w, and t and x are inverses of each other.   **d. and e.** Answers will vary.   **f.** $\{e, u\}$

CHAPTER 7 REVIEW EXERCISES *page 443*

1. 2   [Sec. 7.1]   **2.** 2   [Sec. 7.1]   **3.** 5   [Sec. 7.1]   **4.** 6   [Sec. 7.1]   **5.** 9   [Sec. 7.1]
6. 4   [Sec. 7.1]   **7.** 11   [Sec. 7.1]   **8.** 7   [Sec. 7.1]   **9.** 3   [Sec. 7.1]   **10.** 4   [Sec. 7.1]

11. True   [Sec. 7.1]   **12.** False   [Sec. 7.1]   **13.** False   [Sec. 7.1]   **14.** True   [Sec. 7.1]   **15.** 2   [Sec. 7.1]
16. 4   [Sec. 7.1]   **17.** 0   [Sec. 7.1]   **18.** 3   [Sec. 7.1]   **19.** 8   [Sec. 7.1]   **20.** 3   [Sec. 7.1]
21. 7   [Sec. 7.1]   **22.** 5   [Sec. 7.1]   **23. a.** 2 o'clock   **b.** 6 o'clock   [Sec. 7.1]   **24.** Thursday   [Sec. 7.1]
25. 3, 7, 11, 15, 19, 23, . . .   [Sec. 7.1]   **26.** 7, 16, 25, 34, 43, 52, . . .   [Sec. 7.1]   **27.** 0, 5, 10, 15, 20, 25, 30, . . .   [Sec. 7.1]
28. 4, 15, 26, 37, 48, 59, 70, . . .   [Sec. 7.1]   **29.** 2, 3   [Sec. 7.1]   **30.** 5, 7   [Sec. 7.1]   **31.** 6   [Sec. 7.1]
32. 2   [Sec. 7.1]   **33.** 6   [Sec. 7.2]   **34.** 6   [Sec. 7.2]   **35.** 2   [Sec. 7.2]   **36.** 1   [Sec. 7.2]
37. No   [Sec. 7.2]   **38.** Yes   [Sec. 7.2]   **39.** No   [Sec. 7.2]   **40.** No   [Sec. 7.2]   **41.** THF AOL MVYJL IL
DPAO FVB   [Sec. 7.2]   **42.** NLYNPW LWW AWLYD   [Sec. 7.2]   **43.** GOOD LUCK TOMORROW   [Sec. 7.2]
44. THE DAY HAS ARRIVED   [Sec. 7.2]   **45.** UVR YX NDU PGVU   [Sec. 7.2]   **46.** YOU PASSED
THE TEST   [Sec. 7.2]   **47.** Yes   [Sec. 7.3]   **48.** Yes   [Sec. 7.3]   **49.** No, properties 1, 3, and 4 fail.   [Sec. 7.3]
50. Yes   [Sec. 7.3]   **51.** R_{240} [Sec. 7.3]   **52.** R_l [Sec. 7.3]   **53.** R_r [Sec. 7.3]   **54.** R_{270} [Sec. 7.3]
55. R_{180} [Sec. 7.3]   **56.** R_r [Sec. 7.3]   **57.** There are only four distinct ways to place the rectangle in the

reference rectangle.   [Sec. 7.3]   **58.** $\begin{pmatrix} 1 & 2 & 3 & 4 \\ 1 & 2 & 3 & 4 \end{pmatrix}, \begin{pmatrix} 1 & 2 & 3 & 4 \\ 3 & 4 & 1 & 2 \end{pmatrix}, \begin{pmatrix} 1 & 2 & 3 & 4 \\ 2 & 1 & 4 & 3 \end{pmatrix}, \begin{pmatrix} 1 & 2 & 3 & 4 \\ 4 & 3 & 2 & 1 \end{pmatrix}$ [Sec. 7.3]

59. Yes   [Sec. 7.3]   **60.** Answers will vary.   [Sec. 7.3]   **61.** Answers will vary.   [Sec. 7.3]   **62.** A [Sec. 7.3]

63. A [Sec. 7.3]   **64.** $\begin{pmatrix} 1 & 2 & 3 & 4 & 5 \\ 3 & 4 & 1 & 5 & 2 \end{pmatrix}$ [Sec. 7.3]   **65.** $\begin{pmatrix} 1 & 2 & 3 & 4 & 5 \\ 3 & 5 & 1 & 4 & 2 \end{pmatrix}$ [Sec. 7.3]

CHAPTER 7 TEST   *page 445*

1. a. 3   **b.** 5   [Sec. 7.1]   **2. a.** 0200   **b.** 1300   [Sec. 7.1]   **3. a.** True   **b.** False   [Sec. 7.1]
4. 4   [Sec. 7.1]   **5.** 6   [Sec. 7.1]   **6.** 8   [Sec. 7.1]   **7. a.** 6 o'clock   **b.** 5 o'clock   [Sec. 7.1]
8. 5, 14, 23, 32, 41, 50, . . .   [Sec. 7.1]   **9.** 1, 3, 5, 7, 9, 11, . . .   [Sec. 7.1]   **10.** 4, 2   [Sec. 7.1]   **11.** 5   [Sec. 7.2]
12. 0   [Sec. 7.2]   **13.** Yes   [Sec. 7.2]   **14.** BOZYBD LKMU   [Sec. 7.2]   **15.** NEVER QUIT   [Sec. 7.2]
16. a. Yes   **b.** No   [Sec. 7.3]   **17.** No, Property 4 fails; many elements do not have an inverse.   [Sec. 7.3]

18. a. R_t **b.** R_r [Sec. 7.3]   **19.** $\begin{pmatrix} 1 & 2 & 3 \\ 1 & 3 & 2 \end{pmatrix}$ [Sec. 7.3]   **20.** $\begin{pmatrix} 1 & 2 & 3 & 4 \\ 2 & 4 & 1 & 3 \end{pmatrix}$ [Sec. 7.3]

CHAPTER 8

EXCURSION EXERCISES, SECTION 8.1   *page 461*

1.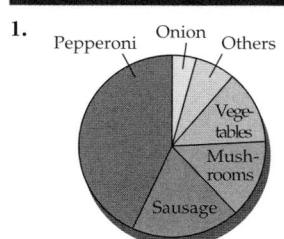

2.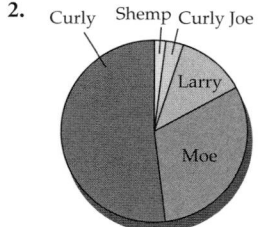

3.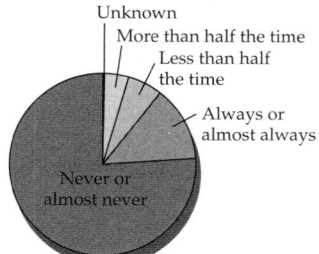

4.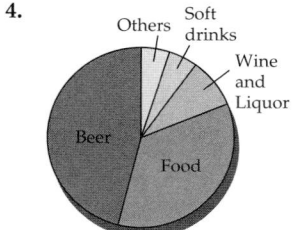

EXERCISE SET 8.1   *page 462*

1. $\angle O$, $\angle AOB$, and $\angle BOA$ 2.   360°; 180°; 90°   3.   40°, acute   4.   130°, obtuse   5.   30°, acute
6.   90°, right   7.   120°, obtuse   8.   45°, acute   9.   Yes   10.   No   11.   No   12.   Yes
13.   A 28° angle   14.   A 59° angle   15.   An 18° angle   16.   A 108° angle   17.   14 cm   18.   19 mm
19.   28 ft   20.   45 in.   21.   30 m   22.   24 cm   23.   86°   24.   47°   25.   71°   26.   63°
27.   30°   28.   18°   29.   36°   30.   33°   31.   127°   32.   53°   33.   116°   34.   121°
35.   20°   36.   15°   37.   20°   38.   18°   39.   20°   40.   45°   41.   141°   42.   128°
43.   106°   44.   49°   45.   11°   46.   12°   47. $m\angle a = 38°$, $m\angle b = 142°$ 48. $m\angle a = 122°$, $m\angle b = 58°$
49. $m\angle a = 47°$, $m\angle b = 133°$ 50. $m\angle a = 44°$, $m\angle b = 136°$ 51.   20°   52.   20°   53.   47°
54.   40°   55. $m\angle x = 155°$, $m\angle y = 70°$ 56. $m\angle x = 160°$, $m\angle y = 145°$ 57. $m\angle a = 45°$, $m\angle b = 135°$
58. $m\angle a = 40°$, $m\angle b = 140°$ 59. $90° - x$ 60. $75° - x$ 61.   60°   62.   45°   63.   35°
64.   73°   65.   102°   66.   43°   67.   The three angles form a straight angle. The sum of the measures of the angles of a
triangle is 180°.   68.   The distances differ because you can travel in a straight line in the air but you cannot travel in a straight line
on the roads. The shortest distance between two points is a straight line.   69.   Zero dimensions; one dimension; one dimension; one
dimension; two dimensions   70.   The line segments are the same length.   71.   360°   72.   The sum of the measures of
the interior angles of a triangle is 180°; therefore, $m\angle a + m\angle b + m\angle c = 180°$. The sum of the measures of an interior and exterior
angle is 180°; therefore, $m\angle c + m\angle x = 180°$. Solving this equation for $m\angle c$, $m\angle c = 180° - m\angle x$. Substitute $180° - m\angle x$ for $m\angle c$
in the equation $m\angle a + m\angle b + m\angle c = 180°$: $m\angle a + m\angle b + 180° - m\angle x = 180°$. Add $m\angle x$ to each side of the equation, and
subtract 180° from each side of the equation: $m\angle a + m\angle b = m\angle x$. The measure of an exterior angle of a triangle is equal to the sum of
the measures of the two opposite interior angles: $m\angle a + m\angle c = m\angle z$.   73. $\angle AOC$ and $\angle BOC$ are supplementary angles;
therefore, $m\angle AOC + m\angle BOC = 180°$. Because $m\angle AOC = m\angle BOC$, by substitution, $m\angle AOC + m\angle AOC = 180°$. Therefore,
$2(m\angle AOC) = 180°$, and $m\angle AOC = 90°$. Hence $\overline{AB} \perp \overline{CD}$.

EXCURSION EXERCISES, SECTION 8.2   *page 483*

1.   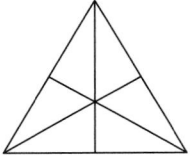   2. a.   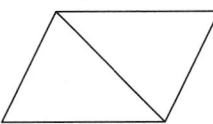   b.   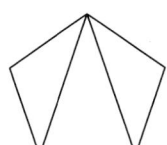   c.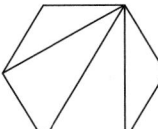

d.   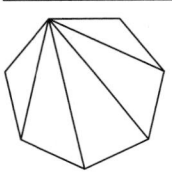   3.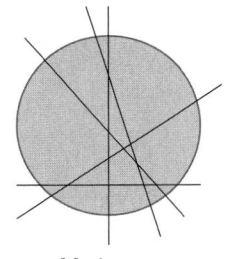

Maximum
number of
pieces with
5 cuts: 16

EXERCISE SET 8.2   *page 484*

1.   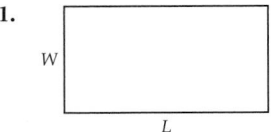   2.   Either of the horizontal line segments should be labeled *b*. Either line segment adjacent to a base
should be labeled *s*. The dashed line segment should be labeled *h*.

3. a.   Perimeter is not measured in square units.   b.   Area is measured in square units.   4.   Hexagon   5.   Heptagon
6.   Pentagon   7.   Quadrilateral   8.   Scalene   9.   Isosceles   10.   Equilateral   11.   Scalene
12.   Obtuse   13.   Right   14.   Acute   15.   Obtuse   16. a.   36 in.   b.   77 in²   17. a.   30 m

b. 50 m² **18. a.** 28 ft **b.** 48 ft² **19. a.** 16 cm **b.** 16 cm² **20. a.** 36 mi **b.** 81 mi²
21. a. 40 km **b.** 100 km² **22. a.** 24 m **b.** 30 m² **23. a.** 40 ft **b.** 72 ft² **24. a.** 68 cm
b. 224 cm² **25. a.** 8π cm; 25.13 cm **b.** 16π cm²; 50.27 cm² **26. a.** 24π m; 75.40 m
b. 144π m²; 452.39 m² **27. a.** 11π mi; 34.56 mi **b.** 30.25π mi²; 95.03 mi² **28. a.** 18π in.; 56.66 in.
b. 81π in²; 254.47 in² **29. a.** 17π ft; 53.41 ft **b.** 72.25π ft²; 226.98 ft² **30. a.** 6.6π km; 20.73 km
b. 10.89π km²; 34.21 km² **31.** $29\frac{1}{2}$ ft **32.** $55\frac{3}{4}$ ft **33.** $10\frac{1}{2}$ mi **34.** $68\frac{1}{6}$ yd **35.** 68 ft
36. 20 m **37.** 20 in. **38.** 50 ft **39.** 214 yd **40.** 54 cm **41.** 10 mi **42.** 560 ft
43. 15 packages **44.** 432 ft² **45.** 144 m² **46.** 8800 yd² **47.** 9 in. **48.** 6 ft **49.** 10 in.
50. 39 ft **51.** 8 m **52.** 10 × 20 unit **53.** 96 m² **54.** 136.5 ft² **55.** 607.5 m² **56.** 160 km²
57. 2 bags **58.** 8 rolls **59.** 2 qt **60.** 7 ft² **61.** 20 tiles **62.** $120 **63.** $40 **64.** $456
65. $34 **66.** 176 m² **67.** 120 ft² **68.** 56.5 ft **69.** 13.19 ft **70.** 9.42 m **71.** 1256.6 ft²
72. 12,064 in. **73.** 94.25 ft **74.** 62.83 ft **75.** 144π in² **76.** 2500π ft² **77.** 113.10 in²
78. 339.29 in² larger; more than twice the size **79.** 266,281 km **80.** 120 ft **81.** $8r^2 - 2\pi r^2$ **82.** $\pi r^2 - 2r^2$
83. 4 times larger **84. a.** Sometimes true **b.** Sometimes true **c.** Always true **85.** $(a + 4)$ by $(a - 4)$
86. $(2x + 3)$ by $(2x - 3)$ **87. a.** 18 cm; 8 cm² **b.** 20 cm; 16 cm² **c.** 32 cm; 64 cm² **88. a.** 12 units
b. 10 units **c.** 14 units **d.** 10 units

EXCURSION EXERCISES, SECTION 8.3 *page 497*

1. c. ray **2.** c. fork **3.** e. T

EXERCISE SET 8.3 *page 497*

1. $\frac{1}{2}$ **2.** $\frac{1}{3}$ **3.** $\frac{3}{4}$ **4.** $\frac{1}{3}$ **5.** 7.2 cm **6.** 13.7 in. **7.** 3.3 m **8.** 4.9 ft
9. 12 m **10.** 38 cm **11.** 12 in. **12.** 45 cm² **13.** 56.3 cm² **14.** 49 m²
15. 18 ft **16.** 22.5 ft **17.** 16 m **18.** 20.8 ft **19.** $14\frac{3}{8}$ ft **20.** 6.25 cm
21. 15 m **22.** 6 in. **23.** 8 ft **24.** 10 ft **25.** 13 cm **26.** 12 m **27.** 35 m
28. 45 ft **29. a.** Always true **b.** Sometimes true **c.** Always true **d.** Always true
30. Triangle *ABC* is similar to triangle *ACD* because both have a right angle and $m \angle A = m \angle A$. The measures of two angles of one triangle are equal to the measures of two angles of the other triangle.
Triangle *ABC* is similar to triangle *CBD* because both have a right angle and $m \angle B = m \angle B$. The measures of two angles of one triangle are equal to the measures of two angles of the other triangle.
Triangles *ACD* and *BCD* are right triangles because both have a right angle. $\angle A$ is complementary to $\angle B$, $\angle ACD$ is complementary to $\angle CBD$, $\angle ACD$ is complementary to $\angle A$, and $\angle BCD$ is complementary to $\angle B$. Therefore, $m \angle A = m \angle BCD$ and $m \angle B = m \angle ACD$. The measures of two angles of one triangle are equal to the measures of two angles of the other triangle. Triangles *ACD* and *CBD* are similar triangles.

EXCURSION EXERCISES, SECTION 8.4 *page 510*

1. 0.21 cm **2.** 1.88 in. **3.** 0.25 lb/in³

EXERCISE SET 8.4 *page 511*

1. 840 in³ **2.** 168π ft³; 527.79 ft³ **3.** 15 ft³ **4.** 421.88 m³ **5.** 4.5π cm³; 14.14 cm³
6. 128π cm³; 402.12 cm³ **7.** 94 m² **8.** 1176 ft² **9.** 56 m² **10.** 4π cm²; 12.57 cm²
11. 96π in²; 301.59 in² **12.** 15.75π ft²; 49.48 ft² **13.** 34 m³ **14.** 20.25 ft³ **15.** 15.625 in³
16. 343 cm³ **17.** 36π ft³ **18.** 7.2 m³ **19.** 8143.01 cm³ **20.** 115.2π m³ **21.** 75π in³
22. 392.70 cm³ **23.** 120 in³ **24.** 216 m³ **25.** 7.80 ft³ **26.** 184 ft² **27.** 6416 cm² **28.** 69.36 m²
29. 13.5 in² **30.** 225π cm² **31.** 50.27 in² **32.** 402.12 in² **33.** 2.88π m² **34.** 6π ft²
35. 874.15 in² **36.** 297 in² **37.** 832 m² **38.** 2.5 ft **39.** 8.5 in. **40.** 3 cm **41.** 3 ft

42. 11 cans   **43.** 3217 ft^2 **44.** 456 in^2 **45.** 881.22 cm^2 **46.** 22.53 cm^2 **47.** 1.67 m^3 **48.** 5 m^3
49. 115.43 cm^3 **50.** 69.12 in^3 **51.** 4580.44 cm^3 **52.** 192 in^3 **53.** 19 m^2 **54.** 208 in^2
55. 622.65 m^2 **56.** 204.57 cm^2 **57.** 19,405.66 m^2 **58.** 165.99 cm^2 **59.** 888.02 ft^3 **60.** $3515.00
61. 95,000 L   **62.** 158 cans   **63.** 79.17 g   **64.** $498.75   **65.** $4860   **66.** 256,000 gal

67. $V = \dfrac{2}{3}\pi r^3$; $SA = 3\pi r^2$ **68. a.** Drawings will vary. For example:

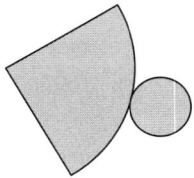

b. Drawings will vary. For example:

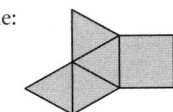

69. Surface area of the sphere $= 4\pi r^2$. Surface area of the side of the cylinder $= 2\pi rh = 2\pi r(2r) = 4\pi r^2$.

70. a. Always true   **b.** Never true   **c.** Sometimes true   **71. a.** Doubled $+ 4WH$ **b.** Quadrupled
c. 8 times larger   **d.** Quadrupled   **72. a.** For example, make a cut perpendicular to the top and bottom faces and parallel to two of the sides.   **b.** For example, beginning at an edge that is perpendicular to the bottom face, cut at an angle through to the bottom face.   **c.** For example, beginning at the top face at a distance d from the vertex, cut at an angle to the bottom face, ending at a distance greater than d from the opposite vertex.   **d.** For example, beginning on the top face at a distance d from a vertex, cut across the cube to a point just above the opposite vertex.

EXCURSION EXERCISES, SECTION 8.5   *page 522*

1. a.

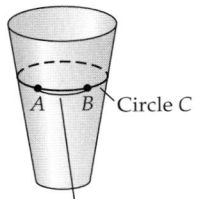

The geodesic from A to B dips slightly below circle C.

b.

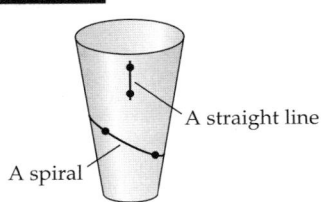

2. A piece of tape placed directly over circle D

does not lie flat. Thus, by the Inverse of the Tape Test Theorem, circle D is not a geodesic of the sphere.   **3.** A piece of tape placed directly over circle E will not lie flat. Thus, by the Inverse of the Tape Test Theorem, circle E is not a geodesic of the figure.
4. d. Alaska   **5. c.** Godthaab, Greenland   **6.** 1800 mi   **7. a.** A piece of tape placed directly over the blue circle will lie flat. A piece of tape placed directly over either of the red circles will not lie flat.   **b.** A piece of tape placed directly over the blue parabola will lie flat. A piece of tape placed directly over either of the red parabolas will not lie flat.

EXERCISE SET 8.5   *page 524*

1. a. Through a given point not on a given line, exactly one line can be drawn parallel to the given line.   **b.** Through a given point not on a given line, there are at least two lines parallel to the given line.   **c.** Through a given point not on a given line, there exist no lines parallel to the given line.   **2.** Nikolai Lobachevsky   **3.** Carl Friedrich Gauss   **4.** Bernhard Riemann
5. a. The sum equals 180°.   **b.** The sum is less than 180°.   **c.** The sum is greater than 180° but less than 540°.
6. a. One   **b.** One   **c.** Three   **7.** Imaginary geometry   **8.** The great circles of a sphere are the geodesics of the sphere, just as lines are the geodesics of a plane.   **9.** A geodesic is a curve on a surface such that for any two points of the curve the portion of the curve between the points is the shortest path on the surface that joins these points.   **10.** Lobachevskian or hyperbolic geometry   **11.** An infinite saddle surface   **12.** Riemannian geometry   **13.** π square units
14. 6,160,000 mi^2 **15.** 1,370,000 mi^2 **16.** 4,110,000 mi^2 **17. a.** 0 mi^2 **b.** The quadrilateral shown by the map is a plane quadrilateral. The spherical polygon area formula can only be applied to spherical polygons.   **18.** The triangle drawn

on the map is a plane triangle. The spherical triangle area formula can only be applied to spherical triangles.   **19. a.** 10
b. 3   **20. a.** 15   **b.** 6   **21.**   Janos Bolyai (1802–1860) learned about the problem of the Parallel Postulate from his father, who told him not to spend any time on the problem because it would not lead to any result and it would ruin his life. Janos ignored his father's wishes and, after many failed attempts to prove the Parallel Postulate, decided (like Gauss) that the postulate was an independent postulate. At this time Janos proposed a geometry that excluded Euclid's Parallel Postulate. Janos proved many theorems in this new geometry, which he referred to as the *absolute science of space.* Janos published his work in 1832. Janos's father sent a copy of this publication to Gauss, who wrote the following response.

> To praise it, would amount to praising myself. For the entire content of the work…coincides almost exactly with my own meditations which have occupied my mind for the past thirty or thirty-five years. (*Source:* School of Mathematics and Statistics University of St Andrews, Scotland. **http://www-gap.dcs.st-and.ac.uk/~history/Mathematicians/Bolyai.html**)

After reading Gauss's response, Janos fell into a period of deep depression. Most mathematicians would have taken the remarks of Gauss as a wonderful compliment, but Janos could only concentrate on the fact that he was not the first to solve the problem of the Parallel Postulate. From that point on, Janos did not publish any more mathematical papers.   **22.**   The Jesuit priest Girolamo Saccheri (1667–1733) tried to establish the Parallel Postulate as a theorem. To produce a proof, Saccheri started with a quadrilateral that has come to be called a *Saccheri quadrilateral.* He assumed that the base angles of the quadrilateral were right angles and that the left and right sides of the quadrilateral were the same length. His plan was to show that the two upper angles (summit angles) could not be acute angles and could not be obtuse angles. If he could do this, then he could conclude that the summit angles must be right angles, and this would establish the Parallel Postulate as a theorem. Saccheri was never able to prove that the summit angles could not be acute angles. Today we know that Saccheri's failure to establish the Parallel Postulate as a theorem was not due to a lack of ingenuity on his part, but rather due to the fact that the Parallel Postulate is truly a postulate.

EXCURSION EXERCISES, SECTION 8.6   *page 537*

1.   The folding procedure would require you to fold the paper 10 times. This is not humanly possible with a small piece of paper. (Try it yourself. Were you able to make more than six folds?)

2. 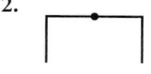   **3.** Left   **4.** Right   **5.** Right

EXERCISE SET 8.6   *page 537*

1.

2.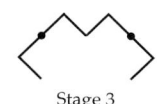

3.

Stage 2

4.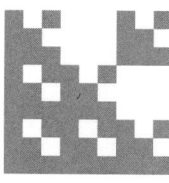

Stage 2

5.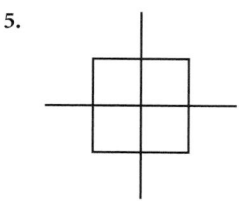

Stage 2

6.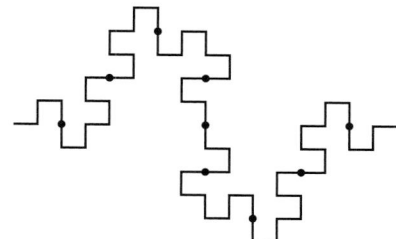

Stage 2

7.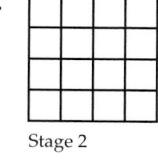

Stage 2

8.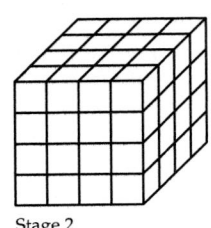

Stage 2

9.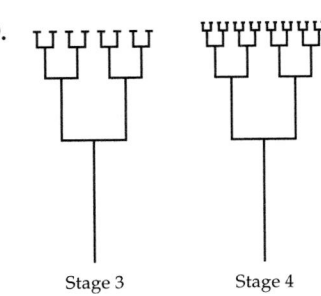

Stage 3   Stage 4

10.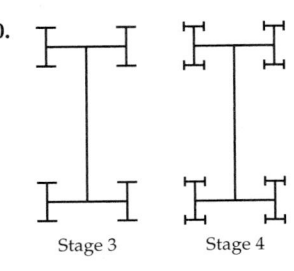

Stage 3   Stage 4

11. 0.631   **12.** 2.000   **13.** 1.465   **14.** 1.771   **15.** 2.000   **16.** 1.500   **17.** 2.000
18. 3.000   **19.** 1.613   **20.** 2.727   **21. a.** Sierpinski carpet, 1.893; Variation 2, 1.771; Variation 1, 1.465
b. The Sierpinski carpet   **22. a.** 2   **b.** The Peano curve completely covers a square region in a plane.
23. The binary tree fractal is not a strictly self-similar fractal.   **24.**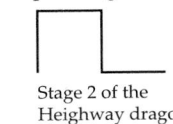

Stage 2 of   Stage 2 of the
Lévy's curve   Heighway dragon

25. Answers will vary.   **26.** The replacement ratio of the Sierpinski pyramid is 4. The scaling ratio is 2. Thus the similarity
dimension of the Sierpinski pyramid is $\frac{\log 4}{\log 2} = 2$.

CHAPTER 8 REVIEW EXERCISES   *page 541*

1. $\angle x = 22°, \angle y = 158°$ [Sec. 8.1]   **2.** 24 in. [Sec. 8.3]   **3.** 240 in³ [Sec. 8.4]   **4.** 68° [Sec. 8.1]
5. 220 ft² [Sec. 8.4]   **6.** 40π m² [Sec. 8.4]   **7.** 44 cm [Sec. 8.1]   **8.** 19° [Sec. 8.1]   **9.** 27 in² [Sec. 8.2]
10. 96 cm³ [Sec. 8.4]   **11.** 14.14 m [Sec. 8.2]   **12.** $\angle a = 138°; \angle b = 42°$ [Sec. 8.1]
13. a 148° angle [Sec. 8.1]   **14.** 39 ft³ [Sec. 8.4]   **15.** 95° [Sec. 8.1]   **16.** 8 cm [Sec. 8.2]
17. 288π mm³ [Sec. 8.4]   **18.** 21.5 cm [Sec. 8.2]   **19.** 4 cans [Sec. 8.4]   **20.** 208 yd [Sec. 8.2]
21. 90.25 m² [Sec. 8.2]   **22.** 276 m² [Sec. 8.2]   **23.** Carl Friedrich Gauss [Sec. 8.5]
24. Nikolai Lobachevsky [Sec. 8.5]   **25.** Spherical geometry or elliptical geometry [Sec. 8.5]
26. Hyperbolic geometry [Sec. 8.5]   **27.** Lobachevskian or hyperbolic geometry [Sec. 8.5]
28. Riemannian or spherical geometry [Sec. 8.5]   **29.** 120π in² [Sec. 8.5]   **30.** $\frac{25\pi}{3}$ ft² [Sec. 8.5]

31. $\frac{338}{9}\pi$ ft² ≈ 117.98 ft² [Sec. 8.5]   **32.** $871{,}200\pi$ mi² $\approx 2{,}736{,}955.5$ mi² [Sec. 8.5]

33.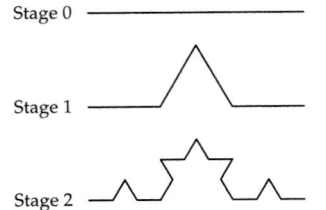

Stage 0

Stage 1

Stage 2   [Sec. 8.6]

34.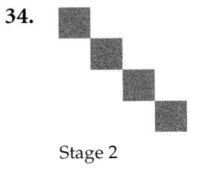

Stage 2   [Sec. 8.6]

35. $\frac{\log 5}{\log 4} \approx 1.161$ [Sec. 8.6]

36. 1 [Sec. 8.6]

CHAPTER 8 TEST   *page 543*

1. 169.65 m³ [Sec. 8.4]   **2.** 6.8 m [Sec. 8.2]   **3.** a 58° angle [Sec. 8.1]   **4.** 3.14 m² [Sec. 8.2]
5. 150° [Sec. 8.1]   **6.** $\angle a = 45°; \angle b = 135°$ [Sec. 8.1]   **7.** 5.0625 ft² [Sec. 8.2]   **8.** 448π cm³ [Sec. 8.4]
9. $1\frac{1}{5}$ ft [Sec. 8.3]   **10.** 90° and 50° [Sec. 8.1]   **11.** 125° [Sec. 8.1]   **12.** 32 m² [Sec. 8.2]
13. 25 ft [Sec. 8.3]   **14.** 113.10 in² [Sec. 8.2]   **15.** 103.87 ft² [Sec. 8.2]   **16.** 780 in³ [Sec. 8.4]
17. a. Through a given point not on a given line, exactly one line can be drawn parallel to the given line.

b. Through a given point not on a given line, there exist no lines parallel to the given line.   [Sec. 8.5]   **18. a.** 1
b. 3   [Sec. 8.5]   **19.** A great circle of a sphere is a circle on the surface of the sphere whose center is at the center of the sphere.
[Sec. 8.5]   **20.** 80π ft$^2 \approx 251.3$ ft^2 [Sec. 8.5]   **21.**

Stage 2
[Sec. 8.6]   **22.**

Stage 2   [Sec. 8.6]

23. Replacement ratio: 2; scale ratio: 2; similarity dimension: 1   [Sec. 8.6]
24. Replacement ratio: 3; scale ratio: 2; similarity dimension: $\frac{\log 3}{\log 2} \approx 1.585$ [Sec. 8.6]

CHAPTER 9

EXCURSION EXERCISES, SECTION 9.1   *page 560*

1. One answer is: 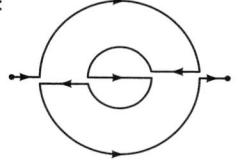   **2.** One answer is: 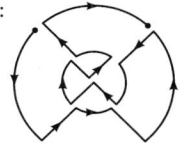   **3.** One answer is: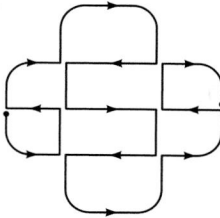

4. One answer is: 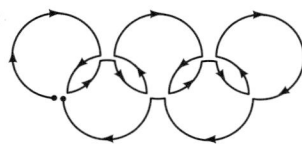   **5.** More than two vertices are of odd degree.

EXERCISE SET 9.1   *page 561*

1. a. 6   **b.** 7   **c.** 6   **d.** Yes   **e.** No   **2. a.** 8   **b.** 6   **c.** 4   **d.** No   **e.** No
3. a. 6   **b.** 4   **c.** 4   **d.** Yes   **e.** Yes   **4. a.** 6   **b.** 2   **c.** 2   **d.** Yes   **e.** No
5. 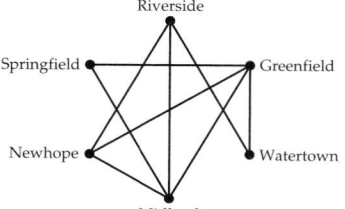   **6.** 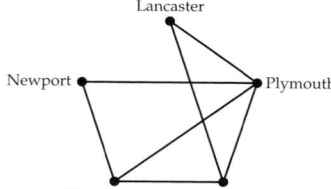   **7.**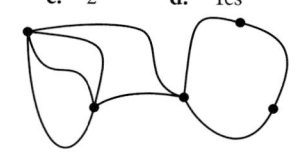

8. 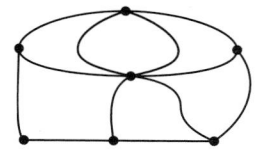   **9. a.** No   **b.** 3   **c.** Ada   **d.** A loop would correspond to a friend speaking to himself or herself.

10. a. Giants   **b.** Titans   **c.** Yes   **11.** Equivalent   **12.** Not equivalent   **13.** Not equivalent
14. Not equivalent   **15.** The graph on the right has a vertex of degree 4 and the graph on the left does not.

16. 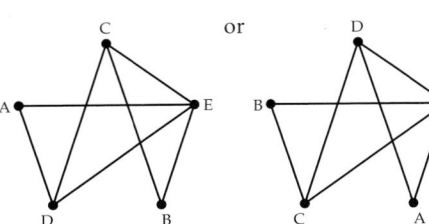 or

17. a. D–A–E–B–D–C–E–D

18. a. Not Eulerian **b.** A–B–D–A–C–E–B–E **19. a.** Not Eulerian **b.** A–E–A–D–E–D–C–E–C–B–E–B
20. a. Not Eulerian **b.** No **21. a.** Not Eulerian **b.** No **22. a.** A–B–A–D–C–D–E–F–E–B–C–A
23. a. Not Eulerian **b.** E–A–D–E–G–D–C–G–F–C–B–F–A–B–E–F
24. a. A–B–C–D–E–A–G–B–H–C–I–D–J–E–F–H–J–G–I–F–A **25. a.** 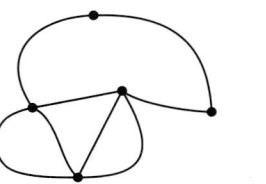 **b.** Yes

26. a. 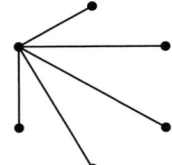 **b.** No; yes **27.** Yes **28.** No **29.** Yes, but the hamster cannot return to its starting point.

30. Yes (exactly two vertices are of odd degree) **31.** Yes; you will always return to the starting room.
32. Not possible **33.** A–B–C–D–E–G–F–A **34.** A–D–B–F–C–E–A **35.** A–B–E–C–H–D–F–G–A
36. A–B–C–D–F–E–A **37.** Springfield–Greenfield–Watertown–Riverside–Newhope–Midland–Springfield
38. Newport–Plymouth–Lancaster–Auburn–Dorset–Newport **39.** A route through the museum that visits each room once
and returns to the starting room without visiting any room twice **40.** A subway trip that visits each train junction once and returns
to the starting point without visiting any junction twice **41.** Euler circuit **42.** Hamiltonian circuit **43.** There is not
an Euler circuit or an Euler walk; there is a Hamiltonian circuit. **44.** There is an Euler walk and a Hamiltonian circuit.

45. a.

(many possible answers)

b. (many possible answers)

c. (many possible answers)

46. a.

b. Not possible; an Euler walk **c.** Venezuela–Guyana–Suriname–French Guiana–Brazil–Suriname–Guyana–Brazil–
Venezuela–Columbia–Brazil–Peru–Columbia–Peru–Bolivia–Brazil–Paraguay–Bolivia–Paraguay–Argentina–Brazil–Uruguay–
Argentina–Bolivia–Chile–Argentina–Chile–Peru–Ecuador–Columbia–Venezuela

---

EXCURSION EXERCISES, SECTION 9.2 *page 578*

1. C–A–B–E–D–C, total weight 18; D–E–B–A–C–D, total weight 18; E–B–A–C–D–E, total weight 18
2. A–F–C–D–E–B–A, total weight 65; B–D–E–F–C–A–B, total weight 66; C–F–A–D–E–B–C, total weight 70;
D–E–B–C–F–A–D, total weight 70; E–D–B–C–F–A–E, total weight 61; F–C–D–E–B–A–F, total weight 65
3. A–F–C–B–D–E–A, total weight 61

EXERCISE SET 9.2   *page 579*

1. A–B–E–D–C–A, total weight 31; A–D–E–B–C–A, total weight 32   **2.** A–F–B–D–C–E–A, total weight 44;
A–D–C–E–B–F–A, total weight 41   **3.** A–D–C–E–B–F–A, total weight 114; A–C–D–E–B–F–A, total weight 158
4. A–G–B–C–D–E–F–A, total weight 339; A–F–G–E–D–C–B–A, total weight 353   **5.** A–D–B–C–F–E–A
6. A–D–F–C–B–E–A   **7.** A–C–E–B–D–A   **8.** A–F–B–E–C–G–D–A   **9.** A–D–B–F–E–C–A
10. A–D–F–C–B–E–A   **11.** A–C–E–B–D–A   **12.** A–F–B–E–C–G–D–A   **13.** Louisville–Evansville–
Bloomington–Indianapolis–Lafayette–Fort Wayne–Louisville   **14.** Toronto–Niagara Falls–Kingston–Ottawa–Windsor–Toronto
15. Louisville–Evansville–Bloomington–Indianapolis–Lafayette–Fort Wayne–Louisville   **16.** Toronto–Kingston–Ottawa–
Windsor–Niagara Falls–Toronto   **17.** Tokyo–Seoul–Beijing–Hong Kong–Bangkok–Tokyo   **18.** Boulder–Denver–
Colorado Springs–Grand Junction–Durango–Boulder   **19.** Tokyo–Seoul–Beijing–Hong Kong–Bangkok–Tokyo
20. Boulder–Denver–Colorado Springs–Durango–Grand Junction–Boulder   **21.** Home–pharmacy–pet store–farmer's
market–shopping mall–home; home–pharmacy–pet store–shopping mall–farmer's market–home   **22.** Courier company–Bank
of America building–Imperial Bank building–Prudential building–GE Tower–Design Center–courier company; courier company–
Bank of America building–Imperial Bank building–Prudential building–Design Center–GE Tower–courier company
23. Home state–task B–task D–task A–task C–home state; Edge–Picking Algorithm gives the same sequence.
24. A–D–E–C–B–F–A   **25.**   **26.**   **27.**

28. Answers will vary.

EXCURSION EXERCISES, SECTION 9.3   *page 592*

1. 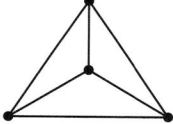   **2.** Octahedron   **3. a.** 3   **b.** 30   **c.** 20

4. a. Not all faces of the graph have the same number of edges; thus not all faces of the corresponding polyhedron are the same shape.
b. 14   **c.** Answers will vary.

EXERCISE SET 9.3   *page 593*

1. 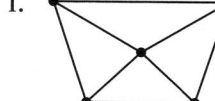   **2.** 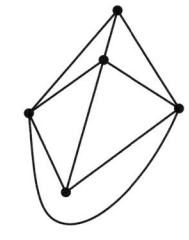   **3.** 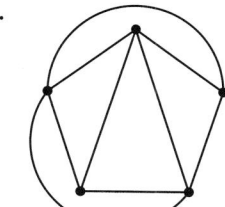   **4.**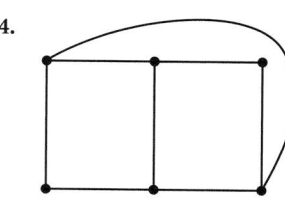

5. 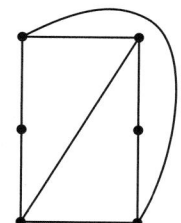   **6.** 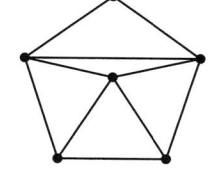   **7.** 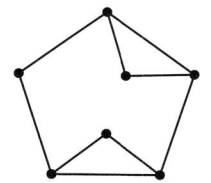   **8.**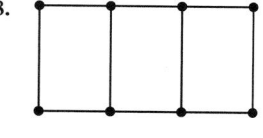

17. 5 faces, 5 vertices, 8 edges   **18.** 4 faces, 6 vertices, 8 edges   **19.** 2 faces, 8 vertices, 8 edges
20. 3 faces, 6 vertices, 7 edges   **21.** 5 faces, 10 vertices, 13 edges   **22.** 7 faces, 7 vertices, 12 edges   **23.** 9

24. 148

25.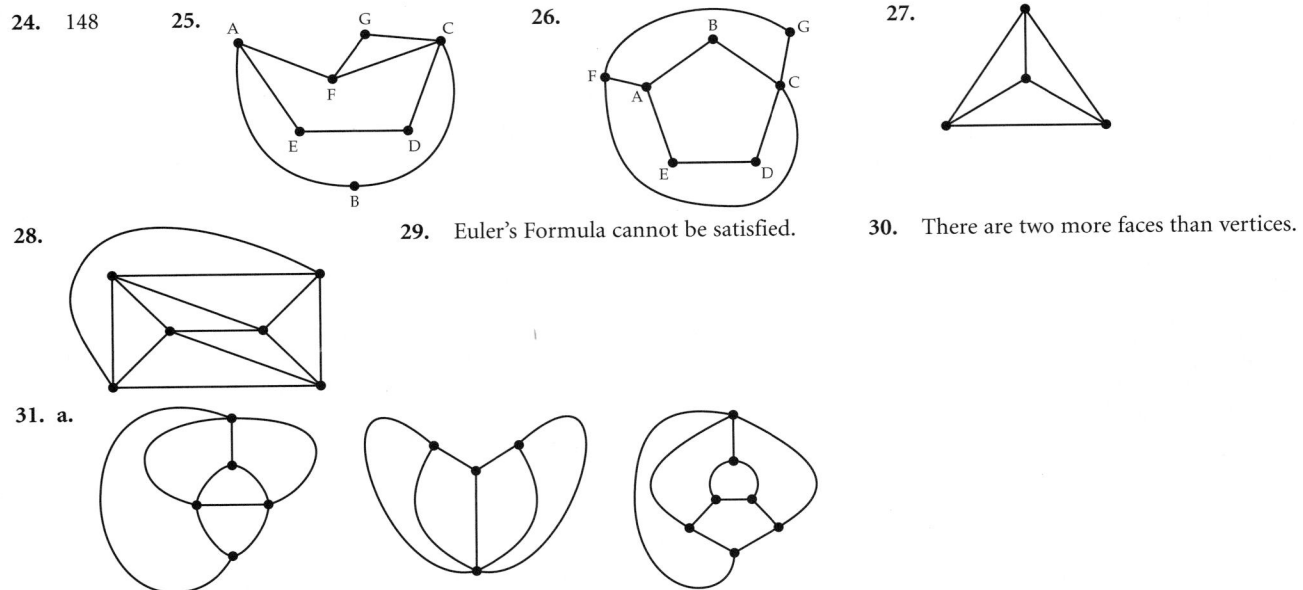

26.

27.

28.

29. Euler's Formula cannot be satisfied.

30. There are two more faces than vertices.

31. a.

b. The number of vertices in the dual graph is equal to the number of faces in the original graph, and the number of faces in the dual graph is equal to the number of vertices in the original graph.   **c.** The dual of the dual is equivalent to the original graph.

EXCURSION EXERCISES, SECTION 9.4   *page 606*

1.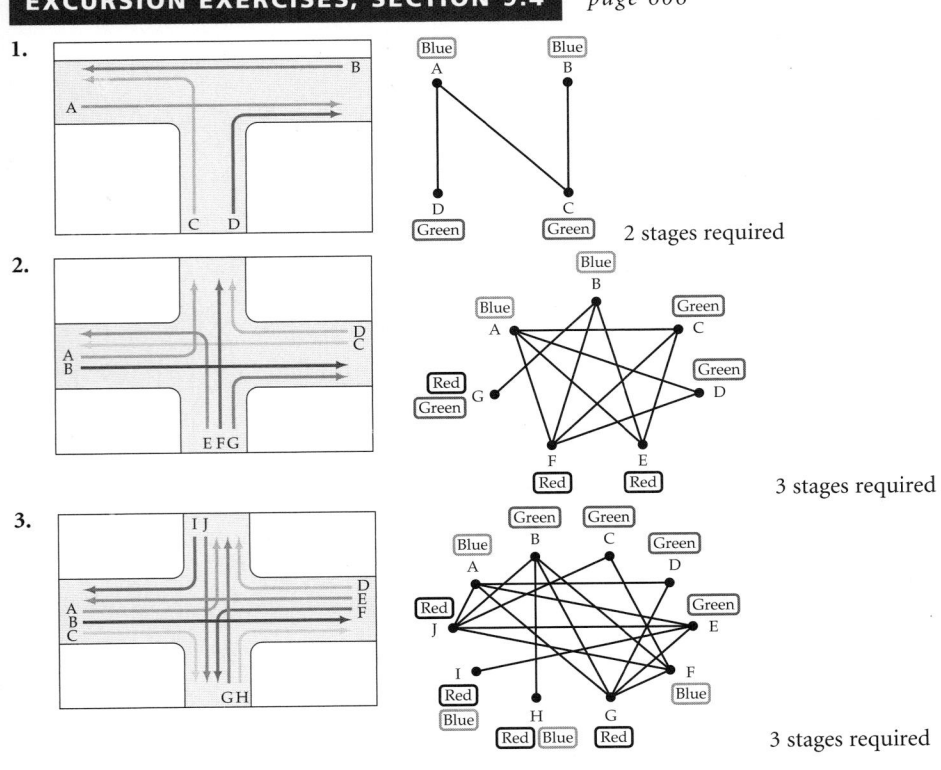

2 stages required

2. 3 stages required

3. 3 stages required

EXERCISE SET 9.4   *page 606*

1.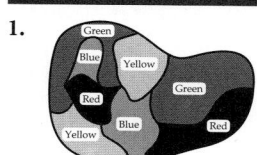

2.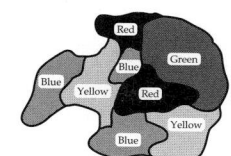

3.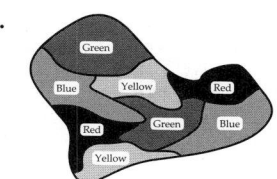

4.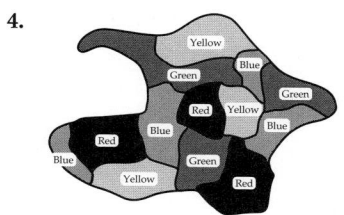

5.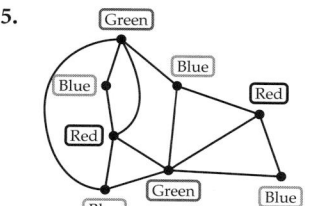

6.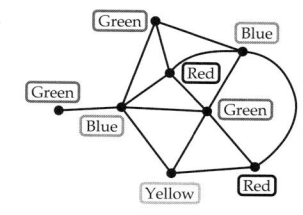

7.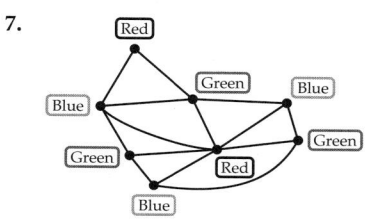

8.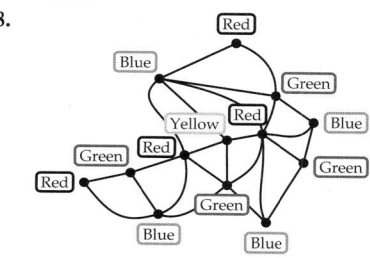

9.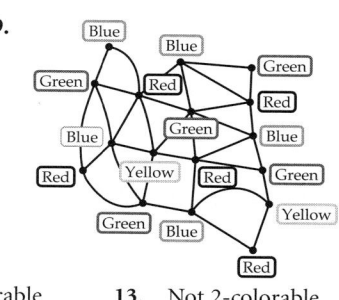

10.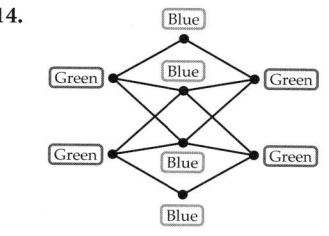

11.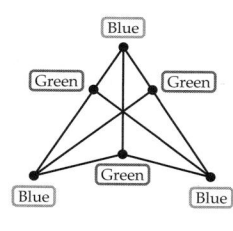

12. Not 2-colorable

13. Not 2-colorable

14.

15.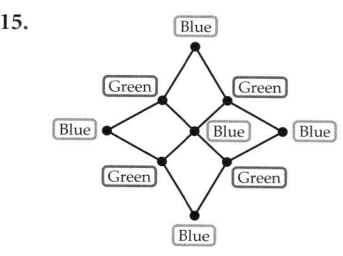

16.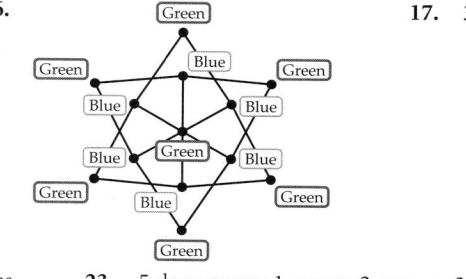

17. 3

18. 4   **19.** 4

and 5, group 4, group 6   **20.** 3   **21.** 2 time slots   **22.** 3 meeting times   **23.** 5 days: group 1, group 2, groups 3

films 3 and 5   **24.** 3 days: charities A and E, charities B and C, charity D

26.    **27.** 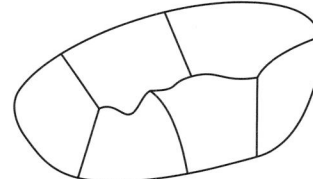   **25.** 3 days: films 1 and 4, films 2 and 6,   **28.**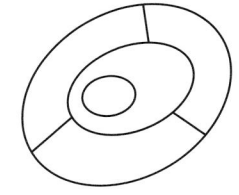

29.

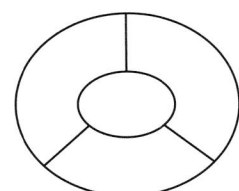

30. Disconnected vertices with no edges

31. a.

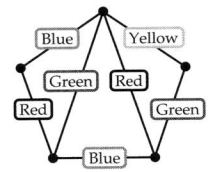

b.

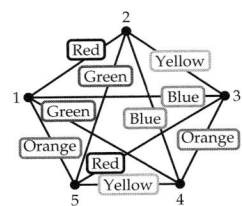

c.

d.

e. Answers will vary.

32.

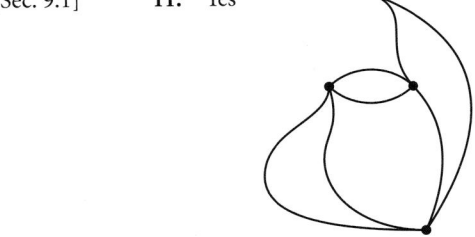

Day 1: 1 vs. 3, 2 vs. 4; day 2: 1 vs. 4, 2 vs. 5; day 3: 1 vs. 2, 3 vs. 5; day 4: 2 vs. 3, 4 vs. 5; day 5: 1 vs. 5, 3 vs. 4

CHAPTER 9 REVIEW EXERCISES   *page 612*

1. a. 8   **b.** 4   **c.** All vertices have degree 4.   **d.** Yes [Sec. 9.1]   **2. a.** 6   **b.** 7
c. 1, 1, 2, 2, 2, 2, 2   **d.** No [Sec. 9.1]   **3.**   **4. a.** No   **b.** 4

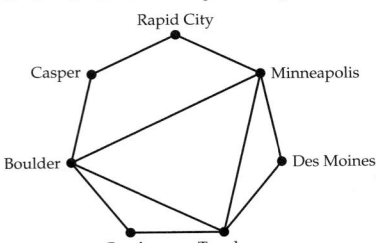

c. 110, 405   **d.** 105 [Sec. 9.1]   **5.** Equivalent [Sec. 9.1]   **6.** Equivalent [Sec. 9.1]
7. a. E–A–B–C–D–B–E–C–A–D   **b.** Not possible [Sec. 9.1]   **8. a.** Not possible   **b.** Not possible [Sec. 9.1]
9. a. and b. F–A–E–C–B–A–D–B–E–D–C–F [Sec. 9.1]   **10. a.** B–A–E–C–A–D–F–C–B–D–E
b. Not possible [Sec. 9.1]   **11.** Yes   **12.** Yes; no [Sec. 9.1]

[Sec. 9.1]

13. A–B–C–E–D–A [Sec. 9.1]   **14.** A–D–F–B–C–E–A [Sec. 9.1]
15.

Casper–Rapid City–Minneapolis–Des Moines–Topeka–Omaha–Boulder–Casper
[Sec. 9.1]
16. Casper–Boulder–Topeka–Minneapolis–Boulder–Omaha–Topeka–Des Moines–Minneapolis–Rapid City–Casper [Sec. 9.1]

17. A–D–F–E–B–C–A   [Sec. 9.2]   **18.** A–B–E–C–D–A   [Sec. 9.2]
19. A–D–F–E–C–B–A   [Sec. 9.2]   **20.** A–B–E–D–C–A   [Sec. 9.2]
21.

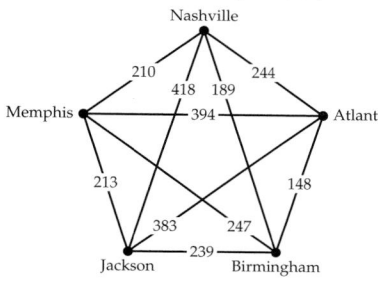

22. A–E–B–C–D–A   [Sec. 9.2]   **23.**

Memphis–Nashville–Birmingham–Atlanta–Jackson–Memphis   [Sec. 9.2]

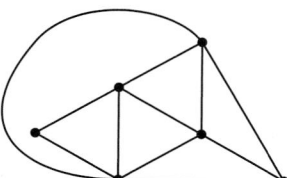

[Sec. 9.3]

27. 5 vertices, 8 edges, 5 faces   [Sec. 9.3]

24.

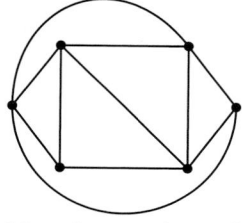

[Sec. 9.3]

28. 14 vertices, 16 edges, 4 faces   [Sec. 9.3]   **29.** Requires 4 colors

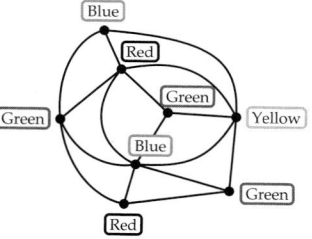

[Sec. 9.4]

30. Requires 4 colors

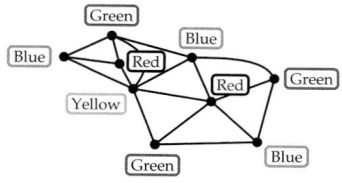

[Sec. 9.4]

31. 2-colorable

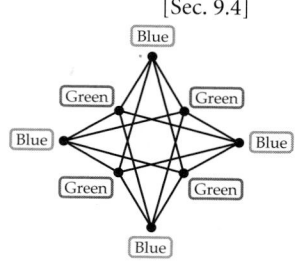

[Sec. 9.4]

32. Not 2-colorable   [Sec. 9.4]   **33.** 3   [Sec. 9.4]   **34.** 5   [Sec. 9.4]
35. 3 time slots

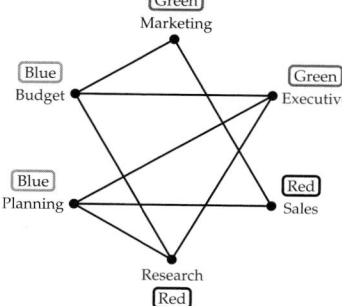

[Sec. 9.4]

CHAPTER 9 TEST   *page 617*

1 a. No   **b.** Monique   **c.** 0   **d.** No   [Sec. 9.1]   **2.** Equivalent   [Sec. 9.1]   **3. a.** No
b. A–B–E–A–F–D–C–F–B–C–E–D   [Sec. 9.1]   **4.** No   **5. a.** See page 557.

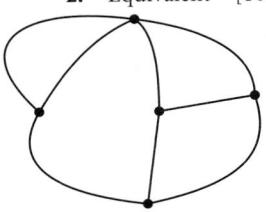

[Sec. 9.1]

b. A–G–C–D–F–B–E–A   [Sec. 9.1]   **6. a.**

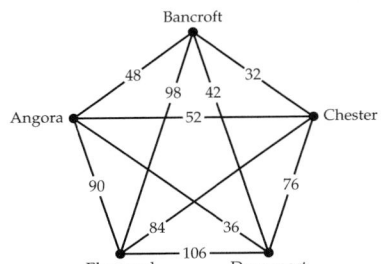

b. Angora–Elmwood–Chester–Bancroft–Davenport–Angora; $284   **c.** Yes   [Sec. 9.2]   **7.** A–E–D–B–C–F–A   [Sec. 9.2]
8.   **10. a.** 6   **b.** 7   **c.** $v = 12, f = 7, e = 17; 12 + 7 = 17 + 2$ [Sec. 9.3]

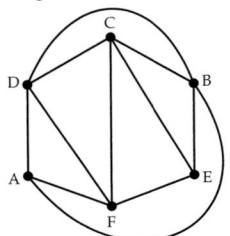

[Sec. 9.3]

11. a.   **b.** 3   **c.**

[Sec. 9.4]

12.

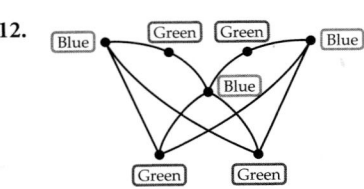

[Sec. 9.4]

CHAPTER 10

EXCURSION EXERCISES, SECTION 10.1   *page 632*

1. $29,663.13   **2.** $19,865.36   **3.** $59,924.80   **4.** $39,819.64

EXERCISE SET 10.1   *page 632*

1. Divide the number of months by 12.   **2.** Using the ordinary method, divide the number of days by 360.
3. *I* is the interest, *P* is the principal, *r* is the interest rate, and *t* is the time period.   **4.** $120   **5.** $560   **6.** $82.50

7. $227.50 **8.** $89.25 **9.** $202.50 **10.** $117 **11.** $16.80 **12.** $30.80 **13.** $159.60
14. $36 **15.** $125 **16.** $252 **17.** $168 **18.** $8789 **19.** $15,667.50 **20.** $4749.50
21. $7390.80 **22.** $13,208 **23.** $2864.40 **24.** 6.25% **25.** 7.5% **26.** 9.5% **27.** 10.5%
28. 8.2% **29.** 9.3% **30.** $152 **31.** $161 **32.** $26.25 **33.** $78 **34.** $2 **35.** $18
36. $145 **37.** $7406 **38.** $7136.25 **39.** $5730.40 **40.** $960.96 **41.** $804.75 **42.** 9%
43. 10.2% **44.** 9.7% **45.** 10% **46.** $1222 **47.** $133.32 **48.** $42.50 **49.** The ordinary
method. The lender benefits. **50.** Approximately 2.5% **51.** 1.25% **52.** Approximately 1.8% **53.** 0.8%
54. $252 **55.** $562.50 **56.** $840 **57.** $2832 **58.** Answers will vary. **59.** There are fewer
days in September than there are in August. **60. a.** $25, $50, $75, $100, $125 **b.** $150 **c.** $175
d. $200 **e.** $225 **f.** Multiply the interest due by 7. **g.** Two times larger **h.** Three times larger
61. a. $5, $10, $15, $20, $25 **b.** $30 **c.** $35 **d.** $40 **e.** $45 **f.** Multiply the interest due by 8.
g. Two times larger **h.** Three times larger

EXCURSION EXERCISES, SECTION 10.2   *page 651*

1. 107.6% **2.** 113.5 **3.** 4.82% **4.** Yes; medical care **5.** Food and beverages, housing, and transportation
6. $123 **7.** $8.71 **8.** 1985

EXERCISE SET 10.2   *page 651*

1. $2739.99 **2.** $10,495.46 **3.** $852.88 **4.** $18,588.48 **5.** $20,836.54 **6.** $7658.76
7. $12,575.23 **8.** $2665.98 **9.** $3532.86 **10.** $6739.77 **11.** $5450.09 **12.** $14,548.16
13. $2213.84 **14.** $14,690.04 **15.** $13,625.23 **16.** $4460.81 **17.** $14,835.46 **18.** $12,463.05
19. $41,210.44 **20.** $10,778.75 **21.** $7641.78 **22.** $11,008.99 **23.** $3182.47 **24.** $402.58
25. $10,094.57 **26.** $20,464.80 **27.** $11,887.58 **28.** $4020.29 **29.** $3583.16 **30.** $1690.74
31. $8188.40 **32.** $17,664.32 **33.** $391.24 **34.** $4700.24 **35.** $18,056.35 **36.** $20,941.99
37. $11,120.58 **38.** $7068.25 **39. a.** $450 **b.** $568.22 **c.** $118.22 **40. a.** $2400
b. $2711.98 **c.** $311.98 **41. a.** $20,528.54 **b.** $20,591.79 **c.** $63.25 **42. a.** $29,775.40
b. $29,851.31 **c.** $75.91 **43. a.** $1698.59 **b.** $1716.59 **c.** $18.00 **44. a.** $6764.51
b. $6878.33 **c.** $113.82 **45. a.** $10,401.63 **b.** $3401.63 **46. a.** $9831.70 **b.** $3831.70
47. $9612.75 **48.** $15,919.00 **49. a.** $67,228.19 **b.** $94,161.98 **50. a.** $39,178.17 **b.** $84,361.02
51. $6789.69 **52.** $20,653.72 **53.** $8773.72 **54.** $4721.16 **55.** $56,102.07 **56.** $89,542.38
57. $12,152.77 **58.** $36,504.48 **59.** $2495.45 **60.** $25,686.16 **61.** 7.40% **62.** 8.67%
63. 7.76% **64.** 7.12% **65.** 8.44% **66.** 6.50% **67.** 6.10% **68.** 6.45%
69. $1.94, $2.60, $4.65 **70.** $3.47, $4.64, $8.31 **71.** $2.93, $3.92, $7.02 **72.** $3.01, $4.03, $7.22
73. $10.71, $14.33, $25.66 **74.** $11.98, $16.03, $28.70 **75.** $234,189.48, $313,398.35, $561,248.71
76. $32,117.41, $42,980.34, $76,971.25 **77.** $25,417.46 **78.** $14,550.23 **79.** $25,841.90 **80.** $10,873.38
81. $53,473.96 **82.** $3684.98 **83. a.** 4.0%, 4.04%, 4.06%, 4.07%, 4.08% **b.** increase **84.** 4.07%
85. 3.04% **86.** 6.6% compounded quarterly **87.** 6.25% compounded semiannually **88.** 7.8% compounded
monthly **89.** 5.8% compounded quarterly **90. a.** $1171.66, $2343.32, $4686.64 **b.** $9373.28 **91.** 10%
92. $1,800,000 **93.** $340.54 **94. a.** $3080.37 **b.** $14,861.21 **c.** $17,515.24 **d.** $4312.52
e. $1049.58 **f.** $4835.32 **g.** $9358.55 **h.** $29,167.36 **i.** $8757.62 **j.** $437.53

EXCURSION EXERCISES, SECTION 10.3   *page 667*

1. $23,227.91 **2.** $449.06 **3.** $26,943.60 **4.** $16,082.85 **5.** $21,635 **6.** $275.15
7. $16,509 **8.** Buying the car **9.** Answers will vary.

EXERCISE SET 10.3   *page 668*

1. $1.48 **2.** $6.85 **3.** $152.32 **4.** $254.34 **5.** $335.87 **6.** $955.20 **7.** $15.34
8. $6.81 **9.** $5.00 **10.** $5.65 **11.** $26.93 **12.** $20.37 **13.** 13.5% **14.** 13.7%
15. 19.2% **16.** 12.9% **17.** $56.21 **18.** $56.03 **19. a.** $696.05 **b.** $174.01 **c.** $88.46
20. a. $19,436.25 **b.** $3887.25 **c.** $470.92 **21. a.** $68,569.73 **b.** $13,713.95 **c.** $641.17
22. a. $2680.82 **b.** $402.12 **c.** $103.23 **23.** $874.88 **24.** $367.15 **25.** $571.31
26. $422.91 **27. a.** $377.06 **b.** $2653.88 **28. a.** $533.28 **b.** $5571.80 **29.** $13,575.25

30. $9990.66   **31.** $3472.57   **32.** $9900.81   **33. a.** $22,740   **b.** $101.90   **c.** $161.25
d. $263.15   **34. a.** $25,165   **b.** $145.83   **c.** $144.42   **d.** $290.25   **35. a.** $21,100   **b.** 0.003375
c. $121.84   **d.** $169.44   **e.** $291.28   **36. a.** $160,000   **b.** 0.0035   **c.** $857.50   **d.** $1250
e. $2107.50   **37.** The monthly payment for the loan is *PMT*. The interest rate per period, *i*, is the annual interest rate divided by 12. The term of the loan, *n*, is the number of years of the loan times 12. Substitute these values into the Payment Formula for an APR Loan and solve for *A*, the selling price of the car. The selling price of the car is $9775.72.   **38. a.** $2704.15
b. $20,938.15   **c.** $515.10   **39. a.** $168.48   **b.** $2669   **c.** $11,391.04   **40. a.** $15.24
b. $20.62   **c.** $5.38   **d.** $5.38   **e.** In part c, the credit card company is charging interest on the additional $299 of unpaid balance. In part d, the interest is earned on the $299 invested.   **41. a.** $336.04   **b.** $339.29   **c.** $1.40
d. $22.63   **42. a.** 92 months   **b.** $1200.66   **c.** Answers will vary.

EXCURSION EXERCISES, SECTION 10.4   *page 682*

1. $600.46   **2.** $1425   **3.** $96,425   **4.** 6.36%   **5.** Option 1: $682.18; Option 2: $665.30
6. Option 1: $1500; Option 2: $2000   **7.** Option 1: $17,872.32; Option 2: $17,967.20   **8.** Option 1: $26,058.48;
Option 2: $25,950.80   **9.** Option 1 is more cost effective if you stay in the home for 2 years or less. Option 2 is more cost effective if you stay in the home for 3 years or more.

EXERCISE SET 10.4   *page 683*

1. $21,500; $64,500   **2.** $12,500; $112,500   **3.** $5625   **4.** $4675   **5.** $35,339   **6.** $19,217
7. $34,289.38   **8.** $34,841   **9.** $974.37   **10.** $1548.57   **11.** $2155.28   **12.** $1300.87
13. a. $1088.95   **b.** $392,022   **c.** $240,022   **14. a.** $780.57   **b.** $234,171   **c.** $135,171
15. $174,606   **16.** $664,141.60   **17.** Interest: $1407.38; principal: $495.89   **18.** Interest: $547.34; principal: $15.83
19. Interest: $619.21; principal: $56.79   **20.** Interest: $986.59; principal: $110.44   **21.** $56,012.75   **22.** $63,437.00
23. $61,039.75   **24.** $96,924.63   **25.** $572.80   **26.** $1026.98   **27.** $2022.50   **28.** $1827.28
29. a. $330.57   **b.** $140,972.40   **30. a.** $201.02   **b.** $81,264.60   **31. a.** $390.62   **b.** $178,273.20
32. a. $343.07   **b.** $188,254.80   **33.** $75,000   **34.** $60,000   **35.** $105,000   **36.** 260th payment
37. No   **38.** Yes. If the interest rate is lower, it will take fewer months.   **39.** You pay less total interest on a 15-year mortgage loan.   **40. a.** $65,641.88   **b.** $138,596.60   **c.** $28,881.52   **d.** 44%   **41. a.** $15,065; yes
b. $431.28   **c.** $646.32   **d.** $1593.37   **e.** 28.9%   **f.** Answers will vary.

CHAPTER 10 REVIEW EXERCISES   *page 687*

1. $61.88 [Sec. 10.1]   **2.** $782 [Sec. 10.1]   **3.** $90 [Sec. 10.1]   **4.** $7218.40 [Sec. 10.1]
5. 7.5% [Sec. 10.1]   **6.** $3654.90 [Sec. 10.2]   **7.** $11,609.72 [Sec. 10.2]   **8.** $7859.52 [Sec. 10.2]
9. $200.23 [Sec. 10.2]   **10.** $10,683.29 [Sec. 10.2]   **11. a.** $11,318.23   **b.** $3318.23 [Sec. 10.2]
12. $19,225.50 [Sec. 10.2]   **13.** 1.1% [Sec. 10.1]   **14.** $9000 [Sec. 10.1]   **15.** $2.31 [Sec. 10.2]
16. $43,650.68 [Sec. 10.2]   **17.** 6.06% [Sec. 10.2]   **18.** 5.4% compounded semiannually [Sec. 10.2]
19. $431.16 [Sec. 10.3]   **20.** $6.12 [Sec. 10.3]   **21. a.** $259.38   **b.** 12.9% [Sec. 10.3]
22. a. $36.03   **b.** 12.9% [Sec. 10.3]   **23.** $45.41 [Sec. 10.3]   **24. a.** $10,092.69   **b.** $2018.54
c. $253.01 [Sec. 10.3]   **25.** $704.85 [Sec. 10.3]   **26. a.** $540.02   **b.** $12,196.80 [Sec. 10.3]
27. a. $29,450   **b.** $181.80   **c.** $224.17   **d.** $436.42 [Sec. 10.3]   **28.** $99,041 [Sec. 10.4]
29. a. $829.56   **b.** $298,641.60   **c.** $170,741.60 [Sec. 10.4]   **30. a.** $1396.69   **b.** $150,665.74 [Sec. 10.4]
31. $2658.53 [Sec. 10.4]

CHAPTER 10 TEST   *page 689*

1. $108.28 [Sec. 10.1]   **2.** $202.50 [Sec. 10.1]   **3.** $8408.89 [Sec. 10.1]   **4.** 9% [Sec. 10.1]
5. $7340.87 [Sec. 10.2]   **6.** $312.03 [Sec. 10.2]   **7. a.** $15,331.03   **b.** $4831.03 [Sec. 10.2]
8. $21,949.06 [Sec. 10.2]   **9.** 1.2% [Sec. 10.1]   **10.** $1900 [Sec. 10.1]   **11.** $397.15 [Sec. 10.2]
12. 6.40% [Sec. 10.2]   **13.** 4.6% compounded semiannually [Sec. 10.2]   **14.** $8.11 [Sec. 10.3]   **15. a.** $48.56
b. 16.6% [Sec. 10.3]   **16.** $60.61 [Sec. 10.3]   **17. a.** $6985.94   **b.** $1397.19   **c.** $174.62 [Sec. 10.3]
18. $37,483.50 [Sec. 10.4]   **19. a.** $1530.69   **b.** $221,546.46 [Sec. 10.4]   **20.** $2595.97 [Sec. 10.4]

CHAPTER 11

EXCURSION EXERCISES, SECTION 11.1 *page 700*

1. Three weighings   **2.** Three weighings

EXERCISE SET 11.1 *page 702*

1. {0, 2, 4, 6, 8}   **2.** {1, 3, 5, 7, 9}   **3.** {Monday, Tuesday, Wednesday, Thursday, Friday, Saturday, Sunday}
4. {January, February, March, April, May, June, July, August, September, October, November, December}
5. {HH, TT, HT, TH}   **6.** {HHH, HHT, HTH, THH, HTT, THT, TTH, TTT}
7. {1H, 2H, 3H, 4H, 5H, 6H, 1T, 2T, 3T, 4T, 5T, 6T}   **8.** {H1, H2, H3, H4, T1, T2, T3, T4}
9. {$S_1E_1D_1$, $S_1E_1D_2$, $S_1E_2D_1$, $S_1E_2D_2$, $S_1E_3D_1$, $S_1E_3D_2$, $S_2E_1D_1$, $S_2E_1D_2$, $S_2E_2D_1$, $S_2E_2D_2$, $S_2E_3D_1$, $S_2E_3D_2$}
10. {$B_1R_1C_1$, $B_1R_1C_2$, $B_1R_2C_1$, $B_1R_2C_2$, $B_2R_1C_1$, $B_2R_1C_2$, $B_2R_2C_1$, $B_2R_2C_2$, $B_3R_1C_1$, $B_3R_1C_2$, $B_3R_2C_1$, $B_3R_2C_2$}
11. {ABCD, ABDC, ACBD, ACDB, ADBC, ADCB}   **12.** {ABCDE, ABCED, ABDCE, ABDEC, ABECD, ABEDC,
ADBCE, ADBEC, ADCBE, ADCEB, ADEBC, ADECB, AEBCD, AEBDC, AECBD, AECDB, AEDBC, AEDCB}
13. {AA BB CC, AA BC CB, AB BA CC, AC BA CB, AC BB CA, AB BC CA}
14. {N3Y, S3Y, E3Y, W3Y, N3N, S3N, E3N, W3N, N4Y, S4Y, E4Y, W4Y, N4N, S4N, E4N, W4N}   **15.** 12   **16.** 16
17. 4^{20} **18.** 2^{25} **19.** 7000   **20.** 125   **21.** 90   **22.** 45   **23.** 18   **24.** 30   **25.** 62
26. 49   **27.** 24   **28.** 9   **29.** 15   **30.** Yes   **31.** 13^4 **32.** 24   **33.** 1   **34.** 17,160
37. 150; 25   **38. a.** 45   **b.** 190   **39.** Answers will vary.   **40.** In the tree diagram, 9–13 signifies moving a
checker from square 9 to square 13.

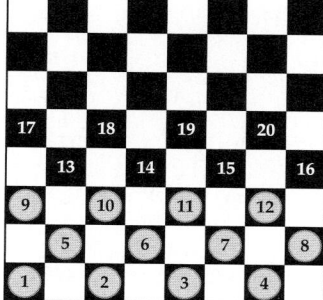

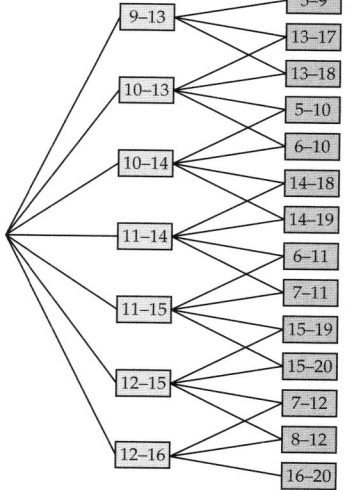

EXCURSION EXERCISES, SECTION 11.2 *page 713*

1. 487,635   **2.** 684,400   **3.** 336,300   **4.** 68,400   **5.** 4845

EXERCISE SET 11.2 *page 714*

1. 40,320   **2.** 120   **3.** 362,760   **4.** 24   **5.** 120   **6.** 40,314   **7.** 40,320   **8.** 3,628,800
9. 120   **10.** 3,628,800   **11.** $n = 7$ **12.** $n = 6$ **13.** 6720   **14.** 42   **15.** 181,440
16. 30,240   **17.** 1   **18.** 1   **19.** 40,320   **20.** 3,628,800   **21.** 3360   **22.** 3   **23.** $\dfrac{1}{720}$
24. 100   **25.** 36   **26.** 28   **27.** 1   **28.** 1   **29.** 525   **30.** 7056   **31.** $\dfrac{60}{143}$ **32.** $\dfrac{20}{63}$

33. $\dfrac{360}{1001}$ **34.** 336   **35.** 2880   **36.** 120   **37.** 21   **38.** 56   **39.** 792   **40.** 36   **41.** 1

42. 1   **43.** No   **44.** No   **45.** 120   **46.** 6840   **47.** 43,680   **48.** 504   **49.** 35

50. 210   **51.** 455   **52.** 3003   **53.** 11,880   **54.** 24   **55.** 28   **56.** 2,118,760   **57.** 30

58. 3136   **59.** 252   **60.** 495   **61.** 21   **62.** 5   **63.** 9   **64.** 35   **65.** 48,620

66. 756,756   **67.** 7,484,400   **68.** 792   **69.** 45,360   **70.** 5040   **71.** 1680

72. 60   **73.** 48   **74.** 1728   **75.** 4512   **76.** 24   **77.** 103,776   **78.** 1584

81. $x^5 + 5x^4y + 10x^3y^2 + 10x^2y^3 + 5xy^4 + y^5$ **82.** $x^6 + 6x^5y + 15x^4y^2 + 20x^3y^3 + 15x^2y^4 + 6xy^5 + y^6$

83. a. 2520   **b.** 60   **c.** Approx. 1929 years   **d.** Approx. 3.23×10^{26} years   **84.** 2,303,805

EXCURSION EXERCISES, SECTION 11.3   *page 726*

1. Answers will vary.   **2.** 0.6366   **3–6.** Answers will vary.

EXERCISE SET 11.3   *page 727*

1. {HHH, HHT, HTH, HTT, THH, THT, TTH, TTT}   **2.** {2, 4, 6, 8, 10}

3. {Nov. 1, Nov. 2, Nov. 3, Nov. 4, Nov. 5, Nov. 6, Nov. 7, Nov. 8, Nov. 9, Nov. 10, Nov. 11, Nov. 12, Nov. 13, Nov. 14}

4. {penny, nickel, dime, quarter, half-dollar, dollar}   **5.** {Alaska, Alabama, Arizona, Arkansas}

6. {April, June, September, November}   **7.** {BBB, BBG, BGB, BGG, GBB, GBG, GGB, GGG}   **8.** {BBG, BGB, GBB}

9. {BGG, GBG, GGB, GGG}   **10.** {BBB}   **11.** {BBG, BGB, BGG, GBB, GBG, GGB, GGG}

12. $\dfrac{3}{8}$ **13.** $\dfrac{1}{2}$ **14.** $\dfrac{1}{8}$ **15.** $\dfrac{7}{8}$ **16.** $\dfrac{3}{8}$ **17.** $\dfrac{1}{4}$ **18.** $\dfrac{1}{16}$ **19.** $\dfrac{1}{8}$ **20.** $\dfrac{5}{16}$

21. $\dfrac{11}{16}$ **22.** $\dfrac{15}{16}$ **23.** $\dfrac{1}{12}$ **24.** $\dfrac{11}{12}$ **25.** $\dfrac{1}{4}$ **26.** $\dfrac{7}{12}$ **27.** $\dfrac{5}{36}$ **28.** $\dfrac{1}{18}$ **29.** $\dfrac{1}{36}$

30. $\dfrac{1}{36}$ **31.** 0   **32.** 0   **33.** $\dfrac{1}{6}$ **34.** $\dfrac{5}{18}$ **35.** $\dfrac{1}{2}$ **36.** $\dfrac{1}{2}$ **37.** $\dfrac{1}{6}$ **38.** $\dfrac{5}{6}$

39. $\dfrac{1}{2}$ **40.** $\dfrac{1}{4}$ **41.** $\dfrac{1}{13}$ **42.** $\dfrac{3}{13}$ **43.** $\dfrac{3}{13}$ **44.** $\dfrac{2}{13}$ **45.** $\dfrac{1267}{3228}$ **46.** $\dfrac{2190}{3228}$ **47.** $\dfrac{804}{3228}$

48. $\dfrac{1651}{3228}$ **49.** $\dfrac{150}{3228}$ **50.** $\dfrac{432}{3228}$ **51.** $\dfrac{26}{425}$ **52.** $\dfrac{137}{425}$ **53.** $\dfrac{18}{425}$ **54.** $\dfrac{103}{850}$ **55.** $\dfrac{58}{293}$

56. $\dfrac{39}{293}$ **57.** $\dfrac{36}{293}$ **58.** $\dfrac{117}{293}$ **59.** $\dfrac{2492}{5111}$ **60.** $\dfrac{253}{556}$ **61.** $\dfrac{1}{4}$ **62.** 0   **63.** 0   **64.** $\dfrac{3}{4}$

67. $\dfrac{1}{3}$ **68.** $\dfrac{1}{5}$ **69.** $\dfrac{3}{10}$ **70.** $\dfrac{3}{8}$ **71.** $\dfrac{8}{13}$ **72.** $\dfrac{11}{20}$ **73.** 1 to 4   **74.** 3 to 2   **75.** 3 to 5

76. 7 to 18   **77.** 11 to 9   **78.** 81 to 19   **79.** 1 to 8   **80.** 1 to 1   **81.** 1 to 3   **82.** 15 to 1

83. $\dfrac{3}{11}$ **84.** $\dfrac{1}{101}$ **85. a.** 7 to 50   **b.** $\dfrac{7}{57}$ **86. a.** 4 to 31   **b.** $\dfrac{4}{35}$ **87.** $\dfrac{9}{24}$ **88.** $\dfrac{63}{256}$

90. $\dfrac{1}{54,145}$ **91.** $\dfrac{1}{108,290}$ **92.** $\dfrac{9}{19}$ **93.** $\dfrac{9}{19}$ **94.** $\dfrac{6}{19}$ **95.** $\dfrac{1}{38}$ **96.** $\dfrac{1}{19}$ **97.** $\dfrac{3}{19}$

EXCURSION EXERCISES, SECTION 11.4   *page 737*

1. 77.3%   **2.** 99.9%   **3.** 0.00000207%   **4.** 26.2%   **5.** Probability of not matching any lucky numbers: 0.12%; Probability of matching five lucky numbers: 23.3%   **6.** 99.9%

EXERCISE SET 11.4   *page 738*

3. $\dfrac{2}{13}$ **4.** $\dfrac{1}{2}$ **5.** $\dfrac{1}{9}$ **6.** $\dfrac{11}{36}$ **7.** 0.6   **8.** 0.8   **9.** 0.2   **10.** 0.5   **11.** $\dfrac{7}{10}$

12. $\dfrac{7}{10}$ **13.** $\dfrac{4}{5}$ **14.** $\dfrac{7}{10}$ **15.** $\dfrac{5}{18}$ **16.** $\dfrac{1}{3}$ **17.** $\dfrac{1}{2}$ **18.** $\dfrac{1}{2}$ **19.** $\dfrac{11}{18}$ **20.** $\dfrac{5}{9}$

21. $\dfrac{4}{13}$ **22.** $\dfrac{7}{13}$ **23.** $\dfrac{3}{13}$ **24.** $\dfrac{8}{13}$ **25.** $\dfrac{3}{4}$ **26.** $\dfrac{3}{4}$ **27.** $\dfrac{1150}{3179}$ **28.** $\dfrac{84}{187}$ **29.** $\dfrac{1170}{3179}$

30. $\dfrac{2644}{3179}$ **31.** 0.96   **32.** 0.62   **33.** 2499 in 2500   **34.** $\dfrac{5}{8}$ **35.** $\dfrac{5}{6}$ **36.** $\dfrac{5}{6}$ **37.** $\dfrac{11}{12}$

38. $\dfrac{35}{36}$ **39.** $\dfrac{12}{13}$ **40.** $\dfrac{10}{13}$ **41.** $\dfrac{15}{16}$ **42.** $\dfrac{11}{16}$ **43.** 42.1%   **44.** 51.8%   **45.** 36.1%

46. 20.4%   **47.** 54.5%   **48.** 27.4%   **49.** 55.3%   **50.** 28.1%   **51.** 88.8%   **52.** 45.9%

53. 20.4%   **54.** 48.7%   **55.** No   **56. a.** $\dfrac{1}{3}$ **b.** $\dfrac{2}{3}$ **c.** $\dfrac{2}{3}$ **57.** 100%

58–59. Answers will vary.   **60. a.** 2,598,960   **b.** 1287   **c.** 0.000495   **d.** 0.000495
e. Yes   **f.** 0.00198

EXCURSION EXERCISES, SECTION 11.5   *page 748*

1. 7.43%   **2.** 25.3%   **3.** 50.7%   **4.** 89.1%

EXERCISE SET 11.5   *page 749*

3. $P(A\,|\,B) = 0.625$; $P(B\,|\,A) = 0.357$ **4.** $P(A\,|\,B) = 0.375$; $P(B\,|\,A) = 0.667$ **5.** $P(A\,|\,B) = 0.389$; $P(B\,|\,A) = 0.115$

6. $P(A\,|\,B) = 0.4$; $P(B\,|\,A) = 1$ **7.** $\dfrac{179}{864}$ **8.** $\dfrac{173}{778}$ **9.** $\dfrac{557}{921}$ **10.** $\dfrac{341}{1642}$ **11.** 0.30   **12.** 0.31

13. 0.15   **14.** 0.11   **15.** 0.28   **16.** 0.17   **17.** 0.20   **18.** 0.24   **19.** 0.050   **20.** 0.059

21. 0.13   **22.** 0.072   **23.** $\dfrac{6}{1045}$ **24.** $\dfrac{13}{5852}$ **25.** $\dfrac{3}{1045}$ **26.** $\dfrac{3}{7315}$ **27.** $\dfrac{13}{102}$ **28.** $\dfrac{13}{850}$

29. $\dfrac{8}{5525}$ **30.** $\dfrac{13}{204}$ **31.** 0.000484   **32.** 0.003916   **33.** 0.001424   **34.** 0.00005324

35. Independent   **36.** Not independent   **37.** Not independent   **38.** Independent   **39.** $\dfrac{25}{1296}$

40. $\dfrac{5}{1296}$ **41.** $\dfrac{1}{72}$ **42.** $\dfrac{1}{36}$ **43.** $\dfrac{1}{4}$ **44.** $\dfrac{1}{36}$ **45.** $\dfrac{1}{16}$ **46.** $\dfrac{1}{216}$ **47.** $\dfrac{1}{216}$ **48.** $\dfrac{1}{1728}$

49. $\dfrac{1}{169}$ **50.** $\dfrac{9}{169}$ **51.** $\dfrac{1}{16}$ **52.** $\dfrac{12}{169}$ **53.** $\dfrac{1}{16}$ **54.** $\dfrac{3}{26}$ **55. a.** $\dfrac{1}{32}$ **b.** $\dfrac{13}{425}$

56. a. $\dfrac{27}{2197}$ **b.** $\dfrac{11}{1105}$ **57. a.** $\dfrac{100}{4913}$ **b.** $\dfrac{1}{51}$ **58. a.** $\dfrac{25}{576}$ **b.** $\dfrac{1}{22}$ **59.** 0.46   **60.** 0.81

61. 0.11   **62.** 0.983   **63. a.** $\dfrac{1}{8}$ **b.** $\dfrac{1}{4}$ **c.** $\dfrac{3}{8}$ **64. a–d.** Answers will vary.   **e.** $\dfrac{4}{7}$

65. a. $\dfrac{1}{2}$ **b.** 0   **c.** 1   **d.** $\dfrac{1}{3}$ **e.** $\dfrac{2}{3}$ **f.** Yes

EXCURSION EXERCISES, SECTION 11.6   *page 756*

1. $\dfrac{1}{216}$ **2.** $\dfrac{25}{72}$ **3.** $\dfrac{5}{72}$ **4.** $\dfrac{125}{216}$ **5.** −5 cents   **6.** 0   **7.** 0   **8.** Under 11 or Over 10;
Numbers bet

EXERCISE SET 11.6   *page 757*

1. 49.5   **2.** 6.27   **3.** −5 cents   **4.** −5 cents   **5.** 52 cents   **6.** −22 cents   **7.** −11 cents
8. −17 cents   **9.** $25   **10.** $35   **11.** −$235   **12.** $95   **13.** More than $600
14. More than $975   **15.** $39,100   **16.** $6000   **17.** $20,250   **18.** $242,500   **19.** −31 cents

20. −78 cents   **21.** 7   **22. a.** {01, 02, 03, 04, 05, 06, 61, 62, 63, 64, 65, 66}   **b.** All sums have a probability of $\dfrac{1}{12}$.

c. 6.5   **23.** 7   **25.** Red dice   **26. a.** −52 cents   **b.** One is almost three times more likely than the other but
the prize is the same.

CHAPTER 11 REVIEW EXERCISES   *page 762*

1. {11, 12, 13, 21, 22, 23, 31, 32, 33} [Sec. 11.1]   **2.** {26, 28, 62, 68, 82, 86} [Sec. 11.1]   **3.** {HHHH, HHHT, HHTH, HHTT, HTHH, HTHT, HTTH, HTTT, THHH, THHT, THTH, THTT, TTHH, TTHT, TTTH, TTTT} [Sec. 11.1]
4. {7A, 8A, 9A, 7B, 8B, 9B} [Sec. 11.1]   **5.** 72 [Sec. 11.1]   **6.** 10,000 [Sec. 11.1]   **7.** 1,423,656 [Sec. 11.1]
8. 64 [Sec. 11.1]   **9.** 5040 [Sec. 11.2]   **10.** 40,296 [Sec. 11.2]   **11.** 1260 [Sec. 11.2]

12. 151,200 [Sec. 11.2]   **13.** 336 [Sec. 11.2]   **14.** $\frac{60}{143}$ [Sec. 11.2]   **15.** 5040 [Sec. 11.2]

16. 5040 [Sec. 11.2]   **17.** 2520 [Sec. 11.2]   **18.** 180 [Sec. 11.2]   **19.** 495 [Sec. 11.2]
20. 60 [Sec. 11.2]   **21.** 3,268,760 [Sec. 11.2]   **22.** 165 [Sec. 11.2]   **23.** 660 [Sec. 11.2]

24. 282,240 [Sec. 11.2]   **25.** 624 [Sec. 11.2]   **26.** $\frac{1}{4}$ [Sec. 11.3]   **27.** $\frac{3}{8}$ [Sec. 11.3]

28. 0.56 [Sec. 11.3]   **29.** 0.37 [Sec. 11.3]   **30.** 0.85 [Sec. 11.3]   **31.** $\frac{1}{9}$ [Sec. 11.3]   **32.** $\frac{17}{18}$ [Sec. 11.4]

33. $\frac{1}{6}$ [Sec. 11.4]   **34.** $\frac{5}{9}$ [Sec. 11.4]   **35.** $\frac{2}{9}$ [Sec. 11.5]   **36.** $\frac{1}{6}$ [Sec. 11.5]   **37.** $\frac{3}{4}$ [Sec. 11.4]

38. $\frac{4}{13}$ [Sec. 11.4]   **39.** $\frac{12}{13}$ [Sec. 11.4]   **40.** $\frac{2}{3}$ [Sec. 11.5]   **41.** 5 to 31 [Sec. 11.3]

42. 1 to 3 [Sec. 11.3]   **43.** $\frac{5}{9}$ [Sec. 11.3]   **44.** $\frac{1}{2}$ [Sec. 11.3]   **45.** .018 [Sec. 11.4]

46. $\frac{173}{1000}$ [Sec. 11.4]   **47.** $\frac{5}{8}$ [Sec. 11.3]   **48.** 1 to 5 [Sec. 11.3]   **49.** $\frac{7}{16}$ [Sec. 11.5]

50. $\frac{646}{1771}$ [Sec. 11.4]   **51.** $\frac{7}{253}$ [Sec. 11.5]   **52.** 0.37 [Sec. 11.3]   **53.** 0.62 [Sec. 11.4]

54. 0.067 [Sec. 11.3]   **55.** 0.47 [Sec. 11.5]   **56.** 0.29 [Sec. 11.5]   **57.** $\frac{1}{216}$ [Sec. 11.5]

58. 0.60 [Sec. 11.5]   **59.** 0.16 [Sec. 11.5]   **60.** $\frac{175}{256}$ [Sec. 11.5]   **61.** 0.648 [Sec. 11.5]

62. 0.029 [Sec. 11.5]   **63.** 0.135 [Sec. 11.4]   **64.** −50 cents [Sec. 11.6]   **65.** 25 cents [Sec. 11.6]
66. −37.5 cents [Sec. 11.6]   **67.** about 5.4 [Sec. 11.6]   **68.** $200 [Sec. 11.6]   **69.** −$80 [Sec. 11.6]
70. $11,000 [Sec. 11.6]

CHAPTER 11 TEST   *page 764*

1. {A2, D2, G2, K2, A3, D3, G3, K3, A4, D4, G4, K4} [Sec. 11.1]   **2.** 72 [Sec. 11.1]   **3.** 5040 [Sec. 11.2]

4. about 1.09×10^{10} [Sec. 11.2]   **5.** 116,280 [Sec. 11.2]   **6.** $\frac{7}{12}$ [Sec. 11.3]   **7.** Yes [Sec. 11.5]

8. $\frac{1}{17}$ [Sec. 11.3]   **9.** 0.72 [Sec. 11.4]   **10.** 1 to 7 [Sec. 11.5]   **11.** 0.635 [Sec. 11.5]
12. 0.466 [Sec. 11.5]   **13.** 50% [Sec. 11.3]   **14.** $38,350 [Sec. 11.6]

CHAPTER 12

EXCURSION EXERCISES, SECTION 12.1   *page 776*

1. 1.9372   **2.** 0.87311   **3.** 1.541   **4.** 2.4   **5.** 315.3 pounds

EXERCISE SET 12.1 *page 777*

1. 7; 7; 7   **2.** 11.6; 9; 3   **3.** 22; 14; no mode   **4.** 73.5; 79.5; 74   **5.** 18.8; 8.1; no mode   **6.** 5; 5; 5
7. 192.4; 190; 178   **8.** 132.2; 118; 118   **9.** 0.1; -3; -5 **10.** 3.1; 4.1; 4.1   **11. a.** Yes. The mean is
computed by using the sum of all the data.   **b.** No. The median is not affected unless the middle value, or one of the two middle
values, in a data set is changed.   **12.** Yes. The mode is the number (or numbers) in a data set that occurs most frequently.
13. $\approx$38.7; 35; 33   **14. a.** $\approx$45.5; 43; 40   **b.** Each result in Exercise 14a is larger than the corresponding result in
Exercise 13.   **15. a.** Answers will vary.   **b.** Answers will vary.   **16.** Answers will vary. However, the median is more
representative of most of the expenditures.   **17.** $\approx$82.9   **18.** 79.4$\overline{6}$ **19.** 82   **20.** 84.5   **21.** 2.5
22. $\approx$2.64   **23.** $\approx$6.1 points; 5 points; 2 points and 5 points   **24.** $\approx$\$12.58; \$11.50; \$8.00   **25.** $\approx$7.2; 7; 7
26. $\approx$10.1 years; 10 years; 10 years   **27.** 64°   **28.** 10°   **29.** -6°F   **30.** 20.5°F   **31.** 92
32. No. Alisa needs a score of 122 to raise her average to 90, but the next test only counts as 100 points.   **33. a.** $\approx$0.275
b. $\approx$0.273   **c.** No   **34.** 48 miles per hour   **35.** 81   **36.** Answers will vary.
37. $d_1 = d_2$

$$r_1 = \frac{d_1}{t_1} \qquad r_2 = \frac{d_2}{t_2} = \frac{d_1}{t_2}$$

$$t_1 = \frac{d_1}{r_1} \qquad t_2 = \frac{d_1}{r_2}$$

$$r = \frac{d_1 + d_2}{t_1 + t_2} = \frac{d_1 + d_1}{t_1 + t_2} = \frac{2d_1}{t_1 + t_2}$$

$$= \frac{2d_1}{\dfrac{d_1}{r_1} + \dfrac{d_1}{r_2}} = \frac{2d_1}{d_1\left(\dfrac{1}{r_1} + \dfrac{1}{r_2}\right)}$$

$$= \frac{2}{\dfrac{1}{r_1} + \dfrac{1}{r_2}} \cdot \left(\frac{r_1 r_2}{r_1 r_2}\right) = \frac{2r_1 r_2}{r_1 + r_2}$$

38. a. Answers will vary.   **b.** The new mean is 12 larger than the mean of the original data.   **c.** The new median is
12 larger than the median of the original data.   **39.** Yes. Joanne has a smaller average for the first month and the second month,
but she has a larger average for both months combined.   **40.** No. Wendy has a larger average in English and in history, and she
also has a larger average for English and history combined.   **41.** Answers will vary.

EXCURSION EXERCISES, SECTION 12.2 *page 789*

1. a. and b.

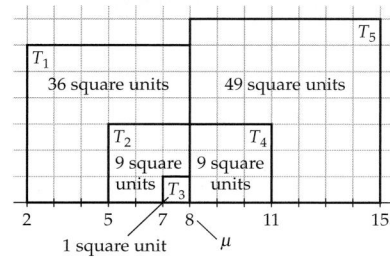

c. 104 square units   **d.** 20.8 square units

e. See the above figure.   **f.** The variance is 20.8. It is the area of the average tile shown above.   **g.** The standard deviation is
$\sqrt{20.8} \approx 4.56$. It is the width of the average tile.   **2. a. and b.**

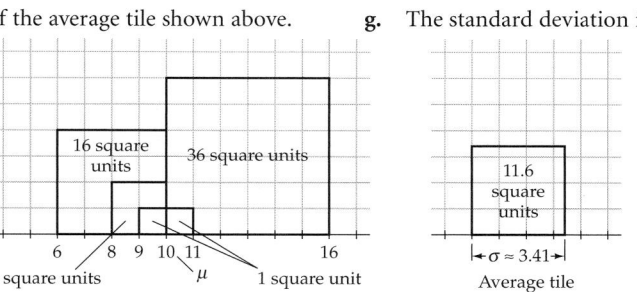

c. 58 square units   **d.** 11.6 square units   **e.** See the above figure.   **f.** The variance is 11.6. It is the area of the
average tile shown above.   **g.** The standard deviation is $\sqrt{11.6} \approx 3.41$. It is the width of the average tile.   **h.** The data set in
Excursion Exercise 2 has the larger mean, but the data set in Excursion Exercise 1 has the larger standard deviation.

1. 84°F **2.** 62°F **3.** 21; 8.2; 67.1 **4.** 13; 4.8; 23.4 **5.** 3.3; 1.3; 1.7 **6.** 15.4; 6.1; 37.7
7. 52; 17.7; 311.6 **8.** 47; 16.9; 284.5 **9.** 0; 0; 0 **10.** 2; 1.0; 1.1 **11.** 23; 8.3; 69.6 **12.** 27; 9.8; 95.6
13. Opinions will vary. However, many climbers would consider rope B to be safer because of its small standard deviation.
14. Opinions will vary. However, many students will choose golfer B because that golfer has a higher standard deviation.
15. Generally, more physical activity will occur during a break period than in most history classes. This increase in physical activity will tend to increase the standard deviation of the pulse rates. **16.** Opinions will vary. However, many students will choose the group of community college students because they would tend to be a more diverse group. **17. a.** 30.8; 10.1 **b.** 14.5; 6.4
c. Winning scores; winning scores **18. a.** 38.9; 12.9 **b.** 45.4; 10.9 **c.** Male actors; female actors
19. a. 44.5; 9.7 **b.** 42.2; 7.6 **c.** National League; National League **20. a.** 9.9 hours; 1.1 hours
b. 8.8 hours; 0.9 hour **c.** Women; women **21.** 54.8 years; 6.2 years **22.** 69.1 years; 11.1 years
23. a. Answers will vary. **b.** The population standard deviation remains the same. **24. a.** Answers will vary.
b. The standard deviation of the new data is k times the standard deviation of the original data. **25. a.** 0 **b.** Yes
c. No **26.** If the variance is 0 or 1 **27. a.** Identical **b.** Identical **c.** Identical **d.** They will be identical. **28.** The standard deviation is the same as in Exercise 27b—about 2.87.

1. *Ages of Customers Who Purchased a Cruise*

Stems	Leaves
2	1 3 8
3	2 3 3 5 8
4	0 1 2 4 5 5 6
5	0 1 1 2 2 5
6	1 2 4 6 8
7	2 7

Legend:
5 | 2 represents 52 years old

2. *Winning Scores vs. Losing Scores*

Winning Scores		Losing Scores
5 2	5	
9 6 2	4	
9 8 8 7 5 5 5 4 4 3 2 1 1 0	3	1
7 7 7 7 6 4 4 3 3 1 0 0	2	6 4 4 1 1 0
6 6 6 4	1	9 9 9 7 7 7 7 6 6 6 4 4 3 3 0 0 0 0 0 0
	0	9 7 7 7 7 7 6 3

Legend:
6 | 2 represents 26 points   2 | 6 represents 26 points
As expected, the winning scores tend to be larger than the losing scores. Also, we can see that most of the winning scores are in the 20s and 30s, whereas most of the losing scores are in the teens.

3. *Ages of Best Actor Award Recipients*

Female		Male
2 5 6 6 9	2	
1 1 2 3 3 3 3 4 4 4 4 5 5 5 5 7 8 9 9	3	0 1 1 5 5 6 7 7 8 8
1 1 2 4 4 9 9	4	0 0 0 2 2 3 3 4 5 8 8
	5	1 2 3 4 6
1	6	0 0 1 2
4	7	6
0	8	

Legend:
5 | 3 represents 35 years old   3 | 5 represents 35 years old
On average, male recipients of the Best Actor award tend to be older than female recipients.

4. *Number of Home Runs by Home Run Leaders per Season*

National League		American League
3 0	7	
5	6	
2 0	5	6 6 2 2 1 0
9 9 8 8 8 7 7 6 4 3 0 0 0 0 0	4	9 8 7 6 6 5 4 3 3 2 1 0 0 0
9 8 8 8 7 7 7 6 6 5 1	3	9 9 9 7 6 6 3 2 2 2
	2	2

Legend:
2|5 represents 52 home runs   5|2 represents 52 home runs
The National League home run leaders tend to hit more home runs than the American League home run leaders. However, the American League home run leaders have more total home runs in the 50s than do the National League leaders.

EXERCISE SET 12.3   *page 802*

1. a. ≈0.87   **b.** ≈1.74   **c.** ≈−2.17   **d.** 0.0   **2. a.** −0.3   **b.** 0.3   **c.** 2.2
d. −2.8   **3. a.** ≈−0.32   **b.** ≈0.21   **c.** ≈1.16   **d.** ≈−0.95   **4. a.** ≈−1.48   **b.** ≈1.65
c. ≈2.52   **d.** ≈0.17   **5. a.** −0.67   **b.** 147.78 mm Hg   **6. a.** ≈−0.73   **b.** ≈10.9 fluid ounces
7. a. 0.72   **b.** ≈112.16 mg/dl   **8. a.** ≈1.07   **b.** 31,200 miles   **9.** The score in part a.
10. The score in part a.   **11.** 59th percentile   **12.** ≈91st percentile   **13.** 6396 students   **14.** 2016 students
15. a. 50%   **b.** 12%   **c.** 38%   **16. a.** 75%   **b.** 50%   **c.** 25%
17. $Q_1 = 5$, $Q_2 = 10$, $Q_3 = 26$ **18.** $Q_1 = 9$, $Q_2 = 11$, $Q_3 = 14$
19.

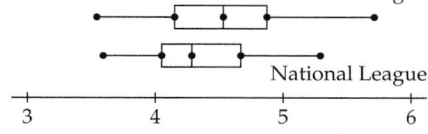

American League / National League

The ERAs in the National League tend to be smaller than the ERAs in the American League. Also, the range of the ERAs is larger for the American League.

20.

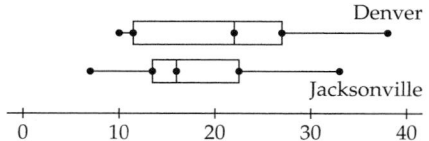

Denver / Jacksonville

The number of points scored by Jacksonville tended to be less than the number of points scored by Denver. The range of the points scored was less for Jacksonville than for Denver.

21.

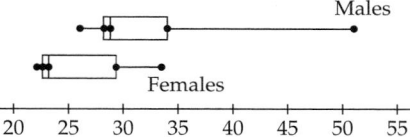

Males / Females

The percent of males who smoked tended to be larger than the percent of females who smoked. Also, the range of the males who smoked was more than double the range of the females who smoked.

22.

The box-and-whisker plots show that the median price of a home tended to be less in the Midwest and the South compared with the Northeast and the West. Also, the range of the median prices in the Midwest was larger than the range in the other areas.
23. a. $\mu = 0, \sigma = 1$ **b.** $\mu = 0, \sigma = 1$ **c.** $\mu = 0, \sigma = 1$ **24. a.** False   **b.** False   **c.** True
d. False   **25.** A value is an outlier for a set of data provided the number is less than $Q_1 - 1.5(Q_3 - Q_1)$ or greater than $Q_3 + 1.5(Q_3 - Q_1)$. For the given data, 85 is the only outlier.

EXCURSION EXERCISES, SECTION 12.4   *page 817*

1. 77.4   **2. a.** At least 124.48   **b.** 86.72   **3.** 64.63 seconds or less

EXERCISE SET 12.4   *page 818*

1. Greater   **2.** Less   **3. a.** 95.4%   **b.** 15.9%   **c.** 81.8%   **4. a.** 99.7%   **b.** 2.3%
c. 97.55%   **5. a.** 81.8%   **b.** 0.15%   **6. a.** 68.2%   **b.** 2.3%   **7. a.** 1272 vehicles   **b.** 12 vehicles
8. a. 682   **b.** 954   **9.** 0.433 square unit   **10.** 0.471 square unit   **11.** 0.468 square unit
12. 0.489 square unit   **13.** 0.130 square unit   **14.** 0.215 square unit   **15.** 0.878 square unit
16. 0.636 square unit   **17.** 0.097 square unit   **18.** 0.027 square unit   **19.** 0.013 square unit
20. 0.352 square unit   **21.** 0.926 square unit   **22.** 0.966 square unit   **23.** 0.997 square unit
24. 0.973 square unit   **25.** $z = 0.84$ **26.** $z = 0.75$ **27.** $z = -0.90$ **28.** $z = -0.44$ **29.** $z = 0.35$
30. $z = -0.15$ **31. a.** 30.9%   **b.** 33.4%   **32. a.** $\approx 62.9\%$ **b.** 38.9%   **33. a.** 13.6%
b. 35.2%   **34. a.** 61.8%   **b.** 72.2%   **35. a.** 10.6%   **b.** 98.8%   **36. a.** $\approx 1.8\%$
b. $\approx 58.8\%$ **37. a.** 0.106   **b.** ≈ 0.460 **38. a.** ≈ 0.053 **b.** ≈ 0.466 **39. a.** ≈ 0.749
b. 0.023   **40. a.** 22.7%   **b.** 61.4%   **41.** Answers will vary.   **42.** Answers will vary.   **43.** True
44. True   **45.** True   **46.** False   **47.** False   **48.** True   **49.** False   **50.** False   **51.** True
52. False   **53. a.** They are identical.   **b.** The area of the region is the same with or without the vertical boundary at $z = 1$.
54. a. and b. Answers will vary.   **55.** -0.84 and 0.84 **56.** ≈ -0.67 and 0.67 **57.** For any positive constant k, the

probability that a random variable will take on a value within k standard deviations of the mean is at least $1 - \dfrac{1}{k^2}$. According to

Chebyshev's Theorem, a minimum of 75% of the data in any data set must lie within 2 standard deviations of the mean.

EXCURSION EXERCISES, SECTION 12.5   *page 830–831*

1. a. $\hat{y} \approx 0.73x - 0.66$ **b.** $r \approx 0.97$ **2.** ≈ 1.7 meters per second   **3.** ≈ 3.5 meters per second

EXERCISE SET 12.5   *page 831*

1. a. b   **b.** c   **2. a.** a   **b.** d   **3. a.**

$\hat{y} \approx -1.12x + 7.40$

b. $n = 5, \Sigma x = 17, \Sigma y = 18, \Sigma x^2 = 75, (\Sigma x)^2 = 289, \Sigma xy = 42$ **c.** $a = -\dfrac{48}{43} \approx -1.12, b = \dfrac{318}{43} \approx 7.40$ **d.** See the graph in part a. **e.** Yes **f.** $y = 3.6$ **g.** $r \approx -0.96$ **4. a.**

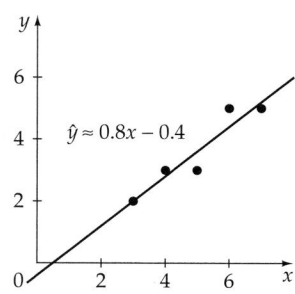

b. $n = 5, \Sigma x = 25, \Sigma y = 18, \Sigma x^2 = 135, (\Sigma x)^2 = 625, \Sigma xy = 98$ **c.** $a = 0.8, b = -0.4$ **d.** See the graph in part a.
e. Yes **f.** $y = 5.44$ **g.** $r \approx 0.94$ **5.** $\hat{y} \approx 2.01x + 0.56; r \approx 0.96$ **6.** $\hat{y} \approx -1.92x + 0.46; r \approx -0.94$
7. $\hat{y} \approx -0.72x + 9.23; r \approx -0.96$ **8.** $\hat{y} \approx 0.66x - 6.66; r \approx 0.99$ **9.** $\hat{y} \approx 1.66x + 2.25; r \approx 0.99$
10. $\hat{y} \approx 2.25x - 0.36; r \approx 0.98$ **11. a.** $\hat{y} \approx -1.15x + 10.45$ **b.** 2.4 students per computer **c.** $r \approx -0.96$
d. As the years go by, the number of students per computer is decreasing. **12. a.** $\hat{y} \approx 23.56x - 24.43$
b. ≈ 1248 centimeters **13. a.** $\hat{y} \approx 0.27x + 1.82$ **b.** ≈ 3.7 million **14. a.** $\hat{y} \approx 1.67x + 16.33$
b. ≈ 46.4 centimeters **15. a.** $r \approx 0.99$ **b.** Yes, at least for the years 1995 to 2000. The correlation coefficient is very close
to 1, which indicates a near perfect linear correlation. **16.** Yes. The correlation coefficient is very close to -1, which indicates a near
perfect linear correlation. **17. a.** $r \approx 0.30$ **b.** No. The median income rose steadily for women but remained near constant
for men. **18. a.** $\hat{y} \approx -0.90x + 78.63$ **b.** ≈ 56 years **19. a.** $r = 1.00$ **b.** The data display a perfect
linear relationship. **c.** $\hat{y} = 1.8x + 32$ **d.** 95°F **e.** Interpolation **20. a.** $r = -1.00$ **b.** The graph
of the regression line passes through all three data points. **c.** $\hat{y} = -0.85x + 187$ **d.** ≈ 126 beats per minute
e. Extrapolation **21.** No. The linear correlation coefficient is $r \approx 0.68$. Thus there is a linear correlation, but not a strong one.
22. a. $\hat{y} \approx 0.12x + 1.80$ **b.** $\hat{y} \approx 0.05x + 3.32$ **c.** No. The linear correlation coefficient is about 0.89, which is slightly
less than 0.9. **23. a.** $\hat{y} \approx 0.26x + 1.83$ **b.** In situations in which one or two of the y data values are suspected of being off
by a considerable amount due to experimental error. **24. a.–d.** Answers will vary.

CHAPTER 12 REVIEW EXERCISES *page 837*

1. 14.7; 14; 12 [Sec. 12.1] **2.** The mode [Sec. 12.1] **3.** Answers will vary. [Sec. 12.1] **4. a.** Median
b. Mode **c.** Mean [Sec. 12.1] **5.** 1331.125 feet; 1223.5 feet; 1200 feet; 462 feet [Sec. 12.1/12.2]
6. 36 miles per hour [Sec. 12.1] **7.** ≈ 3.10 [Sec. 12.1] **8. a.** 1.25 **b.** The 89th percentile [Sec. 12.3]
9. $\approx 3.16; \approx 10.00$ [Sec. 12.2] **10.** $4.51; 4.385; \approx 0.436$ [Sec. 12.1/12.2] **11. a.** The second student's mean is
5 points higher than the first student's mean. **b.** They are the same. [Sec. 12.1/12.2] **12. a.** ≈ -1.73
b. ≈ 0.58 [Sec. 12.3] **13.**

$$\leftmapsto\!\!\!\!\!\!\!\!\text{[box plot]}\!\!\!\!\!\!\!\!\mapsto$$

160 180 200 220 240 260 280 300 320 [Sec. 12.3]

c. 8 [Sec. 12.4] **15. a.** 20% **b.** 0.56 [Sec. 12.4] **16. a.** $r \approx 0.96$ **b.** None
c. Yes [Sec. 12.5] **17.** Answers will vary. [Sec. 12.4] **18.** No. No information is given about how the scores are
distributed below the mean. [Sec. 12.4] **19. a.** 0.933 **b.** 0.309 [Sec. 12.4] **20. a.** 21.2%
b. 95.4% [Sec. 12.4] **21. a.** 6.7% **b.** 64.5% **c.** 78.8% [Sec. 12.4]
22. $\hat{y} \approx 1.57x + 4.96; r \approx 0.88$ [Sec. 12.5] **23. a.**

b. $n = 5, \Sigma x = 67, \Sigma y = 25, \Sigma x^2 = 921, (\Sigma x)^2 = 4489, \Sigma xy = 310$ **c.** $a \approx -1.08, b \approx 19.44$ **d.** See the graph in part a.

e. Yes   **f.** ≈10.8   **g.** $r \approx -0.95$ [Sec. 12.5]   **24. a.** $r \approx 0.999$ **b.** $\hat{y} \approx 0.07x + 0.29$
c. 13.94 inches [Sec. 12.5]   **25. a.** $\hat{y} \approx 0.018x + 0.0005$ **b.** Yes. $r \approx 0.999$, which is very close to 1.
c. 1.80 seconds [Sec. 12.5]   **26. a.** $\hat{y} \approx 13.78x + 109.30, r \approx 0.998$ **b.** 564.0 [Sec. 12.5]

CHAPTER 12 TEST   *page 840*

1. 9.1; 9; 7 [Sec. 12.1]   **2.** 82.2 [Sec. 12.1]   **3.** 24; ≈8.76; ≈76.7 [Sec. 12.2]   **4. a.** ≈1.18
b. ≈−0.49 [Sec. 12.3]   **5.**   **6. a.** 29%

[Sec. 12.3]

b. 40% [Sec. 12.4]   **7. a.** 47.7%   **b.** 15.9% [Sec. 12.4]   **8. a.** 10.6%   **b.** 20.3% [Sec. 12.4]
9. $\hat{y} \approx 1.36x + 3.32, r \approx 0.89$ [Sec. 12.5]   **10. a.** $\hat{y} \approx -7.98x + 767.12$ **b.** 57 calories [Sec. 12.5]

CHAPTER 13

EXCURSION EXERCISES, SECTION 13.1   *page 858*

2. Connecticut: 7; Delaware: 2; Georgia: 2; Maryland: 8; Massachusetts: 14; Kentucky: 2; New Hampshire: 4; Vermont: 2; New York: 10;
New Jersey: 5; Pennsylvania: 13; North Carolina: 10; South Carolina: 6; Virginia: 18; Rhode Island: 2   **3.** Virginia
4. Connecticut: 7; Delaware: 2; Georgia: 2; Maryland: 8; Massachusetts: 14; Kentucky: 2; New Hampshire: 4; Vermont: 3; New York: 10;
New Jersey: 5; Pennsylvania: 12; North Carolina: 10; South Carolina: 6; Virginia: 18; Rhode Island: 2

EXERCISE SET 13.1   *page 858*

1. To calculate the standard divisor, divide the total population p by the number of items to apportion n.   **2. a.** 646,952; There is
one representative for every 646,952 citizens in the U.S.   **b.** Underrepresented; The average constituency is greater than the
standard divisor.   **c.** Overrepresented; The average constituency is less than the standard divisor.   **3.** The standard quota
for a state is the whole number part of the quotient of the state's population divided by the standard divisor.   **4.** The quota rule
stipulates that the number of representatives apportioned to a state must be equal to or one more than the standard quota.
5. a. 8.96; There is one teacher aid for every 8.96 students.   **b.** Kindergarten: 4; first: 4; second: 4; third: 3; fourth: 2; fifth: 4;
sixth: 4   **c.** Kindergarten: 4; first: 5; second: 4; third: 3; fourth: 2; fifth: 3; sixth: 4; no   **d.** Using the Jefferson method, the first
grade receives one more teacher's aid and the fifth grade receives one less teacher's aid when compared to the Hamilton method.
6. a. 87.8; There is one new computer for every 88 students.   **b.** Liberal arts: 39; business: 66; humanities: 22; science: 53
c. Using 86 as a modified standard divisor yields an apportionment of 183, which is greater than the number of available new computers.
Using 87 yields an apportionment of 179, which is less than the number of available new computers. Using 86.5 yields an apportionment
that is equal to the number of available new computers.   **d.** Using a number larger than the standard divisor results in an
apportionment that is less than the number of available new computers.   **e.** Liberal arts: 39; business: 66; humanities: 21;
science: 54; no   **f.** Using the Jefferson method, the humanities division receives one less computer and the sciences division receives
one more computer when compared to using the Hamilton method.   **7. a.** 37.10; There is one new nurse for every 37.10 beds.
b. Sharp: 7; Palomar: 10; Tri-City: 8; Del Raye: 5; Rancho Verde: 7; Bel Aire: 11   **c.** Sharp: 7; Palomar: 10; Tri-City: 8; Del Raye: 5;
Rancho Verde: 7; Bel Aire: 11   **d.** They are identical.   **8.** The Alabama paradox occurs when adding one representative to the
number of representatives to be apportioned results in a state losing a representative, even though the populations do not change.
9. The population paradox occurs when the population of one state is increasing faster than that of another state, yet the first state still
loses a representative.   **10.** The new states paradox occurs when a new state is added along with the appropriate number of
representatives to account for its population, but after reapportionment, another state ends up losing a representative.
11. The Balinski-Young Impossibility Theorem states that any apportionment method will either violate the quota rule or will produce
paradoxes such as the Alabama paradox.   **12. a.** *A*: 4; *B*: 9; *C*: 2; *D*: 13. No. None of the campus locations lost a projector.
b. *A*: 4; *B*: 10; *C*: 1; *D*: 14. Yes. Increasing the number of projectors to be apportioned from 28 to 29 causes campus C to lose a projector.

13. a. Yes   **b.** No   **c.** Yes   **14. a.** A: 4; B: 10; C: 26   **b.** Yes. The enrollment of center A increased at a greater rate than that of center B, yet after reappportionment, center A lost a security person and center B gained one.   **15. a.** Boston: 2; Chicago: 20   **b.** Yes. Chicago lost a vice president while Boston gained one.   **16. a.** Math: 14; physics: 1; chemistry: 4   **b.** Yes. After adding a new department and adjusting the number of clerical assistants to account for the increased total enrollment, reapportionment causes the math department to lose an assistant even though its population has not changed.   **17. a.** Sixth grade   **b.** Sixth grade; same   **18. a.** Valley office   **b.** Valley office; same   **c.** Hill Street office   **19.** Valley   **20.** Utah's 2000 census Huntington-Hill number must be larger than the numbers of all other states because, using the Huntington-Hill apportionment method, a new representative is awarded to the state with the highest Huntington-Hill number.   **21. a.** They are the same.   **b.** Using the Jefferson method, the humanities division gets one less computer and the sciences division gets one more computer compared with using the Webster method.   **22. a.** Using 12,600,000 as a modified standard divisor yields 21 committee members.   **b.** Using 12,700,000 as a modified standard divisor yields 19 committee members.   **c.** Using 12,650,000 as a modified standard divisor yields 20 committee members.   **d.** France: 5; Germany: 6; Italy: 5; Spain: 3; Belgium: 1   **23.** The Jefferson and Webster methods   **24.** The Webster method   **25.** The Huntington-Hill method   **26.** Answers will vary.   **27.** Answers will vary.   **28.** Calculate each state's standard quota using the standard divisor and always round up to the next highest whole number. Modify the standard divisor until the sum of the states' calculated quotas equals the number of representatives to be apportioned. The Adams method violates the quota rule but is not susceptible to any paradox.

29. Del Mar: 3; Wheatly: 7; West: 5; Mountain View: 7   **30.** $\dfrac{(P_B)^2}{b(b+1)} < \dfrac{(P_A)^2}{a(a+1)}$

EXCURSION EXERCISES, SECTION 13.2   *page 876*

1. Lemon, raspberry, plain   **2.** No   **3.** No   **4.** Yes   **5.** Yes   **6.** Yes   **7.** No

EXERCISE SET 13.2   *page 877*

1. A majority means that a choice receives more than 50% of the votes. A plurality means that the choice with the most votes wins. It is possible to have a plurality without a majority when there are more than two choices.   **2.** If there are more than two choices, the winning choice may not have majority support.   **3.** If there are n choices in an election, each voter ranks the choices by giving n points to the voter's first choice, $n - 1$ points to the voter's second choice, and so on, with the voter's least favorite choice receiving 1 point. The choice with the most points is the winner.   **4.** Voters rank their preferences for all choices. The choice with the fewest first-place votes is eliminated and preferences are renumbered assuming that voters do not change their preferences for the remaining choices when one choice is removed. This process of eliminating the choice of lowest preference is continued until the winner is determined.   **5.** In the pairwise comparison voting method, each choice is compared one-on-one with each of the other choices. A choice receives 1 point for a win, 0.5 point for a tie, and 0 points for a loss. The choice with the greatest number of points is the winner.   **6.** The Condorcet criterion states that a candidate who wins all possible head-to-head matchups should win an election when all candidates appear on the ballot.   **7.** No; no   **8.** If there are only two candidates, the plurality winner will have a majority of the first-place votes. Using the Borda Count method, one point is assigned to the first choice and zero points are assigned to the second choice.   **9.** Al Gore; no   **10.** There is a tie between Corn Flakes and Raisin Bran.   **11. a.** 35   **b.** 18   **c.** Scooby Doo   **12.** There is a tie between Italian and Mexican.   **13.** Go to a theater   **14.** Tie between Corn Flakes and Raisin Bran.   **15.** Bugs Bunny   **16.** Mexican   **17.** Elaine Garcia   **18.** Cingular   **19.** Blue and white   **20.** WKLS   **21.** Raymond Lee   **22.** Cingular   **23. a.** Buy new computers for the club.   **b.** Pay for several members to travel to a convention.   **c.** Pay for several members to travel to a convention.   **d.** Answers will vary.   **24. a.** Amusement park   **b.** Amusement park   **c.** Tie between picnic in a park and amusement park   **25.** *Return of the Jedi*   **26.** Grand Canyon   **27.** There is a tie between the panther and the bobcat.   **28.** There is a three-way tie among P. Gibson, R. Allenbaugh, and G. DeWitte.   **29.** Blue and white   **30.** There is a three-way tie among WKLS, WNNX, and WWVV.   **31.** No   **32.** No   **33.** Yes   **34.** No   **35.** No   **36.** No   **37. a.** Stephen Hyde   **b.** Stephen Hyde received the fewest number of first-place votes.   **c.** John Lorenz   **d.** The candidate that wins all head-to-head matches does not win the election.   **e.** John Lorenz   **f.** The candidate winning the original election (Stephen Hyde) did not remain the winner in a recount in which a losing candidate withdrew from the race.   **38. a.** Film A   **b.** Film B   **c.** The monotonicity criterion has not been violated; Film A won the first vote.   **39.** Mendin is president, Jen is vice president, Andrew is secretary, and Hector is treasurer.   **40.** Mendin is president, Hector is

vice president, Andrew is secretary, and Jen is treasurer, no   **41.**   The Borda Count method   **42. a.**   Round Table
b.   Domino's; no   **c.**   Tie between Domino's and Round Table   **d.**   Yes, the Borda Count method
43. a.   Candidate B   **b.**   Candidate A   **44. a.**   There is a tie between *Good Will Hunting* and *Contact*.
b.   Round Table; yes

EXCURSION EXERCISES, SECTION 13.3   *page 895*

1.   {A, B}, {A, D}, {B, D}, {A, B, C}, {A, B, D}, {A, C, D}, {B, C, D}   **2.**   BPI(A) $= \frac{1}{3}$, BPI(B) $= \frac{1}{3}$, BPI(C) $= 0$, BPI(D) $= \frac{1}{3}$

3.   {B, F}, {B, G}, {B, I}, {F, G}, {F, I}, {F, N}, {G, I}, {G, N}, {I, N}, {B, F, G}, {B, F, I}, {B, F, L}, {B, F, N}, {B, G, I},
{B, G, L}, {B, G, N}, {B, I, L}, {B, I, N}, {F, G, I}, {F, G, L}, {F, G, N}, {F, I, L}, {F, I, N}, {F, L, N}, {G, I, L}, {G, I, N}, {G, L, N}, {I, L, N},
{B, F, G, I}, {B, F, G, L}, {B, F, G, N}, {B, F, I, L}, {B, F, I, N}, {B, F, L, N}, {B, G, I, L}, {B, G, I, N}, {B, G, L, N}, {B, I, L, N}, {F, G, I, L},
{F, G, I, N}, {F, G, L, N}, {F, I, L, N}, {G, I, L, N}, {B, F, G, I, L}, {B, F, G, I, N}, {B, F, G, L, N}, {B, F, I, L, N}, {B, G, I, L, N}, {F, G, I, L, N}

4.   BPI(Belgium) $= \frac{1}{7}$; BPI(Luxembourg) $= 0$ **5.**   Answers will vary.   **6.**   Answers will vary.   **7.**   Answers will vary.

EXERCISE SET 13.3   *page 895*

1. a.   6   **b.**   4   **c.**   3   **d.**   6   **e.**   No   **f.**   A and C   **g.**   15   **h.**   6   **2. a.**   16   **b.**   5
c.   4   **d.**   11   **e.**   No   **f.**   A, B, and D   **g.**   31   **h.**   10   **3.**   0.60, 0.20, 0.20   **4.**   0.33, 0.33, 0.33
5.   0.50, 0.30, 0.10, 0.10   **6.**   0.25, 0.25, 0.25, 0.25   **7.**   0.36, 0.28, 0.20, 0.12, 0.04   **8.**   0.25, 0.25, 0.25, 0.25
9.   1.00, 0.00, 0.00, 0.00, 0.00, 0.00   **10.**   0.27, 0.27, 0.18, 0.18, 0.09   **11.**   0.44, 0.20, 0.20, 0.12, 0.04
12.   0.5, 0.3, 0.1, 0.1   **13. a.**   Exercise 9   **b.**   Exercises 3, 5, 6, 9, and 12   **c.**   None   **d.**   Exercise 8
14.   There are no winning coalitions.   **15.**   0.33, 0.33, 0.33   **16.**   0.5, 0.125, 0.125, 0.125, 0.125
17. a.   {12: 1, 1, 1, 1, 1, 1, 1, 1, 1, 1, 1, 1}   **b.**   Yes   **c.**   Yes   **d.**   Divide the vote power, 1, by the quota, 12.
18. a.   {9: 1, 1, 1, 1, 1, 1, 1, 1, 1, 1, 1, 1}   **b.**   Yes   **c.**   No   **d.**   Take the inverse of the total number of voters.
19.   Dictator: A; dummies: B, C, D, E   **20.**   Dummy: D   **21.**   None   **22.**   Dummies: C, D
23. a.   0.60, 0.20, 0.20   **b.**   Answers will vary.   **24. a.**   0.60, 0.20, 0.20   **b.**   They are the same; no
25. a.   0.33, 0.33, 0.33   **b.**   It is a majority system.   **26.**   0.42, 0.25, 0.25, 0.08   **27. a.**   11 and 14   **b.**   15 and 16
c.   No   **28. a.**   Voter D is never a critical voter in a winning coalition.   **b.**   Because D is a dummy voter, BPI(D) $= 0$. The
remaining votes all have equal weight, so their BPIs are the same. The sum of the BPIs must be 1, so each remaining voter's BPI will be 1
divided by 3.   **29.**   Yes   **30.**   No   **31.**   In the weighted system, all voters have equal weight and the quota is equal to
50% of the sum of the weights rounded up to the nearest whole number, or a strict majority.   **32. a.**   B, C, D, and E would gain
power. A would lose power.   **b.**   B, C, D and E would lose power. A would gain power.   **33.**   Answers will vary.
34. a.   BPI(7) $= 0.1669$, BPI(1) $= 0.0165$ **b.**   About 10 times   **c.**   BPI(5) $= 0.142$, BPI(1) $= 0.029$
d.   About 5 times   **35. a.**   BPI(10) $= 0.1116$, BPI(8) $= 0.0924$, BPI(5) $= 0.0587$, BPI(4) $= 0.0479$, BPI(3) $= 0.0359$,
BPI(2) $= 0.0226$ **b.**   1.6

CHAPTER 13 REVIEW EXERCISES   *page 900*

1. a.   Health: 7; business: 18; engineering: 10; science: 15   **b.**   Health: 6; business: 18; engineering: 10; science: 16
c.   Health: 7; business: 18; engineering: 10; science: 15   [Sec. 13.1]   **2. a.**   Newark: 9; Cleveland: 6; Chicago: 11; Philadelphia: 4;
Detroit: 5   **b.**   Newark: 9; Cleveland: 6; Chicago: 11; Philadelphia: 4; Detroit: 5   **c.**   Newark: 9; Cleveland: 6; Chicago: 11;
Philadelphia: 4; Detroit: 5   [Sec. 13.1]   **3. a.**   No. None of the offices loses a new printer.   **b.**   Yes. Office A drops from two
new printers to only one new printer.   [Sec. 13.1]   **4. a.**   *A*: 2; *B*: 5; *C*: 3; *D*: 16; *E*: 2. No. None of the centers lose an automobile.
b.   No. None of the centers lose an automobile.   [Sec. 13.1]   **5. a.**   Los Angeles: 9; Newark: 2   **b.**   Yes. Newark loses a
computer file server.   [Sec. 13.1]   **6. a.**   *A*: 10; *B*: 3; *C*: 21   **b.**   Yes. The population of region B grew at a higher rate than the
population of region A, yet region B lost an inspector to region A.   [Sec. 13.1]   **7.**   Yes   [Sec. 13.1]   **8.**   Yes   [Sec. 13.1]
9. a.   A   **b.**   B   [Sec. 13.1]   **10. a.**   Shannon M.   **b.**   Hannah A.   **c.**   Hannah A. won all head-to-head
comparisons, but lost the overall election.   **d.**   Hannah A.   **e.**   Hannah A. received a majority of the first-place votes,
but lost the overall election.   [Sec. 13.2]   **11. a.**   Hannah A.   **b.**   Cynthia L., a losing candidate, withdrew from the
race and caused a change in the overall winner of the election.   [Sec. 13.2]   **12.**   The monotonicity criterion was violated
because the only change was that the supporter of a losing candidate changed his or her vote to support the original winner, but
the original winner did not win the second vote.   [Sec. 13.2]   **13. a.**   18   **b.**   18   **c.**   Yes   **d.**   A and C

e. 15 **f.** 6 [Sec. 13.3] **14. a.** 35 **b.** 35 **c.** Yes **d.** A **e.** 31 **f.** 10 [Sec. 13.3]
15. 0.60, 0.20, 0.20 [Sec. 13.3] **16.** 0.20, 0.20, 0.20, 0.20, 0.20 [Sec. 13.3] **17.** 0.42, 0.25, 0.25, 0.08 [Sec. 13.3]
18. 0.62, 0.14, 0.14, 0.05, 0.05 [Sec. 13.3] **19.** Dictator: A; dummies: B, C, D, E [Sec. 13.3]
20. Dummy: D [Sec. 13.3] **21.** 0.50, 0.125, 0.125, 0.125, 0.125 [Sec. 13.3] **22. a.** Manuel Ortega
b. No **c.** Crystal Kelley [Sec. 13.2] **23. a.** Vail **b.** Aspen [Sec. 13.2] **24.** A. Kim [Sec. 13.2]
25. Snickers [Sec. 13.2] **26.** A. Kim [Sec. 13.2] **27.** Snickers [Sec. 13.2]

CHAPTER 13 TEST *page 903*

1. a. Sales: 17; advertising: 4; service: 11; manufacturing: 53 **b.** Sales: 17; advertising: 4; service: 10;
manufacturing: 54; no [Sec. 13.1] **2. a.** Cedar Falls ≈ 77,792; Lake View ≈ 70,290 **b.** Cedar Falls [Sec. 13.1]
3. a. 33 **b.** 26 **c.** No **d.** A and C **e.** 31 **f.** 10 [Sec. 13.3] **4. a.** Aquafina
b. No **c.** Evian [Sec. 13.2] **5.** New York [Sec. 13.2] **6. a.** Afternoon **b.** Noon [Sec. 13.2]
7. a. Proposal A **b.** Proposal B **c.** Eliminating a losing choice changed the outcome of the vote. **e.** Proposal B
won all head-to-head comparisons, but lost the vote when all the choices were on the ballot. [Sec. 13.2] **8.** 0.42, 0.25, 0.25, 0.08
[Sec. 13.3] **9.** 0.40, 0.20, 0.20, 0.20 [Sec. 13.3] **10. a.** Dummy: D **b.** Dictator: A; dummies: B, C, D [Sec. 13.3]

APPENDIX

APPENDIX *page 910*

1. Meter, liter, gram **2. a.** Move the decimal point in the number two places to the right. **b.** Move the decimal point in
the number three places to the left. **3.** Kilometer **4.** Kilogram **5.** Centimeter **6.** Milliliter **7.** Gram
8. Kiloliter **9.** Meter **10.** Centimeter **11.** Gram **12.** Kilogram **13.** Milliliter **14.** Liter
15. Gram **16.** Gram **17.** Millimeter **18.** Centimeter **19.** Milligram **20.** Kilogram
21. Gram **22.** Milliliter **23.** Kiloliter **24.** Kilogram **25.** Milliliter **26.** Kilometer
27. a. Column 2: k, c, m; column 3: 10^9, 10^3, 10^2, $\dfrac{1}{10^3}$, $\dfrac{1}{10^{12}}$; column 4: 1 000 000, 10, 0.1, 0.01, 0.000 001, 0.000 000 001

b. Answers will vary. **28.** 420 **29.** 910 **30.** 0.360 **31.** 1.856 **32.** 5.194 **33.** 7.285
34. 2 000 **35.** 8 000 **36.** 0.217 **37.** 0.034 **38.** 4 520 **39.** 29.7 **40.** 8.406 **41.** 7.530
42. 2 400 **43.** 9 200 **44.** 6 180 **45.** 36 **46.** 9 612 **47.** 2 350 **48.** 240 **49.** 83
50. 2.98 **51.** 0.716 **52.** 2.431 **53.** 6.302 **54.** 66 **55.** 458 **56.** 24.3 **57.** 9.2
58. a. 50 kilometers **b.** 10 kilometers **59.** 2 grams **60.** 8 pieces **61.** 24 liters **62.** $110.25
63. 16 servings **64.** 40 grams **65.** 4 tablets **66.** 8 liters **67.** The case containing 12 one-liter bottles
68. 215.5 centimeters **69.** 500 seconds **70.** Answers will vary. **71.** $42.50 **72.** $124.50 **73.** $8715

Chapter 1
p. 2, Mark Antman/The Image Works; p. 3, Courtesy of Marilyn Savant; p. 4, The New Yorker Collection 1978 Warren Miller from Cartoonbank.com. All Rights Reserved; p. 5, Hulton Archive/Getty Images; p. 9, Michael Newman/PhotoEdit; p. 12, Istituto E Museo Di Storia Della Scienza; p. 14, Istituto E. Museo Di Stonria Della Scienza; p. 19, Corbis; p.22, Hulton Archive/Getty Images; p. 26, Science Photo Library/Photo Researchers; p. 26, Science Photo Library/Photo Researchers; p.28, Courtesy of George Polya; p. 34, Margarte Ross/Stock Boston; p. 36, Hulton Archive/Getty Images; p.40, Michael Newman/PhotoEdit.

Chapter 2
p.48, Larry Dale Gordon/Getty Images; p.49, Michael Newman/PhotoEdit; p. 49, Burke/Triolo/Brand X Pictures/PictureQuest; p. 50, The Granger Collection; p. 54, Science Photo Library/Photo Researchers; p.55, Courtesy of Lofti Zadeh; p. 58, Courtesy of Warner Brothers/Getty Images; p. 58, Corbis; p. 64, Coribs Images/PictureQuest; p.67, © IPS/Indes Stock Imagery/PictureQuest; p.68, © Journal-Courier/Steve Warnowski/The Image Works; p.69, Nicholas Devore III/Photographers/Aspen/PictureQuest; p. 70, Bernard Wolff/PhotoEdit; p. 72, The Granger Collection; p. 74, Corbis; p. 75, Charles Gupton/Stock Boston/PictureQuest; p. 81, G.R. Vikki Hart/Brand X Pictures/PictureQuest; p. 83, Corbis; p. 84, Courtesy of Sylvia Wiegand; p. 86, David Young Wolff/PhotoEdit; p. 87, Bob Daemmrich/Stock Boston/PictureQuest; p. 88, The Granger Collection; p. 89, Dale ODell/Stock Connection/PictureQuest; p. 90, M. Bernett/Stock Boston; p. 97, Henry Kaiser/eStock Photography/PictureQuest; p. 102, H. Carol Moran/Indes Stock Imagery/PictureQuest; p. 105, David Young Wolff/PhotoEdit.

Chapter 3
p. 108, Michael De Young/Corbis; p. 109, Jean Marc Gibson/Getty Images; p. 110, Science Photo Library/Photo Researchers; p. 112, The Granger Collection; p. 112, The Granger Collection; p. 117, Hulton Archive/Getty Images; p. 120, Raymond Smullyan the logician; p. 123, © Tribune Media Services, Inc. All Rights Reserved. Reprinted with permission; p. 131, The Everett Collection; p. 133, Michael Newman/PhotoEdit; p. 143, Corbis; p. 143, Don Boroughs/The Image Works; p. 146, The Granger Collection; p. 146, Corbis; p. 151, The Granger Collection; p. 154, c Tribune Media services. All rights reserved. Reprint with permission; p. 159, Hulton Archive/Getty Images; p. 160, Pierre-Auguste Renoir, French, 1841–1919. Dance at Gougival, 1883. Oil on canvas. 181.9 × 98.1 cm (71 5/8 × 38 5/8 in.) Museum of Fine Arts, Boston. picture fund 37.375 c 2002 Museum of Fine Arts, Boston.

Chapter 4
p. 172, Eastcott-Momatuik/The Image Works; p. 173, The Granger Collection; p. 175, Art Resource, NY; p. 178, Image Source/Elektra Vision/PictureQuest; p. 179, The Granger Collection; p.182, Doug Plummer/Stock Connection/PictureQuest; p.201, Michael Newman/PhotoEdit; p. 210, The Fields Medal; p. 219, The Granger Collection; p. 219, AP Photo; p. 226, The Granger Collection; p. 230, Corbis; p. 231, Professor Peter Goddard/PhotoScience Library/Photo Researchers.

Chapter 5
p. 238, Getty Images; p. 240, Hulton Archive/Getty Images; p. 245, Hulton Archive/Getty Images; p. 255, AP Photo/NASA; p. 259, Bob Daemmrich/Stock Boston; p. 260, Mike and Carol Weiner/Stock Boston; p. 263, George Slye; p. 268, Gamma Press; p. 270, Hulton Archive/Getty Images.

Chapter 6
p. 322, Sven Martson/The Image Works; p. 324, Richard Norwitz/Photo Researchers, Inc.; p. 326, The Granger Collection; p. 331, AP Photo/Wide World Photos; p. 343, Hideo Kurihara/Stone/Getty Images; p. 360, Hulton Archive/Getty Images; p. 362, Joe Sohn/Photo Researchers, Inc.; p. 362, Rafael Macia/Photo Researchers, Inc; p. 367, Bill W. Marsh/Photo Researchers, Inc; p. 377, Corbis; p. 380, Photo by Luc Norovitch REUTERS; p. 379, Marie Sklodowska-Curie; p. 391, AP Photo.

Chapter 7
p. 404, Bob Daemmrich/The Image Works; p. 405, AP Photo/Lana Harris; p. 405, Bob Daemmrich/Stock Boston; p. 428, The Granger Collection; p. 424, Hulton Archive/Getty Images; p. 425, The Granger Collection.

Chapter 8
p. 446, Mitch Wejnarowicz/The Image Works; p. 447, The Granger Collection; p. 448, Science Photo Library/Photo Researchers; p. 451, Hideo Kurihara/Stone/Getty Images; p. 468, AP Photo; p. 473, The Granger Collection; p. 472, The Granger Collection; p. 509, Leonard Lee Rue III/Stock Boston; p. 516, The Granger Collection; p. 517, The Granger Collection; p. 256, Phyllis Picardi/Stock Boston; p. 530, Hank Morgan/Photo Researchers; p. 534, Gregory Sams/Photo Researchers, Inc; p. 534, Dr. Fred Espanek/Photo Researchers, Inc; p. 534, Alfred Pasicks/Photo Researchers, Inc; p. 534, Mike and Carol Werner/Stock Boston; p. 535, Andy Ryan; p. 535, Fractal Antena; p. 536, Visuals Unlimited; p. 510, Hulton Archives/Getty Images.

Chapter 9

p. 546, Cindy Charles/PhotoEdit; p. 551, Courtesy of Intentix Software; p. 552, Corbis; p. 573, Courtesy of Princeton Edu; p. 590, Laima Druskis/Stock Boston.

Chapter 10

p. 620, John Henley/Corbis; p. 629, Hulton Archive/Getty Images; p. 637, courtesy of Western Currency Facility; p. 639 Getty Images.

Chapter 11

p. 629, © Topham/The Image Works; p. 693, Visuals Unlimited; p. 693, Visuals Unlimited; p. 694, Mark Burnett/Stock Boston; p. 707, Vaughn Youtz/Newsmakers/Getty Images; p. 715, Darren McCollester/Getty Images; p. 716, Aldo Torelli/Stone/Getty Images; p. 718, Michael Simpson/FPG/Getty Images; p. 718, Richard Kaylin/Stone/Getty Images; p. 722, Topham/The Image Works; p. 724, AP Photo/Doug Mills; p. 728, Visuals Unlimited; p. 728, Najhah Feanny/Stock Boston; p. 730, Richard Epstein/The Image Works; p. 730, Hank de Lespinasse/The Image Bank; p. 740, Tim Boyle/Getty Images; p. 748, John Kelly/Stone/Getty Images; p. 749, Johnny Crawford/The Image Works; p. 750, Michael Newman/PhotoEdit; p. 753, Photo Disc; p. 756, Visuals Unlimited; p. 757, Barbara Alper/Stock Boston; p. 758, Steve Smith/FPG/Getty Images; p. 759, Billy Hustace/Stone/Getty Images.

Chapter 12

p. 766, Richard Pipes/AP Photo; p. 770, Spencer Grant/PhotoEdit; p. 772, AP Photo/Matt York; p. 782, Ali Jareekji/Reuters/Getty Images; p. 790, Tony Gutierrrz/AP Photo; p. 791, Getty Images; p. 798, c Bettmann/Corbis; p. 831, Courtesy Department of Library Services/American Museum of Natural History.

Chapter 13

p. 842, Richard Ellis/The Image Works; p. 843, Hulton Archives/Getty Images; p. 859, David Young Wolff/PhotoEdit; p. 865, Getty Images; p. 869, Getty Images; p. 871, The Granger Collection; p. 873, AP Photo; p. 888 and p.898, AP Photo/Marty Lederhandler; p. 896, © Journal-Courier/Steve Warnowski/The Image Works.

Photo credits for back cover:

Rhind papyrus: Art Resource, NY; Aristotle: Corbis; Euclid: Science Photo Library/Stone/Getty Images; Archimedes: Hulton Archives/Getty Images; Hypatia: © Bettamann/Corbis; Fibonacci: Corbis; Solar System: Science Photo Library/Photo Researchers: Galileo: Hulton Archive/Getty Images; Einstein: Hulton Archive/Getty Images.